INTRODUCTION TO
Genetic Analysis

D0326124

About the Authors

[*Courtesy of Barbara Moon.*]

ANTHONY J. F. GRIFFITHS is a Professor of Botany, Emeritus, at the University of British Columbia. His research focuses on developmental genetics using the model fungus *Neurospora crassa*. He has served as president of the Genetics Society of Canada and two terms as Secretary-General of the International Genetics Federation. He was recently awarded the Fellow Medal of the International Mycological Association.

[*Courtesy of Feng Tian.*]

JOHN DOEBLEY is a Professor of Genetics and Chair of the Department of Genetics at the University of Wisconsin–Madison. He studies the genetics of crop domestication using the methods of population and quantitative genetics. He was elected to the National Academy of Sciences in 2003 and served as the president of the American Genetic Association in 2005. In 2015, he was awarded the Gregor Mendel Medal by the British Genetics Society. He teaches general genetics at the University of Wisconsin.

[*Courtesy of Catherine Peichel and Oliver Moser, Photo Video Zumstein AG.*]

CATHERINE PEICHEL is a Professor in the Institute of Ecology and Evolution at the University of Bern, Switzerland. She studies the genetic, developmental, and genomic mechanisms that underlie evolutionary processes using stickleback fish as a model system. She was named a Fellow of the John Simon Guggenheim Memorial Foundation in 2013 and served as president of the American Genetic Association in 2015. She teaches evolutionary biology and evolutionary genetics at the University of Bern.

[*Becky Katzenberger, courtesy of David Wassarman.*]

DAVID A. WASSARMAN is a Professor of Medical Genetics at the University of Wisconsin–Madison. His research focuses on the genetics of neurodegenerative diseases using *Drosophila melanogaster*. In 1997, he was awarded the Presidential Early Career Award for Scientists and Engineers. He teaches molecular genetics at the University of Wisconsin–Madison.

INTRODUCTION TO

Genetic Analysis

Twelfth Edition

Anthony J. F. Griffiths
University of British Columbia

John Doebley
University of Wisconsin–Madison

Catherine Peichel
University of Bern

David A. Wassarman
University of Wisconsin–Madison

Austin • Boston • New York • Plymouth

Vice President, STEM: Daryl Fox
Executive Program Director: Sandy Lindelof
Executive Marketing Manager: Will Moore
Marketing Assistant: Madeleine Inskeep
Development Editors: Erica Champion, Erica Frost, Michael Zierler
Senior Media Editors: Cassandra Korsvik, Heather Held
Media Editor: Jennifer Compton
Editorial Assistant: Casey Blanchard
Director of Content Management Enhancement: Tracey Kuehn
Senior Managing Editor: Lisa Kinne
Senior Content Project Manager: Harold Chester
Director of Design, Content Management: Diana Blume
Design Services Manager: Natasha Wolfe
Cover Design Manager: John Callahan
Cover Designer: Joseph DePinho
Text Designer: Maureen McCutcheon
Director of Digital Production: Keri deManigold
Media Project Manager: Daniel Comstock
Senior Workflow Project Manager: Paul Rohloff
Production Supervisor: Robert Cherry
Executive Permissions Editor: Robin Fadool
Photo Researcher: Richard Fox, Lumina Datamatics, Inc.
Composition: Lumina Datamatics, Inc.
Printing and Binding: King Printing Co., Inc.
Cover Images: Butterflies photo: Courtesy of Mathieu Joron; DNA art: Emiko Paul

Library of Congress Control Number: 2018968563

Student Edition Hardcover:
ISBN-13: 978-1-319-11478-7
ISBN-10: 1-319-11478-4

Student Edition Loose-leaf:
ISBN-13: 978-1-319-11481-7
ISBN-10: 1-319-11481-4

© 2020, 2015, 2012, 2008 by W. H. Freeman and Company

Printed in the United States of America

2   3   4   5   6   24   23   22   21

Macmillan Learning
One New York Plaza
Suite 4600
New York, NY 10004-1562
www.macmillanlearning.com

w.h.freeman
Macmillan Learning

In 1946, William Freeman founded W. H. Freeman and Company and published Linus Pauling's *General Chemistry*, which revolutionized the chemistry curriculum and established the prototype for a Freeman text. W. H. Freeman quickly became a publishing house where leading researchers can make significant contributions to mathematics and science. In 1996, W. H. Freeman joined Macmillan and we have since proudly continued the legacy of providing revolutionary, quality educational tools for teaching and learning in STEM.

Contents in Brief

Contents in Brief

Contents in Brief

Contents

The Evolution of a Classic

The twelfth edition of *Introduction to Genetic Analysis* takes this corner-stone textbook to the next level. The hallmark focus on genetic analysis, quantitative problem solving, and experimentation continues in this new edition.

The twelfth edition also introduces **SaplingPlus,** the best online resource to teach students the problem-solving skills they need to succeed in genetics. SaplingPlus combines Sapling's acclaimed automatically graded online homework with an extensive suite of engaging multimedia learning resources.

NEW TO THE TWELFTH EDITION

▶ **SaplingPlus** includes tools to help students prepare for class and study for their exams. LearningCurve adaptive quizzing is a great tool to help students learn basic concepts and do assigned readings before coming to lecture. SaplingPlus also includes a wealth of multimedia and problem-solving resources to help students make the most of their study time.

▶ **Chapter Objectives**—New to the twelfth edition, each chapter begins with a two-sentence paragraph describing the goals of the chapter and placing the chapter topic into context of the surrounding chapters. This helps students "see the forest" before stepping into the "trees."

▶ **Core Principles**—*Introduction to Genetic Analysis* divides genetics into three segments: transmission genetics, molecular genetics, and evolutionary genetics. Each part of the text now begins with a three- to five-page introduction outlining the core principles that characterize that segment of genetics. These help orient students by providing an overview of the themes they'll encounter as they read each part.

New Authors and Cutting-Edge Content

The twelfth edition introduces two new co-authors to the team:

[*Becky Katzenberger, courtesy of David Wassarman.*]

Dr. David A. Wassarman is a professor in the Department of Medical Genetics and chair of the Cellular and Molecular Biology Graduate Program at the University of Wisconsin–Madison. His lab uses *Drosophila* as an experimental model to identify genetic modifiers of human neurodegenerative diseases, including Ataxia-telangiectasia and traumatic brain injury. Over the years, he has studied a variety of topics in the fields of molecular, developmental, and transmission genetics. At UW-Madison, he teaches a course on eukaryotic molecular biology to undergraduate and graduate students. His main goal in this course is to teach students how to use what other people have discovered to make their own discoveries.

With this goal in mind, David extensively revised the molecular genetics content of the new edition. He modified, reorganized, and updated the material in Chapters 7–10, 12, and 15, to connect molecular genetics to other fields of genetics and to illuminate the core principles that characterize molecules, molecular processes, and experiments in molecular genetics. Topics that he added to this edition include transcription by the three eukaryotic RNA polymerases and mRNA editing, modification, and decay (Chapter 8); protein modification and decay (Chapter 9); real-time PCR and CRISPR-Cas9 technologies (Chapter 10); and chromatin-mediated regulation of transcription (Chapter 12).

[*Courtesy of Catherine Peichel and Oliver Moser, Photo Video Zumstein AG.*]

Dr. Catherine (Katie) Peichel is an evolutionary geneticist and Professor in the Institute of Ecology and Evolution at the University of Bern, Switzerland. She first fell in love with genetics as an undergraduate at the University of California, Berkeley, where she used the fifth edition of *IGA*. Since then, she has continued to use genetic approaches to study biological processes. Throughout her career in both the US and Switzerland, she has shared her love of genetics and evolutionary biology by teaching undergraduate and graduate students. Katie has also volunteered her time to teach and develop genetics curriculum at medical schools in Nepal.

For this edition of *IGA*, Katie revised Chapter 14 (Genomes and Genomics) to include much-needed updates to next-generation sequencing techniques (e.g., Illumina sequencing) and bioinformatics approaches to analyzing modern and ancient human genomes. She also added a box on "Direct-to-consumer genetic testing" to link these methods to real-world applications.

Modern genetics and genomics techniques have expanded the ability to address important questions in biology using the most appropriate model organisms. Thus, Katie updated Chapters 13, 17, and 20 to include more about genetic analyses in non-traditional model organisms. She also added a new section to the Model Organisms Appendix, entitled "Beyond Model Organisms."

Problem-Solving Skills for Success

Introduction to Genetic Analysis has always been known for its rigorous and powerful problem sets. The twelfth edition expands on this tradition both in the text and online in SaplingPlus.

Working with the Figures

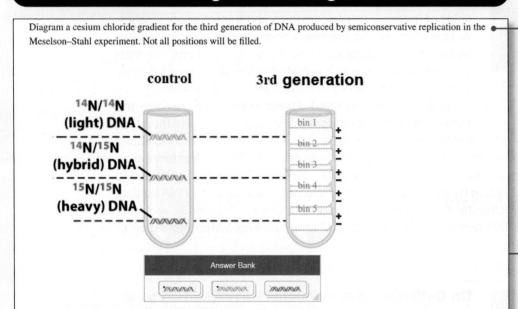

Diagram a cesium chloride gradient for the third generation of DNA produced by semiconservative replication in the Meselson–Stahl experiment. Not all positions will be filled.

Working with the Figures questions prompt students to examine book figures and tease out important information, understand the intricacies of an experimental protocol, relate concepts, or draw a conclusion about what is shown.

NEW   Genetics and Society Questions

NEW Genetics and Society Questions at the end of each chapter ask students to consider the societal relevance of a topic in the chapter; good for initiating classroom discussion.

GENETICS AND SOCIETY

In this chapter, you learned that the shortening of chromosome telomeres due to diminished telomerase activity is associated with aging. This raises the possibility that gene therapy aimed at overexpression of telomerase will increase longevity. Do you think that it is ethical to use this approach to increase the longevity of normal, healthy people? Does your answer change if you consider that there are nongenetic means such as calorie restriction that may increase longevity, or that gene therapy is being pursued to treat numerous diseases?

SaplingPlus introduces a powerful new homework and study engine for students to learn the skills they need to succeed. SaplingPlus gives students the support they need to help them learn from their mistakes, and also provides instructors with an easy way to assign automatically graded homework assignments.

Unpacking the Problem

A man's paternal grandfather had galactosemia. This is a rare disease caused by the inability to process galactose, leading to muscle, nerve, and kidney malfunction. The man married a woman whose sister had galactosemia but whose parents did not. The woman is now pregnant with their first child. What is the probability their first child has galactosemia? If the first child does have galactosemia, what is the probability a second child will have it?

To unpack this problem refer to the list of steps in the Hint.

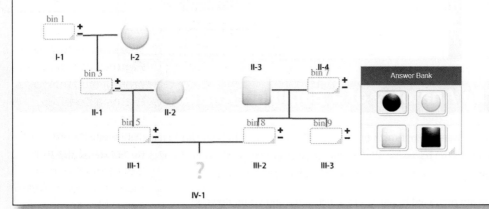

Unpacking the Problem (also available in SaplingPlus) include step-by-step tutorials helping students approach a challenging problem. In the book, this is a list of questions shaped around "what do we know and what are we trying to find out"; in Sapling, these are interactive tutorials that break the problems down into more approachable steps and build to the final answer.

Add ▾   Student Hint

Unpack the Problem: Break this problem into several parts and arrive at a solution using this guided, step-by-step approach.

Part A (step 1): Draw the pedigree as described using the standard pedigree symbols provided by dragging them to the correct places. Individual III-1 is the husband of the pregnant wife, individual III-2. Remember that the trait is rare and thus people coming into the pedigree are not considered carriers or affected unless there is evidence to the contrary.

Part B (steps 2-9): Decide what the probability is that this child will have galact

Step 2: What is the apparent mode of inheritance in this pedigree?

Step 3: What is the genotype of the man's father (II-1)?

Step 4: What is the genotype of the man (III-1)?

Step 5: What is the probability of the man (III-1) carrying the recessive allele fo

Step 6: What is the genotype of the man's wife (III-2)?

Step 7: What is the probability of the man's wife carrying the recessive allele fo

Step 8: What is the combined probability of both the man and his wife being ca

Step 9: What is the probability their first child has galactosemia?

Part C (step 10): Resolve what the probability is that a second child will have ga have it.

The Hint helps students unpack the problem step by step.

Add ▾   ≡ ▾   Solution Explanation

Step 1: Based on the description given, the only affected male is individual I-1 and the only affected female is individual III-3. All other individuals in the pedigree were unaffected.

Step 2: The diagnostic in determining the mode of inheritance in this pedigree is that the trait skips generations and is found in both male and female individuals. Skipping generations is characteristic of recessive traits while appearing in both genders indicates no sex-linkage. The mode is, therefore, autosomal recessive.

Step 3: The genotype of the man's father (II-1) is heterozygous for the disease allele. Because the grandfather (III-1) is affected by the autosomal recessive disease, he must be homozygous for the disease alleles and will pass on one of his affected alleles to his offspring (II-1). Therefore, (I-1) is a heterozygous carrier of the disease allele.

Step 4: The genotype of the man (III-1) is either a heterozygous carrier or he does not carry the disease allele at all depending upon which allele he inherited from his father.

Step 5: The probability of the man (III-1) carrying the recessive allele for galactosemia is 0.5. Because his father (II-1) was heterozygous for the disease allele, his father has a 50% chance of passing on the disease allele to III-1.

Step 6: The genotype of the man's wife (III-2) is either a heterozygous carrier or she does not carry the disease allele at all, depending upon which alleles she inherited from her parents. Because her sister is affected by the disease, both parents must carry the disease allele.

Step 7: The probability of the man's wife (III-2) carrying the recessive allele for galactosemia is 0.5. Because each of her parents carries one disease allele, each parent has a 50% chance to pass on a disease allele to their offspring. Likewise, each parent has a 0.5 probability to pass on a wild-type allele to their offspring. If III-2 is a heterozygous carrier, she could have received the disease allele from her mother and the wild-type allele from her father, or she received a disease allele from her father and the wild-type allele from her mother. The two out of four total possible genotypes means that III-2 has a 0.5

The Solution also follows the unpacked step-by-step breakdown.

SaplingPlus

SaplingPlus combines Sapling's acclaimed automatically graded online homework with a powerful e-book and an extensive suite of engaging multimedia learning resources. Problems feature hints for when students get stuck, answer-specific feedback to help them learn from their mistakes, and solutions to reinforce what they've learned.

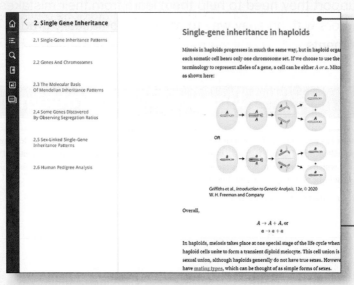

SaplingPlus incorporates a powerful e-book along with all of the plentiful online resources, giving students a single place to read, study, and assess their knowledge.

The Anatomy of a Sapling Problem

Suppose Stephen breeds flowers and wants to optimize production of offspring with both short stems and white flowers, which are coded for by two genes with the recessive alleles *t* and *p*, respectively. In flowers, *T* codes for tall stems and *P* codes for purple flowers. Stephen crosses two heterozygotes that produce 656 offspring.

How many of these 656 offspring are predicted to have both short stems and white flowers?

number of offspring: _____

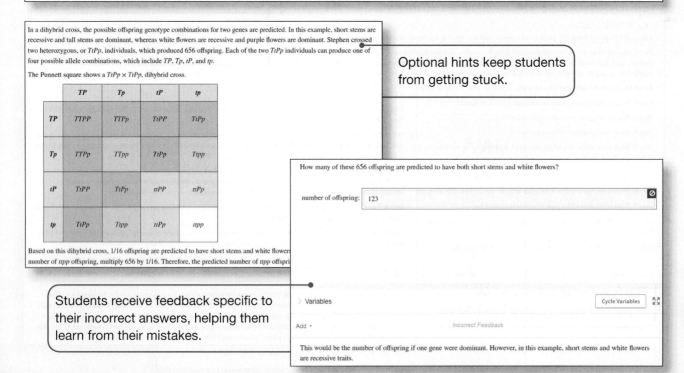

Optional hints keep students from getting stuck.

Students receive feedback specific to their incorrect answers, helping them learn from their mistakes.

SaplingPlus also includes tools to help students prepare for class and study for their exams. **LearningCurve** adaptive quizzing ties back to the e-book and is a great tool to help students learn basic concepts and do assigned readings before coming to lecture. SaplingPlus also includes a wealth of multimedia and problem-solving resources to help students make the most of their study time.

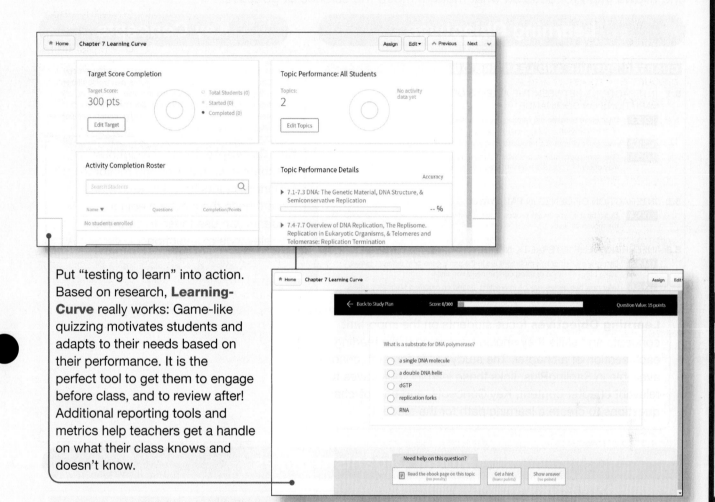

Put "testing to learn" into action. Based on research, **Learning-Curve** really works: Game-like quizzing motivates students and adapts to their needs based on their performance. It is the perfect tool to get them to engage before class, and to review after! Additional reporting tools and metrics help teachers get a handle on what their class knows and doesn't know.

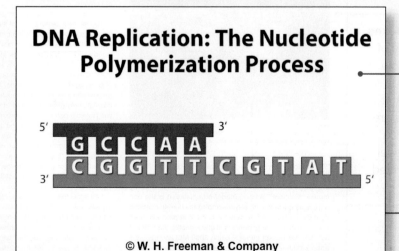

DNA Replication: The Nucleotide Polymerization Process

SaplingPlus also includes powerful multimedia assets such as videos and animations to help students visualize complex processes.

© W. H. Freeman & Company

Powerful Pedagogy

Well-thought-out pedagogy creates a smooth learning path for students and allows them to focus on what matters most: the science of genetics.

Learning Objectives

CHAPTER OUTLINE AND LEARNING OBJECTIVES

5.1 INTERACTIONS BETWEEN THE ALLELES OF A SINGLE GENE: VARIATIONS ON DOMINANCE

LO 5.1 Distinguish between the various types of dominance, based on the phenotypes of heterozygotes.

LO 5.2 Recognize phenotypic ratios diagnostic of the presence of a lethal allele.

LO 5.3 Give some possible reasons why incomplete penetrance and variable expressivity may occur in a population of individuals with identical genotypes at a locus under study.

5.2 INTERACTION OF GENES IN PATHWAYS

LO 5.4 Describe the hypotheses proposed to explain various types of gene interaction at the molecular level.

5.3 INFERRING GENE INTERACTIONS

LO 5.5 Determine whether two mutations are in the same gene or in different genes, using progeny ratios or using complementation tests.

LO 5.6 Infer how two genes may be interacting, based on modified Mendelian ratios.

LO 5.7 For known cases of gene interaction, predict progeny ratios in crosses.

Learning Objectives focus students on the important concepts and skills they should be gaining while reading each section of a chapter. The study guide for each chapter, available in SaplingPlus, links these learning objectives to relevant chapter content, Key Concepts, and end-of-chapter questions to create a learning path for the student.

Key Concepts

KEY CONCEPT For most genes, a single wild-type copy is adequate for full expression (such genes are haplosufficient), and their null mutations are fully recessive. Harmful mutations of haploinsufficient genes are often dominant. Mutations in genes that encode units in homo- or heterodimers can behave as dominant negatives, acting through "spoiler" proteins.

Key Concepts, found throughout the chapter, summarize and reinforce important points from the text. Many more than in the eleventh edition, students can use these to zero in on the relevance of a section.

Model Organism Boxes

Model Organism Boxes describe key features of a model organism: how it is used to study a particular system, what types of experiments are done with it, or why it serves as a good model for the studies described in the main text.

MODEL ORGANISM — *Mus musculus*

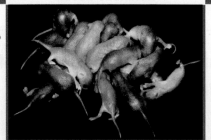

Green-glowing genetically modified mice. The jellyfish gene for green fluorescent protein has been inserted into the chromosomes of the glowing mice. The other mice are normal. [*Eye of Science/Science Source.*]

The laboratory mouse is descended from the house mouse *Mus musculus*. The pure lines used today as standards are derived from mice bred in past centuries by mouse "fanciers." Among model organisms, it is the one whose genome most closely resembles the human genome. Its diploid chromosome number is 40 (compared with 46 in humans), and the genome is slightly smaller than that of humans (the human genome being 3000 Mb) and contains approximately the same number of genes (current estimate 25,000). Furthermore, all mouse genes seem to have counterparts in humans. A large proportion of genes are arranged in blocks in exactly the same positions as those of humans.

Research on the Mendelian genetics of mice began early in the twentieth century. One of the most important early contributions was the elucidation of the genes that control coat color and pattern. Genetic control of the mouse coat has provided a model for all mammals, including cats, dogs, horses, and cattle. A great deal of work was also done on mutations induced by radiation and chemicals. Mouse genetics has been of great significance in medicine. A large proportion of human genetic diseases have mouse counterparts useful for experimental study (they are called "mouse models"). The mouse has played a particularly important role in the development of our current understanding of the genes underlying cancer.

The mouse genome can be modified by the insertion of specific fragments of DNA into a fertilized egg or into somatic cells. The mice in the photograph have received a jellyfish gene for green fluorescent protein (GFP) that makes them glow green under special lights. Gene knockouts and replacements also are possible.

A major limitation of mouse genetics is its cost. Whereas working with a million individuals of *E. coli* or *S. cerevisiae* is a trivial matter, working with a million mice requires a factory-size building. Furthermore, although mice do breed rapidly compared with humans, they cannot compete with microorganisms for speedy life cycle. Hence, the large-scale selections and screens necessary to detect rare genetic events are not possible.

Acknowledgments

We extend our thanks and gratitude to our colleagues who reviewed this edition and whose insights and advice were most helpful:

Rashid Abu-Ghazalah, *McMaster University*

Uduak Afangideh, *Faulkner University*

Faiz Ahmad, *Brandon University*

Shivanthi Anandan, *Drexel University*

Katsura Asano, *Kansas State University*

Rao Ayyagari, *Lindenwood University*

Guy F. Barbato, *Stockton University*

Isabelle H. Barrette-Ng, *University of Calgary*

Bruce Bejcek, *Western Michigan University*

John Belote, *Syracuse University*

Renaud Berlemont, *California State University, Long Beach*

Jaime E. Blair, *Franklin & Marshall College*

Nicole Bournias-Vardiabasis, *California State University, San Bernardino*

Mirjana Milosevic Brockett, *Georgia Institute of Technology*

Judy Brusslan, *California State University, Long Beach*

Patrick Calie, *Eastern Kentucky University*

Jeffrey D. Camper, *Francis Marion University*

John C. Carmen, *Northern Kentucky University*

Steven M. Carr, *Memorial University of Newfoundland*

J. Aaron Cassill, *University of Texas at San Antonio*

Maria V. Cattell, *University of Colorado*

Sarah Certel, *University of Montana*

Henry C. Chang, *Purdue University*

Hexin Chen, *University of South Carolina*

Ian Chin-Sang, *Queens University*

Youngkoo Cho, *Eastern New Mexico University*

Sara G. Cline, *Athens State University*

Craig E. Coleman, *Brigham Young University*

Diane M. Cook, *Louisburg College*

Sarah Adelaide Crawford, *Southern Connecticut State University*

Hongchang Cui, *Florida State University*

Cristina M. Cummings, *Stockton University*

Ann Marie Davison, *Kwantlen Polytechnic University*

Elizabeth A. De Stasio, *Lawrence University*

Matt Dean, *University of Southern California*

Tracie Delgado, *Northwest University*

Alyce DeMarais, *University of Puget Sound*

Teresa Donze-Reiner, *West Chester University*

David Durica, *University of Oklahoma*

Deborah Eastman, *Connecticut College*

Edward Eivers, *California State University, Los Angeles*

Nancy L. Elwess, *State University of New York, Plattsburgh*

Bert Ely, *University of South Carolina*

Yiwen Fang, *Loyola Marymount University*

Robert E. Farrell Jr., *Penn State University*

Steven D. Fenster, *Fort Lewis College*

Victor Fet, *Marshall University*

Christy Fillman, *University of Colorado*

Wayne Forrester, *Indiana University*

Richard D. Gardner, *Southern Virginia University*

Topher Gee, *University of North Carolina, Charlotte*

Vaughn Gehle, *Southwest Minnesota State University*

Matthew Gilg, *University of North Florida*

Michael Gleason, *Georgia College & State University*

Paul Goldstein, *University of Texas at El Paso*

Eli Greenbaum, *University of Texas at El Paso*

Joanna Gress, *Abraham Baldwin Agricultural College*

Chris Griffin, *Ohio University*

Patrick J. Gulick, *Concordia University*

Jody Hall, *Brown University*

Mike Harrington, *University of Alberta*

J. Scott Harrison, *Georgia Southern University*

Elizabeth Hart, *University of Massachusetts, Dartmouth*

George Haughn, *University of British Columbia*

Christopher J. Hickey, *Wilkes Honors College at Florida Atlantic University*

Gregory Hocutt, *Mesa Community College*

Liza Holeski, *Northern Arizona University*

Margaret Hollingsworth, *University at Buffalo*

Adam W. Hrincevich, *Louisiana State University*

Jeffrey A. Hughes, *Millikin University*

Diana S. Ivankovic, *Anderson University*

Varuni Jamburuthugoda, *Fordham University*

Zhenyu Jia, *University of California, Riverside*

Lan Jiang, *Oakland University*

Whitney M. Jones, *North Carolina State University*

Katie Vermillion Kalmon, *University of Wisconsin, Madison*

Kathleen Karrer, *Marquette University*

Christin Mercedes Kastl, *University of Maine at Fort Kent*

Kathrin Schrick, *Kansas State University*

Oliver Kerscher, *The College of William & Mary*

Nobuaki Kikyo, *University of Minnesota*

Miriam K. Konkel, *Clemson University*

Lori Koziol, *New England College*

Brian Kreiser, *University of Southern Mississippi*

Jason N. Kuehner, *Emmanuel College*

Dana Robert Kurpius, *Elgin Community College*

Howard Laten, *Loyola University Chicago*

Jason P. Lee, *Lander University*

John Loike, *Touro College*

Mark S. Longo, *University of Connecticut*

Xu Lu, *University of Findlay*

Bethany Lucas, *Regis University*

Michael Martin, *John Carroll University*

Endre Mathe, *University of Debrecen, Hungary, Vasile Goldis University of Arad, Romania*

P. J. Maughan, *Brigham Young University*

Herman Mays, *University of Cincinnati,*

Andrew McCubbin, *Washington State University*

Virginia McDonough, *Hope College*

Thomas Mennella, *Bay Path University*

Julie E. Minbiole, *Columbia College Chicago*

Ekaterina Mirkin, *Tufts University*

Anni Moore, *Morningside College*

Sarah Mordan-McCombs, *Franklin College of Indiana*

Jeanelle M. Morgan, *University of North Georgia*

Gary Z. Morris, *Glenville State College*

Christopher O'Connor, *Maryville University*

Daniel Odom, *California State University, Northridge*

Greg Odorizzi, *University of Colorado at Boulder*

Maria E. Orive, *University of Kansas*

Pamela Osenkowski, *Loyola University Chicago*

Ana Otero, *Emmanuel College*

Paul Overvoorde, *Macalester College*

Leocadia Paliulis, *Bucknell University*

Holly Paquette, *College of Western Idaho*

Sally G. Pasion, *San Francisco State University*

Thomas R. Peavy, *California State University, Sacramento*

Guy M. L. Perry, *University of Prince Edward Island*

Lynn A. Petrullo, *College of New Rochelle*

Susanne Pfeifer, *Arizona State University, Tempe*

Ruth Phillips, *Syracuse University*

Helen Piontkivska, *Kent State University*

Andres Posso-Terranova, *University of Saskatchewan*

Heather Prior, *The King's University*

Jeffrey L. Reinking, *State University of New York, New Paltz*

Keefe Riedel Reuther, *University of California, San Diego*

Eugenia Ribeiro-Hurley, *Fordham University*

Todd Rimkus, *Marymount University*

Edmund Rucker, *University of Kentucky*

Melanie A. Sacco, *California State University, Fullerton*

Jon Schnorr, *Pacific University*

Aaron Schrey, *Georgia Southern University*

Dana Schroeder, *University of Manitoba*

Sandra Schulze, *Western Washington University*

Bin Shuai, *Wichita State University*

Elaine Sia, *University of Rochester*

Amanda Simons, *Framingham State University*

Elspeth Smith, *University of Guelph*

Marc Spingola, *University of Missouri, St. Louis*

Amy E. Sprowles, *Humboldt State University*

Emily Stowe, *Bucknell University*

Alice Tarun, *Alfred State College*

Michael A. Thomas, *Idaho State University*

Judith M. Thorn, *Knox College*

Douglas Thrower, *University of California, Santa Barbara*

Abe Tucker, *Southern Arkansas University*

Jennifer C. Tudor, *Saint Joseph's University*

L. K. Tuominen, *John Carroll University*

Ludmila Tyler, *University of Massachusetts, Amherst*

Philip Villani, *Butler University*

Darlene Walro, *Walsh University*

Yunqiu Wang, *University of Miami*

Randal Westrick, *Oakland University*

Matt White, *Ohio University*

Daniel Williams, *Coastal Carolina University*

Darla J. Wise, *Concord University*

Donald Withers, *Husson University*

Glenn Yasuda, *Seattle University*

Mary Alice Yund, *University of California, Berkeley Extension*

Xing-Hai Zhang, *Florida Atlantic University, Boca Raton*

Jianmin Zhong, *Humboldt State University*

David S. Zuzga, *La Salle University*

Tony Griffiths would like to acknowledge the pedagogical insights of David Suzuki, who was a co-author of the early editions of this book, and whose teaching in the media is now an inspiration to the general public around the world. Great credit is also due to Jolie Mayer-Smith and Barbara Moon, who introduced Tony to the power of the constructivist approach applied to teaching genetics.

John Doebley would like to thank his University of Wisconsin colleagues Bill Engels, Carter Denniston, and Jim Crow, who shaped his approach to teaching genetics, as well as Jim Birchler, Allen Laughon, and Anna-Lisa Doebley for helpful comments of select chapters.

Katie Peichel would like to thank Jasper Rine for inspiring her love of genetics as an undergraduate, Tom Vogt for teaching her how to actually be a geneticist, and David Kingsley for mentoring throughout her career as a geneticist.

David Wassarman is particularly grateful for the teaching influences of Joe Pelliccia, Tom Wenzel, Joan Steitz, Karen Wassarman, Doug Wassarman, and Kelly Wassarman.

The authors also thank the team at W. H. Freeman for their hard work and patience. In particular we thank our developmental editors, Erica Champion, Erica Frost, and Michael Zierler; program director Sandy Lindelof; senior content project manager Harold Chester; and copy editor Matthew Van Atta. We also thank Paul Rohloff, senior workflow project manager; Natasha Wolfe, design services manager; Matthew McAdams, art manager; Robin Fadool, executive permissions editor; Richard Fox, permissions project manager; Cassandra Korsvik, senior media editor; Jennifer Compton, media editor; and Casey Blanchard, editorial assistant. Finally, we especially appreciate the marketing and sales efforts of Will Moore, executive marketing manager, and the entire sales force.

The Genetics Revolution

DNA (deoxyribonucleic acid) is the molecule that encodes genetic information. The strings of four different chemical bases in DNA store genetic information in much the same way that strings of 0's and 1's store information in computer code. [*Sergey Nivens/Shutterstock.*]

CHAPTER OUTLINE AND LEARNING OBJECTIVES

1.1 THE BIRTH OF GENETICS

LO 1.1 Know the experiments by which genetics developed from Mendel to today.

LO 1.2 Know the molecules involved in storage and expression of genetic information.

1.2 AFTER CRACKING THE CODE

LO 1.3 Know the basic tools for genetic research including model organisms.

1.3 GENETICS TODAY

LO 1.4 Give examples of how genetics has influenced our society.

Genetics is a form of information science. Geneticists seek to understand the rules that govern the transmission of genetic information at three levels—from parent to offspring within families, from DNA to gene action within and between cells, and over many generations within populations of organisms. These three foci of genetics are known as transmission genetics, molecular-developmental genetics, and population-evolutionary genetics. The three parts of this text examine these three foci of genetics.

The science of genetics was born about 120 years ago. Since that time, genetics has profoundly changed our understanding of life, from the level of the individual cell to that of a population of organisms evolving over millions of years. In 1900, William Bateson, a prominent British biologist, wrote presciently that an "exact determination of the laws of heredity will probably work more change in man's outlook on the world, and in his power over nature, than any other advance in natural knowledge that can be foreseen." Throughout this text, you will see the realization of Bateson's prediction. Genetics has driven a revolution in both the biological sciences and society in general.

In this first chapter, we will look back briefly at the history of genetics, and in doing so, we will review some of the basic concepts of genetics that were discovered over the last century. After that, we will look at a few examples of how genetic analysis is being applied to critical problems in biology, agriculture, and human health today. You will see how contemporary research in genetics integrates concepts discovered decades ago with recent technological advances. You will see that genetics today is a dynamic field of investigation in which new discoveries continually advance our understanding of the biological world.

Like begets like

FIGURE 1-1 Family groups in the gray wolf show familial resemblances for coat colors and patterning. [(Top) *DLILLC/Corbis/ VCG/Getty Images;* (bottom) *Bev McConnell/Getty Images.*]

1.1 THE BIRTH OF GENETICS

LO 1.1 Know the experiments by which genetics developed from Mendel to today.

LO 1.2 Know the molecules involved in storage and expression of genetic information.

Throughout recorded history, people around the world have understood that "like begets like." Children resemble their parents, the seed from a tree bearing flavorful fruit will in turn grow into a tree laden with flavorful fruit, and even members of wolf packs show familial resemblances (Figure 1-1). Although people were confident in these observations, they were left to wonder as to the underlying mechanism. The Native American Hopi tribe of the southwestern United States understood that if they planted a red kernel of maize in their fields, it would grow into a plant that also gave red kernels. The same was true for blue, white, or yellow kernels. So they thought of the kernel as a message to the gods in the Earth about the type of maize

the Hopi farmers hoped to harvest. Upon receiving this message, the gods would faithfully return them a plant that produced kernels of the desired color.

In the 1800s in Europe, horticulturalists, animal breeders, and biologists also sought to explain the resemblance between parents and offspring. A commonly held view at that time was the *blending theory* of inheritance, or the belief that inheritance worked like the mixing of fluids such as paints. Red and white paints, when mixed, give pink; and so a child of one tall parent and one short parent could be expected to grow to a middling height. While blending theory works at times, it is also clear that there are exceptions, such as tall children born to parents of average height. Blending theory also provides no mechanism by which the imagined "heredity fluids," once mixed, could be separated—the red and white paints cannot be reconstituted from the pink. Thus, the long-term expectation of blending theory over many generations of intermating among individuals is that all members of the population will come to express the same average value of a trait. Clearly, this is not how nature works. There are people with a range of heights, from short to tall, and we have not all narrowed in on a single average height despite the many generations that humans have dwelled on Earth.

Gregor Mendel—A monk in the garden

While the merits and failings of blending theory were being debated, Gregor Mendel, an Austrian monk, was working to understand the rules that govern the transmission of traits from parent to offspring after hybridization among different varieties of pea plants (**Figure 1-2**). The setting for his work was the monastery garden in the town of Brünn, Austria (Brno, Czech Republic, today). From 1856 to 1863, Mendel cross-pollinated or intermated different varieties of the pea plant. One of his experiments involved crossing a pea variety with purple flowers to one with white flowers (**Figure 1-3**). Mendel recorded that the first hybrid generation of offspring from this cross all had purple flowers, just like one of the parents. There was no blending. Then, Mendel self-pollinated the first-generation hybrid plants and grew a second generation of offspring. Among the progeny, he saw plants with purple flowers as well as plants with white flowers. Of the 929 plants, he recorded 705 with purple flowers and 224 with white flowers (**Figure 1-4**). He observed that there were roughly 3 purple-flowered plants for every 1 white-flowered plant.

How did Mendel explain his results? Clearly, blending theory would not work since that theory predicts a uniform group of first-generation hybrid plants with light purple flowers. So Mendel proposed that the factors that control traits act like *particles* rather than fluids and that these particles do not blend together but are passed intact from one generation to the next. Today, Mendel's particles are known as **genes**.

Mendel proposed that each individual pea plant has two copies of the gene that controls flower color in each of the cells of the plant body (*somatic cells*). However, when the plant forms sex cells, or *gametes* (eggs and sperm), only one copy of the gene enters into these reproductive cells (see Figure 1-3). Then, when egg and sperm unite to start a new individual, once again there will be two copies of the flower color gene in each cell of the plant body.

Mendel had some further insights. He proposed that the gene for flower color comes in two gene variants, or **alleles**—one that conditions purple flowers

FIGURE 1-2 Gregor Mendel was an Austrian monk who discovered the laws of inheritance. [*James King-Holmes/ Science Source.*]

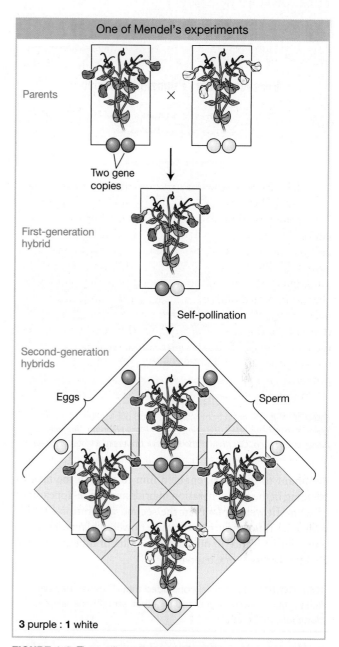

FIGURE 1-3 The mating scheme for Mendel's experiment involving the crossing of purple- and white-flowered varieties of pea plants. The purple and white circles signify the gene variants for purple vs. white flower color. Gametes carry one gene copy; the plants each carry two gene copies. The "×" signifies a cross-pollination between the purple- and white-flowered plants.

ANIMATED ART   **Sapling** Plus

A basic plant cross

and one that conditions white flowers. He proposed that the purple allele of the flower color gene is **dominant** to the white allele such that a plant with one purple allele and one white allele would have purple flowers. Only plants with two white alleles would have white flowers (see Figure 1-3). Mendel's two conclusions, (1) that genes behaved like particles that do not blend together and (2) that one allele is

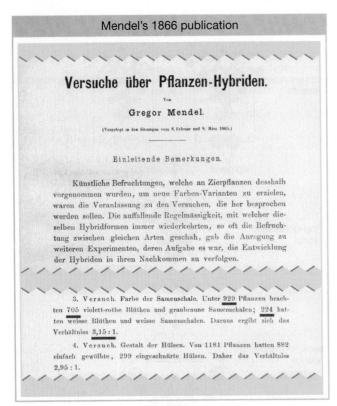

Mendel's 1866 publication

Versuche über Pflanzen-Hybriden.

Von

Gregor Mendel.

(Vorgelegt in den Sitzungen vom 8. Februar und 8. März 1865.)

Einleitende Bemerkungen.

Künstliche Befruchtungen, welche an Zierpflanzen desshalb vorgenommen wurden, um neue Farben-Varianten zu erzielen, waren die Veranlassung zu den Versuchen, die hier besprochen werden sollen. Die auffallende Regelmässigkeit, mit welcher dieselben Hybridformen immer wiederkehrten, so oft die Befruchtung zwischen gleichen Arten geschah, gab die Anregung zu weiteren Experimenten, deren Aufgabe es war, die Entwicklung der Hybriden in ihren Nachkommen zu verfolgen.

3. Versuch. Farbe der Samenschale. Unter 929 Pflanzen brachten 705 violett-rothe Blüthen und graubraune Samenschalen; 224 hatten weisse Blüthen und weisse Samenschalen. Daraus ergibt sich das Verhältniss 3,15 : 1.

4. Versuch. Gestalt der Hülsen. Von 1181 Pflanzen hatten 882 einfach gewölbte, 299 eingeschnürte Hülsen. Daher das Verhältniss 2,95 : 1.

FIGURE 1-4 Excerpts from Mendel's 1866 publication, *Versuche über Pflanzen-Hybriden* (*Experiments on Plant Hybrids*). [*Augustinian Abbey in Old Brno, Courtesy of the Masaryk University, Mendel Museum.*]

dominant to the other, enabled him to explain the lack of blending in the first-generation hybrids and the re-appearance of white-flowered plants in the second-generation hybrids with a 3:1 ratio of purple- to white-flowered plants. This revolutionary advance in our understanding of inheritance will be fully discussed in Chapter 2.

KEY CONCEPT Mendel concluded that (1) genes behave like particles and do not blend together, and (2) one allele is dominant to the other.

How did Mendel get it right when so many others before him were wrong? Mendel chose a good organism and good traits to study. The traits he studied were all controlled by single genes. Traits that are controlled by several genes, as many traits are, would not have allowed him to discover the laws of inheritance so easily. Mendel was also a careful observer, and he kept detailed records of each of his experiments. Finally, Mendel was a creative thinker capable of reasoning well beyond the ideas of his times.

Mendel's particulate theory of inheritance was published in 1866 in the *Proceedings of the Natural History Society of Brünn* (see Figure 1-4). At that time, his work was read by some other biologists, but its implications and importance went unappreciated for almost 40 years. Unlike Charles Darwin, whose theory of evolution by natural selection made him world-renowned virtually overnight, when Mendel died in 1884, he was more or less unknown in the world of science. As biochemist Erwin Chargaff put

it, "There are people who seem to be born in a vanishing cap. Mendel was one of them."

Mendel rediscovered

As the legend goes, when the British biologist William Bateson (**Figure 1-5**) boarded a train bound for a conference in London in 1900, he had no idea how profoundly his world would change during the brief journey. Bateson carried with him a copy of Mendel's 1866 paper on the hybridization of plant varieties. Bateson had recently learned that biologists in Germany, the

William Bateson gave genetics its name

FIGURE 1-5 William Bateson, the British zoologist and evolutionist who introduced the term *genetics* for the study of inheritance and promoted Mendel's work. [*SPL/Science Source.*]

Netherlands, and Austria had each independently reproduced Mendel's 3:1 ratio, and they each cited Mendel's original work. Bateson needed to read Mendel's paper. By the time he stepped off the train, Bateson had a new mission in life. He understood that the mystery of inheritance had been solved. He soon became a relentless apostle of Mendel's laws of inheritance. A few years later in 1905, Bateson coined the term **genetics**—the study of inheritance. The genetics revolution had begun.

When Mendel's laws of inheritance were rediscovered in 1900, a flood of new thinking was unleashed. Mendelism became the organizing principle for much of biology. There were many new questions to be asked about inheritance. **Table 1-1** summarizes the chronology of seminal discoveries made over the coming decades and the chapters of this text that cover each of these topics. Let's look briefly at a few of the questions and their answers that transformed the biological sciences.

Where in the cell are Mendel's genes? The answer came in 1910, when Thomas H. Morgan at Columbia University in New York demonstrated that Mendel's genes are located on chromosomes—he proved the *chromosome theory* of inheritance. The idea was not new. Walter Sutton, who was raised on a farm in Kansas and later served as a surgeon for the U.S. army during WWI had proposed the chromosome theory of inheritance in 1903. Theodor Boveri, a German biologist, independently proposed it at the same time. It was a compelling hypothesis, but there were no experimental data to support it. This changed in 1910, when Morgan proved the chromosome theory of inheritance using the fruit fly as his experimental organism. In Chapter 4, you will retrace Morgan's experiments that proved genes are on chromosomes.

Can Mendelian genes explain the inheritance of continuously variable traits such as human height? While 3:1 segregation ratios could be directly observed for simple

TABLE 1-1	Key Events in the History of Genetics	
Year	**Event**	**Chapters**
1865	Gregor Mendel showed that traits are controlled by discrete factors now known as genes.	2, 3
1903	Walter Sutton and Theodor Boveri hypothesized that chromosomes are the hereditary elements.	4
1905	William Bateson introduced the term *genetics* for the study of inheritance.	2
1908	G. H. Hardy and Wilhelm Weinberg proposed the Hardy–Weinberg law, the foundation for population genetics.	18
1910	Thomas H. Morgan demonstrated that genes are located on chromosomes.	4
1913	Alfred Sturtevant made a genetic linkage map of the *Drosophila X* chromosome, the first genetic map.	4
1918	Ronald Fisher proposed that multiple Mendelian factors can explain continuous variation for traits, founding the field of quantitative genetics.	19
1931	Harriet Creighton and Barbara McClintock showed that crossing over is the cause of recombination.	4, 15
1941	Edward Tatum and George Beadle proposed the one-gene–one-polypeptide hypothesis.	5
1944	Oswald Avery, Colin MacLeod, and Maclyn McCarty provided compelling evidence that DNA is the genetic material in bacterial cells.	7
1946	Joshua Lederberg and Edward Tatum discovered bacterial conjugation.	6
1948	Barbara McClintock discovered mobile elements (transposons) that move from one place to another in the genome.	16
1950	Erwin Chargaff showed DNA composition follows some simple rules for the relative amounts of A, C, G, and T.	7
1952	Alfred Hershey and Martha Chase proved that DNA is the molecule that encodes genetic information.	7
1953	James Watson and Francis Crick, using data produced by Rosalind Franklin and Maurice Wilkins, determined that DNA forms a double helix.	7
1958	Matthew Meselson and Franklin Stahl demonstrated the semiconservative nature of DNA replication.	7
1958	Jérôme Lejeune discovered that Down syndrome resulted from an extra copy of the 21st chromosome.	17
1961	François Jacob and Jacques Monod proposed that enzyme levels in cells are controlled by feedback mechanisms.	11
1961–1967	Marshall Nirenberg, Har Gobind Khorana, Sydney Brenner, and Francis Crick "cracked" the genetic code.	9
1968	Motoo Kimura proposed the neutral theory of molecular evolution.	18, 20
1977	Fred Sanger, Walter Gilbert, and Allan Maxam invented methods for determining the nucleotide sequences of DNA molecules.	10
1980	Christiane Nüsslein-Volhard and Eric F. Wieschaus defined the complex of genes that regulate body plan development in *Drosophila*.	13
1989	Francis Collins and Lap-Chee Tsui discovered the gene causing cystic fibrosis.	4, 10
1995	First genome sequence of a living organism (*Haemophilus influenzae*) published.	14
1998	Andrew Fire and Craig Mello discover a mechanism of gene silencing by double-stranded RNA.	8, 13
1998	First genome sequence of an animal (*Caenorhabditis elegans*) published.	14
2001	The sequence of the human genome is first published.	14
2009	Elizabeth H. Blackburn, Carol W. Greider, and Jack W. Szostak win the Nobel prize for their discovery of how chromosomes are protected by telomeres and the enzyme telomerase.	7
2012	John Gurdon and Shinya Yamanaka win the Nobel Prize for their discovery that just four regulatory genes can convert adult cells into stem cells.	8, 12

Continuous variation for height

| 4:10 | 4:11 | 5:0 | 5:1 | 5:2 | 5:3 | 5:4 | 5:5 | 5:6 | 5:7 | 5:8 | 5:9 | 5:10 | 5:11 | 6:0 | 6:1 | 6:2 |

FIGURE 1-6 Students at the Connecticut Agriculture College in 1914 show a range of heights. Ronald Fisher proposed that continuously variable traits such as human height are controlled by multiple Mendelian genes.

traits such as flower color, many traits show a continuous range of values in second-generation hybrids without simple ratios such as 3:1. In 1918, Ronald Fisher, the British statistician and geneticist, resolved how Mendelian genes explained the inheritance of continuously variable traits such as height in people (**Figure 1-6**). Fisher's core idea was that continuous traits are each controlled by multiple Mendelian genes. Fisher's insight is known as the **multifactorial hypothesis**. In Chapter 19, we will dissect the experimental evidence for Fisher's hypothesis.

KEY CONCEPT The multifactorial hypothesis states that continuously variable traits are each controlled by multiple Mendelian genes.

How do genes function inside cells in a way that enables them to control different states for a trait such as flower color? In 1941, Edward Tatum and George Beadle proposed that genes encode enzymes. Using bread mold (*Neurospora crassa*) as their experimental organism, they demonstrated that genes encode the enzymes that perform metabolic functions within cells (**Figure 1-7**). In the case of the pea plant, there is a gene that encodes an enzyme required to make the purple pigment in the cells of a flower.

Tatum and Beadle's breakthrough became known as the **one-gene–one-enzyme hypothesis**. You will see how they developed this hypothesis in Chapter 5.

What is the physical nature of the gene? Are genes composed of protein, nucleic acid, or some other substance? In 1944, Oswald Avery, Colin MacLeod, and Maclyn McCarty offered the first compelling experimental evidence that genes are made of deoxyribonucleic acid (DNA). They showed that DNA extracted from a virulent strain of bacteria carried the necessary genetic information to transform a nonvirulent strain into a virulent one. Their inference was confirmed in 1952 by Alfred Hersey and Martha Chase. You will learn exactly how they demonstrated this in Chapter 7.

How can DNA molecules store information? In the 1950s, there was something of a race among several groups of scientists to answer this question. In 1953, James Watson and Francis Crick, working at Cambridge University in England, won that race. They determined that the molecular structure of DNA was in the form of a double helix—two strands of DNA wound side-by-side in a spiral. Their structure of the double helix is like a twisted ladder (**Figure 1-8**). The sides of the ladder are made of sugar and phosphate groups. The rungs of the ladder are made of four bases: **adenine (A)**, **thymine (T)**, **guanine (G)**, and

The one-gene–one-enzyme model

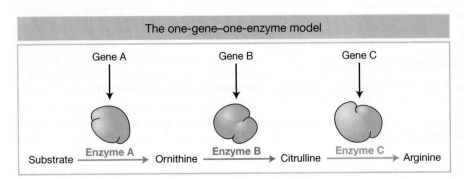

FIGURE 1-7 The one-gene–one-enzyme hypothesis proposed that genes encode enzymes that carry out biochemical functions within cells. Tatum and Beadle proposed this model based on the study of the synthesis of arginine (an amino acid) in the bread mold *Neurospora crassa*.

The structure of DNA

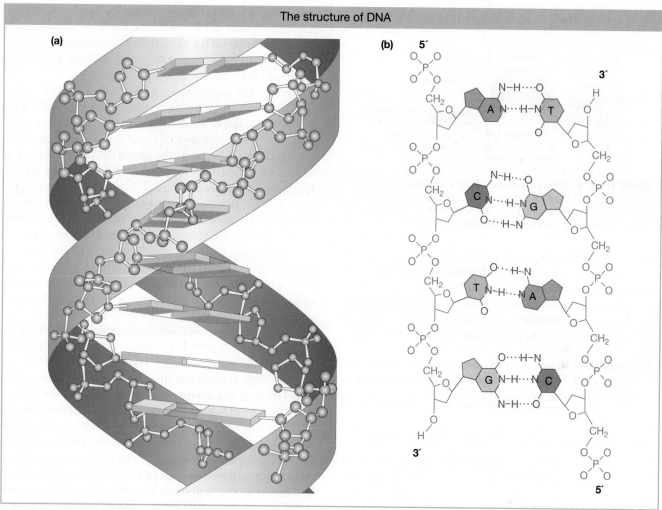

FIGURE 1-8 (a) The double-helical structure of DNA, showing the sugar–phosphate backbone in blue and paired bases in brown. (b) A flattened representation of DNA showing how A always pairs with T, and G always pairs with C. Each row of dots between the bases represents a hydrogen bond.

cytosine (C). The bases face the center, and each base is hydrogen bonded to the base facing it in the opposite strand. Adenine in one strand is always paired with thymine in the other by a *double hydrogen bond*, whereas guanine is always paired with cytosine by a *triple hydrogen bond*. The bonding specificity is based on the **complementary** shapes and charges of the bases. The sequence of A, T, G, and C represents the coded information carried by the DNA molecule. You will learn in Chapter 7 how this was all worked out.

KEY CONCEPT DNA is a double helix in which the nucleotide bases of one strand are paired with those of the other strand. Adenine always pairs with thymine, and guanine always pairs with cytosine.

How are genes regulated? Cells need mechanisms to turn genes on or off in specific cell and tissue types and at specific times during development. In 1961, François Jacob and Jacques Monod made a conceptual breakthrough on this question. Working on the genes necessary to metabolize the sugar lactose in the bacterium *Escherichia coli*, they

demonstrated that genes have **regulatory elements** that control **gene expression**—that is, whether a gene is turned on or off (**Figure 1-9**). The regulatory elements are specific DNA sequences to which a regulatory protein binds and acts as either an activator or repressor of the expression of the gene. In Chapter 11, you will explore the logic behind the experiments of Jacob and Monod with *E. coli*, and in Chapter 12, you will explore the details of gene regulation in eukaryotes.

How is the information stored in DNA decoded to synthesize proteins? While the discovery of the double-helical structure of DNA was a watershed for biology, many details were still unknown. Precisely how information was encoded into DNA and how it was decoded to form the enzymes that Tatum and Beadle had shown to be the workhorses of gene action remained unknown. From 1961 through 1967, teams of geneticists and chemists working in several countries answered these questions when they "cracked the genetic code." What this means is that they deduced how a string of DNA nucleotides, each with one of four different bases (A, T, C, or G), encodes the set of 20 different amino acids

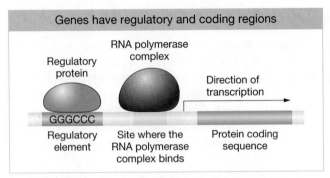

Genes have regulatory and coding regions

FIGURE 1-9 A protein-coding gene includes a regulatory DNA element (GGGCCC) to which a regulatory protein binds, the site where a group of proteins called the RNA polymerase complex binds to initiate transcription, and a protein-coding sequence.

that are the building blocks of proteins. They also discovered that there is a messenger molecule made of ribonucleic acid (RNA) that carries information in the DNA in the nucleus to the cytoplasm where proteins are synthesized. By 1967, the basic flowchart for information transmission in cells was known. This flowchart is called the central dogma of molecular biology.

KEY CONCEPT Genes reside on chromosomes and are made of DNA. Genes encode proteins that conduct the basic enzymatic work within cells.

The central dogma of molecular biology

In 1958, Francis Crick introduced the phrase "central dogma" to represent the flow of genetic information within cells from DNA to RNA to protein, and he drew a simple diagram to summarize these relationships (**Figure 1-10a**).

He curiously used the word *dogma*, "a belief that is to be accepted without doubt," when he intended *hypothesis*, "a testable explanation for an observed phenomenon." Despite this awkward beginning, the phrase had an undeniable power and it has survived.

Figure 1-10b captures much of what was learned about the biochemistry of inheritance from 1905 until 1967. Let's review the wealth of knowledge that this simple figure captures. At the left, you see DNA and a circular arrow representing **DNA replication**, the process by which a copy of the DNA is produced. This process enables each of the two daughter cells that result from cell division to have a complete copy of all the DNA in the parent cell. In Chapter 7, you will explore the details of the structure of DNA and its replication.

Another arrow connects DNA to RNA, symbolizing how the sequence of base pairs in a gene (DNA) is copied to an RNA molecule. The process of RNA synthesis from a DNA template is called **transcription**. One class of RNA molecules made by transcription is **messenger RNA**, or **mRNA** for short. mRNA is the template for protein synthesis. In Chapter 8, you will discover how transcription is accomplished.

The final arrow in Figure 1-10b connects mRNA and protein. This arrow symbolizes protein synthesis, or the **translation** of the information in the specific sequence of bases in the mRNA into the sequence of amino acids that compose a protein. Proteins are the workhorses of cells, comprising enzymes, structural components of the cell, and molecules for cell signaling. The process of translation takes place at the ribosomes in the cytoplasm of each cell. In Chapter 9, you will learn how the genetic code is written in three-letter words called **codons**. A codon is a set of

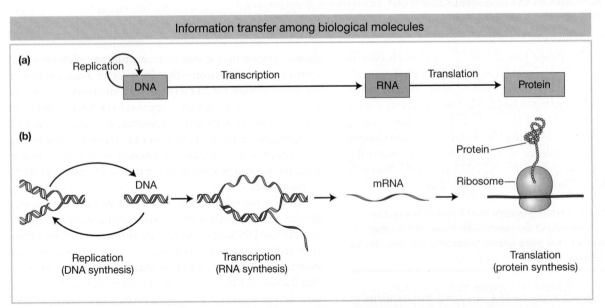

Information transfer among biological molecules

FIGURE 1-10 (a) One version of Francis Crick's sketch of the central dogma, showing information flow between biological molecules. The circular arrow represents DNA replication, the central straight arrow represents the transcription of DNA into RNA, and the right arrow the translation of RNA into protein. (b) More detailed sketch showing how the two strands of the DNA double helix are independently replicated, how the two strands are disassociated for transcription, and how the messenger RNA (mRNA) is translated into protein at the ribosome.

ANIMATED ART SaplingPlus

The central dogma

three consecutive nucleotides in the mRNA that specifies an amino acid in a protein. For example, CGC specifies the amino acid arginine, AGC specifies serine, and so forth.

Since Crick proposed the central dogma, additional pathways of genetic information flow have been discovered. We now know that there are classes of RNA that do not code for proteins, instances in which mRNA is edited after transcription, and cases in which the information in RNA is copied back to DNA (see Chapters 8, 9, and 16).

KEY CONCEPT Genes are made of DNA, which is transcribed to RNA molecules that serve as the template for protein synthesis.

1.2 AFTER CRACKING THE CODE

LO 1.3 Know the basic tools for genetic research including model organisms.

With the basic laws of inheritance largely worked out, the 1970s and beyond witnessed an era of applying genetic analysis to many questions in biology. Much effort has been and continues to be invested in developing tools to address these questions. Geneticists focused their research on a small number of species known as "model organisms" that are well suited for genetic analysis. Then in the late 1990s, the first complete genome sequences were published, launching the genomics era and the ability to study all the genes in the genome simultaneously.

Model organisms

Geneticists make special use of a small set of model organisms for genetic analysis. A **model organism** is a species used in experimental biology with the presumption that what is learned from the analysis of that species will hold true for other species, especially other closely related species. The philosophy underlying the use of model organisms in biology was wryly expressed by Jacques Monod: "Anything found to be true of *E. coli* must also be true of elephants."[1]

As genetics matured and focused on model organisms, Mendel's pea plants fell to the wayside, but Morgan's fruit flies rose to prominence to become one of the most important model organisms for genetic research. New species were added to the list. An inconspicuous little plant that grows as a weed called *Arabidopsis thaliana* became the model plant species, and a minute roundworm called *Caenorhabditis elegans* that lives in compost heaps became a star of genetic analysis in developmental biology (**Figure 1-11**).

KEY CONCEPT Genetic discoveries made in a model organism are often true of related species and may even apply to all forms of life.

[1] F. Jacob and J. Monod, *Cold Spring Harbor Quant. Symp. Biol.* 26, 1963, 393.

What features make a species suitable as a model organism? (1) Small organisms that are easy and inexpensive to maintain are very convenient for research. So fruit flies are good, blue whales not so good. (2) A short generation time is imperative because geneticists, like Mendel, need to cross different strains and then study their first- and second-generation hybrids. The shorter the generation time, the sooner the experiments can be completed. (3) A small genome is useful. As you will learn in Chapter 16, some species have large genomes and others small genomes in terms of the total number of DNA base pairs. Much of the extra size of large genome species is composed of repetitive DNA elements between the genes. If a geneticist is looking for genes, these can be more easily found in organisms with smaller genomes and fewer repetitive elements. (4) Organisms that are easy to cross or mate and that produce large numbers of offspring are best.

As you read this textbook, you will encounter certain organisms over and over. Organisms such as *Escherichia coli* (a bacterium), *Saccharomyces cerevisiae* (baker's yeast), *Caenorhabditis elegans* (nematode or roundworm), *Drosophila melanogaster* (fruit fly), and *Mus musculus* (mice) have been used repeatedly in experiments and revealed much of what we know about how inheritance works. Model organisms can be found on diverse branches of the tree of life (see Figure 1-11), representing bacteria, fungi, algae, plants, and invertebrate and vertebrate animals. This diversity enables each geneticist to use a model best suited to a particular question. Each model organism has a community of scientists working on it who share information and resources, thereby facilitating each other's research. More information on each of the most commonly used model organisms can be found in "A Brief Guide to Model Organisms" at the end of this book.

Mendel's experiments were possible because he had several different varieties of pea plants, each of which carried a different genetic variant for traits such as purple versus white flowers, or tall versus dwarf stems. For each of the model species, geneticists have assembled large numbers of varieties (also called strains or stocks) with special genetic characters that make them useful in research. For example, there are strains of fruit flies that have trait variants such as red versus white eyes. Similarly, there are strains of mice that are prone to develop specific forms of cancer or other diseases such as diabetes. Genetic strains enable geneticists to study how genes influence physiology, development, and disease. The different strains of each model organism are available to researchers through stock centers that maintain and distribute the strains.

KEY CONCEPT Model organisms have features that make them well-suited for genetic studies, such as small size, small genome, large numbers of offspring, and short generation time. Geneticists working with the same model organism share stocks and information with one another.

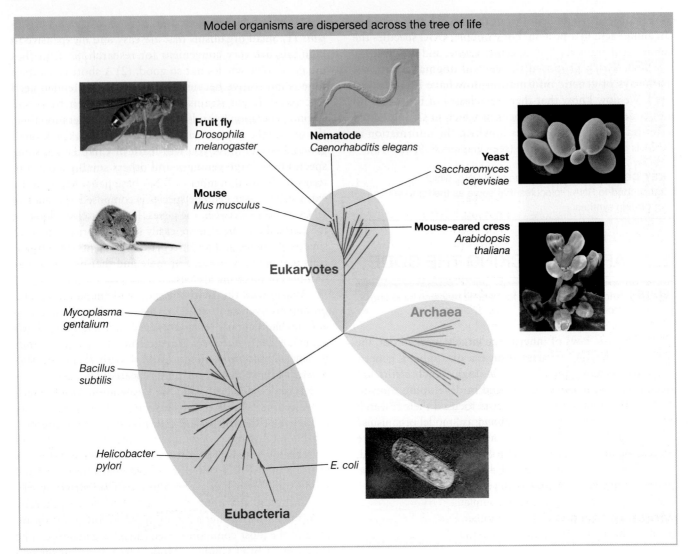

Model organisms are dispersed across the tree of life

Fruit fly
Drosophila melanogaster

Nematode
Caenorhabditis elegans

Yeast
Saccharomyces cerevisiae

Mouse
Mus musculus

Mouse-eared cress
Arabidopsis thaliana

Eukaryotes

Mycoplasma gentalium

Archaea

Bacillus subtilis

Helicobacter pylori

E. coli

Eubacteria

FIGURE 1-11 The tree shows evolutionary relationships among the major groups of organisms: Bacteria, Archaea, and Eukaryota (plants, fungi, and animals). [(*Clockwise, from top, center*) Sinclair Stammers/Science Source; SCIMAT/Science Source; Darwin Dale/Science Source; Biophoto Associates/ Science Source; imageBROKER/Superstock; blickwinkel/Alamy.]

Tools for genetic analysis

Geneticists and biochemists have created an incredible array of tools for characterizing and manipulating DNA, RNA, and proteins. Many of these tools are described in Chapter 10 or in other chapters relevant to a specific tool. There are a few themes to mention here.

First, geneticists have harnessed the cell's own enzymatic machinery for copying, pasting, cutting, and transcribing DNA, enabling researchers to perform these reactions inside test tubes. The enzymes that perform each of these functions in living cells have been purified and are available to researchers: **DNA polymerases** can make a copy of a single DNA strand by synthesizing a matching strand with the complementary sequence of A's, C's, G's, and T's. **Nucleases** can cut DNA molecules in specific locations or degrade an entire DNA molecule into single nucleotides. **Ligases** can join two DNA molecules together end-to-end. Using DNA polymerase

or other enzymes, DNA can also be "labeled" or "tagged" with a fluorescent dye or radioactive element so that the DNA can be detected using a fluorescence or radiation detector.

Second, geneticists have developed methods to *clone DNA* molecules. Here, cloning refers to making many copies (*clones*) of a DNA molecule. The common way of doing this involves isolating a relatively small DNA molecule (up to a few thousand base pairs in length) from an organism of interest. The DNA molecule might be an entire gene or a portion of a gene. The molecule is inserted into a host organism (often *E. coli*) where it is replicated many times by the host's DNA polymerase. Having many copies of a gene is important for a vast array of experiments used to characterize and manipulate it.

Third, geneticists have developed methods to insert foreign DNA molecules into the genomes of many species, including those of all the model organisms (**Figure 1-12**). This process is called **transformation**, and it is possible, for instance, to

Genetically modified
tobacco

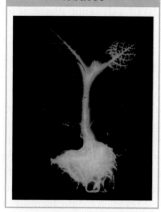

FIGURE 1-12 This genetically modified tobacco plant has a gene from the firefly inserted into its genome, giving it the capability to emit light. [*Republished with permission of the American Association for the Advancement of Science, from D.W. Ow et al., "Transient and Stable Expression of the Firefly Luciferase Gene in Plant Cells and Transgenic Plants" Science 234, 4778: (1986) pp. 856–859, Figure 5. Permission conveyed through Copyright Clearance Center, Inc.*]

transform genes from one species into the genome of another. The recipient species then becomes a **genetically modified organism (GMO)**. In the last few years, geneticists have developed an exciting new method called CRISPR/Cas9 that facilitates editing the genes of an organism and is expected to revolutionize not just laboratory genetics, but also medicine and agriculture (see Chapter 10).

Fourth, geneticists have developed a large set of methods based on hybridizing DNA molecules to one another (or to RNA molecules). The two complementary strands of DNA in the double helix are bound together by hydrogen bonds, either $G \equiv C$ or $A = T$. These bonds can be broken by heat (denatured) in an aqueous solution to give two single-stranded DNA molecules (Figure 1-13a). When the solution is cooled under controlled conditions, DNA molecules with complementary strands will preferentially hybridize with one another. DNA hybridization methods have enabled many discoveries. For example, the cloned DNA of a gene can be tagged with a fluorescent dye and then hybridized to chromosomes fixed on a microscope slide, revealing the chromosome on which the gene is located (**Figure 1-13b**).

Fifth, geneticists and biochemists have developed multiple methods for determining the exact sequence of all the A's, C's, G's, and T's in a DNA molecule. These methods are collectively called **DNA sequencing**, and they have allowed geneticists to read the language of life. Recently, cost-effective, high-through-put methods to sequence both very short (100 bp) and very long (10,000 bp) DNA molecules were developed, enabling sequencing of the complete genomes of thousands of individuals of a single species such as humans (see Chapter 14).

Finally, over the last 20 years, researchers have created molecular and computational tools for analyzing the entire genome of an organism. These efforts gave birth to the field of **genomics**—the study of the structure and function of entire genomes (see Chapter 14). Geneticists and genomicists have assembled mind-boggling amounts of information on model organisms and their genomes, including the complete DNA sequence of their genomes, lists of all their genes, catalogs of variants in these genes, data on the cell and tissue types in which each gene is expressed, and much more. To get an idea of what is available, try browsing Fly Base (http://flybase.org/), the genomics Web site for the fruit fly (see also Appendix B).

KEY CONCEPT Geneticists developed tools to replicate, cut, label, and degrade DNA as well as use it as a template to be transcribed into RNA. These tools allow the assembly of the DNA sequence of whole genomes. Computational tools allow biological questions to be answered by the analysis of genome sequences and associated information.

Strands of nucleic acids hybridize to complementary sequences

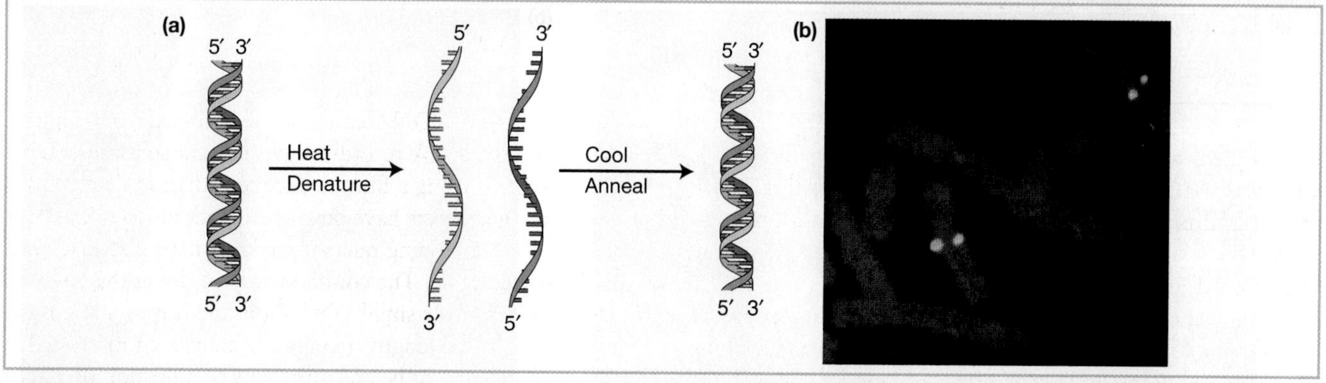

(a)

Heat
Denature

Cool
Anneal

(b)

FIGURE 1-13 (a) The two strands of the DNA double helix can be dissociated by heat in aqueous solutions. Upon cooling under controlled conditions, strands reassociate, or *hybridize,* with their complement. (b) A cloned copy of the human gene for muscle glycogen phosphorylase was tagged with a yellow fluorescent dye. The fluorescent-tagged DNA was then denatured and allowed to hybridize to the chromosomes in a single cell. The fluorescent-tagged clone hybridized to the location on chromosome 11 (yellow fluorescent regions) where the gene is located. [*(b) Republished with permission of the American Association for the Advancement of Science, from P Lichter, CJ Tang, K Call, G Hermanson, GA Evans, D Housman, DC Ward, "High-resolution mapping of human chromosome 11 by in situ hybridization with cosmid clones" Science 05 Jan 1990: Vol. 247, Issue 4938, pp. 64–69, Figure 1B. Permission conveyed through Copyright Clearance Center, Inc.*]

1.3   GENETICS TODAY

LO 1.4 Give examples of how genetics has influenced our society.

In an interview in 2008, geneticist Leonid Kruglyak remarked,

> "You have this clear, tangible phenomenon in which children resemble their parents. Despite what students get told in elementary-school science, we just don't know how that works."
>
> B. Maher, *Nature* 456:18, 6 Nov 2008.

Although Kruglyak's remark might seem disparaging to the progress made in the understanding of inheritance over the last 100 years, this was certainly not his intention. Rather, his remark highlights that despite the paradigm-shifting discoveries of the nineteenth and twentieth centuries, enigmas abound in genetics and the need for new thinking and new technologies remains. Mendel, Morgan, McClintock, Watson, Crick, and many others (see Table 1-1) delimited the foundation of the laws of inheritance, but most of the details that rest atop that foundation remain obscure. The six feet of DNA in the single cell of a human zygote encodes the information needed to transform that cell into an adult, but exactly how this works is not understood.

In this section, we will review some recent advances in genetics—discoveries of enough general interest that they were featured in the popular press. Reading about these discoveries will both reveal the power of genetics to answer critical questions about life and highlight how this knowledge can be applied to addressing problems in society. This textbook and the course of study in which you are engaged should convey a dual message—the science of genetics has profoundly changed our understanding of life, but it is also a youthful field in the midst of a dynamic phase of its development.

From classical genetics to medical genomics

Meet patient VI-1 (**Figure 1-14a**). Her name is Louise Benge, and as a young woman, she developed a crippling illness. Starting in her early 20s, she began to experience excruciating pain in her legs after walking as little as a city block. At first, she ignored the pain, then spoke with her primary care physician, and later visited specialists. She was given a battery of tests and X rays, and these revealed the problem—her arteries from her aorta on down to her legs were calcified, clogged with calcium phosphate deposits (**Figure 1-14b**). It was a disease for which her doctors had no name and no therapy. She had a disease, but not a diagnosis. There was only one thing left to do; her primary care physician referred Benge to the Undiagnosed Diseases Program (UDP) at the National Institutes of Health in Bethesda, Maryland.

The UDP is a group of MDs and scientists that has connections with specialists throughout the National Institutes of Health. This is the team that is asked to tackle the most challenging cases. Working with Benge, the UDP team subjected her to a vast array of tests, and soon they found the underlying defect that caused her disease. Benge had a very low level of an enzyme called CD73. This enzyme is involved in signaling between cells, and specifically it sends a signal that blocks calcification. Now the UDP doctors could give Benge a diagnosis. They named her disease "arterial calcification due to deficiency of CD73," or ACDC.

Louise Benge has an undiagnosed disease

(a)   (b)

FIGURE 1-14 (a) Louise Benge developed an undiagnosed disease as a young woman. (b) An X ray revealed that Louise Benge's disease condition caused calcification of the arteries in her legs. [(a) Jeannine Mjoseth, NHGRI/www.genome.gov; (b) National Human Genome Research Institute (NHGRI).]

What intrigued the UDP team about Benge's case was that she was not alone in having this disease. Benge had two brothers and two sisters, and all of them had arterial calcification. Remarkably, however, Benge's parents were unaffected. Moreover, Benge and her siblings all had children, and none of these children had arterial calcification. This pattern of inheritance suggested that the underlying cause might be genetic. Specifically, it suggested that Benge and all of her siblings inherited two defective copies of either CD73 or a gene that influences CD73 expression—one from their mother and one from their father. A person with one good copy and one defective copy can be normal, but if both of a person's copies are defective, then they lack the function that the gene provides. The situation is just like Mendel's white-flowered pea plants. Since the functional allele is dominant to the dysfunctional allele, ACDC, like white flowers, appears only if an individual carries two defective alleles.

The UDP team delved further into Benge's family history and learned that Benge's parents were third cousins (**Figure 1-15**). This revelation fit well with the idea that the cause was a defective gene. When a husband and wife are close relatives such as third cousins, there is an increased chance that they will both have inherited the same version of a defective gene from their common ancestor and that they will both pass on this defective gene to their children. Children with one copy of a defective gene are often normal, but a child who inherits a defective copy from both parents is likely to have a genetic disorder.

In Figure 1-15, we can see how this works. Benge's mother and father (individuals V-1 and V-2 in the figure) have the same great-great-grandparents (I-1 and I-2). If one of these great-great-grandparents had a mutant gene for CD73, then it could have been passed down over the generations to both Benge's mother and father (follow the red arrows). After that, if Benge received the mutant copy from both her mother and her father, then both of her

copies would be defective. Each of Benge's siblings would also need to have inherited two mutant copies from their parents to explain the fact that they have ACDC. In Chapter 2, you will learn how to calculate the probability of this actually happening.

With this hint from the family history, the UDP team now knew where to look in the genome for the mutant gene. They needed to look for a segment on one of the chromosomes for which the copy that Benge inherited from her mother is identical to the copy she inherited from her father. Moreover, each of Benge's siblings must also have two copies of this segment identical to Benge's. Such regions are very rare in people unless their parents are related, as in the case of Benge since her parents are third cousins. Generally, a segment of a chromosome that is just a few hundred base pairs long will have several differences in the sequence of A's, C's, G's, and T's between the copy we inherited from our mother and the one we inherited from our father. These differences are known as **single nucleotide polymorphisms**, or **SNPs** for short (see **Box 1-1**).

The UDP team used a genomic technology, called a DNA microarray (see Chapter 18), that allowed them to study one million base-pair positions across the genome. At each of these base-pair positions along the chromosomes, the team could see where Benge's two chromosomal segments were identical, and whether all of Benge's siblings also carried two identical copies in this segment. The UDP team found exactly the type of chromosome segment for which they were looking, and furthermore, they discovered that the gene that encodes the CD73 enzyme is located in this segment. This result suggested that Benge and her siblings all had two identical copies of the same defective CD73-encoding gene. The team seemed to have found the needle in a haystack; however, there was one last experiment to perform.

The team needed to identify the specific defect in the defective CD73 gene that Benge and her siblings had

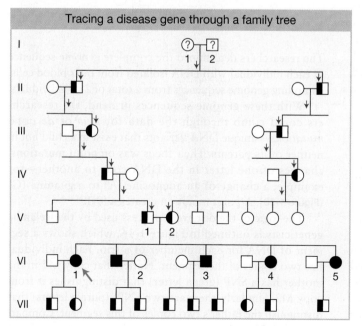

FIGURE 1-15 Family tree or pedigree showing the inheritance of the mutant gene causing arterial calcification due to deficiency of CD73 (ACDC). Squares are males, and circles are females. Horizontal lines connecting a male and female are matings. Vertical lines connect a mating pair to its offspring. Roman numerals designate generations; Arabic numerals designate individuals within generations. Half-filled squares or circles indicate an individual carrying one copy of the mutant gene. Filled squares or circles indicate an individual with two copies of the mutant gene and who have the ACDC disease. Either individual I-1 or individual I-2 must have carried the mutant gene, but which one carried it is uncertain as indicated by the question marks. The blue arrow indicates Louise Benge. The red arrows show the path of the mutant gene through the generations. [Data from C. St. Hilaire et al., New England Journal of Medicine 364, 2011, 432–442.]

BOX 1-1   Single Nucleotide Polymorphisms

Genetic variation is any difference between two copies of the same gene or DNA molecule. The simplest form of genetic variation one might observe at a single nucleotide site is a difference in the nucleotide base present, whether adenine, cytosine, guanine, or thymine. These types of variants are called single nucleotide polymorphisms (SNPs), and they are the most common type of variation in most, if not all, organisms. The figure shows two copies of a DNA molecule from the same region of a chromosome. Notice that the bases are the same in the two molecules except where one molecule has a CG pair and the other a TA pair. If we read strand 1 of the two molecules, then the top molecule has a "G" and the lower molecule an "A" at the SNP site.

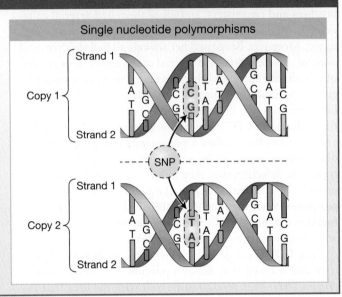

Single nucleotide polymorphisms

inherited. After determining the DNA sequence for the CD73 gene from Benge and her siblings, the team found the defect in the gene—"the smoking gun." The defective gene encoded only a short, or truncated, protein—it did not encode the complete sequence of amino acids. One of the DNA codons with letters TCG that encodes the amino acid serine was mutated to TAG, which signals the end of the protein. The protein made from Benge's version of the CD73 gene was truncated, so it could not signal cells in the arteries to keep the calcification pathway turned off.

Louise Benge's journey from first experiencing pain in her legs to learning that she had a new disease called ACDC was a long one. The diagnosis of her disease was a triumph made possible by the integration of classic transmission genetics and genomics. Knowing the defect underlying the disease ACDC allowed the doctors to try a medication that they would never have considered before they knew that the cause was a defective CD73 enzyme. The medication in question is called etidronate, and it can substitute for CD73 in signaling cells to keep the calcification pathway turned off. Clinical trials with etidronate began in 2012 and are scheduled for completion in 2020.

KEY CONCEPT The integration of classical genetics and genomic technologies allows the causes of inherited diseases to be readily identified and appropriate therapies applied.

Investigating mutation and disease risk

Shortly after the rediscovery of Mendel's work, the German physician Wilhelm Weinberg reported that there appeared to be a higher incidence of short-limbed dwarfism (achondroplasia) among children born last in German families than among those born first. A few decades later, British geneticist J. B. S. Haldane observed another unusual pattern

of inheritance. The genealogies of some British families suggested that new mutations for the blood-clotting disorder hemophilia tended to arise in men more frequently than in women. Taken together, these two observations suggested that the risk of an inherited disorder for a child is greater as the parents age and also that fathers are more likely than mothers to contribute new mutations to their children.

Advances in genomics and DNA sequencing technology (see Chapter 14) allowed new analyses proving that Weinberg's and Haldane's suspicions were correct and providing a very detailed picture of the origin of new mutations within families. Here is how it was accomplished. A team of geneticists in Iceland studied 78 "trios"—a family group of a mother, a father, and their child.

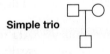

Simple trio

The researchers determined the complete genome sequence of each individual with DNA isolated from their blood cells, compiling genome sequences from a total of 219 individuals.

With these genome sequences in hand, the researchers could comb through the data for *new* or *de novo* *mutations*—unique DNA variants that exist in a child but in neither of its parents. Their focus was on **point mutations**, changes of one letter in the DNA code to another—for example, a change of an adenosine (A) to a guanine (G) (**Figure 1-16**). A point mutation creates a SNP.

The logic of the discovery process used by the Icelandic geneticists is outlined in Figure 1-16, which shows a segment of DNA for each member of a trio. Each individual has two copies of the segment. Notice that copy M1 in the mother has a SNP (green letter) that distinguishes it from copy M2. Similarly, there are two SNPs (purple letters) that distinguish the father's two copies of this segment. Comparing the child to the parents, we see that the child inherited

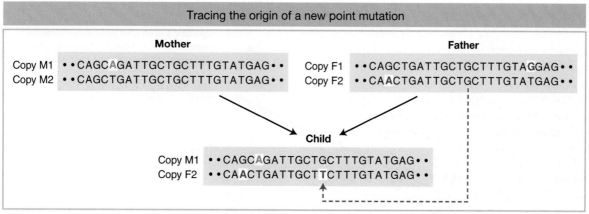

FIGURE 1-16 A short segment of DNA from a specific location in the genome is depicted using the nucleotide base letters of just one strand of the DNA duplex. Each individual has two copies of the DNA segment. In the mother, these are labeled M1 and M2; in the father, F1 and F2. The child inherited copy M1 from its mother and F2 from its father. The version of F2 in the child carries a new point mutation (red, arrow). Single nucleotide polymorphisms (SNPs) that distinguish the different copies are shown in green (mother) and purple (father).

copy M1 from its mother and copy F2 from its father. Look more closely at the child's two copies of the segment, and you will notice a unique variant (red letter) that occurs in the child but in neither of its parents. This is a *de novo* point mutation. It this case, it is a mutation from a guanine (G) to a thymine (T). We can see that the mutation arose in the father because it is on the F2 copy of the segment.

Where and exactly when did the new mutation depicted in Figure 1-16 arise? Most of our bodies are composed of somatic cells that make up everything from our brain to our blood. However, we also have a special lineage of cells called the germline that divide to produce eggs in women and sperm in men. New mutations that arise in somatic cells as they divide during the growth and development of our bodies are not passed on to our offspring. However, a new mutation that occurs in the germline can be transmitted to the offspring. The mutation depicted in Figure 1-16 arose in the germline of the father.

With the genome sequence data for the trios, the Icelandic geneticists made some startling discoveries. First, among the 78 children in the study, they observed a total of 4933 new point mutations. Each child carried about 63 unique mutations that did not exist in its parents. Most of these occurred in parts of the genome where they have only a small chance to pose a health risk, but 62 of the 4933 mutations caused potentially damaging changes to the genes such that they altered the amino acid sequence of the protein encoded. Second, among the mutations that could be assigned a parent of origin, there were on average 55 from the father for every 14 from the mother. The children were inheriting nearly four times as many new mutations from their fathers as their mothers. The Icelandic team had confirmed Haldane's prediction made 90 years earlier.

The genome sequences also allowed the team to test Weinberg's prediction that the frequency of mutation rises with the age of the parents. For each trio, the researchers knew the ages of the mother and the father at the time of conception. When they investigated whether the frequency

of mutation rises with the mother's age when controlling for the age of the father, the team found no evidence that it did. Older mothers did not pass on more new point mutations to their offspring than younger ones. (Older mothers are known to produce more chromosomal aberrations than younger mothers, such as an extra copy of the 21st chromosome that causes Down syndrome; see Chapter 17.) Next, they examined the relationship between mutation and the age of the father when controlling for the age of the mother. Here, they found a powerful relationship. Older fathers produce more new point mutations than young ones (**Figure 1-17**). In fact, for each year of increase in his age, a father will pass on two additional new mutations to his children. A 20-year-old father will pass on about 25 new mutations to each of his children, but a 40-year-old father will pass on about 65 new mutations. Weinberg's observation made 100 years earlier was confirmed.

Why does the age of the father matter, while that of the mother seems to have no effect on the frequency of new point mutations? The answer lies in the different ways by which men and women form gametes. In women, as in the females of other mammals, the process of making eggs takes place largely before a woman is born. Thus, when a woman is born, she possesses in her ovaries a set of egg precursor cells that will mature into egg cells without further rounds of DNA replication. For a woman, from the point when she was conceived until the formation of the egg cells in her ovaries, there are about 23 rounds of cell division with DNA replication and an opportunity for a copying error or mutation. All 23 of these rounds of chromosome replication occur before a woman is born, so there are no additional rounds after her birth and no chance for additional mutations as she ages. Thus, older mothers contribute no more new point mutations to their children than younger mothers.

Sperm production is altogether different. The cell divisions that produce sperm continue throughout a man's life, and there are many more rounds of cell division in sperm formation than in egg formation. Sperm produced

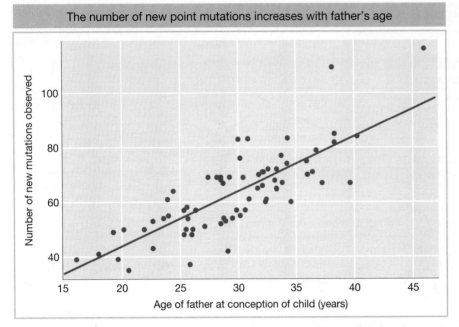

The number of new point mutations increases with father's age

FIGURE 1-17 Plot of the number of new point mutations in each child (y-axis) by the age of the child's father (x-axis). Each dot represents one of the 78 children studied. The diagonal line indicates the rate of increase in new mutations with the father's age. [*Data from A. Kong et al.*, Nature *488, 2012, 471–475.*]

by 20-year-old men will have replicated their DNA about 150 times since the man's conception, compared to only 23 DNA replications for the eggs produced by 20-year-old women. By the time a man is age 40, his sperm will have a history that involves over 25 times as many rounds of DNA replication as for eggs in a woman of the same age. Thus, there is much greater risk of new point mutations occurring during these extra rounds of cell division and DNA replication with the increase in the age of the father.

There is one final twist to the remarkable project performed by the Icelandic geneticists. The 78 trios that they studied were chosen because the children in most of the trios had inherited disorders. These included 44 children with autism spectrum disorder and 21 with schizophrenia. For all these children, there were no other cases of these disorders among their relatives, suggesting that their condition was due to a new mutation. As anticipated, the researchers observed a correlation between the father's age and disease risk—older fathers were more likely to have children with autism and schizophrenia.

Studies such as this can have important implications for individuals and society. Some men who intend to delay parenting until later in life might choose to freeze samples of their sperm while still young. This study also informs us that changes in society can impact the number of new mutations that enter the human gene pool. If men choose to delay fatherhood for postsecondary education or establishing their careers, there will be an associated increase in the number of new mutations among their children. It is common knowledge that infertility rises with age for women—as is often stated, a woman's "biological clock" is ticking once she is past puberty. This work by the Icelandic geneticists informs us that a clock is ticking for men as well.

KEY CONCEPT Mutation is a random process that occurs during DNA replication.

When rice gets its feet a little too wet

Among the cereal crops, rice is unique. Whereas wheat, barley, maize, and the other grain crops grow solely in dry fields, rice is commonly grown in flooded fields called paddies (**Figure 1-18**). The ability of rice to grow in flooded fields offers it an advantage: rice can survive modest flooding (up to 25 cm of standing water) in the paddies, but most weeds cannot. So rice farmers can use flooding to control the weeds in their field while their rice thrives.

The strategy works well where farmers have irrigation systems to control the water levels in their paddies and heavy rains do not exceed their capacity to control these levels. If the water in the paddies gets too deep (greater than 50 cm) for a prolonged period, then the rice plants, like the weeds, can suffer or even die.

Paddy agriculture, as practiced in the lowlands of India, Southeast Asia, and West Africa, relies on natural rainfall, rather than irrigation, to flood the fields. This circumstance poses a risk. When the rains are heavy, water depth in the paddies can exceed 50 cm and completely submerge the plants, causing rice plants to either suffer a loss in yield or simply die. Of the 60 million hectares of rain-fed lowland paddies, one-third experience damaging floods that reduce yield on a regular basis. Since this loss is incurred mostly by the poorest farmers, it can lead to malnourishment and even starvation.

In the early 1990s, David Mackill, a plant geneticist and breeder at the International Rice Research Institute, had an idea about how to improve rice so that it could tolerate being submerged in flood waters. He identified a remarkable variety of rice called FR13A that could survive submergence and even thrive after the plants remained fully submerged in deep water for up to two weeks. Unfortunately, FR13A had a low yield and the quality of its grain was marginal. So Mackill set out to transfer FR13A's

Rice growing in a flooded field or paddy

FIGURE 1-18 Rice is grown in fields with standing water called paddies. Rice is adapted to tolerate modest levels of standing water, but the water suppresses the growth of weeds that could compete with the rice. [*Debasish Banerjee/Dinodia Photo/AGE Fotostock.*]

genetic factor(s) for submergence tolerance into a rice variety with a higher yield and higher grain quality. He first crossed FR13A and a superior variety of rice and then for several generations crossed the hybrid plants back to the superior variety until he had created an improved form of rice that combined submergence tolerance and high yield.

Mackill had achieved his initial goal of transferring submergence tolerance into a superior variety, but the genetic basis for why FR13A was submergence tolerant remained obscure. Was FR13A's submergence tolerance controlled by many genes on multiple chromosomes, or might it be mostly controlled by just one gene? To delve into the genetic basis of submergence tolerance, Mackill and his team conducted a form of genetic analysis called **quantitative trait locus (QTL) mapping** (see Chapter 19). A QTL is a genetic locus that contributes incrementally or quantitatively to variation for a trait. Unlike in Mendel's experiments, where one locus controlled one trait, a QTL is just one of multiple loci that all affect the same trait. Using QTL mapping, Mackill learned that the submergence tolerance trait of FR13A was controlled by several QTL, but one of these had a particularly large effect. He named this large-effect QTL *SUB1* for "submergence tolerant."

To understand molecular nature of *SUB1*, molecular geneticists Pamela Ronald at the University of California, Davis, and Julia Bailey-Serres at the University of California, Riverside, joined Mackill's team. This expanded team determined that *SUB1* is a member of a class of genes called *ethylene response factors (ERFs)*. *ERF* genes encode regulatory proteins that bind to regulatory elements in other genes and thereby regulate their expression. Thus, *SUB1* is a gene that regulates the expression of other genes. Moreover, they determined that the allele of *SUB1* in FR13A is switched on in response to submergence, while the allele of *SUB1* found in submergence-sensitive varieties is not switched on by submergence.

The next question was, how does switching on *SUB1* enable FR13A to survive complete submergence? To answer this question, let's review how ordinary rice plants respond to submergence. When a plant is completely submerged, oxygen levels in its cells drop, and the concentration of ethylene, a plant hormone, in the cells increases. Ethylene signals the plant to escape submergence by elongating its leaves and stems to keep its "head" above water. This *escape strategy* works fine as long as the water is not too deep. If the flood waters are too deep, then the plant cannot grow enough to escape. As a plant in such deeply flooded circumstances grows, it uses up all its energy reserves (carbohydrates), becomes spindly and weak, and eventually dies.

How does the FR13A variety manage to survive submergence while many other types of rice cannot? FR13A has a different strategy that could be called *sit tight*, and *SUB1* acts as the master switch or regulatory gene to activate this strategy. When the flood waters rise and the concentration of ethylene increases, the ethylene turns on *SUB1*, an ethylene response factor. The ERF protein that *SUB1* encodes orchestrates the plant's response by switching on (or off) a battery of genes involved in plant growth and metabolism. In FR13A plants that become submerged, genes involved in stem and leaf elongation as part of the escape strategy are switched off, as are genes involved in mobilizing the energy reserves (carbohydrates) needed to fuel the escape strategy. As a result, the plant prevents itself from burning up all its reserve carbohydrates and becoming weak and spindly. Using the tools of molecular genetics and genomics such as DNA microarrays (see Chapters 10 and 14), the rice team was able to decipher the extensive catalog of genes controlling organ elongation, carbon metabolism, flowering, and photosynthesis that are regulated by *SUB1* to achieve the sit-tight response.

With the basic genetics of *SUB1* elucidated, the rice team could transfer it into a superior variety with surgical precision. This precision is important because it enabled the team to avoid transferring other undesirable genes at the same time. For this project, they worked with a submergence-intolerant, but superior, Indian variety called *Swarna*, which is widely grown and favored by farmers. The new line they created, called *Swarna-Sub1*, has lived up to expectations. Field trials showed a striking difference in plant survival and yield between *Swarna* and *Swarna-Sub1* when there is complete submergence (**Figure 1-19**). As shown in **Figure 1-20**, *Swarna-Sub1* provides higher yield than the original *Swarna* under all different levels of flooding. In various trials, the *SUB1* improved yield between 1 and 3 tons of grain per hectare.

With the support of international research organizations, governmental agencies, and philanthropies, *Swarna-Sub1* and other superior varieties carrying the *SUB1* allele from FR13A have now been distributed to farmers. By 2017, an estimated 10 million farmers were growing *SUB1*-enhanced rice. Although precise data on how this has reduced losses due to flooding are not available, the rapid adoption of *SUB1*-enhanced rice by farmers suggests it is

Flood-intolerant and flood-tolerant rice

FIGURE 1-19 An Indian farmer with rice variety *Swarna* that is not tolerant to flooding (*left*) compared to variety *Swarna-Sub1* that is tolerant (*right*). This field was flooded for 10 days. The photo was taken 27 days after the flood waters receded. [*Republished with permission of Elsevier, from Ismail, Abdelbagi M. et al., "The contribution of submergence-tolerant (Sub1) rice varieties to food security in flood-prone rainfed lowland areas in Asia" Field Crops Research, 2013, October; 152, 83–93, Figure 1. Permission conveyed through Copyright Clearance Center, Inc.*]

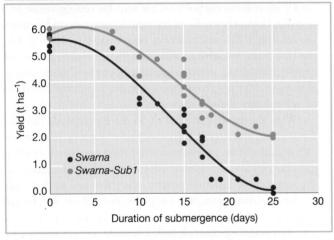

FIGURE 1-20 Yield comparison between variety *Swarna* that is not tolerant to flooding (purple circles) and variety *Swarna-Sub1* that is tolerant (green circles). Yield in tons per hectare (y-axis) versus duration of flooding in days (x-axis). [*Data from Ismail et al., "The contribution of submergence-tolerant (Sub 1) rice varieties to food security in flood-prone rainfed lowland areas in Asia," Field Crops Research 152, 2013, 83–93, © Elsevier.*]

having an impact. Since 2008, when *SUB1*-enhanced rice first appeared, world rice production has climbed from 433 to 486 milled million metric tons of milled grain per year.

The impact of the *SUB1* research may not be limited to rice in the long run. Many crops are subjected to damaging floods that reduce yields or destroy the crop altogether. The genetic research on *SUB1* has provided a deep understanding of the molecular genetics of how plants respond to flooding. With this knowledge, it will be possible to manipulate the genomes of other crop plants so that they too can withstand getting their feet a little too wet.

> **KEY CONCEPT** Genetic analysis enables crop scientists to identify beneficial genes and transfer them from one crop variety to others in order to improve yield, potentially feeding more people.

Recent evolution in humans

One goal of genetics is to understand the rules that govern how genes and the information they encode change over generations. Genes change over time for several different reasons. For example, as we have seen, mutation in the germline can cause a new gene variant or allele to occur in the next generation that was not present in the current generation. Another factor is *natural selection*, which was first described by Charles Darwin. Briefly, if individuals with a certain gene variant contribute more offspring to the next generation than individuals who lack that variant, then the frequency of that variant will rise over time in the population. The last three chapters of the text focus on rules governing the transmission of genes from one generation to the next within populations over long periods of time.

Over the past decade, evolutionary geneticists have described in remarkable detail how genetic changes have enabled human populations to adapt to the conditions of life on different parts of the globe. This work revealed that three factors have been particularly powerful in shaping the types of gene variants that occur in different human populations. These factors are (1) pathogens such as malaria or smallpox; (2) local climatic conditions including solar radiation, temperature, and altitude; and (3) diet, such as the relative amounts of meat, cereals, or dairy products eaten. In Chapter 20, you will learn how a genetic variant in the hemoglobin gene has enabled people in Africa to adapt to the ravages of malaria. Let's look briefly at a case of human adaptation to life at high altitude.

In their effort to colonize the Andes mountains of South America, Spanish colonists established towns high up in the mountains near the settlements of the native peoples. Soon, they realized something was wrong. Spanish parents were not producing children. At Potosí, Bolivia, which is situated 4000 meters above sea level, it was 53 years after the founding of the town before the first child was born to Spanish parents. As noted by the Spanish priest Father Cobo, "The Indians are healthiest and where they multiply the most prolifically is in these same cold air-tempers, which is quite the reverse of what happens to the children of the Spaniards, most of whom when born in such regions do not survive."[2] Unlike the Andean natives, the Spanish were experiencing chronic mountain sickness (CMS), a

[2] V. J. Vitzthum, "The Home Team Advantage: Reproduction in Women Indigenous to High Altitude," *J. Exp. Biol.* 204, 2001, 3141–3150.

condition caused by their inability to obtain enough oxygen from the thin air of the mountains.

Since these and other early observations, geneticists have invested much effort into the study of human adaptation to high altitude in South America, Tibet, and Ethiopia. What enables the natives of these regions to flourish while lowlanders who move to high elevations suffer the grave health consequences of CMS? Let's look at the case in Tibet, where the Tibetan highlanders live at altitudes up to 4000 meters above sea level (**Figure 1-21**). The high Tibetan Plateau was colonized thousands of years ago by people who are closely related to the modern Han Chinese. However, at high altitude, native Tibetans are far less likely than Han Chinese to experience CMS and conditions such as pulmonary hypertension and the associated formation of blood clots that underlie it.

To understand the genetics of how Tibetans adapted to life at high elevation, a research team led by Cynthia Beall of Case Western Reserve University compared Tibetans to Han Chinese at over 500,000 SNPs across the genome. Because Tibetans and Chinese are closely related, she expected each SNP variant to occur at about the same frequency in both groups. If the T variant of a SNP occurs at a frequency of 10 percent in Han Chinese, it should also be at about 10 percent in Tibetans. However, if the variant is associated with improved health at high elevation, its frequency would have risen among Tibetans over the many generations since they colonized the Tibetan Plateau, because Tibetans with this variant would have been healthier and would have had more surviving children than those who lacked it. Charles Darwin's natural selection would be at work.

When the research team analyzed their SNP data, the SNPs in one gene stood out. The gene is called *EPAS1*, and some SNPs in it occur at very different frequencies in Tibetans (87 percent) and Han Chinese (9 percent). Their results are shown in **Figure 1-22**. In this figure, the human chromosomes, numbered 1 through 22, are along the x-axis, and a measure of the difference in SNP variant frequency between Tibetans and Chinese is on the y-axis. Each dot represents a SNP. SNPs that fall above the horizontal red line are those for which the frequency difference between Tibetans and Han Chinese is so large that the gene near these SNPs likely provided some advantage to people who colonized the Tibetan Plateau. The SNPs in *EPAS1* fall above this line.

These results suggest that Tibetans have a special variant of *EPAS1* that helps them adapt to life at high elevation. *EPAS1* regulates the number of red blood cells (RBCs) that our bodies produce in response to the level of oxygen in our tissues. When oxygen levels in our tissues are low, *EPAS1* signals the body to produce more RBCs. The *EPAS1* response to low oxygen may be how our bodies normally respond to anemia (too few RBCs). People with low RBC counts get too little oxygen in their tissues, and so *EPAS1* could signal the body to make more RBCs to correct anemia. This mechanism could explain why people who live at low elevation need the *EPAS1* gene.

Now, let's think about how a person from low elevation would respond if they move to high elevation. Because of the thin air at high elevation, their tissues would get less oxygen. If their bodies interpreted low oxygen due to thin air as a sign of anemia, then *EPAS1* would try to correct the problem by signaling their body to make more RBCs. However, since they are not anemic and already have enough RBCs, their blood would become overloaded with RBCs. Too many RBCs can cause pulmonary hypertension and the formation of blood clots, the conditions underlying CMS.

Finally, how does the Tibetan variant of *EPAS1* help them avoid CMS and adapt to high elevation? The Tibetan version of *EPAS1* is expressed at a lower level than the lowland version, so their bodies are not as stimulated by *EPAS1* to overproduce RBCs at high altitude, and thus they avoid the associated blood clots and pulmonary hypertension. Remarkably, a single SNP in a regulatory element for *EPAS1* seems to be the key genetic variant for this adaptation.

Tibetans are genetically adapted to life at high elevation

China

Tibet

FIGURE 1-21 A young Tibetan woman. Inset shows the location of Tibet in Asia. [*Stefan Auth/imageBROKER/AGE Fotostock; (inset) Planet Observer/UIG/Getty Images.*]

KEY CONCEPT Evolutionary genetics provides the tools to document how gene variants that provide a beneficial effect can rise in frequency in a population and make individuals in the population better adapted to the environment in which they live.

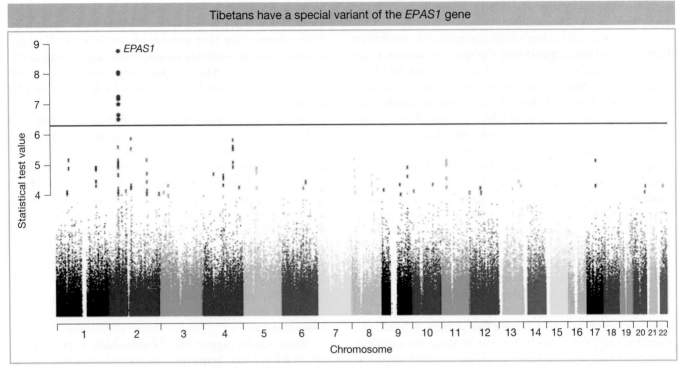

FIGURE 1-22 Twenty-two human chromosomes are arrayed from left to right. The y-axis shows results from a statistical test of whether there is a significant difference in SNP frequency between Tibetans and Han Chinese. Each small dot represents one of the SNPs that was tested. SNPs above the horizontal red line are significantly different. Only the SNPs in the *EPAS1* gene show a significant difference. [*Proceedings of the National Academy of Sciences USA, 107, 25 2010, 11459–11464, Fig. 1.*]

The complex genetics of color blindness

Look at **Figure 1-23a**. Everyone should see the number "12." Now, look at **Figure 1-23b**. Do you see "74" or "21"? If you have normal color vision, you will see "74," but if you have red-green color blindness, you will see "21."

Color blindness is an inherited disorder and a favorite example among genetics instructors for several reasons. (1) The color vision test can be administered in class. (2) Color blindness is common, so a class of 50 students may have at least one affected individual. (3) The genes for the major form of color blindness are on the X chromosome, making it a prime example of sex-linked inheritance. To this list, we can now add a fourth compelling reason to study color blindness: it has recently been corrected by gene therapy.

To understand color vision, let's start with some basics. We perceive light on our retinas where the surface has two types of photoreceptor cells—rod cells, which work in low light conditions, and cone cells, which work in high light conditions and allow us to distinguish colors.

Cone cells come in three forms, depending on which one of three opsin (light-sensitive protein) genes is expressed within them. The major opsin genes are short wavelength (blue), medium wavelength (green), and long wavelength (red). Each cone cell expresses one of these three genes, creating a mosaic of red, green, and blue sensory cells on the retina (**Figure 1-24**). Deficiency of any one or more of the opsin genes can cause a form of color blindness.

The opsin gene for blue light detection is on one of our non-sex chromosomes, or *autosomes*, for which both men and women have two copies—one copy from our mother and one copy from our father. Mutations in the blue opsin are rare, and only 0.01 percent of people carry mutations in both blue opsin alleles, making them unable to distinguish blue and yellow.

Far more common is red-green color blindness that occurs in about 5 percent of people, mostly men. Why is this form of color blindness so much more common? First, the green and red opsin genes are on the X chromosome. Since men have only one X chromosome, they will have red-green color blindness if they have a single mutant allele in either of these two genes. One in 12 men has this condition. Since women carry two X chromosomes, they need two defective alleles of one of these genes, which happens less frequently, so only about 0.6 percent of women are red-green color blind. Second, red-green color blindness can be caused by a mutation in either the red or green opsin gene. Thus, there are two targets for mutation, placing red-green color vision in double jeopardy of being lost.

But something else is up genetically. Notice that the frequency of red-green color blindness in women (0.6 percent) is much higher than the frequency of blue-yellow color blindness in women (0.01 percent). Why is this? The red and green opsin genes are neighbors on the X chromosome.

Red opsin   Green opsin

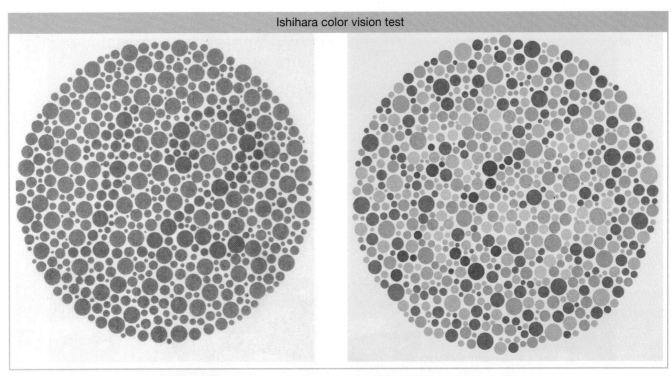

FIGURE 1-23 Plates for testing color vision based on the design of Dr. Shinobu Ishihara of the University of Tokyo. See text for details. [*PRISMA ARCHIVO/Alamy; Phanie/Alamy.*]

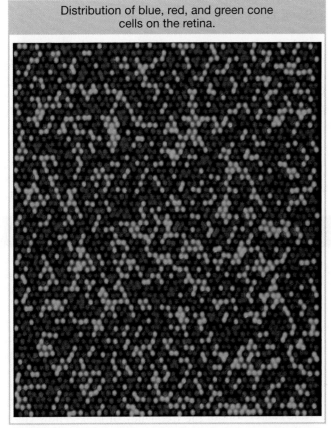

FIGURE 1-24 The blue, red, and green cone cells are arranged randomly on the retina with about equal numbers of red and green cone cells. There are many fewer blue cone cells than red or green due to a developmentally regulated difference. [*Dr. Jay Neitz, Neitz Lab, University of Washington.*]

Because the red and green opsin genes are neighbors, they can undergo a process called **unequal crossing over** that can produce the chromosomes shown below, in which either the green opsin is missing or a hybrid gene is formed. Either of these causes red-green color blindness.

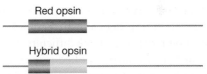

Thus, unequal crossing over provides a second way for these genes to be disrupted, partly explaining why red-green color blindness is more frequent than blue-yellow in women. You will learn about unequal crossing over in Chapter 17.

Gene therapy for color blindness is being developed by Drs. Jay and Maureen Neitz at the University of Washington. Their plan is straightforward. First, they will test red-green color-blind people to determine whether they lack the green or red opsin. Next, they will engineer the appropriate opsin gene into a virus that can insert the gene into their cone cells. The "engineering" is done using the tools mentioned earlier, such as DNA polymerases and ligases, and about which you will learn more in Chapter 10. Once expressed in a cone cell, the inserted gene will shift the spectral sensitivity of the patient, providing improved color vision.

The Neitzes have tested this plan in monkeys (**Figure 1-25**). Male squirrel monkeys lack either the red or green opsin gene, and thus they have red-green color blindness. The Neitzes trained the monkeys to receive a reward (some juice) for distinguishing colors in a test. Before gene therapy, the monkeys could not distinguish green from red,

Dalton the squirrel monkey

FIGURE 1-25 Dalton the squirrel monkey received gene therapy for color vision. (a) How Dalton saw the world through his original red-green color-blind eyes; (b) Dalton enjoying his new color sense after gene therapy. [*Dr. Jay Neitz, Neitz Lab, University of Washington.*]

but after gene therapy, the monkeys could tell green and red apart. Gene therapy worked.

Why should we treat color blindness through gene therapy? This condition does impair people's lives. Red-green color-blind individuals have difficulty distinguishing cooked (brown) from partly cooked (pink) meat or ripe from green fruits. They cannot be commercial pilots or air traffic controllers, and they generally encounter difficulty navigating a world designed for trichromats. Moreover,

gene therapy for color blindness sets the stage for treating more severe eye diseases such as macular degeneration. With advances in technology, we can anticipate a day when gene therapy for visual impairments is no more costly or complicated than putting on a pair of contact lenses.

KEY CONCEPT Genetic tools allow gene therapy to correct some disorders caused by mutant genes.

SUMMARY

As you begin your study of genetics, imagine yourself as a person at halftime during an amazing journey of discovery. The last 100 years have witnessed a remarkable revolution in human knowledge about how biological systems are put together and how they work. Genetics has been at the epicenter of that revolution. Genetic analysis has answered many fundamental questions about the transmission of genetic information within families, inside cells, and over the eons of evolutionary time. Yet, as you will learn, the discovery process in genetics has never been more dynamic and the pace of growth in knowledge never greater. Unanswered questions abound.

- How do all the genes in the genome work together to transform a fertilized egg into an adult organism?

- How do cells manage to seamlessly orchestrate the incredibly complex array of interacting genes and biochemical reactions that are found within them?

- How do genetic variants at hundreds or even thousands of genes control the yield of crop plants?

- How can genetics guide both the prevention and treatment of cancer, autism, and other diseases?

- How do genes give humans the capacity for language and consciousness?

Genetic analysis over the next 100 years promises to help answer many questions such as these.

KEY TERMS

adenine (A) (p. 6)
allele (p. 3)
codon (p. 8)
complementary (base pairs) (p. 7)
cytosine (C) (p. 7)
DNA polymerase (p. 10)
DNA replication (p. 8)
DNA sequencing (p. 11)
dominant (p. 3)
gene (p. 3)
gene expression (p. 7)

genetically modified organism
    (GMO) (p. 11)
genetics (p. 4)
genomics (p. 11)
guanine (G) (p. 6)
ligase (p. 10)
messenger RNA (mRNA) (p. 8)
model organism (p. 9)
multifactorial hypothesis (p. 6)
nuclease (p. 10)
one-gene–one-enzyme hypothesis
    (p. 6)

point mutation (p. 14)
quantitative trait locus (QTL)
    (p. 17)
regulatory element (p. 7)
single nucleotide polymorphism
    (SNP) (p. 13)
thymine (T) (p. 6)
transcription (p. 8)
transformation (p. 10)
translation (p. 8)
unequal crossing over (p. 21)

PROBLEMS

Visit SaplingPlus for supplemental content. Problems with the icon are available for review/grading. Problems with the icon have an Unpacking the Problem exercise.

WORKING WITH THE FIGURES

1. If the white-flowered parental variety in Figure 1-3 were crossed to the first-generation hybrid plant in that figure, what types of progeny would you expect to see, and in what proportions?

2. In Mendel's 1866 publication as shown in Figure 1-4, he reports 705 purple-flowered (violet) offspring and 224 white-flowered offspring. The ratio he obtained is 3.15:1 for purple to white. How do you think he explained the fact that the ratio is not exactly 3:1?

3. Figure 1-7 shows a simplified pathway for arginine synthesis in *Neurospora*. Suppose you have a special strain of *Neurospora* that makes citrulline but not arginine. Which gene(s) are likely mutant or missing in your special strain? You have a second strain of *Neurospora* that makes neither citrulline nor arginine but does make ornithine. Which gene(s) are mutant or missing in this strain?

4. Consider Figure 1-8a.

   a. What do the small, blue spheres represent?

   b. What do the brown slabs represent?

   c. Do you agree with the analogy that DNA is structured like a ladder?

5. In Figure 1-8b, is the number of hydrogen bonds between adenine and thymine the same as that between cytosine and guanine? Do you think that a DNA molecule with a high content of A + T would be more stable than one with high content of G + C?

6. Which of the three major groups (domains) of life in Figure 1-11 is not represented by a model organism?

7. Figure 1-13b shows the human chromosomes in a single cell. The yellow dots show the location of the muscle glycogen phosphorylase gene. Is the cell in this figure a sex cell (gamete)? Explain your answer.

8. Figure 1-15 shows the family tree, or pedigree, for Louise Benge (Individual VI-1) who suffers from the disease ACDC because she has two mutant copies of the CD73 gene. She has four siblings (VI-2, VI-3, VI-4, and VI-5) who have this disease for the same reason. Do all of the 10 children of Louise and her siblings have the same number of mutant copies of the CD73 gene, or might this number be different for some of the 10 children?

BASIC PROBLEMS

9. State four questions about inheritance that arose after Mendel's rules of inheritance were rediscovered.

10. Name four tools or enzymes that molecular geneticists can use to manipulate DNA or RNA molecules.

11. Below is the sequence of a single strand of a short DNA molecule. On a piece of paper, rewrite this sequence and then write the sequence of the complementary strand below it.

    GTTCGCGGCCGCGAAC

    Compare the sequences of the top and bottom strands. What do you notice about the relationship between them?

12. Mendel studied a *tall* variety of pea plants with stems that are 20 cm long and a *dwarf* variety with stems that are only 12 cm long.

a. Under blending theory, how long would you expect the stems of first and second hybrids to be?

b. Under Mendelian rules, and assuming stem length is controlled by a single gene, what would you expect to observe in the second-generation hybrids if all the first-generation hybrids were tall?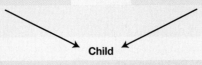

13. If a DNA double helix that is 100 base pairs in length has 32 adenines, how many cytosines, guanines, and thymines must it have?

14. The complementary strands of DNA in the double helix are held together by hydrogen bonds: $G \equiv C$ or $A = T$. These bonds can be broken (denatured) in aqueous solutions by heating to yield two single strands of DNA (see Figure 1-13a). How would you expect the relative amounts of GC versus AT base pairs in a DNA double helix to affect the amount of heat required to denature it? How would you expect the length of a DNA double helix in base pairs to affect the amount of heat required to denature it?

15. The figure below shows the DNA sequence of a portion of one of the chromosomes from a trio (mother, father, and child). Can you spot any new point mutations in the child that are not in either parent? In which parent did the mutation arise?

Mother

Copy M1 ••CAGCAGATTGCTGCTTTGTATGAG••
Copy M2 ••CAGCTGATTGCTGCTTTGTATGAG••

Father

Copy F1 ••CAGCTGATTGCTGCTTTGTAGGAG••
Copy F2 ••CAACTGATTGCTGCTTTGTATGAG••

Child

••CAGCAGATTGCTGCTTTGTCTGAG••
••CAGCTGATTGCTGCTTTGTAGGAG••

CHALLENGING PROBLEMS

16. Fathers contribute more new point mutations to their children than mothers. You may know from general biology that people have sex chromosomes—two X chromosomes in females and an X plus a Y chromosome in males. Both sexes have the autosomes (A's).

a. On which type of chromosome (A, X, or Y) would you expect the genes to have the greatest number of new mutations per base pair over many generations in a population? Why?

b. On which type of chromosome would you expect the least number of new mutations per base pair? Why?

c. Can you calculate the expected number of new mutations per base pair for a gene on the X and Y chromosomes for every one new mutation in a gene on an autosome if the mutation rate in males is twice that in females?

17. For young men of age 20, there have been 150 rounds of DNA replication during sperm production as compared to only 23 rounds for a woman of age 20. That is a 6.5-fold greater number of cell divisions and proportionately greater opportunity for new point mutations. Yet, on average, 20-year-old men contribute only about twice as many new point mutations to their offspring as do women. How can you explain this discrepancy?

18. The human genome is approximately 3 billion base pairs in size.

a. Using standard 8.5" × 11" paper with one-inch margins, a 12-point font size, and single-spaced lines, how many sheets of paper printed on one side would be required to print out the human genome?

b. A ream of 500 sheets of paper is about 5 cm thick. How tall would the stack of paper with the entire human genome be?

c. Would you want a backpack, shopping cart, or a semitrailer truck to haul around this stack?

19. Calculating probabilities is an important skill in genetics. It is used to determine an individual's chance of inheriting genetic diseases, to determine the chances of carrying a disease or genetic trait, as well as for a variety of other tasks, such as predicting the success of a cloning project. When you are trying to calculate the probability of two or more independent events occurring at the same time, use the product rule. An example of this would be the flipping of two coins simultaneously. If we want to calculate the probability of obtaining heads on both flipped coins, we simply take the probability of landing heads on the first coin (1/2) and multiply it by the probability of landing heads on the second coin (1/2). Thus, the probability of landing heads on both coins simultaneously is the product of the individual probabilities: (1/2) × (1/2) = 1/4. Whenever we are calculating the combined probability resulting from more than one independent event, we use the product rule. How many different DNA molecules 10 base pairs long are possible?

GENETICS AND SOCIETY

In this chapter, you read how older fathers contribute more new point mutations to their offspring than younger fathers. Should this information be used to inform public policy or individual decisions? What policy decisions might a society or an individual make and implement in this regard?

PART 1

Core Principles in Transmission Genetics

The subject of this book is the science of *genetics*, which is broadly defined as the study of *heredity*. Heredity in turn is defined as the *transmission* of properties from generation to generation. Since the beginning of time, humans have no doubt wondered about how heredity is possible. Generally speaking, heredity has two equally mysterious components, *constancy* and *variation*.

Constancy is the simple fact that people have human babies, cats have kittens, dogs have puppies, and so on. Thus, the constancy of the species is transmitted down through the generations. Of course, this general observation about heredity raises the question, how is such constancy possible? In other words, what is the mechanism that allows faithful reproduction of a species? If it is assumed that two parents are needed to beget offspring, common sense suggests that some material must be transmitted from each parent to contribute to the offspring. However, for most of the history of mankind, the nature of this material was a mystery.

Variation is the common observation (using humans as an example) that although people beget people, there is considerable difference among the members of a population generally, and also within the progeny of a mating. These differences allow us to recognize individuals in the general pool of the human population. Such variation is concerned with essentially minor aspects of a species. Although the main characteristics that define the species are held constant (we are bipedal, big-brained, have generally hairless bodies, etc.), there are clear differences in properties such as hair color, nose shape, skin color, height, predisposition to certain diseases, and a myriad of other qualities.

We can contrast two types of variation. The first is called *continuous variation*. A good example is height or length, which tends to vary in the population from a low value to a high value and *all* values between. The second is *discontinuous variation*, the existence in a species of individuals with markedly distinct forms of a particular property. In humans, a good example is the presence or absence of a chin dimple. This is not continuous variation; people either have a chin dimple or none. The movie actor Kirk Douglas has a prominent chin dimple, and furthermore his son Michael Douglas (also an actor) has clearly inherited his chin dimple.

Another example from history is the Hapsburg lip, a striking discontinuous variant that was handed down through generations of the Spanish Hapsburg royal family. We see around us many other examples of the clear inheritance of such distinct properties in both animals and plants. A mechanism is needed that explains not only how such variation arises, but also how it is transmitted from one generation to the next, that is, a transmission mechanism.

The inheritance of continuous variation is less obvious, and often no clear pattern of inheritance is immediately obvious. An important complicating factor regarding continuous variation is that environmental effects can have a profound influence. A simple example is weight, which varies continuously. Although in some cases it can seem to be inherited, the weight of an individual is obviously greatly influenced by the availability of food, a crucial component of the environment.

The existence of variation and its inheritance also demands an explanation. How are variants produced, and how are these variants passed on to subsequent generations?

In summary of the above preliminary ideas, we can set down several overarching observations about heredity:

1. *A species always begets progeny of the same species.*
2. *Within a species there are variant properties.*
3. *Variation in a property can be continuous or discontinuous.*
4. *Some variants are passed down through the generations.*
5. *The environment can affect variation.*

Kirk and Michael Douglas

Michael Douglas (*right*), son of Kirk Douglas (*left*), has inherited his father's chin dimple. [*Sunset Boulevard/Getty Images; ScreenProd/Photononstop/Alamy.*]

The Hapsburg Lip

Philip IV (*left*) and his son Charles II (*right*) were the last of the Hapsburgs to rule Spain. Both ruled in the seventeenth century, and both had a Hapsburg lip. [*DEA/G. NIMATALLAH/Getty Images; Heritage Images/Getty Images.*]

These observations must have been apparent to anyone thinking about heredity for as long as humans have existed. However, when we ask about the mechanisms behind these phenomena, history tells us that there were very few productive ideas on the subject until the middle of the nineteenth century. We simply did not know how species and their variants were accurately propagated. This is when the science of genetics as an analytical subject began, when the basic experimental rules for elucidating the mechanisms of inheritance began to be laid down. Then, over the following century and a half, as the experimental approaches gained sophistication, we learned the mechanisms of inheritance as the well-tested principles we embrace today.

This first section of this book is called Transmission Genetics because it covers the mechanisms that allow both constancy and variation to be transmitted from one generation to another. These chapters present the main mechanisms of transmission genetics, which we can call general principles of heredity. This summary of the principles of heredity serves as a general road map to follow through this initial block of chapters.

PRINCIPLES OF HEREDITY

1. DNA is the genetic material that determines the basic properties of an organism.

We now know that each cell of an organism has a fundamental and unique set of DNA, called the *genome*, which has encoded in it the information for building that organism. DNA is a long filamentous molecule made up of many thousands of functional units called genes. The genome of eukaryotic organisms is composed of several DNA molecules, each coiled up as a chromosome, with each chromosome bearing many genes. In most cases, the information in each gene is translated into a protein, and these many proteins are the essential units of form and function in an organism. Hence, overall

$$\underset{information}{\text{Gene (DNA)}} \xrightarrow[translation]{} \underset{form\ or\ function}{\text{Protein}}$$

2. Hereditary constancy is based on DNA replication.

In a cell, DNA can be copied using a process called replication. During DNA replication, each DNA molecule produces two identical "daughter" DNA molecules, destined to carry the essential information of the organism onto the next generation. This is the fundamental chemical process behind all hereditary transmission between cell generations and organismal generations.

$$\text{DNA} \xrightarrow[replication]{} \text{2 identical daughter DNA molecules}$$

3. During cell division, daughter DNA molecules are packaged into the resulting cells.

Somatic cells in eukaryotes divide to produce more cells of the same organism, to increase cell number during growth. During this cell division, the accompanying nuclear division called *mitosis* ensures that each daughter cell receives one daughter of each chromosome, the same number and identity as in the parent cell.

$$\text{somatic cell} \xrightarrow[mitosis]{} \text{2 genetically identical daughter cells}$$

In prokaryotes, an analogous mechanism ensures that each daughter cell has a daughter DNA molecule. Because prokaryotes are single-celled, the daughter cells can be considered offspring.

In the sexual cycle of eukaryotes, specialized cells called *meiocytes* undergo cell division accompanied by the nuclear division called *meiosis*, and the result is four gamete cells (often called eggs and sperm) that each have one single daughter of each chromosome. These four gametes are generally not identical because of recombination (see principle 8).

$$\text{meiocyte} \xrightarrow[meiosis]{} \text{4 non-identical gametes}$$

Hence in such somatic and sexual processes, we see the basis of the constancy, the way in which life propagates itself. DNA is propagated, and the daughter DNA molecules are parceled out in an organized way to the resultant cells depending on the requirements of the somatic and sexual cell cycles.

4. Heritable variation is caused by changes in genes at the DNA level.

Although DNA is generally a stable molecule, it does have an inherent tendency to change at a low rate, in a process known as *mutation*. Some of these changes are essentially caused by random mistakes in the chemistry of DNA. Within a gene, such changes often result in changes in, or lack of, gene function, which in turn can lead to detectable changes in the structure or function of the organism. The changes in the DNA are said to change its *genotype*, while the corresponding changes in the structure or function of an organism change its *phenotype*, the outward manifestation of its genotype.

Gene mutations are transmitted as part of the chromosome. They can be passed down from one cell to descendant cells via mitosis, or from a parental organism to its descendants via meiosis. We see that faithful replication of DNA is the basis of transmission of both constancy and variation.

5. In eukaryotic sexual reproduction, single gene mutations are transmitted to progeny in precise mathematical ratios.

The cells of most animals and plants carry two chromosome sets, arranged in chromosomal pairs. If the members of a pair carry different forms of a gene (such as one normal and one mutation) then meiosis pulls these chromosomes and their genes apart in a process called *segregation* and produces gametes that are 1/2 normal-bearing and 1/2 mutation-bearing. In turn, the same ratio can then be detected in the progeny. This simple mathematical ratio is the basis for several more complex ratios found in the progeny of various types of crosses.

Prokaryotes also have a sex-like cycle and also produce special ratios in the progeny, but these are mathematically more complex.

6. During gamete production, mutations on different chromosomes are transmitted independently.

Because different chromosomes are each manipulated by their own molecular "ropes" during gamete production, the genes on one chromosome are distributed to daughter cells independent of those on other chromosomes.

7. During gamete production, mutations close together on the same chromosome tend to be transmitted together.

Genes located near each other on the same chromosome are physically linked by the segment of chromosome between them, so tend to be inherited as a package.

8. Recombination contributes to variation among progeny.

Both eukaryotes and prokaryotes have mechanisms for recombination, the production of new allelic combinations. In eukaryotes, independent assortment and crossing-over (chromosomal breakage and reunion) at meiosis are the two mechanisms of recombination. In prokaryotes, a crossover-like process occurs when cells fuse and form partial diploids.

9. Mutations of different genes can interact functionally.

Because each biological property of an organism is influenced by an array of several-to-many genes, if several mutations under study are in genes that happen to be part of this collaborative array, the mutations can show interactive effects, some qualitative, some quantitative. These interactions can be detected as modified phenotypic ratios in the progeny of crosses.

10. Gene activity can be influenced by the environment.

A gene, which is a segment of DNA, can do nothing in a test tube by itself. It needs to be enclosed in a cell in order to function. The cell in turn is also not independent because it receives its nutrients and necessary external signals from the environment. Hence, analyses of gene function, genetic variation, and inheritance must always consider and control for variation in the environment.

Viruses are generally not considered to be living, but nevertheless do have nucleic acids as their genetic material, and they demonstrate many of the principles delineated earlier.

The language of these 10 fundamental principles is not particularly difficult to understand or to remember, and you might wonder why the book does not end here. The answer is that understanding in the science of genetics is not about understanding sentences; it is about the ability to make deductions from observed data. In general, a genetic experiment is performed and results obtained that need to be explained. Armed with a set of principles such as those outlined here (and many more are in the other sections), the researcher has to analyze whether or not the results conform to one of the established principles. If not, the results might point to a new principle. At all times we must ask, "How do we know this to be true?" In such analysis knowledge of how principles came to be established in the first place is often very informative. What experiments were done, and how were they analyzed to give credence to a certain principle? Questions of this type are the content of this book. We shall ask many questions such as "How do we know that there are genes?" "How do we know the hereditary material is DNA?" "How would we recognize single gene inheritance?" "How many genes influence this particular characteristic?" "What type of data would allow us to infer that two genes are close together on a chromosome?" That is the stuff of genetic analysis, the subject of this book, and to which we now turn.

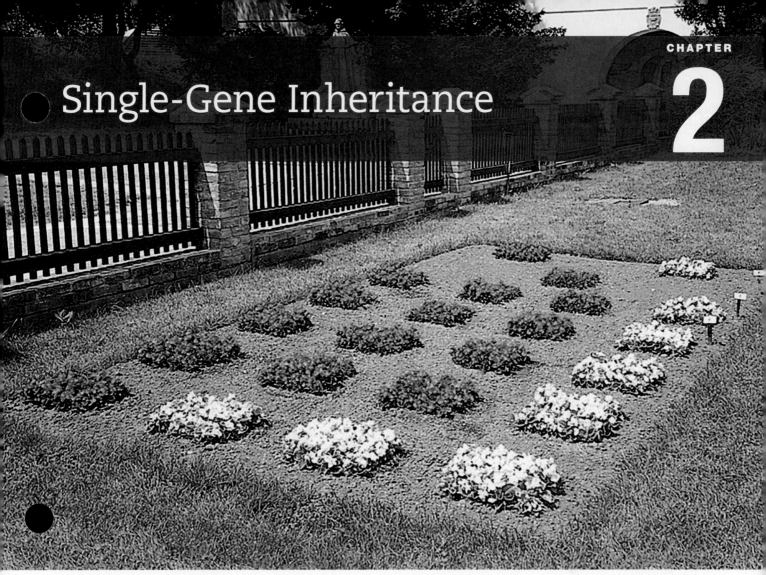

Single-Gene Inheritance

The monastery of the father of genetics, Gregor Mendel. A statue of Mendel is visible in the background. Today, this part of the monastery is a museum, and the curators have planted red and white begonias in a grid that graphically represents the type of inheritance patterns obtained by Mendel with peas. [*Anthony Griffiths.*]

CHAPTER OUTLINE AND LEARNING OBJECTIVES

2.1 SINGLE-GENE INHERITANCE PATTERNS

LO 2.1 In the progeny of controlled crosses, recognize phenotypic ratios diagnostic of single-gene inheritance.

2.2 GENES AND CHROMOSOMES

LO 2.2 Explain single-gene inheritance ratios in terms of chromosome behavior at meiosis.

2.3 THE MOLECULAR BASIS OF MENDELIAN INHERITANCE PATTERNS

LO 2.3 Propose reasonable hypotheses to explain dominance and recessiveness of specific alleles at the molecular level.

2.4 SOME GENES DISCOVERED BY OBSERVING SEGREGATION RATIOS

LO 2.4 Predict phenotypic ratios among descendants from crosses of parents differing at a single gene.

2.5 SEX-LINKED SINGLE-GENE INHERITANCE PATTERNS

LO 2.5 In the progeny of controlled crosses, recognize phenotypic ratios diagnostic of X-linked single-gene inheritance.

2.6 HUMAN PEDIGREE ANALYSIS

LO 2.6 Recognize inheritance patterns diagnostic of autosomal dominant, autosomal recessive, X-linked dominant, X-linked recessive, and Y-linked conditions in human pedigrees.

29

We saw in Chapter 1 that genes are the basic functional units of inheritance. Our broad objective for this chapter is to learn the inheritance patterns in crosses that reveal the existence of individual genes having a measurable effect on the phenotype of a eukaryotic organism. We will see how the pioneering work of Mendel and the principles of genetics he proposed have become valuable analytical tools for geneticists as they dissect biological properties of interest.

What kinds of research do biologists do? One central area of research in the biology of all organisms is the attempt to understand how an organism develops from a fertilized egg into an adult—in other words, what makes an organism the way it is. Usually, this overall goal is broken down into the study of individual biological properties such as the development of plant flower color, or animal locomotion, or nutrient uptake, although biologists also study broader areas such as how cells work. How do geneticists analyze biological properties? In genetics, individual biological properties of a species are referred to as **characters** or **traits**. The genetic approach to understanding any biological trait is to find the subset of genes in the genome that influence it, a process sometimes referred to as **gene discovery**. After these genes have been identified, their cellular functions can be elucidated through further research.

There are several different types of analytical approaches to gene discovery, but one widely used method relies on the detection of *single-gene inheritance patterns*, and that is the topic of this chapter.

All of genetics, in one aspect or another, is based on heritable *variants*: individuals who inherit a trait that expresses differently from some standard form. For example, in regard to the flower color trait in some plant, it might express white in the variant instead of the normal blue. The basic approach of genetics is to compare the properties of variants with the standard and, from these comparisons, to make deductions about genetic function. It is similar to the way in which you could make inferences about how an unfamiliar machine works by changing the composition or positions of the working parts, or even by removing parts one at a time. Each variant represents a "tweak" of the biological machine, from which its function can be deduced.

In genetics, the most common form of any trait of an organism is called the **wild type**, that which is found "in the wild," or in nature. The heritable variants observed in a species that differ from wild type are called **mutants**, individual organisms having some abnormal form of the trait. The alternative forms of a trait are called **phenotypes**, for example, the blue and white phenotypes of the flower color trait. **Figure 2-1** shows examples of the wild-type phenotype and several mutant phenotypes for a given trait in two different model organisms.

Compared to wild type, mutants are rare. We know that they arise from wild types by a process called **mutation**, which results in a heritable change in the DNA of a gene. The changed form of the gene is also called a mutation. Mutations are not always detrimental to an organism; sometimes they can be advantageous, but most often they have no observable effect. A great deal is known about the mechanisms of mutation (see Chapter 15), but generally it can be said that they arise from mistakes in cellular processing of DNA.

Genetic analysis begins with mutants

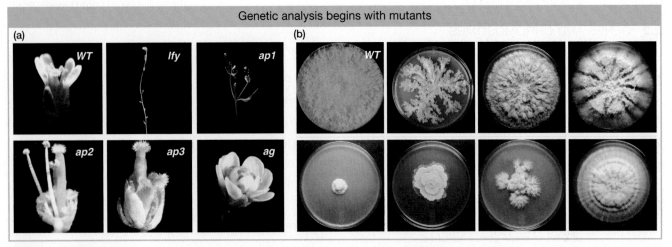

FIGURE 2-1 These photographs show the range of mutant phenotypes typical of those obtained in the genetic dissection of traits. These examples are from the dissection of floral development in *Arabidopsis thaliana*, a plant (a) and hyphal growth in *Neurospora crassa*, a mold (b). *WT* = wild type.
[(a) courtesy of George Haughn; (b) courtesy of Anthony Griffiths/Olivera Gavric.]

Simply stated, the general steps in analyzing a trait through gene discovery are as follows:

1. Amass mutants affecting the trait of interest.
2. **Cross** (mate) mutant individuals to wild-type individuals to see if their descendants show ratios of wild-type-to-mutant phenotypes that are characteristic of single-gene inheritance.
3. Deduce the functions of the gene at the molecular level.
4. Deduce how the gene interacts with other genes to produce the trait in question.

Of these steps, only 1 and 2 will be covered in the present chapter.

Gene discovery starts with a "hunt" to amass mutants in which the biological function under investigation is altered or destroyed. Even though mutants are individually rare, there are ways of enhancing their recovery. One widely used method is to treat the organism with radiation or chemicals that increase the mutation rate. After treatment, the most direct way to identify mutants is to visually *screen* a very large number of individuals, looking for a chance occurrence of mutants in that population. Also, various selection methods can be devised to enrich for the types sought.

Armed with a set of mutants affecting the trait of interest, one hopes that each mutant represents a mutation in one of a set of genes that control the trait, and that a reasonably complete gene pathway or network is represented. However, not all mutants are caused by a single mutation within one gene (some have far more complex determination), so first each mutant has to be tested to see if indeed it is caused by a single-gene mutation.

The test for single-gene inheritance is to mate individuals showing the mutant phenotype with wild-type individuals, and then to analyze the first and second generations of descendants. As an example, a mutant plant with white flowers would be crossed to a wild-type plant showing blue flowers. The progeny of this cross are analyzed, and then they themselves are interbred to produce a second generation of descendants. In each generation, the diagnostic ratios of plants with blue flowers to those with white flowers will reveal whether a single gene controls white versus blue flower color. If so, then by inference, the wild-type phenotype (blue flowers) would be encoded by the wild-type form of the gene, and the mutant phenotype (white flowers) would be encoded by a form of the same gene in which a mutation event has altered the DNA sequence in some way. Other mutations affecting flower color (perhaps mauve, blotched, striped, and so on) would be analyzed in the same way, resulting overall in a set of defined "flower-color genes." Through genetics, the set of gene functions that interact to produce the trait we call flower color can be defined. The use of mutants in this way is sometimes called **genetic dissection,** because the trait in question (flower color in this case) is picked apart to reveal

its underlying genetic program, not with a scalpel but with mutants. Each mutant potentially identifies a separate gene affecting that trait.

After a set of key genes has been defined in this way, several different molecular methods can be used to establish the functions of each of the genes. These methods will be covered in later chapters.

This type of approach to gene discovery is sometimes called **forward genetics,** a strategy to understanding biological function starting with random single-gene mutants and ending with their DNA sequence and biochemical function. (In later chapters, we shall see **reverse genetics** at work. In brief, reverse genetics starts with genomic analysis at the DNA level to identify a set of genes as candidates for encoding the biological trait of interest, then induces mutants targeted specifically to those genes, and then examines the mutant phenotypes to see if they indeed affect the trait under study.)

KEY CONCEPT The genetic approach to understanding a biological trait is to discover the genes that control it. One approach to gene discovery is to isolate mutants and check each one for single-gene inheritance patterns (specific ratios of wild-type and mutant expression of the trait in descendants).

Gene discovery is important not only in experimental organisms, but also in applied studies. One crucial area is agriculture, where gene discovery can be used to understand a desirable commercial property of an organism, such as its protein content. We have already encountered an example of the power of genetics to affect agriculture in Chapter 1, where we saw how genetic analysis facilitated the creation of a flood-resistant rice strain. Human genetics is another important area: to know which gene functions are involved in a specific disease or condition is useful information in finding therapies, such as the gene therapy technique being developed to treat color blindness, as discussed in Chapter 1.

The rules for single-gene inheritance were originally elucidated in the 1860s by the monk Gregor Mendel, who worked in a monastery in the town of Brno, now part of the Czech Republic. Mendel's analysis is the prototype of the experimental approach to single-gene discovery still used today. Indeed, Mendel was the first person to discover any gene! Mendel did not know what genes were, how they influenced traits, or how they were inherited at the cellular level. Now we know that genes either encode proteins or RNA molecules that facilitate or regulate protein expression, a topic that we shall return to in later chapters. We also know that single-gene inheritance patterns are produced because genes are parts of chromosomes, and chromosomes are partitioned very precisely down through the generations, as we shall see later in this chapter.

2.1   SINGLE-GENE INHERITANCE PATTERNS

LO 2.1 In the progeny of controlled crosses, recognize phenotypic ratios diagnostic of single-gene inheritance.

Recall that the first step in genetic dissection is to obtain variants that differ in the trait under scrutiny. With the assumption that we have acquired a collection of relevant mutants, the next question is whether each of the mutations is inherited as a single gene.

Mendel's pioneering experiments

The first-ever analysis of single-gene inheritance as a pathway to gene discovery was carried out by Gregor Mendel. His is the analysis that we shall follow as an example. Mendel chose the garden pea, *Pisum sativum*, as his research organism. The choice of organism for any biological research is crucial, and Mendel's choice proved to be a good one because peas are easy to grow and breed. Note, however, that Mendel did not embark on a hunt for mutants of peas; instead, he made use of mutants that had already been found by others and had been used in horticulture. Moreover, Mendel's work differs from most genetics research undertaken today in that it was not a genetic dissection; he was not interested in the traits of peas themselves, but rather in the way in which the hereditary units that influenced those traits were inherited from generation to generation. Nevertheless, the laws of inheritance deduced by Mendel are exactly those that we use today in modern genetics in identifying single-gene inheritance patterns.

Mendel chose to investigate the inheritance of seven characters (traits) of his chosen pea species: pea color, pea shape, pod color, pod shape, flower color, plant height, and position of the flowering shoot. For each of these seven characters, he obtained from his horticultural supplier two pea plant types that showed distinct and contrasting phenotypes. These contrasting phenotypes are illustrated in **Figure 2-2**. His results were substantially the same for each character, and so we can use one character, pea seed color, as an illustration. All of the plants used by Mendel were from **pure lines**, meaning that, for the phenotype in question, all offspring produced by matings within the members of that line were identical. For example, within the yellow-seeded line, all the progeny of any mating were yellow seeded.

Mendel's analysis of pea heredity made extensive use of crosses. To make a cross in plants such as the pea, pollen is simply transferred from the anthers of one plant to the stigmata of another. A special type of mating is a **self** (self-pollination), which is carried out by allowing pollen from a flower to fall on its own stigma. Crossing and selfing are illustrated in **Figure 2-3**. The first cross made by Mendel mated plants of the yellow-seeded line with plants of the green-seeded line. In his overall breeding program,

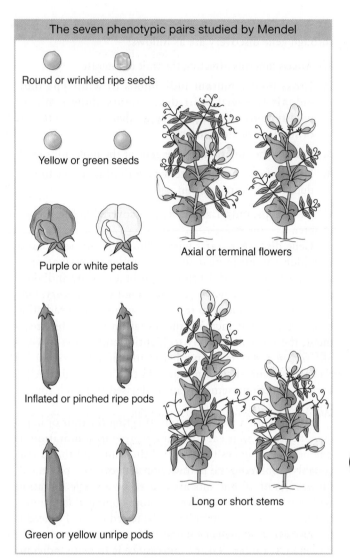

The seven phenotypic pairs studied by Mendel

Round or wrinkled ripe seeds

Yellow or green seeds

Purple or white petals

Axial or terminal flowers

Inflated or pinched ripe pods

Long or short stems

Green or yellow unripe pods

FIGURE 2-2 For each character, Mendel studied two contrasting phenotypes.

these lines constituted the **parental generation**, abbreviated **P**. In *Pisum sativum*, the color of the seed (the pea) is determined by the seed's own genetic makeup; hence, the peas resulting from a cross are effectively progeny and can be conveniently classified for phenotype without the need to grow them into plants. The progeny peas from the cross between the different pure lines were found to be all yellow, no matter which parent (yellow or green) was used as male or female. This progeny generation is called the **first filial generation**, or F_1. The word *filial* comes from the Latin words *filia* (daughter) and *filius* (son). The results of these two reciprocal crosses were as follows, where × represents a cross:

> female from yellow line × male from green line →
>
> F_1 peas all yellow
>
> female from green line × male from yellow line →
>
> F_1 peas all yellow

The results observed in the descendants of both reciprocal crosses were the same, and so we will treat them as one cross.

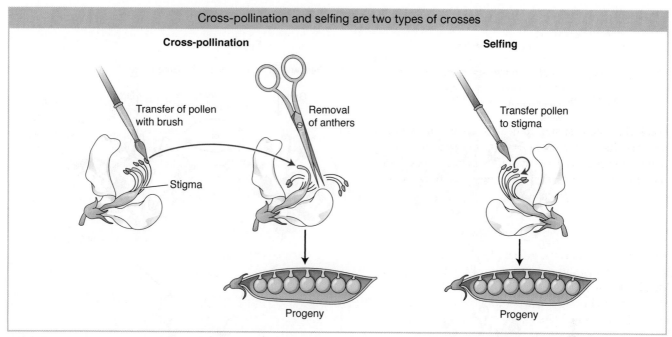

FIGURE 2-3 In a cross of a pea plant (*left*), pollen from the anthers of one plant is transferred to the stigma of another. In a self (*right*), pollen is transferred from the anthers to the stigmata of the same plant.

Mendel grew F_1 peas into plants, and he selfed these plants to obtain the **second filial generation**, or F_2. The F_2 was composed of 6022 yellow peas and 2001 green peas. In summary,

yellow F_1 × yellow F_1 → F_2 comprised of   6022 yellow
                                                    2001 green
                                          Total   8023

Mendel noted that this outcome was very close to a mathematical ratio of three-fourths (75%) yellow and one-fourth (25%) green. A simple calculation shows us that $6022/8023 = 0.751$ or 75.1%, and $2001/8023 = 0.249$ or 24.9%. In other words, there was a 3:1 ratio of yellow to green. Interestingly, the green phenotype, which had disappeared in the F_1, had reappeared in one-fourth of the F_2 individuals, showing that the genetic determinants for green must have been present in the yellow F_1, although unexpressed (not observed in the F_1 phenotype).

To further investigate the nature of the F_2 plants, Mendel selfed plants grown from the F_2 seeds. He found three different types of results. The plants grown from the F_2 green seeds, when selfed, were found to bear only green peas. However, plants grown from the F_2 yellow seeds, when selfed, were found to be of two types: one-third of them were pure breeding for yellow seeds, but two-thirds of them gave mixed progeny: three-fourths yellow seeds and one-fourth green seeds, just as the F_1 plants had. In summary,

$\frac{1}{4}$ of the F_2 were green, which when selfed gave all greens

$\frac{3}{4}$ of the F_2 were yellow;
  of these $\frac{1}{3}$ when selfed gave all yellows
              $\frac{2}{3}$ when selfed gave $\frac{3}{4}$ yellow and $\frac{1}{4}$ green

To put it another way, the F_2 was comprised of

$\frac{1}{4}$ pure-breeding greens
$\frac{1}{2}$ F_1-like yellows (mixed progeny)
$\frac{1}{4}$ pure-breeding yellows

Thus, the 3:1 ratio of phenotypes observed in the F_2 generation, at a more fundamental level, is a 1:2:1 ratio.

Mendel made another informative cross between the F_1 yellow-seeded plants and any green-seeded plant. In this cross, the progeny showed the proportions of one-half yellow and one-half green. In summary,

F_1 yellow × green → $\frac{1}{2}$ yellow
                        $\frac{1}{2}$ green

These two types of matings, the F_1 self and the cross of the F_1 with any green-seeded plant, both gave yellow and green progeny, but in different ratios. These two ratios are represented in **Figure 2-4**. Notice that the ratios are seen only when the peas in several pods are combined.

The 3:1 and 1:1 ratios found for pea color were also found for comparable crosses for the other six characters that Mendel studied. The actual numbers for the 3:1 ratios for those characters are shown in **Table 2-1**.

Mendel's law of equal segregation

Initially, the meaning of these precise and repeatable mathematical ratios must have been unclear to Mendel, but he was able to devise a brilliant model that not only accounted for all the results, but also represented the historical birth of the science of genetics. Mendel's model for

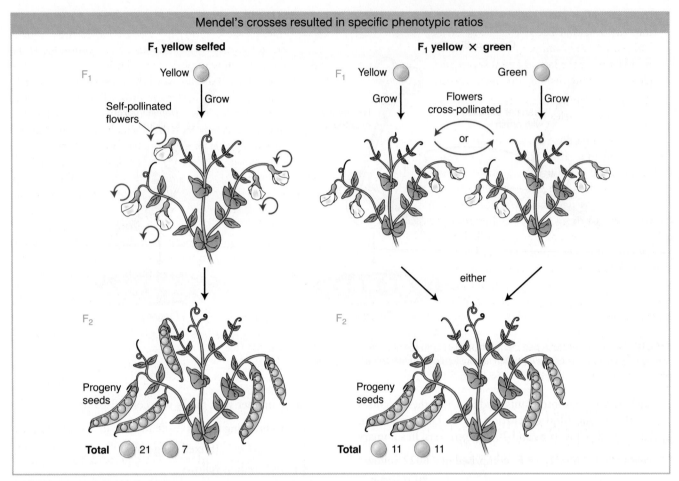

FIGURE 2-4 Mendel obtained a 3:1 phenotypic ratio in his self-pollination of the F₁ (*left*) and a 1:1 phenotypic ratio in his cross of F₁ yellow with green (*right*). Sample sizes are arbitrary.

TABLE 2-1	Results of All Mendel's Crosses in Which Parents Differed in One Character			
Parental phenotypes	**F₁**	**F₂**	**F₂ ratio**	
1. round × wrinkled seeds	All round	5474 round; 1850 wrinkled	2.96:1	
2. yellow × green seeds	All yellow	6022 yellow; 2001 green	3.01:1	
3. purple × white petals	All purple	705 purple; 224 white	3.15:1	
4. inflated × pinched pods	All inflated	882 inflated; 299 pinched	2.95:1	
5. green × yellow pods	All green	428 green; 152 yellow	2.82:1	
6. axial × terminal flowers	All axial	651 axial; 207 terminal	3.14:1	
7. long × short stems	All long	787 long; 277 short	2.84:1	

the pea-color example, translated into modern terms, was as follows:

1. A hereditary factor called a **gene** is necessary for producing pea color.

2. Each plant has a pair of this type of gene.

3. The gene comes in two forms called **alleles.** If the gene is phonetically called a "wye" gene, then the two alleles can be represented by Y (standing for the yellow phenotype) and y (standing for the green phenotype).

4. A plant can be either Y/Y, y/y, or Y/y. The slash shows that the alleles are a pair.

5. The phenotype of Y/y plants is always yellow, even though an allele for the green phenotype is present. One could say that the Y allele "dominates" over the y allele, resulting in the yellow phenotype. The allele whose phenotype is displayed in the Y/y plant, Y, is called the **dominant** allele. The allele whose phenotype is not displayed in the Y/y plant, y, is known as the **recessive** allele.

6. In meiosis, the members of a gene pair separate equally into the cells that become eggs and sperm, the *gametes*. This equal partitioning has become known as **Mendel's first law** or as the **law of equal segregation**. Hence, a single gamete contains only one member of the gene pair.

7. At fertilization, gametes fuse randomly, regardless of which of the alleles they bear.

Here, we introduce some terminology. A fertilized egg, the first cell that develops into a progeny individual, is called a **zygote**. A plant with a pair of identical alleles for a given gene is called a **homozygote** (adjective homozygous), and a plant in which the alleles of the gene pair differ is called a **heterozygote** (adjective heterozygous). Sometimes a heterozygote for one gene is called a **monohybrid**. An individual can be classified as either **homozygous dominant** (such as Y/Y), **heterozygous** (Y/y), or **homozygous recessive** (y/y). In genetics, allelic combinations underlying phenotypes are called **genotypes**. Hence, Y/Y, Y/y, and y/y are all genotypes.

KEY CONCEPT At meiosis, the members of a gene pair segregate equally into the product cells (often sperm or eggs). This is known as Mendel's first law or the law of equal segregation.

Figure 2-5 shows how Mendel's postulates explain the progeny ratios illustrated in Figure 2-4. The pure-breeding lines are homozygous, either Y/Y or y/y. Hence, each line produces only Y gametes or only y gametes and thus can

only breed true. When crossed with each other, the Y/Y and the y/y lines produce an F_1 generation composed of all heterozygous individuals (Y/y). Because Y is dominant, all F_1 individuals are yellow in phenotype. Selfing the F_1 individuals can be thought of as a cross of the type $Y/y \times Y/y$, which is sometimes called a **monohybrid cross**. Equal segregation of the Y and y alleles in the heterozygous F_1 results in gametes, both male and female, half of which are Y and half of which are y. Male and female gametes fuse randomly at fertilization, with the results shown in the grid in Figure 2-5. The composition of the F_2 is three-fourths yellow seeds and one-fourth green, a 3:1 ratio. The one-fourth of the F_2 seeds that are green breed true as expected of the genotype y/y. However, the yellow F_2 seeds (totaling three-fourths) are of two genotypes: two-thirds of them are heterozygotes Y/y, and one-third are homozygous dominant Y/Y. Underlying the 3:1 phenotypic ratio in the F_2 is a 1:2:1 genotypic ratio:

$$\left.\begin{array}{l} \frac{1}{4} \ Y/Y \ \text{yellow} \\ \frac{2}{4} \ Y/y \ \text{yellow} \end{array}\right\} \quad \frac{3}{4} \ \text{yellow} \ (Y/-) $$
$$\frac{1}{4} \ y/y \ \text{green}$$

The general representation of an individual expressing the dominant allele is $Y/-$; the dash represents a slot that can be filled by either another Y or a y. Note that equal segregation is detectable only in the meiosis of a heterozygote; Y/y produces one-half Y gametes and one-half y gametes. Although equal segregation is taking place in homozygotes,

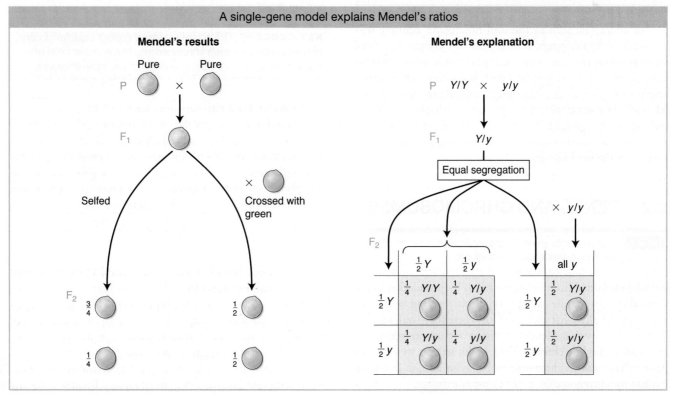

FIGURE 2-5 Mendel's results (*left*) are explained by a single-gene model (*right*) that postulates the equal segregation of the members of a gene pair into gametes.

too, neither segregation $\frac{1}{2}Y : \frac{1}{2}Y$ nor segregation $\frac{1}{2}y : \frac{1}{2}y$ is meaningful or detectable at the genetic level.

We can now also explain results of the cross between the plants grown from F$_1$ yellow seeds (Y/y) and the plants grown from green seeds (y/y). In this case, equal segregation in the yellow heterozygous F$_1$ gives gametes with a $\frac{1}{2}Y : \frac{1}{2}y$ ratio. The y/y parent can make only y gametes, however; so the phenotype of the progeny depends only on which allele they inherit from the Y/y parent. Thus, the $\frac{1}{2}Y : \frac{1}{2}y$ *gametic* ratio from the heterozygote is converted into a $\frac{1}{2}Y/y : \frac{1}{2}y/y$ *genotypic* ratio, which corresponds to a 1:1 *phenotypic* ratio of yellow-seeded to green-seeded plants. This is illustrated in the right-hand panel of Figure 2-5.

KEY CONCEPT Equal segregation of a gene pair during meiosis is observable only in heterozygotes, making them critical tools for genetic analysis. In a cross between a heterozygote and a homozygous recessive, the gametic ratio produced by meiosis in the heterozygote is observed in the phenotypic ratio of the progeny.

Note that, in defining the allele pairs that underlay his phenotypes, Mendel had identified a gene that radically affects pea color. This identification was not his prime interest, but we can see how finding single-gene inheritance patterns is a process of gene discovery, identifying individual genes that influence a biological trait.

KEY CONCEPT All 1:1, 3:1, and 1:2:1 genetic ratios are diagnostic of single-gene inheritance and are based on equal segregation in a heterozygote.

Mendel's research in the mid-nineteenth century was not noticed by the international scientific community until similar observations were independently published by several other researchers in 1900. Soon research in many species of plants, animals, fungi, and algae showed that Mendel's law of equal segregation was applicable to all sexual eukaryotes and, in all cases, was based on the chromosomal segregations taking place in meiosis, a topic that we turn to in the next section.

2.2 GENES AND CHROMOSOMES

LO 2.2 Explain single-gene inheritance ratios in terms of chromosome behavior at meiosis.

Mendel explained his inheritance ratios by postulating hypothetical entities that he called *heritable factors*, but he did not know specifically what they were, or where they were located. Today we know that Mendel's heritable factors are genes, and genes are located on chromosomes. In this section, we correlate gene behavior with chromosome behavior and, in doing so, explain the chromosomal basis of gene inheritance.

In the normal day-to-day chemistry of the cell, genes carry out their activities on chromosomes that are relatively immobile within the nucleus. However, when cells divide, chromosomes are involved in a highly programmed set of moves that partition them into new cells. We must now turn to these movements.

There are two types of cell division in eukaryotes, each with unique consequences. **Somatic cell division** is division of cells of the main body, known as the soma. The products of somatic cell division are exact copies of the parent cell. **Sexual cell division** takes place in sex organs. Specialized cells called **meiocytes** divide to produce sex cells such as sperm and eggs in plants and animals, or sexual spores in fungi or algae. When cells divide, so do their nuclei; somatic cell nuclear division is called **mitosis**, and sexual cell nuclear division is called **meiosis**. The life cycles of some well-known organisms, showing when mitosis and meiosis occur, are found in **Figure 2-6**.

KEY CONCEPT During somatic cell division, the accompanying nuclear division is mitosis. During sexual cell division, the accompanying nuclear division is meiosis.

Notice that in Figure 2-6, in the animal and plant examples, the cells of the adult body are labeled $2n$. In this terminology, $n =$ the number of chromosomes in the genome, and the number 2 indicates that there are two genomes (chromosome sets) per adult cell. $2n$ somatic cells are called **diploid**. A photograph of a diploid cell in a muntjac deer is shown in **Figure 2-7**. Note that in a diploid cell, the chromosomes are in pairs (there are n pairs); the two members of a pair are called *homologous chromosomes*, or *homologs*. The third example in Figure 2-6 shows a **haploid** organism (in this case a fungus) whose somatic cells have just one chromosome set, n. A large proportion of organisms on the planet are haploid.

KEY CONCEPT The somatic cells of diploid organisms contain two copies of each chromosome. The somatic cells of haploid organisms contain one copy of each chromosome.

To understand chromosome segregation, we must first understand and contrast the two types of nuclear divisions that take place in eukaryotic cells. Somatic cell division and the accompanying nuclear division (mitosis) is a programmed stage of all eukaryotic cell-division cycles (**Figure 2-8**). Mitosis can take place in diploid or haploid cells. As a result, one progenitor cell becomes two genetically identical cells. Hence,

$$\text{either } 2n \rightarrow 2n + 2n$$

$$\text{or } n \rightarrow n + n$$

This "trick" of constancy is accomplished when each chromosome replicates to make two identical copies of itself, with underlying DNA replication. The two identical copies of each chromosome are pulled to opposite ends of the cell. When the cell divides, each daughter cell has the same chromosomal set as its progenitor.

In sexual cell division two sequential divisions take place, along with two nuclear divisions (meiosis). Because there are two divisions, four cells are produced from each progenitor cell. Meiosis takes place only in diploid cells, and the resulting gametes are haploid. Hence, the net result of meiosis is

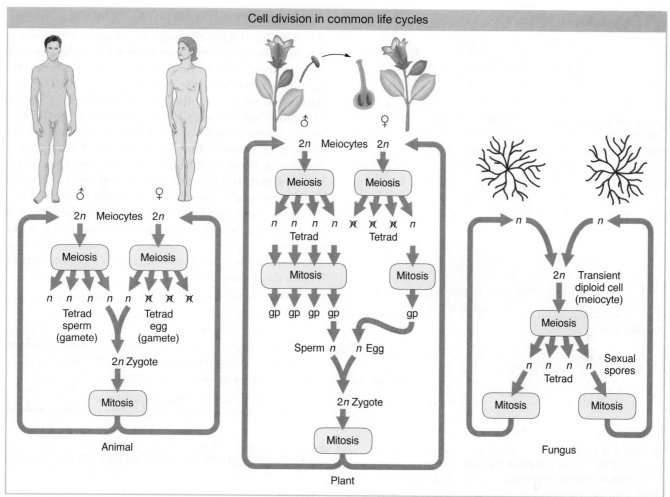

FIGURE 2-6 The life cycles of humans, plants, and fungi, showing the points at which mitosis and meiosis take place. Note that in the females of humans and many plants, three cells of the meiotic tetrad abort; only one develops into an egg. The abbreviation *n* indicates a haploid cell, 2*n* a diploid cell; "gp" stands for gametophyte, the name of the small structure composed of haploid cells that will produce gametes. In many plants such as corn, a nucleus from the male gametophyte fuses with two nuclei from the female gametophyte, giving rise to a triploid (3*n*) cell, which then replicates to form the endosperm, a nutritive tissue that surrounds the embryo (which is derived from the 2*n* zygote).

$$2n \rightarrow n + n + n + n$$

The group of haploid cells is called a tetrad (tetra is Greek for four). This overall halving of chromosome number during meiosis arises because, even though two cell divisions take place, chromosome replication occurs only once. As we will see shortly, the chromosome movements that occur during meiosis ensure that each haploid gamete contains a complete set of chromosomes.

In the sections that follow, we will see how chromosomes and genes are segregated in both diploids and haploids, and at mitosis and meiosis in each.

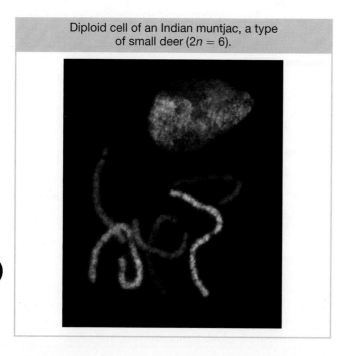

Diploid cell of an Indian muntjac, a type of small deer (2*n* = 6).

FIGURE 2-7 The six visible chromosomes are from a cell caught in the process of nuclear division. The three pairs of chromosomes have been stained with chromosome-specific DNA probes, each tagged with a different fluorescent dye (chromosome paint). A nucleus derived from another cell is at the stage between divisions. [*Republished with permission of Annual Reviews, from Ferguson-Smith, Malcolm A., "Putting Medical Genetics into Practice,"* Annual Review of Genomics and Human Genetics, *2011, September; 12: 1–23, Figure 2. Permission conveyed through Copyright Clearance Center, Inc.*]

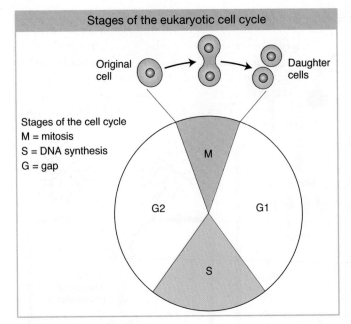

Stages of the eukaryotic cell cycle

Original cell → Daughter cells

Stages of the cell cycle
M = mitosis
S = DNA synthesis
G = gap

M
G2
G1
S

FIGURE 2-8

Single-gene inheritance in diploids

At the onset of mitosis in somatic cells, the chromosomes condense by coiling to become shorter, and they are seen to have duplicated to form daughter units called **chromatids**. At this stage the two sister chromatids remain associated with each other, joined together at a special chromosomal region called the centromere. Each chromatid represents one of two identical DNA molecules formed just before mitosis by DNA replication (the S phase in Figure 2-8). Each pair of chromatids aligns on the equatorial plane of the cell and then, as the cell divides, molecular threads called spindle fibers attach to the centromere and pull one sister chromatid into each daughter cell as the centromere divides. Once in the daughter cells, the chromatids become individual chromosomes in their own right. Mitosis in a diploid heterozygote *Aa* is shown here:

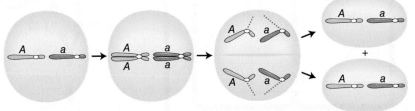

FIGURE 2-9 Simplified representation of mitosis and meiosis in diploid cells (2*n*, diploid; *n*, haploid). (Detailed versions of mitosis and meiosis are shown in Appendix 2-1, page 75, and Appendix 2-2, pages 76–77.)

ANIMATED ART   SaplingPlus
Mitosis

ANIMATED ART   SaplingPlus
Meiosis

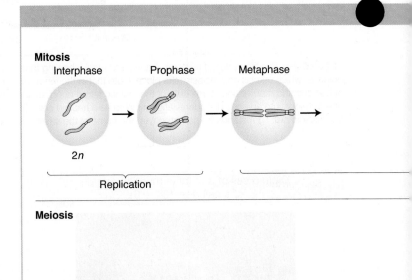

Mitosis

Interphase   Prophase   Metaphase

2*n*

Replication

Meiosis

Interphase   Prophase I   Metaphase I

2*n*

Replication

Pairing

We see that mitosis faithfully produces two cells of the same genotype as the original cell:

$$Aa \rightarrow Aa + Aa$$

Meiosis is also preceded by replication and chromosome condensation, but a key difference is that in this case, the two homologous chromosomes pair to form a group of four chromatids at the equatorial plane. Meiosis takes place over two cell divisions. In the first division, the centromere holding a pair of sister chromatids together does not divide. One pair of chromatids is pulled into each daughter cell by spindles that attach to the undivided centromeres. At the second division of meiosis, the centromeres divide; and now each chromatid is pulled into its own cell, which is now effectively haploid. Meiosis in a diploid heterozygote *Aa* is shown here:

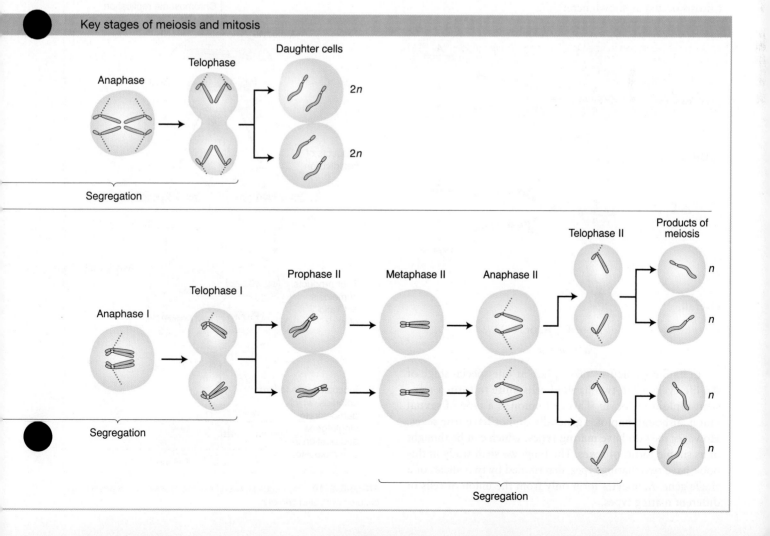

Key stages of meiosis and mitosis

We can see that a single diploid cell of genotype *Aa* produces four haploid cells, two of genotype *A* and two of genotype *a*. Hence, we now see the underlying chromosomal mechanism that produces the 1:1 gametic ratio postulated by Mendel:

$$Aa \rightarrow A + A + a + a$$
$$Aa \rightarrow 1/2 \ A \text{ and } 1/2 \ a$$
$$\text{ratio} = 1A : 1a$$

Figure 2-9 shows the named stages of mitosis and meiosis in a diploid organism. Note the difference in alignment of homologous chromosomes in metaphase of mitosis and metaphase I of meiosis, and the way the chromatids are partitioned during the subsequent anaphase.

> **KEY CONCEPT** The physical separation of chromosome pairs during anaphase I of meiosis is the basis for Mendel's law of equal segregation.

Single-gene inheritance in haploids

Mitosis in haploids progresses in much the same way, but in haploid organisms, each somatic cell bears only one chromosome set. If we choose to use the *A* and *a* terminology to represent alleles of a gene, a cell can be either *A or a*. Mitosis occurs as shown here:

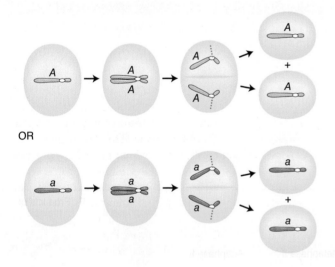

Overall,

$$A \rightarrow A + A, \text{ or}$$
$$a \rightarrow a + a$$

In haploids, meiosis takes place at one special stage of the life cycle when two haploid cells unite to form a transient diploid meiocyte. This cell union is a type of sexual union, although haploids generally do not have true sexes. However they do have **mating types**, which can be thought of as simple forms of sexes. The fungi we shall study in this book have two mating types, determined by two alleles of a single gene. Meiocytes form only from the union of cells of different mating types.

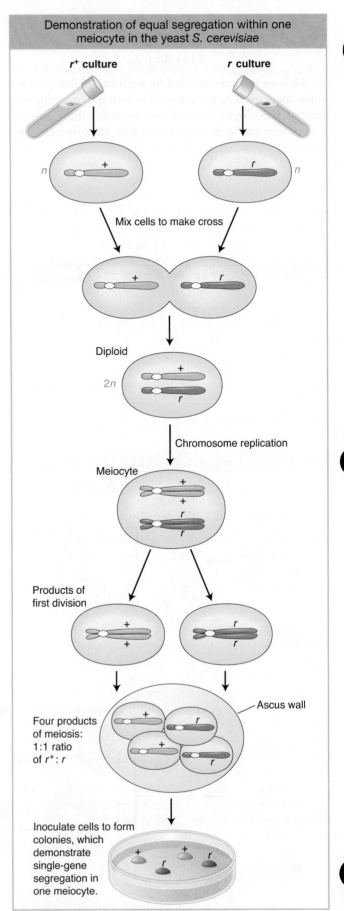

Demonstration of equal segregation within one meiocyte in the yeast *S. cerevisiae*

FIGURE 2-10 One ascus isolated from the cross $r^+ \times r$ leads to two cultures of r^+ and two of r.

Let's look at a cross in the haploid fungus baker's yeast, involving a red mutant, which contrasts with the normal white color of wild type. We will postulate a meiocyte formed by the union of a haploid cell of a red mutant, r, and a haploid cell of wild type r^+ (the + superscript is commonly used to designate wild type alleles). The transient diploid meiocyte is a heterozygote r^+/r. As expected from our previous discussion of a diploid meiosis, the four haploid cells produced are $1/2\ r$ and $1/2\ r^+$, also reflecting Mendel's first law of equal segregation.

It is noteworthy that in many haploid organisms, especially fungi, the four cells that are the products of a single meiosis remain together enclosed in a membranous sac. In yeast, this sac is called an *ascus*. The four haploid nuclear products in the ascus represent the meiotic tetrad. Thus, it is possible to perform analyses of single meioses by using a needle to separate and isolate each of the contents of the sac, and such tetrad analyses have made valuable contributions to research on the analysis of meiosis itself, and on single-gene inheritance. The yeast cross of $r \times r^+$ is shown in detail in **Figure 2-10**.

KEY CONCEPT Mitotic division results in the original chromosome number in each of the two product cells. Meiotic division results in half the original chromosome number in each of the four product cells.

2.3 THE MOLECULAR BASIS OF MENDELIAN INHERITANCE PATTERNS

LO 2.3 Propose reasonable hypotheses to explain dominance and recessiveness of specific alleles at the molecular level.

Of course, Mendel had no idea of the molecular nature of the concepts he was working with. In this section, we can begin putting some of Mendel's concepts into a molecular context. Let's begin with alleles. We have used the concept of *alleles* without defining them at the molecular level. What are the *structural differences* between wild-type and mutant alleles at the DNA level of a gene? What are the *functional differences* at the protein level? Mutant alleles can be used to study single-gene inheritance without needing to understand their structural or functional nature. However, because a primary reason for embarking on single-gene inheritance is ultimately to investigate a gene's function, we must come to grips with the molecular nature of wild-type and mutant alleles at both the structural and the functional level.

Structural differences between alleles at the molecular level

Mendel proposed that genes come in different forms we now call alleles. What are alleles at the molecular level? When alleles such as A and a are examined at the DNA

level by using modern technology, they are generally found to be identical in most of their sequences and differ only at one or several nucleotides of the hundreds or thousands of nucleotides that make up the gene. Therefore, we see that the alleles are truly different versions of the same gene. The following diagram represents the DNA of two alleles of one gene; the letter x represents a difference in the nucleotide sequence:

If the nucleotide sequence of an allele changes as the result of a rare chemical "accident," a new mutant allele is created. Such changes can occur anywhere along the nucleotide sequence of a gene. For example, a mutation could be a change in the identity of a single nucleotide, or the deletion of one or more nucleotides, or even the addition of one or more nucleotides.

A gene can be changed by mutation in many ways. For one thing, the mutational damage can occur at any one of many different sites. We can represent the situation as follows, where dark blue indicates the normal wild-type DNA sequence, and red with the letter x represents the altered sequence:

Molecular aspects of gene transmission

Replication of alleles during the S phase What happens to alleles at the molecular level during cell division? We know that the primary genomic component of each chromosome is a DNA molecule. This DNA molecule is replicated during the S phase, which precedes both mitosis and meiosis. As we will see in Chapter 7, replication is an accurate process, and so all the genetic information is duplicated, whether wild type or mutant. For example, if a mutation is the result of a change in a single nucleotide pair—say, from GC (wild type) to AT (mutant)—then in a heterozygote, replication will be as follows:

$$\text{homolog GC} \rightarrow \text{replication} \rightarrow \begin{array}{l}\text{chromatid GC}\\\text{chromatid GC}\end{array}$$

$$\text{homolog AT} \rightarrow \text{replication} \rightarrow \begin{array}{l}\text{chromatid AT}\\\text{chromatid AT}\end{array}$$

DNA replication before mitosis in a haploid and a diploid are shown in **Figure 2-11**. This type of illustration serves to remind us that, in our considerations of the mechanisms of inheritance, it is essentially DNA molecules that are being moved around in the dividing cells.

Meiosis and mitosis at the molecular level The replication of DNA during the S phase produces two copies of each of the alleles A and a, that can now be segregated into

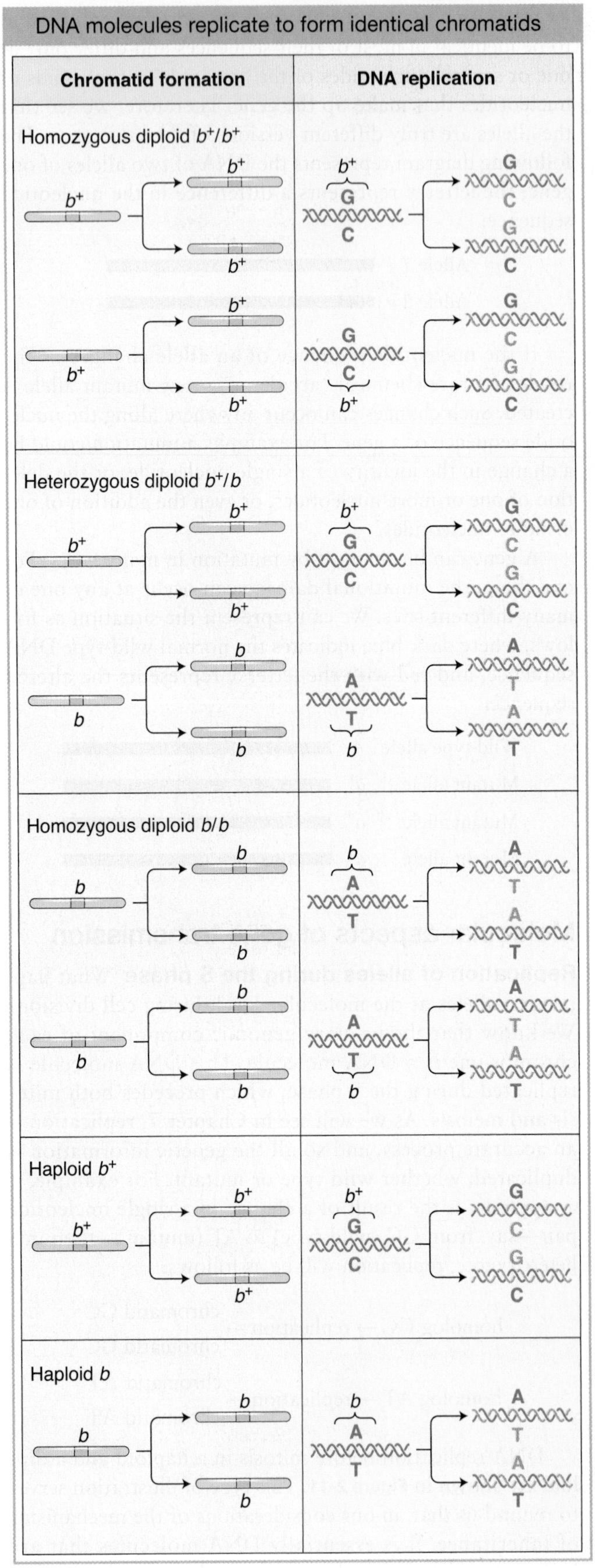

DNA molecules replicate to form identical chromatids

Chromatid formation	DNA replication

Homozygous diploid b^+/b^+

Heterozygous diploid b^+/b

Homozygous diploid b/b

Haploid b^+

Haploid b

separate cells. Nuclear division visualized at the DNA level is shown in **Figure 2-12**.

Demonstrating chromosome segregation at the molecular level

We have interpreted single-gene phenotypic inheritance patterns in relation to the segregation of chromosomal DNA at meiosis. Is there any way to show DNA segregation directly (as opposed to phenotypic segregation)? The most straightforward approach would be to sequence the alleles (say, *A* and *a*) in the parents and the meiotic products: the result would be that one-half of the products would have the *A* DNA sequence and one-half would have the *a* DNA sequence. The same would be true for any DNA sequence that differed in the inherited chromosomes, including regions of DNA found between genes (i.e., not inside alleles correlated with known phenotypes such as red and white flowers). Thus, we see the rules of segregation enunciated by Mendel apply not only to genes, but to any stretch of DNA along a chromosome.

KEY CONCEPT Mendelian inheritance is shown by any segment of DNA on a chromosome: by genes and their alleles and by molecular markers not necessarily associated with any biological function.

Alleles at the molecular level

At the molecular level, the primary phenotype of a gene is the protein it produces. What are the functional differences between proteins that explain the different effects of wild-type and mutant alleles on the traits of an organism?

Let's explore the topic by using the human disease phenylketonuria (PKU). We shall see in a later section on pedigree analysis that the PKU phenotype is inherited as a Mendelian recessive. The disease is caused by a defective allele of the gene that encodes the liver enzyme phenylalanine hydroxylase (PAH). This enzyme normally converts phenylalanine in food into the amino acid tyrosine:

$$\text{phenylalanine} \xrightarrow{\text{phenylalanine hydroxylase}} \text{tyrosine}$$

FIGURE 2-11 Each chromosome divides longitudinally into two chromatids (*left*); at the molecular level (*right*), the single DNA molecule of each chromosome replicates, producing two DNA molecules, one for each chromatid (orange indicates the newly synthesized strand). Also shown are various combinations of a gene with wild-type allele b^+ and mutant form *b*, caused by the change in a single base pair from GC to AT. Notice that, at the DNA level, the two chromatids produced when a chromosome replicates are always identical with each other and with the original chromosome. Also, note that after DNA replication, but before anaphase of mitosis, each pair of sister chromatids is joined at the centromere (not shown).

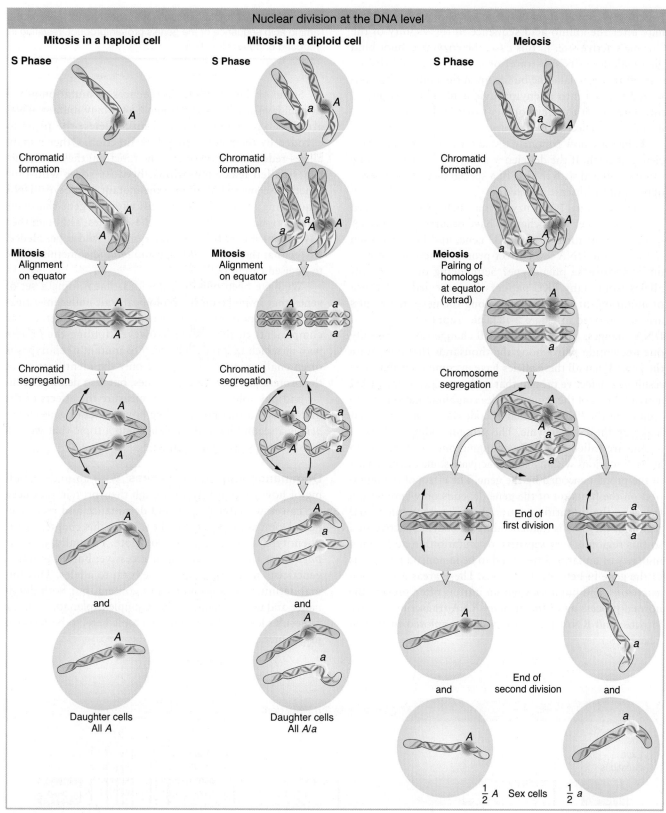

FIGURE 2-12 DNA and gene transmission in mitosis and meiosis in eukaryotes. The S phase and the main stages of mitosis and meiosis are shown. Mitotic divisions (*left and middle*) conserve the genotype of the original cell. At the right, the two successive meiotic divisions that take place during the sexual stage of the life cycle have the net effect of halving the number of chromosomes. The alleles *A* and *a* of one gene are used to show how genotypes are transmitted in cell division.

However, a mutation in the gene encoding this enzyme may alter the amino acid sequence in the vicinity of the enzyme's active site. In this case, the enzyme cannot bind phenylalanine (its substrate) or convert it into tyrosine. Therefore, phenylalanine builds up in the body and is converted instead into phenylpyruvic acid. This compound interferes with the development of the nervous system, leading to intellectual disabilities.

Babies are now routinely tested for this processing deficiency at birth. If the deficiency is detected, phenylalanine can be withheld with the use of a special diet and the development of the disease arrested.

The functional PAH enzyme is a single protein, encoded by a single gene. What changes have occurred in the DNA of the mutant form of the PKU gene, and how can such change at the DNA level affect protein function and produce the disease phenotype? Sequencing of the mutant alleles from many PKU patients has revealed a plethora of mutations at different sites along the gene; the results are summarized in **Figure 2-13**. They represent a range of DNA changes, but most are small changes affecting only one nucleotide pair among the thousands that constitute the gene. What all these alleles have in common is that they result in a defective protein that no longer has normal PAH activity. Most of the mutant alleles contain mutations in the regions of the PKU gene that encode the amino acids that make up the PAH enzyme. The protein-coding regions of a gene are called *exons*. By changing one or more amino acids, mutations within exons inactivate some essential part of the protein encoded by the gene. The effect of the mutation on the function of the gene depends on where within the gene the mutation occurs. An important functional region of the gene is that encoding an enzyme's active site; so this region is very sensitive to mutation. In addition, a minority of mutations are found to be in noncoding regions of the gene in between the exons. These areas are known as *introns*, and mutations within introns often prevent the normal processing of the primary RNA transcript. (Exons, introns, and RNA processing will be explored further in Chapter 8.)

> **KEY CONCEPT** Most mutations that alter phenotype alter the amino acid sequence of the gene's protein product, resulting in reduced or absent function.

Some of the general consequences of mutation at the protein level are shown in **Figure 2-14**. Many mutant alleles are of a type generally called **null alleles**: the proteins encoded by them completely lack function. Other mutant alleles reduce the level of enzyme function; they are sometimes called **leaky mutations,** because some wild-type function seems to "leak" into the mutant phenotype. DNA sequencing often detects changes within a gene that have no functional impact at all, so these alleles, although they have **silent mutations,** are functionally wild type. Hence, we see that the terms *wild type* and *mutant* sometimes have to be used carefully.

We have been pursuing the idea that finding a set of genes that impinge on the biological trait under investigation is an important goal of genetics, because it defines the components of the system. However, finding the *precise* way in which mutant alleles lead to mutant phenotypes is often challenging, requiring not only the identification of the protein products of these genes, but also detailed cellular and physiological studies to measure the effects of the mutations. Furthermore, finding how the set of genes *interacts* is a second level of challenge and a topic that we will pursue later, starting in Chapter 5.

Dominance and recessiveness With an understanding of how genes function through their protein products, we can now better understand dominance and recessiveness. Dominance was defined earlier in this chapter as the phenotype shown by a heterozygote. Formally, it is the *phenotype* that is dominant or recessive; but, in practice, geneticists more often apply the term to alleles. This formal definition has no molecular content, but both dominance and recessiveness can have simple explanations at the molecular level. We introduce the topic here, to be revisited in Chapter 5.

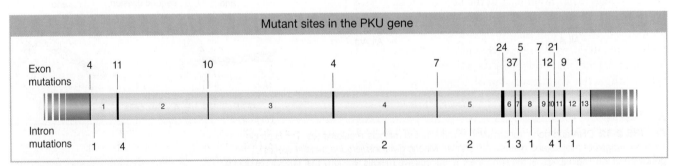

FIGURE 2-13 Many mutations of the human phenylalanine hydroxylase gene that cause enzyme malfunction are known. The number of mutations in the exons, or protein-encoding regions (black), are listed above the gene. The number of mutations in the intron regions (green, numbered 1 through 13) that alter RNA processing are listed below the gene. [*Data from C. R. Scriver,* Ann. Rev. Genet. 28, *1994, 141–165.*]

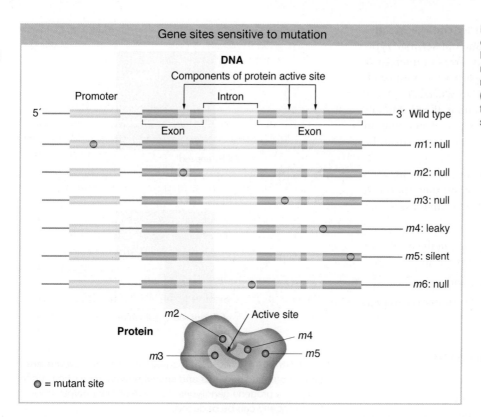

FIGURE 2-14 Mutations in the parts of a gene encoding enzyme active sites lead to enzymes that do not function (null mutations). Mutations elsewhere in the gene may have a partial effect on enzyme function (leaky mutations), or no effect on enzyme function (silent mutations). Promoters are sites important in transcription initiation.

How can alleles be dominant? How can they be recessive? Recessiveness is observed in null mutations in genes that are functionally **haplosufficient,** loosely meaning that one gene copy has enough function to produce a wild-type phenotype. Although a wild-type diploid cell normally has two fully functional copies of a gene, one copy of a haplosufficient gene provides enough gene product (generally a protein) to carry out the normal transactions of the cell. In a heterozygote (say, $+/m$, where m is a null allele), the single functional copy encoded by the $+$ allele provides enough protein product for normal cellular function. In a simple example, assume a cell needs a minimum of 10 protein units to function normally. Each wild-type allele can produce 12 units. Hence, a homozygous wild type $+/+$ will produce 24 units. The heterozygote $+/m$ will produce 12 units, in excess of the 10-unit minimum, and hence the mutant allele is recessive as it has no impact in the heterozygote.

Other genes are **haploinsufficient.** In such cases, a null mutant allele will be dominant because, in a heterozygote $(+/P)$, the single wild-type allele cannot provide enough product for normal function. As another example, let's assume the cell needs a minimum of 20 units of this protein, and the wild-type allele produces only 12 units. A homozygous wild type $+/+$ makes 24 units, which is over the minimum. However, a heterozygote involving a null mutation $(+/P)$ produces only 12; hence, the presence of the mutant allele in the heterozygote results in an inadequate supply of product, and a mutant phenotype ensues. In this situation, the mutant allele P is dominant.

In some cases, mutation results in a *new function* for the gene. Such mutations can be dominant because, in a heterozygote, the wild-type allele cannot mask this new function.

From the above brief considerations, we see that *phenotype*, the description or measurement that we track during Mendelian inheritance, is an emergent property based on the nature of alleles and the way in which the gene functions normally and abnormally. The same can be said for the descriptions "dominant" and "recessive" that we apply to a phenotype.

KEY CONCEPT As a general rule, a null mutation is recessive in a haplosufficient gene, and a null mutation is dominant in a haploinsufficient gene.

2.4 SOME GENES DISCOVERED BY OBSERVING SEGREGATION RATIOS

LO 2.4 Predict phenotypic ratios among descendants from crosses of parents differing at a single gene.

Recall that one general aim of genetic analysis today is to dissect a biological trait by discovering the set of single genes that affect it. We learned that an important way to identify these genes is by the phenotypic segregation ratios generated by their mutations—most often 1:1 and 3:1 ratios, both of which are based on equal segregation as defined by Gregor Mendel.

Let's look at some examples that extend the Mendelian approach into a modern experimental setting. Typically, the researcher is confronted by an array of interesting mutant phenotypes that affect the property of interest (such as those depicted in Figure 2-1) and now needs to know whether they are inherited as single-mutant alleles. Mutant alleles can be either dominant or recessive, depending on their action, so the question of dominance also needs to be considered in the analysis.

The standard procedure is to cross a mutant with wild type. (If the mutant is sterile, then another approach is needed.) First, we will consider three simple cases that cover most of the possible outcomes:

1. A fertile flower mutant with no pigment in the petals (for example, white petaled in contrast with the normal red)

2. A fertile fruit-fly mutant with short wings

3. A fertile mold mutant that produces excess hyphal branches (hyperbranching)

A gene active in the development of flower color

To begin the process, the white-flowered plant is crossed with the normal wild-type red. All the F_1 plants are red flowered, and, of 500 F_2 plants sampled, 378 are red flowered and 122 are white flowered. If we acknowledge the existence of sampling error, these F_2 numbers are very close to a $\frac{3}{4}:\frac{1}{4}$, or 3:1, ratio. Because this ratio indicates single-gene inheritance, we can conclude that the mutant is caused by a recessive alteration in a single gene. According to the general rules of gene nomenclature, the mutant allele for white petals might be called *alb* for *albino* and the wild-type allele would be alb^+ or just +. (The conventions for allele nomenclature vary somewhat among organisms: some of the variations are shown in Appendix A on nomenclature.) We surmise that the wild-type allele plays an essential role in producing the colored petals of the plant, a property that is almost certainly necessary for attracting pollinators to the flower. The gene might be implicated in the biochemical synthesis of the pigment or in the part of the signaling system that tells the cells of the flower to start making pigment or in a number of other possibilities that require further investigation. At the purely genetic level, the crosses made would be represented symbolically as

P	$+/+ \times alb/alb$
F$_1$	all $+/alb$
F$_2$	$\frac{1}{4}\ +/+$
	$\frac{1}{2}\ +/alb$
	$\frac{1}{4}\ alb/alb$

or graphically as in the grids in the next column (see also Figure 2-5). This type of grid showing gametes and gametic fusions is called a *Punnett square*, named after an early geneticist, Reginald C. Punnett. They are useful devices for explaining genetic ratios, and we shall encounter more in later discussions.

All F$_1$ are red

$\frac{3}{4}$ of F$_2$ are red, $\frac{1}{4}$ are white

KEY CONCEPT The Punnett square is a graphical representation of parental gametes and shows how they randomly unite to produce progeny genotypes, from which phenotypic ratios of the progeny can be deduced.

A gene for wing development

In the fruit-fly example, the cross of the mutant short-winged fly with wild-type long-winged stock yielded 788 progeny, classified as follows:

    196 short-winged males
    194 short-winged females
    197 long-winged males
    201 long-winged females

In total, there are 390 short- and 398 long-winged progeny, very close to a 1:1 ratio. The ratio is the same within males and females, again within the bounds of sampling error. Hence, from these results, the "short wings" mutant was very likely produced by a dominant mutation. Note that, for a dominant mutation to be expressed, only a single "dose" of mutant allele is necessary; so, in most cases, when the mutant first shows up in the population, it will be in the heterozygous state. (This is not true for a recessive mutation such as that in the preceding plant example, which must be homozygous to be expressed and must have come from the selfing of an unidentified heterozygous plant in the preceding generation.)

When long-winged progeny were interbred, all of their progeny were long winged, as expected of a recessive wild-type allele. When the short-winged progeny were interbred, their progeny showed a ratio of three-fourths short to one-fourth long.

Dominant mutations are represented by uppercase letters or words: in the present example, the mutant allele might be named *SH*, standing for "short." Then the crosses would be represented symbolically as

$$P \qquad +/+ \times SH/+$$
$$F_1 \qquad \tfrac{1}{2} +/+$$
$$\qquad \tfrac{1}{2} SH/+$$

$$F_1 \qquad +/+ \times +/+$$
$$\qquad \text{all } +/+$$

$$F_1 \qquad SH/+ \times SH/+$$
$$\qquad \tfrac{1}{4} SH/SH$$
$$\qquad \tfrac{1}{2} SH/+$$
$$\qquad \tfrac{1}{4} +/+$$

or graphically as shown in the grids below.

This analysis of the fly mutant identifies a gene that is part of a subset of genes that, in wild-type form, are crucial for the normal development of a wing. Such a result is the starting point of further studies that would focus on the precise developmental and cellular ways in which the growth of the wing is arrested, which, once identified, reveal the time of action of the wild-type allele in the course of development.

P	+	SH
+	+/+	SH/+
+	+/+	SH/+

F_1	+	+
+	+/+	+/+
+	+/+	+/+

F_1	+	SH
+	+/+	SH/+
SH	SH/+	SH/SH

KEY CONCEPT A dominant mutation in the heterozygous state will be expressed. A cross between heterozygous dominant and wild type parents will result in a 1:1 phenotypic ratio in the progeny.

A gene for hyphal branching

A hyperbranching fungal mutant (such as the button-like colony in Figure 2-1) was crossed with a wild-type fungus with normal sparse branching. In a sample of 300 progeny, 152 were wild type and 148 were hyperbranching, very close to a 1:1 ratio. We infer from this single-gene inheritance ratio that the hyperbranching mutation is of a single gene. In haploids, assigning dominance is usually not possible, but, for convenience, we can call the hyperbranching allele *hb* and the wild type hb^+ or +. The cross must have been

P	$+ \times hb$
Diploid meiocyte	$+/hb$
F_1	$\tfrac{1}{2} +$
	$\tfrac{1}{2} hb$

The mutation and inheritance analysis has uncovered a gene whose wild-type allele is essential for normal control of branching, a key function in fungal dispersal and nutrient acquisition. Now the mutant needs to be investigated to see the location in the normal developmental sequence at which the mutant produces a block. This information will reveal the time and place in the cells at which the normal allele acts.

Sometimes, the severity of a mutant phenotype renders the organism sterile, unable to go through the sexual cycle. How can the single-gene inheritance of sterile mutants be demonstrated? In a diploid organism, a sterile recessive mutant can be propagated as a heterozygote, and then the heterozygote can be selfed to produce the expected 25 percent homozygous recessive mutants for study. A sterile dominant mutant is a genetic dead end and cannot be propagated sexually, but, in plants and fungi, such a mutant can be easily propagated asexually.

What if a cross between a mutant and a wild type does not produce a 3:1 or a 1:1 ratio as discussed here, but some other ratio? Such a result can be due to the interactions of several genes or to an environmental effect. Some of these possibilities are discussed in Chapter 5, and environmental effects on phenotype are also considered in Solved Problem 1 at the end of this chapter.

KEY CONCEPT In research on a new mutation affecting a trait of interest, the demonstration of Mendelian single-gene ratios in crossing analysis reveals a gene that is important in the developmental pathways for that trait.

Predicting progeny proportions or parental genotypes by applying the principles of single-gene inheritance

We can summarize the direction of analysis of gene discovery as follows:

Observe phenotypic ratios in progeny →
Deduce genotypes of parents (*A/A*, *A/a*, or *a/a*)

However, the same principle of inheritance (essentially, Mendel's law of equal segregation) can also be used to predict phenotypic ratios in the progeny of parents of *known genotypes*. These parents would be from stocks maintained by the researcher. The types and proportions of the progeny of crosses such as $A/A \times A/A$, $A/A \times a/a$, $A/a \times A/a$, and $A/a \times a/a$ can be easily predicted. In summary,

Cross parents of known genotypes →
   Predict phenotypic ratios in progeny

This type of analysis is used in general breeding to synthesize genotypes for research or for agriculture. It is also useful in predicting likelihoods of various outcomes in human matings in families with histories of single-gene diseases.

After single-gene inheritance has been established, an individual showing the dominant phenotype but of *unknown genotype* can be tested to see if the genotype is homozygous or heterozygous. Such a test can be performed by crossing the individual (of phenotype $A/?$) with a recessive tester strain a/a. If the individual is heterozygous, a 1:1 ratio will result ($\frac{1}{2} A/a$ and $\frac{1}{2} a/a$); if the individual is homozygous, all progeny will show the dominant phenotype (all A/a). In general, the cross of an individual of unknown heterozygosity (for one gene or more) with a fully recessive parent is called a **testcross**, and the recessive individual is called a **tester**. We will encounter testcrosses many times throughout subsequent chapters; they are very useful in deducing the meiotic events taking place in more complex genotypes. The use of a fully recessive tester means that meiosis in the tester parent can be ignored because all of its gametes are recessive and do not contribute to the phenotypes of the progeny. An alternative test for heterozygosity (useful if a recessive tester is not available and the organism can be selfed) is simply to self the unknown: if the organism being tested is heterozygous, a 3:1 ratio will be found in the progeny. Such tests are useful and common in routine genetic analysis.

KEY CONCEPT The principles of inheritance (such as the law of equal segregation) can be applied in two directions: (1) inferring genotypes from phenotypic ratios and (2) predicting phenotypic ratios from parents of known genotypes.

2.5 SEX-LINKED SINGLE-GENE INHERITANCE PATTERNS

LO 2.5 In the progeny of controlled crosses, recognize phenotypic ratios diagnostic of X-linked single-gene inheritance.

The chromosomes that we have analyzed so far are autosomes, the "regular" chromosomes that form most of the genomic set. However, many animals and plants have a special pair of chromosomes associated with sex. The sex chromosomes also segregate equally, but the phenotypic ratios

seen in progeny are often different from the autosomal ratios.

Sex chromosomes

Most animals and many plants show sexual dimorphism; in other words, individuals are either male or female. In most of these cases, sex is determined by a special pair of **sex chromosomes.** Let's look at humans as an example. Human body cells have 46 chromosomes: 22 homologous pairs of autosomes plus 2 sex chromosomes. Females have a pair of identical sex chromosomes called the **X chromosomes.** Males have a nonidentical pair, consisting of one X and one Y. The **Y chromosome** is considerably shorter than the X. Hence, if we let A represent autosomal chromosomes, we can write

$$\text{females} = 44A + XX$$
$$\text{males} = 44A + XY$$

At meiosis in females, the two X chromosomes pair and segregate like autosomes, and so each egg receives one X chromosome. Hence, with regard to sex chromosomes, the gametes are of only one type and the female is said to be the **homogametic sex.** At meiosis in males, the X and the Y chromosomes pair over a short region, which ensures that the X and Y segregate so that there are two types of sperm, half with a single X and the other half with a single Y. Therefore, the male is called the **heterogametic sex.**

KEY CONCEPT Human sex chromosomes, X and Y, contain different sets of genes. Females are the homogametic sex, with a pair of X chromosomes (XX). Males are the heterogametic sex, with a nonidentical pair of sex chromosomes (XY).

The inheritance patterns of genes on the sex chromosomes are different from those of autosomal genes. Sex-chromosome inheritance patterns were first investigated in the early 1900s in the laboratory of the great geneticist Thomas Hunt Morgan, using the fruit fly *Drosophila melanogaster* (see the Model Organism box on page 50). This insect has been one of the most important research organisms in genetics; its short, simple life cycle contributes to its usefulness in this regard. Fruit flies have three pairs of autosomes plus a pair of sex chromosomes, again referred to as X and Y. As in mammals, *Drosophila* females have the constitution XX and males are XY. However, the mechanism of sex determination in *Drosophila* differs from that in mammals. In *Drosophila*, the *number of X chromosomes* in relation to the autosomes determines sex: two X's result in a female, and one X results in a male. In mammals, the *presence of the Y chromosome* determines maleness and the absence of a Y determines femaleness. However, it is important to note that, despite this somewhat different basis for sex determination, the single-gene inheritance patterns of genes on the sex chromosomes are remarkably similar in *Drosophila* and mammals.

Vascular plants show a variety of sexual arrangements. **Dioecious species** are those showing animal-like sexual

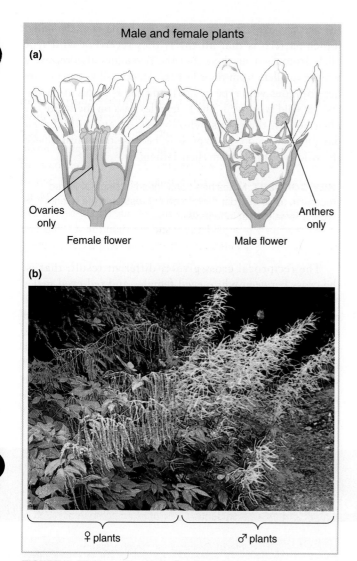

FIGURE 2-15 Examples of two dioecious plant species are (a) *Osmaronia dioica* and (b) *Aruncus dioicus*. [(a) *Leslie Bohm;* (b) *Anthony Griffiths.*]

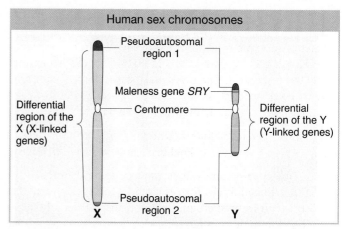

FIGURE 2-16 Human sex chromosomes contain a differential region and two pairing regions. The regions were located by observing where the chromosomes paired up in meiosis and where they did not.

dimorphism, with female plants bearing flowers containing only ovaries and male plants bearing flowers containing only anthers (**Figure 2-15**). Some, but not all, dioecious plants have a nonidentical pair of chromosomes associated with (and almost certainly determining) the sex of the plant. Of the species with nonidentical sex chromosomes, a large proportion have an XY system. For example, the dioecious plant *Melandrium album* has 22 chromosomes per cell: 20 autosomes plus 2 sex chromosomes, with XX females and XY males. Other dioecious plants have no visibly different pair of chromosomes; they may still have sex chromosomes but not visibly distinguishable types.

Sex-linked patterns of inheritance

Cytogeneticists divide the X and Y chromosomes into homologous and differential regions. Again, let's use humans as an example (**Figure 2-16**). The differential regions, which contain most of the genes, have no counterparts on the other sex chromosome. The genes in the differential

regions are said to be **hemizygous** ("half zygous"). The differential region of the X chromosome contains many hundreds of genes; most of these genes do not take part in sexual function, and they influence a great range of human properties. The Y chromosome contains only a few dozen genes. Some of these genes have counterparts on the X chromosome, but most do not. The latter type take part in male sexual function. One of these genes, *SRY*, determines maleness itself. Several other genes are specific for sperm production in males.

In general, genes in the differential regions are said to show inheritance patterns called **sex linkage**. Mutant alleles in the differential region of the X chromosome show a single-gene inheritance pattern called **X linkage**. Mutant alleles of the few genes in the differential region of the Y chromosome show **Y linkage**. A gene that is sex linked can show phenotypic ratios that are different in each sex. In this respect, sex-linked inheritance patterns contrast with the inheritance patterns of genes in the autosomes, which are the same in each sex. If the genomic location of a gene is unknown, a sex-linked inheritance pattern indicates that the gene lies on a sex chromosome.

The human X and Y chromosomes have two short homologous regions, one at each end (see Figure 2-16). In the sense that these regions are homologous, they are autosomal-like, and so they are called **pseudoautosomal regions 1 and 2**. One or both of these regions pairs with the other sex chromosome in meiosis and undergoes crossing over (see Chapter 4 for details of crossing over). For this reason, the X and the Y chromosomes can act as a pair and segregate into equal numbers of sperm.

X-linked inheritance

For our first example of X linkage, we turn to eye color in *Drosophila*. The wild-type eye color of *Drosophila* is dull red, but pure lines with white eyes are available (**Figure 2-17**). This phenotypic difference is determined by two alleles of a gene located on the differential region of the X chromosome. The mutant allele in the present case

White-eyed and red-eyed *Drosophila*

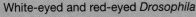

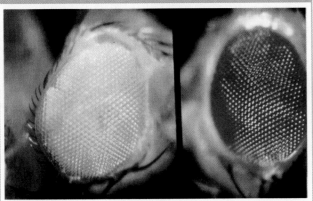

FIGURE 2-17 The red-eyed fly is wild type, and the white-eyed fly is a mutant. [*Science Source/Getty Images.*]

is w for white eyes (the lowercase letter indicates that the allele is recessive), and the corresponding wild-type allele is w^+. When white-eyed males are crossed with red-eyed females, all the F_1 progeny have red eyes, suggesting that the allele for white eyes is recessive. Crossing these red-eyed F_1 males and females produces a 3:1 F_2 ratio of red-eyed to white-eyed flies, but *all the white-eyed flies are males*. This inheritance pattern, which shows a clear difference between the sexes, is explained in **Figure 2-18**. The basis of the inheritance pattern is that all the F_1 flies receive a wild-type allele from their mothers, but the F_1 females also receive a white-eye allele from their fathers. Hence, all F_1 females are heterozygous wild type (w^+/w), and the F_1 males are hemizygous wild type (w^+). The F_1 females pass on the white-eye allele to half their sons, who express it, and to half their daughters, who do not express it, because they must inherit the wild-type allele from their fathers.

KEY CONCEPT Males need only inherit a single X-linked recessive allele in order for it to be expressed in the phenotype; a female must inherit two.

The reciprocal cross gives a different result; that is, the cross between white-eyed females and red-eyed males gives an F_1 in which all the females are red eyed but all the males are white eyed. In this case, every female inherited the dominant w^+ allele from the father's X chromosome, whereas every male inherited the recessive w allele from its mother. The F_2 consists of one-half red-eyed and one-half white-eyed flies of both sexes. Hence, in sex linkage, we see examples not only of different ratios in different sexes, but also of differences between reciprocal crosses.

MODEL ORGANISM   *Drosophila melanogaster*

Drosophila melanogaster, the common fruit fly. [*blickwinkel/Alamy.*]

Time flies like an arrow; fruit flies like a banana.

(Groucho Marx)

Drosophila melanogaster was one of the first model organisms to be used in genetics. It is readily available from ripe fruit, has a short life cycle, and is simple to culture and cross. Sex is determined by X and Y sex chromosomes (XX = female, XY = male), and males and females are easily distinguished. Mutant phenotypes regularly arise in lab populations, and their frequency can be increased by treatment with mutagenic radiation or chemicals. It is a diploid organism, with four pairs of homologous chromosomes ($2n = 8$). In salivary glands and certain other tissues, multiple rounds of DNA replication without chromosomal division result in "giant chromosomes," each with a unique banding pattern that provides geneticists with landmarks for the study of chromosome mapping and rearrangement. It is also noteworthy that there are many species and local races of *Drosophila*, which have been important raw material for the study of evolution.

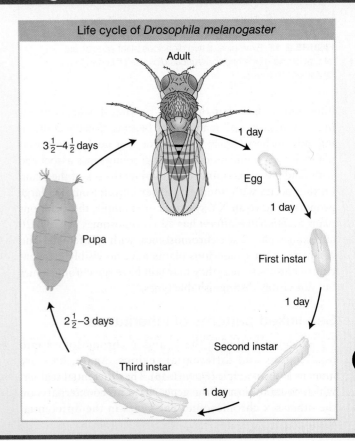

Life cycle of *Drosophila melanogaster*

Adult

$3\frac{1}{2}$–$4\frac{1}{2}$ days

1 day

Egg

1 day

Pupa

First instar

$2\frac{1}{2}$–3 days

1 day

Third instar

Second instar

1 day

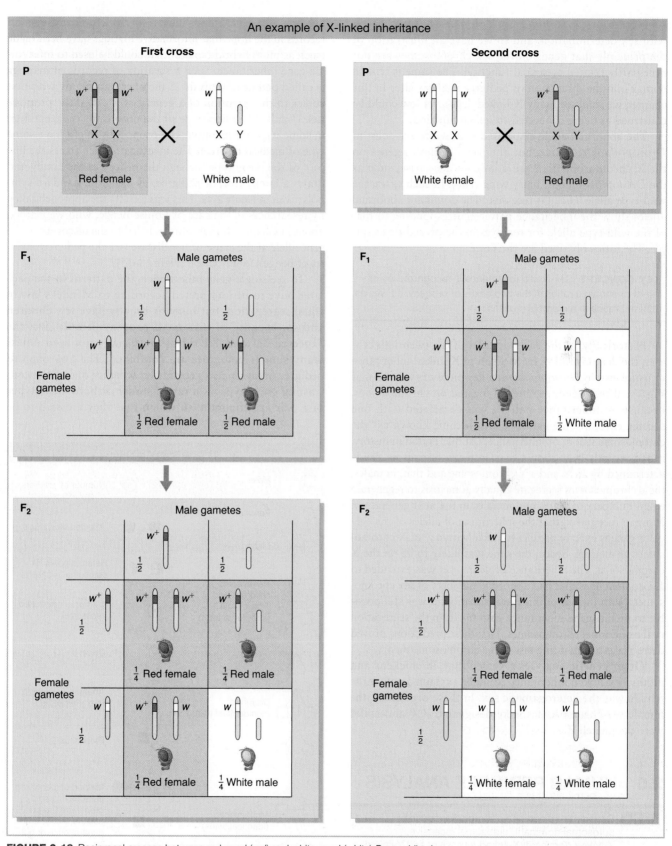

FIGURE 2-18 Reciprocal crosses between red-eyed (red) and white-eyed (white) *Drosophila* give different results. The alleles are X linked, and the inheritance of the X chromosome explains the phenotypic ratios observed, which are different from those of autosomal genes. (In *Drosophila* and many other experimental systems, a superscript plus sign is used to designate the normal, or wild-type, allele. Here, w^+ encodes red eyes and w encodes white eyes.)

ANIMATED ART   🐒 SaplingPlus

X-linked inheritance in flies

Note that *Drosophila* eye color has nothing to do with sex determination, and so we have an illustration of the principle that genes on the sex chromosomes are not necessarily related to sexual function. The same is true in humans: in the discussion of pedigree analysis later in this chapter, we shall see many X-linked genes, yet few could be construed as being connected to sexual function.

The abnormal allele associated with white eye color in *Drosophila* is recessive, but abnormal alleles of genes on the X chromosome that are dominant also arise, such as the *Drosophila* mutant hairy wing (*Hw*). In such cases, the wild-type allele (Hw^+) is recessive. The dominant abnormal alleles show the inheritance pattern corresponding to that of the wild-type allele for red eyes in the preceding example. The ratios obtained are the same.

KEY CONCEPT Sex-linked inheritance is recognized by different phenotypic ratios in the two sexes of progeny, as well as different ratios in reciprocal crosses.

Historically, in the early decades of the twentieth century, the demonstration by Morgan of X-linked inheritance of *white eyes* in *Drosophila* was a key piece of evidence that suggested that genes are indeed located on chromosomes, because an inheritance pattern was correlated with one specific chromosome pair. The idea became known as "the chromosome theory of inheritance." At that period in history, it had recently been shown that, in many organisms, sex is determined by an X and a Y chromosome and that, in males, these chromosomes segregate equally at meiosis to regenerate equal numbers of males and females in the next generation. Morgan recognized that the inheritance of alleles of the eye-color gene is exactly parallel to the inheritance of X chromosomes at meiosis; hence, the gene was likely to be on the X chromosome. The inheritance of *white eyes* was extended to *Drosophila* lines that had abnormal numbers of sex chromosomes. With the use of this novel situation, it was still possible to predict gene-inheritance patterns from the segregation of the abnormal chromosomes. That these predictions proved correct was a convincing test of the chromosome theory.

Other genetic analyses revealed that, in chickens and moths, sex-linked inheritance could be explained only if the female was the heterogametic sex. In these organisms, the female sex chromosomes were designated ZW and males were designated ZZ.

2.6 HUMAN PEDIGREE ANALYSIS

LO 2.6 Recognize inheritance patterns diagnostic of autosomal dominant, autosomal recessive, X-linked dominant, X-linked recessive, and Y-linked conditions in human pedigrees.

Human matings, like those of experimental organisms, provide many examples of single-gene inheritance. However, controlled experimental crosses cannot be made with

humans, and so geneticists must resort to scrutinizing medical records in the hope that informative matings have been made (such as monohybrid crosses) that could be used to infer single-gene inheritance. Such a scrutiny of records of matings is called **pedigree analysis**. A member of a family who first comes to the attention of a geneticist is called the **propositus**. Usually, the phenotype of the propositus is exceptional in some way; for example, the propositus might have some type of medical disorder. The investigator then traces the history of the phenotype through the history of the family and draws a family tree, or pedigree, by using the standard symbols given in **Figure 2-19**. This was exactly the approach taken by researchers in the case of Louise Benge, who we learned about in Chapter 1, that ultimately led to the discovery of the gene allele at the cause of her disease (see the pedigree analysis of Benge's family tree in Figure 1-15).

To see single-gene inheritance, the patterns in the pedigree have to be interpreted according to Mendel's law of equal segregation, but humans usually have few children and so, because of this small progeny sample size, the expected 3:1 and 1:1 ratios are usually not seen unless many similar pedigrees are combined. The approach to pedigree analysis also depends on whether one of the contrasting phenotypes is a rare disorder or both phenotypes of a pair are common (in which case they are said to be

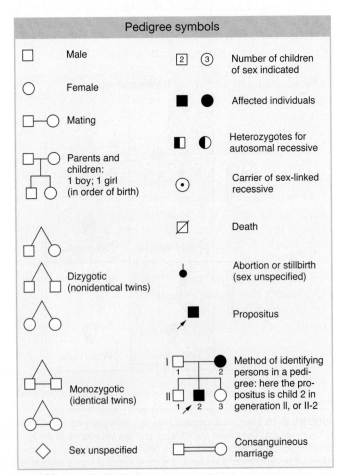

FIGURE 2-19 A variety of symbols are used in human pedigree analysis.

"morphs" of a polymorphism). Most pedigrees are drawn for medical reasons and therefore concern medical disorders that are almost by definition rare. In this case, we have two phenotypes: the presence and the absence of the disorder. Four patterns of single-gene inheritance are revealed in pedigrees. Let's look, first, at recessive disorders caused by recessive alleles of single autosomal genes.

Autosomal recessive disorders

The affected phenotype of an autosomal recessive disorder is inherited as a recessive allele; hence, the corresponding unaffected phenotype must be inherited as the corresponding dominant allele. For example, the human disease phenylketonuria (PKU), discussed earlier, is inherited in a simple Mendelian manner as a recessive phenotype, with PKU determined by the allele *p* and the normal condition determined by *P*. Therefore, people with this disease are of genotype *p/p*, and people who do not have the disease are either *P/P* or *P/p*. Recall that the term *wild type* and its allele symbols are not used in human genetics because wild type is impossible to define.

What patterns in a pedigree would reveal autosomal recessive inheritance? The two key points are that (1) generally the disorder appears in the progeny of unaffected parents, and (2) the affected progeny include both males and females. When we know that both male and female progeny are affected, we can infer that we are most likely dealing with simple Mendelian inheritance of a gene on an autosome, rather than a gene on a sex chromosome. The following typical pedigree illustrates the key point that affected children are born to unaffected parents:

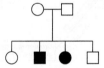

From this pattern, we can deduce a simple monohybrid cross, with the recessive allele responsible for the exceptional phenotype (indicated in black). Both parents must be heterozygotes—say, *A/a*; both must have an *a* allele because each contributed an *a* allele to each affected child, and both must have an *A* allele because they are phenotypically normal. We can identify the genotypes of the children (shown left to right) as *A/−*, *a/a*, *a/a*, and *A/−*. The pedigree can be rewritten as follows:

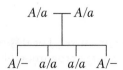

This pedigree does not support the hypothesis of X-linked recessive inheritance, because, under that hypothesis, an affected daughter must have a heterozygous mother (possible) and a hemizygous father, which is clearly impossible because the father would have expressed the phenotype of the disorder.

Notice that, even though Mendelian rules are at work, Mendelian ratios are not necessarily observed in single families because of small sample size, as predicted earlier. In the preceding example, we observe a 1:1 phenotypic ratio in the progeny of a monohybrid cross. If the couple were to have, say, 20 children, the ratio would be something like 15 unaffected children and 5 with PKU (a 3:1 ratio), but, in a small sample of 4 children, any ratio is possible, and all ratios are commonly found.

The family pedigrees of autosomal recessive disorders tend to look rather bare, with few black symbols. A recessive condition shows up in groups of affected siblings, and the people in earlier and later generations tend not to be affected. To understand why this is so, it is important to have some understanding of the genetic structure of populations underlying such rare conditions. By definition, if the condition is rare, most people do not carry the abnormal allele. Furthermore, most of those people who do carry the abnormal allele are heterozygous for it rather than homozygous. The basic reason why heterozygotes are much more common than recessive homozygotes is that to be a recessive homozygote, both parents must have the *a* allele, but to be a heterozygote, only one parent must have it.

The birth of an affected person usually depends on the rare chance union of unrelated heterozygous parents. However, inbreeding (mating between relatives, sometimes referred to as *consanguinity* in humans) increases the chance that two heterozygotes will mate. An example of mating between cousins is shown in **Figure 2-20**. Individuals

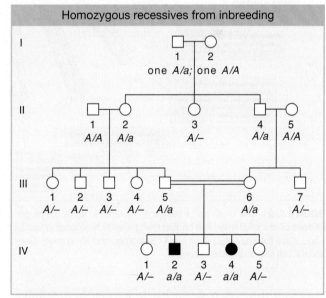

FIGURE 2-20 Pedigree of a rare recessive phenotype determined by a recessive allele *a*. Gene symbols are normally not included in pedigree charts, but genotypes are inserted here for reference. Persons II-1 and II-5 are not related to the bloodline of the family; they are assumed to be normal because the heritable condition under scrutiny is rare. Note also that it is not possible to be certain of the genotype in some persons with normal phenotype; such persons are indicated by *A/−*. Persons III-5 and III-6, who generate the recessives in generation IV, are first cousins. They both obtain their recessive allele from a grandparent, either I-1 or I-2.

Many human diseases are caused by mutations in single genes

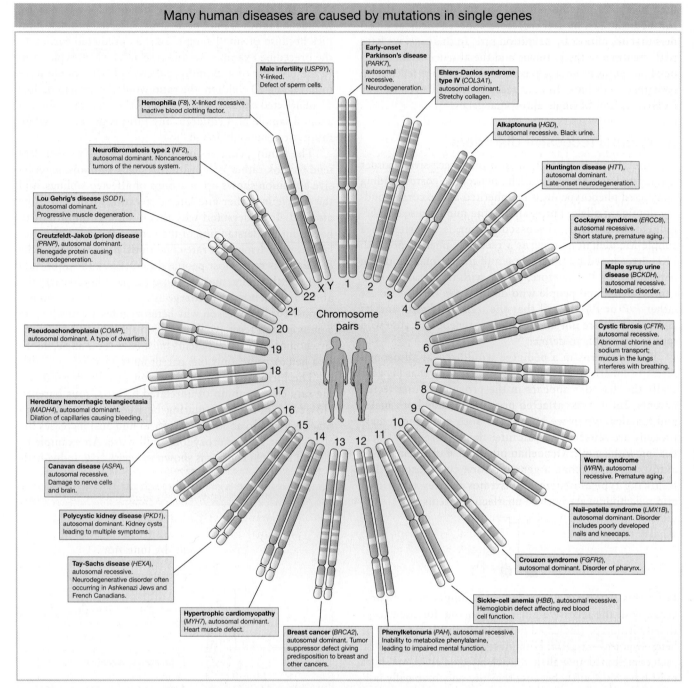

FIGURE 2-21 The positions of the genes mutated in some single-gene diseases, shown in the 23 pairs of chromosomes in a human being. Each chromosome has a characteristic banding pattern. X and Y are the sex chromosomes (XX in women and XY in men). Genes associated with each disease are shown in parentheses.

III-5 and III-6 are first cousins and produce two homozygotes for the rare allele. You can see from Figure 2-20 that an ancestor who is a heterozygote may produce many descendants who also are heterozygotes. Hence, two cousins can carry the *same* rare recessive allele inherited from a common ancestor. For two *unrelated* persons to be heterozygous, they would have to inherit the rare allele from *both* their families. Thus, matings between relatives generally run a higher risk of producing recessive disorders than do

matings between nonrelatives. For this reason, first-cousin matings contribute a large proportion of people with recessive diseases in the population.

Some other examples of human recessive disorders are shown in **Figure 2-21.** Cystic fibrosis is a disease inherited on chromosome 7 according to Mendelian rules as an autosomal recessive phenotype. Its most important symptom is the secretion of large amounts of mucus into the lungs, resulting in death from a combination of effects but usually precipitated by infection of the respiratory tract. The mucus can be dislodged by mechanical chest thumpers, and pulmonary infection can be prevented by antibiotics; thus, with treatment, cystic fibrosis patients can live to adulthood. The cystic fibrosis gene (a mutant allele of the wild-type gene) was one of the first human disease genes to be isolated at the DNA level, in 1989. This line of research eventually revealed that the disorder is caused by a defective protein that normally transports chloride ions across the cell membrane. The resultant alteration of the salt balance changes the constitution of the lung mucus. This new understanding of gene function in affected and unaffected persons has given hope for more effective treatment.

Human albinism also is inherited in the standard autosomal recessive manner. The mutant allele is of a gene that normally synthesizes the brown or black pigment melanin, normally found in skin, hair, and the retina of the eye (**Figure 2-22**).

> **KEY CONCEPT** In human pedigrees, an autosomal recessive disorder is generally revealed by the appearance of the disorder in the male and female progeny of unaffected parents.

Autosomal dominant disorders

What pedigree patterns are expected from autosomal dominant disorders? Here, the normal allele is recessive, and the defective allele is dominant. It may seem paradoxical that a rare disorder can be dominant, but remember that dominance and recessiveness are simply properties of how alleles act in heterozygotes and are not defined in reference to how common they are in the population. A good example of a rare dominant phenotype that shows single-gene inheritance is pseudoachondroplasia, a type of dwarfism (**Figure 2-23**). In regard to this gene, people with normal stature are genotypically *d/d*, and the dwarf phenotype could

A mutant gene causes albinism

FIGURE 2-22 A nonfunctional version of a skin-pigment gene results in lack of pigment. In this case, both members of the gene pair are mutated. [*Friedrich Stark/Alamy.*]

be in principle *D/d* or *D/D*. However, the two "doses" of the *D* allele in the *D/D* genotype are believed to produce such a severe effect that this genotype is lethal. If this belief is generally true, all dwarf individuals are heterozygotes.

In pedigree analysis, the main clues for identifying an autosomal dominant disorder with Mendelian inheritance are that the phenotype tends to appear in every generation of the pedigree and that affected fathers or mothers transmit the phenotype to both sons and daughters. Again, the equal representation of both sexes among the affected offspring rules out inheritance through the sex chromosomes. The phenotype appears in every generation because, generally, the abnormal allele carried by a person must have come from a parent in the preceding generation. (Abnormal alleles can also arise de novo by mutation. This possibility must be kept in mind for disorders that interfere with

Pseudoachondroplasia phenotype

FIGURE 2-23 The human pseudoachondroplasia phenotype is illustrated here by a family of five sisters and two brothers. The phenotype is determined by a dominant allele, which we can call *D*, that interferes with the growth of long bones during development. This photograph was taken when the family arrived in Israel after the end of World War II. [*Bettmann/Getty Images.*]

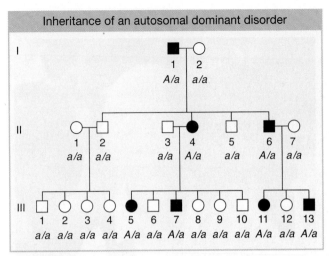

Inheritance of an autosomal dominant disorder

FIGURE 2-24 Pedigree of a dominant phenotype determined by a dominant allele *A*. In this pedigree, all the genotypes have been deduced.

reproduction because, here, the condition is unlikely to have been inherited from an affected parent.) A typical pedigree for a dominant disorder is shown in **Figure 2-24**. Once again, notice that Mendelian ratios are not necessarily observed in families. As with recessive disorders, persons bearing one copy of the rare *A* allele (*A/a*) are much more common than those bearing two copies (*A/A*); so most affected people are heterozygotes, and virtually all matings that produce progeny with dominant disorders are *A/a* × *a/a*. Therefore, if the progeny of such matings are totaled, a 1:1 ratio is expected of unaffected (*a/a*) to affected (*A/a*) persons.

Huntington disease is an example of a disease inherited as a dominant phenotype determined by an allele of a single gene. The phenotype is one of neural degeneration, leading to convulsions and premature death. Folk singer Woody Guthrie suffered from Huntington disease. The disease is rather unusual in that it shows late onset, the symptoms generally not appearing until after the person has reached reproductive age. When the disease has been diagnosed in a parent, each child already born knows that he or she has a 50 percent chance of inheriting the allele and the associated disease. This tragic pattern has inspired a great effort to find ways of identifying people who carry the abnormal allele before they experience the onset of the disease. Now there are molecular diagnostics for identifying people who carry the Huntington allele.

Some other rare dominant conditions are polydactyly (extra digits), shown in **Figure 2-25**, and piebald spotting, shown in **Figure 2-26**.

KEY CONCEPT Pedigrees of Mendelian autosomal dominant disorders show affected males and females in each generation; they also show affected men and women transmitting the condition to equal proportions of their sons and daughters.

Autosomal polymorphisms

Most natural populations also show **polymorphisms**, defined as the coexistence of two or more reasonably common phenotypes of a biological property, such as the occurrence of both red- and orange-fruited plants in a population of wild raspberries. The alternative phenotypes of a polymorphism (the **morphs**) are often inherited as alleles of a single autosomal gene in the standard Mendelian manner. Among the many human examples are the following **dimorphisms** (with two morphs, the simplest polymorphisms):

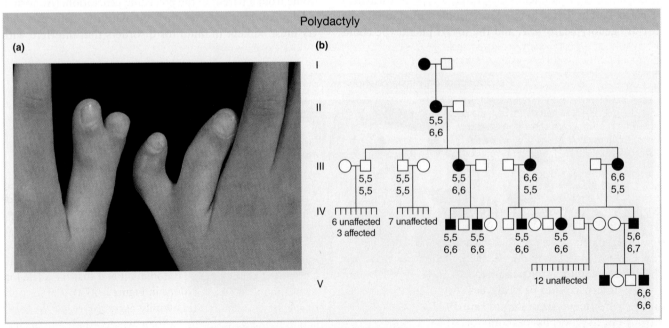

Polydactyly

(a)

(b)

FIGURE 2-25 Polydactyly is a rare dominant phenotype of the human hands and feet. (a) Polydactyly, characterized by extra fingers, toes, or both, is determined by an allele *P*. The numbers in the pedigree (b) give the number of fingers in the upper lines and the number of toes in the lower. (Note the variation in expression of the *P* allele.) [(a) Biophoto Associates/Science Source.]

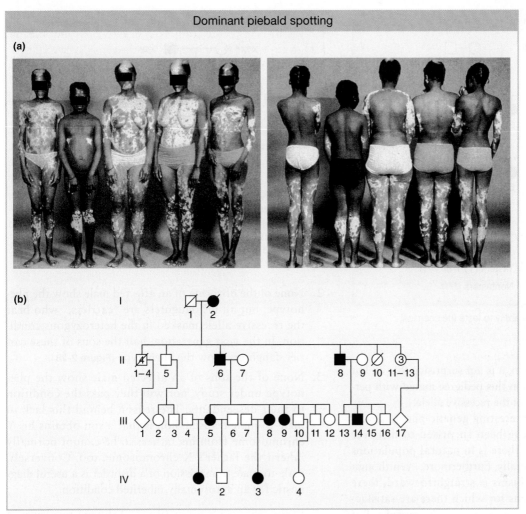

FIGURE 2-26 Piebald spotting is a rare dominant human phenotype. Although the phenotype is encountered sporadically in all races, the patterns show up best in those with dark skin. (a) The photographs show front and back views of affected persons IV-1, IV-3, III-5, III-8, and III-9 from (b) the family pedigree. Notice the variation in expression of the piebald gene among family members. The patterns are believed to be caused by the dominant allele interfering with the migration of melanocytes (melanin-producing cells) from the dorsal to the ventral surface in the course of development. The white forehead blaze is particularly characteristic and is often accompanied by a white forelock in the hair.

Piebaldism is not a form of albinism; the cells in the light patches have the genetic potential to make melanin, but, because they are not melanocytes, they are not developmentally programmed to do so. In true albinism, the cells lack the potential to make melanin. (Piebaldism is caused by mutations in *c-kit*, a type of gene called a *proto-oncogene;* see Chapter 15.) [*Winship, K. Young, R. Martell, R. Ramesar, D. Curtis, and P. Beighton, "Piebaldism: An Autonomous Autosomal Dominant Entity,"* Clinical Genetics 39, 1991, 330. © Reproduced with permission of John Wiley & Sons, Inc.]

brown versus blue eyes, pigmented versus blond hair, ability to smell freesia flowers versus inability, widow's peak versus none, sticky versus dry earwax, and attached versus free earlobes. In each example, the morph determined by the dominant allele is written first.

The interpretation of pedigrees for polymorphisms is somewhat different from that of rare disorders because, by definition, the morphs are common. Let's look at a pedigree for an interesting human case. Most human populations are dimorphic for the ability to taste the chemical phenylthiocarbamide (PTC); that is, people can either detect

it as a foul, bitter taste or—to the great surprise and disbelief of tasters—cannot taste it at all. From the pedigree in **Figure 2-27**, we can see that two tasters sometimes produce nontaster children, which makes it clear that the allele that confers the ability to taste is dominant and that the allele for nontasting is recessive. Notice in Figure 2-27 that almost all people who enter into this family carry the recessive allele either in heterozygous or in homozygous condition. Such a pedigree thus differs from those of rare recessive disorders, for which the conventional assumption is that all outsiders who enter into a family are homozygous normal. Because

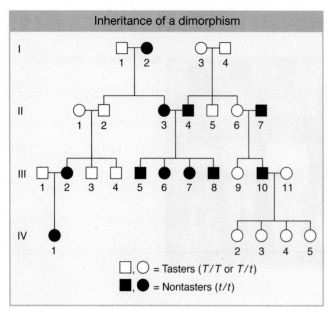

FIGURE 2-27 Pedigree for the ability to taste the chemical phenylthiocarbamide.

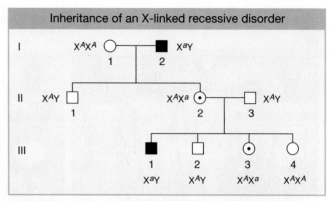

FIGURE 2-28 As is usually the case, expression of the X-linked recessive alleles is only in males. These alleles are carried unexpressed by daughters in the next generation, to be expressed again in sons. Note that III-3 and III-4 cannot be distinguished phenotypically.

both PTC alleles are common, it is not surprising that all but one of the family members in this pedigree mated with persons with at least one copy of the recessive allele.

Polymorphism is an interesting genetic phenomenon. Population geneticists have been surprised to discover how much polymorphism there is in natural populations of plants and animals generally. Furthermore, even though the genetics of polymorphisms is straightforward, there are very few polymorphisms for which there are satisfactory explanations for the coexistence of the morphs. But polymorphism is rampant at every level of genetic analysis, even at the DNA level; indeed, polymorphisms observed at the DNA level have been invaluable as landmarks to help geneticists find their way around the chromosomes of complex organisms, as will be described in Chapter 4. The population and evolutionary genetics of polymorphisms are considered in Chapters 18 and 20.

KEY CONCEPT Populations of plants and animals (including humans) are highly polymorphic. Contrasting morphs are often inherited as alleles of a single gene.

X-linked recessive disorders

Let's look at the pedigrees of disorders caused by rare recessive alleles of genes located on the X chromosome. Such pedigrees typically show the following features:

1. Many more males than females show the rare phenotype under study. The reason is that a female can inherit the genotype only if both her mother *and* her father bear the allele (for example, $X^A X^a \times X^a Y$), whereas a male can inherit the phenotype when *only* the mother carries the allele ($X^A X^a \times X^A Y$). If the recessive allele is very rare, almost all persons showing the phenotype are male.

2. None of the offspring of an affected male show the phenotype, but all his daughters are "carriers," who bear the recessive allele masked in the heterozygous condition. In the next generation, half the sons of these carrier daughters show the phenotype (**Figure 2-28**).

3. None of the sons of an affected male show the phenotype under study, nor will they pass the condition to their descendants. The reason behind this lack of male-to-male transmission is that a son obtains his Y chromosome from his father; so he cannot normally inherit the father's X chromosome, too. Conversely, male-to-male transmission of a disorder is a useful diagnostic for an autosomally inherited condition.

In the pedigree analysis of rare X-linked recessives, a normal female of unknown genotype is assumed to be homozygous unless there is evidence to the contrary.

Perhaps the most familiar example of X-linked recessive inheritance is red–green color blindness. People with this condition are unable to distinguish red from green. The genes for color vision have been characterized at the molecular level. Color vision is based on three different kinds of cone cells in the retina, each sensitive to red, green, or blue wavelengths. The genetic determinants for the red and green cone cells are on the X chromosome. Red–green color-blind people have a mutation in one of these two genes. As with any X-linked recessive disorder, there are many more males with the phenotype than females.

Another familiar example is *hemophilia*, the failure of blood to clot. Many proteins act in sequence to make blood clot. The most common type of hemophilia is caused by the absence or malfunction of one of these clotting proteins, called *factor VIII*. A well-known pedigree of hemophilia is of the interrelated royal families in Europe (**Figure 2-29a**). The original hemophilia allele in the pedigree possibly arose spontaneously as a mutation in the reproductive cells of either Queen Victoria's parents or Queen Victoria herself. However, some have proposed that the origin of the allele was a secret lover of Victoria's mother. Alexis, the son of the last czar of Russia, inherited the hemophilia allele

Inheritance of hemophilia in European royalty

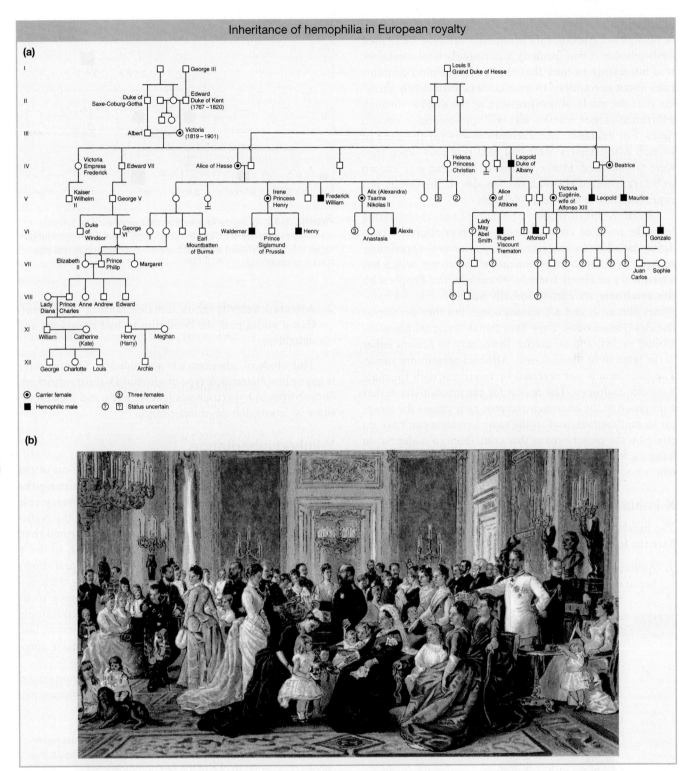

FIGURE 2-29 A pedigree for the X-linked recessive condition hemophilia in the royal families of Europe. A recessive allele causing hemophilia (failure of blood clotting) arose through mutation in the reproductive cells of Queen Victoria or one of her parents. This hemophilia allele spread into other royal families by intermarriage. (a) This partial pedigree shows affected males and carrier females (heterozygotes). Most spouses marrying into the families have been omitted from the pedigree for simplicity. Can you deduce the likelihood of the present British royal family's harboring the recessive allele? (b) A painting showing Queen Victoria surrounded by her numerous descendants. [*(b) Lebrecht Music and Arts Photo Library/Alamy.*]

ultimately from Queen Victoria, who was the grandmother of his mother, Alexandra. Hemophilia can now be treated medically, but it was formerly a potentially fatal condition. It is interesting to note that the Jewish Talmud contains rules about exemptions to male circumcision clearly showing that the mode of transmission of the disease through unaffected carrier females was well understood in ancient times. For example, one exemption was for the sons of women whose sisters' sons had bled profusely when they were circumcised. Hence, abnormal bleeding was known to be transmitted through the females of the family but expressed only in their male children.

A rare X-linked recessive phenotype that is interesting from the point of view of sexual differentiation is a condition called *androgen insensitivity syndrome* (previously referred to as *testicular feminization syndrome*), which has a frequency of about 1 in 65,000 male births. People with this syndrome are chromosomally males, having 44 autosomes plus an X and a Y chromosome, but they develop as females (**Figure 2-30**). They have female external genitalia, a blind vagina, and no uterus. Testes may be present either in the labia or in the abdomen. Afflicted persons are sterile. The condition is not reversed by treatment with the male hormone androgen. The reason for the insensitivity is that a mutation in the androgen-receptor gene causes the receptor to malfunction, and so the male hormone can have no effect on the target organs that contribute to maleness. In humans, femaleness results when the male-determining system is not functional.

X-linked dominant disorders

The inheritance patterns of X-linked dominant disorders have the following characteristics in pedigrees (**Figure 2-31**):

1. Affected males pass the condition to all their daughters but to none of their sons.

FIGURE 2-30 Supermodel Hanne Gaby Odiele, an XY individual with androgen insensitivity syndrome, caused by a recessive X-linked allele. [*Piero Oliosi/Polaris Images/Cap Antibes/France/Newscom.*]

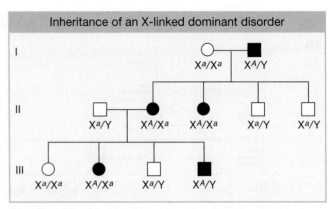

FIGURE 2-31 All the daughters of a male expressing an X-linked dominant phenotype will show the phenotype. Females heterozygous for an X-linked dominant allele will pass the condition on to half their sons and daughters.

2. Affected heterozygous females mating with unaffected males pass the condition to half their sons and daughters.

This mode of inheritance is not common. One example is hypophosphatemia, a type of vitamin D–resistant rickets. Some forms of hypertrichosis (excess body and facial hair) show X-linked dominant inheritance.

Y-linked inheritance

Only males inherit genes in the differential region of the human Y chromosome, with fathers transmitting the genes to their sons. The gene that plays a primary role in maleness is the **SRY gene,** sometimes called the *testis-determining factor.* Genomic analysis has confirmed that, indeed, the SRY gene is in the differential region of the Y chromosome. Hence, maleness itself is Y linked and shows the expected pattern of exclusively male-to-male transmission. Some cases of male sterility have been shown to be caused by deletions of Y-chromosome regions containing sperm-promoting genes. Male sterility is not heritable, but, interestingly, the fathers of these men have normal Y chromosomes, showing that the deletions are new.

There have been no convincing cases of nonsexual phenotypic variants associated with the Y chromosome, although there are cases in other animals.

> **KEY CONCEPT** Inheritance patterns with an unequal representation of phenotypes in males and females can locate the genes concerned to one of the sex chromosomes.

Calculating risks in pedigree analysis

When a disorder with well-documented single-gene inheritance is known to be present in a family, knowledge of transmission patterns can be used to calculate the probability of prospective parents' having a child with the disorder. For example, consider a case in which a new couple find out that each had an uncle with Tay-Sachs disease, a severe autosomal recessive disease caused by malfunction of the enzyme hexosaminidase A. The defect leads to the buildup of fatty deposits in nerve cells, causing paralysis followed by an early death. The pedigree is as follows:

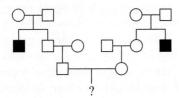

The probability of the couple's first child having Tay-Sachs can be calculated in the following way. Because neither of the couple has the disease, each can only be an unaffected homozygote or heterozygote. If both are heterozygotes, then they each stand a chance of passing the recessive allele on to a child, who would then have Tay-Sachs disease. Hence, we must calculate the probability of their both being heterozygotes, and then, if so, the probability of passing the deleterious allele on to a child.

1. The man's grandparents must have both been heterozygotes (T/t) because they produced a t/t child (the uncle). Therefore, they effectively constituted a monohybrid cross. The man's father could be T/T or T/t, but within the 3/4 of unaffected progeny, we know that the relative probabilities of these genotypes must be 1/4 and 1/2, respectively (the expected progeny ratio in a monohybrid cross is $\frac{1}{4}T/T, \frac{1}{2}T/t, \frac{1}{4}t/t$). Therefore, there is a 2/3 probability that the man's father is a heterozygote (two-thirds is the proportion of unaffected progeny who are heterozygotes: that is, the ratio of 2/4 to 3/4).

2. The man's mother is assumed to be T/T, because she is an outsider who entered into the family and disease alleles are generally rare. Thus, if the father is T/t, then the mating with the mother was a cross $T/t \times T/T$ and the expected proportions in the progeny (which includes the man in question) are $\frac{1}{2}T/T, \frac{1}{2}T/t$.

3. The overall probability of the man being a heterozygote must be calculated with the use of a statistical rule called the **product rule**, which states that

> The probability of two independent events both occurring is the product of their individual probabilities.

Because gene transmissions in different generations are independent events, we can calculate that the probability of the man being a heterozygote is the probability of his father being a heterozygote (2/3) *times* the probability of his father having a heterozygous son (1/2), which is $2/3 \times 1/2 = 1/3$.

4. Likewise, the probability of the woman being heterozygous is also 1/3.

5. If they are both heterozygous (T/t), their mating would be a standard monohybrid cross, and so the probability of their having a t/t child is 1/4.

6. Overall, the probability of the couple having an affected child is the probability of them both being heterozygous and then both transmitting the recessive allele to a child. Again, these events are independent, and so we can calculate the overall probability as $1/3 \times 1/3 \times 1/4 = 1/36$. In other words, there is a 1 in 36 chance of their having a child with Tay-Sachs disease.

In some Jewish communities, the Tay-Sachs allele is not as rare as it is in the general population. In such cases, unaffected people who enter into families with a history of Tay-Sachs cannot be assumed to be T/T. If the frequency of T/t heterozygotes in the community is known, this frequency can be factored into the product-rule calculation. Nowadays, molecular diagnostic tests for Tay-Sachs alleles are available, and the judicious use of these tests has drastically reduced the frequency of the disease in some communities.

SUMMARY

In somatic cell division, the genome is transmitted by mitosis, a nuclear division. In this process, each chromosome replicates into a pair of chromatids, and the chromatids are pulled apart to produce two identical daughter cells. (Mitosis can take place in diploid or haploid cells.) At meiosis, which takes place in the sexual cycle in meiocytes, each chromosome replicates to form a pair of chromatids; then, homologous chromosomes pair up at the equatorial plane of the cell. The homologous chromosomes (each a pair of chromatids) segregate over the course of two cell divisions. The result is four haploid cells, or gametes. Meiosis can take place only in a diploid cell; hence, haploid organisms must temporarily unite to form a diploid meiocyte.

An easy way to remember the main events of meiosis, by using your fingers to represent chromosomes, is shown in **Figure 2-32**.

Genetic dissection of a biological trait begins with a collection of mutants. Each mutant has to be tested to see if it is inherited as a single-gene change. The procedure followed is essentially unchanged from the time of Mendel, who performed the prototypic analysis of this type. The analysis is based on observing specific phenotypic ratios in the progeny of controlled crosses. In a typical case, a cross of $A/A \times a/a$ produces an F_1 that is all A/a. When the F_1 is selfed or intercrossed, a genotypic ratio of $\frac{1}{4}A/A : \frac{1}{2}A/a : \frac{1}{4}a/a$ is produced in the F_2. (At the phenotypic level, this ratio is $\frac{3}{4}A/- : \frac{1}{4}a/a$.)

The main events of mitosis and meiosis

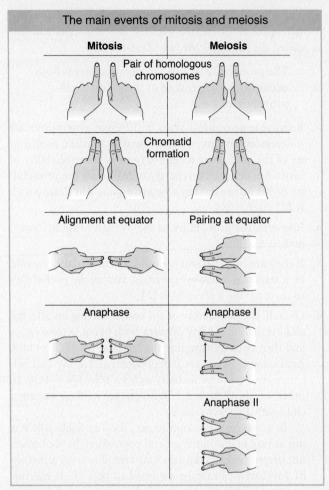

FIGURE 2-32 Using fingers to remember the main events of mitosis and meiosis.

The three single-gene genotypes are homozygous dominant, heterozygous (monohybrid), and homozygous recessive. If an *A/a* individual is crossed with *a/a* (a testcross), a 1:1 ratio is produced in the progeny. The 1:1, 3:1, and 1:2:1 ratios stem from the principle of equal segregation, which is that the haploid products of meiosis from *A/a* will be $\frac{1}{2}A$ and $\frac{1}{2}a$. The cellular basis of the equal segregation of alleles is the segregation of homologous chromosomes at meiosis. Haploid fungi can be used to show equal segregation at the level of a single meiosis (a 1:1 ratio in an ascus).

The molecular basis for chromatid production in meiosis is DNA replication. Segregation at meiosis can be observed directly at the molecular (DNA) level. The molecular force of segregation is the depolymerization and subsequent shortening of spindle fibers that are attached to the centromeres. Recessive mutations are generally in genes that are haplosufficient, whereas dominant mutations are often due to gene haploinsufficiency.

In many organisms, sex is determined chromosomally, and, typically, XX is female and XY is male. Genes on the X chromosome (X-linked genes) typically have no counterparts on the Y chromosome and show a single-gene inheritance pattern that differs in the two sexes, often resulting in different ratios in the male and female progeny.

Mendelian single-gene segregation is useful in identifying mutant alleles underlying many human disorders. Analyses of pedigrees can reveal autosomal or X-linked disorders of both dominant and recessive types. The logic of Mendelian genetics has to be used with caution, taking into account that human progeny sizes are small and phenotypic ratios are not necessarily typical of those expected from larger sample sizes. If a known single-gene disorder is present in a pedigree, Mendelian logic can be used to predict the likelihood of children inheriting the disease.

KEY TERMS

allele (p. 34)
character (p. 30)
chromatid (p. 38)
cross (p. 31)
dimorphism (p. 56)
dioecious species (p. 48)
diploid (p. 36)
dominant (p. 34)
first filial generation (F_1) (p. 32)
forward genetics (p. 31)
gene (p. 34)
gene discovery (p. 30)
genetic dissection (p. 31)
genotype (p. 35)
haploid (p. 36)
haploinsufficient (p. 45)
haplosufficient (p. 45)
hemizygous (p. 49)
heterogametic sex (p. 48)

heterozygote (p. 35)
heterozygous (p. 35)
homogametic sex (p. 48)
homozygote (p. 35)
homozygous dominant (p. 35)
homozygous recessive (p. 35)
law of equal segregation (Mendel's
    first law) (p. 35)
leaky mutation (p. 44)
mating types (p. 40)
meiocyte (p. 36)
meiosis (p. 36)
mitosis (p. 36)
monohybrid (p. 35)
monohybrid cross (p. 35)
morph (p. 56)
mutant (p. 30)
mutation (p. 30)
null allele (p. 44)

parental generation (P) (p. 32)
pedigree analysis (p. 52)
phenotype (p. 30)
polymorphism (p. 56)
product rule (p. 61)
propositus (p. 52)
pseudoautosomal regions 1 and 2
    (p. 49)
pure line (p. 32)
recessive (p. 34)
reverse genetics (p. 31)
second filial generation (F_2) (p. 33)
self (p. 32)
sex chromosome (p. 48)
sex linkage (p. 49)
sexual cell division (p. 36)
silent mutation (p. 44)
somatic cell division (p. 36)
SRY gene (p. 60)

SOLVED PROBLEMS

This section in each chapter contains a few solved problems that show how to approach the problem sets that follow. The purpose of the problem sets is to challenge your understanding of the genetic principles learned in the chapter. The best way to demonstrate an understanding of a subject is to be able to use that knowledge in a real or simulated situation. Be forewarned that there is no machine-like way of solving these problems. The three main resources at your disposal are the genetic principles just learned, logic, and trial and error.

Here is some general advice before beginning. First, for each problem, it is absolutely essential to read and understand the entire problem. Most of the problems use data taken from research that somebody actually carried out: ask yourself why the research might have been initiated and what was the probable goal. Find out exactly what facts are provided, what assumptions have to be made, what clues are given in the problem, and what inferences can be made from the available information. Second, be methodical. Staring at the problem rarely helps. Restate the information in the problem in your own way, preferably using a diagrammatic representation or flowchart to help you think out the problem. Good luck.

SOLVED PROBLEM 1

Crosses were made between two pure lines of rabbits that we can call A and B. A male from line A was mated with a female from line B, and the F_1 rabbits were subsequently intercrossed to produce an F_2. Three-fourths of the F_2 animals were discovered to have white subcutaneous fat, and one-fourth had yellow subcutaneous fat. Later, the F_1 was examined and was found to have white fat. Several years later, an attempt was made to repeat the experiment by using the same male from line A and the same female from line B. This time, the F_1 and all the F_2 (22 animals) had white fat. The only difference between the original experiment and the repeat that seemed relevant was that, in the original, all the animals were fed fresh vegetables, whereas in the repeat, they were fed commercial rabbit chow. Provide an explanation for the difference and a test of your idea.

SOLUTION

The first time that the experiment was done, the breeders would have been perfectly justified in proposing that a pair of alleles determine white versus yellow body fat because the data clearly resemble Mendel's results in peas. White

must be dominant, and so we can represent the white allele as W and the yellow allele as w. The results can then be expressed as follows:

$$
\begin{aligned}
&\text{P} && W/W \times w/w \\
&\text{F}_1 && W/w \\
&\text{F}_2 && \tfrac{1}{4}\,W/W \\
& && \tfrac{1}{2}\,W/w \\
& && \tfrac{1}{4}\,w/w
\end{aligned}
$$

No doubt, if the parental rabbits had been sacrificed, one parent (we cannot tell which) would have been predicted to have white fat and the other yellow. Luckily, the rabbits were not sacrificed, and the same animals were bred again, leading to a very interesting, different result. Often in science, an unexpected observation can lead to a novel principle, and, rather than moving on to something else, it is useful to try to explain the inconsistency. So why did the 3:1 ratio disappear? Here are some possible explanations.

First, perhaps the genotypes of the parental animals had changed. This type of spontaneous change affecting the whole animal, or at least its gonads, is very unlikely, because even common experience tells us that organisms tend to be stable to their type.

Second, in the repeat, the sample of 22 F_2 animals did not contain any yellow fat simply by chance ("bad luck"). This explanation, again, seems unlikely, because the sample was quite large, but it is a definite possibility.

A third explanation draws on the principle that genes do not act in a vacuum; they depend on the environment for their effects. Hence, the formula "genotype + environment = phenotype" is a useful mnemonic. A corollary of this formula is that genes can act differently in different environments; so

genotype 1 + environment 1 = phenotype 1

and

genotype 1 + environment 2 = phenotype 2

In the present problem, the different diets constituted different environments, and so a possible explanation of the results is that the homozygous recessive w/w produces yellow fat only when the diet contains fresh vegetables. This explanation is testable. One way to test it is to repeat the experiment again and use vegetables as food, but the parents might be dead by this time. A more convincing way is to breed several of the white-fatted F_2 rabbits from the second experiment. According to the original interpretation, some of them should be heterozygous, and, if their progeny are raised on vegetables, yellow fat should appear in Mendelian proportions. For example, if a cross happened to be W/w and w/w, the progeny would be $\frac{1}{2}$ white fat and $\frac{1}{2}$ yellow fat.

If this outcome did not happen and no progeny having yellow fat appeared in any of the matings, we would be forced back to the first or second explanation. The second explanation can be tested by using larger numbers, and if this explanation does not work, we are left with the first explanation, which is difficult to test directly.

As you might have guessed, in reality, the diet was the culprit. The specific details illustrate environmental effects beautifully. Fresh vegetables contain yellow substances called xanthophylls, and the dominant allele W gives rabbits the ability to break down these substances to a colorless ("white") form. However, w/w animals lack this ability, and the xanthophylls are deposited in the fat, making it yellow. When no xanthophylls are ingested, both $W/-$ and w/w animals end up with white fat.

SOLVED PROBLEM 2

Phenylketonuria (PKU) is a human hereditary disease resulting from the inability of the body to process the chemical phenylalanine, which is contained in the protein that we eat. PKU is manifested in early infancy and, if it remains untreated, generally leads to intellectual disabilities. PKU is caused by a recessive allele with simple Mendelian inheritance.

A couple intends to have children but consult a genetic counselor because the man has a sister with PKU and the woman has a brother with PKU. There are no other known cases in their families. They ask the genetic counselor to determine the probability that their first child will have PKU. What is this probability?

SOLUTION

What can we deduce? If we let the allele causing the PKU phenotype be p and the respective normal allele be P, then the sister and brother of the man and woman, respectively, must have been p/p. To produce these affected persons, all four grandparents must have been heterozygous normal. The pedigree can be summarized as follows:

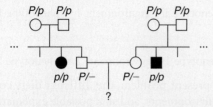

When these inferences have been made, the problem is reduced to an application of the product rule. The only way in which the man and woman can have a PKU child is if both of them are heterozygotes (it is obvious that they themselves do not have the disease). Both the grandparental matings are simple Mendelian monohybrid crosses expected to produce progeny in the following proportions:

$$\left.\begin{array}{l} \frac{1}{4}\ P/P \\[4pt] \frac{1}{2}\ P/p \end{array}\right\} \quad \text{Normal}\left(\tfrac{3}{4}\right)$$

$$\frac{1}{4}\ p/p \qquad \text{PKU}\left(\tfrac{1}{4}\right)$$

We know that the man and the woman are normal, and so the probability of each being a heterozygote is 2/3 because, within the $P/-$ class, 2/3 are P/p and 1/3 are P/P.

The probability of *both* the man and the woman being heterozygotes is $2/3 \times 2/3 = 4/9$. If both are heterozygous, then one-quarter of their children would have PKU, and so the probability that their first child will have PKU is 1/4 and the probability of their being heterozygous *and* of their first child's having PKU is $4/9 \times 1/4 = 4/36 = 1/9$, which is the answer.

SOLVED PROBLEM 3

A rare human disease is found in a family as shown in the accompanying pedigree.

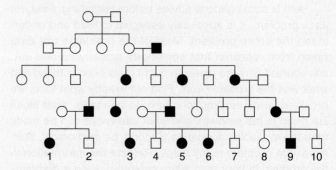

a. Deduce the most likely mode of inheritance.

b. What would be the outcomes of the cousin matings 1×9, 1×4, 2×3, and 2×8?

SOLUTION

a. The most likely mode of inheritance is X-linked dominant. We assume that the disease phenotype is dominant because, after it has been introduced into the pedigree by the male in generation II, it appears in every generation. We assume that the phenotype is X linked because fathers do not transmit it to their sons. If it were autosomal dominant, father-to-son transmission would be common.

In theory, autosomal recessive could work, but it is improbable. In particular, note the matings between affected members of the family and unaffected outsiders. If the condition were autosomal recessive, the only way in which these matings could have affected offspring is if each person entering into the family were a heterozygote; then the matings would be a/a (affected) $\times A/a$ (unaffected). However, we are told that the disease is rare; in such a case, heterozygotes are highly unlikely to be so common. X-linked recessive inheritance is impossible, because a mating of an affected woman with a normal man could not produce affected daughters. So we can let A represent the disease-causing allele and a represent the normal allele.

b. 1×9: Number 1 must be heterozygous A/a because she must have obtained a from her normal mother. Number 9 must be A/Y. Hence, the cross is A/a ♀ $\times A/Y$ ♂.

Female gametes	Male gametes	Progeny
$\frac{1}{2}A$	$\frac{1}{2}A \longrightarrow$	$\frac{1}{2}A/A$ ♀
	$\frac{1}{4}Y \longrightarrow$	$\frac{1}{4}A/Y$ ♂
$\frac{1}{2}a$	$\frac{1}{2}A \longrightarrow$	$\frac{1}{2}A/a$ ♀
	$\frac{1}{4}Y \longrightarrow$	$\frac{1}{4}a/Y$ ♂

1×4: Must be A/a ♀ $\times a/Y$ ♂.

Female gametes	Male gametes	Progeny
$\frac{1}{2}A$	$\frac{1}{2}a \longrightarrow$	$\frac{1}{2}A/a$ ♀
	$\frac{1}{4}Y \longrightarrow$	$\frac{1}{4}A/Y$ ♂
$\frac{1}{2}a$	$\frac{1}{2}a \longrightarrow$	$\frac{1}{2}a/a$ ♀
	$\frac{1}{4}Y \longrightarrow$	$\frac{1}{4}a/Y$ ♂

2×3: Must be a/Y ♂ $\times A/a$ ♀ (same as 1×4).
2×8: Must be a/Y ♂ $\times a/a$ ♀ (all progeny normal).

PROBLEMS

Visit SaplingPlus for supplemental content. Problems with the 🌀 icon are available for review/grading. Problems with the ⬚ icon have a Problem Solving Video. Problems with the 🧬 icon have an Unpacking the Problem exercise.

WORKING WITH THE FIGURES

(The first 25 questions require inspection of text figures.)

1. In Figure 2-1, what seems to be abnormal about the *Arabidopsis* mutants *ap3* and *ag*?

2. In the left-hand part of Figure 2-4, the red arrows show selfing as pollination within single flowers of one F_1 plant. Would the same F_2 results be produced by cross-pollinating two different F_1 plants?

3. In the right-hand part of Figure 2-4, in the plant showing an 11:11 ratio, do you think it would be possible to find a pod with all yellow peas? All green? Explain.

4. In Table 2-1, state the recessive phenotype in each of the seven cases. 🌀

5. In Table 2-1, line 6, if the 651 axial F_1 plants were individually selfed, approximately how many of them would have some terminal progeny?

6. In Figure 2-6, what would you say was the "job" of mitosis in each of the three life cycles?

7. Considering Figure 2-9, is the sequence "pairing → replication → segregation → segregation" a good shorthand description of meiosis?

8. In assessing Figure 2-9, what would you say was the main difference between the metaphase of mitosis and the metaphase II of meiosis?

9. In Figure 2-11, is there any case of a chromosome producing one GC-bearing daughter chromatid and one AT-bearing daughter chromatid?

10. In Figure 2-12, assume (as in corn plants) that allele A encodes an allele that produces starch in pollen and that allele a does not. Iodine solution stains starch black. How would you demonstrate Mendel's first law directly with such a system?

11. Considering Figure 2-14, if you had a homozygous double mutant $m3/m3$ $m5/m5$, would you expect it to be mutant in phenotype? (Note: This line would have two mutant sites in the same coding sequence.)

12. In Figure 2-14, mutant $m5$, whose locus is not in an active site (green area), is silent. Can you think of a way in which a mutation in such a green area could have a phenotypic effect?

13. In which of the stages of the *Drosophila* life cycle (represented in the box on page 50) would you find the products of meiosis? 🌀

14. If you assume Figure 2-16 also applies to mice and you irradiate male sperm with X rays (known to inactivate genes), what phenotype would you look for in progeny in order to find cases of individuals with an inactivated *SRY* gene?

15. In Figure 2-18, how does the 3:1 ratio in the bottom-left-hand grid differ from the 3:1 ratios obtained by Mendel?

16. In Figure 2-18, what progeny would you predict from a cross of a red F_2 male from the first cross and a red F_2 female from the second cross?

17. In Figure 2-20, assume that the pedigree is for mice, in which any chosen cross can be made. If you bred IV-1 with IV-3, what is the probability that the first baby will show the recessive phenotype?

18. In Figure 2-20, (1) can you tell which of the generation I parents is heterozygous in this pedigree? (2) Is generation IV an F_2 Mendelian ratio? Explain.

19. In Figure 2-21, list all the mutations that affect the human nervous system.

20. Which part of the pedigree in Figure 2-24 in your opinion best demonstrates Mendel's first law?

21. Considering all the individuals shown in Figure 2-24, would a cross between any male and female produce progeny all of which are *A/A*?

22. In Figure 2-25b, the first progeny set in generation IV contains three afflicted children. Can you explain why?

23. In Figure 2-27, what are the likely genotypes of II-1 and III-11? 🐟

24. Could the pedigree in Figure 2-31 be explained as an autosomal dominant disorder? Explain.

25. Refer back to Figure 1-15 concerning the condition ACDC and the family tree of Louise Benge.

    a. Assign Mendelian allelic symbols for the affected and unaffected condition and apply them to the individuals in generations V, VI, and VII. Indicate cases where the genotype is in doubt.

    b. Discuss the siblings in generation VI in terms of Mendelian ratios covered in Chapter 2.

    c. In a family with five children born to parents who are both unaffected carriers of the mutant allele, what is the probability that all five children would be affected?

BASIC PROBLEMS

26. Make up a sentence including the words *chromosome*, *genes*, and *genome*.

27. Peas (*Pisum sativum*) are diploid and $2n = 14$. In *Neurospora*, the haploid fungus, $n = 7$. If you were to isolate genomic DNA from both species and use gel electrophoresis to separate DNA molecules by size, how many distinct DNA bands would be visible in each species? (See Section 10.1 and Figure 10.4 for a description of the gel electrophoresis technique.)

28. The broad bean (*Vicia faba*) is diploid and $2n = 18$. Each haploid chromosome set contains approximately 4 m of DNA. The average size of each chromosome during metaphase of mitosis is 13 μm. What is the average packing ratio of DNA at metaphase? (Packing ratio = length of chromosome/length of DNA molecule therein.) How is this packing achieved?

29. If we call the amount of DNA per genome "*x*," then name a situation or situations in diploid organisms in which the amount of DNA per cell is 🐟

    a. *x*

    b. 2*x*

    c. 4*x*

30. Name the key function of mitosis.

31. Name two key functions of meiosis.

32. Design a different nuclear-division system that would achieve the same outcome as that of meiosis.

33. In a possible future scenario, male fertility drops to zero, but, luckily, scientists develop a way for women to produce babies by virgin birth. Meiocytes are converted directly (without undergoing meiosis) into zygotes, which implant in the usual way. What would be the short- and long-term effects in such a society?

34. In what ways does the second division of meiosis differ from mitosis?

35. Make up mnemonics for remembering the five stages of prophase I of meiosis and the four stages of mitosis.

36. In an attempt to simplify meiosis for the benefit of students, mad scientists develop a way of preventing pre-meiotic S phase and making do with having just one division, including pairing, crossing over, and segregation. Would this system work, and would the products of such a system differ from those of the present system?

37. Theodor Boveri said, "The nucleus doesn't divide; it is divided." What was he getting at?

38. Francis Galton, a geneticist of the pre-Mendelian era, devised the principle that half of our genetic makeup is derived from each parent, one-quarter from each grandparent, one-eighth from each great-grandparent, and so forth. Was he right? Explain.

39. If children obtain half their genes from one parent and half from the other parent, why aren't siblings identical?

40. State where cells divide mitotically and where they divide meiotically in a fern, a moss, a flowering plant, a pine tree, a mushroom, a frog, a butterfly, and a snail.

41. Human cells normally have 46 chromosomes. For each of the following stages, state the number of nuclear DNA molecules present in a human cell:

    a. Metaphase of mitosis

    b. Metaphase I of meiosis

    c. After telophase of mitosis

    d. After telophase I of meiosis

    e. After telophase II of meiosis

42. Four of the following events are part of both meiosis and mitosis, but only one is meiotic. Which one? (1) Chromatid formation, (2) spindle formation, (3) chromosome condensation, (4) chromosome movement to poles, (5) chromosome pairing.

43. In corn, the allele *f′* causes floury endosperm, and the allele *f″* causes flinty endosperm. In the cross *f′/f′* ♀ × *f″/f″* ♂, all the progeny endosperms are floury, but, in the reciprocal cross, all the progeny endosperms are flinty. What is a possible explanation? (Check the legend for Figure 2-6.)

44. What is Mendel's first law?

45. If you had a fruit fly (*Drosophila melanogaster*) that was of phenotype *A*, what cross would you make to determine if the fly's genotype was *A/A* or *A/a*?

46. In examining a large sample of yeast colonies on a petri dish, a geneticist finds an abnormal-looking colony that is very small. This small colony was crossed with wild type, and products of meiosis (ascospores) were spread on a plate to produce colonies. In total, there were 188 wild-type (normal-size) colonies and 180 small ones.

    a. What can be deduced from these results regarding the inheritance of the small-colony phenotype? (Invent genetic symbols.)

    b. What would an ascus from this cross look like?

47. Two black guinea pigs were mated and over several years produced 29 black and 9 white offspring. Explain these results, giving the genotypes of parents and progeny.

48. In a fungus with four ascospores, a mutant allele *lys-5* causes the ascospores bearing that allele to be white, whereas the wild-type allele *lys-5$^+$* results in black ascospores. (Ascospores are the spores that constitute the four products of meiosis.) Draw an ascus from each of the following crosses:

    a. *lys-5 × lys-5$^+$*

    b. *lys-5 × lys-5*

    c. *lys-5$^+$ × lys-5$^+$*

49. For a certain gene in a diploid organism, eight units of protein product are needed for normal function. Each wild-type allele produces five units.

    a. If a mutation creates a null allele, do you think this allele will be recessive or dominant?

    b. What assumptions need to be made to answer part *a*?

50. A *Neurospora* colony at the edge of a plate seemed to be sparse (low density) in comparison with the other colonies on the plate. This colony was thought to be a possible mutant, and so it was removed and crossed with a wild type of the opposite mating type. From this cross, 100 ascospore progeny were obtained. None of the colonies from these ascospores was sparse, all appearing to be normal. What is the simplest explanation of this result? How would you test your explanation? (Note: *Neurospora* is haploid.)

51. From a large-scale screen of many plants of *Collinsia grandiflora*, a plant with three cotyledons was discovered (normally, there are two cotyledons). This plant was crossed with a normal pure-breeding wild-type plant, and 600 seeds from this cross were planted. There were 298 plants with two cotyledons and 302 with three cotyledons. What can be deduced about the inheritance of three cotyledons? Invent gene symbols as part of your explanation.

52. In the plant *Arabidopsis thaliana*, a geneticist is interested in the development of trichomes (small projections). A large screen turns up two mutant plants (A and B) that have no trichomes, and these mutants seem to be potentially useful in studying trichome development. (If they were determined by single-gene mutations, then finding the normal and abnormal functions of these genes would be instructive.) Each plant is crossed with wild type; in both cases, the next generation (F$_1$) had normal trichomes. When F$_1$ plants were selfed, the resulting F$_2$ progeny were as follows:

    F$_2$ from mutant A: 602 normal; 198 no trichomes

    F$_2$ from mutant B: 267 normal; 93 no trichomes

    a. What do these results show? Include proposed genotypes of all plants in your answer.

    b. Under your explanation to part *a*, is it possible to confidently predict the F$_1$ from crossing the original mutant A with the original mutant B?

53. You have three dice: one red (R), one green (G), and one blue (B). When all three dice are rolled at the same time, calculate the probability of the following outcomes:

    a. 6 (R), 6 (G), 6 (B)

    b. 6 (R), 5 (G), 6 (B)

    c. 6 (R), 5 (G), 4 (B)

    d. No sixes at all

    e. A different number on all dice

54. In the pedigree below, the black symbols represent individuals with a very rare blood disease.

If you had no other information to go on, would you think it more likely that the disease was dominant or recessive? Give your reasons.

55. a. The ability to taste the chemical phenylthiocarbamide is an autosomal dominant phenotype, and the inability to taste it is recessive. If a taster woman with a nontaster father meets a taster man who in a previous relationship had a nontaster daughter, what is the probability that their first child will be

    1. A nontaster girl

    2. A taster girl

    3. A taster boy

    b. What is the probability that their first two children will be tasters of either sex?

56. John and Martha are contemplating having children, but John's brother has galactosemia (an autosomal recessive disease) and Martha's great-grandmother also had galactosemia. Martha has a sister who has three children, none of whom have galactosemia. What is the probability that John and Martha's first child will have galactosemia?

 UNPACKING PROBLEM 56

Before attempting a solution to this problem, try answering the following questions:

1. Can the problem be restated as a pedigree? If so, write one.

2. Can parts of the problem be restated by using Punnett squares?

3. Can parts of the problem be restated by using branch diagrams?

4. In the pedigree, identify a mating that illustrates Mendel's first law.

5. Define all the scientific terms in the problem, and look up any other terms about which you are uncertain.

6. What assumptions need to be made in answering this problem?

7. Which unmentioned family members must be considered? Why?

8. What statistical rules might be relevant, and in what situations can they be applied? Do such situations exist in this problem?

9. What are two generalities about autosomal recessive diseases in human populations?

10. What is the relevance of the rareness of the phenotype under study in pedigree analysis generally, and what can be inferred in this problem?

11. In this family, whose genotypes are certain, and whose are uncertain?

12. In what way is John's side of the pedigree different from Martha's side? How does this difference affect your calculations?

13. Is there any irrelevant information in the problem as stated?

14. In what way is solving this kind of problem similar to solving problems that you have already successfully solved? In what way is it different?

15. Can you make up a short story based on the human dilemma in this problem?

Now try to solve the problem. If you are unable to do so, try to identify the obstacle and write a sentence or two describing your difficulty. Then go back to the expansion questions and see if any of them relate to your difficulty. If this approach does not work, inspect the Learning Objectives and Key Concepts of this chapter and ask yourself which might be relevant to your difficulty.

57. Holstein cattle are normally black and white. A superb black-and-white bull, Charlie, was purchased by a farmer for $100,000. All the progeny sired by Charlie were normal in appearance. However, certain pairs of his progeny, when interbred, produced red-and-white progeny at a frequency of about 25 percent. Charlie was soon removed from the stud lists of the Holstein breeders. Use symbols to explain precisely why.

58. Suppose that a man and a woman are both heterozygous for a recessive allele for albinism. If they have dizygotic (two-egg) twins, what is the probability that both the twins will have the same phenotype for pigmentation?

59. The plant blue-eyed Mary grows on Vancouver Island and on the lower mainland of British Columbia. The populations are dimorphic for purple blotches on the leaves—some plants have blotches and others do not. Near Nanaimo, one plant in nature had blotched leaves. This plant, which had not yet flowered, was dug up and taken to a laboratory, where it was allowed to self. Seeds were collected and grown into progeny. One randomly selected (but typical) leaf from each of the progeny is shown in the accompanying illustration.

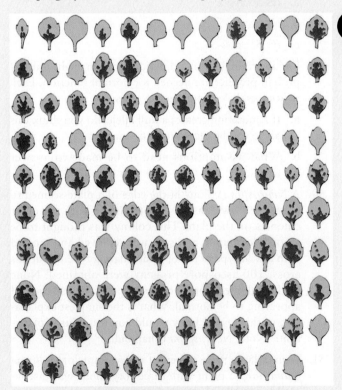

a. Formulate a concise genetic hypothesis to explain these results. Explain all symbols and show all genotypic classes (and the genotype of the original plant).

b. How would you test your hypothesis? Be specific.

60. Can it ever be proven that an animal is not a carrier of a recessive allele (that is, not a heterozygote for a given gene)? Explain.

61. In nature, the plant *Plectritis congesta* is dimorphic for fruit shape; that is, individual plants bear either wingless or winged fruits, as shown in the illustration.

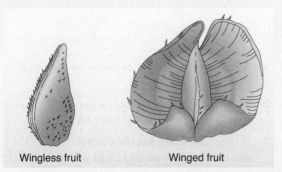

Wingless fruit   Winged fruit

Plants were collected from nature before flowering and were crossed or selfed with the following results:

	Number of progeny	
Pollination	Winged	Wingless
Winged (selfed)	91	1*
Winged (selfed)	90	30
Wingless (selfed)	4*	80
Winged × wingless	161	0
Winged × wingless	29	31
Winged × wingless	46	0
Winged × winged	44	0

*Phenotype probably has a nongenetic explanation.

Interpret these results, and derive the mode of inheritance of these fruit-shaped phenotypes. Use symbols. What do you think is the nongenetic explanation for the phenotypes marked by asterisks in the table?

62. The accompanying pedigree is for a rare, but relatively mild, hereditary disorder of the skin.

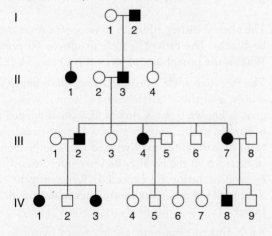

a. How is the disorder inherited? State reasons for your answer.

b. Give genotypes for as many individuals in the pedigree as possible. (Invent your own defined allele symbols.)

c. Consider the four unaffected children of parents III-4 and III-5. In all four-child progenies from parents of these genotypes, what proportion is expected to contain all unaffected children?

63. Four human pedigrees are shown in the accompanying illustration. The black symbols represent an abnormal phenotype inherited in a simple Mendelian manner.

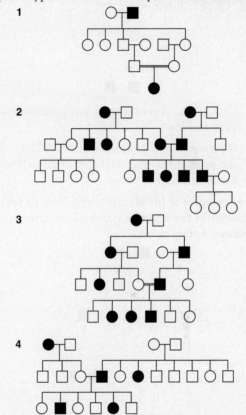

a. For each pedigree, state whether the abnormal condition is dominant or recessive. Try to state the logic behind your answer.

b. For each pedigree, describe the genotypes of as many persons as possible.

64. Tay-Sachs disease is a rare human disease in which toxic substances accumulate in nerve cells. The recessive allele responsible for the disease is inherited in a simple Mendelian manner. For unknown reasons, the allele is more common in populations of Ashkenazi Jews of eastern Europe. A woman is in a relationship with her male first cousin, but the couple discovers that their shared grandfather's sister died in infancy of Tay-Sachs disease.

a. Draw the relevant parts of the pedigree, and show all the genotypes as completely as possible.

b. What is the probability that the cousins' first child will have Tay-Sachs disease, assuming that all outsiders who enter into the family are homozygous normal?

65. The pedigree below was obtained for a rare kidney disease.

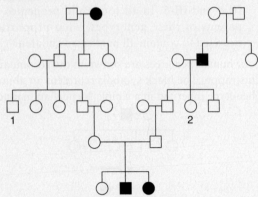

   a. Deduce the inheritance of this condition, stating your reasons.

   b. If persons 1 and 2 decide to have children, what is the probability that their first child will have the kidney disease?

66. This pedigree is for Huntington disease, a late-onset disorder of the nervous system. The slashes indicate deceased family members.

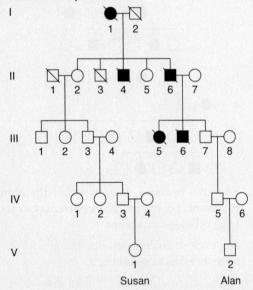

   a. Is this pedigree compatible with the mode of inheritance for Huntington disease mentioned in the chapter?

   b. Consider two newborn children in the two arms of the pedigree, Susan in the left arm and Alan in the right arm. Form an opinion on the likelihood that they will develop Huntington disease. Assume for the sake of the discussion that parents are about 25 years of age when their children are born.

67. Consider the accompanying pedigree of a rare autosomal recessive disease, PKU.

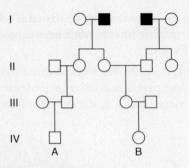

   a. List the genotypes of as many of the family members as possible.

   b. If persons A and B mate, what is the probability that their first child will have PKU?

   c. If their first child has the disease, what is the probability that their second child will be unaffected?

   (Assume that all people entering into the pedigree from the outside lack the abnormal allele.)

68. A man has attached earlobes, whereas his wife has free earlobes. Their first child, a boy, has attached earlobes.

   a. If the phenotypic difference is assumed to be due to two alleles of a single gene, is it possible that the gene is X linked?

   b. Is it possible to decide if attached earlobes is a dominant or recessive trait?

69. A rare recessive allele inherited in a Mendelian manner causes the disease cystic fibrosis. A phenotypically normal man whose father had cystic fibrosis marries a phenotypically normal woman from outside the family, and the couple consider having a child.

   a. Draw the pedigree as far as described.

   b. If the frequency in the population of heterozygotes for cystic fibrosis is 1 in 50, what is the chance that the couple's first child will have cystic fibrosis?

   c. If the first child does have cystic fibrosis, what is the probability that the second child will be normal?

70. The allele c causes albinism in mice (C causes mice to be black). The cross $C/c \times c/c$ produces 10 progeny. What is the probability of all of them being black?

71. The recessive allele s causes *Drosophila* to have small wings, and the s^+ allele causes normal wings. This gene is known to be X linked. If a small-winged male is crossed with a homozygous wild-type female, what ratio of normal to small-winged flies can be expected in each sex in the F_1? If F_1 flies are intercrossed, what F_2 progeny ratios are expected? What progeny ratios are predicted if F_1 females are backcrossed with their father?

72. An X-linked dominant allele causes hypophosphatemia in humans. A man with hypophosphatemia

marries a normal woman. If they have children, what proportion of their sons will have hypophosphatemia?

73. Duchenne muscular dystrophy is sex linked and usually affects only males. Victims of the disease become progressively weaker, starting early in life.

a. What is the probability that a woman whose brother has Duchenne's disease will have an affected child?

b. If your mother's brother (your uncle) had Duchenne's disease, what is the probability that you have received the allele?

c. If your father's brother had the disease, what is the probability that you have received the allele?

74. A man and woman discover that each had an uncle with alkaptonuria (black urine disease), a rare disease caused by an autosomal recessive allele of a single gene. They are about to have their first baby. What is the probability that their child will have alkaptonuria?

75. The accompanying pedigree concerns a rare inherited dental abnormality, amelogenesis imperfecta.

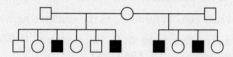

a. What mode of inheritance best accounts for the transmission of this trait?

b. Write the genotypes of all family members according to your hypothesis.

76. A couple learn from studying their family histories that, in *both* their families, their unaffected grandparents had siblings with cystic fibrosis (a rare autosomal recessive disease).

a. If the couple has a child, what is the probability that the child will have cystic fibrosis?

b. If they have four children, what is the chance that the children will have the precise Mendelian ratio of 3:1 for normal:cystic fibrosis?

c. If their first child has cystic fibrosis, what is the probability that their next three children will be normal?

77. A sex-linked recessive allele *c* produces a red–green color blindness in humans. A normal woman whose father was color blind marries a color-blind man.

a. What genotypes are possible for the mother of the color-blind man?

b. If the couple has children, what are the chances that their first child will be a color-blind boy?

c. Of any girls that may be born to these parents, what proportion can be expected to be color blind?

d. Of all the children (sex unspecified) of these parents, what proportion can be expected to have normal color vision?

78. Male house cats are either black or orange; females are black, orange, or calico.

a. If these coat-color phenotypes are governed by a sex-linked gene, how can these observations be explained?

b. Using appropriate symbols, determine the phenotypes expected in the progeny of a cross between an orange female and a black male.

c. Half the females produced by a certain kind of mating are calico, and half are black; half the males are orange, and half are black. What colors are the parental males and females in this kind of mating?

d. Another kind of mating produces progeny in the following proportions: one-fourth orange males, one-fourth orange females, one-fourth black males, and one-fourth calico females. What colors are the parental males and females in this kind of mating?

79. The pedigree below concerns a certain rare disease that is incapacitating but not fatal.

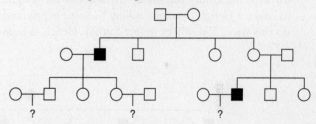

a. Determine the most likely mode of inheritance of this disease.

b. Write the genotype of each family member according to your proposed mode of inheritance.

c. If you were this family's doctor, how would you advise the three couples in the third generation about the likelihood of having an affected child?

80. In corn, the allele *s* causes sugary endosperm, whereas *S* causes starchy. What endosperm genotypes result from each of the following crosses?

a. *s/s* female × *S/S* male

b. *S/S* female × *s/s* male

c. *S/s* female × *S/s* male

81. A plant geneticist has two pure lines, one with purple petals and one with blue. She hypothesizes that the phenotypic difference is due to two alleles of one gene. To test this idea, she aims to look for a 3:1 ratio in the F_2. She crosses the lines and finds that all the F_1 progeny are purple. The F_1 plants are selfed, and 400 F_2 plants are obtained. Of these F_2 plants, 320 are purple

and 80 are blue. Do these results fit her hypothesis well? If not, suggest why.

82. A man's grandfather has galactosemia, a rare autosomal recessive disease caused by the inability to process galactose, leading to muscle, nerve, and kidney malfunction. The man married a woman whose sister had galactosemia. The woman is now pregnant with their first child.

    a. Draw the pedigree as described.

    b. What is the probability that this child will have galactosemia?

    c. If the first child does have galactosemia, what is the probability that a second child will have it?

CHALLENGING PROBLEMS

83. A geneticist working on peas has a single plant monohybrid Y/y (yellow) plant and, from a self of this plant, wants to produce a plant of genotype y/y to use as a tester. How many progeny plants need to be grown to be 95 percent sure of obtaining at least one in the sample?

84. A curious polymorphism in human populations has to do with the ability to curl up the sides of the tongue to make a trough ("tongue rolling"). Some people can do this trick, and others simply cannot. Hence, it is an example of a dimorphism. Its significance is a complete mystery. In one family, a boy was unable to roll his tongue but, to his great chagrin, his sister could. Furthermore, both his parents were rollers, and so were both grandfathers, one paternal uncle, and one paternal aunt. One paternal aunt, one paternal uncle, and one maternal uncle could not roll their tongues.

    a. Draw the pedigree for this family, defining your symbols clearly, and deduce the genotypes of as many individual members as possible.

    b. The pedigree that you drew is typical of the inheritance of tongue rolling and led geneticists to come up with the inheritance mechanism that no doubt you came up with. However, in a study of 33 pairs of identical twins, both members of 18 pairs could roll, neither member of 8 pairs could roll, and one of the twins in 7 pairs could roll but the other could not. Because identical twins are derived from the splitting of one fertilized egg into two embryos, the members of a pair must be genetically identical. How can the existence of the seven discordant pairs be reconciled with your genetic explanation of the pedigree?

85. Red hair runs in families, as the pedigree here shows.

    (Pedigree data from W. R. Singleton and B. Ellis, *Journal of Heredity* 55, 1964, **261**.)

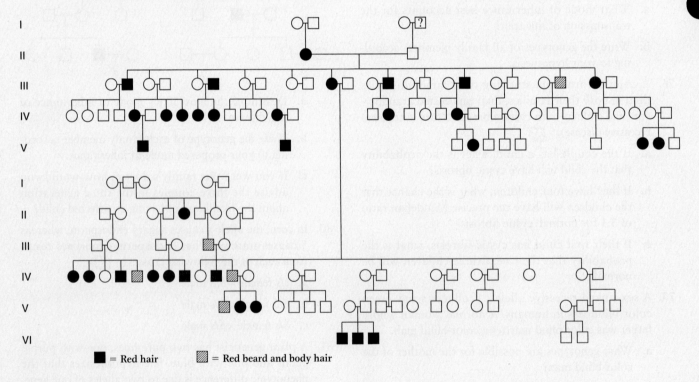

■ = Red hair   ▨ = Red beard and body hair

a. Does the inheritance pattern in this pedigree suggest that red hair could be caused by a dominant or a recessive allele of a gene that is inherited in a simple Mendelian manner?

b. Do you think that the red-hair allele is common or rare in the population as a whole?

86. When many families were tested for the ability to taste the chemical phenylthiocarbamide, the matings were grouped into three types and the progeny were totaled, with the results shown below:

| | Children | | |
Parents	Number of families	Tasters	Non-Parents
Taster × taster	425	929	130
Taster × nontaster	289	483	278
Nontaster × nontaster	86	5	218

With the assumption that PTC tasting is dominant (P) and nontasting is recessive (p), how can the progeny ratios in each of the three types of mating be accounted for?

87. A condition known as icthyosis hystrix gravior appeared in a boy in the early eighteenth century. His skin became very thick and formed loose spines that were sloughed off at intervals. When he grew up, this "porcupine man" married and had six sons, all of whom had this condition, and several daughters, all of whom were normal. For four generations, this condition was passed from father to son. From this evidence, what can you postulate about the location of the gene?

88. The wild-type (W) *Abraxas* moth has large spots on its wings, but the lacticolor (L) form of this species has very small spots. Crosses were made between strains differing in this character, with the following results:

| Cross | Parents | | Progeny | |
	♀	♂	F$_1$	F$_2$
1	L	W	♀ W	♀ $\frac{1}{2}$ L, $\frac{1}{2}$ W
			♂ W	♂ W
2	W	L	♀ L	♀ $\frac{1}{2}$ W, $\frac{1}{2}$ L
			♂ W	♂ $\frac{1}{2}$ W, $\frac{1}{2}$ L

Provide a clear genetic explanation of the results in these two crosses, showing the genotypes of all individual moths.

89. This pedigree shows the inheritance of a rare human disease.

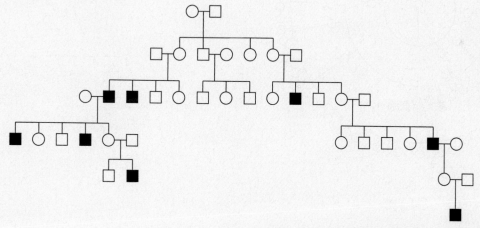

Is the pattern best explained as being caused by an X-linked recessive allele, or by an autosomal dominant allele with expression limited to males? (Pedigree data from J. F. Crow, *Genetics Notes*, 6th ed. Copyright 1967 by Burgess Publishing Co., Minneapolis.)

90. A certain type of deafness in humans is inherited as an X-linked recessive trait. An unaffected woman is expecting a child with a man who has this type of deafness. They find out that they are distantly related. Part of the family tree is shown here.

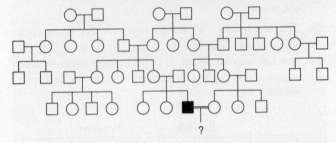

How would you advise the parents about the probability of their child being a deaf boy, a deaf girl, a normal boy, or a normal girl? Be sure to state any assumptions that you make.

91. The accompanying pedigree shows a very unusual inheritance pattern that actually did exist. All progeny are shown, but the fathers in each mating have been omitted to draw attention to the remarkable pattern.

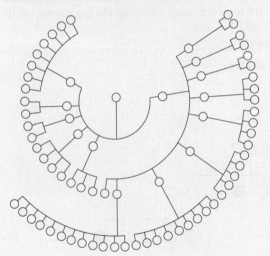

a. Concisely state exactly what is unusual about this pedigree.

b. Can the pattern be explained by Mendelian inheritance?

GENETICS AND SOCIETY

1. A newlywed couple finds out that they are both heterozygous for PKU, but they want to have children. What ethical dilemmas might they have, and what courses of action are available to them?

2. Discuss the genetic issues regarding first-cousin marriages.

3. Most people are heterozygous for several recessive Mendelian alleles causing ill health. Should this be a matter of concern in family planning? How could it be dealt with?

Mitosis usually takes up only a small proportion of the cell cycle, approximately 5 to 10 percent. The remaining time is the interphase, composed of G1, S, and G2 stages. The DNA is replicated during the S phase, although the duplicated DNA does not become visible until later in mitosis. The chromosomes cannot be seen during interphase, mainly because they are in an extended state and are intertwined with one another like a tangle of yarn.

The photographs below show the stages of mitosis in the nuclei of root-tip cells of the royal lily, *Lilium regale*. In each stage, a photograph is shown at the left and an interpretive drawing at the right.

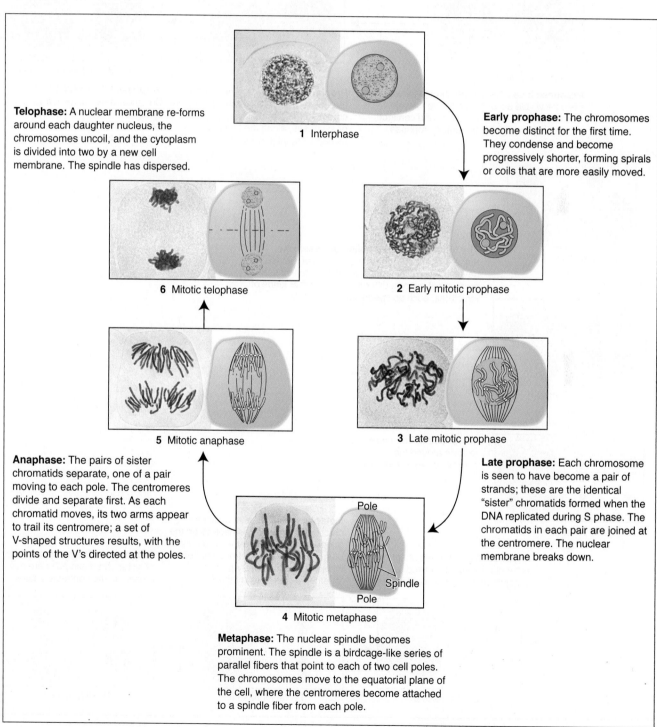

1 Interphase

Early prophase: The chromosomes become distinct for the first time. They condense and become progressively shorter, forming spirals or coils that are more easily moved.

Telophase: A nuclear membrane re-forms around each daughter nucleus, the chromosomes uncoil, and the cytoplasm is divided into two by a new cell membrane. The spindle has dispersed.

6 Mitotic telophase

2 Early mitotic prophase

5 Mitotic anaphase

3 Late mitotic prophase

Anaphase: The pairs of sister chromatids separate, one of a pair moving to each pole. The centromeres divide and separate first. As each chromatid moves, its two arms appear to trail its centromere; a set of V-shaped structures results, with the points of the V's directed at the poles.

Late prophase: Each chromosome is seen to have become a pair of strands; these are the identical "sister" chromatids formed when the DNA replicated during S phase. The chromatids in each pair are joined at the centromere. The nuclear membrane breaks down.

Pole

Spindle

Pole

4 Mitotic metaphase

Metaphase: The nuclear spindle becomes prominent. The spindle is a birdcage-like series of parallel fibers that point to each of two cell poles. The chromosomes move to the equatorial plane of the cell, where the centromeres become attached to a spindle fiber from each pole.

The photographs show mitosis in the nuclei of root-tip cells of *Lilium regale*. [*Republished with permission of Springer Nature, after J. McLeish and B. Snoad,* Looking at Chromosomes, *copyright 1972, St. Martin's, Macmillan, Red Globe Press, permission conveyed through Copyright Clearance Center, Inc.*]

ANIMATED ART

Mitosis

Sapling Plus

Meiosis consists of two nuclear divisions distinguished as meiosis I and meiosis II, which take place in consecutive cell divisions. Each meiotic division is formally divided into prophase, metaphase, anaphase, and telophase. Of these stages, the most complex and lengthy is prophase I, which itself is divided into five stages.

The photographs below show the stages of meiosis in the nuclei of root-tip cells of the royal lily, *Lilium regale*. In each stage, a photograph is shown at the left and an interpretive drawing at the right.

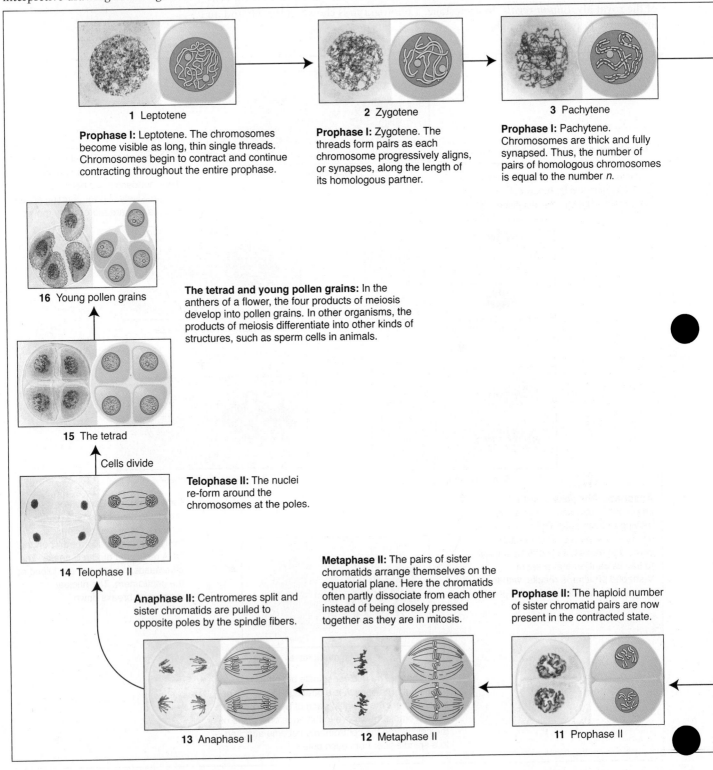

1 Leptotene

Prophase I: Leptotene. The chromosomes become visible as long, thin single threads. Chromosomes begin to contract and continue contracting throughout the entire prophase.

2 Zygotene

Prophase I: Zygotene. The threads form pairs as each chromosome progressively aligns, or synapses, along the length of its homologous partner.

3 Pachytene

Prophase I: Pachytene. Chromosomes are thick and fully synapsed. Thus, the number of pairs of homologous chromosomes is equal to the number *n*.

16 Young pollen grains

The tetrad and young pollen grains: In the anthers of a flower, the four products of meiosis develop into pollen grains. In other organisms, the products of meiosis differentiate into other kinds of structures, such as sperm cells in animals.

15 The tetrad

Cells divide

14 Telophase II

Telophase II: The nuclei re-form around the chromosomes at the poles.

13 Anaphase II

Anaphase II: Centromeres split and sister chromatids are pulled to opposite poles by the spindle fibers.

12 Metaphase II

Metaphase II: The pairs of sister chromatids arrange themselves on the equatorial plane. Here the chromatids often partly dissociate from each other instead of being closely pressed together as they are in mitosis.

11 Prophase II

Prophase II: The haploid number of sister chromatid pairs are now present in the contracted state.

The photographs show meiosis and pollen formation in *Lilium regale*. Note: For simplicity, multiple chiasmata are drawn between only two chromatids; in reality, all four chromatids can take part. [*Republished with permission of Springer Nature, After J. McLeish and B. Snoad, Looking at Chromosomes, Copyright 1972, St. Martin's, Macmillan, Red Globe Press, Permission conveyed through Copyright Clearance Center, Inc.*]

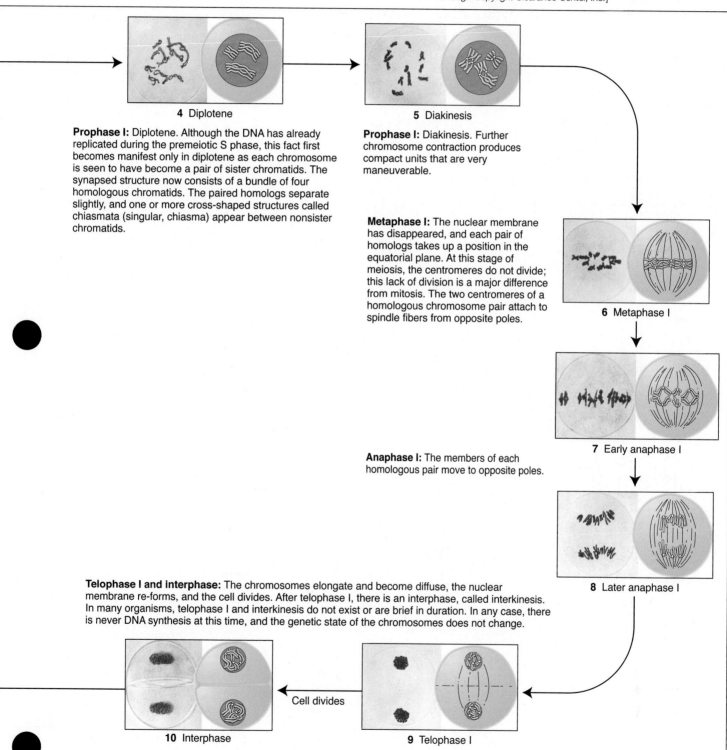

4 Diplotene

5 Diakinesis

6 Metaphase I

7 Early anaphase I

8 Later anaphase I

9 Telophase I

Cell divides

10 Interphase

Prophase I: Diplotene. Although the DNA has already replicated during the premeiotic S phase, this fact first becomes manifest only in diplotene as each chromosome is seen to have become a pair of sister chromatids. The synapsed structure now consists of a bundle of four homologous chromatids. The paired homologs separate slightly, and one or more cross-shaped structures called chiasmata (singular, chiasma) appear between nonsister chromatids.

Prophase I: Diakinesis. Further chromosome contraction produces compact units that are very maneuverable.

Metaphase I: The nuclear membrane has disappeared, and each pair of homologs takes up a position in the equatorial plane. At this stage of meiosis, the centromeres do not divide; this lack of division is a major difference from mitosis. The two centromeres of a homologous chromosome pair attach to spindle fibers from opposite poles.

Anaphase I: The members of each homologous pair move to opposite poles.

Telophase I and interphase: The chromosomes elongate and become diffuse, the nuclear membrane re-forms, and the cell divides. After telophase I, there is an interphase, called interkinesis. In many organisms, telophase I and interkinesis do not exist or are brief in duration. In any case, there is never DNA synthesis at this time, and the genetic state of the chromosomes does not change.

Independent Assortment of Genes

The Green Revolution in agriculture is fostered by the widespread planting of superior lines of crops (such as rice, shown here) made by combining beneficial genetic traits. [*Jorgen Schytte*.]

CHAPTER OUTLINE AND LEARNING OBJECTIVES

3.1 MENDEL'S LAW OF INDEPENDENT ASSORTMENT

LO 3.1 Using standard genetic symbolism, diagram how a dihybrid organism can be constructed starting from two pure parental lines; use the F_1 dihybrid in a self- and testcross to show the expected phenotype frequencies that would result if the two genes are assorting independently.

3.2 WORKING WITH INDEPENDENT ASSORTMENT

LO 3.2 In crosses involving independently assorting dihybrids, predict the genotypic ratios in meiotic products, genotypic ratios in progeny, and phenotypic ratios in progeny.

LO 3.3 Use chi-square analysis to test whether observed phenotypic ratios are an acceptable fit to those predicted by independent assortment.

LO 3.4 In diploids, outline how pure lines homozygous for two or more gene mutations can be created starting from parental lines homozygous for single-gene mutations.

3.3 THE CHROMOSOMAL BASIS OF INDEPENDENT ASSORTMENT

LO 3.5 Explain two-gene independent assortment ratios in terms of chromosome behavior at meiosis, in haploids and in diploids.

3.4 POLYGENIC INHERITANCE

LO 3.6 Extend the principle of independent assortment to multiple genes that each contribute to a phenotype showing continuous distribution.

3.5 ORGANELLE GENES: INHERITANCE INDEPENDENT OF THE NUCLEUS

LO 3.7 Apply the diagnostic criteria for assessing whether a gene of interest resides on a nuclear chromosome or on an organelle chromosome.

We saw from Chapter 2 that individual genes can be analyzed genetically only when they are heterozygous. In this chapter, the broad objective is to extend that analysis to learn how to analyze two or more heterozygous genes if they are inherited independently, most typically when they occupy different chromosomes.

This chapter is about the principles at work when two or more cases of single-gene inheritance are analyzed simultaneously. Nowhere have these principles been more important than in plant and animal breeding in agriculture. For example, between the years 1960 and 2000, the world production of food plants doubled, marking a so-called Green Revolution. What made this Green Revolution possible? In part, it was due to improved agricultural practice, but more important was the development of superior crop genotypes by plant geneticists. These breeders are constantly on the lookout for the chance occurrence of single-gene mutations that significantly increase yield or nutrient value. However, such mutations arise in different lines in different parts of the world. For example, in rice, one of the world's main food crops, the following mutations have been crucial in the Green Revolution:

sd1. This recessive allele results in short stature, making the plant more resistant to "lodging," or falling over, in wind and rain; it also increases the relative amount of the plant's energy that is routed into the seed, the part that we eat.

se1. This recessive allele alters the plant's requirement for a specific daylength, enabling it to be grown at different latitudes.

Xa4. This dominant allele confers resistance to the disease bacterial blight.

bph2. This allele confers resistance to brown plant hoppers (a type of insect).

Snb1. This allele confers tolerance to plant submersion after heavy rains.

To make a truly superior genotype, combining such alleles into one line is clearly desirable. To achieve such a combination, mutant lines must be intercrossed two at a time. For instance, a plant geneticist might start by crossing a strain homozygous for *sd1* to another homozygous for *Xa4*. The F_1 progeny of this cross would carry both mutations, but in a heterozygous state. However, most agriculture uses pure lines, because they can be efficiently propagated and distributed to farmers. To obtain a pure-breeding, doubly mutant *sd1/sd1 · Xa4/Xa4* line, the F_1 would have to be bred further to allow the alleles to "assort" into the desirable combination. Some products of such breeding are shown in **Figure 3-1**. What must take place during gamete formation and fertilization in order for the offspring to obtain the desired combination of alleles? It depends very much on whether the two genes are on the same chromosome pair or on different chromosome pairs. In the latter case, the chromosome pairs act independently at meiosis, and the alleles of two heterozygous gene pairs are said to show **independent assortment.**

This chapter explains how we can recognize independent assortment and how the principle of independent assortment can be used in strain construction, both in agriculture and in basic genetic research. (Chapter 4 covers the analogous principles applicable to heterozygous gene pairs located on the *same* chromosome pair.)

We shall also see that independent assortment of an array of genes is also useful in providing a basic mechanism of inheritance for traits that display continuous phenotypes.

Rice lines

FIGURE 3-1 Superior genotypes of crops such as rice have revolutionized agriculture. This photograph shows some of the key genotypes used in rice breeding programs. [*Bloomberg/Getty Images.*]

These are traits such as height or weight where phenotypes do not fall into distinct categories but are nevertheless often heavily influenced by multiple genes collectively called *polygenes*. We shall examine the role of independent assortment in the inheritance of continuous phenotypes influenced by such polygenes. We will see that independent assortment of polygenes can produce a continuous phenotypic distribution among progeny.

Lastly, we will introduce a different type of independent inheritance, that of genes in the organelles mitochondria and chloroplasts. Unlike nuclear chromosomes, these genes are inherited cytoplasmically and result in different patterns than observed for nuclear genes and chromosomes. However, this cytoplasmic inheritance pattern is independent of genes showing nuclear inheritance, which is why they are included in this chapter.

To begin, we examine the analytical procedures that pertain to the independent assortment of nuclear genes. These were first developed by the father of genetics, Gregor Mendel. So, again, we turn to his work as a prototypic example.

3.1 MENDEL'S LAW OF INDEPENDENT ASSORTMENT

LO 3.1 Using standard genetic symbolism, diagram how a dihybrid organism can be constructed starting from two pure parental lines; use the F₁ dihybrid in a self- and testcross to show the expected phenotype frequencies that would result if the two genes are assorting independently.

In much of his original work on peas, Mendel analyzed the descendants of pure lines that differed in *two* characters. The following general symbolism is used to represent genotypes that include two genes. If two genes are on different chromosomes, the gene pairs are separated by a semicolon—for example, *A/a ; B/b*. If they are on the same chromosome, the alleles on one homolog are written adjacently with no punctuation and are separated from those on the other homolog by a slash—for example, *AB/ab* or *Ab/aB*. An accepted symbolism does not exist for situations in which it is not known whether the genes are on the same chromosome or on different chromosomes. For this situation of unknown position in this book, we will use a dot to separate the genes—for example, *A/a · B/b*. Recall from Chapter 2 that a heterozygote for a *single gene* (such as *A/a*) is sometimes called a monohybrid: accordingly, a *double* heterozygote such as *A/a · B/b* is sometimes called a **dihybrid**. From studying **dihybrid crosses** (*A/a · B/b × A/a · B/b*), Mendel came up with his second important principle of heredity, the **law of independent assortment**, sometimes called **Mendel's second law**.

KEY CONCEPT Dihybrids, organisms heterozygous for two genes (*A/a · B/b*) are the key genotypes for the analysis of independent assortment in this chapter, and departures from it in Chapter 4.

The pair of characters that Mendel began working with were seed shape and seed color. We have already followed the monohybrid cross for seed color (*Y/y × Y/y*), which gave a progeny ratio of 3 yellow:1 green (see Figure 2-5). The seed shape phenotypes (**Figure 3-2**) were round (determined by allele *R*) and wrinkled (determined by allele *r*). The monohybrid cross *R/r × R/r* gave a progeny ratio of 3 round:1 wrinkled as expected (see Table 2-1, page 34). To perform a dihybrid cross, Mendel started with two pure parental lines. One line had wrinkled, yellow seeds. Because Mendel had no concept of the chromosomal location of genes, we must initially use the dot representation to write the combined genotype as *r/r · Y/Y*. The other line had round, green seeds, with genotype *R/R · y/y*. When these two lines were crossed, they must have produced gametes that were *r · Y* and *R · y*, respectively. Hence, the F₁ seeds had to be dihybrid, of genotype *R/r · Y/y*. Mendel discovered that the F₁ seeds were round and yellow. This result showed that the dominance of *R* over *r* and of *Y* over *y* was unaffected by the condition of the other gene pair in the *R/r · Y/y* dihybrid. In other words,

Round and wrinkled phenotypes

FIGURE 3-2 Round (*R/R* or *R/r*) and wrinkled (*r/r*) peas are present in a pod of a selfed heterozygous plant (*R/r*). The phenotypic ratio in this pod happens to be precisely the 3 : 1 ratio expected on average in the progeny of this selfing. [*Madan K. Bhattacharyya.*]

R remained dominant over *r*, regardless of seed color, and *Y* remained dominant over *y*, regardless of seed shape.

Next, Mendel selfed the dihybrid F₁ to obtain the F₂ generation. The F₂ seeds were of four different types in the following proportions:

$$\frac{9}{16} \text{ round, yellow}$$

$$\frac{3}{16} \text{ round, green}$$

$$\frac{3}{16} \text{ wrinkled, yellow}$$

$$\frac{1}{16} \text{ wrinkled, green}$$

The result is illustrated in **Figure 3-3** with the actual numbers obtained by Mendel. This initially unexpected 9:3:3:1 ratio for these two characters seems a lot more complex than the simple 3:1 ratios of the monohybrid crosses. Nevertheless, the 9:3:3:1 ratio proved to be a consistent inheritance pattern in peas. As evidence, Mendel also made dihybrid crosses that included several other combinations of characters and found that *all* of the dihybrid F₁ individuals produced 9:3:3:1 ratios in the F₂. This characteristic ratio was another inheritance pattern that required the development of a new idea to explain it.

First, let's check the actual numbers obtained by Mendel in Figure 3-3 to determine if the monohybrid 3:1 ratios can still be found in the F₂. In regard to seed shape, there are 423 round seeds (315 + 108) and 133 wrinkled seeds

(101 + 32). This result is close to a 3:1 ratio (actually 3.2:1). Next, in regard to seed color, there are 416 yellow seeds (315 + 101) and 140 green seeds (108 + 32), almost exactly a 3:1 ratio. The presence of these two 3:1 ratios hidden in the 9:3:3:1 ratio was undoubtedly a source of the insight that Mendel needed to explain the 9:3:3:1 ratio, because he realized that it was simply two different 3:1 ratios combined at random. One way of visualizing the random combination of these two ratios is with a branch diagram, as follows:

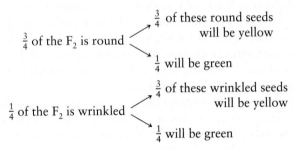

The probabilities of the four possible outcomes are calculated by using the *product rule*, to which we were introduced in Chapter 2 (the probability of two independent events occurring together is the product of their individual probabilities). Hence, we multiply along the branches in the diagram. For example, 3/4 of all seeds will be round, and 3/4 of the round seeds will be yellow, so the probability of a seed being both round and yellow is calculated as 3/4 × 3/4, which equals 9/16. These multiplications give the following four proportions:

$$\frac{3}{4} \times \frac{3}{4} = \frac{9}{16} \text{ round, yellow}$$

$$\frac{3}{4} \times \frac{1}{4} = \frac{3}{16} \text{ round, green}$$

$$\frac{1}{4} \times \frac{3}{4} = \frac{3}{16} \text{ wrinkled, yellow}$$

$$\frac{1}{4} \times \frac{1}{4} = \frac{1}{16} \text{ wrinkled, green}$$

These proportions constitute the 9:3:3:1 ratio that we are trying to explain. However, is this exercise not merely number juggling? What could the combination of the two 3:1 ratios mean biologically? The way that Mendel phrased his explanation does in fact amount to a biological mechanism. In what is now known as the law of independent assortment (Mendel's second law), he concluded that *different gene pairs assort independently during gamete formation.* The consequence is that, for two heterozygous gene pairs *A/a* and *B/b*, the *b* allele is just as likely to end up in a gamete with an *a* allele as with an *A* allele, and likewise for the *B* allele. In hindsight, we now know that, for the most part, this "law" applies to genes on different chromosomes. Genes on the same chromosome generally do not assort independently because they are held together by the chromosome itself.

KEY CONCEPT Mendel's second law (the law of independent assortment) states that the alleles of gene pairs on different chromosome pairs assort independently at meiosis.

Mendel's breeding program that produced a 9:3:3:1 ratio

P   *R/R • y/y* (round, green)   ×   *r/r • Y/Y* (wrinkled, yellow)

Gametes   *R • y*   *r • Y*

F₁   *R/r • Y/y* (round, yellow)

F₁   ×   F₁

F₂			Ratio
315 round, yellow			9
108 round, green			3
101 wrinkled, yellow			3
32 wrinkled, green			1
556 seeds			16

FIGURE 3-3 Mendel created a dihybrid that, when selfed, produced F₂ progeny in the ratio 9:3:3:1.

KEY CONCEPT The 9:3:3:1 phenotypic ratio observed in the progeny of a dihybrid self results from two 3:1 ratios combining at random and is diagnostic of independent assortment of the two genes.

Mendel's original statement of this law was that different genes assort independently because he apparently did not encounter (or he ignored) any exceptions that might have led to the concept of linkage. As we will see in Chapter 4, when two genes are located together on the same chromosome, their alleles do not always assort independently.

We have explained the 9:3:3:1 phenotypic ratio as two randomly combined 3:1 phenotypic ratios. But can we also arrive at the 9:3:3:1 ratio from a consideration of the frequency of gametes, the actual meiotic products? Let's consider the gametes produced by the F_1 dihybrid R/r ; Y/y (the semicolon shows that we are now embracing the idea that the genes are on different chromosomes). Again, we will use the branch diagram to get us started because it visually illustrates independence. Combining Mendel's laws of equal segregation and independent assortment, we can predict that

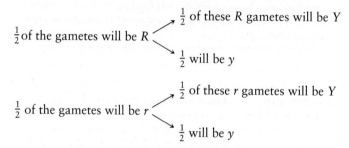

Multiplication along the branches according to the product rule gives us the gamete proportions:

$$\tfrac{1}{4}\,R\,;\,Y$$
$$\tfrac{1}{4}\,R\,;\,y$$
$$\tfrac{1}{4}\,r\,;\,Y$$
$$\tfrac{1}{4}\,r\,;\,y$$

These proportions are a direct result of the application of the two Mendelian laws: of segregation and of independence. However, we still have not arrived at the 9:3:3:1 ratio. The next step is to recognize that, because male and female gametes obey the same laws during formation, both the male and the female gametes will show the same proportions just given. The four female gametic types will be fertilized randomly by the four male gametic types to obtain the F_2. The best graphic way of showing the outcomes of the cross is by using a 4×4 grid called a *Punnett square*, which is depicted in **Figure 3-4**. We first encountered Punnett squares in Chapter 2, and we have already seen that grids are useful in genetics for providing a visual representation of the data. Their usefulness lies in the fact that their proportions can be drawn according to the genetic proportions or ratios under consideration. In the Punnett square in Figure 3-4, for example, four rows and four columns were

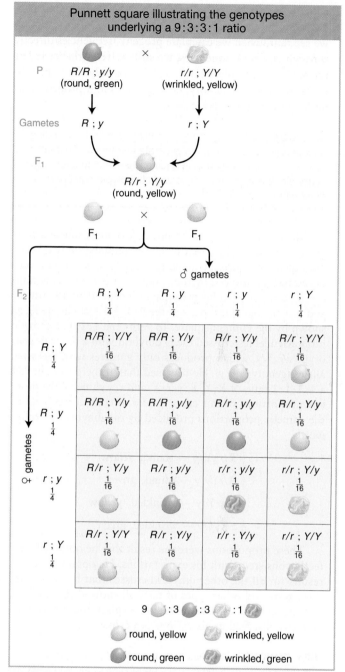

FIGURE 3-4 We can use a Punnett square to predict the result of a dihybrid cross. This Punnett square shows the predicted genotypic and phenotypic constitution of the F_2 generation from a dihybrid cross.

ANIMATED ART 🍁 SaplingPlus

Punnett squares and branch diagrams

drawn to correspond to the four genotypes of female gametes and the four genotypes of male gametes. We see that there are 16 boxes representing the various gametic fusions and that each box is 1/16th of the total area of the grid. In accord with the product rule, each 1/16th is a result of the fertilization of one egg type at frequency 1/4 by one sperm type also at frequency 1/4, giving the probability of that fusion as $(1/4)^2$. As the Punnett square shows, the F_2

contains a variety of genotypes, but there are only four phenotypes, and their proportions are in the 9:3:3:1 ratio. So we see that, when we calculate progeny frequencies directly through gamete frequencies, we still arrive at the 9:3:3:1 ratio. Hence, Mendel's laws explain not only the F_2 phenotypes, but also the genotypes of gametes and progeny that underlie the F_2 phenotypic ratio.

KEY CONCEPT Both male and female dihybrids exhibiting independent assortment will generate four types of gametes, in a 1:1:1:1 ratio. The Punnett square is a useful tool for graphically displaying the genotypic and phenotypic outcomes of their random union.

Mendel went on to test his principle of independent assortment, experimentally confirming it in a number of ways. The most direct way focused on the 1:1:1:1 gametic ratio hypothesized to be produced by the F_1 dihybrid R/r; Y/y. This ratio sprang directly from his principle of independent assortment and was the biological basis of the 9:3:3:1 ratio in the F_2, as shown by the Punnett square. To verify the 1:1:1:1 gametic ratio, Mendel testcrossed the F_1 dihybrid with a tester of genotype r/r; y/y, which produces only gametes with recessive alleles (genotype r; y). He reasoned that, if there were in fact a 1:1:1:1 ratio of R;Y, R;y, r;Y, and r;y gametes, the progeny proportions of this cross should directly correspond to the gametic proportions produced by the dihybrid; in other words,

$$\frac{1}{4} \quad R/r\,;Y/y \rightarrow \text{round, yellow}$$
$$\frac{1}{4} \quad R/r\,;y/y \rightarrow \text{round, green}$$
$$\frac{1}{4} \quad r/r\,;Y/y \rightarrow \text{wrinkled, yellow}$$
$$\frac{1}{4} \quad r/r\,;y/y \rightarrow \text{wrinkled, green}$$

These proportions were the result that he obtained, perfectly consistent with his expectations. He obtained similar results for all the other dihybrid crosses that he made, and these tests and other types of tests all showed that he had, in fact, devised a robust model to explain the inheritance patterns observed in his various pea crosses.

KEY CONCEPT In a testcross, the ratio of phenotypes in the progeny reflects the gametic genotypes of the nontester parent. For a dihybrid exhibiting independent assortment, this ratio is 1:1:1:1. The testcross is a useful tool for diploids, where the phenotypes and genotypes of gametes cannot be observed directly.

In the early 1900s, both of Mendel's laws were tested in a wide spectrum of eukaryotic organisms. The results of these tests showed that Mendelian principles were generally applicable. Mendelian ratios (such as 3:1, 1:1, 9:3:3:1, and 1:1:1:1) were extensively reported, suggesting that equal segregation and independent assortment are fundamental hereditary processes found throughout nature. Mendel's laws are not merely laws about peas; they are laws about the genetics of eukaryotic organisms in general.

As an example of the universal applicability of the principle of independent assortment, we can examine its action in haploids. If the principle of equal segregation is valid across the board, then we should be able to observe it in haploids, given that haploids undergo meiosis. Indeed, independent assortment can be observed in a cross of the type A; $B \times a$; b. Fusion of parental cells results in a transient diploid meiocyte that is a dihybrid A/a; B/b, and the randomly sampled products of meiosis (sexual spores such as ascospores in fungi) will be

$$\frac{1}{4} \quad A\,;B$$
$$\frac{1}{4} \quad A\,;b$$
$$\frac{1}{4} \quad a\,;B$$
$$\frac{1}{4} \quad a\,;b$$

Hence, we see the same ratio as in the dihybrid testcross in a diploid organism; again, the ratio is a random combination of two monohybrid 1:1 ratios because of independent assortment.

KEY CONCEPT In haploids, the genotypes of the products of meiosis (spores) are directly observable in the haploid organisms that arise from each spore through mitotic cell division. A ratio of 1:1:1:1 is diagnostic of independent assortment in a dihybrid meiocyte.

We should pause to compare Mendel's work on one- and two-gene systems. In working with several genes one at a time, he was able to demonstrate the principle of equal segregation of alleles of each gene at meiosis. In two-gene systems, he showed that for the genes at his disposal, the equal segregation principle for each gene was taking place independently of the other genes.

3.2 WORKING WITH INDEPENDENT ASSORTMENT

LO 3.2 In crosses involving independently assorting dihybrids, predict the genotypic ratios in meiotic products, genotypic ratios in progeny, and phenotypic ratios in progeny.

LO 3.3 Use chi-square analysis to test whether observed phenotypic ratios are an acceptable fit to those predicted by independent assortment.

LO 3.4 In diploids, outline how pure lines homozygous for two or more gene mutations can be created starting from parental lines homozygous for single-gene mutations.

In this section, we will examine several analytical procedures that are part of everyday genetic research and are all based on the concept of independent assortment. These procedures are all used to analyze phenotypic ratios.

Predicting progeny ratios

Genetics can work in either of two directions: (1) predicting the unknown genotypes of parents by using phenotype ratios of progeny, or (2) predicting progeny phenotype ratios from parents of known genotype. The latter is an important part of genetics concerned with predicting the types of progeny that emerge from a cross and calculating their expected frequencies—in other words, their probabilities. This is useful not only in research on model organisms but also in predicting outcomes of matings in human genetics; for example, in genetic counseling, people appreciate specific risk estimates. We have already examined two methods for prediction: Punnett squares and branch diagrams. Punnett squares can be used to show hereditary patterns based on one gene pair, two gene pairs, or more. Such grids are good graphic devices for representing progeny, but drawing them can be time consuming. Even the 16-compartment Punnett square that we used to analyze a dihybrid cross takes a long time to write out; but for a trihybrid cross, there are 2^3, or 8, different gamete types, and the Punnett square has 64 compartments. The branch diagram (shown below) is easier to create and is adaptable for phenotypic, genotypic, or gametic proportions, as illustrated for the dihybrid A/a ; B/b.

Progeny genotypes from a self	Progeny phenotypes from a self	Gametes

$\frac{1}{4} A/A$ ⟨ $\frac{1}{4} B/B$, $\frac{1}{2} B/b$, $\frac{1}{4} b/b$

$\frac{3}{4} A/-$ ⟨ $\frac{3}{4} B/-$, $\frac{1}{4} b/b$

$\frac{1}{2} B$ ⟨ $\frac{1}{2} B$, $\frac{1}{2} b$

$\frac{1}{2} A/a$ ⟨ $\frac{1}{4} B/B$, $\frac{1}{2} B/b$, $\frac{1}{4} b/b$

$\frac{1}{4} a/a$ ⟨ $\frac{3}{4} B/-$, $\frac{1}{4} b/b$

$\frac{1}{2} a$ ⟨ $\frac{1}{2} B$, $\frac{1}{2} b$

$\frac{1}{4} a/a$ ⟨ $\frac{1}{4} B/B$, $\frac{1}{2} B/b$, $\frac{1}{4} b/b$

Note, however, that the "tree" of branches for genotypes is quite unwieldy even in this simple case, which uses two gene pairs, because there are $3^2 = 9$ genotypes. For three gene pairs, there are 3^3, or 27, possible genotypes. To simplify this problem, we can use a statistical approach, which constitutes a third method for calculating the probabilities (expected frequencies) of specific phenotypes or genotypes coming from a cross. The two statistical rules needed are the **product rule** (introduced earlier, and also in Chapter 2) and the **sum rule,** which we will now consider together.

The product rule states that the probability of independent events occurring together is the product of their individual probabilities. The possible outcomes from rolling

two dice follow the product rule because the outcome on one die is independent of the other. As an example, let us calculate the probability, p, of rolling a pair of 4's. The probability of a 4 on one die is 1/6 because the die has six sides and only one side carries the number 4. This probability is written as follows:

$$p(\text{rolling one } 4) = \tfrac{1}{6}$$

Therefore, with the use of the product rule, the probability of a 4 appearing on both dice is $1/6 \times 1/6 = 1/36$, which is written:

$$p(\text{rolling two } 4\text{'s}) = \tfrac{1}{6} \times \tfrac{1}{6} = \tfrac{1}{36}$$

> **KEY CONCEPT** The product rule states that the probability of independent events *occurring together* is the product of their individual probabilities.

Now, we turn to the sum rule. The sum rule states that the probability of either one or the other of two mutually exclusive events occurring is the sum of their individual probabilities. Dice can also be used to illustrate the sum rule. We have already calculated that the probability of two 4's is 1/36; clearly, with the use of the same type of calculation, the probability of two 5's will be the same, or 1/36. Now, we can calculate the probability of either two 4's *or* two 5's. Because these outcomes are mutually exclusive, the sum rule can be used to tell us that the answer is $1/36 + 1/36$, which is 1/18. This probability can be written as follows:

$$p(\text{rolling two } 4\text{'s or rolling two } 5\text{'s}) = \tfrac{1}{36} + \tfrac{1}{36} = \tfrac{1}{18}$$

> **KEY CONCEPT** The sum rule states that the probability of *either one or the other* of two mutually exclusive events occurring is the sum of their individual probabilities.

> **KEY CONCEPT** The product rule is used to determine the probability of observing *both* outcome A *and* outcome B. The sum rule is used to determine the probability of observing *either* outcome A *or* outcome B.

What proportion of progeny will be of a specific genotype? Now we can turn to a genetic example. Assume that we have two plants of genotypes

$$A/a\,;b/b\,;C/c\,;D/d\,;E/e$$

and

$$A/a\,;B/b\,;C/c\,;d/d\,;E/e$$

From a cross between these plants, we want to recover a progeny plant of genotype $a/a\,;b/b\,;c/c\,;d/d\,;e/e$ (perhaps for the purpose of acting as the tester strain in a testcross). What proportion of the progeny should we expect to be of that genotype? If we assume that all the gene pairs assort independently, then we can do this calculation easily by using the product rule. The five different gene pairs are considered individually, as if five separate crosses, and then the individual probabilities of obtaining each genotype are multiplied together to arrive at the answer:

From $A/a \times A/a$, one-fourth of the progeny will be a/a.

From $b/b \times B/b$, one-half of the progeny will be b/b.

From $C/c \times C/c$, one-fourth of the progeny will be c/c.

From $D/d \times d/d$, one-half of the progeny will be d/d.

From $E/e \times E/e$, one-fourth of the progeny will be e/e.

Therefore, the overall probability (or expected frequency) of obtaining progeny of genotype a/a; b/b; c/c; d/d; e/e will be $1/4 \times 1/2 \times 1/4 \times 1/2 \times 1/4 = 1/256$. This probability calculation can be extended to predict phenotypic frequencies or gametic frequencies. Indeed, there are many other uses for this method in genetic analysis, and we will encounter some in later chapters.

KEY CONCEPT For independently assorting genes, the probability of a multigene genotype or phenotype can be obtained by multiplying the probabilities of the genotype or phenotype for each of the individual genes.

How many progeny do we need to grow? To take the preceding example a step farther, suppose we need to estimate how many progeny plants need to be grown to stand a reasonable chance of obtaining the desired genotype a/a; b/b; c/c; d/d; e/e. We first calculate the proportion of progeny that is expected to be of that genotype. As just shown, we learn that we need to examine at least 256 progeny to stand an average chance of obtaining one individual plant of the desired genotype.

The probability of actually obtaining one "success" (a fully recessive plant) out of 256 has to be considered more carefully. One in 256 is the *average* probability of success. Unfortunately, if we isolated and tested 256 progeny, we would very likely have no successes at all, simply from bad luck. From a practical point of view, a more meaningful question to ask would be, "What sample size do we need to be *95 percent confident* that we will obtain at least one success?" This 95 percent confidence value is standard in science. The simplest way to perform this calculation is to approach it by considering the probability of complete failure—that is, the probability of obtaining no individuals of the desired genotype. In our example, for every individual isolated, the probability of its *not* being the desired type is $1 - (1/256) = 255/256$. Extending this idea to a sample of size n, we see that the probability of no successes in a sample of n is $(255/256)^n$. (This probability is a simple application of the product rule: $255/256$ multiplied by itself n times.) Hence, the probability of obtaining *at least one success* is the probability of all possible outcomes (this probability is 1) minus the probability of total failure, or $(255/256)^n$. That is, the probability of at least one success is $1 - (255/256)^n$. Remember that we want to calculate the sample size needed to have a 95 percent chance of at least one success. So to satisfy the 95 percent confidence level, we must put this expression equal to 0.95 (the equivalent of 95 percent).

Therefore,

$$1 - (255/256)^n = 0.95$$

Solving this equation for n gives us a value of 765, the number of progeny needed to virtually guarantee success. Notice how different this number is from the naïve expectation of success in 256 progeny. This type of calculation is useful in many applications in genetics and in other situations in which a successful outcome is needed from many trials.

KEY CONCEPT To calculate the progeny sample size needed to be 95 percent certain of obtaining at least one individual of the desired genotype, start by calculating the probability of no successes in a sample size of n. The resulting sample size is always much larger than one calculated using hypothetical expectations.

How many distinct genotypes will a cross produce? The rules of probability can be easily used to predict the number of genotypes or phenotypes in the progeny of complex parental strains. (Such calculations are used routinely in research, in progeny analysis, and in strain building.) For example, in a self of the "tetrahybrid" A/a; B/b; C/c; D/d, there will be three genotypes for each gene pair; for example, for the first gene pair, the three genotypes will be A/a, A/A, and a/a. Because there are four gene pairs in total, there will be $3^4 = 81$ different genotypes. In a testcross of such a tetrahybrid, there will be two genotypes for each gene pair (for example, A/a and a/a) and a total of $2^4 = 16$ genotypes in the progeny. Because we are assuming that all the genes are on different chromosomes, all these testcross genotypes will occur at an equal frequency of 1/16.

Using the chi-square test on monohybrid and dihybrid ratios

In genetics, a researcher is often confronted with results that are close to an expected ratio but not identical to it. Such ratios can be from monohybrids, dihybrids, or more complex genotypes and with independence or not. But how close to an expected result is close enough? A statistical test is needed to check results against expectations, and the **chi-square test**, or χ^2 test, fulfills this role.

In which experimental situations is the χ^2 test applicable? The general situation is one in which observed results are compared with those predicted by a hypothesis. In a simple genetic example, suppose you have bred a plant that you hypothesize on the basis of a preceding analysis to be a heterozygote, A/a. To test this hypothesis, you cross this heterozygote with a tester of genotype a/a and count the numbers of phenotypes with genotypes $A/-$ and a/a in the progeny. Then, you must assess whether the numbers that you obtain constitute the expected 1 : 1 ratio. If there is a close match, then the hypothesis is deemed consistent with the result, whereas if there is a poor match, the hypothesis is rejected. As part of this process, a judgment has to be made about whether the observed numbers are close *enough* to those expected. Very close matches and blatant mismatches generally present no problem, but, inevitably, there are gray areas in which the match is not obvious.

The χ^2 test is simply a way of quantifying the various deviations expected by chance if a hypothesis is true. Take the preceding simple hypothesis predicting a 1 : 1 ratio, for example. Even if the hypothesis were true, we can only rarely expect an exact 1 : 1 ratio. We can model this idea with a barrelful of equal numbers of red and white marbles. If we blindly remove samples of 100 marbles, on the basis of chance we would expect samples to show small deviations such as 52 red : 48 white quite commonly and to show larger deviations such as 60 red : 40 white less commonly. Even 100 red marbles is a possible outcome, at a very low probability of $(1/2)^{100}$. However, if *any* result is possible at some level of probability even if the hypothesis is true, how can we ever reject a hypothesis? A general scientific convention is that a hypothesis will be rejected as false if there is a probability of less than 5 percent of observing a deviation from expectations at least as large as the one actually observed. The hypothesis might still be true, but we have to make a decision somewhere, and 5 percent is the conventional decision line. The implication is that, although results this far from expectations are expected 5 percent of the time even when the hypothesis is true, we will mistakenly reject the hypothesis in only 5 percent of cases, and we are willing to take this chance of error. (This 5 percent is the converse of the 95 percent confidence level used earlier.)

> **KEY CONCEPT** The χ^2 test quantifies the probability of various deviations expected by chance if a hypothesis is true. It is used to decide whether or not an observed experimental deviation is reasonably compatible with a working hypothesis.

Let's look at some real data. We will test our earlier hypothesis that a plant is a heterozygote. We will let *A* stand for red petals and *a* stand for white. Scientists test a hypothesis by making predictions based on the hypothesis. In the present situation, one possibility is to predict the results of a testcross. Assume that we testcross the presumed heterozygote. On the basis of the hypothesis, Mendel's law of equal segregation predicts that we should have 50 percent *A/a* and 50 percent *a/a*. Assume that, in reality, we obtain 120 progeny and find that 55 are red and 65 are white. These numbers differ from the precise expectations, which would have been 60 red and 60 white. The result seems a bit far off the expected ratio, which raises uncertainty; so we need to use the χ^2 test. We calculate χ^2 by using the following formula:

$$\chi^2 = \sum (O - E)^2/E \text{ for all classes}$$

in which *E* is the expected number in a class, *O* is the observed number in a class, and Σ means "sum of." The resulting value, χ^2, will provide a numerical value that estimates the degree of agreement between the expected (hypothesized) and observed (actual) results, with the number growing larger as the agreement increases.

The calculation is most simply performed by using a table:

Class	O	E	$(O - E^2)$	$(O - E^2)/E$
Red	55	60	25	25/60 = 0.42
White	65	60	25	25/60 = 0.42
				Total = χ^2 = 0.84

Now we must look up this χ^2 value in **Table 3-1**, which will give us the probability value (*p*) we seek. The rows in Table 3-1 list different values of *degrees of freedom* (*df*). The number of degrees of freedom is the number of independent variables in the data. In the present context, the number of independent variables is simply the number of phenotypic classes minus 1. In this case, df = 2 − 1 = 1. So we look only at the 1 df line. We see that our χ^2 value of 0.84 lies somewhere between the columns marked 0.5 and 0.1—in other words, a probability of between 50 percent and 10 percent. This probability value is much greater than the cutoff value of 5 percent, and so we accept the observed results as being compatible with the hypothesis.

Some important notes on the application of this test follow:

1. What does the probability value actually mean? It is the probability of observing a deviation from the expected results *at least as large* (not *exactly* this deviation) on the basis of chance if the hypothesis is correct.

2. The fact that our results have "passed" the chi-square test because $p > 0.05$, does not mean that the hypothesis is true; it merely means that the results are compatible with that hypothesis. However, if we had obtained a value of $p < 0.05$, we would have been forced to reject the hypothesis. Science is all about falsifiable hypotheses, not "truth."

3. We must be careful about the wording of the hypothesis because tacit assumptions are often buried within it. The present hypothesis is a case in point; if we were to state it carefully, we would have to say that the "individual under test is a heterozygote *A/a*, these alleles show equal segregation at meiosis, and the *A/a* and *a/a* progeny are of equal viability." We will investigate allele effects on viability in Chapter 5, but, for the time being, we must keep them in mind as a possible complication because differences in survival would affect the sizes of the various classes. The problem is that, if we reject a hypothesis that has hidden components, we do not know which of the components we are rejecting. For example, in the present case, if we were forced to reject the hypothesis as a result of the χ^2 test, we would not know if we were rejecting equal segregation or equal viability, or both.

TABLE 3-1   Critical Values of the χ^2 Distribution

df	0.995	0.975	0.9	0.5	0.1	0.05	0.025	0.01	0.005	df
1	.000	.000	0.016	0.455	2.706	3.841	5.024	6.635	7.879	1
2	0.010	0.051	0.211	1.386	4.605	5.991	7.378	9.210	10.597	2
3	0.072	0.216	0.584	2.366	6.251	7.815	9.348	11.345	12.838	3
4	0.207	0.484	1.064	3.357	7.779	9.488	11.143	13.277	14.860	4
5	0.412	0.831	1.610	4.351	9.236	11.070	12.832	15.086	16.750	5
6	0.676	1.237	2.204	5.348	10.645	12.592	14.449	16.812	18.548	6
7	0.989	1.690	2.833	6.346	12.017	14.067	16.013	18.475	20.278	7
8	1.344	2.180	3.490	7.344	13.362	15.507	17.535	20.090	21.955	8
9	1.735	2.700	4.168	8.343	14.684	16.919	19.023	21.666	23.589	9
10	2.156	3.247	4.865	9.342	15.987	18.307	20.483	23.209	25.188	10
11	2.603	3.816	5.578	10.341	17.275	19.675	21.920	24.725	26.757	11
12	3.074	4.404	6.304	11.340	18.549	21.026	23.337	26.217	28.300	12
13	3.565	5.009	7.042	12.340	19.812	22.362	24.736	27.688	29.819	13
14	4.075	5.629	7.790	13.339	21.064	23.685	26.119	29.141	31.319	14
15	4.601	6.262	8.547	14.339	22.307	24.996	27.488	30.578	32.801	15

4. The outcome of the χ^2 test depends heavily on sample sizes (numbers in the classes). Hence, the test must use *actual numbers*, not proportions or percentages. Additionally, the larger the samples, the more powerful is the test.

Any of the familiar Mendelian ratios considered in this chapter or in Chapter 2 can be tested by using the χ^2 test—for example, 3:1 (1 df), 1:2:1 (2 df), 9:3:3:1 (3 df), and 1:1:1:1 (3 df). We will return to more applications of the χ^2 test in Chapter 4.

KEY CONCEPT In genetics, the χ^2 test is commonly used to assess whether or not the number of observed individuals with certain phenotypes are an acceptable fit to an expected Mendelian ratio.

Synthesizing pure lines

Pure lines are among the essential tools of genetics. For one thing, only these fully homozygous lines will express recessive alleles, but the main need for pure lines is in the maintenance of stocks for research. The members of a pure line can be left to interbreed over time and thereby act as a constant source of the genotype for use in experiments. Hence, for most model organisms, there are international stock centers that are repositories of pure lines for use in research. Similar stock centers provide lines of plants and animals for use in agriculture.

KEY CONCEPT Homozygous pure lines are important research tools that allow geneticists to maintain a source of any given genotype. Recessive alleles can only be expressed in pure lines.

Pure lines of plants or animals are made through repeated generations of selfing. (In animals, selfing is accomplished by mating animals of identical genotype.) Selfing a monohybrid plant shows the principle at work. Suppose we start with a population of individuals that are all *A/a* and allow them to self. We can apply Mendel's first law to predict that, in the next generation, there will be $\frac{1}{4}$ *A/A*, $\frac{1}{2}$ *A/a*, and $\frac{1}{4}$ *a/a*. Note that the *heterozygosity* (the proportion of heterozygotes) has halved, from 1 to $\frac{1}{2}$. If we repeat this process of selfing for another generation, all descendants of homozygotes will be homozygous, but, again, the heterozygotes will halve their proportion to a quarter. The process is shown in the following display:

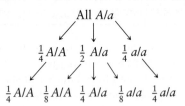

After, say, eight generations of selfing, examination of one gene pair reveals the proportion of heterozygotes in the progeny is reduced to $(1/2)^8$, which is 1/256, or about 0.4 percent. Let's look at this probability value in a slightly different way. Assume that we start such a program with a genotype that is heterozygous at 256 gene pairs. If we also assume independent assortment, then, after selfing for eight generations, we would end up with an average of only one heterozygous gene (that is, 1/256), and the rest will be homozygous. In other words, we are well on our way to creating a pure line.

Let's apply this principle to the selection of agricultural lines, the topic with which we began the chapter.

We can use as our example the selection of Marquis wheat by Charles Saunders in the early part of the twentieth century. Saunders's goal was to develop a productive wheat line that would have a shorter growing season and hence open up large areas of terrain in northern countries such as Canada and Russia for growing wheat, another of the world's staple foods. He crossed a line having excellent grain quality called Red Fife with a line called Hard Red Calcutta, which, although its yield and quality were poor, matured 20 days earlier than Red Fife. The F_1 produced by the cross was presumably heterozygous for multiple genes controlling the wheat qualities. From this F_1, Saunders made selfings and selections that eventually led to a pure line that had the combination of favorable properties needed—good-quality grain and early maturation. This line was called Marquis. It was rapidly adopted in many parts of the world.

A similar approach can be applied to the rice lines with which we began the chapter. All the single-gene mutations are crossed in pairs, and then their F_1 plants are selfed or intercrossed with other F_1 plants. As a demonstration, let's consider just four mutations, *1* through *4*. A breeding program might be as follows, in which the mutant alleles and their wild-type counterparts are always listed in the same order (recall that the + sign designates wild type):

1/1 ; +/+ ; +/+ ; +/+ × +/+ ; *2/2* ; +/+ ; +/+   +/+ ; +/+ ; *3/3* ; +/+ × +/+ ; +/+ ; +/+ ; *4/4*

↓   ↓

F_1 *1/+* ; *2/+* ; +/+ ; +/+ F_1 +/+ ; +/+ ; *3/+* ; *4/+*

↓   ↓

Self   Self

↓   ↓

Select the homozygote *1/1* ; *2/2* ; +/+ ; +/+   Select the homozygote +/+ ; +/+ ; *3/3* ; *4/4*

↘ ↙

Cross these homozygotes

↓

F_1 *1/+* ; *2/+* ; *3/+* ; *4/+*

↓

Self

↓

Select the homozygote *1/1* ; *2/2* ; *3/3* ; *4/4*

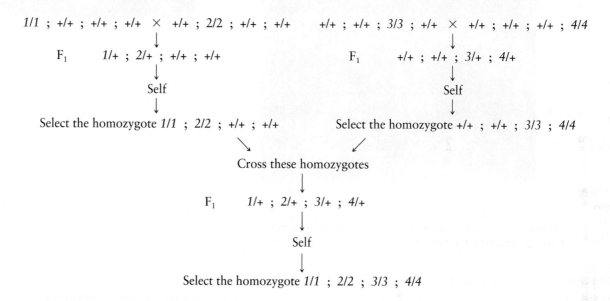

This type of breeding has been applied to many other crop species. The colorful and diverse pure lines of tomatoes used in commerce are shown in **Figure 3-5**.

FIGURE 3-5 Tomato breeding has resulted in a wide range of lines of different genotypes and phenotypes. [*barmalini/Shutterstock.*]

Note that, in general when a multiple heterozygote is selfed, a range of different homozygotes is produced. For example, from *A/a* ; *B/b* ; *C/c*, there are two homozygotes for each gene pair (that is, for the first gene, the homozygotes are *A/A* and *a/a*), and so there are $2^3 = 8$ different homozygotes possible:

A/A ; *B/B* ; *C/C*   *a/a* ; *b/b* ; *c/c*
A/A ; *b/b* ; *C/C*   *a/a* ; *B/B* ; *c/c*
A/A ; *B/B* ; *c/c*   *a/a* ; *b/b* ; *C/C*
A/A ; *b/b* ; *c/c*   *a/a* ; *B/B* ; *C/C*

Each distinct homozygote can be the start of a new pure line.

KEY CONCEPT Pure lines are generated through repeated selfing, which reduces the proportion of heterozygotes by half and results in an increased proportion of homozygotes with each generation.

Hybrid vigor

We have been considering the synthesis of superior pure lines for research and for agriculture. Pure lines are convenient in that propagation of the genotype from year to year is fairly easy. However, a large proportion of commercial seed that farmers (and gardeners) use is called *hybrid seed*. Curiously, in many cases in which two disparate homozygous lines of plants (and animals) are united in an F_1 hybrid (presumed heterozygote), the hybrid shows greater size and vigor than do the two contributing lines (**Figure 3-6**). This general superiority of multiple heterozygotes is called **hybrid vigor**. Seed companies must develop pure lines by the assortment through selfing methods we saw previously, and then cross them every season to generate hybrid seeds for the commercial market.

The molecular reasons for hybrid vigor are mostly unknown and still hotly debated, but the phenomenon is undeniable and has made large contributions to agriculture.

Hybrid vigor in corn

FIGURE 3-6 Multiple heterozygous hybrid flanked by the two pure lines crossed to make it. (a) The plants. (b) Cobs from the same plants. [*(a) Photo courtesy of Jun Cao, Schnable Laboratory, Iowa State University; (b) Deana Namuth-Covert, PhD, Univ of Nebraska, Plant and Soil Sciences eLibrary (http:// passel.unl.edu) hosted at the University of Nebraska, Institute of Agriculture and Natural Resources.*]

A negative aspect of using hybrids is that, every season, the two parental lines must be grown separately and then intercrossed to make hybrid seed for sale. This process is much more inconvenient than maintaining pure lines, which requires only letting plants self; consequently, hybrid seed is more expensive than seed from pure lines.

From the user's perspective, there is another negative aspect of using hybrids. After a hybrid plant has grown and produced its crop for sale, it is not realistic to keep some of the seeds that it produces and expect this seed to be equally vigorous the next year. The reason is that, when the hybrid undergoes meiosis, independent assortment of the various mixed gene pairs will form many different allelic combinations, and very few of these combinations will be that of the original hybrid. For example, a tetrahybrid, when selfed, produces 81 different genotypes, of which only a minority will be tetrahybrid. If we assume independent assortment, then, for each gene pair, selfing will produce one-half heterozygotes $A/a \rightarrow \frac{1}{4}\ A/A$, $\frac{1}{2}\ A/a$, and $\frac{1}{4}\ a/a$. Because there are four gene pairs in this tetrahybrid, the proportion of progeny that will be like the original hybrid $A/a\,;B/b\,;C/c\,;D/d$ will be $(1/2)^4 = 1/16$.

KEY CONCEPT   Some hybrids between genetically different pure lines show hybrid vigor. However, gene assortment when the hybrid undergoes meiosis breaks up the favorable allelic combination, and thus few members of the next generation have it.

3.3 THE CHROMOSOMAL BASIS OF INDEPENDENT ASSORTMENT

LO 3.5 Explain two-gene independent assortment ratios in terms of chromosome behavior at meiosis, in haploids and in diploids.

Like equal segregation, the independent assortment of gene pairs on different chromosomes is explained by the behavior of chromosomes during meiosis. Consider a chromosome that we might call number 1; its two homologs could be named 1′ and 1″. If the chromosomes pair and align on either side of the equator, then 1′ might go "north" and 1″ "south," or vice versa. Similarly, for a chromosome 2 with homologs 2′ and 2″, 2′ might go north and 2″ south, or vice versa. Hence, chromosome 1′ could end up packaged with either chromosome 2′ or 2″, depending on which chromosomes were pulled in the same direction.

Independent assortment is not easy to demonstrate by observing segregating chromosomes under the microscope because homologs such as 1′ and 1″ do not usually look different, although they might carry minor sequence variation. However, independent assortment can be observed in certain specialized cases. One case was instrumental in the historical development of the chromosome theory.

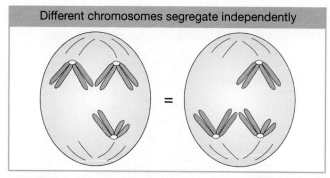

FIGURE 3-7 Carothers observed these two equally frequent patterns by which a heteromorphic pair (brown) and an unpaired chromosome (purple) move into gametes at meiosis.

In 1913, Elinor Carothers found an unusual chromosomal situation in a certain species of grasshopper—a situation that permitted a direct test of whether different chromosome pairs do indeed segregate independently. Studying meioses in the testes of grasshoppers, she found a grasshopper in which one chromosome "pair" had nonidentical members. Such a pair is called a *heteromorphic* pair; presumably, the chromosomes show only partial homology. In addition, the same grasshopper had another chromosome (unrelated to the heteromorphic pair) that had no pairing partner at all. Carothers was able to use these unusual chromosomes as visible cytological markers of the behavior of chromosomes during meiosis. She visually screened many meioses and found that there were two distinct patterns, which are shown in **Figure 3-7**. In addition, she found that the two patterns were equally frequent. To summarize, if we hold the segregation of the heteromorphic pair constant (brown in the figure), then the unpaired (purple) chromosome was found to go to either pole equally frequently, half the time with the long form and half the time with the short form. In other words, the purple and brown sets were segregating independently. Although these are obviously not typical chromosomes, Carothers's results do strongly suggest that different chromosomes assort independently at the first division of meiosis.

Independent assortment in diploid organisms

The chromosomal basis of the law of independent assortment is formally diagrammed in **Figure 3-8**, which illustrates how the separate behavior of two different chromosome pairs gives rise to the 1 : 1 : 1 : 1 Mendelian ratios of gametic types expected from independent assortment. The hypothetical cell has four chromosomes: a pair of homologous long chromosomes (yellow) and a pair of homologous short ones (blue). The genotype of the meiocytes is A/a ; B/b, and the two allelic pairs, A/a and B/b, are shown on two different chromosome pairs. Parts 4 and 4' of Figure 3-8 show the key step in independent assortment: there are two equally frequent allelic segregation patterns, one shown in 4 and the other in 4'. In one case, the A/A and B/B alleles are

pulled together into one cell, and the a/a and b/b are pulled into the other cell. In the other case, the alleles A/A and b/b are united in the same cell and the alleles a/a and B/B are united in the other cell. The two patterns result from two equally frequent spindle attachments to the centromeres in the first anaphase. Meiosis then produces four cells of the indicated genotypes from each of these segregation patterns. Because segregation patterns 4 and 4' are equally common, the meiotic product cells of genotypes A ; B, a ; b, A ; b, and a ; B are produced in equal frequencies. In other words, the frequency of each of the four genotypes is 1/4. This gametic distribution is that postulated by Mendel for a dihybrid, and it is the one that we inserted along one edge of the Punnett square in Figure 3-4. The random fusion of these gametes results in the 9 : 3 : 3 : 1 F_2 phenotypic ratio.

> **KEY CONCEPT** The mechanical basis of equal segregation and independent assortment of alleles is the anaphase segregation of chromosomes at meiosis. Segregation of a pair of homologs by spindle attachment from each pole accounts for Mendel's first law. The randomness of spindle attachment throughout the chromosome set accounts for Mendel's second law.

Independent assortment in haploid organisms

In the ascomycete fungi, we can actually inspect the products of a single meiocyte to show independent assortment directly. Let's use the filamentous fungus *Neurospora crassa* to illustrate this point. As we have seen from earlier fungal examples in Chapter 2, a cross in *Neurospora* is made by mixing two parental haploid strains of opposite mating type. In a manner similar to that of yeast, mating type is determined by two "alleles" of one gene—in this species, called MAT-A and MAT-a.

The products of meiosis in fungi are sexual spores. Recall that the *ascomycetes* (which include *Neurospora* and *Saccharomyces*) are unique in that, for any given meiocyte, the spores are held together in a membranous sac called an *ascus*. Thus, for these organisms, the products of a single meiosis can be recovered and tested. In the orange bread mold *Neurospora*, the nuclear divisions of meioses I and II take place along the linear axis of the ascus and do not overlap, and so the four products of a single meiocyte lie in a straight row (**Figure 3-9a**). Furthermore, for some reason not understood, there is a *postmeiotic mitosis*, which also shows no spindle overlap. Hence, meiosis and the extra mitosis result in a linear ascus containing eight *ascospores*, or an *octad*. In a heterozygous meiocyte A/a, if there are no crossovers between the gene and its centromere (a possibility we will explore in Chapter 4), then there will be two adjacent blocks of ascospores, four of A and four of a (Figure 3-9b).

Now we can examine a dihybrid. Let's make a cross between two distinct mutants having mutations in different

FIGURE 3-8 Meiosis in a diploid cell of genotype *A/a ; B/b*. The diagram shows how the segregation and assortment of different chromosome pairs give rise to the 1 : 1 : 1 : 1 Mendelian gametic ratio.

ANIMATED ART 🐿 SaplingPlus

Meiotic recombination between unlinked genes by independent assortment

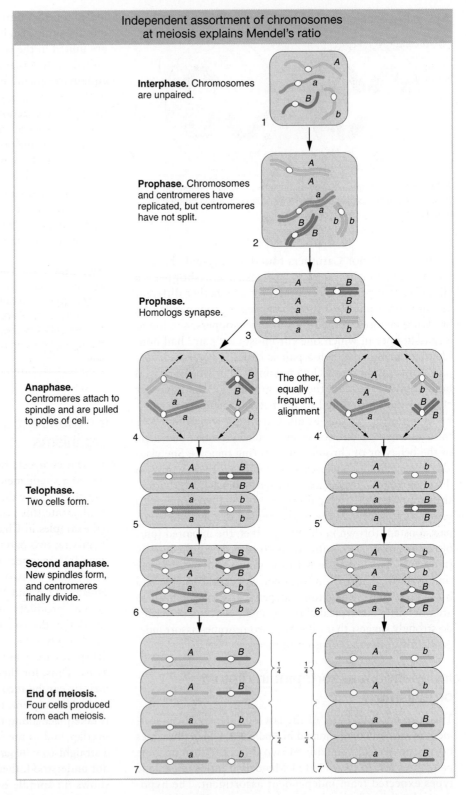

Independent assortment of chromosomes at meiosis explains Mendel's ratio

Interphase. Chromosomes are unpaired.

Prophase. Chromosomes and centromeres have replicated, but centromeres have not split.

Prophase. Homologs synapse.

Anaphase. Centromeres attach to spindle and are pulled to poles of cell.

The other, equally frequent, alignment

Telophase. Two cells form.

Second anaphase. New spindles form, and centromeres finally divide.

End of meiosis. Four cells produced from each meiosis.

genes on different chromosomes. By assuming that the loci of the mutated genes are both very close to their respective centromeres, we avoid complications due to crossing over between the loci and the centromeres (again, we will see examples of such scenarios in Chapter 4). The first mutant is albino (*a*), contrasting with the normal pink wild type (*a⁺*). The second mutant is biscuit (*b*), which has a very compact colony shaped like a biscuit in contrast with the

flat, spreading colony of wild type (*b⁺*). We will assume that the two mutants are of opposite mating type. Hence, the cross is

$$a ; b^+ \times a^+ ; b$$

Because of random spindle attachment, the two octad types will be produced with equal frequency. (Inspect Figure 3-8, which shows the mechanisms behind this result.)

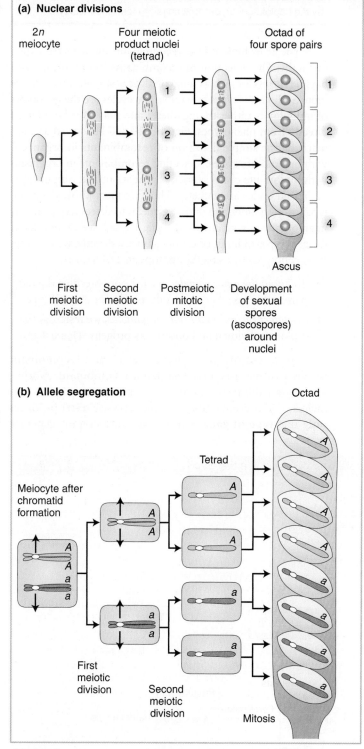

FIGURE 3-9 *Neurospora* is an ideal model system for studying allelic segregation at meiosis. (a) The four products of meiosis (tetrad) undergo mitosis to produce an octad. The products are contained within an ascus. (b) An *A/a* meiocyte undergoes meiosis followed by mitosis, resulting in equal numbers of *A* and *a* products and demonstrating the principle of equal segregation.

$a^+;b$	$a;b$
$a^+;b$	$a;b$
$a^+;b$	$a;b$
$a^+;b$	$a;b$
$a;b^+$	$a^+;b^+$
$a;b^+$	$a^+;b^+$
$a;b^+$	$a^+;b^+$
$a;b^+$	$a^+;b^+$
50%	**50%**

The equal frequency of these two types is a convincing demonstration of independent assortment occurring in individual meiocytes.

Recombination

The independent assortment of genes at meiosis is one of the main ways by which an organism produces new combinations of alleles. The production of new allele combinations is formally called **recombination.**

There is general agreement that the evolutionary advantage of producing new combinations of alleles is that it provides variation as the raw material for natural selection. Recombination is a crucial principle in genetics, partly because of its relevance to evolution but also because of its use in genetic analysis. It is particularly useful for analyzing inheritance patterns of multigene genotypes. In this section, we define recombination in such a way that we would recognize it in experimental results, and we lay out the way in which recombination is analyzed and interpreted.

Recombination is observed in a variety of biological situations, but, for the present, we define it in relation to meiosis.

> **Meiotic recombination** is any meiotic process that generates a haploid product with new combinations of the alleles carried by the haploid genotypes that united to form the meiocyte.

This seemingly wordy definition is actually quite simple; it makes the important point that we detect recombination by comparing the *inputs* into meiosis with the *outputs* (**Figure 3-10**). The inputs are the two haploid genotypes that combine to form the meiocyte, the diploid cell that undergoes meiosis. For humans, the inputs are the parental egg and sperm. They unite to form a diploid zygote, which divides to yield all the body cells, including the meiocytes that are set aside within the gonads. The output genotypes are the haploid products of meiosis. In humans, these haploid products are a person's own eggs or sperm. Any meiotic product that has a new combination of the alleles provided by the two input genotypes is by definition a **recombinant.**

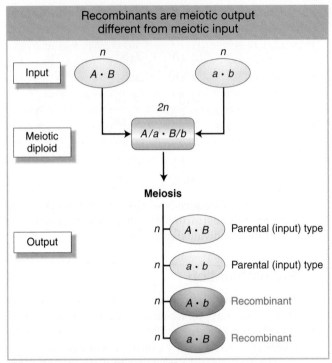

Recombinants are meiotic output different from meiotic input

FIGURE 3-10 Recombinants (blue) are those products of meiosis with allele combinations different from those of the haploid cells that formed the meiotic diploid (yellow). Note that genes *A/a* and *B/b* are shown separated by a dot because they may be on the same chromosome or on different chromosomes.

KEY CONCEPT Meiosis generates recombinants, which are haploid meiotic products with new combinations of the alleles carried by the haploid genotypes that united to form the meiocyte.

First, let us look at how recombinants are detected experimentally. The detection of recombinants in organisms with haploid life cycles such as fungi or algae is straightforward. The input and output types in haploid life cycles are the genotypes of individuals rather than gametes and may thus be inferred directly from phenotypes. Figure 3-10 can be viewed as summarizing the simple detection of recombinants in organisms with haploid life cycles. Detecting recombinants in organisms with diploid life cycles is trickier. The input and output types in diploid cycles are gametes. Thus, we must know the genotypes of both input and output gametes to detect recombinants in an organism with a diploid cycle. Though we cannot detect the genotypes of input or output gametes directly, we can infer these genotypes by using the appropriate techniques:

- *To know the input gametes*, we use pure-breeding diploid parents because they can produce only one gametic type.
- *To detect recombinant output gametes*, we testcross the diploid individual and observe its progeny (**Figure 3-11**).

A testcross offspring that arises from a recombinant product of meiosis also is called a recombinant. Notice, again, that the testcross allows us to concentrate on *one* meiosis and prevent ambiguity (the recessive tester produces only one type of gamete in meiosis and cannot generate

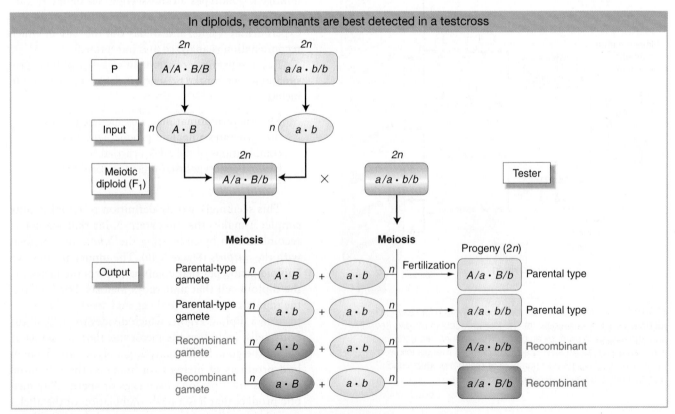

In diploids, recombinants are best detected in a testcross

FIGURE 3-11 Recombinant products of a diploid meiosis are most readily detected in a cross of a heterozygote and a recessive tester. Note that Figure 3-10 is repeated as part of this diagram.

recombinant gametes). From a *self* of the F₁ in Figure 3-11, for example, a recombinant *A/A · B/b* offspring could not be distinguished from *A/A · B/B* without further crosses.

A central part of recombination analysis is recombinant frequency. One reason for focusing on recombinant frequency is that its numerical value is a convenient test for whether two genes are on different chromosomes. Recombinants are produced by two different cellular processes: the independent assortment of genes on different chromosomes (this chapter) and crossing over between genes on the same chromosome (discussed in Chapter 4). The *proportion* of recombinants is the key idea here because the diagnostic value can tell us whether genes are on different chromosomes. We will deal with independent assortment here.

For genes on separate chromosomes, recombinants are produced by independent assortment, as shown in **Figure 3-12**. Again, we see the 1:1:1:1 ratio that we have seen before, but now the progeny of the testcross are classified as either recombinant or resembling the P (parental) input types. Set up in this way, the proportion of recombinants is clearly $\frac{1}{4} + \frac{1}{4} = \frac{1}{2}$, or

50 percent of the total progeny. Hence, we see that independent assortment at meiosis produces a recombinant frequency of 50 percent. If we observe a recombinant frequency of 50 percent in a testcross, we can infer that the two genes under study assort independently. The simplest and most likely interpretation of independent assortment is that the two genes are on separate chromosome pairs. (However, we must note that genes that are very far apart on the *same* chromosome pair can assort virtually independently and produce the same result, as we will see in Chapter 4.)

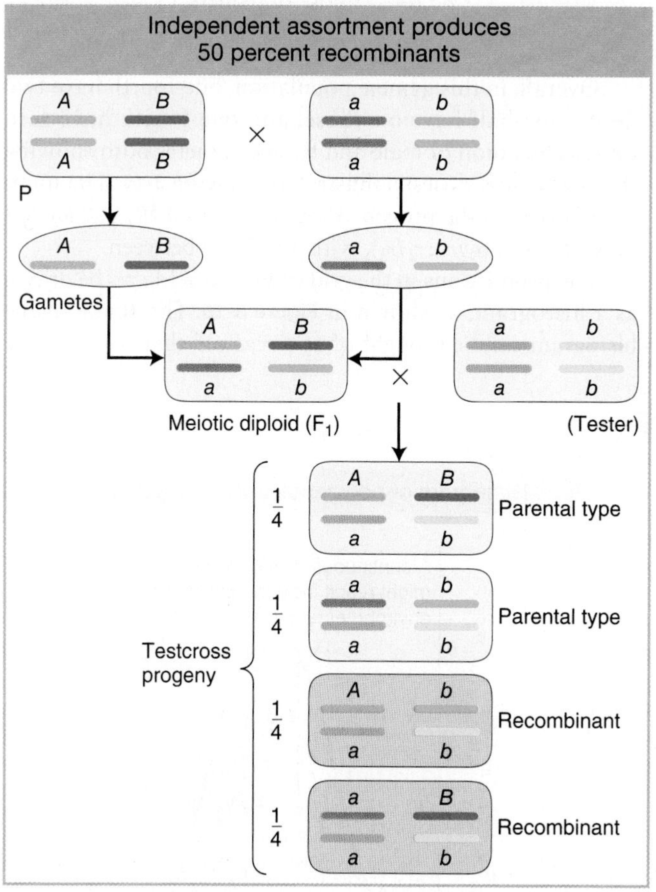

KEY CONCEPT A recombinant frequency of 50 percent indicates that the genes are independently assorting and are most likely on different chromosomes.

3.4 POLYGENIC INHERITANCE

LO 3.6 Extend the principle of independent assortment to multiple genes that each contribute to a phenotype showing continuous distribution.

So far, our analysis in this book has focused on single-gene differences, with the use of sharply contrasting phenotypes such as red versus white petals, smooth versus wrinkled seeds, and long- versus vestigial-winged *Drosophila*. However, much of the variation found in nature is *continuous*, in which a phenotype can take any measurable value between two extremes. Height, weight, and color intensity are examples of such *metric*, or *quantitative*, *phenotypes* (see Figure 1-6). Typically, when the metric value is plotted against frequency in a natural population, the distribution curve is shaped like a bell (**Figure 3-13**). The bell shape is due to the fact that average values in the middle are the most common, whereas extreme values are rare. At first, it is difficult to see how continuous distributions can be influenced by genes inherited in a Mendelian manner; after all, Mendelian analysis is facilitated by using clearly distinguishable

Independent assortment produces 50 percent recombinants

P

A   B   ×   a   b
A   B   a   b

Gametes

A   B   a   b

Meiotic diploid (F₁)   (Tester)
A   B   ×   a   b
a   b   a   b

Testcross progeny

$\frac{1}{4}$ A   B   Parental type
        a   b

$\frac{1}{4}$ a   b   Parental type
        a   b

$\frac{1}{4}$ A   b   Recombinant
        a   b

$\frac{1}{4}$ a   B   Recombinant
        a   b

FIGURE 3-12 This diagram shows two chromosome pairs of a diploid organism with *A* and *a* on one pair and *B* and *b* on the other. Independent assortment produces a recombinant frequency of 50 percent. Note that we could represent the haploid situation by removing the parental (P) generations and the tester.

ANIMATED ART 🌎 Sapling Plus

Meiotic recombination between unlinked genes by independent assortment

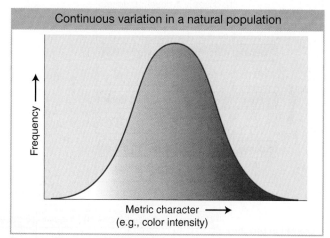

Continuous variation in a natural population

Frequency →

Metric character →
(e.g., color intensity)

FIGURE 3-13 In a population, a metric character such as color intensity can take on many values. Hence, the distribution is in the form of a smooth curve, with the most common values representing the high point of the curve. If the curve is symmetrical, it is bell shaped, as shown.

categories. However, it soon became clear that the interaction of several genes affecting a metric trait could produce a bell-shaped curve, a concept that became known as the *multifactorial hypothesis*. In this section, we shall see how the multifactorial hypothesis works. The topic is explored at length in Chapter 19.

> **KEY CONCEPT** Natural populations may show continuous variation of metric traits such as height or weight. Often the distribution of measurements is in the form of a bell-shaped curve.

Of course, many cases of continuous variation have a purely environmental basis, little affected by genetics. For example, a population of genetically homozygous plants grown in a plot of ground often show a bell-shaped curve for height, with the smaller plants around the edges of the plot and the larger plants in the middle. This variation can be explained only by environmental factors such as moisture and amount of fertilizer applied. However, many cases of continuous variation do have a genetic basis. Human skin color is an example: all degrees of skin darkness can be observed in populations from different parts of the world, and this variation clearly has a genetic component. In such cases, from several to many alleles interact with a more or less additive effect. The interacting genes underlying hereditary continuous variation are called **polygenes** or **quantitative trait loci (QTLs)**. (The term *quantitative trait locus* needs some definition: *quantitative* is more or less synonymous with continuous; *trait* is more or less synonymous with character or property; *locus*, which literally means place on a chromosome, is more or less synonymous with gene.) The polygenes, or QTLs, for the same trait are distributed throughout the genome; in many cases, they are on different chromosomes and show independent assortment.

> **KEY CONCEPT** Both environment and genotype can contribute to continuous variation.

Let's see how the inheritance of several heterozygous polygenes (even as few as two) can generate a bell-shaped distribution curve. We can consider a simple model that was originally used to explain continuous variation in the degree of redness in wheat seeds. The work was done by Hermann Nilsson-Ehle in the early twentieth century. We will assume two independently assorting gene pairs R_1/r_1 and R_2/r_2. Both R_1 and R_2 contribute to wheat-seed redness. Each "dose" of an R allele of either gene is additive, meaning that it increases the degree of redness proportionately. An illustrative cross is a self of a dihybrid $R_1/r_1 ; R_2/r_2$. Both male and female gametes will show the genotypic proportions as follows:

$$R_1 ; R_2 \quad \text{2 doses of redness}$$
$$R_1 ; r_2 \quad \text{1 dose of redness}$$
$$r_1 ; R_2 \quad \text{1 dose of redness}$$
$$r_1 ; r_2 \quad \text{0 doses of redness}$$

Overall, in this gamete population, one-fourth have two doses, one-half have one dose, and one-fourth have zero doses. The union of male and female gametes both showing this array of R doses is illustrated in **Figure 3-14**. The number of doses in the progeny ranges from four ($R_1/R_1 ; R_2/R_2$) down to zero ($r_1/r_2 ; r_2/r_2$), with all values between.

The proportions in the grid of Figure 3-14 can be drawn as a histogram, as shown in **Figure 3-15**. The shape of the histogram can be thought of as a scaffold that could be the

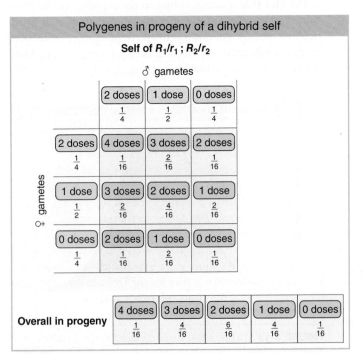

FIGURE 3-14 The progeny of a dihybrid self for two polygenes can be expressed as numbers of additive allelic "doses."

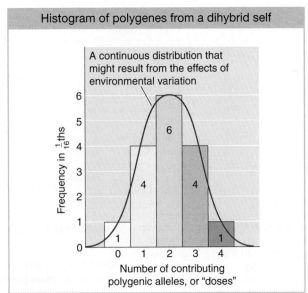

FIGURE 3-15 The progeny shown in Figure 3-14 can be represented as a frequency histogram of contributing polygenic alleles ("doses").

underlying basis for a bell-shaped distribution curve. When this analysis of redness in wheat seeds was originally done, variation was found within each class that allegedly represented one polygene "dose" level. Presumably, this variation within a class is the result of environmental differences. Hence, the environment can be seen to contribute in a way that rounds off the sharp shoulders of the histogram bars, resulting in a smooth bell-shaped curve (the red line in the histogram). If the number of polygenes is increased, the histogram more closely approximates a smooth continuous distribution. For example, for a characteristic determined by three polygenes, the histogram is as shown in **Figure 3-16**.

KEY CONCEPT The interaction of several additive heterozygous genes can by themselves result in a bell-shaped curve, their alleles acting as metric "doses."

In our illustration, we used a dihybrid self to show how the histogram is produced. But how is our example relevant to what is going on in natural populations? After all, not all crosses could be of this type. Nevertheless, if the alleles at each gene pair are approximately equal in frequency in the population (for example, R_1 is about as common as r_1), then the dihybrid cross can be said to represent an average cross for a population in which two polygenes are segregating.

Identifying polygenes and understanding how they act and interact are important challenges for geneticists in the twenty-first century. Identifying polygenes will be especially important in medicine. Many common human diseases such as atherosclerosis (hardening of the arteries) and hypertension (high blood pressure) are thought to have a polygenic component. If so, a full understanding of these conditions, which affect large proportions of human populations, requires an understanding of these polygenes, their inheritance, and their function. Today, several molecular approaches can be applied to the job of finding polygenes, and we will consider some in

Chapter 19. Note that polygenes are not considered a special functional class of genes. They are identified as a group only in the sense that they have alleles that contribute to continuous variation of a trait.

3.5 ORGANELLE GENES: INHERITANCE INDEPENDENT OF THE NUCLEUS

LO 3.7 Apply the diagnostic criteria for assessing whether a gene of interest resides on a nuclear chromosome or on an organelle chromosome.

So far, we have considered how nuclear genes assort independently by virtue of their loci on different chromosomes. However, although the nucleus contains most of a eukaryotic organism's genes, a distinct and specialized subset of the genome is found in the mitochondria, and, in plants, also in the chloroplasts. These subsets are inherited independently of the nuclear genome, and so they constitute a special case of independent inheritance, sometimes called extranuclear inheritance.

Mitochondria and chloroplasts are specialized organelles located in the cytoplasm. They contain small circular chromosomes that carry a defined subset of the total cell genome. Mitochondrial genes are concerned with the mitochondrion's task of energy production, whereas chloroplast genes are needed for the chloroplast to carry out its function of photosynthesis. However, neither organelle is functionally autonomous because each relies to a large extent on nuclear genes for its function. Why some of the necessary genes are in the organelles themselves and others are in the nucleus is still something of a mystery, which will not be addressed here.

Another peculiarity of organelle genes is the large number of copies present in a cell. Each organelle is present in many copies per cell, and, furthermore, each organelle contains many copies of its chromosome. Hence, a eukaryotic cell can contain hundreds or thousands of organelle chromosomes. Consider chloroplasts, for example. Any green cell of a plant has many chloroplasts, and each chloroplast contains many identical circular DNA molecules, the so-called chloroplast chromosomes. Hence, the number of chloroplast chromosomes per cell can be in the thousands, and the number can even vary somewhat from cell to cell. The DNA is sometimes seen to be packaged into suborganellar structures called *nucleoids*, which become visible if stained with a DNA-binding dye. The DNA is folded within the nucleoid but does not have the type of histone-associated coiling shown by nuclear chromosomes. The same arrangement is true for the DNA in mitochondria (**Figure 3-17**). For the time being, we will assume that all copies of an organelle chromosome within a cell are identical, but we will have to relax this assumption later.

Many organelle chromosomes have now been sequenced. Examples of relative gene size and spacing in **mitochondrial DNA (mtDNA)** and **chloroplast DNA (cpDNA)** are shown

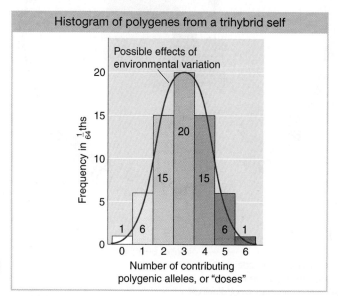

FIGURE 3-16 The progeny of a polygene trihybrid can be graphed as a frequency histogram of contributing polygenic alleles ("doses").

Cell showing nucleoids within mitochondria

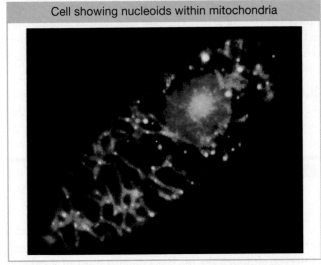

FIGURE 3-17 Fluorescent staining of a cell of *Euglena gracilis*. With the dyes used, the nucleus appears red because of the fluorescence of large amounts of nuclear DNA. The mitochondria fluoresce green, and, within mitochondria, the concentrations of mitochondrial DNA (nucleoids) fluoresce yellow. [*Republished with permission from The Company of Biologists Ltd., Y. Hayashi and K. Ueda, "The shape of mitochondria and the number of mitochondrial nucleoids during the cell cycle of Euglena gracilis," J. Cell Sci. 93, 1989, 565. Permission conveyed through Copyright Clearance Center, Inc.*]

in **Figure 3-18**. Organelle genes are very closely spaced, and, in some organisms, organelle genes can contain untranslated segments called introns. Note how most genes concern the chemical reactions taking place within the organelle itself: photosynthesis in chloroplasts and oxidative phosphorylation in mitochondria.

Patterns of inheritance in organelles

Organelle genes show their own special mode of inheritance called **uniparental inheritance**: progeny inherit organelle genes exclusively from one parent but not the other. In most cases, that parent is the mother, a pattern called **maternal inheritance**. Why only the mother? The answer lies in the fact that the organelle chromosomes are located in the cytoplasm and the male and female gametes do not contribute cytoplasm equally to the zygote. In regard to nuclear genes, both parents contribute equally to the zygote. However, the egg contributes the bulk of the cytoplasm, whereas the sperm contributes essentially none. Therefore, because organelles reside in the cytoplasm, the female parent contributes the organelles along with the cytoplasm, and essentially none of the organelle DNA in the zygote is from the male parent.

Some phenotypic variants are caused by a mutant allele of an organelle gene, and we can use these mutants to track

Organelle genomes

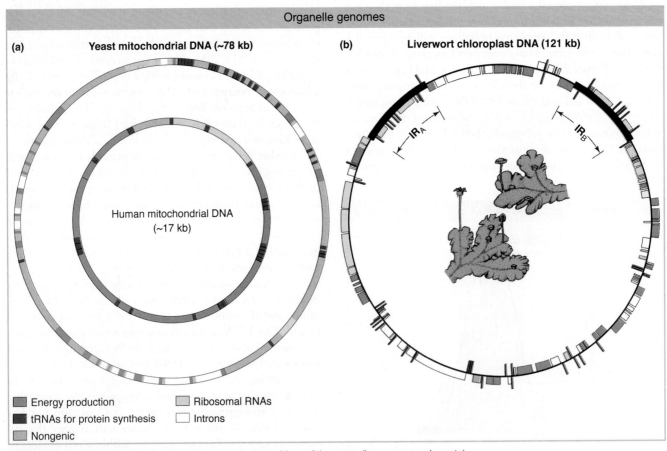

(a) **Yeast mitochondrial DNA (~78 kb)**

Human mitochondrial DNA (~17 kb)

(b) **Liverwort chloroplast DNA (121 kb)**

IR$_A$ IR$_B$

Energy production

tRNAs for protein synthesis

Nongenic

Ribosomal RNAs

Introns

FIGURE 3-18 DNA maps for mitochondria and chloroplasts. Many of the organelle genes encode proteins that carry out the energy-producing functions of these organelles (green), whereas others (red and orange) function in protein synthesis. (a) Maps of yeast and human mtDNAs. (Note that the human map is not drawn at the same scale as the yeast map.) (b) The 121-kb chloroplast genome of the liverwort *Marchantia polymorpha*. Genes shown inside the map are transcribed clockwise, and those outside are transcribed counterclockwise. IR$_A$ and IR$_B$ indicate inverted repeats. The upper drawing in the center of the map depicts a male *Marchantia* plant; the lower drawing depicts a female. [*Data from K. Umesono and H. Ozeki,* Trends Genet. *3, 1987.*]

patterns of organelle inheritance. We will temporarily assume that the mutant allele is present in all copies of the organelle chromosome, a situation that is indeed often found. In a cross, the variant phenotype will be transmitted to progeny if the variant used is the female parent, but not if it is the male parent. Generally, cytoplasmic inheritance shows the following pattern:

mutant female × wild-type male → progeny all mutant

wild-type female × mutant male → progeny all wild type

Indeed, this inheritance pattern is diagnostic of organelle inheritance in cases in which the genomic location of a mutant allele is not known.

Maternal inheritance can be clearly demonstrated in certain mutants of fungi. For example, in the fungus *Neurospora*, a mutant called *poky* has a slow-growth phenotype. *Neurospora* can be crossed in such a way that one parent acts as the maternal parent, contributing the cytoplasm. The results of the following reciprocal crosses suggest that the mutant gene resides in the mitochondria (fungi have no chloroplasts):

poky female × wild-type male → progeny all poky

wild-type female × poky male → progeny all wild type

Sequencing has shown that the poky phenotype is caused by a mutation of a ribosomal RNA gene in mtDNA. Its inheritance is shown diagrammatically in **Figure 3-19**. The cross includes an allelic difference (*ad* and *ad*⁺) in a nuclear gene in addition to *poky*; notice how the Mendelian inheritance of the nuclear gene is independent of the maternal inheritance of the poky phenotype.

> **KEY CONCEPT** Variant phenotypes caused by mutations in cytoplasmic organelle DNA are generally inherited maternally and independent of the Mendelian patterns shown by nuclear genes.

Cytoplasmic segregation

In some cases, cells contain mixtures of mutant and normal organelles. These cells are called *cytohets*, or *heteroplasmons*. In these mixtures, a type of **cytoplasmic segregation** can be detected, in which the two types apportion themselves into different daughter cells. The process most likely stems from chance partitioning of the multiple organelles over the course of multiple rounds of cell division. Plants provide a good example. Many cases of white leaves are caused by mutations in chloroplast genes that control the production and deposition of the green pigment chlorophyll. Because chlorophyll is necessary for a plant to live, this type of mutation is lethal, and white-leaved plants cannot be obtained for experimental crosses. However, some plants are variegated, bearing both green and white patches, and these plants are viable. Thus, variegated plants provide a way of demonstrating cytoplasmic segregation.

The four-o'clock plant in **Figure 3-20** shows a commonly observed variegated leaf and branch phenotype that demonstrates the inheritance of a mutant allele of a chloroplast gene. The mutant allele causes chloroplasts to be white; in turn, the color of the chloroplasts determines the color of cells and hence the color of the branches composed of those cells. Variegated branches are mosaics of all-green and all-white cells. Flowers can develop on green, white, or variegated branches, and the chloroplast genes of a flower's cells are those of the branch on which it grows. Hence, in a cross (**Figure 3-21**), the maternal gamete within the flower (the egg cell) determines the color of the leaves and branches of the progeny plant. For example, if an egg cell is from a flower on a green branch, all the progeny will be green, regardless of the origin of the pollen. A white branch will have white chloroplasts, and the resulting progeny plants will be white. (Because of

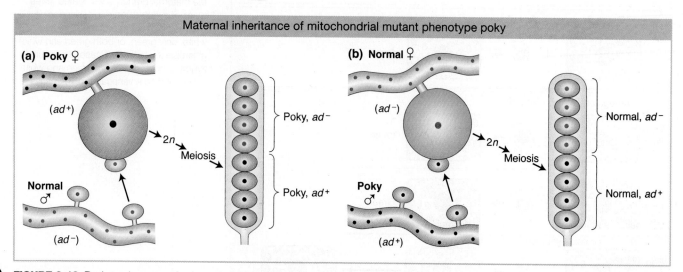

Maternal inheritance of mitochondrial mutant phenotype poky

(a) Poky ♀

(*ad*⁺)

Normal ♂

(*ad*⁻)

2*n* → Meiosis

Poky, *ad*⁻

Poky, *ad*⁺

(b) Normal ♀

(*ad*⁻)

Poky ♂

(*ad*⁺)

2*n* → Meiosis

Normal, *ad*⁻

Normal, *ad*⁺

FIGURE 3-19 Reciprocal crosses of poky and wild-type *Neurospora* produce different results because a different parent contributes the cytoplasm. The female parent contributes most of the cytoplasm of the progeny cells. Brown shading represents cytoplasm with mitochondria containing the *poky* mutation, and green shading represents cytoplasm with wild-type mitochondria. Note that all the progeny in part *a* are poky, whereas all the progeny in part *b* are normal. Hence, both crosses show maternal inheritance. The nuclear gene with the alleles *ad*⁺ (black) and *ad*⁻ (red) is used to illustrate the segregation of the nuclear genes in the 1:1 Mendelian ratio expected for this haploid organism.

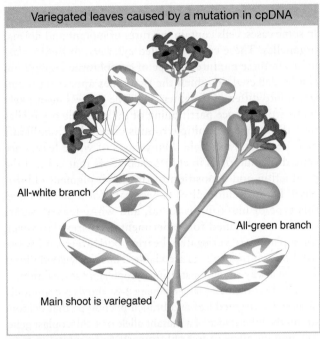

FIGURE 3-20 Leaf variegation in *Mirabilis jalapa*, the four-o'clock plant. Flowers can form on any branch (variegated, green, or white), and these flowers can be used in crosses.

lethality, white descendants would not live beyond the seedling stage.)

The variegated zygotes (bottom of Figure 3-21) demonstrate cytoplasmic segregation. These variegated progeny come from eggs that are cytohets. Interestingly, when such a zygote divides, the white and green chloroplasts often segregate over successive cell divisions; that is, they sort themselves into separate cells, yielding the distinct green and white sectors that cause the variegation in the branches. Here, then, is a direct demonstration of cytoplasmic segregation.

Given that a cell is a population of organelle molecules, how is it ever possible to obtain a "pure" mutant cell, containing only mutant chromosomes? Most likely, pure mutants are created in asexual cells as follows. The variants arise by mutation of a single gene in a single chromosome. Then, in some cases, the mutation-bearing chromosome may by chance increase in frequency in the population within the cell. This process is called *random genetic drift* and is discussed more fully in Chapter 18. A cell that is a cytohet may have, say, 60 percent *A* chromosomes and 40 percent *a* chromosomes. When this cell divides, sometimes all the *A* chromosomes go into one daughter, and all the *a* chromosomes into the other (again, by chance). More often, this partitioning requires several subsequent generations of cell division to be complete. Hence, as a result of

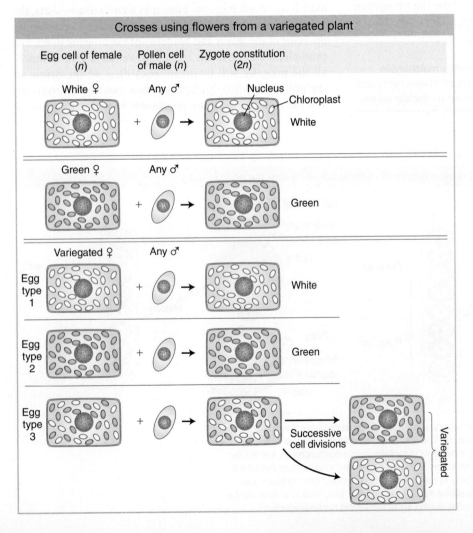

FIGURE 3-21 The results of the *Mirabilis jalapa* crosses can be explained by autonomous chloroplast inheritance. The large, dark spheres represent nuclei. The smaller bodies represent chloroplasts, either green or white. Each egg cell is assumed to contain many chloroplasts, and each pollen cell is assumed to contain no chloroplasts. The first two crosses exhibit strict maternal inheritance. If, however, the maternal branch is variegated, three types of zygotes can result, depending on whether the egg cell contains only white, only green, or both green and white chloroplasts. In the last case, the resulting zygote can produce both green and white tissue, and so a variegated plant results.

these chance events, the two alleles are expressed in different daughter cells, and this separation will continue through the descendants of these cells. Note that cytoplasmic segregation is not a mitotic process; it does take place in dividing asexual cells, but it is unrelated to mitosis. In chloroplasts, cytoplasmic segregation is a common mechanism for producing variegated (green-and-white) plants, as already mentioned. In fungal mutants such as the *poky* mutant of *Neurospora*, the original mutation in one mtDNA molecule must have accumulated and undergone cytoplasmic segregation to produce the strain expressing the poky symptoms.

KEY CONCEPT Organelle populations that contain mixtures of two genetically distinct chromosomes often show segregation of the two types into the daughter cells after one or more cell divisions. This process is called cytoplasmic segregation.

KEY CONCEPT Alleles on organelle chromosomes

1. in sexual crosses are inherited from one parent only (generally the maternal parent) and hence show no segregation ratios of the type nuclear genes do.
2. in asexual cells can show cytoplasmic segregation.

Cytoplasmic mutations in humans

Are there cytoplasmic mutations in humans? Some human pedigrees show the transmission of rare disorders only through females and never through males. This pattern strongly suggests cytoplasmic inheritance and points to a mutation in mtDNA as the reason for the phenotype. The disease MERRF (myoclonic epilepsy with ragged red fibers) is such a phenotype, resulting from a single base change in mtDNA. It is a disease that affects muscles, but the symptoms also include eye and hearing disorders. Another example is Kearns–Sayre syndrome, a constellation of symptoms affecting the eyes, heart, muscles, and brain that is caused by the loss of part of the mtDNA. In some of these cases, the cells of those affected contain mixtures of normal and mutant chromosomes, and the proportions of each passed on to progeny can vary as a result of cytoplasmic segregation. The proportions in one person can also vary in different tissues or over time. The accumulation of certain types of mitochondrial mutations over time has been proposed as a possible cause of aging.

Figure 3-22 shows some of the mutations in human mitochondrial genes that can lead to disease when, by random

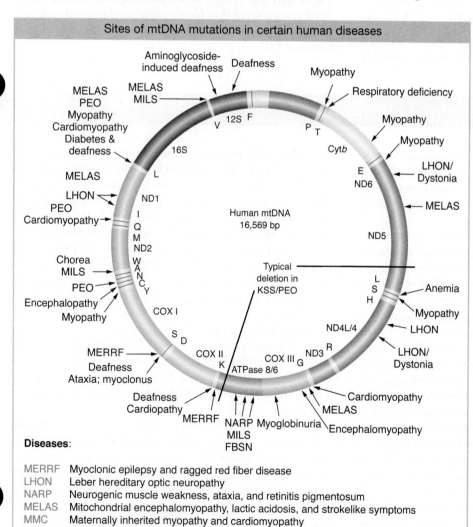

Sites of mtDNA mutations in certain human diseases

Human mtDNA
16,569 bp

Typical deletion in KSS/PEO

Diseases:

MERRF	Myoclonic epilepsy and ragged red fiber disease
LHON	Leber hereditary optic neuropathy
NARP	Neurogenic muscle weakness, ataxia, and retinitis pigmentosum
MELAS	Mitochondrial encephalomyopathy, lactic acidosis, and strokelike symptoms
MMC	Maternally inherited myopathy and cardiomyopathy
PEO	Progressive external opthalmoplegia
KSS	Kearns–Sayre syndrome
MILS	Maternally inherited Leigh syndrome

FIGURE 3-22 This map of human mtDNA shows loci of mutations leading to cytopathies. Gene labels are on the inside of the chromosome, and disorders are labeled around the outside of the chromosome. The transfer RNA genes are represented by single-letter amino acid abbreviations. ND = NADH dehydrogenase; COX = cytochrome C oxidase; and 12S and 16S refer to ribosomal RNAs. [*Data from S. DiMauro et al., "Mitochondria in Neuromuscular Disorders," Biochim. Biophys. Acta 1366, 1998, 199–210.*]

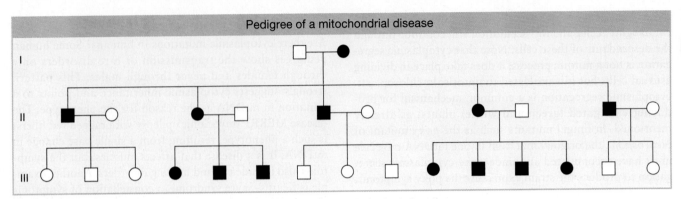

FIGURE 3-23 This pedigree shows that a human mitochondrial disease is inherited only from the mother.

drift and cytoplasmic segregation, they rise in frequency to such an extent that cell function is impaired. The inheritance of a human mitochondrial disease is shown in **Figure 3-23**. Note that the condition is always passed to offspring by mothers and never fathers. Occasionally, a mother will produce an unaffected child (not shown), probably owing to cytoplasmic segregation in the gamete-forming tissue.

mtDNA in evolutionary studies

Differences and similarities of homologous mtDNA sequences between species have been used extensively to construct evolutionary trees. Furthermore, it has been possible to introduce some extinct organisms into evolutionary trees using mtDNA sequences obtained from their remains, such as skins and bones in museums. mtDNA evolves relatively rapidly, so this approach has been most useful in plotting recent evolution such as the evolution of humans and other primates. One key finding is that the "root" of the human mtDNA tree is in Africa, suggesting that *Homo sapiens* originated in Africa and from there dispersed throughout the world (see Chapter 18).

SUMMARY

Genetic research and plant and animal breeding often necessitate the synthesis of genotypes that are complex combinations of alleles from different genes. Such genes can be on the same chromosome or on different chromosomes; the latter is the main subject of this chapter.

In the simplest case—a dihybrid for which the two gene pairs are on different chromosome pairs—each individual gene pair shows equal segregation at meiosis as predicted by Mendel's first law. Because nuclear spindle fibers attach randomly to centromeres at meiosis, the two gene pairs are partitioned independently into the meiotic products. This principle of independent assortment is called Mendel's second law because Mendel was the first to observe it. From a dihybrid A/a; B/b, four genotypes of meiotic products are produced, A; B, A; b, a; B, and a; b, all at an equal frequency of 25 percent each. Hence, in a testcross of a dihybrid with a double recessive, the phenotypic proportions of the progeny also are 25 percent (a 1:1:1:1 ratio). If such a dihybrid is selfed, the phenotypic classes in the progeny are $\frac{9}{16}$ $A/-$; $B/-$, $\frac{3}{16}$ $A/-$; b/b, $\frac{3}{16}$ a/a; $B/-$, and $\frac{1}{16}$ a/a; b/b. The 1:1:1:1 and 9:3:3:1 ratios are both diagnostic of independent assortment.

More complex genotypes composed of independently assorting genes can be treated as extensions of the case for single-gene segregation. Overall genotypic, phenotypic, or gametic ratios are calculated by applying the product rule—that is, by multiplying the proportions relevant to the individual genes. The probability of the occurrence of any of several categories of progeny is calculated by applying the sum rule—that is, by adding their individual probabilities. In mnemonic form, the product rule deals with "A AND B," whereas the sum rule deals with "A' OR A"." The χ^2 test can be used to test whether the observed proportions of classes in genetic analysis conform to the expectations of a genetic hypothesis, such as a hypothesis of single- or two-gene inheritance. If a probability value of less than 5 percent is calculated, the hypothesis must be rejected.

Sequential generations of selfing increase the proportions of homozygotes, according to the principles of equal segregation and independent assortment (if the genes are on different chromosomes). Hence, selfing is used to create complex pure lines with combinations of desirable mutations.

The independent assortment of chromosomes at meiosis can be observed cytologically by using heteromorphic chromosome pairs (those that show a structural difference). The X and Y chromosomes are one such case, but other, rarer cases can be found and used for this demonstration. The independent assortment of genes at the level of single meiocytes can be observed in the ascomycete fungi, because the asci show the two alternative types of segregations at equal frequencies.

One of the main functions of meiosis is to produce recombinants, new combinations of alleles of the haploid genotypes that united to form the meiocyte. Independent assortment is the main source of recombinants. In a dihybrid testcross showing independent assortment, the recombinant frequency will be 50 percent.

Metric characters such as color intensity show a continuous distribution in a population. Continuous distributions can be based on environmental variation or on variant alleles of multiple genes, or on a combination of both. A simple genetic model proposes that the active alleles of several genes (called polygenes) contribute more or less additively to the metric character. In an analysis of the progeny from the self of a multiply heterozygous individual, the histogram showing the proportion of each phenotype approximates a bell-shaped curve typical of continuous variation.

The small subsets of the genome found in mitochondria and chloroplasts are inherited independently of the nuclear genome. Mutants in these organelle genes often show maternal inheritance, along with the cytoplasm, which is the location of these organelles. In genetically mixed cytoplasms (cytohets), the two genotypes (say, wild type and mutant) often sort themselves out into different daughter cells by a poorly understood process called cytoplasmic segregation. Mitochondrial mutation in humans results in diseases that show cytoplasmic segregation in body tissues and maternal inheritance in a mating.

KEY TERMS

chi-square test (p. 86)
chloroplast DNA (cpDNA) (p. 97)
cytoplasmic segregation (p. 99)
dihybrid (p. 81)
dihybrid cross (p. 81)
hybrid vigor (p. 89)
independent assortment (p. 80)

law of independent assortment (Mendel's second law) (p. 81)
maternal inheritance (p. 98)
meiotic recombination (p. 93)
mitochondrial DNA (mtDNA) (p. 97)
polygene (quantitative trait locus) (p. 96)

product rule (p. 85)
quantitative trait locus (QTL) (p. 96)
recombinant (p. 93)
recombination (p. 93)
sum rule (p. 85)
uniparental inheritance (p. 98)

SOLVED PROBLEMS

SOLVED PROBLEM 1

Two *Drosophila* flies that had normal (transparent, long) wings were mated. In the progeny, two new phenotypes appeared: dusky wings (having a semi-opaque appearance) and clipped wings (with squared ends). The progeny were as follows:

Females	Males
179 transparent, long	92 transparent, long
58 transparent, clipped	89 dusky, long
	28 transparent, clipped
	31 dusky, clipped

a. Provide a chromosomal explanation for these results, showing chromosomal genotypes of parents and of all progeny classes under your model.

b. Design a test for your model.

SOLUTION

a. The first step is to state any interesting features of the data. The first striking feature is the appearance of two new phenotypes. We encountered the phenomenon in Chapter 2, where it was explained as recessive alleles being masked by their dominant counterparts. So, first, we might suppose that one or both parental flies have recessive alleles of two different genes. This inference is strengthened by the observation that some progeny express only one of the new phenotypes. If the new phenotypes always appeared together, we might suppose that the same recessive allele determines both.

However, the other striking feature of the data, which we cannot explain by using the Mendelian principles from Chapter 2, is the obvious difference between the sexes: although there are approximately equal numbers of males and females, the males fall into four phenotypic classes, but the females constitute only two. This fact should immediately suggest some kind of sex-linked inheritance. When we study the data, we see that the long and clipped phenotypes are segregating in both males and females, but only males have the dusky phenotype. This observation suggests that the inheritance of wing transparency differs from the inheritance of wing shape. First, long and clipped are found in a 3 : 1 ratio in both males and females. This ratio can be explained if both parents are heterozygous for an autosomal gene; we can represent them as L/l, where L stands for long and l stands for clipped.

Having done this partial analysis, we see that only the inheritance of wing transparency is associated with sex. The most obvious possibility is that the alleles for transparent (D) and dusky (d) are on the X chromosome, because we have seen in Chapter 2 that gene location on this chromosome gives inheritance patterns correlated with sex. If this suggestion is true, then the parental female must be the one sheltering the d allele, because, if the male had the d, he would have been dusky, whereas we were told that he had transparent wings. Therefore, the female parent would be D/d and the male D. Let's see if this suggestion works: if it is true, all female progeny would inherit the D allele from their father, and so all would be transparent winged, as was observed. Half the sons would be D (transparent) and half d (dusky), which also was observed.

So, overall, we can represent the female parent as $D/d\,;L/l$ and the male parent as $D\,;L/l$. Then the progeny would be

Females

$$\frac{1}{2}\,D/D\left<\begin{array}{l}\frac{3}{4}\,L/-\longrightarrow\frac{3}{8}\,D/D\,;L/-\\\frac{1}{4}\,l/l\longrightarrow\frac{1}{8}\,D/D\,;l/l\end{array}\right.$$

$$\frac{1}{2}\,D/d\left<\begin{array}{l}\frac{3}{4}\,L/-\longrightarrow\frac{3}{8}\,D/d\,;L/-\\\frac{1}{4}\,l/l\longrightarrow\frac{1}{8}\,D/d\,;l/l\end{array}\right.$$

$\begin{array}{l}\frac{3}{4}\text{ transparent, long}\\\frac{1}{4}\text{ transparent, clipped}\end{array}$

Males

$$\frac{1}{2}\,D\left<\begin{array}{l}\frac{3}{4}\,L/-\longrightarrow\frac{3}{8}\,D\,;L/-\quad\text{transparent, long}\\\frac{1}{4}\,l/l\longrightarrow\frac{1}{8}\,D\,;l/l\quad\text{transparent, clipped}\end{array}\right.$$

$$\frac{1}{2}\,d\left<\begin{array}{l}\frac{3}{4}\,L/-\longrightarrow\frac{3}{8}\,d\,;L/-\quad\text{dusky, long}\\\frac{1}{4}\,l/l\longrightarrow\frac{1}{8}\,d\,;l/l\quad\text{dusky, clipped}\end{array}\right.$$

b. Generally, a good way to test such a model is to make a cross and predict the outcome. But which cross? We have to predict some kind of ratio in the progeny, and so it is important to make a cross from which a unique phenotypic ratio can be expected. Notice that using one of the female progeny as a parent would not serve our needs: we cannot say from observing the phenotype of any one of these females what her genotype is. A female with transparent wings could be D/D or D/d, and one with long wings could be L/L or L/l. It would be good to cross the parental female of the original cross with a dusky, clipped son, because the full genotypes of both are specified under the model that we have created. According to our model, this cross is

$$D/d\,;L/l\times d\,;l/l$$

From this cross, we predict

Females

$$\frac{1}{2}\,D/d\left<\begin{array}{l}\frac{1}{2}\,L/l\longrightarrow\frac{1}{4}\,D/d\,;L/l\\\frac{1}{2}\,l/l\longrightarrow\frac{1}{4}\,D/d\,;l/l\end{array}\right.$$

$$\frac{1}{2}\,d/d\left<\begin{array}{l}\frac{1}{2}\,L/l\longrightarrow\frac{1}{4}\,d/d\,;L/l\\\frac{1}{2}\,l/l\longrightarrow\frac{1}{4}\,d/d\,;l/l\end{array}\right.$$

Males

$$\frac{1}{2}\,D\left<\begin{array}{l}\frac{1}{2}\,L/l\longrightarrow\frac{1}{4}\,D\,;L/l\\\frac{1}{2}\,l/l\longrightarrow\frac{1}{4}\,D\,;l/l\end{array}\right.$$

$$\frac{1}{2}\,d\left<\begin{array}{l}\frac{1}{2}\,L/l\longrightarrow\frac{1}{4}\,d\,;L/l\\\frac{1}{2}\,l/l\longrightarrow\frac{1}{4}\,d\,;l/l\end{array}\right.$$

SOLVED PROBLEM 2   **ANIMATED ART** 🍁 SaplingPlus

Analyzing a cross

Consider three yellow, round peas, labeled A, B, and C. Each was grown into a plant and crossed with a plant grown from a green wrinkled pea. Exactly 100 peas issuing from each cross were sorted into phenotypic classes as follows:

A:	51 yellow, round	
	49 green, round	
B:	100 yellow, round	
C:	24 yellow, round	
	26 yellow, wrinkled	
	25 green, round	
	25 green, wrinkled	

What were the genotypes of A, B, and C? (Use gene symbols of your own choosing; be sure to define each one.)

SOLUTION

Notice that each of the crosses is

yellow, round × green, wrinkled → progeny

Because A, B, and C were all crossed with the same plant, all the differences between the three progeny populations must be attributable to differences in the underlying genotypes of A, B, and C.

You might remember a lot about these analyses from the chapter, which is fine, but let's see how much we can deduce from the data. What about dominance? The key cross for deducing dominance is B. Here, the inheritance pattern is

yellow, round × green, wrinkled → all yellow, round

So yellow and round must be dominant phenotypes because dominance is literally defined by the phenotype of a hybrid. Now we know that the green, wrinkled parent used in each cross must be fully recessive; we have a very convenient situation because it means that each cross is a test-cross, which is generally the most informative type of cross.

Turning to the progeny of A, we see a 1:1 ratio for yellow to green. This ratio is a demonstration of Mendel's first law (equal segregation) and shows that, for the character of color, the cross must have been heterozygote × homozygous recessive. Letting Y represent yellow and y represent green, we have

$$Y/y \times y/y \rightarrow \tfrac{1}{2} \; Y/y \text{ (yellow)} \rightarrow \tfrac{1}{2} \; y/y \text{ (green)}$$

For the character of shape, because all the progeny are round, the cross must have been homozygous dominant × homozygous recessive. Letting R represent round and r represent wrinkled, we have

$$R/R \times r/r \rightarrow R/r \text{ (round)}$$

Combining the two characters, we have

$$Y/y \,;\, R/R \times y/y \,;\, r/r \rightarrow \tfrac{1}{2} \; Y/y \,;\, R/r \;\; \tfrac{1}{2} \; y/y \,;\, R/r$$

Now cross B becomes crystal clear and must have been

$$Y/Y \,;\, R/R \times y/y \,;\, r/r \rightarrow Y/y \,;\, r/r$$

because any heterozygosity in pea B would have given rise to several progeny phenotypes, not just one.

What about C? Here, we see a ratio of 50 yellow : 50 green (1:1) and a ratio of 49 round : 51 wrinkled (also 1:1). So both genes in pea C must have been heterozygous, and cross C was

$$Y/y \,;\, R/r \times y/y \,;\, r/r$$

which is a good demonstration of Mendel's second law (independent assortment of different genes).

How would a geneticist have analyzed these crosses? Basically, the same way that we just did but with fewer intervening steps. Possibly something like this: "yellow and round dominant; single-gene segregation in A; B homozygous dominant; independent two-gene segregation in C."

PROBLEMS

Visit SaplingPlus for supplemental content. Problems with the  icon are available for review/grading. Problems with the  icon have a problem Solving Video.

WORKING WITH THE FIGURES

(The first 23 questions require inspection of text figures.)

1. Using Table 3-1, answer the following questions about probability values (see p. 88):

   a. If χ^2 is calculated to be 17 with 9 df, what is the approximate probability value?

   b. If χ^2 is 17 with 6 df, what is the probability value?

   c. What trend ("rule") do you see in the previous two calculations?

2. In Figure 3-3, what are the expected numbers from a 9:3:3:1 ratio with a total of 556 seeds? Compare with the observed numbers and comment on possible reasons for the differences.

3. Redraw the F_2 in Figure 3-4 using the gamete order $R;Y, R;y, r;Y, r;y$. Which is clearer, this one or the one in Figure 3-4?

4. In Figure 3-6, part b, can you see any evidence of dominance?

5. What stage of meiosis is illustrated in Figure 3-7?

6. Inspect Figure 3-8: which meiotic stage is responsible for generating Mendel's second law?

7. In Figure 3-8, to get from step 5 to step 6, what un-shown steps are needed?

8. Inspect Figure 3-9: what would be the outcome in the octad if on rare occasions a nucleus from the postmeiotic mitotic division of nucleus 2 slipped past a nucleus from the postmeiotic mitotic division of nucleus 3? How could you measure the frequency of such a rare event?

9. Regarding Figure 3-9, some fungi do not have a post-meiotic mitosis; what would their linear asci look like regarding the alleles A and a?

10. In Figure 3-10, if the input genotypes were $a \cdot B$ and $A \cdot b$, what would be the genotypes colored blue?

11. In the crossing sequence in Figure 3-11, would a pure-breeding tester of genotype $A/A \cdot B/B$ be of any use?

12. In the progeny seen in Figure 3-12, what are the origins of the chromosomes colored dark blue, light blue, and very light blue?

13. In Figure 3-12, the legend refers to a comparable analysis of a haploid: draw it.

14. In Figure 3-13, would the top part of the curve correspond to a frequency of 100%?

15. In Figure 3-14, in the central box of the 3×3 grid, how is the value of 4/16 obtained?

16. In Figure 3-15, in the central bar of the histogram, how is the number 6 derived?

17. In Figure 3-16, in which bar of the histogram would the genotype $R_1/r_1 \cdot R_2/R_2 \cdot r_3/r_3$ be found?

18. In Figure 3-16, how is the value of 1/64 for the outer bars calculated?

19. Regarding Figure 3-18:

    a. Do you think the tRNAs encoded by the three different organelles participate in protein synthesis in the cytosol?

b. Which organelle has the largest proportion of non-genic DNA?

c. What do you think is the main reason for the difference in size of yeast and human mtDNA?

20. Regarding Figure 3-19:

   a. Do you think it possible that some paternal mitochondria leak in the meiocyte?

   b. How would you test that possibility?

   c. What color is used to denote cytoplasm containing wild-type mitochondria?

21. In Figure 3-20, what would be the leaf types of progeny of the apical (top) flower?

22. In Figure 3-22, count the sites where mutations cause myopathy. The word myopathy means disease of the muscles; why should mitochondrial mutation so often cause this phenotype?

23. From the pedigree in Figure 3-23, what principle can you deduce about the inheritance of mitochondrial disease from affected fathers?

BASIC PROBLEMS

24. Assume independent assortment and start with a plant that is dihybrid A/a; B/b:

   a. What phenotypic ratio is produced from selfing it?

   b. What genotypic ratio is produced from selfing it?

   c. What phenotypic ratio is produced from testcrossing it?

   d. What genotypic ratio is produced from testcrossing it?

25. Normal mitosis takes place in a diploid cell of genotype A/a; B/b. Which of the following genotypes might represent possible daughter cells?

   a. A; B b. a; b c. A; b

   d. a; B e. A/A; B/B f. A/a; B/b

   g. a/a; b/b

26. In a diploid organism of $2n = 10$, assume that you can label all the centromeres derived from its female parent and all the centromeres derived from its male parent. When this organism produces gametes, how many male- and female-labeled centromere combinations are possible in the gametes?

27. It has been shown that when a thin beam of light is aimed at a nucleus, the amount of light absorbed is proportional to the cell's DNA content. Using this method, the DNA in the nuclei of several different types of cells in a corn plant were compared. The following numbers represent the relative amounts of DNA in these different types of cells:

   0.7, 1.4, 2.1, 2.8, and 4.2

Which cells could have been used for these measurements? (Note: In plants, the endosperm part of the seed is often triploid, $3n$.)

28. Draw a haploid mitosis of the genotype a^+; b.

29. In moss, the genes A and B are expressed only in the gametophyte. A sporophyte of genotype A/a; B/b is allowed to produce gametophytes.

   a. What proportion of the gametophytes will be A; B?

   b. If fertilization is random, what proportion of sporophytes in the next generation will be A/a; B/b?

30. When a cell of genotype A/a; B/b; C/c having all the genes on separate chromosome pairs divides mitotically, what are the genotypes of the daughter cells?

31. In the haploid yeast *Saccharomyces cerevisiae*, the two mating types are known as **a** and **α**. Mating type is determined by two alleles of a single gene, *MATa* and *MATα*. You cross a purple (ad^-) strain of mating type *a* and a white (ad^+) strain of mating type α. If ad^- and ad^+ are alleles of one gene, and *MATa* and *MATα* are alleles of an independently inherited gene on a separate chromosome pair, what progeny do you expect to obtain? In what proportions?

32. In mice, dwarfism is caused by an X-linked recessive allele, and pink coat is caused by an autosomal dominant allele (coats are normally brownish). If a dwarf female from a pure line is crossed with a pink male from a pure line, what will be the phenotypic ratios in the F_1 and F_2 in each sex? (Invent and define your own gene symbols.)

33. Suppose you discover two interesting *rare* cytological abnormalities in the karyotype of a human male. (A karyotype is the total visible chromosome complement.) There is an extra piece (satellite) on *one* of the chromosomes of pair 4, and there is an abnormal pattern of staining on one of the chromosomes of pair 7. With the assumption that all the gametes of this male are equally viable, what proportion of his children will have the same karyotype that he has?

34. Suppose that meiosis occurs in the transient diploid stage of the cycle of a haploid organism of chromosome number n. What is the probability that an individual haploid cell resulting from the meiotic division will have a complete parental set of centromeres (that is, a set all from one parent or all from the other parent)?

35. Pretend that the year is 1868. You are a skilled young lens maker working in Vienna. With your superior new lenses, you have just built a microscope that has better resolution than any others available. In your testing of this microscope, you have been observing the cells in the testes of grasshoppers and have been fascinated by the behavior of strange elongated structures that you have seen within the dividing cells. One day, in the

library, you read a recent journal paper by G. Mendel on hypothetical "factors" that he claims explain the results of certain crosses in peas. In a flash of revelation, you are struck by the parallels between your grasshopper studies and Mendel's pea studies, and you resolve to write him a letter. What do you write? (Problem 35 is based on an idea by Ernest Kroeker.)

36. From a presumed testcross $A/a \times a/a$, in which A represents red and a represents white, use the χ^2 test to find out which of the following possible results would fit the expectations:

   a. 120 red, 100 white

   b. 5000 red, 5400 white

   c. 500 red, 540 white

   d. 50 red, 54 white

37. Look at the Punnett square in Figure 3-4. 🌀

   a. How many different genotypes are shown in the 16 squares of the grid?

   b. What is the genotypic ratio underlying the $9:3:3:1$ phenotypic ratio?

   c. Can you devise a simple formula for the calculation of the number of progeny genotypes in dihybrid, trihybrid, and so forth crosses? Repeat for phenotypes.

   d. Mendel predicted that, within all but one of the phenotypic classes in the Punnett square, there should be several different genotypes. In particular, he performed many crosses to identify the underlying genotypes of the round, yellow phenotype. Show two different ways that could be used to identify the various genotypes underlying the round, yellow phenotype. (Remember, all the round, yellow peas look identical.)

38. Assuming independent assortment of all genes, develop formulas that show the number of phenotypic classes and the number of genotypic classes from selfing a plant heterozygous for n gene pairs.

39. Note: The first part of this problem was introduced in Chapter 2. The line of logic is extended here.

   In the plant *Arabidopsis thaliana*, a geneticist is interested in the development of trichomes (small projections) on the leaves. A large screen turns up two mutant plants (A and B) that have no trichomes, and these mutants seem to be potentially useful in studying trichome development. (If they are determined by single-gene mutations, then finding the normal and abnormal function of these genes will be instructive.) Each plant was crossed with wild type; in both cases, the next generation (F_1) had normal trichomes. When F_1 plants were selfed, the resulting F_2's were as follows:

   F_2 from mutant A: 602 normal; 198 no trichomes

   F_2 from mutant B: 267 normal; 93 no trichomes

   a. What do these results show? Include proposed genotypes of all plants in your answer.

   b. Assume that the genes are located on separate chromosomes. An F_1 is produced by crossing the original mutant A with the original mutant B. This F_1 is testcrossed: What proportion of testcross progeny will have no trichomes?

40. In dogs, dark coat color is dominant over albino, and short hair is dominant over long hair. Assume that these effects are caused by two independently assorting genes. Seven crosses were done as shown below, in which D and A stand for the dark and albino phenotypes, respectively, and S and L stand for the short-hair and long-hair phenotypes.

	Number of progeny			
Parental phenotypes	D, S	D, L	A, S	A, L
a. D, S × D, S	88	31	29	12
b. D, S × D, L	19	18	0	0
c. D, S × A, S	21	0	20	0
d. A, S × A, S	0	0	29	9
e. D, L × D, L	0	31	0	11
f. D, S × D, S	45	16	0	0
g. D, S × D, L	31	30	10	10

Write the genotypes of the parents in each cross. Use the symbols C and c for the dark and albino coat-color alleles and the symbols H and h for the short-hair and long-hair alleles, respectively. Assume parents are homozygous unless there is evidence otherwise.

41. In tomatoes, one gene determines whether the plant has purple (P) or green (G) stems, and a separate, independent gene determines whether the leaves are "cut" (C) or "potato" (Po). Five matings of tomato-plant phenotypes give the following results:

	Parental	Number of progeny			
Mating	phenotypes	P, C	P, Po	G, C	G, Po
1	P, C × G, C	323	102	309	106
2	P, C × P, Po	220	206	65	72
3	P, C × G, C	723	229	0	0
4	P, C × G, Po	405	0	389	0
5	P, Po × G, C	71	90	85	78

   a. Which alleles are dominant?

   b. What are the most probable genotypes for the parents in each cross? 🖷

42. A mutant allele in mice causes a bent tail. Six pairs of mice were crossed. Their phenotypes and those of their progeny are given in the following table. N is normal phenotype; B is bent phenotype. Deduce the mode of inheritance of bent tail. 🌀

	Parents		Progeny	
Cross	♀	♂	♀	♂
1	N	B	All B	All N
2	B	N	$\frac{1}{2}$ B, $\frac{1}{2}$ N	$\frac{1}{2}$ B, $\frac{1}{2}$ N
3	B	N	All B	All B
4	N	N	All N	All N
5	B	B	All B	All B
6	B	B	All B	$\frac{1}{2}$ B, $\frac{1}{2}$ N

a. Is it recessive or dominant?

b. Is it autosomal or sex-linked?

c. What are the genotypes of all parents and progeny?

43. The normal eye color of *Drosophila* is red, but strains in which all flies have brown eyes are available. Similarly, wings are normally long, but there are strains with short wings. A female from a pure line with brown eyes and short wings is crossed with a male from a normal pure line. The F$_1$ consists of normal females and short-winged males. An F$_2$ is then produced by intercrossing the F$_1$. *Both* sexes of F$_2$ flies show phenotypes as follows:

$$\frac{3}{8} \text{ red eyes, long wings}$$

$$\frac{3}{8} \text{ red eyes, short wings}$$

$$\frac{1}{8} \text{ brown eyes, long wings}$$

$$\frac{1}{8} \text{ brown eyes, short wings}$$

Deduce the inheritance of these phenotypes; use clearly defined genetic symbols of your own invention. State the genotypes of all three generations and the genotypic proportions of the F$_1$ and F$_2$.

 UNPACKING PROBLEM 43

Before attempting a solution to this problem, try answering the following questions:

1. What does the word *normal* mean in this problem?

2. The words *line* and *strain* are used in this problem. What do they mean, and are they interchangeable?

3. Draw a simple sketch of the two parental flies showing their eyes, wings, and sexual differences.

4. How many different characters are there in this problem?

5. How many phenotypes are there in this problem, and which phenotypes go with which characters?

6. What is the full phenotype of the F$_1$ females called "normal"?

7. What is the full phenotype of the F$_1$ males called "short winged"?

8. List the F$_2$ phenotypic ratios for each character that you came up with in answer to question 4.

9. What do the F$_2$ phenotypic ratios tell you?

10. What major inheritance pattern distinguishes sex-linked inheritance from autosomal inheritance?

11. Do the F$_2$ data show such a distinguishing criterion?

12. Do the F$_1$ data show such a distinguishing criterion?

13. What can you learn about dominance in the F$_1$? The F$_2$?

14. What rules about wild-type symbolism can you use in deciding which allelic symbols to invent for these crosses?

15. What does "deduce the inheritance of these phenotypes" mean?

Now try to solve the problem. If you are unable to do so, try to identify the obstacle and write a sentence or two describing your difficulty. Then go back to the expansion questions and see if any of them relate to your difficulty. If this approach doesn't work, inspect the Learning Objectives and Key Concepts of this chapter and ask yourself which might be relevant to your difficulty.

44. In a natural population of annual plants, a single plant is found that is sickly looking and has yellowish leaves. The plant is dug up and brought back to the laboratory. Photosynthesis rates are found to be very low. Pollen from a normal dark-green-leaved plant is used to fertilize emasculated flowers of the yellowish plant. A hundred seeds result, of which only 60 germinate. All the resulting plants are sickly yellow in appearance.

a. Propose a genetic explanation for the inheritance pattern.

b. Suggest a simple test for your model.

c. Account for the reduced photosynthesis, sickliness, and yellowish appearance.

45. What is the basis for the green-and-white color variegation in the leaves of *Mirabilis*? If the following cross is made,

variegated ♀ × green ♂

what progeny types can be predicted? What about the reciprocal cross?

46. In *Neurospora*, the mutant *stp* exhibits erratic stop-and-start growth. The mutant site is known to be in the mtDNA. If an *stp* strain is used as the female parent in a cross with a normal strain acting as the male, what type of progeny can be expected? What about the progeny from the reciprocal cross?

47. Two corn plants are studied. One is resistant (R) and the other is susceptible (S) to a certain pathogenic fungus. The following crosses are made, with the results shown:

$S♀ \times R♂ \rightarrow$ all progeny S

$R♀ \times S♂ \rightarrow$ all progeny R

What can you conclude about the location of the genetic determinants of R and S?

48. A presumed dihybrid in *Drosophila*, *B/b* ; *F/f*, is test-crossed with *b/b* ; *f/f*. (*B* = black body ; *b* = brown body; *F* = forked bristles; *f* = unforked bristles.) The results are

black, forked	230	brown, forked	240
black, unforked	210	brown, unforked	250

Use the χ^2 test to determine if these results fit the results expected from testcrossing the hypothesized dihybrid. 📚

49. A plant geneticist has two pure lines, one with purple petals and one with blue. She hypothesizes that the phenotypic difference is due to two alleles of one gene. To test this idea, she aims to look for a 3:1 ratio in the F_2. She crosses the lines and finds that all the F_1 progeny are purple. The F_1 plants are selfed, and 400 F_2 plants are obtained. Of these F_2 plants, 320 are purple and 80 are blue. Do these results fit her hypothesis well? If not, suggest why.

50. Are the following progeny numbers consistent with the results expected from selfing a plant presumed to be a dihybrid of two independently assorting genes, *H/h* ; *R/r*? (*H* = hairy leaves; *h* = smooth leaves; *R* = round ovary; *r* = elongated ovary.) Explain your answer.

hairy, round	178	smooth, round	56
hairy, elongated	62	smooth, elongated	24

51. A dark female moth is crossed with a dark male. All the male progeny are dark, but half the female progeny are light and the rest are dark. Propose an explanation for this pattern of inheritance.

52. In *Neurospora*, a mutant strain called stopper (*stp*) arose spontaneously. Stopper showed erratic "stop and start" growth, compared with the uninterrupted growth of wild-type strains. In crosses, the following results were found:

♀ stopper × ♂ wild type → progeny all stopper

♀ wild type × ♂ stopper → progeny all wild type

a. What do these results suggest regarding the location of the stopper mutation in the genome?

b. According to your model for part *a*, what progeny and proportions are predicted in octads from the following cross, including a mutation *nic3* located on chromosome VI?

♀ *stp · nic3* × wild type ♂

53. In polygenic systems, how many phenotypic classes corresponding to number of polygene "doses" are expected in selfs

a. of strains with four heterozygous polygenes?

b. of strains with six heterozygous polygenes?

54. In the self of a polygenic trihybrid R_1/r_1 ; R_2/r_2 ; R_3/r_3, use the product and sum rules to calculate the proportion of progeny with just one polygene "dose."

55. Reciprocal crosses and selfs were performed between the two moss species *Funaria mediterranea* and *F. hygrometrica*. The sporophytes and the leaves of the gametophytes are shown in the accompanying diagram.

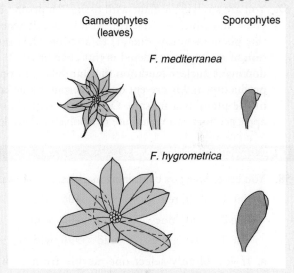

Gametophytes (leaves)   Sporophytes

F. mediterranea

F. hygrometrica

The crosses are written with the female parent first.

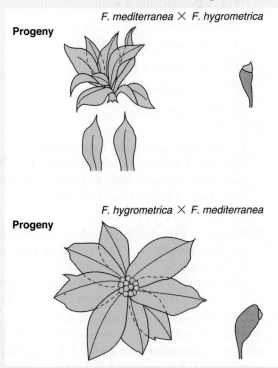

F. mediterranea × *F. hygrometrica*

Progeny

F. hygrometrica × *F. mediterranea*

Progeny

a. Describe the results presented, summarizing the main findings.

b. Propose an explanation of the results.

c. Show how you would test your explanation; be sure to show how it could be distinguished from other explanations.

56. Assume that diploid plant A has a cytoplasm genetically different from that of plant B. To study nuclear–cytoplasmic relations, you wish to obtain a plant with the cytoplasm of plant A and the nuclear genome predominantly of plant B. How would you go about producing such a plant?

57. You are studying a plant with tissue comprising both green and white sectors. You wish to decide whether this phenomenon is due (1) to a chloroplast mutation of the type considered in this chapter or (2) to a dominant nuclear mutation that inhibits chlorophyll production and is present only in certain tissue layers of the plant as a mosaic. Outline the experimental approach that you would use to resolve this problem.

CHALLENGING PROBLEMS

58. You have three jars containing marbles, as follows:

jar 1	600	red	and	400	white
jar 2	900	blue	and	100	white
jar 3	10	green	and	990	white

a. If you blindly select one marble from each jar, calculate the probability of obtaining

(1) a red, a blue, and a green.

(2) three whites.

(3) a red, a green, and a white.

(4) a red and two whites.

(5) a color and two whites.

(6) at least one white.

b. In a certain plant, R = red and r = white. You self a red R/r heterozygote with the express purpose of obtaining a white plant for an experiment. What minimum number of seeds do you have to grow to be at least 95 percent certain of obtaining at least one white individual?

c. When a woman is injected with an egg fertilized in vitro, the probability of its implanting successfully is 20 percent. If a woman is injected with five eggs simultaneously, what is the probability that she will become pregnant? (Part c is from Margaret Holm.)

59. In tomatoes, red fruit is dominant over yellow, two-loculed fruit is dominant over many-loculed fruit, and tall vine is dominant over dwarf. A breeder has two pure lines: (1) red, two-loculed, dwarf and (2) yellow, many-loculed, tall. From these two lines, he wants to

produce a new pure line for trade that is yellow, two-loculed, and tall. How exactly should he go about doing so? Show not only which crosses to make, but also how many progeny should be sampled in each case.

60. We have dealt mainly with only two genes, but the same principles hold for more than two genes. Consider the following cross:

$A/a\,;B/b\,;C/c\,;D/d\,;E/e \times a/a\,;B/b\,;c/c\,;D/d\,;e/e$

a. What proportion of progeny will *phenotypically* resemble (1) the first parent, (2) the second parent, (3) either parent, and (4) neither parent?

b. What proportion of progeny will be *genotypically* the same as (1) the first parent, (2) the second parent, (3) either parent, and (4) neither parent?

Assume independent assortment.

61. The accompanying pedigree shows the pattern of transmission of two rare human phenotypes: cataract and pituitary dwarfism. Family members with cataract are shown with a solid left half of the symbol; those with pituitary dwarfism are indicated by a solid right half.

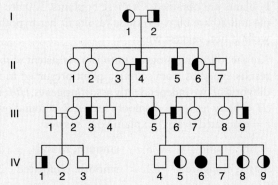

a. What is the most likely mode of inheritance of each of these phenotypes? Explain.

b. List the genotypes of all members in generation III as far as possible.

c. If a hypothetical mating took place between IV-1 and IV-5, what is the probability of the first child's being a dwarf with cataracts? A phenotypically normal child?

(Problem 61 is adapted from J. Kuspira and R. Bhambhani, *Compendium of Problems in Genetics.* Copyright 1994 by Wm. C. Brown.)

62. A corn geneticist has three pure lines of genotypes $a/a\,;B/B\,;C/C$, $A/A\,;b/b\,;C/C$, and $A/A\,;B/B\,;c/c$. All the phenotypes determined by a, b, and c will increase the market value of the corn; so, naturally, he wants to combine them all in one pure line of genotype $a/a\,;b/b\,;c/c$.

a. Outline an effective crossing program that can be used to obtain the $a/a\,;b/b\,;c/c$ pure line.

b. At each stage, state exactly which phenotypes will be selected and give their expected frequencies.

c. Is there more than one way to obtain the desired genotype? Which is the best way?

Assume independent assortment of the three gene pairs. (Note: Corn will self- or cross-pollinate easily.)

63. In humans, color vision depends on genes encoding three pigments. The *R* (red pigment) and *G* (green pigment) genes are close together on the X chromosome, whereas the *B* (blue pigment) gene is autosomal. A recessive mutation in any one of these genes can cause color blindness. Suppose that a color-blind man married a woman with normal color vision. The four sons from this marriage were color-blind, and the five daughters were normal. Specify the most likely genotypes of both parents and their children, explaining your reasoning. (A pedigree drawing will probably be helpful.) (Problem 63 is by Rosemary Redfield.)

64. Consider the accompanying pedigree for a rare human muscle disease.

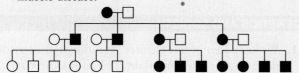

a. What unusual feature distinguishes this pedigree from those studied earlier in this chapter?

b. Where do you think the mutant DNA responsible for this phenotype resides in the cell?

65. The plant *Haplopappus gracilis* has a 2*n* of 4. A diploid cell culture was established and, at premitotic S phase, a radioactive nucleotide was added and was incorporated into newly synthesized DNA. The cells were then removed from the radioactivity, washed, and allowed to proceed through mitosis. Radioactive chromosomes or chromatids can be detected by placing photographic emulsion on the cells; radioactive chromosomes or chromatids appeared covered with spots of silver from the emulsion. (The chromosomes "take their own photograph.") Draw the chromosomes at prophase and telophase of the first and second mitotic divisions after the radioactive treatment. If they are radioactive, show it in your diagram. If there are several possibilities, show them, too.

66. In the species of Problem 65, you can introduce radioactivity by injection into the anthers at the S phase before meiosis. Draw the four products of meiosis with their chromosomes, and show which are radioactive.

67. The plant *Haplopappus gracilis* is diploid and 2*n* = 4. There are one long pair and one short pair of chromosomes. The diagrams below (numbered 1 through 12) represent anaphases ("pulling apart" stages) of individual cells in meiosis or mitosis in a plant that is genetically a dihybrid (*A/a* ; *B/b*) for genes on different chromosomes. The lines represent chromosomes or chromatids, and the points of the V's represent centromeres. In each case, indicate if the diagram represents a cell in meiosis I, meiosis II, or mitosis. If a diagram shows an impossible situation, say so.

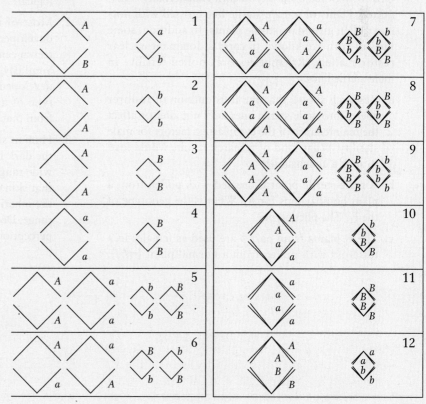

68. The pedigree below shows the recurrence of a rare neurological disease (large black symbols) and spontaneous fetal abortion (small black symbols) in one family. (A slash means that the individual is deceased.) Provide an explanation for this pedigree in regard to the cytoplasmic segregation of defective mitochondria.

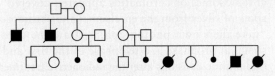

69. A man is brachydactylous (very short fingers; rare autosomal dominant), and his wife is not. Both can taste the chemical phenylthiocarbamide (autosomal dominant; common allele), but their mothers could not.

    a. Give the genotypes of the couple.

    If the genes assort independently and the couple has four children, what is the probability of

    b. all of them being brachydactylous?

    c. none being brachydactylous?

    d. all of them being tasters?

    e. all of them being nontasters?

    f. all of them being brachydactylous tasters?

    g. none being brachydactylous tasters?

    h. at least one being a brachydactylous taster?

70. One form of male sterility in corn is maternally transmitted. Plants of a male-sterile line crossed with normal pollen give male-sterile plants. In addition, some lines of corn are known to carry a dominant nuclear restorer allele (*Rf*) that restores pollen fertility in male-sterile lines.

    a. Research shows that the introduction of restorer alleles into male-sterile lines does not alter or affect the maintenance of the cytoplasmic factors for male sterility. What kind of research results would lead to such a conclusion?

    b. A male-sterile plant is crossed with pollen from a plant homozygous for *Rf*. What is the genotype of the F₁? The phenotype?

    c. The F₁ plants from part *b* are used as females in a testcross with pollen from a normal plant (*rf/rf*).

What are the results of this testcross? Give genotypes and phenotypes, and designate the kind of cytoplasm.

    d. The restorer allele already described can be called *Rf-1*. Another dominant restorer, *Rf-2*, has been found. *Rf-1* and *Rf-2* are located on different chromosomes. Either or both of the restorer alleles will give pollen fertility. With the use of a male-sterile plant as a tester, what will be the result of a cross in which the male parent is

       (1) heterozygous at both restorer loci?

       (2) homozygous dominant at one restorer locus and homozygous recessive at the other?

       (3) heterozygous at one restorer locus and homozygous recessive at the other?

       (4) heterozygous at one restorer locus and homozygous dominant at the other?

GENETICS AND SOCIETY

1. We have seen in this chapter that independent assortment can be used by plant and animal breeders to develop new lines that have a combination of favorable mutant alleles making them suitable for commerce. This type of "genetic engineering" by making crosses and selfs is deemed to be appropriate by the general public, whereas genetic engineering by DNA manipulation (details in later chapters) is not. Why do you think there is this perceived distinction, and is it logical?

2. Most of the dog breeds familiar to us today were developed by dog fanciers in Victorian times (nineteenth century) before the discovery of Mendel's Laws. Similarly, many breeds of agricultural animals were developed pre-Mendel. In what way do you think that type of animal breeding might have been different from that which is used today?

3. Human skin shows great variation in the amount of the dark brown pigment called melanin, resulting in a wide range of skin colors. It is a common observation that skin color is heritable. Devise a genetic model for the inheritance of skin color, which explains this wide range. Does your model shine any light on the popular perceptions of race and racism?

Mapping Eukaryote Chromosomes by Recombination

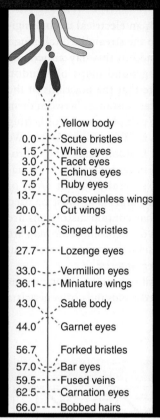

	Yellow body
0.0	Scute bristles
1.5	White eyes
3.0	Facet eyes
5.5	Echinus eyes
7.5	Ruby eyes
13.7	Crossveinless wings
20.0	Cut wings
21.0	Singed bristles
27.7	Lozenge eyes
33.0	Vermillion eyes
36.1	Miniature wings
43.0	Sable body
44.0	Garnet eyes
56.7	Forked bristles
57.0	Bar eyes
59.5	Fused veins
62.5	Carnation eyes
66.0	Bobbed hairs

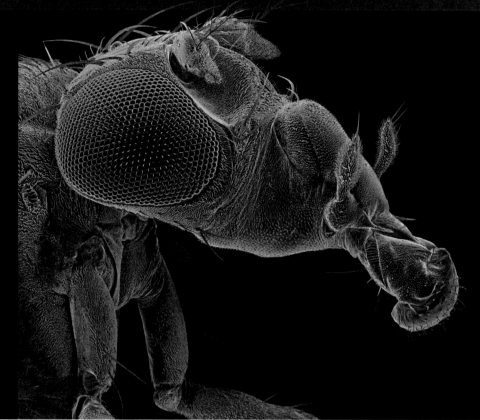

At the left is a recombination-based map of one of the chromosomes of *Drosophila* (the organism in the image above), showing the loci of genes whose mutations produce known phenotypes. [*David Scharf/Science Source.*]

CHAPTER OUTLINE AND LEARNING OBJECTIVES

4.1 DIAGNOSTICS OF LINKAGE

LO 4.1 Use a recombinant frequency of 50% or <50% in the progeny of a dihybrid testcross as a key diagnostic to show whether or not two genes are linked on the same chromosome.

4.2 MAPPING BY RECOMBINANT FREQUENCY

LO 4.2 Show the impact of one or more crossovers in producing recombinants between linked genes.

LO 4.3 Use recombinant frequency in a testcross to calculate the chromosomal distance between two genes in map units.

LO 4.4 Extend the same type of analysis to a three-gene ("three-point") testcross to assess whether or not three genes are linked, and if so, their relative order and map distances.

4.3 MAPPING WITH MOLECULAR MARKERS

4.4 USING THE CHI-SQUARE TEST TO INFER LINKAGE

LO 4.5 Apply the chi-square test to statistically assess whether or not the recombinant frequency is 50% in a particular data set.

4.5 THE MOLECULAR MECHANISM OF CROSSING OVER

LO 4.6 Diagram how double-strand DNA breaks and heteroduplex formation can lead to a crossover at the molecular level.

4.6 USING RECOMBINATION-BASED MAPS IN CONJUNCTION WITH PHYSICAL MAPS

Having analyzed two or more independently inherited genes in Chapter 3, our broad objective for this chapter is to learn what inheritance patterns are produced by two or more heterozygous genes located on the same chromosome.

Some of the questions that geneticists want to answer about the genome are: *What genes* are present in the genome? *What functions* do they have? *What positions* do they occupy on the chromosomes? Their pursuit of the third question is broadly called mapping. Mapping is the main focus of this chapter, but all three questions are interrelated, as we will see.

We all have an everyday feeling for the importance of maps in general, and, indeed, we have all used them at some time in our lives to find our way around. Relevant to the focus of this chapter is that, in some situations, several maps need to be used simultaneously. A good example in everyday life is in navigating the dense array of streets and buildings of a city such as London, England. A street map that shows the general layout is one necessity. However, the street map is used by tourists and Londoners alike in conjunction with another map, that of the underground railway system. The underground system is so complex

and spaghetti-like that, in 1933, an electrical circuit engineer named Harry Beck drew up the streamlined (although distorted) map that has remained to this day an icon of London. The street and underground maps of London are compared in **Figure 4-1**. Note that the positions of the underground stations and the exact distances between them are of no interest in themselves, except as a way of getting to a destination of interest such as Westminster Abbey. We will see three parallels with the London maps when chromosome maps are used to zero in on individual "destinations," or specific genes. First, several different types of chromosome maps are often necessary and must be used in conjunction; second, maps that contain distortions are still useful; and third, many sites on a chromosome map are charted only because they are useful in trying to zero in on other sites that are the ones of real interest.

Obtaining a map of gene positions on the chromosomes is an endeavor that has occupied thousands of geneticists

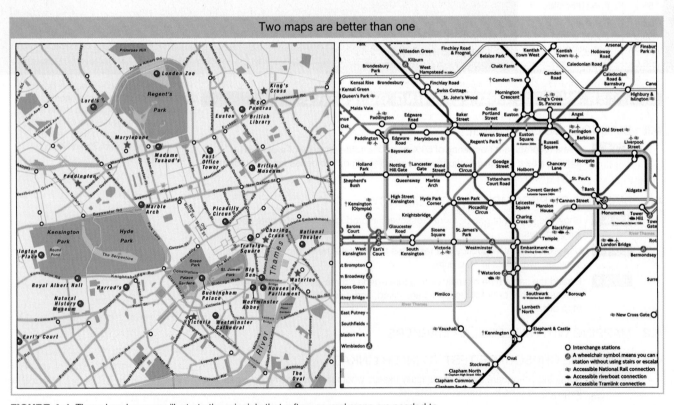

Two maps are better than one

FIGURE 4-1 These London maps illustrate the principle that, often, several maps are needed to get to a destination of interest. The map of the underground railway ("the Tube") is used to get to a destination of interest such as a street address, shown on the street map. In genetics, two different kinds of genome maps are often useful in locating a gene, leading to an understanding of its structure and function. [*(right) Transport for London.*]

for the past 100 years or so. Why is it so important? There are several reasons:

1. Gene position is crucial information needed to build *complex genotypes* required for experimental purposes or for commercial applications. For example, in Chapter 5, we will see cases in which special allelic combinations must be put together to explore gene interaction.

2. Knowing the position occupied by a gene provides a way of isolating it and discovering its *structure and function*. A gene's position can be used to define it at the DNA level. In turn, the DNA sequence of a wild-type gene or its mutant allele is a necessary part of deducing its underlying function.

3. The genes present and their arrangement on chromosomes are often slightly different in related species. For example, the rather long human chromosome number 2 is split into two shorter chromosomes in the great apes. By comparing such differences, geneticists can deduce the *evolutionary genetic mechanisms* through which these genomes diverged. Hence, chromosome maps are useful in interpreting mechanisms of evolution.

The arrangement of genes on chromosomes is represented diagrammatically as a unidimensional **chromosome map**, showing gene positions known as **loci** (sing., **locus**), and the distances between the loci based on some kind of scale. Two basic types of chromosome maps are currently used in genetics; they are assembled by quite different procedures yet are used in a complementary way. *Recombination-based maps*, which are the topic of this chapter, map the loci of genes that have been identified by mutant phenotypes showing single-gene inheritance. *Physical maps* (see Chapter 14) show the genes as segments arranged along the long DNA molecule that constitutes a chromosome. These maps show different views of the genome, but, just like the maps of London, they can be used together to arrive at an understanding of what a gene's function is at the molecular level and how that function influences phenotype.

KEY CONCEPT Genetic maps are useful for strain building, for interpreting evolutionary mechanisms, and for discovering a gene's unknown function. Discovering a gene's function is facilitated by integrating information on recombination-based and physical maps.

4.1 DIAGNOSTICS OF LINKAGE

LO 4.1 Use a recombinant frequency of 50% or <50% in the progeny of a dihybrid testcross as a key diagnostic to show whether or not two genes are linked on the same chromosome.

Recombination maps of chromosomes are usually assembled two or three genes at a time, with the use of a method called *linkage analysis*. When geneticists say that two genes

are **linked,** they mean that the loci of those genes are on the same chromosome and, hence, the alleles on any one homolog are physically joined (linked) by the DNA between them. The way in which early geneticists deduced linkage is a useful means of introducing most of the key ideas and procedures in the analysis.

Using recombinant frequency to recognize linkage

In the early 1900s, William Bateson and R. C. Punnett (for whom the Punnett square was named) were studying the inheritance of two genes in sweet peas. In a standard self of a dihybrid F_1, the F_2 did not show the $9:3:3:1$ ratio predicted by the principle of independent assortment. In fact, Bateson and Punnett noted that certain combinations of alleles showed up more often than expected, almost as though they were physically attached in some way. However, they had no explanation for this discovery.

Later, Thomas Hunt Morgan found a similar deviation from Mendel's second law while studying two autosomal genes in *Drosophila*. Morgan proposed linkage as a hypothesis to explain the phenomenon of apparent allele association.

Let's look at some of Morgan's data. One of the genes affected eye color (pr, purple, and pr^+, red), and the other gene affected wing length (vg, vestigial, and vg^+, normal). (Vestigial wings are very small compared to wild type.) The wild-type alleles of both genes are dominant. Morgan performed a cross to obtain dihybrids and then followed with a testcross:

P $\qquad\qquad pr/pr \cdot vg/vg \times pr^+/pr^+ \cdot vg^+/vg^+$

$$\downarrow$$

Gametes $\qquad\qquad pr \cdot vg \quad pr^+ \cdot vg^+$

$$\downarrow$$

F_1 dihybrid $\qquad\qquad pr^+/pr \cdot vg^+/vg$

Testcross:

$\qquad pr^+/pr \cdot vg^+/vg\,♀ \quad \times \quad pr/pr \cdot vg/vg\,♂$

$\qquad\qquad F_1$ dihybrid female $\qquad\qquad$ Tester male

Morgan's use of the testcross is important. Because the tester parent contributes gametes carrying only recessive alleles, the phenotypes of the offspring directly reveal the alleles contributed by the gametes of the dihybrid parent, as described in Chapters 2 and 3. Hence, the analyst can concentrate on meiosis in one parent (the dihybrid) and essentially forget about meiosis in the other (the homozygous recessive tester). In contrast, from an F_1 *self*, there are two sets of meioses to consider in the analysis of progeny: one in the male parent and the other in the female.

Morgan's testcross results were as follows (listed as the gametic classes from the dihybrid):

$pr^+ \cdot vg^+$	1339
$pr \cdot vg$	1195
$pr^+ \cdot vg$	151
$pr \cdot vg^+$	154
	2839

Obviously, these numbers deviate drastically from the Mendelian prediction of a 1:1:1:1 ratio expected from independent assortment (approximately 710 in each of the four classes). In Morgan's results, we see that the first two allele combinations are in the great majority, clearly indicating that they are associated, or "linked."

Another useful way of assessing the testcross results is by considering the percentage of *recombinants* in the progeny. In Chapter 3, we learned that we detect recombination by comparing the *inputs* into meiosis with the *outputs*, and a recombinant is any meiotic product that has a new combination of the alleles provided by the two input genotypes (see Figure 3-11). The recombinants in the present cross are the two types $pr^+ \cdot vg$ and $pr \cdot vg^+$ because they are clearly not the two input genotypes contributed to the F_1 dihybrid by the original homozygous parental flies (more precisely, by their gametes). We see that the two recombinant types are approximately equal in frequency ($151 \approx 154$). Their total is 305, which is a frequency of $(305/2839) \times 100$, or 10.7 percent. This is significantly less than the 50 percent frequency of recombinants that we would expect if the genes were on different chromosomes and assorting independently (see Chapter 3). We can make sense of these data, as Morgan did, by postulating that the genes were linked on the same chromosome, and so the parental allelic combinations are held together in the majority of progeny. In the dihybrid, the allelic conformation must have been as follows:

$$\frac{pr^+ \qquad vg^+}{pr \qquad vg}$$

The tendency of linked alleles to be inherited as a package is illustrated in **Figure 4-2.**

Now let's look at another cross that Morgan made with the use of the same alleles but in a different combination. In this cross, each parent is homozygous for the wild-type allele of one gene and the mutant allele of the other. Again, F_1 females were testcrossed:

P $pr^+/pr^+ \cdot vg/vg \times pr/pr \cdot vg^+/vg^+$

↓

Gametes $pr^+ \cdot vg$ $pr \cdot vg^+$

↓

F_1 dihybrid $pr^+/pr \cdot vg^+/vg$

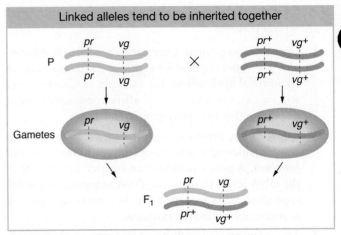

FIGURE 4-2 Simple inheritance of two genes located on the same chromosome pair. The same genes are present together on a chromosome in both parents and progeny.

Testcross:

$pr^+/pr \cdot vg^+/vg\,♀$ × $pr/pr \cdot vg/vg\,♂$

F_1 dihybrid female   Tester male

The following progeny were obtained from the testcross:

$pr^+ \cdot vg^+$	157
$pr \cdot vg$	146
$pr^+ \cdot vg$	965
$pr \cdot vg^+$	1067
	2335

Again, these results are not even close to a 1:1:1:1 Mendelian ratio. Now, however, the recombinant classes, $pr^+ \cdot vg^+$ and $pr \cdot vg$, are the converse of those in the first analysis. But notice that their frequency is approximately the same: $(157 + 146)/2335 \times 100 = 12.9$ percent of the progeny. Again, linkage is suggested, but, in this case, the F_1 dihybrid must have been as follows:

$$\frac{pr^+ \qquad vg}{pr \qquad vg^+}$$

Dihybrid testcross results like those just presented are commonly encountered in genetics. They follow the general pattern:

Two equally frequent nonrecombinant classes totaling
in excess of **50 percent**

Two equally frequent recombinant classes totaling
less than **50 percent**

KEY CONCEPT When two genes are close together on the same chromosome pair (that is, when they are linked), they do not assort independently but produce a recombinant frequency of less than 50 percent. Hence, a recombinant frequency of less than 50 percent is a diagnostic for linkage.

How crossovers produce recombinants for linked genes

The linkage hypothesis explains why allele combinations from the parental generations remain together: the genes are physically attached by the segment of chromosome between them. But exactly how are *any* recombinants produced when genes are linked? Morgan suggested that, when homologous chromosomes pair at meiosis, the chromosomes occasionally break and exchange parts in a process called **crossing over.** Figure 4-3 illustrates this physical exchange of chromosome segments. The two new combinations are called **crossover products.**

Is there any microscopically observable process that could account for crossing over? At meiosis, when duplicated homologous chromosomes pair with each other, a cross-shaped structure called a *chiasma* (pl., chiasmata) often forms between two nonsister chromatids. Chiasmata are shown in Figure 4-4. To Morgan, the appearance of the chiasmata visually corroborated the concept of crossing over. (Note that the chiasmata seem to indicate that *chromatids*, not unduplicated chromosomes, participate in a crossover. We will return to this point later.)

> **KEY CONCEPT** For linked genes, recombinants are produced by crossovers between nonsister chromatids during meiosis. Chiasmata are the visible manifestations of crossovers.

Linkage symbolism and terminology

The work of Morgan showed that linked genes in a dihybrid may be present in one of two basic conformations. In one, the two dominant, or wild-type, alleles are present on the same homolog (as in Figure 4-3); this arrangement is called a **cis conformation** (*cis* means "adjacent"). In the other, they

are on different homologs, in what is called a **trans conformation** (*trans* means "opposite"). The two conformations are written as follows:

Cis AB/ab or $++/ab$

Trans Ab/aB or $+b/a+$

Note the following conventions that pertain to linkage symbolism:

1. Alleles on the same homolog have no punctuation between them.
2. A slash symbolically separates the two homologs.
3. Alleles are always written in the same order on each homolog.
4. As in earlier chapters, genes known to be on different chromosomes (unlinked genes) are shown separated by a semicolon—for example, A/a ; C/c.
5. In this book, genes of *unknown* linkage are shown separated by a dot, $A/a \cdot D/d$.

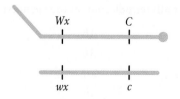

Chiasmata are the sites of crossing over

FIGURE 4-4 Several chiasmata appear in this photograph taken in the course of meiosis in a grasshopper testis. [*Republished with permission of Elsevier, Gareth H. Jones, F. Chris H. Franklin, "Meiotic Crossingover: Obligation and Interference," Cell, 2006, July; 126 (2): 246–248, Figure 1. Permission conveyed through Copyright Clearance Center, Inc.*]

Evidence that crossing over is a breakage-and-rejoining process

The idea that recombinants are produced by some kind of exchange of material between homologous chromosomes was a compelling one. But experimentation was necessary to test this hypothesis. A first step was to find a case in which the exchange of parts between chromosomes would be visible under the microscope. Several investigators approached this problem in the same way, and one of their analyses follows.

In 1931, Harriet Creighton and Barbara McClintock were studying two genes in corn that they knew were both located on chromosome 9. One affected seed color (C, colored; c, colorless), and the other affected endosperm composition (Wx, waxy; wx, starchy). The plant was a dihybrid in cis conformation. However, in one plant, the chromosome 9 carrying the alleles C and Wx was unusual in that it also carried a large, densely staining element (called a *knob*) on the C end and a longer piece of chromosome on the Wx end; thus, the heterozygote was

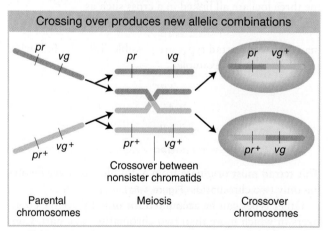

FIGURE 4-3 The exchange of parts by crossing over may produce gametic chromosomes whose allelic combinations differ from the parental combinations.

ANIMATED ART SaplingPlus

Crossing over produces new allelic combinations

In the progeny of a testcross of this plant, they examined the chromosomes of the recombinants and parental genotypes. They found that all the recombinants inherited one or the other of the two following chromosomes, depending on their recombinant makeup:

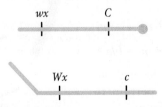

Thus, there was a precise correlation between the *genetic* event of the appearance of recombinants and the *chromosomal* event of crossing over. Consequently, the chiasmata appeared to be the sites of exchange.

What can we say about the molecular mechanism of chromosome exchange in a crossover event? The short answer is that a crossover results from the breakage and reunion of DNA. Two parental chromosomes break at the same position, and then each piece joins up with the neighboring piece from the *other* chromosome. In Section 4.5, we will see a model of the molecular processes that allow DNA to break and rejoin in a precise manner such that no genetic material is lost or gained.

KEY CONCEPT A crossover is the breakage of two DNA molecules at the same position and their rejoining in two reciprocal recombinant combinations.

Evidence that crossing over takes place at the four-chromatid stage

As already noted, the diagrammatic representation of crossing over in Figure 4-3 shows a crossover taking place at the four-chromatid stage of meiosis; in other words, crossovers are between *nonsister chromatids*. However, it was *theoretically* possible that crossing over took place before replication, at the *two-chromosome* stage. This uncertainty was resolved through the genetic analysis of organisms whose four products of meiosis remain together in groups of four called *tetrads*. These organisms, which we met in Chapters 2 and 3, are fungi and unicellular algae. The products of meiosis of a single tetrad can be isolated, which is equivalent to isolating all four chromatids from a single meiocyte. Tetrad analyses of crosses *in which genes are linked* show many tetrads that contain four different allele combinations. For example, from the cross

$$AB \times ab$$

some (but not all) tetrads contain four genotypes:

$$AB$$
$$Ab$$
$$aB$$
$$ab$$

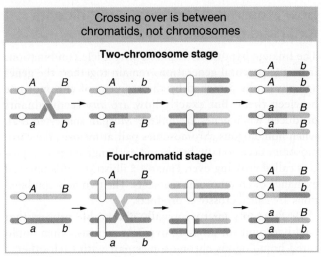

FIGURE 4-5 Crossing over takes place at the four-chromatid stage. Because more than two different products of a single meiosis can be seen in some tetrads, crossing over cannot take place at the two-strand stage (before DNA replication). The white circle designates the position of the centromere. When sister chromatids are visible, the centromere appears unreplicated.

This result can be explained only if crossovers take place at the four-chromatid stage because, if crossovers took place at the two-chromosome stage, there could only ever be a maximum of two different genotypes in an individual tetrad. This reasoning is illustrated in **Figure 4-5**.

Multiple crossovers can include two or more than two chromatids

Tetrad analysis can also show two other important features of crossing over. First, within one meiocyte, several crossovers can occur along a chromosome pair. Second, these multiple crossovers can exchange material between two or more chromatids. To think about this matter, we need to look at the simplest case: double crossovers. To study double crossovers, we need three linked genes. For example, if the three loci are all linked in a cross such as

$$ABC \times abc$$

many different tetrad types are possible. Take the following tetrad as an initial example:

$$ABC$$
$$AbC$$
$$aBc$$
$$abc$$

This tetrad must originate from a double crossover involving only two chromatids (**Figure 4-6a**).

Other types can be accounted for only by double crossovers in which more than two chromatids take part. Consider the following tetrad as an example:

$$ABc$$
$$AbC$$
$$aBC$$
$$abc$$

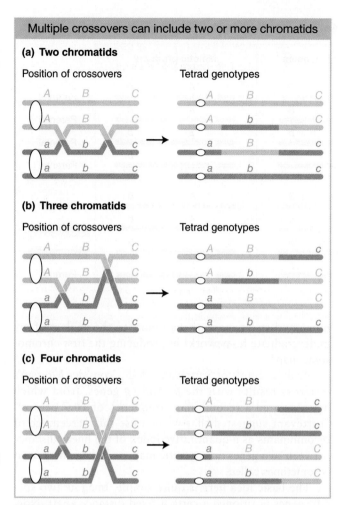

Multiple crossovers can include two or more chromatids

(a) Two chromatids

Position of crossovers   Tetrad genotypes

(b) Three chromatids

Position of crossovers   Tetrad genotypes

(c) Four chromatids

Position of crossovers   Tetrad genotypes

FIGURE 4-6 Multiple crossovers can involve (a) two chromatids, (b) three chromatids, or (c) four chromatids. This figure shows double crossovers, the simplest case of multiple crossovers.

This tetrad can only be explained by two crossovers in which *three* chromatids take part, as shown in Figure 4-6b. Furthermore, the following type of tetrad shows that all *four* chromatids can participate in crossing over in the same meiosis (Figure 4-6c):

$$ABc$$
$$Abc$$
$$aBC$$
$$abC$$

Therefore, for any pair of homologous chromosomes, two, three, or four chromatids can take part in crossing over events in a single meiocyte. Note, however, that any single crossover is between just two chromatids.

You might be wondering about crossovers between *sister* chromatids. They do occur but are rare. They do not produce new allele combinations and so are not usually considered.

KEY CONCEPT Multiple (two or more) crossovers can produce both recombinant and parental chromatids.

4.2   MAPPING BY RECOMBINANT FREQUENCY

LO 4.2 Show the impact of one or more crossovers in producing recombinants between linked genes.

LO 4.3 Use recombinant frequency in a testcross to calculate the chromosomal distance between two genes in map units.

LO 4.4 Extend the same type of analysis to a three-gene ("three-point") testcross to assess whether or not three genes are linked, and if so, their relative order and map distances.

The frequency of recombinants produced by crossing over is the key to chromosome mapping. Fungal tetrad analysis has shown that, for any two specific linked genes, crossovers take place between them in some, but not all, meiocytes (**Figure 4-7**). The farther apart the genes are, the more likely that a crossover will take place and the higher the proportion of recombinant products will be. Thus, the proportion of recombinants is a clue to the distance separating two gene loci on a chromosome map.

As stated earlier in regard to Morgan's data, the recombinant frequency was significantly less than 50 percent, specifically 10.7 percent. **Figure 4-8** shows the general situation for linkage in which recombinants are less than 50 percent. Recombinant frequencies for different linked genes range from 0 to 50 percent, depending on their closeness. The farther apart genes are, the more closely their recombinant frequencies approach 50 percent, and, in such cases, one cannot immediately discern whether genes are linked or are on different chromosomes. What about recombinant frequencies greater than 50 percent? The answer is that such frequencies are *never* observed (a mathematical explanation of this phenomenon appears later in the chapter).

Note in Figure 4-7 that a single crossover generates two reciprocal recombinant products, which explains why the reciprocal recombinant classes are generally approximately equal in frequency. The corollary of this point is that the two parental nonrecombinant types also must be equal in frequency, as also observed by Morgan.

Map units

The basic method of mapping genes with the use of recombinant frequencies was devised by a student of Morgan's. As Morgan studied more and more linked genes, he saw that the proportion of recombinant progeny varied considerably, depending on which linked genes were being studied, and he thought that such variation in recombinant frequency might somehow indicate the actual distances separating genes on the chromosomes. Morgan assigned the quantification of this process to an undergraduate student, Alfred Sturtevant, who also became one of the great geneticists. Morgan asked Sturtevant to try to make some sense

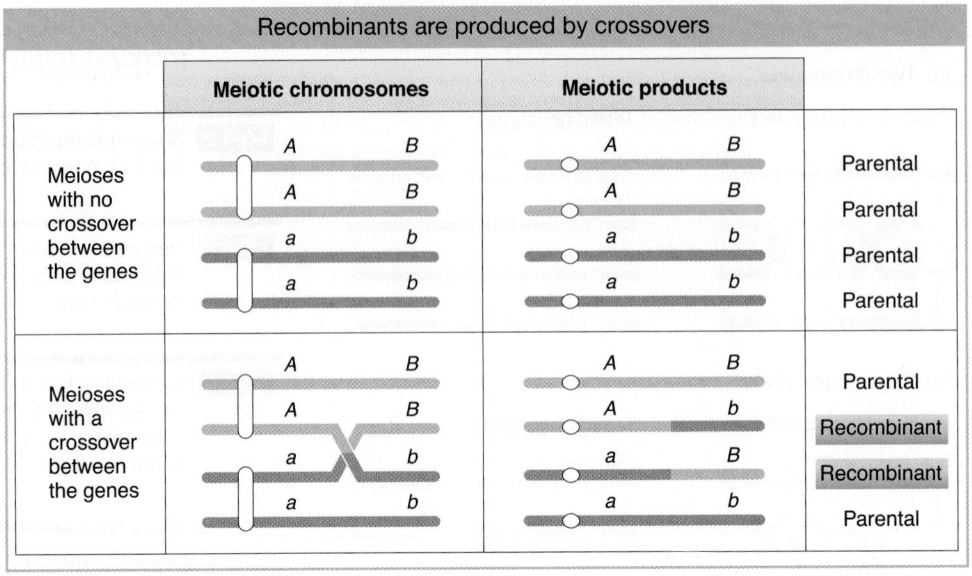

FIGURE 4-7 Recombinants arise from meioses in which a crossover takes place between nonsister chromatids.

**ANIMATED Sapling Plus
ART**

Meiotic recombination between linked genes by crossing over

Recombinants are produced by crossovers		
	Meiotic chromosomes	**Meiotic products**
Meioses with no crossover between the genes	A B A B a b a b	A B Parental A B Parental a b Parental a b Parental
Meioses with a crossover between the genes	A B A B a b a b	A B Parental A b Recombinant a B Recombinant a b Parental

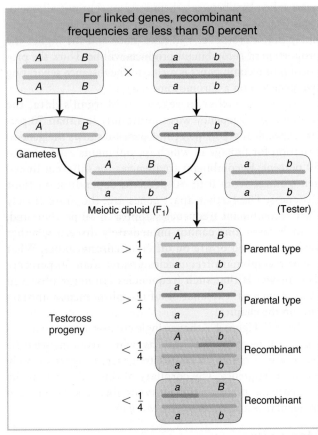

For linked genes, recombinant frequencies are less than 50 percent

P

A B × a b
A B   a b

Gametes

A B   a b

Meiotic diploid (F₁)

A B × a b
a b   a b

(Tester)

Testcross progeny

$> \frac{1}{4}$ A B Parental type a b

$> \frac{1}{4}$ a b Parental type a b

$< \frac{1}{4}$ A b Recombinant a b

$< \frac{1}{4}$ a B Recombinant a b

FIGURE 4-8 A testcross reveals that the frequencies of recombinants arising from crossovers between linked genes are less than 50 percent.

of the data on crossing over between different linked genes. In one evening, Sturtevant developed a method for mapping genes that is still used today. In Sturtevant's own words, "In the latter part of 1911, in conversation with Morgan, I suddenly realized that the variations in strength of linkage, already attributed by Morgan to differences in the spatial separation of genes, offered the possibility of determining sequences in the linear dimension of a chromosome. I went

home and spent most of the night (to the neglect of my undergraduate homework) in producing the first chromosome map."

As an example of Sturtevant's logic, consider Morgan's testcross results with the *pr* and *vg* genes, from which he calculated a recombinant frequency of 10.7 percent. Sturtevant suggested that we can use this percentage of recombinants as a quantitative index of the linear distance between two genes on a genetic map, or **linkage map**, as it is sometimes called.

The basic idea here is quite simple. Imagine two specific genes positioned a certain fixed distance apart. Now imagine random crossing over along the paired homologs. In some meioses, nonsister chromatids cross over by chance in the chromosomal region between these genes; from these meioses, recombinants are produced. In other meiotic divisions, there are no crossovers between these genes; no recombinants result from these meioses. (See Figure 4-7 for a diagrammatic illustration.) Sturtevant postulated a rough proportionality: the greater the distance between the linked genes, the greater the chance of crossovers in the region between the genes and, hence, the greater the proportion of recombinants that would be produced. Thus, by determining the frequency of recombinants, we can obtain a measure of the map distance between the genes. In fact, Sturtevant defined one **genetic map unit (m.u.)** as that distance between genes for which 1 product of meiosis in 100 is recombinant. For example, the **recombinant frequency (RF)** of 10.7 percent obtained by Morgan is defined as 10.7 m.u. A map unit is sometimes referred to as a **centimorgan (cM)** in honor of Thomas Hunt Morgan.

Does this method produce a linear map corresponding to chromosome linearity? Sturtevant predicted that, on a linear map, if 5 map units (5 m.u.) separate genes A and B, and 3 m.u. separate genes A and C, then the distance separating B and C should be either 8 or 2 m.u. (**Figure 4-9**). Sturtevant found his prediction to be the case. In other words, his analysis strongly suggested that genes are

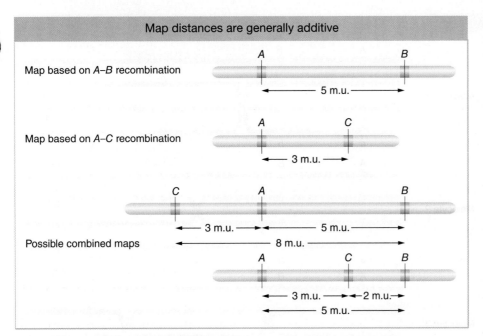

Map distances are generally additive

Map based on A–B recombination

5 m.u.

Map based on A–C recombination

3 m.u.

Possible combined maps

3 m.u. ——— 5 m.u.

8 m.u.

3 m.u. ——— 2 m.u.

5 m.u.

FIGURE 4-9 A chromosome region containing three linked genes. Because map distances are additive, calculation of A–B and A–C distances leaves us with the two possibilities shown for the B–C distance.

arranged in some linear order, making map distances additive. (There are some minor but not insignificant exceptions, as we will see later.) Since we now know from molecular analysis that a chromosome is a single DNA molecule with the genes arranged along it, it is no surprise for us today to learn that recombination-based maps are linear because they reflect a linear array of genes.

How is a map represented? As an example, in *Drosophila*, the locus of the eye-color gene and the locus of the wing-length gene are approximately 11 m.u. apart, as mentioned earlier. The relation is usually diagrammed in the following way:

$$pr \quad\quad 11.0 \text{ m.u.} \quad\quad vg$$

Generally, we refer to the locus of this eye-color gene in shorthand as the "*pr* locus," after the first discovered mutant allele, but we mean the place on the chromosome where *any* allele of this gene will be found, mutant or wild type.

As stated in Chapters 2 and 3, genetic analysis can be applied in two opposite directions. This principle is applicable to recombinant frequencies. In one direction, recombinant frequencies can be used to make maps. In the other direction, when given an established map with genetic distance in map units, we can predict the frequencies of progeny in different classes. For example, the genetic distance between the *pr* and *vg* loci in *Drosophila* is approximately 11 m.u. So, knowing this value, we know that there will be 11 percent recombinants in the progeny from a testcross of a female dihybrid heterozygote in cis conformation (*pr vg/pr⁺ vg⁺*). These recombinants will consist of two reciprocal recombinants of equal frequency: thus, 5.5 percent will be *pr vg⁺* and 5.5 percent will be *pr⁺ vg*. We also

know that $100 - 11 = 89$ percent will be nonrecombinant in two equal classes, 44.5 percent *pr⁺ vg⁺* and 44.5 percent *pr vg*. (Note that the tester contribution *pr vg* was ignored in writing out these genotypes.)

There is a strong implication that the "distance" on a linkage map is a physical distance along a chromosome, and Morgan and Sturtevant certainly intended to imply just that. But we should realize that the linkage map is a *hypothetical* entity constructed from a purely genetic analysis. The linkage map could have been derived without even knowing that chromosomes existed. Furthermore, at this point in our discussion, we cannot say whether the "genetic distances" calculated by means of recombinant frequencies in any way represent actual physical distances on chromosomes. However, physical mapping has shown that genetic distances are, in fact, roughly proportional to recombination-based distances. There are exceptions caused by recombination hotspots, places in the genome where crossing over takes place more frequently than usual. The presence of hotspots causes proportional expansion of some regions of the map. Recombination blocks, which have the opposite effect, also are known.

A summary of the way in which recombinants from crossing over are used in mapping is shown in **Figure 4-10**. Crossovers occur more or less randomly along the chromosome pair. In general, in longer regions, the average number of crossovers is higher and, accordingly, recombinants are more frequently obtained, translating into a longer map distance.

KEY CONCEPT Recombination between linked genes can be used to map their distance apart on a chromosome. The unit of mapping (1 m.u.) is defined as a recombinant frequency of 1 percent. Map distances are roughly additive.

FIGURE 4-10 Crossovers produce recombinant chromatids whose frequency can be used to map genes on a chromosome. Longer regions produce more crossovers. Brown shows recombinants for that interval.

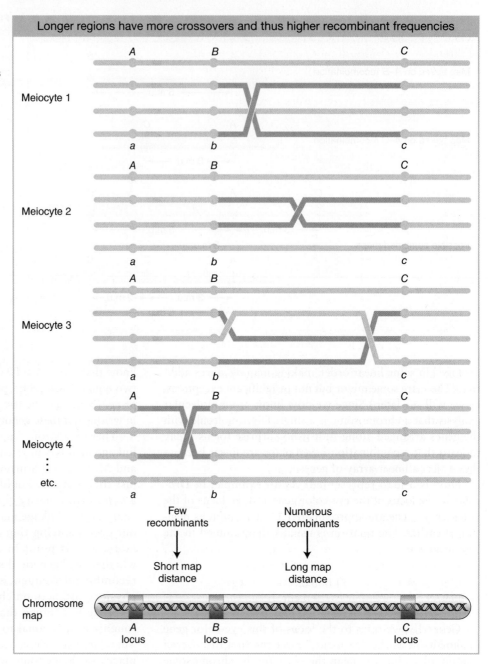

Three-point testcross

So far, we have looked at linkage in crosses of dihybrids (double heterozygotes) with doubly recessive testers. The next level of complexity is a cross of a trihybrid (triple heterozygote) with a triply recessive tester. This kind of cross, called a **three-point testcross** or a **three-factor cross**, is commonly used in linkage analysis. The goal is to deduce whether the three genes are linked and, if they are, to deduce their order and the map distances between them.

Let's look at an example, also from *Drosophila*. In our example, the mutant alleles are *v* (vermilion eyes), *cv* (crossveinless, or absence of a crossvein on the wing), and *ct* (cut, or snipped, wing edges). The analysis is carried out by performing the following crosses:

P $v^+/v^+ \cdot cv/cv \cdot ct/ct \times v/v \cdot cv^+/cv^+ \cdot ct^+/ct^+$

↓

Gametes $v^+ \cdot cv \cdot ct$ $v \cdot cv^+ \cdot ct^+$

F₁ trihybrid $v/v^+ \cdot cv/cv^+ \cdot ct/ct^+$

Trihybrid females are testcrossed with triple recessive males:

$v/v^+ \cdot cv/cv^+ \cdot ct/ct^+ ♀ \times v/v \cdot cv/cv \cdot ct/ct ♂$

F₁ trihybrid female   Tester male

From any trihybrid, only $2 \times 2 \times 2 = 8$ gamete genotypes are possible. They are the genotypes seen in the testcross progeny. The following chart shows the number of each of the eight gametic genotypes in a sample of 1448 progeny flies. The columns alongside show which genotypes are recombinant (R) for the loci taken two at a time. We must be careful in our classification of parental and recombinant types. Note that the parental input genotypes for the triple heterozygotes are $v^+ \cdot cv \cdot ct$ and $v \cdot cv^+ \cdot ct^+$; any combination other than these two constitutes a recombinant.

		Recombinant for loci		
Gametes		v and cv	v and ct	cv and ct
$v \cdot cv^+ \cdot ct^+$	580			
$v^+ \cdot cv \cdot ct$	592			
$v \cdot cv \cdot ct^+$	45	R		R
$v^+ \cdot cv^+ \cdot ct$	40	R		R
$v \cdot cv \cdot ct$	89	R	R	
$v^+ \cdot cv^+ \cdot ct^+$	94	R	R	
$v \cdot cv^+ \cdot ct$	3		R	R
$v^+ \cdot cv \cdot ct^+$	5		R	R
	1448	268	191	93

Let's analyze the loci two at a time, starting with the v and cv loci. In other words, we look at just the first two columns under "Gametes" and cover up the third one. Because the parentals for this pair of loci are $v^+ \cdot cv$ and $v \cdot cv^+$, we know that the recombinants are by definition $v \cdot cv$ and $v^+ \cdot cv^+$. There are $45 + 40 + 89 + 94 = 268$ of these recombinants. Of a total of 1448 flies, this number gives an RF of 18.5 percent.

For the v and ct loci, the recombinants are $v \cdot ct$ and $v^+ \cdot ct^+$. There are $89 + 94 + 3 + 5 = 191$ of these recombinants among 1448 flies, and so the RF = 13.2 percent.

For ct and cv, the recombinants are $cv \cdot ct^+$ and $cv^+ \cdot ct$. There are $45 + 40 + 3 + 5 = 93$ of these recombinants among the 1448, and so the RF = 6.4 percent.

Clearly, all the loci are linked, because the RF values are all considerably less than 50 percent. Because the v and cv loci have the largest RF value, they must be farthest apart; therefore, the ct locus must lie between them. A map can be drawn as follows:

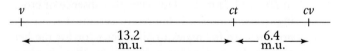

The testcross can be rewritten as follows, now that we know the linkage arrangement:

$$v^+ct\ cv/v\ ct^+cv^+ \ \times \ v\ ct\ cv/v\ ct\ cv$$

Note several important points here. First, we have deduced a gene order that differs from that used in our list of the progeny genotypes. Because the point of the exercise was to determine the linkage relation of these genes, the original listing was of necessity arbitrary; the order was simply not known before the data were analyzed. Henceforth, the genes must be written in correct order.

Second, we have definitely established that ct is between v and cv. In the diagram, we have arbitrarily placed v to the

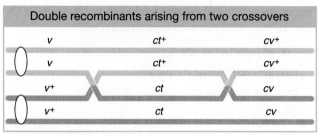

FIGURE 4-11 Example of a double crossover between two chromatids. Notice that a double crossover produces double-recombinant chromatids that have the parental allele combinations at the outer loci. The position of the centromere cannot be determined from the data. It has been added for completeness.

left and cv to the right, but the map could equally well be drawn with the placement of these loci inverted.

Third, note that linkage maps merely map the loci in relation to one another, with the use of standard map units. We do not know where the loci are on a chromosome—or even which specific chromosome they are on. In subsequent analyses, as more loci are mapped in relation to these three, the complete chromosome map would become "fleshed out."

KEY CONCEPT Three-point (and higher) testcrosses enable geneticists to evaluate linkage between three (or more) genes and to determine gene order, all in one cross.

A final point to note is that the two smaller map distances, 13.2 m.u. and 6.4 m.u., add up to 19.6 m.u., which is greater than 18.5 m.u., the distance calculated for v and cv. Why? The answer to this question lies in the way in which we have treated the two rarest classes of progeny (totaling 8) with respect to the recombination of v and cv. Now that we have the map, we can see that these two rare classes are in fact double recombinants, arising from two crossovers (**Figure 4-11**). However, when we calculated the RF value for v and cv, we did not count the $v\ ct\ cv^+$ and v^+ct^+cv genotypes; after all, with regard to v and cv, they are parental combinations ($v\ cv^+$ and $v^+\ cv$). In light of our map, however, we see that this oversight led us to underestimate the distance between the v and the cv loci. Not only should we have counted the two rarest classes, we should have counted each of them *twice* because each represents double recombinants. Hence, we can correct the value by adding the numbers $45 + 40 + 89 + 94 + 3 + 3 + 5 + 5 = 284$. Of the total of 1448, this number is exactly 19.6 percent, which is identical with the sum of the two component values. (In practice, we do not need to do this calculation, because the sum of the two shorter distances gives us the best estimate of the overall distance.)

Deducing gene order by inspection

Now that we have had some experience with the three-point testcross, we can look back at the progeny listing and see that, for trihybrids of linked genes, *gene order* can usually be deduced by inspection, without a recombinant frequency analysis. Typically, for linked genes, we have the eight genotypes at the following frequencies:

two at high frequency

two at intermediate frequency

two at a different intermediate frequency

two rare

Only three gene orders are possible, each with a different gene in the middle position. It is generally true that the double-recombinant classes are the smallest ones. Only one order is compatible with the smallest classes ($v\ ct\ cv^+$ and $v^+\ ct^+\ cv$) having been formed by double crossovers, as shown in **Figure 4-12**. A simple rule of thumb for deducing the gene in the middle is to compare the gametes in the parental classes with the gametes in the double-recombinant classes, and to identify the allele pair that has moved in relation to the other two allele pairs that remain in the original parental arrangement. In other words, the gene in the middle is the allele pair that has "flipped" position in the double-recombinant classes.

Interference

Knowing the existence of double crossovers permits us to ask questions about their possible interdependence. We can ask, are the crossovers in adjacent chromosome regions independent events, or does a crossover in one region affect the likelihood of there being a crossover in an adjacent region? The answer is that, generally, crossovers inhibit each other somewhat in an interaction called **interference**. Double-recombinant classes can be used to deduce the extent of this interference.

Interference can be measured in the following way. If the crossovers in the two regions are independent, we can use the product rule (see Chapter 3) to predict the frequency of double recombinants: that frequency would equal the product of the recombinant frequencies in the adjacent regions. In the v-ct-cv recombination data, the v-ct RF value is 0.132 and the ct-cv value is 0.064; so, if there is no interference, double recombinants might be expected at the frequency $0.132 \times 0.064 = 0.0084$ (0.84 percent). In the sample of 1448 flies, $0.0084 \times 1448 = 12$ double recombinants are expected. But the data show that only 8 double recombinants were actually observed. If this deficiency of double recombinants were consistently observed, it would show us that the two regions are not independent and suggest that the distribution of crossovers favors singles at the expense of doubles. In other words, there is some kind of interference: a crossover does reduce the probability of a crossover in an adjacent region.

Interference is quantified by first calculating a term called the **coefficient of coincidence** (c.o.c.), which is the ratio of observed to expected double recombinants. Interference (I) is defined as $1 - $ c.o.c. Hence,

$$I = 1 - \frac{\text{observed frequency or number of double recombinants}}{\text{expected frequency or number of double recombinants}}$$

In our example,

$$I = 1 - \tfrac{8}{12} = \tfrac{4}{12} = \tfrac{1}{3}, \text{ or 33 percent}$$

In some regions, there are never any observed double recombinants. In these cases, c.o.c. = 0, and so I = 1 and interference is complete. Interference values anywhere between 0 and 1 are found in different regions and in different organisms.

A statistical approach to correcting map distances for unseen multiple crossovers is shown in **Box 4-1**.

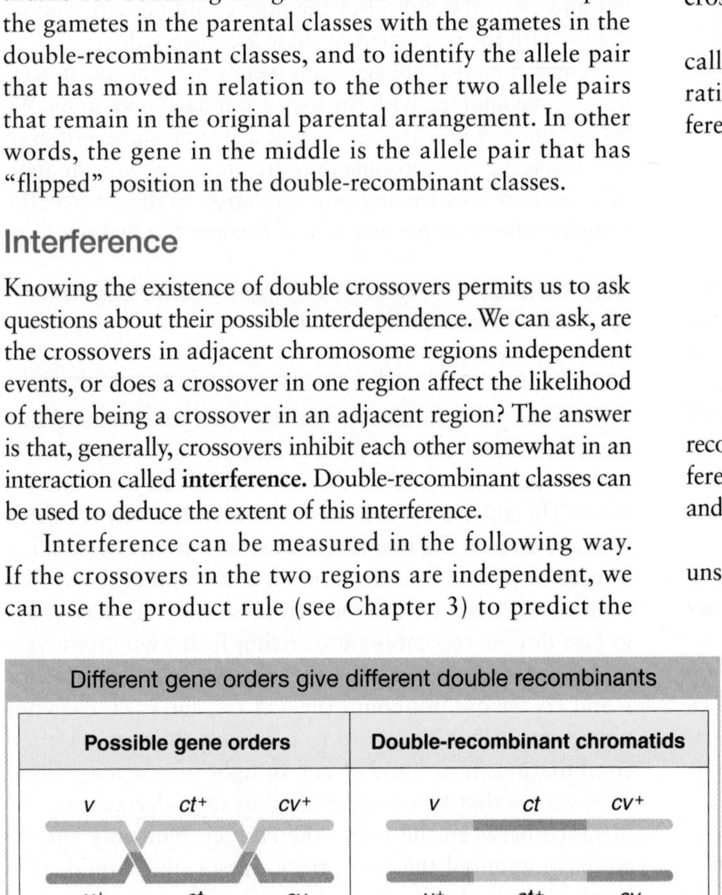

You may have wondered why we always use heterozygous females for testcrosses in *Drosophila*. The explanation lies in an unusual feature of *Drosophila* males. When, for example, $pr\ vg/pr^+\ vg^+$ males are crossed with $pr\ vg/pr\ vg$ females, only $pr\ vg/pr^+\ vg^+$ and $pr\ vg/pr\ vg$ progeny are recovered. This result shows that there is no crossing over in *Drosophila* males. However, this absence of crossing over in one sex is limited to certain species; it is not the case for males of all species (or for the heterogametic sex). In other organisms, there is crossing over in XY males and in WZ females. The reason for the absence of crossing over in *Drosophila* males is that they have an unusual prophase I, with no synaptonemal complexes (the molecular assemblages generally visible between paired chromosomes). Incidentally, there is a recombination difference between human sexes as well. Women show higher recombinant frequencies for the same autosomal loci than do men.

With the use of a reiteration of the preceding recombination-based techniques, maps have been produced of thousands of genes for which variant (mutant) phenotypes have been identified. A simple illustrative example from the tomato is shown in **Figure 4-13**. The tomato chromosomes are shown

Different gene orders give different double recombinants	
Possible gene orders	**Double-recombinant chromatids**

FIGURE 4-12 The three possible gene orders shown on the left yield the six products of a double crossover shown on the right. Only the first possibility is compatible with the data in the text. Note that only the nonsister chromatids taking part in the double crossover are shown.

BOX 4-1   Accounting for Unseen Multiple Crossovers

In the discussion of the three-point testcross, some parental (nonrecombinant) chromatids resulted from *double* crossovers. These crossovers initially could not be counted in the recombinant frequency, skewing the results. This situation leads to the worrisome notion that *all* map distances based on recombinant frequency might be underestimations of physical distances because undetected multiple crossovers might have occurred, some of whose products would not be recombinant. Several creative mathematical approaches have been designed to get around the multiple-crossover problem. We will look at two methods. First, we examine a method originally worked out by J. B. S. Haldane in the early years of genetics.

A mapping function

The approach worked out by Haldane was to devise a **mapping function,** a formula that relates an observed recombinant-frequency value to a map distance corrected for multiple crossovers. The approach works by relating RF to the mean number of crossovers, *m*, that must have taken place in that chromosomal segment per meiosis and then deducing what map distance this *m* value *should* have produced.

To find the relation of RF to *m*, we must first think about outcomes of the various crossover possibilities. In any chromosomal region, we might expect meioses with 0, 1, 2, 3, 4, or more crossovers. Surprisingly, the only class that is really crucial is the zero class. To see why, consider the following. It is a curious but nonintuitive fact that *any number* of crossovers produces a frequency of 50 percent recombinants *within those meioses.* The diagram proves this statement for single and double crossovers as examples, but it is true for any number of crossovers. Hence, the true determinant of RF is the relative sizes of the classes with no crossovers (the zero class) compared with the classes with any nonzero number of crossovers.

Now the task is to calculate the size of the zero class. The occurrence of crossovers in a specific chromosomal region is well described by a statistical distribution called the **Poisson distribution.** The Poisson formula in general describes the distribution of "successes" in samples when the average probability of successes is low. An illustrative example is to dip a child's net into a pond of fish: most dips will produce no fish, a smaller proportion will produce one fish, an even smaller proportion two, and so on. This analogy can be directly applied to a chromosomal region, which will have 0, 1, 2, and so forth, crossover "successes" in different meioses. The Poisson formula, given here, will tell us the proportion of the classes with different numbers of crossovers:

$$f_i = (e^{-m}m^i)/i\,!$$

The terms in the formula have the following meanings:

e = the base of natural logarithms (approximately 2.7)

m = the mean number of successes in a defined sample size

i = the actual number of successes in a sample of that size

f_i = the frequency of samples with i successes in them

$!$ = the factorial symbol (for example, $5! = 5 \times 4 \times 3 \times 2 \times 1$)

The Poisson distribution tells us that the frequency of the $i = 0$ class (the key one) is

$$e^{-m}\frac{m^0}{0!}$$

Because m^0 and $0!$ both equal 1, the formula reduces to e^{-m}.

Now we can write a function that relates RF to m. The frequency of the class with any nonzero number of crossovers will be $1 - e^{-m}$, and, in these meioses, 50 percent (1/2) of the products will be recombinant; so

$$\text{RF} = \tfrac{1}{2}(1 - e^{-m})$$

and this formula is the mapping function that we have been seeking.

Let's look at an example in which RF is converted into a map distance corrected for multiple crossovers. Assume that, in one testcross, we obtain an RF value of 27.5 percent (0.275). Plugging this into the function allows us to solve for m:

$$0.275 = \tfrac{1}{2}(1 - e^{-m})$$

so

$$e^{-m} = 1 - (2 \times 0.275) = 0.45$$

By using a calculator to find the natural logarithm (ln) of 0.45, we can deduce that $m = 0.8$. That is, on average, there are 0.8 crossovers per meiosis in that chromosomal region.

The final step is to convert this measure of crossover frequency to give a "corrected" map distance. All that we have to do to convert into corrected map units is to multiply the calculated average crossover frequency by 50 because, on average, a crossover produces a recombinant frequency of 50 percent. Hence, in the preceding numerical example, the m value of 0.8 can be converted into a corrected recombinant fraction of $0.8 \times 50 = 40$ corrected m.u. We see that, indeed, this value is substantially larger than the 27.5 m.u. that we would have deduced from the observed RF.

Note that the mapping function neatly explains why the maximum RF value for linked genes is 50 percent. As m gets very large, e^{-m} tends to zero and the RF tends to 1/2, or 50 percent.

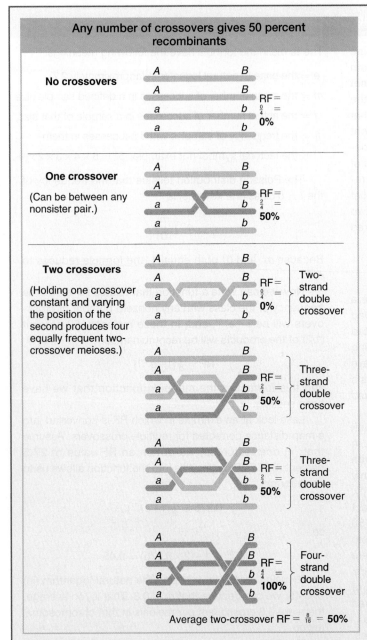

Any number of crossovers gives 50 percent recombinants

No crossovers

$RF = \frac{0}{4} = 0\%$

One crossover

(Can be between any nonsister pair.)

$RF = \frac{2}{4} = 50\%$

Two crossovers

(Holding one crossover constant and varying the position of the second produces four equally frequent two-crossover meioses.)

$RF = \frac{0}{4} = 0\%$ Two-strand double crossover

$RF = \frac{2}{4} = 50\%$ Three-strand double crossover

$RF = \frac{2}{4} = 50\%$ Three-strand double crossover

$RF = \frac{4}{4} = 100\%$ Four-strand double crossover

Average two-crossover RF = $\frac{8}{16}$ = **50%**

Demonstration that the average RF is 50 percent for meioses in which the number of crossovers is not zero. Recombinant chromatids are brown. Two-strand double crossovers produce all parental types; so all the chromatids are orange. Note that all crossovers are between nonsister chromatids. Try the triple crossover class yourself.

The Perkins formula

For fungi and other tetrad-producing organisms, there is another way of compensating for multiple crossovers—specifically, double crossovers (the most common type expected). In tetrad analysis of "dihybrids" generally, only three types of tetrads are possible, when classified on the basis of the presence of parental and recombinant genotypes in the products. The classification of tetrads is based on whether there are two genotypes present (ditype) or four (tetratype). Within ditypes there are two classes: parental (showing two parental genotypes) and nonparental

(showing two nonparental genotypes). From a cross $AB \times ab$, they are:

Parental ditype (PD)	Tetratype (T)	Nonparental ditype (NPD)
$A \cdot B$	$A \cdot B$	$A \cdot b$
$A \cdot B$	$A \cdot b$	$A \cdot b$
$a \cdot b$	$a \cdot B$	$a \cdot B$
$a \cdot b$	$a \cdot b$	$a \cdot B$

The recombinant genotypes are shown in red. If the genes are linked, a simple approach to mapping their distance apart might be to use the following formula:

$$\text{map distance} = RF = 100 \left(NPD + \tfrac{1}{2}T\right)$$

because this formula gives the percentage of all recombinants. However, in the 1960s, David Perkins developed a formula that compensates for the effects of double crossovers. The Perkins formula thus provides a more accurate estimate of map distance:

$$\text{corrected map distance} = 50 \left(T + 6\,NPD\right)$$

We will not go through the derivation of this formula other than to say that it is based on the totals of the PD, T, and NPD classes expected from meioses with 0, 1, and 2 crossovers (it assumes that higher numbers are vanishingly rare). Let's look at an example of its use. In our hypothetical cross of $A\,B \times a\,b$, the observed frequencies of the tetrad classes are 0.56 PD, 0.41 T, and 0.03 NPD. By using the Perkins formula, we find the corrected map distance between the a and b loci to be

$$50\,[0.41 + (6 \times 0.03)] = 50\,(0.59) = 29.5 \text{ m.u.}$$

Let us compare this value with the uncorrected value obtained directly from the RF. By using the same data, we find

$$\text{uncorrected map distance} = 100\left(\tfrac{1}{2}T + NPD\right)$$
$$= 100(0.205 + 0.03)$$
$$= 23.5 \text{ m.u.}$$

This distance is 6 m.u. less than the estimate that we obtained by using the Perkins formula because we did not correct for double crossovers.

As an aside, what PD, NPD, and T values are expected when dealing with *unlinked* genes? The sizes of the PD and NPD classes will be equal as a result of independent assortment. The T class can be produced only from a crossover between either of the two loci and their respective centromeres and, therefore, the size of the T class will depend on the total size of the two regions lying between locus and centromere. However, the formula $\tfrac{1}{2}T + NPD$ should always yield 0.50, reflecting independent assortment.

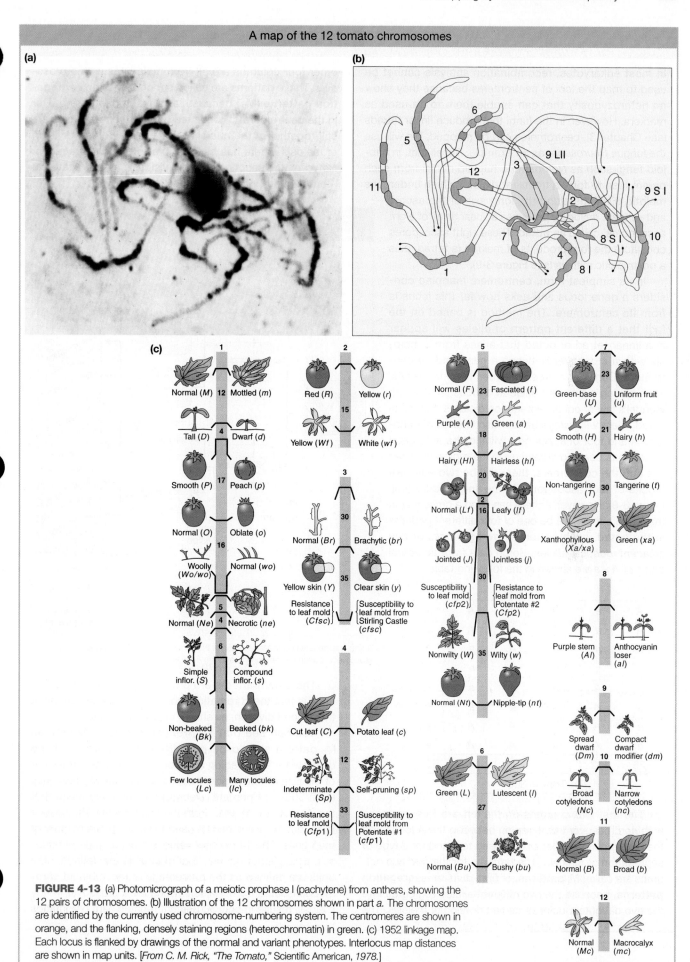

FIGURE 4-13 (a) Photomicrograph of a meiotic prophase I (pachytene) from anthers, showing the 12 pairs of chromosomes. (b) Illustration of the 12 chromosomes shown in part a. The chromosomes are identified by the currently used chromosome-numbering system. The centromeres are shown in orange, and the flanking, densely staining regions (heterochromatin) in green. (c) 1952 linkage map. Each locus is flanked by drawings of the normal and variant phenotypes. Interlocus map distances are shown in map units. [From C. M. Rick, "The Tomato," Scientific American, 1978.]

BOX 4-2 Centromere Mapping with Linear Tetrads

In most eukaryotes, recombination analysis cannot be used to map the loci of centromeres because they show no heterozygosity that can enable them to be used as markers. However, in the fungi that produce linear tetrads (see Chapter 3), centromeres *can* be mapped. We will use the fungus *Neurospora* as an example. Recall that, in haploid fungi such as *Neurospora*, haploid nuclei from each parent fuse to form a transient diploid, which undergoes meiotic divisions along the long axis of the ascus, and so each meiocyte produces a linear array of eight ascospores, called an **octad.** These eight ascospores constitute the four products of meiosis (a tetrad) plus a postmeiotic mitosis (see Figure 3-9).

In its simplest form, centromere mapping considers a gene locus and asks how far this locus is from its centromere. The method is based on the fact that a different pattern of alleles will appear in a linear tetrad or octad that arises from a meiosis with a crossover between a gene and its centromere. Consider a cross between two individuals, each having a different allele at a locus (say, *A* × *a*). Mendel's law of equal segregation dictates that, in an octad, there will always be four ascospores of genotype *A* and four of *a*, but how will they be arranged? If there has been no crossover in the region between *A*/*a* and the centromere, there will be two adjacent blocks of four ascospores in the linear octad (see Figure 3-9). However, if there has been a crossover in that region, there will be one of four different patterns in the octad, each pattern showing *blocks of two adjacent identical alleles.* Some data from an actual cross of *A* × *a* are shown in the following table.

Octads

A	*a*	*A*	*a*	*A*	*a*
A	*a*	*A*	*a*	*A*	*a*
A	*a*	*a*	*A*	*a*	*A*
A	*a*	*a*	*A*	*a*	*A*
a	*A*	*A*	*a*	*a*	*A*
a	*A*	*A*	*a*	*a*	*A*
a	*A*	*a*	*A*	*A*	*a*
a	*A*	*a*	*A*	*A*	*a*
126	132	9	11	10	12

Total = 300

The first two columns on the left are from meioses with *no* crossover in the region between the *A* locus and the centromere. The letter M is used to stand for a type of segregation at meiosis. The patterns for the first two columns are called **M$_I$ patterns,** or **first-division segregation patterns,** because the two different alleles segregate into the two daughter nuclei at the first division of meiosis. The

other four columns are all from meiocytes *with* a crossover. These patterns are called **second-division segregation patterns (M$_{II}$)** because, as a result of crossing over in the centromere-to-locus region, the *A* and *a* alleles are still together in the nuclei at the end of the first division of meiosis. There has been no first-division segregation. However, the second meiotic division does segregate the *A* and *a* alleles into separate nuclei:

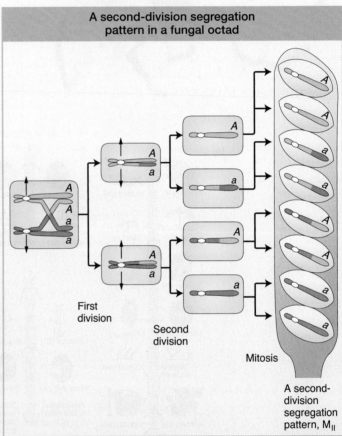

A second-division segregation pattern in a fungal octad

First division

Second division

Mitosis

A second-division segregation pattern, M$_{II}$

A and *a* segregate into separate nuclei at the second meiotic division when there is a crossover between the centromere and the *A* locus.

The other patterns are produced similarly; the difference is that the chromatids move in different directions at the second division, as shown on the next page.

You can see that the frequency of octads with an M$_{II}$ pattern should be proportional to the size of the centromere—*A*/*a* region and could be used as a measure of the size of that region. In our example, the M$_{II}$ frequency is 42/300 = 14 percent. Does this percentage mean that the *A*/*a* locus is 14 m.u. from the centromere? The answer is no, but this value can be used to calculate the number of map units. The 14 percent value is a percentage of *meioses*, which is not the way that map units are defined. Map units are defined as the percentage of recombinant *chromatids* issuing from meiosis. Because a crossover in any meiosis results in only 50 percent recombinant chromatids

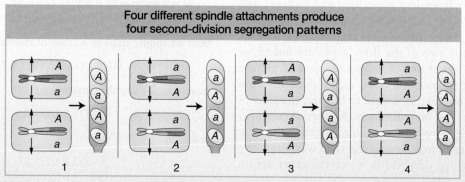

Four different spindle attachments produce
four second-division segregation patterns

In the second meiotic division, the centromeres attach to the spindle at random, producing the four arrangements shown. The four arrangements are equally frequent.

(four out of eight), we must divide the 14 percent by 2 to convert the M_{II} frequency (a frequency of *meioses*) into map units (a frequency of recombinant *chromatids*). Hence, this region must be 7 m.u. in length, and this measurement can be introduced into the map of that chromosome.

in Figure 4-13a, their numbering in Figure 4-13b, and recombination-based gene maps in Figure 4-13c. The chromosomes are shown as they appear under the microscope, together with chromosome maps based on linkage analysis of various allelic pairs shown with their phenotypes.

In some organisms, recombination-based techniques can be used to map the positions of centromeres. Centromeres are not genes, but they are regions of DNA on which the orderly reproduction of living organisms absolutely depends and are therefore of great interest in genetics (**Box 4-2**).

Using ratios as diagnostics

The analysis of ratios is one of the pillars of genetics. In the text so far, we have encountered many different ratios whose derivations are spread out over several chapters. Because recognizing ratios and using them in diagnosis of the genetic system under study are part of everyday genetics, let's review the main ratios that we have covered so far. They are shown in **Figure 4-14**. You can read the ratios from the relative widths of the colored boxes in a row. Figure 4-14 deals with selfs and testcrosses of monohybrids,

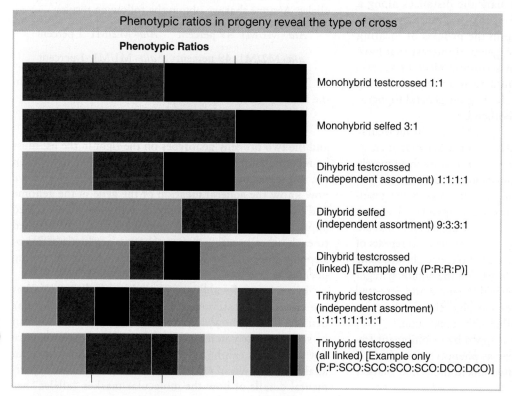

Phenotypic ratios in progeny reveal the type of cross

Phenotypic Ratios

Monohybrid testcrossed 1:1

Monohybrid selfed 3:1

Dihybrid testcrossed (independent assortment) 1:1:1:1

Dihybrid selfed (independent assortment) 9:3:3:1

Dihybrid testcrossed (linked) [Example only (P:R:R:P)]

Trihybrid testcrossed (independent assortment) 1:1:1:1:1:1:1:1

Trihybrid testcrossed (all linked) [Example only (P:P:SCO:SCO:SCO:SCO:DCO:DCO)]

FIGURE 4-14 P = parental, R = recombinant, SCO = single crossover, DCO = double crossover.

dihybrids (with independent assortment and linkage), and trihybrids (also with independent assortment and linkage of all genes). One situation not represented is a trihybrid in which only two of the three genes are linked; as an exercise, you might like to deduce the general pattern that would have to be included in such a diagram from this situation. Note that, in regard to linkage, the sizes of the classes depend on map distances. A geneticist deduces unknown genetic states in something like the following way: "a 9:3:3:1 ratio tells me that this ratio was very likely produced by a selfed dihybrid in which the genes are on different chromosomes."

> **KEY CONCEPT** Single gene inheritance and two-gene inheritance (linked and unlinked) can be inferred from diagnostic phenotypic ratios in both selfing and testcrossing.

4.3 MAPPING WITH MOLECULAR MARKERS

So far in this chapter we have mapped loci of genes with alleles that result in visible phenotypic differences. One might call these loci *phenotypic markers* as they mark certain points on the chromosome that can produce visibly different phenotypes in the outward appearance of progeny. However, molecular studies such as DNA sequencing have revealed many DNA differences between chromosomes that are neutral; that is, they do not seem to produce visible phenotypic effects. These are generally called **molecular markers**. Because they are so numerous, they are useful in "fleshing out" the chromosome map. They can be compared to milestones that mark the distances along a well-traveled road; in themselves they are not particularly interesting, but they can help in the process of locating important towns—in this case, genes of interest that have been shown to be relevant in a specific piece of genetic research. Locating a gene of interest is an important step toward isolating the gene and charting the general topography of its chromosomal neighborhood.

Here, we will focus on two general types of molecular markers, and will revisit the subject in more detail in later chapters. (Hence, for the time being, we are treating them merely as milestones and ignoring their origin and detection.) First, there are molecular markers that are merely loci showing neutral simple DNA sequence differences, perhaps a G-C base pair replaced by a T-A base pair. Second, there are markers that are loci showing variable numbers of tandem (adjacent) repeats of short, simple DNA sequences; for example, one chromosome might have five repeats at that locus, and the homolog might have eight. Both simple sequence differences and repeated DNA differences are highly polymorphic; that is, there are often many "alleles" of each marker in the population.

Molecular markers can be mapped by recombinant frequency in exactly the same way as phenotypic markers. If we use D and R to represent two linked molecular marker loci, where

D is a locus with a simple DNA sequence difference, and R is a locus showing variable numbers of tandem repeats, then we might analyze a cross of the following type:

$$D1\ R1/D2\ R2 \times D3\ R3/D3\ R3$$

from which recombinants would be

$$D1\ R2/D3\ R3$$

and

$$D2\ R1/D3\ R3$$

The frequency of recombinants would then be used to calculate the map distance between the two loci.

Thus, molecular markers can be mapped in relation to each other in this way, but a molecular marker can also be mapped in relation to a phenotypic marker. The location of the gene for the human disease cystic fibrosis was originally discovered in this way through its linkage to molecular markers known to be located on chromosome 7. This discovery led to the isolation and sequencing of the gene, resulting in the further discovery that it encodes the protein now called *cystic fibrosis transmembrane conductance regulator* (CFTR). The gene for Huntington disease was also located in this way, leading to the discovery that it encodes a muscle protein now called *huntingtin*. The general experimental procedure might be as follows. Let A and a be the disease-gene alleles and M1 and M2 be alleles of a specific molecular-marker locus. Assume that the cross is $A/a \cdot M1/M2 \times a/a \cdot M1/M1$, a kind of testcross. Progeny would be first scored for the A and a phenotypes, and then DNA would be extracted from each individual and sequenced or otherwise assessed to determine the molecular alleles. Assume that we obtain the following results:

$A/a \cdot M1/M1$ 49 percent $A/a \cdot M2/M1$ 1 percent

$a/a \cdot M2/M1$ 49 percent $a/a \cdot M1/M1$ 1 percent

These results tell us that the testcross must have been in the following conformation:

$$A\ M1/a\ M2 \times a\ M1/a\ M1$$

and the two progeny genotypes on the right in the list must be recombinants, giving a map distance of 2 m.u. between the A/a locus and the molecular locus M1/M2. Hence, we now know the general location of the gene in the genome and can narrow its location down with more finely scaled approaches. Additional molecular markers can be mapped to each other, creating a map that can act like a series of stepping stones on the way to some gene with an interesting phenotype.

In our examples above, we have been dealing with what are effectively testcrosses. However, because molecular markers do not show dominance or recessiveness, and can be scored directly in molecular tests, often crosses that are not testcrosses can be assessed for recombinants. Such an analysis is shown in **Figure 4-15**. The figure shows experimental details of how the molecular marker "alleles" are

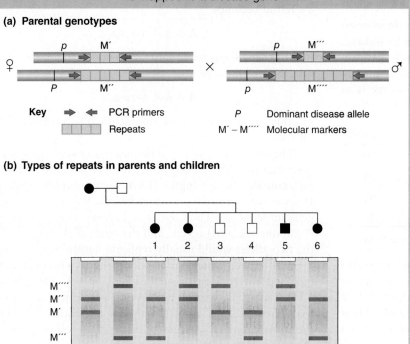

A molecular marker locus with variable number of short sequence repeats is mapped to a disease gene

(a) Parental genotypes

Key
→ ← PCR primers
▭▭▭ Repeats

P Dominant disease allele
M′ – *M″″* Molecular markers

(b) Types of repeats in parents and children

PCR products

FIGURE 4-15 A PCR banding pattern is shown for a family with six children, and this pattern is interpreted at the top of the illustration with the use of four differently sized molecular "alleles," M′ through M″″. One of these markers (M″) is probably linked in cis configuration to the disease allele *P*. (Note: This mating is not a testcross, yet it is informative about linkage.)

detected using a technique called the polymerase chain reaction (PCR). For now, simply recognize that PCR allows different numbers of tandem repeats to be detected as differently sized products on an electrophoretic gel (the details of both PCR and gel electrophoresis are discussed in Chapter 10).

KEY CONCEPT Loci of any DNA heterozygosity can be mapped and used as molecular chromosome markers or milestones.

4.4 USING THE CHI-SQUARE TEST TO INFER LINKAGE

LO 4.5 Apply the chi-square test to statistically assess whether or not the recombinant frequency is 50% in a particular data set.

The standard genetic test for linkage is a dihybrid testcross. Consider a general cross of that type, in which it is not known if the genes are linked or not:

$$A/a \cdot B/b \times a/a \cdot b/b$$

If there is *no* linkage—that is, if the genes assort independently—we have seen from the discussions in this chapter and Chapter 3 that the following phenotypic proportions are expected in progeny:

A B	0.25
A b	0.25
a B	0.25
a b	0.25

A cross of this type was made and the following phenotypes obtained in a progeny sample of 200.

A B	60
A b	37
a B	41
a b	62

There is clearly a deviation from the prediction of no linkage (which would have given the progeny numbers 50:50:50:50). The results suggest that the dihybrid was a cis configuration of linked genes, *A B/a b*, because the progeny *A B* and *a b* are in the majority. The recombinant frequency would be $(37 + 41)/200 = 78/200 = 39$ percent, or 39 m.u.

However, we know that chance deviations due to sampling error can provide results that resemble those produced by genetic processes; hence, we need the χ^2 (chi-square) test to help us calculate the probability of a chance deviation of this magnitude from a 1:1:1:1 ratio.

First, let us examine the allele ratios for both loci. These are 97:103 for *A:a*, and 101:99 for *B:b*. Such numbers are close to the 1:1 allele ratios expected from Mendel's first law, so skewed allele ratios cannot be responsible for the quite large deviations from the expected numbers of 50:50:50:50.

We must apply the χ^2 analysis to test a hypothesis of *no linkage*. If that hypothesis is rejected, we can infer linkage. (We cannot test a hypothesis of linkage directly because we have no way of predicting what recombinant frequency to test.) The calculation for testing lack of linkage is as follows:

Observed (O)	Expected (E)	O − E	(O − E)²	(O − E)²/E
60	50	10	100	2.00
37	50	−13	169	3.38
41	50	−9	81	1.62
62	50	12	144	2.88

$$\chi^2 = \sum (O - E)^2 / E \text{ for all classes} = 9.88$$

Since there are four genotypic classes, we must use $4 - 1 = 3$ degrees of freedom. Consulting the chi-square table in Chapter 3 (Table 3-1), we see our values of 9.88 and 3 df give a *p* value of ~ 0.025, or 2.5 percent. This is less than the standard cut-off value of 5 percent, so we can reject the hypothesis of no linkage. Hence, we are left with the conclusion that the genes are very likely linked, approximately 39 m.u. apart.

Notice, in retrospect, that it was important to make sure alleles were segregating 1:1 to avoid a *compound* hypothesis of 1:1 allele ratios *and* no linkage. If we rejected such a compound hypothesis, we would not know which part of it was responsible for the rejection.

KEY CONCEPT The chi-square test is useful in testing the significance of deviations from a 1:1:1:1 ratio in deducing linkage between two genes.

4.5 THE MOLECULAR MECHANISM OF CROSSING OVER

LO 4.6 Diagram how double-strand DNA breaks and heteroduplex formation can lead to a crossover at the molecular level.

In this chapter we have analyzed the genetic consequences of the cytologically visible process of crossing over without worrying about the mechanism of crossing over. However, crossing over is remarkable in itself as a molecular process: how can two large coiled molecules of DNA exchange segments with a precision so exact that no nucleotides are lost or gained?

Studies on fungal octads gave a clue. Although most octads show the expected 4:4 segregation of alleles such as 4A:4a, some rare octads show aberrant ratios. There are several types, but as an example we will use 5:3 octads (either 5A:3a or 5a:3A). Two things are peculiar about this ratio. First, there is one too many spores of one allele and one too few of the other. Second, there is a

nonidentical sister-spore pair. Normally, postmeiotic replication gives identical sister-spore pairs as follows: the *A A a a* tetrad becomes

A-A A-A a-a a-a

(the hyphens show sister spores). In contrast, an aberrant 5*A*:3*a* octad must be

A-A A-A A-a a-a

In other words, there is one nonidentical sister-spore pair (in red).

The observation of a nonidentical sister-spore pair suggests that the DNA of one of the final four meiotic homologs contains **heteroduplex DNA**. Heteroduplex DNA is DNA in which there is a mismatched nucleotide pair in the gene under study. The logic is as follows. If in a cross of $A \times a$, one allele (*A*) is G:C and the other allele (*a*) is A:T, the two alleles would usually replicate faithfully. However, a heteroduplex, which forms only rarely, would be a mismatched nucleotide pair such as G:T or A:C (effectively, a DNA molecule bearing both *A* and *a* information). Note that a heteroduplex involves only one nucleotide position: the surrounding DNA segment might be as follows, where the heteroduplex site is shown in red:

GCTAATGTTATTAG

CGATTATAATAATC

At replication to form an octad, a G:T heteroduplex would pull apart and replicate faithfully, with G bonding to C and A bonding to T. The result would be a nonidentical spore pair of G:C (allele *A*) and A:T (allele *a*).

Nonidentical sister spores (and aberrant octads generally) were found to be statistically correlated with crossing over in the region of the gene concerned, providing an important clue that crossing over might be based on the formation of heteroduplex DNA.

In the currently accepted model (follow it in **Figure 4-16**), the heteroduplex DNA and a crossover are both produced by a **double-stranded break** in the DNA of one of the chromatids participating in the crossover. Let's see how that works. Molecular studies show that broken ends of DNA will promote recombination between different chromatids. In step 1, both strands of a chromatid break in the same location. From the break, DNA is eroded at the 5' end of each broken strand, leaving both 3' ends single stranded (step 2). One of the single strands "invades" the DNA of the other participating chromatid; that is, it enters the center of the helix and base-pairs with its homologous sequence (step 3), displacing the other strand. Then the tip of the invading strand uses the adjacent sequence as a template for new polymerization, which proceeds by forcing the two resident strands of the helix apart (step 4). The displaced single-stranded loop hydrogen bonds with the other single strand (the blue one in Figure 4-16). If the invasion and strand displacement spans a site of heterozygosity (such as *A/a*), then a region of heteroduplex DNA

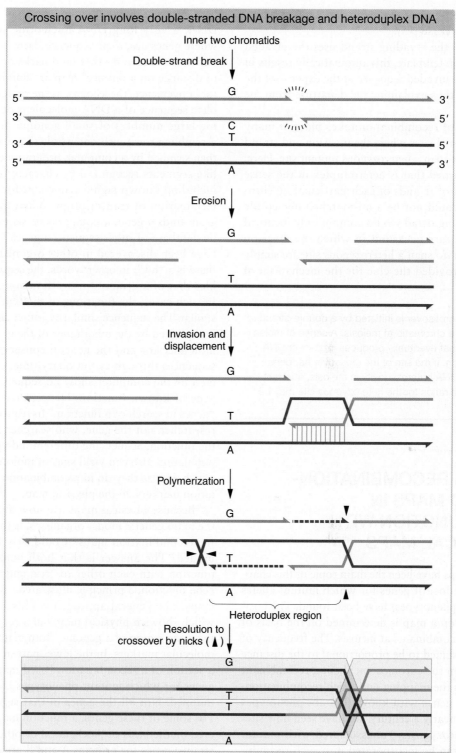

FIGURE 4-16 A molecular model of crossing over. Only the two chromatids (blue and red) participating in the crossover are shown. The 3′-to-5′ strand is placed on the inside of both for clarity. The chromatids differ at one site, GC, in one allele (perhaps allele *A*) and AT in the other (perhaps *a*). Only the outcome with mispaired heteroduplex DNA and a crossover are shown. The final crossover products are shaded in yellow and blue.

ANIMATED ART 🌀 SaplingPlus

A molecular model of crossing over

is formed. Replication also takes place from the other single-stranded end to fill the gap left by the invading strand (also shown on the upper blue strand in step 4 of Figure 4-16). The replicated ends are sealed, and the net result is a strange structure with two single-stranded junctions called *Holliday junctions* after their original proposer, Robin Holliday. These junctions are potential sites of single-strand breakage and reunion; two such events, shown by the darts in the figure, then lead to a complete double-stranded crossover (step 5).

Note that when the invading strand uses the invaded DNA as a replication template, this automatically results in an extra copy of the invaded sequence at the expense of the invading sequence, thus explaining the departure from the expected 4:4 ratio.

This same sort of recombination takes place at many different chromosomal sites where the invasion and strand displacement do *not* span a heterozygous mutant site. Here, DNA would be formed that is heteroduplex in the sense that it is composed of strands of each participating chromatid, but there would not be a mismatched nucleotide pair, and the resulting octad would contain only identical spore pairs. Those rare occasions in which the invasion and polymerization *do* span a heterozygous site are simply lucky cases that provided the clue for the mechanism of crossing over.

> **KEY CONCEPT** A crossover is initiated by a double-stranded break in the DNA of a chromatid at meiosis. A series of molecular events ensues that eventually produces crossover DNA molecules. In addition, if the site of the crossover happens to be near a site of DNA heterozygosity in meiosis, aberrant non-Mendelian allele ratios for the heterozygous site may be produced.

4.6 USING RECOMBINATION-BASED MAPS IN CONJUNCTION WITH PHYSICAL MAPS

Recombination maps have been the main topic of this chapter. They show the loci of genes for which mutant alleles (and their mutant phenotypes) have been found. The positions of these loci on a map is determined on the basis of the frequency of recombinants at meiosis. The frequency of recombinants is assumed to be proportional to the distance apart of two loci on the chromosome; hence, recombinant frequency becomes the mapping unit. Such recombination-based mapping of genes with known mutant phenotypes has been done for nearly a century. We have seen how sites of molecular heterozygosity (unassociated with mutant phenotypes) also can be incorporated into such recombination maps. Like any heterozygous site, these molecular markers are mapped by recombination and then used to navigate toward a gene of biological interest. We make the perfectly reasonable assumption that a recombination map represents the arrangement of genes on chromosomes, but, as stated earlier, these maps are really hypothetical constructs. In contrast, physical maps are as close to the real genome map as science can get.

The topic of **physical maps** will be examined more closely in Chapters 10 and 14, but we can foreshadow it

here. A physical map is simply a map of the actual genomic DNA, a very long DNA nucleotide sequence, showing where genes are, their sequence, how big they are, what is between them, and other landmarks of interest. The units of distance on a physical map are numbers of DNA bases; for convenience, the kilobase is the preferred unit. The complete sequence of a DNA molecule is obtained by sequencing large numbers of small genomic fragments and then assembling them into one whole sequence. The sequence is then scanned by a computer, programmed to look for gene-like segments recognized by characteristic base sequences including known signal sequences for the initiation and termination of transcription. When the computer's program finds a gene, it compares its sequence with the public database of other sequenced genes for which functions have been discovered in other organisms. In many cases, there is a "hit"; in other words, the sequence closely resembles that of a gene of known function in another species. In such cases, the functions of the two genes also may be similar. The sequence similarity (often close to 100 percent) is explained by the inheritance of the gene from some common ancestor and the general conservation of functional sequences through evolutionary time. Other genes discovered by the computer show no sequence similarity to any gene of known function. Hence, they can be considered "genes in search of a function." In reality, of course, it is the researcher, not the gene, who searches and who must find the function. Sequencing different individual members of a population also can yield sites of molecular heterozygosity, which, just as they do in recombination maps, act as orientation markers on the physical map.

Because physical maps are now available for most of the main genetic model organisms, is there really any need for recombination maps? Could they be considered outmoded? The answer is that both maps are used in conjunction with each other to "triangulate" in determining gene function, a principle illustrated earlier by the London maps. The general approach is illustrated in **Figure 4-17**, which shows a physical map and a recombination map of the same region of a genome. Both maps contain genes and molecular markers. In the lower part of Figure 4-17, we see a section of a recombination-based map, with positions of genes for which mutant phenotypes have been found and mapped. Not all the genes in that segment are included. For some of these genes, a function may have been discovered on the basis of biochemical or other studies of mutant strains; genes for proteins A and B are examples. The gene in the middle is a "gene of interest" that a researcher has found to affect the aspect of development being studied. To determine its function, the physical map can be useful. The genes in the physical map that are in the general region of the gene of interest on the recombination map become *candidate genes*, any one of which could be the gene of interest. Further studies are needed to narrow the choice to one. If that single case is a gene whose function is known for other organisms, then a function for the gene of interest is suggested. In this way, the phenotype mapped on

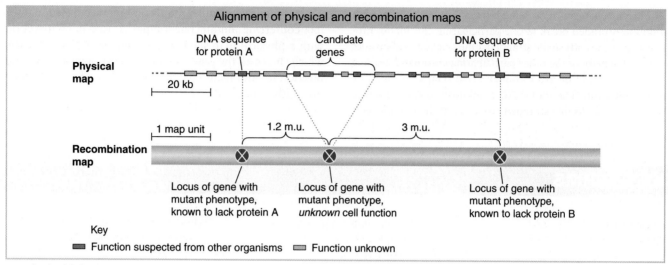

FIGURE 4-17 Comparison of relative positions on physical and recombination maps can connect phenotype with an unknown gene function.

the recombination map can be tied to a function deduced from the physical map. Molecular markers on both maps (not shown in Figure 4-17) can be aligned to help in the zeroing-in process. Hence, we see that both maps contain elements of function: the physical map shows a gene's possible action at the cellular level, whereas the recombination map contains information related to the effect of the gene at the phenotypic level. At some stage, the two have to be melded to understand the gene's contribution to the development of the organism.

There are several other genetic-mapping techniques, some of which we will encounter in Chapters 6, 18, and 19.

KEY CONCEPT The union of recombination and physical maps can ascribe biochemical function to a gene identified by its mutant phenotype.

SUMMARY

In a dihybrid testcross in *Drosophila*, Thomas Hunt Morgan found a deviation from Mendel's law of independent assortment. He postulated that the two genes were located on the same pair of homologous chromosomes. This relation is called linkage.

Linkage explains why the parental gene combinations stay together, but not how the recombinant (nonparental) combinations arise. Morgan postulated that, in meiosis, there may be a physical exchange of chromosome parts by a process now called crossing over. A result of the physical breakage and reunion of chromosome parts, crossing over takes place at the four-chromatid stage of meiosis. Thus, there are two types of meiotic recombination. Recombination by Mendelian independent assortment results in a recombinant frequency of 50 percent. Crossing over results in a recombinant frequency (RF) of generally less than 50 percent.

As Morgan studied more linked genes, he discovered many different values for recombinant frequency and wondered if these values corresponded to the actual distances between genes on a chromosome. Alfred Sturtevant, a student of Morgan's, developed a method of determining the distance between genes on a linkage map, based on the RF. The easiest way to measure RF is with a testcross of a dihybrid or trihybrid. RF values calculated as percentages can be used as map units to construct a chromosomal map showing the loci of the genes analyzed. In ascomycete fungi, centromeres also can be located on the map by measuring second-division segregation frequencies.

Single nucleotide differences in sequences, and differences in the number of repeating units, can be used as molecular markers for mapping genes.

Although the basic test for linkage is deviation from independent assortment, such a deviation may not be obvious in a testcross, and a statistical test is needed. The χ^2 test, which tells how often observations deviate from expectations purely by chance, is particularly useful in determining whether loci are linked.

The mechanism of crossing over is thought to start with a double-stranded break in one participating chromatid. Erosion leaves the ends single stranded. One single strand invades the double helix of the other participating chromatid, leading to the formation of heteroduplex DNA. Gaps are filled by polymerization. The molecular resolution of this structure becomes a full double-stranded crossover at the DNA level.

In genetics generally, the recombination-based map of loci conferring mutant phenotypes is used in conjunction with a physical map such as the complete DNA sequence, which shows all the gene-like sequences. Knowledge of gene position in both maps enables the melding of cellular function with a gene's effect on phenotype.

KEY TERMS

centimorgan (cM) (p. 120)
chromosome map (p. 115)
cis conformation (p. 117)
coefficient of coincidence (c.o.c.) (p. 124)
crossing over (p. 117)
crossover product (p. 117)
double-stranded break (p. 132)
first-division segregation pattern (M_I pattern) (p. 128)

genetic map unit (m.u.) (p. 120)
heteroduplex DNA (p. 132)
interference (p. 124)
linkage map (p. 120)
linked (p. 115)
locus (p. 115)
mapping function (p. 125)
molecular marker (p. 130)
octad (p. 128)
physical map (p. 134)

Poisson distribution (p. 125)
recombinant frequency (RF) (p. 120)
recombination map (p. 115)
second-division segregation pattern (M_{II} pattern) (p. 128)
three-point testcross (three factor-testcross) (p. 122)
trans conformation (p. 117)

SOLVED PROBLEMS

SOLVED PROBLEM 1

A human pedigree shows people affected with the rare nail–patella syndrome (misshapen nails and kneecaps) and gives the ABO blood-group genotype of each person (I^A, I^B, and i are alleles; I^A determines blood group A, I^B group B, and i group O). Both loci concerned are autosomal. Study the pedigree below.

a. Is the nail–patella syndrome a dominant or recessive phenotype? Give reasons to support your answer.

b. Is there evidence of linkage between the nail–patella gene and the gene for ABO blood type, as judged from this pedigree? Why or why not?

c. If there is evidence of linkage, then draw the alleles on the relevant homologs of the grandparents. If there is no evidence of linkage, draw the alleles on two homologous pairs.

d. According to your model, which generation II descendants are recombinants?

e. What is the best estimate of RF?

f. If man III-1 mates with a normal woman of blood type O, what is the probability that their first child will be blood type B with nail–patella syndrome?

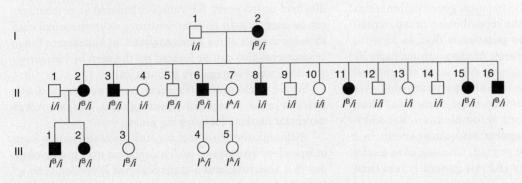

SOLUTION

a. Nail–patella syndrome is most likely dominant. We are told that it is a rare abnormality, and so the unaffected people marrying into the family are unlikely to carry a presumptive recessive allele for nail–patella syndrome. Let N be the causative allele. Then all people with the syndrome are heterozygotes N/n because all (probably including the grandmother) result from matings with n/n normal people. Notice that the syndrome appears in all three generations—another indication of dominant inheritance.

b. There is evidence of linkage. Notice that most of the affected people—those who carry the N allele—also carry the I^B allele; most likely, these alleles are linked on the same chromosome.

c.
$$\frac{n \qquad i}{n \qquad i} \times \frac{N \qquad I^B}{n \qquad i}$$

(The grandmother must carry both recessive alleles to produce offspring of genotype i/i and n/n.)

d. Notice that the grandparental mating is equivalent to a testcross; so the recombinants in generation II are

$$\text{II-5}: n\, I^B\, /n\, i \text{ and } \text{II-8}: N\, i/n\, i$$

whereas all others are nonrecombinants, being either $N\, I^B\, /n\, i$ or $n\, i/n\, i$.

e. Notice that the grandparental cross and the first two crosses in generation II are identical and are testcrosses. Three of the total 16 progeny are recombinant (II-5, II-8, and III-3). The cross of II-6 with II-7 is not a testcross, but the chromosomes donated from II-6 can be deduced to be nonrecombinant. Thus, RF = 3/18, which is 17 percent.

f. (III-1♂)
$$\frac{N \qquad I^B}{n \qquad i} \times \frac{n \qquad i}{n \qquad i} \begin{array}{l}(\text{normal} \\ \text{type O♀})\end{array}$$
$$\downarrow$$

Gametes

$$83.0\% \begin{cases} N\, I^B & 41.5\% \longleftarrow \text{nail–patella} \\ n\, i & 41.5\% \qquad\quad \text{blood type B} \end{cases}$$

$$17.0\% \begin{cases} N\, i & 8.5\% \\ n\, I^B & 8.5\% \end{cases}$$

The two parental classes are always equal, and so are the two recombinant classes. Hence, the probability that the first child will have nail–patella syndrome and blood type B is 41.5 percent.

SOLVED PROBLEM 2	**ANIMATED ART** SaplingPlus

Mapping a three point cross

The allele b gives *Drosophila* flies a black body, and b^+ gives brown, the wild-type phenotype. The allele wx of a separate gene gives waxy wings, and wx^+ gives nonwaxy, the wild-type phenotype. The allele cn of a third gene gives cinnabar eyes, and cn^+ gives red, the wild-

type phenotype. A female heterozygous for these three genes is testcrossed, and 1000 progeny are classified as follows: 5 wild type; 6 black, waxy, cinnabar; 69 waxy, cinnabar; 67 black; 382 cinnabar; 379 black, waxy; 48 waxy; and 44 black, cinnabar. Note that a progeny group may be specified by listing only the mutant phenotypes.

a. Explain these numbers.

b. Draw the alleles in their proper positions on the chromosomes of the triple heterozygote.

c. If appropriate according to your explanation, calculate interference.

SOLUTION

a. A general piece of advice is to be methodical. Here, it is a good idea to write out the genotypes that may be inferred from the phenotypes. The cross is a testcross of type

$$b^+/b \cdot wx^+/wx \cdot cn^+/cn \times b/b \cdot wx/wx \cdot cn/cn$$

Notice that there are distinct pairs of progeny classes in regard to frequency. Already, we can guess that the two largest classes represent parental chromosomes, that the two classes of about 68 represent single crossovers in one region, that the two classes of about 45 represent single crossovers in the other region, and that the two classes of about 5 represent double crossovers. We can write out the progeny as classes derived from the female's gametes, grouped as follows:

$b^+ \cdot wx^+ \cdot cn$	382
$b \cdot wx \cdot cn^+$	379
$b^+ \cdot wx \cdot cn$	69
$b \cdot wx^+ \cdot cn^+$	67
$b^+ \cdot wx \cdot cn^+$	48
$b \cdot wx^+ \cdot cn$	44
$b \cdot wx \cdot cn$	6
$b^+ \cdot wx^+ \cdot cn^+$	5
	1000

Listing the classes in this way confirms that the pairs of classes are in fact reciprocal genotypes arising from zero, one, or two crossovers.

At first, because we do not know the parents of the triple heterozygous female, it looks as if we cannot apply the definition of recombination in which gametic genotypes are compared with the two parental genotypes that form an individual fly. But, on reflection, the only parental types that make sense in regard to the data presented are $b^+/b^+ \cdot wx^+/wx^+ \cdot cn/cn$ and $b/b \cdot wx/wx \cdot cn^+/cn^+$ because these types represent the most common gametic classes.

Now, we can calculate the recombinant frequencies. For *b–wx*,

$$RF = \frac{69 + 67 + 48 + 44}{1000} = 22.8\%$$

for *b–cn*,

$$RF = \frac{48 + 44 + 6 + 5}{1000} = 10.3\%$$

and for *wx–cn*,

$$RF = \frac{69 + 67 + 6 + 5}{1000} = 14.7\%$$

The map is therefore

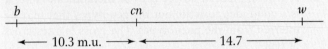

b. The parental chromosomes in the triple heterozygote are

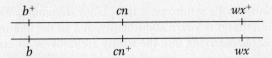

c. The expected number of double recombinants is $0.103 \times 0.147 \times 1000 = 15.141$. The observed number is $6 + 5 = 11$, and so interference can be calculated as

$$I = 1 - (11/15.141) = 1 - 0.726 = 0.274 = 27.4\%$$

SOLVED PROBLEM 3

A cross is made between a haploid strain of *Neurospora* of genotype *nic*$^+$ *ad* and another haploid strain of genotype *nic ad*$^+$. From this cross, a total of 1000 linear asci are isolated and categorized as in the table below. Map the *ad* and *nic* loci in relation to centromeres and to each other. (Note: Information from Box 4-2 is required for this problem.)

1	2	3	4	5	6	7
nic$^+$ · *ad*	*nic*$^+$ · *ad*$^+$	*nic*$^+$ · *ad*$^+$	*nic*$^+$ · *ad*	*nic*$^+$ · *ad*	*nic*$^+$ · *ad*$^+$	*nic*$^+$ · *ad*$^+$
nic$^+$ · *ad*	*nic*$^+$ · *ad*$^+$	*nic*$^+$ · *ad*$^+$	*nic*$^+$ · *ad*	*nic*$^+$ · *ad*	*nic*$^+$ · *ad*$^+$	*nic*$^+$ · *ad*$^+$
nic$^+$ · *ad*	*nic*$^+$ · *ad*$^+$	*nic*$^+$ · *ad*	*nic* · *ad*	*nic* · *ad*$^+$	*nic* · *ad*	*nic* · *ad*
nic$^+$ · *ad*	*nic*$^+$ · *ad*$^+$	*nic*$^+$ · *ad*	*nic* · *ad*	*nic* · *ad*$^+$	*nic* · *ad*	*nic* · *ad*
nic · *ad*$^+$	*nic* · *ad*	*nic* · *ad*$^+$	*nic*$^+$ · *ad*$^+$	*nic*$^+$ · *ad*	*nic*$^+$ · *ad*$^+$	*nic*$^+$ · *ad*
nic · *ad*$^+$	*nic* · *ad*	*nic* · *ad*$^+$	*nic*$^+$ · *ad*$^+$	*nic*$^+$ · *ad*	*nic*$^+$ · *ad*$^+$	*nic*$^+$ · *ad*
nic · *ad*$^+$	*nic* · *ad*	*nic* · *ad*	*nic* · *ad*$^+$	*nic* · *ad*$^+$	*nic* · *ad*	*nic*$^+$ · *ad*
nic · *ad*$^+$	*nic* · *ad*	*nic* · *ad*	*nic* · *ad*$^+$	*nic* · *ad*$^+$	*nic* · *ad*	*nic*$^+$ · *ad*
808	1	90	5	90	1	5

SOLUTION

What principles can we draw on to solve this problem? It is a good idea to begin by doing something straightforward, which is to calculate the two locus-to-centromere distances. We do not know if the *ad* and the *nic* loci are linked, but we do not need to know. The frequencies of the M$_{II}$ patterns for each locus give the distance from locus to centromere. (We can worry about whether it is the same centromere later.)

Remember that an M$_{II}$ pattern is any pattern that is not two blocks of four. Let's start with the distance between the *nic* locus and the centromere. All we have to do is add the ascus types 4, 5, 6, and 7, because all of them are M$_{II}$ patterns for the *nic* locus. The total is $5 + 90 + 1 + 5 = 101$ of 1000, or 10.1 percent. In this chapter, we have seen that to convert this percentage into map units, we must divide by 2, which gives 5.05 m.u.

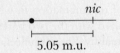

We do the same for the *ad* locus. Here, the total of the M$_{II}$ patterns is given by types 3, 5, 6, and 7 and is $90 + 90 + 1 + 5 = 186$ of 1000, or 18.6 percent, which is 9.3 m.u.

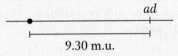

Now we have to put the two together and decide between the following alternatives, all of which are compatible with the preceding locus-to-centromere distances:

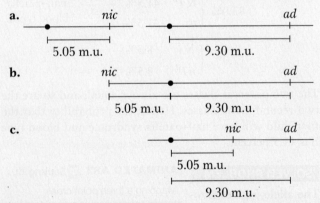

Here, a combination of common sense and simple analysis tells us which alternative is correct. First, an inspection of the asci reveals that the most common single type is the one labeled 1, which contains more than 80 percent of all the asci. This type contains only $nic^+ \cdot ad$ and $nic \cdot ad^+$ genotypes, and they are *parental* genotypes. So we know that recombination is quite low and the loci are certainly linked. This rules out alternative **a**.

Now consider alternative **c**. If this alternative were correct, a crossover between the centromere and the *nic* locus would generate not only an M_{II} pattern for that locus, but also an M_{II} pattern for the *ad* locus, because it is farther from the centromere than *nic* is. The ascus pattern produced by a crossover between *nic* and the centromere in alternative **c** should be

$$
\begin{array}{ll}
nic^+ & ad \\
nic^+ & ad \\
nic & ad^+ \\
nic & ad^+ \\
nic^+ & ad \\
nic^+ & ad \\
nic & ad^+ \\
nic & ad^+ \\
\end{array}
$$

Remember that the *nic* locus shows M_{II} patterns in asci types 4, 5, 6, and 7 (a total of 101 asci); of them, type 5 is the very one that we are talking about and contains 90 asci. Therefore, alternative **c** appears to be correct because ascus type 5 comprises about 90 percent of the M_{II} asci for the *nic* locus. This relation would not hold if alternative **b** were correct because crossovers on either side of the centromere would generate the M_{II} patterns for the *nic* and the *ad* loci independently.

Is the map distance from *nic* to *ad* simply $9.30 - 5.05 = 4.25$ m.u.? Close, but not quite. The best way of calculating map distances between loci is always by measuring the recombinant frequency. We could go through the asci and count all the recombinant ascospores, but using the formula $RF = T + NPD$ is simpler. The T asci are classes 3, 4, and 7, and the NPD asci are classes 2 and 6. Hence, $RF = [(100) + 2]/1000 = 5.2$ percent, or 5.2 m.u., and a better map is

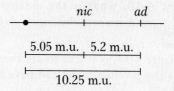

The reason for the underestimation of the *ad*-to-centromere distance calculated from the M_{II} frequency is the occurrence of double crossovers, which can produce an M_I pattern for *ad*, as in ascus type 4:

$$
\begin{array}{ll}
nic^+ & ad \\
nic^+ & ad \\
nic & ad \\
nic & ad \\
nic^+ & ad^+ \\
nic^+ & ad^+ \\
nic & ad^+ \\
nic & ad^+ \\
\end{array}
$$

PROBLEMS

Visit SaplingPlus for supplemental content. Problems with the 🐟 icon are available for review/grading. Problems with the ▭ icon have a Problem Solving Video. Problems with the 🧊 icon have an Unpacking the Problem exercise.

WORKING WITH THE FIGURES

(The first 20 questions require inspection of text figures.)

1. (i) In Figure 4-1, under which street do you think the red underground line lies?

   (ii) Using the underground map, determine where Oxford Circus is on the street map.

2. Redraw Figure 4-2 with the heterozygous alleles in the trans configuration.

3. In Figure 4-3, would there be any meiotic products that did not undergo a crossover in the meiosis illustrated? If so, what colors would they be in the color convention used?

4. In Figure 4-4, can you tell if the three crossovers on the long chromosome all involve the same two chromatids?

5. a. Redraw Figure 4-5 with the heterozygous alleles in trans configuration.

   b. Is the outcome from the upper case possible from a meiosis in which crossing over takes place at the four strand stage?

6. Redraw Figure 4-6 part *b* with one crossover occurring between chromatids 2 and 3, and the other between 2 and 4, and list the meiotic products.

7. Redraw Figure 4-7 with the heterozygous alleles in trans configuration, and state genotypes of the recombinants.

8. a. In Figure 4-8, some meiotic products are labeled parental. Which parent is being referred to in this terminology?

   b. Redraw the figure with the heterozygous alleles in trans configuration and then compare the recombinant genotypes with those from this figure.

9. a. In Figure 4-9, why is only locus *A* shown in a constant position?

   b. In an actual experiment 50 progeny are analyzed and 2 are found to be recombinant. What is the likely gene order?

10. In Figure 4-10, what is the mean frequency of crossovers per meiosis in the region *A–B*? The region *B–C*? 🔗

11. a. In Figure 4-11, is it true to say that from such a cross the product *v cv*$^+$ can have two different origins?

   b. Redraw the figure showing a double recombinant arising from a three-chromatid double crossover.

12. Redraw Figure 4-12 using parental chromosomes *A B C* and *a b c*.

13. Point to the approximate location of the wooly locus in Figure 4-13b.

14. a. In Figure 4-14, in the bottom row, four colors are labeled SCO. Why are they not all the same size (frequency)?

   b. Consider the three cases in which there are four progeny types and in words state the different diagnostics they represent.

   c. Consider the two cases in which there are eight progeny types, and in words state the different diagnostics they represent.

15. a. Using the conventions of Figure 4-15, draw parents and progeny classes from a cross

   $$P\ M'''/p\ M' \times p\ M'/p\ M''''$$

   b. Why are there no M'M'' individuals in the progeny?

   c. Are there any exceptions to the rule that P is associated with M''?

   d. In the gel, why are there only two DNA bands shown for each progeny individual?

16. a. For the figure in Box 4-1, what would be the RF between *A/a* and *B/b* in a cross in which purely by chance all meioses had four-strand double crossovers in that region?

   b. Redraw the two-crossover section, making the constant crossover between chromatids 1 and 3. Does this affect the main conclusion of that part of the figure?

17. For the first figure in Box 4-2, draw the arrangements of alleles in an octad from a similar meiosis in which the upper product of the first division is segregated in an upside-down manner at the second division.

18. For the second figure in Box 4-2, draw the outcome if each pair of arrow is "flipped." Does this change significantly affect the outcomes?

19. a. In Figure 4-16, let GC = *A* and AT = *a*, then draw the fungal octad that would result from the final structure (5).

   b. (Challenging) Insert some closely linked flanking markers into the diagram, say *P/p* to the left and *Q/q* to the right (assume either cis or trans arrangements). Assume neither of these loci show non-Mendelian segregation. Then draw the final octad based on the structure in part 5.

20. In Figure 4-17, suggest a way in which you might be able to decide between which of the five or six candidate genes that might actually be the central locus on the recombination map.

BASIC PROBLEMS

21. A plant of genotype

$$\frac{A \qquad B}{a \qquad b}$$

   is testcrossed with

$$\frac{a \qquad b}{a \qquad b}$$

   If the two loci are 10 m.u. apart, what proportion of progeny will be *AB/ab*? 🔗

22. The *A* locus and the *D* locus are so tightly linked that no recombination is ever observed between them. If *Ad/Ad* is crossed with *aD/aD* and the F$_1$ is intercrossed, what phenotypes will be seen in the F$_2$, and in what proportions?

23. The *R* and *S* loci are 35 m.u. apart. If a plant of genotype

$$\frac{R \qquad S}{r \qquad s}$$

   is selfed, what progeny phenotypes will be seen, and in what proportions?

24. The cross *E/E · F/F × e/e · f/f* is made, and the F$_1$ is then backcrossed with the recessive parent. The progeny genotypes are inferred from the phenotypes. The progeny genotypes, written as the gametic contributions of the heterozygous parent, are in the following proportions:

$$E \cdot F\ \tfrac{2}{6}$$
$$E \cdot f\ \tfrac{1}{6}$$
$$e \cdot F\ \tfrac{1}{6}$$
$$e \cdot f\ \tfrac{2}{6}$$

Explain these results.

25. A strain of *Neurospora* with the genotype $H \cdot I$ is crossed with a strain with the genotype $h \cdot i$. Half the progeny are $H \cdot I$, and the other half are $h \cdot i$. Explain how this outcome is possible.

26. A female animal with genotype $A/a \cdot B/b$ is crossed with a double-recessive male ($a/a \cdot b/b$). Their progeny include 442 $A/a \cdot B/b$, 458 $a/a \cdot b/b$, 46 $A/a \cdot b/b$, and 54 $a/a \cdot B/b$. Explain these results.

27. If $A/A \cdot B/B$ is crossed with $a/a \cdot b/b$ and the F_1 is testcrossed, what percentage of the testcross progeny will be $a/a \cdot b/b$ if the two genes are (a) unlinked; (b) completely linked (no crossing over at all); (c) 10 m.u. apart; (d) 24 m.u. apart?

28. In a haploid organism, the C and D loci are 8 m.u. apart. From a cross $C \, d \times c \, D$, give the proportion of each of the following progeny classes: (a) $C \, D$; (b) $c \, d$; (c) $C \, d$; (d) all recombinants combined.

29. A fruit fly of genotype $B \, R/b \, r$ is testcrossed with $b \, r/b \, r$. In 84 percent of the meioses, there are no chiasmata between the linked genes; in 16 percent of the meioses, there is one chiasma between the genes. What proportion of the progeny will be $B \, r/b \, r$?

30. A three-point testcross was made in corn. The results and a recombination analysis are shown in the display below, which is typical of three-point testcrosses (p = purple leaves, + = green; v = virus-resistant seedlings, + = sensitive; b = brown midriff to seed, + = plain). Study the display, and answer parts *a* through *c*.

P $\qquad$ $+/+ \cdot +/+ \cdot +/+ \times p/p \cdot v/v \cdot b/b$

Gametes $\qquad$ $+ \cdot + \cdot +$ $\qquad$ $p \cdot v \cdot b$

a. Determine which genes are linked.

b. Draw a map that shows distances in map units.

c. Calculate interference, if appropriate.

				Recombinant for		
Class	Progeny phenotypes	F_1 gametes	Numbers	*p-b*	*p-v*	*v-b*
1	gre sen pla	$+ \cdot + \cdot +$	3210			
2	pur res bro	$p \cdot v \cdot b$	3222			
3	gre res pla	$+ \cdot v \cdot +$	1024		R	R
4	pur sen bro	$p \cdot + \cdot b$	1044		R	R
5	pur res pla	$p \cdot v \cdot +$	690	R		R
6	gre sen bro	$+ \cdot + \cdot b$	678	R		R
7	gre res bro	$+ \cdot v \cdot b$	72	R	R	
8	pur sen pla	$p \cdot + \cdot +$	60	R	R	
		Total	10,000	1,500	2,200	3,436

UNPACKING PROBLEM 30

Before attempting a solution to this problem, try answering the following questions:

1. Sketch cartoon drawings of the P, F_1, and tester corn plants, and use arrows to show exactly how you would perform this experiment. Show where seeds are obtained.

2. Why do all the +'s look the same, even for different genes? Why does this not cause confusion?

3. How can a phenotype be purple and brown, for example, at the same time?

4. Is it significant that the genes are written in the order *p-v-b* in the problem?

5. What is a tester, and why is it used in this analysis?

6. What does the column marked "Progeny phenotypes" represent? In class 1, for example, state exactly what "gre sen pla" means.

7. What does the line marked "Gametes" represent, and how is it different from the column marked "F_1 gametes"? In what way is comparison of these two types of gametes relevant to recombination?

8. Which meiosis is the main focus of study? Label it on your drawing.

9. Why are the gametes from the tester not shown?

10. Why are there only eight phenotypic classes? Are there any classes missing?

11. What classes (and in what proportions) would be expected if all the genes are on separate chromosomes?

12. To what do the four pairs of class sizes (very big, two intermediates, very small) correspond?

13. What can you tell about gene order simply by inspecting the phenotypic classes and their frequencies?

14. What will be the expected phenotypic class distribution if only two genes are linked?

15. What does the word "point" refer to in a three-point testcross? Does this word usage imply linkage? What would a four-point testcross be like?

16. What is the definition of *recombinant*, and how is it applied here?

17. What do the "Recombinant for" columns mean?

18. Why are there only three "Recombinant for" columns?

19. What do the R's mean, and how are they determined?

20. What do the column totals signify? How are they used?

21. What is the diagnostic test for linkage?

22. What is a map unit? Is it the same as a centimorgan?

23. In a three-point testcross such as this one, why are the F_1 and the tester not considered to be parental in calculating recombination? (They *are* parents in one sense.)

24. What is the formula for interference? How are the "expected" frequencies calculated in the coefficient-of-coincidence formula?

25. Why does part *c* of the problem say "if appropriate"?

26. How much work is it to obtain such a large progeny size in corn? Which of the three genes would take the most work to score? Approximately how many progeny are represented by one corncob?

Now try to solve the problem. If you are unable to do so, try to identify the obstacle and write a sentence or two describing your difficulty. Then go back to the expansion questions and see if any of them relate to your difficulty. If this approach does not work, inspect the Learning Objectives and Key Concepts of this chapter and ask yourself which might be relevant to your difficulty.

31. You have a *Drosophila* line that is homozygous for autosomal recessive alleles *a*, *b*, and *c*, linked in that order. You cross females of this line with males homozygous for the corresponding wild-type alleles. You then cross the F_1 heterozygous males with their heterozygous sisters. You obtain the following F_2 phenotypes (where letters denote recessive phenotypes and pluses denote wild-type phenotypes): $1364 + + +, 365\ a\ b\ c, 87\ a\ b +, 84 + + c, 47\ a + +, 44 + b\ c, 5\ a + c$, and $4 + b +$.

    a. What is the recombinant frequency between *a* and *b*? Between *b* and *c*? (Remember, there is no crossing over in Drosophila males.)

    b. What is the coefficient of coincidence?

32. R. A. Emerson crossed two different pure-breeding lines of corn and obtained a phenotypically wild-type F_1 that was heterozygous for three alleles that determine recessive phenotypes: *an* determines anther; *br*, brachytic; and *f*, fine. He testcrossed the F_1 with a tester that was homozygous recessive for the three genes and obtained these progeny phenotypes: 355 anther; 339 brachytic, fine; 88 completely wild type; 55 anther, brachytic, fine; 21 fine; 17 anther, brachytic; 2 brachytic; 2 anther, fine.

    a. What were the genotypes of the parental lines?

    b. Draw a linkage map for the three genes (include map distances).

    c. Calculate the interference value.

33. Chromosome 3 of corn carries three loci (*b* for plant-color booster, *v* for virescent, and *lg* for liguleless). A testcross of triple recessives with F_1 plants heterozygous for the three genes yields progeny having the following genotypes: $305 + v\ lg, 275\ b + +, 128\ b + lg, 112 + v +, 74 + + lg, 66\ b\ v +, 22 + + +$, and $18\ b\ v\ lg$. Give the gene sequence on the chromosome, the map distances between genes, and the coefficient of coincidence.

34. Groodies are useful (but fictional) haploid organisms that are pure genetic tools. A wild-type groody has a fat body, a long tail, and flagella. Mutant lines are known that have thin bodies, are tailless, or do not have flagella. Groodies can mate with one another (although they are so shy that we do not know how) and produce recombinants. A wild-type groody mates with a thin-bodied groody lacking both tail and flagella. The 1000 baby groodies produced are classified as shown in the illustration here. Assign genotypes, and map the three genes. (Problem 34 is from Burton S. Guttman.)

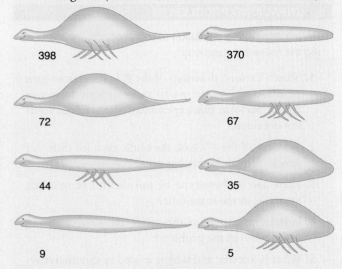

398   370

72   67

44   35

9   5

35. In *Drosophila*, the allele dp^+ determines long wings and *dp* determines short ("dumpy") wings. At a separate

locus, e^+ determines gray body and e determines ebony body. Both loci are autosomal. The following crosses were made, starting with pure-breeding parents:

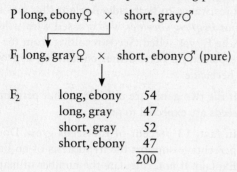

P long, ebony♀ × short, gray♂

F₁ long, gray♀ × short, ebony♂ (pure)

F₂		
long, ebony	54	
long, gray	47	
short, gray	52	
short, ebony	47	
	200	

Use the χ^2 test to determine if these loci are linked. In doing so, indicate (**a**) the hypothesis, (**b**) calculation of χ^2, (**c**) p value, (**d**) what the p value means, (**e**) your conclusion, and (**f**) the inferred chromosomal constitutions of parents, F₁, tester, and progeny.

36. The mother of a family with 10 children has blood type Rh⁺. She also has a very rare condition (elliptocytosis, phenotype E) that causes red blood cells to be oval rather than round in shape but that produces no adverse clinical effects. The father is Rh⁻ (lacks the Rh⁺ antigen) and has normal red blood cells (phenotype e). The children are 1 Rh⁺ e, 4 Rh⁺ E, and 5 Rh⁻ e. Information is available on the mother's parents, who are Rh⁺ E and Rh⁻ e. One of the 10 children (who is Rh⁺ E) marries someone who is Rh⁺ e, and they have an Rh⁺ E child.

a. Draw the pedigree of this whole family.

b. Is the pedigree in agreement with the hypothesis that the Rh⁺ allele is dominant and Rh⁻ is recessive?

c. What is the mechanism of transmission of elliptocytosis?

d. Could the genes governing the E and Rh phenotypes be on the same chromosome? If so, estimate the map distance between them, and comment on your result.

37. From several crosses of the general type $A/A \cdot B/B \times a/a \cdot b/b$, the F₁ individuals of type $A/a \cdot B/b$ were testcrossed with $a/a \cdot b/b$. The results are as follows:

	Testcross progeny			
Testcross of F₁ from cross	$A/a \cdot B/b$	$a/a \cdot b/b$	$A/a \cdot b/b$	$a/a \cdot B/b$
1	310	315	287	288
2	36	38	23	23
3	360	380	230	230
4	74	72	50	44

For each set of progeny, use the χ^2 test to decide if there is evidence of linkage.

38. In the two pedigrees diagrammed here, a vertical bar in a symbol stands for steroid sulfatase deficiency, and a horizontal bar stands for ornithine transcarbamylase deficiency.

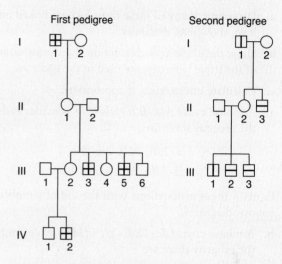

a. Is there any evidence in these pedigrees that the genes determining the deficiencies are linked?

b. If the genes are linked, is there any evidence in the pedigree of crossing over between them?

c. Assign genotypes of these individuals as far as possible.

39. In the accompanying pedigree, the vertical lines stand for protan color blindness, and the horizontal lines stand for deutan color blindness. These are separate conditions causing different misperceptions of colors; each is determined by a separate gene.

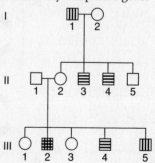

a. Does the pedigree show any evidence that the genes are linked?

b. If there is linkage, does the pedigree show any evidence of crossing over?

Explain your answers to parts *a* and *b* with the aid of the diagram.

c. Can you calculate a value for the recombination between these genes? Is this recombination by independent assortment or by crossing over?

40. In corn, a triple heterozygote was obtained carrying the mutant alleles *s* (shrunken), *w* (white aleurone), and *y* (waxy endosperm), all paired with their normal wild-type

alleles. This triple heterozygote was testcrossed, and the progeny contained 116 shrunken, white; 4 fully wild type; 2538 shrunken; 601 shrunken, waxy; 626 white; 2708 white, waxy; 2 shrunken, white, waxy; and 113 waxy.

a. Determine if any of these three loci are linked and, if so, show map distances.

b. Show the allele arrangement on the chromosomes of the triple heterozygote used in the testcross.

c. Calculate interference, if appropriate.

41. a. A mouse cross $A/a \cdot B/b \times a/a \cdot b/b$ is made, and in the progeny there are

$$25\%\ A/a \cdot B/b,\quad 25\%\ a/a \cdot b/b,$$
$$25\%\ A/a \cdot b/b,\quad 25\%\ a/a \cdot B/b$$

Explain these proportions with the aid of simplified meiosis diagrams.

b. A mouse cross $C/c \cdot D/d \times c/c \cdot d/d$ is made, and in the progeny there are

$$45\%\ C/c \cdot d/d,\quad 45\%\ c/c \cdot D/d,$$
$$5\%\ c/c \cdot d/d,\quad 5\%\ C/c \cdot D/d$$

Explain these proportions with the aid of simplified meiosis diagrams.

42. In the tiny model plant *Arabidopsis*, the recessive allele *hyg* confers seed resistance to the drug hygromycin, and *her*, a recessive allele of a different gene, confers seed resistance to herbicide. A plant that was homozygous $hyg/hyg \cdot her/her$ was crossed with wild type, and the F_1 was selfed. Seeds resulting from the F_1 self were placed on petri dishes containing hygromycin and herbicide.

a. If the two genes are unlinked, what percentage of seeds are expected to grow?

b. In fact, 13 percent of the seeds grew. Does this percentage support the hypothesis of no linkage? Explain. If not, calculate the number of map units between the loci.

c. Under your hypothesis, if the F_1 is testcrossed, what proportion of seeds will grow on the medium containing hygromycin and herbicide?

43. In a diploid organism of genotype $A/a\,;B/b\,;D/d$, the allele pairs are all on different chromosome pairs. The diagrams below purport to show anaphases ("pulling apart" stages) in individual cells. State whether each drawing represents mitosis, meiosis I, or meiosis II or is impossible for this particular genotype.

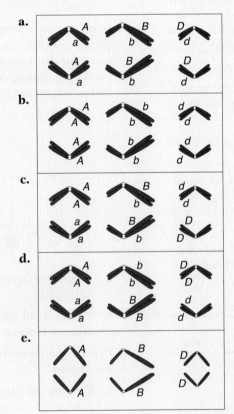

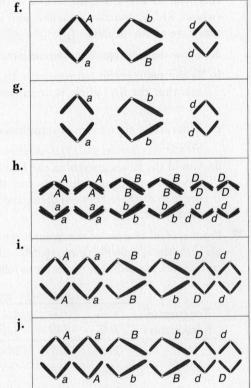

44. The *Neurospora* cross *al-2⁺* × *al-2* is made. A linear tetrad analysis reveals that the second-division segregation frequency is 8 percent.

    a. Draw two examples of second-division segregation patterns in this cross.

    b. What can be calculated by using the 8 percent value?

(Note: Information from Box 4-2 is required for this problem.)

45. From the fungal cross *arg-6 · al-2* × *arg-6⁺ · al-2⁺*, what will the spore genotypes be in unordered tetrads that are **(a)** parental ditypes? **(b)** tetratypes? **(c)** nonparental ditypes? (Note: Information from Box 4-2 is required for this problem.)

46. For a certain chromosomal region, the mean number of crossovers at meiosis is calculated to be two per meiosis. In that region, what proportion of meioses are predicted to have **(a)** no crossovers? **(b)** one crossover? **(c)** two crossovers? (Note: Information from Box 4-1 is required for this problem.)

47. A *Neurospora* cross was made between a strain that carried the mating-type allele *A* and the mutant allele *arg-1* and another strain that carried the mating-type allele *a* and the wild-type allele for *arg-1*(+). Four hundred linear octads were isolated, and they fell into the seven classes given in the table below. For simplicity, they are shown as tetrads. (Note: Information from Box 4-2 is required for this problem.)

    a. Deduce the linkage arrangement of the mating-type locus and the *arg-1* locus. Include the centromere or centromeres on any map that you draw. Label *all* intervals in map units.

    b. Diagram the meiotic divisions that led to class 6. Label clearly.

1	2	3	4	5	6	7
A · arg	*A · +*	*A · arg*	*A · arg*	*A · arg*	*A · +*	*A · +*
A · arg	*A · +*	*A · +*	*a · arg*	*a · +*	*a · arg*	*a · arg*
a · +	*a · arg*	*a · arg*	*A · +*	*A · arg*	*A · +*	*A · arg*
a · +	*a · arg*	*a · +*	*a · +*	*a · +*	*a · arg*	*a · +*
127	125	100	36	2	4	6

UNPACKING PROBLEM 47

Before attempting a solution to this problem, try answering the following questions:

1. Are fungi generally haploid or diploid?

2. How many ascospores are in the ascus of *Neurospora*? Does your answer match the number presented in this problem? Explain any discrepancy.

3. What is mating type in fungi? How do you think it is determined experimentally?

4. Do the symbols *A* and *a* have anything to do with dominance and recessiveness?

5. What does the symbol *arg-1* mean? How would you test for this genotype?

6. How does the *arg-1* symbol relate to the symbol +?

7. What does the expression *wild type* mean?

8. What does the word *mutant* mean?

9. Does the biological function of the alleles shown have anything to do with the solution of this problem?

10. What does the expression *linear octad analysis* mean?

11. In general, what more can be learned from linear tetrad analysis that cannot be learned from unordered tetrad analysis?

12. How is a cross made in a fungus such as *Neurospora*? Explain how to isolate asci and individual ascospores. How does the term *tetrad* relate to the terms *ascus* and *octad*?

13. Where does meiosis take place in the *Neurospora* life cycle? (Show it on a diagram of the life cycle.)

14. What does Problem 47 have to do with meiosis?

15. Can you write out the genotypes of the two parental strains?

16. Why are only four genotypes shown in each class?

17. Why are there only seven classes? How many ways have you learned for classifying tetrads generally? Which of these classifications can be applied to both linear and unordered tetrads? Can you apply these classifications to the tetrads in this problem? (Classify each class in as many ways as possible.) Can you think of more possibilities in this cross? If so, why are they not shown?

18. Do you think there are several different spore orders within each class? Why would these different spore orders not change the class?

19. Why is the following class not listed?

        a · +   *A · arg*

        a · +   *A · arg*

20. What does the expression *linkage arrangement* mean?

21. What is a genetic *interval*?

22. Why does the problem state "centromere or centromeres" and not just "centromere"? What is the general method for mapping centromeres in tetrad analysis?

23. What is the total frequency of $A \cdot +$ ascospores? (Did you calculate this frequency by using a formula or by inspection? Is this a recombinant genotype? If so, is it the only recombinant genotype?)

24. The first two classes are the most common and are approximately equal in frequency. What does this information tell you? What is their content of parental and recombinant genotypes?

Now try to solve the problem. If you are unable to do so, try to identify the obstacle and write a sentence or two describing your difficulty. Then go back to the expansion questions and see if any of them relate to your difficulty. If this approach does not work, inspect the Learning Objectives and Key Concepts of this chapter and ask yourself which might be relevant to your difficulty.

48. A geneticist studies 11 different pairs of *Neurospora* loci by making crosses of the type $a \cdot b \times a^+ \cdot b^+$ and then analyzing 100 linear asci from each cross. For the convenience of making a table, the geneticist organizes the data as if all 11 pairs of genes had the same designation—*a* and *b*—as shown below. For each cross, map the loci in relation to each other and to centromeres. (Note: Information from Box 4-2 is required for this problem.)

49. Three different crosses in *Neurospora* are analyzed on the basis of unordered tetrads. Each cross combines a different pair of linked genes. The results are shown in the following table:

Cross	Parents (%)	Parental ditypes (%)	Tetra-types (%)	Non-parental ditypes (%)
1	$a \cdot b^+ \times a^+ \cdot b$	51	45	4
2	$c \cdot d^+ \times c^+ \cdot d$	64	34	2
3	$e \cdot f^+ \times e^+ \cdot f$	45	50	5

For each cross, calculate

a. the frequency of recombinants (RF).

b. the uncorrected map distance, based on RF.

c. the corrected map distance, based on tetrad frequencies.

d. the corrected map distance, based on the mapping function.

(Note: Information from both Box 4-1 and Box 4-2 is required for this problem.)

50. On *Neurospora* chromosome 4, the *leu3* gene is just to the left of the centromere and always segregates at the first division, whereas the *cys2* gene is to the right of the centromere and shows a second-division segregation frequency of 16 percent. In a cross between a *leu3* strain and a *cys2* strain, calculate the predicted frequencies of the seven classes of linear tetrads shown on the next page, where $l = leu3$ and $c = cys2$. Ignore double and other multiple crossovers. (Note: Information from Box 4-2 is required for this problem.)

Number of asci of type

Cross	$a \cdot b$ / $a \cdot b$ / $a^+ \cdot b^+$ / $a^+ \cdot b^+$	$a \cdot b^+$ / $a \cdot b^+$ / $a^+ \cdot b$ / $a^+ \cdot b$	$a \cdot b$ / $a \cdot b^+$ / $a^+ \cdot b^+$ / $a^+ \cdot b$	$a \cdot b$ / $a^+ \cdot b$ / $a^+ \cdot b^+$ / $a \cdot b^+$	$a \cdot b$ / $a^+ \cdot b^+$ / $a^+ \cdot b^+$ / $a \cdot b$	$a \cdot b^+$ / $a^+ \cdot b$ / $a^+ \cdot b$ / $a \cdot b^+$	$a \cdot b^+$ / $a^+ \cdot b$ / $a^+ \cdot b^+$ / $a \cdot b$
1	34	34	32	0	0	0	0
2	84	1	15	0	0	0	0
3	55	3	40	0	2	0	0
4	71	1	18	1	8	0	1
5	9	6	24	22	8	10	20
6	31	0	1	3	61	0	4
7	95	0	3	2	0	0	0
8	6	7	20	22	12	11	22
9	69	0	10	18	0	1	2
10	16	14	2	60	1	2	5
11	51	49	0	0	0	0	0

(i) *l c* (ii) *l* + (iii) *l c* (iv) *l c* (v) *l c* (vi) *l* + (vii) *l* +

l c	*l* +	*l* +	+*c*	++	+*c*	+*c*
++	+*c*	++	++	++	+*c*	++
++	+*c*	+*c*	*l* +	*l c*	*l* +	*l c*

51. A rice breeder obtained a triple heterozygote carrying the three recessive alleles for albino flowers (*al*), brown awns (*b*), and fuzzy leaves (*fu*), all paired with their normal wild-type alleles. This triple heterozygote was testcrossed. The progeny phenotypes were

170	wild type	710	albino
150	albino, brown, fuzzy	698	brown, fuzzy
5	brown	42	fuzzy
3	albino, fuzzy	38	albino, brown

a. Are any of the genes linked? If so, draw a map labeled with map distances. (Do not bother with a correction for multiple crossovers.)

b. The triple heterozygote was originally made by crossing two pure lines. What were their genotypes?

52. In a fungus, a proline mutant (*pro*) was crossed with a histidine mutant (*his*). A nonlinear tetrad analysis gave the following results:

+	+	+	+	+	*his*
+	+	+	*his*	+	*his*
pro	*his*	*pro*	+	*pro*	+
pro	*his*	*pro*	*his*	*pro*	+
	6		82		112

a. Are the genes linked or not?

b. Draw a map (if linked) or two maps (if not linked), showing map distances based on straightforward recombinant frequency where appropriate.

c. If there is linkage, correct the map distances for multiple crossovers. Choose one approach only. (Note: Information from Box 4-1 is required for this part of the problem.)

53. In the fungus *Neurospora*, a strain that is auxotrophic for thiamine (mutant allele *t*) was crossed with a strain that is auxotrophic for methionine (mutant allele *m*). Linear asci were isolated and classified into the following groups:

Spore pair	Ascus types					
1 and 2	*t* +	*t* +	*t* +	*t* +	*t m*	*t m*
3 and 4	*t* +	*t m*	+*m*	++	*t m*	++
5 and 6	+*m*	++	*t* +	*t m*	++	*t* +
7 and 8	+*m*	+*m*	+*m*	+*m*	++	+*m*
Number	260	76	4	54	1	5

a. Determine the linkage relations of these two genes to their centromere(s) and to each other. Specify distances in map units.

b. Draw a diagram to show the origin of the ascus type with only one single representative (second from right).

(Note: Information from Box 4-2 is required for this problem.)

54. A corn geneticist wants to obtain a corn plant that has the three dominant phenotypes: anthocyanin (*A*), long tassels (*L*), and dwarf plant (*D*). In her collection of pure lines, the only lines that bear these alleles are *AA LL dd* and *aa ll DD*. She also has the fully recessive line *aa ll dd*. She decides to intercross the first two and testcross the resulting hybrid to obtain in the progeny a plant of the desired phenotype (which would have to be *Aa Ll Dd* in this case). She knows that the three genes are linked in the order written, that the distance between the *A/a* and the *L/l* loci is 16 m.u., and that the distance between the *L/l* and the *D/d* loci is 24 m.u.

a. Draw a diagram of the chromosomes of the parents, the hybrid, and the tester.

b. Draw a diagram of the crossover(s) necessary to produce the desired genotype.

c. What percentage of the testcross progeny will be of the phenotype that she needs?

d. What assumptions did you make (if any)?

55. In the model plant *Arabidopsis thaliana*, the following alleles were used in a cross:

T = presence of trichomes	*t* = absence of trichomes
D = tall plants	*d* = dwarf plants
W = waxy cuticle	*w* = nonwaxy
A = presence of purple anthocyanin pigment	*a* = absence (white)

The *T/t* and *D/d* loci are linked 26 m.u. apart on chromosome 1, whereas the *W/w* and *A/a* loci are linked 8 m.u. apart on chromosome 2.

A pure-breeding double-homozygous recessive trichomeless nonwaxy plant is crossed with another pure-breeding double-homozygous recessive dwarf white plant.

a. What will be the appearance of the F_1?

b. Sketch the chromosomes 1 and 2 of the parents and the F_1, showing the arrangement of the alleles.

c. If the F_1 is testcrossed, what proportion of the progeny will have all four recessive phenotypes?

56. In corn, the cross *WW ee FF × ww EE ff* is made. The three loci are linked as follows:

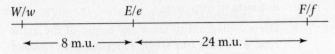

W/w ————————————— E/e ——————————————— F/f

◄—— 8 m.u. ——► ◄————— 24 m.u. —————►

Assume no interference.

a. If the F₁ is testcrossed, what proportion of progeny will be *ww ee ff*?

b. If the F₁ is selfed, what proportion of progeny will be *ww ee ff*?

57. The fungal cross + · + × c · m was made, and *nonlinear* (unordered) tetrads were collected. The results were

+ +	+ +	+ m
+ +	+ m	+ m
c m	c +	c +
c m	c m	c +
Total 112	82	6

a. From these results, calculate a simple recombinant frequency.

b. Compare the Haldane mapping function and the Perkins formula in their conversions of the RF value into a "corrected" map distance.

c. In the derivation of the Perkins formula, only the possibility of meioses with zero, one, and two crossovers was considered. Could this limit explain any discrepancy in your calculated values? Explain briefly (no calculation needed).

(Note: Information from Box 4-1 is required for this problem.)

58. In mice, the following alleles were used in a cross:

W = waltzing gait   w = nonwaltzing gait

G = normal gray color   g = albino

B = bent tail   b = straight tail

A waltzing gray bent-tailed mouse is crossed with a nonwaltzing albino straight-tailed mouse and, over several years, the following progeny totals are obtained:

waltzing	gray	bent	18
waltzing	albino	bent	21
nonwaltzing	gray	straight	19
nonwaltzing	albino	straight	22
waltzing	gray	straight	4
waltzing	albino	straight	5
nonwaltzing	gray	bent	5
nonwaltzing	albino	bent	6
Total			100

a. What were the genotypes of the two parental mice in the cross?

b. Draw the chromosomes of the parents.

c. If you deduced linkage, state the map unit value or values and show how they were obtained.

59. Consider the *Neurospora* cross +; + × f; p

It is known that the +/f locus is very close to the centromere on chromosome 7—in fact, so close that there are never any second-division segregations. It is also known that the +/p locus is on chromosome 5, at such a distance that there is usually an average of 12 percent second-division segregations. With this information, what will be the proportion of octads that are

a. parental ditypes showing M₁ patterns for both loci?

b. nonparental ditypes showing M₁ patterns for both loci?

c. tetratypes showing an M₁ pattern for +/f and an M₁₁ pattern for +/p?

d. tetratypes showing an M₁₁ pattern for +/f and an M₁ pattern for +/p?

(Note: Information from Box 4-2 is required for this problem.)

60. In a haploid fungus, the genes *al-2* and *arg-6* are 30 m.u. apart on chromosome 1, and the genes *lys-5* and *met-1* are 20 m.u. apart on chromosome 6. In a cross

al-2 +; + met-1 × + arg-6; lys-5 +

what proportion of progeny would be prototrophic + +; + +?

61. The recessive alleles *k* (kidney-shaped eyes instead of wild-type round), *c* (cardinal-colored eyes instead of wild-type red), and *e* (ebony body instead of wild-type gray) identify three genes on chromosome 3 of *Drosophila*. Females with kidney-shaped, cardinal-colored eyes were mated with ebony males. The F₁ was wild type. When F₁ females were testcrossed with *kk cc ee* males, the following progeny phenotypes were obtained:

k	c	e	3
k	c	+	876
k	+	e	67
k	+	+	49
+	c	e	44
+	c	+	58
+	+	e	899
+	+	+	4
Total			2000

a. Determine the order of the genes and the map distances between them.

b. Draw the chromosomes of the parents and the F_1.

c. Calculate interference and say what you think of its significance.

62. From parents of genotypes $A/A \cdot B/B$ and $a/a \cdot b/b$, a dihybrid was produced. In a testcross of the dihybrid, the following seven progenies were obtained:

$$A/a \cdot B/b, a/a \cdot b/b, A/a \cdot B/b, A/a \cdot b/b,$$
$$a/a \cdot b/b, A/a \cdot B/b, \text{ and } a/a \cdot B/b$$

Do these results provide convincing evidence of linkage?

CHALLENGING PROBLEMS

63. Use the Haldane map function to calculate the corrected map distance in cases where the measured RF = 5%, 10%, 20%, 30%, and 40%. Sketch a graph of RF against corrected map distance, and use it to answer the question, When should one use a map function? (Note: Information from Box 4-1 is required for this problem.)

64. An individual heterozygous for four genes, $A/a \cdot B/b \cdot C/c \cdot D/d$, is testcrossed with $a/a \cdot b/b \cdot c/c \cdot d/d$, and 1000 progeny are classified by the gametic contribution of the heterozygous parent as follows:

$a \cdot B \cdot C \cdot D$	42
$A \cdot b \cdot c \cdot d$	43
$A \cdot B \cdot C \cdot d$	140
$a \cdot b \cdot c \cdot D$	145
$a \cdot B \cdot c \cdot D$	6
$A \cdot b \cdot C \cdot d$	9
$A \cdot B \cdot c \cdot d$	305
$a \cdot b \cdot C \cdot D$	310

a. Which genes are linked?

b. If two pure-breeding lines had been crossed to produce the heterozygous individual, what would their genotypes have been?

c. Draw a linkage map of the linked genes, showing the order and the distances in map units.

d. Calculate an interference value, if appropriate.

65. An autosomal allele N in humans causes abnormalities in nails and patellae (kneecaps) called the nail–patella syndrome. Consider matings in which one partner has the nail–patella syndrome and blood type A and the other partner has normal nails and patellae and blood type O. These matings produce some children who have both the nail–patella syndrome and blood type A. Assume that unrelated children from this phenotypic group mature, mate, and have children. Four phenotypes are observed in the following percentages in this second generation:

nail–patella syndrome, blood type A	66%
normal nails and patellae, blood type O	16%
normal nails and patellae, blood type A	9%
nail–patella syndrome, blood type O	9%

Fully analyze these data, explaining the relative frequencies of the four phenotypes. (See pages 156–157 for the genetic basis of these blood types.)

66. Assume that three pairs of alleles are found in *Drosophila*: x^+ and x, y^+ and y, and z^+ and z. As shown by the symbols, each non-wild-type allele is recessive to its wild-type allele. A cross between females heterozygous at these three loci and wild-type males yields progeny having the following genotypes: 1010 $x^+ \cdot y^+ \cdot z^+$ females, 430 $x \cdot y^+ \cdot z$ males, 441 $x^+ \cdot y \cdot z^+$ males, 39 $x \cdot y \cdot z$ males, 32 $x^+ \cdot y^+ \cdot z$ males, 30 $x^+ \cdot y^+ \cdot z^+$ males, 27 $x \cdot y \cdot z^+$ males, 1 $x^+ \cdot y \cdot z$ male, and 0 $x \cdot y^+ \cdot z^+$ males.

a. On what chromosome of *Drosophila* are the genes carried?

b. Draw the relevant chromosomes in the heterozygous female parent, showing the arrangement of the alleles.

c. Calculate the map distances between the genes and the coefficient of coincidence.

67. The five sets of data given in the following table represent the results of testcrosses using parents with the same alleles but in different combinations. Determine the order of genes by inspection—that is, without calculating recombination values. Recessive phenotypes are symbolized by lowercase letters and dominant phenotypes by pluses.

Phenotypes observed in 3-point testcross	Data sets				
	1	2	3	4	5
+ + +	317	1	30	40	305
+ + c	58	4	6	232	0
+ b +	10	31	339	84	28
+ b c	2	77	137	201	107
a + +	0	77	142	194	124
a + c	21	31	291	77	30
a b +	72	4	3	235	1
a b c	203	1	34	46	265

68. From the phenotype data given in the following table for two 3-point testcrosses for (1) a, b, and c and (2) b, c, and d, determine the sequence of the four genes a, b, c, and d and the three map distances between them. Recessive phenotypes are symbolized by lowercase letters and dominant phenotypes by pluses.

1			2	
+ + +	669		b c d	8
a b +	139		b + +	441
a + +	3		b + d	90
+ + c	121		+ c d	376
+ b c	2		+ + +	14
a + c	2280		+ + d	153
a b c	653		+ c +	65
+ b +	2215		b c +	141

69. Vulcans have pointed ears (determined by allele *P*), absent adrenals (determined by *A*), and a right-sided heart (determined by *R*). All these alleles are dominant to normal Earth alleles: rounded ears (*p*), present adrenals (*a*), and a left-sided heart (*r*). The three loci are autosomal and linked as shown in this linkage map:

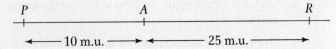

Mr. Spock, first officer of the starship *Enterprise*, has a Vulcan father and an Earthling mother. If Mr. Spock mates with an Earth woman and there is no (genetic) interference, what proportion of their children will have

a. Vulcan phenotypes for all three characters?

b. Earth phenotypes for all three characters?

c. Vulcan ears and heart but Earth adrenals?

d. Vulcan ears but Earth heart and adrenals?

70. In a certain diploid plant, the three loci *A*, *B*, and *C* are linked as follows:

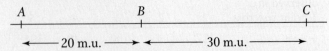

One plant is available to you (call it the parental plant). It has the constitution *A b c/a B C*.

a. With the assumption of no interference, if the plant is selfed, what proportion of the progeny will be of the genotype *a b c/a b c*?

b. Again, with the assumption of no interference, if the parental plant is crossed with the *a b c/a b c* plant, what genotypic classes will be found in the progeny? What will be their frequencies if there are 1000 progeny?

c. Repeat part *b*, this time assuming 20 percent interference between the regions.

71. The following pedigree shows a family with two rare abnormal phenotypes: blue sclerotic (a brittle-bone defect), represented by a black-bordered symbol, and hemophilia, represented by a black center in a symbol. Members represented by completely black symbols have both disorders. The numbers in some symbols are the numbers of individuals with those types.

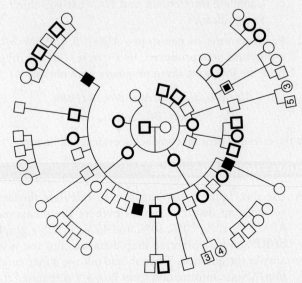

a. What pattern of inheritance is shown by each condition in this pedigree?

b. Provide the genotypes of as many family members as possible.

c. Is there evidence of linkage?

d. Is there evidence of independent assortment?

e. Can any of the members be judged as recombinants (that is, formed from at least one recombinant gamete)?

72. The human genes for color blindness and for hemophilia are both on the X chromosome, and they show a recombinant frequency of about 10 percent. The linkage of a pathological gene to a relatively harmless one can be used for genetic prognosis. Shown here is part of a bigger pedigree. Blackened symbols indicate that the subjects had hemophilia, and crosses indicate color blindness. What information could be given to women III-4 and III-5 about the likelihood of their having sons with hemophilia?

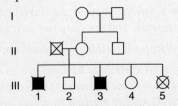

(Problem 72 is adapted from J. F. Crow, *Genetics Notes: An Introduction to Genetics*. Burgess, 1983.)

73. A geneticist mapping the genes *A*, *B*, *C*, *D*, and *E* makes two 3-point testcrosses. The first cross of pure lines is

$$A/A \cdot B/B \cdot C/C \cdot D/D \cdot E/E \times a/a \cdot b/b \cdot C/C \cdot d/d \cdot E/E$$

The geneticist crosses the F_1 with a recessive tester and classifies the progeny by the gametic contribution of the F_1:

$A \cdot B \cdot C \cdot D \cdot E$	316
$a \cdot b \cdot C \cdot d \cdot E$	314
$A \cdot B \cdot C \cdot d \cdot E$	31
$a \cdot b \cdot C \cdot D \cdot E$	39
$A \cdot b \cdot C \cdot d \cdot E$	130
$a \cdot B \cdot C \cdot D \cdot E$	140
$A \cdot b \cdot C \cdot D \cdot E$	17
$a \cdot B \cdot C \cdot d \cdot E$	13
	1000

The second cross of pure lines is $A/A \cdot B/B \cdot C/C \cdot D/D \cdot E/E \times a/a \cdot B/B \cdot c/c \cdot D/D \cdot e/e$.

The geneticist crosses the F_1 from this cross with a recessive tester and obtains

$A \cdot B \cdot C \cdot D \cdot E$	243
$a \cdot B \cdot c \cdot D \cdot e$	237
$A \cdot B \cdot c \cdot D \cdot e$	62
$a \cdot B \cdot C \cdot D \cdot E$	58
$A \cdot B \cdot C \cdot D \cdot e$	155
$a \cdot B \cdot c \cdot D \cdot E$	165
$a \cdot B \cdot C \cdot D \cdot e$	46
$A \cdot B \cdot c \cdot D \cdot E$	34
	1000

The geneticist also knows that genes D and E assort independently.

a. Draw a map of these genes, showing distances in map units wherever possible.

b. Is there any evidence of interference?

74. In the plant *Arabidopsis*, the loci for pod length (L, long; l, short) and fruit hairs (H, hairy; h, smooth) are linked 16 m.u. apart on the same chromosome. The following crosses were made: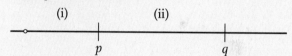

(i) $L\ H/L\ H \times l\ h/l\ h \rightarrow F_1$

(ii) $L\ h/L\ h \times l\ H/l\ H \rightarrow F_1$

If the F_1's from cross i and cross ii are crossed,

a. what proportion of the progeny are expected to be $l\ h/l\ h$?

b. what proportion of the progeny are expected to be $L\ h/l\ h$?

75. In corn (*Zea mays*), the genetic map of part of chromosome 4 is as follows, where w, s, and e represent recessive mutant alleles affecting the color and shape of the pollen:

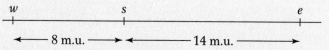

If the following cross is made

$$+++/+++\times w\ s\ e/w\ s\ e$$

and the F_1 is testcrossed with $w\ s\ e/w\ s\ e$, and it is assumed that there is no interference on this region of the chromosome, what proportion of progeny will be of the following genotypes?

a.	+	+	+		**e.**	+	+	e
b.	w	s	e		**f.**	w	s	+
c.	+	s	e		**g.**	w	+	e
d.	w	+	+		**h.**	+	s	+

76. Every Friday night, genetics student Jean Allele, exhausted by her studies, goes to the student union's bowling lane to relax. But, even there, she is haunted by her genetic studies. The rather modest bowling lane has only four bowling balls: two red and two blue. They are bowled at the pins and are then collected and returned down the chute in random order, coming to rest at the end stop. As the evening passes, Jean notices familiar patterns of the four balls as they come to rest at the stop. Compulsively, she counts the different patterns. What patterns did she see, what were their frequencies, and what is the relevance of this matter to genetics?

77. In a tetrad analysis, the linkage arrangement of the p and q loci is as follows:

Assume that

- in region i, there is no crossover in 88 percent of meioses, and there is a single crossover in 12 percent of meioses;

- in region ii, there is no crossover in 80 percent of meioses, and there is a single crossover in 20 percent of meioses; and

- there is no interference (in other words, the situation in one region does not affect what is going on in the other region).

What proportions of tetrads will be of the following types? (a) $M_I M_I$, PD; (b) $M_I M_I$, NPD; (c) $M_I M_{II}$, T; (d) $M_{II} M_I$, T; (e) $M_{II} M_{II}$, PD; (f) $M_{II} M_{II}$, NPD; (g) $M_{II} M_{II}$, T. (Note: Here the M pattern written first is the one that pertains to the p locus.) Hint: The easiest way to do this problem is to start by calculating the frequencies of asci with crossovers in both regions, region i, region ii, and neither region. Then determine what M_I and M_{II} patterns result. (Note: Information from Box 4-2 is required for this problem.)

78. For an experiment with haploid yeast, you have two different cultures. Each will grow on minimal medium to which arginine has been added, but neither will grow on minimal medium alone. (Minimal medium is inorganic salts plus sugar.) Using appropriate methods,

you induce the two cultures to mate. The diploid cells then divide meiotically and form unordered tetrads. Some of the ascospores will grow on minimal medium. You classify a large number of these tetrads for the phenotypes ARG⁻ (arginine requiring) and ARG⁺ (arginine independent) and record the following data:

Segregation of ARG⁻ : ARG⁺	Frequency (%)
4 : 0	40
3 : 1	20
2 : 2	40

a. Using symbols of your own choosing, assign genotypes to the two parental cultures. For each of the three kinds of segregation, assign genotypes to the segregants.

b. If there is more than one locus governing arginine requirement, are these loci linked?

79. A molecular analysis of two pure lines $A/A \cdot B/B$ and $a/a \cdot b/b$ showed that the former was homozygous for

a long tandem repeat (l) and the latter for a short tandem repeat (s). The two were crossed to form an F_1, which was then backcrossed to the second pure line. A thousand progeny were scored as follows:

Aa Bb ss	9		Aa bb ss	43
Aa Bb ls	362		Aa bb ls	93
aa bb ls	11		aa Bb ls	37
aa bb ss	358		aa Bb ss	87

a. What do these results tell us about linkage?

b. Draw a map if appropriate.

c. Incorporate the variable tandem repeat locus into your map.

GENETICS AND SOCIETY

Mapping the human chromosomes by the analysis of recombinant frequencies between phenotypic markers was of only limited success. What might be some possible reasons for this?

Gene Interaction

The colors of peppers are determined by the interaction of several genes. An allele *Y* promotes the early elimination of chlorophyll (a green pigment), whereas *y* does not. Allele *R* determines red and *r* determines yellow carotenoid pigments. Alleles *c1* and *c2* of two different genes down-regulate the amounts of carotenoids, causing the lighter shades. Orange is down-regulated red. Brown is green plus red. Pale yellow is down-regulated yellow. [*Anthony Griffiths.*]

CHAPTER OUTLINE AND LEARNING OBJECTIVES

5.1 INTERACTIONS BETWEEN THE ALLELES OF A SINGLE GENE: VARIATIONS ON DOMINANCE

LO 5.1 Distinguish between the various types of dominance, based on the phenotypes of heterozygotes.

LO 5.2 Recognize phenotypic ratios diagnostic of the presence of a lethal allele.

LO 5.3 Give some possible reasons why incomplete penetrance and variable expressivity may occur in a population of individuals with identical genotypes at a locus under study.

5.2 INTERACTION OF GENES IN PATHWAYS

LO 5.4 Describe the hypotheses proposed to explain various types of gene interaction at the molecular level.

5.3 INFERRING GENE INTERACTIONS

LO 5.5 Determine whether two mutations are in the same gene or in different genes, using progeny ratios or using complementation tests.

LO 5.6 Infer how two genes may be interacting, based on modified Mendelian ratios.

LO 5.7 For known cases of gene interaction, predict progeny ratios in crosses.

CHAPTER OBJECTIVE	Reflecting the fact that the thousands of genes in the genome must clearly interact at the cellular level, many genes are observed to interact at the phenotypic level, resulting in modified inheritance ratios. Our broad objective in this chapter is to catalog the inheritance patterns that reveal various types of gene interaction.

The thrust of our presentation in the book so far has been to show how geneticists identify a gene that affects some biological property of interest. We have seen how the approaches of forward genetics can be used to identify individual genes. The researcher begins with a set of mutants and then crosses each mutant with the wild type to see if the mutant shows single-gene inheritance. The cumulative data from such a research program would reveal a set of genes that all have roles in the development of the property under investigation. In some cases, the researcher may be able to identify specific biochemical functions for many of the genes by comparing gene sequences with those of other organisms. The next step, which is a greater challenge, is to deduce how the genes in a set interact to influence phenotype.

How are the gene interactions underlying a property deduced? One molecular approach is to analyze protein interactions directly in vitro by using one protein as "bait" and observing which other cellular proteins attach to it. Proteins that are found to bind to the bait are candidates for interaction in the living cell. Another molecular approach is to analyze mRNA transcripts. The genes that collaborate in some specific developmental process can be defined by the set of RNA transcripts present when that process is going on, a type of analysis now carried out with a technique called *RNA-seq*, as we will see in Chapter 14. Finally, gene interactions and their significance in shaping phenotype can be deduced by *genetic analysis*, which is the focus of this chapter.

Gene interactions can be classified broadly into two categories. The first category consists of interactions between alleles of a single gene (a single locus). These types of interactions can be thought of, broadly speaking, as variations on dominance. In earlier chapters we dealt with alleles displaying full dominance or full recessiveness, but as we shall see in this chapter, there are other types of dominance, each with its own underlying cell biology. Although this information does not address the range of genes affecting a function, a great deal can be learned of a gene's role by considering allelic interactions. The second category consists of interactions between two or more loci. These interactions reveal the number and types of genes in the overall program underlying a particular biological function.

5.1 INTERACTIONS BETWEEN THE ALLELES OF A SINGLE GENE: VARIATIONS ON DOMINANCE

LO 5.1 Distinguish between the various types of dominance, based on the phenotypes of heterozygotes.

LO 5.2 Recognize phenotypic ratios diagnostic of the presence of a lethal allele.

LO 5.3 Give some possible reasons why incomplete penetrance and variable expressivity may occur in a population of individuals with identical genotypes at a locus under study.

There are thousands of different ways to alter the sequence of a gene, each producing a mutant allele, although only some of these mutant alleles will appear in a real population. The known mutant alleles of a gene and its wild-type allele are referred to as **multiple alleles,** or an **allelic series.**

One of the tests routinely performed on a new mutant allele is to see if it is dominant or recessive. Basic information about dominance and recessiveness is useful in working with the new mutation and can be a source of insight into the way the gene functions, as we will see in the examples. Dominance is a manifestation of how *the alleles of a single gene* interact in a heterozygote. In any experiment, the interacting pair of alleles may be wild-type and mutant alleles (+/*m*), or two different mutant alleles (*m₁*/*m₂*). Several types of dominance have been discovered, each representing a different type of interaction between a pair of alleles.

Complete dominance and recessiveness

The simplest type of dominance is **full dominance,** also called **complete dominance,** which we examined in Chapter 2. The phenotype of a fully dominant allele will be displayed when only one copy is present, such as in a heterozygote individual; in a heterozygote, the other allele whose phenotype is not displayed is the fully recessive allele. In full dominance, the homozygous dominant cannot be distinguished from the heterozygote; that is, at the phenotypic level, genotype *A/A* cannot be distinguished from genotype

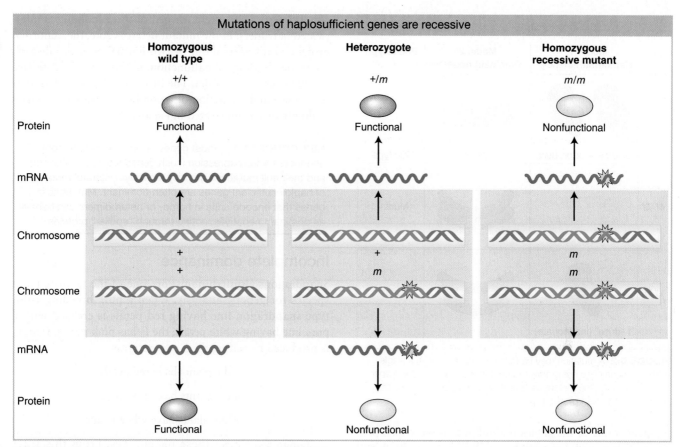

FIGURE 5-1 In the heterozygote, even though the mutated copy of the gene produces nonfunctional protein, the wild-type copy generates enough functional protein to produce the wild-type phenotype.

A/a. As we saw in Chapter 2, the alleles that result in phenylketonuria (PKU) and many other single-gene human diseases are fully recessive, whereas their wild-type alleles are dominant. Other single-gene diseases such as pseudo-achondroplasia result from alleles that are fully dominant, whereas, in those cases, the wild-type allele is recessive. How can these dominance relations be interpreted at the cellular level?

The disease PKU is a good general model for recessive mutations. Recall from Chapter 2 that PKU is caused by a defective allele of the gene encoding the enzyme phenylalanine hydroxylase (PAH). In the absence of normal PAH, the phenylalanine entering the body in food is not broken down and hence accumulates. Under such conditions, phenylalanine is converted into phenylpyruvic acid, which is transported to the brain through the bloodstream and there impedes normal development, leading to intellectual disabilities. The reason that the defective allele is recessive is that just one copy of the wild-type allele *P* produces enough

PAH to break down the phenylalanine entering the body. One "dose" of functional PAH protein, produced by one *P* allele, results in the wild-type phenotype. Thus, the PAH wild-type allele is said to be *haplosufficient. Haplo* means a haploid dose (one), and *sufficient* refers to the ability of that single dose to produce the wild-type phenotype. Hence, both *P/P* (two doses) and *P/p* (one dose) have enough PAH activity to result in the normal cellular chemistry. People with *p/p* have zero doses of PAH activity. **Figure 5-1** illustrates this general notion.

How can we explain fully dominant mutations? There are several molecular mechanisms for dominance. A regularly encountered mechanism is that the wild-type allele of a gene is *haploinsufficient.* In haploinsufficiency, one wild-type dose is *not* enough to achieve normal levels of function. Assume that 16 units of a gene's product are needed for normal chemistry and that each wild-type allele can make 10 units. Two wild-type alleles will produce 20 units of product, well over the minimum. But consider what

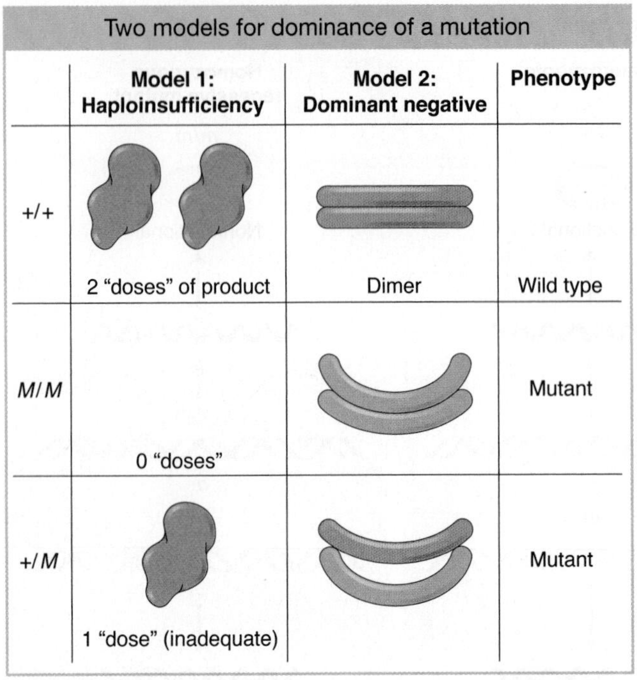

Two models for dominance of a mutation		
Model 1: Haploinsufficiency	**Model 2:** Dominant negative	**Phenotype**
+/+ 2 "doses" of product	Dimer	Wild type
M/M 0 "doses"		Mutant
+/M 1 "dose" (inadequate)		Mutant

FIGURE 5-2 A mutation may be dominant because (*left*) a single wild-type gene does not produce enough protein product for proper function or (*right*) the mutant allele acts as a dominant negative that produces a "spoiler" protein product.

happens if one of the alleles is a **null mutation,** which produces a nonfunctional protein (or no protein at all). A null mutation in combination with a single wild-type allele would produce $10 + 0 = 10$ units, well below the minimum. Hence, the heterozygote (wild type/null) is mutant, and the mutant allele is, by definition, dominant. In mice, the gene *Tbx1* is haploinsufficient. This gene encodes a transcription-regulating protein (a *transcription factor*) that acts on genes responsible for the development of the pharynx. A knockout of one wild-type allele results in an inadequate concentration of the regulatory protein, which results in defects in the development of the pharyngeal arteries. The same haploinsufficiency is thought to be responsible for DiGeorge syndrome in humans, a condition with cardiovascular and craniofacial abnormalities.

Another important type of dominant mutation is called a **dominant negative.** Polypeptides with this type of mutation act as "spoilers" or "rogues." In some cases, the gene product is a unit of a *homodimeric* protein, a protein composed of two units of the same type. In the heterozygote (+/M), the mutant polypeptide binds to the wild-type polypeptide and acts as a spoiler by distorting it or otherwise interfering with its function. The same type of spoiling can also hinder the functioning of a *heterodimer* composed of polypeptides from different genes. In other cases, the gene product is a monomer, and, in these situations, the mutant protein binds the substrate, and it acts as a spoiler by hindering the ability of the wild-type protein to bind to the substrate.

An example of a mutation that can act as a dominant negative is found in the gene for collagen protein. Some mutations in this gene give rise to the human phenotype osteogenesis imperfecta (brittle-bone disease). Collagen is a connective-tissue protein formed of three monomers intertwined (a trimer). In the mutant heterozygote, the abnormal protein wraps around one or two normal ones and distorts the trimer, leading to malfunction. In this way, the defective collagen acts as a spoiler. The difference between haploinsufficiency and the action of a dominant negative as causes of dominance is illustrated in **Figure 5-2**.

KEY CONCEPT For most genes, a single wild-type copy is adequate for full expression (such genes are haplosufficient), and their null mutations are fully recessive. Harmful mutations of haploinsufficient genes are often dominant. Mutations in genes that encode units in homo- or heterodimers can behave as dominant negatives, acting through "spoiler" proteins.

Incomplete dominance

Snapdragons (Antirrhinums) are one of the favorite plant species for genetic analysis. When a pure-breeding wild-type snapdragon line having red petals is crossed with a pure line having white petals, the F_1 has pink petals. If an F_2 is produced by selfing the F_1, the result is

$\frac{1}{4}$ of the plants have red petals

$\frac{1}{2}$ of the plants have pink petals

$\frac{1}{4}$ of the plants have white petals

Figure 5-3 shows these phenotypes. From this 1 : 2 : 1 ratio in the F_2, we can deduce that the inheritance pattern is based on two alleles of a single gene. However, the heterozygotes (the F_1 and half the F_2) are intermediate in phenotype. By inventing allele symbols, we can list the genotypes in this experiment as c^+/c^+ (red), c/c (white), and c^+/c (pink). The occurrence of the intermediate phenotype suggests an **incomplete dominance,** the term used to describe the general case in which the phenotype of a heterozygote is intermediate between those of the two homozygotes, on some quantitative scale of measurement.

How do we explain incomplete dominance at the molecular level? In incomplete dominance, each wild-type allele generally produces a set dose of its protein product. The number of doses of a wild-type allele determines the concentration of a chemical made by the protein, such as pigment. In the four-o'clock plant, two doses (c^+/c^+) produce the most copies of transcript, thus producing the greatest amount of protein and, hence, the greatest amount of pigment, enough to make the flower petals red. One dose (c^+/c) produces less pigment, and so the petals are pink. A zero dose (c/c) produces no pigment.

Codominance

Another variation on the theme of dominance is **codominance,** the expression of both alleles of a heterozygote. A clear example is seen in the human ABO blood groups, where there is codominance of antigen alleles. The ABO blood groups are determined by three alleles of one gene. These three alleles interact in several ways to produce the four blood types of the ABO system. The three most important

Incomplete dominance

FIGURE 5-3 In snapdragons, a heterozygote is pink (*right*), intermediate between the two homozygotes red (*middle*) and white (*left*). The pink heterozygote demonstrates incomplete dominance. [*John Kaprielian/Science Source.*]

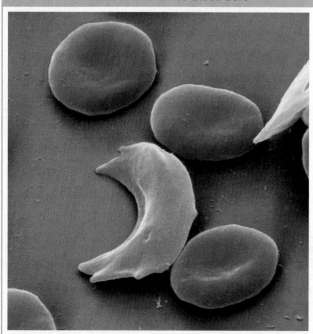

Sickled and normal red blood cells

FIGURE 5-4 The sickle-shaped cell is caused by a single mutation in the gene for hemoglobin. [*Eye of Science/Science Source.*]

alleles are i, I^A, and I^B, but a person can have only two of the three alleles or two copies of one of them. The pairwise combinations result in six different genotypes: the three homozygotes and three different types of heterozygotes, as follows:

Genotype	Blood type
I^A/I^A, I^A/i	A
I^B/I^B, I^B/i	B
I^A/I^B	AB
i/i	O

In this allelic series, the alleles determine the presence and form of a complex sugar molecule present on the surface of red blood cells. This sugar molecule is an antigen, a cell-surface molecule that can be recognized by the immune system. The alleles I^A and I^B determine two different forms of this cell-surface molecule. However, the allele i results in no cell-surface molecule of this type (it is a null allele). In the genotypes I^A/i and I^B/i, the alleles I^A and I^B are fully dominant over i. However, in the genotype I^A/I^B, each of the alleles produces its own form of the cell-surface molecule, and so the A and B alleles are codominant.

The human disease sickle-cell anemia illustrates the somewhat arbitrary ways in which we classify dominance. The gene concerned encodes the molecule hemoglobin, which is responsible for transporting oxygen in blood vessels and is the major constituent of red blood cells. There are two main alleles, Hb^A and Hb^S, and the three possible genotypes have different phenotypes, as follows:

Hb^A/Hb^A: normal; red blood cells never sickle

Hb^S/Hb^S: severe, often fatal anemia; abnormal hemoglobin causes red blood cells to have sickle shape

Hb^A/Hb^S: no anemia; red blood cells sickle only under low oxygen concentrations

Figure 5-4 shows an electron micrograph of blood cells including some sickled cells. In regard to the presence or absence of anemia, the Hb^A allele is dominant. In the heterozygote, a single Hb^A allele produces enough functioning hemoglobin to prevent anemia. In regard to blood-cell shape, however, there is incomplete dominance, as shown by the fact that, in the heterozygote, many of the cells have a slight sickle shape. Finally, in regard to hemoglobin itself, there is codominance. The alleles Hb^A and Hb^S encode two different forms of hemoglobin that differ by a single amino acid, and both forms are synthesized in the heterozygote. The A and S forms of hemoglobin can be separated by electrophoresis because it happens that they have different charges (**Figure 5-5**). We see that homozygous Hb^A/Hb^A people have one type of hemoglobin (A), and anemics have another (type S), which moves more slowly in the electric field. The heterozygotes have both types, A and S. In other words, there is codominance at the molecular level. The fascinating population genetics of the Hb^A and Hb^S alleles will be considered in Chapter 20.

Sickle-cell anemia illustrates the arbitrariness of the terms *dominance, incomplete dominance,* and *codominance*. The type of dominance inferred depends on the phenotypic level at which the assay is made—organismal, cellular, or molecular. Indeed, caution should be applied to many of the categories that scientists use to classify structures and processes; these categories are devised by humans for the convenience of analysis.

KEY CONCEPT In general, three main types of dominance can be distinguished: full dominance, incomplete dominance, and codominance. The type of dominance is determined by the molecular functions of the alleles of a gene and by the investigative level of analysis.

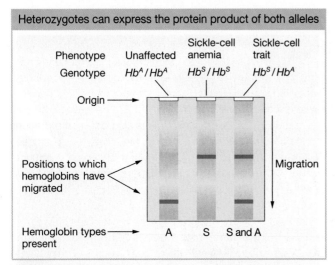

Heterozygotes can express the protein product of both alleles

FIGURE 5-5 The electrophoresis of normal and mutant hemoglobins. Shown are results produced by hemoglobin from a person with sickle-cell trait (a heterozygote), a person with sickle-cell anemia, and a healthy person. The green bands show the positions to which the hemoglobins migrate on the starch gel.

The leaves of clover plants show several variations on the dominance theme. Clover is the common name for plants of the genus *Trifolium*. There are many species. Some are native to North America, whereas others grow there as introduced weeds. Much genetic research has been done with white clover, which shows considerable variation among individual plants in the curious V, or chevron, pattern on the leaves. The different chevron forms (and the absence of chevrons) are determined by a series of seven alleles, as seen in **Figure 5-6**, which shows the many different types of interactions possible for even one allele. In most practical cases, many alleles of a gene can be found together in a population, constituting an allelic series. The phenotypes shown by the allelic combinations are many and varied, reflecting the relative nature of dominance: an allele can show dominance with one partner but not with another. Hence, the complexity illustrated by the ABO blood type system is small compared with that in a case such as clover chevrons.

KEY CONCEPT A gene can take on many forms, called alleles, each caused by various mutations of the DNA sequence. Some mutant alleles have phenotypic impact; others do not.

Recessive lethal alleles

An allele that is capable of causing the death of an organism is called a **lethal allele.** In the characterization of a set of newly discovered mutant alleles, a recessive mutation (a mutation in the homozygous state) is sometimes found to be lethal. This information is potentially useful in that it shows that the newly discovered gene (of yet unknown

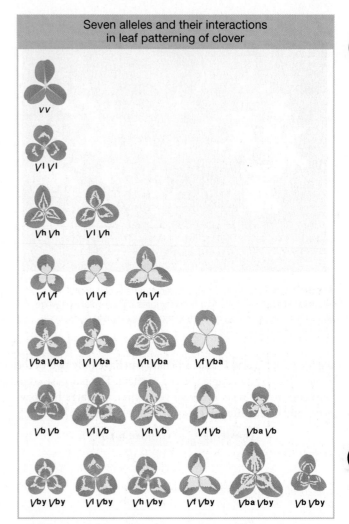

Seven alleles and their interactions in leaf patterning of clover

FIGURE 5-6 Multiple alleles determine the chevron pattern on the leaves of white clover. The genotype of each plant is shown below it. There is a variety of dominance interactions. [*Research by W. Ellis Davies.*]

function) is essential to the organism's operation. **Essential genes** are those without which an organism dies. (An example of an essential gene might be a ribosomal gene without which no protein would be made.) Indeed, with the use of modern DNA technology, a null mutant allele of a gene of interest can now be made intentionally and made homozygous to see if it is lethal and under which environmental conditions. Lethal alleles are also useful in determining the developmental stage at which the gene normally acts. In this case, geneticists look for whether death from a lethal mutant allele occurs early or late in the development of a zygote. The phenotype associated with death can also be informative in regard to gene function; for example, if a certain organ appears to be abnormal, the gene is likely to be expressed in that organ.

What is the diagnostic test for lethality? The test is well illustrated by one of the prototypic examples of a lethal allele, a coat-color allele in mice (see the Model Organism box on page 159). Normal wild-type mice have coats with a

Green-glowing genetically modified mice. The jellyfish gene for green fluorescent protein has been inserted into the chromosomes of the glowing mice. The other mice are normal. [*Eye of Science/Science Source.*]

arranged in blocks in exactly the same positions as those of humans.

Research on the Mendelian genetics of mice began early in the twentieth century. One of the most important early contributions was the elucidation of the genes that control coat color and pattern. Genetic control of the mouse coat has provided a model for all mammals, including cats, dogs, horses, and cattle. A great deal of work was also done on mutations induced by radiation and chemicals. Mouse genetics has been of great significance in medicine. A large proportion of human genetic diseases have mouse counterparts useful for experimental study (they are called "mouse models"). The mouse has played a particularly important role in the development of our current understanding of the genes underlying cancer.

The mouse genome can be modified by the insertion of specific fragments of DNA into a fertilized egg or into somatic cells. The mice in the photograph have received a jellyfish gene for green fluorescent protein (GFP) that makes them glow green under special lights. Gene knockouts and replacements also are possible.

A major limitation of mouse genetics is its cost. Whereas working with a million individuals of *E. coli* or *S. cerevisiae* is a trivial matter, working with a million mice requires a factory-size building. Furthermore, although mice do breed rapidly compared with humans, they cannot compete with microorganisms for speedy life cycle. Hence, the large-scale selections and screens necessary to detect rare genetic events are not possible.

The laboratory mouse is descended from the house mouse *Mus musculus*. The pure lines used today as standards are derived from mice bred in past centuries by mouse "fanciers." Among model organisms, it is the one whose genome most closely resembles the human genome. Its diploid chromosome number is 40 (compared with 46 in humans), and the genome is slightly smaller than that of humans (the human genome being 3000 Mb) and contains approximately the same number of genes (current estimate 25,000). Furthermore, all mouse genes seem to have counterparts in humans. A large proportion of genes are

rather dark overall pigmentation. A mutation called *yellow* (a lighter coat color) shows a curious inheritance pattern. If any yellow mouse is mated with a homozygous wild-type mouse, a 1:1 ratio of yellow to wild-type mice is always observed in the progeny. This result suggests that a yellow mouse is always heterozygous for the yellow allele and that the yellow allele is dominant over wild type. However, if any two yellow mice are crossed with each other, the result is always as follows:

$$\text{yellow} \times \text{yellow} \rightarrow \tfrac{2}{3} \text{ yellow}, \ \tfrac{1}{3} \text{ wild type}$$

Figure 5-7 shows a typical litter from a cross between yellow mice.

How can the 2:1 ratio be explained? The results make sense if the yellow allele is assumed to be lethal when homozygous. The yellow allele is known to be of a coat-color gene called *A*. Let's call it A^Y. Hence, the results of crossing two yellow mice are

$$A^Y/A \times A^Y/A$$

Progeny		
$\tfrac{1}{4}$	A^Y/A^Y	lethal
$\tfrac{1}{2}$	A^Y/A	yellow
$\tfrac{1}{4}$	A/A	wild type

The expected monohybrid ratio of 1:2:1 would be found among the zygotes, but it is altered to a 2:1 ratio in the progeny actually seen at birth because zygotes with a lethal A^Y/A^Y genotype do not survive to be counted. This hypothesis is supported by the removal of uteri from pregnant females of the yellow × yellow cross; one-fourth of the embryos are found to be dead.

A recessive lethal allele, yellow coat

FIGURE 5-7 A litter from a cross between two mice heterozygous for the dominant yellow coat-color allele. The allele is lethal in a double dose. Not all progeny are visible. [*Anthony Griffiths.*]

159

Tailless, a recessive lethal allele in cats

FIGURE 5-8 A Manx cat. A dominant allele causing taillessness is lethal in the homozygous state. The phenotype of two eye colors is unrelated to taillessness. [*Gerard Lacz/NHPA/Photoshot.*]

The A^Y allele produces effects on two characters: coat color and survival. In general, the term **pleiotropic** is used for any allele that affects several properties of an organism.

The tailless Manx phenotype in cats (**Figure 5-8**) also is produced by an allele that is lethal in the homozygous state. A single copy of the Manx allele, M^L, severely interferes with normal spinal development, resulting in the absence of a tail in the M^L/M heterozygote. But in the M^L/M^L homozygote, two copies of the Manx allele produces such an extreme abnormality in spinal development that the embryo does not survive.

KEY CONCEPT Some mutant alleles are lethal; that is, they can result in the death of the organism. Lethality is most often recessive.

The *yellow* and M^L alleles have their own phenotypes in a heterozygote, but most recessive lethals are silent in the heterozygote. In such a situation, recessive lethality is diagnosed by observing the death of 25 percent of the progeny at some stage of development.

Whether an allele is lethal or not often depends on the environment in which the organism develops. Whereas certain alleles are lethal in virtually any environment, others are viable in one environment but lethal in another. Human hereditary diseases provide some examples. Cystic fibrosis and sickle-cell anemia are diseases that would be lethal without treatment. Furthermore, many of the alleles favored and selected by animal and plant breeders would almost certainly be eliminated in nature as a result of competition with the members of the natural population. The dwarf mutant varieties of grain, which are very high yielding, provide good examples; only careful nurturing by farmers has maintained such alleles for our benefit.

Geneticists commonly encounter situations in which expected phenotypic ratios are consistently skewed in one direction because a mutant allele reduces viability. For example, in the cross $A/a \times a/a$, we predict a progeny ratio of 50 percent A/a and 50 percent a/a, but we might consistently observe a ratio such as 55 percent:45 percent or 60 percent:40 percent. In such a case, the recessive allele is said to be *sublethal* because the lethality is expressed in only some but not all of the homozygous individuals. Thus, lethality may range from 0 to 100 percent, depending on the gene itself, the rest of the genome, and the environment.

We have seen that lethal alleles are useful in diagnosing the time at which a gene acts and the nature of the phenotypic defect that kills. However, maintaining stocks bearing lethal alleles for laboratory use is a challenge. In diploids, recessive lethal alleles can be maintained as heterozygotes. In haploids, heat-sensitive lethal alleles are useful. They are members of a general class of **temperature-sensitive (ts) mutations.** Their phenotype is wild type at the **permissive temperature** (often room temperature) but mutant at some higher **restrictive temperature.** Temperature-sensitive alleles are thought to be caused by mutations that make the protein prone to twist or bend its shape to an inactive conformation at the restrictive temperature. Research stocks can be maintained easily under permissive conditions, and the mutant phenotype can be assayed in a subset of individuals by a switch to the restrictive conditions. In diploids, temperature-sensitive dominant lethal mutations also are useful. This type of mutation is lethal even when present in a single dose, but only when the experimenter switches the organism to the restrictive temperature.

Null alleles for genes identified through genomic sequencing can be made by using a variety of "reverse genetic" procedures that specifically knock out the function of that gene. These will be described in Chapter 14.

KEY CONCEPT To see if a gene is essential, a null allele is tested for lethality.

Penetrance and expressivity

In the analysis of single-gene inheritance, there is a natural tendency to choose mutants that produce clear Mendelian ratios. In such cases, we can use the phenotype to distinguish mutant and wild-type genotypes with almost 100 percent certainty. In these cases, we say that the mutation is 100 percent *penetrant* into the phenotype. However, many mutations show *incomplete* penetrance; that is, not every individual with the genotype expresses the corresponding phenotype. Thus, **penetrance** is defined as the percentage of *individuals* with a given allele who exhibit the phenotype associated with that allele.

Why would an organism have a particular genotype and yet not express the corresponding phenotype? There are several possible reasons:

1. *The influence of the environment.* Individuals with the same genotype may show a range of phenotypes,

depending on the environment. The range of phenotypes for mutant and wild-type individuals may overlap: the phenotype of a mutant individual raised in one set of circumstances may match the phenotype of a wild-type individual raised in a different set of circumstances. Should this matching happen, the mutant cannot be distinguished from the wild type.

2. *The influence of other interacting genes.* Uncharacterized modifiers, epistatic genes, or suppressors in the rest of the genome (all discussed shortly) may act to prevent the expression of the typical phenotype.

3. *The subtlety of the mutant phenotype.* The subtle effects brought about by the absence of a gene function may be difficult to measure in a laboratory situation.

A typical encounter with incomplete penetrance is shown in **Figure 5-9**. In this human pedigree, we see a normally dominantly inherited phenotype disappearing in the second generation only to reappear in the next.

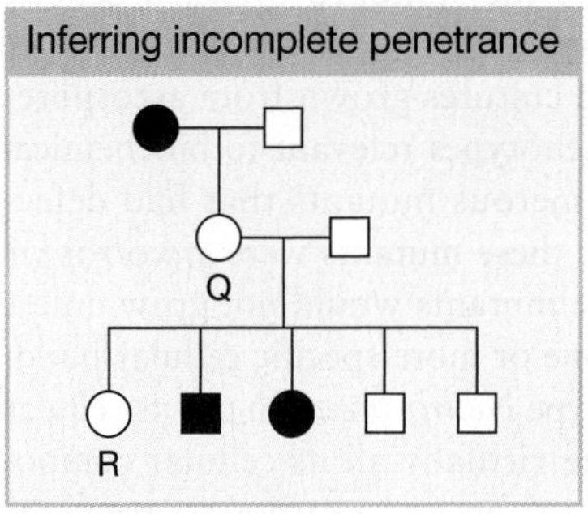

Another measure for describing the range of phenotypic expression is called **expressivity**. Expressivity measures the degree to which a given allele is expressed at the phenotypic level; that is, expressivity measures the intensity of the phenotype. For example, "brown" animals

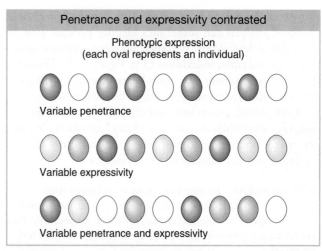

FIGURE 5-10 Assume that all the individuals shown have the same pigment allele (*P*) and possess the same potential to produce pigment. Effects from the rest of the genome and the environment may suppress or modify pigment production in any one individual. The color indicates the level of expression.

(genotype *b/b*) from different stocks might show very different intensities of brown pigment from light to dark. As for penetrance, variable expressivity may be due to variation in the allelic constitution of the rest of the genome or to environmental factors. **Figure 5-10** illustrates the distinction between penetrance and expressivity. An example of variable expressivity in dogs is found in **Figure 5-11**.

The phenomena of incomplete penetrance and variable expressivity can make any kind of genetic analysis substantially more difficult, including human pedigree analysis and predictions in genetic counseling. For example, it is often the case that a disease-causing allele is not fully penetrant. Thus, someone could have the allele but not show any signs of the disease. If that is the case, it is difficult to give a clean genetic bill of health to any person in a disease pedigree

FIGURE 5-9 In this human pedigree of a dominant allele that is not fully penetrant, person Q does not display the phenotype but passed the dominant allele to at least two progeny. Because the allele is not fully penetrant, the other progeny (for example, R) may or may not have inherited the dominant allele.

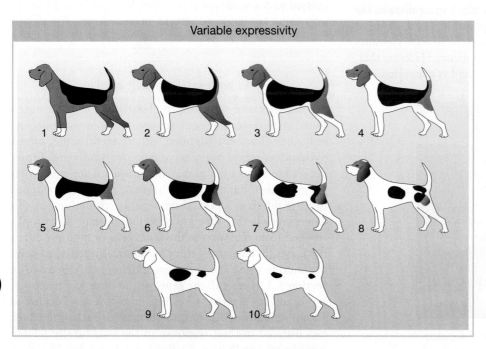

FIGURE 5-11 Ten grades of piebald spotting in beagles. Each of these dogs has the allele S^P, the allele responsible for piebald spots in dogs. The variation is caused by variation at other loci.

(for example, person R in Figure 5-9). On the other hand, pedigree analysis can sometimes identify persons who do not express but almost certainly do have a disease genotype (for example, individual Q in Figure 5-9). Similarly, variable expressivity can complicate counseling because persons with low expressivity might be misdiagnosed.

Even though penetrance and expressivity can be quantified, they nevertheless represent "fuzzy" situations because rarely is it possible to identify the specific factors causing variation without substantial extra research.

> **KEY CONCEPT** The terms *penetrance* and *expressivity* quantify the modification of a gene's effect by varying environment and genetic background; they measure, respectively, the percentage of cases in which the phenotype is observed and its severity.

We now turn to the approaches that can be used to detect the interaction between two or more loci.

5.2 INTERACTION OF GENES IN PATHWAYS

> **LO 5.4** Describe the hypotheses proposed to explain various types of gene interaction at the molecular level.

Genes act by controlling cellular chemistry. Early in the twentieth century, Archibald Garrod, an English physician (**Figure 5-12**), made the first observation supporting this insight. Garrod noted that several recessive human diseases show defects in what is called metabolism, the general set of chemical reactions taking place in an organism. This observation led to the notion that such genetic diseases are "inborn errors of metabolism." Garrod worked on a disease called alkaptonuria (AKU), or black urine disease. He discovered that the substance responsible for black urine was homogentisic acid, which is present in high amounts and secreted into the urine in AKU patients. He knew that, in unaffected people, homogentisic acid is converted into

Discoverer of inborn errors of metabolism

FIGURE 5-12 British physician Archibald Garrod (1857–1936). [*SPL/Science Source.*]

maleylacetoacetic acid; so he proposed that, in AKU, there is a defect in this conversion. Consequently, homogentisic acid builds up and is excreted. Garrod's observations raised the possibility that the cell's chemical pathways were under the control of a large set of interacting genes. However, the direct demonstration of this control was provided by the later work of Beadle and Tatum on the fungus *Neurospora*.

Biosynthetic pathways in *Neurospora*

The landmark study by George Beadle and Edward Tatum in the 1940s not only clarified the role of genes but also demonstrated the interaction of genes in biochemical pathways. They later received a Nobel Prize for their study, which marks the beginning of all molecular biology. Beadle and Tatum did their work on the haploid fungus *Neurospora*, which we have met in earlier chapters. Their plan was to investigate the genetic control of cellular chemistry. In what has become the standard forward genetic approach, they first irradiated *Neurospora* cells to produce mutations and then tested cultures grown from ascospores for interesting mutant phenotypes relevant to biochemical function. They found numerous mutants that had defective nutrition. Specifically, these mutants were *auxotrophic* mutants, meaning that the mutants would not grow unless their medium contained one or more specific cellular building blocks. Whereas wild-type *Neurospora* can use its cellular biochemistry to synthesize virtually all its cellular components from the inorganic nutrients and a carbon source in the medium, auxotrophic mutants cannot. In order to grow, such mutants require a nutrient to be supplied (a nutrient that a wild-type fungus is able to synthesize for itself), suggesting that the mutant is defective for some normal synthetic step.

As their first step, Beadle and Tatum confirmed that each mutation that generated a nutrient requirement was inherited as a single-gene mutation because each gave a 1:1 ratio when crossed with a wild type (remember, *Neurospora* is a haploid organism). Letting *aux* represent an auxotrophic mutation,

$$+ \times aux$$
$$\downarrow$$
$$\text{progeny: } \tfrac{1}{2} + \text{ and } \tfrac{1}{2} \ aux$$

Their second step was to classify the specific nutritional requirement of each auxotroph. Some would grow only if proline was supplied, others methionine, others pyridoxine, others arginine, and so on. Beadle and Tatum decided to focus on arginine auxotrophs. They found that the genes that mutated to give arginine auxotrophs mapped to three different loci on three separate chromosomes. (To determine whether their collection of arginine auxotrophs resulted from mutation in the same gene or multiple different genes, Beadle and Tatum methodically analyzed pairs of mutants using the *complementation test*, discussed in Section 5.3.) Let's call the genes at the three loci the *arg-1*, *arg-2*, and *arg-3* genes. A key breakthrough was Beadle and Tatum's discovery that the auxotrophs for each of the three loci differed in their response to the structurally related compounds ornithine and citrulline (**Figure 5-13**). The *arg-1*

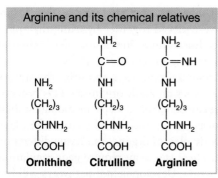

| Arginine and its chemical relatives |

FIGURE 5-13 The chemical structures of arginine and the structurally related compounds citrulline and ornithine.

mutants grew when supplied with any one of the chemicals ornithine, citrulline, or arginine. The *arg-2* mutants grew when given arginine or citrulline but not ornithine. The *arg-3* mutants grew only when arginine was supplied. These results are summarized in Table 5-1.

Cellular enzymes were already known to interconvert such related compounds. On the basis of the properties of the *arg* mutants, Beadle and Tatum and their colleagues proposed a biochemical pathway for such conversions in *Neurospora*:

precursor $\xrightarrow{\text{enzyme X}}$ ornithine $\xrightarrow{\text{enzyme Y}}$

citrulline $\xrightarrow{\text{enzyme Z}}$ arginine

This pathway nicely explains the three classes of mutants shown in Table 5-1. Under the model, the *arg-1* mutants have a defective enzyme X, and so they are unable to convert the precursor into ornithine as the first step in producing arginine. However, they have normal enzymes Y and Z, and so the *arg-1* mutants are able to produce arginine if supplied with either ornithine or citrulline. Similarly, the *arg-2* mutants lack enzyme Y, and the *arg-3* mutants lack enzyme Z. Thus, a mutation at a particular gene is assumed to interfere with the production of a single enzyme. The defective enzyme creates a block in some biosynthetic pathway. The block can be circumvented by supplying to the cells any compound that normally comes after the block in the pathway.

We can now diagram a more complete biochemical model:

$$arg\text{-}1^+ \qquad\qquad arg\text{-}2^+$$
$$\downarrow \qquad\qquad\qquad \downarrow$$
precursor $\xrightarrow{\text{enzyme X}}$ ornithine $\xrightarrow{\text{enzyme Y}}$

$$arg\text{-}3^+$$
$$\downarrow$$
citrulline $\xrightarrow{\text{enzyme Z}}$ arginine

TABLE 5-1	Growth of *arg* Mutants in Response to Supplements		
	Supplements		
Mutant	**Ornithine**	**Citrulline**	**Arginine**
arg-1	+	+	+
arg-2	−	+	+
arg-3	−	−	+

Note: A plus sign means growth; a minus sign means no growth.

This brilliant model, which was initially known as the *one-gene–one-enzyme hypothesis*, was the source of the first exciting insight into the functions of genes: genes somehow were responsible for the function of enzymes, and each gene apparently controlled one specific enzyme in a series of interconnected steps in a biochemical pathway. Other researchers obtained similar results for other biosynthetic pathways, and the hypothesis soon achieved general acceptance. All proteins, whether or not they are enzymes, also were found to be encoded by genes, and so the phrase was refined to become the **one-gene–one-polypeptide hypothesis**. (Recall that a polypeptide is the simplest type of protein, a single chain of amino acids.) It soon became clear that a gene encodes the *physical structure* of a protein, which in turn dictates its function. Beadle and Tatum's hypothesis became one of the great unifying concepts in biology because it provided a bridge that brought together the two major research areas of genetics and biochemistry.

We must add parenthetically that, although the great majority of genes encode proteins, some are known to encode RNAs that have special functions. All genes are transcribed to make RNA. Protein-encoding genes are transcribed to messenger RNA (mRNA), which is then translated into protein. However, the RNA encoded by a minority of genes is never translated into protein because the RNA itself has a unique function. These are called **functional RNAs**. Some examples are transfer RNAs, ribosomal RNAs, and small cytoplasmic RNAs; more about them will be covered in later chapters.

KEY CONCEPT Chemical synthesis in cells is by pathways of sequential steps catalyzed by enzymes. The genes encoding the enzymes of a specific pathway constitute a functionally interacting subset of the genome.

Gene interaction in other types of pathways

The notion that genes interact through pathways is a powerful one that finds application in all organisms. The *Neurospora* arginine pathway is an example of a synthetic pathway, a chain of enzymatic conversions that synthesizes essential molecules. We can extend the idea again to a human case already introduced, the disease phenylketonuria (PKU), which is caused by an autosomal recessive allele. This disease results from an inability to convert phenylalanine into tyrosine. As a result of the block, phenylalanine accumulates and is spontaneously converted into a toxic compound, phenylpyruvic acid. The PKU gene is part of a metabolic pathway like the *Neurospora* arginine pathway, a section of which is shown in Figure 5-14. The illustration includes several other diseases caused by blockages in steps in this pathway (including alkaptonuria, the disease investigated by Garrod).

Another type of pathway is a *signal-transduction pathway*. This type of pathway is a chain of complex signals, from the environment to the internal components

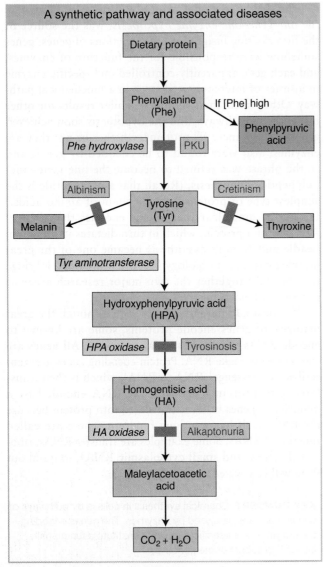

A synthetic pathway and associated diseases

FIGURE 5-14 A section of the phenylalanine metabolic pathway in humans, including diseases associated with enzyme blockages. The disease PKU is produced when the enzyme phenylalanine hydroxylase malfunctions. Accumulation of phenylalanine results in an increase in phenylpyruvic acid, which interferes with the development of the nervous system.

of the cell, that result in activation of cellular responses. These pathways are crucial to the proper function of an organism. One of the best-understood signal-transduction pathways was worked out from a genetic analysis of the mating response in baker's yeast. Two mating types, determined by the alleles MATa and MATα, are necessary for yeast mating to occur. When a cell is in the presence of another cell of opposite mating type, it undergoes a series of changes in shape and behavior to prepare for mating. This mating response is triggered when a mating pheromone (hormone) is released by a cell of the opposite mating type and binds to a membrane receptor on the receiving cell. This signal promotes the sequential action of a set of genes, which ultimately activates the transcription of mating-specific genes that enable the cell to mate. This set of genes was discovered through a standard interaction

analysis of mutants with aberrant mating response, and the steps were pieced together by using the approaches in the next section. A mutation at any one of these steps may disrupt the mating process.

Developmental pathways comprise the steps by which a zygote becomes an adult organism. This process involves many genetically controlled steps, including establishment of the anterior-posterior and dorsal-ventral axes, laying down the basic body plan of organs, and tissue differentiation and movement. These steps can require gene regulation and signal transduction. Developmental pathways will be taken up in detail in Chapter 13, but the interaction of genes in these pathways is analyzed in the same way, as we will see next.

KEY CONCEPT Gene interaction occurs in any cellular pathway, such as biosynthetic, signal transduction, and developmental.

5.3 INFERRING GENE INTERACTIONS

LO 5.2 Recognize phenotypic ratios diagnostic of the presence of a lethal allele.

LO 5.5 Determine whether two mutations are in the same gene or in different genes, using progeny ratios or using complementation tests.

LO 5.6 Infer how two genes may be interacting, based on modified Mendelian ratios.

LO 5.7 For known cases of gene interaction, predict progeny ratios in crosses.

The genetic approach that reveals the interacting genes for a particular biological property is briefly as follows:

Step 1. Obtain many single-gene mutants and test for dominance.

Step 2. Test the mutants for allelism—are they at one or several loci?

Step 3. Combine the mutants in pairs to form **double mutants** to see if the genes interact.

Gene interaction is inferred from the phenotype of the double mutant: if the genes interact, then the phenotype differs from the simple combination of both single-gene mutant phenotypes. If mutant alleles from different genes interact, then we infer that the wild-type genes interact normally as well. In cases in which the two mutants interact, a modified 9:3:3:1 Mendelian ratio will often result.

A procedure that must be carried out before testing interactions is to determine whether each mutation is of a different locus (step 2 above). The mutant screen could have unintentionally favored certain genes. Thus, the set of gene loci needs to be defined, as shown in the next section.

KEY CONCEPT The complementation test is a standard way of determining whether or not two recessive mutations are in the same gene. The mutations are united in one cell, and if the cell shows the wild-type phenotype, the mutations have complemented and must be in different genes.

Sorting mutants using the complementation test

How is it possible to decide whether two mutations belong to the same gene? There are several ways. First, each mutant allele could be mapped. Then, if two mutations map to two different chromosomal loci, they are likely of different genes. However, this approach is time consuming on a large set of mutations. A quicker approach often used is the **complementation test.**

In a diploid, the complementation test is performed by intercrossing two individuals that are homozygous for different recessive mutations. The next step is to observe whether the progeny have the wild-type phenotype. If the progeny are wild type, the two recessive mutations must be in *different* genes because the respective wild-type alleles provide wild-type function. In this case, the two mutations are said to have *complemented.* Consider two genes *a1* and *a2*, named after their mutant alleles. We can represent the heterozygotes as follows, depending on whether the genes are on the same chromosome or are on different chromosomes:

Same chromosome:

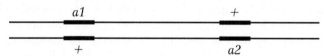

Different chromosomes:

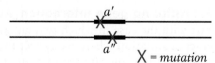

You can see that each locus has one wild-type allele to provide wild-type function, resulting in wild-type progeny.

However, if the progeny are *not* wild type, then the recessive mutations must be alleles of the *same* gene. Because both alleles of the gene are mutants, there is no wild-type allele to provide wild-type function. These alleles could have different mutant sites within the same gene, but they would both be nonfunctional. Consider two recessive mutations, a' and a'', of a gene whose wild type allele is a^+. The heterozygote a'/a'' would be

$$X = mutation$$

Since there is no a^+ allele to provide wild-type function, the progeny will not be wild type.

At the operational level, **complementation** is defined as the production of a wild-type phenotype when two haploid genomes bearing different recessive mutations are united in the same cell.

Let's illustrate the complementation test with an example from harebell plants (genus *Campanula*). The wild-type flower color of this plant is blue. Let's assume that, from a mutant hunt, we have obtained three white-petaled mutants and that they are available as homozygous pure-breeding strains. They all look the same, and so we do not know a priori whether they are genetically identical. We will call the mutant strains $, £, and ¥ to avoid any symbolism using letters, which might imply dominance. When crossed with wild type, each mutant gives the same results in the F_1 and F_2 as follows:

white $ × blue → F_1, all blue → F_2, $\frac{3}{4}$ blue, $\frac{1}{4}$ white

white £ × blue → F_1, all blue → F_2, $\frac{3}{4}$ blue, $\frac{1}{4}$ white

white ¥ × blue → F_1, all blue → F_2, $\frac{3}{4}$ blue, $\frac{1}{4}$ white

In each case, the results show that the mutant condition is determined by the recessive allele of a single gene. However, are they three alleles of one gene, of two genes, or of three genes? Because the mutants are recessive, the question can be answered by the complementation test, which asks if the mutants *complement* one another.

Let us intercross the mutants to test for complementation. Assume that the results of intercrossing mutants $, £, and ¥ are as follows:

white $ × white £ → F_1, all white

white $ × white ¥ → F_1, all blue

white £ × white ¥ → F_1, all blue

From this set of results, we can conclude that mutants $ and £ must be caused by alleles of one gene (say, $w1$) because they do not complement, but ¥ must be caused by a mutant allele of another gene ($w2$) because ¥ complements both $ and £.

KEY CONCEPT When two independently derived recessive mutant alleles producing similar recessive phenotypes fail to complement, they must be alleles of the same gene.

How does complementation work at the molecular level? The normal blue color of the harebell flower is caused by a blue pigment called *anthocyanin*. Pigments are chemicals that absorb certain colors of light; in regard to the

Harebell plant

Flowers of the harebell plant (Campanula species).
[*Gregory G. Dimijian, M.D./Science Source.*]

harebell, the anthocyanin absorbs all wavelengths except blue, which is reflected into the eye of the observer. However, this anthocyanin is made from chemical precursors that are not pigments; that is, they do not absorb light of any specific wavelength and simply reflect back the white light of the sun to the observer, giving a white appearance. The blue pigment is the end product of a series of biochemical conversions of nonpigments. Each step is catalyzed by a specific enzyme encoded by a specific gene. We can explain the results with a pathway as follows:

$$\begin{array}{ccc}
\text{gene } w1^+ & & \text{gene } w2^+ \\
\downarrow & & \downarrow \\
\text{enzyme 1} & & \text{enzyme 2}
\end{array}$$

$$\longrightarrow \text{precursor 1} \longrightarrow \text{precursor 2} \longrightarrow \text{blue anthocyanin}$$

A homozygous mutation in either of the genes will lead to the accumulation of a precursor that will simply make the plant white. Now the mutant designations could be written as follows:

$$\begin{array}{ll}
\$ & w1_\$/w1_\$ \cdot w2^+/w2^+ \\
\pounds & w1_\pounds/w1_\pounds \cdot w2^+/w2^+ \\
\yen & w1^+/w1^+ \cdot w2_\yen/w2_\yen
\end{array}$$

However, in practice, the subscript symbols would be dropped and the genotypes would be written as follows:

$$\begin{array}{ll}
\$ & w1/w1 \cdot w2^+/w2^+ \\
\pounds & w1/w1 \cdot w2^+/w2^+ \\
\yen & w1^+/w1^+ \cdot w2/w2
\end{array}$$

Hence, an F_1 from $\$ \times \pounds$ will be

$$w1/w1 \cdot w2^+/w2^+$$

These F_1 plants will have two defective alleles for $w1$ and will therefore be blocked at step 1. Even though enzyme 2 is fully functional, it has no substrate on which to act; so no blue pigment will be produced, and the phenotype will be white.

The F_1 plants from the other crosses, however, will have the wild-type alleles for both of the enzymes needed to take the intermediates to the final blue product. Their genotypes will be

$$w1^+/w1 \cdot w2^+/w2$$

Hence, we see that complementation is actually a result of the cooperative interaction of the *wild-type* alleles of the two genes. **Figure 5-15** summarizes the interaction of the complementing and noncomplementing white mutants at the genetic and cellular levels.

In a haploid organism, the complementation test cannot be performed by intercrossing. In fungi, an alternative method brings mutant alleles together to test complementation: fusion resulting in a **heterokaryon** (**Figure 5-16**). Fungal cells fuse readily. When two different strains fuse, the haploid nuclei from the different strains occupy one cell, which is the *heterokaryon* (Greek; different kernels).

The nuclei in a heterokaryon do not generally fuse. In one sense, this condition is a "mimic" diploid.

Assume that, in different strains, there are mutations in two different genes conferring the same mutant phenotype—for example, an arginine requirement. We will call these genes *arg-1* and *arg-2*. The genotypes of the two strains can be represented as *arg-1 · arg-2⁺* and *arg-1⁺ · arg-2*. These two strains can be fused to form a heterokaryon with the two nuclei in a shared cytoplasm:

$$\text{Nucleus 1 is } arg\text{-}1 \cdot arg\text{-}2^+$$
$$\text{Nucleus 2 is } arg\text{-}1^+ \cdot arg\text{-}2$$

Because gene products are made in a common cytoplasm, the two wild-type alleles can exert their dominant effect and cooperate to produce a heterokaryon of wild-type phenotype. In other words, the two mutations complement, just as they would in a diploid. If the mutations had been alleles of the same gene, there would have been no complementation.

Analyzing double mutants of random mutations

Recall that, to learn whether two genes interact, we need to assess the phenotype of the double mutant to see if it is different from the combination of both single mutations. The double mutant is obtained by intercrossing. The F_1 is obtained as part of the complementation test; so with the assumption that complementation has been observed, suggesting different genes, the F_1 is selfed or intercrossed to obtain an F_2 homozygous for both mutations. This double mutant may then be identified by looking for Mendelian ratios. For example, if a standard 9:3:3:1 Mendelian ratio is obtained, the phenotype present in only 1/16 of the progeny represents the double mutant (the "1" in 9:3:3:1). In cases of gene interaction, however, the phenotype of the double mutant may not be distinct but will match that of one of the single mutants. In this case, a modified Mendelian ratio will result, such as 9:3:4 or 9:7.

The standard 9:3:3:1 Mendelian ratio is the simplest case, expected if there is no gene interaction and if the two mutations under test are on different chromosomes. This 9:3:3:1 ratio is the null hypothesis: any modified Mendelian ratio representing a departure from this null hypothesis would be informative, as the following examples will show.

KEY CONCEPT A range of modified 9:3:3:1 F_1 ratios can reveal specific types of gene interaction.

The 9:3:3:1 ratio: no gene interaction As a baseline, let's start with the case in which two mutated genes do not interact, a situation where we expect the 9:3:3:1 ratio. Let's look at the inheritance of skin coloration in corn snakes. The snake's natural color is a repeating black-and-orange camouflage pattern, as shown in **Figure 5-17a**. The phenotype is produced by two separate pigments, both of which are under genetic control. One gene determines the orange pigment, and the alleles that

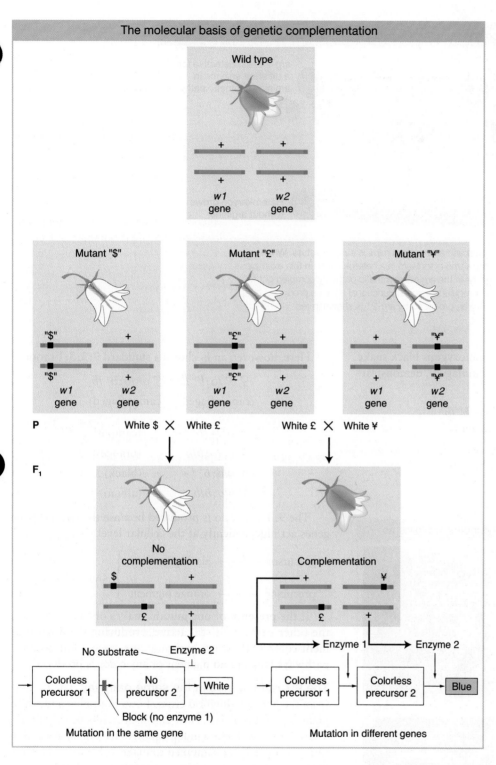

FIGURE 5-15 Three phenotypically identical white harebell mutants— $, £, and ¥—are intercrossed. Mutations in the same gene (such as $ and £) cannot complement because the F₁ has one gene with two mutant alleles. The pathway is blocked and the flowers are white. When the mutations are in different genes (such as £ and ¥), there is complementation by the wild-type alleles of each gene in the F₁ heterozygote. Pigment is synthesized and the flowers are blue.

we will consider are o^+ (presence of orange pigment) and o (absence of orange pigment). Another gene determines the black pigment, and its alleles are b^+ (presence of black pigment) and b (absence of black pigment). These two genes are unlinked. The natural pattern is produced by the genotype $o^+/-;b^+/-$. (The dash represents the presence of either allele.) A snake that is $o/o;b^+/-$ is black because it lacks the orange pigment (Figure 5-17b), and a snake that is $o^+/-;b/b$ is orange because it lacks the

black pigment (Figure 5-17c). The double homozygous recessive $o/o;b/b$ is albino (Figure 5-17d). Notice, however, that the faint pink color of the albino is from yet another pigment, the hemoglobin of the blood that is visible through this snake's skin when the other pigments are absent. The albino snake also clearly shows that there is another element to the skin-pigmentation pattern in addition to pigment: the repeating motif in and around which pigment is deposited.

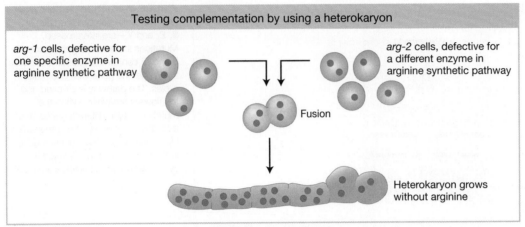

Testing complementation by using a heterokaryon

arg-1 cells, defective for one specific enzyme in arginine synthetic pathway

arg-2 cells, defective for a different enzyme in arginine synthetic pathway

Fusion

Heterokaryon grows without arginine

FIGURE 5-16 A heterokaryon of *Neurospora* and similar fungi mimics a diploid state. When vegetative cells fuse, haploid nuclei share the same cytoplasm in a heterokaryon. In this example, haploid nuclei with mutations in different genes in the arginine synthetic pathway complement to produce a *Neurospora* culture that no longer requires arginine. Functional enzyme produced by *arg-1$^+$* is shown in purple, and functional enzyme produced by *arg-2$^+$* is shown in red.

If a homozygous orange and a homozygous black snake are crossed, the F$_1$ is wild type (camouflaged), demonstrating complementation:

$$♀ \ o^+/o^+ \ ; b/b \times ♂ \ o/o ; b^+/b^+$$
(orange)   (black)
↓
$$F_1 \quad o^+/o ; b^+/b$$
(camouflaged)

Independently synthesized and inherited pigments

(a)   (b)

(c)   (d)

FIGURE 5-17 In corn snakes, combinations of orange and black pigments determine the four phenotypes shown. (a) A wild-type black-and-orange camouflaged snake synthesizes both black and orange pigments. (b) A black snake does not synthesize orange pigment. (c) An orange snake does not synthesize black pigment. (d) An albino snake synthesizes neither black nor orange pigment. [*Anthony Griffiths.*]

Here, however, an F$_2$ shows a standard 9:3:3:1 ratio:

$$♀ \ o^+/o ; b^+/b \times ♂ \ o^+/o ; b^+/b$$
(camouflaged)   (camouflaged)
↓

F$_2$	9	$o^+/- ; b^+/-$	(camouflaged)
	3	$o^+/- ; b/b$	(orange)
	3	$o/o ; b^+/-$	(black)
	1	$o/o ; b/b$	(albino)

The 9:3:3:1 ratio is produced because the two pigment genes act independently at the cellular level.

$$\text{precursor} \xrightarrow{b^+} \text{black pigment}$$
$$\text{precursor} \xrightarrow{o^+} \text{orange pigment}$$
} camouflaged

If the presence of one mutant makes one pathway fail, the other pathway is still active, producing the other pigment color. Only when both mutants are present do both pathways fail, and no pigment of any color is produced.

The 9:7 ratio: genes in the same pathway The F$_2$ ratio from the harebell dihybrid cross shows both blue and white plants in a ratio of 9:7. How can such results be explained? The 9:7 ratio is clearly a modification of the dihybrid 9:3:3:1 ratio with the 3:3:1 combined to make 7; hence, some kind of interaction is inferred. The cross of the two white lines and subsequent generations can be represented as follows:

$$w1/w1 ; w2^+/w2^+ \text{(white)} \times w1^+/w1^+ ; w2/w2 \text{(white)}$$
↓
$$F_1 \qquad w1^+/w1 ; w2^+/w2 \text{ (blue)}$$
$$w1^+/w1 ; w2^+/w2 \times w1^+/w1 ; w2^+/w2$$
↓

F$_2$	9	$w1^+/- ; w2^+/-$	(blue)	9
	3	$w1^+/- ; w2/w2$	(white)	
	3	$w1/w1 ; w2^+/-$	(white)	7
	1	$w1/w1 ; w2/w2$	(white)	

Clearly, in this case, the only way in which a 9:7 ratio is possible is if the double mutant has the same phenotype as the two single mutants. Hence, the modified ratio constitutes a way of identifying the double mutant's phenotype. Furthermore, the identical phenotypes of the single and double mutants suggest that each mutant allele controls a different step in the *same* pathway. The results show that a plant will have white petals if it is homozygous for the recessive mutant allele of *either* gene or *both* genes. To have the blue phenotype, a plant must have at least one copy of the dominant, wild-type allele of both genes because both are needed to complete the sequential steps in the pathway. No matter which is absent, the same pathway fails, producing the same phenotype. Thus, three of the genotypic classes will produce the same phenotype, and so, overall, only two phenotypes result.

The example in harebells entailed different steps in a synthetic pathway. Similar results can come from gene regulation. A regulatory gene often functions by producing a protein that binds to a regulatory site upstream of a target gene, facilitating the transcription of the gene (**Figure 5-18**). In the absence of the regulatory protein, the target gene would be transcribed at very low levels, inadequate for cellular needs. Let's cross a pure line r/r defective for the regulatory protein to a pure line a/a defective for the target protein. The cross is $r/r\,;a^+/a^+ \times r^+/r^+\,;a/a$. The $r^+/r\,;a^+/a$ dihybrid will show complementation between the mutant genotypes because both r^+ and a^+ are present, permitting normal transcription of the wild-type allele. When selfed,

the F_1 dihybrid will also result in a 9:7 phenotypic ratio in the F_2:

Proportion	Genotype	Functional a^+ protein	Ratio
$\frac{9}{16}$	$r^+/-\,;a^+/-$	Yes	9
$\frac{3}{16}$	$r^+/-\,;a/a$	No	
$\frac{3}{16}$	$r/r\,;a^+/-$	No	7
$\frac{1}{16}$	$r/r\,;a/a$	No	

KEY CONCEPT A 9:7 F_2 ratio suggests interacting genes in the same pathway; absence of either gene function leads to absence of the end product of the pathway.

The 9:3:4 ratio: recessive epistasis A 9:3:4 ratio in the F_2 suggests a type of gene interaction called **epistasis**. This word means "stand upon," referring to the situation in which a double mutant shows the phenotype of one mutation but not the other. The overriding mutation is *epistatic*, whereas the overridden one is *hypostatic*. Epistasis also results from genes being in the same pathway. In a simple synthetic pathway, the epistatic mutation is carried by a gene that is farther upstream (earlier in the pathway) than the gene of the overridden mutation (**Figure 5-19**). The mutant phenotype of the upstream gene takes precedence, no matter what is taking place later in the pathway.

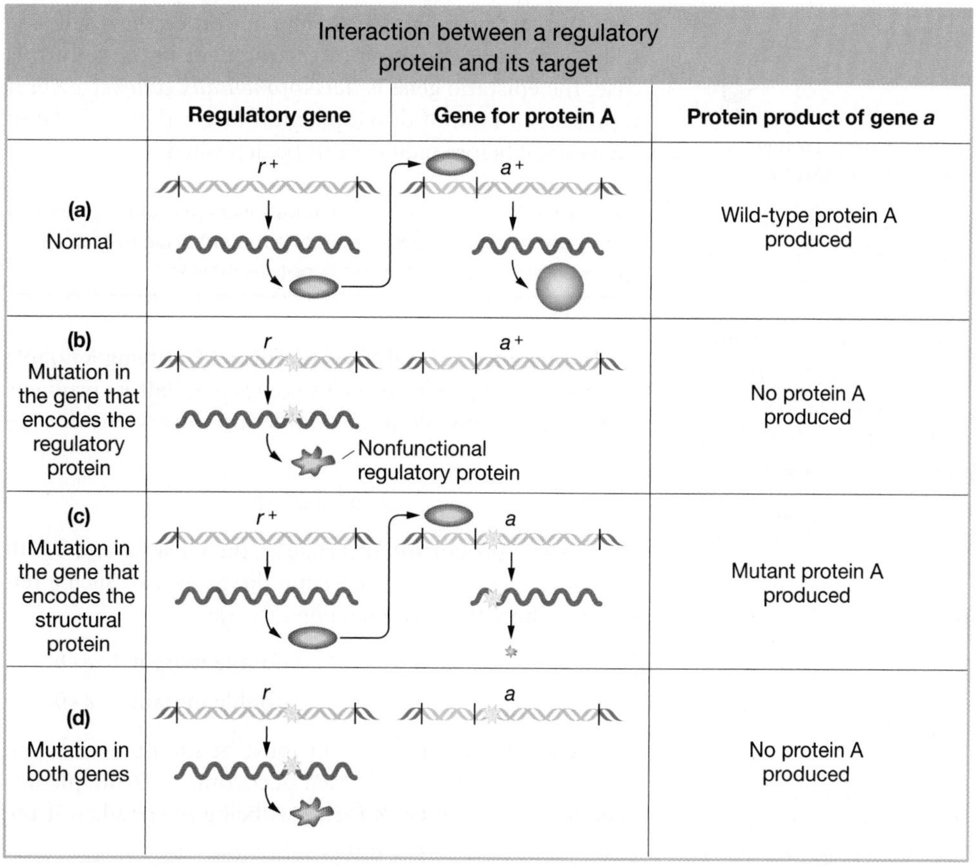

FIGURE 5-18 The r^+ gene encodes a regulatory protein, and the a^+ gene encodes a structural protein. Both must be normal for a functional ("active") structural protein to be synthesized.

Interaction between a regulatory protein and its target

	Regulatory gene	Gene for protein A	Protein product of gene a
(a) Normal	r^+	a^+	Wild-type protein A produced
(b) Mutation in the gene that encodes the regulatory protein	r — Nonfunctional regulatory protein	a^+	No protein A produced
(c) Mutation in the gene that encodes the structural protein	r^+	a	Mutant protein A produced
(d) Mutation in both genes	r	a	No protein A produced

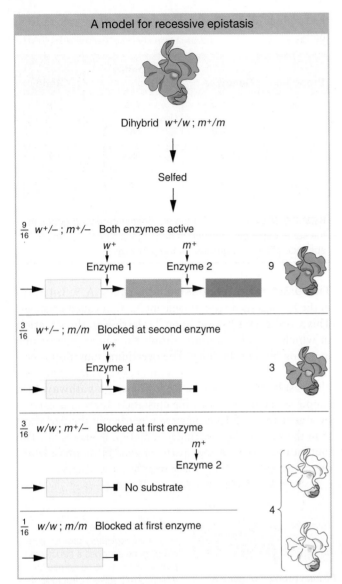

A model for recessive epistasis

Dihybrid w^+/w ; m^+/m

↓

Selfed

↓

$\frac{9}{16}$ $w^+/-$; $m^+/-$ Both enzymes active

w^+ m^+

Enzyme 1   Enzyme 2   9

$\frac{3}{16}$ $w^+/-$; m/m Blocked at second enzyme

w^+

Enzyme 1   3

$\frac{3}{16}$ w/w ; $m^+/-$ Blocked at first enzyme

m^+

Enzyme 2

No substrate

4

$\frac{1}{16}$ w/w ; m/m Blocked at first enzyme

FIGURE 5-19 Wild-type alleles of two genes (w^+ and m^+) encode enzymes catalyzing successive steps in the synthesis of a blue petal pigment. Homozygous m/m plants produce magenta flowers, and homozygous w/w plants produce white flowers. The double mutant w/w ; m/m also produces white flowers, indicating that white is epistatic to magenta.

Let's look at an example concerning petal-pigment synthesis in the plant blue-eyed Mary (*Collinsia parviflora*). From the blue wild type, we'll start with two pure mutant lines, one with white (w/w) and the other with magenta petals (m/m). The w and m genes are not linked. The F_1 and F_2 are as follows:

w/w ; m^+/m^+ (white) × w^+/w^+ ; m/m (magenta)

F_1 $\qquad$ w^+/w ; m^+/m (blue)

↓

w^+/w ; m^+/m × w^+/w ; m^+/m

↓

F_2 $\qquad$ 9 $w^+/-$; $m^+/-$ (blue)   9

3 $w^+/-$; m/m (magenta)   3

3 w/w ; $m^+/-$ (white) ⎫
                                 ⎬ 4
1 w/w ; m/m (white) ⎭

In the F_2, the 9:3:4 phenotypic ratio is diagnostic of recessive epistasis. As in the preceding case, we see, again, that the ratio tells us what the phenotype of the double must be, because the $\frac{4}{16}$ component of the ratio must be a grouping of one single mutant class $\left(\frac{3}{16}\right)$ plus the double mutant class $\left(\frac{1}{16}\right)$. Hence, the double mutant expresses only one of the two mutant phenotypes; so, by definition, white must be epistatic to magenta. (To find the double mutant within the group, white F_2 plants would have to be individually testcrossed.) This interaction is called recessive epistasis because a recessive phenotype (white) overrides the other phenotype. Dominant epistasis will be considered in the next section.

At the cellular level, we can account for the recessive epistasis in *Collinsia* by the following type of pathway (see also Figure 5-19).

$$\text{colorless} \xrightarrow{\text{gene } w^+} \text{magenta} \xrightarrow{\text{gene } m^+} \text{blue}$$

Notice that the epistatic mutation occurs in a step in the pathway leading to blue pigment; this step is upstream of the step that is blocked by the masked mutation.

Another informative case of recessive epistasis is the yellow coat color of some Labrador retriever dogs. Two alleles, B and b, stand for black and brown coats, respectively. The two alleles produce black and brown melanin. The allele e of another gene is epistatic on these alleles, giving a yellow coat (**Figure 5-20**). Therefore, the genotypes $B/-$; e/e and b/b ; e/e both produce a yellow phenotype, whereas $B/-$; $E/-$ and b/b ; $E/-$ are black and brown, respectively. This case of epistasis is *not* caused by an upstream block in a pathway leading to dark pigment. Yellow dogs can make black or brown pigment, as can be seen in their noses and lips. The action of the allele e is to prevent the deposition of the pigment in hairs. In this case, the epistatic gene is *developmentally downstream*; it represents a kind of developmental target that must be of E genotype before pigment can be deposited.

> **KEY CONCEPT** Epistasis is inferred when a mutant allele of one gene masks the expression of a mutant allele of another gene and expresses its own phenotype instead.

In fungi, tetrad analysis is useful in identifying a double mutant. For example, an ascus containing half its products as wild type must contain double mutants. Consider the cross

$$a \cdot b^+ \times a^+ \cdot b$$

In some proportion of progeny, the alleles a and b will segregate together (a nonparental ditype ascus). Such a tetrad will show the following phenotypes:

| wild type | $a^+ \cdot b^+$ | double mutant | $a \cdot b$ |
| wild type | $a^+ \cdot b^+$ | double mutant | $a \cdot b$ |

Hence, the double mutant must be the non-wild-type genotype and can be assessed accordingly. If the phenotype is the a phenotype, then b is being overridden; if the

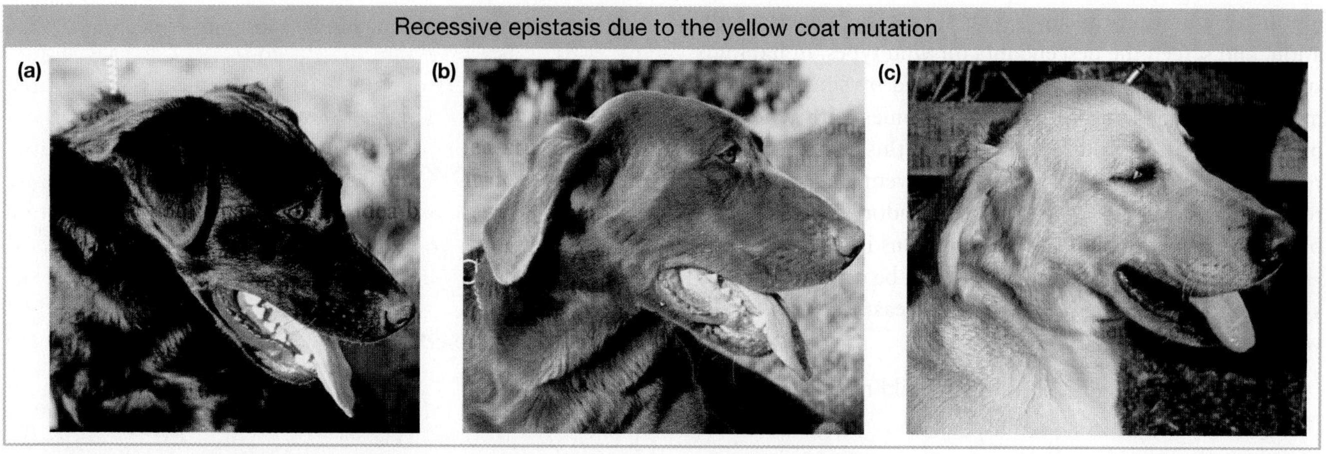

Recessive epistasis due to the yellow coat mutation

(a)　(b)　(c)

FIGURE 5-20 Three different coat colors in Labrador retrievers. Two alleles *B* and *b* of a pigment gene determine (a) black and (b) brown, respectively. At a separate gene, *E* allows color deposition in the coat, and *e/e* prevents deposition, resulting in (c) the gold phenotype. Part c illustrates recessive epistasis. [*Anthony Griffiths.*]

phenotype is the *b* phenotype, then *a* is being overridden. If both phenotypes are present, then there is no epistasis.

The 12:3:1 ratio: dominant epistasis In foxgloves (*Digitalis purpurea*), two genes interact in the pathway that determines petal coloration. The two genes are unlinked. One gene affects the intensity of the red pigment in the petal; allele *d* results in the light red color seen in natural populations of foxgloves, whereas *D* is a mutant allele that produces dark red color (**Figure 5-21**). The other gene determines in which cells the pigment is synthesized: allele *w* allows synthesis of the pigment throughout the petals as in the wild type, but the mutant allele *W* confines pigment synthesis to the small throat spots. If we self a dihybrid *D/d*; *W/w*, then the F_2 ratio is as follows:

$$\left.\begin{array}{l} 9 \quad D/-;W/- \text{ (white with spots)} \\ 3 \quad d/d;W/- \text{ (white with spots)} \end{array}\right\} 12$$

$$3 \quad D/-;w/w \text{ (dark red)} \qquad 3$$

$$1 \quad d/d;w/w \text{ (light red)} \qquad 1$$

The ratio tells us that the dominant allele *W* is epistatic, producing the 12:3:1 ratio. The $\frac{12}{16}$ component of the ratio must include the double mutant class $\left(\frac{9}{16}\right)$, which is clearly white in phenotype, establishing the epistasis of the dominant allele *W*. The two genes act in a common developmental pathway: *W* prevents the synthesis of red pigment but only in a special class of cells constituting the main area of the petal; synthesis is allowed in the throat spots. When synthesis is allowed, the pigment can be produced in either high or low concentrations.

KEY CONCEPT Genetic analysis of gene interaction works in both directions. (1) Specific progeny ratios can be used to infer gene interaction. (2) In a known case of gene interaction, ensuing progeny ratios can be predicted.

Suppressors It is not easy to specifically select or screen for epistatic interactions, and cases of epistasis have to be

Dominant epistasis due to a white mutation

FIGURE 5-21 In foxgloves, *D* and *d* cause dark and light pigments, respectively, whereas the epistatic *W* restricts pigment to the throat spots. [*Anthony Griffiths.*]

built up by the laborious combination of candidate mutations two at a time. However, for our next type of gene interaction, the experimenter can readily select interesting mutant alleles. A **suppressor** is a mutant allele of a gene that reverses the effect of a mutation of another gene, resulting in a wild-type or near-wild-type phenotype. Suppression implies that the target gene and the suppressor gene normally interact at some functional level in their wild-type states. For example, assume that an allele a^+ produces the normal phenotype, whereas a recessive mutant allele *a* results in abnormality. A recessive mutant allele *s* at another gene suppresses the effect of *a*, and so the genotype *a/a · s/s* will have the wild-type (a^+-like) phenotype. Suppressor alleles sometimes have no effect in the absence of the other mutation; in such a case, the phenotype of $a^+/a^+ \cdot s/s$ would be wild type. In other cases, the suppressor allele produces its own abnormal phenotype.

Screening for suppressors is quite straightforward. Start with a mutant in some process of interest, expose this mutant to mutation-causing agents such as high-energy radiation, and screen the descendants for wild types. In haploids such as fungi, screening is accomplished by simply plating mutagenized cells and looking for colonies with wild-type phenotypes. Most wild types arising in this way are merely reversals of the original mutational event and are called **revertants.** However, some will be "pseudorevertants," double mutants in which one of the mutations is a suppressor.

Revertant and suppressed states can be distinguished by appropriate crossing. For example, in yeast, the two results would be distinguished as follows:

true revertant a^+ × standard wild-type a^+

↓

Progeny   all a^+

suppressed mutant $a \cdot s$ × standard wild-type $a^+ \cdot s^+$

↓

Progeny $a^+ \cdot s^+$ wild type
             $a^+ \cdot s$ wild type
             $a \cdot s^+$ original mutant
             $a \cdot s$ wild type (suppressed)

The appearance of the original mutant phenotype identifies the parent as a suppressed mutant.

In diploids, suppressors produce various modified F_2 ratios, which are useful in confirming suppression. Let's look at a real-life example from *Drosophila*. The recessive allele *pd* results in purple eye color when unsuppressed. A recessive allele *su* has no detectable phenotype itself but suppresses the unlinked recessive allele *pd*. Hence, *pd/pd*; *su/su* is wild type in appearance and has red eyes. The following analysis illustrates the inheritance pattern. A homozygous purple-eyed fly is crossed with a homozygous red-eyed stock carrying the suppressor.

$$pd/pd \, ; su^+/su^+ \text{ (purple)} \times pd^+/pd^+ \, ; su/su \text{ (red)}$$

↓

F_1 all $pd^+/pd \, ; su^+/su$ (red)

Self $pd^+/pd \, ; su^+/su$ (red) $\times pd^+/pd \, ; su^+/su$ (red)

↓

F_2 9 $pd^+/- \, ; su^+/-$ red ⎫
         3 $pd^+/- \, ; su/su$ red ⎬ 13
         1 $pd/pd \, ; su/su$ red ⎭
         3 $pd/pd \, ; su^+/-$ purple   3

The overall ratio in the F_2 is 13 red:3 purple. The $\frac{13}{16}$ component must include the double mutant, which is clearly wild type in phenotype. This ratio is expected from a recessive suppressor that itself has no detectable phenotype.

Suppression is sometimes confused with epistasis. However, the key difference is that a suppressor cancels the expression of a mutant allele and restores the corresponding wild-type phenotype. Furthermore, often only two phenotypes segregate (as in the preceding examples) rather than three, as in epistasis.

How do suppressors work at the molecular level? There are many possible mechanisms. A particularly useful type of

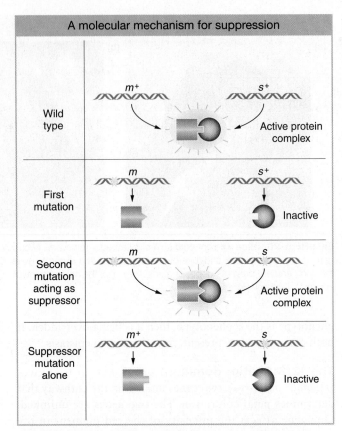

A molecular mechanism for suppression

FIGURE 5-22 A first mutation alters the binding site of one protein so that it can no longer bind to a partner. A suppressor mutation in the partner alters the binding site so that both proteins are able to bind once again.

suppression is based on the physical binding of gene products in the cell—for example, protein–protein binding. Assume that two proteins normally fit together to provide some type of cellular function. When a mutation causes a shape change in one protein, it no longer fits together with the other; hence, the function is lost (**Figure 5-22**). However, a suppressor mutation that causes a compensatory shape change in the second protein can restore fit and hence normal function. In this figure, if the genotypes were diploids representing an F_2 from a dihybrid, then a 14:2 ratio would result because the only mutant genotypes would be $m/m \cdot s^+/s^+$ (1/16) and $m^+/m^+ \cdot s/s$ (1/16), totaling (2/16). If this were a haploid dihybrid cross (such as $m^+ s^+ \times m \, s$), a 1:1 ratio would result. From suppressor ratios generally, interacting proteins often can be deduced.

Alternatively, in situations in which a mutation causes a block in a metabolic pathway, the suppressor finds some way of bypassing the block—for example, by rerouting into the blocked pathway intermediates similar to those beyond the block. In the following example, the suppressor provides an intermediate B to circumvent the block.

No suppressor

A ⟶⊣ ✗ ⟶ ~~product~~

With suppressor

A ⟶⊣ B ⟶ product
  B⟍

In several organisms, *nonsense suppressors* have been found—mutations in tRNA genes resulting in an anticodon that will bind to a premature stop codon within a mutant coding sequence. Hence, the suppressor allows translation to proceed past the former block and make a complete protein rather than a truncated one. Such suppressor mutations often have little effect on the phenotype other than in suppression.

KEY CONCEPT Mutant alleles called suppressors cancel the effect of a mutant allele of another gene, resulting in wild-type phenotype.

Modifiers As the name suggests, a **modifier** mutation at a second locus changes the degree of expression of a mutated gene at the first locus. Regulatory genes provide a simple illustration. As in an earlier example, regulatory proteins bind to the sequence of the DNA upstream of the start site for transcription. These proteins regulate the level of transcription. In the discussion of complementation, we considered a null mutation of a regulatory gene that almost completely prevented transcription. However, some regulatory mutations change the level of transcription of the target gene so that either more or less protein is produced. In other words, a mutation in a regulatory protein can down-regulate or up-regulate the transcribed gene. Let's look at an example using a down-regulating regulatory mutation b, affecting a gene A in a fungus such as yeast. We look at the effect of b on a *leaky* mutation of gene A. A leaky mutation is one with some low level of gene function. We cross a leaky mutation a with the regulatory mutation b:

leaky mutant $a \cdot b^+ \times$ inefficient regulator $a^+ \cdot b$

Progeny	Phenotype
$a^+ \cdot b^+$	wild type
$a^+ \cdot b$	defective (low transcription)
$a \cdot b^+$	defective (defective protein A)
$a \cdot b$	extremely defective (low transcription of defective protein)

Hence, the action of the modifier is seen in the appearance of two grades of mutant phenotypes *within* the a progeny.

Synthetic lethals In some cases, when two viable single mutants are intercrossed, the resulting double mutants are lethal. In a diploid F_2, this result would be manifested as a 9:3:3 ratio because the double mutant (which would be the "1" component of the ratio) would be absent. These **synthetic lethals** can be considered a special category of gene interaction. They can point to specific types of interactions of gene products. For instance, genome analysis has revealed that evolution has produced many duplicate systems within the cell. One advantage of these duplicates might be to provide "backups." If there are null mutations in genes in both duplicate systems, then a faulty system will

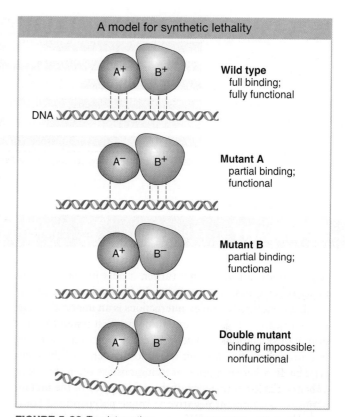

FIGURE 5-23 Two interacting proteins perform some essential function on some substrate such as DNA but must first bind to it. Reduced binding of either protein allows some functions to remain, but reduced binding of both is lethal.

ANIMATED ART Sapling Plus

A model for synthetic lethality

have no backup, and the individual will lack essential function and die. In another instance, a leaky mutation in one step of a pathway may cause the pathway to slow down, but leave enough function for life. However, if double mutants combine, each with a leaky mutation in a different step, the whole pathway grinds to a halt. One version of the latter interaction is two mutations in a protein machine, as shown in **Figure 5-23**.

In the earlier discussions of modified Mendelian ratios, all the crosses were dihybrid selfs. As an exercise, you might want to calculate the ratios that would be produced in the same systems if testcrosses were made instead of selfs.

KEY CONCEPT Two mutations that are individually benign can become lethal when united in the same genotype. Such synthetic lethals can point to some type of normal gene interaction in the wild type.

KEY CONCEPT Genetic analysis of gene interaction makes use of mutant alleles, but the gene interaction revealed is one that is taking place normally in the wild type.

A summary of some of the ratios that reveal gene interaction is shown in **Table 5-2**.

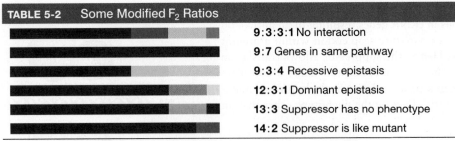

TABLE 5-2   Some Modified F₂ Ratios

9:3:3:1 No interaction
9:7 Genes in same pathway
9:3:4 Recessive epistasis
12:3:1 Dominant epistasis
13:3 Suppressor has no phenotype
14:2 Suppressor is like mutant

Note: Some of these ratios can be produced with other mechanisms of interaction.

SUMMARY

A gene does not act alone; rather, it acts in concert with many other genes in the genome. In forward genetic analysis, deducing these complex interactions is an important stage of the research. Individual mutations are first tested for their dominance relations, a type of allelic interaction. Recessive mutations are often a result of haplosufficiency of the wild-type allele, whereas dominant mutations are often the result either of haploinsufficiency of the wild type or of the mutant acting as a dominant negative (a rogue polypeptide). Some mutations cause severe effects or even death (lethal mutations). Lethality of a homozygous recessive mutation is a way to assess if a gene is essential in the genome.

The interaction of different genes is a result of their participation in the same or connecting pathways of various kinds—synthetic, signal transduction, or developmental. Genetic dissection of gene interactions begins by the experimenter amassing mutants affecting a character of interest. The complementation test determines whether two distinct recessive mutations are of one gene or of two different genes. The mutant genotypes are brought together in an F₁ individual, and if the phenotype is mutant, then no complementation has occurred and the two alleles must be of the same gene. If the phenotype is wild type, then complementation has occurred, and the alleles must be of different genes.

The interaction of different genes can be detected by testing double mutants because allele interaction implies interaction of gene products at the functional level. Some key types of interaction are epistasis, suppression, and synthetic lethality. Epistasis is the replacement of a mutant phenotype produced by one mutation with a mutant phenotype produced by mutation of another gene. The observation of epistasis suggests a common developmental or chemical pathway. A suppressor is a mutation of one gene that can restore wild-type phenotype to a mutation at another gene. Suppressors often reveal physically interacting proteins or nucleic acids. Some combinations of viable mutants are lethal, a result known as synthetic lethality. Synthetic lethals can reveal a variety of interactions, depending on the nature of the mutations.

The different types of gene interactions produce F₂ dihybrid ratios that are modifications of the standard 9:3:3:1. For example, recessive epistasis results in a 9:3:4 ratio.

In more general terms, gene interaction and gene-environment interaction are revealed by incomplete penetrance (the ability of a genotype to express itself in the phenotype) and variable expressivity (the quantitative degree of phenotypic manifestation of a genotype).

KEY TERMS

allelic series (multiple alleles) (p. 154)
codominance (p. 156)
complementation (p. 165)
complementation test (p. 165)
dominant negative mutation (p. 156)
double mutants (p. 164)
epistasis (p. 169)
essential gene (p. 158)
expressivity (p. 161)
full (complete) dominance (p. 154)

functional RNA (p. 163)
heterokaryon (p. 166)
incomplete dominance (p. 156)
lethal allele (p. 158)
modifier (p. 173)
multiple alleles (p. 154)
null mutation (p. 156)
one-gene–one-polypeptide hypothesis (p. 163)
penetrance (p. 160)

permissive temperature (p. 160)
pleiotropic allele (p. 160)
restrictive temperature (p. 160)
revertant (p. 172)
suppressor (p. 171)
synthetic lethal (p. 173)
temperature-sensitive (ts) mutations (p. 160)

SOLVED PROBLEMS

SOLVED PROBLEM 1

Most pedigrees show polydactyly (see Figure 2-25) inherited as a rare autosomal dominant, but the pedigrees of some families do not fully conform to the patterns expected for such inheritance. Such a pedigree is shown here. (The unshaded diamonds stand for the specified number of unaffected persons of unknown sex.)

a. What irregularity does this pedigree show?

b. What genetic phenomenon does this pedigree illustrate?

c. Suggest a specific gene-interaction mechanism that could produce such a pedigree, showing genotypes of pertinent family members.

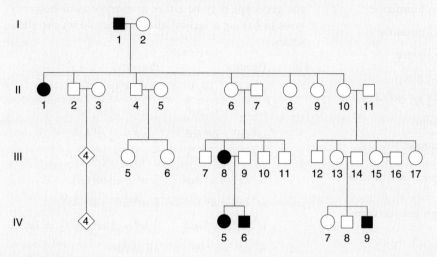

SOLUTION

a. The normal expectation for an autosomal dominant is for each affected individual to have an affected parent, but this expectation is not seen in this pedigree, which constitutes the irregularity. What are some possible explanations?

Could some cases of polydactyly be caused by a different gene, one that is an X-linked dominant gene? This suggestion is not useful, because we still have to explain the absence of the condition in persons II-6 and II-10. Furthermore, postulating recessive inheritance, whether autosomal or sex-linked, requires many people in the pedigree to be heterozygotes, which is inappropriate because polydactyly is a rare condition.

b. Thus, we are left with the conclusion that polydactyly must sometimes be incompletely penetrant. As described in this chapter, some individuals who have the genotype for a particular phenotype do not express it. In this pedigree, II-6 and II-10 seem to belong in this category; they must carry the polydactyly gene inherited from I-1 because they transmit it to their progeny.

c. As discussed in this chapter, environmental suppression of gene expression can cause incomplete penetrance, as can suppression by another gene. To give the requested genetic explanation, we must come up with a genetic hypothesis. What do we need to explain? The key is that I-1 passes the mutation on to two types of progeny, represented by II-1, who expresses the mutant phenotype, and by II-6 and II-10, who do not. (From the pedigree, we cannot tell whether the other children of I-1 have the mutant allele.) Is genetic suppression at work? I-1 does not have a suppressor allele because he expresses

polydactyly. So the only person from whom a suppressor could come is I-2. Furthermore, I-2 must be heterozygous for the suppressor allele because at least one of her children does express polydactyly. Therefore, the suppressor allele must be dominant. We have thus formulated the hypothesis that the mating in generation I must have been

$$(\text{I-1}) \; P/p \cdot s/s \times (\text{I-2}) \; p/p \cdot S/s$$

where S is the suppressor and P is the allele responsible for polydactyly. From this hypothesis, we predict that the progeny will comprise the following four types if the genes assort:

Genotype	Phenotype	Example
$P/p \cdot S/s$	normal (suppressed)	II-6, II-10
$P/p \cdot s/s$	polydactylous	II-1
$p/p \cdot S/s$	normal	
$p/p \cdot s/s$	normal	

If S is rare, the progeny of II-6 and II-10 are:

Progeny genotype	Example
$P/p \cdot S/s$	III-13
$P/p \cdot s/s$	III-8
$p/p \cdot S/s$	
$p/p \cdot s/s$	

We cannot rule out the possibilities that II-2 and II-4 have the genotype $P/p \cdot S/s$ and that by chance none of their descendants are affected.

Beetles of a certain species may have green, blue, or turquoise wing covers. Virgin beetles were selected from a polymorphic laboratory population and mated to determine the inheritance of wing-cover color. The crosses and results were as given in the following table:

Cross	Parents	Progeny
1	blue × green	all blue
2	blue × blue	$\frac{3}{4}$ blue : $\frac{1}{4}$ turquoise
3	green × green	$\frac{3}{4}$ green : $\frac{1}{4}$ turquoise
4	blue × turquoise	$\frac{1}{2}$ blue : $\frac{1}{2}$ turquoise
5	blue × blue	$\frac{3}{4}$ blue : $\frac{1}{4}$ green
6	blue × green	$\frac{1}{2}$ blue : $\frac{1}{2}$ green
7	blue × green	$\frac{1}{2}$ blue : $\frac{1}{4}$ green : $\frac{1}{4}$ turquoise
8	turquoise × turquoise	all turquoise

a. Deduce the genetic basis of wing-cover color in this species.

b. Write the genotypes of all parents and progeny as completely as possible.

SOLUTION

a. These data seem complex at first, but the inheritance pattern becomes clear if we consider the crosses one at a time. A general principle of solving such problems, as we have seen, is to begin by looking over all the crosses and by grouping the data to bring out the patterns.

One clue that emerges from an overview of the data is that all the ratios are one-gene ratios: there is no evidence of two separate genes taking part at all. How can such variation be explained with a single gene? The answer is that there is variation for the single gene itself—that is, multiple allelism. Perhaps there are three alleles of one gene; let's call the gene w (for wing-cover color) and represent the alleles as w^g, w^b, and w^t. Now we have an additional problem, which is to determine the dominance of these alleles.

Cross 1 tells us something about dominance because all of the progeny of a blue × green cross are blue; hence, blue appears to be dominant over green. This conclusion is supported by cross 5, because the green determinant must have been present in the parental stock to appear in the progeny. Cross 3 informs us about the turquoise determinants, which must have been present, although unexpressed, in the parental stock because there are turquoise wing covers in the progeny. So green must be dominant over turquoise. Hence, we have formed a model in which the dominance is $w^b > w^g > w^t$. Indeed, the inferred position of the w^t allele at the bottom of the dominance series is supported by the results of cross 7, where turquoise shows up in the progeny of a blue × green cross.

b. Now it is just a matter of deducing the specific genotypes. Notice that the question states that the parents were taken from a polymorphic population, which means that they could be either homozygous or heterozygous. A parent with blue wing covers, for example, might be homozygous (w^b/w^b) or heterozygous (w^b/w^g or w^b/w^t). Here, a little trial and error and common sense are called for, but, by this stage, the question has essentially been answered, and all that remains is to "cross the t's and dot the i's." The following genotypes explain the results. A dash indicates that the genotype may be either homozygous or heterozygous in having a second allele farther down the allelic series.

Cross	Parents	Progeny
1	$w^b/w^b \times w^g/\text{–}$	w^b/w^g or $w^b/\text{–}$
2	$w^b/w^t \times w^b/w^t$	$\frac{3}{4} w^b/\text{–} : \frac{1}{4} w^t/w^t$
3	$w^g/w^t \times w^g/w^t$	$\frac{3}{4} w^g/\text{–} : \frac{1}{4} w^t/w^t$
4	$w^b/w^t \times w^t/w^t$	$\frac{1}{2} w^b/w^t : \frac{1}{2} w^t/w^t$
5	$w^b/w^g \times w^b/w^g$	$\frac{3}{4} w^b/\text{–} : \frac{1}{4} w^g/w^g$
6	$w^b/w^g \times w^g/w^g$	$\frac{1}{2} w^b/w^g : \frac{1}{2} w^g/w^g$
7	$w^b/w^t \times w^g/w^t$	$\frac{1}{2} w^b/\text{–} : \frac{1}{4} w^g/w^t : \frac{1}{4} w^t/w^t$
8	$w^t/w^t \times w^t/w^t$	all w^t/w^t

The leaves of pineapples can be classified into three types: spiny (S), spiny tip (ST), and piping (nonspiny; P). In crosses between pure strains followed by intercrosses of the F_1, the following results appeared:

		Phenotypes	
Cross	Parental	F_1	F_2
1	ST × S	ST	99 ST : 34 S
2	P × ST	P	120 P : 39 ST
3	P × S	P	95 P : 25 ST : 8 S

a. Assign gene symbols. Explain these results in regard to the genotypes produced and their ratios.

b. Using the model from part *a*, give the phenotypic ratios that you would expect if you crossed (1) the F_1 progeny from piping × spiny with the spiny parental stock and (2) the F_1 progeny of piping × spiny with the F_1 progeny of spiny × spiny tip.

SOLUTION

a. First, let's look at the F_2 ratios. We have clear 3:1 ratios in crosses 1 and 2, indicating single-gene segregations. Cross 3, however, shows a ratio that is almost certainly a 12:3:1 ratio. How do we know this ratio? Well, there are simply not that many complex ratios in genetics, and trial and error brings us to the 12:3:1 quite quickly. In the 128 progeny total, the numbers of

96:24:8 are expected, but the actual numbers fit these expectations remarkably well.

One of the principles of this chapter is that modified Mendelian ratios reveal gene interactions. Cross 3 gives F_2 numbers appropriate for a modified dihybrid Mendelian ratio, and so it looks as if we are dealing with a two-gene interaction. It seems the most promising place to start; we can return to crosses 1 and 2 and try to fit them in later.

Any dihybrid ratio is based on the phenotypic proportions 9:3:3:1. Our observed modification groups them as follows:

$$\left.\begin{array}{l} 9\ A/-;B/- \\ 3\ A/-;b/b \end{array}\right\}\ 12\ \text{piping}$$

$$3\ a/a;B/-\qquad 3\ \text{spiny tip}$$

$$1\ a/a;b/b\qquad 1\ \text{spiny}$$

So, without worrying about the name of the type of gene interaction (we are not asked to supply this anyway), we can already define our three pineapple-leaf phenotypes in relation to the proposed allelic pairs A/a and B/b:

$$\text{piping} = A/-\ (B/b\ \text{irrelevant})$$
$$\text{spiny tip} = a/a;B/-$$
$$\text{spiny} = a/a;b/b$$

What about the parents of cross 3? The spiny parent must be $a/a;b/b$, and, because the B gene is needed to produce F_2 spiny-tip leaves, the piping parent must be $A/A;B/B$. (Note that we are told that all parents are pure, or homozygous.) The F_1 must therefore be $A/a;B/b$.

Without further thought, we can write out cross 1 as follows:

$$a/a\ ;\ B/B\ \times\ a/a\ ;\ b/b\ \longrightarrow$$

$$a/a\ ;\ B/b\ \left\langle\begin{array}{l}\frac{3}{4}\ a/a\ ;\ B/- \\ \frac{1}{4}\ a/a\ ;\ b/b \end{array}\right.$$

Cross 2 can be partly written out without further thought by using our arbitrary gene symbols:

$$A/A\ ;\ -/-\ \times\ a/a\ ;\ B/B\ \longrightarrow$$

$$A/a\ ;\ B/-\ \left\langle\begin{array}{l}\frac{3}{4}\ A/-\ ;\ -/- \\ \frac{1}{4}\ a/a\ ;\ B/- \end{array}\right.$$

We know that the F_2 of cross 2 shows single-gene segregation, and it seems certain now that the A/a allelic pair has a role. But the B allele is needed to produce the spiny-tip phenotype, and so all plants must be homozygous B/B:

$$A/A\ ;\ B/B\ \times\ a/a\ ;\ B/B\ \longrightarrow$$

$$A/a\ ;\ B/B\ \left\langle\begin{array}{l}\frac{3}{4}\ A/-\ ;\ B/B \\ \frac{1}{4}\ a/a\ ;\ B/B \end{array}\right.$$

Notice that the two single-gene segregations in crosses 1 and 2 do not show that the genes are *not* interacting. What is shown is that the two-gene interaction is not *revealed* by these crosses—only by cross 3, in which the F_1 is heterozygous for both genes.

b. Now it is simply a matter of using Mendel's laws to predict cross outcomes:

(1) $A/a;B/b \times a/a;b/b \longrightarrow$

(independent assortment in a standard testcross)

$$\left.\begin{array}{l}\frac{1}{4}\ A/a;B/b \\ \frac{1}{4}\ A/a;b/b \end{array}\right\}\ \text{piping}$$
$$\frac{1}{4}\ a/a;B/b\qquad \text{spiny tip}$$
$$\frac{1}{4}\ a/a;b/b\qquad \text{spiny}$$

(2) $A/a;B/b \times a/a;B/b \longrightarrow$

$$\frac{1}{2}\ A/a\ \left\langle\begin{array}{ll}\frac{3}{4}\ B/- \longrightarrow \frac{3}{8} \\ \frac{1}{4}\ b/b \longrightarrow \frac{1}{8} \end{array}\right\}\ \frac{1}{2}\ \text{piping}$$

$$\frac{1}{2}\ a/a\ \left\langle\begin{array}{ll}\frac{3}{4}\ B/- \longrightarrow \frac{3}{8}\quad \text{spiny tip} \\ \frac{1}{4}\ b/b \longrightarrow \frac{1}{8}\quad \text{spiny} \end{array}\right.$$

PROBLEMS

Visit SaplingPlus for supplemental content. Problems with the ⊠ icon are available for review/grading. Problems with the ▭ icon have a Problem Solving Video.

WORKING WITH THE FIGURES

(The first 19 questions require inspection of text figures.)

1. a. In Figure 5-1, what do the yellow stars represent?

   b. Explain in your own words why the heterozygote is functionally wild type.

   c. In the system defined in this figure, if we assume the gene codes for an enzyme catalyzing the synthesis of a black pigment, what would be the phenotype of the heterozygote?

d. At the structural level, what might be the difference between the proteins represented by the colors orange and yellow?

2. a. In Figure 5-2, explain how the mutant polypeptide acts as a spoiler and what its net effect on phenotype is.

   b. What might cause a bend in the mutant protein?

   c. In Model 1, what can you say about the possibility of up-regulation of the protein in the heterozygote?

3. In Figure 5-4, does the photo show the blood of one individual (if so, which one), or a mixture of bloods (if so, which ones)?

4. a. In Figure 5-5, what is the object represented by the color blue?

   b. Is it true to say that the sickle-cell hemoglobin migrates faster than normal hemoglobin?

   c. What might cause the different migration rates?

5. a. In Figure 5-6, assess the allele V^f with respect to the V^{by} allele. Is it dominant? Recessive? Codominant? Incompletely dominant?

   b. In this figure, is there any case in which the heterozygote has a truly new phenotype?

   c. Predict the phenotype of the heterozygotes of v combined with the other alleles.

6. In Figure 5-7, if you assume that all the progeny are visible, is the observed color ratio the one expected?

7. In Figure 5-9, propose a specific genetic explanation for individual Q (give a possible genotype, defining the alleles).

8. In Figure 5-10, point to the individuals that show full expressivity. 🐟

9. Speculate logically on the minimum number of modifying genes and alleles that could produce the variation shown by the allele S^P in Figure 5-11.

10. From a knowledge of the structures shown in Figure 5-13, do you think Beadle and Tatum might have had a clue about the sequential steps in the synthetic pathway before doing their genetic tests?

11. a. In Figure 5-14, in view of the position of HPA oxidase earlier in the pathway compared to that of HA oxidase, would you expect people with tyrosinosis to show symptoms of alkaptonuria?

    b. If a double mutant could be found, would you expect tyrosinosis to be epistatic to alkaptonuria?

    c. Do you think it might be possible to cure the symptoms of PKU by ingesting tyrosine? (Research this possibility yourself.)

    d. How might you treat cretinism? (Research this possibility yourself.)

e. Do you think it would be possible to treat albinism by ingesting melanin? (Research this possibility yourself.)

12. a. In Figure 5-15, what do the dollar, pound, and yen symbols represent?

    b. Why can't the left-hand F_1 heterozygote synthesize blue pigment?

    c. Draw out the results of crossing the $ and ¥ lines.

    d. Write out all the genotypes and phenotypes of progeny from a self of the blue F_1.

    e. Write out all the genotypes and phenotypes produced by crossing the blue F_1 to the white F_1.

13. a. In Figure 5-16, explain at the protein level why this heterokaryon can grow on minimal medium.

    b. A heterokaryon produces spores by pinching off cells that contain a few nuclei. Will any of these spores be arginine-dependent? Explain.

14. a. In Figure 5-17, write possible genotypes for each of the four snakes illustrated.

    b. Explain in short sentences the meaning of the header for this figure.

15. a. In Figure 5-18, which panel represents the double mutant?

    b. State the function of the regulatory gene.

    c. In the situation in panel b, would protein from the active protein gene be made?

    d. What is the function of the pale green region?

    e. What is the element represented in yellow?

    f. Panels b and d have the same outcome: state the two different mechanisms that produce this.

16. a. In Figure 5-19, if you selfed 10 different F_2 pink plants, would you expect to find any white-flowered plants among the offspring? Any blue-flowered plants?

    b. Some white F_2 plants have a functional enzyme 2: Since enzyme 2 produces the blue pigment, why are these plants not blue?

17. In Figure 5-21, write down possible genotypes for each of the three petals. 🐟

18. a. In Figure 5-22, what do the square/triangular pegs and holes represent?

    b. Is the suppressor mutation alone wild type in phenotype?

    c. Would it be reasonable to call the s gene a suppressor gene in a wild-type cell?

19. a. In Figure 5-23, explain why the interacting alleles are called synthetic lethals.

    b. For the model to work, is it essential that the red and blue proteins bind to each other?

BASIC PROBLEMS

20. In humans, the disease galactosemia causes intellectual disabilities at an early age. Lactose (milk sugar) is broken down to galactose plus glucose. Normally, galactose is broken down further by the enzyme galactose-1-phosphate uridyltransferase (GALT). However, in patients with galactosemia, GALT is inactive, leading to a buildup of high levels of galactose, which, in the brain, causes intellectual disabilities. How would you provide a secondary cure for galactosemia? Would you expect this disease phenotype to be dominant or recessive?

21. In humans, PKU (phenylketonuria) is a recessive disease caused by an enzyme inefficiency at step A in the following simplified reaction sequence, and AKU (alkaptonuria) is another recessive disease due to an enzyme inefficiency in one of the steps summarized as step B here:

$$\text{phenylalanine} \xrightarrow{A} \text{tyrosine} \xrightarrow{B} CO_2 + H_2O$$

A person with PKU marries a person with AKU. What phenotypes do you expect for their children? All normal, all having PKU only, all having AKU only, all having both PKU and AKU, or some having AKU and some having PKU?

22. In *Drosophila*, the autosomal recessive *bw* causes a dark brown eye, and the unlinked autosomal recessive *st* causes a bright scarlet eye. A homozygote for both genes has a white eye. Thus, we have the following correspondences between genotypes and phenotypes:

$st^+/st^+ ; bw^+/bw^+$ = red eye (wild type)

$st^+/st^+ ; bw/bw$ = brown eye

$st/st ; bw^+/bw^+$ = scarlet eye

$st/st ; bw/bw$ = white eye

Construct a hypothetical biosynthetic pathway showing how the gene products interact and why the different mutant combinations have different phenotypes.

23. Several mutants are isolated, all of which require compound G for growth. The compounds (A to E) in the biosynthetic pathway to G are known, but their order in the pathway is not known. Each compound is tested for its ability to support the growth of each mutant (1 to 5). In the following table, a plus sign indicates growth and a minus sign indicates no growth.

		Compound tested					
		A	B	C	D	E	G
Mutant	1	–	–	–	+	–	+
	2	–	+	–	+	–	+
	3	–	–	–	–	–	+
	4	–	+	+	+	–	+
	5	+	+	+	+	–	+

a. What is the order of compounds A to E in the pathway?

b. At which point in the pathway is each mutant blocked?

c. Would a heterokaryon composed of double mutants 1,3 and 2,4 grow on a minimal medium? Would 1,3 and 3,4? Would 1,2 and 2,4 and 1,4?

24. In a certain plant, the flower petals are normally purple. Two recessive mutations arise in separate plants and are found to be on different chromosomes. Mutation 1 (m_1) gives blue petals when homozygous (m_1/m_1). Mutation 2 (m_2) gives red petals when homozygous (m_2/m_2). Biochemists working on the synthesis of flower pigments in this species have already described the following pathway:

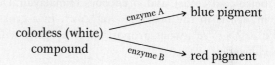

a. Which mutant would you expect to be deficient in enzyme A activity?

b. A plant has the genotype M_1/m_1; M_2/m_2. What would you expect its phenotype to be?

c. If the plant in part *b* is selfed, what colors of progeny would you expect, and in what proportions?

d. Why are these mutants recessive?

25. In sweet peas, the synthesis of purple anthocyanin pigment in the petals is controlled by two genes, *B* and *D*. The pathway is

$$\text{white intermediate} \xrightarrow[\text{enzyme}]{\text{gene } B} \text{blue intermediate} \xrightarrow[\text{enzyme}]{\text{gene } D} \text{anthocyanin (purple)}$$

a. What color petals would you expect in a pure-breeding plant unable to catalyze the first reaction?

b. What color petals would you expect in a pure-breeding plant unable to catalyze the second reaction?

c. If the plants in parts *a* and *b* are crossed, what color petals will the F_1 plants have?

d. What ratio of purple:blue:white plants would you expect in the F_2?

26. If a man of blood-group AB marries a woman of blood-group A whose father was of blood-group O, to what different blood groups can this man and woman expect their children to belong?

27. Most of the feathers of erminette fowl are light colored, with an occasional black one, giving a flecked appearance. A cross of two erminettes produced a total of 48 progeny, consisting of 22 erminettes, 14 blacks, and 12 pure whites. What genetic basis of the erminette pattern is suggested? How would you test your hypotheses?

28. Radishes may be long, round, or oval, and they may be red, white, or purple. You cross a long, white variety with a round, red one and obtain an oval, purple F_1. The F_2 shows nine phenotypic classes as follows: 9 long, red; 15 long, purple; 19 oval, red; 32 oval, purple; 8 long, white; 16 round, purple; 8 round, white; 16 oval, white; and 9 round, red.

    a. Provide a genetic explanation of these results. Be sure to define the genotypes and show the constitution of the parents, the F_1, and the F_2.

    b. Predict the genotypic and phenotypic proportions in the progeny of a cross between a long, purple radish and an oval, purple one.

29. In the multiple-allele series that determines coat color in rabbits, c^+ encodes agouti, c^{ch} encodes chinchilla (a beige coat color), and c^h encodes Himalayan. Dominance is in the order $c^+ > c^{ch} > c^h$. In a cross of $c^+/c^{ch} \times c^{ch}/c^h$, what proportion of progeny will be chinchilla?

30. Black, sepia, cream, and albino are coat colors of guinea pigs. Individual animals (not necessarily from pure lines) showing these colors were intercrossed; the results are tabulated as follows, where the abbreviations A (albino), B (black), C (cream), and S (sepia) represent the phenotypes:

Cross	Parental phenotypes	Phenotypes of progeny			
		B	S	C	A
1	B × B	22	0	0	7
2	B × A	10	9	0	0
3	C × C	0	0	34	11
4	S × C	0	24	11	12
5	B × A	13	0	12	0
6	B × C	19	20	0	0
7	B × S	18	20	0	0
8	B × S	14	8	6	0
9	S × S	0	26	9	0
10	C × A	0	0	15	17

    a. Deduce the inheritance of these coat colors, and use gene symbols of your own choosing. Show all parent and progeny genotypes.

    b. If the black animals in crosses 7 and 8 are crossed, what progeny proportions can you predict by using your model?

31. In a maternity ward, four babies become accidentally mixed up. The ABO types of the four babies are known to be O, A, B, and AB. The ABO types of the four sets of parents are determined. Indicate which baby belongs to each set of parents: (a) AB × O, (b) A × O, (c) A × AB, (d) O × O.

32. Consider two blood polymorphisms that humans have in addition to the ABO system. Two alleles L^M and L^N determine the M, N, and MN blood groups. The dominant allele R of a different gene causes a person to have the Rh$^+$(rhesus positive) phenotype, whereas the homozygote for r is Rh$^-$ (rhesus negative). Two men took a paternity dispute to court, each claiming three children to be his own. The blood groups of the men, the children, and their mother were as follows:

Person	Blood group		
husband	O	M	Rh$^+$
wife's lover	AB	MN	Rh$^-$
wife	A	N	Rh$^+$
child 1	O	MN	Rh$^+$
child 2	A	N	Rh$^+$
child 3	A	MN	Rh$^-$

From this evidence, can the paternity of the children be established?

33. On a fox ranch in Wisconsin, a mutation arose that gave a "platinum" coat color. The platinum color proved very popular with buyers of fox coats, but the breeders could not develop a pure-breeding platinum strain. Every time two platinums were crossed, some normal foxes appeared in the progeny. For example, the repeated matings of the same pair of platinums produced 82 platinum and 38 normal progeny. All other such matings gave similar progeny ratios. State a concise genetic hypothesis that accounts for these results.

34. For several years, Hans Nachtsheim investigated an inherited anomaly of the white blood cells of rabbits. This anomaly, termed the *Pelger anomaly*, is the arrest of the segmentation of the nuclei of certain white cells. This anomaly does not appear to seriously burden the rabbits.

    a. When rabbits showing the Pelger anomaly were mated with rabbits from a true-breeding normal stock, Nachtsheim counted 217 offspring showing the Pelger anomaly and 237 normal progeny. What is the genetic basis of the Pelger anomaly?

    b. When rabbits with the Pelger anomaly were mated with each other, Nachtsheim found 223 normal progeny, 439 with the Pelger anomaly, and 39 extremely abnormal progeny. These very abnormal progeny not only had defective white blood cells, but also showed severe deformities of the skeletal system; almost all of them died soon after birth. In genetic terms, what do you suppose these extremely defective rabbits represented? Why were there only 39 of them?

c. What additional experimental evidence might you collect to test your hypothesis in part *b*?

d. In Berlin, about 1 human in 1000 shows a Pelger anomaly of white blood cells very similar to that described for rabbits. The anomaly is inherited as a simple dominant, but the homozygous type has not been observed in humans. Based on the condition in rabbits, why do you suppose the human homozygous has not been observed?

e. Again by analogy with rabbits, what phenotypes and genotypes would you expect among the children of a man and woman who both show the Pelger anomaly?

(Data from A. M. Srb, R. D. Owen, and R. S. Edgar, *General Genetics*, 2nd ed. W. H. Freeman and Company, 1965.)

35. Two normal-looking fruit flies were crossed, and, in the progeny, there were 202 females and 98 males.

a. What is unusual about this result?

b. Provide a genetic explanation for this anomaly.

c. Provide a test of your hypothesis.

36. You have been given a virgin *Drosophila* female. You notice that the bristles on her thorax are much shorter than normal. You mate her with a normal male (with long bristles) and obtain the following F$_1$ progeny: $\frac{1}{3}$ short-bristled females, $\frac{1}{3}$ long-bristled females, and $\frac{1}{3}$ long-bristled males. A cross of the F$_1$ long-bristled females with their brothers gives only long-bristled F$_2$. A cross of short-bristled females with their brothers gives $\frac{1}{3}$ short-bristled females, $\frac{1}{3}$ long-bristled females, and $\frac{1}{3}$ long-bristled males. Provide a genetic hypothesis to account for all these results, showing genotypes in every cross.

37. A dominant allele *H* reduces the number of body bristles that *Drosophila* flies have, giving rise to a "hairless" phenotype. In the homozygous condition, *H* is lethal. An independently assorting dominant allele *S* has no effect on bristle number except in the presence of *H*, in which case a single dose of *S* suppresses the hairless phenotype, thus restoring the hairy phenotype. However, *S* also is lethal in the homozygous (*S/S*) condition.

a. What ratio of hairy to hairless flies would you find in the live progeny of a cross between two hairy flies both carrying *H* in the suppressed condition?

b. When the hairless progeny are backcrossed with a parental hairy fly, what phenotypic ratio would you expect to find among their live progeny?

38. After irradiating wild-type cells of *Neurospora* (a haploid fungus), a geneticist finds two leucine-requiring auxotrophic mutants. He combines the two mutants in a heterokaryon and discovers that the heterokaryon is prototrophic.

a. Were the mutations in the two auxotrophs in the *same* gene in the pathway for synthesizing leucine, or in two *different* genes in that pathway? Explain.

b. Write the genotype of the two strains according to your model.

c. What progeny, and in what proportions, would you predict from crossing the two auxotrophic mutants? (Assume independent assortment.)

39. A yeast geneticist irradiates haploid cells of a strain that is an adenine-requiring auxotrophic mutant, caused by mutation of the gene *ade1*. Millions of the irradiated cells are plated on minimal medium, and a small number of cells divide and produce prototrophic colonies. These colonies are crossed individually with a wild-type strain. Two types of results are obtained:

1. prototroph × wild type : progeny all prototrophic

2. prototroph × wild type : progeny 75% prototrophic, 25% adenine-requiring auxotrophs

a. Explain the difference between these two types of results.

b. Write the genotypes of the prototrophs in each case.

c. What progeny phenotypes and ratios do you predict from crossing a prototroph of type 2 by the original *ade1* auxotroph?

40. In roses, the synthesis of red pigment is by two steps in a pathway, as follows:

colorless intermediate $\xrightarrow{\text{gene } P}$

magenta intermediate $\xrightarrow{\text{gene } Q}$ red pigment

a. What would the phenotype be of a plant homozygous for a null mutation of gene *P*?

b. What would the phenotype be of a plant homozygous for a null mutation of gene *Q*?

c. What would the phenotype be of a plant homozygous for null mutations of genes *P* and *Q*?

d. Write the genotypes of the three strains in parts *a*, *b*, and *c*.

e. What F$_2$ ratio is expected from crossing plants from parts *a* and *b*? (Assume independent assortment.)

41. Because snapdragons (*Antirrhinum*) possess the pigment anthocyanin, they have reddish purple petals. Two pure anthocyaninless lines of *Antirrhinum* were developed, one in California and one in Holland. They looked identical in having no red pigment at all, manifested as white (albino) flowers. However, when petals from the two lines were ground up together in buffer in the same test tube, the solution, which appeared colorless at first, gradually turned red.

a. What control experiments should an investigator conduct before proceeding with further analysis?

b. What could account for the production of the red color in the test tube?

c. According to your explanation for part *b*, what would be the genotypes of the two lines?

d. If the two white lines were crossed, what would you predict the phenotypes of the F_1 and F_2 to be?

42. The frizzle fowl is much admired by poultry fanciers. It gets its name from the unusual way that its feathers curl up, giving the impression that it has been (in the memorable words of animal geneticist F. B. Hutt) "pulled backwards through a knothole." Unfortunately, frizzle fowl do not breed true: when two frizzles are intercrossed, they always produce 50 percent frizzles, 25 percent normal, and 25 percent with peculiar woolly feathers that soon fall out, leaving the birds naked.

a. Give a genetic explanation for these results, showing genotypes of all phenotypes, and provide a statement of how your explanation works.

b. If you wanted to mass-produce frizzle fowl for sale, which types would be best to use as a breeding pair?

43. The petals of the plant *Collinsia parviflora* are normally blue, giving the species its common name, blue-eyed Mary. Two pure-breeding lines were obtained from color variants found in nature; the first line had pink petals, and the second line had white petals. The following crosses were made between pure lines, with the results shown:

Parents	F_1	F_2
blue × white	blue	101 blue, 33 white
blue × pink	blue	192 blue, 63 pink
pink × white	blue	272 blue, 121 white, 89 pink

a. Explain these results genetically. Define the allele symbols that you use, and show the genetic constitution of the parents, the F_1, and the F_2 in each cross.

b. A cross between a certain blue F_2 plant and a certain white F_2 plant gave progeny of which $\frac{3}{8}$ were blue, $\frac{1}{8}$ were pink, and $\frac{1}{2}$ were white. What must the genotypes of these two F_2 plants have been?

UNPACKING PROBLEM 43

Before attempting a solution to this problem, try answering the following questions:

1. What is the character being studied?

2. What is the wild-type phenotype?

3. What is a variant?

4. What are the variants in this problem?

5. What does "in nature" mean?

6. In what way would the variants have been found in nature? (Describe the scene.)

7. At which stages in the experiments would seeds be used?

8. Would the way of writing a cross "blue × white," for example, mean the same as "white × blue"? Would you expect similar results? Why or why not?

9. In what way do the first two rows in the table differ from the third row?

10. Which phenotypes are dominant?

11. What is complementation?

12. Where does the blueness come from in the progeny of the pink × white cross?

13. What genetic phenomenon does the production of a blue F_1 from pink and white parents represent?

14. List any ratios that you can see.

15. Are there any monohybrid ratios?

16. Are there any dihybrid ratios?

17. What does observing monohybrid and dihybrid ratios tell you?

18. List four modified Mendelian ratios that you can think of.

19. Are there any modified Mendelian ratios in the problem?

20. What do modified Mendelian ratios indicate generally?

21. What is indicated by the specific modified ratio or ratios in this problem?

22. Draw chromosomes representing the meioses in the parents in the cross blue × white and representing meiosis in the F_1.

23. Repeat step 22 for the cross blue × pink.

Now try to solve the problem. If you are unable to do so, try to identify the obstacle and write a sentence or two describing your difficulty. Then go back to the expansion questions and see if any of them relate to your difficulty. If this approach does not work, inspect the Learning Objectives and Key Concepts of this chapter and ask yourself which might be relevant to your difficulty.

44. A woman who owned a purebred albino poodle (an autosomal recessive phenotype) wanted white puppies. She took the dog to a breeder, who said he would mate the female with an albino stud male, also from a pure stock. When six puppies were born, all of them were black; so the woman sued the breeder, claiming that he replaced the stud male with a black dog, giving her six unwanted puppies. You are called in as an expert witness, and the defense asks you if it is possible

to produce black offspring from two pure-breeding recessive albino parents. What testimony do you give?

45. A snapdragon plant that bred true for white petals was crossed with a plant that bred true for purple petals, and all the F_1 had white petals. The F_1 was selfed. Among the F_2, three phenotypes were observed in the following numbers:

white	240
solid purple	61
spotted purple	19
Total	320

   a. Propose an explanation for these results, showing genotypes of all generations (make up and explain your symbols).

   b. A white F_2 plant was crossed with a solid purple F_2 plant, and the progeny were

white	50%
solid purple	25%
spotted purple	25%

   What were the genotypes of the F_2 plants crossed?

46. Most flour beetles are black, but several color variants are known. Crosses of pure-breeding parents produced the following results (see table) in the F_1 generation, and intercrossing the F_1 from each cross gave the ratios shown for the F_2 generation. The phenotypes are abbreviated Bl, black; Br, brown; Y, yellow; and W, white.

Cross	Parents	F_1	F_2
1	Br × Y	Br	3 Br:1 Y
2	Bl × Br	Bl	3 Bl:1 Br
3	Bl × Y	Bl	3 Bl:1 Y
4	W × Y	Bl	9 Bl:3 Y:4 W
5	W × Br	Bl	9 Bl:3 Br:4 W
6	Bl × W	Bl	9 Bl:3 Y:4 W

   a. From these results, deduce and explain the inheritance of these colors.

   b. Write the genotypes of each of the parents, the F_1, and the F_2 in all crosses.

47. Two albinos marry and have four normal children. How is this possible?

48. Consider the production of flower color in the Japanese morning glory (*Pharbitis nil*). Dominant alleles of either of two separate genes ($A/-$ · b/b or a/a · $B/-$) produce purple petals. $A/-$ · $B/-$ produces blue petals, and a/a · b/b produces scarlet petals. Deduce the genotypes of parents and progeny in the following crosses:

Cross	Parents	Progeny
1	blue × scarlet	$\frac{1}{4}$ blue: $\frac{1}{2}$ purple: $\frac{1}{4}$ scarlet
2	purple × purple	$\frac{1}{4}$ blue: $\frac{1}{2}$ purple: $\frac{1}{4}$ scarlet
3	blue × blue	$\frac{3}{4}$ blue: $\frac{1}{4}$ purple
4	blue × purple	$\frac{3}{8}$ blue: $\frac{4}{8}$ purple: $\frac{1}{8}$ scarlet
5	purple × scarlet	$\frac{1}{2}$ purple: $\frac{1}{2}$ scarlet

49. Corn breeders obtained pure lines whose kernels turn sun red, pink, scarlet, or orange when exposed to sunlight (normal kernels remain yellow in sunlight). Some crosses between these lines produced the following results. The phenotypes are abbreviated O, orange; P, pink; Sc, scarlet; and SR, sun red.

		Phenotypes	
Cross	Parents	F_1	F_2
1	SR × P	all SR	66 SR:20 P
2	O × SR	all SR	998 SR:314 O
3	O × P	all O	1300 O:429 P
4	O × Sc	all Y	182 Y:80 O:58 Sc

   Analyze the results of each cross, and provide a unifying hypothesis to account for *all* the results. (Explain all symbols that you use.)

50. Many kinds of wild animals have the agouti coloring pattern, in which each hair has a yellow band around it.

   a. Black mice and other black animals do not have the yellow band; each of their hairs is all black. This absence of wild agouti pattern is called *nonagouti*. When mice of a true-breeding agouti line are crossed with nonagoutis, the F_1 is all agouti and the F_2 has a 3:1 ratio of agoutis to nonagoutis. Diagram this cross, letting *A* represent the allele responsible for the agouti phenotype and *a*, nonagouti. Show the phenotypes and genotypes of the parents, their gametes, the F_1, their gametes, and the F_2.

   b. Another inherited color deviation in mice substitutes brown for the black color in the wild-type hair. Such brown-agouti mice are called *cinnamons*. When wild-type mice are crossed with cinnamons, all of the F_1 are wild type and the F_2 has a 3:1 ratio of wild type to cinnamon. Diagram this cross as in part *a*, letting *B* stand for the wild-type black allele and *b* stand for the cinnamon brown allele.

   c. When mice of a true-breeding cinnamon line are crossed with mice of a true-breeding nonagouti (black) line, all of the F_1 are wild type. Use a genetic diagram to explain this result.

d. In the F_2 of the cross in part *c*, a fourth color called *chocolate* appears in addition to the parental cinnamon and nonagouti and the wild type of the F_1. Chocolate mice have a solid, rich brown color. What is the genetic constitution of the chocolates?

e. Assuming that the *A/a* and *B/b* allelic pairs assort independently of each other, what do you expect to be the relative frequencies of the four color types in the F_2 described in part *d*? Diagram the cross of parts *c* and *d*, showing phenotypes and genotypes (including gametes).

f. What phenotypes would be observed in what proportions in the progeny of a backcross of F_1 mice from part *c* with the cinnamon parental stock? With the nonagouti (black) parental stock? Diagram these backcrosses.

g. Diagram a testcross for the F_1 of part *c*. What colors would result, and in what proportions?

h. Albino (pink-eyed white) mice are homozygous for the recessive member of an allelic pair *C/c*, which assorts independently of the *A/a* and *B/b* pairs. Suppose that you have four different highly inbred (and therefore presumably homozygous) albino lines. You cross each of these lines with a true-breeding wild-type line, and you raise a large F_2 progeny from each cross. What genotypes for the albino lines can you deduce from the following F_2 phenotypes?

		Phenotypes of progeny			
F_2 of line	Wild type	Black	Cinnamon	Chocolate	Albino
1	87	0	32	0	39
2	62	0	0	0	18
3	96	30	0	0	41
4	287	86	92	29	164

(Adapted from A. M. Srb, R. D. Owen, and R. S. Edgar, *General Genetics*, 2nd ed. W. H. Freeman and Company, 1965.)

51. An allele *A* that is not lethal when homozygous causes rats to have yellow coats. The allele *R* of a separate gene that assorts independently produces a black coat. Together, *A* and *R* produce a grayish coat, whereas *a* and *r* produce a white coat. A gray male is crossed with a yellow female, and the F_1 is $\frac{3}{8}$ yellow, $\frac{3}{8}$ gray, $\frac{1}{8}$ black, and $\frac{1}{8}$ white. Determine the genotypes of the parents.

52. The genotype *r/r* ; *p/p* gives fowl a single comb, *R/* – ; *P/* – gives a walnut comb, *r/r* ; *P/* – gives a pea comb, and *R/* – ; *p/p* gives a rose comb (see the illustrations). Assume independent assortment.

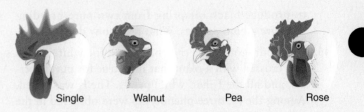

Single   Walnut   Pea   Rose

a. What comb types will appear in the F_1 and in the F_2 and in what proportions if single-combed birds are crossed with birds of a true-breeding walnut strain?

b. What are the genotypes of the parents in a walnut × rose mating from which the progeny are $\frac{3}{8}$ rose, $\frac{3}{8}$ walnut, $\frac{1}{8}$ pea, and $\frac{1}{8}$ single?

c. What are the genotypes of the parents in a walnut × rose mating from which all the progeny are walnut?

d. How many genotypes produce a walnut phenotype? Write them out.

53. The production of eye-color pigment in *Drosophila* requires the dominant allele *A*. The dominant allele *P* of a second independent gene turns the pigment to purple, but its recessive allele leaves it red. A fly producing no pigment has white eyes. Two pure lines were crossed with the following results:

P   red-eyed female × white-eyed male

$\downarrow$

F_1 purple-eyed females
        red-eyed males
        $F_1 \times F_1$

$\downarrow$

F_2 both males and females: $\frac{3}{8}$ purple eyed

        $\frac{3}{8}$ red eyed

        $\frac{2}{8}$ white eyed

Explain this mode of inheritance, and show the genotypes of the parents, the F_1, and the F_2.

54. When true-breeding brown dogs are mated with certain true-breeding white dogs, all the F_1 pups are white. The F_2 progeny from some $F_1 \times F_1$ crosses were 118 white, 32 black, and 10 brown pups. What is the genetic basis for these results?

55. Wild-type strains of the haploid fungus *Neurospora* can make their own tryptophan. An abnormal allele *td* renders the fungus incapable of making its own tryptophan. An individual of genotype *td* grows only when its medium supplies tryptophan. The allele *su* assorts independently of *td*; its only known effect is to suppress the *td* phenotype. Therefore, strains carrying both *td* and *su* do not require tryptophan for growth.

a. If a *td* ; *su* strain is crossed with a genotypically wild-type strain, what genotypes are expected in the progeny, and in what proportions?

b. What will be the ratio of tryptophan-dependent to tryptophan-independent progeny in the cross of part *a*?

56. Mice of the genotypes A/A; B/B; C/C; D/D; S/S and a/a; b/b; c/c; d/d; s/s are crossed. The progeny are intercrossed. What phenotypes will be produced in the F$_2$, and in what proportions? [The allele symbols stand for the following: A = agouti, a = solid (nonagouti); B = black pigment, b = brown; C = pigmented, c = albino; D = nondilution, d = dilution (milky color); S = unspotted, s = pigmented spots on white background.]

57. Consider the genotypes of two lines of chickens: the pure-line mottled Honduran is i/i; D/D; M/M; W/W, and the pure-line leghorn is I/I; d/d; m/m; w/w, where

    I = white feathers, i = colored feathers

    D = duplex comb, d = simplex comb

    M = bearded, m = beardless

    W = white skin, w = yellow skin

    These four genes assort independently. Starting with these two pure lines, what is the fastest and most convenient way of generating a pure line that has colored feathers, has a simplex comb, is beardless, and has yellow skin? Make sure that you show

    a. the breeding pedigree.

    b. the genotype of each animal represented.

    c. how many eggs to hatch in each cross, and why this number.

    d. why your scheme is the fastest and the most convenient.

58. The following pedigree is for a dominant phenotype governed by an autosomal allele. What does this pedigree suggest about the phenotype, and what can you deduce about the genotype of individual A?

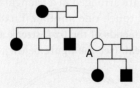

59. Petal coloration in foxgloves is determined by three genes. *M* encodes an enzyme that synthesizes anthocyanin, the purple pigment seen in these petals; *m/m* produces no pigment, resulting in the phenotype albino with yellowish spots. *D* is an enhancer of anthocyanin, resulting in a darker pigment; *d/d* does not enhance. At the third locus, *w/w* allows pigment deposition in petals, but *W* prevents pigment deposition except in the spots and so results in the white, spotted phenotype. Consider the following two crosses:

Cross	Parents		Progeny
1	dark purple	× white with yellow spots	$\frac{1}{2}$ dark purple: $\frac{1}{2}$ light purple
2	white with yellow spots	× light purple	$\frac{1}{2}$ white with purple spots: $\frac{1}{4}$ dark purple: $\frac{1}{4}$ light purple

In each case, give the genotypes of parents and progeny with respect to the three genes.

60. In one species of *Drosophila*, the wings are normally round in shape, but you have obtained two pure lines, one of which has oval wings and the other sickle-shaped wings. Crosses between pure lines reveal the following results:

Parents		F1	
Female	Male	Female	Male
sickle	round	sickle	sickle
round	sickle	sickle	round
sickle	oval	oval	sickle

a. Provide a genetic explanation of these results, defining all allele symbols.

b. If the F$_1$ oval females from cross 3 are crossed with the F$_1$ round males from cross 2, what phenotypic proportions are expected for each sex in the progeny?

61. Mice normally have one yellow band on each hair, but variants with two or three bands are known. A female mouse having one band was crossed with a male having three bands. (Neither animal was from a pure line.) The progeny were

    Females $\frac{1}{2}$ one band   Males $\frac{1}{2}$ one band
    $\frac{1}{2}$ three bands $\frac{1}{2}$ two bands

a. Provide a clear explanation of the inheritance of these phenotypes.

b. In accord with your model, what would be the outcome of a cross between a three-banded daughter and a one-banded son?

62. In minks, wild types have an almost black coat. Breeders have developed many pure lines of color variants for the mink-coat industry. Two such pure lines are platinum (blue gray) and aleutian (steel gray). These lines were used in crosses, with the following results:

Cross	Parents	F$_1$	F$_2$
1	wild × platinum	wild	18 wild, 5 platinum
2	wild × aleutian	wild	27 wild, 10 aleutian
3	platinum × aleutian	wild	133 wild
			41 platinum
			46 aleutian
			17 sapphire (new)

a. Devise a genetic explanation of these three crosses. Show genotypes for the parents, the F_1, and the F_2 in the three crosses, and make sure that you show the alleles of each gene that you hypothesize for every mink.

b. Predict the F_1 and F_2 phenotypic ratios from crossing sapphire with platinum and with aleutian pure lines.

63. In *Drosophila*, an autosomal gene determines the shape of the hair, with *B* giving straight and *b* giving bent hairs. On another autosome, there is a gene of which a dominant allele *I* inhibits hair formation so that the fly is hairless (*i* has no known phenotypic effect).

a. If a straight-haired fly from a pure line is crossed with a fly from a pure-breeding hairless line known to be an inhibited bent genotype, what will the genotypes and phenotypes of the F_1 and the F_2 be?

b. What cross would give the ratio 4 hairless: 3 straight:1 bent?

64. The following pedigree concerns eye phenotypes in *Tribolium* beetles. The solid symbols represent black eyes, the open symbols represent brown eyes, and the cross symbols (X) represent the "eyeless" phenotype, in which eyes are totally absent.

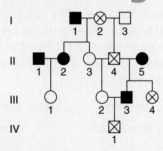

a. From these data, deduce the mode of inheritance of these three phenotypes.

b. Using defined gene symbols, show the genotype of beetle II-3.

65. A plant believed to be heterozygous for a pair of alleles *B/b* (where *B* encodes yellow and *b* encodes bronze) was selfed, and, in the progeny, there were 280 yellow and 120 bronze plants. Do these results support the hypothesis that the plant is *B/b*?

66. A plant thought to be heterozygous for two independently assorting genes (*P/p*;*Q/q*) was selfed, and the progeny were

88 *P/–*;*Q/–*   25 *p/p*;*Q/–*
32 *P/–*;*q/q*   14 *p/p*;*q/q*

Do these results support the hypothesis that the original plant was *P/p*;*Q/q*?

67. A plant of phenotype 1 was selfed, and, in the progeny, there were 100 plants of phenotype 1 and 60 plants of an alternative phenotype 2. Are these numbers compatible with expected ratios of 9:7, 13:3, and 3:1? Formulate a genetic hypothesis on the basis of your calculations.

68. Four homozygous recessive mutant lines of *Drosophila melanogaster* (labeled 1 through 4) showed abnormal leg coordination, which made their walking highly erratic. These lines were intercrossed; the phenotypes of the F_1 flies are shown in the following grid, in which "+" represents wild-type walking and "−" represents abnormal walking:

	1	2	3	4
1	−	+	+	+
2	+	−	−	+
3	+	−	−	+
4	+	+	+	−

a. What type of test does this analysis represent?

b. How many different genes were mutated in creating these four lines?

c. Invent wild-type and mutant symbols, and write out full genotypes for all four lines and for the F_1 flies.

d. Do these data tell us which genes are linked? If not, how could linkage be tested?

e. Do these data tell us the total number of genes taking part in leg coordination in this animal?

69. Three independently isolated tryptophan-requiring mutants of haploid yeast are called *trpB*, *trpD*, and *trpE*. Cell suspensions of each are streaked on a plate of nutritional medium supplemented with just enough tryptophan to permit weak growth for a *trp* strain. The streaks are arranged in a triangular pattern so that they do not touch one another. Luxuriant growth is noted at both ends of the *trpE* streak and at one end of the *trpD* streak (see the figure below).

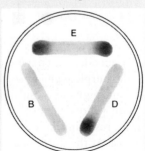

a. Do you think complementation has a role?

b. Briefly explain the pattern of luxuriant growth.

c. Draw the enzymatic steps that are defective in mutants *trpB*, *trpD*, and *trpE* in order in the tryptophan-synthesizing pathway.

d. Why was it necessary to add a small amount of tryptophan to the medium to demonstrate such a growth pattern?

CHALLENGING PROBLEMS

70. A pure-breeding strain of squash that produced disk-shaped fruits (see the accompanying illustration) was crossed with a pure-breeding strain having long fruits. The F₁ had disk fruits, but the F₂ showed a new phenotype, sphere, and was composed of the following proportions, shown in the next column:

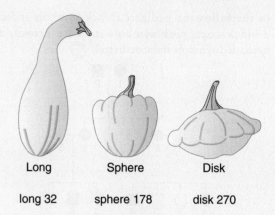

Long	Sphere	Disk
long 32	sphere 178	disk 270

Propose an explanation for these results, and show the genotypes of the P, F₁, and F₂ generations.

71. Marfan's syndrome is a disorder of the fibrous connective tissue, characterized by many symptoms, including long, thin digits; eye defects; heart disease; and long limbs. (Flo Hyman, the American volleyball star, suffered from Marfan's syndrome. She died from a ruptured aorta.)

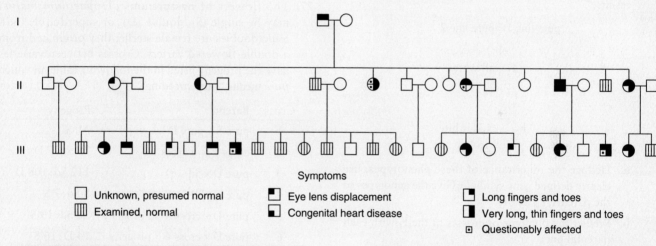

Symptoms

- ☐ Unknown, presumed normal
- ⊞ Examined, normal
- ◼ Eye lens displacement
- ◼ Congenital heart disease
- ◼ Long fingers and toes
- ◼ Very long, thin fingers and toes
- ⊡ Questionably affected

a. Use the pedigree above to propose a mode of inheritance for Marfan's syndrome.

b. What genetic phenomenon is shown by this pedigree?

c. Speculate on a reason for such a phenomenon.

(Data from J. V. Neel and W. J. Schull, *Human Heredity.* University of Chicago Press, 1954.)

72. In corn, three dominant alleles, called *A*, *C*, and *R*, must be present to produce colored seeds. Genotype *A/−*; *C/−*; *R/−* is colored; all others are colorless. A colored plant is crossed with three tester plants of known genotype. With tester *a/a*; *c/c*; *R/R*, the colored plant produces 50 percent colored seeds; with *a/a*; *C/C*; *r/r*, it produces 25 percent colored; and with *A/A*; *c/c*; *r/r*, it produces 50 percent colored. What is the genotype of the colored plant?

73. The production of pigment in the outer layer of seeds of corn requires each of the three independently assorting genes *A*, *C*, and *R* to be represented by at least one

dominant allele, as specified in Problem 72. The dominant allele *Pr* of a fourth independently assorting gene is required to convert the biochemical precursor into a purple pigment, and its recessive allele *pr* makes the pigment red. Plants that do not produce pigment have yellow seeds. Consider a cross of a strain of genotype *A/A*; *C/C*; *R/R*; *pr/pr* with a strain of genotype *a/a*; *c/c*; *r/r*; *Pr/Pr*.

a. What are the phenotypes of the parents?

b. What will be the phenotype of the F₁?

c. What phenotypes, and in what proportions, will appear in the progeny of a selfed F₁?

d. What progeny proportions do you predict from the testcross of an F₁?

74. The allele *B* gives mice a black coat, and *b* gives a brown one. The genotype *e/e* of another, independently assorting gene prevents the expression of *B* and *b*, making the coat color beige, whereas *E/−* permits the expression of *B* and *b*. Both genes are autosomal.

In the following pedigree, black symbols indicate a black coat, pink symbols indicate brown, and unshaded symbols indicate beige.

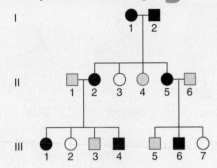

a. What is the name given to the type of gene interaction in this example?

b. What are the genotypes of the individual mice in the pedigree? (If there are alternative possibilities, state them.)

75. A researcher crosses two white-flowered lines of *Antirrhinum* plants as follows and obtains the following results:

pure line 1 × pure line 2

↓

F_1 all white

$F_1 \times F_1$

↓

F_2 131 white

29 red

a. Deduce the inheritance of these phenotypes; use clearly defined gene symbols. Give the genotypes of the parents, F_1, and F_2.

b. Predict the outcome of crosses of the F_1 with each parental line.

76. Assume that two pigments, red and blue, mix to give the normal purple color of petunia petals. Separate biochemical pathways synthesize the two pigments, as shown in the top two rows of the accompanying diagram. "White" refers to compounds that are not pigments. (Total lack of pigment results in a white petal.) Red pigment forms from a yellow intermediate that is normally at a concentration too low to color petals.

pathway I $\cdots \longrightarrow$ white$_1$ $\xrightarrow{\text{E}}$ blue

pathway II $\cdots \longrightarrow$ white$_2$ $\xrightarrow{\text{A}}$ yellow $\xrightarrow{\text{B}}$ red

$\text{C} \vdots$

pathway III $\cdots \to$ white$_3$ $\xrightarrow{\text{D}}$ white$_4$

A third pathway, whose compounds do not contribute pigment to petals, normally does not affect the blue and red pathways, but, if one of its intermediates (white$_3$) should build up in concentration, it can be converted into the yellow intermediate of the red pathway.

In the diagram, the letters A through E represent enzymes; their corresponding genes, all of which are unlinked, may be symbolized by the same letters.

Assume that wild-type alleles are dominant and encode enzyme function and that recessive alleles result in a lack of enzyme function. Deduce which combinations of true-breeding parental genotypes could be crossed to produce F_2 progeny in the following ratios:

a. 9 purple : 3 green : 4 blue

b. 9 purple : 3 red : 3 blue : 1 white

c. 13 purple : 3 blue

d. 9 purple : 3 red : 3 green : 1 yellow

(Note: Blue mixed with yellow makes green; assume that no mutations are lethal.)

77. The flowers of nasturtiums (*Tropaeolum majus*) may be single (S), double (D), or superdouble (Sd). Superdoubles are female sterile; they originated from a double-flowered variety. Crosses between varieties gave the progeny listed in the following table, in which *pure* means "pure breeding."

Cross	Parents	Progeny
1	pure S × pure D	All S
2	cross 1 F_1 × cross 1 F_1	78 S : 27 D
3	pure D × Sd	112 Sd : 108 D
4	pure S × Sd	8 Sd : 7 S
5	pure D × cross 4 Sd progeny	18 Sd : 19 S
6	pure D × cross 4 S progeny	14 D : 16 S

Using your own genetic symbols, propose an explanation for these results, showing

a. all the genotypes in each of the six rows.

b. the proposed origin of the superdouble.

78. In a certain species of fly, the normal eye color is red (R). Four abnormal phenotypes for eye color were found: two were yellow (Y1 and Y2), one was brown (B), and one was orange (O). A pure line was established for each phenotype, and all possible combinations of the pure lines were crossed. Flies of each F_1 were intercrossed to produce an F_2. The F_1 and the F_2 flies are shown within the following square; the pure lines are given at the top and at the left-hand side.

		Y1	Y2	B	O
Y1	F_1	all Y	all R	all R	all R
	F_2	all Y	9 R	9 R	9 R
		7 Y	4 Y	4 O
			3 B	3 Y
Y2	F_1		all Y	all R	all R
	F_2		all Y	9 R	9 R
				4 Y	4 Y
				3 B	3 O
B	F_1			all B	all R
	F_2			all B	9 R
					4 O
					3 B
O	F_1				all O
	F_2				all O

a. Define your own symbols, and list the genotypes of all four pure lines.

b. Show how the F_1 phenotypes and the F_2 ratios are produced.

c. Show a biochemical pathway that explains the genetic results, indicating which gene controls which enzyme.

79. In common wheat, *Triticum aestivum*, kernel color is determined by multiply duplicated genes, each with an *R* and an *r* allele. Any number of *R* alleles will give red, and a complete lack of *R* alleles will give the white phenotype. In one cross between a red pure line and a white pure line, the F_2 was $\frac{63}{64}$ red and $\frac{1}{64}$ white.

a. How many R genes are segregating in this system?

b. Show the genotypes of the parents, the F_1, and the F_2.

c. Different F_2 plants are backcrossed with the white parent. Give examples of genotypes that would give the following progeny ratios in such backcrosses: (1) 1 red:1 white, (2) 3 red:1 white, (3) 7 red:1 white.

d. What is the formula that generally relates the number of segregating genes to the proportion of red individuals in the F_2 in such systems?

80. The following pedigree shows the inheritance of deaf-mutism.

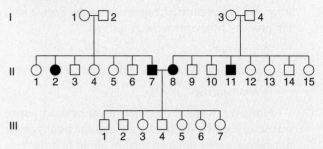

a. Provide an explanation for the inheritance of this rare condition in the two families in generations I and II, showing the genotypes of as many persons as possible; use symbols of your own choosing.

b. Provide an explanation for the production of only normal persons in generation III, making sure that your explanation is compatible with the answer to part *a*.

81. The pedigree below is for blue sclera (bluish thin outer wall of the eye) and brittle bones.

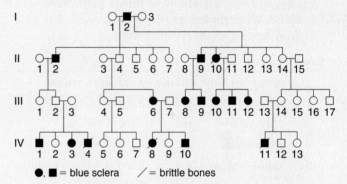

●, ■ = blue sclera   / = brittle bones

a. Are these two abnormalities caused by the same gene or by separate genes? State your reasons clearly.

b. Is the gene (or genes) autosomal or sex-linked?

c. Does the pedigree show any evidence of incomplete penetrance or expressivity? If so, make the best calculations that you can of these measures.

82. Workers of the honeybee line known as *Brown* (nothing to do with color) show what is called "hygienic behavior"; that is, they uncap hive compartments containing dead pupae and then remove the dead pupae. This behavior prevents the spread of infectious bacteria through the colony. Workers of the *Van Scoy* line, however, do not perform these actions, and therefore this line is said to be "nonhygienic." When

a queen from the *Brown* line was mated with *Van Scoy* drones, all the F_1 were nonhygienic. When drones from this F_1 inseminated a queen from the *Brown* line, the progeny behaviors were as follows:

$\frac{1}{4}$ hygienic

$\frac{1}{4}$ uncapping but no removing of pupae

$\frac{1}{2}$ nonhygienic

However, when the compartment of dead pupae was uncapped by the beekeeper and the nonhygienic honeybees were examined further, about half the bees were found to remove the dead pupae, but the other half did not.

a. Propose a genetic hypothesis to explain these behavioral patterns.

b. Discuss the data in relation to epistasis, dominance, and environmental interaction.

(Note: Workers are sterile, and all bees from one line carry the same alleles.)

83. The normal color of snapdragons is red. Some pure lines showing variations of flower color have been found. When these pure lines were crossed, they gave the following results (see the table):

Cross	Parents	F_1	F_2
1	orange × yellow	orange	3 orange : 1 yellow
2	red × orange	red	3 red : 1 orange
3	red × yellow	red	3 red : 1 yellow
4	red × white	red	3 red : 1 white
5	yellow × white	red	9 red : 3 yellow : 4 white
6	orange × white	red	9 red : 3 orange : 4 white
7	red × white	red	9 red : 3 yellow : 4 white

a. Explain the inheritance of these colors.

b. Write the genotypes of the parents, the F_1, and the F_2.

84. Consider the following F_1 individuals in different species and the F_2 ratios produced by selfing:

F_1	Phenotypic ratio in the F_2			
1 cream	$\frac{12}{16}$ cream	$\frac{3}{16}$ black	$\frac{1}{16}$ gray	
2 orange	$\frac{9}{16}$ orange	$\frac{7}{16}$ yellow		
3 black	$\frac{13}{16}$ black	$\frac{3}{16}$ white		
4 solid red	$\frac{9}{16}$ solid red	$\frac{3}{16}$ mottled red	$\frac{4}{16}$ small red dots	

If each F_1 were testcrossed, what phenotypic ratios would result in the progeny of the testcross?

85. To understand the genetic basis of locomotion in the diploid nematode *Caenorhabditis elegans*, recessive mutations were obtained, all making the worm "wiggle" ineffectually instead of moving with its usual smooth gliding motion. These mutations presumably affect the nervous or muscle systems. Twelve homozygous mutants were intercrossed, and the F_1 hybrids were examined to see if they wiggled. The results were as follows, where a plus sign means that the F_1 hybrid was wild type (gliding) and "w" means that the hybrid wiggled.

	1	2	3	4	5	6	7	8	9	10	11	12
1	w	+	+	+	w	+	+	+	+	+	+	+
2		w	+	+	+	w	+	w	+	w	+	+
3			w	w	+	+	+	+	+	+	+	+
4				w	+	+	+	+	+	+	+	+
5					w	+	+	+	+	+	+	+
6						w	+	w	+	w	+	+
7							w	+	+	+	w	w
8								w	+	w	+	+
9									w	+	+	+
10										w	+	+
11											w	w
12												w

a. Explain what this experiment was designed to test.

b. Use this reasoning to assign genotypes to all 12 mutants.

c. Explain why the phenotype of the F_1 hybrids between mutants 1 and 2 differed from that of the hybrids between mutants 1 and 5.

86. A geneticist working on a haploid fungus makes a cross between two slow-growing mutants called *mossy* and *spider* (referring to the abnormal appearance of the colonies). Tetrads from the cross are of three types (A, B, C), but two of them contain spores that do not germinate.

Spore	A	B	C
1	wild type	wild type	spider
2	wild type	spider	spider
3	no germination	mossy	mossy
4	no germination	no germination	mossy

Devise a model to explain these genetic results, and propose a molecular basis for your model.

87. In the nematode *C. elegans*, some worms have blistered cuticles due to a recessive mutation in one of the *bli* genes. Someone studying a suppressor mutation that suppressed *bli-3* mutations wanted to know if it would also suppress mutations in *bli-4*. They had a strain that was homozygous for this recessive suppressor mutation, and its phenotype was wild type.

a. How would they determine whether this recessive suppressor mutation would suppress mutations in *bli-4*? In other words, what is the genotype of the worms required to answer the question?

b. What cross(es) would they do to make these worms?

c. What results would they expect in the F_2 if
   (1) it did act as a suppressor of *bli-4*?
   (2) it did not act as a suppressor of *bli-4*?

88. Six proline-requiring auxotrophic mutants were obtained in the haploid fungus *Saccharomyces cerevisiae* (yeast). Each was crossed to wild type in order to obtain the mutants in each mating type ("sex"), then all combinations were crossed and the resultant random ascospores were plated onto minimal medium. In some cases, proline-independent colonies were obtained (numbers are not shown); but in other cases, none of the ascospores grew. The results are summarized in the following table, in which + indicates presence of colonies, and 0 indicates no colonies. Formulate a hypothesis that explains the results of each cross.

	1	2	3	4	5	6
1	0	+	+	0	+	0
2	+	0	0	+	+	+
3	+	0	0	+	+	+
4	0	+	+	0	+	0
5	+	+	+	+	0	+
6	0	+	+	0	+	0

GENETICS AND SOCIETY

1. How might a recessive lethal allele reveal itself in a human pedigree?

2. How has Beadle and Tatum's pioneering work influenced the therapy of human disease?

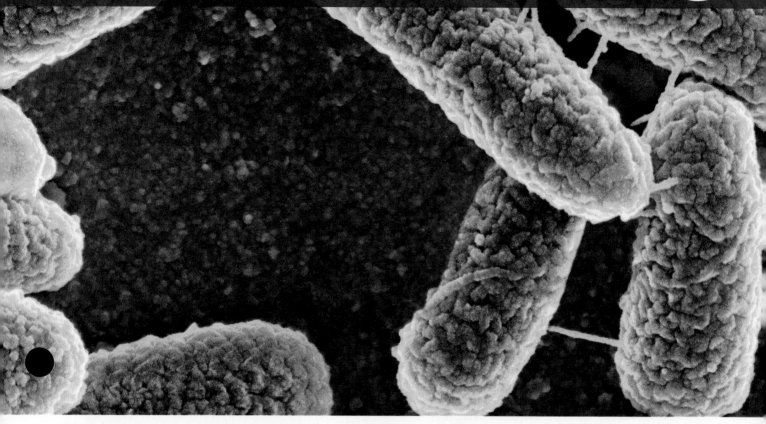

The Genetics of Bacteria and Their Viruses

CHAPTER 6

CHAPTER 6

E. coli cells connected by pili. [*Science Photo Library/Getty Images.*]

CHAPTER OUTLINE AND LEARNING OBJECTIVES

6.1 WORKING WITH MICROORGANISMS

6.2 BACTERIAL CONJUGATION

LO 6.1 Distinguish between the three main ways by which bacteria exchange genes, and describe the experimental procedures that reveal them.

LO 6.2 Map bacterial genomes using interrupted conjugation.

LO 6.3 Map bacterial genomes using recombinant frequency.

LO 6.4 In bacterial crosses, predict the inheritance of genetic elements carried on plasmids.

6.3 BACTERIAL TRANSFORMATION

LO 6.5 Assess the outcome of double transformation experiments in terms of linkage.

6.4 BACTERIOPHAGE GENETICS

LO 6.6 Map phage genomes by recombination in double infections of bacteria.

6.5 TRANSDUCTION

LO 6.7 Predict the outcomes of transduction experiments using phages capable of generalized or specialized transduction.

6.6 PHYSICAL MAPS AND LINKAGE MAPS COMPARED

LO 6.8 Explain how transposons can be used in insertional mutagenesis to create and map new mutations.

193

So far, the book has focused on inheritance patterns produced by heterozygous genes in eukaryotes. Our broad objective in this chapter is to learn the inheritance patterns produced by heterozygous genes in crosses of bacteria and viruses.

DNA technology is responsible for the rapid advances being made in the genetics of all model organisms. It is also a topic of considerable interest in the public domain. Examples are the highly publicized announcement of the full genome sequences of humans and chimpanzees in recent years and the popularity of DNA-based forensic analysis in television shows and movies (**Figure 6-1**). Indeed, improvements in technology have led to the sequencing of the genomes of many hundreds of species. Such dramatic results, whether in humans, fish, insects, plants, or fungi, are all based on the use of methods that permit small pieces of DNA to be isolated, carried from cell to cell, and amplified into large pure samples. The sophisticated systems that permit these manipulations of the DNA of any organism are almost all derived from bacteria and their viruses. Hence, the advance of modern genetics to its present state of understanding was entirely dependent on the development of bacterial genetics, the topic of this chapter.

However, the goal of bacterial genetics has never been to facilitate eukaryotic molecular genetics. Bacteria are biologically important in their own right. They are the most numerous organisms on our planet. They contribute to the recycling of nutrients such as nitrogen, sulfur, and carbon in ecosystems. Some are agents of human, animal, and plant disease. Others live symbiotically inside our mouths and intestines. In addition, many types of bacteria are useful for the industrial synthesis of a wide range of organic products. Hence, the impetus for the genetic dissection of bacteria has been the same as that for multicellular organisms—to understand their biological function.

Bacteria belong to a class of organisms known as **prokaryotes**, which also includes the blue-green algae (classified as *cyanobacteria*). A key defining feature of prokaryotes is that their DNA is not enclosed in a membrane-bounded nucleus. Like higher organisms, bacteria have genes composed of DNA arranged in a long series on a chromosome. However, the organization of their genetic material is unique in several respects. The genome of most bacteria is a single molecule of double-stranded DNA in the form of a closed circle. In addition, bacteria in nature often contain extra autonomous DNA elements called *plasmids*. Most plasmids are also DNA circles but are much smaller than the main bacterial genome.

Bacteria can be parasitized by specific **viruses** called **bacteriophages** or, simply, **phages**. Phages and other viruses are very different from the organisms that we have been studying so far. Viruses have some properties in common with organisms; for example, their genetic material can be DNA or RNA, constituting a short "chromosome." However, most biologists regard viruses as nonliving because they are not cells and they have no metabolism of their own. Hence, for the study of their genetics, viruses must be propagated in the cells of their host organisms.

When scientists began studying bacteria and phages, they were naturally curious about their hereditary systems. Clearly, bacteria and phages must have hereditary systems because they show a constant appearance and function from one generation to the next (they are true to type). But how do these hereditary systems work? Bacteria, like unicellular eukaryotic organisms, reproduce asexually by cell growth and division, one cell becoming two. This asexual reproduction is quite easy to demonstrate experimentally. However, is there ever a union of different types for the purpose of sexual reproduction? Furthermore, how do the much smaller phages reproduce? Do they ever unite for a sex-like cycle? These questions are pursued in this chapter.

We will see that there is a variety of hereditary processes in bacteria and phages. These processes are interesting because of the basic biology of these forms, but they also act as models—as sources of insight into genetic processes at work in *all* organisms. For a geneticist, the attraction of bacteria and phages is that they can be cultured in very large numbers because they are so small. Consequently, it is possible to detect and study very rare genetic events that are difficult or impossible to study in eukaryotes.

The fruits of DNA technology, made possible by bacterial genetics

FIGURE 6-1 The dramatic results of modern DNA technology, such as sequencing the human genome, were possible only because bacterial genetics led to the invention of efficient DNA manipulation vectors. [*Republished with permission of the American Association for the Advancement of Science, from* Science *vol. 291 16 February 2001 no. 5507 p. 1145–1434/image by Ann E. Cutting. Permission conveyed through Copyright Clearance Center, Inc.*]

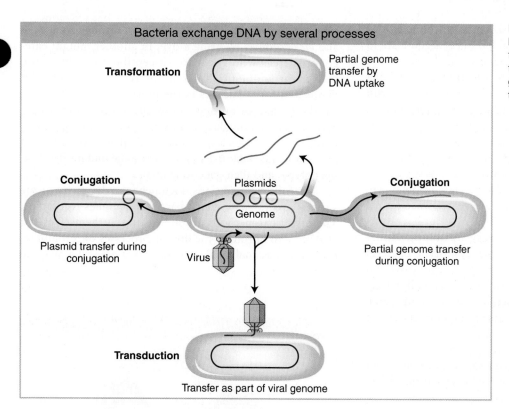

FIGURE 6-2 Bacterial DNA can be transferred from cell to cell in four ways: conjugation with plasmid transfer, conjugation with partial genome transfer, transformation, and transduction.

What hereditary processes are observed in bacteria? They can undergo both asexual and sexual reproduction. Mutation occurs in asexual cells in much the same way as it does in eukaryotes, and mutant alleles can be followed through both these processes in an approach analogous to that used in eukaryotes. We shall follow alleles in this way in the chapter ahead.

When bacterial cells reproduce asexually, their genomic DNA replicates and is partitioned into daughter cells, but the partitioning method is quite different from mitosis.

In sexual reproduction, two DNA molecules from different sources are brought together. However, an important difference from eukaryotes is that, in bacteria, rarely are two complete chromosomes brought together; usually, the union is of one complete chromosome plus a fragment of another. The ways in which bacteria exchange DNA are outlined in **Figure 6-2**.

The first process of gene exchange to be examined will be *conjugation*, which is the contact and fusion of two different bacterial cells. After fusion, one cell, called a donor, sometimes transfers genomic DNA to the other cell. This transferred DNA may be part or (rarely) all of the bacterial genome. In some cases, one or more plasmids, if present, are transferred. Some plasmids are capable of carrying genomic DNA into the recipient cell. Any genomic fragment, transferred by whatever route, may recombine with the recipient's chromosome after entry.

A bacterial cell can also take up a piece of DNA from the external environment and incorporate this DNA into its own chromosome, a process called *transformation*. In addition, certain phages can pick up a piece of DNA from one bacterial cell and inject it into another, where it can be incorporated into the chromosome, in a process known as *transduction*.

DNA transfer by conjugation, transformation, or transduction constitutes a process known as **horizontal transmission**, a type of gene transmission without the need for cell division. This term distinguishes this type of DNA transfer from that during **vertical transmission**, the passage of DNA down thorough the bacterial generations. Horizontal transmission can spread DNA rapidly through a bacterial population by contact in much the same way that a disease spreads. For bacteria, horizontal transmission provides a powerful method by which they can adapt rapidly to changing environmental conditions.

Phages themselves can undergo recombination when two different genotypes both infect the same bacterial cell (**phage recombination**, not shown in Figure 6-2).

Before we analyze these modes of genetic exchange, let's consider the practical ways of handling bacteria, which are much different from those used in handling multicellular organisms.

6.1 WORKING WITH MICROORGANISMS

Bacteria are fast-dividing and take up little space; so they are very convenient to use as genetic model organisms. They can be cultured in a liquid medium or on a solid surface such as an agar gel, as long as basic nutrients are supplied. Each bacterial cell divides asexually from

$1 \rightarrow 2 \rightarrow 4 \rightarrow 8 \rightarrow 16$ cells, and so on, until the nutrients are exhausted or until toxic waste products accumulate to levels that halt the population growth. A small amount of a liquid culture can be pipetted onto a petri plate containing solid agar medium and spread evenly on the surface with a sterile spreader, in a process called **plating** (Figure 6-3). The cells divide, but, because they cannot travel far on the surface of the gel, all the cells remain together in a clump. When this mass reaches more than 10^7 cells, it becomes visible to the naked eye as a **colony**. Each distinct colony on the plate has been derived from a single original cell. Members of a colony that have a single genetic ancestor are known as **cell clones.**

Bacterial mutants are quite easy to obtain. Nutritional mutants are a good example. Wild-type bacteria are **prototrophic,** which means that they can grow and divide on **minimal medium**—a substrate containing only inorganic salts, a carbon source for energy, and water. From a prototrophic culture, **auxotrophic** mutants can be obtained: these mutants are cells that will not grow unless the medium contains one or more specific cellular building blocks such as adenine, threonine, or biotin. Another type of useful mutant differs from wild type in the ability to use a specific energy source; for example, the wild type (lac^+) can use lactose and grow, whereas a mutant (lac^-)

cannot. **Figure 6-4** shows another way of distinguishing lac^+ and lac^- colonies by using a dye. In another mutant category, whereas wild types are susceptible to an inhibitor, such as the antibiotic streptomycin, **resistant mutants** can divide and form colonies in the presence of the inhibitor. All these types of mutants allow the geneticist to distinguish different individual strains, thereby providing **genetic markers** (marker alleles) to keep track of genomes and cells in experiments. **Table 6-1** summarizes some mutant bacterial phenotypes and their genetic symbols.

The following sections document the discovery of the various processes by which bacterial genomes recombine. The historical methods are interesting in themselves but also serve to introduce the diverse processes of recombination as well as analytical techniques that are still applicable today.

Distinguishing lac^+ and lac^- by using a red dye

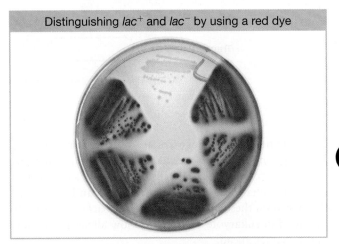

FIGURE 6-4 Wild-type bacteria able to use lactose as an energy source (lac^+) stain red in the presence of this indicator dye. The unstained cells are mutants unable to use lactose (lac^-). [*Jeffrey H. Miller.*]

Bacterial colonies, each derived from a single cell

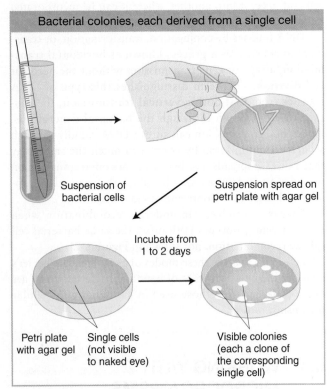

Suspension of bacterial cells

Suspension spread on petri plate with agar gel

Incubate from 1 to 2 days

Petri plate with agar gel

Single cells (not visible to naked eye)

Visible colonies (each a clone of the corresponding single cell)

FIGURE 6-3 Bacterial phenotypes can be assessed in their colonies. A stock of bacterial cells can be grown in a liquid medium containing nutrients, and then a small number of bacteria from the liquid suspension can be spread on solid agar medium. Each cell will give rise to a colony. All cells in a colony have the same genotype and phenotype.

TABLE 6-1	Some Genotypic Symbols Used in Bacterial Genetics
Symbol	**Character or phenotype associated with symbol**
bio^-	Requires biotin added as a supplement to minimal medium
arg^-	Requires arginine added as a supplement to minimal medium
met^-	Requires methionine added as a supplement to minimal medium
lac^-	Cannot utilize lactose as a carbon source
gal^-	Cannot utilize galactose as a carbon source
str^r	Resistant to the antibiotic streptomycin
str^s	Sensitive to the antibiotic streptomycin

Note: Minimal medium is the basic synthetic medium for bacterial growth without nutrient supplements.

6.2 BACTERIAL CONJUGATION

LO 6.1 Distinguish between the three main ways by which bacteria exchange genes, and describe the experimental procedures that reveal them.

LO 6.2 Map bacterial genomes using interrupted conjugation.

LO 6.3 Map bacterial genomes using recombinant frequency.

LO 6.4 In bacterial crosses, predict the inheritance of genetic elements carried on plasmids.

The earliest studies in bacterial genetics revealed the unexpected process of cell conjugation.

Discovery of conjugation

Do bacteria possess any processes similar to sexual reproduction and recombination? The question was answered by the elegantly simple experimental work of Joshua Lederberg and Edward Tatum, who in 1946 discovered a sex-like process in what became the main model for bacterial genetics, *Escherichia coli* (see the Model Organism box on page 198). They were studying two strains of *E. coli* with different sets of auxotrophic mutations. Strain A⁻ would grow only if the medium were supplemented with methionine and biotin; strain B⁻ would grow only if it were supplemented with threonine, leucine, and thiamine. Thus, we can designate the strains as

$$\text{strain A}^-: \quad met^-\ bio^-\ thr^+\ leu^+\ thi^+$$
$$\text{strain B}^-: \quad met^+\ bio^+\ thr^-\ leu^-\ thi^-$$

Figure 6-5a displays in simplified form the design of their experiment. Strains A⁻ and B⁻ were mixed together, incubated for a while, and then plated on minimal medium,

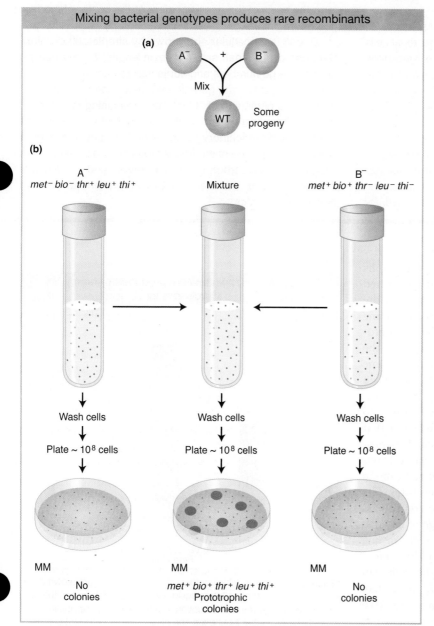

FIGURE 6-5 With the use of this method, Lederberg and Tatum demonstrated that genetic recombination between bacterial genotypes is possible. (a) The basic concept: two auxotrophic cultures (A⁻ and B⁻) are mixed, yielding prototrophic wild types (WT). (b) Cells of type A⁻ (red dots) or type B⁻ (blue dots) cannot grow on an unsupplemented (minimal) medium (MM) because A⁻ and B⁻ each carry mutations that cause the inability to synthesize constituents needed for cell growth. All A⁻ or B⁻ cells deposited on minimal medium plates will die and not form colonies. When A⁻ and B⁻ are mixed for a few hours and then plated, however, a few colonies (shown here in purple) appear on the agar plate. These colonies derive from single cells in which genetic material has been exchanged; they are therefore capable of synthesizing all the required constituents of metabolism.

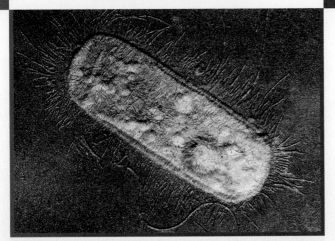

An electron micrograph of an *E. coli* cell showing long flagella, used for locomotion, and fimbriae, proteinaceous hairs that are important in anchoring the cells to animal tissues. (Sex pili are not shown in this micrograph.) [*Biophoto Associates/Science Source.*]

The seventeenth-century microscopist Antony van Leeuwenhoek was probably the first to see bacterial cells and to recognize their small size: "There are more living in the scum on the teeth in a man's mouth than there are men in the whole kingdom." However, bacteriology did not begin in earnest until the nineteenth century. In the 1940s, Joshua Lederberg and Edward Tatum made the discovery that launched bacteriology into the burgeoning field of genetics: they discovered that, in a certain bacterium, there was a type of sexual cycle including a crossing-over-like process. The organism that they chose for this experiment has become the model not only for bacterial genetics but, in a sense, for

all of genetics. The organism was *Escherichia coli*, a bacterium named after its discoverer, the nineteenth-century German bacteriologist Theodore Escherich.

The choice of *E. coli* was fortunate because it has proved to have many features suitable for genetic research, not the least of which is that it is easily obtained, given that it lives in the gut of humans and other animals. In the gut, it is a benign symbiont, but it occasionally causes urinary tract infections and diarrhea.

E. coli has a single circular chromosome 4.6 Mb in length. Of its 4000 intron-free genes, about 35 percent are of unknown function. The sexual cycle is made possible by the action of an extragenomic plasmid called F, which confers a type of "maleness." Other plasmids carry genes whose functions equip the cell for life in specific environments, such as drug-resistance genes. These plasmids have been adapted as gene *vectors*, which are gene carriers that form the basis of the gene transfers at the center of modern genetic engineering.

E. coli is unicellular and grows by simple cell division. Because of its small size (~1 μm in length), *E. coli* can be grown in large numbers and subjected to intensive selection and screening for rare genetic events. *E. coli* research represents the beginning of "black box" reasoning in genetics: through the selection and analysis of mutants, the workings of the genetic machinery could be deduced even though it was too small to be seen. Phenotypes such as colony size, drug resistance, carbon-source utilization, and colored-dye production took the place of the visible phenotypes of eukaryotic genetics.

on which neither auxotroph could grow. A small minority of the cells (1 in 10^7) was found to grow as prototrophs and, hence, must have been wild type, having regained the ability to grow without added nutrients. Some of the dishes were plated only with strain A⁻ bacteria and some only with strain B⁻ bacteria to act as controls, but no prototrophs arose from these platings. Figure 6-5b illustrates the experiment in more detail. These results suggested that some form of recombination of genes had taken place between the genomes of the two strains to produce the prototrophs.

It could be argued that the cells of the two strains do not really exchange genes but instead leak substances that the other cells can absorb and use for growing. This possibility of "cross-feeding" was ruled out by Bernard Davis in the following way. He constructed a U-shaped tube in which the two arms were separated by a fine filter. The pores of the filter were too small to allow bacteria to pass through but large enough to allow easy passage of any dissolved substances (**Figure 6-6**). Strain A⁻ was put in one arm, strain B⁻ in the other. After the strains had been incubated for a while, Davis tested the contents of each arm to see if there were any prototrophic cells, but none were found. In other words, *physical contact* between the two strains was needed for wild-type cells to form. It looked as though some kind of genome union had taken place, and

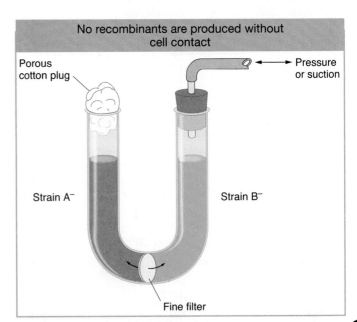

FIGURE 6-6 Auxotrophic bacterial strains A⁻ and B⁻ are grown on either side of a U-shaped tube. Liquid may be passed between the arms by applying pressure or suction, but the bacterial cells cannot pass through the filter. After incubation and plating, no recombinant colonies grow on minimal medium.

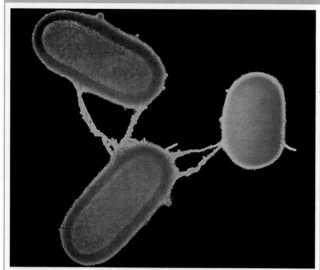

Bacteria conjugate by using pili

FIGURE 6-7 A donor cell extends one or more projections, or pili, that attach to a recipient cell and pull the two bacteria together. [*Dr. L. Caro/Science Source.*]

genuine recombinants had been produced. The physical union of bacterial cells can be confirmed under an electron microscope and is now called **conjugation** (Figure 6-7).

Discovery of the fertility factor (F)

In 1953, William Hayes discovered that, in the types of "crosses" just described here, the conjugating parents acted *unequally* (later, we will see ways to demonstrate this unequal participation). One parent (and *only* that parent) seemed to transfer some or all of its genome into another cell. Hence, one cell acts as a **donor,** and the other cell acts as a **recipient.** This "cross" is quite different from eukaryotic crosses in which parents contribute nuclear genomes equally to a progeny individual.

KEY CONCEPT The transfer of genetic material in *E. coli* conjugation is not reciprocal. One cell, the donor, transfers part of its genome to the other cell, which acts as the recipient.

By accident, Hayes discovered a variant of his original donor strain that would not produce recombinants on crossing with the recipient strain. Apparently, the donor-type strain had lost the ability to transfer genetic material and had changed into a recipient-type strain. In working with this "sterile" donor variant, Hayes found that it could regain the ability to act as a donor by association with other donor strains. Indeed, the donor ability was transmitted rapidly and effectively between strains during conjugation. A kind of "infectious transfer" of some factor seemed to be taking place. He suggested that the ability to be a donor is itself a hereditary state, imposed by a **fertility factor (F)**. Strains that carry F can donate and are designated F$^+$. Strains that lack F cannot donate and are recipients, designated F$^-$.

We now know much more about F. It is a type of small, nonessential circular DNA molecule called a **plasmid** that can replicate in the cytoplasm independent of the host chromosome. **Figure 6-8** shows how bacteria can transfer plasmids such as F. The F plasmid directs the synthesis of pili (sing., pilus), projections that initiate contact with a recipient (see Figures 6-7 and 6-8) and draw it closer. The F plasmid DNA in the donor cell makes a single-stranded version of itself in a peculiar mechanism called **rolling circle replication**. The circular plasmid "rolls," and as it turns, it reels out a newly synthesized, single-stranded "fishing line." This single strand of DNA passes through a pore into the recipient cell, where the other DNA strand is synthesized, forming a double helix. Hence, a copy of F remains in the donor and another appears in the recipient, as shown in Figure 6-8. Note that the *E. coli* genome is depicted as a single circular chromosome in Figure 6-8. We will examine the evidence for it later. Most bacterial genomes are circular, a feature quite different from eukaryotic nuclear chromosomes. We will see that this feature leads to the many idiosyncrasies of bacterial genetics.

KEY CONCEPT F plasmids from F$^+$ donor cells are transmitted rapidly to F$^-$ recipient cells by rolling circle replication, but the bacterial chromosome is not transferred.

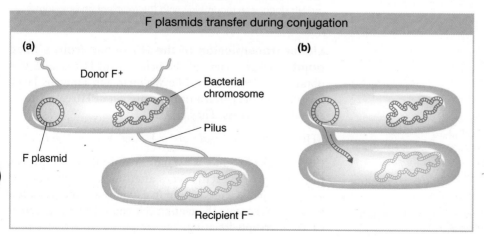

F plasmids transfer during conjugation

(a) Donor F+ · Bacterial chromosome · Pilus · F plasmid · Recipient F−

(b)

FIGURE 6-8 (a) During conjugation, the pilus pulls two bacteria together. (b) Next, a pore forms between the two cells. A single-stranded copy of plasmid DNA is produced in the donor cell and then passes into the recipient bacterium, where the single strand, serving as a template, is converted into the double-stranded helix.

Hfr strains

An important breakthrough came when Luca Cavalli-Sforza discovered a derivative of an F$^+$ strain with two unusual properties:

1. On crossing with F$^-$ strains, this new strain produced 1000 times as many recombinants as a normal F$^+$ strain. Cavalli-Sforza designated this derivative an **Hfr** strain to symbolize its ability to promote a *high frequency* of *recombination*.

2. In Hfr × F$^-$ crosses, virtually none of the F$^-$ parents were converted into F$^+$ or into Hfr. This result is in contrast with F$^+$ × F$^-$ crosses, in which, as we have seen, infectious transfer of F results in a large proportion of the F$^-$ parents being converted into F$^+$.

It became apparent that an Hfr strain results from the integration of the F factor into the chromosome, as pictured in **Figure 6-9**. We can now explain the first unusual property of Hfr strains. During conjugation, the F factor inserted in the chromosome efficiently drives part or all of that chromosome into the F$^-$ cell. The chromosomal fragment can then engage in recombination with the recipient chromosome. The rare recombinants observed by Lederberg and Tatum in F$^+$ × F$^-$ crosses were due to the spontaneous, but rare, formation of Hfr cells in the F$^+$ culture. Cavalli-Sforza isolated examples of these rare cells from F$^+$ cultures and found that, indeed, they now acted as true Hfrs.

Does an Hfr cell die after donating its chromosomal material to an F$^-$ cell? The answer is no. Just like the F plasmid, the Hfr chromosome replicates and transfers a single strand to the F$^-$ cell during conjugation. That the transferred DNA is a single strand can be demonstrated visually with the use of special strains and antibodies, as shown in **Figure 6-10**. The replication of the chromosome ensures a complete chromosome for the donor cell after mating. The transferred strand is converted into a double helix in the recipient cell, and donor genes may become incorporated in the recipient's chromosome through crossovers, creating

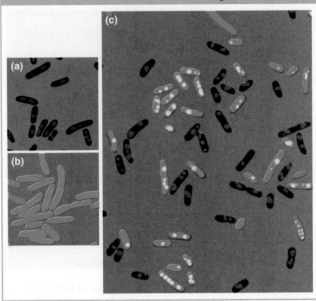

FIGURE 6-10 The photographs show a visualization of single-stranded DNA transfer in conjugating *E. coli* cells, with the use of special fluorescent antibodies. Parental Hfr strains (a) are black with red DNA. The red is from the binding of an antibody to a protein normally attached to DNA. The recipient F$^-$ cells (b) are green due to the presence of the gene for a jellyfish protein that fluoresces green, and, because they are mutant for a certain gene, their DNA protein does not bind to the fluorescent antibody. When Hfr donor single-stranded DNA enters the recipient, it promotes atypical binding of the DNA protein. Bound antibody fluoresces yellow in this background. Part c shows Hfrs (unchanged) and exconjugants (cells that have undergone conjugation) with yellow transferred single-stranded DNA. A few unmated F$^-$ cells are visible. [*Republished with permission of the American Association for the Advancement of Science, from M. Kohiyama, S. Hiraga, I. Matic, and M. Radman, "Bacterial Sex: Playing Voyeurs 50 Years Later," Science 301, 2003, p. 803, Fig 1. Permission conveyed through Copyright Clearance Center, Inc.*]

a recombinant cell (**Figure 6-11**). If there is no recombination, the transferred fragments of DNA are simply lost in the course of cell division.

KEY CONCEPT An Hfr strain is a strain in which the F plasmid has inserted into the bacterial chromosome.

KEY CONCEPT A DNA fragment entering an F$^-$ recipient from an Hfr donor can recombine with the recipient chromosome.

Linear transmission of the Hfr genes from a fixed point A clearer view of the behavior of Hfr strains was obtained in 1957, when Elie Wollman and François Jacob investigated the pattern of transmission of Hfr genes to F$^-$ cells during a cross. They crossed

Hfr *azi*^{r} *ton*^{r} *lac*$^+$ *gal*$^+$ *str*^{s} × F$^-$ *azi*^{s} *ton*^{s} *lac*$^-$ *gal*$^-$ *str*^{r}

(Superscripts "r" and "s" stand for resistant and sensitive, respectively.) At specific times after mixing, they removed samples, which were each put in a kitchen blender for a few seconds to separate the mating cell pairs. This procedure is called **interrupted mating**. The sample was then plated onto

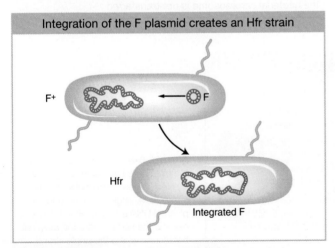

Integration of the F plasmid creates an Hfr strain

F$^+$ ⟵ ⊙ F

Hfr

Integrated F

FIGURE 6-9 In an F$^+$ strain, the free F plasmid occasionally integrates into the *E. coli* chromosome, creating an Hfr strain.

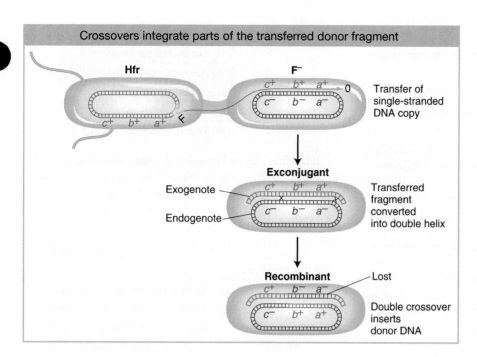

Crossovers integrate parts of the transferred donor fragment

Hfr   F⁻

c^+ b^+ a^+ 0   Transfer of single-stranded DNA copy

c^- b^- a^-

c^+ b^+ a^+

Exconjugant

Exogenote c^+ b^+ a^+ Transferred fragment converted into double helix

Endogenote c^- b^- a^-

Recombinant   Lost

c^+ b^- a^-

c^- b^+ a^+ Double crossover inserts donor DNA

FIGURE 6-11 After conjugation, crossovers are needed to integrate genes from the donor fragment into the recipient's chromosome and, hence, become a stable part of its genome. An even number of crossovers is required (a minimum of two).

ANIMATED ART   📶 Sapling Plus

Bacterial conjugation and mapping by recombination

a medium containing streptomycin to kill the Hfr donor cells, which bore the sensitivity allele str^s. The surviving str^r cells then were tested for the presence of alleles from the donor Hfr genome. Any str^r cell bearing a donor allele must have taken part in conjugation; such cells are called **exconjugants.** The results are plotted in **Figure 6-12a**, showing a time course of entry of each donor allele azi^r, ton^r, lac^+, and gal^+. Figure 6-12b portrays the transfer of Hfr alleles.

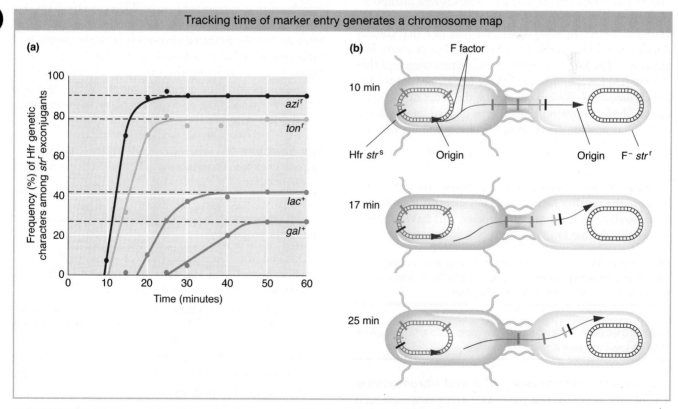

Tracking time of marker entry generates a chromosome map

(a)

Frequency (%) of Hfr genetic characters among str^r exconjugants

azi^r
ton^r
lac^+
gal^+

Time (minutes)

(b)   F factor

10 min

Hfr str^s Origin   Origin   F⁻ str^r

17 min

25 min

FIGURE 6-12 In this interrupted-mating conjugation experiment, F⁻ streptomycin-resistant cells with mutations in *azi*, *ton*, *lac*, and *gal* are incubated for varying times with Hfr cells that are sensitive to streptomycin and carry wild-type alleles for these genes. (a) A plot of the frequency of donor alleles in exconjugants as a function of time after mating. (b) A schematic view of the transfer of markers (shown in different colors) with the passage of time. [(a) Data from E. L. Wollman, F. Jacob, and W. Hayes, Cold Spring Harbor Symp. Quant. Biol. 21, 1956, 141.]

The key elements in these results are

1. Each donor allele first appears in the F⁻ recipients at a specific time after mating began.

2. The donor alleles appear in a specific sequence.

3. Later donor alleles are present in fewer recipient cells.

Putting all these observations together, Wollman and Jacob deduced that, in the conjugating Hfr, single-stranded DNA transfer begins from a fixed point on the donor chromosome, termed the **origin (O)**, and continues in a linear fashion. The point O is now known to be the site at which the F plasmid is inserted. The farther a gene is from O, the later it is transferred to the F⁻. The transfer process will generally stop before the farthermost genes are transferred, and, as a result, these genes are included in fewer exconjugants. Note that a type of chromosome map can be produced in units of minutes, based on time of entry of marked genes. In the example in Figure 6-12, the map would be:

$$azi^r \quad ton^r \quad lac^+ \quad gal^+$$
$$0 \quad 10 \quad 12 \quad 17 \quad 25$$
$$\leftarrow \quad 10 \quad 2 \quad 5 \quad 8$$

How can we explain the second unusual property of Hfr crosses, that F⁻ exconjugants are rarely converted into Hfr or F⁺? When Wollman and Jacob allowed Hfr × F⁻ crosses to continue for as long as 2 hours before disruption, they found that in fact a few of the exconjugants were converted into Hfrs. In other words, the part of F that confers donor ability was eventually transmitted but at a very low frequency. The rareness of Hfr exconjugants suggested that the inserted F was transmitted as the *last* element of the linear chromosome. We can summarize the order of transmission with the following general type of map, in which the arrow indicates the direction of transfer, beginning with O:

$$O \quad a \quad b \quad c \quad F$$
$$\leftarrow$$

Thus, almost none of the F⁻ recipients are converted, because the fertility factor is the last element transmitted and usually the transmission process will have stopped before getting that far.

KEY CONCEPT The Hfr chromosome, originally circular, unwinds a copy of itself that is transferred to the F⁻ cell in a linear fashion, with the F factor entering last.

KEY CONCEPT Time of entry of Hfr alleles into an F⁻ recipient can be used to make a chromosome map.

Inferring integration sites of F and chromosome circularity Wollman and Jacob went on to shed more light on how and where the F plasmid integrates to form an Hfr cell and, in doing so, deduced that the chromosome is circular. They performed interrupted-mating experiments with different, separately derived Hfr strains. Significantly, the order of transmission of the alleles differed from strain to strain, as in the following examples:

Hfr strain	
H	O thr pro lac pur gal his gly thi F
1	O thr thi gly his gal pur lac pro F
2	O pro thr thi gly his gal pur lac F
3	O pur lac pro thr thi gly his gal F
AB 312	O thi thr pro lac pur gal his gly F

Each line can be considered a map showing the order of alleles on the chromosome. At first glance, there seems to be a random shuffling of genes. However, when some of the Hfr maps are inverted, the relation of the sequences becomes clear.

H (written backward)	F thi gly his gal pur lac pro thr O
1	O thr thi gly his gal pur lac pro F
2	O pro thr thi gly his gal pur lac F
3	O pur lac pro thr thi gly his gal F
AB 312 (written backward)	F gly his gal pur lac pro thr thi O

The relation of the sequences to one another is explained if each map is the segment of a circle. It was the first indication that bacterial chromosomes are circular. Furthermore, Allan Campbell proposed a startling hypothesis that accounted for the different Hfr maps. He proposed that, if F is a ring, then the mechanism for its insertion into the bacterial chromosome might simply be a single crossover between F and the bacterial chromosome (**Figure 6-13**). That being the case, any of the linear Hfr chromosomes could be generated simply by the insertion of F into the ring in the appropriate place and orientation (**Figure 6-14**).

KEY CONCEPT The insertion of a free F plasmid into the bacterial chromosome occurs by a single crossover at a region of DNA homology.

Several hypotheses—later supported—followed from Campbell's proposal.

1. One end of the integrated F factor would be the origin, where transfer of the Hfr chromosome begins. The **terminus** would be at the other end of F.

2. The orientation in which F is inserted would determine the order of entry of donor alleles. If the circle contains genes *A*, *B*, *C*, and *D*, then insertion between *A* and *D* would give the order *ABCD* or *DCBA*, depending on orientation. Check the different orientations of the insertions in Figure 6-14.

How is it possible for F to integrate at different sites and in different orientations? If F DNA had a region homologous to any of several regions on the bacterial chromosome, any one of them could act as a pairing region at which pairing could be followed by a crossover. These regions of homology are now known to be mainly segments

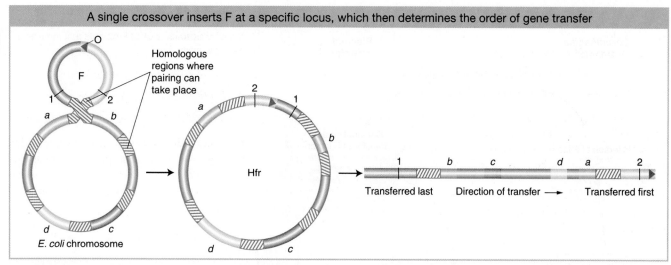

FIGURE 6-13 The insertion of F creates an Hfr cell. Hypothetical markers 1 and 2 are shown on F to depict the direction of insertion. The origin (O) is the mobilization point where insertion into the *E. coli* chromosome occurs; the pairing region is homologous with a region on the *E. coli* chromosome; *a* through *d* are representative genes in the *E. coli* chromosome. Pairing regions (hatched) are identical in plasmid and chromosome. They are derived from mobile elements called *insertion sequences* (see Chapter 16). In this example, the Hfr cell created by the insertion of F would transfer its genes in the order *a, d, c, b*.

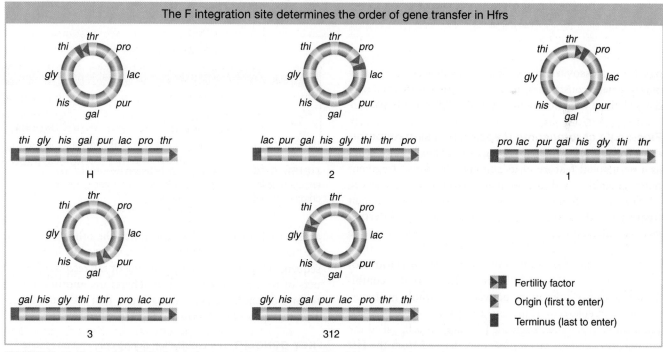

FIGURE 6-14 The five *E. coli* Hfr strains shown each have different F plasmid insertion points and orientations. All strains have the same order of genes on the *E. coli* chromosome. The orientation of the F factor determines which gene enters the recipient cell first. The gene closest to the terminus enters last.

of transposable elements called *insertion sequences*. For a full explanation of insertion sequences, see Chapter 16.

The fertility factor thus exists in two states:

1. **The plasmid state:** As a free cytoplasmic element, F is easily transferred to F⁻ recipients.

2. **The integrated state:** As a contiguous part of a circular chromosome, F is transmitted only very late in conjugation.

The *E. coli* conjugation cycle is summarized in **Figure 6-15**.

Mapping of bacterial chromosomes

Broad-scale chromosome mapping by using time of entry Wollman and Jacob realized that the construction of linkage maps from the interrupted-mating results would be easy by using as a measure of "distance" the times at which the donor alleles first appear after mating. The units of map distance in this case are minutes. Thus, if b^+ begins to enter the F⁻ cell 10 minutes after a^+ begins to enter, then a^+ and b^+ are 10 units apart (see map on p. 202). Like eukaryotic maps

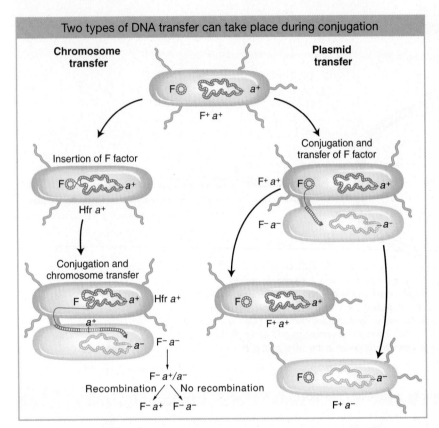

Two types of DNA transfer can take place during conjugation

Chromosome transfer

Plasmid transfer

F+ a+

Insertion of F factor

Hfr a+

Conjugation and chromosome transfer

Hfr a+

F− a−

F− a+/a−

Recombination / \ No recombination

F− a+   F− a−

Conjugation and transfer of F factor

F+ a+

F− a−

F+ a+

F+ a−

FIGURE 6-15 Conjugation can take place by partial transfer of a chromosome containing the F factor or by transfer of an F plasmid that remains a separate entity.

based on crossovers, these linkage maps were originally purely genetic constructions. At the time they were originally devised, there was no way of testing their physical basis.

Fine-scale chromosome mapping by using recombinant frequency For an exconjugant to acquire donor genes as a permanent feature of its genome, the donor fragment must recombine with the recipient chromosome. However, note that time-of-entry mapping is not based on recombinant frequency. Indeed, the units are minutes, not RF. Nevertheless, recombinant frequency can be used for a more fine-scale type of mapping in bacteria, a method to which we now turn.

First, we need to understand some special features of the recombination event in bacteria. Recall that recombination does not take place between two whole genomes, as it does in eukaryotes. In contrast, it takes place between one *complete* genome, from the F⁻ recipient cell, called the **endogenote**, and an *incomplete* one, derived from the Hfr donor cell and called the **exogenote**. The cell at this stage has two copies of one segment of DNA: one copy is part of the endogenote and the other copy is part of the exogenote. Thus, at this stage, the cell is a *partial* diploid, called a **merozygote**. Bacterial genetics is merozygote genetics. A single crossover in a merozygote would break the ring and thus not produce viable recombinants, as shown in **Figure 6-16**. To keep the circle intact, there must be an even number of crossovers. An even number of crossovers produces a circular, intact chromosome and a fragment. Although such recombination events are represented in a shorthand way as double crossovers, the actual molecular mechanism is somewhat different, more like an invasion of the endogenote by

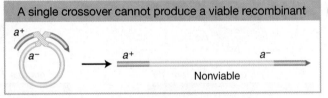

A single crossover cannot produce a viable recombinant

a+

a−

a+   a−

Nonviable

FIGURE 6-16 A single crossover between exogenote and endogenote in a merozygote would lead to a linear, partly diploid chromosome that would not survive.

an internal section of the exogenote. The other product of the "double crossover," the fragment, is generally lost in subsequent cell growth. Hence, only one of the reciprocal products of recombination survives. Therefore, another unique feature of bacterial recombination is that we must forget about reciprocal exchange products in most cases.

KEY CONCEPT Recombination during conjugation results from a double-crossover-like event, which gives rise to reciprocal recombinants of which only one survives.

With this understanding, we can examine recombination mapping. Suppose that we want to calculate map distances separating three close loci: *met*, *arg*, and *leu*. To examine the recombination of these genes, we need "trihybrids," exconjugants that have received all three donor markers. Assume that an interrupted-mating experiment has shown that the order is *met*, *arg*, *leu*, with *met* transferred first and *leu* last. To obtain a trihybrid, we need the merozygote diagrammed here:

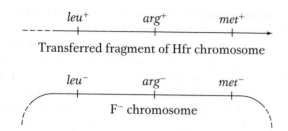

To obtain this merozygote, we must first select stable exconjugants bearing the *last* donor allele, which, in this case, is *leu⁺*. Why? In *leu⁺* exconjugants, we know all three markers were transferred into the recipient because *leu* is the last donor allele. We also know that at least the *leu⁺* marker was integrated into the endogenote. We want to know how often the other two markers were also integrated so that we can determine the number of recombination events in which *arg⁺* or *met⁺* was omitted due to double crossover.

The goal now is to count the frequencies of crossovers at different locations. Note that we now have a different situation from the analysis of interrupted conjugation. In mapping by interrupted conjugation, we measure the time of entry of individual loci; to be stably inherited, each marker has to recombine into the recipient chromosome by a double crossover spanning it. However, in the recombinant frequency analysis, we have specifically selected trihybrids as a starting point, and now we have to consider the various possible combinations of the three donor alleles that can be inserted by double crossing over in the various intervals. We know that *leu⁺* must have entered and inserted because we selected it, but the *leu⁺* recombinants that we select may or may not have incorporated the other donor markers, depending on where the double crossover took place. Hence, the procedure is to first select *leu⁺* exconjugants and then isolate and test a large sample of them to see which of the other markers were integrated.

Let's look at an example. In the cross Hfr *met⁺ arg⁺ leu⁺ str*ˢ × F⁻ *met⁻ arg⁻ leu⁻ str*ʳ, we would select *leu⁺* recombinants and then examine them for the *arg⁺* and *met⁺* alleles, called the **unselected markers**. **Figure 6-17** depicts the types

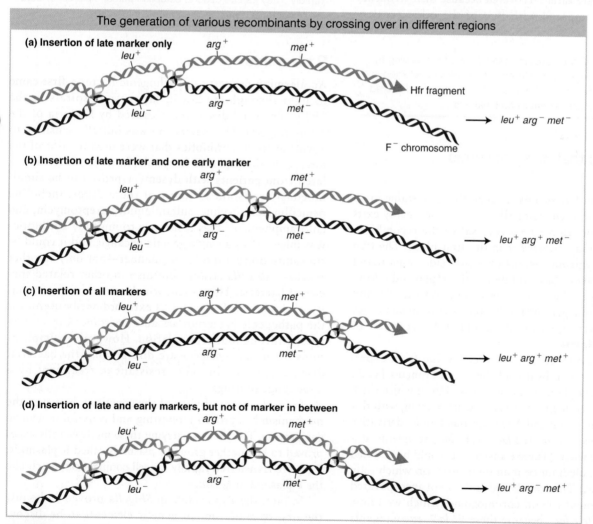

FIGURE 6-17 The diagram shows how genes can be mapped by recombination in *E. coli*. In exconjugants, selection is made for merozygotes bearing the *leu⁺* marker, which is donated late. The early markers (*arg⁺* and *met⁺*) may or may not be inserted, depending on the site where recombination between the Hfr fragment and the F⁻ chromosome takes place. The frequencies of events diagrammed in (a) and (b) are used to obtain the relative sizes of the *leu–arg* and *arg–met* regions. Note that, in each case, only the DNA inserted into the F⁻ chromosome survives; the other fragment is lost.

ANIMATED ART   Sapling Plus

Bacterial conjugation and mapping by recombination

of double-crossover events expected. One crossover must be on the left side of the *leu* marker and the other must be on the right side. Let's assume that the *leu*⁺ exconjugants are of the following types and frequencies:

$$
\begin{array}{ll}
leu^+\ arg^-\ met^- & 4\% \\
leu^+\ arg^+\ met^- & 9\% \\
leu^+\ arg^+\ met^+ & 87\%
\end{array}
$$

The double crossovers needed to produce these genotypes are shown in Figure 6-17a, b, and c. The first two types are the key because they require a crossover between *leu* and *arg* in the first case and between *arg* and *met* in the second. Hence, the relative frequencies of these types correspond to the sizes of these two regions between the genes. We would conclude that the *leu–arg* region is 4 m.u. and that the *arg–met* region is 9 m.u.

In a cross such as the one just described, one type of potential recombinants of genotype *leu*⁺ *arg*⁻ *met*⁺ requires four crossovers instead of two (see Figure 6-17d). These recombinants are rarely recovered because their frequency is very low compared with that of the other types of recombinants.

KEY CONCEPT Merozygotes can be used for mapping by recombinant frequency. A late-entering marker is selected and these cells are tested for the unselected markers and scored for recombinant and parental combinations.

F plasmids that carry genomic fragments

The F factor in Hfr strains is generally quite stable in its inserted position. Occasionally, an F factor cleanly exits from the chromosome by a reversal of the recombination process that inserted it in the first place. The two homologous pairing regions on either side re-pair, and a crossover takes place to liberate the F plasmid. However, sometimes the exit is not clean, and the plasmid carries with it a part of the bacterial chromosome. An F plasmid carrying bacterial genomic DNA is called an **F′ (F prime) plasmid.**

The first evidence of this process came from experiments in 1959 by Edward Adelberg and François Jacob. One of their key observations was of an Hfr in which the F factor was integrated near the *lac*⁺ locus. Starting with this Hfr *lac*⁺ strain, Jacob and Adelberg found an F⁺ derivative that, in crosses, transferred *lac*⁺ to F⁻ *lac*⁻ recipients at a very high frequency. (These transferrants could be detected by plating on medium containing lactose, on which only *lac*⁺ can grow.) The transferred *lac*⁺ is not incorporated into the recipient's main chromosome, which we know retains the allele *lac*⁻ because these F⁺ *lac*⁺ exconjugants occasionally gave rise to F⁻ *lac*⁻ daughter cells, at a frequency of 1×10^{-3}. Thus, the genotype of these recipients appeared to be F′ *lac*⁺/F⁻ *lac*⁻. In other words, the *lac*⁺

exconjugants seemed to carry an F′ plasmid with a piece of the donor chromosome incorporated. The origin of this F′ plasmid is shown in **Figure 6-18**. Note that the faulty excision occurs because there is another homologous region nearby that pairs with the original. The F′ in our example is called F′ *lac* because the piece of host chromosome that it picked up has the *lac* gene on it. F′ factors have been found carrying many different chromosomal genes and have been named accordingly. For example, F′ factors carrying *gal* or *trp* are called F′ *gal* and F′ *trp*, respectively. Because F′ *lac*⁺/F⁻ *lac*⁻ cells are *lac*⁺ in phenotype, we know that *lac*⁺ is dominant over *lac*⁻.

Partial diploids made with the use of F′ strains are useful for some aspects of routine bacterial genetics, such as the study of dominance or of allele interaction. Some F′ strains can carry very large parts (as much as one-quarter) of the bacterial chromosome.

KEY CONCEPT The DNA of an F′ plasmid is part F factor and part bacterial genome. Like F plasmids, F′ plasmids transfer rapidly. They can be used to establish partial diploids for studies of bacterial dominance and allele interaction.

R plasmids

An alarming property of pathogenic bacteria first came to light through studies in Japanese hospitals in the 1950s. Bacterial dysentery is caused by bacteria of the genus *Shigella*. This bacterium was initially sensitive to a wide array of antibiotics that were used to control the disease. In the Japanese hospitals, however, *Shigella* isolated from patients with dysentery proved to be simultaneously resistant to many of these drugs, including penicillin, tetracycline, sulfanilamide, streptomycin, and chloramphenicol. This resistance to multiple antibiotics was inherited as a single genetic package, and it could be transmitted in an infectious manner—not only to other sensitive *Shigella* strains, but also to other related species of bacteria. This talent, which resembles the mobility of the *E. coli* F plasmid, is extraordinarily useful for the pathogenic bacterium because resistance can rapidly spread throughout a population. However, its implications for medical science are dire because the bacterial disease suddenly becomes resistant to treatment by a large range of drugs.

From the point of view of the geneticist, however, the mechanism has proved interesting and is useful in genetic engineering. The vectors carrying these multiple resistances proved to be another group of plasmids called **R plasmids.** They are transferred rapidly on cell conjugation, much like the F plasmid in *E. coli*.

In fact, the R plasmids in *Shigella* proved to be just the first of many similar genetic elements to be discovered. All exist in the plasmid state in the cytoplasm. These elements have been found to carry many different kinds of genes in bacteria. **Table 6-2** shows some of the

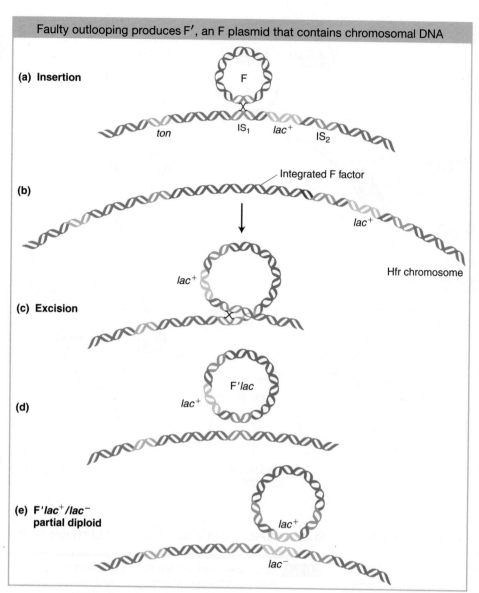

Faulty outlooping produces F′, an F plasmid that contains chromosomal DNA

(a) **Insertion**

F

ton   IS₁   lac⁺   IS₂

(b)

Integrated F factor

lac⁺

Hfr chromosome

(c) **Excision**

lac⁺

lac⁺

(d)

F′lac

lac⁺

(e) **F′lac⁺/lac⁻ partial diploid**

lac⁺

lac⁻

FIGURE 6-18 An F plasmid can pick up chromosomal DNA as it exits a chromosome. (a) F is inserted in an Hfr strain at a repetitive element identified as IS₁ (insertion sequence 1) between the *ton* and *lac⁺* alleles. (b) The integrated F factor. (c) Abnormal "outlooping" by crossing over with a different element, IS₂, to include the *lac* locus. (d) The resulting F′ *lac⁺* plasmid. (e) F′ *lac⁺* /F⁻ *lac⁻* partial diploid produced by the transfer of the F′ *lac⁺* plasmid to an F⁻ *lac⁻* recipient. [*Data from G. S. Stent and R. Calendar, Molecular Genetics, 2nd ed.*]

characteristics that can be borne by plasmids. **Figure 6-19** shows an example of a well-traveled plasmid isolated from the dairy industry.

Engineered derivatives of R plasmids, such as pBR322 and pUC (see Chapter 10), have become the preferred vectors for the molecular cloning of the DNA of all organisms. The genes on an R plasmid that confer resistance can be used as markers to keep track of the movement of the vectors between cells.

On R plasmids, the alleles for antibiotic resistance are often contained within a unit called a *transposon* (**Figure 6-20**). Transposons are unique segments of DNA that can move around to different sites in the genome, a process called transposition. (The mechanisms for transposition, which occurs in most species studied, will be detailed in Chapter 16.) When a transposon in the genome moves to a new location, it can occasionally embrace between its ends various types of genes, including alleles for drug resistance, and carry them along to their new locations as passengers. Sometimes, a transposon carries a drug-resistance allele to a plasmid, creating an R plasmid. Like F plasmids, many R plasmids are conjugative; in other words, they are effectively transmitted to a recipient cell during conjugation. Even R plasmids that are not conjugative and never leave their own cells can donate their R alleles to a conjugative plasmid by transposition. Hence, through plasmids,

TABLE 6-2	Genetic Determinants Borne by Plasmids
Characteristic	**Plasmid examples**
Fertility	F, R1, Col
Bacteriocin production	Col E1
Heavy-metal resistance	R6
Enterotoxin production	Ent
Metabolism of camphor	Cam
Tumorigenicity in plants	T1 (in *Agrobacterium tumefaciens*)

FIGURE 6-19 The diagram shows the origins of genes of the *Lactococcus lactis* plasmid pK214. The genes are from many different bacteria. [*Data from Table 1 in V. Perreten, F. Schwarz, L. Cresta, M. Boeglin, G. Dasen, and M. Teuber,* Nature *389, 1997, 801–802.*]

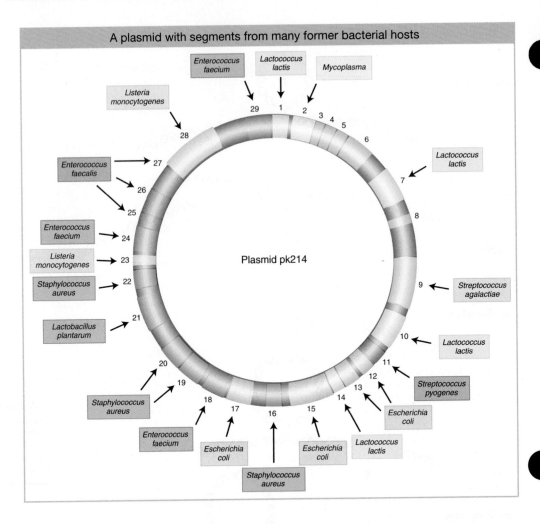

A plasmid with segments from many former bacterial hosts

Plasmid pk214

antibiotic-resistance alleles can spread rapidly throughout a population of bacteria. Although the spread of R plasmids is an effective strategy for the survival of bacteria, it presents a major problem for medical practice, as mentioned earlier, because bacterial populations rapidly become resistant to any new antibiotic drug that is invented and applied to humans.

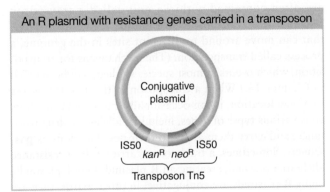

An R plasmid with resistance genes carried in a transposon

Conjugative plasmid

IS50   IS50
kan^R *neo*^R
Transposon Tn5

FIGURE 6-20 A transposon such as Tn5 can acquire several drug-resistance genes (in this case, those for resistance to the drugs kanamycin and neomycin) and transmit them rapidly on a plasmid, leading to the infectious transfer of resistance genes as a package. Insertion sequence 50 (IS50) forms the flanks of Tn5.

6.3 BACTERIAL TRANSFORMATION

LO 6.1 Distinguish between the three main ways by which bacteria exchange genes, and describe the experimental procedures that reveal them.

LO 6.5 Assess the outcome of double transformation experiments in terms of linkage.

Some bacteria can take up fragments of DNA or intact plasmids from the external medium, and such uptake constitutes another way in which bacteria can exchange their genes. The source of the DNA can be other cells of the same species or cells of other species. In some cases, the DNA has been released from dead cells; in other cases, the DNA has been secreted from live bacterial cells. The DNA taken up can integrate into the recipient's chromosome (circular plasmids remain extrachromosomal). If this DNA is of a different genotype from that of the recipient, the genotype of the recipient can become permanently changed, a process aptly termed **transformation**.

The nature of transformation

Transformation was discovered in the bacterium *Streptococcus pneumoniae* in 1928 by Frederick Griffith. Later, in 1944, Oswald T. Avery, Colin M. MacLeod, and Maclyn

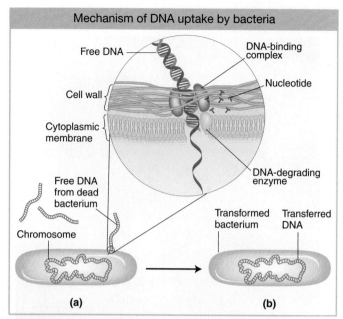

FIGURE 6-21 A bacterium undergoing transformation (a) picks up free DNA released from a dead bacterial cell. As DNA-binding complexes on the bacterial surface take up the DNA (*inset*), enzymes break down one strand into nucleotides; a derivative of the other strand may integrate into the bacterium's chromosome (b).

McCarty demonstrated that the "transforming principle" was DNA. Both results are milestones in the elucidation of the molecular nature of genes. We consider this work in more detail in Chapter 7.

The transforming DNA is incorporated into the bacterial chromosome by a process analogous to the double-recombination events observed in Hfr × F⁻ crosses. Note, however, that, in *conjugation*, DNA is transferred from one living cell to another through close contact, whereas in *transformation*, isolated pieces of external DNA are taken up by a cell through the cell wall and plasma membrane. **Figure 6-21** shows one way in which this process can take place.

Transformation has been a handy tool in several areas of bacterial research because the genotype of a strain can be deliberately changed in a very specific way by transforming with an appropriate DNA fragment or plasmid. For example, transformation is used widely in genetic engineering. It has been found that even eukaryotic cells can be transformed, by using quite similar procedures, and this technique has been invaluable for modifying eukaryotic cells (see Chapter 10).

Chromosome mapping using transformation

Transformation can be used to measure how closely two genes are linked on a bacterial chromosome. When DNA (the bacterial chromosome) is extracted for transformation experiments, some breakage into smaller pieces is inevitable. If two donor genes are located close together on the chromosome, there is a good chance that sometimes they will

be carried on the same piece of transforming DNA. Hence, both will be taken up, causing a **double transformation.** Conversely, if genes are widely separated on the chromosome, they will most likely be carried on separate transforming segments. A genome could possibly take up both segments independently, creating a double transformant, but that outcome is not likely. Hence, in widely separated genes, the frequency of double transformants will equal the product of the single-transformant frequencies. Therefore, testing for close linkage by testing for a departure from the product rule should be possible. In other words, if genes are linked, then the proportion of double transformants will be greater than the product of single-transformant frequencies.

Unfortunately, the situation is made more complex by several factors—the most important being that not all cells in a population of bacteria are competent to be transformed. Nevertheless, at the end of this chapter, you can sharpen your skills in transformation analysis in one of the problems, which assumes that 100 percent of the recipient cells are competent.

KEY CONCEPT Bacteria can take up DNA fragments from the surrounding medium. Inside the cell, these fragments can integrate into the chromosome.

6.4 BACTERIOPHAGE GENETICS

LO 6.6 Map phage genomes by recombination in double infections of bacteria.

The word *bacteriophage*, which is a name for bacterial viruses, means "eater of bacteria." These viruses parasitize and kill bacteria. Pioneering work on the genetics of bacteriophages in the middle of the twentieth century formed the foundation of more recent research on tumor-causing viruses and other kinds of animal and plant viruses. In this way, bacterial viruses have provided an important model system.

These viruses can be used in two different types of genetic analysis. First, two distinct phage genotypes can be crossed to measure recombination and hence map the viral genome. Mapping of the viral genome by this method is the topic of this section. Second, bacteriophages can be used as a way of bringing bacterial genes together for linkage and other genetic studies. We will study the use of phages in bacterial studies in Section 6.5. In addition, as we will see in Chapter 10, phages are used in DNA technology as vectors, or carriers, of foreign DNA. Before we can understand phage genetics, we must first examine the infection cycle of phages.

Infection of bacteria by phages

Most bacteria are susceptible to attack by bacteriophages. A phage consists of a nucleic acid "chromosome" (DNA or RNA) surrounded by a coat of protein molecules. Phage types are identified not by species names but by symbols—for example, phage T4, phage λ, and so forth. **Figures 6-22** and **6-23** show the structure of phage T4. During infection, a phage attaches to a bacterium and injects its genetic material into the bacterial cytoplasm, as diagrammed in Figure 6-22. An electron micrograph of the process is shown in **Figure 6-24**. The phage genetic information then takes over the machinery of the bacterial cell by turning off the synthesis of bacterial components and redirecting the bacterial synthetic machinery to make phage components. Newly made phage heads are individually stuffed with replicates of the phage chromosome. Ultimately, many phage descendants are made and are released when the bacterial cell wall breaks open. This breaking-open process is called **lysis**. The population of phage progeny is called the phage **lysate**.

How can we study inheritance in phages when they are so small that they are visible only under the electron microscope? In this case, we cannot produce a visible colony by plating, but we can produce a visible manifestation of a phage by taking advantage of several phage characters.

Let's look at the consequences of a phage infecting a single bacterial cell. **Figure 6-25** shows the sequence of events in the infectious cycle that leads to the release of progeny phages from the lysed cell. After lysis, the progeny phages

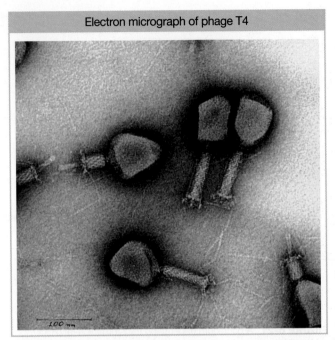

Electron micrograph of phage T4

FIGURE 6-23 Enlargement of the *E. coli* phage T4 reveals details of head, tail, and tail fibers. [*Omikron/Science Source.*]

infect neighboring bacteria. This cycle is repeated through progressive rounds of infection, and, as these cycles repeat, the number of lysed cells increases exponentially. Within 15 hours after one single phage particle infects a single bacterial cell, the effects are visible to the naked eye as a clear area, or **plaque**, in the opaque lawn of bacteria covering the surface of a plate of solid medium (**Figure 6-26**). Such plaques can be large or small, fuzzy or sharp, and so forth, depending on the phage genotype. Thus, *plaque morphology* is a phage character that can be analyzed at the genetic level. Another phage phenotype that we can analyze genetically is *host range*, because phages may differ in the spectra of bacterial strains that they can infect and lyse.

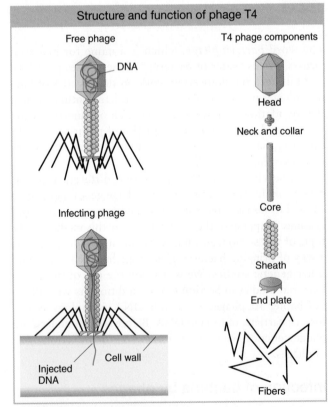

Structure and function of phage T4

FIGURE 6-22 An infecting phage injects DNA through its core structure into the cell. *Left:* Bacteriophage T4 is shown as a free phage and then in the process of infecting an *E. coli* cell. *Right:* The major structural components of T4.

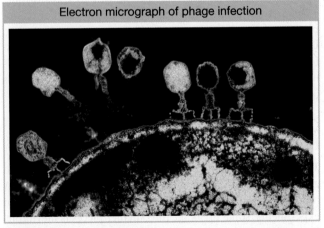

Electron micrograph of phage infection

FIGURE 6-24 Bacteriophages are shown in several stages of the infection process, which includes attachment and DNA injection. [*Eye of Science/Science Source.*]

Cycle of phage that lyses the host cells

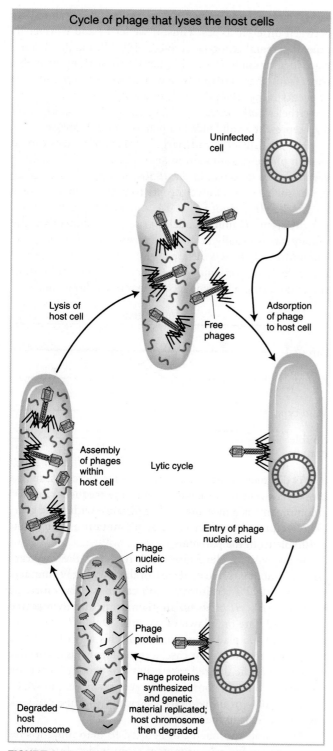

Uninfected cell

Lysis of host cell

Free phages

Adsorption of phage to host cell

Assembly of phages within host cell

Lytic cycle

Entry of phage nucleic acid

Phage nucleic acid

Phage protein

Phage proteins synthesized and genetic material replicated; host chromosome then degraded

Degraded host chromosome

FIGURE 6-25 Infection by a single phage redirects the cell's machinery into making progeny phages, which are released at lysis.

For example, a specific strain of bacteria might be immune to phage 1 but susceptible to phage 2.

KEY CONCEPT A phage inserts its genome into a bacterial cell where it directs the cellular machinery to make many copies of phage DNA and sheath components, which then assemble into progeny phages.

A plaque is a clear area in which all bacteria have been lysed by phages

Clear areas, or plaques

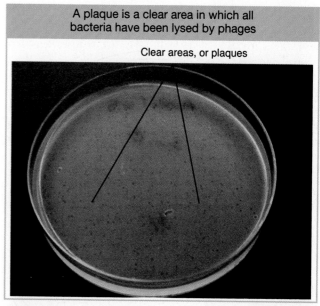

FIGURE 6-26 Through repeated infection and production of progeny phage, a single phage produces a clear area, or plaque, on the opaque lawn of bacterial cells. [*D. Sue Katz, Rogers State University, Claremore, OK.*]

Mapping phage chromosomes by using phage crosses

Two phage genotypes can be crossed in much the same way that we cross organisms. A phage cross can be illustrated by a cross of T2 phages originally studied by Alfred Hershey. The genotypes of the two parental strains in Hershey's cross were $h^- r^+ \times h^+ r^-$. The alleles correspond to the following phenotypes:

> h^-: can infect two different *E. coli* strains (which we can call strains 1 and 2)
>
> h^+: can infect only strain 1
>
> r^-: rapidly lyses cells, thereby producing large plaques
>
> r^+: slowly lyses cells, producing small plaques

To make the cross, *E. coli* strain 1 is infected with both parental T2 phage genotypes. This kind of infection is called a **mixed infection** or a **double infection** (**Figure 6-27**).

A phage cross made by doubly infecting the host cell with parental phages

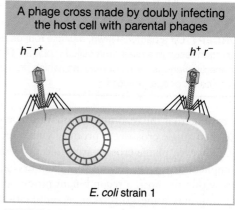

$h^- r^+$ $h^+ r^-$

E. coli strain 1

FIGURE 6-27

Plaques from recombinant and
parental phage progeny

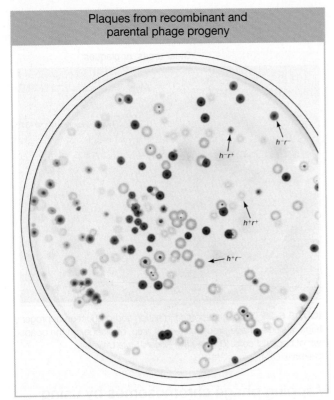

FIGURE 6-28 These plaque phenotypes were produced by progeny of the cross $h^- r^+ \times h^+ r^-$. Four plaque phenotypes can be differentiated, representing two parental types and two recombinants. [*From G. S. Stent,* Molecular Biology of Bacterial Viruses. *©1963 by W.H. Freeman & Company.*]

After an appropriate incubation period, the phage lysate (containing the progeny phages) is analyzed by spreading it onto a bacterial lawn composed of a mixture of *E. coli* strains 1 and 2. Four plaque types are then distinguishable (**Figure 6-28**). Large plaques indicate rapid lysis (r^-), and small plaques indicate slow lysis (r^+). Phage plaques with the allele h^- will infect both hosts, forming a clear plaque, whereas phage plaques with the allele h^+ will infect only one host, forming a cloudy plaque. Thus, the four genotypes can be easily classified as parental ($h^- r^+$ and $h^+ r^-$) and recombinant ($h^+ r^+$ and $h^- r^-$), and a recombinant frequency can be calculated as follows:

$$RF = \frac{(h^+ r^+) + (h^- r^-)}{\text{total plaques}}$$

KEY CONCEPT On a bacterial lawn, multiple rounds of phage infection result in a clear area called a plaque. Different phages generate various forms of plaques, which provide useful phenotypes for phage genetics.

If we assume that the recombining phage chromosomes are linear, then single crossovers produce viable reciprocal products. However, phage crosses are subject to some analytical complications. First, several rounds of exchange can take place within the host: a recombinant produced shortly after infection may undergo further recombination in the same cell or in later infection cycles. Second, recombination

can take place between genetically similar phages as well as between different types. Thus, if we let P_1 and P_2 refer to general parental genotypes, crosses of $P_1 \times P_1$ and $P_2 \times P_2$ take place in addition to $P_1 \times P_2$. For both these reasons, recombinants from phage crosses are a consequence of a *population* of events rather than defined, single-step exchange events. Nevertheless, *all other things being equal*, the RF calculation does represent a valid index of map distance in phages.

Because astronomically large numbers of phages can be used in phage-recombination analyses, very rare crossover events can be detected. In the 1950s, Seymour Benzer made use of such rare crossover events to map the mutant sites *within* the *rII* gene of phage T4, a gene that controls lysis. For different *rII* mutant alleles arising spontaneously, the mutant site is usually at different positions within the gene. Therefore, when two different *rII* mutants are crossed, a few rare crossovers may take place between the mutant sites, producing wild-type recombinants, as shown here:

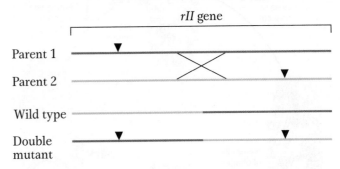

As distance between two mutant sites increases, such a crossover event is more likely. Thus, the frequency of rII^+ recombinants is a measure of that distance within the gene. (The reciprocal product is a double mutant and indistinguishable from the parentals.)

Benzer used a clever approach to detect the very rare rII^+ recombinants. He made use of the fact that *rII* mutants will not infect a strain of *E. coli* called K. Therefore, he made the $rII \times rII$ cross on another strain and then plated the phage lysate on a lawn of strain K. Only rII^+ recombinants will form plaques on this lawn. This way of finding a rare genetic event (in this case, a recombinant) is a **selective system:** *only* the desired rare event can produce a certain visible outcome. In contrast, a **screen** is a system in which large numbers of individuals are visually scanned to seek the rare "needle in the haystack."

This same approach can be used to map mutant sites within genes for any organism from which large numbers of cells can be obtained and for which wild-type and mutant phenotypes can be distinguished. However, this sort of intragenic mapping has been largely superseded by the advent of inexpensive chemical methods for DNA sequencing, which identify the positions of mutant sites directly.

KEY CONCEPT Recombination between phage chromosomes can be studied by bringing the parental chromosomes together in one host cell through mixed infection. Progeny phages can be examined for both parental and recombinant genotypes.

6.5 TRANSDUCTION

LO 6.1 Distinguish between the three main ways by which bacteria exchange genes, and describe the experimental procedures that reveal them.

LO 6.7 Predict the outcomes of transduction experiments using phages capable of generalized or specialized transduction.

Some phages are able to pick up bacterial genes and carry them from one bacterial cell to another, a process known as **transduction**. Thus, transduction joins the battery of modes of transfer of genomic material between bacteria—along with Hfr chromosome transfer, F' plasmid transfer, and transformation.

Discovery of transduction

In 1951, Joshua Lederberg and Norton Zinder were testing for recombination in the bacterium *Salmonella typhimurium* by using the techniques that had been successful with *E. coli*. The researchers used two different strains: one was *phe⁻ trp⁻ tyr⁻*, and the other was *met⁻ his⁻*. We won't worry about the nature of these alleles except to note that all are auxotrophic. When either strain was plated on a minimal medium, no wild-type cells were observed. However, after the two strains were mixed, wild-type prototrophs appeared at a frequency of about 1 in 10^5. Thus far, the situation seems similar to that for recombination in *E. coli*.

However, in this case, the researchers also recovered recombinants from a U-tube experiment, in which conjugation was prevented by a filter separating the two arms (recall Figure 6-6). They hypothesized that some agent was carrying genes from one bacterium to another. By varying the size of the pores in the filter, they found that the agent responsible for gene transfer was the same size as a known phage of *Salmonella*, called phage P22. Furthermore, the filterable agent and P22 were identical in sensitivity to antiserum and in immunity to hydrolytic enzymes. Thus, Lederberg and Zinder had discovered a new type of gene transfer, mediated by a virus. They were the first to call this process *transduction*. As a rarity in the lytic cycle, virus particles sometimes pick up bacterial genes and transfer them when they infect another host. Transduction has subsequently been demonstrated in many bacteria.

To understand the process of transduction, we need to distinguish two types of phage cycle. **Virulent phages** are those that immediately lyse and kill the host. **Temperate phages** can remain within the host cell for a period without killing it. Their DNA either integrates into the host chromosome, to replicate with it, or replicates separately in the cytoplasm, as does a plasmid. A phage integrated into the bacterial genome is called a **prophage**. A bacterium harboring a quiescent phage is described as **lysogenic** and is itself called a **lysogen**. Occasionally, the quiescent phage in a lysogenic bacterium becomes active, replicates itself, and causes the spontaneous lysis of its host cell. A resident temperate phage confers resistance to infection by other phages of that type.

KEY CONCEPT Virulent phages cannot become prophages; they replicate and lyse a cell immediately. Temperate phages can exist within the bacterial cell as prophages, allowing their hosts to survive as lysogenic bacteria; they are also capable of occasional bacterial lysis.

There are two kinds of transduction: generalized and specialized. *Generalized* transducing phages can carry any part of the bacterial chromosome, whereas *specialized* transducing phages carry only certain specific parts.

Generalized transduction

By what mechanisms can a phage carry out **generalized transduction**? In 1965, H. Ikeda and J. Tomizawa threw light on this question in some experiments on the *E. coli* phage P1. They found that, when a donor cell is lysed by P1, the bacterial chromosome is broken up into small pieces. Occasionally, the newly forming phage particles mistakenly incorporate a piece of the bacterial DNA into a phage head in place of phage DNA. This event is the origin of the transducing phage.

A phage carrying bacterial DNA can infect another cell. That bacterial DNA fragment can then be incorporated into the recipient cell's chromosome by recombination (**Figure 6-29**). Because genes on any of the cut-up parts of the host genome can be transduced, this type of transduction is by necessity of the generalized type.

Phages P1 and P22 both belong to a phage group that shows generalized transduction. P22 DNA inserts into the host chromosome, whereas P1 DNA remains free, like a large plasmid. However, both transduce by faulty head stuffing.

KEY CONCEPT When a bacterial cell harboring an inserted phage occasionally lyses, some of the phage progeny carry fragments of bacterial DNA, and these phages can transform the genotypes of recipient bacterial cells.

Generalized transduction can be used to obtain bacterial linkage information when genes are close enough that the phage can pick them up and transduce them in a single piece of DNA. For example, suppose that we wanted to find the linkage distance between *met* and *arg* in *E. coli*. We could grow phage P1 on a donor *met⁺ arg⁺* strain and then allow P1 phages from lysis of this strain to infect a *met⁻ arg⁻* strain. First, one donor allele is selected, say, *met⁺*. Then the percentage of *met⁺* colonies that are also *arg⁺* is measured. Strains transduced to both *met⁺* and *arg⁺* are called **cotransductants**. The *greater* the cotransduction frequency, the *closer* two genetic markers must be (the opposite of most mapping measurements). Linkage values are usually expressed as cotransduction frequencies (**Figure 6-30**).

By using an extension of this approach, we can estimate the *size* of the piece of host chromosome that a phage can

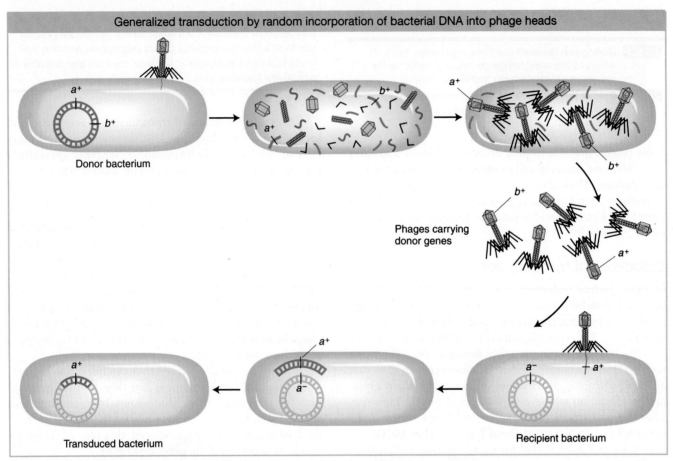

FIGURE 6-29 A newly forming phage may pick up DNA from its host cell's chromosome (*top*) and then inject it into a new cell (*bottom right*). The injected DNA may insert into the new host's chromosome by recombination (*bottom left*). In reality, only a very small minority of phage progeny (1 in 10,000) carry donor genes.

pick up, as in the following type of experiment, which uses P1 phage:

donor *leu*+ *thr*+ *azi*ʳ → recipient *leu*− *thr*− *azi*ˢ

In this experiment, P1 phage grown on the *leu*+ *thr*+ *azi*ʳ donor strain infect the *leu*− *thr*− *azi*ˢ recipient strain. The strategy is to select one or more donor alleles in the recipient and then test these transductants for the presence of the unselected alleles. The results are outlined in **Table 6-3**.

TABLE 6-3	Accompanying Markers in Specific P1 Transductions	
Experiment	**Selected marker**	**Unselected markers**
1	*leu*+	50% are *azi*ʳ; 2% are *thr*+
2	*thr*+	3% are *leu*+; 0% are *azi*ʳ
3	*leu*+ and *thr*+	0% are *azi*ʳ

FIGURE 6-30 The diagram shows a genetic map of the *purB*-to-*cysB* region of *E. coli* determined by P1 cotransduction. The numbers given are the averages in percent for cotransduction frequencies obtained in several experiments. The values in parentheses are considered unreliable. [*Data from J. R. Guest, Mol. Gen. Genet. 105, 1969, p. 285.*]

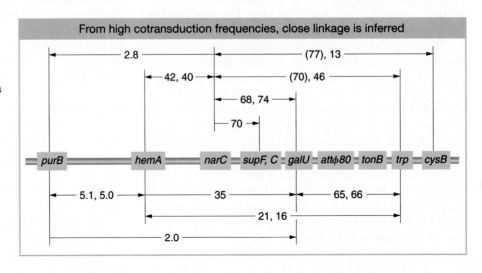

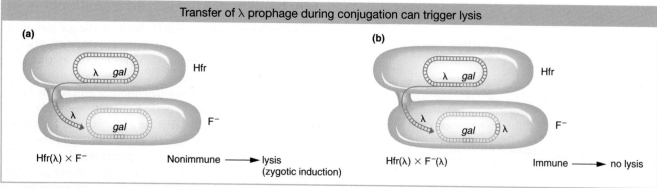

FIGURE 6-31 A λ prophage can be transferred to a recipient during conjugation, but the prophage triggers lysis, a process called zygotic induction, only if the recipient has no prophage already—that is, in the case shown in part *a* but not in part *b*.

Experiment 1 in Table 6-3 tells us that *leu* is relatively close to *azi* and distant from *thr*, leaving us with two possibilities:

$$\frac{thr \qquad leu \quad azi}{}$$

or

$$\frac{thr \qquad azi \quad leu}{}$$

Experiment 2 tells us that *leu* is closer to *thr* than *azi* is, and so the map must be

$$\frac{thr \qquad leu \quad azi}{}$$

By selecting for *thr⁺* and *leu⁺* together in the transducing phages in experiment 3, we see that the transduced piece of genetic material never includes the *azi* locus because the phage head cannot carry a fragment of DNA that big. P1 can only cotransduce genes less than approximately 1.5 minutes apart on the *E. coli* chromosome map.

Specialized transduction

A generalized transducer, such as phage P22, picks up fragments of broken host DNA at random. How are other phages, which act as specialized transducers, able to carry only certain host genes to recipient cells? The short answer is that a specialized transducer inserts into the bacterial chromosome at one position only. When it exits, a faulty outlooping occurs (similar to the type that produces F′ plasmids). Hence, it can pick up and transduce only genes that are close by.

The prototype of **specialized transduction** was provided by studies undertaken by Joshua and Esther Lederberg on a temperate *E. coli* phage called *lambda* (λ). Phage λ has become the most intensively studied and best-characterized phage.

Behavior of the prophage Phage λ has unusual effects when cells lysogenic for it are used in crosses. In the cross of an uninfected Hfr with a lysogenic F⁻ recipient [Hfr × F⁻ (λ)], lysogenic F⁻ exconjugants with Hfr genes are readily recovered, as expected. However, in the reciprocal cross Hfr(λ) × F⁻, the *early* genes from the Hfr chromosome are recovered among the exconjugants, but recombinants for *late* genes are not recovered. Furthermore, lysogenic F⁻ exconjugants are

almost never recovered from this reciprocal cross. What is the explanation? The observations make sense if the λ prophage is behaving as a bacterial gene locus behaves (that is, as part of the bacterial chromosome). Thus, in the Hfr(λ) × F⁻ cross, the prophage would enter the F⁻ cell at a specific time corresponding to its position in the chromosome. Earlier genes are recovered because they enter before the prophage. Later genes are not recovered because lysis destroys the recipient cell. In interrupted-mating experiments, the λ prophage does in fact always enter the F⁻ cell at a specific time, closely linked to the *gal* locus.

In an Hfr(λ) × F⁻ cross, the entry of the λ prophage into the cell immediately triggers the prophage into a lytic cycle; this process is called **zygotic induction** **(Figure 6-31)**. However, in the cross of *two* lysogenic cells Hfr(λ) × F⁻(λ), there is no zygotic induction. The presence of any prophage prevents another infecting virus from causing lysis. This is because the prophage produces a cytoplasmic factor that represses the multiplication of the virus. (The phage-directed cytoplasmic repressor nicely explains the immunity of the lysogenic bacteria, because a phage would immediately encounter a repressor and be inactivated.)

λ insertion The interrupted-mating experiments heretofore described showed that the λ prophage is part of the lysogenic bacterium's chromosome. How is the λ prophage inserted into the bacterial genome? In 1962, Allan Campbell proposed that it inserts by a single crossover between a circular λ phage chromosome and the circular *E. coli* chromosome, as shown in **Figure 6-32**. The crossover point would be between a specific site in λ, the **λ attachment site**, and an attachment site in the bacterial chromosome located between the genes *gal* and *bio*, because λ integrates at that position in the *E. coli* chromosome.

An attraction of Campbell's proposal is that from it follow predictions that geneticists can test. For example, integration of the prophage into the *E. coli* chromosome should increase the genetic distance between flanking bacterial genes, as can be seen in Figure 6-32 for *gal* and *bio*. In fact, studies show that lysogeny *does* increase time-of-entry or recombination distances between the bacterial genes. This unique location of λ accounts for its specialized transduction.

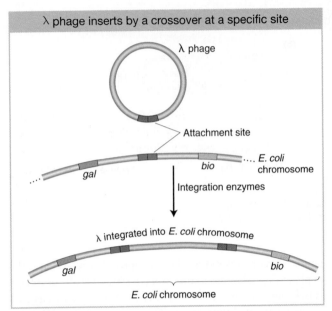

λ phage inserts by a crossover at a specific site

FIGURE 6-32 Reciprocal recombination takes place between a specific attachment site on the circular DNA and a specific region called the attachment site on the *E. coli* chromosome between the *gal* and *bio* genes.

FIGURE 6-33 The diagram shows how specialized transduction operates in phage λ. (a) A crossover at the specialized attachment site produces a lysogenic bacterium. (b) The lysogenic bacterium can produce a normal λ (i) or, rarely, λdgal (ii), a transducing particle containing the *gal* gene. (c) *gal*⁺ transductants can be produced by either (i) the co-incorporation of λdgal and λ (acting as a helper) or (ii) crossovers flanking the *gal* gene, a rare event. The blue double boxes are the bacterial attachment site, the purple double boxes are the λ attachment site, and the pairs of blue and purple boxes are hybrid integration sites, derived partly from *E. coli* and partly from λ.

Mechanism of specialized transduction

As a prophage, λ always inserts between the *gal* region and the *bio* region of the host chromosome (**Figure 6-33**), and, in transduction experiments, as expected, λ can transduce only the *gal* and *bio genes.*

How does λ carry away neighboring genes? The explanation lies, again, in an imperfect reversal of the Campbell insertion mechanism, like that for F′ formation. The recombination event between specific regions of λ and the bacterial chromosome is catalyzed by a specialized phage-encoded enzyme system that uses the λ attachment site as a substrate. The enzyme system dictates that λ integrates only at a specific point between *gal* and *bio* in the chromosome (see Figure 6-33a). Furthermore, during lysis, the λ prophage normally excises at precisely the correct point to produce a normal circular λ chromosome, as seen in Figure 6-33b(i). Very rarely, excision is abnormal owing to faulty outlooping. In this case, the outlooping phage DNA can pick up a nearby gene and leave behind some phage genes, as seen in Figure 6-33b(ii). The resulting phage genome is defective because of the genes left behind, but it has also gained

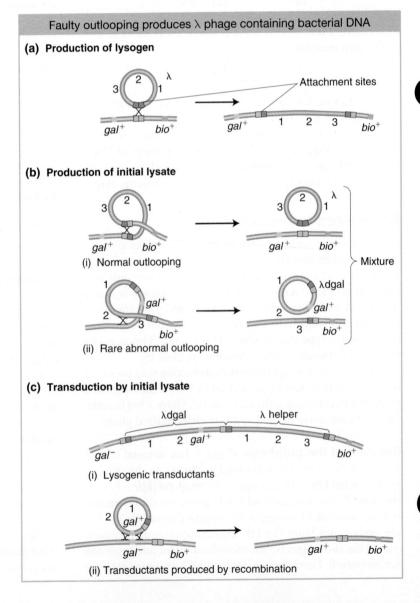

Faulty outlooping produces λ phage containing bacterial DNA

(a) Production of lysogen

Attachment sites

(b) Production of initial lysate

(i) Normal outlooping

(ii) Rare abnormal outlooping

Mixture

(c) Transduction by initial lysate

λdgal   λ helper

(i) Lysogenic transductants

(ii) Transductants produced by recombination

a bacterial gene, *gal* or *bio*. The abnormal DNA carrying nearby genes can be packaged into phage heads to produce phage particles that can infect other bacteria. These phages are referred to as λdgal (λ-defective *gal*) or λdbio. In the presence of a second, normal phage particle in a double infection, the λdgal can integrate into the chromosome at the λ attachment site (Figure 6-33c). In this manner, the *gal* genes in this case are transduced into the second host.

KEY CONCEPT Transduction occurs when newly forming phages acquire host genes and transfer them to other bacterial cells. *Generalized transduction* can transfer any host gene. It occurs when phage packaging accidentally incorporates bacterial DNA instead of phage DNA. *Specialized transduction* is due to faulty outlooping of the prophage from the bacterial chromosome, and so the new phage includes both phage and bacterial genes. The transducing phage can transfer only specific host genes.

6.6 PHYSICAL MAPS AND LINKAGE MAPS COMPARED

LO 6.8 Explain how transposons can be used in insertional mutagenesis to create and map new mutations.

Some very detailed chromosomal maps for bacteria have been obtained by combining the mapping techniques of interrupted mating, recombination mapping, transformation, and transduction. Today, new genetic markers are typically mapped first into a segment of about 10 to 15 map minutes by using interrupted mating. Then additional, closely linked markers can be mapped in a more fine-scale analysis with the use of P1 cotransduction or recombination.

By 1963, the *E. coli* map (**Figure 6-34**) already detailed the positions of approximately 100 genes. After 27 years of further refinement, the 1990 map depicted the positions

A map of the *E. coli* genome obtained genetically

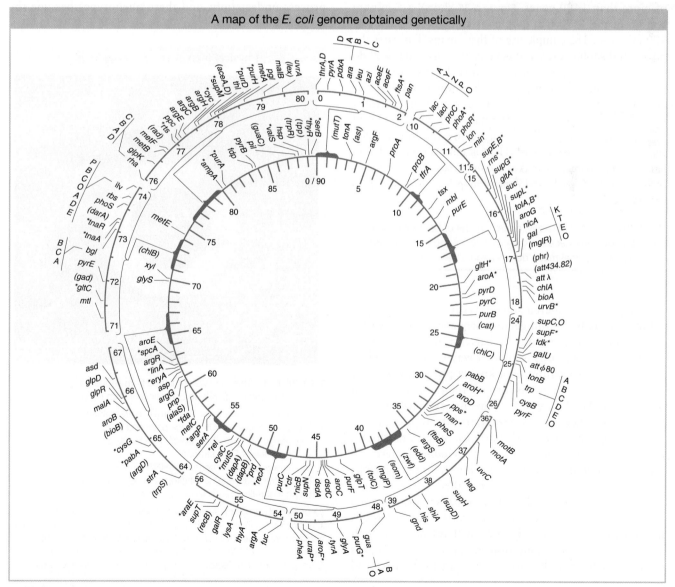

FIGURE 6-34 The 1963 genetic map of *E. coli* genes with mutant phenotypes. Units are minutes, based on interrupted-mating and recombination experiments. Asterisks refer to map positions that are not as precise as the other positions. [*Data from* G. S. Stent, *Molecular Biology of Bacterial Viruses*.]

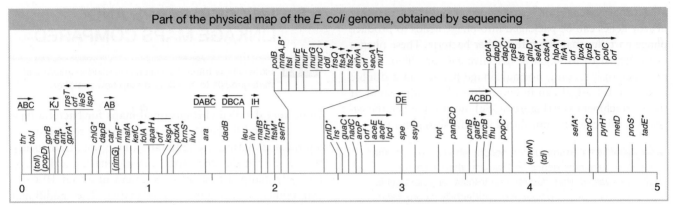

FIGURE 6-35 A linear scale drawing of a sequenced 5-minute section of the 100-minute 1990 *E. coli* linkage map. The parentheses and asterisks indicate markers for which the exact location was unknown at the time of publication. Arrows above genes and groups of genes indicate the direction of transcription. [*Data from B. J. Bachmann, "Linkage Map of Escherichia coli K-12, Edition 8,"* Microbiol. Rev. *54, 1990, 130–197.*]

of more than 1400 genes. **Figure 6-35** shows a 5-minute section of the 1990 map (which is adjusted to a scale of 100 minutes). The complexity of these maps illustrates the power and sophistication of genetic analysis. How well do these maps correspond to physical reality? In 1997, the DNA sequence of the entire *E. coli* genome of 4,632,221 base pairs was completed, allowing us to compare the exact position of genes on the genetic map with the position of the corresponding coding sequence on the linear DNA sequence (the physical map). The full map is represented in **Figure 6-36. Figure 6-37** makes a comparison for a segment of both maps. Clearly, the genetic map is a close match to the physical map.

KEY CONCEPT Generally, genetic maps of bacterial chromosomes show their genes in the same order and relative position as physical bacterial maps derived from DNA sequencing.

Chapter 4 considered some ways in which the physical map (usually the full genome sequence) can be useful in mapping new mutations. In bacteria, the technique of **insertional mutagenesis** is another way to zero in rapidly on a mutation's position on a known physical map. The technique causes mutations through the random insertion of "foreign" DNA fragments. The inserts inactivate any gene in which they land by interrupting the transcriptional unit. Transposons are particularly useful inserts for this purpose in several model organisms, including bacteria. To map a new mutation, the procedure is as follows. The DNA of a transposon carrying a resistance allele or other selectable marker is introduced by transformation into bacterial recipients that have no active transposons. The transposons insert more or less randomly, and any that land in the middle of a gene cause a mutation. A subset of all mutants obtained will have phenotypes relevant to the bacterial

process under study, and these phenotypes become the focus of the analysis.

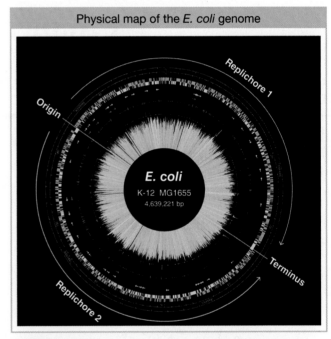

FIGURE 6-36 This map was obtained from sequencing DNA and plotting gene positions. Key to components from the outside in:

- The DNA replication origin and terminus are marked.
- The two scales are in DNA base pairs and in minutes.
- The orange and yellow histograms show the distribution of genes on the two different DNA strands.
- The arrows represent genes for rRNA (red) and tRNA (green).
- The central "starburst" is a histogram of each gene with lines of length that reflect predicted level of transcription.

[*Republished with permission of the American Association for the Advancement of Science, from F. R. Blattner et al., "The Complete Genome Sequence of Escherichia coli K-12,"* Science *277, 1997, 1453–1462, Figure 1. DOI: 10.1126/science.277.5331. Permission conveyed through Copyright Clearance Center, Inc. Image courtesy of Dr. Guy Plunkett III.*]

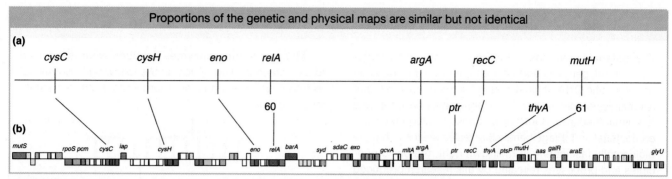

FIGURE 6-37 An alignment of the genetic and physical maps. (a) Markers on the 1990 genetic map in the region near 60 and 61 minutes. (b) The exact positions of every gene, based on the complete sequence of the *E. coli* genome. (Not every gene is named in this map, for simplicity.) The elongated boxes are genes and putative genes. Each color represents a different type of function. For example, red denotes regulatory functions, and dark blue denotes functions in DNA replication, recombination, and repair. Lines between the maps in parts *a* and *b* connect the same gene in each map. [Data from F. R. Blattner et al., The Complete Science *277, 1997, 1453–1462.*]

The beauty of inserting transposons is that, because their sequence is known, the mutant gene can be located and sequenced. DNA replication primers are created that match the known sequence of the transposon (see Chapter 10). These primers are used to initiate a sequencing analysis that proceeds *outward* from the transposon into the surrounding gene. The short sequence obtained can then be fed into a computer and compared with the complete genome sequence. From this analysis, the position of the gene and its full sequence are obtained. The function of a homolog of this gene might already have been deduced in other organisms. Hence, you can see that this approach (like that introduced in Chapter 4) is another way of uniting mutant phenotype with map position and potential function. **Figure 6-38** summarizes the approach.

As an aside in closing, it is interesting that many of the historical experiments revealing the circularity of bacterial and plasmid genomes coincided with the publication and popularization of J. R. R. Tolkien's *The Lord of the Rings*. Consequently, a review of bacterial genetics at that time led off with the following quotation from the trilogy:

One Ring to rule them all, One Ring to find them,
One Ring to bring them all and in the darkness
bind them.

Indeed, plasmid and bacterial chromosome rings turned out to be even more powerful than originally realized. Discoveries made in bacteria were critical in furthering the research of higher organisms and in the development of sophisticated genetic engineering techniques. **Box 6-1** discusses some of the bacterial processes that paved the way for these advancements.

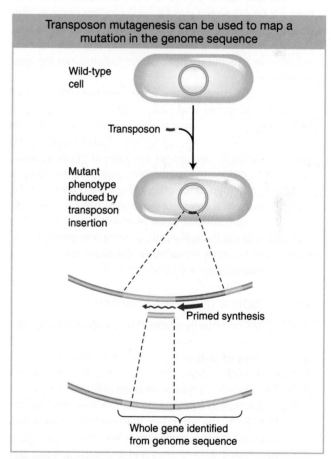

FIGURE 6-38 The insertion of a transposon inserts a mutation into a gene of unknown position and function. The segment next to the transposon is replicated, sequenced, and matched to a segment in the complete genome sequence.

BOX 6-1   Bacterial and Phage Genetics Harnessed for Manipulation of Eukaryotic DNA

The material in this box is a bridge between this chapter on bacterial genetics and later chapters on manipulation of eukaryotic DNA. It illustrates a historical principle that seemingly esoteric research can sometimes pave the way to unimagined important applications. Although research on bacterial and phage genetics revealed systems that are biologically fascinating in their own right, the basic principles and processes revealed by this research produced unexpected spinoffs that have revolutionized modern genetic analysis. Some of the genetic elements and processes discovered in these simpler life forms have provided powerful approaches to the elucidation and genetic manipulation of the more complex eukaryotic genomes, including the genome of our own species. Hence, the reader should treat this section as a relaxing appetizer for what is to come; that is, all the techniques mentioned here will be expanded into full analytical coverage in later chapters (beginning with Chapter 10), which are where the key concepts are delineated in full. The material here will provide a useful springboard for jumping into the later treatments.

Eukaryotic genomes are large, composed of tens of thousands of genes and billions of nucleotide pairs of DNA. Hence, a direct frontal attack on the study of such genomes is difficult to impossible. Consequently, the general approach devised for this type of genomic analysis was to cut up the eukaryotic genome into defined fragments and then, once characterized, the parts can be reassembled into a complete genome or can be used individually for more specific types of intervention. Hence, the first step is to cut the eukaryotic genome into defined fragments, and bacteria provided an excellent way of doing this, as shown in the next paragraph.

Restriction enzymes in bacteria

Bacteria are constantly under threat from parasitic elements, especially viruses. Therefore, they have evolved several types of defense systems, and one of these is *restriction modification*. It was found in the 1950s that phages produced in a lysate are often defective as a result of their DNA having been cut by the previous host bacterium. This led to the discovery that bacteria have genes encoding DNA-cutting endonucleases called *restriction enzymes*. Restriction enzymes find and cut a specific target sequence in the viral DNA; often the target sequence is around 6 to 10 base pairs long. These target sequences are not necessarily in functional regions, but nevertheless, the likelihood of a genome containing such target sequences by chance is high. (The target sequences in the bacterial genome are protected by the addition of methyl groups.)

Most restriction enzymes produce what is referred to as a *staggered cut*, such as the following hypothetical example where each arrow represents a cut in a single strand of DNA:

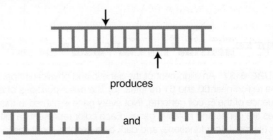

produces

These single-stranded tails with their hydrogen-bonding sites exposed are called "sticky ends." This stickiness is important, as we shall see later.

However, the key element for now is that restriction enzymes can be extracted from bacteria, purified, and used to cut eukaryotic DNA because their target sites will also undoubtedly be present by chance in that DNA. Furthermore, because the target sites are part of the genomic sequence for that particular eukaryote, the cuts will all be in the same positions in any homologous DNA molecule. Hence, starting with a sample of extracted eukaryotic DNA (which in a test tube resembles an unappetizing blob of mucus), the bacterial restriction enzymes cut it into defined segments, which can be sorted, identified, and manipulated in various ways. This represents the beginning stage of eukaryotic genomics.

Bacterial vectors and hosts for eukaryotic DNA

For detailed study, *individual* segments of restriction-digested eukaryotic DNA need to be amplified to convert them into pure samples that are effectively chemical reagents. Here, the sticky tails left by the restriction enzymes are particularly useful because if the DNA of a bacterial plasmid or a phage is cut with the same restriction enzyme and the two DNAs mixed, they join by virtue of complementary hydrogen bonding of their identical sticky ends, and hence, we get replicating bacterial molecules with defined eukaryotic inserts. These hybrid molecules, which are called *recombinant DNAs*, can be inserted into a bacterial host where they will replicate and produce a large pure sample of recombinant DNA (including the eukaryotic insert) for subsequent study. The replicating molecules in such as a study are called *vectors* (carriers). In this way, eukaryotic DNA can be prepared in easily characterizable units called DNA *clones*.

The clones can be used in a number of ways. For example, the insert DNAs can be sequenced and assembled to obtain a full sequence of the eukaryotic genome. Another approach is to use the clones to modify eukaryotic cells. For example, a wild-type recombinant DNA insert can be used to "correct" or reverse a mutation in a eukaryotic recipient; it happens that many eukaryotic cells can be transformed in a way similar to bacteria, so the entry of the corrective fragment is facilitated. Once inside their eukaryotic host, transforming fragments often go through homologous double recombination with host DNA, thus replacing the resident sequence.

DNA clones can also be used to tailor the DNA of the eukaryotic recipient in a highly specific way, and an example of this is shown in the following section.

Bacterial CRISPR systems for engineering eukaryotic DNA

Many species of bacteria and archaea have an immune system that (in contrast to the *general* defensive action of restriction enzymes) protects them against *specific* infectious viruses and plasmids. The basis of the system is loci composed of <u>C</u>lustered <u>R</u>egularly <u>I</u>nterspersed <u>S</u>hort <u>P</u>alindromic <u>R</u>epeats. These are generally referred to as *CRISPRs* for short (and pronounced "crispers"). In language, palindromes are words that have the same spelling whether read forward or backward; for example, RADAR. In DNA, a palindromic sequence is one where the 5′-to-3′ sequence of one strand is identical to the 5′-to-3′ sequence of the complementary strand, such as

5′ AAGGCCTT 3′

3′ TTCCGGAA 5′

At CRISPR loci, the short palindromic repeats are separated by several different unique sequences that initially seemed mysterious until it was discovered that they were in fact non-genomic sequences derived from various phages or plasmids. This observation led to the idea that they were part of an immune system against invasive DNAs. At one end of each repeated sequence, there is a sequence encoding one to several proteins called CRISPR-associated proteins or *Cas proteins,* which are DNA-cutting nucleases.

The antiviral mechanism works in the following way. When viral DNA enters the cell, either it can either kill the cell or, in some cases, a fragment of its DNA can be inserted into the CRISPR array, where it acts as a heritable immune protectant against that specific virus. If the cell is later infected by this virus, the CRISPR array is transcribed and cut into short specific RNAs, each of which are hooked up to Cas proteins. One of the RNAs will be homologous to a region of the infecting virus DNA and will bind through base pairing. At that point, the Cas protein cuts the viral DNA with a double-strand break, thus rendering it incapable of encoding for phage propagation. **Figure 1** summarizes the process.

The components of the bacterial CRISPR systems have been harnessed to design very effective genetic engineering methods for eukaryotes. Remarkably, the CRISPR system works in eukaryotic cells. One commonly used system uses Cas9, derived from a species of *Streptococcus*. DNA constructs bearing Cas9 plus a "guide RNA" homologous to a target gene to be modified are inserted into the eukaryotic cell. The guide RNA (which is substituting for the phage immunity fragment) finds the target gene by base homology, and Cas9 cuts it, resulting in a double-strand break. At this point, the eukaryotic cellular repair mechanisms take over and mend the break. Such systems are inherently error-prone, and the repair can sometimes result in a faulty sequence leading to a random mutation. A random mutation in a specific target gene might in itself be very useful; however, the versatility of the gene modification process can be greatly expanded by adding a tailored piece of DNA flanked by sequences homologous to the target gene. In such a situation, the repair system is tricked into inserting the tailored sequence into the target gene at the location of the double-strand break. Hence, the CRISPR-based technology has efficiently inserted an extra piece of specially tailored DNA into the eukaryotic gene of interest. This provides immense scope for genetic modification of the eukaryotic genome, for example in correcting for deleterious mutations or adding new functions.

The molecular details of all these types of techniques and their applications to genetic engineering will be covered in Chapter 10.

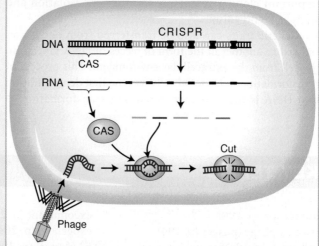

How bacterial CRISPR loci act as agents of immunity

FIGURE 1 The black regions represent repetitive DNA and the colored regions are fragments derived from parasitic elements such as phage. The net outcome is that the Cas protein cuts the invading DNA recognized by the red RNA, rendering it inactive.

SUMMARY

Advances in bacterial and phage genetics within the past 50 years have provided the foundation for molecular biology and cloning (discussed in later chapters). Early in this period, gene transfer and recombination were found to take place between different strains of bacteria. In bacteria, however, genetic material is passed in only one direction—for example, in *Escherichia coli*, from a donor cell (F^+ or Hfr) to a recipient cell (F^-). Donor ability is determined by the presence in the cell of a fertility factor (F), a type of plasmid. On occasion, the F factor present in the free state in F^+ cells can integrate into the *E. coli* chromosome and form an Hfr cell. When this occurs, a fragment of donor chromosome can transfer into a recipient cell and subsequently recombine with the recipient chromosome. Because the F factor can insert at different places on the host chromosome, early investigators were able to piece the transferred fragments together to show that the *E. coli* chromosome is a single circle, or ring. Interruption of the transfer at different times has provided geneticists with an unconventional method (interrupted mating) for constructing a linkage map of the single chromosome of *E. coli* and other similar bacteria, in which the map unit is a unit of time (minutes). In an extension of this technique, the frequency of recombinants between markers known to have entered the recipient can provide a finer-scale map distance.

Several types of plasmids other than F can be found. R plasmids carry antibiotic-resistance alleles, often within a mobile element called a transposon. Rapid plasmid spread causes population-wide resistance to medically important drugs. Derivatives of such natural plasmids have become important cloning vectors, useful for gene isolation and study in all organisms.

Genetic traits can also be transferred from one bacterial cell to another in the form of pieces of DNA taken into the cell from the extracellular environment. This process of transformation in bacterial cells was the first demonstration that DNA is the genetic material. For transformation to occur, DNA must be taken into a recipient cell, and recombination must then take place between a recipient chromosome and the incorporated DNA.

Bacteria can be infected by viruses called bacteriophages. In one method of infection, the phage chromosome may enter the bacterial cell and, by using the bacterial metabolic machinery, produce progeny phages that burst the host bacterium. The new phages can then infect other cells. If two phages of different genotypes infect the same host, recombination between their chromosomes can take place.

In another mode of infection, lysogeny, the injected phage lies dormant in the bacterial cell. In many cases, this dormant phage (the prophage) incorporates into the host chromosome and replicates with it. Either spontaneously or under appropriate stimulation, the prophage can leave its dormant state and lyse the bacterial host cell.

A phage can carry bacterial genes from a donor to a recipient. In generalized transduction, random host DNA is incorporated alone into the phage head during lysis. In specialized transduction, faulty excision of the prophage from a unique chromosomal locus results in the inclusion of specific host genes as well as phage DNA in the phage head.

Today, a physical map in the form of the complete genome sequence is available for many bacterial species. With the use of this physical genome map, the map position of a mutation of interest can be precisely located. First, appropriate mutations are produced by the insertion of transposons (insertional mutagenesis). Then, the DNA sequence surrounding the inserted transposon is obtained and matched to a sequence in the physical map. This technique provides the locus, the sequence, and possibly the function of the gene of interest.

As we will see in subsequent chapters, many of the bacterial genetic mechanisms discovered historically have proved to be the basis for powerful approaches to engineering DNA in general.

KEY TERMS

auxotroph (p. 196)
bacteriophage (phage) (p. 194)
cell clone (p. 196)
colony (p. 196)
conjugation (p. 199)
cotransductant (p. 213)
donor (p. 199)
double (mixed) infection (p. 211)
double transformation (p. 209)
endogenote (p. 204)

exconjugant (p. 201)
exogenote (p. 204)
F^+ (donor) (p. 199)
F^- (recipient) (p. 199)
F' plasmid (p. 206)
fertility factor (F)
   (p. 199)
generalized transduction
   (p. 213)
genetic marker (p. 196)

Hfr (high frequency of recombination) (p. 200)
horizontal transmission (p. 195)
insertional mutagenesis (p. 218)
interrupted mating (p. 200)
λ attachment site (p. 215)
lysate (p. 210)
lysis (p. 210)
lysogen (lysogenic bacterium)
   (p. 213)

SOLVED PROBLEMS

SOLVED PROBLEM 1

Suppose that a bacterial cell were unable to carry out generalized recombination (*rec⁻*). How would this cell behave as a recipient in generalized and in specialized transduction? First, compare each type of transduction, and then determine the effect of the *rec⁻* mutation on the inheritance of genes by each process.

SOLUTION

Generalized transduction entails the incorporation of chromosomal fragments into phage heads, which then infect recipient strains. Fragments of the chromosome are incorporated randomly into phage heads, and so any marker on the bacterial host chromosome can be transduced to another strain by generalized transduction. In contrast, specialized transduction entails the integration of the phage at a specific point on the chromosome and the rare incorporation of chromosomal markers near the integration site into the phage genome. Therefore, only those markers that are near the specific integration site of the phage on the host chromosome can be transduced.

Markers are inherited by different routes in generalized and specialized transduction. A generalized transducing phage injects a fragment of the donor chromosome into the recipient. This fragment must be incorporated into the recipient's chromosome by recombination, with the use of the recipient's recombination system. Therefore, a *rec⁻* recipient will not be able to incorporate fragments of DNA and cannot inherit markers by generalized transduction. On the other hand, the major route for the inheritance of markers by specialized transduction is by integration of the specialized transducing particle into the host chromosome at the specific phage integration site. This integration, which sometimes requires an additional wild-type (helper) phage, is mediated by a phage-specific enzyme system that is independent of the normal recombination enzymes. Therefore, a *rec⁻* recipient can still inherit genetic markers by specialized transduction.

SOLVED PROBLEM 2

In *E. coli*, four Hfr strains donate the following genetic markers, shown in the order donated:

Strain 1:	Q	W	D	M	T
Strain 2:	A	X	P	T	M
Strain 3:	B	N	C	A	X
Strain 4:	B	Q	W	D	M

All these Hfr strains are derived from the same F⁺ strain. What is the order of these markers on the circular chromosome of the original F⁺?

SOLUTION

A two-step approach works well: (1) determine the underlying principle, and (2) draw a diagram. Here, the principle is clearly that each Hfr strain donates genetic markers from a fixed point on the circular chromosome and that the earliest markers are donated with the highest frequency. Because not all markers are donated by each Hfr, only the early markers must be donated for each Hfr. Each strain allows us to draw the following circles:

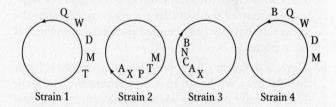

| Strain 1 | Strain 2 | Strain 3 | Strain 4 |

From this information, we can consolidate each circle into one circular linkage map of the order Q, W, D, M, T, P, X, A, C, N, B, Q.

SOLVED PROBLEM 3

In an Hfr × F⁻ cross, *leu⁺* enters as the first marker, but the order of the other markers is unknown. If the Hfr is wild type and the F⁻ is auxotrophic for each marker in question, what is the order of the markers in a cross where *leu⁺* recombinants are selected if 27 percent are *ile⁺*, 13 percent are *mal⁺*, 82 percent are *thr⁺*, and 1 percent are *trp⁺*?

SOLUTION

Recall that spontaneous breakage creates a natural gradient of transfer, which makes it less and less likely for a recipient to receive later and later markers. Because we have selected for the earliest marker in this cross, the frequency of recombinants is a function of the order of entry for each marker. Therefore, we can immediately determine the order of the genetic markers simply by looking at the percentage of recombinants for any marker among the *leu*⁺ recombinants. Because the inheritance of *thr*⁺ is the highest, *thr*⁺ must be the first marker to enter after *leu*. The complete order is *leu, thr, ile, mal, trp*.

SOLVED PROBLEM 4

A cross is made between an Hfr that is *met*⁺ *thi*⁺ *pur*⁺ and an F⁻ that is *met*⁻ *thi*⁻ *pur*⁻. Interrupted-mating studies show that *met*⁺ enters the recipient last, and so *met*⁺ recombinants are selected on a medium containing supplements that satisfy only the *pur* and *thi* requirements. These recombinants are tested for the presence of the *thi*⁺ and *pur*⁺ alleles. The following numbers of individuals are found for each genotype:

met⁺ *thi*⁺ *pur*⁺	280
met⁺ *thi*⁺ *pur*⁻	0
met⁺ *thi*⁻ *pur*⁺	6
met⁺ *thi*⁻ *pur*⁻	52

a. Why was methionine (Met) left out of the selection medium?

b. What is the gene order?

c. What are the map distances in recombination units?

SOLUTION

a. Methionine was left out of the medium to allow selection for *met*⁺ recombinants because *met*⁺ is the last marker to enter the recipient. The selection for *met*⁺ ensures that all the loci that we are considering in the cross will have already entered each recombinant that we analyze.

b. Here, a diagram of the possible gene orders is helpful. Because we know that *met* enters the recipient last, there are only two possible gene orders if the first marker enters on the right: *met, thi, pur* or *met, pur, thi*. How can we distinguish between these two orders? Fortunately, one of the four possible classes of recombinants requires two additional crossovers. Each possible order predicts a different class that arises by four crossovers rather than two. For instance, if the order was *met, thi, pur*, then *met*⁺ *thi*⁻ *pur*⁺ recombinants would be very rare. On the other hand, if the order was *met, pur, thi*, then the four-crossover class would be *met*⁺ *pur*⁻ *thi*⁺. From the information given in the table, the *met*⁺ *pur*⁻ *thi*⁺ class is clearly the four-crossover class and therefore the gene order *met, pur, thi* is correct.

c. Refer to the following diagram:

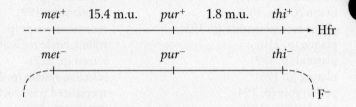

To compute the distance between *met* and *pur*, we compute the percentage of *met*⁺ *pur*⁻ *thi*⁻, which is 52/338 = 15.4 m.u. Similarly, the distance between *pur* and *thi* is 6/338 = 1.8 m.u.

SOLVED PROBLEM 5

Compare the mechanism of transfer and inheritance of the *lac*⁺ genes in crosses with Hfr, F⁺, and F′ *lac*⁺ strains. How would an F⁻ cell that cannot undergo normal homologous recombination (*rec*⁻) behave in crosses with each of these three strains? Would the cell be able to inherit the *lac*⁺ gene?

SOLUTION

Each of these three strains donates genes by conjugation. In the Hfr and F⁺ strains, the *lac*⁺ genes on the host chromosome are donated. In the Hfr strain, the F factor is integrated into the chromosome in every cell, and so chromosomal markers can be efficiently donated, particularly if a marker is near the integration site of F and is donated early. The F⁺ cell population contains a small percentage of Hfr cells, in which F is integrated into the chromosome. These cells are responsible for the gene transfer displayed by cultures of F⁺ cells. In the Hfr and F⁺-mediated gene transfer, inheritance requires the incorporation of a transferred fragment by recombination (recall that two crossovers are needed) into the F⁻ chromosome. Therefore, an F⁻ strain that cannot undergo recombination cannot inherit donor chromosomal markers even though they are transferred by Hfr strains or Hfr cells in F⁺ strains. The fragment cannot be incorporated into the chromosome by recombination. Because these fragments do not possess the ability to replicate within the F⁻ cell, they are rapidly diluted out during cell division.

Unlike Hfr cells, F′ cells transfer genes carried on the F′ factor, a process that does not require chromosome transfer. In this case, the *lac*⁺ genes are linked to the F′ factor and are transferred with it at a high efficiency. In the F⁻ cell, no recombination is required because the F′ *lac*⁺ strain can replicate and be maintained in the dividing F⁻ cell population. Therefore, the *lac*⁺ genes are inherited even in a *rec*⁻ strain.

PROBLEMS

Visit SaplingPlus for supplemental content. Problems with the 🔁 icon are available for review/grading. Problems with the 🖵 icon have a Problem Solving Video. Problems with the 🝾 icon have an Unpacking the Problem exercise.

WORKING WITH THE FIGURES

(The first 31 questions require inspection of text figures.)

1. a. In Figure 6-2, in which of the four processes shown can a complete bacterial genome be transferred from one cell to another?

   b. Which of the four methods shown are forms of horizontal transmission? 🔁

2. a. In Figure 6-3, if the concentration of bacterial cells in the original suspension is 200/ml and 0.2 ml is plated onto each of 100 petri dishes, what is the expected average number of colonies per plate?

   b. If colonies are found after plating, how can you rule out the possibility of their being the result of bacterial cells in the air?

3. In Table 6-1, distinguish two different ways in which a superscript "−" is used.

4. In Figure 6-5,

   a. Why do A^- and B^- cells, by themselves, not form colonies on the plating medium?

   b. What genetic event do the purple colonies in the middle plate represent?

   c. If prototrophs are inoculated onto various other media, will they grow on a medium containing only methionine?

5. In the experiment described in Figure 6-6, a few very rare prototrophic colonies are observed. What might be their origin?

6. In Figure 6-8, draw the next stage that would be expected after (b).

7. In Figure 6-9, if Hfr cells are obtained and subsequently analyzed, how would you look for the possibility of a rare Hfr in which F has exited?

8. a. In Figure 6-10c, what do the yellow dots represent?

   b. In part *c*, are there any pure green cells? What is their origin?

9. a. In Figure 6-11, which donor alleles become part of the recombinant genome produced?

   b. Draw a crossover diagram that would show the integration of a^+ and c^+ but not b^+.

10. In Figure 6-12,

    a. Which Hfr gene enters the recipient last? (Which diagram shows it actually entering?)

    b. What is the maximum percentage of cases of transfer of this gene?

    c. Which genes have entered at 25 minutes? Could they all become part of a stable exconjugant genome?

    d. Redraw part *b* for an Hfr with F inserted in the opposite direction but at the same locus. 🔁

11. In Figure 6-13, re-draw the diagram to show integration between *c* and *d*.

12. a. In Figure 6-14, which is the last gene to be transferred into the F^- from each of the five Hfr strains?

    b. Redraw each diagram with F oriented in the Hfr in the opposite direction.

13. In Figure 6-15, how are each of the following genotypes produced?

    a. $F^+ a^-$ c. $F^- a^+$

    b. $F^- a^-$ d. $F^+ a^+$

14. Redraw Figure 6-16 showing how a viable a^+ recombinant could be produced.

15. a. In Figure 6-17, how many crossovers are required to produce a completely prototrophic exconjugant?

    b. Draw a diagram to show production of a $leu^- arg^+ met^+$ recombinant.

16. a. In Figure 6-18c, why is the crossover shown occurring in the orange segments of DNA?

    b. Redraw the diagram to show how an F' *ton* plasmid could be formed

17. In Figure 6-19, how many different bacterial species are shown as having contributed DNA to the plasmid pk214?

18. Referring to Figure 6-20, draw a diagram to show how Tn5 might become incorporated into a bacterial chromosome.

19. In Figure 6-21, draw a diagram to show how the transferred fragment could become integrated.

20. In Figure 6-25, can you point to any phage progeny that could transduce? 🔁

21. Regarding Figure 6-27, outline an experimental protocol to set up double infection in the lab.

22. In Figure 6-28, what are the physical features of the plaques of recombinant phages?

23. a. Outline an experimental lab protocol that would allow the experiment in Figure 6-29 to be accomplished.

   b. In Figure 6-29, do you think that b^+ could be transduced instead of a^+? As well as a^+?

24. a. In Figure 6-30, which genes show the highest frequencies of cotransduction?

   b. What is the cotransduction frequency of *narC* and *purB*?

25. In Figure 6-31, what essential event on a petri dish is observed in one experiment and not in the other?

26. a. In Figure 6-32, what do the half-red, half-blue segments represent?

   b. Would you say that integration is by one or by two crossovers?

27. a. In Figure 6-33, which is the rarest λ genotype produced in the initial lysate?

   b. Draw a diagram to show how the λdgal and λ helper lysate could have been produced.

28. In Figure 6-34, if F is inserted at minute 45, which two genes would tell you in which orientation it was inserted and in what kind of experiment?

29. Look at Figures 6-34 and 6-35, find the *ara* gene in each, and compare the gene landscape in its vicinity. Why is there a difference?

30. Looking at Figure 6-37, which region would you say shows the best proportionality between the DNA and the recombination maps?

31. In Figure 6-38, precisely which gene is eventually identified from the genome sequence?

BASIC PROBLEMS

32. Describe the state of the F factor in an Hfr, F^+, and F^- strain.

33. How does a culture of F^+ cells transfer markers from the host chromosome to a recipient?

34. With respect to gene transfer and the integration of the transferred gene into the recipient genome, compare

   a. Hfr crosses by conjugation and generalized transduction.

   b. F′ derivatives such as F′ *lac* and specialized transduction.

35. Why is generalized transduction able to transfer any gene, but specialized transduction is restricted to only a small set?

36. A microbial geneticist isolates a new mutation in *E. coli* and wishes to map its chromosomal location. She uses interrupted-mating experiments with Hfr strains and generalized-transduction experiments with phage P1. Explain why each technique, by itself, is insufficient for accurate mapping.

37. In *E. coli*, four Hfr strains donate the following markers, shown in the order donated:

Strain 1:	M	Z	X	W	C
Strain 2:	L	A	N	C	W
Strain 3:	A	L	B	R	U
Strain 4:	Z	M	U	R	B

All these Hfr strains are derived from the same F^+ strain. What is the order of these markers on the circular chromosome of the original F^+?

38. You are given two strains of *E. coli*. The Hfr strain is arg^+ ala^+ glu^+ pro^+ leu^+ T^s; the F^- strain is arg^- ala^- glu^- pro^- leu^- T^r. All the markers are nutritional except T, which determines sensitivity or resistance to phage T1. The order of entry is as given, with arg^+ entering the recipient first and T^s last. You find that the F^- strain dies when exposed to penicillin (pen^s), but the Hfr strain does not (pen^r). How would you locate the locus for *pen* on the bacterial chromosome with respect to *arg*, *ala*, *glu*, *pro*, and *leu*? Formulate your answer in logical, well-explained steps, and draw explicit diagrams where possible.

39. A cross is made between two *E. coli* strains: Hfr arg^+ bio^+ leu^+ × F^- arg^- bio^- leu^-. Interrupted mating studies show that arg^+ enters the recipient last, and so arg^+ recombinants are selected on a medium containing *bio* and *leu* only. These recombinants are tested for the presence of bio^+ and leu^+. The following numbers of individuals are found for each genotype:

arg^+ bio^+ leu^+	320	arg^+ bio^- leu^+	0
arg^+ bio^+ leu^-	8	arg^+ bio^- leu^-	48

   a. What is the gene order?

   b. What are the map distances in recombination percentages?

40. Linkage maps in an Hfr bacterial strain are calculated in units of minutes (the number of minutes between genes indicates the length of time that it takes for the second gene to follow the first in conjugation). In making such maps, microbial geneticists assume that the bacterial chromosome is transferred from Hfr to F^- at a constant rate. Thus, two genes separated by 10 minutes near the origin end are assumed to be the same physical distance apart as two genes separated by 10 minutes near the F^- attachment end. Suggest a critical experiment to test the validity of this assumption.

41. A particular Hfr strain normally transmits the pro^+ marker as the last one in conjugation. In a cross of

this strain with an F⁻ strain, some *pro⁺* recombinants are recovered early in the mating process. When these *pro⁺* cells are mixed with F⁻ cells, the majority of the F⁻ cells are converted into *pro⁺* cells that also carry the F factor. Explain these results.

42. F′ strains in *E. coli* are derived from Hfr strains. In some cases, these F′ strains show a high rate of integration back into the bacterial chromosome of a second strain. Furthermore, the site of integration is often the site occupied by the fertility factor in the original Hfr strain (before production of the F′ strains). Explain these results.

43. You have two *E. coli* strains, F⁻ *str^s ala⁻* and Hfr *str^s ala⁺*, in which the F factor is inserted close to *ala⁺*. Devise a screening test to detect strains carrying F′ *ala⁺*.

44. Five Hfr strains A through E are derived from a single F⁺ strain of *E. coli*. The following chart shows the entry times of the first five markers into an F⁻ strain when each is used in an interrupted-conjugation experiment:

A	B	C	D	E
mal⁺ (1)	*ade⁺* (13)	*pro⁺* (3)	*pro⁺* (10)	*his⁺* (7)
str^s (11)	*his⁺* (28)	*met⁺* (29)	*gal⁺* (16)	*gal⁺* (17)
ser⁺ (16)	*gal⁺* (38)	*xyl⁺* (32)	*his⁺* (26)	*pro⁺* (23)
ade⁺ (36)	*pro⁺* (44)	*mal⁺* (37)	*ade⁺* (41)	*met⁺* (49)
his⁺ (51)	*met⁺* (70)	*str^s* (47)	*ser⁺* (61)	*xyl⁺* (52)

a. Draw a map of the F⁺ strain, indicating the positions of all genes and their distances apart in minutes.

b. Show the insertion point and orientation of the F plasmid in each Hfr strain.

c. In the use of each of these Hfr strains, state which allele you would select to obtain the highest proportion of Hfr exconjugants.

45. *Streptococcus pneumoniae* cells of genotype *str^s mtl⁻* are transformed by donor DNA of genotype *str^r mtl⁺* and (in a separate experiment) by a mixture of two DNAs with genotypes *str^r mtl⁻* and *str^s mtl⁺*. The accompanying table shows the results.

Transforming DNA	*Percentage of cells transformed into*		
	str^r mtl⁻	*str^s mtl⁺*	*str^r mtl⁺*
str^r mtl⁺	4.3	0.40	0.17
str^r mtl⁻ + *str^s mtl⁺*	2.8	0.85	0.0066

a. What does the first row of the table tell you? Why?

b. What does the second row of the table tell you? Why?

46. Recall that, in Chapter 4, we considered the possibility that a crossover event may affect the likelihood of another crossover. In the bacteriophage T4, gene *a* is 1.0 m.u. from gene *b*, which is 0.2 m.u. from gene *c*. The gene order is *a*, *b*, *c*. In a recombination experiment, you recover five double crossovers between *a* and *c* from 100,000 progeny viruses. Is it correct to conclude that interference is negative? Explain your answer.

47. *E. coli* cells were infected with two strains of T4 virus. One strain is minute (*m*), rapid lysis (*r*), and turbid (*t*); the other is wild type for all three markers. The lytic products of this infection were plated and classified. The resulting 10,342 plaques were distributed among eight genotypes as follows:

m r t	3469	*m + +*	521
+ + +	3727	*+ r t*	475
m r +	854	*+ r +*	171
m + t	163	*+ + t*	963

a. What are the linkage distances between *m* and *r*, between *r* and *t*, and between *m* and *t*.

b. Determine the linkage order for the three genes.

c. What is the coefficient of coincidence (see Chapter 4) in this cross? What does it signify?

48. With the use of P22 as a generalized transducing phage grown on a *pur⁺ pro⁺ his⁺* bacterial donor, a recipient strain of genotype *pur⁻ pro⁻ his⁻* was infected and incubated. Afterward, transductants for *pur⁺*, *pro⁺*, and *his⁺* were selected individually in experiments I, II, and III, respectively.

a. What medium is used in each of these selection experiments?

b. The transductants were examined for the presence of unselected donor markers, with the following results:

I		II		III	
pro⁻ his⁻	86%	*pur⁻ his⁻*	44%	*pur⁻ pro⁻*	20%
pro⁺ his⁻	0%	*pur⁺ his⁻*	0%	*pur⁺ pro⁻*	14%
pro⁻ his⁺	10%	*pur⁻ his⁺*	54%	*pur⁻ pro⁺*	61%
pro⁺ his⁺	4%	*pur⁺ his⁺*	2%	*pur⁺ pro⁺*	5%

What is the order of the bacterial genes?

c. Which two genes are closest together?

d. Based on your answer to part *c*, explain the relative proportions of genotypes observed in experiment II.

49. Although most λ-mediated *gal⁺* transductants are inducible lysogens, a small percentage of these transductants in fact are not lysogens (that is, they contain no integrated λ). Control experiments show that these transductants are not produced by mutation. What is the likely origin of these types?

50. An $ade^+ arg^+ cys^+ his^+ leu^+ pro^+$ bacterial strain is known to be lysogenic for a newly discovered phage, but the site of the prophage is not known. The bacterial map is

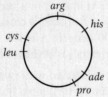

The lysogenic strain is used as a source of the phage, and the phages are added to a bacterial strain of genotype $ade^- arg^- cys^- his^- leu^- pro^-$. After a short incubation, samples of these bacteria are plated on six different media, with the supplementations indicated in the following table. The table also shows whether colonies were observed on the various media.

| Medium | \multicolumn{6}{c}{Nutrient supplementation in medium} | Presence of colonies |
	Ade	Arg	Cys	His	Leu	Pro	
1	−	+	+	+	+	+	N
2	+	−	+	+	+	+	N
3	+	+	−	+	+	+	C
4	+	+	+	−	+	+	N
5	+	+	+	+	−	+	C
6	+	+	+	+	+	−	N

(In this table, a plus sign indicates the presence of a nutrient supplement, a minus sign indicates that a supplement is not present, N indicates no colonies, and C indicates colonies present.)

a. What genetic process is at work here?

b. What is the approximate locus of the prophage?

51. In a generalized-transduction system using P1 phage, the donor is $pur^+ nad^+ pdx^-$ and the recipient is $pur^- nad^- pdx^+$. The donor allele pur^+ is initially selected after transduction, and 50 pur^+ transductants are then scored for the other alleles present. Here are the results:

Genotype	Number of colonies
$nad^+ pdx^+$	3
$nad^+ pdx^-$	10
$nad^- pdx^+$	24
$nad^- pdx^-$	13
	50

a. What is the cotransduction frequency for pur and nad?

b. What is the cotransduction frequency for pur and pdx?

c. Which of the unselected loci is closest to pur?

d. Are nad and pdx on the same side or on opposite sides of pur? Explain.

(Draw the exchanges needed to produce the various transformant classes under either order to see which requires the minimum number to produce the results obtained.)

52. In a generalized-transduction experiment, phages are collected from an *E. coli* donor strain of genotype $cys^+ leu^+ thr^+$ and used to transduce a recipient of genotype $cys^- leu^- thr^-$. Initially, the treated recipient population is plated on a minimal medium supplemented with leucine and threonine. Many colonies are obtained.

a. What are the possible genotypes of these colonies?

b. These colonies are then replica plated onto three different media: (1) minimal plus threonine only; (2) minimal plus leucine only; and (3) minimal. What genotypes could, in theory, grow on these three media?

c. Of the original colonies, 56 percent are observed to grow on medium 1, 5 percent on medium 2, and no colonies on medium 3. What are the actual genotypes of the colonies on media 1, 2, and 3?

d. Draw a map showing the order of the three genes and which of the two outer genes is closer to the middle gene.

53. Deduce the genotypes of the following *E. coli* strains 1 through 4:

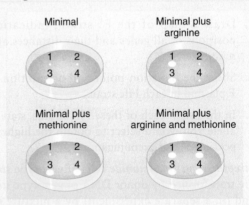

54. In an interrupted-conjugation experiment in *E. coli*, the pro gene enters after the thi gene. A $pro^+ thi^+$ Hfr is crossed with a $pro^- thi^-$ F$^-$ strain, and exconjugants are plated on medium containing thiamine but no proline. A total of 360 colonies are observed, and they are isolated and cultured on fully supplemented medium. These cultures are then tested for their ability to grow on medium containing no proline or thiamine (minimal medium), and 320 of the cultures are found to be able to grow but the remainder cannot.

a. Deduce the genotypes of the two types of cultures.

b. Draw the crossover events required to produce these genotypes.

c. Calculate the distance between the *pro* and *thi* genes in recombination units.

 UNPACKING PROBLEM 54

Before attempting a solution to this problem, try answering the following questions:

1. What type of organism is *E. coli*?

2. What does a culture of *E. coli* look like?

3. On what sort of substrates does *E. coli* generally grow in its natural habitat?

4. What are the minimal requirements for *E. coli* cells to divide?

5. Define the terms *prototroph* and *auxotroph*.

6. Which cultures in this experiment are prototrophic, and which are auxotrophic?

7. Given some strains of unknown genotype regarding thiamine and proline, how would you test their genotypes? Give precise experimental details, including equipment.

8. What kinds of chemicals are proline and thiamine? Does it matter in this experiment?

9. Draw a diagram showing the full set of manipulations performed in the experiment.

10. Why do you think the experiment was done?

11. How was it established that *pro* enters after *thi*? Give precise experimental steps.

12. In what way does an interrupted-mating experiment differ from the experiment described in this problem?

13. What is an exconjugant? How do you think that exconjugants were obtained? (It might include genes not described in this problem.)

14. When the *pro* gene is said to enter after *thi*, does it mean the *pro* allele, the *pro*⁺ allele, either, or both?

15. What is "fully supplemented medium" in the context of this question?

16. Some exconjugants did not grow on minimal medium. On what medium would they grow?

17. State the types of crossovers that take part in Hfr × F⁻ recombination. How do these crossovers differ from crossovers in eukaryotes?

18. What is a recombination unit in the context of the present analysis? How does it differ from the map units used in eukaryote genetics?

Now try to solve the problem. If you are unable to do so, try to identify the obstacle and write a sentence or two describing your difficulty. Then go back to the expansion questions and see if any of them relate to your difficulty. If this approach does not work, inspect the Learning

Objectives and Key Concepts of this chapter and ask yourself which might be relevant to your difficulty.

55. A generalized transduction experiment uses a $metE^+$ $pyrD^+$ strain as donor and $metE^-$ $pyrD^-$ as recipient. $metE^+$ transductants are selected and then tested for the $pyrD^+$ allele. The following numbers were obtained:

$$metE^+ \ pyrD^- \quad 857$$
$$metE^+ \ pyrD^+ \quad 1$$

Do these results suggest that these loci are closely linked? What other explanations are there for the lone "double"?

56. An $argC^-$ strain was infected with transducing phage, and the lysate was used to transduce $metF^-$ recipients on medium containing arginine but no methionine. The $metF^+$ transductants were then tested for arginine requirement: most were $argC^+$, but a small percentage were found to be $argC^-$. Draw diagrams to show the likely origin of the $argC^+$ and $argC^-$ strains.

CHALLENGING PROBLEMS

57. Four *E. coli* strains of genotype $a^+ \ b^-$ are labeled 1, 2, 3, and 4. Four strains of genotype $a^- \ b^+$ are labeled 5, 6, 7, and 8. The two genotypes are mixed in all possible combinations and (after incubation) are plated to determine the frequency of $a^+ \ b^+$ recombinants. The following results are obtained, where M = many recombinants, L = low numbers of recombinants, and 0 = no recombinants:

	1	2	3	4
5	0	M	M	0
6	0	M	M	0
7	L	0	0	M
8	0	L	L	0

On the basis of these results, assign a sex type (either Hfr, F⁺, or F⁻) to each strain.

58. An Hfr strain of genotype $a^+ \ b^+ \ c^+ \ d^- \ str^s$ is mated with a female strain of genotype $a^- \ b^- \ c^- \ d^+ \ str^r$. At various times, mating pairs are separated by vigorously shaking the culture. The cells are then plated on three types of agar, as shown in the accompanying table, where nutrient A allows the growth of a^- cells; nutrient B, of b^- cells; nutrient C, of c^- cells; and nutrient D, of d^- cells. (A plus indicates the presence of streptomycin or a nutrient, and a minus indicates its absence.)

Agar type	Str	A	B	C	D
1	+	+	+	−	+
2	+	−	+	+	+
3	+	+	−	+	+

a. What donor genes are being selected on each type of agar?

b. The following table shows the number of colonies on each type of agar for samples taken at various times after the strains are mixed. Use this information to determine the order of genes *a*, *b*, and *c*.

Time of sampling (minutes)	Number of colonies on agar of type		
	1	2	3
0	0	0	0
5	0	0	0
7.5	102	0	0
10	202	0	0
12.5	301	0	74
15	400	0	151
17.5	404	49	225
20	401	101	253
25	398	103	252

c. From each of the 25-minute plates, 100 colonies are picked and transferred to a petri dish containing agar with all the nutrients except D. The numbers of colonies that grow on this medium are 90 for the sample from agar type 1, 52 for the sample from agar type 2, and 9 for the sample from agar type 3. Using these data, fit gene *d* into the sequence of *a*, *b*, and *c*.

d. At what sampling time would you expect colonies to first appear on agar containing C and streptomycin but no A or B?

59. In the cross Hfr *aro⁺ arg⁺ eryʳ strˢ* × F⁻ *aro⁻ arg⁻ eryˢ strʳ*, the markers are transferred in the order given (with *aro⁺* entering first), but the first three genes are very close together. Exconjugants are plated on a medium containing Str (streptomycin, to kill Hfr cells), Ery (erythromycin), Arg (arginine), and Aro (aromatic amino acids). The following results are obtained for 300 colonies isolated from these plates and tested for growth on various media: on Ery only, 263 strains grow; on Ery + Arg, 264 strains grow; on Ery + Aro, 290 strains grow; on Ery + Arg + Aro, 300 strains grow.

a. Draw up a list of genotypes, and indicate the number of individuals in each genotype.

b. Calculate the recombination frequencies.

c. Calculate the ratio of the size of the *arg*-to-*aro* region to the size of the *ery*-to-*arg* region.

60. A bacterial transformation is performed with a donor strain that is resistant to four drugs, A, B, C, and D, and a recipient strain that is sensitive to all four drugs. The resulting recipient cell population is divided and plated on media containing various combinations of the drugs. The following table shows the results.

Drugs added	Number of colonies	Drugs added	Number of colonies
None	10,000	BC	50
A	1155	BD	48
B	1147	CD	785
C	1162	ABC	31
D	1140	ABD	43
AB	47	ACD	631
AC	641	BCD	35
AD	941	ABCD	29

a. One of the genes is distant from the other three, which appear to be closely linked. Which is the distant gene?

b. What is the likely order of the three closely linked genes?

61. You have two strains of λ that can lysogenize *E. coli*; their linkage maps are as follows:

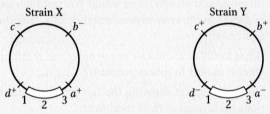

The segment shown at the bottom of the chromosome, designated 1–2–3, is the region responsible for pairing and crossing over with the *E. coli* chromosome. (Keep the markers on all your drawings.)

a. Diagram the way in which λ strain X is inserted into the *E. coli* chromosome (so that the *E. coli* is lysogenized).

b. The bacteria that are lysogenic for strain X can be superinfected by using strain Y. A certain percentage of these superinfected bacteria become "doubly" lysogenic (that is, lysogenic for both strains). Diagram how it will take place. (Don't worry about how double lysogens are detected.)

c. Diagram how the two λ prophages can pair.

d. Crossover products between the two prophages can be recovered. Diagram a crossover event and the consequences.

62. You have three strains of *E. coli*. Strain A is F′ *cys⁺ trp1/cys⁺ trp1* (that is, both the F′ and the chromosome carry *cys⁺* and *trp1*, an allele for tryptophan requirement). Strain B is F⁻ *cys⁻ trp2* Z (this strain requires cysteine for growth and carries *trp2*, another allele causing a tryptophan requirement; strain B is lysogenic for the generalized transducing phage Z). Strain C is F⁻ *cys⁺ trp1* (it is an F⁻ derivative of strain A that has lost the F′). How would you determine whether *trp1*

and *trp2* are alleles of the same locus? (Describe the crosses and the results expected.)

63. A generalized transducing phage is used to transduce an $a^- b^- c^- d^- e^-$ recipient strain of *E. coli* with an $a^+ b^+ c^+ d^+ e^+$ donor. The recipient culture is plated on various media with the results shown in the following table. (Note that a^- indicates a requirement for A as a nutrient, and so forth.) What can you conclude about the linkage and order of the genes?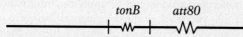

Compounds added to minimal medium	Presence (+) or absence (−) of colonies
CDE	−
BDE	−
BCE	+
BCD	+
ADE	−
ACE	−
ACD	−
ABE	−
ABD	+
ABC	−

64. In 1965, Jon Beckwith and Ethan Signer devised a method of obtaining specialized transducing phages carrying the *lac* region. They knew that the integration site, designated *att80*, for the temperate phage ϕ80 (a relative of phage λ) was located near *tonB*, a gene that confers resistance to the virulent phage T1:

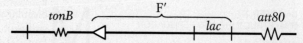

They used an F′ *lac*⁺ plasmid that could not replicate at high temperatures in a strain carrying a deletion of the *lac* genes. By forcing the cell to remain *lac*⁺ at high temperatures, the researchers could select strains in which the plasmid had integrated into the chromosome, thereby allowing the F′ *lac* to be maintained at high temperatures. By combining this selection with a simultaneous selection for resistance to T1 phage infection, they found that the only survivors were cells in which the F′ *lac* had integrated into the *tonB* locus, as shown here:

This result placed the *lac* region near the integration site for phage ϕ80. Describe the subsequent steps that the researchers must have followed to isolate the specialized transducing particles of phage ϕ80 that carried the *lac* region.

65. Wild-type *E. coli* takes up and concentrates a certain red food dye, making the colonies blood red. Transposon mutagenesis was used, and the cells were plated on food dye. Most colonies were red, but some colonies did not take up dye and appeared white. In one white colony, the DNA surrounding the transposon insert was sequenced, with the use of a DNA replication primer identical with part of the end of the transposon sequence, and the sequence adjacent to the transposon was found to correspond to a gene of unknown function called *atoE*, spanning positions 2.322 through 2.324 Mb on the map (numbered from an arbitrary position zero). Propose a function for *atoE*. What biological process could be investigated in this way, and what other types of white colonies might be expected?

GENETICS AND SOCIETY

The methods of DNA transfer in bacteria have been applied more recently to the genetic modification of eukaryotic DNA in general, and the genomes of plants, animals, and humans have all been modified or manipulated using these techniques. This has led to the criticism that modern genetics is akin to "playing God." Do you think this criticism is justified?

PART 2

Core Principles in Molecular and Developmental Genetics

Molecular genetics, also called molecular biology, is the study of how DNA, RNA, and protein molecules store, transmit, and exchange the information that determines the phenotypes of organisms. The foundation of molecular biology is the central dogma (DNA makes RNA makes protein) and its integral molecular processes (DNA replication, transcription, and translation) (see Figure 1-10). Molecular biology provides mechanistic explanations for why mutations in the sequence of a gene alter its expression and function and why cells that contain the same genome sequence express different genes in response to developmental and environmental signals. This mechanistic understanding can be used in innumerable ways, including the treatment of diseases and the improvement of crop plants. As an example, in Chapter 1 (pages 16–18), you read about a quantitative trait locus (QTL) that allows rice plants to survive submergence in deep water for up to two weeks, making the plants tolerant to floods. Researchers used molecular biology principles and techniques to determine that a gene called *submergence tolerant (SUB1)* contains the information that confers flood tolerance and that the encoded *SUB1* protein functions by regulating the expression of other genes. Researchers were then able to use this understanding to increase the yield of rice and other plants.

The large number of molecules and processes in molecular biology is related to the huge variety of forms and functions of organisms. As an example, in eukaryotic but not prokaryotic cells, the physical separation of DNA from ribosomes by the nuclear membrane necessitates mechanisms that transport RNAs and proteins between the nucleus and the cytoplasm. Despite differences among organisms, there are principles in molecular biology that apply to all organisms from single-celled bacteria and yeasts to multi-celled plants and animals because all cells are largely the same; they grow, divide, and respond to developmental and environmental signals.

The following catalog of core principles in molecular and developmental genetics will help guide your learning of the molecules and molecular processes that are described in Chapters 7–14. Knowing these principles should make it easier for you to understand the unique details of each molecule and molecular process and to apply the information to solve present-day problems. We suggest that you think of the principles as shelves to organize molecular biology information. As you work through the chapters, return to the principles and place the information that you learn on the shelves.

CORE PRINCIPLES ABOUT MOLECULES

1. **The ability of molecules (DNA, RNAs, and proteins) to properly function in cells is regulated by molecular processes that control their:**

   - **Synthesis**—making a molecule
   - **Decay**—destroying a molecule
   - **Interactions**—physical contacts with other molecules
   - **Localization**—the location of a molecule in a cell
   - **Folding**—generating the three-dimensional structure of a molecule
   - **Modification**—altering the chemical structure of a molecule

   For example, the function of a protein, such as an enzyme, in a cell depends on its abundance, which is determined by regulatory factors that control its rate of *synthesis* by transcription and translation and its rate of *decay* by proteases. The function of an enzyme is also determined by regulatory factors that control its ability to *interact* with its substrates, to *localize* in a cell where its substrates reside, to *fold* into a three-dimensional structure that is capable of catalytic activity, and to be *modified* to activate its catalytic activity.

2. Information is stored in nucleic acids.

DNA sequences store two types of information, (1) coding information that determines the sequence of RNAs and proteins, and (2) noncoding information that regulates the production of DNA, RNAs, and proteins. Noncoding DNA sequences of varying length (five to several hundred base pairs) contain the information that regulates molecular processes such as DNA replication and transcription. In addition, noncoding information in short (5–20 base pairs) DNA sequences within the coding region of genes is transferred to RNAs by transcription and regulates molecular processes such as translation. Regulatory sequences within DNA and RNA are located in specific places, and they are bound by RNAs or proteins that directly catalyze molecular processes or serve as scaffolds for the binding of other proteins that catalyze molecular processes. Thus, DNA coding sequences tell RNAs and proteins *how to function* by determining their sequence, which controls the ways that RNAs and proteins fold, interact with other molecules, localize in cells, and carry out enzymatic reactions. Whereas DNA noncoding sequences act as signposts that tell RNAs and proteins involved in molecular processes such as DNA replication, transcription, and translation *where to function*.

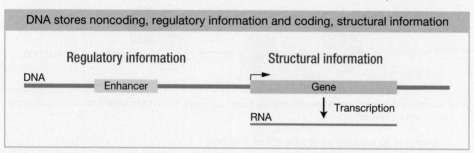

DNA stores noncoding, regulatory information and coding, structural information

In this example, a noncoding DNA sequence called an enhancer contains information that regulates transcription of the coding DNA sequence of a gene that contains information to make an RNA.

3. Information is transferred between nucleic acids by base pairing.

All organisms use DNA replication, transcription, and translation to transmit information from DNA to DNA, from DNA to RNA, and from RNA to protein, respectively. Furthermore, while DNA replication, transcription, and translation involve different molecules and have different outcomes, they are fundamentally similar because they transfer information through complementary base pairing between nucleic acids. Thus, the principle of complementary base pairing makes it immediately apparent how some viruses transmit information from RNA to RNA in the process of RNA replication or from RNA to DNA by reverse transcription.

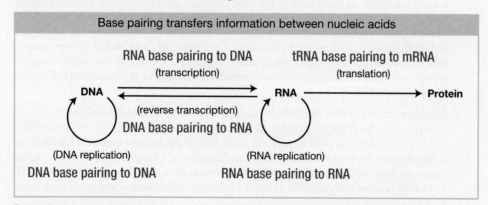

Base pairing transfers information between nucleic acids

Each molecular process in the central dogma of molecular biology (indicated in parentheses) involves DNA-DNA, RNA-DNA, or RNA-RNA base pairing.

4. Structure determines function.

The structure of DNA, RNAs, and proteins influences the roles that they play in cells. As a consequence, processes that change the structure of molecules play a major role in regulating biological phenomena.

Changes to higher-order structures affect function.

The chemical identities and sequences of nucleotides or amino acids determine the overall three-dimensional structures of entire nucleic acids and proteins, and these structures determine function by dictating what other molecules they can bind. Nucleic acids bind to other nucleic acids by complementary base pairing (DNA-DNA, RNA-DNA, and RNA-RNA interactions), proteins bind to nucleic acids through nucleic acid binding domains (protein-DNA and protein-RNA interactions), and proteins bind to other proteins by covalent and non-covalent interactions (stable protein-protein interactions and temporary protein-protein interactions). Nucleic acid and protein structures are governed by fundamental chemical principles, including covalent bonds, bond rotations, and hydrogen bonds and other non-covalent interactions, and they are dynamic. For example, enzymes called helicases alter the secondary structure of DNA by breaking hydrogen bonds between strands of DNA, exposing single-strand sequences for interactions with RNAs and proteins.

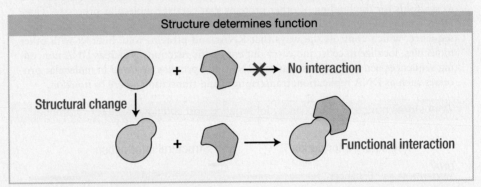

Structure determines function

In this example, a change in the structure of the circular protein converts it from nonfunctional to functional by enabling it to interact with the hexagon protein.

Changes to primary structure affect function.

In some organisms, nucleotide sequences of DNA and RNAs and amino acid sequences of proteins are changed after DNA replication, transcription, and translation, respectively. For example, in eukaryotic organisms, some RNAs undergo splicing, which precisely removes large regions of nucleotides called introns and links together remaining nucleotide regions called exons. Structural changes that result from splicing can affect the sequence and function of the encoded proteins.

Structural changes due to chemical modification affect function.

In all organisms, specific nucleotides in DNA and RNAs and amino acids in proteins are modified by the addition of chemical groups such as methyl groups (CH_3). These chemical groups often affect interactions with other molecules. For example, addition of methyl groups to DNA in bacteria prevents restriction enzymes from binding and cutting the DNA. In eukaryotic organisms, chemical modifications of histone proteins that package DNA serve as binding sites for molecules that regulate transcription. Three types of proteins are involved in chemical modifications: *writers* that add chemical modifications, such as methyltransferase enzymes that add methyl groups; *erasers* that remove chemical modifications, such as demethylase enzymes that remove methyl groups; and *readers* that bind chemical modifications, such as proteins that bind methyl groups.

Structural changes that result from nucleotide hydrolysis affect function.

Some proteins bind the nucleotides ATP or GTP and use the energy produced by their hydrolysis to ADP and GDP to perform some type of mechanical work. GTP hydrolysis is mostly used by proteins to induce a conformational change that controls progress from one step to the next in a molecular process or a signaling pathway, and ATP hydrolysis is commonly used by proteins to change conformation and generate a force.

5. Molecular outcomes are reversible.

Molecules are synthesized and destroyed, folded and unfolded, and localized and dispersed; molecular interactions are formed and disrupted; and molecular modifications are added and removed. Reversibility means that two processes determine a single molecular outcome. For example, the balance of RNA synthesis and destruction determines the abundance of RNAs in a cell. In almost all cases, opposing molecular processes involve different molecules and mechanisms. Chapters 7–14 focus on the molecules and mechanisms involved in synthesis, forming interactions, and adding modifications; however, the principles underlying these forward processes also apply to the reverse processes.

RNA synthesis
(RNA polymerase)

A
U G → AACUGCACCGU
C
(RNA nucleases)
RNA decay

In this example, an RNA is synthesized from nucleotides (A, C, G, and U) by RNA polymerase in the process of transcription, and, in the reverse reaction, the RNA is taken apart into nucleotides by RNA nucleases.

CORE PRINCIPLES ABOUT MOLECULAR PROCESSES

1. Molecular processes are made up of distinct stages.

The occurrence and timing of each stage in a molecular process is regulated to ensure that the products are accurate. Typically, the stages in a molecular process are:

- **Repression**—keeping a process off
- **Activation/initiation**—starting a process
- **Maintenance**—keeping a process going once it is started
- **Termination**—stopping a process

For example, there are regulatory mechanisms that *repress* the transcription of particular genes in cells where the function of the genes is not needed or would be harmful. Furthermore, in those same cells, the transcription of genes that are needed is not uniformly turned on. Instead, the transcription of each gene is regulated at three main stages: *initiation*, which involves defining where in a gene RNA polymerase will start transcription and synthesize a short RNA molecule; *elongation*, which involves *maintaining* RNA synthesis through the whole gene; and *termination*, which involves defining where in a gene RNA synthesis stops and RNA polymerase dissociates from DNA.

2. Signals regulate molecular processes.

Cells respond to physical and chemical signals during development and in their environment by altering the activity of molecular processes, including DNA replication, transcription, and translation. Signals are most often conveyed in cells by chemical modifications of DNA, RNAs, or proteins. For example, cells grow in response to signals from nutrients in the environment by adding a phosphate group onto specific proteins, enabling them to bind DNA regulatory sequences and turn on the transcription of genes involved in cell growth. Similarly, developmental signals initiated by receptor-ligand interactions at the surface of cells trigger chemical modifications that affect molecular processes that alter gene expression. Therefore, despite the fact that cells in multi-celled organisms have the same information stored in their genomic DNA sequence, they express different genes and have different phenotypes because they receive different signals.

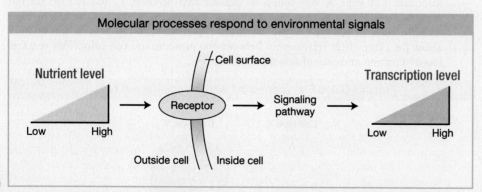

Molecular processes respond to environmental signals

In this example, the amount of nutrients outside of a cell (i.e., an environmental signal) is sensed by a receptor on the surface of the cell. The receptor then turns on a signaling pathway that activates the transcription of genes whose function is to respond to the amount of nutrients.

3. Combinatorial control mechanisms determine the specificity and accuracy of molecular processes.

Molecular processes are regulated by the collective effects of multiple pieces of information. A particularly illustrative example is transcription regulation in eukaryotic organisms. The information that controls transcription comes from short sequences in DNA; however, one short sequence is not specific enough to control the transcription of one gene or even a small set of genes. A sequence of 8 base pairs randomly occurs every 65,536 base pairs (4^8 base pairs), which means that in the human genome of about 3 billion base pairs, the 8-base-pair sequence appears about 45,000 times. In contrast, the number of times that different 8-base-pair sequences are located near one another in the genome is considerably lower. By analogy, many sentences in this book have the words "the," "to," or "of," but very few sentences have all three words. Therefore, multiple regulatory sequences working in combination provide information that is specific enough to regulate the transcription of one gene or a small set of genes. In addition to providing specificity, combinatorial control also ensures that molecular processes are accurate. For example, in the process of translation, the enzymes that attach amino acids to tRNAs check in two different ways that the molecules are correctly paired. This is similar to the "measure twice, cut once" system, which ensures accuracy in carpentry.

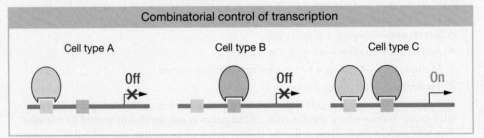

Combinatorial control of transcription

In this example, the gene is transcribed in cell type C but not in A and B because the *combined* activities of two transcription factors (i.e., binding of the green and purple proteins to the green and purple regulatory sequences in the gene) is required to activate transcription. In cell types A and B, only one of the transcription factors is expressed, so transcription is not activated.

CORE PRINCIPLES ABOUT MOLECULAR EXPERIMENTS

1. There are three basic types of molecular biology experiments that have different purposes.

a. Discovery/observation experiments

Discovery/observation experiments are used to identify molecules that *may be involved in* a molecular process or phenotype. This type of experiment provides information that is descriptive rather than mechanistic. Included in this category are genomic experiments that determine the sequence of an organism's genome and the RNAs and proteins that are expressed in cells of an organism (termed the transcriptome and proteome, respectively). In the example that follows, to determine the molecular mechanism that makes cell types X and Y different colors, the proteins expressed in the cell types were identified. Cell type X was found to express two proteins, C and F, that are not expressed in cell type Y. Thus, the experiment showed that proteins C and F *correlate* with the color of cell type X. However, the experiment did not provide information about the cause-effect relationship between the proteins and cell color. This requires loss-of-function and gain-of-function experiments.

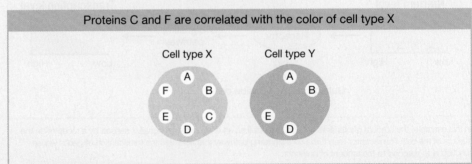

Proteins C and F are correlated with the color of cell type X

b. Loss-of-function experiments

Loss-of-function experiments are used to determine whether molecules are *necessary* for a molecular process or phenotype to occur. A common type of loss-of-function experiment is a gene knockout in a whole organism. If a molecular process or phenotype differs between knockout and wild-type organisms, it can be concluded that the gene is necessary for the event. In the following example, knocking out gene C changed the color of cell type X into that of cell type Y, but knocking out gene F had no effect on the cell color. Thus, gene C is necessary for the molecular mechanism that generates the color of cell type X, but gene F is not.

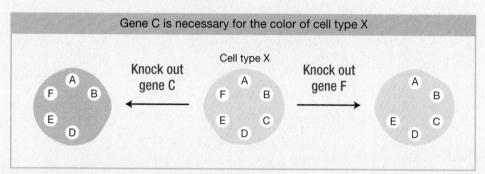

c. Gain-of-function experiments

Gain-of-function experiments are used to determine whether molecules are *sufficient* for a molecular process or phenotype to occur. A common type of gain-of-function experiment is to express a gene in a cell that normally does not express the gene. If a process or phenotype differs between cells that misexpress the gene and wild-type cells, it can be concluded that the gene is sufficient for these events. In the following example, misexpression of gene C changed the color of cell type Y, but misexpression of gene F had no effect on the cell color. Thus, gene C is sufficient to activate the mechanism that generates the color of cell type X, but gene F is not.

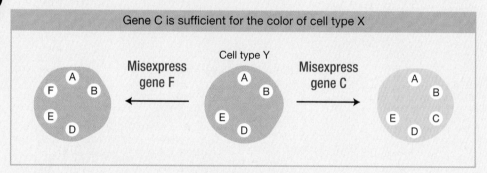

2. Molecular biology experiments are carried out using whole organisms (in vivo) or isolated molecules (in vitro).

A major advantage of in vivo experiments is that all molecules and molecular processes are intact and at physiological levels, so the experimental findings are automatically biologically relevant. However, a major disadvantage of in vivo experiments is that, due to the extraordinary complexity of molecules and molecular processes in whole organisms, it is very difficult to determine detailed molecular mechanisms and whether biological outcomes are due to direct or indirect effects. In contrast, in vitro experiments allow tremendous simplification and control of molecules and molecular processes, allowing investigators to work out detailed molecular mechanisms. However, because the experimental systems are artificial, the findings may not reflect what happens in whole organisms. Therefore, both in vivo and in vitro experiments are needed to gain a complete understanding of molecules and molecular processes. There are benefits and drawbacks to every experimental system and model organism (see the Index to Model Organisms at the end of the book). Consequently, researchers have to balance the ease of experimentation and the ability to get clear results with the physiological relevance of experimental results.

DNA: Structure and Replication

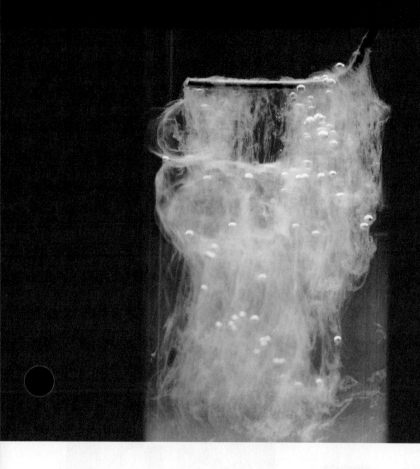

Spooling of DNA. [*TED M. KINSMAN/ Science Source.*]

CHAPTER OUTLINE AND LEARNING OBJECTIVES

7.1 DNA IS THE GENETIC MATERIAL

LO 7.1 Describe evidence demonstrating that DNA is the genetic material.

7.2 DNA STRUCTURE

LO 7.2 Describe evidence used to construct the double helix model of DNA.

LO 7.3 Draw the chemical structure of a DNA double helix.

7.3 DNA REPLICATION IS SEMICONSERVATIVE

LO 7.4 Describe evidence used to support semiconservative DNA replication.

7.4 DNA REPLICATION IN BACTERIA

LO 7.5 Outline the factors and events involved in DNA replication.

7.5 DNA REPLICATION IN EUKARYOTES

LO 7.6 Explain why and how DNA replication differs between bacteria and eukaryotes.

Building upon the finding of early twentieth-century geneticists that heredity involves chromosomes, the broad objective for this chapter is to understand how the structure of DNA, with its base-paired strands, (1) holds the genetic information that determines how organisms are built, and (2) is accurately copied to make it possible for genetic information to be inherited when cells divide and organisms reproduce.

In this chapter, we describe the structure of DNA and the process of DNA replication, which makes an identical copy of DNA every time a cell divides. The story begins in the early 1900s, when results of several experiments led scientists to conclude that DNA, rather than another biological molecule such as carbohydrate, protein, or lipid, is the genetic material. DNA is a simple molecule made up of only four building blocks called **nucleotides.** It was thus necessary to understand how this very simple molecule could be the blueprint for the incredible diversity of organisms on Earth.

A large part of this understanding came from the structure of DNA, which was determined in 1953 by James Watson and Francis Crick through modeling based on the data of others. Their model of the structure of DNA was revolutionary because it defined genes in chemical terms and, in doing so, paved the way for understanding gene action and heredity at a molecular level. A measure of the importance of their discovery is that the double-helical structure of DNA has become a cultural icon that is seen more and more frequently in various forms of art (**Figure 7-1**).

The model of DNA proposed by Watson and Crick was built upon the results of scientists before them. They relied on earlier discoveries of the chemical composition of DNA and the ratios of its nucleotide bases. In addition, pictures of DNA fibers produced by X-ray diffraction revealed to the trained eye that DNA is a helix of precise dimensions. Watson and Crick concluded that DNA is a **double helix** composed of two strands of linked nucleotides that wind around each other.

The proposed structure of DNA immediately suggested that the sequence of nucleotides composing the two DNA strands of the helix could serve as a blueprint for constructing an organism. In addition, the structure hinted at how the blueprint could be copied into all cells in an organism. Because of the rules of base complementarity discovered by Watson and Crick, the sequence of one strand determines the sequence of the other strand. In this way, genetic information in the sequence of DNA can be passed from a mother cell to each daughter cell by having each of the separated strands of DNA serve as a **template** for producing new copies of double-stranded DNA.

In summary, this chapter focuses on the structure of DNA and the molecules and mechanisms that produce

An artistic representation of DNA

FIGURE 7-1 A 15-meter-tall sculpture of DNA is housed at the Prince Felipe Museum of Science in Valencia, Spain. [*Peter Blixt/Alamy.*]

DNA copies in a process called **DNA replication.** This information is essential for understanding the molecular basis of genes and genetic inheritance. Precisely how DNA is replicated is still an active area of research more than 65 years after the discovery of the double-helix structure. Our current understanding of the mechanism of replication gives a central role to a protein machine called the **replisome.** This complex of proteins coordinates numerous reactions that are necessary for rapid and accurate replication of DNA.

7.1 DNA IS THE GENETIC MATERIAL

LO 7.1 Describe evidence demonstrating that DNA is the genetic material.

Before we see how Watson and Crick solved the structure of DNA, let's review what was known about genes and DNA at the time they began their historic collaboration:

1. Genes—the hereditary "factors" described by Mendel—were known to be associated with specific traits, but their physical nature was not understood. Similarly, mutations were known to alter gene function, but the precise chemical nature of a mutation was not understood.

2. The one-gene–one-enzyme hypothesis (described in Chapter 5) postulated that genes determine the structure of proteins.

3. Genes were known to be carried on chromosomes.

4. Chromosomes were known to consist of DNA and protein.

5. As described next, experiments beginning in the 1920s revealed that DNA is the genetic material.

The discovery of bacterial transformation: the Griffith experiment

In 1928, Frederick Griffith made the puzzling observation that the genotype and phenotype of a live bacterial strain can be changed, that is, "transformed," by mixing it with a different, heat-killed bacterial strain. His studies used the bacterium *Streptococcus pneumoniae*, which causes pneumonia in humans and is normally lethal in mice. However, some strains of this bacterial species have evolved to be less virulent (less able to cause disease or death). In experiments summarized in **Figure 7-2**, Griffith used two strains that are distinguishable by the appearance of their colonies when grown in laboratory cultures. One strain was a normal virulent type, deadly to most laboratory animals. The cells of this strain are enclosed in a polysaccharide capsule, giving colonies a smooth appearance; hence, this strain is identified as S. Griffith's other strain was a mutant, nonvirulent type that grows in mice but is not lethal. In this strain, the polysaccharide coat is absent, giving colonies a rough appearance; this strain is called R.

Griffith killed some virulent S cells by boiling them. He then injected the heat-killed cells into mice. The mice survived, showing that the carcasses of the cells do not cause death. However, mice injected with a mixture of heat-killed virulent S cells and live nonvirulent R cells did die. Furthermore, live cells could be recovered from the dead mice; these cells gave smooth colonies and were virulent on subsequent injection. Somehow, the cell debris of the boiled S cells converted some of the live R cells into live S cells. That is, the live R cells were transformed into S cells by picking up some chemical component of the dead S cells. The process, already discussed in Chapter 6, is called *transformation*.

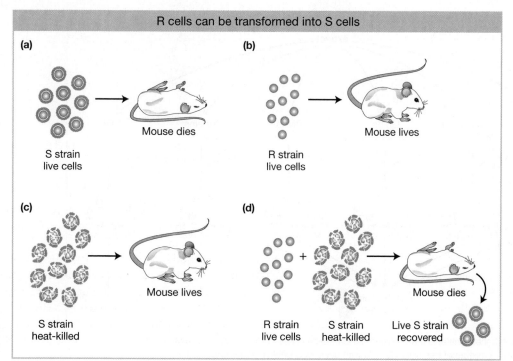

R cells can be transformed into S cells

(a) S strain live cells → Mouse dies

(b) R strain live cells → Mouse lives

(c) S strain heat-killed → Mouse lives

(d) R strain live cells + S strain heat-killed → Mouse dies → Live S strain recovered

FIGURE 7-2 The presence of heat-killed S cells transforms live R cells into live S cells. (a) Mice die after injection with virulent S cells. (b) Mice survive after injection with R cells. (c) Mice survive after injection with heat-killed S cells. (d) Mice die after injection with a mixture of heat-killed S cells and live R cells. Live S cells were isolated from dead mice, indicating that heat-killed S cells somehow transform nonvirulent R cells into virulent S cells.

Evidence that DNA is the genetic material in bacteria: the Avery, Macleod, and McCarty experiments

The next step was to determine which chemical component of dead S cells caused transformation. This molecule had changed the genotype of the recipient strain and therefore was a candidate for the hereditary material. The problem was solved by experiments conducted in 1944 by Oswald Avery, Colin MacLeod, and Maclyn McCarty (**Figure 7-3**). Their approach to the problem was to destroy all of the major categories of chemicals in an extract of dead S cells one at a time, and to find out if the extract had lost the ability to transform. Virulent S cells had a smooth polysaccharide coat, whereas nonvirulent R cells did not; hence, polysaccharides were an obvious candidate for the transforming agent. However, when polysaccharides were destroyed, the mixture could still transform. Lipids, RNAs, and proteins were all similarly shown not to be the transforming agent. In contrast, the mixture lost its transforming ability when the donor mixture was treated with the enzyme deoxyribonuclease (DNase), which destroys DNA. These results strongly implicated DNA as the genetic material. It is now known that fragments of the transforming DNA that confer virulence enter the bacterial chromosome and replace their counterparts that confer nonvirulence.

KEY CONCEPT The demonstration that DNA is the transforming agent was the first evidence that genes (the hereditary material) are composed of DNA.

Evidence that DNA is the genetic material in phage: the Hershey–Chase experiment

The experiments conducted by Avery and his colleagues were definitive, but many scientists were reluctant to accept DNA (rather than proteins) as the genetic material. After all, how could such a low-complexity molecule as DNA encode the diversity of all living things? In 1952, Alfred Hershey and Martha Chase provided additional evidence in an experiment that made use of bacteriophage T2 (or phage T2 for short), a virus that infects bacteria. They reasoned that the infecting phage must inject into the bacterium the specific information that directs the production of new viral particles. If they could find out what material the phage was injecting into the bacterial host, they would have determined the genetic material of phages.

The phage is relatively simple in molecular composition. The T2 structure is similar to T4 shown in Figures 6-22 to 6-24. Most of its structure is protein, with DNA contained inside the protein sheath of its "head." Hershey and Chase used radioisotopes to give DNA and protein distinct labels

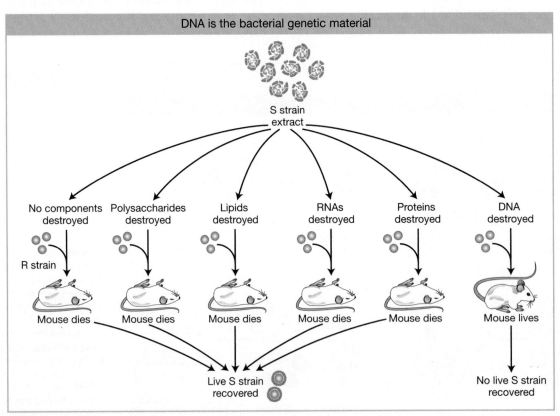

FIGURE 7-3 DNA is the genetic material that transforms nonvirulent R cells into virulent S cells. Mice survive when injected with a mixture of heat-killed S cells with destroyed DNA and live nonvirulent R cells. However, destroying polysaccharides, lipids, RNAs, or proteins does not allow the mice to survive. Thus, DNA, but not polysaccharide, lipid, RNA, or protein, is necessary for transformation.

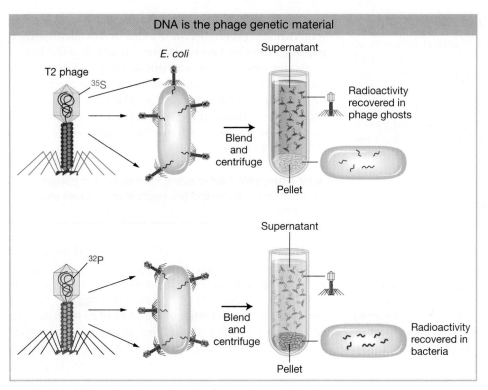

DNA is the phage genetic material

FIGURE 7-4 The Hershey–Chase experiment demonstrates that the genetic material of phages is DNA, not protein. The experiment uses two sets of T2 phage. In one set, the protein coat is labeled with radioactive sulfur (^{35}S), not found in DNA. In the other set, the DNA is labeled with radioactive phosphorus (^{32}P), not found in protein. Only ^{32}P is recovered from *E. coli* and phage progeny, indicating that DNA is the genetic material necessary for production of new phages.

that they could track during infection. Phosphorus is not found in the amino acid building blocks of proteins but is found in DNA; conversely, sulfur is not in the nucleotide building blocks of DNA but is in proteins. Hershey and Chase incorporated a radioisotope of phosphorus (^{32}P) into DNA and that of sulfur (^{35}S) into proteins of separate phage cultures. **Radioisotopes** are unstable (i.e., radioactive) isotopes of an element that emit radiation to transform into a more stable form. Emitted radiation can be measured using instruments such as a scintillation counter or a Geiger counter or by autoradiography (described in Section 7.3).

As shown in **Figure 7-4**, after labeling the phage DNA and proteins, they then infected two *E. coli* cultures with many phage particles per cell: one *E. coli* culture received phage labeled with ^{32}P, and the other received phage labeled with ^{35}S. After allowing sufficient time for infection to take place, they sheared the empty phage carcasses (called *ghosts*) off the bacterial cells in a kitchen blender. They separated the bacterial cells from the phage ghosts in a centrifuge and then measured the radioactivity in the solid pellet of bacteria and the liquid supernatant of phage ghosts. When the ^{32}P-labeled phages were used to infect *E. coli*, the radioactivity ended up inside the bacterial cells, indicating that phage DNA entered the cells. In contrast, when ^{35}S-labeled phages were used, the radioactive material ended up in the phage ghosts, indicating that phage proteins did not enter the bacterial cell. In addition, the progeny of ^{32}P-labeled phages remained labeled, but the progeny of ^{35}S-labeled phages were not labeled. These data once again indicated that DNA is the hereditary material. The phage proteins are mere structural

packaging that is discarded after delivering the viral DNA to the bacterial cell.

7.2 DNA STRUCTURE

LO 7.2 Describe evidence used to construct the double helix model of DNA.

LO 7.3 Draw the chemical structure of a DNA double helix.

Even before the structure of DNA was elucidated, genetic studies indicated that the hereditary material must have three key properties:

1. Because essentially every cell in the body of an organism has the same genetic makeup, accurate replication of the genetic material at every cell division is crucial. Thus, structural features of the genetic material *must allow accurate replication*. The structural features of DNA will be covered in this section of the chapter.

2. Because it must encode the collection of proteins expressed by an organism, structural features of the genetic material *must have informational content*. How information coded in DNA is deciphered to produce proteins is the subject of Chapters 8 and 9.

3. Because hereditary changes, called mutations, provide the raw material for evolutionary selection, genetic material *must be able to change* on rare occasion. Nevertheless, the structure of the genetic material must be stable enough for an organism to rely on its encoded information. The mechanisms of DNA mutations will be covered in Chapter 16.

DNA structure before Watson and Crick

Consider the discovery of the double-helical structure of DNA by Watson and Crick as the solution to a complicated three-dimensional puzzle. To solve this puzzle, Watson and Crick used a process called "model building," in which they assembled the results of earlier and ongoing experiments (the puzzle pieces) to form the three-dimensional puzzle (the double helix model). To understand how they built the DNA model, we first need to know what pieces of the puzzle were available to them.

The building blocks of DNA The first piece of the puzzle was knowledge of the basic building blocks of DNA. As a chemical, DNA is quite simple. It contains three components: (1) **phosphate**, (2) a sugar called **deoxyribose**, and (3) four nitrogenous **bases—adenine, guanine, cytosine,** and **thymine** (**Figure 7-5**). The sugar in DNA is called "deoxyribose" because it contains ribose sugars that are missing an oxygen atom. Deoxyribose has a hydrogen atom (H) at the 2'-carbon atom, unlike **ribose** (a component of RNA), which has a hydroxyl (OH) group at that position.

Two of the bases, adenine and guanine, have a double-ring structure characteristic of a class of chemicals called **purines.** The other two bases, cytosine and thymine, have a single-ring structure characteristic of another class of chemicals called **pyrimidines.** Carbon and nitrogen atoms in the rings of the bases are assigned numbers for ease of reference. Carbon atoms in the sugar group also are assigned numbers—in this case, each number is followed by a prime (1', 2', and so forth).

> **KEY CONCEPT** DNA contains four bases—two purines (adenine and guanine) and two pyrimidines (cytosine and thymine).

The chemical subunits of DNA are nucleotides or, more specifically, **deoxynucleotides,** each composed of a phosphate group, a deoxyribose sugar molecule, and one of the four bases (Figure 7-5). It is convenient to refer to each nucleotide by the first letter of the name of its base: A, G, C, or T. The nucleotide with the adenine base is called deoxyadenosine 5'-monophosphate and abbreviated dAMP,

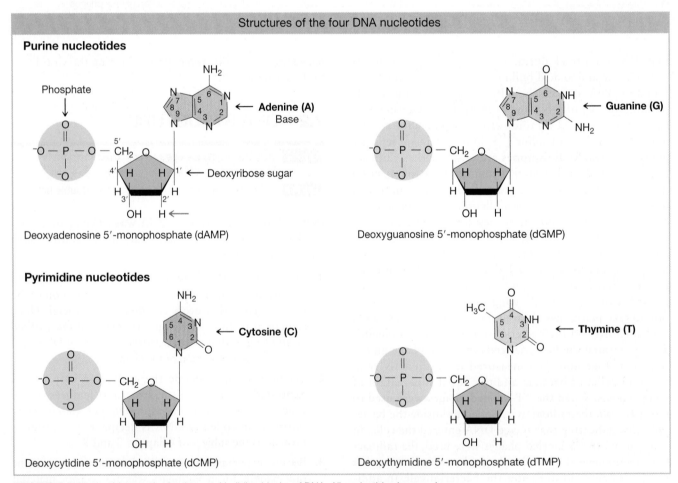

Structures of the four DNA nucleotides

Purine nucleotides

Deoxyadenosine 5'-monophosphate (dAMP)

Deoxyguanosine 5'-monophosphate (dGMP)

Pyrimidine nucleotides

Deoxycytidine 5'-monophosphate (dCMP)

Deoxythymidine 5'-monophosphate (dTMP)

FIGURE 7-5 Nucleotides are the fundamental building blocks of DNA. All nucleotides have a phosphate, a sugar, and a base. The sugar is called deoxyribose because it is a variant of ribose that lacks an oxygen atom, indicated by the red arrow. There are two purine bases (adenine and guanine) and two pyrimidine bases (cytosine and thymine). Note that each of the bases contains nitrogen atoms.

where the 5′ refers to the position of the carbon atom in the sugar ring to which the single (mono) phosphate group is attached. The other nucleotides are named using the same convention.

KEY CONCEPT DNA nucleotides are known as deoxynucle-otides and are composed of a phosphate, a deoxyribose, and a purine or pyrimidine base.

Chargaff's rules of base composition The second piece of the puzzle used by Watson and Crick came from work done several years earlier by Erwin Chargaff. Studying a large selection of DNAs from different organisms (**Table 7-1**), Chargaff established certain empirical rules about the amounts of each type of nucleotide found in DNA:

1. The total amount of purine nucleotides (A + G) always equals the total amount of pyrimidine nucleotides (T + C).

2. The amount of A always equals the amount of T, and the amount of G always equals the amount of C; that is, A/T and G/C is close to 1.0, regardless of the source of DNA (Table 7-1).

3. The amount of A + T is not necessarily equal to the amount of G + C, as can be seen in the last column of Table 7-1. The (A + T)/(G + C) ratio varies among different organisms. For example, in sea urchins, the ratio is 1.85, indicating that the sea urchin genome has almost twice as much A + T than G + C; it is said to be AT-rich. In contrast, the *Mycobacterium tuberculosis* genome is GC-rich, with about twice as much G + C than A + T. However, the ratio is virtually the same in different tissues of the same organism (as seen for human tissues in the last three rows of Table 7-1), supporting the idea that all cells of an organism have the same genomic DNA sequence.

KEY CONCEPT DNA contains an equal amount of A and T nucleotides and G and C nucleotides. Organisms vary in the relative amount of A + T versus G + C, but different tissues in the same organism have the same relative amount of A + T versus G + C.

X-ray diffraction analysis of DNA: Rosalind Franklin

The third piece of the puzzle came from the X-ray diffraction pattern of DNA fibers (**Figure 7-6a**) that was collected by Rosalind Franklin (Figure 7-6b). In this experiment, X rays were fired at DNA fibers that were collected from cells, as shown in the opening photograph of this chapter. The scatter of the X rays from the fibers is detected as spots on photographic film (Figure 7-6a). The angle of scatter represented by each spot on the film gives information about the position of an atom or certain groups of atoms in DNA. Darker spots are where the film was hit multiple times by X rays from repeated parts of DNA such as nucleotide bases. This procedure is not simple to carry out (or to explain), and interpretation of the spot patterns requires complex mathematical treatment that is beyond the scope of this text. The available data suggested that DNA is long and skinny and that it has two similar parts that are parallel to each other and run along the length of the molecule. The X-ray data showed that DNA is helical, like a spiral staircase. Unknown to Franklin, her best X-ray picture (Figure 7-6a) was shown to Watson and Crick, and this crucial piece of the puzzle allowed them to deduce the three-dimensional structure of DNA (**Figure 7-7**).

KEY CONCEPT The X-ray diffraction pattern of DNA showed that it is a long and skinny, two-stranded helix (that is, a double helix).

TABLE 7-1	Molar Properties of Bases* in DNAs from Various Sources					
Organism	**Tissue**	**Adenine**	**Thymine**	**Guanine**	**Cytosine**	$\dfrac{\text{A}+\text{T}}{\text{G}+\text{C}}$
E. coli (K12)	—	26.0	23.9	24.9	25.2	1.00
D. pneumoniae	—	29.8	31.6	20.5	18.0	1.59
M. tuberculosis	—	15.1	14.6	34.9	35.4	0.42
Yeast	—	31.3	32.9	18.7	17.1	1.79
Sea urchin	Sperm	32.8	32.1	17.7	18.4	1.85
Herring	Sperm	27.8	27.5	22.2	22.6	1.23
Rat	Bone marrow	28.6	28.4	21.4	21.5	1.33
Human	Thymus	30.9	29.4	19.9	19.8	1.52
Human	Liver	30.3	30.3	19.5	19.9	1.53
Human	Sperm	30.7	31.2	19.3	18.8	1.62

*Defined as moles of nitrogenous constituents per 100 g-atoms phosphate in hydrolysate.
Source: Data from E. Chargaff and J. Davidson, eds., *The Nucleic Acids*. Academic Press, 1955.

Rosalind Franklin's critical experimental result

(a)

(b)

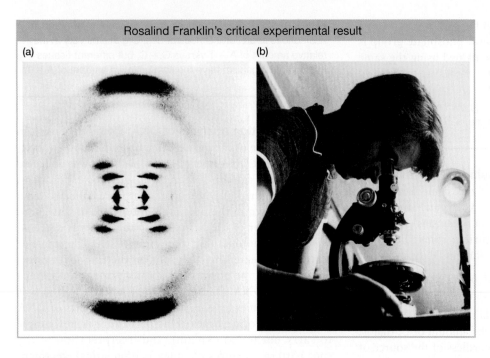

FIGURE 7-6 (a) The X-ray diffraction pattern of DNA determined by (b) Rosalind Franklin. [(a) OMIKRON/ Science Source; (b) © World History Archive/Alamy Stock Photo.]

The DNA double helix structure: Watson and Crick

A 1953 paper by Watson and Crick in the journal *Nature* began with two sentences that ushered in a new age of biology: "We wish to suggest a structure for the salt of deoxyribose nucleic acid (D.N.A.). This structure has novel features which are of considerable biological interest."[1] The structure of DNA had been a subject of great debate since the experiments of Avery and co-workers in 1944. The general composition of DNA was known, but how the parts fit together was not known. The structure had to fulfill the main requirements for a hereditary molecule: the ability to store information, the ability to be replicated, and the ability to mutate.

By studying models that they made of the structure, Watson and Crick realized that the observed diameter of the double helix (known from the X-ray data) would be explained if a purine base always pairs (by hydrogen bonding) with a pyrimidine base (**Figure 7-8**). Such pairing would account for the $(A + G) = (T + C)$ regularity observed by Chargaff, but it would predict four possible pairings: A-T, G-T, A-C, and G-C. However, Chargaff's data indicate that G pairs only with C, and A pairs only with T. Watson and Crick concluded that each base pair consists of one purine

Watson and Crick's DNA model

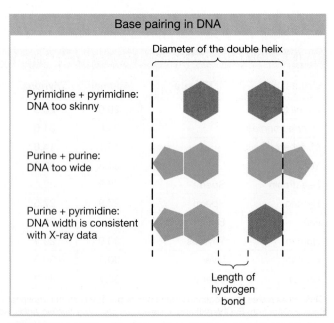

FIGURE 7-7 James Watson (*left*) and Francis Crick (*right*) with their three-dimensional DNA model and a two-dimensional drawing of DNA on the wall. [*BARRINGTON BROWN/Science Source.*]

Base pairing in DNA

Diameter of the double helix

Pyrimidine + pyrimidine: DNA too skinny

Purine + purine: DNA too wide

Purine + pyrimidine: DNA width is consistent with X-ray data

Length of hydrogen bond

FIGURE 7-8 Pairing of purines with pyrimidines accounts exactly for the diameter of the DNA double helix determined from X-ray data.

base and one pyrimidine base, paired according to the following rule: G pairs with C (G-C), and A pairs with T (A-T). These are called **complementary bases**. The double helix accounted nicely for Franklin's X-ray data as well as Chargaff's base composition data.

> **KEY CONCEPT** The two strands of DNA contain complementary base pairs—G base pairs with C and A base pairs with T.

The three-dimensional structure derived by Watson and Crick is composed of two side-by-side chains ("strands") of nucleotides twisted into the shape of a double helix with 10 base pairs in each complete turn of the helix (**Figure 7-9a**). DNA is a right-handed helix; in other words, it has the same structure as that of a screw that would be screwed into place using a clockwise turning motion. The two strands are held together by hydrogen bonds between purine and pyrimidine bases of each strand, forming the stairs of a spiral staircase. On the outside of the double helix, the backbone of each

strand is formed by alternating phosphate and deoxyribose sugar units that are connected by phosphodiester linkages (Figure 7-9b). These linkages are used to describe how a nucleotide chain is organized. As already mentioned, the carbon atoms of the sugar groups are numbered 1′ through 5′. A phosphodiester linkage connects the 5′-carbon atom of one deoxyribose to the 3′-carbon atom of the adjacent deoxyribose. Thus, each sugar–phosphate backbone is said to have a 5′-to-3′ polarity, or direction. Understanding this polarity is essential in understanding how DNA fulfills its roles. In double-stranded DNA, the two backbones are in opposite, or **antiparallel,** orientation; one is oriented 5′-to-3′ and the other is oriented 3′-to-5′ (Figure 7-9b).

> **KEY CONCEPT** The base-paired strands of DNA are oriented antiparallel to one another—one strand is oriented in the 5′-to-3′ direction and the other strand is oriented in the 3′-to-5′ direction.

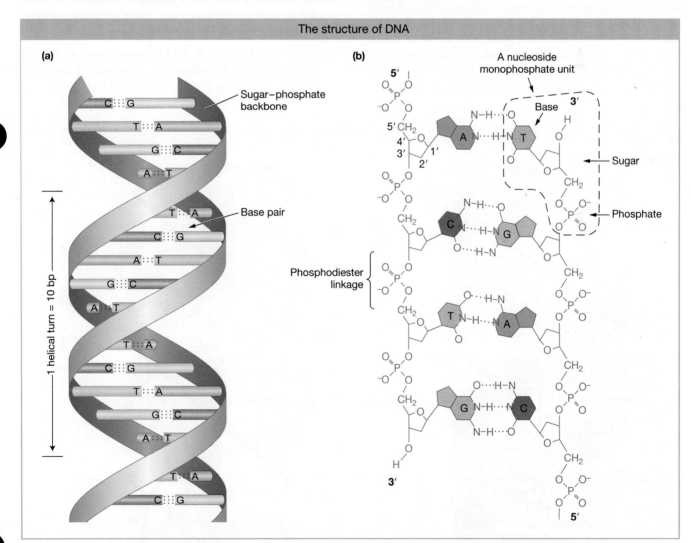

The structure of DNA

FIGURE 7-9 (a) A simplified model showing the helical structure of DNA. Horizontal sticks represent base pairs; blue ribbons represent the sugar–phosphate backbones of the two antiparallel DNA chains. (b) A chemical diagram of the DNA double helix, unrolled to show the sugar–phosphate backbones (blue) and base-pair rungs (purple,

orange). The backbones run in opposite directions. The strand on the left is oriented 5′-to-3′ from top to bottom, and the strand on the right is oriented 3′-to-5′ from top to bottom. Each base pair has one purine base, adenine (A) or guanine (G), and one pyrimidine base, thymine (T) or cytosine (C), connected by hydrogen bonds (red dots).

Each base is attached to the 1'-carbon atom of a deoxyribose sugar in the backbone of each strand and faces inward toward a base on the other strand. Hydrogen bonds between pairs of purine and pyrimidine bases (indicated by dots in Figure 7-9b) hold the two strands of DNA together. Note that G–C base pairs have three hydrogen bonds, whereas A–T base pairs have only two. We would predict that DNA containing many G–C base pairs would be more stable than DNA containing many A–T base pairs. In fact, this prediction is confirmed. Heat causes the two strands of the DNA double helix to separate (a process called DNA melting or DNA denaturation); DNAs with higher G + C content require higher temperatures to melt because of greater attraction of G–C base pairs.

KEY CONCEPT A–T base pairs have two hydrogen bonds, and G–C base pairs have three.

The two complementary nucleotide strands paired in an antiparallel manner automatically assume a double-helical conformation (**Figure 7-10**), mainly through the interaction of base pairs. Base pairs, which are flat planar structures, stack on top of one another at the center of the double helix (Figure 7-10a). Stacking adds to the stability of DNA by excluding water molecules from spaces between the base pairs. A single strand of nucleotides is not helical; the helical shape of DNA depends entirely on pairing and stacking of bases in the antiparallel strands. The most stable form that results from base stacking is a double helix with two distinct sizes of grooves running in a spiral: shallow **major grooves** occur where the sugar–phosphate backbones are far apart, and deep **minor grooves** occur where they are close together. Both types of grooves can be seen in ribbon (Figure 7-10a) and space-filling (Figure 7-10b) models. Proteins that bind DNA interact specifically with either major or minor grooves.

KEY CONCEPT The geometry of base pairs creates shallow, wide major grooves and narrow, deep minor grooves along the DNA helix; features that are recognized for protein binding.

The structure of DNA is considered by some to be the most important biological discovery of the twentieth century. The reason that this discovery is considered so important is that the double helix model, in addition to being consistent with earlier data about DNA structure, fulfilled the three requirements for a hereditary substance:

1. The double-helical structure suggested how the genetic material might determine the structure of proteins. Perhaps the *sequence* of nucleotides in DNA dictates the

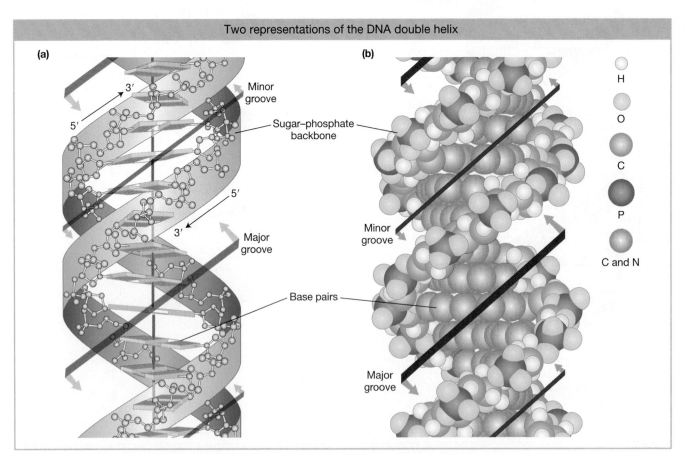

Two representations of the DNA double helix

(a)

3'
5'
Minor groove
Sugar–phosphate backbone
5'
3'
Major groove
Base pairs

(b)

Minor groove
Major groove

H
O
C
P
C and N

FIGURE 7-10 (a) The ribbon diagram highlights the stacking of base pairs, whereas (b) the space-filling model emphasizes the major and minor grooves. In both models, the sugar–phosphate backbone is blue and the bases are gold.

sequence of amino acids in the protein specified by that gene. In other words, some sort of **genetic code** may write information in DNA as a sequence of nucleotides and then translate it into a different language of amino acid sequence in protein. Just how this is done is the subject of Chapter 9.

2. As Watson and Crick stated in the concluding words of their 1953 *Nature* paper that reported the double-helical structure of DNA: "It has not escaped our notice that the specific pairing we have postulated immediately suggests a possible copying mechanism for the genetic material."[2] To geneticists at the time, this statement proposed that DNA is replicated by a semiconservative mechanism, as described in the next section.

3. If the nucleotide sequence of DNA specifies the amino acid sequence, mutations are possible by the substitution of one nucleotide for another at one or more positions. Mutations will be discussed in Chapter 16.

7.3 DNA REPLICATION IS SEMICONSERVATIVE

LO 7.4 Describe evidence used to support semiconservative DNA replication.

In **semiconservative replication** hypothesized by Watson and Crick, the double helix is unwound and each DNA strand acts as a template to direct assembly of complementary bases following the A–T and G–C base-pairing rules to create two double helices that are identical to the original. This mode of replication is called semiconservative because each of the new helices conserves one of the original strands (that is, the **parental molecule**) and the other strand (that is, the **daughter molecule**) is new (**Figure 7-11a**). However, two other modes of replication were also hypothesized. In **conservative replication,** the parent DNA double helix is conserved, and a daughter double helix is produced consisting of two newly synthesized strands (Figure 7-11b). In **dispersive replication,** two new DNA double helices are produced, with each strand containing segments of *both* parental DNA and newly synthesized daughter DNA (Figure 7-11c).

FIGURE 7-11 Three mechanisms were hypothesized for how DNA is replicated: (a) semiconservative, (b) conservative, and (c) dispersive. The Meselson–Stahl experiment demonstrates that DNA is copied by semiconservative replication. DNA centrifuged in a cesium chloride (CsCl) gradient will form bands according to its density. (a) In accord with semiconservative replication, when cells grown in heavy ^{15}N are transferred to light ^{14}N medium, the first generation produces a single DNA band of intermediate density and the second generation produces two bands: one intermediate and one light. (b and c) In contrast, the data do not match results predicted for conservative and dispersive replication.

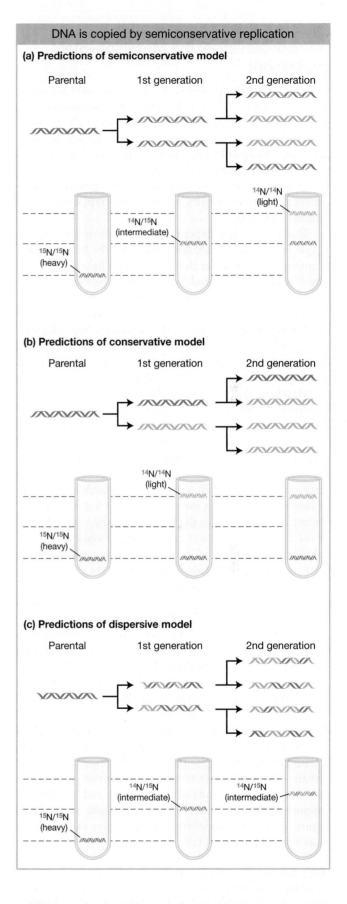

DNA is copied by semiconservative replication

(a) Predictions of semiconservative model

Parental   1st generation   2nd generation

$^{14}N/^{14}N$ (light)
$^{14}N/^{15}N$ (intermediate)
$^{15}N/^{15}N$ (heavy)

(b) Predictions of conservative model

Parental   1st generation   2nd generation

$^{14}N/^{14}N$ (light)
$^{15}N/^{15}N$ (heavy)

(c) Predictions of dispersive model

Parental   1st generation   2nd generation

$^{14}N/^{15}N$ (intermediate) $^{14}N/^{15}N$ (intermediate)
$^{15}N/^{15}N$ (heavy)

[2] J. Watson and F. Crick, *Nature* 171:737, 1953.

Evidence that DNA replication is semiconservative: the Meselson–Stahl experiment

In 1958, Matthew Meselson and Franklin Stahl set out to discover whether DNA replication was semiconservative, conservative, or dispersive. Their idea was to allow parental DNA containing nucleotides of one density to replicate using nucleotides of a different density. They realized that after two rounds of DNA replication, the three proposed replication mechanisms could be distinguished by differences in the density of the newly replicated DNA (Figure 7-11, left).

To carry out their experiment, Meselson and Stahl grew *E. coli* cells in liquid medium containing the heavy isotope of nitrogen (^{15}N) rather than the normal light (^{14}N) form. The ^{15}N isotope was used by the cells to synthetize nitrogenous bases, which then were incorporated into newly synthesized DNA strands. After many cell divisions in ^{15}N, the DNA was almost completely labeled with the heavy isotope. Cells were then removed from the ^{15}N medium and placed into ^{14}N medium; after one and two cell divisions, DNA was isolated from each sample and analyzed.

Meselson and Stahl were able to distinguish DNA of different densities using a separation procedure called *cesium chloride gradient centrifugation*. If cesium chloride (CsCl) is spun in a centrifuge at a tremendously high speed (50,000 revolutions per minute, or rpm) for many hours, the cesium and chloride ions are pushed by centrifugal force toward the bottom of the tube. Ultimately, a gradient of ions is established in the tube, with the highest ion concentration, or density, at the bottom and the lowest density at the top. When DNA is centrifuged along with cesium chloride, it forms a band in the gradient at a position identical to its density (Figure 7-11, right). DNA of different densities will form bands at different places in the gradient. Cells initially grown in the heavy isotope ^{15}N showed DNA of high density. This DNA is shown in blue in the left-most tube of Figure 7-11. After growing these cells in the light isotope ^{14}N for one generation, they found that the DNA was of intermediate density, shown in half blue (^{15}N) and half gold (^{14}N) in the middle tube of Figure 7-11. This banding pattern supported the semiconservative (Figure 7-11a) and dispersive (Figure 7-11c) models and disproved the conservative (Figure 7-11b) model. Meselson and Stahl continued the experiment through a second *E. coli* generation so that they could distinguish semiconservative from dispersive replication. After two generations, they observed two bands of intermediate and low density, supporting the semiconservative model (right-most tube of Figure 7-11a) and disproving the dispersive model (right-most tube of Figure 7-11c).

KEY CONCEPT DNA is replicated semiconservatively by unwinding the two strands of the double helix and building a new complementary strand on each of the separated strands of the original double helix.

Evidence for a replication fork: the Cairns experiment

The next problem was to determine where replication initiates on the chromosome. The possibilities were that replication could initiate at one site or many sites, and that the sites could be random or defined. In 1963, John Cairns addressed this problem by allowing replicating DNA in bacterial cells to incorporate tritiated thymidine ([^{3}H]thymidine)—a **nucleoside** (a base linked to a sugar) labeled with a radioactive hydrogen isotope called tritium. In the cells, the nucleoside was converted to a nucleotide by phosphorylation and incorporated into newly replicated DNA. After varying the number of replication cycles in the presence of tritiated thymidine, Cairns carefully isolated the DNA and covered it with photographic emulsion for several weeks. This procedure, called autoradiography, allowed Cairns to develop a picture of the location of ^{3}H in the DNA. As ^{3}H decays, it emits a beta particle (an energetic electron). A photograph of black spots results from a chemical reaction that occurs wherever a beta particle strikes the emulsion. Hence, each tritiated thymidine incorporated into the DNA appears as a black spot on the photograph.

Since DNA is replicated semiconservatively, after one round of DNA replication, each newly synthesized daughter chromosome should contain one radioactive ("hot") strand (with ^{3}H) that is detected in the autoradiograph, and another nonradioactive ("cold") strand that is not detected. Indeed, after one replication cycle in [^{3}H]thymidine, a ring of black spots appeared in the autoradiograph. Cairns interpreted this ring as a newly formed radioactive strand in a circular daughter DNA molecule, as shown in **Figure 7-12a**. It is thus apparent that the bacterial chromosome is circular—a fact that also emerged from genetic analysis described earlier (see Chapter 6). Furthermore, Cairns found that chromosomes captured in the middle of a second replication cycle formed a structure that resembled the Greek letter theta (θ), with a thin circle of dots consisting of a single radioactive strand and a thick curve of dots cutting through the interior of the circle of DNA consisting of two radioactive strands (Figure 7-12b). Thus, this type of replication is often called theta replication. The ends of the thick curve of dots defined two sites of ongoing DNA replication and are referred to as **replication forks.** Cairns saw all sizes of theta autoradiographic patterns, suggesting that replication begins at one place and the replication forks progressively move around the ring. Other experiments showed that DNA replication initiates at a single, specific DNA sequence and spreads bidirectionally (that is, in opposite directions) from this site, and both DNA strands are simultaneously replicated.

KEY CONCEPT The Cairns experiment provided additional evidence for semiconservative replication and also demonstrated that replication in bacteria begins at one site in the genome and spreads bidirectionally by means of two replication forks.

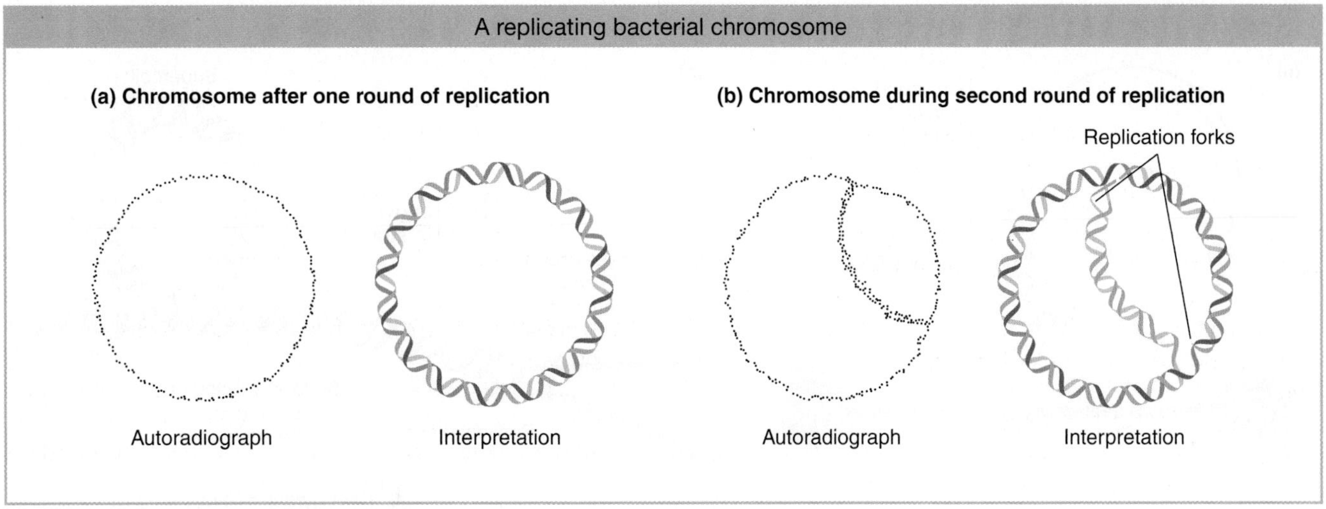

A replicating bacterial chromosome

(a) Chromosome after one round of replication

(b) Chromosome during second round of replication

Replication forks

Autoradiograph   Interpretation   Autoradiograph   Interpretation

FIGURE 7-12 A replicating bacterial chromosome has two replication forks. (a) *Left:* An autoradiograph of a bacterial chromosome after one generation of replication in tritiated thymidine. *Right:* An interpretation of the autoradiograph. The gold helix represents the tritiated strand. In agreement with the semiconservative model of replication, one of the two strands should be radioactive. (b) *Left:* An autogradiograph of a bacterial chromosome in the second generation of replication in tritiated thymidine. *Right:* An interpretation of the autoradiograph. Replication forks occur at the two sites of ongoing, bidirectional DNA replication. Once again, in accord with the semiconservative model of replication, the newly replicated double helix that crosses the circle consists of two radioactive strands (if the parental strand was the radioactive one).

7.4 DNA REPLICATION IN BACTERIA

LO 7.5 Outline the factors and events involved in DNA replication.

In this section, we walk through the steps of DNA replication in bacteria, emphasizing the activities of enzymes in the replisome, the multi-protein molecular machine that carries out DNA replication. Similar steps occur in eukaryotes, and they are carried out by analogous enzymes (**Table 7-2**).

Unwinding the DNA double helix

When the double helix was proposed in 1953, a major objection was that replication of such a structure would require unwinding the double helix and breaking hydrogen bonds that hold the strands together. How could DNA be unwound so rapidly and, even if it could, wouldn't that overwind the DNA behind the fork and make it hopelessly tangled? The problem can be envisioned by thinking of two strands of a rope that are separated at one end while the other end is held stationary (**Figure 7-13a**). We now know that the replisome contains proteins that open the helix and

TABLE 7-2	Analogous DNA Replication Factors in Bacteria and Eukaryotes	
Function	**Bacteria (*E. coli*)**	**Eukaryotes (humans)**
Recognizes origins	DnaA	ORC (origin recognition complex)
Unwinds double-stranded DNA	DnaB helicase	MCM2-7 (minichromosome maintenance 2-7) helicase
Assists helicase binding	DnaC	Cdc6 and ORC
Stabilizes single-stranded DNA	SSB	RPA (replication factor A)
Removes twists and supercoils	Gyrase	Topoisomerases
Synthesizes RNA primers	Primase	DNA pol α-primase complex
Elongates DNA	DNA pol III	DNA pol ε (leading strand) and δ (lagging strand)
Sliding clamp	β-clamp	PCNA (proliferating cell nuclear antigen)
Clamp loader	τ complex	RFC (replication factor C)
Removes RNA primers	DNA pol I	FEN1
Replaces RNA primers with DNA	DNA pol I	DNA pol ε
Ligates Okazaki fragments	DNA ligase	DNA ligase I

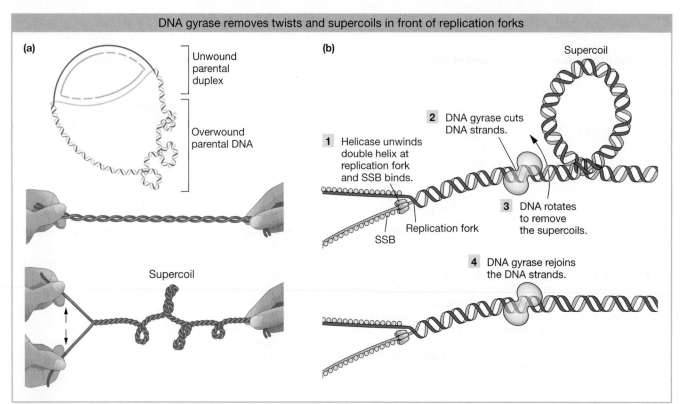

DNA gyrase removes twists and supercoils in front of replication forks

(a)
Unwound parental duplex

Overwound parental DNA

Supercoil

(b)
1 Helicase unwinds double helix at replication fork and SSB binds.

2 DNA gyrase cuts DNA strands.

3 DNA rotates to remove the supercoils.

Supercoil

Replication fork

SSB

4 DNA gyrase rejoins the DNA strands.

FIGURE 7-13 (a) Twisted and supercoiled regions accumulate ahead of the replication fork as parental DNA strands separate for replication, analogous to what happens when a rope is separated into individual strands. (b) Helicase encircles one DNA strand at each replication fork and uses the energy of ATP hydrolysis to break hydrogen bonds between bases, unzipping the two strands and causing twisting and supercoiling. SSB molecules bind the single-stranded DNA to prevent their reannealing. A topoisomerase called DNA gyrase counteracts twisting and supercoiling by cutting the DNA strands, allowing them to rotate, and then rejoining the strands.

prevent overwinding: they are **helicases** and **topoisomerases,** respectively.

Helicases are enzymes that disrupt hydrogen bonds that hold the two strands of the double helix together. The DNA replication helicase is a ring-shaped homohexamer of DnaB proteins (that is, a complex of six copies of DnaB) that encircles single-stranded DNA at the replication forks. From this position, helicases use energy from ATP hydrolysis to rapidly unzip the double helix ahead of DNA synthesis (Figure 7-13b, step 1). Unwound DNA is stabilized by **single-strand DNA-binding (SSB) proteins,** which bind to single-stranded DNA and prevent the duplex from re-forming.

Unwinding of DNA by helicases causes extra twisting to occur ahead of replication forks, and supercoils form to release the strain of the extra twisting (Figure 7-13b, steps 1). Twists and supercoils must be removed (commonly called *relaxed*) to allow replication to continue. This is done by enzymes termed topoisomerases, of which an example is **DNA gyrase** (Figure 7-13b, steps 2 and 3). Topoisomerases relax supercoiled DNA by breaking either a single DNA strand or both strands, which allows DNA to rotate into a relaxed molecule. Topoisomerases finish their job by religating the strands of the now-relaxed DNA (Figure 7-13b, step 4).

KEY CONCEPT Helicases, topoisomerases, and single-strand-binding proteins generate and maintain single-stranded DNA that is used as a template for DNA replication.

Assembling the replisome: replication initiation

Assembly of the replisome is an orderly process that begins at a precise site on the chromosome called the **origin of replication,** or simply origin. *E. coli* replication begins from a single origin (a locus called *oriC*) and then proceeds in both directions (with moving forks at both ends, as shown in Figure 7-12b) until the forks merge. *OriC* is 245 base pairs long and contains five copies of 9-base-pair sequences called DnaA boxes and an adjacent DNA unwinding element that is AT-rich (**Figure 7-14**). The first step in replisome assembly is binding of a protein called DnaA to the DnaA boxes, which helps other copies of DnaA bind at the origin in a process called oligomerization. Subsequent binding of DnaA to the AT-rich region promotes unwinding to form a single-stranded DNA bubble. Recall that A–T base pairs are held together with only two hydrogen bonds, whereas G–C base pairs are held

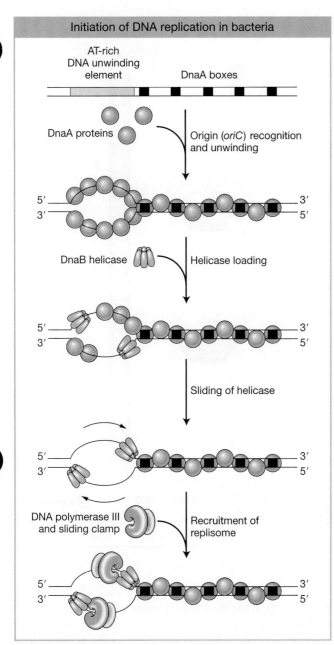

Initiation of DNA replication in bacteria

AT-rich
DNA unwinding
element

DnaA boxes

DnaA proteins

Origin (*oriC*) recognition
and unwinding

5′   3′
3′   5′

DnaB helicase   Helicase loading

5′   3′
3′   5′

Sliding of helicase

5′   3′
3′   5′

DNA polymerase III
and sliding clamp   Recruitment of
replisome

5′   3′
3′   5′

FIGURE 7-14 DNA synthesis is initiated at the origin of replication (*oriC*) in bacteria. DnaA proteins (pink) first bind DnaA boxes and then oligomerize throughout the origin. At the AT-rich DNA unwinding element, DnaA separates the two strands of the double helix and recruits the DnaB helicase (green) and other replisome components (blue) to the two replication forks.

together with three. Thus, it is easier to separate (melt) the double helix at stretches of DNA that are enriched in A–T base pairs.

After unwinding begins, additional DnaA proteins bind to the newly unwound single-stranded regions. With DnaA coating the origin, two DnaB helicases now bind and slide in a 5′-to-3′ direction to begin unzipping the helix at the replication forks. Although DnaA is necessary for the assembly of the replisome, it is not part of the replication machinery. Rather, its job is solely to bring the replisome to the correct place in the circular chromosome

for the initiation of replication. As replication progresses through the origin, the replisome displaces DnaA from the DNA.

KEY CONCEPT Where and when replication takes place are carefully controlled by ordered assembly of the replisome at a precise site called the origin.

DNA polymerases catalyze DNA chain elongation

Although scientists suspected that enzymes play a role in synthesizing DNA, that possibility was not verified until 1959, when Arthur Kornberg isolated a DNA polymerase from *E. coli* and demonstrated its enzymatic activity in vitro. This enzyme adds deoxyribonucleotides to the 3′ end of a growing nucleotide chain, using for its template a single strand of DNA that has been exposed by localized unwinding of the double helix (**Figure 7-15**). The substrates for DNA polymerases are the triphosphate forms of the deoxyribonucleotides, dATP, dGTP, dCTP, and dTTP (dNTP is used to refer to any of four <u>d</u>eoxy<u>n</u>ucleoside <u>t</u>riphosphates). Addition of each base to the growing polymer is accompanied by removal of two of the three phosphates in the form of pyrophosphate (PP_i). Energy produced by cleaving this bond and the subsequent hydrolysis of pyrophosphate to two inorganic phosphate molecules help drive the process of building a DNA polymer.

Five DNA polymerases are now known in *E. coli*. The enzyme that Kornberg purified is called **DNA polymerase I** or **DNA pol I**. This enzyme has three activities, which appear to be located in different parts of the molecule: (1) a polymerase activity that catalyzes DNA chain growth in the 5′-to-3′ direction, (2) a 3′-to-5′ exonuclease activity that removes mismatched nucleotides, and (3) a 5′-to-3′ exonuclease activity that degrades single strands of DNA or RNA. We will return to the significance of the two exonuclease activities later in this chapter.

Although DNA pol I has a role in DNA replication (see the next section), some scientists suspected that it was not responsible for the majority of DNA synthesis because it was too slow (~20 nucleotides/second; at this rate it would take ~30 hours to replicate the *E. coli* genome) and too abundant (~400 molecules/cell, which is more than needed for the two replication forks), and it was not processive (it dissociated from DNA after incorporating only 20 to 50 nucleotides). In 1969, John Cairns and Paula DeLucia settled this matter when they demonstrated that an *E. coli* strain harboring a mutation in the DNA pol I gene that had less than 1% of DNA pol I activity was still able to grow normally and replicate its DNA. They concluded that another DNA polymerase catalyzes DNA synthesis at the replication fork. This enzyme was later shown to be **DNA polymerase III (DNA pol III)**.

KEY CONCEPT DNA polymerases synthesize DNA in the 5′-to-3′ direction using single-stranded DNA as a template.

Reaction catalyzed by DNA polymerases

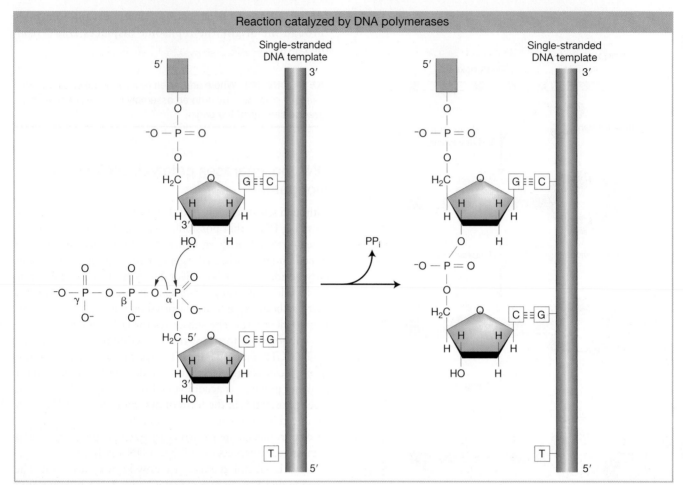

FIGURE 7-15 DNA polymerases catalyze the chain-elongation reaction. A dNTP, in this case dCTP, base pairs to the single-stranded DNA template, the free 3′-OH group at the end of the growing DNA chain is activated to attack the alpha (α) phosphate of the dNTP, resulting in attachment of dNMP to the 3′ end and release of linked β and γ phosphates (PP$_i$). Thus, the DNA chain is elongated in the 5′-to-3′ direction by DNA polymerases.

ANIMATED ART 🅱 SaplingPlus

The nucleotide polymerization process

DNA replication is semidiscontinuous

Another problem in DNA replication arises because DNA polymerases can extend a chain but cannot start a chain. Therefore, synthesis must be initiated by a **primer**, a short chain of nucleotides that forms a segment of duplex nucleic acid (**Figure 7-16**). Primers are synthesized by a set of proteins called a **primosome**, of which a central component is an RNA polymerase called **primase** (**Figure 7-17**). Primase copies the template DNA in the 5′-to-3′ direction, producing a short RNA of about 11 nucleotides. DNA pol III then takes over and continues to copy the template DNA, extending off the 3′ end of the RNA primer.

Because DNA polymerases synthesize DNA only in the 5′-to-3′ direction, only one of the two DNA template strands can serve as a template for replication in the same direction as movement of each replication fork (see Figure 7-16). For this strand, called the **leading strand**, synthesis takes place in a smooth continuous manner. Synthesis

on the other template is also in the 5′-to-3′ direction, but because it is in the direction opposite to that of replication fork movement it must be in short segments. Reinitiation of synthesis occurs for each segment as the growing fork exposes new DNA template. The 1000–2000 nucleotide stretches of newly synthesized DNA are called **Okazaki fragments**, in honor of their discoverer Reiji Okazaki. As with synthesis of the leading strand, each Okazaki fragment is primed at the 5′ end by an RNA primer synthesized by primase. Thus, for this strand, called the **lagging strand**, synthesis takes place in a discontinuous manner. Because DNA replication is continuous for the leading strand and discontinuous for the lagging strand, the overall process is described as **semidiscontinuous**.

KEY CONCEPT DNA replication is described as semidiscontinuous because one DNA template strand is synthesized continuously and the other strand is synthesized as a series of discontinuous fragments.

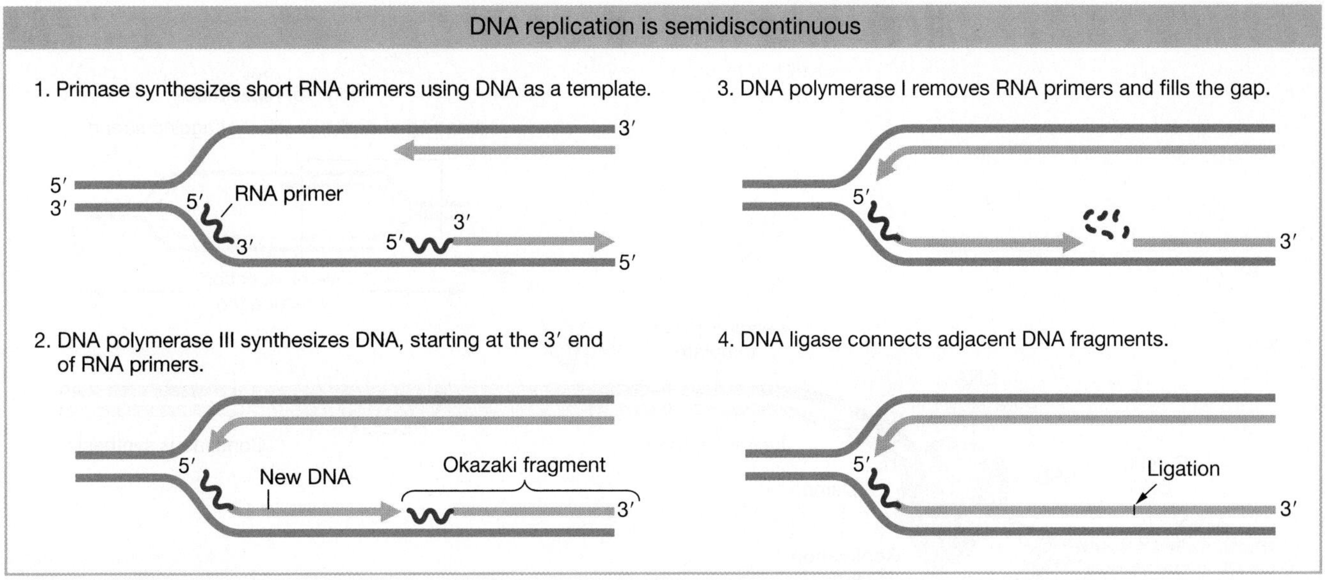

DNA replication is semidiscontinuous

1. Primase synthesizes short RNA primers using DNA as a template.

3. DNA polymerase I removes RNA primers and fills the gap.

2. DNA polymerase III synthesizes DNA, starting at the 3′ end of RNA primers.

New DNA

Okazaki fragment

4. DNA ligase connects adjacent DNA fragments.

Ligation

FIGURE 7-16 DNA replication takes place at the replication fork, where the double helix is unwound and the two strands are separated. DNA replication proceeds continuously in the direction of the unwinding replication fork for the leading strand. In contrast, DNA is synthesized in short segments in the direction away from the replication fork for the lagging strand. DNA polymerase requires a primer, a short chain of nucleotides, to be in place to begin synthesis. Additional details are provided in the text.

A different DNA polymerase, DNA pol I, removes the RNA primers with its 5′-to-3′ exonuclease activity and fills in the gaps with its 5′-to-3′ polymerase activity (Figure 7-17). As mentioned earlier, DNA pol I is the enzyme originally purified by Kornberg. Another enzyme, **DNA ligase,** joins the 3′ end of the gap-filling DNA to the 5′ end of the downstream Okazaki fragment. In general, DNA ligases join broken pieces of DNA by catalyzing formation of a phosphodiester bond between a 5′ phosphate of one fragment and a 3′ OH group of an adjacent fragment.

> **KEY CONCEPT** DNA synthesis by DNA polymerase III requires an RNA primer, synthesized by the primase enzyme, an RNA polymerase.

DNA replication is accurate and rapid

A hallmark of DNA replication is its accuracy, also called fidelity: overall, less than one error occurs every 10^{10} nucleotides. Part of the reason for the accuracy of DNA replication is that both DNA pol I and DNA pol III possess a 3′-to-5′ exonuclease activity, which serves a "proofreading" function by excising incorrectly inserted mismatched bases (**Figure 7-18**). Once the mismatched base is removed, the polymerase has another chance to add the correct complementary base.

As you would expect, mutant strains lacking a functional 3′-to-5′ exonuclease activity have a higher rate of mutation. In addition, because primase lacks a proofreading function,

the RNA primer is more likely than DNA to contain errors. The need to maintain the high fidelity of replication is one reason why RNA primers at the ends of Okazaki fragments must be removed and replaced with DNA. Only after the RNA primer is gone does DNA pol I catalyze DNA synthesis to replace the primer. Mismatches that escape proofreading are corrected by DNA repair mechanisms that will be covered in detail in Chapter 15.

> **KEY CONCEPT** DNA polymerases I and III have proofreading activity but primase does not.

Another hallmark of DNA replication is speed. It takes *E. coli* about 40 minutes to replicate its chromosome. Therefore, its genome of about 5 million base pairs must be copied at a rate of about *2000 nucleotides per second.* From the experiment of Cairns, we know that *E. coli* uses only two replication forks to copy its entire genome. Thus, each fork must be able to move at a rate of about *1000 nucleotides per second.* What is remarkable about the entire process of DNA replication is that it does not sacrifice speed for accuracy. How can it maintain both speed and accuracy, given the complexity of the reactions at the replication fork? The answer is that DNA polymerase is part of a large complex that coordinates the activities at the replication fork. This complex, the replisome, is an example of a "molecular machine." You will encounter other examples in later chapters. The discovery that most major functions

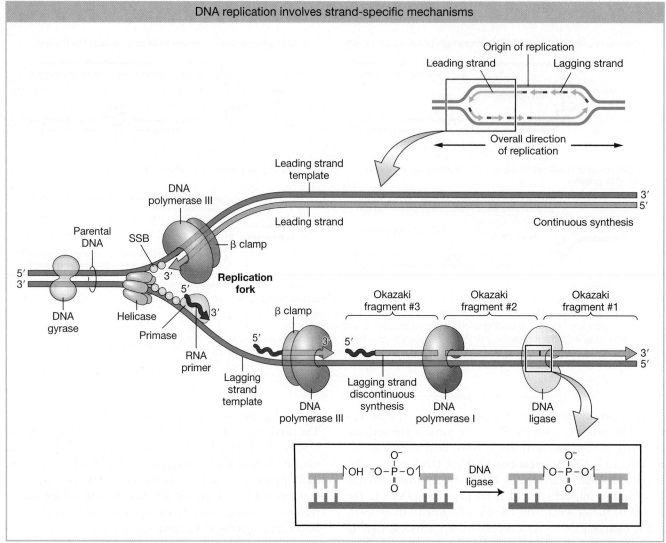

FIGURE 7-17 DNA replication occurs in both directions. The same protein factors are required for synthesis of both the leading and lagging strands, but synthesis occurs in a single continuous stretch for the leading strand and in short stretches for the lagging strand. The β clamp is required for processive DNA synthesis. Helicase separates the strands of parental DNA and gyrase (a topoisomerase) removes the twists and supercoils in DNA. Single-strand DNA-binding proteins (SSBs) prevent the separated DNA strands from reannealing. Primase makes short RNA primers for continuous synthesis by DNA pol III of the leading strand as well as discontinuous synthesis of the lagging strand as Okazaki fragments. DNA pol I removes RNA primers and fills in the resulting gaps. Lastly, DNA ligase joins the DNA fragments together.

ANIMATED ART Sapling Plus

Leading and lagging strand synthesis

of cells—replication, transcription, and translation, for example—are carried out by large multisubunit complexes has changed the way that we think about cells. To begin to understand why, let's look at the replisome more closely.

Some of the interacting components of the replisome in *E. coli* are shown in **Figure 7-19.** At the replication fork, the catalytic core of DNA pol III is part of a much larger complex, called the **DNA pol III holoenzyme,** which consists of two catalytic cores and several accessory proteins. One of the catalytic cores handles synthesis of the leading strand while the other handles the lagging strand. The lagging strand is shown looping around so that the replisome can coordinate the synthesis of both strands and move in the direction of the replication fork. Some of the accessory

proteins (not visible in Figure 7-19) form a connection that bridges the two catalytic cores, thus coordinating synthesis of the leading and lagging strands.

KEY CONCEPT A molecular machine called the replisome carries out DNA synthesis. It includes two DNA polymerase units to handle synthesis on each strand, and it coordinates the activity of accessory proteins required for unwinding the double helix, stabilizing the single strands and processing RNA primers.

Attachment of DNA pol III to the DNA template is maintained by other accessory proteins, the β **clamp** (also known as the **sliding clamp**), which encircles the DNA like a

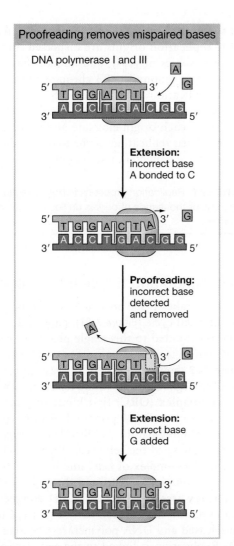

Proofreading removes mispaired bases

DNA polymerase I and III

Extension:
incorrect base
A bonded to C

Proofreading:
incorrect base
detected
and removed

Extension:
correct base
G added

FIGURE 7-18 DNA polymerases I and III use their 3′-to-5′ exonuclease activity to remove the A–C mismatch.

donut, and the clamp loader (also called τ complex), which assembles β clamps onto DNA. The β clamp transforms DNA pol III from an enzyme that can add only 10 nucleotides before falling off the template (termed a **distributive enzyme**) into an enzyme that stays at the moving fork and adds tens of thousands of nucleotides (termed a **processive enzyme**). In sum, through the action of accessory proteins, synthesis of both the leading and lagging strands is rapid and highly coordinated.

Note that primase, the enzyme that synthesizes the RNA primer, is not touching the clamp protein. Therefore, primase acts as a distributive enzyme—it adds only a few ribonucleotides before dissociating from the template. This mode of action makes sense because primers only need to be long enough to form a suitable duplex starting point for DNA pol III.

KEY CONCEPT The β clamp converts DNA polymerase III from a distributive to a processive enzyme.

FIGURE 7-19 A dimer of DNA pol III enzymes coordinates replication of the leading and lagging DNA strands. Looping of the template for the lagging strand orients it for synthesis by DNA pol III in the 5′-to-3′ direction. DNA pol III releases the lagging strand template after synthesizing 1000–2000 nucleotides, a new loop is formed, and primase synthesizes an RNA primer to initiate another Okazaki fragment.

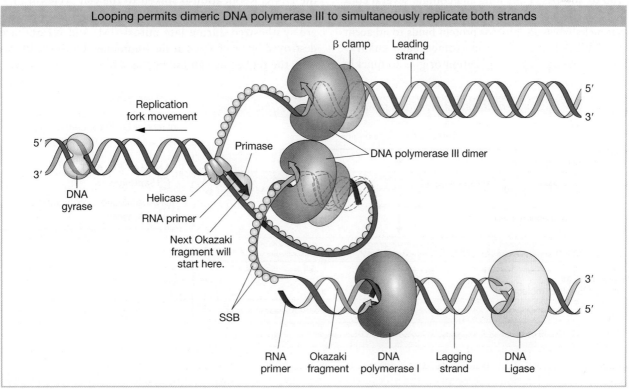

Looping permits dimeric DNA polymerase III to simultaneously replicate both strands

7.5 DNA REPLICATION IN EUKARYOTES

LO 7.6 Explain why and how DNA replication differs between bacteria and eukaryotes.

DNA replication in both bacteria and eukaryotes uses a semiconservative mechanism and employs leading and lagging strand synthesis. For this reason, it should not come as a surprise that components of the bacterial and eukaryotic replisomes are very similar (see Table 7-2). However, because eukaryotic genomes are larger and have linear, not circular, chromosomes, there are additional mechanistic complexities and associated factors. Bacteria such as *E. coli* usually complete replication in about 40 minutes, but in eukaryotes, the amount of time to complete replication can vary from a few minutes to many hours depending on many features, including the size of the genome, the number of origins, and the type of cell. Eukaryotes also have to solve the problem of coordinating the replication of more than one chromosome.

Eukaryotic origins of replication

To understand eukaryotic replication origins, we will first turn our attention to the simple eukaryote yeast (*Saccharomyces cerevisiae*). Many eukaryotic proteins having roles at replication origins were first identified in yeast because of the ease of genetic analysis (see the yeast Model Organism box in Chapter 12). Origins of replication in yeast are referred to as autonomously replicating sequences (ARSs) and are very much like *oriC* in *E. coli*. ARSs are about 100 to 200 base pairs long and contain several conserved DNA sequence elements, including an AT-rich element that melts when an initiator protein binds to adjacent elements. Unlike bacterial chromosomes, each eukaryotic chromosome has many replication origins to quickly

replicate the much larger eukaryotic genomes. Approximately 400 replication origins are dispersed throughout the 16 chromosomes of yeast, and humans have 40,000 to 80,000 origins among the 23 chromosomes. Thus, in eukaryotes, replication proceeds in both directions from multiple points of origin (**Figure 7-20**). Double helices that are produced at each origin elongate and eventually join one another. When replication of the two strands is complete, two identical daughter molecules of DNA result.

KEY CONCEPT Replication proceeds in both directions from hundreds or thousands of origins on linear eukaryotic chromosomes.

DNA replication and the yeast cell cycle

DNA synthesis takes place only in S (synthesis) phase of the eukaryotic **cell cycle** (**Figure 7-21**). How is the onset of DNA synthesis limited to this single phase? In yeast, the method of control is to link replisome assembly to the cell cycle. **Figure 7-22** shows the process. In yeast, three proteins are required to begin assembly of the replisome. The **origin recognition complex (ORC)** first binds to sequences in yeast origins, much as DnaA protein does in *E. coli*. ORC then acts as a landing pad to recruit Cdc6 to origins early in gap 1 (G1) phase of the cell cycle. Together, ORC and Cdc6 then load a complex of Cdt1 and helicase. A second helicase-Cdt1 complex is recruited through association with the already assembled helicase-Cdt1 complex. Once the helicases are on the DNA in early S phase, Cdc6 and Cdt1 are released and DNA polymerases are loaded onto the DNA. Replication is linked to the cell cycle through the availability of Cdc6 and Cdt1. In yeast, these proteins are synthesized during late mitosis (M) and G1 and are destroyed by proteolysis at the beginning of S phase. In this way, the replisome can only be assembled before S phase.

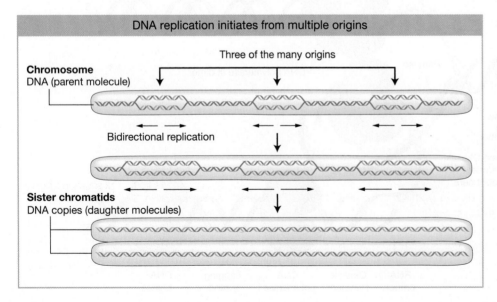

FIGURE 7-20 DNA replication proceeds in both directions from an origin of replication. Three origins of replication are shown in this example.

ANIMATED ART
SaplingPlus

DNA replication: replication of a chromosome

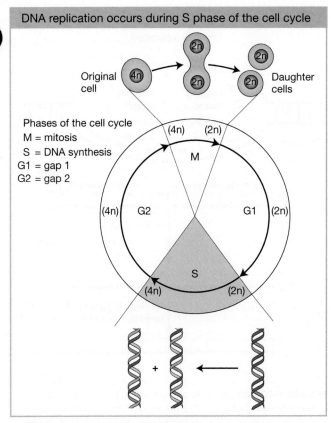

DNA replication occurs during S phase of the cell cycle

Phases of the cell cycle
M = mitosis
S = DNA synthesis
G1 = gap 1
G2 = gap 2

FIGURE 7-21 The cell cycle is composed of G1, S, G2, and M phases. DNA replication occurs during S phase, and chromosome segregation and cell division occur during M phase. Numbers in parentheses indicate the ploidy (that is, the number of sets of chromosomes) at each phase of the cell cycle in a diploid organism.

Once replication has begun, new replisomes cannot form at origins, because Cdc6 and Cdt1 are no longer available.

KEY CONCEPT DNA replication in eukaryotes requires Cdc6 and Cdt1, proteins that are only available during late mitosis (M) and G1 phase, ensuring that the genome is only replicated once per cell cycle.

Replication origins in higher eukaryotes

As already stated, most of the approximately 400 origins of replication in yeast are composed of similar DNA sequence motifs (100–200 base pairs in length) that are recognized by ORC subunits. Although all characterized eukaryotes have similar ORC proteins, origins of replication in higher eukaryotes such as humans are much longer, possibly as long as tens of thousands or hundreds of thousands of base pairs. Significantly, they have limited sequence similarity. Thus, although yeast ORC recognizes specific DNA sequences in yeast chromosomes, what the related ORCs of higher eukaryotes recognize is not clear at this time, but the feature recognized is probably not a specific DNA sequence. In practical terms, this uncertainty means

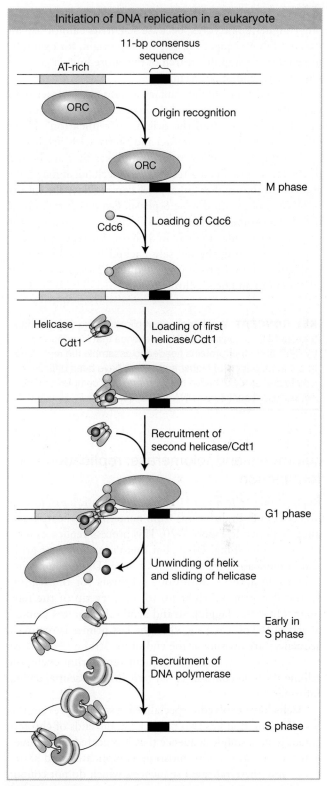

Initiation of DNA replication in a eukaryote

FIGURE 7-22 This eukaryotic example from yeast shows initiation of DNA synthesis at an origin of replication. Similar to initiation in bacteria (see Figure 7-14), sequences at the origin of replication are bound by a protein complex, in this case the origin recognition complex (ORC), which recruits helicases to separate the two strands of the double helix. The cell cycle-regulated factors Cdc6 and Cdt1 are essential for recruitment of the helicases, thereby linking DNA replication to the cell cycle.

that it is much harder to isolate origins from humans and other higher eukaryotes, because scientists cannot use an isolated DNA sequence of one human origin, for example, to perform a computer search of the entire human genome sequence to find other origins.

If the ORCs of higher eukaryotes do not interact with a specific sequence scattered throughout the chromosomes, then how do they find the origins of replication? These ORCs are thought to interact indirectly with origins by associating with other protein complexes that are bound to chromosomes. Such a recognition mechanism may have evolved so that higher eukaryotes can regulate the timing of DNA replication during S phase. Gene-rich regions of the chromosome (the euchromatin) have been known for some time to replicate early in S phase, whereas gene-poor regions, including the densely packed heterochromatin, replicate late in S phase (see Chapter 12 for more about euchromatin and heterochromatin).

KEY CONCEPT Yeast origins of replication, like origin in bacteria, contain a conserved DNA sequence that is recognized by ORC and other proteins needed to assemble the replisome. In contrast, origins of higher eukaryotes have been difficult to isolate and study because they are long and complex and do not contain a conserved DNA sequence.

Telomeres and telomerase: replication termination

Replication of the linear DNA in a eukaryotic chromosome proceeds in both directions from numerous replication origins, as shown in Figure 7-20. This process replicates most of the chromosomal DNA, but there is an inherent problem in replicating the two ends of linear DNA, the regions called **telomeres**. Continuous synthesis for the leading strand can proceed right up to the very tip of the template. However, lagging strand synthesis requires primers ahead of the process; so, when the last primer is removed, sequences are missing at the end of the strand (**Figure 7-23**, terminal gap). At each subsequent replication cycle, the telomere would continue to shorten, losing essential coding information.

Cells have evolved a specialized system to prevent this loss. The solution has two parts. First, the ends of chromosomes have a simple sequence that is repeated many times. Thus, every time a chromosome is replicated and shortened, only these repeated sequences, which do not contain protein-coding information, are lost. Second, an enzyme called **telomerase** adds these repeated sequences back to the chromosome ends.

KEY CONCEPT Telomeres stabilize chromosomes by preventing loss of genomic information after each round of DNA replication.

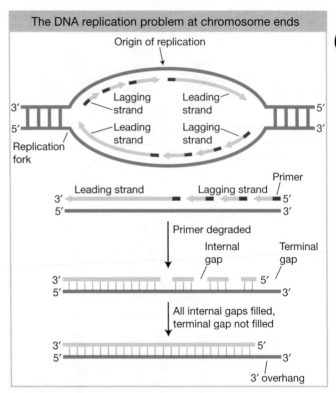

The DNA replication problem at chromosome ends

FIGURE 7-23 *Top:* The replication of each Okazaki fragment for the lagging strand begins with a primer. *Bottom:* The fate of the bottom strand in the transcription bubble. When the primer for the last Okazaki fragment of the lagging strand is removed, there is no way to fill the gap by conventional replication. A shortened chromosome would result when the chromosome containing the gap is replicated.

The discovery that the ends of chromosomes are made up of sequences repeated in tandem was made in 1978 by Elizabeth Blackburn and Joe Gall, who were studying DNA in the unusual macronucleus of the single-celled ciliate *Tetrahymena*. Like other ciliates, *Tetrahymena* has a conventional micronucleus and an unusual macronucleus in which the chromosomes are fragmented into thousands of gene-size pieces with new ends added to each piece. With so many chromosome ends, *Tetrahymena* has about 40,000 telomeres and, as such, was the perfect choice to determine telomere composition. Blackburn and Gall were able to isolate fragments containing the genes for ribosomal RNA (fragments called rDNA; see Chapter 9 for more on ribosomes) by using CsCl gradient centrifugation, the technique developed by Meselson and Stahl to study newly replicated *E. coli* DNA. The ends of rDNA fragments contained tandem arrays of the sequence TTGGGG (that is, TTGGGGTTGGGGTTGGGG . . .). We now know that virtually all eukaryotes have short tandem repeats at their chromosome ends; however, the sequence is not exactly the same. Human chromosomes, for example, end in about 10 to 15 kb of tandem repeats of the sequence TTAGGG.

The question of how these repeats are added to chromosome ends after each round of replication was addressed by

FIGURE 7-24 (a) Telomerase carries a small RNA (red letters) that acts as a template for addition of a complementary DNA sequence, which is added to the 3' overhang (blue letters). To add another repeat, telomerase translocates to the end of the repeat that it just added. (b) The extended 3' overhang then serves as template for conventional DNA replication.

Elizabeth Blackburn and Carol Grieder. They hypothesized that an enzyme catalyzed the process. Working again with extracts from the *Tetrahymena* macronucleus, they identified the telomerase enzyme, which adds the short repeats to the 3' ends of DNA. Telomerase is an RNA-protein complex, also called a ribonucleoprotein (RNP) complex. The protein component of the telomerase complex is a special type of DNA polymerase known as **reverse transcriptase** that uses RNA as a template to synthesize DNA. The RNA component of the telomerase complex varies in length from 159 nucleotides in *Tetrahymena* to 450 nucleotides in humans and about 1300 nucleotides in the yeast *Saccharomyces cerevisiae*. In all vertebrates, including humans, a region in the telomerase RNA contains the sequence 3'-AAUCCC-5' that serves as the template for synthesis of the 5'-TTAGGG-3' repeat unit by the mechanism shown in **Figure 7-24**. Briefly, the telomerase RNA first anneals to the 3'-end DNA overhang, which is then extended with the use of the telomerase's two components: the RNA and the reverse transcriptase protein. After addition of a repeat to the 3' end, the telomerase RNA moves along the DNA so that the 3' end can be further extended by its polymerase activity. The 3' end continues to be extended by repeated movement of the telomerase RNA. Primase and DNA polymerases then use the very long 3' overhang as a template to fill in the end of the other DNA strand. Working with Blackburn, a third researcher, Jack Szostak, went on to show that telomeres also exist in the less unusual eukaryote yeast. For contributing to the discovery of how telomerase protects chromosomes from shortening, Blackburn, Grieder, and Szostak were awarded the 2009 Nobel Prize in Medicine or Physiology.

KEY CONCEPT Telomeres are specialized structures at the ends of linear chromosomes that contain tandem repeats of a short DNA sequence that is added to the 3' end by the enzyme telomerase.

In addition to preventing the erosion of genetic material after each round of replication, telomeres preserve chromosomal integrity by associating with proteins such as WRN, TRF1, and TRF2, to form a protective structure called a **telomeric loop** (**t-loop**) (**Figure 7-25**). These structures sequester the 3' single-stranded overhang, which can be as much as 100 nucleotides long. Without t-loops, the ends of chromosomes would be mistaken for double-strand breaks by the cell and dealt with accordingly. As you will see in Chapter 15, double-strand breaks are potentially very dangerous because they can result in chromosomal instability that can lead to cancer and a variety of

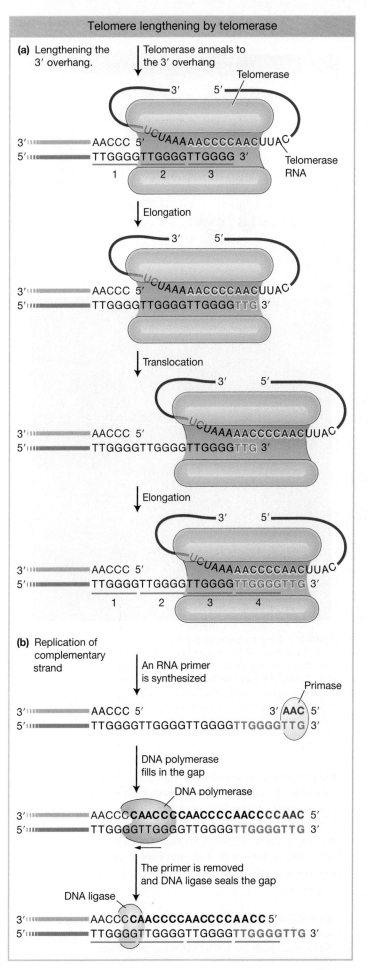

Telomere lengthening by telomerase

(a) Lengthening the 3' overhang.

Telomerase anneals to the 3' overhang

Telomerase

Telomerase RNA

Elongation

Translocation

Elongation

(b) Replication of complementary strand

An RNA primer is synthesized

Primase

DNA polymerase fills in the gap

DNA polymerase

The primer is removed and DNA ligase seals the gap

DNA ligase

The telomeric loop structure

(a)

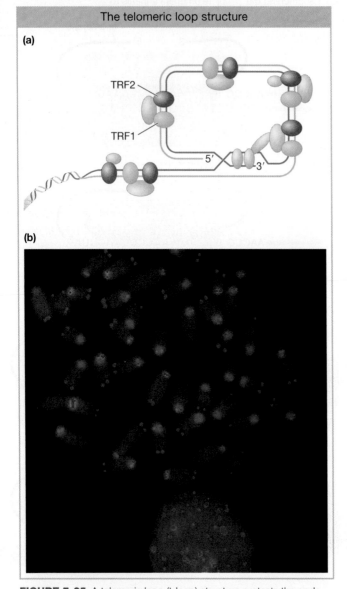

(b)

FIGURE 7-25 A telomeric loop (t-loop) structure protects the ends of chromosomes. (a) The 3′ overhang is "hidden" when it displaces a DNA strand in a region where telomeric repeats are double-stranded. The proteins TRF1 and TRF2 bind to telomeric repeats, while other proteins, including WRN, bind to TRF1 and TRF2, thus forming the protective structure. (b) Visualized by immumofluorescence microscopy in pink are telomeres at the ends of chromosomes, shown in blue. [SCIENCE SOURCE/Science Source.]

Werner syndrome causes premature aging

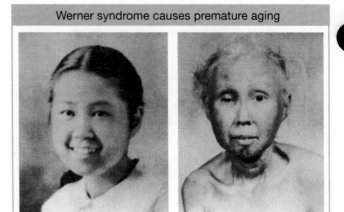

FIGURE 7-26 A woman with Werner syndrome at ages 15 (*left*) and 48 (*right*). [*International Registry of Werner Syndrome, www.wernersyndrome.org.*]

phenotypes associated with aging. For this reason, when a double-strand break is detected, cells respond in a variety of ways, depending, in part, on the cell type and the extent of the damage. For example, a double-strand break can be fused to another break, or the cell can limit damage to the organism by stopping further cell division (called senescence) or by initiating a cell-death pathway (called apoptosis).

KEY CONCEPT Telomeres stabilize chromosomes by associating with proteins to form a structure that "hides" chromosome ends from the cell's DNA repair machinery.

Surprisingly, although most germ cells have ample telomerase, somatic cells produce very little or no telomerase. For this reason, chromosomes of proliferating somatic cells get progressively shorter with each cell division until the cell stops all divisions and enters a senescence phase. This observation led many investigators to suspect that there was a link between telomere shortening and aging. Geneticists studying human diseases that lead to a premature-aging phenotype have uncovered evidence that supports such a connection. People with Werner syndrome experience early onset of many age-related events, including wrinkling of skin, cataracts, osteoporosis, graying of hair, and cardiovascular disease (**Figure 7-26**). Genetic and biochemical studies have found that afflicted people have shorter telomeres than those of normal people owing to a mutation in a gene called WRN (conferring Werner syndrome), which encodes a helicase that associates with proteins that comprise the telomeric loop (TRF2, Figure 7-25). This mutation is hypothesized to disrupt the normal telomere, resulting in chromosomal instability and premature aging. Patients with another premature-aging syndrome called dyskeratosis congenita also have shorter telomeres than those of healthy people of the same age, and they harbor mutations in genes required for telomerase activity.

Geneticists are also very interested in connections between telomeres and cancer. Unlike normal somatic cells, about 80% of cancer cells have telomerase activity. The ability to maintain functional telomeres may be one reason why cancer cells, but not normal cells, can grow in cell culture for decades and are considered to be immortal. As such, many pharmaceutical companies are seeking to capitalize on this difference between cancerous and normal cells by developing drugs that selectively target cancer cells by inhibiting telomerase activity.

KEY CONCEPT Telomeres and telomerase are associated with aging and cancer.

SUMMARY

Experimental work on the molecular nature of hereditary material has demonstrated conclusively that DNA (not protein, lipids, or carbohydrates) is the genetic material. Using data obtained by others, Watson and Crick deduced a double-helical model with two DNA strands, wound around each other, running in antiparallel fashion. Binding of the two strands together is based on the fit of adenine (A) to thymine (T) and guanine (G) to cytosine (C). The former pair is held by two hydrogen bonds; the latter, by three.

The Watson–Crick model shows how DNA can be replicated in an orderly fashion—a prime requirement for genetic material. Replication is accomplished semiconservatively in both bacteria and eukaryotes. One double helix is replicated to form two identical helices, each with their nucleotides in the identical linear order; each of the two new double helices is composed of one old and one newly polymerized strand of DNA.

The DNA double helix is unwound at a replication fork, and the two single strands serve as templates for polymerization of free nucleotides. Nucleotides are polymerized by the enzyme DNA polymerase, which adds new nucleotides only to the 3′ end of a growing DNA chain. Because addition is only at 3′ ends, polymerization on one template is continuous, producing the leading strand; and on the other, it is discontinuous in short stretches (Okazaki fragments), producing the lagging strand. Synthesis of the leading strand and of every Okazaki fragment is primed by a short RNA primer (synthesized by primase) that provides a 3′ end for deoxyribonucleotide addition.

The multiple events that have to occur accurately and rapidly at the replication fork are carried out by a biological machine called the replisome. This protein complex includes two DNA polymerase units, one to produce the leading strand and the other to produce the lagging strand. In this way, the more complex synthesis and joining of Okazaki fragments into a continuous strand can be temporally coordinated with the less complex synthesis of the leading strand. Where and when replication takes place is carefully controlled by the ordered assembly of the replisome at certain sites on chromosomes called origins. Eukaryotic genomes can have tens of thousands of origins. Assembly of replisomes at origins takes place only at a specific time in the cell cycle.

The ends of linear chromosomes (telomeres) in eukaryotes present a problem for the replication system because there is always a short stretch on one strand that cannot be primed. The enzyme telomerase adds numerous short, repetitive sequences to maintain the length of telomeres. Telomerase carries a small RNA that acts as the template for synthesis of telomeric repeats. These noncoding telomeric repeats associate with proteins to form a telomeric loop that protects against DNA damage. Telomeres shorten with age because telomerase is not produced in somatic cells. Individuals who have defective telomeres experience premature aging.

KEY TERMS

adenine (p. 244)
antiparallel (p. 247)
bases (p. 244)
β-clamp (sliding clamp) (p. 256)
cell cycle (p. 258)
complementary bases (p. 247)
conservative replication (p. 249)
cytosine (p. 244)
daughter molecule (p. 249)
deoxynucleotide (p. 244)
deoxyribose (p. 244)
dispersive replication (p. 249)
distributive enzyme (p. 257)
DNA gyrase (p. 252)
DNA ligase (p. 255)
DNA polymerase I (DNA pol I) (p. 253)
DNA polymerase III (DNA pol III) (p. 253)

DNA pol III holoenzyme (p. 256)
DNA replication (p. 240)
double helix (p. 240)
genetic code (p. 249)
guanine (p. 244)
helicase (p. 252)
lagging strand (p. 254)
leading strand (p. 254)
major groove (p. 248)
minor groove (p. 248)
nucleoside (p. 250)
nucleotide (p. 240)
Okazaki fragment (p. 254)
origin of replication (origin) (p. 252)
Origin recognition complex (ORC) (p. 258)
parental molecule (p. 249)
phosphate (p. 244)
primase (p. 254)

primer (p. 254)
primosome (p. 254)
processive enzyme (p. 257)
purine (p. 244)
pyrimidine (p. 244)
radioisotope (p. 243)
replication fork (p. 250)
replisome (p. 240)
reverse transcriptase (p. 261)
ribose (p. 244)
semiconservative replication (p. 249)
semidiscontinuous (p. 254)
single-strand DNA-binding (SSB) protein (p. 252)
telomerase (p. 260)
telomere (p. 260)
telomeric loop (t-loop) (p. 261)
template (p. 240)
thymine (p. 244)
topoisomerase (p. 252)

PROBLEMS

Visit SaplingPlus for supplemental content. Problems with the icon are available for review/grading. Problems with the icon have an Unpacking the Problem exercise.

WORKING WITH THE FIGURES

(The first 27 questions require inspection of text figures.)

1. In Table 7-1, complete the table for a genome that is 20 percent adenine.

2. In Figure 7-1, what features of the sculpture of DNA are correct or incorrect?

3. In Figure 7-2, speculate as to why Griffith did not conduct the experiment the other way around, that is, with heat-killed R cells and live S cells.

4. In Figure 7-3, what types of enzymes could Avery, Macleod, and McCarty have used to destroy proteins and RNAs?

5. In Figure 7-4, what part of the DNA structure is labeled by ^{32}P?

6. In Figure 7-5, draw 7-methylguanine and 5-methylcytosine. A methyl group is CH_3.

7. In Figure 7-6, what information did Rosalind Franklin's X-ray diffraction data provide that was key to determining the structure of DNA?

8. In Figure 7-7, why do you think that Watson and Crick built a three-dimensional model of DNA rather than only a two-dimensional model?

9. In Figure 7-8, would the diameter of DNA change if the pyrimidine was on the left and the purine on the right? Justify your answer.

10. In Figure 7-9a, why are there two rows of dots between A–T base pairs but three rows between G–C base pairs?

11. In Figure 7-10a, is a purine or a pyrimidine on the left in the bottom base pair?

12. In Figure 7-11, draw cesium chloride gradients for a Meselson–Stahl experiment in which cells are first grown in ^{14}N and then in ^{15}N for two generations.

13. In Figure 7-12, draw an autoradiograph for a chromosome during the second round of replication in which the DNA that crosses the circle has one blue parental strand.

14. In Figure 7-13a, what would happen in the rope demonstration if you cut one of the two strands in the supercoiled region.

15. In Figure 7-14, in the second to last diagram, why do the arrows show the two helicase molecules moving in opposite directions?

16. In Figure 7-15, draw the chemical reaction that occurs to add the next nucleotide in the DNA chain.

17. In Figure 7-16, draw an analogous diagram for the other replication fork.

18. In Figure 7-17, which factors are involved in lagging strand synthesis but not leading strand synthesis?

19. In Figure 7-18, draw the phosphodiester linkage between T and the misincorporated A in the strand being synthesized, and place an arrow at the bond that is broken by the 3'-to-5' exonuclease activity of DNA polymerase.

20. In Figure 7-19, why is the DNA looped for one strand but not for the other strand when they are both serving as templates for DNA synthesis by the DNA polymerase III dimer?

21. In Figure 7-20, could the spacing of origins affect the amount of time that it takes to replicate a chromosome?

22. In Figure 7-21, how much DNA would a cell contain if it went through two cell cycles that did not include an M phase?

23. In Figure 7-22, why does replication not initiate at origins in G2 phase?

24. In Figure 7-23, analogous to the last diagram in the figure, draw the top DNA strand in the bubble after the primers are degraded and the gaps are filled. Based on this drawing, is telomerase required for the replication of both ends of chromosomes?

25. In Figure 7-24a, the telomerase RNA template contains one and a half copies of the repeat sequence. Circle the full copy and put a box around the half copy.

26. In Figure 7-25b, fluorescent antibodies to what protein may have been used to detect the telomeres?

27. In Figure 7-26, if Figure 7-25b represents chromosomes from this individual at a young age, how might the image in Figure 7-25b differ at the older age?

BASIC PROBLEMS

28. Does the Hershey-Chase experiment definitively demonstrate that DNA is the genetic material, or just that it is consistent with being the genetic material? Justify your answer.

29. Does the Avery, MacLeod, and McCarty experiment definitively demonstrate that DNA is the genetic material or just that it is consistent with being the genetic material? Justify your answer.

30. Draw a cesium chloride gradient for the third generation of DNA produced by semiconservative replication in the Meselson–Stahl experiment.

31. Write the sequence of the telomerase RNA that serves as a template for the telomere repeat sequence 5′-TTAGGG-3′.

32. Why might Werner syndrome increase the chances of getting cancer?

33. Draw 2′,3′ dideoxyadenosine and predict what would happen if this nucleotide was incorporated into the growing DNA chain during replication.

34. Explain how DNA fulfills the three main requirements for a hereditary molecule: (1) the ability to store information, (2) the ability to be replicated, and (3) the ability to mutate.

35. Match the protein with its function.

    A. DNA polymerase   creates RNA primers
    B. Helicase   links short DNA chains
    C. Ligase   helps hold polymerase on DNA
    D. Primase   separates DNA strands
    E. Gyrase   prevents reannealing of DNA
    F. Sliding clamp   extends DNA strand
    G. SSB   removes supercoils in DNA

36. Why is telomerase not required for replication of the bacterial genome?

37. Explain what is meant by the terms *conservative* and *semiconservative replication*.

38. Describe two pieces of evidence indicating that DNA polymerase I is not the chromosomal replicase.

39. What is meant by a *primer*, and why are primers necessary for DNA replication?

40. A molecule of composition

    5′-AAAAAAAAAAAAAA-3′

    3′-TTTTTTTTTTTTTT-5′

    is replicated in a solution containing unlabeled (not radioactive) dGTP, dCTP, and dTTP plus dATP with all its phosphorus atoms in the form of the radioactive isotope ^{32}P. Will both daughter molecules be radioactive? Explain. Then repeat the question for the molecule

    5′-ATATATATATATAT-3′

    3′-TATATATATATATA-5′

41. Why is DNA synthesis continuous on one strand and discontinuous on the opposite strand?

42. Explain why cutting one strand of supercoiled DNA removes the supercoiling.

43. Describe how the enzymatic activities of DNA polymerases I and III are similar and different.

44. If the GC content of a DNA is 48 percent, what are the percentages of the four bases (A, T, G, and C) in this molecule?

45. Would the Meselson–Stahl experiment have worked if diploid eukaryotic cells had been used instead?

46. Consider the following segment of DNA, which is part of a much longer molecule constituting a chromosome:

    5′ . . . ATTCGTACGATCGACTGACTGACAGTC . . . 3′

    3′ . . . TAAGCATGCTAGCTGACTGACTGTCAG . . . 5′

    If the DNA polymerase starts replicating this segment from the right,

    a. which will be the template for the leading strand?

    b. draw the molecule when the DNA polymerase is halfway along this segment.

    c. draw the two complete daughter molecules.

47. The DNA polymerases are positioned over the following DNA segment (which is part of a much larger molecule) and moving from right to left. If we assume that an Okazaki fragment is made from this segment, what will be the fragment's sequence? Label its 5′ and 3′ ends.

    5′ . . . CCTTAAGACTAACTACTTACTGGGATC . . . 3′

    3′ . . . GGAATTCTGATTGATGAATGACCCTAG . . . 5′

CHALLENGING PROBLEMS

48. If you extract the DNA of the coliphage φX174, you will find that its composition is 25 percent A, 33 percent T, 24 percent G, and 18 percent C. Does this composition make sense in regard to Chargaff's rules? How would you interpret this result? How might such a phage replicate its DNA?

49. Given what you know about the structure and function of telomerase, provide a plausible model to explain how a species could exist with a combination of two different repeats (for example, TTAGGG and TTGTGG) on each of their telomeres.

50. Why is it unlikely that continuous replication of both DNA strands occurs but is yet to be discovered?

GENETICS AND SOCIETY

In this chapter, you learned that the shortening of chromosome telomeres due to diminished telomerase activity is associated with aging. This raises the possibility that gene therapy aimed at overexpression of telomerase will increase longevity. Do you think that it is ethical to use this approach to increase the longevity of normal, healthy people? Does your answer change if you consider that there are nongenetic means such as calorie restriction that may increase longevity, or that gene therapy is being pursued to treat numerous diseases?

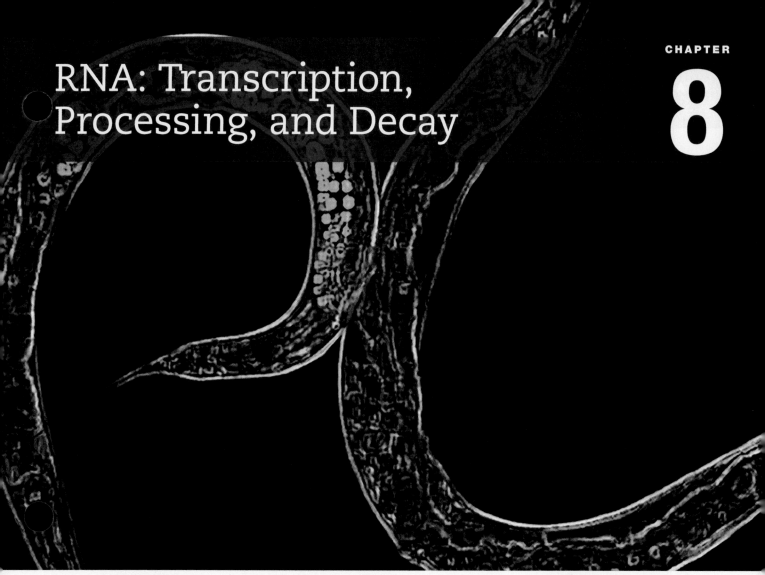

RNA: Transcription, Processing, and Decay

Knowledge of the molecular mechanisms that synthesize and destroy RNAs in cells has led to technologies that allow researchers to control gene expression in precise ways. For example, on the left is a *C. elegans* worm that has been manipulated to express a gene encoding the green fluorescent protein (GFP) in specific cells, and on the right is a genetically identical worm in which GFP expression is silenced. [*Jessica Vasale/Laboratory of Craig Mello.*]

CHAPTER OUTLINE AND LEARNING OBJECTIVES

In this chapter, we describe how the information stored in DNA is transferred to RNA. The key event in this transfer is **transcription,** which copies the information from one strand of DNA into a strand of RNA. In bacteria, the information in protein-coding RNAs is almost immediately converted into protein by a process called translation (the focus of Chapter 9). In contrast, in eukaryotes, transcription and translation are spatially separated: transcription takes place in the nucleus and translation in the cytoplasm. Furthermore, in eukaryotes, before RNAs are ready to be exported to the cytoplasm for translation, they undergo extensive processing, including deletion of internal nucleotides and addition of special nucleotide structures to the 5′ and 3′ ends.

Both bacteria and eukaryotes also produce other types of RNA that are not translated into protein but instead perform a variety of roles in cells by base pairing to other RNAs, binding proteins, and performing enzymatic reactions. Lastly, the chapter describes how RNAs are eliminated from cells by decay mechanisms that disassemble RNAs into individual nucleotides. **Figure 8-1** provides an overview of the chapter by illustrating the timeline of events that occur in the life cycle of protein-coding RNAs (mRNAs) in bacteria and eukaryotes. Every process described in this chapter relies on molecular interactions that are specified by nucleic acid sequences. RNAs interact with DNA and other RNAs by base pairing of complementary sequences, and proteins interact with DNA and RNA by binding specific sequences. Therefore, mutations in DNA and RNA that disrupt molecular interactions can affect the expression of proteins.

In summary, this chapter focuses on the molecules and mechanisms that produce and destroy RNAs. The molecules and mechanisms are important to geneticists because mutations that affect them change which proteins are expressed, their sequence, and their abundance, and lead to altered phenotypes.

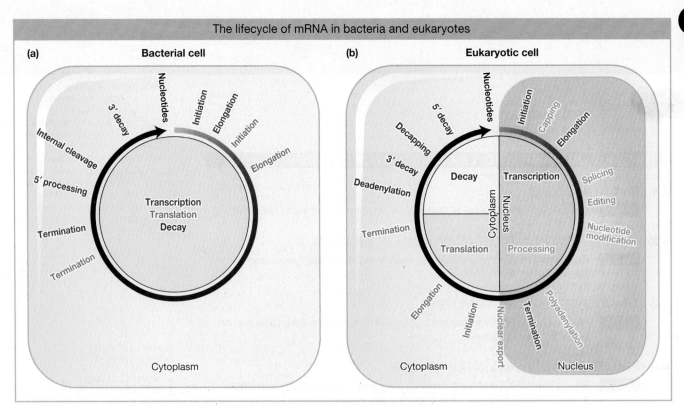

FIGURE 8-1 In both (a) bacteria and (b) eukaryotes, the transfer of information from DNA to protein involves a sequential series of molecular events dedicated to mRNA. Note the similarities and differences in the life cycles of mRNA in bacteria and eukaryotes. Transcription, translation, and decay of mRNA occur in both bacteria and eukaryotes. In contrast, transcription, translation, and decay occur concurrently only in bacteria, and various mRNA processing events and nuclear export are unique to eukaryotes.

8.1 RNA STRUCTURE

LO 8.1 Describe how the structure of RNA enables it to function differently from DNA.

RNAs carry out an amazing variety of biological functions, including providing the information for making proteins, regulating translation, processing RNA, and maintaining chromosome ends. The versatility of RNA relative to DNA is due to the ability of single-stranded RNA to form an immense variety of elaborate three-dimensional structures that scaffold the binding of proteins, base pair with other RNAs, and carry out enzymatic reactions. Furthermore, the versatility of RNA as a biomolecule is enhanced by the variety of ways in which RNA function can be regulated, including mechanisms that alter RNA structure, abundance, and cellular localization.

RNA is the information-carrying intermediate between DNA and proteins

Early investigators had good reason for thinking that information is not transferred directly from DNA to protein. In eukaryotic cells, DNA is located in the nucleus, which physically separates it from the protein synthesis machinery in the cytoplasm. Thus, an intermediate is needed that carries the DNA sequence information from the nucleus to the cytoplasm. That intermediate is RNA.

In 1957, Elliot Volkin and Lawrence Astrachan made an observation suggesting that RNA was the intermediate molecule. They found that one of the most striking molecular changes that takes place when *E. coli* is infected with the bacteriophage T2 is a rapid burst of RNA synthesis. Furthermore, this bacteriophage-induced RNA "turns over"; that is, the amount of time it spends in the cell is brief, on the order of minutes. Its rapid appearance and disappearance suggested that RNA might play some role in the synthesis of more T2 phage particles.

Volkin and Astrachan demonstrated the rapid turnover of RNA using a protocol called a pulse–chase experiment. To conduct a pulse–chase experiment, the infected bacteria are first fed (pulsed with) radioactive uracil, a molecule needed for the synthesis of RNA but not DNA. Any RNA synthesized in the bacteria from then on is "labeled" with the readily detectable radioactive uracil. After a short period of incubation, the radioactive uracil is washed away and replaced (chased) by uracil that is not radioactive. This procedure "chases" the label out of RNA because, as the pulse-labeled RNA breaks down, only the unlabeled uracil is available to synthesize new RNA molecules. Volkin and Astrachan found that the RNA recovered shortly after the pulse was labeled, but RNA recovered just a few minutes later was unlabeled, indicating that the RNA has a very short lifetime in bacteria.

A similar experiment can be done with eukaryotic cells. Cells are first pulsed with radioactive uracil and, after a short time, they are transferred to medium (the liquid they grow in) with unlabeled uracil. In samples taken immediately after the pulse, most of the labeled RNA is in the nucleus. However, in samples taken a few minutes later, the labeled RNA is also found in the cytoplasm (**Figure 8-2**). This indicates that, in eukaryotes, RNA is synthesized in the nucleus and then moves to the cytoplasm, where proteins are synthesized. These data along with other data led to the conclusion that RNA is the information-transfer intermediary between DNA and protein.

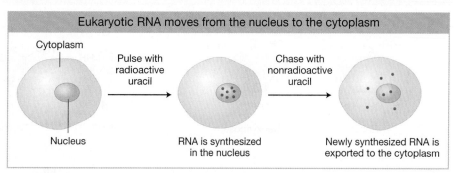

Eukaryotic RNA moves from the nucleus to the cytoplasm

Cytoplasm — Nucleus

Pulse with radioactive uracil

Chase with nonradioactive uracil

RNA is synthesized in the nucleus

Newly synthesized RNA is exported to the cytoplasm

FIGURE 8-2 The pulse–chase experiment showed that RNA moves from the nucleus to the cytoplasm in eukaryotic cells. Cells are grown briefly in medium with radioactive uracil to label newly synthesized RNA (pulse). Cells are washed to remove the radioactive uracil and then grown in medium with nonradioactive uracil (chase). The red dots indicate the location of RNAs containing radioactive uracil over time. The location of radioactive RNAs was not determined by microscopy; rather, it was inferred using a molecular approach. Cells were collected immediately after the pulse as well as after the chase and broken open, nuclei were separated from the cytoplasm by centrifugation, RNA was purified from the nuclear and cytoplasmic fractions (leaving behind radioactive uracil that was not incorporated into RNA), and the amount of radioactivity in the nuclear and cytoplasmic fractions was determined using a scintillation counter (an instrument that measures ionizing radiation).

Consequences of the distinct chemical properties of RNA

Although both RNA and DNA are nucleic acids, the building blocks of RNA differ from those of DNA in three important ways that allow RNA to have greater structural and functional diversity.

1. RNA has **ribose** sugar in its nucleotides, rather than the **deoxyribose** found in DNA (**Figure 8-3a**). As the names suggest, the sugars differ in the presence or absence of just one oxygen atom at the 2′ carbon. The 2′-OH in RNA reduces the stability of single-stranded RNA relative to single-stranded DNA. RNA cleavage can occur when a deprotonated 2′-OH acts as a nucleophile at a nearby phosphodiester bond. Furthermore, the 2′-OH provides an extra site for hydrogen bonding as well as an extra site for chemical modification such as methylation. Properties of the 2′-OH contribute to the ability of RNAs, which are usually single-stranded, to fold into complex three-dimensional structures.

2. RNA contains the pyrimidine base **uracil (U)**, instead of thymine (T) found in DNA (Figure 8-3b). Uracil forms two hydrogen bonds with adenine just like thymine does in DNA (Figure 8-3c). In addition, uracil is capable of base pairing with G in helices of a folded RNA or between two separate RNAs, but not with G in DNA during transcription. The ability of U to base pair with both A and G is a major reason why RNA can form intricate structures.

    Throughout this chapter, we revisit the chemical and structural properties of the ribose 2′-OH and the uracil base because they are critical to the folding, function, and recognition of RNAs in RNA processing events.

3. RNA is usually single-stranded, not double-stranded like DNA. As a consequence, RNA is much more flexible than DNA and can form a greater variety of three-dimensional structures. Base pairing between regions within an RNA (i.e., intramolecular base pairing) is an important determinant of RNA structure. For example, the stem-loop is the fundamental structural element of RNA (Figure 8-3d). Stem-loops are made up of a double-stranded stem of complementary regions of an RNA and a single-stranded loop at the end of the stem.

> **KEY CONCEPT** Unlike DNA, RNA contains 2′-hydroxyls on the ribose sugars, uracil replaces thymine, and it is single-stranded but base pairs to itself to form double-stranded regions.

As exemplified by the terms ribose sugar and uracil base, there is specific terminology for describing the building blocks of RNA. There are terms for each of the four RNA nucleobases (i.e., the base itself), the nucleosides (i.e., the base with a ribose sugar), and the nucleotides (i.e., the base with a ribose sugar and one, two, or three phosphates) (**Figure 8-4**). For example, uracil is a nucleobase, uridine is a nucleoside, and uridine triphosphate is a nucleotide.

Classes of RNA

RNA molecules can be grouped into two general classes. One class of RNA is **messenger RNA (mRNA)** because, like a messenger, it serves as an intermediary that carries information. The information from DNA is transferred to mRNA through the process of transcription, and mRNA passes that information on to proteins through the process of translation. The other class of RNA is **noncoding RNA (ncRNA)** because it does not encode proteins. Instead, the ncRNA is the final product whose function is determined by its sequence and three-dimensional structure.

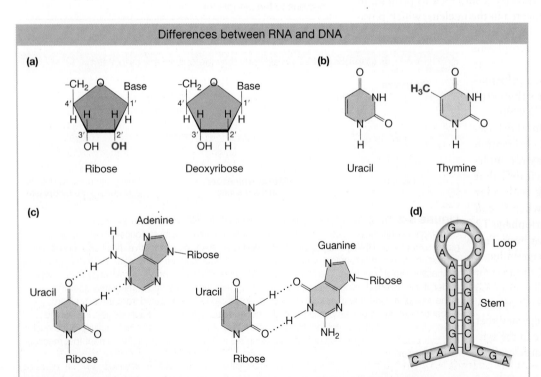

Differences between RNA and DNA

(a)   Ribose   Deoxyribose

(b)   Uracil   Thymine

(c)   Adenine   Guanine   Uracil   Uracil

(d)   Loop   Stem

FIGURE 8-3 (a) The 5-carbon sugar in RNA nucleosides (*left*) versus DNA nucleosides (*right*). Ribose carries a hydroxyl group, indicated in red, at the 2′ carbon instead of a hydrogen atom in deoxyribose. (b) The pyrimidine base uracil (*left*) replaces thymine (*right*) in RNA versus DNA, respectively. Uracil differs from thymine by a methyl group, indicated in red. (c) Uracil base pairs via two hydrogen bonds with adenine (*left*) and guanine (*right*). (d) Stem-loops are basic structural features of RNA. Note that U base pairs with both A and G.

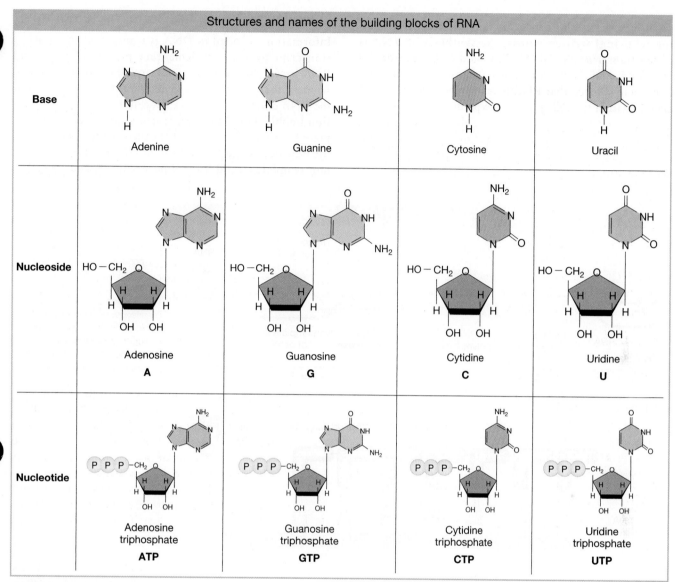

Base	Adenine	Guanine	Cytosine	Uracil
Nucleoside	Adenosine **A**	Guanosine **G**	Cytidine **C**	Uridine **U**
Nucleotide	Adenosine triphosphate **ATP**	Guanosine triphosphate **GTP**	Cytidine triphosphate **CTP**	Uridine triphosphate **UTP**

FIGURE 8-4 Each of the four building blocks of RNA has a distinct name for its nucleobase, nucleoside, and nucleotide.

ncRNAs vary in length and function, although they often act similarly by base pairing with another RNA, by serving as a scaffold for binding proteins, or by catalyzing a chemical reaction. In both bacteria and eukaryotes, some ncRNA such as **transfer RNAs (tRNAs)** and **ribosomal RNAs (rRNAs)** function in translation. Bacteria and eukaryotes also express a variety of ncRNAs that are about 50–300 nucleotides in length and function at multiple steps in gene expression. In eukaryotes, ncRNAs are categorized based on their location in the cell: **small *nuclear* RNAs (snRNAs),** small *nucleolar* RNAs (snoRNAs), and small *cytoplasmic* RNAs (scRNAs). The nucleolus is a non-membrane-bound region in the nucleus where ribosomes are produced. Eukaryotes also express **long noncoding RNAs (lncRNAs)** that are typically greater than 300 nucleotides in length. Thousands of lncRNAs have been identified in humans, but only a few have been assigned functions, and these are mostly regulators of gene expression. Some eukaryotes also encode **microRNAs (miRNAs)** and generate **small interfering RNAs (siRNAs)** and other very small

RNAs, about 21 nucleotides in length, that suppress the expression of genes and help maintain genome stability.

KEY CONCEPT There are two general classes of RNAs, those that encode proteins (mRNA) and those that do not encode proteins (ncRNA). ncRNAs participate in a variety of cellular processes, including protein synthesis (tRNA and rRNA), RNA processing (snRNA), the regulation of gene expression (siRNA and miRNA), and genome defense (siRNA).

8.2 TRANSCRIPTION AND DECAY OF mRNA IN BACTERIA

LO 8.2 Explain how RNA polymerases are directed to begin and end transcription at specific places in genomes.

The first step in the transfer of information from DNA to protein is to produce an RNA strand whose nucleotide sequence matches the nucleotide sequence of a DNA segment. Because this process is reminiscent of transcribing (copying) written words, the synthesis of RNA is called *transcription*. The DNA is said to be transcribed into RNA, and the RNA is called a **transcript**. Volkin and Astrachan showed that RNA is transcribed and degraded rapidly within the cell, and later experiments showed that the abundance of a given RNA is regulated by controlling its rate of transcription and its rate of decay. These processes, though chemically simple, are controlled by a variety of factors.

Overview: DNA as transcription template

Information encoded in DNA is transferred to the RNA transcript by the complementary pairing of DNA and RNA bases. Consider the transcription of a chromosomal segment that constitutes a gene. First, the two strands of the DNA double helix separate locally to form a **transcription bubble**. One of the separated strands acts as a template for RNA synthesis and is called the **template strand** (or noncoding strand) and the other strand is called the **non-template strand** (or coding strand) (**Figure 8-5a**). The

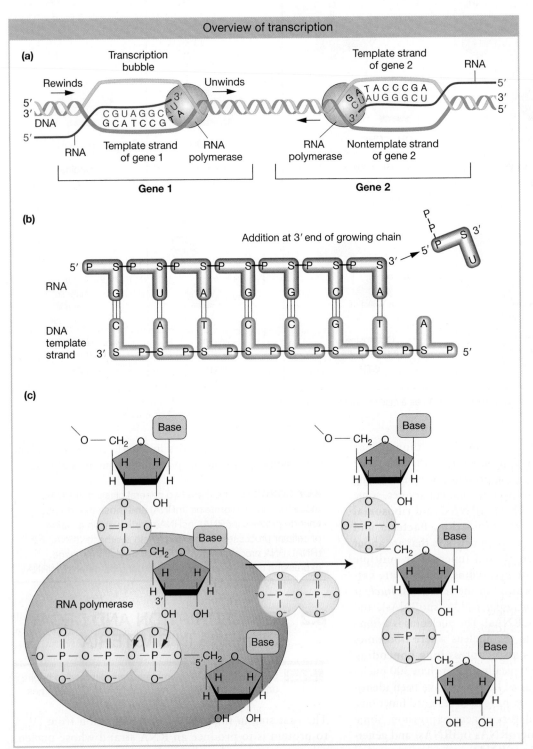

FIGURE 8-5 (a) Only one strand of DNA is the template for gene transcription, but the strand can vary with the gene. RNA is transcribed in the 5'-to-3' direction using DNA oriented in the 3'-to-5' direction as a template. Hence, genes transcribed in different directions use opposite strands of DNA as templates. (b) As a gene is transcribed, the 3'-hydroxyl group on the sugar (S) at the end of the growing RNA strand attaches to the 5'-phosphate group (P) on the entering ribonucleotide (A, C, G, or U) that base pairs with the DNA template nucleotide. (c) To form a phosphodiester bond, the 3' hydroxyl is deprotonated and acts as a nucleophile at the α-phosphate of the entering nucleotide, breaking the bond between the α- and β-phosphates and producing the energy needed to form the new phosphodiester bond.

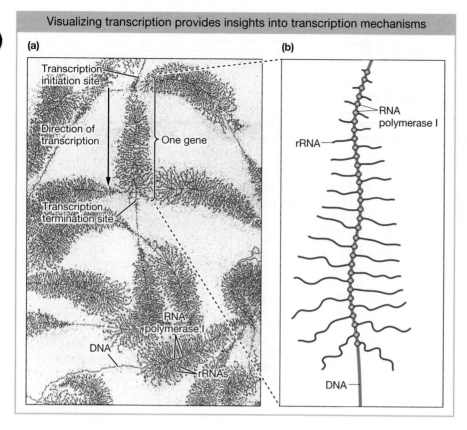

Visualizing transcription provides insights into transcription mechanisms

(a)

Transcription
initiation site

Direction of
transcription

One gene

Transcription
termination site

RNA
polymerase I

DNA

rRNAs

(b)

RNA
polymerase I

rRNA

DNA

FIGURE 8-6 This picture from an electron microscope shows the transcription of tandemly repeated rRNA genes in the nuclear genome of a newt oocyte (an immature egg). Along each gene, many RNA polymerases (in this case, eukaryotic RNA polymerase I) are transcribing in one direction. Growing rRNA transcripts appear as threads extending outward from the DNA. Transcription starts (initiates) and ends (terminates) at specific places. Shorter transcripts are closer to the start of the gene; longer ones are closer to the end of the gene. Based on their appearance, these structures are called "Christmas trees." They are also called "Miller spreads" after their discoverer Oscar Miller [*W Fawcett Don/Getty Images.*]

resulting RNA sequence is complementary to the template strand and identical (except for the use of uracil in place of thymine) to the non-template strand. When DNA sequence is cited in the scientific literature, the sequence of the non-template strand is almost always given because it is the same as the RNA sequence.

KEY CONCEPT RNA sequence is complementary to the template strand and is the same as the coding (non-template) strand, except it contains U in place of T.

Across the genome, both DNA strands may be used as templates, but in any one gene, only one strand is used (Figure 8-5a). Starting at the 3′ end of the template strand, ribonucleotides form base pairs by hydrogen bonding with their complementary DNA nucleotides. The ribonucleotide A pairs with T in the DNA, C with G, G with C, and U with A. Each ribonucleotide is positioned opposite its complementary nucleotide by the enzyme **RNA polymerase**. This enzyme moves along the DNA template strand in the 3′-to-5′ direction forming **phosphodiester bonds** that covalently link aligned ribonucleotides to build RNA in the 5′-to-3′ direction, as shown in Figures 8-5b and c. As the RNA strand is progressively lengthened, the 5′ end is displaced from the DNA template and the transcription bubble closes behind RNA polymerase. Multiple RNA polymerases, each synthesizing an RNA molecule, can move along a gene at the same time (**Figure 8-6**). Hence, we already see two fundamental mechanisms that bring about the transfer of information

from DNA to RNA: base complementarity and protein-nucleic acid interactions.

KEY CONCEPT RNA is transcribed in the 5′-to-3′ direction using a single-stranded DNA template oriented in the 3′-to-5′ direction. Thus, RNAs start with a 5′-triphosphate (5′-ppp) and end with a 3′ hydroxyl (3′-OH).

KEY CONCEPT One gene can be transcribed by multiple RNA polymerase molecules at the same time.

Stages of transcription

Genes are segments of DNA embedded in extremely long DNA molecules (chromosomes). How, then, is a gene accurately transcribed into RNA with a specific beginning and end? Because the DNA of a chromosome is a continuous unit, the transcriptional machinery must be directed to the start of a gene to begin transcription, continue transcribing the length of the gene, and finally stop transcribing at the end of the gene. These three distinct stages of transcription are called **initiation, elongation,** and **termination,** respectively. Although the overall process of transcription is remarkably similar in bacteria and eukaryotes, there are important differences. For this reason, we will follow the three stages first in bacteria by using the gut bacterium *E. coli* as an example, and then we will repeat the process in eukaryotes, emphasizing the similarities and differences between bacteria and eukaryotes.

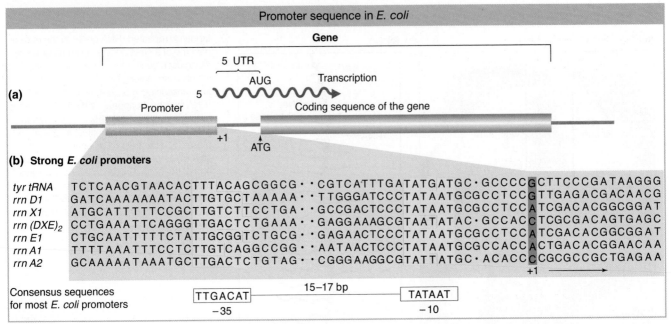

FIGURE 8-7 (a) The promoter lies "upstream" (i.e., 5′) of the transcription start site (+1) and protein-coding sequences. (b) Promoters have regions of similar sequences, as indicated by the yellow shading in seven different promoter sequences in *E. coli*. Spaces (dots) are inserted in the sequence to optimize the alignment of common sequences. Numbers refer to the number of bases before (−) or after (+) the transcription start site.

ANIMATED ART Sapling Plus
Transcription in bacteria

Transcription initiation in bacteria How does RNA polymerase find the correct starting point for transcription? In bacteria, RNA polymerase usually binds to a specific DNA sequence called a **promoter**, located close to the start of the transcribed region. Promoters are an important part of the transcriptional regulatory region of a gene (**Figure 8-7**). The first transcribed base is called the *initiation site* or the **transcription start site**. The promoter is referred to as **upstream** of the initiation site because it is located ahead of the initiation site (5′ of the gene). A **downstream** site is located later in the direction of transcription. Nucleotide positions upstream of the initiation site are indicated by a negative (−) sign and those downstream by a positive (+) sign. By convention, the first DNA base to be transcribed is numbered +1.

KEY CONCEPT Nucleotides in genes are numbered relative to the transcription start site; those before the start site have negative numbers and are called upstream and those after the start site have positive numbers and are called downstream.

Figure 8-7 shows the promoter sequences of seven different genes in *E. coli*. Because the same RNA polymerase binds to the promoter sequences of these different genes, similarities among the promoters are not surprising. In particular, two regions of great similarity appear in virtually every case. These regions have been termed the −35 (minus 35) and −10 (minus 10) regions because they are located 35 base pairs and 10 base pairs, respectively, upstream of the first transcribed base. They are shown in yellow in Figure 8-7. As you can see, the −35 and −10 regions from

different genes do not have to be identical to perform a similar function. Nonetheless, it is possible to arrive at a sequence of nucleotides, called a **consensus sequence**, that is in agreement with most sequences. The *E. coli* promoter consensus sequence is shown at the bottom of Figure 8-7. An RNA polymerase holoenzyme (see the next paragraph) binds to the DNA at this point, then unwinds the DNA double helix and begins the synthesis of an RNA molecule. Note in Figure 8-7 that the protein-coding part of the gene usually begins at an AUG sequence in the mRNA (i.e., ATG in DNA), but the transcription start site, where transcription begins, is usually well upstream of this sequence. The region between the start of transcription and the start of translation is referred to as the **5′ untranslated region (5′ UTR)**.

The bacterial RNA polymerase that scans the DNA for a promoter sequence is called the **RNA polymerase holoenzyme** (**Figure 8-8a**). This multi-protein complex is composed of a five-subunit **core enzyme** (two subunits of α, one of β, one of β′, and one of ω) plus a subunit called **sigma factor (σ)**. The two α subunits help assemble the enzyme and promote interactions with regulatory proteins, the β subunit is active in catalysis, the β′ subunit binds DNA, and the ω subunit has roles in assembly of the holoenzyme and the regulation of gene expression. The σ subunit binds to the −10 and −35 regions, thus positioning the holoenzyme to initiate transcription correctly at the start site (Figure 8-8a). The σ subunit also has a role in separating the DNA strands around the −10 region so that the core enzyme can bind tightly to the DNA in preparation for RNA synthesis. After the holoenzyme is bound, transcription initiates and the σ subunit dissociates (Figure 8-8b). The core enzyme then elongates through the gene (Figure 8-8c).

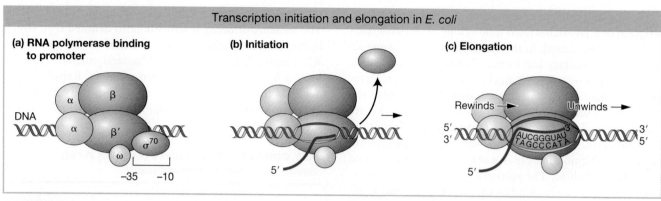

Transcription initiation and elongation in *E. coli*

(a) RNA polymerase binding to promoter

β
α
α   β′
ω   σ⁷⁰
DNA
−35   −10

(b) Initiation

5′

(c) Elongation

Rewinds →   Unwinds →
5′   3′
3′   AUCGGGUAU   5′
TAGCCCATA
5′

FIGURE 8-8 (a) Binding of the σ subunit to the −10 and −35 regions positions the RNA polymerase holoenzyme for correct initiation. (b) Shortly after RNA synthesis begins, the σ subunit dissociates from the core enzyme, which continues transcription. (c) Synthesis of an RNA strand complementary to the single-strand region of the DNA template is in the 5′-to-3′ direction. DNA that is unwound ahead of RNA polymerase is rewound after it is transcribed.

ANIMATED ART   SaplingPlus

Transcription in bacteria

E. coli, like most other bacteria, has several different σ factors. One, called σ⁷⁰ because its mass in kilodaltons (kDa) is 70, is the primary σ factor used to initiate transcription of the vast majority of *E. coli* genes. Alternative σ factors recognize promoters with different consensus sequences. Thus, by associating with different σ factors, the same core RNA polymerase enzyme can transcribe different genes to respond to stresses, changes in cell shape, and nitrogen uptake. This is discussed in greater detail in Chapter 11.

KEY CONCEPT A sequence called a promoter controls where RNA polymerase begins transcription. In bacteria, promoters are bound by RNA polymerase σ factors.

Transcription elongation in bacteria As the RNA polymerase moves along the DNA, it unwinds the DNA ahead of it and rewinds the DNA that has already been transcribed (Figure 8-8c). In this way, it maintains a region of single-stranded DNA, called a transcription bubble, within which the template strand is exposed. In the bubble, RNA polymerase monitors the binding of a free ribonucleoside triphosphate to the next exposed base on the DNA template and, if there is a complementary match, adds it to the chain (Figure 8-5b). Energy for the addition of a nucleotide is derived from breaking a phosphate bond. RNA polymerase synthesizes RNA at a rate of 50 to 90 nucleotides per second. Within this range, slower rates of synthesis may provide time for the RNA to fold properly and for translation to synchronize with transcription.

Inside the transcription bubble, the last 8 to 9 nucleotides added to the RNA chain form an RNA–DNA hybrid by complementary base pairing with the template strand. As the RNA chain lengthens at its 3′ end, the 5′ end is further extruded from the polymerase. The complementary base pairs are broken at the point of exit, leaving the extruded region of RNA single-stranded.

Transcription termination in bacteria Transcription continues beyond the protein-coding segment of a gene, creating a **3′ untranslated region (3′ UTR)** at the end of

the transcript. Elongation proceeds until RNA polymerase recognizes special nucleotide sequences that act as a signal to stop transcription and release RNA polymerase and the nascent (i.e., newly synthesized) RNA from the template. There are two major types of termination mechanisms in *E. coli* (and other bacteria), **factor-independent termination** (also called intrinsic or rho-independent) and **Rho-dependent termination** (also called factor-dependent) (**Figure 8-9**).

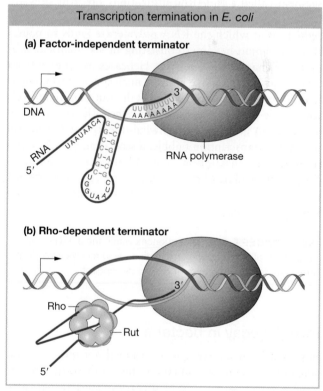

Transcription termination in *E. coli*

(a) Factor-independent terminator

DNA
RNA
5′
UAAUAACA
G–C
G–C
C–G
C–G
U–A
G–C
U–C–G
U   G   C
G   G   U
G   U   A   A
UUUUUUUU
AAAAAAAA
3′
RNA polymerase

(b) Rho-dependent terminator

Rho
Rut
3′
5′

FIGURE 8-9 Transcription termination occurs by two mechanisms that involve different sequences in the RNA. (a) Factor-independent termination depends on a terminator signal consisting of a GC-rich stem-loop structure followed by a stretch of U's. (b) Rho-dependent termination depends on a C-rich binding site for the protein Rho (called a Rut site) that is located upstream of the termination site.

Factor-independent termination occurs after the transcription of a GC-rich stretch followed by an A-rich stretch in the template strand. In the RNA, the GC-rich sequence is self-complementary and forms a 7- to 20-base-pair stem-loop followed by a 7- to 8-nucleotide U-rich tract. Stem-loops are also called hairpins because they resemble the metal clips used to hold a person's hair in place (Figure 8-9a). Normally, in the course of transcription elongation, RNA polymerase will pause if the short RNA–DNA hybrid in the transcription bubble is weak and will backtrack to stabilize the hybrid. The strength of the hybrid is determined by the relative number of 3-hydrogen bond G–C and C–G base pairs compared with 2-hydrogen bond A–T and U–A base pairs. In the factor-independent mechanism, the polymerase is believed to pause after synthesizing the U's (U–A forms a weak RNA–DNA hybrid). However, the backtracking polymerase encounters the hairpin. This roadblock sets off the release of RNA from the polymerase and the polymerase from the DNA template.

In the Rho-dependent termination mechanism, a protein called Rho factor recognizes nucleotide sequences in the RNA that act as a termination signal for RNA polymerase. RNAs with Rho-dependent termination signals do not have the string of U residues at their 3′ end and usually do not have a hairpin (Figure 8-9b). Instead, they have a sequence of about 50–90 nucleotides that is rich in C residues and poor in G residues and includes an upstream segment called a Rut (Rho utilization) site. Rut sites are located upstream (recall that upstream means 5′ of) from sequences at which the RNA polymerase tends to pause. Rho is a homo-hexamer consisting of six identical subunits that has **helicase** activity. Helicases use energy from ATP hydrolysis to move along a nucleic acid and unwind nucleic acid helices. Once bound at the Rut site, Rho travels toward the 3′ end of the transcript. When Rho encounters a paused RNA polymerase, it unwinds the RNA-DNA hybrid within the transcription bubble, dissociating the RNA and terminating transcription. Thus, Rho-dependent termination entails binding of Rho to the Rut site, pausing of RNA polymerase, and Rho-mediated release of the RNA from RNA polymerase.

KEY CONCEPT Special sequences within the 3′ UTR of an mRNA direct transcription termination using mechanisms that are either factor-independent or Rho-dependent.

mRNA decay in bacteria

RNA abundance in cells is determined not only by transcription mechanisms that control the synthesis of RNA by RNA polymerase, but also by decay mechanisms that carry out the destruction of RNAs. RNA destruction, which is commonly called RNA degradation or **decay,** is carried out by ribonucleases. Bacteria have about 25 different ribonucleases. Some are involved in decay and others function in the precise processing of RNA precursors such as cutting

the long rRNA precursor into individual functional rRNAs. The measure of decay is **half-life** (also denoted $t_{1/2}$), which is the amount of time it takes for half of the pool of an RNA molecule to be decayed. In bacteria, mRNAs typically have a half-life of less than two minutes. Rapid mRNA decay is thought to allow bacteria to quickly alter gene expression in response to changing nutritional and environmental conditions.

KEY CONCEPT The abundance of an RNA in cells is determined by transcription and decay.

Decay of mRNA in bacteria is commonly initiated by an **endonuclease** that cuts the mRNA into pieces, followed by digestion of the RNA pieces by **exonucleases** that remove nucleotides one at a time from the 3′ end (**Figure 8-10**). In *E. coli,* decay often begins with conversion of the triphosphate at the 5′ end of the RNA to a monophosphate via removal of pyrophosphate (PP_i) by an RNA pyrophosphohydrolase. The 5′-monophosphate serves as a binding site for the main endonuclease RNase E, which cuts single-stranded RNA. The RNase E products are then digested by 3′-to-5′ exonucleases. Because access to the mRNA by RNase E is critical to decay, the presence of ribosomes can affect the half-life of an mRNA. Remember that translation in bacteria occurs while mRNAs are being transcribed. Inefficient translation initiation presumably increases the distance between translating ribosomes, providing greater opportunity for cleavage by RNase E and decreasing the half-life of an mRNA.

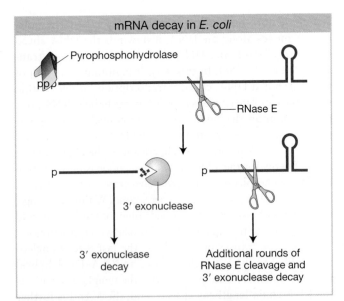

FIGURE 8-10 mRNA decay in *E. coli* is carried out by the sequential activities of an endonuclease (RNase E), which generates RNA fragments, and 3′-to-5′ exonucleases. RNase E recognizes a 5′-monophosphate, which is generated by pyrophosphohydrolase. Exonucleases are commonly drawn as the Pac-Man video game character, since exonucleases eat nucleotides from the ends of RNAs analogous to the way Pac-Man eats through a maze of Pac-Dots.

KEY CONCEPT The first step of decay in bacteria is carried out by an endonuclease, which cuts an RNA into two pieces, and the next step is carried out by exonucleases, which digest the RNA pieces into single nucleotides starting at their 3′ end.

8.3 TRANSCRIPTION IN EUKARYOTES

LO 8.3 Describe how mRNA transcription and decay mechanisms in eukaryotes are similar to those in bacteria.

Transcription in eukaryotic organisms, including humans, is similar to transcription in bacteria in that they retain many of the events associated with initiation, elongation, and termination. However, transcription in eukaryotes is more complex in four main ways.

1. **Eukaryotes have many genes that are spaced far apart.** The larger eukaryotic genomes have many more genes to be recognized and transcribed. Whereas bacteria usually have a few thousand genes, eukaryotes can have tens of thousands of genes. Furthermore, there is much more non-transcribed DNA in eukaryotes. Non-transcribed DNA originates by a variety of mechanisms that will be discussed in Chapter 16. So, even though eukaryotes have more genes than bacteria, their genes are, on average, farther apart. For example, whereas the average gene density in the bacteria *E. coli* is 1 gene per 1400 base pairs, that number drops in eukaryotic organisms to 1 gene per 9000 base pairs for the fruit fly *Drosophila melanogaster* and 1 gene per 100,000 base pairs for humans.

2. **Eukaryotes have three RNA polymerases.** In contrast to bacteria that have one RNA polymerase that transcribes all genes in the cellular genome, eukaryotes have three RNA polymerases that work with distinct initiation, elongation, and termination factors. **RNA polymerase I** transcribes rRNA, excluding 5S rRNA. **RNA polymerase II** transcribes all mRNAs and some ncRNAs, including miRNAs and some snRNAs. **RNA polymerase III** transcribes a variety of ncRNAs, including tRNAs, 5S rRNA, and some snRNAs. Each RNA polymerase is a multisubunit enzyme composed of about 12 proteins, some of which are identical or similar among the polymerases and others of which are unique to one polymerase.

   Unique features of the RNA polymerases carry out polymerase-specific functions. For example, the largest protein in RNA polymerase II contains a unique sequence called the **carboxy-terminal domain (CTD)** that helps coordinate RNA processing events that are exclusive to its transcripts. In addition, since eukaryotic RNA polymerases cannot bind promoters or initiate transcription on their own, each polymerase functions along with its own set of **general transcription factors (GTFs)** that are required to bind promoters and recruit the RNA polymerase to the transcription start site. The roles of GTFs and their interactions with RNA polymerases will be described in the section on eukaryotic transcription initiation.

3. **Transcription in eukaryotes takes place in the nucleus.** An important cellular difference between bacteria and eukaryotes is the presence of a nucleus in eukaryotes (see Figure 8-2). Because bacteria lack a nucleus, the information in RNA is almost immediately translated into protein, as described in Chapter 9. In eukaryotes, the nuclear membrane spatially separates transcription and translation—transcription takes place in the nucleus and translation in the cytoplasm. It also means that mechanisms exist to export RNAs from the nucleus to the cytoplasm. Additionally, before RNAs leave the nucleus, they are modified in several ways. Both ends of an mRNA are chemically modified to protect against degradation: capping at the 5′ end and polyadenylation at the 3′ end. These modifications and others are collectively referred to as **RNA processing.** Newly synthesized RNAs that are not yet processed are called primary transcripts or **precursor RNAs (pre-RNAs)**, for example, pre-mRNA and pre-rRNA. RNA processing often occurs co-transcriptionally, that is, while the RNA is being transcribed. Thus, RNA polymerases synthesize RNA while simultaneously coordinating a variety of processing events.

4. **DNA in eukaryotes is packaged with proteins into chromatin.** The template for transcription, genomic DNA, is tightly wrapped around proteins to form chromatin in eukaryotes, whereas DNA is less compacted in bacteria. The structure of chromatin can affect transcription initiation, elongation, and termination by all three RNA polymerases as well as the processing of their transcripts. These chromatin-based mechanisms will be covered in Chapter 12.

KEY CONCEPT Differences in transcription between eukaryotes and bacteria are related to (1) larger eukaryotic genomes with genes that are spaced further apart, (2) the division of transcription in eukaryotes among three RNA polymerases, (3) the nuclear membrane in eukaryotic cells that decouples transcription and translation and necessitates nuclear RNA export, and (4) the tight packaging of eukaryotic genomic DNA into chromatin.

Transcription initiation in eukaryotes

RNA polymerases I, II, and III cannot recognize promoter sequences on their own. However, unlike bacteria, where promoters are recognized by σ factor as an integral part of the RNA polymerase holoenzyme, eukaryotic promoters are recognized by GTFs that first bind specific sequences in the promoter and then bind the RNA polymerase. Nevertheless, the mechanisms in bacteria and eukaryotes are conceptually similar. In both cases, the information that defines a promoter is provided by short DNA sequences located near the transcription start site, and

KEY CONCEPT RNA polymerase I, II, and III genes have unique promoters that direct transcription initiation. Promoters are first recognized by RNA polymerase-specific general transcription factors (GTFs). One of the main functions of GTFs is to recruit a specific RNA polymerase and position it to begin RNA synthesis at the transcription start site.

the sequences are bound by proteins that associate with RNA polymerase and position it at the correct site to start transcription.

RNA polymerase I promoters and GTFs In eukaryotic organisms, ribosomal RNA (rRNA) is transcribed by RNA polymerase I from hundreds of near-identical copies of rDNA genes that are tandemly repeated in the genome and reside in the **nucleolus**, a non–membrane-bound region in the nucleus where rRNA transcripts are synthesized, processed, and assembled with proteins into ribosomes (see Figure 8.6). Each rDNA gene encodes a single rRNA transcript that contains 18S, 5.8S, and 28S rRNAs along with an external transcribed spacer (ETS) and internal transcribed spacers (ITSs) (**Figure 8-11**, top). After transcription, spacer regions are removed by processing enzymes, 18S rRNA is assembled with ribosomal proteins to form the 40S small ribosomal subunit, and 5.8S, 28S, and 5S rRNAs are assembled with ribosomal proteins to form the 60S ribosomal subunit. RNA polymerase III transcribes 5S rRNA from tandem arrays of hundreds of gene copies located at different places in the genome from those of the RNA polymerase I genes. Note that the "S" in 18S, 5.8S, 28S, 5S, 40S, and 60S stands for Svedberg units, which is a measure of

a molecule's size that is based on its rate of sedimentation upon centrifugation.

Between the tandemly repeated rDNA genes are intergenic spacers (IGSs) that contains two promoter elements important for transcription initiation. A Core element is located at the transcription start site, and an Upstream Control Element (UCE) is located 100 to 150 base pairs upstream of the transcription start site (Figure 8-11, bottom). In humans, the Core element is bound by a multi-protein complex called Selectivity Factor 1 (SL1), which contains TATA-binding protein (TBP), and the UCE is bound by Upstream Binding Factor (UBF). In addition to SL1 and UBF, the protein TIF-1A (Transcription Initiation Factor 1A) is also required for recruitment of RNA polymerase I to the transcription start site of rDNA genes. Unlike SL1 and UBF, TIF-1A does not bind DNA, but instead it functions through protein-protein interactions and forms a bridge between SL1 and RNA polymerase I.

Every time a cell divides, the number of ribosomes must be doubled to keep the number of ribosomes in the two daughter cells equal to that in the parent cell. Mammalian cells have one to two million ribosomes. Therefore, every cell division involves the production of one to two million rRNA transcripts by RNA polymerase I. Consequently, the RNA polymerase I transcription mechanism has evolved to be exquisitely sensitive to environmental conditions that promote or inhibit cell proliferation (i.e., an increase in the number of cells). Conditions that affect cell proliferation can act on transcription initiation by RNA polymerase I by altering the activities of SLI, UBF, and TIF-1A. For example, conditions that promote cell proliferation lead to phosphorylation of TIF-1A on a specific serine residue, increasing its ability to recruit RNA polymerase I and trigger transcription initiation.

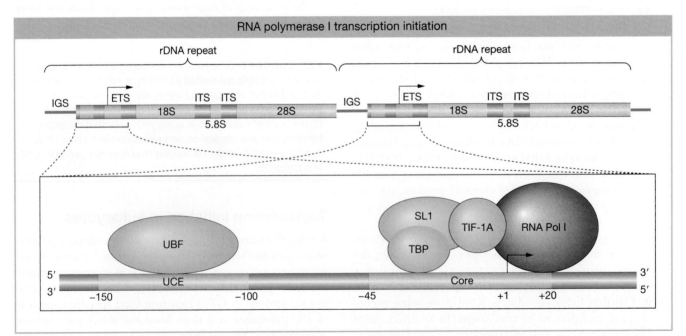

FIGURE 8-11 The information that controls transcription initiation by RNA polymerase I is contained in rDNA promoter sequences (UCE and Core) that are located upstream of the transcription start site (+1) and are bound by protein factors (UBF and SL1). A third factor, TIF-1A, does not directly bind DNA but is important for the recruitment and function of RNA polymerase I. UCE = Upstream Control Element, UBF = Upstream Binding Factor, TBP = TATA Binding Protein, SL1 = Selectivity Factor 1, TIF-1A = Transcription Initiation Factor 1A.

RNA polymerase II promoters and GTFs RNA polymerase II is responsible for transcribing all mRNAs as well as numerous ncRNAs, including snRNAs involved in splicing and miRNAs involved in mRNA decay and inhibition of translation. Transcription by RNA polymerase II is the most complex transcriptional system in eukaryotes because of the large number of gene targets with unique expression patterns.

RNA polymerase II promoters, which are somewhat arbitrarily defined as sequences located within 100 base pairs of the transcription start site, contain a variety of promoter elements, a few of which are relatively common (**Figure 8-12**). About 25 percent of promoters in yeast and humans contain

a **TATA box**, an sequence element so-named because the nucleotide sequence TATA appears in the consensus sequence <u>TATA</u>AAA. In animals, the TATA box is located about 30 nucleotides upstream of the transcription start site (i.e., −30), but in yeast its location is more variable (between −50 and −125). Another common promoter element is the initiator (Inr), which is located right at the transcription start site in about 40 percent of genes.

Collectively, only about 50 percent of RNA polymerase II genes contain a TATA box and/or an Inr. This predicts the existence of other promoter elements. Computational analyses that searched for common sequences surrounding the transcription start sites of RNA polymerase II genes have identified additional promoter elements, including the downstream promoter element (DPE), which is located at about +25, and the TFIIB recognition element (BRE), which is located at about −40. However, since many genes lack all of the known promoter elements, it is likely that promoter elements remain to be discovered.

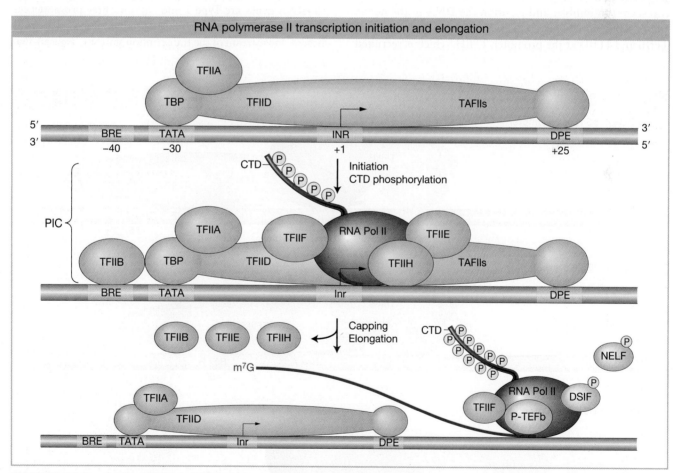

FIGURE 8-12 Transcription initiation by RNA polymerase II is directed by a variety of promoter elements, including the BRE, TATA box, Inr, and DPE, located within 100 base pairs upstream or downstream of the transcription start site (+1). Assembly of the PIC occurs in a sequential manner, starting with TFIID, which contains proteins (TBP and TAFs) that bind promoter elements. Assembly of TFIID leads to recruitment of the other GTFs and RNA polymerase II. TFIIH is required during initiation for opening the transcription bubble and phosphorylation of the CTD. Shortly after initiation, the pre-mRNA is capped and elongation is promoted by the P-TEFb kinase, which phosphorylates the RNA polymerase II CTD, DSIF, and NELF.

ANIMATED ART SaplingPlus
Transcription in eukaryotes

All of the proteins that bind RNA polymerase II promoter elements are subunits of GTFs. Transcription factor IIB (TFIIB) binds the BRE, and TFIID binds the other promoter elements. The TFIID complex contains TBP (the same protein involved in RNA polymerase I transcription) and about 15 TBP-associated factors (TBP). TBP binds the TATA box, and TAFs bind the Inr and DPE. Binding of TFIID at a promoter is the first step in the sequential assembly of other GTFs and RNA polymerase II. TFIID binding instructs assembly of TFIIA and TFIIB, followed by TFIIF and RNA polymerase II as a pre-assembled complex and ending with addition of TFIIE and TFIIH. The assemblage of GTFs and RNA polymerase II constitutes the **preinitiation complex (PIC)**, which serves to position RNA polymerase II at the transcription start site, generate the transcription bubble, and position the DNA in the active site of RNA polymerase II. TFIIA stabilizes the binding of TFIIB and TFIID at the promoter. TFIIH, which is recruited

to the promoter by TFIIE, contains proteins with helicase activity that unwind the DNA into two strands to form the transcription bubble. Lastly, TFIIF places the promoter DNA in a position in RNA polymerase II that is appropriate for DNA unwinding and initiation of transcription at the start site. After transcription has been initiated, RNA polymerase II dissociates from most of the GTFs to elongate the RNA transcript. Some GTFs, including TFIID, remain at the promoter to attract the next RNA polymerase II. In this way, multiple RNA polymerase II molecules can simultaneously synthesize transcripts from a single gene.

RNA polymerase III promoters and GTFs RNA polymerase III transcribes noncoding RNAs (ncRNAs) shorter than 300 nucleotides. The RNA polymerase III gene targets are classified into three types based on their promoter elements. The 5S rRNA genes are Type 1 and contain three promoter elements, Box A, intermediate element (IE), and Box C, that are all located downstream of the transcription start site (**Figure 8-13a**).

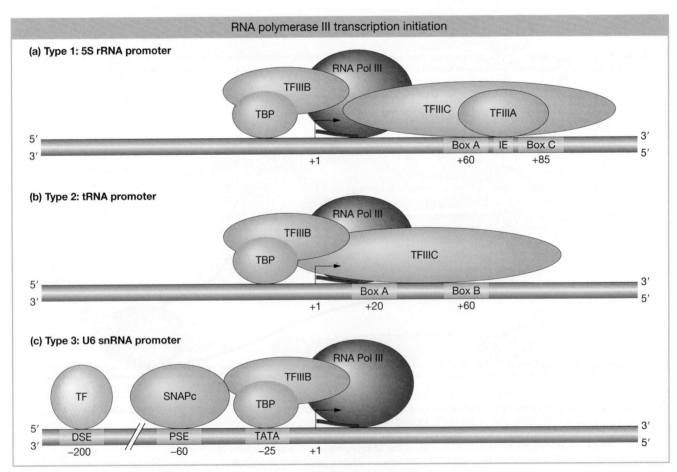

FIGURE 8-13 RNA polymerase III transcribes genes with three types of promoters, Types 1, 2, and 3 (a–c). TFIIIA, TFIIIC, and SNAP$_c$ bind promoter elements (IE, Box B, and PSE, respectively) that are unique to each type of RNA polymerase III gene and can be viewed as the specificity factors for the gene type. The main function of these factors is to recruit TFIIIB to the promoter, which then leads to recruitment of RNA polymerase III and transcription initiation.

tRNA genes are Type 2 and contain the same Box A element as 5S genes, but instead of a Box C they have a Box B (Figure 8-13b). Lastly, Type 3 genes such as the U6 snRNA gene that is involved in pre-mRNA splicing contain promoter elements that are located upstream of the transcription start site and look very similar to promoter elements found in RNA polymerase II genes, including a TATA box (Figure 8-13c).

The GTFs for RNA polymerase III transcription are designated TFIIIA, TFIIIB, TFIIIC (transcription factor for RNA polymerase III), and SNAP$_c$ (snRNA activating protein complex). TFIIIB is required for the transcription of all three types of RNA polymerase III genes and functions to recruit RNA polymerase III as well as open the transcription bubble. TFIIIA binds promoter elements in Type 1 genes and helps recruit TFIIIC, which binds promoter elements in Type 1 and 2 genes. Note that since TFIIIA and TFIIIC bind downstream of the transcription start site, they need to be temporarily displaced as RNA polymerase III transcribes through the promoter DNA. For Type 3 genes, recruitment of RNA polymerase III is assisted by binding of both the TBP subunit of TFIIIB to the TATA box and of SNAP$_c$ to the proximal sequence element (PSE). Binding of SNAP$_c$ to the PSE is helped by transcription factors that bind to the distal sequence element (DSE). Transcription factors are discussed in detail in Chapter 12.

KEY CONCEPT Genes transcribed by RNA polymerase III are divided into three types, based on their promoter elements. Type 1 and Type 2 have promoter elements downstream of the transcription start site. Type 3 promoter elements include a TATA box and are positioned upstream of the transcription start site.

RNA polymerase II transcription elongation

Shortly after transcription initiation, phosphorylation of RNA polymerase II by a protein kinase in TFIIH helps coordinate the processing of mRNAs as they are being transcribed. The carboxy-terminal domain (CTD) of the largest subunit of RNA polymerase II contains the sequence YSPTSPS (tyrosine-serine-proline-threonine-serine-proline-serine) tandemly repeated 26 times in yeast and 52 times in humans. Phosphorylation of the serine in position 5 of the repeat (S5) by TFIIH serves as a signal for the binding of enzymes that cap the 5′ end of the mRNA (discussed in the next section) (**Figure 8-14**). The CTD is located near the site where nascent RNA emerges from RNA polymerase II, so it is in an ideal place to orchestrate the binding and release of proteins needed to process the nascent transcript while RNA synthesis continues. Post-translational modification of S5 and other amino acids in the CTD change as RNA polymerase II transcribes through a gene, creating different binding sites for other processing factors as well as factors that regulate transcription elongation and termination.

KEY CONCEPT During elongation, the CTD of RNA polymerase II is chemically modified to serve as a binding site for other proteins involved in transcription and RNA processing.

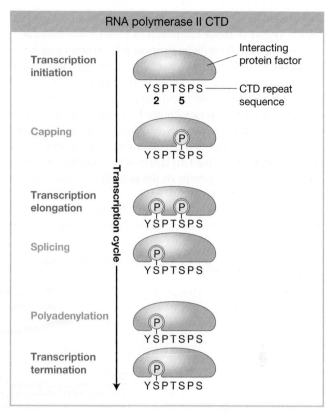

FIGURE 8-14 The pattern of amino acid modifications within the RNA polymerase II CTD changes as the polymerase transcribes through a gene. The modifications, including phosphorylation of serines 2 and 5 by kinases and dephosphorylation by phosphatases, create a code that directs the binding of factors at specific stages of transcription. Factors that bind the CTD include regulators of transcription elongation and termination as well as RNA processing events, including capping, splicing, and cleavage and polyadenylation.

Transcription initiation is not a green light that liberates RNA polymerase II to completely transcribe a gene unimpeded. In fact, transcription continues to be regulated all along the length of a gene. For example, for a large fraction of human genes, transcription elongation is temporarily stopped (i.e., paused) about 50 base pairs downstream of the transcription start site. Pausing of this type is caused by the protein factors NELF (negative elongation factor) and DSIF (DRB sensitivity-inducing factor) and relieved by P-TEFb (positive transcription elongation factor b) (see Figure 8-12). To release paused RNA polymerase II into productive elongation, P-TEFb phosphorylates NELF and DSIF. NELF dissociates from the elongation complex, and DSIF travels along with RNA polymerase II and functions as a positive elongation factor. P-TEFb also phosphorylates the RNA polymerase II CTD on serine 2 (S2) within the YSPTSPS repeats (see Figure 8-14), which serves as a signal for the binding of factors involved in processing of the pre-mRNA and transcription termination.

Transcription termination in eukaryotes

Transcription termination for the three RNA polymerases occurs by different mechanisms. Elongating RNA polymerase I is stopped by protein factors bound at specific DNA sequences called terminator elements and is released from DNA by other factors. In contrast, RNA polymerase III terminates elongation and dissociates from DNA after the synthesis of a poly(U) stretch, similar to factor-independent termination in bacteria.

Two models have been proposed for transcription termination by RNA polymerase II—the torpedo model and the allosteric model. The models are conceptually similar to Rho-dependent and factor-independent mechanisms, respectively, in *E. coli* (see Figure 8-9), but different factors are involved. Both RNA polymerase II termination models couple 3′-end formation to termination. As described below in the section on polyadenylation, the 3′ ends of mRNAs are determined by cleavage of the pre-mRNA and addition of a poly(A) tail to the new 3′ end.

In the **torpedo termination model** (Figure 8-15a), RNA polymerase II continues to transcribe past the site of cleavage, the pre-mRNA is cleaved, and the new 5′-mono-phosphorylated end that is formed is a substrate for a 5′-to-3′ exonuclease called Xrn2, which digests the RNA one nucleotide at a time until eventually reaching RNA polymerase II and causing it to dissociate from DNA. Xrn2 is positioned to act in termination through its association with the CTD phosphorylated on serine 2 (see Figure 8-14).

In the **allosteric termination model** (Figure 8-15b), transcription through the site of cleavage causes elongation factors to dissociate, leading to a conformational change within the active site of RNA polymerase II and its release from DNA. In this model, it remains to be determined how RNA polymerase II senses passage through the site of cleavage and how this leads to dissociation of elongation factors.

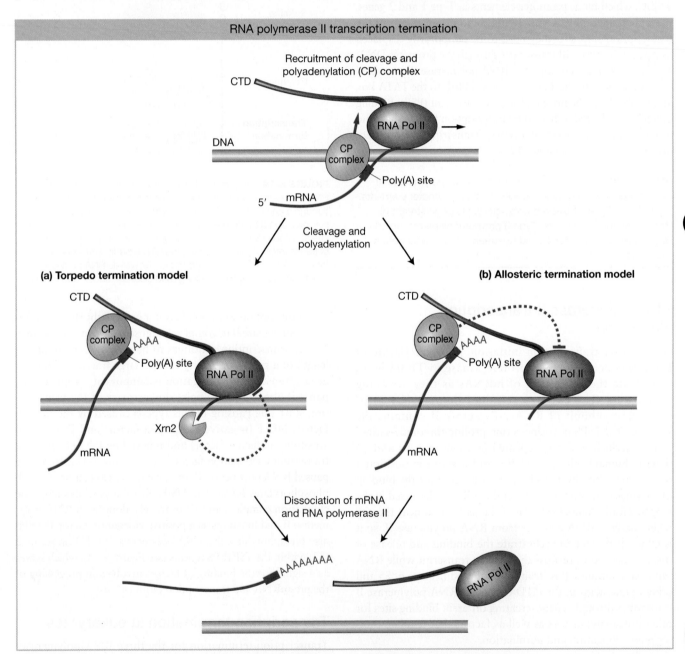

FIGURE 8-15 Current data support two models for how transcription of mRNAs by RNA polymerase II is terminated. (a) In the torpedo model, the piece of RNA that stays associated with RNA polymerase II and continues to be synthesized after cleavage is a substrate for the 5′-to-3′ exonuclease Xrn2 that degrades the RNA to elicit termination. (b) In the allosteric model, upon encountering cleavage and polyadenylation signals, RNA polymerase II undergoes a conformation change that commits it to termination.

(i.e., co-transcriptionally) and are coordinated with transcription initiation, elongation, or termination as well as with one another. Some themes emerge upon comparison of the processing events: (1) sequence elements within mRNAs often direct where processing occurs, and (2) sequence elements are bound by proteins or ncRNAs that are themselves enzymes or that recruit enzymes to carry out the processing.

8.4 PROCESSING OF mRNA IN EUKARYOTES

LO 8.4 Explain how mRNA processing, editing, and modification occur and can affect the abundance and sequence of proteins in eukaryotes.

Unlike bacterial mRNAs, eukaryotic mRNAs undergo numerous processing events that affect their structure and function. Many of these events occur at the same time as transcription

Capping

RNAs synthesized by RNA polymerase II, including mRNAs and snRNAs, are modified at their 5′ end by addition of a methylated guanine nucleotide, **7-methylguanosine** (m^7G), more commonly referred to as a **cap** because it covers the "head" of the RNA (**Figure 8-16a**). The 5′ cap is added during transcription when the RNA is about 25 nucleotides long and has just emerged from the exit channel of RNA polymerase II. The process of adding a cap (capping) involves the sequential action of three enzymes

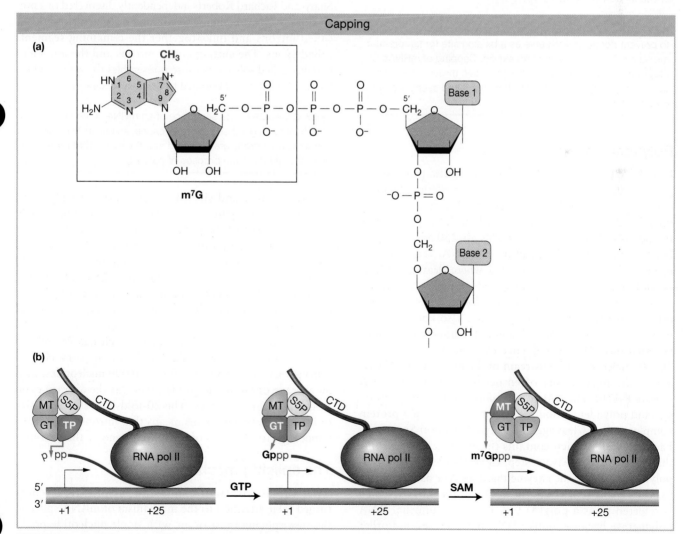

Capping

(a)

m⁷G

(b)

FIGURE 8-16 (a) The first nucleotide in an mRNA is modified by addition of an m^7G cap. (b) Capping is initiated by a triphosphatase (TP) enzyme that removes the terminal phosphate from the first nucleotide. Then a guanylyltransferase (GT) adds a guanine monophosphate nucleotide, and a methyltransferase (MT) adds a methyl group (CH_3).

All three enzymes are positioned to act on the mRNA shortly after it emerges from the exit tunnel of RNA polymerase II because of their association with the CTD that is phosphorylated on serine 5 (S5P) of repeat sequences.

ANIMATED ART 🅟 Sapling Plus   Transcription in eukaryotes

(Figure 8-16b): RNA triphosphatase removes the gamma phosphate from the first nucleotide in the RNA chain, guanylyltransferase uses GTP as a substrate and links GMP to the first nucleotide by an unusual 5′,5′-triphosphate linkage (GpppN, where N is the first nucleotide in the mRNA), and 7-methyltransferase adds a methyl group (CH_3) from *S*-adenosylmethionine (SAM) to the N-7 position of the newly added guanine base. Through their interaction with the CTD of RNA polymerase II, these enzymes are in position to act on RNAs early in the transcription process (see Figure 8-14).

Caps serve multiple functions. They protect RNAs from decay by exonucleases, which often require 5′-phosphates to recognize their substrates. Caps on mRNAs also serve as a binding site for proteins such as the cap binding complex (CBC) that mediate subsequent events, including splicing, polyadenylation, and nuclear export, by interacting with processing and export factors. The CBC is also critical for the first round of translation, while another cap binding protein, eIF4E, is required for subsequent rounds of translation, as described in Chapter 9.

> **KEY CONCEPT** The 5′ end of a eukaryotic mRNA is modified to prevent decay and to serve as a binding site for factors that mediate mRNA processing and export. Capping of mRNAs is programmed to occur early in transcription through the association of capping enzymes with phosphorylated serine 5 on the CTD of RNA polymerase II.

Polyadenylation

Like the 5′ end, the 3′ end of an mRNA is also processed to protect it from decay and to promote translation. Processing at the 3′ end consists of two events: cleavage, which cuts the mRNA away from the transcribing RNA polymerase II, and **polyadenylation**, which adds 50–250 adenosine (A) residues to the end of the cleaved mRNA. Sequence elements within the 3′ UTR determine where cleavage occurs.

In humans, the highly conserved six-nucleotide (hexanucleotide) sequence AAUAAA is located 10–30 nucleotides upstream of the cleavage site, also known as the poly(A) site (**Figure 8-17a**). A less well conserved U-rich or GU-rich downstream sequence element (DSE) is located 20–40 nucleotides downstream of the poly(A) site. Cleavage at the poly(A) site often occurs after a CA or UA (Figure 8-17b). The AAUAAA is important for both cleavage and polyadenylation because it is bound by a protein complex called Cleavage and Polyadenylation Specificity Factor (CPSF), which contains the endonuclease enzyme that executes the cleavage step. CPSF also recruits **poly(A) polymerase (PAP)**, the enzyme that uses ATP as a substrate to add a string of A's onto the 3′-OH of the mRNA, which is referred to as a **poly(A) tail**. PAP is an unusual RNA polymerase because, unlike DNA polymerases and other RNA polymerases, it does not copy a nucleic acid template. In addition to CPSF, the cleavage step involves CstF (Cleavage Stimulatory Factor), which binds the DSE and helps determine the site of cleavage that is bound by Cleavage Factors I and II (CFI and CFII). During its synthesis, the poly(A) tail is bound by poly(A) binding protein (PABP), which in the cytoplasm protects the mRNA from decay by exonucleases and promotes translation by interacting with the translation machinery.

> **KEY CONCEPT** The 3′ end of mRNAs is modified by addition of a long stretch of adenosine nucleotides, which protects the mRNA from decay and supports translation. The poly(A) tail is added by a special type of RNA polymerase following mRNA cleavage at a site that is determined by protein factors that bind sequence elements in the mRNA.

The discovery of splicing

Transcription copies the DNA sequence of protein-coding genes into mRNA, yet sequence comparison of most pairs of human mRNAs and genes shows that they are different: large stretches of DNA sequence are transcribed into RNA and later removed from the RNA. In 1977, the laboratories of Philip Sharp and Richard Roberts independently discovered this process of mRNA **splicing**, which removes segments of mRNA called **introns** and links together the remaining segments called **exons**. The cutting out of introns and the joining of exons is called splicing because it resembles the way in which movie film is cut and rejoined to delete a specific segment.

> **KEY CONCEPT** The sequence of an mRNA is not always identical to its gene sequence because, as pre-mRNAs are transcribed, introns are removed and the exons that remain are joined together in the process of splicing.

The number and size of introns varies from gene to gene and from organism to organism. For example, only about 5 percent of the genes in yeast (*S. cerevisiae*) have introns. Intron-containing genes in yeast almost always have a single intron that ranges in length from 50 to 1000 nucleotides, with an average length of 250 nucleotides. In contrast, 85 percent of human genes have at least one intron, and an average human gene has eight introns and nine exons. An extreme example is the gene that is mutated in Duchenne muscular dystrophy, which has 78 introns and 79 exons spread across 2.3 million base pairs. Human introns vary in length from 50 to 300,000 nucleotides, with an average of 6000 nucleotides, whereas the average exon length is 300 nucleotides. The 20-fold-larger average size of introns relative to exons means that introns account for a much greater fraction of the human genome than exons.

The splicing mechanism

After the discovery of exons and introns, researchers turned their attention to the mechanism of mRNA splicing. Because splicing must occur with single nucleotide precision to maintain the information that directs translation, the intron-containing mRNA precursor (pre-mRNA) must hold the information that points the splicing machine called

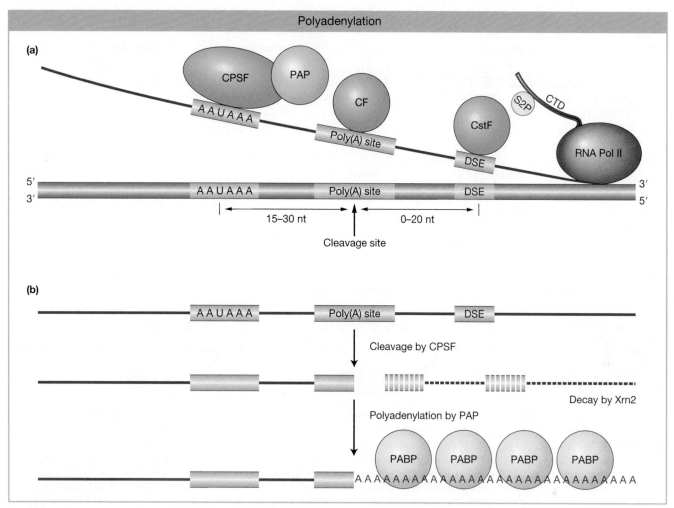

FIGURE 8-17 The 3′ end of an mRNA is generated by consecutive cleavage and polyadenylation reactions. (a) In humans, the site of cleavage is directed by three sequences in the 3′ UTR of an mRNA (i.e., AAUAAA, poly(A) site, and DSE), each of which is bound by protein factors (i.e., CPSF, CF, and CstF, respectively) that are recruited to the ends of genes by association with the RNA polymerase II CTD phosphorylated on serine 2 (S2P) of repeat sequences. (b) Cleavage at the poly(A) site by CPSF is followed by addition of a poly(A) tail to the new 3′ end by PAP and by association of PABP with the poly(A) tail.

the **spliceosome** where to act. Researchers hypothesized that the information would be provided by sequences at the boundaries between exons and introns. In fact, alignments of boundary sequences for many pre-mRNAs revealed that almost all introns begin with GU and end with AG (**Figure 8-18a**). In addition, there is high conservation of intron and exon nucleotides adjacent to the GU and AG. The GU and AG sequence elements define the **5′ splice site** and **3′ splice site,** respectively, where cuts are made by the spliceosome to remove the intron. In addition, a third conserved sequence called the **branch point** is located 15–45 nucleotides upstream of the 3′ splice site. An invariant adenosine within the branch point participates in the first catalytic step of splicing. The existence of conserved nucleotide sequences at splice sites and the branch point suggested that components of the spliceosome are directed to act at specific places in pre-mRNAs by binding to these sequences.

A serendipitous finding in the laboratory of Joan Steitz led to the discovery of components of the spliceosome.

Patients with a variety of autoimmune diseases, including systemic lupus erythematosus, produce antibodies against their own proteins. In the course of analyzing blood samples from patients with lupus, Steitz and colleagues identified antibodies that bound nuclear RNA-protein complexes called **small nuclear ribonucleoproteins (snRNPs),** pronounced "snurps," that are comprised of a small nuclear RNA (snRNA) 100–200 nucleotides in length that serves as a scaffold for binding several proteins. They observed that the sequence at the 5′ end of the snRNA named U1 has extensive complementarity to the sequence at 5′ splice sites, suggesting that the U1 snRNA identifies 5′ splice sites by base pairing (Figure 8-18b). To test this hypothesis, the laboratory of Alan Weiner performed a mutational analysis. They found that splicing was dramatically reduced by mutations in a 5′ splice site sequence that partially disrupted base pairing with the U1 snRNA (Figure 8-18c). Moreover, they found that splicing of the mutant pre-mRNA was recovered by mutations in the 5′ end of the U1 snRNA that restored base pairing. This

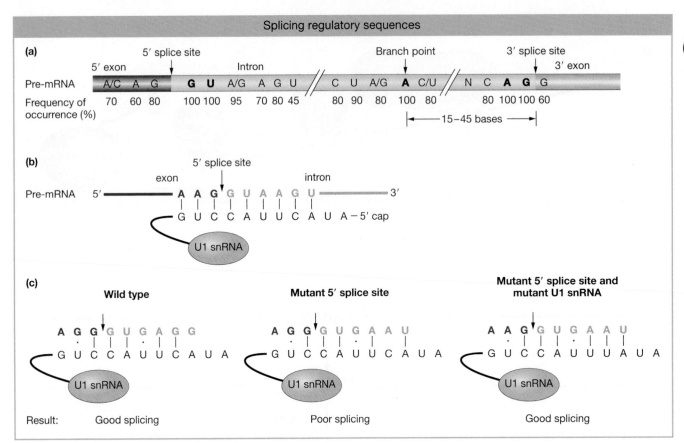

FIGURE 8-18 (a) Conserved nucleotide sequences are present at the junctions of exons and introns (i.e., 5′ and 3′ splice sites) as well as at the branch point, which is near the 3′ splice site. Invariant nucleotides (GU at the 5′ splice site, A at the branch point, and AG at the 3′ splice site) are shown in bold font, and N represents any base. (b) The U1 snRNA recognizes 5′ splice sites by base pairing. Sequences near the 5′ end of the U1 snRNA form eight consecutive base pairs with a consensus 5′ splice site. (c) The efficiency of splicing is affected by the strength of U1 snRNA base pairing at the 5′ splice site. Mutations in the 5′ splice site that reduce the number of hydrogen bonds lead to a decrease in the efficiency of splicing (compare *left* and *middle*); however, splicing efficiency can be restored by compensatory mutations in the U1 snRNA (*right*).

"compensatory mutation" analysis demonstrated that base pairing between snRNAs and the pre-mRNA is important for the selection of splice sites.

KEY CONCEPT snRNAs facilitate splicing by base pairing with conserved sequences in the pre-mRNA.

In addition to the U1 snRNP, the spliceosome contains U2, U4, U5, and U6 snRNPs as well as many proteins that have conserved functions in eukaryotes from yeast to humans. The splicing reaction begins with stepwise recognition of pre-mRNA sequence elements (**Figure 8-19a**). First, U1 binds the 5′ splice site and U2 binds the branch point, with the U2 snRNA base pairing to nucleotides across the branch point, except for the key adenosine. Spliceosome assembly is completed by entry of the U4, U5, and U6 snRNPs as a preassembled tri-snRNP complex. At this point, the spliceosome undergoes several conformational changes to become catalytically active. The U1 and U4 snRNPs are released from the spliceosome, the U6 snRNP base pairs to the 5′ splice site, and the U5 snRNP base pairs to both exon sequences, placing the splice sites in close proximity.

Splicing, then, takes place by means of two transesterification reactions (Figure 8-19b). The first step of the reaction involves nucleophilic attack by the 2′-OH of the unpaired branch point adenosine at the phosphodiester bond at the 5′ splice site, which cuts the pre-mRNA between the 5′ exon and the intron and produces an intron with a loop structure called a lariat because it resembles the shape of a cowboy's lariat (lasso). The second step of the reaction involves nucleophilic attack by the 3′-OH of the 5′ exon at the phosphodiester bond at the 3′ splice site, which covalently links together the 5′ and 3′ exons and frees the intron as a lariat. Lastly, the U2, U5, and U6 snRNPs are released from the excised lariat and participate in another cycle of splicing along with previously released U1 and U4 snRNPs. This process is repeated for each intron in a pre-mRNA.

KEY CONCEPT Splicing is a two-step reaction. The first step is cleavage at the 5′ splice, and the second step is cleavage at the 3′ splice site, which results in removal of the intron and joining of the exons.

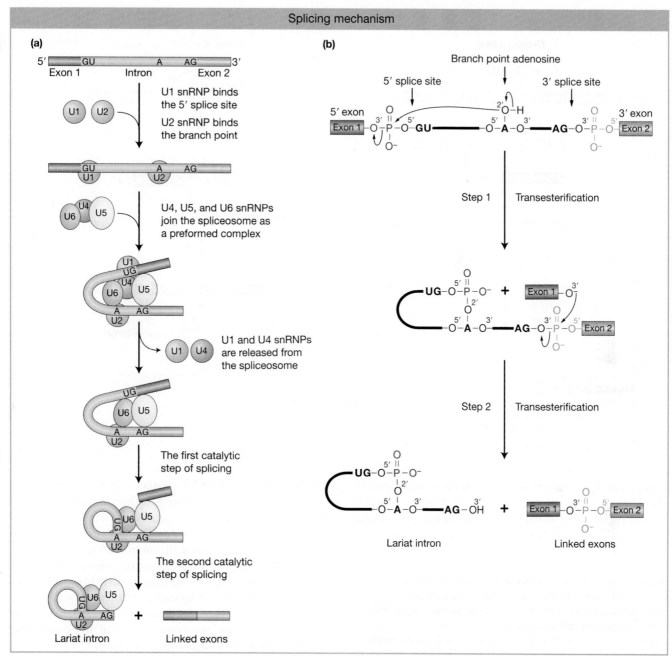

FIGURE 8-19 (a) The spliceosome is formed by sequential assembly of five snRNPs (U1, U2, U4, U5, and U6) onto an intron. Conformational changes that lead to release of the U1 and U4 snRNPs position the U2 and U6 snRNAs and the branch point adenosine near the 5' splice site for the first catalytic step of splicing. (b) In the first step of splicing, the 2'-hydroxyl of the branch point adenosine attacks the phosphodiester bond at the 5' splice site. This produces two intermediates, a 5' exon and an intron–3' exon in a lariat structure. In the second step of splicing, the 3'-hydroxyl of the 5' exon attacks the phosphodiester bond at the 3' splice site, producing linked exons and the released lariat intron.

ANIMATED ART 🖉 SaplingPlus   Mechanism of mRNA splicing

snRNAs in the spliceosome may carry out the catalytic steps of splicing

Researchers initially assumed that proteins in the spliceosome carry out the catalytic reaction, but in 1981 studies by Thomas Cech's laboratory raised the possibility that pre-mRNA splicing by the spliceosome is catalyzed by the snRNAs. Cech and co-workers reported that precursor ribosomal RNA (pre-rRNA) from the ciliated protozoan

Tetrahymena thermophila could splice a 413-nucleotide intron from itself without the help of proteins, thus demonstrating that RNA can function as an enzyme, a **ribozyme**. There are two distinct classes of self-splicing introns, called Group I and II, that are found in bacteria and bacterial viruses as well as some nuclear-encoded mitochondrial and chloroplast genes in fungi, algae, and plants. Since the structure of base-paired pre-mRNA, U2 snRNA, and U6 snRNA in the active site of the spliceosome is similar to the secondary

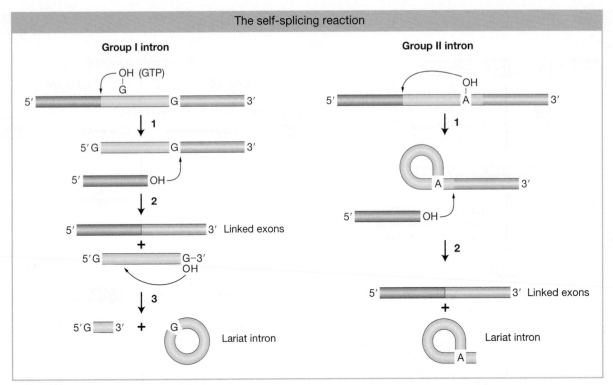

FIGURE 8-20 Similar to splicing by the spliceosome, RNA-catalyzed self-splicing of Group I and Group II introns involves two transesterification reactions. In Group I splicing, the first reaction is carried out by GTP, whereas in Group II splicing and spliceosomal splicing. The first reaction is carried out by the branch point adenosine.

structure of Group II introns and the chemistry of splicing by the spliceosome is the same as that carried out by Group II introns (**Figure 8-20**), it has been hypothesized that spliceosomal and Group II splicing mechanisms are evolutionarily related and that the spliceosome is a ribozyme.

> **KEY CONCEPT** Intron removal and exon joining are catalyzed by RNA molecules. In eukaryotes, the snRNAs of the spliceosome catalyze the removal of introns from pre-mRNA. Some introns are self-splicing; in these cases, the intron catalyzes its own removal. RNAs capable of catalysis are called ribozymes.

Alternative splicing can expand the proteome

A major rationale for having genes with introns is that introns provide a mechanism to encode different proteins (called protein **isoforms**) from a single gene. Through the process of **alternative splicing**, exons in a pre-mRNA can be joined together in different combinations to produce different mature mRNAs that encode protein isoforms. An extreme example is the *Dscam* gene in *Drosophila* that via alternative splicing can produce 38,016 different Dscam proteins.

Alternative splicing can produce protein isoforms with different functional domains. This is illustrated by *FGFR2*, a human gene that encodes a receptor that binds fibroblast growth factors and then transduces a signal inside the cell

(**Figure 8-21**). The FGFR2 protein is made up of several domains, including an extracellular ligand-binding domain. Alternative splicing results in two protein isoforms that differ in their extracellular domains. Because of this difference, each isoform binds different growth factors. In addition, as illustrated by the α-*tropomyosin* gene (**Figure 8-22**), protein isoforms can be produced in particular cells by cell type-specific alternative splicing, and they can also be produced at different stages of development. Thus, alternative splicing expands the **proteome** (the set of all proteins that can be expressed) of eukaryotic organisms.

In humans, about 95 percent of intron-containing genes undergo alternative splicing to encode two or more protein isoforms. There are four general types of alternative splicing, the most common of which is exon skipping, where an exon is either included or excluded in the mature mRNA (**Figure 8-23**). The other types of alternative splicing are alternative 3′ splice sites (i.e., one 5′ splice site and a choice of two 3′ splice sites), alternative 5′ splice sites (i.e., one 3′ splice site and a choice of two 5′ splice sites), and mutually exclusive exons (i.e., only one of several exons is included in the mature mRNA, as illustrated by *FGFR2* in Figure 8-21).

A key feature of alternative splicing mechanisms is that 5′ and 3′ splice site sequences differ among exons. Alternative exons tend to have weak splice site sequences that have lower affinity for spliceosome components than splice sites associated with constitutive exons (i.e., exons that are always spliced into the mature mRNA). Weak splice sites are subject

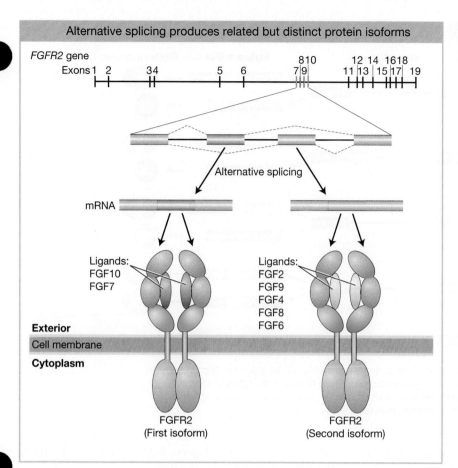

FIGURE 8-21 Alternative splicing of mutually exclusive exons in the *FGFR2* pre-mRNA produces two protein isoforms that bind different FGF proteins.

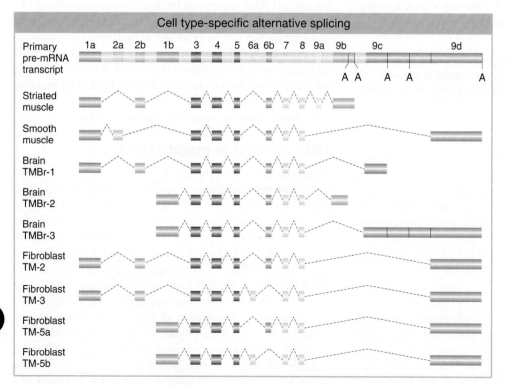

FIGURE 8-22 The rat *α-tropomyosin* gene is alternatively spliced in different patterns in different cell types. Light blue boxes represent introns; other colors represent exons. Note that in addition to alternative splicing, the *α-tropomyosin* gene undergoes alternative transcription initiation (starting transcription at the beginning of either the peach or the green exons) and alternative polyadenylation (occurring at five sites indicated by A's). Dashed lines indicate introns that have been removed by splicing. TM, tropomyosin.

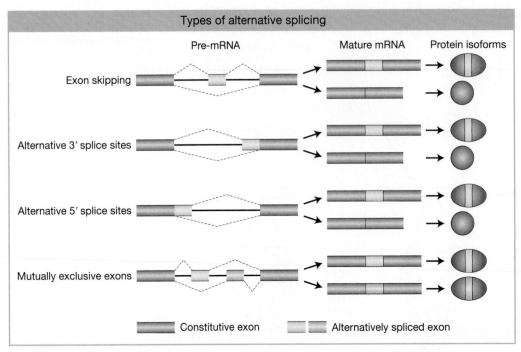

FIGURE 8-23 A single pre-mRNA that contains exons and introns can be spliced in different patterns to produce mature mRNAs that encode different proteins. There are four major types of alternative splicing. Dotted lines on the top and bottom of each pre-mRNA show how the 5′ and 3′ splice sites can be alternatively joined to produce different mature mRNAs.

to alternative splicing because their use by the spliceosome is not optimal and can thus be enhanced or suppressed by regulatory factors. Alternative splicing can also be affected by the rate of transcription elongation. Central to this mechanism is the fact that most splicing events take place during transcription. Spliceosomes assemble on introns as soon as they are transcribed. Slow elongation by RNA polymerase II provides a longer window of opportunity for the use of an alternative weak splice site before a strong splice site is transcribed.

> **KEY CONCEPT** The joining of exons in different patterns via alternative splicing greatly expands the number of proteins encoded in the human genome and other eukaryotic genomes.

RNA editing

RNA sequences encoded in eukaryotic genomes are not only changed by RNA processing events such as splicing but also by RNA editing. **RNA editing** is a general term that describes molecular processes through which nucleotide sequences in RNAs are changed after transcription. Editing events include insertion and deletion of nucleotides as well as base substitution. Many types of RNA, including tRNAs, rRNAs, mRNAs, and snRNAs, are edited.

In animals, the most common type of editing is adenosine-to-inosine (A-to-I) editing, which converts adenosine to inosine by deamination. A-to-I editing is catalyzed by double-strand RNA-binding enzymes called adenosine deaminase acting on RNAs (ADARs) (**Figure 8-24a**). Inosine is a non-canonical nucleoside that can base pair with cytidine, so during translation inosine is read as a guanosine,

rather than adenosine (Figures 8-24b and 8-24c), changing the amino acid sequence of a protein.

A-to-I editing can also affect regulatory elements in RNAs that function by base pairing to another RNA or are bound by a protein. For example, conversion of a stable A-U base pair into a less stable I-U base pair can alter splicing by affecting base pairing between snRNAs and pre-mRNAs. High-throughput RNA sequencing methods have identified over two million A-to-I edited sites in the human **transcriptome** (the set of all RNAs that can be expressed in humans). The physiological consequence of A-to-I editing at most of these sites is yet to be determined, but global effects on A-to-I editing caused by mutation of ADARs leads to behavioral and locomotion abnormalities in *Drosophila* and seizures and early death in mice, highlighting the importance of A-to-I editing.

RNA nucleotide modification

The structure and function of RNAs can be altered by post-transcriptional chemical modifications. Most RNA modifications consist of the addition of a methyl group (CH_3) to a nucleoside base such as N^6-methyladenosine (m^6A) (Figure 8-24d), to a ribose sugar such as 2′-O-methyladenosine (Am) (Figure 8-24e), or to both such as N^6, 2′-O-dimethyladenosine (m^6Am). More than 100 different chemical modifications of RNA have been identified. Each modification can have distinct effects on RNA structure and interactions with other RNAs and proteins, which can affect all aspects of RNA metabolism, including the processing, stability, and translation of mRNAs.

m^6A is the most common modification in human mRNAs. Of the more than 20,000 m^6A sites that have been identified in humans, 70 percent occur in the last exon of a

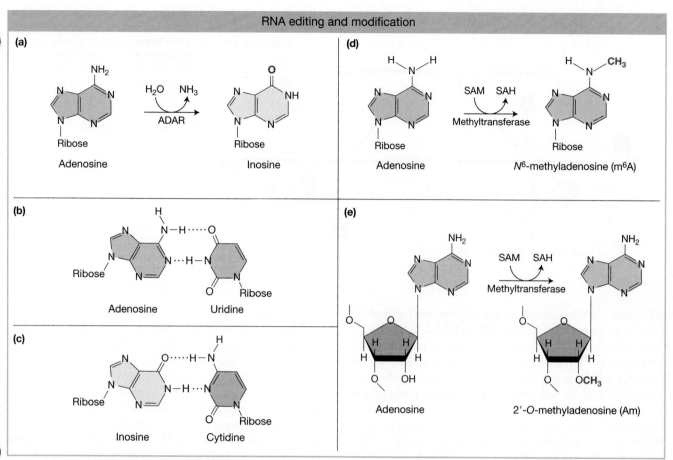

FIGURE 8-24 The information contained in RNA can be changed by editing and modification. (a) A-to-I editing is carried out by ADAR enzymes. (b) Adenosine base pairs with uridine, but (c) inosine base pairs with cytidine. (d) The chemical structure of ribonucleoside bases and (e) ribose sugars can be changed by methylation. These reactions are catalyzed by different methyltransferases, but both reactions use SAM as the methyl-group donor.

transcript and nearly half occur in the 3′ UTR. m⁶A modifications are added during transcription and prior to the completion of splicing. A major function of m⁶A is to destabilize mRNAs, as demonstrated by the finding that knocking out the methyltransferase that writes m⁶A into mRNA results in longer half-lives of m⁶A-containing mRNAs. Newly developed technologies that map the sites of chemical modifications in RNAs are making it possible for researchers to identify the writers, readers, and erasers of modifications as well as determine their molecular and biological functions.

KEY CONCEPT RNAs are subject to editing and modification. Editing can change the protein sequence encoded by an mRNA, and both editing and modification can create new signals in mRNAs and ncRNAs that change their structure, function, and stability.

RNA export from the nucleus

Many eukaryotic RNAs that are transcribed in the nucleus from the nuclear genome spend some part of their life in the cytoplasm. For example, mRNAs are exported from the nucleus to the cytoplasm where they are translated into proteins, and snRNAs involved in splicing are produced in the nucleus, exported to the cytoplasm for assembly with

proteins, and then returned to the nucleus. Export of mRNAs and snRNAs from the nucleus occurs by different mechanisms, but both mechanisms involve adaptor proteins that bind the RNAs early in their biogenesis and escort them to the cytoplasm through channels in the nuclear membrane called nuclear pores. In human cells, mRNAs are transported out of the nucleus by the TREX (transcription export) complex, whereas snRNA are transported by PHAX (phosphorylated adaptor for RNA export). Both TREX and PHAX interact with their RNA cargo during transcription through binding to the cap binding complex (CBC).

KEY CONCEPT Mechanisms exist in eukaryotic organisms to transport and localize RNAs to particular places in cells.

8.5 DECAY OF mRNA IN EUKARYOTES

LO 8.5 Describe how siRNAs regulate the abundance of specific RNAs and play a role in maintaining genome integrity in eukaryotes.

As in bacteria, decay counterbalances transcription to regulate the abundance of mRNAs in eukaryotes. The half-life of

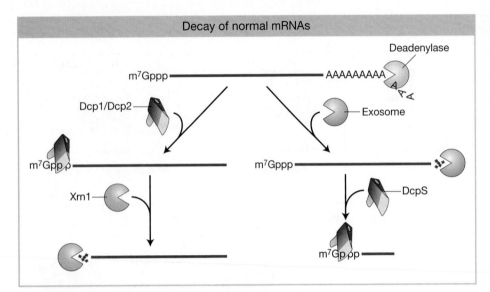

Decay of normal mRNAs

FIGURE 8-25 mRNA decay is normally initiated by removal of the poly(A) tail by a deadenylase enzyme. Deadenylation is followed either by decapping by Dcp1/Dcp2 and 5′-to-3′ decay by the exonuclease Xrn1 or by 3′-to-5′ decay by the exosome and decapping by DcpS.

eukaryotic mRNAs can vary widely. For example, the *β-globin* mRNA, which encodes a subunit of hemoglobin in blood cells, has a half-life of 20–24 hours. In contrast, the *c-Myc* mRNA, which encodes a transcription factor that regulates the cell cycle, has a half-life of 20–30 minutes. Mechanisms that control mRNA half-life help ensure that proteins are only present in cells when they are needed. mRNAs with a long half-life, like *β-globin*, tend to encode proteins with structural or metabolic functions, and mRNAs with a short half-life, like *c-Myc*, tend to encode proteins with regulatory functions.

The half-life of an mRNA is experimentally determined by turning off all RNA polymerase II transcription and then measuring how long it takes for half of the existing mRNA molecules to be degraded. The most common method involves treating cells with Actinomycin D, an inhibitor of RNA polymerase II elongation, and quantifying mRNA abundance by Northern blot or reverse transcription–PCR analysis (discussed in Chapter 10).

mRNA decay mechanisms

mRNA decay typically occurs in the cytoplasm and requires translation. Most mRNA decay occurs by two general pathways that are both initiated via removal of the poly(A) tail by a **deadenylase**, a special type of exonuclease that specifically cleaves phosphodiester bonds between adenosine nucleotides one at a time in the 3′-to-5′ direction (**Figure 8-25**). Deadenylation is sometimes followed by removal of the 5′ m⁷G cap by the **decapping enzyme** Dcp1/Dcp2. This enzyme cuts off 5′-m⁷Gpp (see Figure 8-16a), leaving behind an mRNA with a 5′-monophosphate that serves as a substrate for complete digestion by a 5′-to-3′ exonuclease Xrn1. Alternatively, deadenylation is followed by digestion by a 3′-to-5′ exonuclease called the exosome. Following decay of the mRNA body by the exosome, a different decapping enzyme called the scavenger decapping enzyme (DcpS) catalyzes the hydrolysis of the 5′ m⁷G cap, releasing 5′-m⁷Gp. The major difference between decay in bacteria and decay in eukaryotes is that most decay in eukaryotes is initiated by an exonuclease, not an endonuclease (compare Figures 8-10 and 8-25).

The efficiency of decay pathways can be enhanced or suppressed by RNAs and proteins that bind specific sequences within 3′ UTRs of mRNAs and affect the recruitment of decay factors. For example, base pairing of miRNAs to sequences within the 3′ UTR not only inhibits translation (discussed in Chapter 9) but also enhances decay by recruiting deadenylase and decapping enzymes. Similarly, sequences in the 3′ UTR that are rich in adenosine and uridine known as AU-rich elements (AREs) serve as binding sites for RNA-binding proteins that enhance or suppress decay by affecting recruitment of decay factors.

Related but distinct pathways detect and rapidly decay particular types of abnormal mRNAs during their translation and prevent the production of truncated or erroneous proteins. Nonsense-mediated decay (NMD) detects mRNAs that have a premature translation stop site, non-stop decay (NSD) detects mRNAs that lack a translation stop site, and no-go decay (NGD) detects mRNAs that contain sequences or structures such as strong stem-loops that stall translation elongation. Decay of abnormal mRNAs involves many of the same enzymes as decay of normal mRNAs, including decapping enzymes, deadenylases, and exonucleases, but it can also involve endonucleases.

KEY CONCEPT After removal of the poly(A) tail, mRNA decay by specialized enzymes occurs in both the 5′-to-3′ and 3′-to-5′ direction.

The discovery of RNA interference (RNAi)

In 2002, one of the leading science journals, *Science* magazine, named "Small RNA" as their Breakthrough of the Year. The RNAs to which they were referring were not the previously described small RNAs such as snRNAs or tRNAs, which are considered to have housekeeping roles and, as such, are synthesized all the time (i.e., constitutively). Instead, these other small RNAs are synthesized in response to changes in a cell's developmental state or its surroundings. We now know that small RNAs are critically important for

Three experiments demonstrating gene silencing

(a) Jorgensen: insertion of transgene

1. Pigment transgene inserted into the petunia genome.

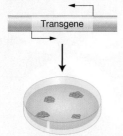

Cells from deep violet plants

2. Transgenic plants have white sectors in flowers.

Conclusion: dsRNAs produced from the transgene silence expression from the transgenic and endogenous pigment genes.

(b) Fire/Mello: injection of dsRNA

1. ssRNA and dsRNA synthesized in the lab.

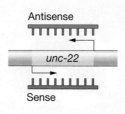

2. ssRNA and dsRNA injected into adult worms.

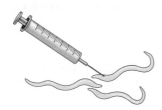

3. F1 progeny display muscle defects.

Conclusion: injected dsRNAs silence *unc-22* expression.

(c) Baulcombe: insertion of viral gene

1. Viral gene inserted into tobacco plant.

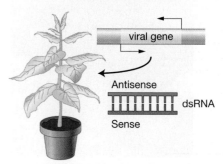

2. Plant exposed to virus but remains healthy.

Conclusion: dsRNAs produced from the inserted viral gene silence expression of the gene from invading viruses.

FIGURE 8-26 Three experiments reveal key features of gene silencing. (a) Jorgensen discovered that a transgene can silence an endogenous petunia gene necessary for floral color. (b) Fire and Mello demonstrated that dsRNA can selectively silence genes in *C. elegans*. (c) Baulcombe showed that plants with a copy of a viral transgene were resistant to viral infection and produced siRNAs complementary to the viral genome.

the regulation of gene expression and the maintenance of a stable genome. Studies that led to the discovery of one class of small RNA called small interfering RNA (siRNA) were preceded by several reports of unanticipated changes in gene expression when RNAs were injected into an organism or expressed from a **transgene** (a gene introduced by researchers into the chromosomes of an organism).

One of the greatest joys of doing scientific research is observing a completely unexpected result. In 1990, this is precisely what occurred for Richard Jorgensen in his studies of plant coloration. To increase pigmentation in petunia plants that had pale pink flowers, Jorgensen inserted a petunia gene that encodes an enzyme necessary for the synthesis of purple-blue flower pigment (**Figure 8-26a**). As a control, he inserted the same gene into plants with deep violet flowers. He expected that relative to the parental plants, the flower color of the pale pink transgenic plants would be darker, but the deep violet transgenic plants would be unchanged. However, none of the transgenic plants had darker flowers. In fact, all the transgenic plants produced flowers that were either pure white or a variety of white patterns (**Figure 8-27**).

In a totally unexpected outcome, Jorgensen found that the transgene suppressed the expression of its own mRNA as well as the mRNA produced from the endogenous pigment gene (i.e., the one that is normally in the petunia genome). Therefore, he called the phenomenon co-suppression.

The mechanism underlying co-suppression remained a mystery until 1998, when Andrew Fire and Craig Mello used the roundworm *C. elegans* to test the hypothesis that **double-stranded RNA (dsRNA)** was the agent that triggered co-suppression. They injected *C. elegans* with RNAs that were identical in sequence to an endogenous gene that when mutated causes adult worms to twitch (Figure 8-26b). If the injected RNAs triggered co-suppression, they expected to see the twitcher phenotype even though the endogenous gene was intact. Indeed, they found that, relative to injection of single-stranded RNA, injection of dsRNA caused a much stronger twitcher phenotype, demonstrating that dsRNA mediates suppression of endogenous gene expression in a process that is now called **RNA interference (RNAi)**. The study by Fire and Mello also uncovered other remarkable features of RNAi: (1) RNAi is very specific, and only RNAs

Petunia flowers demonstrating co-suppression

Wild type   **Transgenic**   **Transgenic**

(a)   (b)   (c)

FIGURE 8-27 (a) The wild-type (no transgene) phenotype. (b and c) So-called co-suppression phenotypes resulting from insertion of a transgene that controls pigmentation into the genome of a wild-type petunia. In white regions of the petals, both the transgene and the endogenous chromosomal copy of the same gene have been inactivated. [*Richard Jorgensen, Department of Plant Biology, Carnegie Institution for Science.*]

with perfect complementarity to the dsRNA are affected; (2) RNAi is extremely potent, as only a few dsRNA molecules are required per cell to inhibit expression of the targeted gene, indicating that the process is catalytic; (3) RNAi can affect cells and tissues that are far removed from the site of introduction, indicating that there is an RNA transport mechanism; and (4) RNAi affects the progeny of injected animals, indicating that the targeting information is heritable.

Many labs continue to study the mechanism underlying RNAi, which is discussed in the next section. Nevertheless, even without a complete understanding of the mechanism, RNAi has had a tremendous impact on almost all fields of biology research through its use as a tool to perform loss-of-function experiments. Researchers have developed creative methods to introduce or express dsRNAs in cells and whole organisms to reduce the expression of a specific gene and determine its necessity for molecular, cellular, and organismal processes. The terms *knockdown* and *silencing* are used in conjunction with RNAi because the reduction in the abundance of targeted RNAs is rarely complete; instead, it is knocked down or silenced.

RNAi technologies have also been developed to perform genome-wide screens for genes involved in cellular processes. In this approach, libraries of dsRNAs are generated that target all of the protein-coding genes in an organism, and screens are performed to identify the few dsRNAs that produce a desired phenotype. Thus, RNAi has made it much easier to perform genetic studies in many organisms and cell culture systems for which there had been no simple method to manipulate gene expression.

siRNA-mediated RNA decay and transcriptional silencing

RNAi silences gene expression by targeting RNAs for decay in the cytoplasm of cells. The RNAi decay mechanism involves three main components: (1) small interfering RNAs (siRNAs) that provide the specificity of RNAi by base pairing to target RNAs, (2) Dicer, an endonuclease that precisely cuts dsRNAs into siRNAs, and (3) Argonaute (Ago), an RNA endonuclease that is programmed to cut RNAs that base pair to bound siRNAs.

siRNAs are approximately 21-nucleotide dsRNAs. Each strand of an siRNA has a 5'-monophosphate, a 3'-hydroxyl, and a two-nucleotide 3' overhang beyond the core base-paired region of 19 nucleotides (**Figure 8-28**). These features of siRNAs are important for their recognition by proteins that carry out RNAi. Dicer uses its PAZ domain to bind the 3' overhang and generate an siRNA from a hairpin or long dsRNA via its two endonuclease domains (**Figure 8-29**). Once bound to the end of a dsRNA, the endonuclease domains are positioned to make cuts in the strands that are 21 nucleotides away and staggered by two nucleotides. Dicer can repeat this process, producing multiple siRNAs from a single dsRNA. One of the 3' overhangs of siRNAs is then bound by the PAZ domain of Ago, which is part of a multi-protein complex called

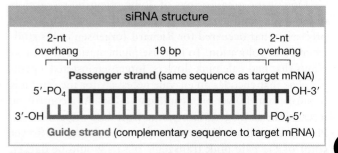

siRNA structure

2-nt overhang   19 bp   2-nt overhang

Passenger strand (same sequence as target mRNA)

5'-PO$_4$ OH-3'

3'-OH   PO$_4$-5'

Guide strand (complementary sequence to target mRNA)

FIGURE 8-28 Small interfering RNAs (siRNAs) that are produced from hairpin or long dsRNAs by Dicer have specific features that are important for recognition by Ago. siRNAs are 19–21 base pairs in length and each strand has a 5'-phosphate and a 2-nucleotide 3' overhang with a 3'-hydroxyl.

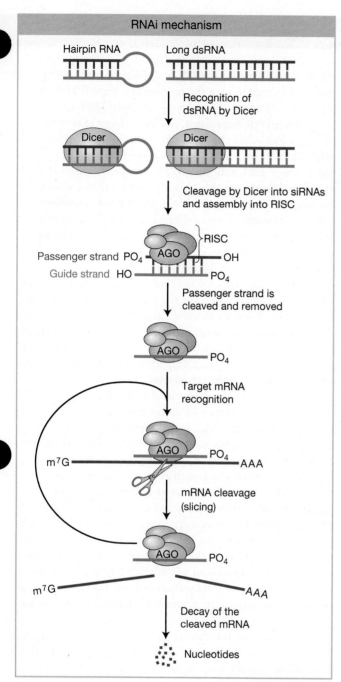

FIGURE 8-29 In the RNAi pathway, Dicer produces siRNAs from dsRNA, and siRNAs are bound by Ago-containing RISC. The siRNA guide strand targets RISC to specific RNAs by base pairing. The endonuclease activity of Ago cuts the target RNA into two fragments that are decayed to nucleotides by normal decay pathways (see Figure 8-25). RNAi results in silencing (i.e., knock down) of the expression of targeted genes.

RNA-induced silencing complex (RISC). Ago also uses its PIWI endonuclease domain to cut and displace one of the siRNA strands called the passenger strand, leaving behind a single-stranded siRNA guide strand. Perfect base pairing between an siRNA and a target mRNA stimulates cleavage of the target by the Ago PIWI domain. The resulting pieces of target RNA are decayed by the normal pathways,

involving deadenylation, decapping, and exonucleolytic cleavage from the 5′ and 3′ ends.

> **KEY CONCEPT** Dicer cuts dsRNA to produce 21-bp siRNAs with 2-nt overhangs on each end. siRNAs are bound by RISC, which contains Ago, an endonuclease that cuts the passenger strand, leaving the guide strand intact. When the guide strand base pairs to a complementary segment of mRNA, Ago cleaves the mRNA, triggering its degradation.

In some cases, siRNAs can enter the nucleus and block transcription of target genes by inducing heterochromatin formation (discussed in Chapter 12). For example, in *C. elegans,* siRNAs are transported from the cytoplasm to the nucleus by an Ago protein that lacks endonuclease activity. The Ago-siRNA complex localizes to a specific gene, presumably by base pairing of the siRNA to an mRNA during its synthesis. This localization leads to recruitment of a methyltransferase enzyme that methylates histone H3 on lysine 9 (H3K9me) to generate heterochromatin and turn off transcription of the gene.

RNAi protects the genome from foreign DNA

The function of RNAi is clearly not to shut off genes at the whim of scientists. In fact, in plants, RNAi is a form of antiviral defense. This discovery came from experiments conducted by David Baulcombe and co-workers who had engineered the genome of tobacco plants to express a viral gene (see Figure 8-26c). He found that plants engineered with a viral transgene but not plants that lacked the transgene were resistant to subsequent infection by the virus. Furthermore, he found that that resistant plants, and only the resistant plants, produced large amounts of siRNAs complementary to the viral genome. Thus, RNAi serves as an antiviral defense system.

RNAi also silences the expression of endogenous repeated sequences such as transposons, which make up a large part of many eukaryotic genomes (Chapter 16). When translated into proteins, transposons facilitate insertion of their DNA into new sites in the genome. Transposon mobilization can disrupt host genes and promote chromosomal rearrangements, leading to diseases such as cancer. However, some transposons contain inverted repeat sequence or antisense promoters that can produce dsRNA and trigger RNAi (see Figure 16-28). Thus, one of the normal functions of RNAi is to protect against invading sources of nucleic acids such as viruses and transposons that threaten the integrity of the host genome.

> **KEY CONCEPT** Many eukaryotic organisms use siRNA-mediated RNAi to silence the expression of foreign genes. Researchers have taken advantage of the endogenous RNAi machinery to knock down the expression of a specific gene by introducing into cells dsRNA that is identical in sequence to the target gene.

SUMMARY

RNAs serve numerous and varied purposes in bacteria and eukaryotic cells. mRNA garners much of the attention in this book and elsewhere because it is the template for the synthesis of proteins, which perform the vast majority of structural and enzymatic roles in cells. Nevertheless, ncRNAs are also important. For example, the translation of mRNA into proteins cannot take place without tRNAs and rRNAs, and ncRNAs function at many steps in gene expression. ncRNAs operate in three general ways: they interact with other RNAs and DNA by base pairing, they serve as scaffolds for the assembly of proteins, and they carry out enzymatic reactions. The activities of ncRNAs are integral to transcription and RNA processing as well as other events in gene expression, including translation (Chapter 9). For example, in ribosomes (the protein synthesis machines), sequences in tRNAs base pair to sequences in mRNA, rRNAs serve as scaffolds for assembly of ribosomal proteins, and rRNAs catalyze peptide bond formation between amino acids. The variety of RNA functions is possible because of RNA's unique chemical features.

In bacteria, all RNAs are synthesized by a single RNA polymerase, whereas in eukaryotic cells there is a division of labor among three RNA polymerases (I, II, and III). Regardless of the organism or the RNA polymerase, the sites of transcription initiation are marked by conserved promoter elements that are located near the transcription start site and are bound by proteins. In bacteria, a subunit of RNA polymerase binds promoter elements, but in eukaryotes, GTFs bind promoter elements and recruit a specific RNA polymerase. After RNA polymerase recruitment, the DNA is locally unwound and RNA polymerase begins incorporating ribonucleotides in the 5′-to-3′ direction that are complementary to the template DNA strand. As transcription transitions from the initiation to the elongation phase, the factors associated with RNA polymerase change. This is exemplified by sigma factors in bacteria and factors that associate with different modified forms of the RNA polymerase II CTD in eukaryotes. In the case of the CTD, associated factors are involved in processing of the pre-mRNA and can alter the elongation rate of RNA polymerase II, including causing RNA polymerase II to pause at particular sites. In both bacteria and eukaryotes, RNA polymerase terminates transcription after transcribing termination signals into a nascent mRNA, and multiple mechanisms are involved in recognizing the termination signals and releasing the mRNA and RNA polymerase from DNA.

In eukaryotic organisms, RNAs undergo extensive processing, often while they are being transcribed. For example, mRNAs are modified with a cap at the 5′ end and a poly(A) tail at the 3′ end, and introns are removed. Modifications at the ends increase the stability of an mRNA and assist translation. Sequences within a pre-mRNA together with snRNA or protein factors that bind them define sequences as intron or exon for splicing as well as dictate the site of cleavage and polyadenylation. Additional factors that bind other RNA sequence elements can enhance or suppress the use of particular sites for splicing. This leads to alternative splicing, which can increase the types of proteins encoded by a gene from one to, in some cases, thousands. In bacteria, ribosomes associate with mRNAs as they are being transcribed, whereas in eukaryotes ribosome association and translation can take place only after mRNAs are exported from the nucleus to the cytoplasm.

The last step in the life cycle of an RNA is decay. mRNA decay in both bacteria and eukaryotes occurs via defined pathways that begin with recruitment of specific enzymes. In bacteria an endonuclease is recruited by interacting with a 5′-monophosphate, and in eukaryotes an exonuclease is recruited by interacting with proteins that associate with the 3′ UTR. The initiating steps of decay generate recognition sites for further decay by other enzymes. Additionally, in some eukaryotes, very short RNAs such as siRNAs base pair to mRNAs and bring along an endonuclease that initiates decay. One of the normal functions of siRNAs is to silence the expression of repetitive genes in genomes such as transposons. Researchers have taken advantage of this decay activity to perform targeted loss-of-function experiments.

The recent development of technologies that detect low-abundance RNAs has led to the discovery of thousands of RNAs of unknown function. However, researchers have a head start in figuring out how these RNAs are transcribed, processed, transported, and decayed as well as how they function, because it is likely that aspects of the mechanisms are shared with mRNAs, tRNAs, rRNAs, snRNAs, or siRNAs.

KEY TERMS

3′ splice site (p. 285)
3′ untranslated region (3′ UTR) (p. 275)
7-methylguanosine (m^{7}G) (p. 283)
5′ splice site (p. 285)
5′ untranslated region (5′ UTR) (p. 274)
allosteric termination model (p. 282)

alternative splicing (p. 288)
branch point (p. 285)
cap (p. 283)
carboxy-terminal domain (CTD) (p. 277)
consensus sequence (p. 274)

deadenylase (p. 292)
decapping enzyme (p. 292)
decay (p. 276)
deoxyribose (p. 270)
double-stranded RNA (dsRNA) (p. 293)

downstream (p. 274)
elongation (p. 273)
endonuclease (p. 276)
exon (p. 284)
exonuclease (p. 276)
factor-independent termination (p. 275)
general transcription factor (GTF)
    (p. 277)
half-life (p. 276)
helicase (p. 276)
initiation (p. 273)
intron (p. 284)
isoform (p. 288)
long noncoding RNA (lncRNA)
    (p. 271)
messenger RNA (mRNA) (p. 270)
microRNA (miRNA) (p. 271)
noncoding RNA (ncRNA) (p. 270)
non-template strand (coding strand)
    (p. 272)
nucleolus (p. 278)

phosphodiester bond (p. 273)
poly(A) polymerase (PAP) (p. 284)
poly(A) tail (p. 284)
polyadenylation (p. 284)
precursor RNA (pre-RNA) (p. 277)
preinitiation complex (PIC) (p. 280)
promoter (p. 274)
proteome (p. 288)
Rho-dependent termination (p. 275)
ribose (p. 270)
ribosomal RNA (rRNA) (p. 271)
ribozyme (p. 287)
RNA editing (p. 290)
RNA interference (RNAi) (p. 293)
RNA polymerase (p. 273)
RNA polymerase I (p. 277)
RNA polymerase II (p. 277)
RNA polymerase III (p. 277)
RNA polymerase core enzyme (p. 274)
RNA polymerase holoenzyme (p. 274)
RNA processing (p. 277)

sigma factor (σ) (p. 274)
small interfering RNA (siRNA) (p. 271)
small nuclear ribonucleoprotein
    (snRNP) (p. 285)
small nuclear RNA (snRNA) (p. 271)
spliceosome (p. 285)
splicing (p. 284)
TATA box (p. 279)
template strand (noncoding strand)
    (p. 272)
termination (p. 273)
torpedo termination model (p. 282)
transcript (p. 272)
transcription (p. 268)
transcription bubble (p. 272)
transcription start site (p. 274)
transcriptome (p. 290)
transfer RNA (tRNA) (p. 271)
transgene (p. 293)
upstream (p. 274)
uracil (U) (p. 270)

PROBLEMS

Visit SaplingPlus for supplemental content. Problems with the icon are available for review/grading.

WORKING WITH THE FIGURES

(The first 30 questions require inspection of text figures.)

1. In Figures 8-1a and b, draw a generic mRNA at each stage in the life cycle.

2. In Figure 8-2, if the chase was continued for a longer period of time, how would the distribution of radioactive RNAs change, and why?

3. In Figure 8-3, draw the ribonucleotide uridine-5′-monophosphate (UMP) base paired to adenosine-5′-monophosphate (AMP).

4. In Figure 8-4, what would have to happen in a cell to convert uracil into uridine triphosphate?

5. In Figure 8-5a, place an arrow at the location of the transcriptional promoter for each gene.

6. In Figure 8-5c, put a circle around 5′ ribose carbons and a square around 3′ ribose carbons.

7. In Figure 8-6, are adjacent rDNA genes transcribed in the same direction, or in different directions?

8. In Figure 8-7b, write the sequence of the first 10 nucleotides of the *rrn D1* transcript.

9. In Figure 8-8, what change occurs to RNA polymerase as it initiates transcription?

10. In Figure 8-9, write the sequence of the DNA template and non-template strands that encode the factor-independent mRNA stem-loop termination signal.

11. In Figure 8-10, why is pyrophosphohydrolase required only for the first endonucleolytic cleavage by RNase E and not for subsequent cleavage events?

12. In Figure 8-11, why are RNA polymerase II or III not recruited to rDNA promoters?

13. In Figure 8-12, what GTF is most likely to recognize promoters that lack a BRE, TATA box, Inr, and DPE? Provide a rationale for your answer.

14. In Figure 8-13, which GTF could be mutated to block transcription of all three types of RNA polymerase III genes?

15. In Figure 8-14, what happens to the CTD between the third and fourth steps in the model? What type of enzyme carries out this reaction?

16. In Figure 8-15, 3′-end formation is "coupled" to transcription termination. Explain what this means in the context of the allosteric and torpedo models.

17. In Figure 8-16a, does the identity of the first nucleotide in the RNA chain (A, C, G, or U) affect capping? Why or why not?

18. In Figure 8-17, which of the sequence elements that regulate cleavage and polyadenylation are retained in the mRNA after the reaction is complete?

19. In Figure 8-18, mutations of which five intron nucleotides is most likely to block splicing? What data support your hypothesis?

20. In Figure 8-19a, what snRNPs are in the spliceosome when the catalytic steps of splicing occur?

21. In Figure 8-20, what is the primary difference between Group I and Group II self-splicing?

22. In Figure 8-21, what other mRNAs could be produced by alternative splicing of the blue and green exons?

23. In Figure 8-22, which exons are spliced by a mutually exclusive mechanism?

24. In Figure 8-23, for each mechanism, which alternatively spliced product is expected to be more frequently produced if transcription is slow?

25. In Figure 8-24, draw N^6, $2'$-O-dimethyladenosine (m^6Am).

26. In Figure 8-25, is the $5'$-to-$3'$ decay pathway on the left or right?

27. In Figure 8-26, what is the source of the foreign dsRNA in each of the experiments?

28. In Figures 8-27b and c, in what part of the flower is co-suppression (RNAi) occurring?

29. In Figure 8-28, draw the location of the Ago PAZ and PIWI domains on the siRNA.

30. In Figure 8-29, why are only a few siRNA molecules needed to knock down hundreds or even thousands of copies of an mRNA?

BASIC PROBLEMS

31. Draw the longest continuous base-pairing interaction between the following RNAs:

$5'$-AAUGCCGGUAACGAUUAACGCCCGAUAUCCG-$3'$

$5'$-GAGCUUCCAUAUCGGGCGUUGGUGAUUCGAA-$3'$

32. What role does the branch point ribose $2'$-OH play in the splicing reaction?

33. How are the ends of an mRNA protected to prevent decay?

34. Why might a mutation in a $3'$ UTR affect the rate of decay of an mRNA?

35. How does Rho in bacteria function similarly to Xrn2 in eukaryotes to terminate transcription?

36. What problem is encountered by $3'$-to-$5'$ exonucleases that might block the complete decay of excised introns?

37. What is the primary function of the sigma factor in bacteria? Is there a factor in eukaryotes that is functionally analogous to the sigma factor?

38. Write the sequence of the template and non-template strands of DNA that encode the following fragment of a bacterial mRNA:

pppGUUCACUGGGACUAAAGCCCGGGAACUAGG

39. Write the sequence of the template and non-template strands of DNA that encode the following eukaryotic mRNA, where the underlined sequence is the poly(A) tail:

m^7GpppGUUCACUGGGACUGAAUAAAGGGAAC-UAGGA<u>AAAAAAAAAAAAA</u>$_{(n = 150)}$

40. Draw the possible alternative splicing products of the following pre-mRNA, where the white boxes are constitutive exons and the shaded boxes are alternative exons:

41. Develop a consensus sequence for the following six RNA sequences:

UCGGUAGAUCCC
CCGCGAGGUUCC
CCGAAAGACCCC
UCGCGAGACUCC
UCGACAGGCUCC
CCGUAAGGUCCC

42. Draw the base-pairing interaction between the following $5'$ splice site (the exon is underlined and the intron is not underlined) and the U1 snRNA, and count the number of hydrogen bonds:

$5'$-<u>CAG</u>GUGACU-$3'$

43. A researcher repeated the pulse–chase experiment shown in Figure 8-2 with UTP that was radioactively labeled on the gamma (γ) phosphate and were unable to detect radioactive RNA in the cells. Why?

44. In addition to phosphorylation of serines 2 and 5 in the CTD repeats of RNA polymerase II, phosphorylation also occurs on serine 7. What additional "codes" are possible for factor recruitment?

45. How often does a random 21-nucleotide siRNA sequence appear in the human genome? How might the answer explain the specificity of RNAi?

46. Can a bacterial promoter direct transcription initiation in a eukaryotic cell? Why or why not?

47. A researcher found that the abundance of an mRNA increased between normal and stressed conditions. What two processes might be affected by the stress?

48. If you knew the sequence of an mRNA and its genes, how would you determine where the introns were located in the pre-mRNA?

49. If you isolated an mRNA from a eukaryotic cell, what features would it have at its $5'$ and $3'$ ends if it is full length?

50. If you had the sequence of an mRNA and the genome of a new organism, how would you determine the location in the genome of the transcription start site of the gene that encodes the mRNA?

51. Describe two functions for ATP and GTP in the production of RNA. 🐾

52. Draw base-paired passenger and guide strands of an siRNA that could be used to knock down the following mRNA:

5′-AAGUCCGGCAAUGCGACCAAGUCGUAAGCU-
UUAGGCGUCUUGGCAAAGA-3′

53. In bacteria and eukaryotes, describe what else is happening to an mRNA while RNA polymerase is synthesizing it from the DNA template.

54. Based on the experiment used to test the requirement for base pairing between the U1 snRNA and the 5′ splice site (Figure 8-18c), how would you test the requirement of base pairing in the stem-loop structure for Rho-dependent termination (Figure 8-9a)?

55. In Figure 8-29, propose how the 5′ mRNA fragment produced by RNAi is decayed to nucleotides.

56. What makes poly(A) polymerase an unusual nucleic acid polymerase?

57. A researcher sequenced an RNA and found that there was a G in a position where there was an A in the genome. What is likely to have happened to the RNA?

58. Which of the types of RNA polymerase III genes is most likely to also be transcribed by RNA polymerase II, and why?

59. Why is single-stranded RNA less stable in a test tube than single-stranded DNA?

60. List four similarities and four differences between eukaryotic mRNAs and ncRNAs. 🐾

CHALLENGING PROBLEMS

61. What information argues for and against the possibility that DNA is directly used as a template for translation? 🐾

62. The following data represent the base compositions of double-stranded DNA from two different bacterial species and their RNA products obtained in experiments conducted in vitro:

Species	$\frac{(A + T)}{(G + C)}$	$\frac{(A + U)}{(G + C)}$	$\frac{(A + G)}{(U + C)}$
Bacillus subtilis	1.36	1.30	1.02
E. coli	1.00	0.98	0.80

a. From these data, determine whether the RNA of these species is copied from a single strand or from both strands of the DNA. Draw a diagram to show how you solve this problem.

b. How can you tell if the RNA itself is single-stranded or double-stranded?

63. Researchers performed a genetic screen for genes that increase the lifespan of *C. elegans*. They sequenced the complete genome of a mutant with a longer lifespan and found a single A-to-T base change. List the possible ways in which the A-to-T change could alter gene expression to produce the longer lifespan phenotype. For example, the A-to-T could change the amino acid sequence of a protein. 🐾

GENETICS AND SOCIETY

In 2018, the first therapy based on RNAi (RNA interference) was approved by the US Food and Drug Administration (FDA). A pharmaceutical company used RNAi to treat hereditary transthyretin amyloidosis, a progressive and often fatal disease caused by an autosomal dominant mutation in the *transthyretin* gene that makes a toxic form of the transthyretin protein. The drug silences expression of the *transthyretin* gene using an siRNA (small interfering RNA) that targets the transthyretin mRNA for destruction. Based on your knowledge of how RNAi works, why do you think that this first RNAi drug has provided great optimism that RNAi-based therapeutics will become a widespread approach to address genetic diseases?

Proteins and Their Synthesis

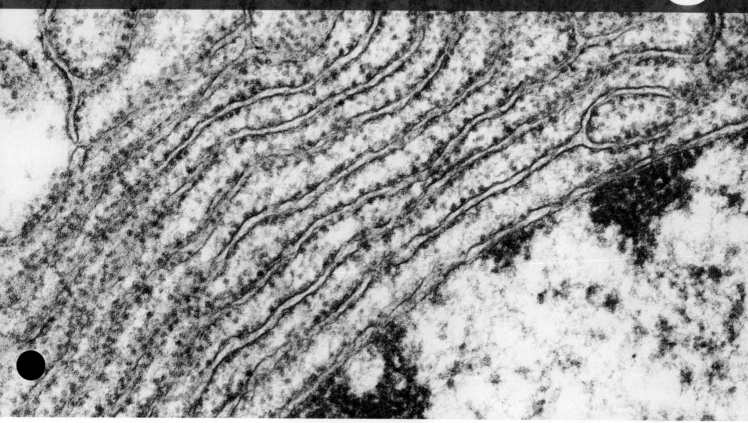

Ribosomes are RNA-protein machines that synthesize proteins in bacteria and eukaryotes. Many ribosomes (rows of black dots) associate with the endoplasmic reticulum in the cytoplasm of eukaryotic cells. [*JOSEPH F. GENNARO JR./Science Source.*]

CHAPTER OUTLINE AND LEARNING OBJECTIVES

Chapters 7 and 8 describe the first two stages of information transfer shown in Figure 1-10: replication (the synthesis of a DNA copy of DNA) and transcription (the synthesis of an RNA copy of a segment of DNA). In this chapter, you will learn about the final stage of information transfer: translation (the synthesis of a protein copy of an RNA).

RNAs play several important roles in translation. As you learned in Chapter 8, RNAs are classified as either messenger RNA (mRNA) or non-coding RNA (ncRNA). The majority of genes encode mRNAs, whose function is to serve as an intermediate in the synthesis of proteins. In contrast, ncRNAs are active as functional RNAs; they are never translated into proteins. ncRNAs involved in protein synthesis include transfer RNAs and ribosomal RNAs. **Transfer RNAs (tRNAs)** carry out the decoding work of translation, associating three-nucleotide sequences in an mRNA with their corresponding amino acids. This decoding occurs inside **ribosomes**, which are composed of several types of **ribosomal RNAs (rRNAs)** and many different proteins (**Figure 9-1**). Ribosomes assemble on mRNAs and catalyze protein synthesis by chemically binding together the amino acids brought to the ribosome by tRNAs. Like tRNAs, ribosomes are general in function, in the sense that they can translate any mRNA.

Although most genes encode mRNAs, ncRNAs make up the largest fraction of total cellular RNA. In a typical actively dividing eukaryotic cell, rRNA and tRNA account for almost 95 percent of the total RNA, whereas mRNA accounts for only about 5 percent. Two factors explain the abundance of rRNAs

and tRNAs. First, they are much more stable than mRNAs, so they remain intact much longer. Second, ribosomes are an abundant component of cells. There are tens of thousands of ribosomes in bacterial cells, about 200,000 ribosomes in yeast cells, and several million ribosomes in mammalian cells.

Components of the translational machinery and the process of translation are very similar in bacteria and eukaryotes. In addition to ribosomes, tRNAs, and mRNAs, each phase of translation involves a distinct set of protein regulatory factors; **initiation factors (IFs)** start translation at the beginning of the mRNA open reading frame (ORF), **elongation factors (EFs)** maintain translation through the ORF, and **termination factors**, also called **release factors (RFs)**, stop translation at the end of the ORF.

The major feature that distinguishes translation in bacteria from that in eukaryotes is the location where transcription and translation take place in the cell: the two processes take place in the same compartment in bacteria, whereas they are physically separated in eukaryotes by the nuclear membrane. After processing, eukaryotic mRNAs are exported from the nucleus for translation by ribosomes that reside in the cytoplasm. In contrast, transcription and translation are coupled

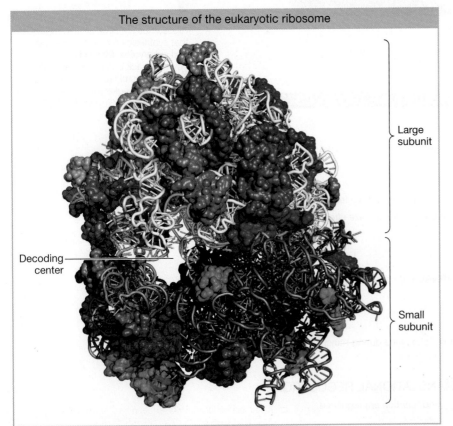

The structure of the eukaryotic ribosome

Decoding center

Large subunit

Small subunit

FIGURE 9-1 In the yeast ribosome (shown), there are four individual rRNAs and about 80 proteins. RNAs are shown in white and gray in the large and small subunits, respectively. Proteins are shown in shades of red and shades of blue in the large and small subunits, respectively. [PDB ID 4V7R.]

in bacteria: translation of an RNA begins at its 5′ end while the rest of the mRNA is still being transcribed.

KEY CONCEPT Translation occurs within ribosomes and requires three types of RNAs: mRNAs carry the sequence information from DNA to ribosomes, tRNAs decode mRNA nucleotide sequences into amino acids, and rRNAs are structural and functional components of ribosomes.

9.1 PROTEIN STRUCTURE

LO 9.1 Explain how the interactions of amino acids determine the structure of proteins.

Before considering how proteins are made, let's start with a discussion of the structure of proteins. Proteins are polymers

composed of building blocks called **amino acids.** In other words, a protein is a chain of amino acids. Because amino acids were once called peptides, a chain is sometimes referred to as a **polypeptide.** Amino acids have the general formula

All amino acids have two functional groups (an amino group and a carboxyl group, shown above) bonded to the same carbon atom (called the α carbon). Also attached to the α carbon is a hydrogen (H) atom and a **side chain,** known as an **R (reactive) group.** There are 20 common amino acids that can make up proteins, each amino acid having a different R group that gives it unique properties (**Figure 9-2**). The side chains are

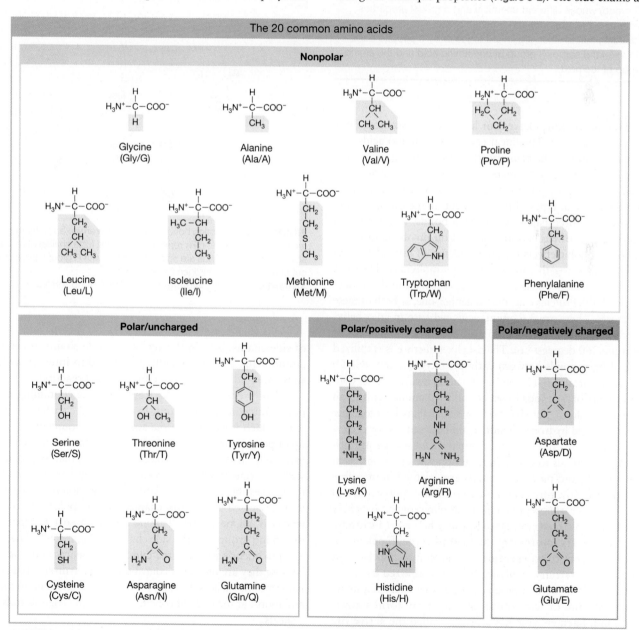

FIGURE 9-2 The chemical structures of the side chains (R groups) of the 20 common amino acids are grouped based on their polarity and charge. Each amino acid is labeled with its full name, abbreviation, and single letter designation such as Glycine, Gly, and G, respectively.

categorized into four groups based on their chemical properties: nonpolar, polar/uncharged, polar/positively charged, and polar/negatively charged. As described in this chapter, the chemical properties of side chains play a role in determining the folded structures of proteins.

In proteins, amino acids are linked together by covalent bonds called **peptide bonds**. A peptide bond is formed by linkage of the **amino group** (NH_3^+) of one amino acid with the **carboxyl group** (COO^-) of another amino acid (**Figure 9-3**). One water molecule is removed during the reaction. Because of the way in which peptide bonds form, a polypeptide chain always has an amino end (N-terminal end) and a carboxyl end (C-terminal end).

KEY CONCEPT Each of the 20 common amino acids has an amino group and a carboxyl group as well as a side chain (R group) whose different chemical and physical properties determine the structure and function of proteins. In polypeptides, amino acids are linked together by a peptide bond between the carboxyl group of one amino acid and the amino group of the next.

Protein structures have four levels of organization, illustrated in **Figure 9-4**. The linear sequence of amino acids in a protein constitutes the **primary structure**. Local regions of the protein fold into specific shapes, called **secondary structures**. Each shape arises from bonding forces between amino acids, including several types of non-covalent interactions, notably electrostatic forces such as hydrogen bonds, van der Waals forces (a type of electrostatic interaction involving dipoles), and hydrophobic effects (i.e., the tendency of nonpolar molecules to gather together to exclude water molecules). The most common secondary structures are the α-helix and the β-sheet (Figure 9-4b). Proteins can contain neither, one, or both of these structures. There are 3.6 amino acids per turn in an α-helix, which means that each amino acid occupies 100 degrees of rotation (360 degrees/3.6). The α-helix structure is stabilized by hydrogen bonds between carbonyl oxygen atoms (C=O) and amide groups (NH) four amino acids away. β-sheets consist of pairs of β-strands (stretches of 3–10 amino acids in an extended conformation) lying side by side that are kept together by inter-strand hydrogen bonds, again between carbonyl oxygen atoms and amide groups. Antiparallel β-strands are oriented (N-terminus to C-terminus) in opposite directions, as in Figure 9-4b, and parallel β-strands are oriented in the same direction. Lastly, turns composed of a few amino acids and loops of longer stretches of amino acids often connect α-helices to α-helices, β-strands to β-strands, and α-helices to β-strands.

Tertiary structures are the overall three-dimensional shape of an entire polypeptide. In addition to noncovalent interactions, tertiary structures can be stabilized by covalent disulfide bridges between cysteine side chains. Cysteine is the only amino acid whose side chain can form a covalent bond. Each enzyme has a pocket called the **active site** into which its substrate or substrates fit. The active sites of enzymes are good illustrations of the precise interactions

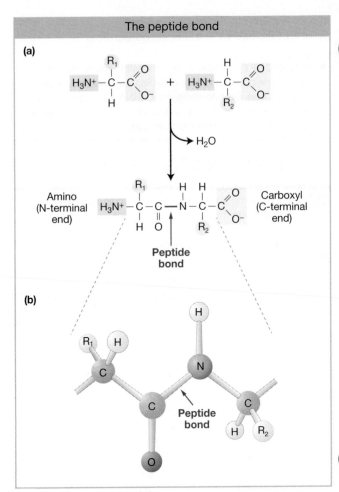

FIGURE 9-3 (a) Peptide bonds form between amino acids by the removal of water. R_1 and R_2 represent R groups (side chains) of amino acids. (b) Peptide bonds are rigid planar units with R groups projecting out from the carbon-nitrogen (C–N) backbone.
ANIMATED ART 🐾 SaplingPlus Peptide bond formation

of side chains. Within the active site, side chains of certain amino acids are strategically positioned to interact with a substrate and catalyze a specific chemical reaction.

Most proteins have a hydrophobic core containing nonpolar amino acid side chains. In contrast, the surface of proteins, which is exposed to the aqueous environment, is made up of polar amino acids, including those that are positively or negatively charged. The surface location of polar amino acids such as serine, threonine, tyrosine, lysine, and arginine makes them accessible to post-translation modification by enzymes, a topic discussed later in this chapter. The nonpolar amino acid proline is unique among all amino acids because it incorporates the amino group into the side chain (see Figure 9-2). Prolines are infrequently found in the middle of α-helices and β-sheets because they are unable to contribute to the hydrogen bonding pattern of the helices. Instead, prolines are often found in turns and loops, at the ends of α-helices, and in the edge strands of β-sheets.

The folding of polypeptides into their correct conformation will be discussed at the end of this chapter. At present,

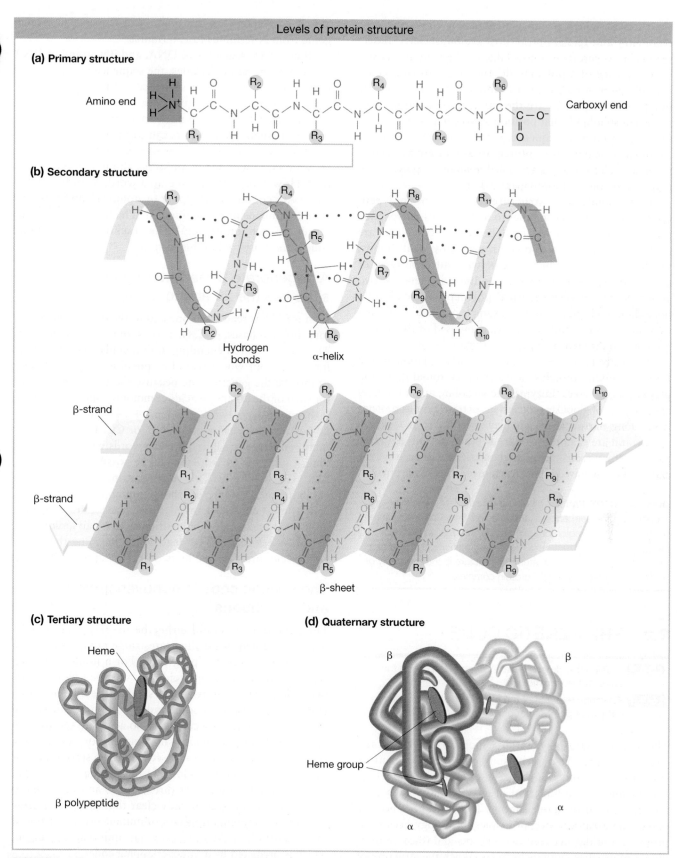

Levels of protein structure

(a) Primary structure

Amino end

Carboxyl end

(b) Secondary structure

Hydrogen bonds

α-helix

β-strand

β-strand

β-sheet

(c) Tertiary structure

Heme

β polypeptide

(d) Quaternary structure

β

β

Heme group

α

α

FIGURE 9-4 Proteins have four levels of structure. (a) Primary structure: the sequence of amino acids. (b) Secondary structure: the three-dimensional form of regions of a protein. Examples are α-helices and β-sheets. Antiparallel β-sheets have two polypeptide segments arranged in opposite polarity, as indicated by the arrows. (c) Tertiary structure: the overall three-dimensional shape of a polypeptide. In hemoglobin, heme is a non-protein molecule. (d) Quaternary structure: the arrangement of polypeptides in a protein complex. As an example, hemoglobin is composed of four subunits, two α subunits and two β subunits.

the rules by which primary structure is converted into secondary and tertiary structures are imperfectly understood. However, from knowledge of the primary amino acid sequence of a polypeptide, the functions of specific regions can be predicted. For example, some characteristic polypeptide sequences are the contact points with membrane phospholipids that position a protein in a membrane. Other characteristic sequences act to bind DNA or RNA. Amino acid sequences or protein folds that are associated with particular functions are called **domains**. A polypeptide may contain one or more separate domains.

Lastly, **quaternary structure** refers to how polypeptides interact with one another to form a multi-polypeptide protein complex. Individual polypeptides in complexes are called **subunits** and are joined together by weak bonds. Quaternary associations can be between different types of polypeptides (resulting in a heterodimer, if there are two subunits) or between identical polypeptides (making a homodimer). Hemoglobin is an example of a heterotetramer (tetramer meaning four subunits), composed of two copies each of two different polypeptides (Figure 9-4d).

There are two general types of proteins, globular and fibrous. **Globular proteins** have a compact, round shape and play functional roles. Enzymes, hemoglobin, and antibodies are examples of globular proteins. In contrast, **fibrous proteins** have a long, narrow shape and play structural roles. Collagen and keratin are examples of fibrous proteins. Collagen is the main structural protein found in connective tissues such as skin, and keratin is involved in the structure of hair and fingernails.

KEY CONCEPT Proteins have four levels of structure. Primary structure is the sequence of amino acids. Secondary structure is the shape of a region of amino acids, such as α-helices or β-sheets. Tertiary structure is the three-dimensional shape of a whole polypeptide, and quaternary structure is the assembly of multiple polypeptides into a protein complex.

9.2 THE GENETIC CODE

LO 9.2 Outline the experimental evidence supporting the rules of the genetic code.

LO 9.3 Describe features of the genetic code that minimize effects of point mutations on protein function.

The one-gene–one-polypeptide hypothesis of Beadle and Tatum (Chapter 5) was the source of the first exciting insight into the functions of genes: genes were somehow responsible for the function of enzymes, and each gene apparently controlled one enzyme. This hypothesis became one of the great unifying principles in biology because it provided a bridge between the concepts and research techniques of genetics and biochemistry. When the structure of DNA was deduced in 1953, it seemed likely that there was a linear correspondence between the nucleotide sequence in DNA and the amino acid sequence in a protein. It was soon deduced that the nucleotide sequence in mRNA going from

5′ to 3′ corresponds to the amino acid sequence in protein going from N-terminus to C-terminus.

If genes are segments of DNA, and if a strand of DNA is just a string of nucleotides, the sequence of nucleotides must somehow dictate the sequence of amino acids in proteins. How does DNA sequence dictate protein sequence? Simple logic tells us that, if nucleotides are the "letters" in a code, a combination of letters can form "words" representing different amino acids. However, in the 1960s, researchers were faced with many questions about how the code is read. How many letters make up a word, or **codon**, in the code? Are codons overlapping or nonoverlapping? Is the code continuous or discontinuous? Which codon or codons represent each amino acid? The cracking of the **genetic code** is the story told in this section.

A degenerate three-letter genetic code specifies the 20 amino acids

If an mRNA is read from one end to the other, only one of four different bases, A, C, G, or U, is found at each position. Thus, if the words encoding amino acids were one letter long, only four words would be possible. This vocabulary cannot be the genetic code because there must be a word for each of the 20 amino acids commonly found in proteins. If the words were two letters long, $4 \times 4 = 16$ words would be possible; for example, AU, CU, or CC. This vocabulary is still not large enough. But if the words were three letters long, $4 \times 4 \times 4 = 64$ words would be possible; for example, AUU, GCG, or UGC. This vocabulary provides more than enough words to describe the 20 amino acids. Therefore, codons must consist of at least three nucleotides. However, if all three-nucleotide combinations specify an amino acid, the genetic code must be **degenerate**, meaning that some amino acids are specified by two or more different triplets.

The genetic code is nonoverlapping and continuous

The genetic code could either be overlapping or nonoverlapping. **Figure 9-5** illustrates these possibilities for a three-nucleotide, or **triplet**, code. For a nonoverlapping code, consecutive amino acids are specified by consecutive code words (codons), and a single nucleotide mutation would only alter one codon and one amino acid. For an overlapping code, consecutive amino acids are specified by codons that have nucleotides in common; for example, the third nucleotide in one codon could be the second or first nucleotide in adjacent codons. In this case, a single nucleotide mutation would alter three codons and three amino acids. By 1961, it was already clear that the genetic code was nonoverlapping. Analyses of mutant proteins showed that almost all of the time, only one amino acid changed, which is predicted by a nonoverlapping code.

The genetic code could be continuous or discontinuous. In a continuous code, codons are arranged side by side with no gaps, whereas in a discontinuous code, codons are separated by one or more nucleotides that act to pause

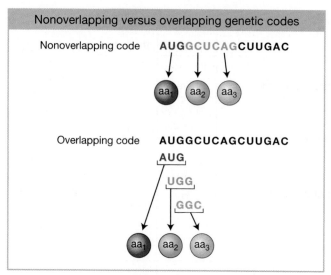

FIGURE 9-5 Nonoverlapping and overlapping genetic codes would translate into different amino acid sequences. The example uses codons with three nucleotides (a triplet code). (Top) In a nonoverlapping code, a protein is translated by reading nucleotides sequentially in sets of three. Each nucleotide is found in only one codon. In this example, the third nucleotide (G) in the RNA is only in the first codon. (Bottom) In contrast, in an overlapping code, each nucleotide occupies positions in multiple codons. In this case, the third nucleotide (G) is found in all three codons.

translation and restart anew at the next codon. If the code was continuous, which turns out to be the case, insertion or deletion of a single nucleotide would cause a shift in the reading frame starting at the site of the mutation and continuing to the end of the open reading frame (**Figures 9-6b and c**). In contrast, if the code was discontinuous, insertion or deletion of a single nucleotide would only affect one codon, and this error would not be propagated through the rest of the open reading frame.

Convincing proof that the genetic code is continuous came from genetic experiments first reported in 1961 by Francis Crick, Sidney Brenner, and their co-workers. These experiments used mutants in the *rII* locus of T4 phage. The use of *rII* mutations in recombination analyses was discussed in Chapter 6. Phage T4 is usually able to grow on two different *E. coli* strains, called B and K. However, mutations in the *rII* gene change the host range of the phage: mutant phages can still grow on an *E. coli* B host, but they cannot grow on an *E. coli* K host. Mutations causing the rII⁻ phenotype were induced using a chemical called proflavin, which causes insertion or deletion of single base pairs in DNA. Starting with one particular proflavin-induced mutation called FCO, Crick and his colleagues found "reversions" (reversals of the mutation) that were able to grow on *E. coli* strain K. Genetic analysis of these phages revealed that the "revertants" were not identical to true wild types. In fact, reversions were found to be due to the presence of a second mutation at a different site from that of FCO, although in the same gene. This second mutation "suppressed" mutant expression of the original FCO. Recall from Chapter 5 that a suppressor mutation counteracts the

effects of another mutation so that the bacterium is more like wild type.

How can these results be explained? If we assume that the gene is read from one end only, the original insertion or deletion induced by proflavin could interrupt the normal reading mechanism that establishes the grouping of bases to be read as words. For example, if each group of three bases in an mRNA makes a word, the **reading frame** might be established by taking the first three bases from the end as the first word, the next three as the second word, and so forth (Figure 9-6a). In that case, a proflavin-induced insertion of a single base pair in the DNA would shift the reading frame on the mRNA from that point on, causing all following words to be misread (Figure 9-6b). Such **frameshift mutations** could reduce most of the genetic message to garbage. However, the proper reading frame could be restored by a compensatory deletion somewhere else, limiting the garbage to the segment between the two mutations (Figure 9-6d). We have assumed here that the original frameshift mutation was an insertion, but the explanation works just as well if the original FCO mutation was a deletion and the suppressor was an insertion (Figures 9-6c and d). The few wrong words in the suppressed genotype could account for the fact that the revertants (suppressed phenotypes) did not look exactly like true wild types. These data demonstrated that the genetic code is continuous.

Crick and his colleagues also found that insertions or deletions of two bases produced the mutant phenotype (Figures 9-6e and f); however, a third mutation of the same type restored a wild-type phenotype because it corrected the reading frame (Figures 9-6g and h). This observation provided the first experimental evidence that a word in the genetic code consists of three successive nucleotides, or a triplet.

KEY CONCEPT The genetic code has the following features:

1. The linear sequence of nucleotides in a gene determines the linear sequence of amino acids in the encoded protein.
2. A codon of three nucleotides specifies an amino acid.
3. The genetic code is degenerate; more than one codon can specify the same amino acid.
4. The genetic code is nonoverlapping; each nucleotide is part of only one codon.
5. The genetic code is continuous; it is read from a fixed starting point and continues uninterrupted to the end of the open reading frame.

Cracking the code

Deciphering the genetic code—determining the amino acid specified by each triplet—is one of the most exciting genetic breakthroughs to occur since elucidation of the structure of DNA. After the necessary experimental techniques became available, the genetic code was cracked quickly.

One technical breakthrough was the discovery that single-stranded RNA can be synthesized in vitro by the enzyme polynucleotide phosphorylase. Unlike transcription, no

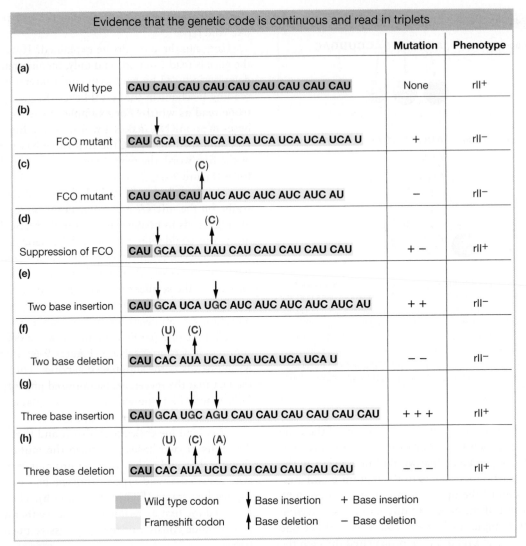

		Mutation	Phenotype
(a) Wild type	CAU CAU CAU CAU CAU CAU CAU CAU CAU	None	rII⁺
(b) FCO mutant	CAU GCA UCA UCA UCA UCA UCA UCA UCA U	+	rII⁻
(c) FCO mutant	CAU CAU CAU AUC AUC AUC AUC AUC AU	−	rII⁻
(d) Suppression of FCO	CAU GCA UCA UAU CAU CAU CAU CAU CAU	+ −	rII⁺
(e) Two base insertion	CAU GCA UCA UGC AUC AUC AUC AUC AUC AU	+ +	rII⁻
(f) Two base deletion	CAU CAC AUA UCA UCA UCA UCA UCA U	− −	rII⁻
(g) Three base insertion	CAU GCA UGC AGU CAU CAU CAU CAU CAU CAU	+ + +	rII⁺
(h) Three base deletion	CAU CAC AUA UCU CAU CAU CAU CAU CAU	− − −	rII⁺

Evidence that the genetic code is continuous and read in triplets

Wild type codon   ↓ Base insertion   + Base insertion
Frameshift codon   ↑ Base deletion   − Base deletion

FIGURE 9-6 Examination of single, double, and triple mutations by Crick and colleagues demonstrated that the genetic code is continuous and read in triplets. A wild-type sequence (a) produces the wild-type rII⁺ phenotype. A single base insertion (b) or deletion (c) causes the rII⁻ phenotype. The phenotype is suppressed by combining these mutations (d), but not by a second insertion (e) or deletion (f). Suppression restores the normal reading frame and indicates that the genetic code is continuous. Furthermore, a double insertion or deletion is suppressed by a third insertion (g) or deletion (h), indicating that the genetic code is read in triplets.

DNA template is needed for this synthesis, so the nucleotides are incorporated at random. The ability to enzymatically synthesize RNA offered the exciting prospect of creating specific RNA sequences and then seeing which amino acids they would specify. The first synthetic RNA was made using only uracil nucleotides, producing . . . UUUU . . . [poly(U)]. In 1961, Marshall Nirenberg and Heinrich Matthaei mixed poly(U) with the protein-synthesizing machinery of *E. coli* in vitro and observed the formation of a protein. The main excitement centered on the question of the amino acid sequence of this protein. It proved to be polyphenylalanine—a string of phenylalanine (Phe) amino acids (**Figure 9-7a**). Thus, the triplet UUU codes for phenylalanine. Expanding this approach to other single nucleotides and combinations of nucleotides led to assignment of about 40 codons to particular amino acids (Figures 9-7b and c). The technical breakthrough that led to

assignment of 61 of the 64 codons was the development by H. Gobind Khorana of methods to chemically synthesize RNAs of defined sequences. In 1968, Nirenberg and Khorana were awarded the Nobel Prize for deciphering the genetic code.

Virtually all organisms use the same genetic code (**Figure 9-8**). There are just a few exceptions in which a small number of codons have different meanings—for example, in mitochondrial and nuclear genomes of ciliates and some other protozoans. How is it that genomes can have very different base compositions but use the same code? Part of the answer is that degeneracy of the code permits the DNA base composition of genomes to vary over a wide range and still encode all 20 amino acids. For example, the G + C content of the coding regions of bacterial genomes ranges from about 20 percent to 70 percent but encodes proteins with very similar sequences. Furthermore, not all codons

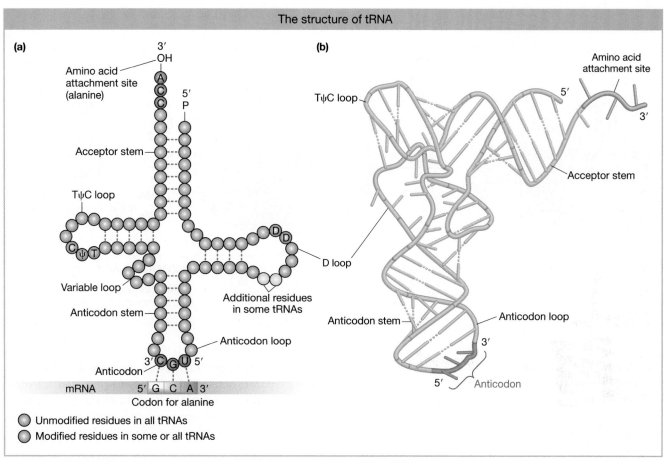

The structure of tRNA

FIGURE 9-11 (a) The structure of yeast alanine tRNA. The labeled features of the structure are important for tRNA function and are discussed in detail in the text. (b) Diagram of the three-dimensional structure of yeast alanine tRNA. [*Based on PDB ID 3WQY.*]

2. Numerous nucleotides in all tRNAs undergo post-transcriptional modification; that is, tRNAs are transcribed with the standard four nucleotides, which are then altered by enzymes. In yeast, each tRNA contains between 7 and 17 modified nucleotides. For example, the D loop contains the nucleotide dihydrouridine (D), which has two extra hydrogen atoms on the uracil base; the TψC loop contains pseudouridine (ψ), which has uracil attached to the ribose sugar at a carbon rather than a nitrogen; and the anticodon can contain the base inosine (I), which is structurally similar to guanosine (G). Nucleotide modifications in the anticodon affect base pairing with the codon (see the discussion about wobble in the next section), and nucleotide modifications at other sites affect tRNA recognition, folding, and stability. However, the precise function of many modifications is yet to be determined.

3. A tRNA normally exists as an inverted L-shaped three-dimensional structure, as shown in Figure 9-11b, rather than the flattened cloverleaf shown in Figure 9-11a. Although tRNAs differ in their primary nucleotide sequence, all tRNAs fold into virtually the same L-shaped conformation, indicating that the shape of a tRNA is important for its function.

KEY CONCEPT tRNAs have four important structural features: (1) the sequence CCA at the 3′ end; (2) modified nucleotides such as dihydrouridine, pseudouridine, and inosine; (3) an overall inverted L shape; and (4) an anticodon.

Amino acids are attached to tRNAs by enzymes called **aminoacyl-tRNA synthetases**. The tRNA with an attached amino acid is said to be **charged tRNA**. There are 20 synthetases, one for each of the 20 amino acids. Because the code is degenerate, some synthetases act on multiple tRNAs. Charging by aminoacyl-tRNA synthetases occurs in two steps (**Figure 9-12**). In the first step, the carboxyl group of the amino acid reacts with the α-phosphate of ATP to form 5′-aminoacyl-AMP and release pyrophosphate (PP_i). 5′-aminoacyl-AMP is referred to as an activated amino acid. In the second step of charging, the amino acid is transferred to the adenosine (A) of the invariant CCA sequence at the 3′ end of the tRNA, and AMP is released as a by-product.

What would happen if the wrong amino acid was covalently attached to a tRNA? A convincing experiment answered this question. The experiment used tRNA^{Cys}, the tRNA specific for cysteine. This tRNA was charged with cysteine, meaning that cysteine was attached to the tRNA.

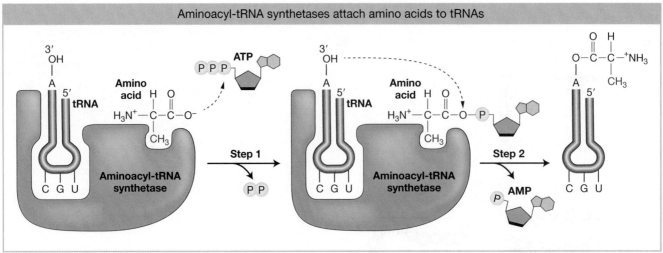

FIGURE 9-12 In a two-step reaction, aminoacyl-tRNA synthetases charge tRNAs with the correct amino acid. In the first step, the amino acid is linked to AMP to form aminoacyl-AMP, with PP_i as a by-product. In the second step, the tRNA replaces AMP to form aminoacyl-tRNA, with AMP as a by-product.

ANIMATED ART Sapling Plus
tRNA charging

The charged tRNA was treated with nickel hydride, which converted the cysteine (while still bound to tRNA^{Cys}) into another amino acid, alanine, without affecting the tRNA:

$$\text{cysteine-tRNA}^{Cys} \xrightarrow{\text{nickel hydride}} \text{alanine-tRNA}^{Cys}$$

Proteins synthesized with alanine-tRNA^{Cys} had alanine wherever cysteine was coded. The experiment demonstrated that amino acids are "illiterate"; they are inserted at the proper position because the tRNA adaptors recognize the mRNA codons. Thus, attachment of the correct amino acid to its corresponding tRNA (also called its cognate tRNA) by an aminoacyl-tRNA synthetase is the critical step in ensuring that the mRNA code is translated correctly. If the wrong amino acid is attached, there is no way to prevent it from being incorporated into a growing polypeptide chain.

Correct charging of tRNAs depends on selection of appropriately paired tRNAs and amino acids by aminoacyl-tRNA synthetases. These enzymes are good at recognizing the correct tRNA because tRNAs have numerous distinguishing structural features, including nucleotide sequence and nucleotide modifications. However, the only distinguishing feature of amino acids is their side chain, which can be very similar. Therefore, to prevent mistakes, aminoacyl-tRNA synthetases have a two-step mechanism that discriminates between chemically similar amino acids such as valine (Val) and isoleucine (Ile), which differ by only a single CH_2 group (see Figure 9-2). The first discrimination step occurs in the activation site of the enzyme where the amino acid is bound and activated to form aminoacyl-AMP. This step rejects amino acids that do not fit into the activation site because they are too large. So, Val-tRNA synthetase will reject Ile because it is too large. In contrast, Ile-tRNA synthetase will sometimes charge tRNA^{Ile} with valine to produce Val-tRNA^{Ile}. But, in the second

discrimination step, Val-tRNA^{Ile}, but not Ile-tRNA^{Ile}, fits into a separate active site of the synthetase and is hydrolyzed to valine and tRNA^{Ile}. Because of this proofreading mechanism, the error rate of protein synthesis is very low, in the range of 1 in 10^4–10^5 amino acids incorporated.

KEY CONCEPT tRNAs are charged by tRNA synthetases in a two-step reaction requiring ATP. There is a different synthetase for each amino acid. Two proofreading steps ensure that synthetases charge tRNAs with the correct amino acid.

Wobble base pairing allows tRNAs to recognize more than one codon

If perfect Watson-Crick base pairing between tRNA anticodons and mRNA codons was required to recognize all of the codons, there would need to be 61 different tRNAs. However, this is not the case; some tRNAs can recognize multiple codons through a different kind of base pairing at the third position of a codon, termed the **wobble** position. For example, the charged tRNA^{Ser} can form either a normal Watson-Crick G–C base pair or an unusual G–U wobble base pair with serine codons (**Figure 9-13**), so one tRNA can be used for both serine codons. In addition, inosine (I), a rare modified base in tRNA, can base pair to C, U, and A (**Table 9-1**). Therefore, because of wobble base pairing, cells require fewer than 61 tRNAs to read all of the codons.

KEY CONCEPT The genetic code is called degenerate because, in many cases, more than one codon is assigned to a single amino acid; in addition, wobble base pairing allows the anticodon of some tRNAs to pair with more than one codon.

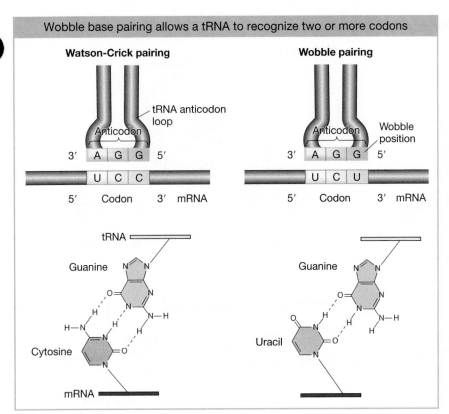

Wobble base pairing allows a tRNA to recognize two or more codons

Watson-Crick pairing

Wobble pairing

FIGURE 9-13 The third position in a codon is often called the wobble position because multiple nucleotides in this position can base pair with same 5′ nucleotide in a tRNA anticodon. In this case, both C and U in the third position base pair with a G in the anticodon. Table 9-1 lists all of the wobble base pairs. (Top) Normal (Watson-Crick) and wobble base pairing interactions of tRNA anticodons and codons. (Bottom) Hydrogen bonds formed in Watson-Crick and wobble base pairs.

TABLE 9-1	Wobble Base-pairing Rules
5′ end of anticodon	**3′ end of codon**
A	U
C	G
G	C or U
U	A or G
I (inosine)	A, C, or U

Wobble base pairs are shown in red.

Ribosome structure and function

Ribosomes are made up of two subunits that were originally characterized by their rate of sedimentation when spun in an ultracentrifuge. Therefore, their names are derived from their sedimentation coefficients in Svedberg (S) units, which is an indication of molecular size. In bacteria, the small and large subunits are called 30S and 50S, respectively, and they associate to form a 70S particle (**Figure 9-14a**). The eukaryotic subunits are called 40S and 60S, and the complete eukaryotic ribosome is called 80S (Figure 9-14b).

Although bacterial and eukaryotic ribosomes differ in size and composition, the steps in protein synthesis are similar overall. The similarities clearly indicate that translation is an ancient process that originated in the common ancestor of bacteria and eukaryotes. On the other hand, because of differences between bacterial and eukaryotic ribosomes, antibiotics are able to inactivate bacterial ribosomes but

leave eukaryotic ribosomes untouched. More than half of all antibiotics currently in use target the bacterial ribosome, including penicillin, tetracycline, ampicillin, and chloramphenicol.

When ribosomes were first studied, the fact that almost two-thirds of their mass is RNA and only one-third is protein was surprising. For decades, rRNAs had been assumed to function as a scaffold for assembly of ribosomal proteins. That role seemed logical because rRNAs fold up by intramolecular base pairing into stable secondary structures (**Figure 9-15**). According to this model, the ribosomal proteins catalyzed protein synthesis. This view changed with the discovery in the 1980s of catalytic RNAs (see Chapter 8). As you will see, there is now considerable evidence that rRNAs, assisted by the ribosomal proteins, catalyze protein synthesis.

Ribosomes bring together the important players in protein synthesis—charged tRNA and mRNA—to translate the nucleotide sequence of an mRNA into the amino acid sequence of a protein. tRNAs and mRNAs are positioned in the ribosome so that codons of the mRNA can interact with anticodons of tRNAs. Key sites of interaction are illustrated in **Figure 9-16**. The binding site for mRNA is completely within the small subunit. There are three binding sites for tRNA molecules. Each bound tRNA bridges the 30S and 50S subunits, positioned with its anticodon end in the 30S subunit and its aminoacyl end (carrying the amino acid) in the 50S subunit. The **A site** (for **aminoacyl-tRNA binding site**) binds an incoming aminoacyl-tRNA whose anticodon is complementary to the mRNA codon in the

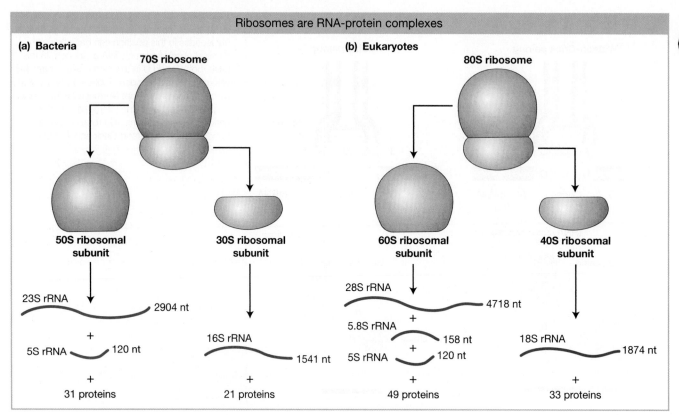

FIGURE 9-14 Ribosomes contain a large and a small subunit. Each subunit contains one large RNA and a set of proteins. In addition, the large subunit of bacterial ribosomes contains one small RNA, 5S rRNA, whereas (b) the large subunit of eukaryotic ribosomes contains two small rRNAs, 5S rRNA and 5.8S rRNA.

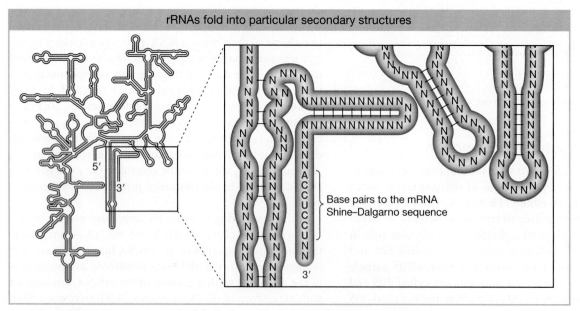

FIGURE 9-15 The folded structure of bacterial 16S rRNA. The magnified area shows the details of the complex secondary structure of 16S rRNA and the sequence at the 3′ end of 16S that binds an mRNA Shine–Dalgarno sequence.

A site of the 30S subunit. Proceeding in the 5′ direction on the mRNA, the next codon interacts with the anticodon of the tRNA in the **P site** (for **peptidyl site**) of the 30S subunit. The P and A sites are situated to facilitate formation of a peptide bond between their amino acids, disconnecting the P site amino acid from its tRNA. The growing peptide chain fits into a tunnel-like structure in the 50S subunit. The **E site** (for **exit site**) contains a deacylated tRNA (it no

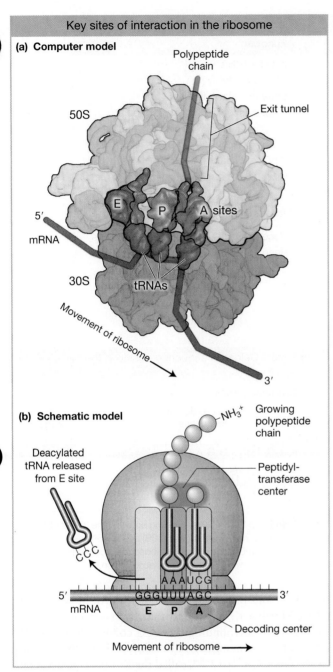

Two additional regions in the ribosome are critical for protein synthesis. The **decoding center** in the 30S subunit ensures that only tRNAs carrying anticodons that match the codon will be accepted into the A site. The **peptidyl-transferase center** in the 50S subunit is the site where peptide bond formation is catalyzed. The Nobel Prize in Chemistry was awarded in 2009 to Thomas Steitz, Venkatraman Ramakrishnan, and Ada Yonath for their laboratories' work using X-ray crystallography to determine the structure of the ribosome at the atomic level. The results of their elegant studies clearly show that both the decoding and peptidyltransferase centers are composed entirely of rRNA. Therefore, peptide bond formation is thought to be catalyzed by an active site in the rRNA and assisted only by ribosomal proteins. In other words, the large ribosomal subunit functions as a ribozyme to catalyze peptide bond formation.

KEY CONCEPT In all organisms, ribosomes have large and small subunits, each containing rRNA and proteins. Two key sites in ribosomes, the peptidyltransferase center in the large subunit that carries out peptide bond formation and the decoding center in the small subunit that accepts the correct tRNA, are built of rRNA.

9.4 TRANSLATION

LO 9.5 Outline the molecular events that take place during translation initiation, elongation, and termination.

Translation is carried out by ribosomes moving along mRNA in the 5′-to-3′ direction. tRNAs bring amino acids to the ribosome, and their anticodons base pair to mRNA codons. An incoming amino acid becomes bonded to the amino end of the growing polypeptide chain in the ribosome. The process of translation can be divided into three phases: initiation, elongation, and termination. Aside from the ribosome, mRNA, and tRNAs, other proteins (factors) are required for each phase (**Table 9-2**). Because certain steps in initiation differ significantly in bacteria and eukaryotes, initiation is described separately for the two groups. The elongation and termination phases are described largely as they take place in bacteria.

Translation initiation

The main task of initiation is to place the first aminoacyl-tRNA in the P site of the ribosome and, in this way, establish the correct reading frame of the mRNA. In most bacteria and all eukaryotes, the first amino acid in any newly synthesized polypeptide is methionine (Met), specified by the **initiation codon** AUG. In bacteria, there are two tRNAs for methionine. tRNA^{Met} is used at AUGs in internal positions in mRNA and an **initiator tRNA**, tRNA^{fMet}, is used at AUG initiation codons. tRNA^{fMet} is charged with methionine to form Met-tRNA^{fMet} and then a formyl (f) group is added to methionine to generate *N*-formylmethionyl-tRNA^{fMet} (fMet-tRNA^{fMet}). The formyl group on fMet-tRNA^{fMet} is

FIGURE 9-16 Ribosome interactions during the elongation phase of translation. (a) A computer model of the three-dimensional structure of the ribosome with mRNA, tRNAs, and the nascent polypeptide chain as it emerges from the large ribosomal subunit. (b) A schematic model of the ribosome during translation elongation. [*Part (a) drawn using PDB IDs 1VSA, 2OW8, and 1GIX.*]

longer carries an amino acid) that is ready to be released from the ribosome. Whether codon–anticodon interactions also take place between the tRNA and the mRNA in the E site is not yet clear.

KEY CONCEPT mRNA base pairs with tRNA in the small subunit of the ribosome, while tRNAs fit into sites that span both subunits. tRNAs begin in the A site, peptide bond formation occurs in the P site, and tRNAs exit from the E site.

TABLE 9-2	Translation Factors		
	Bacteria	**Eukaryotes**	**Function**
Initiation	fMET - tRNA^{fMET}	Met-tRNA$_i^{Met}$	Initiator tRNA
	IF1	eIF1A	Blocks A site
	IF2	eIF2, eIF5B	Entry of initiator tRNA
	IF3	eIF3, eIF1	Blocks association of large subunit
		eIF4F complex	
		eIF4A	Unwinds mRNA
		eIF4E	Binds m^{7}G cap
		eIF4G	Binds PABP to circularize mRNA
Elongation	EF-Tu	eEF1α	Delivers aminoacyl tRNA
	EF-G	eEF2	Translocates ribosome
Termination	RF1	eRF1	Recognizes UAA and UAG stop codons
	RF2	eRF1	Recognizes UAA and UGA stop codons
	RF3	eRF3	Stimulates peptide release

removed during or shortly after synthesis of the polypeptide. Eukaryotes also use distinct methionine tRNAs for internal and initiation AUG codons, called tRNA^{Met} and tRNA$_i^{Met}$, respectively. The use of Met-tRNA$_i^{Met}$ rather than Met-tRNA^{Met} for initiation is specified by interactions with translation initiation factors.

How does the translation machinery know where to begin? In other words, how is the initiation AUG codon selected from among the many AUG codons in an mRNA? Recall that, in both bacteria and eukaryotes, mRNA has a 5' untranslated region (UTR) consisting of the sequence between the transcription start site and the translational start site (see Figure 8-7). As you will see below, the nucleotide sequence of the 5' UTR adjacent to the AUG initiation codon is critical for ribosome binding in bacteria, and the 5' cap is critical for ribosome binding and scanning for the AUG initiation codon in eukaryotes.

In bacteria, AUG initiation codons in mRNA are preceded by a special sequence called the **Shine–Dalgarno sequence**, also known as the **ribosome-binding site (RBS)**, that base pairs with the 3' end of 16S rRNA in the 30S ribosomal subunit (**Figure 9-17**). This base pairing correctly positions the AUG in the P site where the initiator tRNA will bind. The mRNA can interact only with a 30S subunit that is dissociated from the rest of the ribosome. Note again that rRNA performs the key function in ensuring that the ribosome is at the right place to start translation. Once the initiator tRNA is bound, the 50S ribosomal subunit binds to form the 70S initiation complex.

In eukaryotes, translation initiation involves binding of the 40S ribosomal subunit to the capped 5' end of an mRNA, followed by scanning of the 5' UTR for an AUG initiation codon. The 5' cap (m^{7}G) that is added to mRNAs during transcription (described in Chapter 8) is directly bound by a translation initiation factor, which in turn binds other initiation factors to recruit the small ribosomal subunit to the mRNA. The ribosome subsequently scans the mRNA in the 5'-to-3' direction until it encounters the first AUG codon. In 5 to 10 percent of cases, the ribosome will bypass the

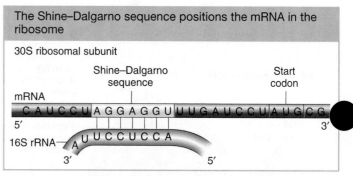

The Shine–Dalgarno sequence positions the mRNA in the ribosome

30S ribosomal subunit

FIGURE 9-17 In bacteria, base pairing between the 3' end of the 16S rRNA of the small ribosomal subunit (30S) and the Shine–Dalgarno sequence of the mRNA positions the ribosome to correctly initiate translation at the downstream AUG initiation codon.

first AUG and initiate at the second, third, or subsequent AUG. Marilyn Kozak found that bypass occurs because the sequence surrounding the AUG codon affects the efficiency of its use in initiation. Kozak found that the sequence CC(A^{-3}/G)CC$\underline{A}^{+1}$U$\underline{G}$G^{+4}, called the **Kozak sequence**, commonly surrounds initiating AUGs, and mutagenesis studies showed that the A at −3 and the G at +4 are particularly important for specifying the AUG initiation codon.

KEY CONCEPT Translation initiation begins when a charged initiator tRNA anticodon and an mRNA AUG initiation codon assemble in the P site of a ribosomal small subunit. In bacteria, the Shine–Dalgarno sequence in the 5' UTR of the mRNA base pairs with the 16S rRNA to position the start codon in the P site. In eukaryotes, the 5' cap of the mRNA is bound by an initiation factor that recruits the ribosomal small subunit. Once bound, the ribosome scans the mRNA for the initiation codon within the Kozak sequence.

In bacteria, three proteins—IF1, IF2, and IF3 (for initiation factor)—are required to assemble an active 70S ribosome (**Figure 9-18a**). Assembly begins by positioning the

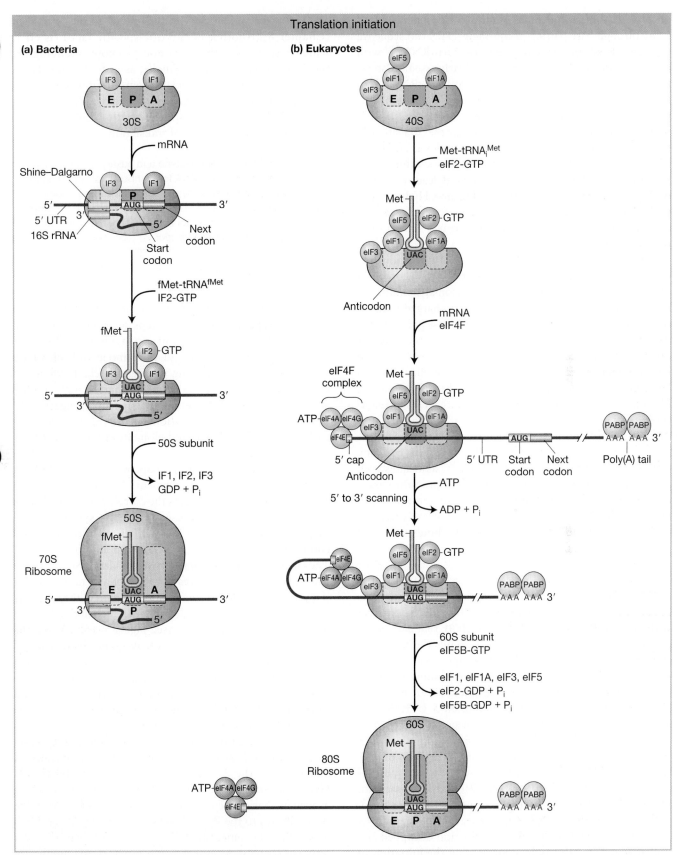

FIGURE 9-18 Initiation factors assist assembly of the ribosome at the translation start site and then dissociate before translation elongation. (a) In bacteria, three initiation factors (IF1, IF2, and IF3) position fMet-tRNA^fMet in the P site of the 70S ribosome. (b) In eukaryotes, many eIFs are required to position Met-tRNA_i^Met in the P site of the 80S ribosome. Factors that serve similar functions in bacteria and eukaryotes are colored the same.

ANIMATED ART   SaplingPlus

Translation

mRNA AUG initiation codon in the 30S ribosomal subunit P site through base pairing of the Shine–Dalgarno sequence with 16S rRNA and by base pairing of fMet-tRNA^{fMet} with the AUG initiation codon. During these early assembly steps, IF2 bound to GTP promotes binding of fMet-tRNA^{fMet} to the P site. In addition, binding of IF1 in the A site blocks tRNA binding to the second codon, and binding of IF3, with help from IF1, blocks association of the 50S subunit. Once formation of the 30S initiation complex is complete, IF3 and IF1 are released, which enables association of the 50S subunit, and hydrolysis of GTP to GDP leads to release of IF2 to generate a functional 70S ribosome. Hydrolysis of GTP and sometimes ATP provides energy for conformation changes that are required to progress through several steps of translation.

Like bacteria, eukaryotes use a set of initiation factors (IFs) to assemble the 80S translation initiation complex containing the small and large ribosomal subunits, an mRNA bound to the small subunit, and a charged initiator tRNA (Met-tRNA$_i^{Met}$) in the P site that is base paired to the AUG initiation codon (Figure 9-18b). The names of eukaryotic factors all begin with an "e" to distinguish them from bacterial factors. Four eIFs have functions similar to IFs in bacteria. Like bacterial IF1, eIF1A blocks tRNA binding in the A site of the small subunit (40S) as well as premature association of the large subunit (60S) with the small subunit. Eukaryotes have two factors, eIF2 and eIF5B, that have functions similar to bacterial IF2. A ternary complex composed of eIF2, GTP, and Met-tRNA$_i^{Met}$ associates with the small ribosomal subunit, while eIF5B promotes association of the small and large subunits. In addition, both eIF2 and eIF5B use hydrolysis of GTP to GDP to carry out their jobs. Lastly, like bacterial IF3, eIF3 binds the small ribosomal subunit and blocks association of the large subunit.

Because of differences between eukaryotic and bacterial mRNAs, there are other eIFs that do not have counterparts in bacteria. Three eIFs—eIF4A, eIF4E, and eIF4G—interact with one another in a complex called eIF4F that carries out activities unique to eukaryotes. eIF4E binds the cap structure at the 5′ end of an mRNA. m^{7}G caps are present only on mRNAs, so the requirement for cap binding by eIF4E ensures that only mRNAs are translated. eIF4A has RNA helicase activity that unwinds regions in the 5′ UTR that are double-stranded RNA due to intramolecular base pairing. This allow ribosomes to scan along single-stranded RNA in search of the AUG initiation codon. Hydrolysis of ATP to ADP by eIF4A is required for its activity. Lastly, eIF4G binds poly(A) binding proteins (PABPs) that associate with the mRNA poly(A) tail, thereby bringing together the 5′ and 3′ ends of the mRNA. Circularization of the mRNA is thought to enhance the rate of translation of capped and polyadenylated mRNAs by coordinating the initiation of ribosomes that have recently terminated translation.

In the first round of translation, the function of eIF4F is carried out by the cap binding complex (CBC), which associates with newly synthesized mRNAs in the nucleus and is transported along with the mRNA to the cytoplasm (Chapter 8). Another eukaryotic-specific factor is eIF1, which associates with eIF4G and promotes scanning for the AUG initiation codon. Finally, as in bacteria, eukaryotic initiation factors dissociate from the ribosome before the elongation phase of translation begins. The exception is eIF4F, which remains associated with PABP, setting the stage for additional rounds of translation.

> **KEY CONCEPT** In both bacteria and eukaryotes, initiation factors bring the initiator tRNA to the ribosomal small subunit and prevent premature binding of the large subunit. Some initiation factors hydrolyze GTP to proceed through the steps of initiation. In eukaryotes, additional initiation factors facilitate mRNA scanning and circularization.

Translation elongation

During translation elongation, the ribosome functions as a factory, repeating the same steps over and over again. The mRNA acts as a blueprint specifying the delivery of tRNAs, each carrying as cargo an amino acid. Each amino acid is added to the growing polypeptide chain, while the deacylated tRNA is recycled by being charged with another amino acid. **Figure 9-19** details the steps in elongation. In bacteria, two protein elongation factors (EFs) called elongation factor Tu (EF-Tu) and elongation factor G (EF-G) assist the elongation process.

As described earlier in this chapter, an aminoacyl-tRNA is formed by the covalent attachment of an amino acid to the 3′ end of a tRNA that contains the correct anticodon. Before aminoacyl-tRNAs can be used in protein synthesis, they associate with the protein factor EF-Tu to form a ternary complex composed of EF-Tu, GTP, and aminoacyl-tRNA. The elongation cycle commences with fMet-tRNA^{fMet} in the P site and with the A site ready to accept a ternary complex (Figure 9-19). Codon–anticodon recognition in the decoding center of the small subunit determines which of the different ternary complexes to accept (see Figure 9-16b). When the correct match has been made, the ribosome changes shape, EF-Tu hydrolyzes GTP to GDP and leaves the ternary complex, and the two amino acids are juxtaposed in the peptidyl-transferase center of the large subunit (see Figure 9-16b). There, a peptide bond is formed with transfer of fMet in the P site to the amino acid in the A site.

At this point, the second elongation factor, EF-G, plays its part. EF-G is structurally similar to a ternary complex and fits into the A site, displacing the peptidyl-tRNA. Hydrolysis of GTP to GDP by EF-G changes its structure as well as that of the ribosome and shifts the tRNAs in the A and P sites to the P and E sites, respectively. When EF-G leaves the ribosome, the A site is open to accept the next ternary complex. As elongation progresses, the number of amino acids on the peptidyl-tRNA (at the P site) increases. Eventually, the amino-terminal end of the growing polypeptide emerges from the tunnel in the 50S subunit and protrudes from the ribosome.

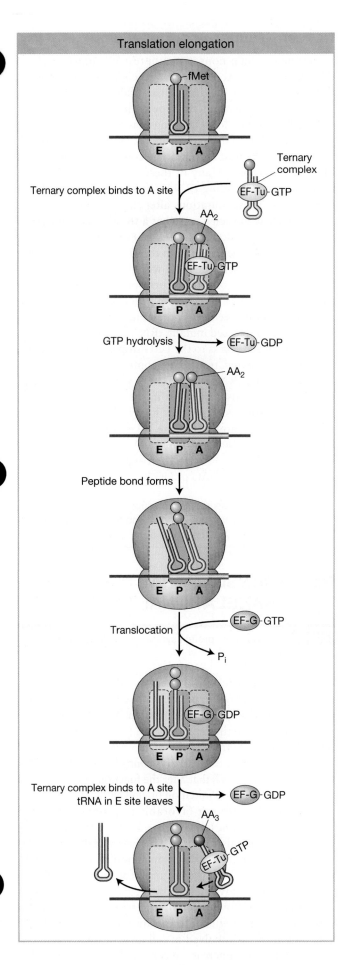

Translation elongation

Ternary complex binds to A site

EF-Tu-GTP — Ternary complex

GTP hydrolysis → EF-Tu-GDP

Peptide bond forms

Translocation — EF-G-GTP → P_i

EF-G-GDP

Ternary complex binds to A site
tRNA in E site leaves → EF-G-GDP

EF-Tu-GTP

Translation elongation in eukaryotes is very similar. Eukaryotic elongation factor 1α (eEF1α) functions similarly to EF-Tu, and eEF2 functions similarly to EF-G.

KEY CONCEPT During translation elongation, two protein elongation factors function repeatedly to grow the polypeptide chain. In bacteria, one elongation factor, EF-Tu, brings charged tRNAs to the ribosome A site to be joined to the polypeptide chain in the P site. The other elongation factor, EF-G, binds in the A site and promotes the translocation of tRNAs from the P and A sites to the E and P sites, respectively. In eukaryotes, the analogous elongation factors are eEF1α and eEF2.

Translation termination

The elongation cycle continues until the codon in the A site is one of the three stop codons: UGA, UAA, or UAG. tRNAs do not recognize these codons. Instead, proteins called release factors (RFs) recognize stop codons (**Figure 9-20**). Just like the structure of EF-G mimics the structure of a ternary complex, RF1 and RF2 mimic the structure of a tRNA. In bacteria, RF1 recognizes UAA or UAG, whereas RF2 recognizes UAA or UGA. The interaction between RF1 or RF2 and the A site differs from that of the ternary complex in two important ways. First, the stop codons are recognized by tripeptides in the RF proteins, not by an anticodon. Second, RFs fit into the A site of the 30S subunit but do not participate in peptide bond formation. Instead, a water molecule gets into the peptidyltransferase center, and its presence leads to release of the polypeptide from the tRNA in the P site. Following release of the peptide chain, RF3 promotes release of RF1 or RF2 from the ribosome. Hydrolysis of GTP to GDP is involved in releasing RF3 from the ribosome. In eukaryotes, translation termination is very similar; eRF1 recognizes stop codons, and eRF3 stimulates peptide chain release by eRF1. However, unlike bacteria, eRF1 recognizes all three stop codons (UAA, UAG, and UGA).

To prepare for a new round of translation, the ribosome recycling factor (RRF) disassembles the post-termination complex (Figure 9-20). With the help of EF-G and IF-3, RRF binds in the A site, translocates to the P site, releases deacylated tRNAs from the E and P sites, and dissociates the small and large ribosome subunits from each other and from the mRNA. IF-3 bound to the small subunit is now ready to initiate translation (see Figure 9-18a).

FIGURE 9-19 In bacteria, two elongation factors, EF-Tu and EF-G, perform repetitive functions for each amino acid that is added to the growing polypeptide. EF-Tu escorts charged tRNAs to the A site and positions them for peptide bond formation with the peptide attached to the tRNA in the P site. The protein factor EF-G then drives the repositioning of tRNAs from the P and A sites into the E and P sites, respectively.

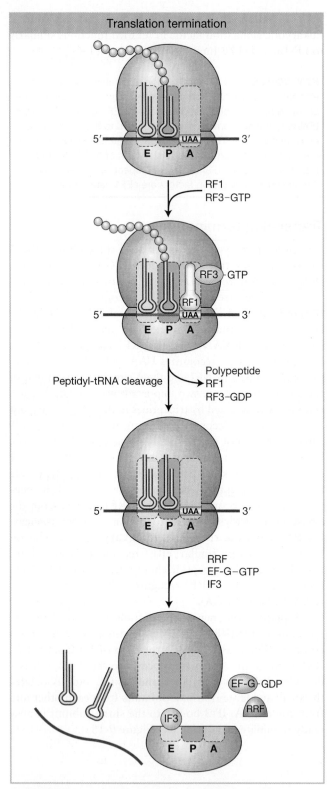

Translation termination

FIGURE 9-20 In bacteria, translation is terminated when a release factor (RF1 or RF2) recognizes a stop codon in the A site of the ribosome and liberates the polypeptide chain. RF3 then promotes the release of RF1 or RF2 (RF1, in this example). To get ready for another round of translation, other factors (RRF, EF-G, and IF3) displace the uncharged tRNAs and dissociate the ribosome subunits from one another and from the mRNA.

KEY CONCEPT During translation termination, release factors (proteins, not tRNAs) bind stop codons and release the polypeptide chain from the tRNA in the P site. Other factors recycle the ribosome to begin another round of translation.

Nonsense suppressor mutations

Experiments using nonsense suppressor mutations show that the ribosome relies on the anticodon–codon interaction in its decoding center to ensure fidelity in translation. In these experiments, wild-type codons in phages were mutated to stop codons, creating nonsense mutations that resulted in truncated phage proteins. However, suppressor mutations in the host chromosome counteracted the effects of these mutations. Many of these suppressors are mutations in genes encoding tRNAs and are known as tRNA suppressors. These mutations alter the anticodon loops of specific tRNAs in such a way that a tRNA becomes able to recognize a stop codon in mRNA. In **Figure 9-21**, a mutation replaces a wild-type codon with the chain-terminating stop codon UAG. By itself, the UAG would cause the protein to be prematurely cut off at the corresponding position. The suppressor mutation in this case produces a tRNA^{Tyr} with an anticodon that recognizes the mutant UAG stop codon. Thus, in the suppressed mutant, tRNA^{Tyr} competes with the release factor for access to the UAG stop codon. As a result, if tyrosine is inserted, translation continues past that triplet.

Presumably, tRNA suppressors also bind to normal termination signals and result in the synthesis of abnormally long proteins. Now that many genomes have been sequenced, it is known that the UAA stop codon is used much more often than UAG or UGA to terminate protein synthesis. As such, cells with UAA suppressors are usually sicker than cells with UAG or UGA suppressor mutations.

KEY CONCEPT Experiments with suppressor mutations show that the ribosome cannot proofread the match between the tRNA anticodon and the amino acid.

9.5 TRANSLATIONAL AND POST-TRANSLATIONAL REGULATION

LO 9.6 Describe how protein synthesis and function are regulated.

Translation is a regulated process in eukaryotic cells. As a result, the quantity of an mRNA is not always representative of the quantity of its encoded protein. For example, some mRNAs localize to the cell cytoplasm but are not assembled with ribosomes and translated until the cell receives a particular signal. In general, translation is controlled by signals from outside and inside a cell that alter the function of both general and specific translation initiation or elongation factors.

Stresses that are not favorable to growth, including nutrient deprivation, temperature shock, and DNA damage, produce signals that cause a general halt to translation and allow only selective translation of a few mRNAs encoding proteins that are required for responding to the stress. In contrast, favorable growth conditions that are rich in nutrients and growth stimuli lead to a global increase in translation and stimulated translation of specific mRNAs encoding proteins involved in cell growth (increase in cell size), proliferation (increase in cell number), and survival.

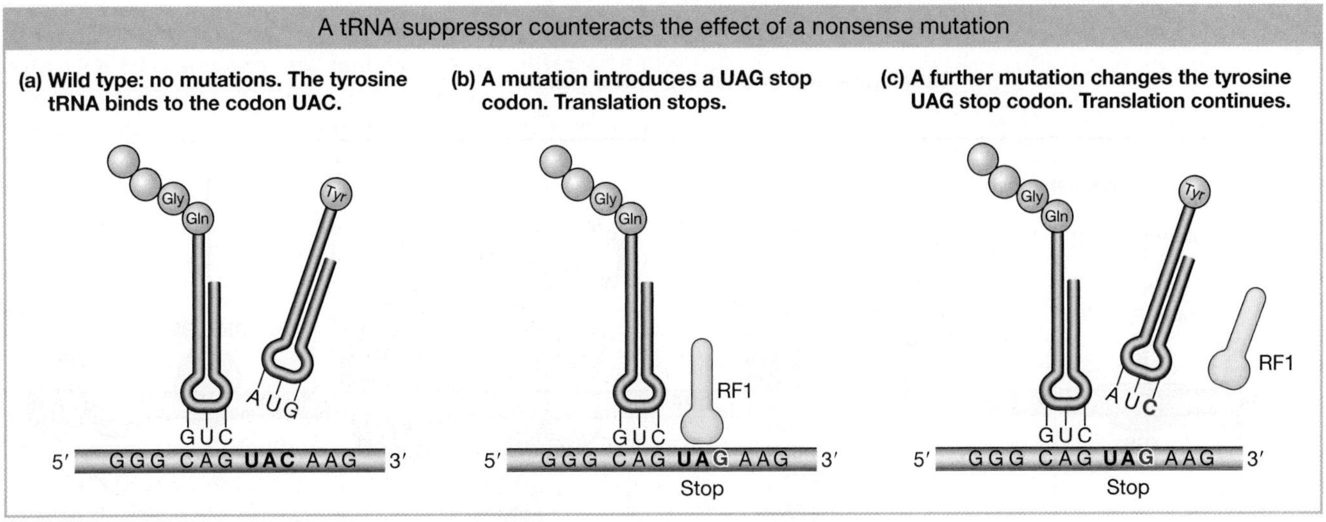

FIGURE 9-21 A suppressor allows translation to continue past a nonsense mutation. (a) In a wild-type mRNA, a Tyr-tRNA^{Tyr} reads the codon UAC, and translation elongation continues. (b) Mutation of UAC to the stop codon UAG terminates translation through the RF1 mechanism. (c) A mutation in the anticodon of Tyr-tRNA^{Tyr} allows the tRNA to read the UAG codon, add a tyrosine to the polypeptide chain, and permit translation elongation to continue.

ANIMATED ART Sapling Plus

Nonsense suppression at the molecular level

Stress → decrease in general transcription

increase in transcription of stress-response proteins

Growth stimuli → permissive to general transcription

increase in transcription of cell growth proteins

In eukaryotes, signals that inhibit the translation of many mRNAs often act by directly altering the ability of eIF4E to bind mRNA 5′ caps and assemble eIF4F to initiate translation (see Figure 9-18). On the other hand, signals that inhibit the translation of particular mRNAs act through microRNAs (miRNAs) or RNA-binding proteins (RNA-BPs) that bind specific sequences or structural motifs in an mRNA's 5′ or 3′ UTR (**Figure 9-22**). Bound factors in turn block translation through multiple mechanisms, including inhibiting ribosome assembly and elongation, inhibiting mRNA circularization, and promoting cleavage of the polypeptide chain.

miRNAs are small RNAs (~21 nucleotides) that bind with imperfect complementarity to their target mRNAs. Humans express over 2,500 miRNAs, and about 50% of human mRNAs are subject to regulation by miRNAs. So, most biological processes, including cell differentiation, growth, and proliferation, are regulated by miRNAs. Single-stranded miRNAs associate with Argonaute (Ago) proteins and other proteins to form an RNA-induced silencing complex (RISC). Guided to specific mRNAs by sequence complementary between the miRNA and the target mRNA, RISC interacts with other proteins to inhibit translation as well as promote mRNA decay.

Most newly synthesized proteins are unable to function until regulatory mechanisms alter their structure and cellular location. To become functional, proteins need to be folded correctly, the amino acids of some proteins need to be chemically modified, and proteins need to be transported to their sites of action within or outside the cell. Some protein folding, modification, and targeting takes place co-translationally (while the protein is being synthesized), and the rest takes place post-translationally (after synthesis is complete).

KEY CONCEPT Developmental and environmental signals regulate the translation of many mRNAs at the same time by altering the function of translation initiation factors and of selected mRNAs via RNA-binding factors (proteins and RNAs) that act on initiation and elongation factors.

Protein folding

Protein folding is the process by which proteins attain their functional tertiary structure. A protein that is folded correctly is said to be in its native conformation (in contrast with an unfolded or misfolded protein that is nonnative). Folding involves the stepwise formation of secondary structures such as α-helices and β-sheets that are stabilized by non-covalent hydrogen bonds. Folded regions then guide and stabilize subsequent folding to progressively build the tertiary structure. The distinct three-dimensional structures of proteins are essential for their enzymatic activity, for their ability to bind to other molecules, and for their structural roles in the cell. Although it has been known since the 1950s that the amino acid sequence of a protein determines its three-dimensional structure, it is also known that the aqueous environment inside the cell does not favor the correct folding of most proteins. Given that proteins do in fact fold correctly in the cell, a long-standing question has been, how is correct folding accomplished?

The answer seems to be that folding often begins co-translationally and is helped by **chaperones**—a class of proteins found in all organisms from bacteria to plants to humans. Chaperones typically bind hydrophobic regions of

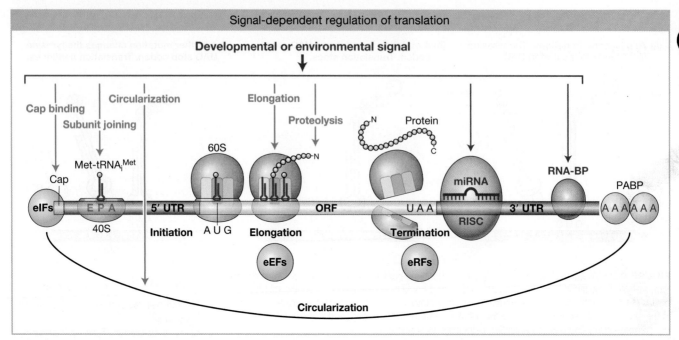

FIGURE 9-22 In eukaryotes, translation is regulated by developmental and environmental signals (red arrow) that affect the function of initiation and elongation factors (blue and purple arrows). Signals elicit their effects on translation by inhibiting or promoting mRNA cap binding by eIF4E, joining of the small and large ribosomal subunits, elongation by the ribosome, proteolysis of the nascent polypeptide, and circularization of the mRNA through the interaction of eIF4G and poly(A)-binding protein (PABP) (blue arrows). Signal-dependent inhibition of translation through these mechanisms is sometimes mediated by microRNAs (miRNAs) and RNA-binding proteins (RNA-BPs) that bind sequences in the mRNA 3′ UTR (purple arrows). AUG is the translation start site, UAA is the translation stop site, AAAAAA is the poly(A) tail, UTRs are untranslated regions, and ribosomal subunits are indicated in gray.

incorrectly or incompletely folded proteins to arrest folding or promote unfolding and then release them to undergo spontaneous refolding. Multiple rounds of binding and release that are driven by ATP hydrolysis occur until the protein is properly folded. Chaperones exist in all cell compartments. Some chaperones are expressed all the time in cells, while others are upregulated by heat shock and other stresses that increase protein misfolding. The latter class of chaperones is classified as stress or heat shock proteins.

> **KEY CONCEPT** The folding of newly synthesized proteins into precise three-dimensional structures is determined by the primary amino acid sequence and assisted by a class of proteins called chaperones.

Post-translational modification of amino acid side chains

Chemical modifications of amino acids greatly increase the functionality of proteins. More than 200 different types of amino acid modifications have been identified, many of which occur post-translationally. Amino acids that undergo post-translational modification often have a functional group that acts as a nucleophile in the enzymatic modification reaction. Examples of nucleophiles include the hydroxyl group of serine, threonine, and tyrosine; the amine group of lysine, arginine, and histidine, the thiol group of cysteine; and the carboxylate group of aspartic acid and glutamic acid (see Figure 9-2). The modifications themselves fall into

five broad categories: the addition of chemical groups, complex molecules, or polypeptides to amino acids; the modification of amino acids; and the cleavage of peptide bonds between amino acids to convert inactive precursor proteins into smaller, active proteins (**Figure 9-23**).

> **KEY CONCEPT** Proteins can undergo five types of post-translational modification:
>
> 1. Addition of chemical groups to amino acids (e.g., phosphorylation, methylation, acetylation, hydroxylation)
> 2. Addition of complex molecules to amino acids (e.g., glycosylation, lipidation)
> 3. Addition of polypeptides to amino acids (e.g., ubiquitination, sumoylation)
> 4. Modification of amino acids (e.g., deamidation, disulfide bond formation)
> 5. Cleavage of peptide bonds to convert inactive precursor proteins into smaller, active proteins (proteolysis)

Three types of proteins are involved in chemical modifications: enzymes called *writers* that add chemical modifications, enzymes called *erasers* that remove chemical modifications, and structural proteins called *readers* that bind chemical modifications. Sometimes, a single protein or a protein complex can be both a writer or eraser and a reader.

Addition and removal of chemical modifications serves as a reversible switch to control the traits of proteins, including increasing or decreasing their biological

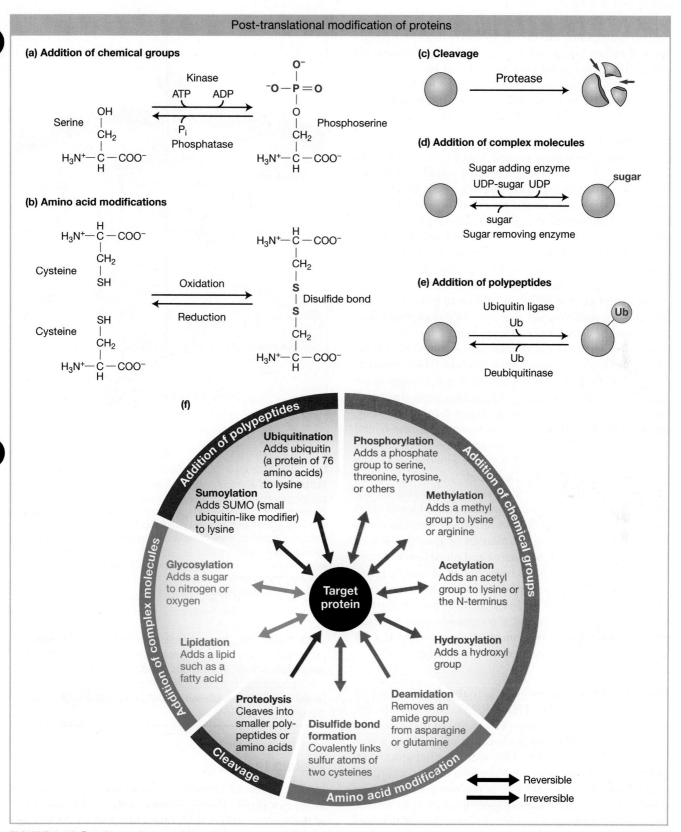

FIGURE 9-23 Proteins can be reversibly and irreversibly modified in numerous ways. (a–e) Examples of the five types of post-translational modifications, shown in red. (f) A larger list of examples of the five types of post-translational modifications.

activity, promoting or inhibiting their transport between cell compartments, increasing or decreasing their stability, and enhancing or suppressing their interactions with DNA, RNA, and other proteins. Two of the more commonly encountered post-translational modifications—phosphorylation and ubiquitination—are considered next.

Phosphorylation Enzymes called **kinases** (i.e., writers) catalyze **phosphorylation**, the addition of a phosphate group from ATP to the hydroxyl group of the amino acids serine, threonine, or tyrosine, whereas enzymes called **phosphatases** (i.e., erasers) catalyze dephosphorylation, resulting in the removal of phosphate groups. Protein phosphorylation is an important regulatory mechanism in eukaryotic cells. For example, phosphorylation and other modifications of the C-terminal domain (CTD) of RNA polymerase II regulate the processing of nascent mRNAs (see Figure 8-13). It is estimated that more than 30 percent of human proteins are regulated by phosphorylation, and abnormal phosphorylation is the cause or consequence of many human diseases.

Post-translational modifications are also used by eukaryotic cells to rapidly convert signals from the cellular environment into changes in gene expression; that is, changes in the transcription and translation of particular genes that allow the cell to respond to the signal (**Figure 9-24**). Commonly, signaling pathways begin when a plasma membrane-bound receptor such as a receptor tyrosine kinase is activated by binding a ligand (the signal from the environment). The activated receptor phosphorylates itself (i.e., autophosphorylation), which creates a binding site to recruit and activate other kinases in the cell cytoplasm. Then, these kinases phosphorylate other kinases that in turn phosphorylate translation factors to alter the translation of specific mRNAs, or they translocate to the nucleus and phosphorylate transcription factors to alter the transcription of specific genes. The transfer of information from kinase to kinase serves to amplify the strength of the initial signal. Thus, in response to signals, sequential cascades of post-translational modifications transmit information from one place to another in a cell and often culminate in the regulation of gene expression.

Ubiquitination Amino acids can also be modified by small polypeptides (see Figure 9-23). For example, the ε-amine of lysines in proteins can be linked to a glycine in the 76-amino-acid polypeptide called **ubiquitin** in a process called **ubiquitination**. This addition of ubiquitin targets proteins for decay by a multiprotein protease called the **proteasome**. Two broad classes of proteins are targeted for destruction by ubiquitination: short-lived proteins such as cell cycle regulators, and proteins that have become damaged or mutated. Ubiquitin can also be covalently linked to itself in many different conformations to form polyubiquitin chains. Both monoubiquitination and polyubiquitination target proteins for decay by the proteasome. Monoubiquitination can also localize proteins to specific cellular compartments and regulate the formation of protein complexes.

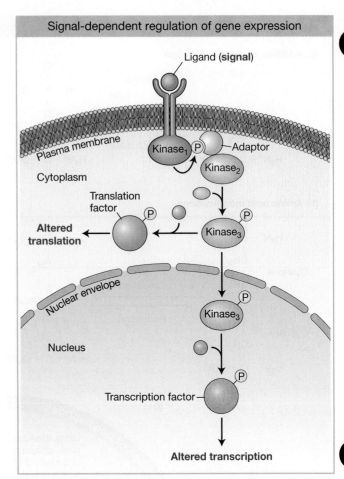

Signal-dependent regulation of gene expression

FIGURE 9-24 Extracellular signals alter gene expression via signaling pathways that use post-translational modifications to alter the function of translation and transcription factors. Signaling begins with a receptor-ligand interaction at the cell surface, proceeds through a kinase cascade in the cell cytoplasm, and culminates in the phosphorylation of translation factors in the cytoplasm or transcription factors in the nucleus that alter gene expression. Colored circles indicate different proteins whose function is labeled, and circled P's indicate a phosphorylated amino acid.

KEY CONCEPT Post-translational modification of amino acids alters the structure of proteins with consequent effects on protein activity, interactions, localization, and stability. Phosphorylation is a common regulatory mechanism used in signaling cascades. Ubiquitination is often used to target proteins for degradation.

Protein targeting

In eukaryotes, all proteins are synthesized on ribosomes in the cytoplasm. However, some of these proteins end up in the nucleus, others in the mitochondria, and still others anchored in a membrane or secreted from the cell. How do these proteins "know" where they are supposed to go? The answer to this seemingly complex problem is actually quite simple: a newly synthesized protein contains a short sequence that targets the protein to the correct place or cellular compartment. For example, a newly synthesized

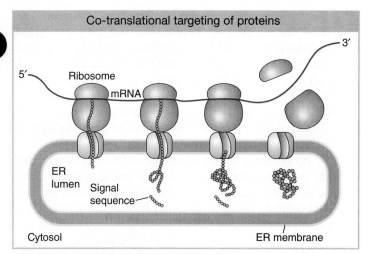

Co-translational targeting of proteins

FIGURE 9-25 Proteins destined to be secreted from the cell have an amino-terminal sequence that is rich in hydrophobic residues. This signal sequence binds to proteins in the endoplasmic reticulum (ER) membrane that draw the remainder of the protein through the lipid bilayer. The signal sequence is cleaved from the protein in this process by an enzyme called signal peptidase (not shown). Once inside the endoplasmic reticulum, the protein is directed to the cell membrane, from which it will be secreted.

membrane protein or a protein destined for an organelle has a short leader peptide, called a **signal sequence**, at its amino-terminal end. For membrane proteins, this stretch of 15 to 25 amino acids directs the protein to channels in the endoplasmic reticulum (ER) membrane where the signal sequence is cleaved by a peptidase (**Figure 9-25**). From the ER, the protein is directed to its ultimate destination. A similar phenomenon exists for certain bacterial proteins that are secreted.

Proteins destined for the nucleus include RNA and DNA polymerases and transcription factors. Amino acid sequences embedded in the interior of such nucleus-bound proteins are necessary for transport from the cytoplasm into the nucleus. These **nuclear localization sequences (NLSs)** are recognized by cytoplasmic receptor proteins that transport newly synthesized proteins through nuclear pores—sites in the membrane through which large molecules are able to pass into and out of the nucleus. A protein not normally found in the nucleus will be directed to the nucleus if an NLS is artificially attached to it.

Why are signal sequences cleaved during targeting, whereas an NLS, located in a protein's interior, remains after the protein moves into the nucleus? One explanation might be that, in the nuclear dissolution that accompanies mitosis (see Chapter 2), proteins localized to the nucleus may find themselves in the cytoplasm. Because such a protein contains an NLS, it can relocate to the nucleus of a daughter cell that results from mitosis.

KEY CONCEPT Signal sequences target proteins co-translationally or post-translationally to the inner space of organelles, to organelle or plasma membranes, or to the exterior of a cell.

SUMMARY

This chapter has described translation, the process by which the nucleotide sequence of an mRNA is converted into the amino acid sequence of a protein. Translation is the last step in the transfer of information from DNA to proteins. It occurs in three phases—initiation, elongation, and termination—that are regulated by distinct factors—IFs, EFs, and RFs, respectively. In addition, initiation, elongation, and termination involve conformational changes driven by the hydrolysis of GTP, and elongation and termination involve protein factors that mimic the function of tRNAs. Some parts of the initiation mechanism are similar between bacteria and eukaryotes, and others are different because bacterial translation occurs co-transcriptionally, whereas eukaryotic transcription and translation are physically separated into the nucleus and cytoplasm, respectively. In addition, eukaryotic mRNAs have 5′ caps and 3′ poly(A) tails that play critical roles in translation. On the other hand, translation elongation and termination mechanisms are largely conserved between bacteria and eukaryotes.

Ribosomes, tRNAs charged with amino acids, and mRNAs are central players in translation. Ribosomes are complex structures made up of small and large subunits, each containing rRNAs and proteins. rRNAs occupy important sites in ribosomes such as the decoding center, which has E, A, and P binding sites for tRNAs, and the peptidyltransferase center, where peptide bonds are formed between amino acids. The accuracy of translation depends on the enzymatic linkage of an amino acid with its cognate tRNA, generating a charged tRNA molecule. As adaptors, tRNAs decode the information in mRNA by base pairing between their anticodons and an mRNA codon, a triplet of nucleotides. Some tRNAs recognize more than one codon by forming both Watson-Crick and wobble base pairs. RNA base pairing also occurs between the mRNA Shine–Dalgarno sequence and the 3′ end of 16S rRNA to position bacterial ribosomes at translation initiation codons. Translation begins at a fixed starting point (an AUG initiation codon that codes for methionine) and continues uninterrupted to the end of the open reading frame (an in-frame stop codon that does not code for an amino acid).

Just because an mRNA is present in a cell does not mean that it is translated. Translation is regulated, and this regulation largely occurs at the initiation and elongation phases.

Some signals globally affect translation by altering the activity of initiation factors that act on most mRNAs, while other signals affect the translation of particular mRNAs via factors that bind specific sequences or structures in mRNA untranslated regions and influence multiple steps in translation initiation and elongation.

The genetic code that deciphers mRNA sequence into amino acids is almost universal among organisms. It contains 64 triplet codons, 61 that code for amino acids and 3 that code for stop codons. The code is degenerate, nonoverlapping, continuous, and organized in a non-random pattern. The organization minimizes effects of nonsynonymous mutations on the functionality of proteins by having codons that differ by one nucleotide code for amino acids with similar polarity, charge, and size.

Proteins are polymers of amino acids linked together by peptide bonds. The 20 common amino acids all have an amino group, a carboxyl group, and a unique side chain. Differences in the polarity, charge, and size of amino acid side chains affect the structure and function of proteins. With the assistance of chaperones, proteins fold into secondary structures such as α-helices and β-sheets as well as tertiary structures. Proteins also assemble into stable multiprotein complexes called quaternary structures. Proteins are the enzymes responsible for cell metabolism, including DNA and RNA synthesis; they are regulatory factors required for expression of the genetic program, and they are structural factors that confer stiffness to otherwise flexible biological machines. Lastly, proteins contain short sequence elements that act as zip codes to target them to distinct places in cells such as inside organelles, organelle or cell membranes, or the exterior of a cell.

An immense variety of amino acid modifications affect the structure, function, and localization of proteins. Many modifications are reversible, so they serve as on/off switches that control molecular processes such as transcription and translation in response to signals. Modifications fall into five broad categories: cleavage of the peptide backbone, modifications of amino acids, or the addition of complex molecules, polypeptides, or chemical groups to amino acids. The process of post-translational modifications involves enzymes that add modifications (writer), enzymes that remove modifications (erasers), and proteins that interpret modifications (readers). In summary, the versatility of proteins as biological molecules is manifested in the diversity of shapes that they can assume.

KEY TERMS

active site (p. 304)

amino acid (p. 303)

aminoacyl tRNA-binding site (A site) (p. 313)

aminoacyl-tRNA synthetase (p. 311)

amino group (p. 304)

anticodon (p. 310)

carboxyl group (p. 304)

chaperone (p. 321)

charged tRNA (p. 311)

codon (p. 306)

decoding center (p. 315)

degenerate code (p. 306)

domain (p. 306)

elongation factor (EF) (p. 302)

exit site (E site) (p. 314)

fibrous protein (p. 306)

frameshift mutation (p. 307)

genetic code (p. 306)

globular protein (p. 306)

initiation codon (p. 315)

initiation factor (IF) (p. 302)

initiator tRNA (p. 315)

kinase (p. 324)

Kozak sequence (p. 316)

large ribosomal subunit (p. 310)

nonsense codon (p. 309)

nuclear localization sequence (NLS) (p. 325)

peptide bond (p. 304)

peptidyl site (P site) (p. 314)

peptidyltransferase center (p. 315)

phosphatase (p. 324)

phosphorylation (p. 324)

polypeptide (p. 303)

primary structure (of a protein) (p. 304)

proteasome (p. 324)

quaternary structure (of a protein) (p. 306)

reactive group (R group), side chain (p. 303)

reading frame (p. 307)

release factor (RF) (p. 302)

ribosomal RNA (rRNA) (p. 302)

ribosome (p. 302)

ribosome-binding site (RBS) (p. 316)

secondary structure (of a protein) (p. 304)

Shine–Dalgarno sequence (p. 316)

signal sequence (p. 325)

small ribosomal subunit (p. 310)

stop codon (termination codon) (p. 309)

subunit (p. 306)

synonymous codon (p. 309)

termination factor (p. 302)

tertiary structure (of a protein) (p. 304)

transfer RNA (tRNA) (p. 302)

triplet (p. 306)

ubiquitin (p. 324)

ubiquitination (p. 324)

wobble (p. 312)

SOLVED PROBLEMS

SOLVED PROBLEM 1

Using Figure 9-8, show the consequences on subsequent translation of the addition of an adenine nucleotide to the beginning of the following coding sequence:

Ⓐ

–CGA–UCG–GAA–CCA–CGU–GAU–AAG–CAU–
– Arg – Ser – Glu – Pro – Arg – Asp – Lys – His –

SOLUTION

With the addition of A at the beginning of the coding sequence, the reading frame shifts, and a different set of amino acids is specified by the sequence, as shown here (note that a set of nonsense codons is encountered, which results in chain termination):

–ACG–AUC–GGA–ACC–ACG–UGA–UAA–GCA–
– Thr – Ile – Gly – Thr – Thr – stop

SOLVED PROBLEM 2

A single nucleotide addition followed by a single nucleotide deletion approximately 20 bp apart in DNA causes a change in the protein sequence from

–His–Thr–Glu–Asp–Trp–Leu–His–Gln–Asp–

to

–His–Asp–Arg–Gly–Leu–Ala–Thr–Ser–Asp–

Which nucleotide has been added, and which nucleotide has been deleted? What are the original and the new mRNA sequences? (*Hint:* Consult Figure 9-8.)

SOLUTION

We can draw the mRNA sequence for the original protein sequence (with the inherent ambiguities at this stage):

– His – Thr – Glu – Asp – Trp – Leu– His – Gln – Asp

$$-CA\,{}^{U}_{C}-ACC-GA\,{}^{A}_{G}-GA\,{}^{U}_{C}-UGG-CUC-CA\,{}^{U}_{C}-CA\,{}^{A}_{G}-GA\,{}^{U}_{C}$$

(with additional options: C, A, G under first; U, A, G under fifth; UUA, G)

Because the protein-sequence change given to us at the beginning of the problem begins after the first amino acid (His) owing to a single nucleotide addition, we can deduce that a Thr codon must change to an Asp codon. This change must result from the addition of a G directly before the Thr codon (indicated by a box), which shifts the reading frame, as shown here:

$$-CA\,{}^{U}_{C}-\boxed{G}AC-UGA-{}^{A}_{G}GA-\widehat{U}UG-G\widehat{C}U-UCA-\widehat{U}CA\!\uparrow-GA\,{}^{U}_{C}-$$

– His – Asp – Arg – Gly – Leu – Ala – Thr – Ser – Asp –

Additionally, because a deletion of a nucleotide must restore the final Asp codon to the correct reading frame, an A or G must have been deleted from the end of the original next-to-last codon, as shown by the arrow. The original protein sequence permits us to draw the mRNA with a number of ambiguities. However, the protein sequence resulting from the frameshift allows us to determine which nucleotide was in the original mRNA at most of these points of ambiguity. Nucleotides that could have appeared in the original sequence are circled. In only a few cases does the ambiguity remain.

PROBLEMS

Visit SaplingPlus for supplemental content. Problems with the icon are available for review/grading. Problems with the ⧉ icon have an Unpacking the Problem exercise.

WORKING WITH THE FIGURES

(The first 25 questions require inspection of text figures.)

1. In Figure 9-1, circle a protein α-helix, an RNA stem, and an RNA loop.

2. In Figure 9-2, for which amino acids does the single-letter abbreviation not match the first letter of the amino acid name?

3. In Figure 9-3a, draw the reaction that takes place to add a third amino acid to the chain.

4. In Figure 9-4c, where are hydrophobic and hydrophilic amino acids most likely located?

5. In Figure 9-5, what is the sequence of the fourth codon in the nonoverlapping mechanism?

6. In Figure 9-6, what phenotype (rII^+ or rII^-) would be expected for two insertions and one deletion?

7. In Figure 9-7, what amino acids are encoded by a repeat of the sequence CCA?

8. In Figure 9-8, list the amino acids that are coded for by 1, 2, 3, 4, and 6 codons.

9. In Figure 9-9, create an analogous figure for the UAA stop codon.

10. In Figure 9-10, in general, what effect does a mutation of a purine to a purine or a pyrimidine to a pyrimidine in the first nucleotide of a codon have on the polarity and charge of an encoded amino acid?

11. In Figure 9-11, draw the secondary structure of tRNA^{Trp}, include the sequence of the anticodon and label the 5′ and 3′ ends as well as the location of amino acid attachment.

12. In Figure 9-12, draw 5′-aminoacyl-AMP for proline.

13. In Figure 9-13 and Table 9-1, what codons are recognized by tRNA^{Val} with an anticodon sequence 5′-UAC-3′?

14. In Figure 9-14, in eukaryotes, what RNA polymerases are required to transcribe rRNAs and ribosomal protein genes?

15. In Figure 9-15, how many nucleotides are commonly found in the loop region of stem-loops?

16. In Figure 9-16, which subunit of the ribosome binds the mRNA, and which subunit carries out peptide bond formation?

17. Using Figure 9-17, circle the Shine–Dalgarno sequence in Figure 9-15.

18. In Figure 9-18, describe three mechanistic differences between translation initiation in bacteria and eukaryotes.

19. In Figure 9-19, draw the next step in elongation.

20. In Figure 9-20, is RF1 a tRNA or a protein?

21. In Figure 9-21, what mutation in tRNA^{Tyr} would suppress a UAA nonsense mutation?

22. In Figure 9-22, why do miRNAs and RNA-BPs affect only the translation of specific mRNAs?

23. In Figure 9-23, draw phosphotyrosine.

24. In Figure 9-24, how might this pathway get turned off?

25. In Figure 9-25, how does this diagram provide insight into the picture of ribosomes shown on the first page of the chapter?

BASIC PROBLEMS

26. **a.** Use the codon table in Figure 9-8 to complete the following table. Assume that reading is from left to right and that the columns represent transcriptional and translational alignments.

				T	G	A				DNA double helix
C										
	C	A			U					mRNA transcribed
						G	C	A		Appropriate tRNA anticodon
		Trp								Amino acids incorporated into protein

b. Label the 5′ and 3′ ends of DNA and RNA, as well as the amino and carboxyl ends of the protein.

27. Consider the following segment of DNA:

5′ GCTTCCCAA 3′

3′ CGAAGGGTT 5′

Assume that the bottom strand is the template strand used by RNA polymerase.

a. Draw the RNA transcribed.

b. Label its 5′ and 3′ ends.

c. Draw the corresponding amino acid chain, assuming that the reading frame starts at the first nucleotide.

d. Label its amino and carboxyl ends.

Repeat parts *a* through *d*, assuming that the top strand is the template strand.

28. A mutational event inserts an extra base pair into DNA. Which of the following outcomes do you expect? (1) No protein at all; (2) a protein in which one amino acid is changed; (3) a protein in which three amino acids are changed; (4) a protein in which two amino acids are changed; (5) a protein in which most amino acids after the site of the insertion are changed.

29. **a.** In how many cases in the genetic code would you fail to know the amino acid specified by a codon if you knew only the first two nucleotides of the codon?

b. In how many cases would you fail to know the first two nucleotides of the codon if you knew which amino acid is specified by it?

30. If a polyribonucleotide contains equal amounts of randomly positioned adenine and uracil bases, what proportion of its triplets will encode (a) phenylalanine, (b) isoleucine, (c) leucine, (d) tyrosine?

31. In the fungus *Neurospora*, some mutants were obtained that lacked activity for a certain enzyme. The mutations were found, by mapping, to be in either of two unlinked genes. Provide a possible explanation in reference to quaternary protein structure.

32. What is meant by the statement "The genetic code is universal"? What is the significance of this finding?

33. A mutant has no activity for the enzyme isocitrate lyase. Does this result prove that the mutation is in the gene encoding isocitrate lyase? Why? 🌐

34. A certain nonsense suppressor corrects a non-growing mutant to a state that is near, but not exactly, wild type (it has abnormal growth). Suggest a possible reason why the reversion is not a full correction.

35. In bacterial genes, as soon as a partial mRNA transcript is produced by RNA polymerase, the ribosome assembles on it and starts translating. Draw a diagram of this process, identifying 5′ and 3′ ends of mRNA, the amino and carboxyl ends of the protein, the RNA polymerase, and at least one ribosome. Why couldn't this system work in eukaryotes?

36. Researchers have found that aspartic acid and glutamic acid can sometimes mimic the function of phosphoserine and phosphothreonine. Why might this be?

37. Why might a mutation in the untranslated region of a bacterial mRNA affect translation? How about for a eukaryotic mRNA?

38. In vitro translation systems have been developed in which a specific mRNA can be added to a test tube containing a bacterial cell extract that includes all the components needed for translation (ribosomes, tRNAs, and amino acids). If a radioactively labeled amino acid is included, any protein translated from that mRNA can be detected on a gel. If a eukaryotic mRNA is added to the in vitro system, would radioactive protein be produced? Explain why or why not.

39. An in vitro translation system contains a eukaryotic cell extract that includes all the components needed for translation (ribosomes, tRNAs, and amino acids). If bacterial RNA is added to the test tube, would a protein be produced? Explain why or why not.

40. Would a chimeric translation system containing the large ribosomal subunit from *E. coli* and the small ribosomal subunit from yeast (a unicellular eukaryote) be able to function in protein synthesis? Explain why or why not.

41. Mutations that change a single amino acid in the active site of an enzyme can result in the synthesis of wild-type amounts of an inactive enzyme. In what other regions of a protein might a single amino acid change have the same result?

42. What evidence supports the view that ribosomes are ribozymes? 🌐

43. Explain why antibiotics, such as erythromycin and Zithromax, that bind the large ribosomal subunit do not harm us.

44. Our immune system makes many different proteins that protect us from viral and bacterial infection. Biotechnology companies must produce large quantities of these immune proteins for human testing and eventual sale to the public. To this end, their scientists engineer bacterial or human cell cultures to express these immune proteins. Explain why proteins isolated from bacterial cultures are often inactive, whereas the same proteins isolated from human cell cultures are active (functional).

45. Would you expect to find nuclear localization sequences (NLSs) in the proteins that make up bacterial and eukaryotic DNA and RNA polymerases? Explain why or why not.

CHALLENGING PROBLEMS

46. Draw the structure and hydrogen bonding of a parallel β-sheet.

47. How were synthetic RNAs such as poly(U) that lacked a Shine–Dalgarno sequence translated in an *E. coli* extract?

48. A single nucleotide addition and a single nucleotide deletion approximately 15 bases apart in the DNA cause a protein change in sequence from

    Phe–Ser–Pro–Arg–Leu–Asn–Ala–Val–Lys

    to

    Phe–Val–His–Ala–Leu–Met–Ala–Val–Lys

    a. What are the old and new mRNA nucleotide sequences? Use the codon table in Figure 9-8.

    b. Which nucleotide has been added? Which has been deleted? 🌐

49. You are studying an *E. coli* gene that specifies a protein. A part of its sequence is

    –Ala–Pro–Trp–Ser–Glu–Lys–Cys–His–

    You recover a series of mutants for this gene that show no enzymatic activity. By isolating the mutant enzyme products, you find the following sequences:

    Mutant 1:

    –Ala–Pro–Trp–Arg–Glu–Lys–Cys–His–

    Mutant 2:

    –Ala–Pro–

    Mutant 3:

    –Ala–Pro–Gly–Val–Lys–Asn–Cys–His–

    Mutant 4:

    –Ala–Pro–Trp–Phe–Phe–Thr–Cys–His–

    What is the molecular basis for each mutation? What is the DNA sequence that specifies this part of the protein? 🌐

50. What structural features are shared by spliceosomes (see Figure 8-19) and ribosomes? Why are both structures used to support the RNA World theory?

51. A double-stranded DNA molecule with the sequence shown here produces, in vivo, a polypeptide that is five amino acids long.

TACATGATCATTTCACGGAATTTCTAGCATGTA

ATGTACTAGTAAAGTGCCTTAAAGATCGTACAT

   a. Which strand of DNA is the template strand, and in which direction is it transcribed?

   b. Label the 5′ and 3′ end of each strand.

   c. If an inversion occurs between the second and the third triplets from the left and right ends, respectively, and the same strand of DNA is transcribed, how long will the resultant polypeptide be?

   d. Assume that the original molecule is intact and that the bottom strand is transcribed from left to right. Give the RNA base sequence and label the 5′ and 3′ ends of the anticodon that inserts the *fourth* amino acid into the nascent polypeptide. What is this amino acid? 📖

52. One of the techniques Khorana used to decipher the genetic code was to synthesize polypeptides in vitro, using synthetic mRNA with various repeating base sequences. For example, (AGA)$_n$, which can be written out as AGAAGAAGAAGA. . . . Sometimes the resulting polypeptide contained just one amino acid (a homopolymer), and sometimes it contained more than one amino acid (a heteropolymer), depending on the repeating sequence used. Khorana found that sometimes different polypeptides were made from the same synthetic mRNA, suggesting that the initiation of protein synthesis in the system in vitro does not always start at the first nucleotide of the messenger. For example, from (CAA)$_n$, three polypeptides may have been made: aa$_1$ homopolymer (abbreviated aa$_1$-aa$_1$), aa$_2$ homopolymer (aa$_2$-aa$_2$), and aa$_3$ homopolymer (aa$_3$-aa$_3$). These polypeptides probably correspond to the following readings derived by starting at different places in the sequence:

CAA CAA CAA CAA . . .

ACA ACA ACA ACA . . .

AAC AAC AAC AAC . . .

The following table shows the results of Khorana's experiment.

Synthetic mRNA	Polypeptide(s) synthesized
(UC)$_n$	(Ser–Leu)
(UG)$_n$	(Val–Cys)
(AC)$_n$	(Thr–His)
(AG)$_n$	(Arg–Glu)
(UUC)$_n$	(Ser–Ser) and (Leu–Leu) and (Phe–Phe)
(UUG)$_n$	(Leu–Leu) and (Val–Val) and (Cys–Cys)
(AAG)$_n$	(Arg–Arg) and (Lys–Lys) and (Glu–Glu)
(CAA)$_n$	(Thr–Thr) and (Asn–Asn) and (Gln–Gln)
(UAC)$_n$	(Thr–Thr) and (Leu–Leu) and (Tyr–Tyr)
(AUC)$_n$	(Ile–Ile) and (Ser–Ser) and (His–His)
(GUA)$_n$	(Ser–Ser) and (Val–Val)
(GAU)$_n$	(Asp–Asp) and (Met–Met)
(UAUC)$_n$	(Tyr–Leu–Ser–Ile)
(UUAC)$_n$	(Leu–Leu–Thr–Tyr)
(GAUA)$_n$	None
(GUAA)$_n$	None

Note: The order in which the polypeptides or amino acids are listed in the table is not significant except for (UAUC)$_n$ and (UAUC)$_n$.

   a. Why do (GUA)$_n$ and (GAU)$_n$ each encode only two homopolypeptides?

   b. Why do (GAUA)$_n$ and (GUAA)$_n$ fail to stimulate synthesis?

   c. Using Khorana's results, assign an amino acid to each triplet in the following list. Remember that there are often several codons for a single amino acid and that the first two letters in a codon are usually the important ones (but that the third letter is occasionally significant). Also keep in mind that some very different-looking codons sometimes encode the same amino acid. Try to solve this problem without consulting Figure 9-8.

GUA	GAU	UUG	AAC
GUG	UUC	UUA	GAA
GUU	AGU	UAU	AGA
AUG	CUU	AUC	GAG
UGU	CUA	UAC	CAA
ACA	UCU	AAG	UAG
CAC	CUC	ACU	UGA

Solving this problem requires both logic and trial and error. Don't be disheartened: Khorana received a Nobel Prize for doing it. Good luck!

(Data from J. Kuspira and G. W. Walker, *Genetics: Questions and Problems.* McGraw-Hill, 1973.)

GENETICS AND SOCIETY

If life were found on another planet, do you think that it would have the same genetic code? Justify your answer.

Gene Isolation and Manipulation

Agarose gel electrophoresis is used to separate DNA fragments based on their size. [*SPL/Science Source.*]

CHAPTER OUTLINE AND LEARNING OBJECTIVES

In prior chapters, we saw that the genome contains the information needed to build and maintain an organism. Researchers are often interested in studying the function of one or a few genes in a genome. The main objective of this chapter is to present methods that are used to do this, including methods to detect and quantify RNAs, proteins, and specific regions of DNA as well as methods to alter the sequence and amounts of these molecules.

In this chapter, we describe experimental techniques used to isolate and manipulate genes and their products, RNAs and proteins. There are many good reasons for including this information in an introductory genetics textbook. To develop these techniques, researchers relied on knowledge of the chemical and functional properties of DNA, RNAs, and proteins as well as the mechanisms underlying fundamental molecular genetic processes such as DNA replication, transcription, and translation. Thus, an understanding of experimental techniques will reinforce principles that are presented in other chapters. It will also aid in the comprehension and evaluation of the primary research literature, as well as the design of experiments to address yet-to-be-solved genetic problems. Lastly, whether or not a problem can be solved is often determined by the techniques that are available. Consequently, some major advances in genetics have become possible only because of the development of a new technique. Standouts include techniques for isolating and manipulating fragments of DNA (DNA cloning), amplifying DNA (PCR), sequencing DNA (dideoxy sequencing), and introducing DNA into an organism (transgenesis). This chapter ends with recently developed techniques such as CRISPR-Cas9 for the precise engineering of genomes. New genome engineering techniques have made possible reverse genetic studies that aim to understand the function of a gene by analyzing the phenotypic consequences of altering the gene sequence or its expression.

Genes are the central focus of genetics, and so, clearly, it is desirable to isolate a gene of interest (or any DNA region) from the genome to study it. Isolating individual genes and producing enough copies of them to analyze can be a daunting task because a single gene is a tiny fraction of an entire genome. For example, the haploid human genome contains over 3 billion base pairs, whereas the coding region of an average gene contains only a few thousand base pairs. How do scientists find the proverbial needle in the haystack—the gene—and then produce suitable quantities of it for analysis?

Many investigations in genetics begin with the desire to study a trait or a disease. In Chapter 2, we described forward genetic approaches to search for mutants that exhibit an altered phenotype and crosses or pedigree analysis to determine whether that phenotype is determined by a single gene. In Chapter 4, we discussed how mapping by recombination helps locate the gene at the DNA level. In this chapter, we continue by presenting molecular methods for identifying a gene of interest and studying its molecular function.

The first step in studying gene function is to isolate its DNA and reproduce it in quantities suitable for study. Just like a construction worker, a genetic engineer needs tools.

Most toolboxes that we are familiar with are filled with tools like hammers, screwdrivers, and wrenches that are designed by people and manufactured in factories. In contrast, the tools of the genetic engineer are molecules isolated from cells. Most of these tools were the product of scientific discovery—where the objective was to answer a biological question. Only later did some scientists appreciate the potential practical value of these molecules and invent ways to put them to use with the goal of isolating and amplifying DNA fragments. As an example, one way to separate our gene of interest from the rest of the genome is to cut the genome with "molecular scissors" and isolate the small fragment containing the gene. Werner Arber discovered these molecular scissors, and for this discovery he was awarded the Nobel Prize in Physiology or Medicine in 1978. However, Arber was not looking for a tool to cut DNA precisely. Rather, he was trying to understand why some bacteria are resistant to infection by bacterial viruses. By answering this biological question, he discovered that resistant bacteria possess a previously unknown type of enzyme—a restriction endonuclease—that cuts DNA at specific sequences. As we will see in this chapter, restriction enzymes are one of the cornerstones of genetic engineering and a common tool found in the genetic engineer's toolbox.

As another example, it is unlikely that anyone would have predicted that DNA polymerase, discovered by Arthur Kornberg, could be fashioned into two powerful tools for DNA isolation and analysis. To this day, many of the techniques used to determine the nucleotide sequence of DNA rely on synthesizing it with DNA polymerase. Similarly, most of the protocols used to isolate and amplify specific regions of DNA from sources as disparate as a crime scene or a fossil embedded in amber rely on the activity of DNA polymerase.

DNA technologies are the collective techniques for obtaining, amplifying, and manipulating specific DNA fragments. Since the mid-1970s, the development of DNA technologies has revolutionized the study of biology, opening many areas of research to molecular investigation. **Genetic engineering**, the application of DNA technologies to specific biological, medical, or agricultural problems, is now a well-established branch of technology. **Genomics** is the ultimate extension of the technologies to the global analysis of the nucleic acids present in a nucleus, a cell, an organism, or a group of related species (see Chapter 14).

In this chapter, we will illustrate gene isolation and manipulation techniques through their application to the insulin gene. The insulin gene encodes a protein that functions to maintain normal blood glucose levels by promoting the uptake of glucose from blood into cells and by regulating the metabolism of carbohydrates, lipids, and proteins. In mammals, including humans and mice, the insulin gene is present in the genome of all cells, but it is expressed (i.e., transcribed and translated) only in beta (β) cells in the pancreas. Diabetes is a human disease in which blood glucose levels are abnormally high, either because β cells do not produce enough insulin (type I diabetes) or because cells are unable to respond to insulin (type II diabetes). Mild forms of type I diabetes can be treated by dietary restrictions, but for many patients, daily insulin injections are necessary.

Until about 35 years ago, cows were the major source of insulin protein. The protein was harvested from pancreases of animals processed in meatpacking plants and purified on a large scale to eliminate the majority of proteins and other contaminants in pancreas extracts. It took about 8,000 pounds of pancreas from 23,500 animals to purify one pound of insulin. Then, in 1982, the first recombinant human insulin came on the market. Because it could be produced on an industrial scale in bacteria by recombinant DNA techniques using the human gene sequence, insulin could be made in a purer form and at a lower cost than the previous method. We will use the generation of recombinant human insulin as an example of the general steps necessary for making any recombinant DNA molecule. These steps are summarized in **Figure 10-1**. The uses

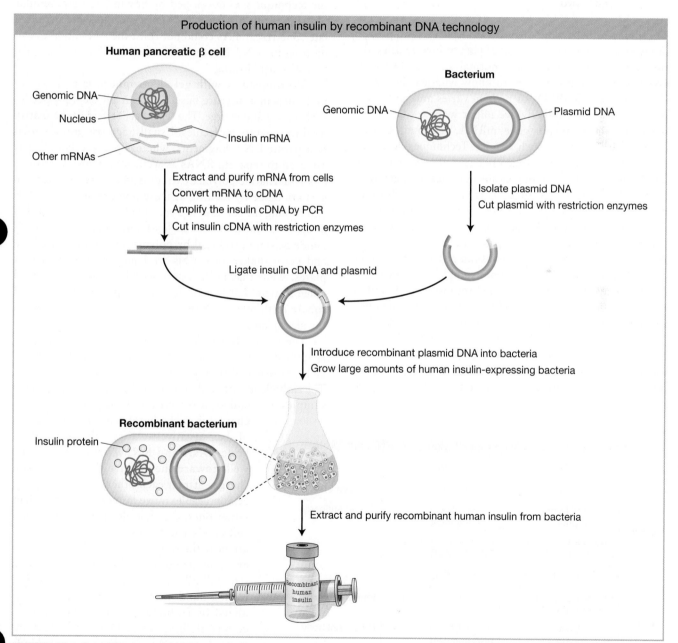

FIGURE 10-1 Recombinant human insulin is currently being produced in bacteria. The first step in this process is the construction of an expression plasmid that contains a human insulin cDNA insert. The plasmid is then transformed into bacteria, and the bacteria are grown in large quantities. As the recombinant bacteria grow, they transcribe insulin mRNA and translate the mRNA into insulin protein. Finally, the bacteria are harvested, and the insulin protein is extracted and purified for therapeutic use in humans.

of recombinant DNA technologies are quite broad, ranging from gene isolation for basic biological research to gene therapy to treat human disease to the production of herbicides and pesticides for crop plants.

10.1 DETECTING AND QUANTIFYING DNA, RNA, AND PROTEIN

LO 10.1 Describe methods for detecting and quantifying specific nucleic acid and protein molecules in vitro and in vivo.

The ability to detect and quantify DNA, RNA, and protein in vivo and in vitro is an essential part of investigating the function of these molecules in normal and disease states. For example, we might want to know whether the size of the insulin gene, mRNA, or protein varies among human populations. Alternatively, we might want to determine whether a similar insulin gene, mRNA, or protein is present in other organisms such as mice. Techniques described in this section and summarized in **Figure 10-2** were used to determine that the human genome contains a single insulin gene (*Ins*) on chromosome 11. Transcription of *Ins* produces a 1431-nucleotide pre-mRNA that contains three exons and two introns (**Figure 10-3a**, top). Translation of the spliced *Ins* mRNA produces a protein of 110 amino acids. In addition, other animals have insulin genes that are similar in sequence to human *Ins*. For example, mice have two insulin genes, *Ins1* and *Ins2*, on chromosomes 19 and 7, respectively (Figure 10-3a, middle and bottom). The open reading frame of the human *Ins* mRNA is 81 percent identical in nucleotide sequence to the mouse *Ins1* mRNA and 83 percent to *Ins2*, and the human Ins protein is 78 percent identical in amino acid sequence to the mouse Ins1 protein and 82 percent to Ins2. As we move through the chapter, consider how the methods presented could be used to isolate and manipulate insulin genes in other animals such as the fruit fly *Drosophila melanogaster* that lack a pancreas and β cells.

Detecting and quantifying molecules by Southern, Northern, and Western blot analysis

Blotting is a commonly used in vitro method to detect and quantify a specific DNA, RNA, or protein molecule within a mixture of many different DNA, RNA, or protein molecules. Blotting for DNA is called **Southern blotting** because the technique was developed by Edwin Southern. Similar blotting techniques for RNA and protein were invented later, and researchers could not resist the temptation to call blotting for RNA **Northern blotting** and blotting for protein **Western blotting**.

Blotting starts with **gel electrophoresis** to separate molecules in a mixture based on their physical properties such as size and charge. The term "gel" refers to the matrix used to separate molecules. Usually, agarose gels are used to separate DNA fragments, whereas polyacrylamide gels are used to separate RNAs as well as proteins. Agarose is a polysaccharide polymer extracted from seaweed. Agarose gels are made by melting agarose powder in a hot buffer and cooling the solution in a rectangular tray to form a slab of agarose that is similar to Jell-O. In contrast, polyacrylamide gels are produced by polymerization of acrylamide and a cross-linker such as bis-acrylamide between two glass plates. Wells that hold experimental samples are formed when agarose hardens or acrylamide polymerizes around square teeth of a comb that is set into the tray or between the glass plates.

The term "electrophoresis" refers to the voltage that is applied to gels that are submerged in a buffer solution. Gels are oriented with electrodes at the top and bottom. The cathode (negative charge) is at the top of the gel, where samples are loaded into wells, and the anode (positive charge) is at the bottom of the gel. Because of their negatively charged phosphate backbone, DNA and RNA migrate out of the wells toward the positive charge (opposite charges attract each other). Smaller nucleic acid molecules move faster through gels than larger ones, so, after electrophoresis, molecules are separated by size; larger molecules are near the top of the gel and smaller ones are near the bottom. Molecules of the same size will all migrate the same distance and form a band in the gel. Bands can be visualized by staining gels with dyes such as coomassie blue for proteins and ethidium bromide for DNA. Ethidium bromide binds DNA by intercalating between base pairs, and enables DNA to fluoresce when exposed to ultraviolet (UV) light (**Figure 10-4**). The

Methods for detecting and quantifying DNA, RNA, and protein			
	DNA	**RNA**	**Protein**
In vitro	Southern blot	Northern blot	Western blot
	Probe: DNA or RNA fragment	Probe: DNA or RNA fragment	Probe: Antibody
	Polymerase chain reaction (PCR)	Reverse transcription-PCR (RT-PCR)	
	Probe: DNA primers	Probe: DNA primers	
In vivo	Fluorescence in situ hybridization (FISH)	In situ hybridization	Immunofluorescence
	Probe: DNA or RNA fragment	Probe: DNA or RNA fragment	Probe: Antibody

FIGURE 10-2 A summary of the main methods used to detect and quantify specific DNA regions, RNAs, and proteins in vitro (i.e., after purification from cells) and in vivo (i.e., in cells, tissues, and whole organisms).

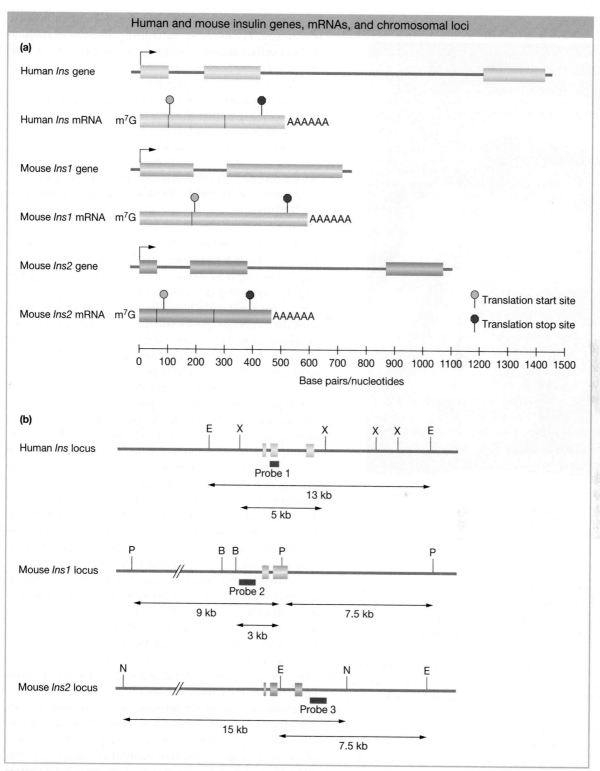

FIGURE 10-3 (a) Structures of insulin genes and mRNAs in humans and mice. Boxes represent exons. The mRNAs are spliced, capped at the 5′ end with m⁷G, and polyadenylated at the 3′ end (Chapter 8). (b) Structures of insulin gene loci in humans and mice. Red lines indicate regions used as probes for Southern and Northern blot analyses. Results for probe 1 are shown in Figure 10-6, and questions related to probes 2 and 3 are in the Working with the Figures section at the end of the chapter. Restriction enzymes used for Southern blot analysis are indicated by single letters: B = *Bam*HI, E = *Eco*RI, N = *Nsi*I, P = *Pvu*II, and X = *Xho*I (restriction sites for these enzymes are shown in Table 10-1). Lines with arrows indicate the sizes of some restriction fragments.

Gel electrophoresis

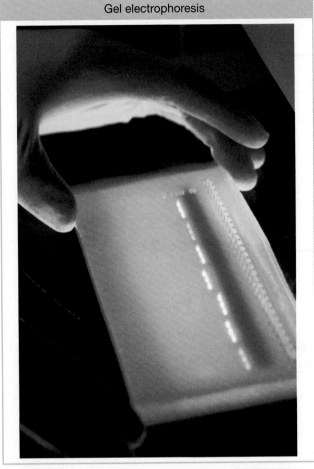

FIGURE 10-4 Agarose gel electrophoresis is used to separate DNA fragments based on size. After the DNA fragments are separated, they are stained with ethidium bromide and visualized with ultraviolet (UV) light. The visualized pink/white lines are bands of DNA of a particular size. [*SPL/Science Source.*]

higher-order structures and allow molecules to migrate true to size. For example, denaturing SDS-PAGE (sodium dodecyl sulfate-polyacrylamide gel electrophoresis) is typically used to separate proteins. SDS is a detergent that not only disrupts protein folding, but also uniformly coats proteins with a negative charge, blocking the charges on amino acid R-groups, so that the rate of migration of a protein toward the positive anode depends on its molecular weight, and not on its amino acid composition.

KEY CONCEPT Gel electrophoresis separates complex mixtures of DNA fragments, RNAs, or proteins, based on size and charge.

After gel electrophoresis is used to separate complex mixtures of DNA fragments, RNAs, or proteins, the next step in blotting is to transfer the molecules from the gel to a special type of paper called a membrane that has high affinity for these molecules. The transfer procedure maintains the molecules' positions relative to each other on the membrane, just as they were in the gel. Transfer is carried out either by capillary action, as illustrated in **Figure 10-5**, or by electrophoresis. At this point, the transferred molecules on the membrane are invisible to the naked eye. The last step in blotting is to use a **probe** to visualize a specific molecule on the membrane. In Southern and Northern blotting, probes are radioactive (^{32}P-labeled) single-stranded nucleic acids that are complementary to the nucleic acid of interest. When the membrane is incubated with a solution containing the probe, the probe anneals with complementary nucleic acid sequences bound to the membrane. This annealing process is commonly called **hybridization**. Unbound probe is washed away, and places where the probe has hybridized are revealed by **autoradiography**, that is, exposing the membrane to X-ray film. Because hybridization requires single-stranded molecules, Southern blotting has an extra step in which the gel is soaked in an alkaline solution such as NaOH (sodium hydroxide) to denature the double-stranded DNA into single-stranded DNA prior to the membrane transfer step.

In Western blotting, probes are **antibodies** that bind specific proteins and are detected in a variety of ways, including light emitting chemiluminescence or fluorescence. Antibodies are proteins made by the immune system of some animals that bind foreign substances called antigens with high affinity. One way to produce an antibody is to inject a large amount of an antigen (e.g., a protein of interest) into an animal (usually rabbits or chickens, but sometimes larger animals such as goats), allow time for the animal to raise an immune response to the antigen, and then, from the blood of the animal, collect serum that contains the antibody.

KEY CONCEPT In Southern blotting, the material that is transferred to the membrane is DNA, while in Northern blotting it is RNA, and in Western blotting it is protein.

size of the molecules within each band in the gel can be determined by comparing a band's migration distance with a set of standard molecules of known sizes (also known as size markers). If the bands are well separated, an individual band can be cut from the gel, and the DNA sample can be purified from the gel matrix. Therefore, DNA electrophoresis can be either diagnostic (showing sizes and relative amounts of DNA fragments present) or preparative (useful in isolating specific DNA fragments).

Gel electrophoresis can be carried out under non-denaturing or denaturing conditions. Non-denaturing conditions maintain the higher-order structures of molecules, including base pairing between DNA strands, base pairing in folded single-stranded RNAs, secondary and tertiary structures in folded proteins, and interactions between molecules. Typically, agarose gel electrophoresis of DNA molecules is carried out under non-denaturing conditions to maintain the double-stranded structure of DNA, whereas polyacrylamide gel electrophoresis of RNAs and proteins is carried out under denaturing conditions to eliminate

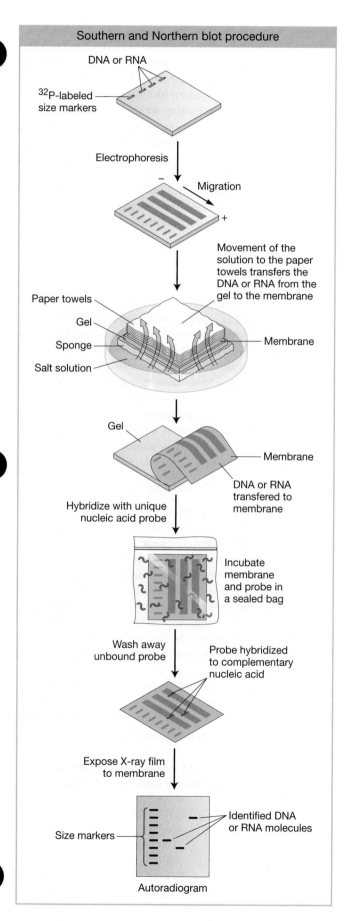

Southern and Northern blot procedure

DNA or RNA

^{32}P-labeled size markers

Electrophoresis

– Migration +

Movement of the solution to the paper towels transfers the DNA or RNA from the gel to the membrane

Paper towels
Gel
Sponge
Salt solution
Membrane

Gel
Membrane
DNA or RNA transfered to membrane

Hybridize with unique nucleic acid probe

Incubate membrane and probe in a sealed bag

Wash away unbound probe

Probe hybridized to complementary nucleic acid

Expose X-ray film to membrane

Identified DNA or RNA molecules

Size markers

Autoradiogram

Techniques analogous to Southern, Northern, and Western blotting are used to detect DNA, RNAs, and proteins in vivo in cultured cells or in whole organisms. Detection of DNA and RNA in vivo is carried out by hybridization with radioactive single-stranded nucleic acid probes, followed by autoradiography. This technique is called **in situ hybridization (ISH)**. A modified form of ISH that uses fluorescently labeled probes rather than radioactive probes is called **fluorescence in situ hybridization (FISH)** (for example, see Figure 13-6). Proteins are detected in vivo by **immunofluorescence**, which uses antibody probes (for example, see Figure 13-15). A fluorescence microscope is used to reveal the location and abundance of the fluorescence signal in FISH and immunofluorescence.

KEY CONCEPT Specific DNA fragments or RNAs are detected in vitro and in vivo by hybridization with nucleic acid probes, and specific proteins are detected by interaction with antibody probes.

The expected results of Northern and Western blot analyses for insulin are shown in **Figures 10-6a** and 10-6b, respectively. Northern blot analysis of the thousands of different mRNAs expressed in human cells with a probe complementary to the human *Ins* mRNA should detect one band, that is, a single type of insulin mRNA in β cells but not in kidney cells (Figure 10-6a, lanes 2 and 3). Because the nucleotide sequence of the insulin mRNA is very similar between humans and mice, the human probe should also detect the two insulin mRNAs (*Ins1* and *Ins2*) in mouse β cells (Figure 10-6a, lane 5). Standardly, blots are probed not only for a molecule of interest, but also for other molecules that serve to confirm that the experiment worked as expected or that serve as the basis for comparison within or between samples, which is referred to as a loading, normalization, or specificity control. In this case, analysis of *ribosomal protein S7* (*RpS7*) mRNA, a transcript that is present in all cells at similar levels, shows that the failure to detect insulin mRNA in the kidney cell samples (Figure 10-6a, lanes 2 and 4) was not due to a problem with the experiment. Furthermore, the intensity of the bands provides information about the abundance of the mRNAs. The data show that, in human β cells, *Ins* mRNA is less abundant than *RpS7* mRNA, and, in mouse β cells, *Ins2* is more abundant than *Ins1*. Western blot analysis of

FIGURE 10-5 Southern and Northern blotting procedures are similar. The key difference is that in Southern blotting, DNA is transferred to the membrane, while in Northern blotting, RNA is transferred to the membrane. Western blotting for proteins is carried out by a comparable procedure, except that the probe is an antibody rather than a radioactive nucleic acid, and the size markers are proteins of different sizes.

ANIMATED ART SaplingPlus   Northern blot analysis

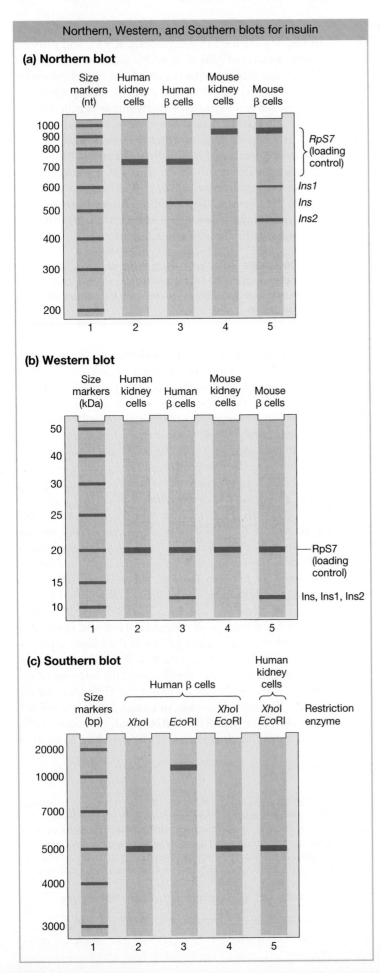

Northern, Western, and Southern blots for insulin

(a) Northern blot

(b) Western blot

(c) Southern blot

proteins from the same types of cells with an antibody to human insulin shows that cell type-specific expression of the insulin protein is similar to that of the insulin mRNA (Figure 10-6b). The expected size of a protein on a Western blot can be estimated based on the average molecular weight of an amino acid, 110 Daltons (Da). One Da is one gram per mole, and one kilodalton (kDa) is 1000 grams per mole. Thus, the 110-amino-acid insulin protein is about 12 kDa (110 amino acids × 110 Da/amino acid). Note that only one insulin protein band should be detected in mouse β cells because the Ins1 and Ins2 proteins are the same size and, thus, should migrate to the same position upon gel electrophoresis.

Sometimes, the starting material for Southern blot analysis is chromosome-sized DNA molecules of genomic DNA. Such large molecules can be analyzed more precisely when cut into fragments of much smaller size. Cutting is carried out by bacterial **restriction enzymes**. These enzymes are endonucleases that cleave phosphodiester bonds between nucleotides at specific DNA sequences, called **restriction sites**, that are usually 4 to 8 base pairs long. **Table 10-1** shows several examples of restriction enzymes and their restriction sites. Restriction sites are **palindromic**, which means that both strands have the same nucleotide sequence but in antiparallel orientation (the same sequence reads 5′ to 3′ on each strand). For example, the restriction enzyme

*Eco*RI recognizes the sequence $\begin{smallmatrix} \downarrow \\ 5'-\text{GAATTC}-3' \\ 3'-\text{CTTAAG}-5' \\ \uparrow \end{smallmatrix}$ and cleaves

the bond between G and A on each strand. Restriction enzyme names are based on the organism in which they were discovered. For example, the enzyme *Eco*RI was discovered in *E. coli*, which also explains why the first part of a restriction enzyme name is italicized. To date, approximately 3000 restriction enzymes have been identified that recognize over 230 restriction sites. Some restriction enzymes cut in the same position on each strand, leaving *blunt* ends (e.g., *Msp*I; see Table 10-1), while others make cuts that are offset, producing *staggered* ends (e.g., *Eco*RI). Thus, a restriction enzyme will cut the DNA from any organism into a set of **restriction fragments** determined by the locations of restriction sites in the DNA, and will produce the same pattern of fragments every time that DNA is cut.

KEY CONCEPT Restriction enzymes cut DNA at specific sequences, producing fragments with staggered or blunt ends.

FIGURE 10-6 (a) Northern, (b) Western, (c) and Southern blot analyses for insulin in different human and mouse tissues. The Southern and Northern blots were analyzed with probe 1 to exon 2 of the human *Ins* gene (see Figure 10-3b). The Northern blot was also analyzed with a probe to the *RpS7* mRNA. The Western blot was probed with antibodies to human insulin and RpS7 proteins.

ANIMATED ART 🌐 SaplingPlus   Western blot analysis

TABLE 10-1	Restriction enzymes			
Restriction enzyme	Source bacterium	Restriction site	Length (bp)	Staggered (S) Blunt (B)
BamHI	B. amyloliquefaciens	5′-G↓GATCC-3′ 3′-CCTAG↑G-5′	6	S
EcoRI	E. coli	5′-G↓AATTC-3′ 3′-CTTAA↑G-5′	6	S
MspI	Moraxella sp.	5′-C↓CGG-3′ 3′-GGC↑C-5′	4	B
NotI	N. otitdis	5′-GC↓GGCCGC-3′ 3′-CGCCGG↑CG-5′	8	S
NsiI	N. sicca	5′-ATGCA↓T-3′ 3′-T↑ACGTA-5′	6	S
PvuII	P. vularis	5′-CAG↓CTG-3′ 3′-GTC↑GAC-5′	6	B
XhoI	X. holcicola	5′-C↓TCGAG-3′ 3′-GAGCT↑C-5′	6	S

Arrows indicate sites of cleavage.

The expected results of Southern blot analysis for the insulin gene are shown in Figure 10-6c. Each sample on the blot contains genomic DNA that was digested with restriction enzymes. If the sequence of the 3×10^9 base pair human genome was completely random, the 6-base pair restriction site for EcoRI should occur every 4096 base pairs (4^6, the number of possible base pairs at each position in the restriction site^the number of base pairs in the restriction site), which means that digestion of the human genome would produce about 730,000 EcoRI fragments ($3 \times 10^9/4096$). Based on **restriction maps** of genomic DNA, as shown in Figure 10-3b, the probe complementary to the insulin gene should detect a single EcoRI band out of the estimated 730,000 EcoRI bands (Figure 10-6c, lane 3). The same probe should detect a smaller band when the DNA is digested with the restriction enzyme XhoI or with both EcoRI and XhoI (Figure 10-6c, lanes 2 and 4), because the XhoI fragment is located within the EcoRI fragment. Southern blot analysis of DNA from β cells and kidney cells produces the same result (Figure 10-6c, lanes 4 and 5) because all cells in an organism have the same genomic DNA. Whereas Northern and Western blot analysis of β cells and kidney cells produces different results (Figures 10-6a and 10-6b), because insulin is expressed only in β cells.

Detecting and amplifying DNA by the polymerase chain reaction

In 1985, the ability of researchers to analyze and manipulate DNA was transformed by the invention of the **polymerase chain reaction (PCR)** by Kary Mullis. PCR makes it possible to produce billions of copies of a specific DNA sequence starting with only one copy. This is called **DNA amplification**. To develop PCR, Mullis brought together several pieces of information obtained through basic research. First, he knew from studies of DNA replication that DNA polymerases copy a single-stranded DNA template by extending off the 3′ end of an annealed primer. Second, he knew from techniques such as Southern blotting that, in solution, a short piece of single-stranded DNA called an oligonucleotide will specifically anneal to DNA sequences that are perfectly complementary. Third, he knew that high temperatures disrupt hydrogen bonds between bases in double-stranded DNA to produce single-stranded DNA. Lastly, he knew that about 20 years earlier, the microbiologist Thomas Brock was able to culture a bacterium *Thermus aquaticus* that grows at high temperatures in the hot springs of Yellowstone National Park. The DNA polymerase from this bacterium, called *Taq* polymerase, not only is active at high temperatures but also remains active over many cycles of heating and cooling. By putting these pieces of information together, Mullis constructed a simple method to amplify any DNA sequence in vitro.

The basic strategy of PCR is outlined in **Figure 10-7**. The process uses a pair of chemically synthesized oligonucleotide DNA primers that are each about 20 nucleotides long. Each primer is designed to base pair to one end of the target gene or region to be amplified, such that the primers base pair to opposite DNA strands with their 3′ ends pointing toward each other. The primers are added to a solution containing the DNA template (e.g., genomic DNA), the four deoxyribonucleoside triphosphates (dATP, dCTP, dGTP, and dTTP) required for DNA synthesis (Figure 7-5), and the heat-stable *Taq* DNA polymerase.

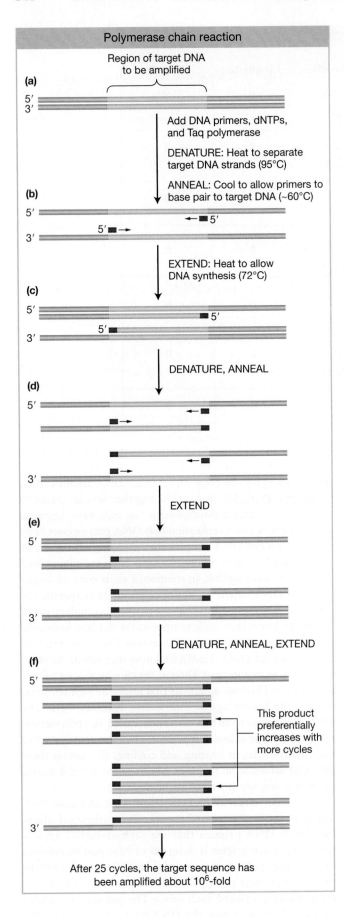

Polymerase chain reaction

Region of target DNA to be amplified

(a)

5′
3′

Add DNA primers, dNTPs, and Taq polymerase

DENATURE: Heat to separate target DNA strands (95°C)

ANNEAL: Cool to allow primers to base pair to target DNA (~60°C)

(b)

5′

5′ →

3′   ← 5′

EXTEND: Heat to allow DNA synthesis (72°C)

(c)

5′

5′   ■ 5′

3′

DENATURE, ANNEAL

(d)

5′

← ■

■ →

3′

■ →   ← ■

EXTEND

(e)

5′

3′

DENATURE, ANNEAL, EXTEND

(f)

5′

3′

This product preferentially increases with more cycles

After 25 cycles, the target sequence has been amplified about 10^6-fold

The DNA template is denatured by heat (95°C), resulting in single-stranded DNA molecules. Upon cooling the reaction to between 50°C and 65°C, the primers anneal to their complementary sequences in the single-stranded DNA template. After the temperature is raised to 72°C, *Taq* polymerase replicates the single-stranded DNA segments by extending from the annealed primers. Complementary new strands are synthesized, as in DNA replication in cells, forming two double-stranded DNA molecules identical to the single parental double-stranded DNA molecule. Thus, one cycle of PCR consists of three main steps (denaturing, annealing, and extending) and results in doubling the starting amount of the target sequence. Subsequent cycles of denaturing, annealing, and extending also double the amount of the target sequence, resulting in an exponential increase ($2^{\text{number of cycles}}$) in the number of copies of DNA. Thus, a typical PCR with 30 cycles of five minutes each will amplify the target DNA about one billion-fold (2^{30}) in 2.5 hours.

PCR is a powerful technique that is routinely used to isolate specific regions of DNA when there is prior knowledge of the sequence to be amplified. What makes PCR so powerful is that only small amounts of starting material are needed, which makes it possible to work with DNA samples that are difficult to obtain, such as from a small number of tumor cells. There are applications for PCR in genotyping, sequencing, cloning, paternity testing, forensics, molecular archeology, detection of infectious diseases, and many other clinical and basic research efforts that involve DNA. In recognition of the importance of PCR, Kary Mullis was awarded the Nobel Prize in Chemistry in 1993.

KEY CONCEPT The polymerase chain reaction uses specially designed primers to amplify specific regions of DNA in a test tube.

FIGURE 10-7 The polymerase chain reaction rapidly synthesizes many copies of a target DNA sequence. (a) Double-stranded DNA (blue) containing the target sequence (orange). (b) Addition of *Taq* polymerase, deoxyribonucleotides, and two DNA primers that have sequences complementary to the 3′ ends of the two strands of the target DNA. The strands are denatured (separated) by heating and then cooled to allow the primers to anneal to the target DNA. (c) After the temperature is raised, *Taq* polymerase synthesizes the first set of complementary strands. These first two strands are of varying length because the template extends beyond the site of binding by the other primer. (d) The two duplexes are heated again, exposing four binding sites, two for each primer. After cooling, the two primers again bind to their respective strands at the 3′ ends of the target region. (e) After the temperature is raised, *Taq* polymerase synthesizes four complementary strands. Although the template strands at this stage are variable in length, two of the four strands just synthesized from them are precisely the length of the target sequence desired. This precise length is achieved because each of these strands begins at the primer-binding site, at one end of the target sequence, and proceeds until it runs out of template, at the other end of the sequence. (f) The process is repeated for many cycles, each time creating twice as many double-stranded DNA molecules that are identical to the target sequence.

ANIMATED ART SaplingPlus Polymerase chain reaction

Quantifying DNA by real-time PCR Since the amount of DNA produced by PCR doubles with each cycle, it is possible to calculate the amount of DNA in a sample based on the amount of DNA produced after a given number of PCR cycles. Quantification of DNA by PCR, called **quantitative PCR (qPCR)**, is automated by real-time PCR instruments that measure the amount of DNA product in "real-time" during each PCR cycle. To be more exact, these instruments measure the intensity of a fluorescent signal generated by a dye called SYBR green, which, like ethidium bromide, intercalates into double-stranded DNA. As the amount of DNA increases with each PCR cycle, the fluorescent signal increases (**Figure 10-8**). The number of PCR cycles that it takes for the fluorescent signal to be detected above background is called the C_T (cycle threshold) value. If two samples have C_T values of 16 and 24, it would mean that the second sample had 256-fold less DNA than the first sample because eight more cycles were required to attain a fluorescent signal above background. In other words, the fold difference in the amount of DNA in the two samples is equal to $2^{-\Delta C_T}$. So, in this case,

$$2^{-(16-24)} = 2^8 = 256$$

qPCR is often used to compare the relative amounts of different DNA molecules in a single sample. For example, to diagnose cancer due a somatic mutation, qPCR is used to determine the fraction of cells in a tumor sample that contain the mutant gene versus the wild-type gene. qPCR is also used to compare the relative amount of the same DNA molecule in different samples. For example, to determine the rate of progression of a viral infection, qPCR is used to compare the amount of viral DNA in blood samples that are collected at different times.

KEY CONCEPT Quantitative PCR (qPCR) is a method that uses a real-time PCR instrument to determine the amount of a specific DNA molecule in a sample.

Detecting and quantifying mRNA by reverse-transcription PCR PCR can also be used to detect, amplify, and quantify mRNA; however, single-stranded RNA must first be converted into double-stranded DNA. **Complementary DNA (cDNA)** is a double-stranded DNA version of an mRNA molecule. cDNA is made from mRNA in vitro by a special enzyme called **reverse transcriptase**, originally isolated from retroviruses (see Chapter 16). Reverse transcriptase is a type of DNA polymerase that synthesizes a DNA strand complementary to an RNA template. Retroviruses such as human immuno-deficiency virus (HIV) use reverse transcriptase to convert their RNA genomes into DNA as part of their replication cycle.

cDNA synthesis begins with the purification of mRNA from a tissue or specific cells such as the pancreas or β cells. Purification is necessary because mRNA accounts for only about 5 percent of the total amount of cellular RNA, with rRNA accounting for about 80 percent and tRNA for 15 percent. mRNAs from eukaryotic cells are commonly purified using affinity methods that target the unique features of mRNA relative to other types of RNA, that is, the $5'$-m^7G cap and the $3'$ poly(A) tail. Next, purified mRNA is incubated with reverse transcriptase, the four dNTPs, and an oligo-dT primer (an oligonucleotide of about 20 T residues) (**Figure 10-9**). The oligo-dT primer anneals to the poly(A) tail of the mRNA, and, using the mRNA as a template, reverse transcriptase synthesizes single-stranded DNA starting from the oligo-dT primer and ending at the m^7G cap. This is called first-strand cDNA synthesis. The RNA strand of the RNA-DNA hybrid is then removed by alkaline hydrolysis (as described in Chapter 8, RNA is susceptible to base-catalyzed hydrolysis because of the hydroxyl group at the ribose sugar $2'$ position) or by RNase H, an enzyme that cuts the RNA strand of an RNA-DNA hybrid. Second-strand cDNA synthesis is carried out by *E. coli* DNA polymerase I and primed by ligation of an oligonucleotide of known sequence to the $3'$ end of the first-strand cDNA.

KEY CONCEPT Reverse transcriptase synthesizes DNA using an RNA template and can be used to create cDNA, a double-stranded DNA copy of an mRNA molecule.

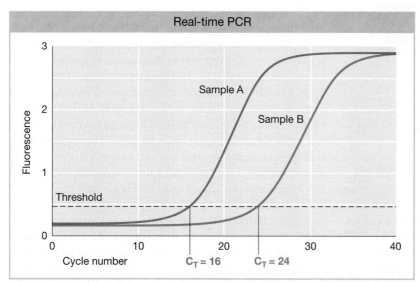

FIGURE 10-8 The amount of a specific DNA molecule in a sample can be quantified by real-time PCR. A real-time PCR instrument measures the fluorescence signal at each PCR cycle. In this example, the same DNA molecule was analyzed in two samples, Sample A and Sample B. The cycle threshold (C_T) is the PCR cycle at which the fluorescence reached a threshold. The C_T values for the two samples are used in the formula $2^{-\Delta C_T}$ to calculate the relative amount of the specific DNA molecule in the samples.

ANIMATED ART *SaplingPlus* Real-time qPCR

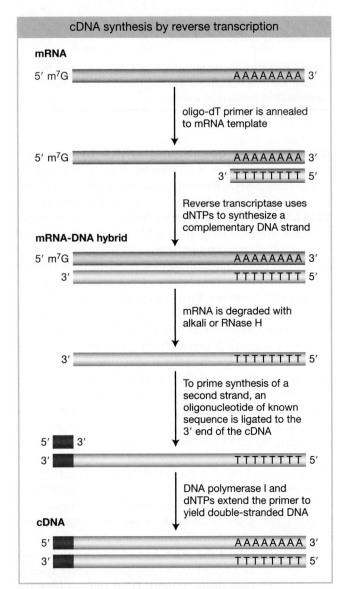

FIGURE 10-9 mRNA is converted to cDNA by the enzymes reverse transcriptase and DNA polymerase I. Reverse transcriptase first synthesizes a single-stranded DNA molecule using the mRNA as a template. Then DNA polymerase I synthesizes a double-stranded DNA molecule (cDNA) using the single-stranded DNA as a template.

This method produces double-stranded cDNA copies of all of the mRNAs that were in the source cells, tissue, or organism. Thus, the collection of cDNAs can be used to amplify by PCR any gene that is transcribed in the cells, tissue, or organism. This process is called **reverse transcription-PCR (RT-PCR)**. Furthermore, for mRNA transcribed from a given gene, the number of copies of cDNA is equal to the number of copies of mRNA. Thus, using cDNA as a template, a real-time PCR instrument can quantify mRNA levels in tissues and cells. For example, to determine if reduced transcription of the insulin gene is the cause of type I diabetes in an individual, insulin mRNA can be quantified by real-time PCR analysis of cDNA generated from β cells of the individual.

KEY CONCEPT Conversion of mRNA into cDNA makes it possible to use PCR approaches to amplify and quantify specific mRNAs.

10.2 GENERATING RECOMBINANT DNA

LO 10.2 Describe the functional components of vectors that are useful for cloning DNA.

LO 10.3 Describe methods for generating and isolating recombinant DNA molecules.

To investigate the function of genes and their products, it is helpful to be able to manipulate DNA sequences. For example, manipulation of the insulin gene has made it possible to carry out experiments that explain the effect of heritable mutations in the insulin gene on insulin protein expression and function in type I diabetes. A general approach to manipulate DNA is **DNA cloning**, which involves isolating a specific piece of DNA called **donor DNA**, or, more informally, **insert DNA**, and combining it with **vector** DNA to form a **recombinant DNA** molecule. Cloning vectors are naturally occurring DNA molecules that serve as vehicles to carry foreign DNA into a cell. Host cells increase the amount of a recombinant DNA molecule by DNA replication. Thus, the term DNA cloning refers to the process by which many identical copies of a piece of DNA, a clone, are produced.

DNA cloning

In DNA cloning, restriction enzymes and DNA ligase are used to combine insert and vector DNAs into a single molecule. Recall that restriction enzymes cut DNA at specific sequences, producing DNA fragments with staggered or blunt ends (see Table 10-1). On the other hand, **DNA ligase** joins two DNA fragments together by catalyzing the formation of phosphodiester bonds. **Figure 10-10** illustrates the basic steps in producing a recombinant DNA molecule. In this example, the restriction enzyme *Eco*RI is used to make a staggered double-strand cut at a single site in a circular vector such as a bacterial plasmid, converting the circular DNA into a single linear molecule with half of an *Eco*RI site at each end. Plasmids can be engineered to contain a **multiple cloning site (MCS)** or **polylinker** that contains restriction enzyme recognition sites that do not occur elsewhere in the plasmid. Thus, cleavage at any one of these sites linearizes the plasmid rather than cutting it into multiple pieces. In the case of the insert, *Eco*RI digestion of a linear piece of DNA at two sites produces a DNA fragment with half of an *Eco*RI site at each end. Mixing the linearized vector with the linear insert allows the "sticky" ends of the vector and insert to hybridize and form a recombinant molecule. DNA ligase finishes the job by creating phosphodiester linkages at the junctions between vector and insert sequences. If a single restriction enzyme is used for cloning, or if two restriction enzymes are used that both create blunt

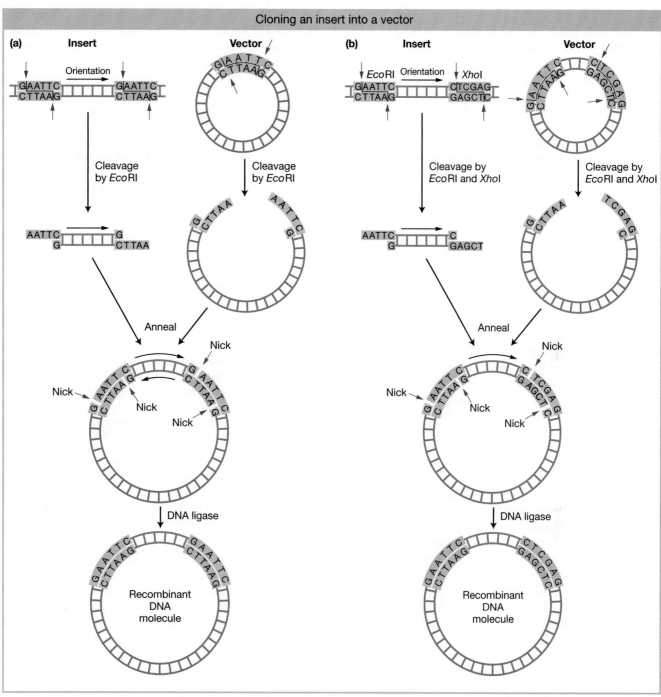

FIGURE 10-10 To form a recombinant DNA molecule, restriction enzymes are used to cut a vector and an insert. Because of sequence complementarity at the ends of the vector and insert, the vector and insert anneal. DNA ligase then permanently links the vector and insert together. Depending on the restriction enzymes used, the insert can hybridize with the vector in (a) two orientations or (b) only one orientation.

ANIMATED ART   SaplingPlus

Plasmid cloning

ends, the insert can hybridize in two orientations relative to the vector (Figure 10-10a); whereas cloning with two different restriction enzymes (at least one of which produces a staggered cut) limits hybridization of the insert and vector to one orientation (Figure 10-10b).

KEY CONCEPT Insert and vector DNAs with the same sticky ends or with blunt ends can be joined efficiently and ligated.

Choice of cloning vector Numerous cloning vectors that meet a wide range of experimental needs are in current use. Vectors are used mainly to express a specific RNA or protein or to increase the amount of a specific DNA molecule so that the DNA can be sequenced or further cloned. As described earlier, all vectors must have convenient restriction sites at which the DNA to be cloned can be inserted. Other important features of vectors are ways to quickly identify the desired recombinant vector as well as to express and purify the recombinant protein. Some general classes of cloning vectors follow.

Plasmid vectors Bacterial **plasmids**, which we first encountered in Chapter 6, are small circular DNA molecules that are replicated independently of the bacterial chromosome because they have an origin of replication (ori). Plasmids routinely used as vectors carry a gene for drug resistance and a gene to distinguish plasmids with and without DNA inserts. Genes that confer resistance to antibiotics such as ampicillin (amp^R gene), tetracycline (tet^R gene), and chloramphenicol (cam^R gene) provide a convenient way to select for bacterial cells transformed by plasmids: those cells still alive after exposure to the drug must carry the plasmid vector. However, because not all plasmids in transformed cells will contain DNA inserts, some plasmid vectors also have a system that allows researchers to identify bacterial colonies with plasmids containing DNA inserts. Such a feature is part of the pUC18 plasmid vector shown in **Figure 10-11a**; DNA inserts disrupt a gene (*lacZ*) in the plasmid that encodes an enzyme (β-galactosidase) capable of cleaving a compound added to the bacterial culture plate (X-gal) so that it produces a blue pigment. Thus, colonies that contain plasmids with an insert will be white rather than blue (i.e., they cannot cleave X-gal because they cannot produce β-galactosidase).

Plasmids called expression plasmids contain sequence that control the transcription and translation of the inserted DNA, often a gene in the form of a cDNA. Expression plasmids can drive transcription of the inserted gene constitutively (i.e., all the time) or inducibly (i.e., only in response to a signal). For example, some pET plasmids are used for constitutive expression of recombinant proteins in *E. coli* (Figure 10-11b). These plasmids contain the bacteriophage T7 promoter that drives transcription of the inserted gene in *E. coli* that express bacteriophage T7 RNA polymerase. Another type of the pET plasmid uses components of the *lac* operon (described in Chapter 11) to inducibly express the inserted gene in *E. coli*. This type of plasmid contains the *lac* operator site near the T7 promoter and also contains the *lacI* gene that encodes the Lac repressor protein. In uninduced cells, the Lac repressor binds the *lac* operator and represses transcription of the inserted gene by T7 polymerase. However, when the compound IPTG (isopropyl β-D-1 thiogalactopyranoside) is added to the growth media, the Lac repressor is inactivated and T7 polymerase can transcribe the inserted gene. Therefore, the recombinant protein is expressed only in the presence of IPTG. Inducible expression is helpful in cases where constitutive expression of the recombinant protein produces a large amount of protein that is toxic to *E. coli*, or it makes the protein insoluble.

A final feature of bacterial expression plasmids, including pET vectors, is a sequence that encodes an epitope tag that can be used to purify recombinant proteins. Epitope tags are short protein sequences that are translated in-frame, often at the N- or C-terminus of a recombinant protein. pET vectors contain an epitope tag called a His-tag that consists of six histidine amino acids (6X-His-tag). Since the tag is small, it typically does not affect the structure or function of the recombinant protein. Purification of His-tagged proteins is based on the affinity of histidine for metal ions such as nickel (Ni^{2+}). As shown in Figure 10-11b, His-tagged recombinant proteins are purified by Ni^{2+} affinity chromatography from an *E. coli* extract, a solution of *E. coli* cells that are broken open to release the recombinant protein as well as *E. coli* proteins. The *E. coli* extract is mixed with inert beads that have Ni^{2+} immobilized on their surface. The His-tagged recombinant protein binds tightly to the beads. The beads are washed several times to remove non-specifically bound *E. coli* proteins, leaving only His-tagged recombinant proteins bound to the beads. The bound proteins are then released (i.e., eluted) from the beads in a pure form by adding a chemical called imidazole that competes with the His-tag for binding to the beads. Bacterial expression plasmids and methods of this kind are used to synthesize and purify human insulin protein.

KEY CONCEPT The essential features of plasmids for cloning are an origin of replication so that the plasmid is replicated when bacteria divide, a drug resistance gene so that bacteria containing the plasmid can be identified, and a polylinker so that DNA can be inserted with restriction enzymes.

KEY CONCEPT Non-essential but useful features of plasmids include sequence elements for the identification of plasmids that contain inserts, the constitutive or inducible expression of inserted genes, and the addition of an epitope tag onto a recombinant protein.

Bacteriophage vectors A bacteriophage vector harbors DNA as an insert packaged inside the phage particle. Different classes of bacteriophage vectors can carry different sizes of insert DNA. Bacteriophage λ (lambda; discussed in Chapters 6 and 11) is an effective cloning vector for double-stranded DNA inserts as long as 15 kb. The central part of the phage genome is not required for replication or packaging of λ DNA molecules in *E. coli* and so can be cut out by restriction enzymes and discarded. The deleted central part is then replaced by insert DNA.

Vectors for larger DNA inserts The standard plasmid and phage λ vectors just described can accept inserts as large as 15 kb. However, many experiments require inserts well in excess of this size. To meet these needs, special vectors that require more sophisticated methods for transferring DNA into the host cell have been engineered. In each case, the recombinant DNAs replicate as large plasmids after they have been delivered into the bacterium.

Fosmids are vectors that can carry 35- to 45-kb inserts (**Figure 10-12**). They are engineered hybrids of λ phage DNA and bacterial F plasmid DNA (Chapter 6). Because of their cos sites from λ phage, fosmids are packaged into λ phage

Practical features of plasmid vectors

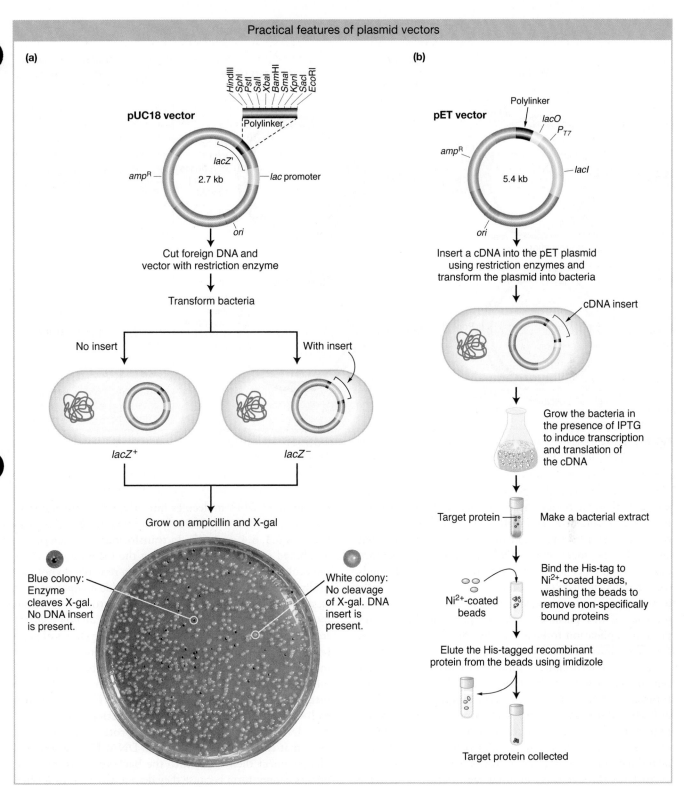

FIGURE 10-11 (a) The pUC18 plasmid has been designed for use as a vector for DNA cloning. The polylinker has multiple restriction sites into which donor DNA can be inserted. Insertion of DNA into pUC18 is detected by inactivation of the β-galactosidase function of *lacZ*, resulting in an inability to convert the artificial substrate X-gal into a blue dye. (b) The pET plasmid has been designed for expression and purification of recombinant proteins. Inducible expression of a recombinant protein is controlled by three pET plasmid elements, the lac operator (*lacO*), the *lacI* gene that encodes the Lac repressor protein, and the T7 polymerase promoter. The plasmid also contains a sequence that encodes a His-tag that is translated in-frame with the recombinant protein. IPTG induces expression of the His-tagged recombinant protein, which is then purified based on the affinity of the His-tag for Ni²⁺-coated beads. [(a) Dr. James M. Burnette III and Dr. Leslie Bañuelos.]

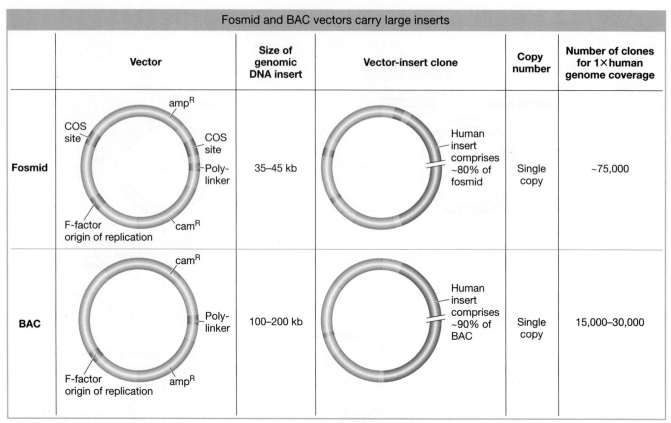

	Vector	Size of genomic DNA insert	Vector-insert clone	Copy number	Number of clones for 1× human genome coverage
Fosmid	amp^R, COS site, COS site, Poly-linker, F-factor origin of replication, cam^R	35–45 kb	Human insert comprises ~80% of fosmid	Single copy	~75,000
BAC	cam^R, Poly-linker, F-factor origin of replication, amp^R	100–200 kb	Human insert comprises ~90% of BAC	Single copy	15,000–30,000

FIGURE 10-12 Features of some large-insert cloning vectors. The number of clones needed to cover the human genome once (1 ×) is based on a genome size of 3000 Mb (3 billion base pairs).

particles that introduce these big pieces of recombinant DNA into recipient *E. coli* cells. Cos (an abbreviation for cohesive) sites are 12 base pair overlapping sticky ends that circularize the linear phage DNA through complementary base pairing. Once introduced into the bacterium, fosmids form circular molecules that replicate extrachromosomally in a manner similar to plasmids. However, because of the presence of an F plasmid origin of replication that couples plasmid replication to host cell chromosome duplication, very few copies accumulate in a cell.

Bacterial artificial chromosomes (BACs) are another type of vector for carrying large inserts. Derived from the F plasmid, BACs can carry inserts ranging from 100 to 200 kb, although the vector itself is only about 7 kb (see Figure 10-12). The DNA to be cloned is inserted into the plasmid, and this large circular recombinant DNA is introduced into the bacterium. BACs were the workhorse vectors for the extensive cloning required by large-scale genome-sequencing projects, including the public project to sequence the human genome (Chapter 14).

KEY CONCEPT Cloning vectors accept inserts of small sizes for plasmids, to medium sizes for bacteriophage, to large sizes for fosmids and BACs.

Entry of recombinant DNA molecules into bacterial cells Three methods are used to introduce

recombinant DNA molecules into bacterial cells: transformation, transduction, and infection (**Figure 10-13**; see also Sections 6.3, 6.4, and 6.5). In **transformation**, bacteria are incubated in a solution containing the recombinant DNA molecule. Because bacterial cells used in research do not naturally take up plasmids, they must be made *competent* (that is, able to take up the DNA from the surrounding media) by either incubation in a calcium solution (*calcium chloride transformation*) or exposure to a high-voltage electrical pulse (*electroporation*). After entering a competent cell through membrane pores, the recombinant molecule becomes a plasmid chromosome (Figure 10-13a). Electroporation is the method of choice for introducing especially large DNAs such as BACs into bacterial cells. In **transduction**, the recombinant DNA is combined with phage proteins to produce a virus that contains largely non-viral DNA. These engineered phages inject their DNA into the bacterial cells, but new phages cannot form because they do not carry the viral genes necessary for phage replication. Fosmids are introduced into cells by transduction (Figure 10-13b). In contrast to transduction, which produces plasmids and bacterial colonies but not new viruses, **infection** of bacteria produces recombinant phage particles (Figure 10-13c). Through repeated rounds of infection, a plaque full of λ phage particles forms from each initial bacterium that was infected. Each phage particle in a plaque contains not only the recombinant DNA, but also viral genes needed to create new infective phage particles.

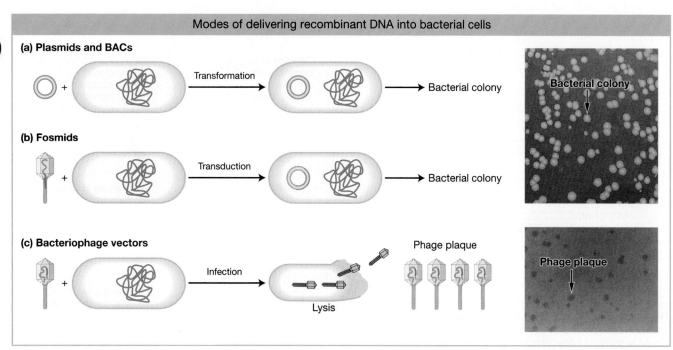

FIGURE 10-13 Recombinant DNA can be delivered into bacterial cells by transformation, transduction, or infection. (a) Plasmid and BAC vectors are delivered by transformation of purified DNA. (b) Certain vectors such as fosmids are delivered within bacteriophage heads (transduction); however, after having been injected into the bacterium, they form circles and replicate as large plasmids.

(c) Bacteriophage vectors such as phage λ infect and lyse bacteria, forming a plaque that contains progeny phages all carrying the identical recombinant DNA molecule within the phage genome. Examples of bacterial colonies and phage plaques are shown on the right. In the bottom picture, uninfected bacteria are found in regions surrounding plaques. [(a) JelenaMV/Medical Images; (b) Branko Velebit/Medical Images.]

DNA libraries

To clone specific genes and mRNAs such as those for insulin, researchers have put into practice the information presented thus far in this chapter. A common cloning approach is to generate a collection of recombinant DNA molecules called a library, and to fish out the molecule of interest. For example, in 1982 the human insulin gene was identified from a library of human genome fragments. To create a **genomic library**, restriction enzymes or physical methods are used to break human genomic DNA into fragments of appropriate size for a cloning vector, and each fragment is inserted into a different copy of the vector. If fosmids that accept ~40 kb inserts are used as the cloning vector, ~75,000 independent clones would be required to represent one human genome's worth of DNA (3×10^9 bp in the human genome/4×10^4 bp per fosmid). To ensure that all regions of the genome are included, genomic libraries aim to have each DNA fragment represented an average of five times. So, in this example there would need to be 375,000 independent clones in the genomic library ($5 \times 75,000$).

To create a **cDNA library**, mRNA is purified from a cellular source, converted into cDNA, and inserted into a vector. One method for preparing cDNAs for insertion into a cloning vector is to add restriction sites to both ends of each cDNA (**Figure 10-14**). To do this, DNA ligase is used to link short double-stranded oligonucleotides called **DNA linkers** or **DNA adapters**, which contain a restriction site, to cDNAs. After ligation, the cDNAs are digested with the

selected restriction enzyme to generate staggered ends for cloning into a vector that is digested with the same restriction enzyme. cDNA libraries require tens or hundreds of thousands of independent cDNA clones to completely represent the set of expressed genes in a particular cellular source. Suppose we want to identify cDNAs corresponding to insulin mRNAs. Since β cells of the pancreas are the most abundant source of insulin, mRNAs from pancreas are the appropriate source for a cDNA library. To completely represent all the mRNAs expressed by an organism, many cDNA libraries from sources such as different tissues, developmental stages, and environmental conditions are needed.

> **KEY CONCEPT** The task of isolating a clone of a specific gene can begin with making a library of genomic DNA or cDNA.

> **KEY CONCEPT** Genomic libraries represent all of the genes in an organism, while cDNA libraries represent only those genes that were expressed in the cells that were the source of mRNA.

Identifying a clone of interest from a genomic or cDNA library

After generating a genomic or cDNA library, the next task in finding a particular clone is to screen the library. Such screening is accomplished by a procedure called colony or

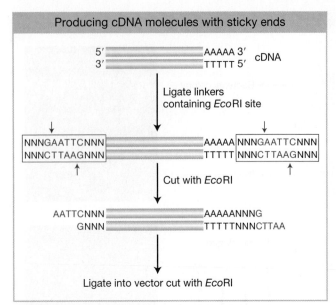

Producing cDNA molecules with sticky ends

FIGURE 10-14 Adding *EcoRI* sites to the ends of cDNA molecules. The cDNA molecules come from the last step in Figure 10-9. Adapters (boxed regions) are added at both ends of the cDNA molecules. These adapters are double-stranded oligonucleotides that contain a restriction site (*EcoRI* is shown in red) and random DNA sequence at both ends (represented by N). Note that in the example shown, any cDNAs that contain an internal *EcoRI* site will be cut into pieces, so some clones will not contain full-length cDNAs.

plaque hybridization, which is similar to Southern blotting; but in this case, the DNA being analyzed comes from bacterial colonies or phage plaques. The procedure shown in **Figure 10-15** is for a library cloned into a fosmid vector, but the steps are similar for libraries of plasmids, BACs, or phages. First, colonies of the library on a petri dish are transferred to a membrane by laying the membrane onto the colonies. The membrane is peeled off, colonies clinging to the surface are lysed in place on the membrane, and the DNA is simultaneously denatured so that it is single-stranded. Second, the membrane is incubated in a solution of a single-stranded probe that is specific for the DNA sequence being sought. Generally, the probe is itself a cloned piece of DNA whose sequence is complementary to that of the desired gene. Since the probe is labeled with either a radioactive isotope or a fluorescent dye, the position of the radioactive or fluorescent label will indicate the position of positive clones. Radioactive probes are detected by autoradiography using X-ray film, and fluorescent probes are detected by photographing the membrane after exposure to a wavelength of light that activates the dye's fluorescence.

Genomic and cDNA clones are used in different ways

Genomic and cDNA clones of a given eukaryotic gene contain different sequences, which dictates how the clones can be used for gene expression. For illustrative purposes, we will use genomic and cDNA clones of the mouse *Ins1* insulin gene as examples (see Figure 10-3a). A genomic clone of *Ins1* can contain all of the regulatory sequences that are

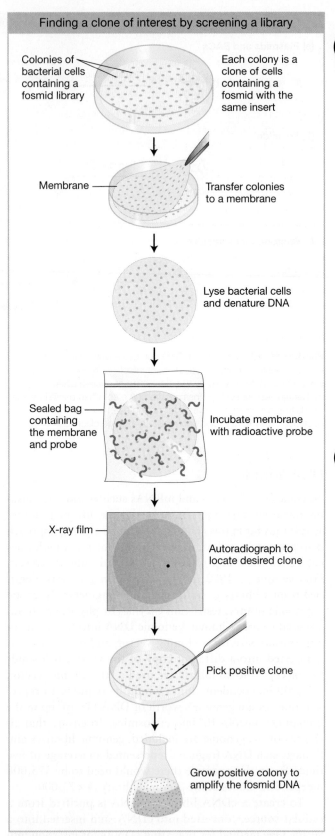

Finding a clone of interest by screening a library

FIGURE 10-15 The clone carrying a gene of interest is identified by probing a genomic library, in this case made by cloning genes in a fosmid vector, with DNA or RNA that has sequence complementary to the desired gene. A radioactive probe hybridizes with any recombinant DNA molecule containing a matching DNA sequence, and the position of the clone having the DNA is revealed by autoradiography. Now the desired clone can be selected from the corresponding spot on the petri dish and grown to high levels in a liquid bacterial culture.

needed to direct the normal expression pattern of *Ins1* in mice. This includes transcriptional regulatory sequences as well as RNA processing sequences for splicing of mRNA introns and polyadenylation of mRNA 3′ ends. However, the genomic clone of *Ins1* cannot be expressed in bacteria because bacterial proteins do not recognize eukaryotic transcriptional regulatory sequences and bacteria do not carry out splicing. In contrast, a cDNA clone of *Ins1* cannot be expressed in mice because it lacks regulatory sequences for transcription and polyadenylation, since they are not transcribed into mRNA. Nevertheless, it is possible to express the *Ins1* cDNA in mice using vectors that contain these regulatory sequences from either the *Ins1* gene or another mouse gene. Likewise, the *Ins1* cDNA cannot be expressed in bacteria because it lacks bacterial transcriptional regulatory sequences; but these sequences can be provided by the vector, as described in Figure 10-11b. Furthermore, the lack of splicing in bacteria is not an issue because splicing has already taken place. The same considerations apply to genomic and cDNA clones of the human insulin gene, which is why a cDNA clone is used to produce recombinant human insulin in bacteria (see Figure 10-1).

KEY CONCEPT Genomic and cDNA clones of a gene are not functionally interchangeable. Both can be used for gene expression, but under different conditions.

Cloning by PCR

Since the widespread use of PCR in the 1990s, PCR, rather than screening a library, is routinely used to construct a particular genomic or cDNA clone. For example, to clone the human insulin cDNA, oligonucleotide primers are designed that are complementary to the 5′ and 3′ ends of the insulin cDNA, and PCR is carried out using cDNA generated from pancreas mRNA as the template. To enable cloning of a PCR product, a common approach is to use PCR primers that have restriction sites at their 5′ end (**Figure 10-16**). Thus, after digestion with a restriction enzyme, the PCR product can be ligated into a vector that is linearized with the same restriction enzyme. A problem with this approach is that the length of PCR products is limited to about 2 kb. To circumvent this problem, cDNAs and genes larger than 2 kb can be cloned by stitching together multiple PCR products with restriction sites at their ends that direct the order in which they are assembled (**Figure 10-17a**).

As an alternative to restriction enzyme cloning, researchers have developed other **DNA assembly** methods that can be used to construct large genomic regions and cDNAs and even whole chromosomes and genomes. As an example, *Gibson assembly* can piece together multiple linear DNA fragments that have 15- to 40-bp regions of sequence similarity, often referred to as homology regions, at their ends (Figure 10-17b). The fragments can be produced by PCR or by chemical synthesis, which can generate oligonucleotides of up to 200 nucleotides in length. Assembly is achieved by incubating the fragments with three enzymes: (1) an exonuclease that chews back the 5′ ends of each fragment, producing long single-stranded regions with sequence complementarity

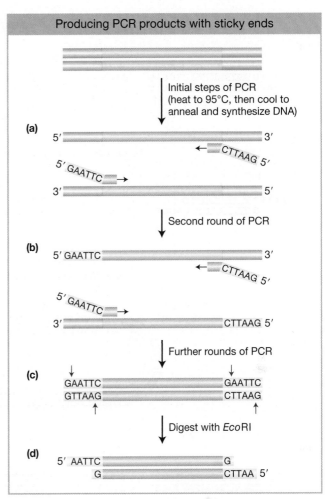

FIGURE 10-16 Adding *Eco*RI sites to the ends of PCR products. (a) A pair of PCR primers is designed so that their 3′ ends anneal to the target sequence, while their 5′ ends contain sequences encoding a restriction site (*Eco*RI in this case). The target DNA is denatured, and 5′ ends with the restriction sites remain single stranded while the rest of each primer anneals and is extended by *Taq* DNA polymerase. (b) In the second round of PCR—only the newly synthesized strands are shown—the DNA primers anneal again, and this time DNA synthesis produces double-stranded DNA molecules with restriction sites at one end. (c) The products of all subsequent rounds have *Eco*RI sites at both ends. (d) Sticky ends are produced when these PCR products are cut with *Eco*RI.

between fragments, (2) a DNA polymerase to fill in the gaps between annealed fragments, and (3) a DNA ligase to form the final phosphodiester bonds between the fragments. The main advantage of assembly over standard restriction enzyme-based cloning is that assembly allows the joining of any two DNA fragments at any position, whereas restriction enzyme-based cloning is limited to positions of natural or engineered restriction sites. In addition, assembly is faster than standard restriction enzyme-based cloning because a greater number of DNA fragments can be efficiently joined in a single reaction.

Assembly methods can also be used to put together parts from different genes. For example, researchers commonly combine the transcriptional regulatory region of one gene with the cDNA of another gene. Returning to the insulin gene, this approach could be used to identify the β cells in a mouse pancreas. A researcher could put together the transcriptional regulatory region of a mouse insulin

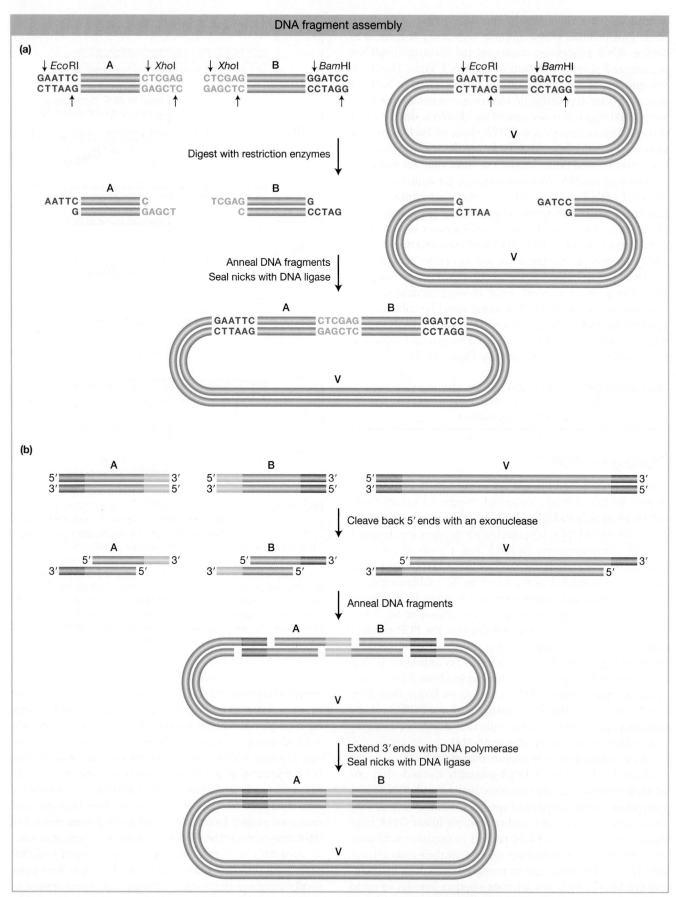

FIGURE 10-17 Recombinant DNA molecules with large inserts can be constructed by assembly methods involving multiple DNA fragments (A and B) and a vector (V) with (a) compatible restriction sites at their ends or, as in Gibson assembly, (b) regions of sequence similarity (homology regions) at their ends. Both restriction sites and regions of sequence similarity can be built into the DNA fragments by the appropriate design of PCR primers.

gene, which is only expressed in β cells, and a cDNA for a reporter gene that encodes a protein that is easy to detect. The green fluorescent protein (GFP) gene from a jellyfish is commonly used as a reporter because it encodes a small protein (238 amino acids) that exhibits bright green fluorescence when exposed to ultraviolet light of a particular wavelength. In mice that have this engineered gene inserted into their genome (using techniques described later in this chapter), β cells in the pancreas can be identified because they will be the only cells that express GFP and glow green. Alternatively, a researcher could construct a gene that expresses a single protein, called a *fusion protein*, that is comprised of both insulin and GFP amino acid sequences. This could be done by assembling an insulin gene along with the GFP cDNA such that the insulin and GFP protein-coding regions are translationally in-frame. In this scenario, the insulin protein is said to be "tagged" with GFP. Another way to tag a protein is to append to it a few amino acids that can be recognized by an antibody in vitro by Western blot analysis and in vivo by immunofluorescence microscopy. Like His-tags that are used for protein purification, such tags are called **epitope tags**. Commonly used epitope tags for this purpose include the 7-amino-acid FLAG tag (DYKDDDK), the 9-amino-acid HA tag (YPYDVPDYA), and the 10-amino-acid Myc tag (NNKLISEEDL). The advantage of epitope tags over GFP is that, because of their small size, epitope tags are less likely to alter the structure and function of the protein to which they are fused.

KEY CONCEPT Assembly methods make it relatively easy to modify the sequence of genes to create tools that are useful for research, including reporter genes that reveal the expression pattern of transcriptional regulatory elements and epitope-tagged genes that enable the detection and purification of recombinant fusion proteins.

10.3 SEQUENCING DNA

LO 10.4 Diagram the steps of dideoxy DNA sequencing.

As described in many chapters in this text, the regulatory and coding information in DNA is determined by its nucleotide sequence. To reveal this information, public and private institutions have invested heavily in the sequencing of DNA genomes, including the human genome, which was completed in 2001. Furthermore, since mRNA can be converted to cDNA (see Figure 10-9), the same technologies used to sequence genomes have been extensively applied to sequence cDNAs and to discover the information that flows from DNA to RNA by transcription in cells, tissues, and organisms. Lastly, sequencing of DNA is a common activity in individual laboratories for purposes such as identifying specific DNA lesions in mutant alleles and confirming the sequence of recombinant DNA molecules and PCR products.

Since the late 1970s, researchers have put considerable effort into developing techniques to sequence DNA. Currently, the most commonly used technique for small scale sequencing is called **dideoxy sequencing** or **Sanger sequencing**, after its inventor Fred Sanger. However, as we will see in Chapter 14, other sequencing technologies have largely supplanted this technique when the goal is to determine the sequence of an entire genome. The term *dideoxy* comes from a modified nucleotide, called a dideoxynucleoside triphosphate (ddNTP). This modified nucleotide is key to the Sanger technique because of its ability to be added to a growing DNA chain but to block continued DNA synthesis. A dideoxynucleotide lacks the ribose sugar 3′-hydroxyl group as well as the 2′-hydroxyl group that is absent in a regular deoxynucleotide (**Figure 10-18a**). For DNA synthesis to take place, DNA polymerase must catalyze phosphodiester bond formation between the 3′-hydroxyl group of the last nucleotide in the growing chain and the α-phosphate of the nucleotide to be added. Because a dideoxynucleotide lacks the 3′-hydroxyl group, this reaction cannot take place, and DNA synthesis terminates.

KEY CONCEPT Dideoxynucleotides (ddNTPs) cause chain termination because they lack the 3′-hydroxyl group on the ribose sugar that is essential for phosphodiester bond formation.

DNA sequencing requires four separate reactions, each containing the DNA segment (e.g., a cloned plasmid insert or a PCR product), a radioactive DNA primer that will hybridize to exactly one location on the DNA segment, DNA polymerase, and the four deoxynucleoside triphosphates (dNTPs: dATP, dCTP, dGTP, and dTTP). In addition, each reaction receives a small amount of a different dideoxynucleoside triphosphate (ddNTP: ddATP, ddCTP, ddGTP, or ddTTP). As in DNA replication, the DNA polymerase will add deoxynucleotides to the 3′ end of the primer, with the identity of the added deoxynucleotide being determined by base pairing complementarity to the template strand (Figure 10-18b). Since the dNTPs and each ddNTP are present in a ratio of about 300:1, most of the time, DNA polymerase will add a dNTP and continue synthesis. However, every once in a while, DNA polymerase will incorporate a ddNTP into the new strand, which will terminate synthesis of that DNA strand. Hence, in the tube that contains ddATP, new DNA being synthesized will terminate when a ddATP is added to the strand, thus marking locations of T nucleotides in the DNA segment being sequenced. When complete, the reaction in the ddATP tube results in a collection of radiolabeled single-stranded DNA fragments of different length, each ending with an A residue. This process is repeated for the reactions with ddCTP, ddGTP, and ddTTP. The DNA fragments in the four reactions are separated by polyacrylamide gel electrophoresis and visualized by autoradiography (Figure 10-18c). Since polyacrylamide gels can resolve fragments of DNA that

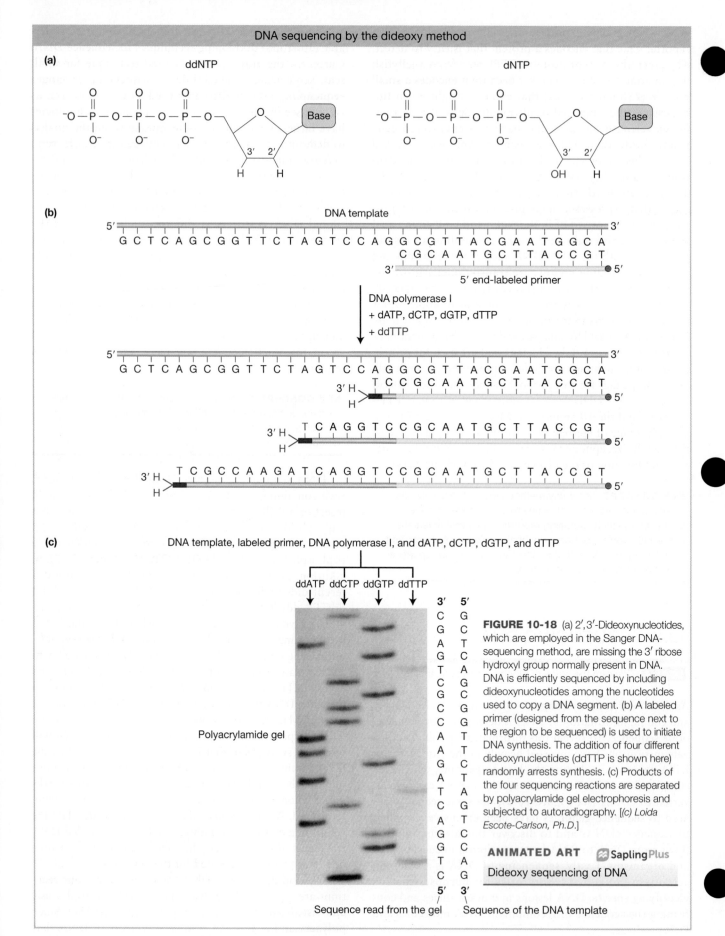

FIGURE 10-18 (a) 2′,3′-Dideoxynucleotides, which are employed in the Sanger DNA-sequencing method, are missing the 3′ ribose hydroxyl group normally present in DNA. DNA is efficiently sequenced by including dideoxynucleotides among the nucleotides used to copy a DNA segment. (b) A labeled primer (designed from the sequence next to the region to be sequenced) is used to initiate DNA synthesis. The addition of four different dideoxynucleotides (ddTTP is shown here) randomly arrests synthesis. (c) Products of the four sequencing reactions are separated by polyacrylamide gel electrophoresis and subjected to autoradiography. [(c) Loida Escote-Carlson, Ph.D.]

ANIMATED ART   Sapling Plus

Dideoxy sequencing of DNA

vary by only one nucleotide in length, the fragments in the gel are separated and ordered by size, with the lengths increasing by one base at a time. Shorter DNA fragments migrate fastest in the gel, so bands at the bottom of the gel represent the sequence closest to the primer. Therefore, the sequence is read in the 5′-to-3′ direction from the bottom to the top of the gel and is complementary to the DNA strand being sequenced.

KEY CONCEPT DNA sequencing by the dideoxy (Sanger) method uses dideoxynucleotides to terminate synthesis by DNA polymerase from a DNA template, producing DNA fragments of different lengths that end at each nucleotide position in the template.

In 1986, a modified dideoxy sequencing method was developed that uses an automated electrophoresis system and labels synthesized DNA strands with fluorescent dideoxynucleotides rather than a radioactive primer. The automated method is superior to the prior method because more samples can be sequenced at the same time, and the length of sequence reads is increased from about 200 to 1000 base pairs. Automated sequencing is carried out in a single reaction containing an unlabeled primer and all four dideoxynucleotides, each labeled with a differently colored fluorescent dye. So, synthesized fragments are not fluorescently labeled until they terminate, but, once they are labeled, the color of the fluorescence indicates the nucleotide at the 3′ end of the fragment. The synthesized fragments are separated by size by capillary gel electrophoresis, in which the gel matrix is contained within a thin tube rather than between glass plates. As in polyacrylamide gel electrophoresis, all of the fragments of the same size migrate as a single band. As the bands reach the bottom of the capillary tube, the fluorescence is detected with a laser beam. The intensity of light in each band is depicted as peak in the computer output, as shown in **Figure 10-19**. This figure shows the sequence of a DNA fragment that was amplified by PCR from the genomic DNA of an individual. At most positions, there is a single peak with green=A, blue=C, black=G,

and red=T. The sequence also contains a single nucleotide polymorphism (SNP) at position 144. This nucleotide is read as both a T and a C, which means that the diploid individual has one allele with a T–A base pair and the other with a C–G base pair. This automated technology was used to sequence the human genome as well as the genomes of many other organisms, but more cost-effective and faster technologies are currently in use for large sequencing projects (discussed in Chapter 14).

KEY CONCEPT Automated sequencing is superior to the original sequencing method that uses a radioactive primer because it involves only one reaction, not four, and it produces longer sequencing reads.

10.4 ENGINEERING GENOMES

LO 10.5 Describe methods for generating transgenic organisms.
LO 10.6 Describe the CRISPR-Cas9 technique for precise engineering of genomes.

Thanks to recombinant DNA technologies, genes can be isolated and characterized as specific nucleotide sequences. But even this achievement is not the end of the story. We will see next that knowledge of a sequence is often the beginning of a fresh round of genetic manipulation. When characterized, a sequence can be manipulated to alter an organism's genotype. The introduction of an altered gene into an organism has become central to basic genetic research, but it also finds wide commercial application. Three examples of the latter are (1) goats that are modified to secrete into their milk human antithrombin protein, which is used to treat a rare human blood clotting disorder, (2) rice that is modified to produce beta-carotene, the precursor to vitamin A, the deficiency of which causes health problems for millions of people around the world, and (3) plants that are modified to kept from freezing by incorporation of arctic-fish "antifreeze" genes into their genomes. The use of recombinant DNA techniques to alter an

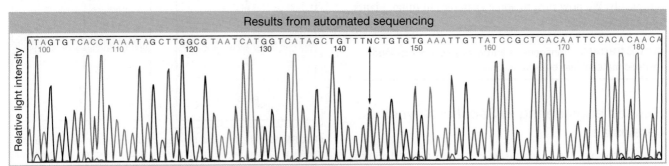

FIGURE 10-19 Printout from an automatic sequencer that uses fluorescent dyes. Each of the four colors represents a different base (A-green, C-blue, G-black, and T-red). The letter N represents a base that cannot be assigned because there are overlapping peaks for two bases. This is diagnostic of a single nucleotide polymorphism (SNP). Note that, if this was a sequencing gel, as in Figure 10-18c, each of the peaks would correspond to one of the dark bands on the gel; in other words, the colored peaks represent a different readout of the same data produced on a sequencing gel.

ANIMATED ART SaplingPlus
Dideoxy sequencing using fluorescent nucleotides

organism's genotype and phenotype is termed *genetic engineering*, and its application for practical purposes is called *biotechnology*.

The techniques of genetic engineering described in the first part of this chapter were originally developed in bacteria. These techniques have been extended to model eukaryotes, which constitute a large proportion of organisms used for research. Eukaryotic genes are still typically cloned into bacterial vectors, but eventually they are introduced into a eukaryote, either the original donor species or a completely different one. The gene transferred is called a **transgene**, and the engineered product is called a **transgenic organism**.

Transgenes are introduced into eukaryotic cells by chemical, physical, and biological methods (**Figure 10-20**). Chemical methods are based on the principle that DNA co-precipitated with minerals such as calcium phosphate or packaged inside tiny phospholipid vesicles can be taken up into cells by endocytosis, a natural process by which cells take in molecules from the environment by engulfing them. Physical methods include electroporation, biolistic particle delivery, and microinjection. Electroporation involves applying an electrical field to cells for a short period of time to create microscopic holes in the plasma membrane through which DNA can enter. Biolistic particle delivery systems, also known as gene guns, bombard cells with DNA-coated metal particles that are small enough to enter cells but not destroy them. The last physical method, microinjection, directly delivers DNA into cells through a fine-point needle. Biological methods use bacteria or viruses to transfer DNA into cells. For example, as described shortly, the bacterium *Agrobacterium tumefaciens* can transfer to a plant genome part of its own genome that carries a gene of interest, and viruses can transfer into animal cells their genome that is engineered to include a gene of interest.

When a transgene enters a cell, it travels to the nucleus, where it becomes a stable part of the genome by either inserting into a chromosome or (in a few species) replicating as part of a plasmid. If insertion occurs, the transgene can either replace the resident gene by homologous recombination or insert **ectopically**—that is, at other locations in the genome. Transgenes from other species typically insert ectopically. We now turn to some examples in fungi, plants, and animals.

KEY CONCEPT Transgenesis introduces new or modified genetic material into eukaryotic cells.

Genetic engineering in *Saccharomyces cerevisiae*

It is fair to say that *S. cerevisiae* is the most easily manipulated eukaryotic genetic model. Most of the techniques typically used for eukaryotic genetic engineering were developed in yeast; so let's consider the general routes for transgenesis in yeast.

The simplest yeast vectors are yeast integrative plasmids (YIps), derivatives of bacterial plasmids into which yeast

Methods of introducing a transgene

(a) Lipid vesicle

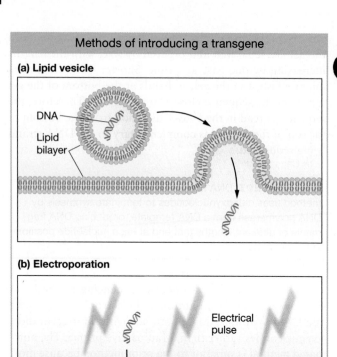

(b) Electroporation

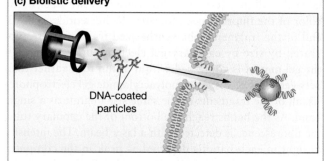

(c) Biolistic delivery

(d) Microinjection

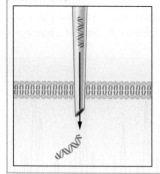

(e) Virus infection

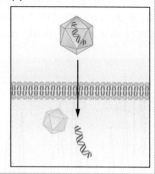

FIGURE 10-20 Chemical, physical, and biological methods are used to deliver transgenes (i.e., recombinant DNA molecules) into eukaryotic cells. Examples of chemical methods include (a) lipid vesicles; physical methods include (b) electroporation, (c) biolistic delivery, and (d) microinjection; and biological methods include (e) virus infection.

DNA of interest has been inserted. When transformed into yeast cells, these plasmids insert into yeast chromosomes, generally by homologous recombination with the resident

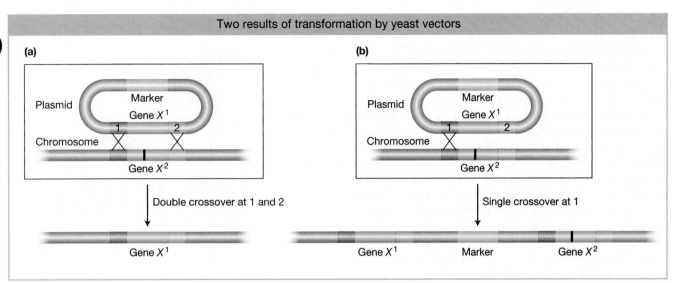

Two results of transformation by yeast vectors

(a)

Plasmid
Marker
Gene X^1

Chromosome
1   2

Gene X^2

Double crossover at 1 and 2

Gene X^1

(b)

Plasmid
Marker
Gene X^1

Chromosome
1   2

Gene X^2

Single crossover at 1

Gene X^1 Marker   Gene X^2

FIGURE 10-21 A plasmid bearing a functional allele (Gene X^1) inserts into a recipient yeast strain bearing a defective allele (Gene X^2) by homologous recombination. The result can be (a) replacement of the X^2 by X^1 or (b) retention of X^2 and concurrent addition of X^1. The mutant site of Gene X^2 is represented as a vertical black bar. Single crossovers at position 2 also are possible but are not shown. The bacterial origin of replication is different from eukaryotic origins, so bacterial plasmids do not replicate in yeast. Therefore, the only way in which such vectors can generate a stable modified genotype is if they are integrated into the yeast chromosome.

gene, by either a single or a double crossover (**Figure 10-21**). As a result, either the entire plasmid is inserted, or the targeted allele is replaced by the allele on the plasmid. The latter is an example of *gene replacement*—in this case, the substitution of an engineered gene for the gene originally in the yeast cell. Gene replacement can be used to delete a gene or substitute a mutant allele for its wild-type counterpart or, conversely, to substitute a wild-type allele for a mutant.

KEY CONCEPT Transgenic yeast cells are generated by homologous recombination between a yeast chromosome and a plasmid that is transformed into yeast and carries a gene of interest.

Genetic engineering in plants

Recombinant DNA technologies have introduced a new dimension to the effort to develop improved crop varieties. No longer is genetic diversity achieved solely by selecting variants within a given species. DNA can now be introduced from other species of plants, animals, or even bacteria, producing **genetically modified organisms (GMOs)**. Genome modifications made possible by this technology are almost limitless. In response to new possibilities, a sector of the public has expressed concern that introduction of GMOs into the food supply may produce unexpected health problems. The concern about GMOs is one facet of an ongoing public debate about complex public health, safety, ethical, and educational issues raised by new genetic technologies.

A vector routinely used to produce transgenic plants is derived from the **Ti plasmid**, a natural plasmid from a soil bacterium called *Agrobacterium tumefaciens*. This

bacterium causes what is known as *crown gall disease*, in which the infected plant produces uncontrolled growths called tumors or galls. The key to tumor production is a large (200 kb) circular DNA plasmid—the *Ti (tumor-inducing) plasmid* (**Figure 10-22**). When the bacterium infects a plant cell, a part of the Ti plasmid is transferred and inserted, apparently more or less at random, into the genome of the host plant. The region of the Ti plasmid that inserts into the host plant is called *T-DNA*, for transfer DNA. The genes whose products catalyze this T-DNA

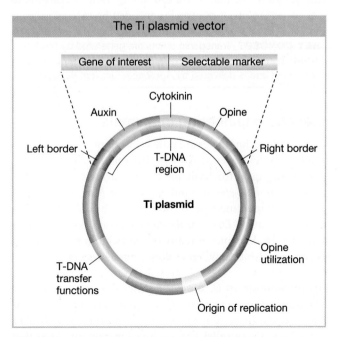

The Ti plasmid vector

Gene of interest   Selectable marker

Cytokinin

Auxin   Opine

Left border   Right border

T-DNA region

Ti plasmid

Opine utilization

T-DNA transfer functions

Origin of replication

FIGURE 10-22 Simplified representation of the major regions of the Ti plasmid of *A. tumefaciens* containing an engineered T-DNA.

transfer reside in a region of the Ti plasmid separate from the T-DNA region itself.

The natural behavior of the Ti plasmid makes it well suited to the role of a vector for plant genetic engineering. In particular, any DNA that is inserted between the left and right T-DNA border sequences (24 base-pair ends) can be mobilized by other functions provided by the Ti plasmid and inserted into plant chromosomes. Thus, scientists are able to eliminate all of the T-DNA sequence between the borders (including the tumor-causing genes) and replace it with a gene of interest and a selectable marker (for example, kanamycin resistance). One method of introducing the T-DNA into the plant genome is shown in **Figure 10-23**. Bacteria containing an engineered Ti plasmid are used to infect cut segments of plant tissue, such as punched-out leaf discs. If the leaf disks are placed on a medium containing kanamycin, only the plant cells that have acquired the kan^R gene engineered into the T-DNA will undergo cell division. Transformed cells grow into a clump, or callus, that can be induced to form shoots and roots. These calli are transferred to soil, where they develop into transgenic plants. Typically, only a single copy of the T-DNA region inserts into a given plant genome, where it segregates at meiosis like a regular Mendelian allele. Therefore, one-quarter of the progeny from crossing of the original transgenic plants will get two copies of the T-DNA. The presence of the insert can be verified by Southern blot analysis of purified DNA with a T-DNA probe or by PCR using primers specific for the T-DNA.

Transgenic plants carrying any one of a variety of foreign genes are in current use, including crop plants carrying genes that confer resistance to certain bacterial or fungal pests, and many more are in development. Not only are the qualities of plants themselves being manipulated, but, like microorganisms, plants are also being used as convenient "factories" to produce proteins encoded by foreign genes.

KEY CONCEPT Transgenic plants are generated by random insertion into a chromosome of a Ti plasmid that carries a gene of interest and is delivered by *Agrobacterium tumefaciens*.

Genetic engineering in animals

Transgenic technologies are now being employed with many animal model systems. We will focus on two animal models heavily used for basic genetic research: the nematode *Caenorhabditis elegans* and the mouse *Mus musculus*. A commonly used method to transform a third model organism, the fruit fly *Drosophila melanogaster*, is described in Chapter 16. Versions of many of the techniques considered so far can also be applied in these animal systems.

Transgenesis in *C. elegans* Microinjection is used to introduce transgenes into *C. elegans*. Transgenic DNAs are injected directly into the organism, typically as plasmids, fosmids, or other DNAs cloned in bacteria. The injection strategy is determined by the worm's reproductive

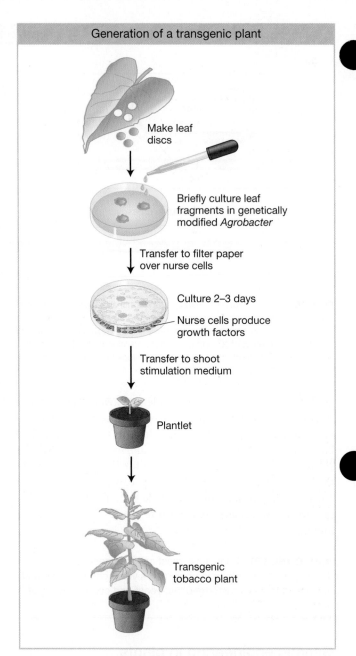

Generation of a transgenic plant

Make leaf discs

Briefly culture leaf fragments in genetically modified *Agrobacter*

Transfer to filter paper over nurse cells

Culture 2–3 days
Nurse cells produce growth factors

Transfer to shoot stimulation medium

Plantlet

Transgenic tobacco plant

FIGURE 10-23 Insertion of T-DNA into plant chromosomes. Incubation of tobacco leaf discs with the bacterium *A. tumefaciens* containing an engineered T-DNA leads to leaf cells with the T-DNA in their genome, which are able to grow on plates with growth factors and can be coaxed to differentiate into transgenic tobacco plants.

biology of the hermaphrodite gonad. The gonads are syncytial, meaning that there are many nuclei within the same gonadal cell. One syncytial cell is a large proportion of one arm of the gonad, and the other syncytial cell is the bulk of the other arm (**Figure 10-24a**). These nuclei do not form individual cells until meiosis, when they begin their transformation into individual eggs or sperm. A solution of DNA is injected into the syncytial region of one of the arms, thereby exposing more than 100 nuclei to the transgenic DNA. By chance, a few of these nuclei will incorporate the DNA (remember, the nuclear membrane breaks down in the course of division, and so the cytoplasm into

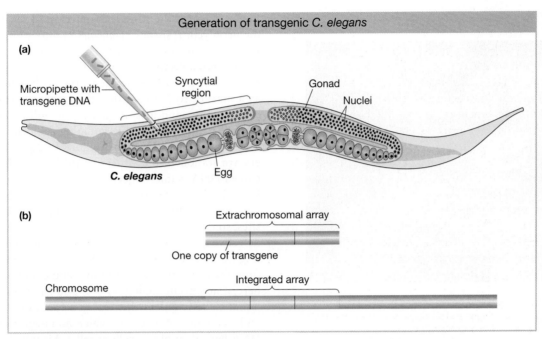

FIGURE 10-24 Transgenic *C. elegans* are created by injecting transgene DNA directly into a gonad. (a) The method of injection. (b) The two main types of transgenic results: extrachromosomal arrays and arrays integrated in ectopic chromosomal locations.

which the DNA is injected becomes continuous with the nucleoplasm). Typically, the transgenic DNA forms multi-copy *extrachromosomal arrays* (Figure 10-24b) that exist as independent units outside the chromosomes. The arrays are stably inherited, but not with the same efficiency as chromosomes. More rarely, the transgenes will become integrated into an ectopic position in a chromosome, still as a multicopy array.

KEY CONCEPT Transgenic worms are generated by injection of a plasmid containing a gene of interest into the gonad. The plasmid is typically stably inherited as a multicopy extrachromosomal array.

Transgenesis in *M. musculus* Mice are a very important model for mammalian genetics because they are relatively easy to breed and genetically manipulate. Furthermore, many of the technologies developed in mice and biological insights gained from studies of mice are potentially applicable to humans. There are two strategies for transgenesis in mice, each having its advantages and disadvantages:

- *Ectopic insertions.* Transgenes are inserted randomly in the genome, usually as multicopy arrays. Mice generated with an ectopic insertion are called *transgenic* mice.
- *Gene targeting.* Transgenes are inserted into a location occupied by a homologous sequence in the genome. That is, the transgene replaces its normal homologous counterpart. Mice generated by gene targeting are called *knock-in* or *knockout* mice. For knock-in mice, new DNA either is added to the targeted gene or is substituted

for DNA sequences at the targeted gene. For knockout mice, part or all of the targeted gene is deleted, or a DNA sequence is inserted into the targeted gene to disrupt its expression, thereby creating a loss-of-function mutation.

Ectopic insertions To insert transgenes in random locations, a solution of bacterially cloned DNA is injected into either the male or female pronucleus of a fertilized egg (**Figure 10-25**). Several injected eggs are inserted into the oviduct of a recipient mouse. Progeny are analyzed for integration of the transgene. Typically, DNA extracted from a piece of the tail is used for Southern blot analysis or PCR analysis for the transgene. Occasionally, mice are mosaic; that is, not every cell contains the transgene because DNA integration occurred at a two-cell or later stage of embryogenesis. Positive mice are subsequently mated, their offspring are analyzed for transgene expression, and positive mice are used to establish transgenic mouse lines with stable integration and expression of the transgene. The technique gives rise to some problems: (1) the expression pattern of the randomly inserted genes may be abnormal due to **position effects** from the local chromatin environment (see Chapter 12 for more on position effects), and (2) DNA rearrangements can occur inside the multicopy arrays (in essence, mutating the sequences). Nonetheless, this technique is much more efficient and less laborious than gene targeting. Because of the ease of generating mice with ectopic insertions, it has been used to produce human antibodies for use as therapeutics.

Gene targeting Gene targeting enables researchers to eliminate a gene or modify its function. In one application,

Generation of transgenic mice

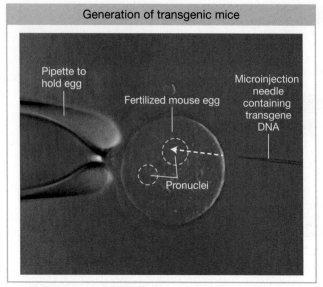

Pipette to hold egg

Fertilized mouse egg

Microinjection needle containing transgene DNA

Pronuclei

FIGURE 10-25 Transgenic mice are created by injection of cloned DNA into fertilized eggs and subsequent insertion in ectopic chromosomal locations. [*RAPHO AGENCE/Science Source.*]

ANIMATED ART   SaplingPlus
Creating a transgenic mouse

called **gene replacement,** a mutant allele can be repaired by substituting a wild-type allele in its normal chromosomal location. Gene replacement avoids both position effects and DNA rearrangements associated with ectopic insertion, because a single copy of the gene is inserted in its normal chromosomal environment. Conversely, a gene may be inactivated by substituting an inactive gene for the normal gene. Such a targeted inactivation is called a **gene knockout.**

Gene targeting in mice is carried out in cultured embryonic stem cells (ES cells). In general, a stem cell is an undifferentiated cell in a given tissue or organ that divides asymmetrically to produce a progeny stem cell and a cell that will differentiate into a terminal cell type. ES cells are special stem cells called pluripotent stem cells that can differentiate to form any cell type in the body—including, most important, the germ line.

To illustrate the process of gene targeting, we look at how it achieves one of its typical outcomes—namely, the substitution of an inactive gene for the normal gene, or gene knockout. The process requires three stages:

1. An inactive gene is targeted to replace the functioning gene in a culture of ES cells, producing ES cells containing a gene knockout (**Figure 10-26**).

2. ES cells containing the inactive gene are transferred to mice embryos (**Figure 10-27a**).

3. Knockout mice are identified and bred to produce mice of known genotype (Figure 10-27b).

Stage 1: A recombinant DNA molecule is generated that disrupts a gene of interest. In the example shown in

Figure 10-26a, the gene was inactivated by insertion of the neomycin-resistance gene (neo^R) into a protein-coding region (exon 2) of the gene. The mutant gene was then cloned into a vector containing the herpes virus *thymidine kinase* (*tk*) gene. In later steps, the neo^R gene will serve as a marker to indicate that the transgene inserted in a chromosome, and loss of the *tk* gene will ensure that the transgene is inserted at the homologous locus rather than randomly in a chromosome (Figure 10-26b). These markers are standard, but others could be used instead. The cloned DNA is microinjected into the nucleus of cultured ES cells. The defective gene inserts far more frequently into nonhomologous (ectopic) sites than into homologous sites, so the next step is to select the rare cells in which the defective gene has replaced the functioning gene as desired (Figure 10-26c). To isolate cells carrying a targeted mutation, the cells are cultured in medium containing drugs—here, a neomycin analog (G418) and ganciclovir. G418 is lethal to cells unless they carry a functional neo^R gene, and so it eliminates cells in which no integration of vector DNA has taken place (yellow cells). Meanwhile, ganciclovir kills any cells that harbor the *tk* gene, thereby eliminating cells bearing a randomly integrated vector (red cells). Consequently, the only cells that survive and proliferate are those harboring the targeted insertion (green cells).

Stage 2: ES cells that contain one copy of the disrupted gene of interest, that is, a gene knockout, are injected into a blastocyst-stage embryo, which is then implanted in a surrogate mother (Figure 10-27a). Some of the ES cells may become incorporated into the host embryo, and if that happens, the mouse that develops will be **chimeric**—that is, it will contain cells from two different mouse strains.

Stage 3: When the chimeric mouse reaches adulthood, it is mated with a normal mouse. If the chimeric mouse contained germ-line cells that were derived from the ES cells (with the knockout gene), some of the resulting offspring will inherit the gene knockout in all their cells. Sibling mice that are identified by Southern blot or PCR analysis as being heterozygous for the knockout version of the gene of interest are then mated to produce mice that are homozygous for the knockout allele. If the gene is essential, homozygotes will be lethal, and none will be obtained from this cross (Figure 10-27b).

KEY CONCEPT Knock-in and knockout mice are generated by homologous recombination in ES cells between a mouse chromosome and a plasmid. Gene-targeted ES are then injected into embryos to generate chimeric mice that are crossed to assess heritable germ-line transmission.

KEY CONCEPT Germ-line transgenic techniques have been developed for all well-studied eukaryotic species. These techniques depend on an understanding of the reproductive biology of the recipient species.

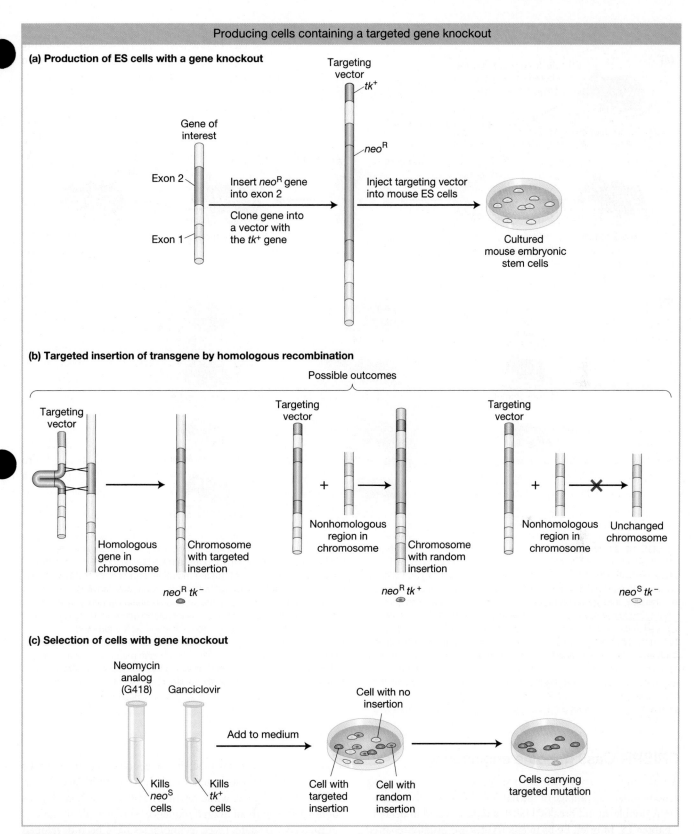

FIGURE 10-26 Producing cells that contain a mutation in one specific gene, known as a targeted mutation or a gene knockout. (a) The gene of interest (cloned gene) is inactivated by insertion of the *neo*R gene in exon 2 and cloned into a targeting vector containing the *tk* gene. Copies of a cloned gene are altered in vitro to produce the targeting vector. The vector is then injected into ES cells. (b) When homologous recombination occurs (*left*), the homologous regions on the vector, together with any DNA in between but excluding the marker at the tip, take the place of the original gene. This event is important because the vector sequences serve as a useful tag for detecting the presence of this mutant gene. In many cells, though, the full vector (complete with the extra marker at the tip) inserts ectopically (*middle*) or does not become integrated at all (*right*). (c) To isolate cells carrying a targeted mutation, all of the cells are cultured in media containing drugs to select for cells containing the targeted insertion.

Producing a mouse containing a targeted gene knockout

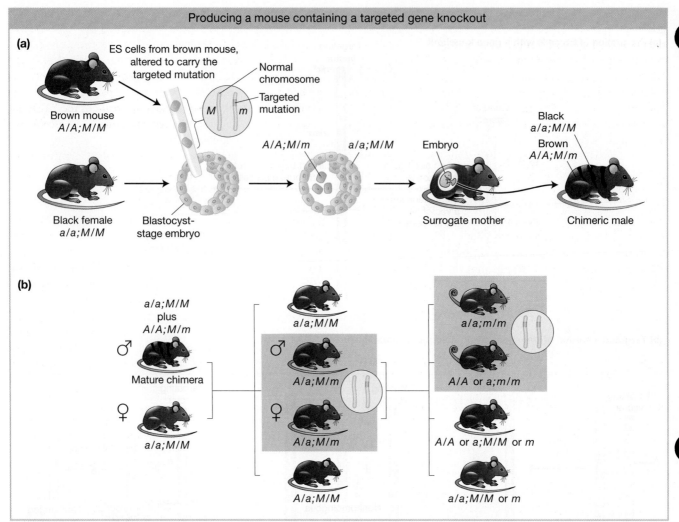

FIGURE 10-27 A knockout mouse is produced by inserting embryonic stem (ES) cells carrying the targeted mutation into an embryo. (a) ES cells are isolated from an agouti (brown) mouse strain (*A/A*) and altered to carry a targeted mutation (*m*) in one chromosome. The ES cells are then inserted into young embryos, one of which is shown. The coat color of the future newborns is a guide to whether the ES cells have survived in the embryo. Hence, ES cells are typically put into embryos that, in the absence of the ES cells, would acquire a totally black coat. Such embryos are obtained from a black strain that lacks the dominant agouti allele (*a/a*). Embryos containing the ES cells grow to term in surrogate mothers. Agouti shading intermixed with black indicates those newborns in which the ES cells have survived and proliferated. Such mice are called *chimeras* because they contain cells derived from two different strains of mice. Solid black coloring, in contrast, indicates that the ES cells have perished, and these mice are excluded. *A* represents agouti; *a*, black; *m* is the targeted mutation; and *M* is its wild-type allele. (b) Chimeric males are mated with black (nonagouti) females. Progeny are screened for evidence of the targeted mutation (green in inset) in the gene of interest. Direct examination of the genes in the agouti mice reveals which of those animals (*boxed*) inherited the targeted mutation. Males and females carrying the mutation are mated with one another to produce mice whose cells carry the chosen mutation in both copies of the target gene (*inset*) and thus lack a functional gene. Such animals (*boxed*) are identified definitively by direct analysis of their DNA. The knockout in this case results in a curly-tail phenotype.

CRISPR-Cas9 genome engineering

An alternative approach to transgenesis for engineering genomes takes advantage of the natural ability of cells to repair DNA double-strand breaks (DSBs), a topic covered in detail in Chapter 15. In brief, DSBs in eukaryotes are usually repaired by **nonhomologous end joining (NHEJ)**, a mechanism that reattaches the two chromosomal pieces but in a sloppy fashion, causing nucleotides to be inserted or deleted at the site of the DSB. Alternatively, DSBs are repaired by **homologous recombination (HR)**, which fixes the break without errors using a homologous donor DNA (e.g., a sister chromatid or a plasmid). Therefore, if DSBs could somehow be directed to occur at a particular place in the genome, repair by NHEJ would create mutations that are likely to result in inactivation of the targeted gene, while repair by HR using a homologous donor DNA with an altered sequence would create mutations in the gene. To date, three technologies have been developed that create site-specific DSBs: zinc-finger nucleases (ZFNs), transcription activator-like effector nucleases (TALENs), and CRISPR RNA-guided Cas nucleases (CRISPR-Cas). ZFNs and TALENs are proteins that contain two functional domains: a domain with DNA-binding activity that is designed to bind a specific DNA sequence, and a domain with non-specific DNA endonuclease activity that produces DSBs. When expressed in cells, ZNFs and TALENs bind their targeted sequence in genomic DNA and generate a DSB at a nearby sequence that is then repaired by NHEJ or HR. In contrast, in the CRISPR-Cas system, base pairing between a noncoding RNA and genomic DNA targets the Cas endonuclease to generate a DSB at a specific place in the genome. A major technical advantage of CRISPR-Cas over ZNFs and TALENs is that it is much easier to produce RNAs than DNA-binding domains that bind a unique sequence in a DNA genome.

The **CRISPR-Cas (clustered, regularly interspaced short palindromic repeats-CRISPR-associated protein)** technology is derived from a bacterial immune system that protects bacteria against foreign plasmid and bacteriophage DNA (as discussed in Chapter 6). For instance, the bacterium *Streptococcus pyogenes*, which contains a relatively simple CRISPR-Cas system, stores the memory of encounters with foreign DNA by integrating 20-nucleotide sequences from the foreign DNA into a particular place in the bacterial genome called a CRISPR array. When *S. pyogenes* is attacked for a second time by the same foreign DNA, guide RNAs (gRNAs) that contain the 20-nucleotide sequences are produced by cutting apart a long RNA transcript from the CRISPR array. A sequence common to all gRNAs base pairs with another noncoding RNA called a trans-activating CRISPR RNA (tracrRNA), and the dual RNAs form a complex with a Cas protein called Cas9. The gRNA then directs Cas9 to produce a DSB at a location in the foreign DNA that contains the complementary 20-nucleotide target sequence located next to a trinucleotide NGG protospacer adjacent motif (PAM). Cas9 has two separate endonuclease domains that cut three nucleotides upstream of the PAM; one domain cuts the target strand that base pairs to the gRNA, and the other cuts the nontarget strand.

In 2012, the laboratories of Emmanuelle Charpentier and Jennifer Doudna demonstrated that when a gRNA and a tracrRNA are modified to be in the same transcript, called a **single guide RNA (sgRNA)**, the sgRNA retains the ability of the separate RNAs to assemble with Cas9, base pair to targeted sequences in DNA, and activate the endonuclease activities of Cas9. This simplification made it practical to employ the CRISPR-Cas9 system in the laboratory to make modifications to eukaryotic genomes with high efficiency and specificity. Application of the CRISPR-Cas9 technology involves two plasmids: one expresses Cas9 protein, and the other expresses an sgRNA with a 20-nucleotide guide sequence that is designed by a researcher to be complementary to a specific genomic site adjacent to a PAM (**Figure 10-28**). After both plasmids are

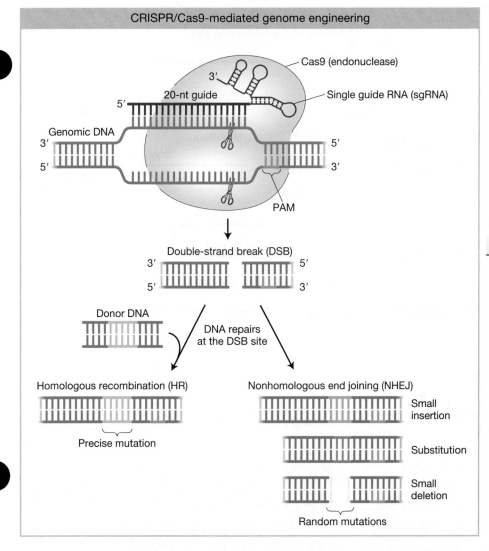

CRISPR/Cas9-mediated genome engineering

FIGURE 10-28 To target a DSB in a genome, the sequence of an sgRNA is designed to base pair to a target site, which is chosen in part because of its proximity to a PAM (orange DNA). A complex consisting of the sgRNA and Cas9 binds the target site, and the endonuclease domains of Cas9 produce a DSB (scissors). The DSB is then repaired either by homologous recombination using a supplied donor DNA, which creates precise mutations, or by nonhomologous end joining, which generates small insertions, base substitutions, or small deletions.

ANIMATED ART　📱 SaplingPlus

CRISPR

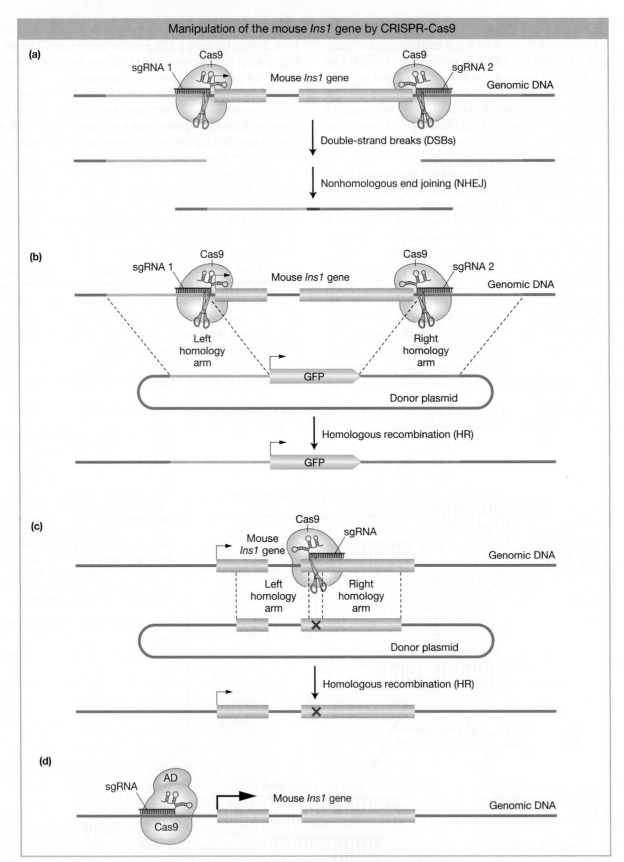

FIGURE 10-29 Four examples of the use of CRISPR-Cas9 to manipulate the mouse *Ins1* gene. (a) Two sgRNAs are designed to target DSBs to the ends of the *Ins1* gene. The chromosome break is then repaired by nonhomologous end joining, which deletes the *Ins1* gene and may introduce other sequence changes, indicated by the red line. (b) Addition of a donor plasmid triggers repair by homologous recombination, which, in this case, replaces the *Ins1* gene with the GFP gene. (c) Repair by homologous recombination of a single DSB in the *Ins1* gene using an altered *Ins1* donor DNA can generate missense and nonsense mutations as well as insertions and deletions in the *Ins1* gene. (d) sgRNA-mediated targeting of a Cas9 protein with inactivated endonuclease domains fused to a transcriptional activation domain (AD) yields enhanced transcription of the *Ins1* gene (bold arrow).

ANIMATED ART 🍁 Sapling Plus   CRISPR

introduced into cells or organisms, Cas9 protein and sgRNA are expressed and form a complex that produces a DSB in the targeted gene. Inaccurate repair of the DSB by NHEJ causes gene inactivation. In contrast, specific mutations in a gene can be introduced by inclusion of a third plasmid, a donor plasmid, that is used for repair of the DSB by HR because it contains sequences identical to those that flank the site of cleavage in addition to the specified mutations.

CRISPR-Cas9 gene editing occurs in reproductive cells of many organisms, making targeted gene modifications heritable. Therefore, CRISPR-Cas9 can be used for essentially all of the same reverse genetic purposes as transgenesis and even more. For example, to study the phenotypic consequences of loss of one of the two insulin genes in mice, a null allele of *Ins1* can be generated by CRISPR-Cas9-mediated deletion of the *Ins1* gene. This is accomplished by expressing two sgRNAs that create DSBs at the 5′ and 3′ ends of the *Ins1* gene, which will cut out the *Ins* gene and trigger repair of the broken chromosome by NHEJ (**Figure 10-29a**). Alternatively, the *Ins1* gene can be replaced by a reporter gene such as GFP. This is done by expressing the same two sgRNAs as well as a donor plasmid containing the GFP gene flanked by homology arms, sequences identical to those found upstream and downstream of the *Ins1* gene (Figure 10-29b). Repair of the DSBs by HR using the donor plasmid will replace the *Ins1* gene with the GFP gene. An advantage of this approach is that mice carrying null alleles of *Ins1* can be easily identified because they will express GFP in β cells under control of *Ins1* transcriptional regulatory sequences. Lastly, a donor plasmid could contain a missense or nonsense mutation that inactivates or alters the function of the *Ins1* gene (Figure 10-29c).

Researchers have also modified the CRISPR-Cas9 system to manipulate gene expression in specific ways. The basic idea behind these technologies is that, when complexed with an sgRNA, a mutant Cas9 protein lacking its endonuclease activity can transport any protein or protein domain to a specific place in the genome. As shown in Figure 10-29d, transcription of the mouse *Ins1* gene can be activated by sgRNA-mediated targeting to the *Ins1* gene of a Cas9 protein that is converted into a transcription factor by the addition of a transcriptional activation domain. Conversely, *Ins1* transcription can be turned off by fusion of a transcriptional repression domain to Cas9. One can imagine that someday CRISPR-Cas9 technologies will be used to treat individuals with type I diabetes by manipulating the sequence or expression of genes.

KEY CONCEPT The CRISPR-Cas9 system efficiently and specifically changes the sequence of targeted genes in an organism, and modified versions of the system alter gene expression without changing gene sequences.

SUMMARY

Recombinant DNA is constructed in the laboratory to allow researchers to manipulate and analyze DNA segments (donor DNA) from any genome or a DNA copy of mRNA. Three sources of donor DNA are (1) genomes digested with restriction enzymes, (2) PCR products of specific DNA regions, and (3) cDNA copies of mRNAs. Sequencing of DNA by the dideoxy (Sanger) method is used to confirm the accuracy of recombinant DNA molecules and also to discover information stored in genomic DNA and mutant genes.

The polymerase chain reaction is a powerful method for direct amplification of a small sequence of DNA from within a complex mixture of DNA, without the need for a host cell or very much starting material. The key is to have primers that are complementary to flanking regions on each of the two DNA strands. These regions act as sites for polymerization. Multiple rounds of denaturation, annealing, and extension amplify the sequence of interest exponentially.

To insert donor DNA into vectors, donor and vector DNA are cut by the same restriction endonuclease, joined by annealing the sticky ends that result from digestion, and ligated to covalently join the molecules. PCR products and cDNA molecules are inserted into vectors by first adding restriction sites to the 5′ end of PCR primers or by ligating short adapters containing restriction sites to their ends before insertion into the vector. Assembly methods that do not require restriction sites have made the construction of recombinant DNA molecules more flexible and efficient.

There are a wide variety of bacterial vectors. The choice of vector depends largely on the size of DNA fragment to be cloned. Plasmids are used to clone small genomic DNA fragments, PCR products, or cDNAs. Intermediate-size fragments, such as those resulting from digestion of genomic DNA, can be cloned into modified versions of λ bacteriophage (for inserts of 10–15 kb) or into phage–plasmid hybrids called fosmids (for inserts of 35–45 kb). Finally, bacterial artificial chromosomes (BACs) are used routinely to clone very large genomic fragments (~100–200 kb). A variety of plasmids have been developed that contain features that make it easier to clone DNA fragments and to control the expression of constituent genes in different organisms.

The vector-donor DNA construct is amplified inside bacterial host cells as extrachromosomal molecules that are replicated when the host is replicating its genome. Amplification of plasmids, phages, and BACs results in clones containing multiple copies of each recombinant DNA construct. In contrast, only a single fosmid is present in each bacterial cell.

Often, finding a specific clone with a gene of interest requires the screening of a genomic library, a set of clones, ligated in the same vector, that together represent all regions of the genome of the organism in question. The number of clones that constitute a genomic library depends on (1) the size of the genome in question, and (2) the insert size tolerated by the particular cloning-vector system. Similarly, a cDNA library is a representation of the total mRNA set produced by a tissue or developmental stage in a given organism.

Hybridization with single-stranded nucleic acid probes is fundamental to both in vitro and in vivo methods for identifying DNA fragments or RNAs of interest. These methods include Southern blotting for DNA, Northern blotting for RNA, and screening of genomic and cDNA libraries. In contrast, labeled antibodies are probes for identifying specific proteins from complex mixtures in Western blotting or immunofluorescence.

Transgenes are engineered DNA molecules that are introduced and expressed in eukaryotic cells. They can be used to engineer a novel mutation or to study the regulatory sequences that constitute part of a gene. Transgenes can be introduced as extrachromosomal molecules, or they can be integrated into a chromosome, either in random (ectopic) locations or in place of the homologous gene, depending on the system. Typically, the mechanisms used to introduce a transgene depend on an understanding and exploitation of the reproductive biology of the organism. New genome engineering methods like the CRISPR-Cas9 system are being developed whose defining features are the creation and repair of site-specific DNA double-strand breaks. These methods have opened the door to new and exciting reverse genetic studies in a wide variety of eukaryotic organisms, and they could potentially be used for gene-editing therapies for patients with serious diseases.

KEY TERMS

antibody (p. 336)
autoradiography (p. 336)
bacterial artificial chromosome (BAC) (p. 346)
cDNA library (p. 347)
chimera (chimeric) (p. 358)
clustered, regularly interspaced short palindromic repeats (CRISPR) (p. 361)
complementary DNA (cDNA) (p. 341)
CRISPR-associated protein (Cas) (p. 361)
dideoxy (Sanger) sequencing (p. 351)
DNA amplification (p. 339)
DNA assembly (p. 349)
DNA cloning (p. 342)
DNA ligase (p. 342)
DNA linker (DNA adapter) (p. 347)
DNA technologies (p. 332)
donor DNA (insert DNA) (p. 342)
ectopic (ectopically) (p. 354)
epitope tag (p. 351)

fluorescence in situ hybridization (FISH) (p. 337)
fosmid (p. 344)
gel electrophoresis (p. 334)
gene knockout (p. 358)
gene replacement (p. 358)
genetically modified organism (GMO) (p. 355)
genetic engineering (p. 332)
genomic library (p. 347)
genomics (p. 332)
homologous recombination (HR) (p. 360)
hybridization (p. 336)
immunofluorescence (p. 337)
infection (p. 346)
in situ hybridization (ISH) (p. 337)
multiple cloning site (MCS) (polylinker) (p. 342)
nonhomologous end joining (NHEJ) (p. 360)
Northern blotting (p. 334)
palindrome (palindromic) (p. 338)
plasmid (p. 344)

polymerase chain reaction (PCR) (p. 339)
position effect (p. 357)
probe (p. 336)
quantitative PCR (qPCR) (p. 341)
recombinant DNA (p. 342)
restriction enzyme (p. 338)
restriction fragment (p. 338)
restriction map (p. 339)
restriction site (p. 338)
reverse transcriptase (p. 341)
reverse transcription-PCR (RT-PCR) (p. 342)
single guide RNA (sgRNA) (p. 361)
Southern blotting (p. 334)
Ti plasmid (p. 355)
transduction (p. 346)
transformation (p. 346)
transgene (p. 354)
transgenic organism (p. 354)
vector (p. 342)
Western blotting (p. 334)

SOLVED PROBLEMS

SOLVED PROBLEM 1

In Chapter 9, we studied the structure of tRNA molecules. Suppose that you want to clone a fungal gene that encodes a certain tRNA. You have a sample of the purified tRNA and an *E. coli* plasmid that contains a single *Eco*RI cutting site in a *tet*R (tetracycline-resistance) gene, as well as a gene for resistance to ampicillin (*amp*R). How can you clone the gene of interest?

SOLUTION

You can use the tRNA itself or a cloned cDNA copy of it to probe for the DNA containing the gene. One method is to digest the genomic DNA with *Eco*RI and then mix it with the plasmid, which you also have cut with *Eco*RI. After transformation of an *amp*S *tet*S recipient, select AmpR colonies, indicating successful transformation. Of these AmpR colonies, select the colonies that are TetS. These TetS colonies will contain vectors with inserts in the *tet*R gene, and a great number of them are needed to make the library. Test the library by using the tRNA as the probe. Those clones that hybridize to the probe will contain the gene of interest. Alternatively, you can subject *Eco*RI-digested genomic DNA to gel electrophoresis and then identify the correct band by probing with the tRNA. This region of the gel can be cut out and used as a source of enriched DNA to clone into the plasmid cut with *Eco*RI. You then probe these clones with the tRNA to confirm that these clones contain the gene of interest.

PROBLEMS

Visit SaplingPlus for supplemental content. Problems with the icon are available for review/grading.

WORKING WITH THE FIGURES

(The first 33 questions require inspection of text figures.)

1. In the opening figure, what would happen if the cathode and anode were switched during gel electrophoresis?

2. In Figure 10-1, by what methods could the plasmid be introduced into bacteria?

3. In Figure 10-2, why can both DNA and RNA be used as probes for both Southern and Northern blot analysis?

4. In Figure 10-3a, what changes occur to the human *Ins* pre-mRNA to produce the mature *Ins* mRNA?

5. In Figure 10-3b, what size fragment would probe 2 detect in Southern blot analysis of mouse genomic DNA digested with *Pvu*II, and what site fragment would probe 3 detect in an *Nsi*I digest?

6. In Figure 10-4, which of the seven lanes contains the smallest piece of DNA?

7. In Figure 10-5, what is the purpose behind transferring the nucleic acid from a gel to a membrane?

8. In Figure 10-6c, what size band would be detected by probe 1 in Southern blot analysis of human smooth muscle cell genomic DNA digested with *Eco*RI?

9. Examine Table 10-1. Draw the staggered ends produced by digestion with *Not*I and the blunt ends produced by *Msp*I.

10. In Figure 10-7, why would PCR not work if *Taq* DNA polymerase was replaced with a DNA polymerase from human cells?

11. In Figure 10-8, if Sample A had a C_T value of 24 and Sample B had a C_T value of 27, which sample had more DNA and how much more?

12. In Figure 10-9, explain why the polymerase used for synthesizing DNA from an RNA template is called reverse transcriptase.

13. In Figure 10-10a, in the second step of the procedure, label the 5′ and 3′ ends of the linearized vector and the *Eco*RI insert.

14. In Figure 10-11a, which colonies (blue or white) contain plasmids with a DNA insert?

15. In Figure 10-12, determine approximately how many BAC clones are needed to provide 1× coverage of
   a. the yeast genome (12 Mbp).
   b. the *E. coli* genome (4.1 Mbp).
   c. the fruit fly genome (130 Mbp).

16. In Figure 10-13, what is the difference between plasmid transformation and fosmid transduction?

17. In Figure 10-14, is it possible for more than one insert to ligate into a single vector? Why or why not?

18. In Figure 10-15, how is screening a genomic library similar to Southern blot analysis?

19. In Figure 10-16, how would you modify the restriction sites in the primers so that PCR products only insert into a vector in one orientation?

20. In Figure 10-17a, what enzymes would you use to cut a full-length insert back out of the vector?

21. In Figure 10-18a, draw a ribonucleotide that would act as a chain terminator during transcription.

22. In Figure 10-18b, analogous to the drawing for the sequencing reaction that contains ddTTP, write the sequence of the first three termination products of the sequencing reaction that contains ddCTP. 🐾

23. In Figure 10-19, what would happen to the height of the peak at the SNP position in an individual that did not contain a SNP?

24. In Figure 10-20, why are multiple procedures needed for introducing DNA into cells?

25. In Figure 10-21, how can the marker be used to determine if a single or double crossover event occurred?

26. In Figure 10-22, what is the purpose of the selectable marker? Provide two examples of selectable markers. 🐾

27. In Figure 10-23, do all of the cells of a transgenic plant grown from one clump of cells contain T-DNA? Justify your answer.

28. In Figure 10-24, what is distinctive about the syncytial region that makes it a good place to inject DNA?

29. In Figure 10-25, why do the fertilized eggs have two nuclei? What is the ploidy of each nucleus?

30. In Figure 10-26c, does the selection procedure distinguish whether the targeting vector inserted in one or both copies of the homologous gene in the diploid ES cells?

31. In Figure 10-27a, why are chimeric males, rather than females, used in the mating crosses to generate a homozygous mutant mouse line?

32. In Figure 10-28, how would you determine the sequence change resulting from nonhomologous end joining (NHEJ) at a double-strand break site created by CRISPR-Cas9? 🐾

33. In Figure 10-29b, draw an analogous figure that shows a scheme for replacing the second exon in *Ins1* with the GFP gene.

BASIC PROBLEMS

34. Why is a range of temperatures indicated in the annealing step of PCR?

35. In the PCR process, if we assume that each cycle takes 5 minutes, what fold amplification would be accomplished in 1 hour? 🐾

36. Would the blot be called a Southern, Northern, or Western blot if RNA was on the membrane and the probe was single-stranded DNA?

37. How can genomic and cDNA sequences be used to determine where introns are located in genes? Use Figure 10-3a to illustrate your answer.

38. In Figure 10-10, it is possible that two or more copies, rather than one copy, of the insert ligated into the vector. How would you test this possibility (a) using restriction enzymes, (b) by PCR, and (c) by Southern blot? **Hints:** The complete sequence of the vector is known, and the *Eco*RI site is one of many restriction sites in the polylinker of the vector.

39. Write out the sequence of 20-nucleotide primers to be used for PCR to amplify the region of interest in the following piece of DNA: 🐾

Region of interest

5′–CCGTAACACGTCAGGGCCTAACAGG————————————TTGACAATGCCTGGAATTCTGTAAC–3′
3′–GGCATTGTGCAGTCCCGGATTGTCC————————————AACTGTTACGGACCTTAAGACATTG–5′

40. Draw a diagram that explains how automated dideoxy sequencing reactions are analyzed by capillary gel electrophoresis and laser detection.

41. Explain why dideoxynucleosides, which are converted to dideoxynucleotides in human cells, are effective drugs to block replication of the human immunodeficiency virus (HIV) genome by the HIV reverse transcriptase.

42. Compare and contrast the use of the word *recombinant* as used in the phrases (a) "recombinant DNA" and (b) "recombinant frequency."

43. Why is DNA ligase needed to make recombinant DNA? What would be the immediate consequence in the cloning process if DNA ligase was not included in the reaction?

44. In Figure 10-26, describe how positive-negative selection is used to find rare homologous recombination-mediated gene targeting events.

45. In T-DNA transformation of a plant with a transgene from a fungus (not found in plants), the presumptive transgenic plant does not display the expected phenotype of the transgene. How would you determine whether the transgene is in fact inserted in the plant genome? How would you determine whether the transgene mRNA and protein are expressed in the plant? 🐾

46. Why was cDNA and not genomic DNA used to express human insulin in *E. coli*?

47. Based on the information presented in Figures 10-26 and 10-29, explain how CRISPR-Cas9 could be used to knock out the *Ins1* gene in mice. In particular, what

RNAs and proteins would be expressed from plasmids that are injected into fertilized mouse eggs?

CHALLENGING PROBLEMS

48. Using the information in Figures 10-3 and 10-26a, draw a targeting vector that could be used to tag the human insulin protein with GFP at the N-terminus.

49. Diagram how Gibson assembly could be used to construct the targeting vector for question 48.

50. A cloned fragment of DNA was sequenced by using the dideoxy method. A part of the autoradiogram of the sequencing gel is represented here.

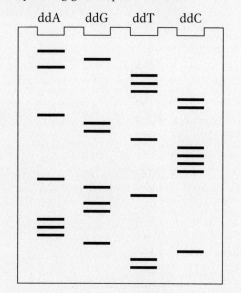

a. Write out the nucleotide sequence of the DNA molecule synthesized from the primer. Label the 5′ and 3′ ends.

b. Write out the nucleotide sequence of the DNA molecule used as the template strand. Label the 5′ and 3′ ends.

c. Write out the nucleotide sequence of the DNA double helix. Label the 5′ and 3′ ends.

51. Transgenic tobacco plants were obtained in which the vector Ti plasmid was designed to insert the gene of interest plus an adjacent kanamycin-resistance gene. The inheritance of chromosomal insertion was followed by testing progeny for kanamycin resistance. Two plants typified the results obtained generally. When plant 1 was backcrossed with wild-type tobacco, 50 percent of the progeny were kanamycin resistant and 50 percent were sensitive. When plant 2 was backcrossed with the wild type, 75 percent of the progeny were kanamycin resistant and 25 percent were sensitive. What must have been the difference between the two transgenic plants? What would you predict about the situation regarding the gene of interest?

52. The sequence of the *actin* gene in the haploid fungus *Neurospora* is known from the complete genome sequence. If you had a slow-growing mutant that you

suspected of being an *actin* mutant, how would you use (a) restriction enzyme cloning and sequencing, (b) PCR and sequencing, and (c) restriction enzyme cloning and functional complementation (rescue) to determine whether your suspicion is correct?

53. Bacterial glucuronidase converts a colorless substance called X-Gluc into a bright blue indigo pigment. The gene for glucuronidase also works in plants if given a plant promoter region. How would you use this gene as a reporter gene to find the tissues in which a plant gene that you have just cloned is normally active? (Assume that X-Gluc is easily taken up by the plant tissues.)

54. The plant *Arabidopsis thaliana* was transformed using the Ti plasmid into which a kanamycin-resistance gene had been inserted in the T-DNA region. Two kanamycin-resistant colonies (A and B) were selected, and plants were regenerated from them. The plants were allowed to self-pollinate, and the results were as follows:

Plant A selfed → $\frac{3}{4}$ progeny resistant to kanamycin

$\frac{1}{4}$ progeny sensitive to kanamycin

Plant B selfed → $\frac{15}{16}$ progeny resistant to kanamycin

$\frac{1}{16}$ progeny sensitive to kanamycin

a. Draw the relevant plant chromosomes in both plants.

b. Explain the two different ratios.

GENETICS AND SOCIETY

In 2018, a researcher claimed to have used the CRISPR-Cas9 genome editing technique to produce the world's first gene-edited babies. The researcher announced that they edited the *CCR5* gene in two embryos, which were then implanted in a woman. *CCR5* encodes a receptor that is expressed on the surface of white blood cells and other cells, where it coordinates immune responses. CCR5 is also the main receptor used by the human immunodeficiency virus (HIV) to gain entry into cells, which is necessary for its replication. Genetic variations in the *CCR5* gene have been identified in the human population that confer natural resistance to HIV infection. This includes a *CCR5* allele called delta-32 that is missing 32 base pairs from the coding region of the gene, causing a deletion and a frameshift in the encoded CCR5 protein that blocks its expression on the cell surface. Therefore, to create babies that were resistant to HIV infection, the researcher used the CRISPR-Cas9 editing technique to produce a deletion in the *CCR5* gene that was similar to delta-32. This use of the CRISPR-Cas9 technique has prompted a great deal of discussion about the scientific and ethical implications of making heritable changes to the human genome. Given your newfound insights into the CRISPR-Cas9 technique and genetic phenomena, what are your concerns?

CHAPTER 11

Regulation of Gene Expression in Bacteria and Their Viruses

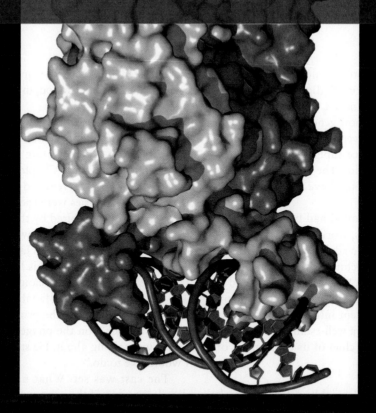

The control of gene expression is governed primarily by DNA-binding proteins that recognize specific control sequences of genes. Here, the binding of the Lac repressor protein to the lac operator DNA is modeled.

CHAPTER OUTLINE AND LEARNING OBJECTIVES

11.1 GENE REGULATION

> **LO 11.1** Illustrate how both positive and negative regulation control the activity of the *lac* operon.

11.2 DISCOVERY OF THE *LAC* SYSTEM: NEGATIVE REGULATION

> **LO 11.2** Infer the components of genetic switches from experimental data and predict the effect of mutations in the different components on gene expression.

11.3 CATABOLITE REPRESSION OF THE *LAC* OPERON: POSITIVE REGULATION

11.4 DUAL POSITIVE AND NEGATIVE REGULATION: THE ARABINOSE OPERON

> **LO 11.3** Illustrate and compare the mechanisms that coordinate expression of sets of genes in bacteria and bacteriophage.

11.5 METABOLIC PATHWAYS AND ADDITIONAL LEVELS OF REGULATION: ATTENUATION

11.6 BACTERIOPHAGE LIFE CYCLES: MORE REGULATORS, COMPLEX OPERONS

> **LO 11.4** Explain the roles of sequence-specific DNA-binding proteins and DNA regulatory sequences in coordinating the expression of sets of genes in bacteria and bacteriophage.

11.7 ALTERNATIVE SIGMA FACTORS REGULATE LARGE SETS OF GENES

369

Bacteria and their viruses use a straightforward logic of positive and negative regulation to coordinately control the expression of genes in response to environmental conditions. The broad objective for this chapter is to learn how this regulatory logic can be uncovered using genetic approaches in bacteria and their viruses.

In December 1965, the king of Sweden presented the Nobel Prize in Physiology or Medicine to François Jacob, Jacques Monod, and André Lwoff of the Pasteur Institute for their discoveries of how gene expression is regulated (**Figure 11-1**). The prizes were the fruit of an exceptional collaboration among three superb scientists. They were also triumphs over great odds. The chances were slim that each of these three men would have lived to see that day, let alone earn such honors.

Twenty-five years earlier, Monod had been a doctoral student at the Sorbonne in Paris, working on a phenomenon in bacteria called "enzymatic adaptation" that seemed so obscure to some that the director of the zoological laboratory where he worked stated, "What Jacques Monod is doing is of no interest whatever to the Sorbonne." Jacob was a 19-year-old medical student intent on becoming a surgeon. Lwoff was by that time a well-established member of the Pasteur Institute in Paris, chief of its department of microbial physiology.

Then came World War II.

As France was invaded and quickly defeated, Jacob raced for the coast to join the Free French forces assembling in England. He served as a medic in North Africa and in Normandy until badly wounded. Monod joined the French Resistance while continuing his work. After a Gestapo raid on his Sorbonne laboratory, Monod decided that working there was too dangerous (his predecessor in the Resistance was arrested and executed), and André Lwoff offered him space at the Pasteur. Monod, in turn, connected Lwoff with the Resistance.

After the liberation of Paris, Monod served in the French army and happened on an article by Oswald Avery and colleagues demonstrating that DNA is the hereditary material in bacteria (see Chapter 7). His interest in genetics was rekindled, and he rejoined Lwoff after the war. Meanwhile, Jacob's injuries were too severe for him to pursue a career in surgery. Inspired by the enormous impact of antibiotics introduced late in the war, Jacob eventually decided to pursue scientific research. Jacob approached Lwoff several times for a position in his laboratory but was declined. He made one last try and caught Lwoff in a jovial mood. The senior scientist told Jacob, "You know, we have just found the induction of the prophage. Would you be interested in working on the phage?" Jacob had no idea what Lwoff was talking about. He stammered, "That's just what I would like to do."

The cast was set. What unfolded in the subsequent decade was one of the most creative and productive collaborations in the history of genetics, whose discoveries still reverberate throughout biology today.

One of the most important insights arrived not in the laboratory but in a movie theater. Struggling with a lecture that he had to prepare, Jacob opted instead to take his wife, Lise, to a Sunday matinee. Bored and daydreaming, Jacob drew a connection between the work he had been doing on the induction of prophage and that of Monod on the induction of enzyme synthesis. Jacob became "involved by a sudden excitement mixed with a vague pleasure. . . . Both experiments . . . on the phage . . . and that done with Pardee and Monod on the lactose system . . . are the same! Same situation. Same result . . . In both cases, a gene governs the formation . . . of a repressor blocking the expression of other genes and so preventing either the synthesis of the galactosidase or the multiplication of the virus. . . . Where can the repressor act to stop everything at once? The only simple answer . . . is on the DNA itself!"[1]

And so was born the concept of a repressor acting on DNA to repress the induction of genes. It would take many years before the hypothesized repressors were isolated and characterized biochemically. The concepts worked out by Jacob and Monod and explained in this chapter—messenger RNA, promoters, operators, regulatory genes,

Pioneers of gene regulation

FIGURE 11-1 François Jacob, Jacques Monod, and André Lwoff were awarded the 1965 Nobel Prize in Physiology or Medicine for their pioneering work on how gene expression is regulated. [*The Pasteur Institute.*]

[1] F. Jacob, *The Statue Within: An Autobiography*, 1988.

operons, and allosteric proteins—were deduced entirely from genetic evidence, and these concepts shaped the future field of molecular genetics.

Walter Gilbert, who isolated the first repressor and was later awarded a Nobel Prize in Chemistry for co-inventing a method of sequencing DNA, explained the effect of Jacob and Monod's work at that time: "Most of the crucial discoveries in science are of such a simplifying nature that they are very hard even to conceive without actually having gone through the experience involved in the discovery. . . . Jacob's and Monod's suggestion made things that were utterly dark, very simple."[2]

The concepts that Jacob and Monod illuminated went far beyond bacterial enzymes and viruses. They understood, and were able to articulate with exceptional eloquence, how their discoveries about gene regulation pertained to the general mysteries of cell differentiation and embryonic development in animals. The two men once quipped, "anything found to be true of *E. coli* must also be true of Elephants."[3] In the next three chapters, we will see to what degree that assertion is true. We'll start in this chapter with bacterial examples that illustrate key themes and mechanisms in the regulation of gene expression. We will largely focus on single regulatory proteins and the genetic "switches" on which they act. Then, in Chapter 12, we'll tackle gene regulation in eukaryotic cells, which entails more complex biochemical and genetic machinery. Finally, in Chapter 13, we'll examine the role of gene regulation in the development of multicellular animals. There we will see how sets of regulatory proteins act on arrays of genetic switches to control gene expression in time and space and choreograph the building of bodies and body parts.

11.1 GENE REGULATION

LO 11.1 Illustrate how both positive and negative regulation control the activity of the *lac* operon.

Despite their simplicity of form, bacteria have in common with larger and more complex organisms the need to regulate expression of their genes. One of the main reasons is that they are nutritional opportunists. Consider how bacteria obtain the many important compounds, such as sugars, amino acids, and nucleotides, needed for metabolism. Bacteria swim in a sea of potential nutrients. They can either acquire the compounds that they need from the environment or synthesize them by enzymatic pathways. But synthesizing these compounds also requires expending energy and cellular resources to produce the necessary enzymes for these pathways. Thus, given the choice, bacteria will

take compounds from the environment instead. Natural selection favors efficiency and selects against the waste of resources and energy. To be economical, bacteria will synthesize the enzymes necessary to produce compounds only when there is no other option—in other words, when compounds are unavailable in their local environment.

Bacteria have evolved regulatory systems that couple the expression of gene products to sensor systems that detect the relevant compound in a bacterium's local environment. The regulation of enzymes taking part in sugar metabolism provides an example. Sugar molecules can be broken down to provide energy, or they can be used as building blocks for a great range of organic compounds. However, there are many different types of sugar that bacteria could use, including lactose, glucose, galactose, and xylose. A different import protein is required to allow each of these sugars to enter the cell. Further, a different set of enzymes is required to process each of the sugars. If a cell were to simultaneously synthesize all the enzymes that it might possibly need, the cell would expend much more energy and materials to produce the enzymes than it could ever derive from breaking down prospective carbon sources. The cell has devised mechanisms to shut down (repress) the transcription of all genes encoding enzymes that are not needed at a given time and to turn on (activate) those genes encoding enzymes that are needed. For example, if only lactose is in the environment, the cell will shut down the transcription of the genes encoding enzymes needed for the import and metabolism of glucose, galactose, xylose, and other sugars. Conversely, *E. coli* will initiate the transcription of the genes encoding enzymes needed for the import and metabolism of lactose. In sum, cells need mechanisms that fulfill two criteria:

1. They must be able to recognize environmental conditions in which they should activate or repress the transcription of the relevant genes.

2. They must be able to toggle on or off, like a switch, the transcription of each specific gene or group of genes.

KEY CONCEPT Cells must be able both to recognize environmental conditions and to respond to those conditions by activating or repressing particular genes.

Let's preview the current model for bacterial transcriptional regulation and then use a well-understood example—the regulation of the genes in the metabolism of the sugar lactose—to examine it in detail. In particular, we will focus on how this regulatory system was dissected with the use of the tools of classical genetics and molecular biology.

The basics of bacterial transcriptional regulation: genetic switches

The regulation of transcription depends mainly on two types of protein–DNA interactions. Both take place near the site at which gene transcription begins.

[2] H. F. Judson, *The Eighth Day of Creation: Makers of the Revolution in Biology*, 1979.

[3] F. Jacob and J. Monod, *Cold Spring Harbor Quant. Symp. Biol.* 26, 1963, 393.

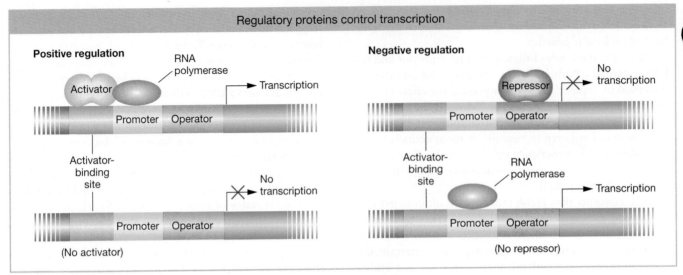

FIGURE 11-2 The binding of regulatory proteins can either activate or block transcription.

One of these DNA–protein interactions determines where transcription begins. The DNA that participates in this interaction is a DNA segment called the **promoter** (Chapter 8, Section 8.2), and the protein that binds to this site is RNA polymerase. When RNA polymerase binds to the promoter DNA, transcription can start a few bases away from the promoter site. Every gene must have a promoter or it cannot be transcribed.

The other type of DNA–protein interaction determines whether promoter-driven transcription takes place. DNA segments near the promoter serve as binding sites for sequence-specific regulatory proteins called **activators** and **repressors**. In bacteria, most binding sites for repressors are termed **operators**. For some genes, an activator protein must bind to its target DNA site as a necessary prerequisite for transcription to begin. Such instances are sometimes referred to as **positive regulation** because the *presence* of the bound protein is required for transcription (**Figure 11-2**). For other genes, a repressor protein must be prevented from binding to its target site as a necessary prerequisite for transcription to begin. Such cases are sometimes termed **negative regulation** because the *absence* of the bound repressor allows transcription to begin.

How do activators and repressors regulate transcription? Often, a DNA-bound activator protein physically helps tether RNA polymerase to its nearby promoter so that polymerase may begin transcribing. A DNA-bound repressor protein typically acts either by physically interfering with the binding of RNA polymerase to its promoter (blocking transcription initiation) or by impeding the movement of RNA polymerase along the DNA chain (blocking transcription). Together, these regulatory proteins and their binding sites constitute **genetic switches** that control the efficient changes in gene expression that occur in response to environmental conditions.

KEY CONCEPT Genetic switches are proteins and DNA sequences that control gene transcription. Activator or repressor proteins bind to operator sequences in the vicinity of the promoter to control its accessibility to RNA polymerase.

Both activator and repressor proteins must be able to recognize when environmental conditions are appropriate for their actions and act accordingly. Thus, for activator or repressor proteins to do their job, each must be able to exist in two states: one that can bind its DNA targets and another that cannot. The binding state must be appropriate to the set of physiological conditions present in the cell and its environment. For many regulatory proteins, DNA binding is effected through the interaction of two different sites in the three-dimensional structure of the protein. One site is the **DNA-binding domain**. The other site, the **allosteric site**, acts as a sensor that sets the DNA-binding domain in one of two modes: functional or nonfunctional. The allosteric site interacts with small molecules called **allosteric effectors**.

In lactose metabolism, it is actually an isomer of the sugar lactose (called allolactose) that is an allosteric effector: the sugar binds to a regulatory protein that inhibits the expression of genes needed for lactose metabolism. In general, an allosteric effector binds to the allosteric site of the regulatory protein in such a way as to change its activity. In this case, allolactose changes the shape and structure of the DNA-binding domain of a regulatory protein. Some activator or repressor proteins must bind to their allosteric effectors before they can bind DNA. Others can bind DNA only in the absence of their allosteric effectors. Two of these situations are shown in **Figure 11-3**.

KEY CONCEPT Allosteric effectors are small molecules that bind to activator or repressor proteins and control their ability to bind to their DNA target sites.

A first look at the *lac* regulatory circuit

The pioneering work of François Jacob and Jacques Monod in the 1950s showed how lactose metabolism is genetically regulated. Let's examine the system under two conditions: the presence and the absence of lactose. **Figure 11-4** is a simplified view of the components of this system. The cast of characters for *lac* operon regulation includes protein-coding genes and sites on the DNA that are targets for DNA-binding proteins.

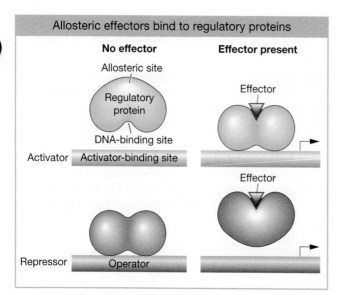

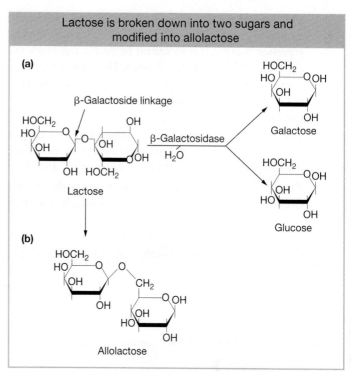

Allosteric effectors bind to regulatory proteins

FIGURE 11-3 Allosteric effectors influence the DNA-binding activities of activators and repressors.

Lactose is broken down into two sugars and modified into allolactose

FIGURE 11-5 The metabolism of lactose. (a) The enzyme β-galactosidase catalyzes a reaction in which water is added to the β-galactoside linkage to break lactose into separate molecules of glucose and galactose. (b) The enzyme also modifies a smaller proportion of lactose into allolactose, which acts as an inducer of the *lac* operon.

The *lac* structural genes The metabolism of lactose requires two enzymes: (1) a permease to transport lactose into the cell, and (2) β-galactosidase to modify lactose into allolactose and to cleave the lactose molecule to yield glucose and galactose (**Figure 11-5**). The structures of the β-galactosidase and permease proteins are encoded by two adjacent sequences, *Z* and *Y*, respectively. A third contiguous sequence, *A*, encodes an additional enzyme, termed *transacetylase*, which is not required for lactose metabolism. We will call *Z*, *Y*, and *A* *structural genes*—in other words, segments encoding proteins—while reserving judgment on this categorization until later. We will focus mainly on the *Z* and *Y* genes. All three genes are transcribed into a single messenger RNA molecule. Regulation of the production of this mRNA coordinates the synthesis of all three enzymes. That is, either all or none of the three enzymes are synthesized. Genes whose transcription is controlled by a common means are said to be **coordinately controlled genes**.

KEY CONCEPT If the genes encoding proteins constitute a single transcription unit, the expression of all these genes will be coordinately regulated.

Regulatory components of the *lac* system Key regulatory components of the lactose metabolic system include a gene encoding a transcription regulatory protein and two binding sites on DNA: one site for the regulatory protein and another site for RNA polymerase.

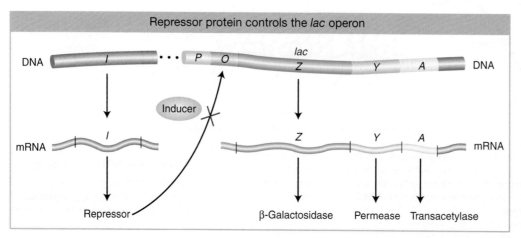

Repressor protein controls the *lac* operon

FIGURE 11-4 A simplified *lac* operon model. Coordinate expression of the *Z, Y,* and *A* genes is under negative regulation by the product of the *I* gene, the repressor. When the inducer binds the repressor, the operon is fully expressed.

1. *The gene for the Lac repressor.* A fourth gene (besides the structural genes *Z*, *Y*, and *A*), the *I* gene, encodes the Lac repressor protein. It is so named because it can block the expression of the *Z*, *Y*, and *A* genes. The *I* gene happens to map close to the *Z*, *Y*, and *A* genes, but this proximity is not important to its function because it encodes a diffusible protein.

2. *The lac promoter site.* The promoter (*P*) is the site on the DNA to which RNA polymerase binds to initiate transcription of the *lac* structural genes (*Z*, *Y*, and *A*).

3. *The lac operator site.* The operator (*O*) is the site on the DNA to which the Lac repressor binds. It is located

between the promoter and the *Z* gene near the point at which transcription of the multigenic mRNA begins.

The induction of the *lac* system   The *P*, *O*, *Z*, *Y*, and *A* segments (shown in **Figure 11-6**) together constitute an **operon**, defined as a segment of DNA that encodes a multigenic mRNA as well as an adjacent common promoter and regulatory region. The *lacI* gene, encoding the Lac repressor, is *not* considered part of the *lac* operon itself, but the interaction between the Lac repressor and the *lac* operator site is crucial to proper regulation of the *lac* operon. The Lac repressor has a *DNA-binding* site that can recognize

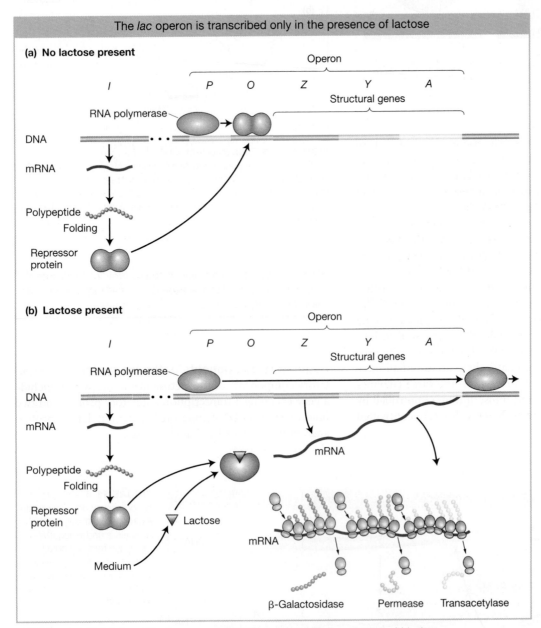

FIGURE 11-6 Regulation of the *lac* operon. The *I* gene continually makes repressor. (a) In the absence of lactose, the repressor binds to the *O* (operator) region and blocks transcription. (b) The binding of lactose changes the shape of the repressor so that the repressor no longer binds to *O* and falls off the DNA. The RNA polymerase is then able to transcribe the *Z*, *Y*, and *A* structural genes, and so the three enzymes are produced.

ANIMATED ART   Sapling Plus

Assaying lactose presence or absence through the Lac repressor the operator DNA sequence and an *allosteric site* that binds allolactose or analogs of lactose that are useful experimentally. The repressor will bind tightly only to the O site on the DNA near the genes that it is controlling and not to other sequences distributed throughout the chromosome. By binding to the operator, the repressor prevents transcription by RNA polymerase that has bound to the adjacent promoter site; the *lac* operon is switched "off."

When allolactose or its analogs bind to the repressor protein, the protein undergoes an **allosteric transition**, a change in shape. This slight alteration in shape in turn alters the DNA-binding site so that the repressor no longer has high affinity for the operator. Thus, in response to binding allolactose, the repressor falls off the DNA, allowing RNA polymerase to proceed (transcribe the gene): the *lac* operon is switched "on." The repressor's response to allolactose satisfies one requirement for such a control system—that the presence of lactose stimulates the synthesis of genes needed for its processing. The relief of repression for systems such as *lac* is termed **induction**. Allolactose and its analogs that allosterically inactivate the repressor, leading to the expression of the *lac* genes, are termed **inducers**.

Let's summarize how the *lac* switch works (Figure 11-6). In the absence of an inducer (allolactose or an analog), the Lac repressor binds to the *lac* operator site and prevents transcription of the *lac* operon by blocking the movement of RNA polymerase. In this sense, the Lac repressor acts as a roadblock on the DNA. Consequently, all the structural genes of the *lac* operon (the Z, Y, and A genes) are repressed, and there are very few molecules of β-galactosidase, permease, or transacetylase in the cell. In contrast, when an inducer is present, it binds to the allosteric site of each Lac repressor subunit, thereby inactivating the site that binds to the operator. The Lac repressor falls off the DNA, allowing the transcription of the structural genes of the *lac* operon to begin. The enzymes β-galactosidase, permease, and transacetylase now appear in the cell in a coordinated fashion. So, when lactose is present in the environment of a bacterial cell, the cell produces the enzymes needed to metabolize it. But when no lactose is present, resources are not wasted.

11.2 DISCOVERY OF THE *LAC* SYSTEM: NEGATIVE REGULATION

| LO 11.1 | Illustrate how both positive and negative regulation control the activity of the *lac* operon. |
| LO 11.2 | Infer the components of genetic switches from experimental data and predict the effect of mutations in the different components on gene expression. |

To study gene regulation, ideally we need three ingredients: a biochemical assay that lets us measure the amount of mRNA or expressed protein or both, reliable conditions in which the levels of expression differ in a wild-type genotype, and genetic mutations that perturb the levels of expression. In other words, we need a way of describing wild-type gene regulation, and we need mutations that can disrupt the wild-type regulatory process. With these elements in hand, we can analyze the expression in mutant genotypes, treating the mutations singly and in combination, to unravel any kind of gene-regulation event. The classical application of this approach was used by Jacob and Monod, who performed the definitive studies of bacterial gene regulation.

Jacob and Monod used the lactose metabolism system of *E. coli* (see Figure 11-4) to genetically dissect the process of enzyme induction—that is, the appearance of a specific enzyme only in the presence of its substrates. This phenomenon had been observed in bacteria for many years, but how could a cell possibly "know" precisely which enzymes to synthesize? How could a particular substrate induce the appearance of a specific enzyme?

In the *lac* system, the presence of lactose causes cells to produce more than 1000 times as much of the enzyme β-galactosidase as they produced when grown in the absence of lactose. What role did lactose play in the induction phenomenon? When Monod and co-workers followed the fate of radioactively labeled amino acids added to growing cells either before or after the addition of an inducer, they found that induction resulted in the synthesis of new enzyme molecules, as indicated by the presence of the radioactive amino acids in the enzymes. These new molecules could be detected as early as three minutes after the addition of an inducer. Additionally, withdrawal of lactose brought about an abrupt halt in the synthesis of the new enzyme. Therefore, it became clear that the cell has a rapid and effective mechanism for turning gene expression on and off in response to environmental signals.

Genes controlled together

When Jacob and Monod induced β-galactosidase, they found that they also induced the enzyme permease, which is required to transport lactose into the cell. The analysis of mutants indicated that each enzyme was encoded by a different gene. The enzyme transacetylase (with a dispensable and as yet unknown function) also was induced together with β-galactosidase and permease and was later shown to be encoded by a separate gene. Therefore, Jacob and Monod could identify three coordinately controlled genes. Recombination mapping showed that the Z, Y, and A genes were very closely linked on the chromosome (see Section 6.2).

Genetic evidence for the operator and repressor

Now we come to the heart of Jacob and Monod's work: How did they deduce the mechanisms of gene regulation in the *lac* system? Their strategy was a classic genetic approach: to examine the physiological consequences

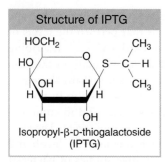

Structure of IPTG

Isopropyl-β-D-thiogalactoside
(IPTG)

FIGURE 11-7 IPTG is an inducer of the *lac* operon.

of mutations. Thus, they induced mutations in the structural genes and regulatory elements of the *lac* operon. As we will see, the properties of mutations in these different components of the *lac* operon are quite different, providing important clues for Jacob and Monod.

Natural inducers, such as allolactose, are not optimal for these experiments because they are broken down by β-galactosidase. The inducer concentration decreases during the experiment, and so the measurements of enzyme induction become quite complicated. Instead, for such experiments, Jacob and Monod used synthetic inducers, such as isopropyl-β-D-thiogalactoside (IPTG; **Figure 11-7**). IPTG is not hydrolyzed by β-galactosidase, but it still induces β-galactosidase enzyme expression.

Jacob and Monod found that several different classes of mutations can alter the expression of the structural genes of the *lac* operon. They were interested in assessing the interactions between the new alleles, such as which alleles exhibited dominance. But to perform such tests, one needs diploids, and bacteria are haploid. However, Jacob and Monod were able to produce bacteria that are partially diploid by inserting F′ factors carrying the *lac* region of the genome. (An F′ factor is a plasmid that carries one or more bacterial genes and that can be transferred from one bacteria to another through a process known as conjugation; see Section 6.2.) They could then create strains that were heterozygous for selected *lac* mutations, but still haploid for the rest of the genome. These **partial diploids** allowed Jacob and Monod to distinguish mutations in the regulatory DNA site (the *lac* operator) from mutations in the regulatory protein (the Lac repressor encoded by the *I* gene).

We begin by examining mutations that inactivate the structural genes for β-galactosidase and permease (designated Z^- and Y^-, respectively). The first thing that we learn is that Z^- and Y^- are recessive to their respective wild-type alleles (Z^+ and Y^+). For example, strain 2 in **Table 11-1** can be induced to synthesize β-galactosidase (like the wild-type

haploid strain 1 in this table), even though it is heterozygous for mutant and wild-type Z alleles. This demonstrates that the Z^+ allele is dominant over its Z^- counterpart.

Jacob and Monod first identified two classes of regulatory mutations, called O^C and I^-. These were called **constitutive mutations** because they caused the *lac* operon structural genes to be expressed regardless of whether inducer was present. Jacob and Monod identified the existence of the operator on the basis of their analysis of the O^C mutations. These mutations make the operator incapable of binding to repressor; they damage the switch such that the operon is always "on" (Table 11-1, strain 3). Importantly, the constitutive effects of O^C mutations were restricted solely to those *lac* structural genes *on the same chromosome* as the O^C mutation. For this reason, the operator mutant was said to be **cis-acting**, as demonstrated by the phenotype of strain 4 in Table 11-1. Here, because the wild-type permease (Y^+) gene is cis to the wild-type operator, permease is expressed only when lactose or an analog is present. In contrast, the wild-type β-galactosidase (Z^+) gene is cis to the O^C mutant operator; hence, β-galactosidase is expressed constitutively. This unusual property of cis action suggested that the operator is a segment of DNA that influences only the expression of the structural genes linked to it (**Figure 11-8**). The operator thus acts simply as a protein-binding site and makes *no* gene product.

Jacob and Monod did comparable genetic tests with the I^- mutations (**Table 11-2**). A comparison of the inducible wild-type I^+ (strain 1) with I^- strains shows that I^- mutations are constitutive (strain 2). That is, they cause the structural genes to be expressed at all times. Strain 3 demonstrates that the inducible phenotype of I^+ is dominant over the constitutive phenotype of I^-. This finding showed Jacob and Monod that the amount of wild-type protein encoded by one copy of the gene is sufficient to regulate both copies of the operator in a diploid cell. Most significantly, strain 4 showed them that the I^+ gene product is **trans-acting**, meaning that the gene product can regulate *all* structural *lac* operon genes, whether residing on the same DNA molecule or on different ones (in cis or in trans, respectively). Unlike the operator, the *I* gene behaves like a standard protein-coding gene. The protein product of the *I* gene is able to diffuse throughout a cell and act on both operators in the partial diploid (**Figure 11-9**).

TABLE 11-1	Synthesis of β-Galactosidase and Permease in Haploid and Heterozygous Diploid Operator Mutants					
		β—*Galactosidase (Z)*		*Permease (Y)*		
Strain	Genotype	Noninduced	Induced	Noninduced	Induced	Conclusion
1	$O^+\ Z^+\ Y^+$	−	+	−	+	Wild type is inducible
2	$O^+\ Z^+\ Y^+/F'\ O^+\ Z^-\ Y^+$	+	+	−	+	Z^+ is dominant to Z^-
3	$O^C\ Z^+\ Y^+$	+	+	+	+	O^C is constitutive
4	$O^+\ Z^-\ Y^+/F'\ O^C\ Z^+\ Y^-$	+	+	−	+	Operator is cis-acting

Note: Bacteria were grown in glycerol (no glucose present) with and without the inducer IPTG. Expression of maximal enzyme levels is indicated by +. Absence or very low levels of enzyme activity is indicated by −. All strains are I^+.

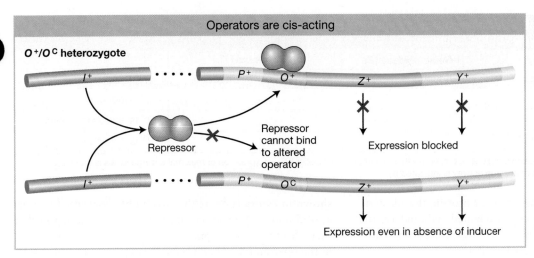

Operators are cis-acting

O^+/O^C **heterozygote**

Repressor

Repressor cannot bind to altered operator

Expression blocked

Expression even in absence of inducer

FIGURE 11-8 O^+/O^C heterozygotes demonstrate that operators are cis-acting. Because a repressor cannot bind to O^C operators, the *lac* structural genes linked to an O^C operator are expressed even in the absence of an inducer. However, the *lac* genes adjacent to an O^+ operator are still subject to repression.

ANIMATED ART 🌊 Sapling Plus

O^C *lac* operator mutations

TABLE 11-2 Synthesis of β-Galactosidase and Permease in Haploid and Heterozygous Diploid Strains Carrying I^+ and I^-

Strain	Genotype	β-*Galactosidase* (Z)		Permease (Y)		Conclusion
		Noninduced	Induced	Noninduced	Induced	
1	$I^+ Z^+ Y^+$	−	+	−	+	I^+ is inducible
2	$I^- Z^+ Y^+$	+	+	+	+	I^- is constitutive
3	$I^+ Z^- Y^+/F' I^- Z^+ Y^+$	−	+	−	+	I^+ is dominant to I^-
4	$I^- Z^- Y^+/F' I^+ Z^+ Y^-$	−	+	−	+	I^+ is trans-acting

Note: Bacteria were grown in glycerol (no glucose present) with and without the inducer IPTG. Expression of maximal enzyme levels is indicated by +. Absence or very low levels of enzyme activity is indicated by −. All strains are O^+.

KEY CONCEPT Operator mutations reveal that such a site is cis-acting; that is, it regulates the expression of an adjacent transcription unit on the same DNA molecule. In contrast, mutations in the gene encoding a repressor protein reveal that this protein is trans-acting; that is, it can act on any copy of the target DNA.

Genetic evidence for allostery

Finally, Jacob and Monod were able to demonstrate allostery through the analysis of another class of repressor mutations. Recall that the Lac repressor inhibits transcription of the *lac* operon in the absence of an inducer but permits transcription when the inducer is present. This regulation is accomplished

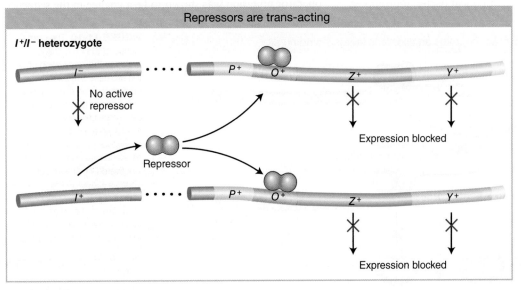

Repressors are trans-acting

I^+/I^- **heterozygote**

No active repressor

Repressor

Expression blocked

Expression blocked

Expression blocked

FIGURE 11-9 The recessive nature of I^- mutations demonstrates that the repressor is trans-acting. Although no active repressor is synthesized from the I^- gene, the wild-type (I^+) gene provides a functional repressor that binds to both operators in a diploid cell and blocks *lac* operon expression (in the absence of an inducer).

ANIMATED ART 🌊 Sapling Plus

I^- Lac repressor mutations

TABLE 11-3   Synthesis of β-Galactosidase and Permease by the Wild Type and by Strains Carrying Different Alleles of the *I* Gene

Strain	Genotype	β-*Galactosidase* (*Z*)		Permease (Y)		Conclusion
		Noninduced	Induced	Noninduced	Induced	
1	$I^+ Z^+ Y^+$	−	+	−	+	I^+ is inducible
2	$I^S Z^+ Y^+$	−	−	−	−	I^S is always repressed
3	$I^S Z^+ Y^+ / F' I^+ Z^+ Y^+$	−	−	−	−	I^S is dominant to I^+

Note: Bacteria were grown in glycerol (no glucose present) with and without the inducer IPTG. Expression of maximal enzyme levels is indicated by +. Absence or very low levels of enzyme activity is indicated by −.

through a second site on the repressor protein, the allosteric site, which binds to the inducer. When bound to the inducer, the repressor undergoes a change in overall structure such that its DNA-binding site can no longer function.

Jacob and Monod isolated another class of repressor mutation, called superrepressor (I^S) mutations. I^S mutations cause repression to persist even in the presence of an inducer (compare strain 2 in **Table 11-3** with the inducible wild-type strain 1). Unlike I^- mutations, I^S mutations are dominant over I^+ (see Table 11-3, strain 3). This key observation led Jacob and Monod to speculate that I^S mutations alter the allosteric site so that it can no longer bind to an inducer. As a consequence, I^S-encoded repressor protein continually binds to the operator—preventing transcription of the *lac* operon even when the inducer is present in the cell. On this basis, we can see why I^S is dominant over I^+. Mutant I^S protein will bind to both copies of the operator in the partial diploid cell, even in the presence of an inducer and regardless of the fact that I^+-encoded protein may be present in the same cell (**Figure 11-10**).

Genetic analysis of the *lac* promoter

Mutational analysis also demonstrated that an element essential for *lac* transcription is located between the gene for the repressor *I* and the operator site *O*. This element, termed the *promoter* (*P*), serves as the initiation site for transcription by RNA polymerase (see Chapter 8). There are two binding regions for RNA polymerase in a typical bacterial promoter,

shown in **Figure 11-11** as the two highly conserved regions at −35 and −10. Promoter mutations are cis-acting in that they affect the transcription of all adjacent structural genes in the operon. Like operators and other cis-acting elements, promoters are sites on the DNA molecule that are bound by proteins and themselves produce no protein product.

Molecular characterization of the Lac repressor and the *lac* operator

Walter Gilbert and Benno Müller-Hill provided a decisive demonstration of the *lac* system in 1966 by monitoring the binding of the radioactively labeled inducer IPTG to purified repressor protein. They showed that in the test tube, repressor protein binds to DNA containing the operator and comes off the DNA in the presence of IPTG. (A more detailed description of how the repressor and other DNA-binding proteins work is given later, at the end of Section 11.6.)

Gilbert and his co-workers showed that the repressor can protect specific bases in the operator from chemical reagents. This information allowed them to isolate the DNA segment constituting the operator and to determine its sequence. They took operon DNA to which repressor was bound and treated it with the enzyme DNase, which breaks up DNA. They were able to recover short DNA strands that had been shielded from the enzyme activity by the repressor molecule. These short strands presumably constituted the operator sequence. The base sequence of each strand was determined, and each operator mutation was shown to be a change in the sequence

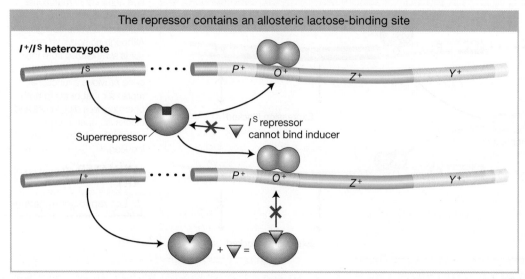

The repressor contains an allosteric lactose-binding site

I^+/I^S heterozygote

I^S

P^+ O^+ Z^+ Y^+

Superrepressor

I^S repressor cannot bind inducer

I^+

P^+ O^+ Z^+ Y^+

FIGURE 11-10 The dominance of the I^S mutation is due to the inactivation of the allosteric site on the Lac repressor. In an I^S/I^+ diploid cell, none of the *lac* structural genes are transcribed. The I^S repressor lacks a functional allolactose-binding site (the allosteric site) and thus is not inactivated by an inducer. Therefore, even in the presence of an inducer, the I^S repressor binds irreversibly to all operators in a cell, thereby blocking transcription of the *lac* operon.

ANIMATED ART   SaplingPlus

I^S Lac superrepressor mutations

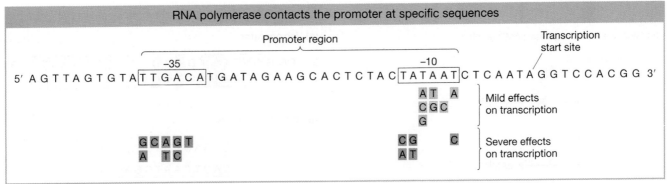

FIGURE 11-11 Specific DNA sequences are important for the efficient transcription of *E. coli* genes by RNA polymerase. Only the coding strand (non-template strand) is shown here (see Figure 8-5). Transcription would proceed from left to right (5′ to 3′), and the mRNA transcript would be homologous to the sequence shown. The boxed sequences are highly conserved in all *E. coli* promoters, an indication of their role as contact sites on the DNA for RNA polymerase binding. Mutations in these regions have mild (gold) and severe (brown) effects on transcription. [*Data from J. D. Watson, M. Gilman, J. Witkowski, and M. Zoller, Recombinant DNA, 2nd ed.*]

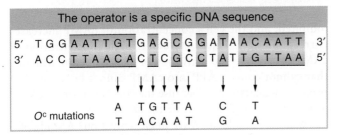

FIGURE 11-12 The DNA base sequence of the lactose operator and the base changes associated with eight O^C mutations. Regions of twofold rotational symmetry are indicated by color and by a dot at their axis of symmetry. [*Data from W. Gilbert, A. Maxam, and A. Mirzabekov, in N. O. Kjeldgaard and O. Malløe, eds., Control of Ribosome Synthesis. Academic Press, 1976.*]

(Figure 11-12). These results showed that the operator locus is a specific sequence of 17 to 25 nucleotides situated just before (5′ to) the structural *Z* gene. They also showed the incredible specificity of repressor–operator recognition, which can be disrupted by a single base substitution. When the sequence of bases in the *lac* mRNA (transcribed from the *lac* operon) was determined, the first 21 bases on the 5′ initiation end proved to be complementary to the operator sequence that Gilbert had determined, showing that the operator sequence is transcribed.

The results of these experiments provided crucial confirmation of the mechanism of repressor action formulated by Jacob and Monod.

> **KEY CONCEPT** The function of cis-acting elements such as promoters and operators is determined by their DNA sequence.

11.3 CATABOLITE REPRESSION OF THE *LAC* OPERON: POSITIVE REGULATION

> **LO 11.1** Illustrate how both positive and negative regulation control the activity of the *lac* operon.

Through a long evolutionary process, the existing *lac* system has been selected to operate for the optimal energy efficiency of the bacterial cell. Presumably to maximize energy efficiency, two environmental conditions have to be satisfied for the lactose metabolic enzymes to be expressed.

One condition is that lactose must be present in the environment. This condition makes sense because it would be inefficient for the cell to produce the lactose metabolic enzymes if there is no lactose to metabolize. We have already seen that the cell is able to respond to the presence of lactose through the action of a repressor protein.

The other condition is that glucose cannot be present in the cell's environment. Because the cell can capture more energy from the breakdown of glucose than it can from the breakdown of other sugars, it is more efficient for the cell to metabolize glucose rather than lactose. Thus, mechanisms have evolved that prevent the cell from synthesizing the enzymes for lactose metabolism when both lactose and glucose are present together. The repression of the transcription of lactose-metabolizing genes in the presence of glucose is an example of **catabolite repression** (glucose is a breakdown product, or a **catabolite**, of lactose). The transcription of genes encoding proteins necessary for the metabolism of many different sugars is similarly repressed in the presence of glucose. We will see that catabolite repression works through an *activator protein*.

The basics of *lac* catabolite repression: choosing the best sugar to metabolize

If both lactose and glucose are present, the synthesis of β-galactosidase is not induced until all the glucose has been metabolized. Thus, the cell conserves its energy by metabolizing any existing glucose before going through the energy-expensive process of creating new machinery to metabolize lactose. There are multiple mechanisms that bacteria have evolved to ensure the preferential use of a carbon source and optimal growth. One mechanism is to exclude lactose from the cell. A second mechanism is to regulate operon expression via catabolites.

The results of studies indicate that a breakdown product of glucose prevents activation of the *lac* operon by

Glucose levels control the *lac* operon

(a) Glucose levels regulate cAMP levels

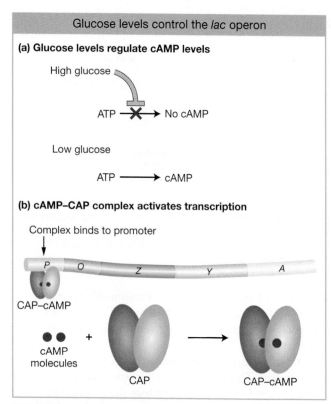

(b) cAMP–CAP complex activates transcription

Complex binds to promoter

FIGURE 11-13 Catabolite control of the *lac* operon. (a) Only under conditions of low glucose is cAMP (cyclic adenosine monophosphate) formed from ATP. (b) When cAMP is present, it forms a complex with CAP (catabolite activator protein) that activates transcription by binding to a region within the *lac* promoter.

Many DNA binding sites are symmetrical

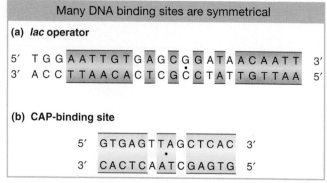

(a) *lac* operator

```
5′  T G G A A T T G T G A G C G G A T A A C A A T T  3′
3′  A C C T T A A C A C T C G C C T A T T G T T A A  5′
```

(b) CAP-binding site

```
5′  G T G A G T T A G C T C A C  3′
3′  C A C T C A A T C G A G T G  5′
```

FIGURE 11-14 The DNA base sequences of (a) the *lac* operator, to which the Lac repressor binds, and (b) the CAP-binding site, to which the CAP–cAMP complex binds. Sequences exhibiting twofold rotational symmetry are indicated by the colored boxes and by a dot at the center point of symmetry. [(a) Data from W. Gilbert, A. Maxam, and A. Mirzabekov, in N. O. Kjeldgaard and O. Malløe, eds., Control of Ribosome Synthesis. *Academic Press, 1976.*]

inhibited, and the cell's concentration of cAMP increases correspondingly (**Figure 11-13a**). A high concentration of cAMP is necessary for activation of the *lac* operon. Mutants that cannot convert ATP into cAMP cannot be induced to produce β-galactosidase because the concentration of cAMP is not great enough to activate the *lac* operon.

What is the role of cAMP in *lac* activation? A study of a different set of mutants provided an answer. These mutants make cAMP but cannot activate the Lac enzymes because they lack yet another protein, called **catabolite activator protein (CAP)**, encoded by the *crp* gene. CAP binds to a specific DNA sequence of the *lac* operon (the CAP-binding site; see **Figure 11-14b**). The DNA-bound CAP is then able to interact physically with RNA polymerase and increases that enzyme's affinity for the *lac* promoter. By itself, CAP cannot bind to the CAP-binding site of the *lac* operon. However, by binding to cAMP, its allosteric effector, CAP is able to bind to the CAP-binding site and activate transcription by RNA polymerase (Figure 11-13b). By inhibiting CAP when glucose is available, the catabolite-repression system ensures that the *lac* operon will be activated only when glucose is scarce.

lactose—the catabolite repression just mentioned. The glucose breakdown product is known to modulate the level of an important cellular constituent—**cyclic adenosine monophosphate (cAMP)**, which is synthesized from the major energy source within the cell: adenosine triphosphate (ATP). When glucose is present in high concentrations, it inhibits the conversion of ATP to cAMP, so the cell's cAMP concentration is low. As the glucose concentration decreases, the conversion of ATP to cAMP is no longer

CAP and RNA polymerase bind next to each other

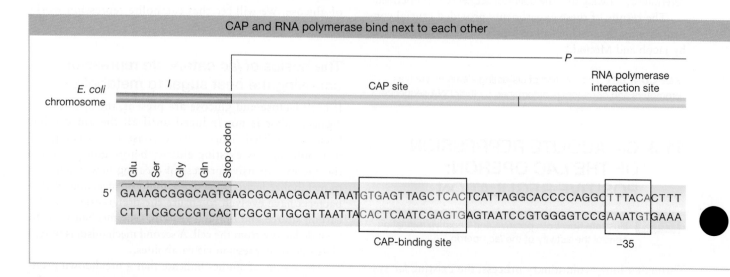

KEY CONCEPT Operons that metabolize a nutrient, such as the *lac* operon, often have an added level of control so that the operon is inactive in the presence of its catabolic breakdown product (i.e., glucose) even if the nutrient (i.e., lactose) is present.

The structures of target DNA sites

The DNA sequences to which the CAP–cAMP complex binds (see Figure 11-14) are different from the sequences to which the Lac repressor binds. These differences underlie the specificity of DNA binding by these very different regulatory proteins. One property that these sequences do have in common and that is common to many other DNA-binding sites is rotational twofold symmetry. In other words, if we rotate the DNA sequence shown in Figure 11-14 by 180 degrees within the plane of the page, the sequence of the highlighted bases of the binding sites will be identical. The highlighted bases are thought to constitute the important contact sites for protein–DNA interactions. This rotational symmetry corresponds to symmetries within the DNA-binding proteins, many of which are composed of two or four identical subunits. We will consider the structures of some DNA-binding proteins later in the chapter.

How does the binding of the cAMP–CAP complex to the operon further the binding of RNA polymerase to the *lac* promoter? In **Figure 11-15**, the DNA is shown as being bent when CAP is bound. This bending of DNA may aid the binding of RNA polymerase to the promoter. There is also evidence that CAP makes direct contact with RNA polymerase. The base sequence shows that CAP and RNA polymerase bind directly adjacent to each other on the *lac* promoter (**Figure 11-16**).

KEY CONCEPT Generalizing from the *lac* operon model, regulatory proteins bind to DNA in the operator sites in the operons that they control. The exact pattern of binding in an operon will depend on physiological signals and whether activators or repressors regulate particular operons.

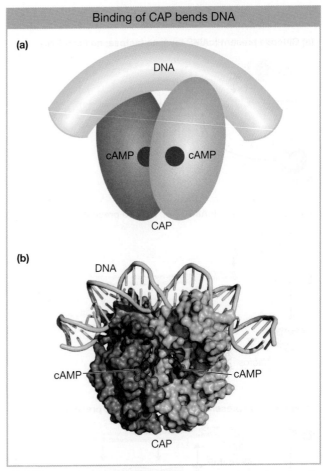

Binding of CAP bends DNA

(a)

(b)

FIGURE 11-15 (a) When CAP binds the promoter, it creates a bend greater than 90 degrees in the DNA. (b) Image derived from the structural analysis of two subunits of CAP bound to the CAP-binding site. [*(b)* PDB ID *1cgp*.]

FIGURE 11-16 The control region of the *lac* operon. The base sequence and the genetic boundaries of the control region of the *lac* operon, with partial sequences for the structural genes. Note that the *lac* operon promoter sequences at the −35 and the −10 sites differ from the consensus in Figure 11-11. [*Data from R. C. Dickson, J. Abelson, W. M. Barnes, and W. S. Reznikoff, "Genetic Regulation: The Lac Control Region," Science 187, 1975, 27.*]

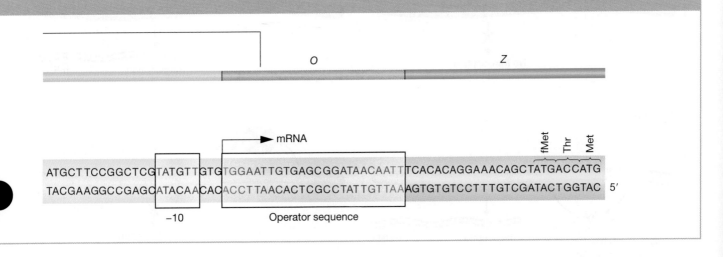

Negative and positive regulation of the *lac* operon

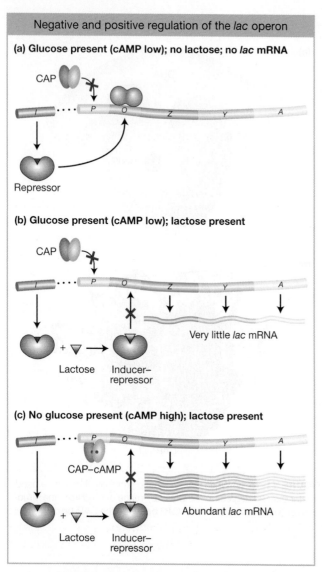

(a) Glucose present (cAMP low); no lactose; no *lac* mRNA

CAP

P   O   Z   Y   A

Repressor

(b) Glucose present (cAMP low); lactose present

CAP

I   P   O   Z   Y   A

Very little *lac* mRNA

Lactose   Inducer–repressor

(c) No glucose present (cAMP high); lactose present

I   P   O   Z   Y   A

CAP–cAMP

Abundant *lac* mRNA

Lactose   Inducer–repressor

FIGURE 11-17 The *lac* operon is controlled jointly by the Lac repressor (negative regulator) and the catabolite activator protein (CAP; positive regulator). Large amounts of mRNA are produced only when lactose is present to inactivate the repressor, and low glucose levels promote the formation of the CAP–cAMP complex, which positively regulates transcription.

A summary of the *lac* operon

We can now fit the CAP–cAMP- and RNA-polymerase-binding sites into the detailed model of the *lac* operon, as shown in **Figure 11-17**. The presence of glucose prevents lactose metabolism because a glucose breakdown product inhibits maintenance of the high cAMP levels necessary for formation of the CAP–cAMP complex, which in turn is required for the RNA polymerase to attach at the *lac* promoter site (see Figure 11-17a, b). Even when there is a shortage of glucose catabolites and CAP–cAMP forms, the mechanism for lactose metabolism will be implemented only if lactose is present (see Figure 11-17c). Only two or three molecules of β-galactosidase are present per cell in the absence of lactose or in the presence of lactose and glucose. These few molecules of β-galactosidase are likely due to a very low level of spurious transcription that results because the repressor can briefly dissociate from the DNA.

FIGURE 11-18 (a) In repression, an active repressor (encoded by the *R* gene in this example) blocks expression of the *A, B, C* operon by binding to an operator site (*O*). (b) In activation, a functional activator is required for gene expression. A nonfunctional activator results in no expression of genes *X, Y, Z*. Small molecules can convert a nonfunctional activator into a functional one that then binds to the control region of the operon, termed *I* in this case. The positions of both *O* and *I* with respect to the promoter *P* in the two examples are arbitrarily drawn, because their positions differ in different operons.

Repression and activation compared

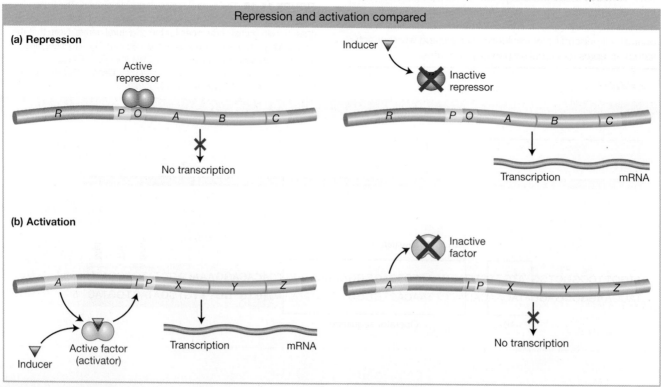

(a) Repression

Active repressor

R   P   O   A   B   C

No transcription

Inducer

Inactive repressor

R   P   O   A   B   C

Transcription   mRNA

(b) Activation

A   I   P   X   Y   Z

Transcription   mRNA

Inducer   Active factor (activator)

Inactive factor

A   I   P   X   Y   Z

No transcription

This increases to approximately 3000 molecules of enzyme when lactose is present and glucose is absent. Thus, the cell conserves its energy and resources by producing the lactose-metabolizing enzymes only when they are both needed and useful.

Inducer–repressor control of the *lac* operon is an example of repression, or negative regulation, in which expression is normally blocked. In contrast, the CAP–cAMP system is an example of activation, or positive regulation, because it acts as a signal that activates expression—in this case, the activating signal is the interaction of the CAP–cAMP complex with the CAP-binding site on DNA. **Figure 11-18** outlines these two basic types of control systems.

KEY CONCEPT Negative regulation promotes gene expression in the absence of the repressor, and positive regulation promotes gene expression in the presence of an activator.

11.4 DUAL POSITIVE AND NEGATIVE REGULATION: THE ARABINOSE OPERON

LO 11.3 Illustrate and compare the mechanisms that coordinate expression of sets of genes in bacteria and bacteriophage.

As with the *lac* system, the control of transcription in bacteria is neither purely positive nor purely negative; rather, both positive and negative regulation may govern individual operons. The regulation of the arabinose operon provides an example in which a single DNA-binding protein may act as *either* a repressor *or* an activator—a twist on the general theme of transcriptional regulation by DNA-binding proteins.

The structural genes *araB*, *araA*, and *araD* encode the metabolic enzymes that break down the sugar arabinose. The three genes are transcribed in a unit as a single mRNA. **Figure 11-19** shows a map of the *ara* operon. Transcription is activated at *araI*, the **initiator** region, which contains a

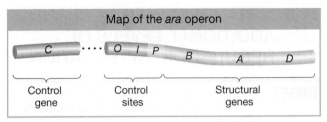

Map of the *ara* operon

FIGURE 11-19 The *B*, *A*, and *D* genes together with the *I* and *O* sites constitute the *ara* operon. *O* is *araO* and *I* is *araI*.

binding site for an activator protein. The *araC* gene, which maps nearby, encodes an activator protein. When bound to arabinose, this protein binds to the *araI* site and activates transcription of the *ara* operon, perhaps by helping RNA polymerase bind to the promoter. In addition, the same CAP–cAMP catabolite repression system that prevents *lac* operon expression in the presence of glucose also prevents expression of the *ara* operon.

In the presence of arabinose, both the CAP–cAMP complex and the AraC–arabinose complex must bind to *araI* in order for RNA polymerase to bind to the promoter and transcribe the *ara* operon (**Figure 11-20a**). In the absence of arabinose, the AraC protein assumes a different conformation and represses the *ara* operon by binding both to *araI* and to a second distant site, *araO*, thereby forming a loop (Figure 11-20b) that prevents transcription. Thus, the AraC protein has two conformations, one that acts as an activator and another that acts as a repressor. The on/off switch of the operon is "thrown" by arabinose. The two conformations, dependent on whether the allosteric effector arabinose has bound to the protein, differ in their abilities to bind a specific target site in the *araO* region of the operon.

KEY CONCEPT Operon transcription is commonly regulated by both activation and repression. However, the specific mechanisms regulating the expression of operons that control the metabolism of similar compounds, such as sugars, can be quite different.

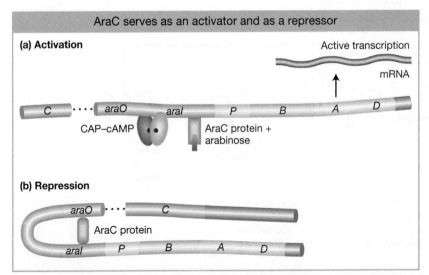

AraC serves as an activator and as a repressor

(a) Activation

(b) Repression

FIGURE 11-20 Dual control of the *ara* operon. (a) In the presence of arabinose, the AraC protein binds to the *araI* region. The CAP–cAMP complex binds to a site adjacent to *araI*. This binding stimulates the transcription of the *araB*, *araA*, and *araD* genes. (b) In the absence of arabinose, the AraC protein binds to both the *araI* and the *araO* regions, forming a DNA loop. This binding prevents transcription of the *ara* operon.

11.5 METABOLIC PATHWAYS AND ADDITIONAL LEVELS OF REGULATION: ATTENUATION

LO 11.3 Illustrate and compare the mechanisms that coordinate expression of sets of genes in bacteria and bacteriophage.

Coordinate control of genes in bacteria is widespread. Just as we saw in the preceding sections, there is a need for the cell to regulate pathways for the breakdown of specific sugars, depending upon the availability of that sugar. Similarly, the pathways that synthesize essential molecules, like amino acids, must be regulated so that the enzymes needed for their synthesis are produced by the bacteria only when amino acids are not available from the environment. In pathways that synthesize essential molecules, the genes that encode the enzymes are also organized into operons, complete with multigenic mRNAs. Furthermore, in cases for which the sequence of catalytic activity is known, there is a remarkable congruence between the order of operon genes on the chromosome and the order in which their products act in the metabolic pathway. This congruence is strikingly illustrated by the organization of the tryptophan operon in *E. coli* (**Figure 11-21**). The tryptophan operon contains five genes (*trpE*, *trpD*, *trpC*, *trpB*, *trpA*) that encode enzymes that contribute to the synthesis of the amino acid tryptophan.

KEY CONCEPT In bacteria, genes that encode enzymes that are in the same metabolic pathways are generally organized into operons.

There are two mechanisms for regulating transcription of the tryptophan operon and some other operons functioning in amino acid biosynthesis. One provides global control of operon mRNA expression, and the other provides fine-tuned control.

The level of *trp* operon gene expression is governed by the level of tryptophan. When tryptophan is absent from the growth medium, *trp* gene expression is high; when levels of tryptophan are high, the *trp* operon is repressed. One mechanism for controlling the transcription of the *trp* operon is similar to the mechanism of negative regulation that we have already seen controls the *lac* operon: a repressor protein binds an operator, preventing the initiation of transcription. This repressor is the Trp repressor, the product of the *trpR* gene. The Trp repressor binds tryptophan when adequate levels of the amino acid are present, and only after binding tryptophan will the Trp repressor bind to the operator and switch off transcription of the operon. This simple mechanism ensures that the cell does not waste energy producing tryptophan when the amino acid is sufficiently abundant in the environment. *E. coli* strains with mutations in *trpR* continue to express the *trp* mRNA and thus continue to produce tryptophan when the amino acid is abundant.

In studying these *trpR* mutant strains, Charles Yanofsky discovered that, when tryptophan was removed from the medium, the production of *trp* mRNA further increased several-fold. This finding was evidence that, in addition to the Trp repressor, a second control mechanism existed to negatively regulate transcription. This mechanism is called **attenuation** because mRNA production is normally *attenuated*, meaning "decreased," when tryptophan is plentiful. Unlike the other bacterial control mechanisms described thus far, attenuation acts at a step *after* transcription initiation.

The mechanisms governing attenuation were discovered by identifying mutations that reduced or abolished attenuation. Strains with these mutations produce *trp* mRNA at maximal levels even in the presence of tryptophan. Yanofsky mapped the mutations to a region between the *trp* operator and the *trpE* gene; this region, termed the **leader sequence**, is at the 5′ end of the *trp* operon mRNA before the first codon of the *trpE* gene (**Figure 11-22**). The *trp* leader sequence is unusually long for a bacterial mRNA, 160 bases, and detailed analyses have revealed how a part of this sequence works as an **attenuator** that governs the further transcription of *trp* mRNA.

The key observations are that, in the absence of the TrpR repressor protein, the presence of tryptophan halts

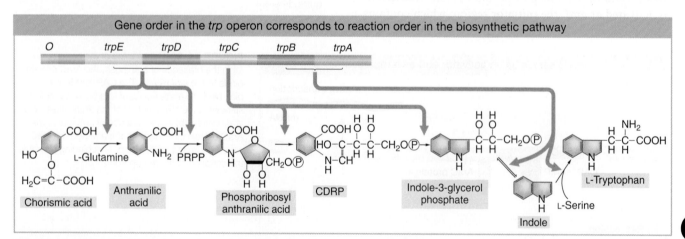

FIGURE 11-21 The chromosomal order of genes in the *trp* operon of *E. coli* and the sequence of reactions catalyzed by the enzyme products of the *trp* structural genes. The products of genes *trpD* and *trpE* form a complex that catalyzes specific steps, as do the products of genes *trpB* and *trpA*. Abbreviations: PRPP, phosphoribosylpyrophosphate; CDRP, 1-(*o*-carboxyphenylamino)-1-deoxyribulose 5-phosphate.

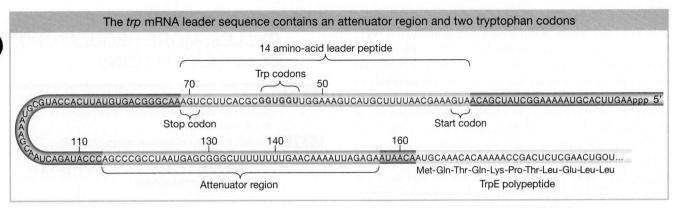

FIGURE 11-22 In the *trp* mRNA leader sequence, the attenuator region precedes the *trpE* coding sequence. Farther upstream, at bases 54 through 59, are the two tryptophan codons (shown in red) found in the 14-amino-acid leader peptide.

transcription after the first 140 bases or so, whereas, in the absence of tryptophan, transcription of the operon continues. The mechanism for terminating or continuing transcription consists of two key elements. First, the *trp* mRNA leader sequence encodes a short, 14-amino-acid peptide that includes two adjacent tryptophan codons. Tryptophan is one of the least abundant amino acids in proteins, and it is encoded by a single codon. This pair of tryptophan codons is therefore an unusual feature. Second, the *trp* mRNA leader sequence consists of four segments that form stem-and-loop RNA structures that are able to alternate between two conformations. One of these conformations favors the termination of transcription, while the other favors the continuation of transcription (Figure 11-23).

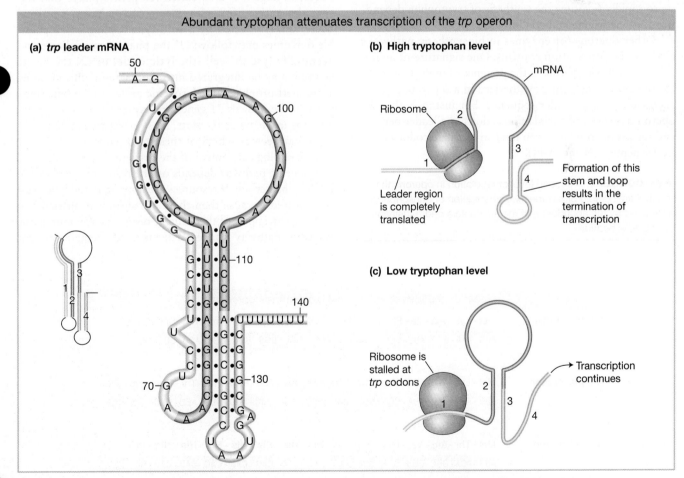

FIGURE 11-23 (a) Proposed secondary structures in the conformation of *trp* leader mRNA that favors termination of transcription. Four regions can base pair to form three stem-and-loop structures, but only two regions base pair with one another at a given time. (b) When tryptophan is abundant, segment 1 of the *trp* mRNA is translated. Segment 2 enters the ribosome, enabling segments 3 and 4 to base pair to form a stem-loop that causes RNA polymerase to terminate transcription. (c) In contrast, when tryptophan is scarce, the ribosome is stalled at the codons of segment 1. Thus, segment 2 can interact with segment 3, and so segments 3 and 4 cannot pair. Consequently, transcription continues. [*Data from D. L. Oxender, G. Zurawski, and C. Yanofsky, Proc. Natl. Acad. Sci. USA 76, 1979, 5524.*]

The regulatory logic of the operon pivots on the abundance of tryptophan. When tryptophan is abundant, there is a sufficient supply of aminoacyl-tRNA^{Trp} to allow translation of the 14-amino-acid leader peptide. Recall that transcription and translation in bacteria are coupled; so ribosomes can engage mRNA transcripts and initiate translation before transcription is complete. The engagement of the ribosome alters *trp* mRNA conformation to the form that favors termination of transcription. Because segment 1 of the *trp* leader mRNA is translated in the presence of tryptophan, segment 2 of the leader mRNA will enter the ribosome. This allows base pairing between the attenuator region found in segments 3 and 4 of the leader mRNA, which leads to the termination of transcription by RNA polymerase (Figure 11-23b). However, when tryptophan is scarce, the ribosome is stalled at the tryptophan codons in segment 1, such that segments 2 and 3 base pair, and transcription is able to continue (Figure 11-23c). This mechanism is exquisitely sensitive to the levels of tryptophan in the environment because the number of transcripts produced will be directly related to the number of stalled ribosomes, which in turn is related to the amount of tryptophan in the cell. Thus, attenuation provides a way for the bacteria to rapidly fine-tune the synthesis of tryptophan depending upon environmental conditions.

Other operons for enzymes in biosynthetic pathways have similar attenuation controls. One signature of amino acid biosynthesis operons is the presence of multiple codons for the amino acid being synthesized in a separate peptide encoded by the 5′ leader sequence. For instance, the *phe* operon has seven phenylalanine codons in a leader peptide and the *his* operon has seven tandem histidine codons in its leader peptide (Figure 11-24).

KEY CONCEPT A second level of regulation in operons that control amino acid biosynthesis is attenuation of transcription mediated by the abundance of the amino acid and translation of a leader peptide.

11.6 BACTERIOPHAGE LIFE CYCLES: MORE REGULATORS, COMPLEX OPERONS

LO 11.3 Illustrate and compare the mechanisms that coordinate expression of sets of genes in bacteria and bacteriophage.

LO 11.4 Explain the roles of sequence-specific DNA-binding proteins and DNA regulatory sequences in coordinating the expression of sets of genes in bacteria and bacteriophage.

In that Paris movie theater, François Jacob had a flash of insight that the phenomenon of prophage induction might be closely analogous to the induction of β-galactosidase synthesis. He was right. Here, we are going to see how the life cycle of the bacteriophage λ is regulated. Although its regulation is more complex than that of individual operons, it is controlled by now-familiar modes of gene regulation.

Regulation of the bacteriophage λ life cycle

Bacteriophage λ is a so-called temperate phage that has two alternative life cycles (**Figure 11-25**). When a normal bacterium is infected by a wild-type λ phage, two possible outcomes may follow: (1) the phage may replicate and eventually lyse the cell (the **lytic cycle**) or (2) the phage genome may be integrated into the bacterial chromosome as an inert prophage (the **lysogenic cycle**). In the lytic state, most of the phage's 71 genes are expressed at some point, whereas in the lysogenic state, most genes are inactive.

What decides which of these two pathways is taken? The physiological control of the decision between the lytic or lysogenic pathway depends on the resources available in the host bacterium. If resources are abundant, the lytic cycle is preferred because then there are sufficient nutrients to make many copies of the virus. If resources are limited, the lysogenic pathway is taken. The virus then remains present

Leader peptides of amino acid biosynthesis operons
(a) *trp* operon   Met - Lys - Ala - Ile - Phe - Val - Leu - Lys - Gly - **Trp** - **Trp** - Arg - Thr - Ser - Stop
5′ AUG - AAA - GCA - AUU - UUC - GUA - CUG - AAA - GGU - UGG - UGG - CGC - ACU - UCC - UGA   3′
(b) *phe* operon   Met - Lys - His - Ile - Pro - **Phe** - **Phe** - **Phe** - Ala - **Phe** - **Phe** - **Phe** - Thr - **Phe** - Pro - Stop
5′ AUG - AAA - CAC - AUA - CCG - UUU - UUU - UUC - GCA - UUC - UUU - UUU - ACC - UUC - CCC - UGA   3′
(c) *his* operon   Met - Thr - Arg - Val - Gln - Phe - Lys - **His** - **His** - **His** - **His** - **His** - **His** - **His** - Pro - Asp
5′ AUG - ACA - CGC - GUU - CAA - UUU - AAA - CAC - CAC - CAU - CAU - CAC - CAU - CAU - CCU - GAC   3′

FIGURE 11-24 (a) The translated part of the *trp* leader region contains two consecutive tryptophan codons, (b) the *phe* leader sequence contains seven phenylalanine codons, and (c) the *his* leader sequence contains seven consecutive histidine codons.

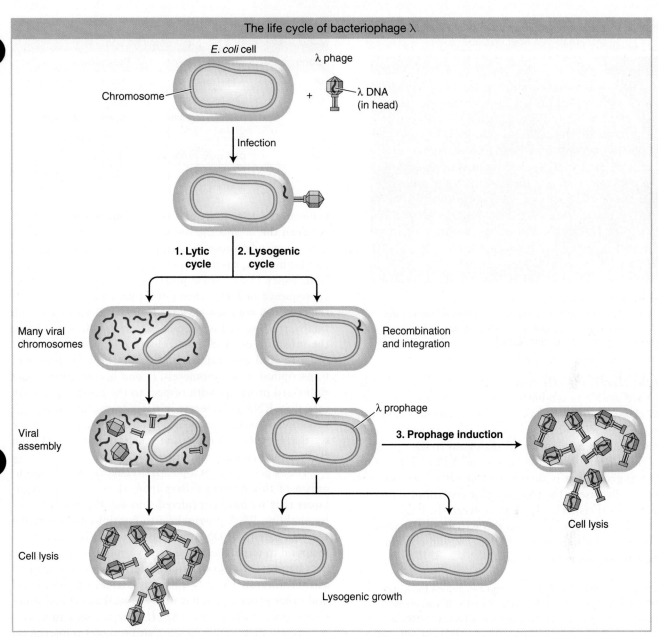

FIGURE 11-25 Whether bacteriophage λ enters the lytic cycle immediately or enters the lysogenic cycle depends on the availability of resources. The lysogenic virus inserts its genome into the bacterial chromosome, where it remains quiescent until conditions are favorable.

as a *prophage* until conditions improve. The inert prophage can be induced by ultraviolet light to enter the lytic cycle—the phenomenon studied by Jacob. The lytic and lysogenic states are characterized by very distinct programs of gene expression that must be regulated. Which alternative state is selected is determined by a complex genetic switch comprising several DNA-binding regulatory proteins and a set of operator sites.

Just as they were for the *lac* and other regulatory systems, genetic analyses of mutants were sources of crucial insights into the components and logic of the λ genetic switch. Jacob used simple phenotypic screens to isolate mutants that were defective in either the lytic or the lysogenic pathway. Mutants of each type could be recognized by the appearance of infected plaques on a lawn of bacteria. When wild-type phage particles are placed on a lawn of sensitive bacteria, clearings (called "plaques") appear where bacteria are infected and lysed, but these plaques are cloudy because bacteria that are lysogenized grow within them (**Figure 11-26**). Mutant phages that are unable to lysogenize cells form clear plaques.

Such *clear* mutants (designated by *c*) turn out to be analogous to the *I* and *O* mutants of the *lac* system. These mutants were often isolated as temperature-sensitive mutants that had *clear* phenotypes at higher temperatures but wild-type phenotypes at lower temperatures. Three classes of mutants led to the identification of the key regulatory features of phage λ. In the first class, mutants

Clear and cloudy bacteriophage plaques on a lawn
of *E. coli* host bacteria

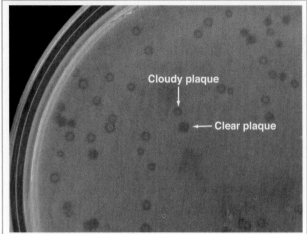

Cloudy plaque

Clear plaque

FIGURE 11-26 Plaques are clear where host cell lysis has occurred; they are cloudy where cells have survived infection and continued to grow as a lysogen. [*From* Microbiology: An Evolving Science 1e, Figure 10.22. © John Foster.]

TABLE 11-4	Major Regulators of Bacteriophage λ Life Cycle	
Gene	**Protein**	**Promotes**
cI	λ repressor	lysogenic pathway
cro	Cro repressor	lytic pathway
N	positive regulator	cII, cIII expression
cII	activator	cI expression
cIII	protease inhibitor	cII activity

for the *cI*, *cII*, and *cIII* genes form clear plaques; that is, they are unable to establish lysogeny. Mutants in the second class were isolated that do not lysogenize cells but can replicate and enter the lytic cycle in a lysogenized cell. These mutants turn out to be analogous to the operator-constitutive mutants of the *lac* system. A third key mutant can lysogenize but is unable to lyse cells. The mutated gene in this case is the *cro* gene (for *c*ontrol of *r*epressor and *o*ther things). The decision between the lytic and the lysogenic pathways hinges on the activity of the proteins encoded by the four genes *cI*, *cII*, *cIII*, and *cro*, three of which are DNA-binding proteins.

KEY CONCEPT Wild-type λ phage can induce the lysogenic cycle in *E. coli*, resulting in cloudy plaques. Because clear plaques indicate that only the lytic cycle has occurred, a genetic screen for clear plaques can be used to identify mutations in genes required for the lysogenic cycle of λ phage.

We will first focus on the two genes *cI* and *cro* and the proteins that they encode (**Table 11-4**). The *cI* gene encodes a repressor, often referred to as λ repressor, which represses lytic growth and promotes lysogeny. The *cro* gene encodes a repressor that represses lysogeny, thereby permitting lytic growth. The genetic switch controlling the two λ phage life cycles has two states: in the lysogenic state, *cI* is on but *cro* is off, and in the lytic cycle, *cro* is on but *cI* is off. Therefore, λ repressor and Cro are in competition, and whichever repressor prevails determines the state of the switch and of the expression of the λ genome.

The race between λ repressor and Cro is initiated when phage λ infects a normal bacterium. The sequence of events in the race is critically determined by the organization of

genes in the λ genome and of promoters and operators between the *cI* and the *cro* genes. The roughly 50-kb λ genome encodes proteins having roles in DNA replication, recombination, assembly of the phage particle, and cell lysis (**Figure 11-27**). These proteins are expressed in a logical sequence such that copies of the genome are made first, these copies are then packaged into viral particles, and, finally, the host cell is lysed to release the virus and begin the infection of other host cells (see Figure 11-25). The order of viral gene expression flows from the initiation of transcription at two promoters, P_L and P_R (for *l*eftward and *r*ightward promoter with respect to the genetic map). On infection, RNA polymerase initiates transcription at both promoters. Looking at the genetic map (Figure 11-27), we see that from P_R, *cro* is the first gene transcribed, and from P_L, *N* is the first gene transcribed.

The *N* gene encodes a positive regulator, but the mechanism of this protein differs from those of other regulators that we have considered thus far. Protein N works by enabling RNA polymerase to continue to transcribe through regions of DNA that would otherwise cause transcription to terminate. A regulatory protein such as N that acts by preventing transcription termination is called an **antiterminator**. Thus, N allows the transcription of *cIII* and other genes to the left of *N*, as well as *cII* and other genes to the right of *cro*. The *cII* gene encodes an activator protein that binds to a site that promotes transcription leftward from a different promoter, P_{RE} (for *p*romoter of *r*epressor *e*stablishment), which activates transcription of the *cI* gene. Recall that the *cI* gene encodes λ repressor, which will prevent lytic growth.

Before the expression of the rest of the viral genes takes place, a "decision" must be made—whether to continue with viral-gene expression and lyse the cell, or to repress that pathway and lysogenize the cell (**Figure 11-28**). The decision whether to lyse or lysogenize a cell pivots on the activity of the cII protein. The cII protein is unstable because it is sensitive to bacterial proteases—enzymes that degrade proteins. These proteases respond to environmental conditions: they are more active when resources are abundant, but less active when cells are starved.

When resources are abundant, cII is degraded and little λ repressor is produced. The genes transcribed from P_L and P_R continue to be expressed, and the lytic cycle prevails.

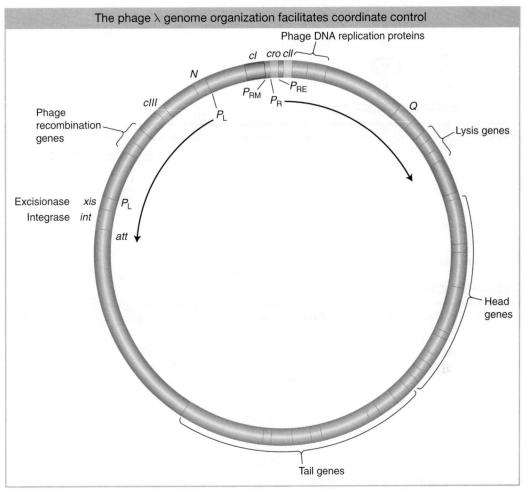

FIGURE 11-27 Map of phage λ in the circular form. The genes for recombination, integration and excision, replication, head and tail assembly, and cell lysis are clustered together and coordinately regulated. Transcription of the right side of the genome begins at P_R, and that of the leftward genes begins at P_L. Key regulatory interactions governing the lysogenic-versus-lytic decision take place at operators between the *cro* and the *cI* genes.

However, if resources are limited, cII is more active and more λ repressor is produced. In this case, the genes transcribed from P_L and P_R are repressed by the λ repressor and the lysogenic cycle is entered. The cII protein is also responsible for activating the transcription of *int*, a gene that encodes an additional protein required for lysogeny—an integrase required for the λ genome to integrate into the host chromosome. The cIII protein shields cII from degradation; so it, too, contributes to the lysogenic decision.

Molecular anatomy of the genetic switch

To see how the decision is executed at the molecular level, let's turn to the activities of λ repressor and Cro. The O_R operator lies between the two genes encoding these proteins and contains three sites, O_{R1}, O_{R2}, and O_{R3}, that overlap two opposing promoters: P_R, which promotes transcription of lytic genes, and P_{RM} (for *repressor maintenance*), which directs transcription of the *cI* gene (see Figures 11-27 and

11-28). Recall that the *cI* gene encodes the λ repressor. The three operator sites are similar but not identical in sequence, and although Cro and λ repressor can each bind to any one of the operators, they do so with different affinities: λ repressor binds to O_{R1} with the highest affinity, whereas Cro binds to O_{R3} with the highest affinity. The λ repressor's occupation of O_{R1} blocks transcription from P_R and thus blocks the transcription of genes for the lytic cycle. Cro's occupation of O_{R3} blocks transcription from P_{RM} and thus blocks maintenance of *cI* transcription. Hence, no λ repressor is produced, and transcription of genes for the lytic cycle can continue. The occupation of the operator sites therefore determines the lytic-versus-lysogenic patterns of λ gene expression (**Figure 11-29**).

After a lysogen has been established, it is generally stable. But the lysogen can be induced to enter the lytic cycle by various environmental changes. Ultraviolet light induces the expression of host genes. One of the host genes encodes a protein, RecA, that stimulates cleavage of the

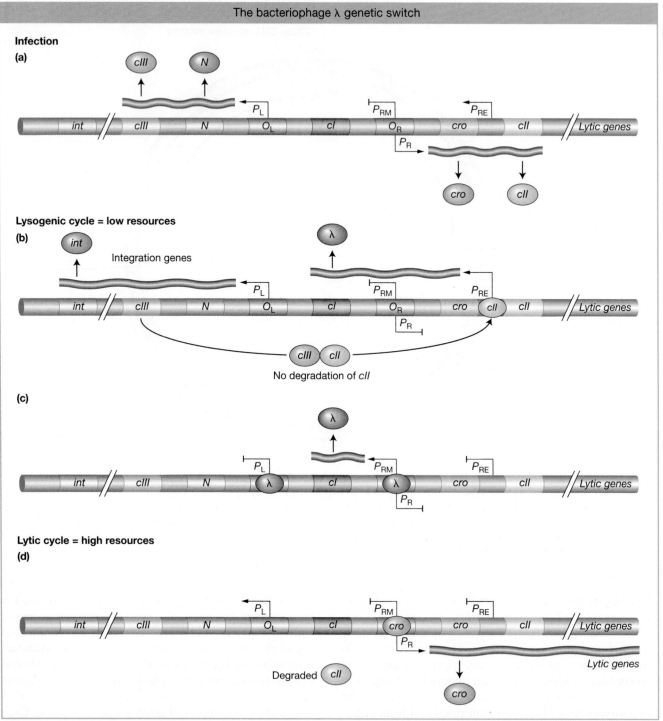

FIGURE 11-28 On infection (a), host RNA polymerase initiates transcription at P_L and P_R, expressing the N and cro genes, respectively. Antiterminator protein N enables transcription of the $cIII$ gene and recombination genes (see Figure 11-27, left), and the cII gene and other genes. Next, (b) the cII protein, protected by the cIII protein, turns on cI by activating transcription at P_{RE} and also activates the transcription of int. If resources and proteases are not abundant, cII remains active, cI transcription proceeds at a high level, and the Int protein integrates the phage chromosome. Eventually (c), the cI protein (λ repressor) shuts off all genes except itself. The phage will then remain in the lysogenic state. However, if resources and proteases are abundant (d), the cII protein is degraded, Cro represses transcription of cI from P_{RM} and activates transcription of Cro and lytic genes from P_L and P_R, and the lytic cycle continues.

λ repressor, thus crippling maintenance of lysogeny and resulting in lytic growth. Prophage induction, just as Jacob and Monod surmised, requires the release of a repressor from DNA. The physiological role of ultraviolet light in lysogen induction makes sense in that this type of radiation damages host DNA and stresses the bacteria; the phage replicates and leaves the damaged, stressed cell for another host.

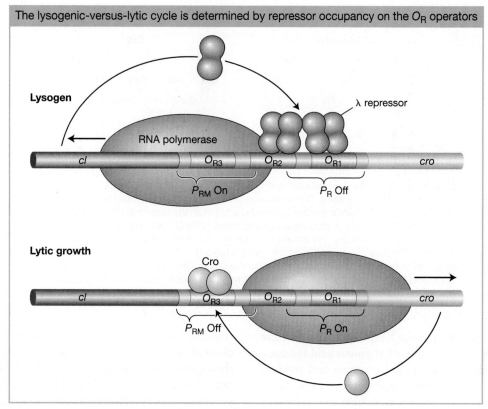

FIGURE 11-29 Lysogeny is promoted by λ repressor binding to O_{R1} and O_{R2}, which prevents transcription from P_R. On induction or in the lytic cycle, the binding of Cro to O_{R3} prevents transcription of the *cl* gene from P_{RM}. [*Data from M. Ptashne and A. Gann,* Genes and Signals, *p. 30, Fig. 1-13.*]

KEY CONCEPT The phage λ genetic switch illustrates that the regulatory logic that underlies the control of gene expression in response to physiological signals is conserved. Just as in the *lac, ara, trp,* and other systems, the alternative states of gene expression in λ phage are determined by the interaction of a few key DNA-binding regulatory proteins with control sites on the DNA. The order and orientation of these genetic elements in the genome is important to the switch function in all of these systems.

Sequence-specific binding of regulatory proteins to DNA

How do λ repressor and Cro recognize different operators with different affinities? This question directs our attention to a fundamental principle in the control of gene transcription—the regulatory proteins bind to specific DNA sequences. For individual proteins to bind to certain sequences and not others requires specificity in the interactions between the side chains of the protein's amino acids and the chemical groups of DNA bases. Detailed structural studies of λ repressor, Cro, and other bacterial regulators have revealed how the three-dimensional structures of regulators and DNA interact, and how the arrangement of particular amino acids enables them to recognize specific base sequences.

Crystallographic analysis has identified a common structural feature of the DNA-binding domains of λ and Cro. Both proteins make contact with DNA through a *helix-turn-helix* domain that consists of two helices joined by a short flexible linker region (**Figure 11-30**). One helix, the recognition helix, fits into the major groove of DNA. In that position, amino acids on the helix's outer face are able to interact with chemical groups on the DNA bases. The specific amino acids in the recognition helix determine the affinity of a protein for a specific DNA sequence.

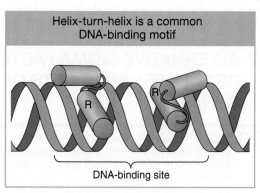

FIGURE 11-30 The binding of a helix-turn-helix motif to DNA. The purple cylinders are alpha helices. Many regulatory proteins bind as dimers to DNA. In each monomer, the recognition helix (R) makes contact with bases in the major groove of DNA.

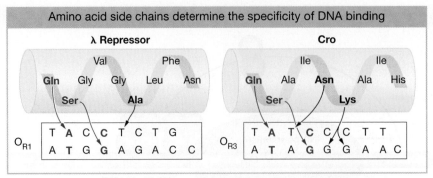

FIGURE 11-31 Interactions between amino acids and bases determine the specificity and affinity of DNA-binding proteins. The amino acid sequences of the recognition helices of the λ repressor and Cro proteins are shown. Interactions between the glutamine (Gln), serine (Ser), and alanine (Ala) residues of the λ repressor and bases in the O_{R1} operator determine the strength of binding. Similarly, interactions between the glutamine, serine, asparagine (Asn), and lysine (Lys) residues of the Cro protein mediate binding to the O_{R3} operator. Each DNA sequence shown is that bound by an individual monomer of the respective repressor; it is half of the operator site occupied by the repressor dimer. [*Data from M. Ptashne*, A Genetic Switch: Phage l and Higher Organisms, *2nd ed.*]

The recognition helices of the λ repressor and Cro have similar structures and some identical amino acid residues. Differences between the helices in key amino acid residues determine their DNA-binding properties. For example, in the λ repressor and Cro proteins, glutamine and serine side chains contact the same bases, but an alanine residue in the λ repressor and lysine and asparagine residues in the Cro protein impart different binding affinities for sequences in O_{R1} and O_{R3} (**Figure 11-31**).

The Lac and TrpR repressors, as well as the AraC activator and many other proteins, also bind to DNA through helix-turn-helix motifs of differing specificities, depending on the primary amino acid sequences of their recognition helices. In general, other domains of these proteins, such as those that bind their respective allosteric effectors, are dissimilar.

KEY CONCEPT The biological specificity of gene regulation is due to the chemical specificity of amino acid–base interactions between individual regulatory proteins and discrete DNA sequences.

11.7 ALTERNATIVE SIGMA FACTORS REGULATE LARGE SETS OF GENES

LO 11.4 Explain the roles of sequence-specific DNA-binding proteins and DNA regulatory sequences in coordinating the expression of sets of genes in bacteria and bacteriophage.

Thus far, we have seen how single switches can control the expression of single operons or two operons containing as many as a couple of dozen genes. Some physiological responses to changes in the environment require the coordinated expression of large sets of unlinked genes located throughout the genome to bring about dramatic physiological and even morphological changes. Analyses of these processes have revealed another twist in bacterial gene regulation: the control of large numbers of genes by alternative sigma (σ) factors of RNA polymerase. One such example, the process of sporulation in *Bacillus subtilis*, has been analyzed in great detail in the past few decades. Under stress, the bacterium forms spores that are remarkably resistant to heat and desiccation.

Early in the process of sporulation, the bacterium divides asymmetrically, generating two components of unequal size that have very different fates. The smaller compartment, the forespore, develops into the spore. The larger compartment, the mother cell, nurtures the developing spore and lyses when spore morphogenesis is complete to liberate the spore (**Figure 11-32a**). Genetic dissection of this process has entailed the isolation of many mutants that cannot sporulate. Detailed investigations have led to the characterization of several key regulatory proteins that directly regulate programs of gene expression that are specific to either the forespore or the mother cell. Four of these proteins are alternative σ factors.

Recall from Chapter 8 that transcription initiation in bacteria includes the binding of the σ subunit of RNA polymerase to the −35 and −11 regions of gene promoters. The σ factor disassociates from the complex when transcription begins and is recycled. In *B. subtilis*, two σ factors, σ^A and σ^H, are active in vegetative cells. During sporulation, a different σ factor, σ^F, becomes active in the forespore and activates a group of more than 40 genes. One gene activated by σ^F is a secreted protein that in turn triggers the proteolytic processing of the inactive precursor pro-σ^E, a distinct σ factor in the mother cell. The σ^E factor is required to activate sets of genes in the mother cell. Two additional σ factors, σ^K

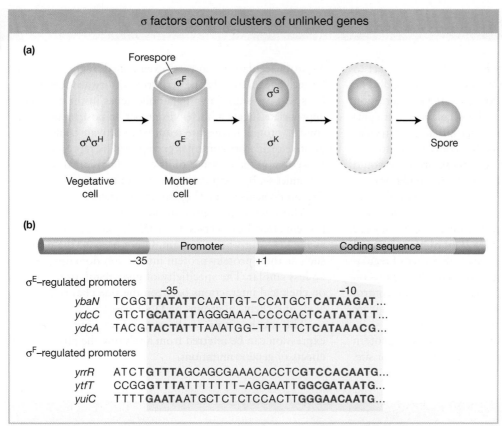

σ factors control clusters of unlinked genes

(a)

Forespore

σ^F

σ^A σ^H   Vegetative cell

σ^E   Mother cell

σ^G

σ^K

Spore

(b)

Promoter   Coding sequence

−35   +1

σ^E–regulated promoters

−35   −10

ybaN   TCGG**TTATATT**CAATTGT–CCATGCT**CATAAGAT**...

ydcC   GTCT**GCATATT**AGGGAAA–CCCCACT**CATATATT**...

ydcA   TACG**TACTATT**TAAATGG–TTTTTCT**CATAAACG**...

σ^F–regulated promoters

yrrR   ATCT**GTTTA**GCAGCGAAACACCTC**GTCCACAATG**...

ytfT   CCGGG**TTTA**TTTTTTT–AGGAATT**GGCGATAATG**...

yuiC   TTTT**GAATAAT**GCTCTCTCCACTT**GGGAACAATG**...

FIGURE 11-32 Sporulation in *Bacillus subtilis* is regulated by cascades of σ factors. (a) In vegetative cells, σ^A and σ^H are active. On initiation of sporulation, σ^F is active in the forespore and σ^E is active in the mother cell. These σ factors are then superseded by σ^G and σ^K, respectively. The mother cell eventually lyses and releases the mature spore. (b) Factors σ^E and σ^F control the regulons of many genes (*ybaN*, and so forth, in this illustration). Three examples of the large number of promoters regulated by each σ factor are shown. Each σ factor has a distinct sequence-specific binding preference at the −35 and −10 sequences of target promoters. [*Data from* P. Eichenberger et al., J. Mol. Biol. *327, 2003, 945–972; and S. Wang et al., J. Mol. Biol.* 358, 2006, 16–37.]

and σ^G, are subsequently activated in the mother cell and forespore, respectively (Figure 11-32a). The expression of distinct σ factors allows for the coordinated transcription of different sets of genes, or **regulons**, by a single RNA polymerase.

New approaches for characterizing the expression of all genes in a genome (see Section 14.7) have made it possible to monitor the transcription of each *B. subtilis* gene during vegetative growth and spore formation and in different compartments of the spore. Several hundred genes have been identified in this fashion that are transcriptionally activated or repressed during spore formation.

How are the different sets of genes controlled by each σ factor? Each σ factor has different sequence-specific DNA-binding properties. The operons or individual genes regulated by particular σ factors have characteristic sequences in the −35 and −11 regions of their promoters that are bound by one σ factor and not others (Figure 11-32b). For example, σ^E binds to at least 121 promoters, within 34 operons and 87 individual genes, to

regulate more than 250 genes, and σ^F binds to at least 36 promoters to regulate 48 genes.

KEY CONCEPT Sequential expression of alternative σ factors that recognize alternative promoter sequences enables the coordinated expression of large numbers of independent operons and unlinked genes during the developmental program of sporulation.

Alternative σ factors also play important roles in the virulence of human pathogens. For example, bacteria of the genus *Clostridium* produce potent toxins that are responsible for severe diseases such as botulism, tetanus, and gangrene. Key toxin genes of *C. botulinum*, *C. tetani*, and *C. perfringens* have recently been discovered to be controlled by related, alternative σ factors that recognize similar sequences in the −35 and −10 regions of the toxin genes. Understanding the mechanisms of toxin-gene regulation may lead to new means of disease prevention and therapy.

SUMMARY

Gene regulation is often mediated by proteins that react to physiological signals from within and around the cell. The proteins respond by raising or lowering the transcription rates of specific genes. The logic of this regulation is straightforward. For regulation to operate appropriately, the regulatory proteins have built-in sensors that continually monitor cellular conditions. The activities of these proteins would then depend on the right set of environmental conditions.

In bacteria and their viruses, the control of several structural genes may be coordinated by clustering the genes together into operons on the chromosome so that they are transcribed into multigenic mRNAs. Coordinated control simplifies the task for bacteria because one cluster of regulatory sites per operon is sufficient to regulate the expression of all the operon's genes. Alternatively, coordinate control can also be achieved through discrete σ factors that regulate dozens of independent promoters simultaneously.

In negative regulatory control, a repressor protein blocks transcription by binding to DNA at the operator site.

Negative regulatory control is exemplified by the *lac* system. Negative regulation is one very straightforward way for the *lac* system to shut down metabolic genes in the absence of appropriate sugars in the environment. In positive regulatory control, protein factors are required to activate transcription. Positive regulatory control is exemplified by repression of the *lac* system in the presence of its catabolite breakdown product, glucose. By contrast, repression of operons that synthesize amino acids is often controlled by attenuation.

Many regulatory proteins are members of families of proteins that have very similar DNA-binding motifs, such as the helix-turn-helix domain. Other parts of the proteins, such as their protein–protein interaction domains, tend to be less similar. The specificity of gene regulation depends on chemical interactions between the side chains of amino acids and chemical groups on DNA bases.

The mechanisms of the regulatory control of gene expression can be inferred from analyzing the physiological effects of genetic mutations.

KEY TERMS

activator (p. 372)
allosteric effector (p. 372)
allosteric site (p. 372)
allosteric transition (p. 375)
antiterminator (p. 388)
attenuation (p. 384)
attenuator (p. 384)
catabolite (p. 379)
catabolite activator protein (CAP) (p. 380)
catabolite repression (p. 379)

cis-acting (p. 376)
constitutive mutation (p. 376)
coordinately controlled genes (p. 373)
cyclic adenosine monophosphate (cAMP) (p. 380)
DNA-binding domain (p. 372)
genetic switch (p. 372)
inducer (p. 375)
induction (p. 375)
initiator (p. 383)
leader sequence (p. 384)

lysogenic cycle (p. 386)
lytic cycle (p. 386)
negative regulation (p. 372)
operator (p. 372)
operon (p. 374)
partial diploid (p. 376)
positive regulation (p. 372)
promoter (p. 372)
regulon (p. 393)
repressor (p. 372)
trans-acting (p. 376)

SOLVED PROBLEMS

This set of four solved problems, which are similar to Problem 15 in the Basic Problems at the end of this chapter, is designed to test understanding of the operon model. Here, we are given several diploids and are asked to determine whether *Z* and *Y* gene products are made in the presence or absence of an inducer. Use a table similar to the one in Problem 15 as a basis for your answers, except that the column headings will be as follows:

Genotype	*Z* gene		*Y* gene	
	No inducer	Inducer	No inducer	Inducer

SOLVED PROBLEM 1

$$\frac{I^- P^- O^C Z^+ Y^+}{I^+ P^+ O^+ Z^- Y^-}$$

SOLUTION

One way to approach these problems is first to consider each chromosome separately and then to construct a diagram. The following illustration diagrams this diploid:

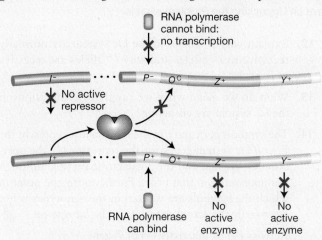

The first chromosome is P^-, and so transcription is blocked and no Lac enzyme can be synthesized from it. The second chromosome (P^+) can be transcribed, and thus transcription is repressible (O^+). However, the structural genes linked to the good promoter are defective; thus, no active Z product or Y product can be generated. The symbols to add to your table are "$-, -, -, -$."

SOLVED PROBLEM 2

$$\frac{I^+ P^- O^+ Z^+ Y^+}{I^- P^+ O^+ Z^+ Y^-}$$

SOLUTION

The first chromosome is P^-, and so no enzyme can be synthesized from it. The second chromosome is O^+, and so transcription is repressed by the repressor supplied from the first chromosome, which can act in trans through the cytoplasm. However, only the Z gene from this chromosome is intact. Therefore, in the absence of an inducer, no enzyme is made; in the presence of an inducer, only the Z gene product, β-galactosidase, is generated. The symbols to add to the table are "$-, +, -, -$."

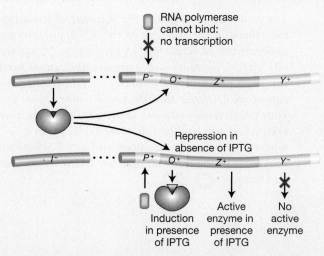

SOLVED PROBLEM 3

$$\frac{I^+ P^+ O^C Z^- Y^+}{I^+ P^- O^+ Z^+ Y^-}$$

SOLUTION

Because the second chromosome is P^-, we need consider only the first chromosome. This chromosome is O^C, and so enzyme is made in the absence of an inducer, although, because of the Z^- mutation, only active permease (Y) is generated. The entries in the table should be "$-, -, +, +$."

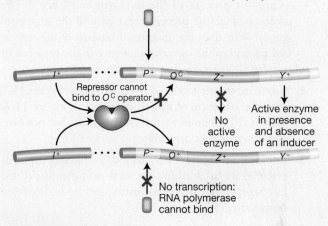

SOLVED PROBLEM 4

$$\frac{I^S P^+ O^+ Z^+ Y^-}{I^- P^+ O^C Z^- Y^+}$$

SOLUTION

In the presence of an I^S repressor, all wild-type operators are shut off, both with and without an inducer. Therefore, the first chromosome is unable to produce any enzyme. However, the second chromosome has an altered (O^C) operator and can produce enzyme in both the absence and the presence of an inducer. Only the Y gene is wild type on the O^C chromosome, and so only permease is produced constitutively. The entries in the table should be "$-, -, +, +$."

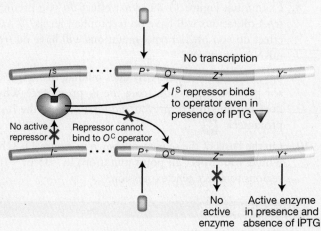

PROBLEMS

Visit SaplingPlus for supplemental content. Problems with the [icon] icon are available for review/grading. Problems with the [icon] icon have a Problem Solving Video. Problems with the [icon] icon have an Unpacking the Problem exercise.

WORKING WITH THE FIGURES

(The first 10 questions require inspection of text figures.)

1. Based on Figure 11-6, why does the binding of the repressor protein to the operator sequence in the absence of lactose prevent expression of the structural genes? Why does the absence of binding of the repressor protein to the operator sequence in the presence of lactose allow expression of the structural genes?

2. Compare the structure of IPTG shown in Figure 11-7 with the structure of galactose shown in Figure 11-5. Why is IPTG bound by the Lac repressor but not broken down by β-galactosidase?

3. Looking at Figure 11-9, why were partial diploids essential for establishing the trans-acting nature of the Lac repressor? Could one distinguish cis-acting from trans-acting genes in haploids?

4. Why do promoter mutations cluster at positions −10 and −35 as shown in Figure 11-11? Which protein-DNA interaction is disrupted by these mutations?

5. Comparing Figures 11-14, 11-15, and 11-30, why are many regulatory DNA-binding sites symmetrical?

6. Looking at Figure 11-16, note the large overlap between the operator and the region of the *lac* operon that is transcribed. Which protein binds specifically to this overlapping sequence, and what effect does it have on transcription?

7. Looking at Figure 11-20b, why do you think binding of the AraC protein to the *araO* and *araI* DNA sites leads to repression of the *ara* operon?

8. Examining Figure 11-21, what effect do you predict *trpA* mutations will have on tryptophan levels? What effect do you predict *trpA* mutations will have on *trp* mRNA expression?

9. Based on Figure 11-29, why does binding of λ repressor prevent expression from the P_R promoter? Why does binding of Cro prevent expression from the P_{RM} promoter? [icon]

10. On the basis of the sequences shown in Figure 11-32b, would you expect all point mutations in −35 or −10 regions to affect gene expression?

BASIC PROBLEMS

11. Which of the following molecules is an inducer of the *lac* operon:

   a. Galactose   c. Allolactose

   b. Glucose   d. Isothiocyanate

12. Explain why I^- alleles in the *lac* system are normally recessive to I^+ alleles and why I^+ alleles are recessive to I^S alleles.

13. What do we mean when we say that O^C mutations in the *lac* system are cis-acting?

14. The symbols *a*, *b*, and *c* in the table below represent the *E. coli lac* system genes for the repressor (*I*), the operator (*O*) region, and the β-galactosidase (*Z*), although not necessarily in that order. Furthermore, the order in which the symbols are written in the genotypes is not necessarily the actual sequence in the *lac* operon. [icon]

Activity (+) or inactivity (−) of Z gene

Genotype	Inducer absent	Inducer present
$a^- \ b^+ \ c^+$	+	+
$a^+ \ b^+ \ c^-$	+	+
$a^+ \ b^- \ c^-$	−	−
$a^+ \ b^- \ c^+ / a^- \ b^+ \ c^-$	+	+
$a^+ \ b^+ \ c^+ / a^- \ b^- \ c^-$	−	+
$a^+ \ b^+ \ c^- / a^- \ b^- \ c^+$	−	+
$a^- \ b^+ \ c^+ / a^+ \ b^- \ c^-$	+	+

   a. Which symbol (*a*, *b*, or *c*) represents each of the *lac* genes *I*, *O*, and *Z*?

   b. In the table, a superscript minus sign on a gene symbol merely indicates a mutant, but some mutant behaviors in this system are given special mutant designations. Using the conventional gene symbols for the *lac* operon, designate each genotype in the table.

15. The map of the *lac* operon is

   POZY

   The promoter (*P*) region is the start site of transcription through the binding of the RNA polymerase molecule before actual mRNA production. Mutationally altered promoters (*P⁻*) apparently cannot bind the RNA polymerase molecule. Certain predictions can be made about the effect of *P⁻* mutations. Use your predictions and your knowledge of the lactose system to complete the following table. Insert a "+" where an enzyme is produced and a "−" where no enzyme is produced. The first one has been done as an example. [icon] [icon]

Genotype	β-Galactosidase		Permease	
	No lactose	Lactose	No lactose	Lactose
$I^+ \ P^+ \ O^+ \ Z^+ \ Y^+/I^+ \ P^+ \ O^+ \ Z^+ \ Y^+$	−	+	−	+
a. $I^- \ P^+ \ O^C \ Z^+ \ Y^-/I^+ \ P^+ \ O^+ \ Z^- \ Y^+$				
b. $I^+ \ P^- \ O^C \ Z^- \ Y^+/I^- \ P^+ \ O^C \ Z^+ \ Y^-$				
c. $I^S \ P^+ \ O^+ \ Z^+ \ Y^-/I^+ \ P^+ \ O^+ \ Z^- \ Y^+$				
d. $I^S \ P^+ \ O^+ \ Z^+ \ Y^+/I^- \ P^+ \ O^+ \ Z^+ \ Y^+$				
e. $I^- \ P^+ \ O^C \ Z^+ \ Y^-/I^- \ P^+ \ O^+ \ Z^- \ Y^+$				
f. $I^- \ P^- \ O^+ \ Z^+ \ Y^+/I^- \ P^+ \ O^C \ Z^+ \ Y^-$				
g. $I^+ \ P^+ \ O^+ \ Z^- \ Y^+/I^- \ P^+ \ O^+ \ Z^+ \ Y^-$				

16. Explain why it makes sense for the cell to synthesize β-galactosidase only when levels of lactose are high and levels of glucose are low.

17. Explain the fundamental differences between negative regulation and positive regulation of transcription in bacteria.

18. Which molecule regulates both the *lac* operon and the *ara* operon?

19. Compare the mechanisms of negative and positive regulation in the *lac* operon with those in the *ara* operon.

20. Mutants that are *lacY*⁻ retain the capacity to synthesize β-galactosidase. However, even though the *lacI* gene is still intact, β-galactosidase can no longer be induced by adding lactose to the medium. Explain.

21. What is the function of the two tryptophan codons in the 14-amino-acid leader peptide in the regulation of *trp* operon?

22. Could the attenuation mechanism found in the *trp* operon regulate gene expression in eukaryotic cells?

23. What are the similarities between the mechanisms controlling the *lac* operon and those controlling bacteriophage λ genetic switches?

24. Compare the arrangement of cis-acting sites in the control regions of the *lac* operon and bacteriophage λ.

25. Which regulatory protein induces the lytic phase genes of the bacteriophage λ life cycle?

   a. cI

   b. Cro

   c. Int

   d. cIII

26. What protein in bacteriophage λ serves as a readout of the level of resources in the cell?

27. What is the function of the cIII protein in the bacteriophage λ genetic switch?

28. Predict the effect of a mutation that eliminates the DNA-binding activity of the σ^E protein on spore formation in *Bacillus subtilis*.

CHALLENGING PROBLEMS

29. An interesting mutation in *lacI* results in repressors with 110-fold increased binding to both operator and nonoperator DNA. These repressors display a "reverse" induction curve, allowing β-galactosidase synthesis in the absence of an inducer (IPTG) but partly repressing β-galactosidase expression in the presence of IPTG. How can you explain this? (Note that, when IPTG binds a repressor, it does not completely destroy operator affinity; rather, it reduces affinity 110-fold. Additionally, as cells divide and new operators are generated by the synthesis of daughter strands, the repressor must find the new operators by searching along the DNA, rapidly binding to nonoperator sequences and dissociating from them.)

30. Certain *lacI* mutations eliminate operator binding by the Lac repressor but do not affect the aggregation of subunits to make a tetramer, the active form of the repressor. These mutations are partly dominant over wild type. Can you explain the partly dominant I^- phenotype of the I^-/I^+ heterodiploids?

31. You are examining the regulation of the lactose operon in the bacterium *Escherichia coli*. You isolate seven new independent mutant strains that lack the products of all three structural genes. You suspect that some of these mutations are *lacI*^{S} mutations and that other mutations are alterations that prevent the binding of RNA polymerase to the promoter region. Using whatever haploid and partial diploid genotypes that you think are necessary, describe a set of genotypes that will permit you to distinguish between the *lacI* and *lacP* classes of uninducible mutations.

32. You are studying the properties of a new kind of regulatory mutation of the lactose operon. This mutation, called *S*, leads to the complete repression of the *lacZ*, *lacY*, and *lacA* genes, regardless of whether lactose is present. The results of studies of this mutation

in partial diploids demonstrate that this mutation is completely dominant over wild type. When you treat bacteria of the S mutant strain with a mutagen and select for mutant bacteria that can express the enzymes encoded by *lacZ*, *lacY*, and *lacA* genes in the presence of lactose, some of the mutations map to the *lac* operator region and others to the *lac* repressor gene. On the basis of your knowledge of the lactose operon, provide a molecular genetic explanation for all these properties of the S mutation. Include an explanation of the constitutive nature of the "reverse mutations."

33. The *trp* operon in *E. coli* encodes enzymes essential for the biosynthesis of tryptophan. The general mechanism for controlling the *trp* operon is similar to that observed with the *lac* operon: when the repressor binds to the operator, transcription is prevented; when the repressor does not bind to the operator, transcription proceeds. The regulation of the *trp* operon differs from the regulation of the *lac* operon in the following way: the enzymes encoded by the *trp* operon are synthesized not when tryptophan is present, but rather when it is absent. In the *trp* operon, the repressor has two binding sites: one for DNA, and the other for the effector molecule, tryptophan. The *trp* repressor must first bind to a molecule of tryptophan before it can bind effectively to the *trp* operator.

a. Draw a map of the tryptophan operon, indicating the promoter (P), the operator (O), and the first structural gene of the tryptophan operon (*trpA*). In your drawing, indicate where on the DNA the repressor protein binds when it is bound to tryptophan.

b. The *trpR* gene encodes the repressor; *trpO* is the operator; *trpA* encodes the enzyme tryptophan synthetase. A *trpR⁻* repressor cannot bind tryptophan, a *trpO⁻* operator cannot be bound by the repressor, and the enzyme encoded by a *trpA⁻* mutant gene is completely inactive. Do you expect to find active tryptophan synthetase in each of the

following mutant strains when the cells are grown in the presence of tryptophan? In its absence?

1. $R^+ \, O^+ \, A^+$ (wild type)
2. $R^- \, O^+ \, A^+/R^+ \, O^+ \, A^-$
3. $R^+ \, O^- \, A^+/R^+ \, O^+ \, A^-$

34. The activity of the enzyme β-galactosidase produced by wild-type cells grown in media supplemented with different carbon sources is measured. In relative units, the following levels of activity are found:

Glucose	Lactose	Lactose + glucose
0	100	1

Predict the relative levels of β-galactosidase activity in cells grown under similar conditions when the cells are $lacI^-$, $lacI^S$, $lacO^C$, and crp^-.

35. A bacteriophage λ is found that is able to lysogenize its *E. coli* host at 30°C but not at 42°C. What genes may be mutant in this phage?

36. What would happen to the ability of bacteriophage λ to lyse a host cell if it acquired a mutation in the O_R binding site for the Cro protein? Why?

37. Sketch the effects of exposure of host cells to UV on the bacteriophage λ genetic switch.

38. Contrast the effects of mutations in genes encoding sporulation-specific σ factors with mutations in the −35 and −10 regions of the promoters of genes in their regulons. Would functional mutations in the σ-factor genes or in the individual promoters have the greater effect on sporulation?

GENETICS AND SOCIETY

How might an understanding of the regulation of gene expression in bacteria be important for the treatment or prevention of human disease?

Regulation of Transcription in Eukaryotes

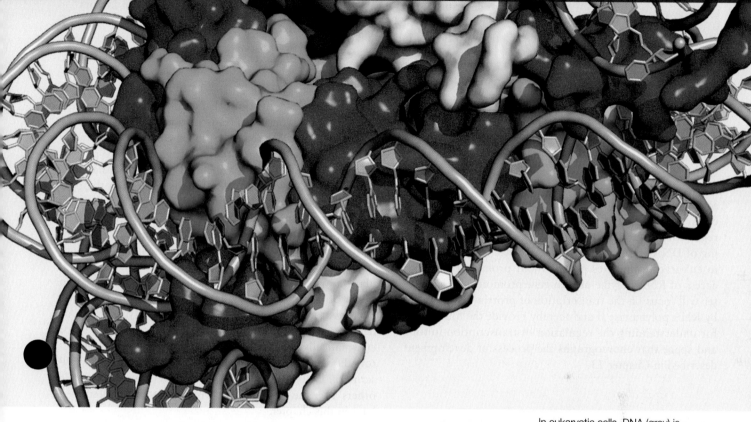

In eukaryotic cells, DNA (gray) is wrapped around histone proteins (blue, green, red, and yellow), which affects access of the transcription machinery to DNA. [*PDB ID 5y0c.*]

CHAPTER OUTLINE AND LEARNING OBJECTIVES

12.1 TRANSCRIPTION FACTORS REGULATE TRANSCRIPTION

LO 12.1 Diagram how transcription factors and DNA enhancer elements control the transcription of individual genes.

12.2 CHROMATIN STRUCTURE

LO 12.2 Draw a segment of chromatin, labeling each histone, a nucleosome, and the structural features that are important to their function in transcription.

12.3 CHROMATIN REGULATES TRANSCRIPTION

LO 12.3 Compare and contrast how chromatin modifying and chromatin remodeling mechanisms contribute to gene-specific transcription regulation.

12.4 CHROMATIN IN EPIGENETIC REGULATION

LO 12.4 Provide examples of the chromatin-based mechanisms that maintain gene expression over cellular or organismal generations.

The process of transcription in bacterial and eukaryotic cells is carried out by fundamentally similar mechanisms, as described in Chapter 8. In contrast, the regulation of transcription in eukaryotic cells is more complex than in bacterial cells. This increased complexity is mediated by a larger variety of regulatory DNA sequence elements and protein factors, including proteins that package DNA to fit into the nucleus. The main goal of this chapter is to describe how the variety of DNA sequence elements and protein factors in eukaryotic cells function in different combinations to precisely control the transcription of individual genes.

In Chapters 8 and 11, you learned that transcription in bacteria is often regulated by single activator or repressor proteins that directly bind DNA. Initial expectations were that eukaryotic transcription would be regulated by similar means. However, in most eukaryotes, multiple proteins and DNA sequences work together to control transcription. A key additional difference between bacteria and eukaryotes is that in eukaryotes access to transcription regulatory sequences in DNA is restricted by the packaging of DNA with proteins in the nucleus. Gene regulation in eukaryotes involves proteins that promote or restrict access of RNA polymerases to gene promoters. This chapter will focus on the transcription of protein-coding genes by RNA polymerase II and thereby provide the foundation for understanding the regulation of transcription in time and space that choreographs the process of development described in Chapter 13.

12.1 TRANSCRIPTION FACTORS REGULATE TRANSCRIPTION

LO 12.1 Diagram how transcription factors and DNA enhancer elements control the transcription of individual genes.

The machinery required for generating the distinct patterns of gene transcription that occur in eukaryotic cells has many components, including trans-acting regulatory proteins and cis-acting regulatory DNA sequences. The regulatory proteins can be divided into two sets, those that

directly bind DNA and those that do not. The first set of regulatory proteins consists of **transcription factors** that directly bind regulatory DNA sequences called **enhancers**. Enhancers that are located close to the core promoter are part of **proximal promoters** and are called **proximal enhancers**, and those that are a considerable distance from the promoter are part of **distal enhancers** and are called enhancers (**Figure 12-1**). In addition, some **general transcription factors (GTFs)** directly bind DNA regulatory sequences within **core promoters** that surround transcription start sites.

The second set of regulatory proteins consists of coregulators, which do not directly bind DNA. There are two types of coregulators: **coactivators** and **corepressors**. Coactivators and corepressors, respectively, increase or decrease the amount of transcription through binding or enzymatically modifying other transcription regulatory factors. For example, some coactivators serve to bridge transcription factors and RNA polymerase II (**Figure 12-2**), while others alter the structure of chromatin, which is described later in this chapter.

KEY CONCEPT Distal and proximal enhancers are DNA sequences that regulate the transcription of genes. Coregulators, which bind transcription factors, control the recruitment and access to DNA of general transcription factors and RNA polymerase II.

Eukaryotic transcription regulatory mechanisms have been discovered through both biochemical and genetic approaches. The latter has been advanced in particular by studies of the single-celled yeast *Saccharomyces cerevisiae* (see the Model Organism box on page 404). This organism,

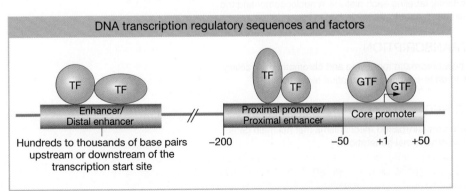

FIGURE 12-1 Transcription of eukaryotic genes is regulated by transcription factors (TF) that bind distal and proximal enhancers and by general transcription factors (GTF) that bind core promoters.

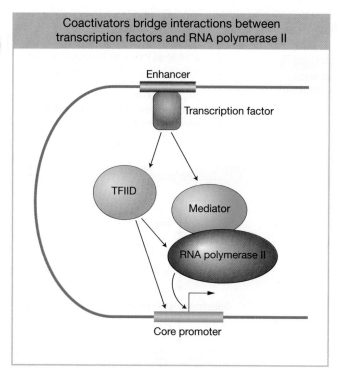

FIGURE 12-2 Transcription factors physically interact with coactivators such as TFIID and mediator that recruit RNA polymerase II to a gene's transcription start site.

which has played a key role in wine making, beer making, and baking for many centuries, has been a passport to understanding much of eukaryotic molecular biology. Several decades of research have produced many fundamental insights into general principles of how eukaryotic transcription regulatory proteins work.

Transcription factors bind distal and proximal enhancers

Mutagenesis studies have revealed the importance of proximal enhancers. As shown in **Figure 12-3**, point mutations in proximal enhancers as well as core promoters reduce transcription of the β-globin gene. This example reveals general features of enhancers: they contain short sequence elements (6–10 base pairs), and multiple elements are often clustered together. Enhancer elements frequently occur as inverted repeats of the same DNA sequence for binding of two similar or identical transcription factors, reminiscent of the DNA sequences controlling the *lac* operon in bacteria (Figure 11-14). Because enhancer elements are short, they randomly occur many times in genomes. However, they are not all bound by transcription factors because binding often requires interactions with partner transcription factors bound to other nearby enhancers.

In addition to binding DNA enhancer elements, transcription factors bind other proteins (**Figure 12-4**). This is exemplified by C/EBP (CCAAT/enhancer-binding protein), the transcription factor that binds one of the proximal enhancer elements in the β-globin gene, the CCAAT box (pronounced "cat" box). C/EBP is characterized by a **DNA-binding domain** and a **dimerization domain**, the latter of which facilitates the formation of homodimers (binding between two C/EBP proteins) and heterodimers (binding between different C/EBP family members). C/EBP also contains an **activation domain** that interacts with other components of the transcription machinery to turn on transcription. Other transcription factors have **repression domains** that use similar mechanisms to turn

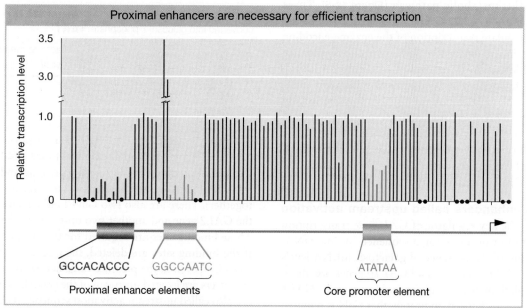

FIGURE 12-3 Point mutations throughout the proximal enhancer and core promoter regions of the β-globin gene were analyzed for their effects on transcription level. The height of each line represents the transcription level relative to a wild-type gene (set to 1.0). Only the base substitutions that lie within the three labeled elements changed the level of transcription. Wild-type sequences are shown for the proximal enhancer and core promoter elements. Positions with black dots were not tested. [Data from T. Maniatis, S. Goodbourn, and J. A. Fischer, "Regulation of Inducible and Tissue-Specific Gene Expression," Science 236, 1987, 1237.]

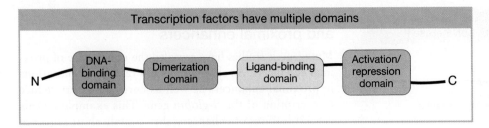

Transcription factors have multiple domains

off transcription. Furthermore, some transcription factors include a **ligand-binding domain** that binds a ligand such as a hormone or a vitamin, changing the structure of the transcription factor and activating it. As an example, binding of the hormone estrogen by a transcription factor called Estrogen receptor in the cytoplasm leads to its dimerization, nuclear localization, and binding to enhancers elements called Estrogen Response Elements. All transcription factors contain a DNA-binding domain and an activation/repression domain, but only some transcription factors contain dimerization and ligand-binding domains.

KEY CONCEPT Transcription factors use their DNA-binding, activation/repression, dimerization, and ligand-binding domains to activate or repress gene transcription.

Transcription factors: lessons from the yeast GAL system

Yeast make use of extracellular galactose (gal) by importing and converting it into a form of glucose that can be metabolized. Several genes—*GAL1*, *GAL2*, *GAL7*, and *GAL10*—in the yeast genome encode enzymes that catalyze steps in this metabolic pathway (**Figure 12-5**). Three additional genes—*GAL3*, *GAL4*, and *GAL80*—encode proteins that regulate transcription of the enzyme-encoding genes. Just as in the *lac* system of *E. coli*, the abundance of the sugar determines the level of transcription in the metabolic pathway. In yeast cells growing in media lacking galactose, the *GAL* genes are largely transcriptionally silent. But, in the presence of galactose (and the absence of glucose), the *GAL* genes are transcriptionally induced. Just as for the *lac* operon, genetic and molecular analyses of mutants have been key to understanding how transcription of genes in the galactose pathway is controlled.

Gal4 binds enhancers called upstream activation sequences The key regulator of *GAL* gene transcription is the Gal4 transcription factor, a sequence-specific DNA-binding protein. In the presence of galactose, mRNA levels for the *GAL1*, *GAL2*, *GAL7*, and *GAL10* genes are about 1000-fold higher than in its absence. However, in *GAL4* mutants they are unchanged, indicating that Gal4 is required for transcription of these genes. Each of the four genes has two or more Gal4-binding sites (i.e., enhancers) located at some distance 5′ (upstream) of its promoter (**Figure 12-6**). Consider the *GAL10* and *GAL1* genes, which are adjacent to each other and transcribed in opposite directions.

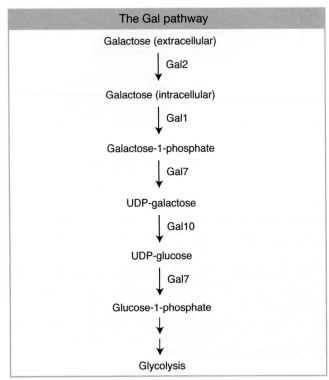

The Gal pathway

Galactose (extracellular)

↓ Gal2

Galactose (intracellular)

↓ Gal1

Galactose-1-phosphate

↓ Gal7

UDP-galactose

↓ Gal10

UDP-glucose

↓ Gal7

Glucose-1-phosphate

↓

Glycolysis

FIGURE 12-5 Galactose is converted into glucose-1-phosphate in a series of biochemical steps. These steps are catalyzed by the enzymes Gal1, Gal2, Gal7, and Gal10, which are encoded by the genes *GAL1*, *GAL2*, *GAL7*, and *GAL10*, respectively.

ANIMATED ART ⊠ SaplingPlus

Galactose metabolism in yeast

Between the *GAL1* and *GAL10* transcription start sites is a 118-base-pair region that contains four Gal4-binding sites. Each Gal4-binding site is 17 base pairs long and is bound by a homodimer of Gal4 proteins (two Gal4 proteins bound together). There are also two Gal4-binding sites upstream of the *GAL2* gene and another two upstream of the *GAL7* gene. These binding sites are required for transcription activation. If the binding sites are deleted, the genes are transcriptionally silent, even in the presence of galactose. Because the Gal4 enhancers are located upstream of the genes they regulate, they are also called **upstream activation sequence (UAS)** elements.

KEY CONCEPT Transcription factors coordinately regulate the transcription of multiple genes involved in the same biological process by binding enhancers that are common to the genes.

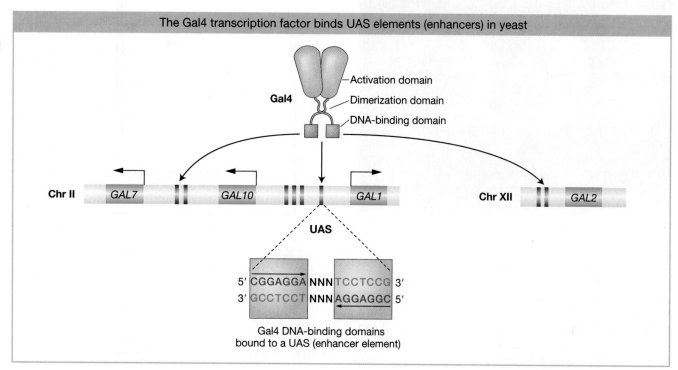

FIGURE 12-6 Gal4 activates the transcription of target genes by binding upstream activation sequence (UAS) elements (red rectangles). The Gal4 protein has three functional domains. The protein binds as a dimer to inverted repeats of the same sequence (arrows show directionality of the sequence) located upstream of the promoters of Gal-pathway genes. Some of the *GAL* genes are on the same chromosome (*GAL7*, *GAL10*, *GAL1*), whereas others are on different chromosomes (*GAL2*).

ANIMATED ART 🔗 SaplingPlus

Gal4 regulation of galactose-metabolizing enzymes

In addition to its action in yeast cells, Gal4 has been shown to activate transcription of UAS-containing genes when they are introduced into insect cells, human cells, and many other eukaryotic organisms. This versatility suggests that transcription machineries and mechanisms of gene activation are common to a broad array of eukaryotes, and that features revealed in yeast are generally present in other eukaryotes and vice versa. Furthermore, because of their versatility, Gal4 and its UAS elements have become favored tools for manipulating gene expression in a wide variety of model organisms.

KEY CONCEPT The ability of Gal4 to function in a variety of eukaryotes indicates that eukaryotes generally have common transcription regulatory machineries and mechanisms.

Gal4 domains function independently of one another

In addition to the DNA-binding domain and dimerization domain, Gal4 has an activation domain. A series of simple, elegant experiments demonstrated that the DNA-binding and activation domains of Gal4 as well as other transcription factors are modular; that is, they function independently of one another (**Figure 12-7**). In this study, researchers fused the Gal4 activation domains to the DNA-binding domain from the *E. coli* transcription factor LexA. Transcription was measured using reporter genes (see Chapter 10) that contained Gal4-binding sites (i.e., UAS) or LexA-binding sites (i.e., LexA site) upstream of a promoter and the *E. coli lacZ* gene coding region. The level of transcription of *lacZ* in yeast cells was determined by measuring the level of its encoded protein product β-galactosidase. Full-length Gal4 activated transcription when bound to the UAS (Figure 12-7a) but the Gal4 DNA-binding domain lacking the activation domain did not (Figure 12-7b). Similarly, the LexA DNA-binding domain did not activate transcription from LexA sites (Figure 12-7c), but a protein fusion of the Gal4 activation domain and the LexA DNA-binding domain did (Figure 12-7d). Likewise, a fusion of the Gal4 binding domains to another activation domain was able to activate transcription (not shown).

Researchers have used the modularity of transcription factors to develop technologies such as the yeast two-hybrid system that is used to detect protein-protein interactions in vivo (Chapter 14). The modularity of transcription factors is also the cause of some cancers such as acute promyelocytic leukemia (APL), a cancer of early blood-forming cells. In almost all cases of APL, a chromosome translocation creates a gene fusion between the activation domain of PML (promyelocytic leukemia) and the DNA-binding and ligand binding domains of RARA (retinoic acid receptor α). The fusion protein assembles with corepressor proteins, instead of coactivator proteins, to block transcription of normal RARA gene targets that control the differentiation of myeloid (blood) cells, which leads to uncontrolled proliferation of these cells.

Yeast

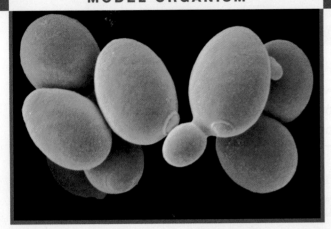

Electron micrograph of budding yeast cells. [SCIMAT/Science Source.]

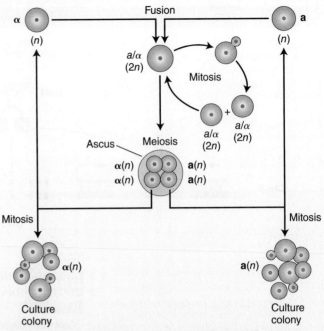

The life cycle of baker's yeast. The nuclear alleles *MAT***a** and *MAT*α determine mating type. Ploidies, *n* and 2*n*, are indicated in parentheses.

Saccharomyces cerevisiae, or budding yeast, is a pre-mier eukaryotic genetic system. Humans have grown yeast for centuries because it is an essential component of beer, bread, and wine. Yeast has many features that make it an ideal model organism. As a unicellular eukaryote, it can be grown on agar plates and, with yeast's life cycle of just 90 minutes, large quantities can be cultured in liquid media. It has a very compact genome with only about 12 megabase pairs of DNA (compared with almost 3000 megabase pairs for humans) containing approximately 6000 genes that are distributed on 16 chromosomes. It was the first eukaryote to have its genome sequenced.

The yeast life cycle makes it very versatile for laboratory studies. Cells can be grown as either diploids or haploids. In both cases, the mother cell produces a bud containing an identical daughter cell. Diploid cells either continue to grow by budding or are induced to undergo meiosis, which pro-duces four haploid spores held together in an ascus (also called a tetrad). Haploid spores of opposite mating type (**a** or α) will fuse and form a diploid. Spores of the same mating type will continue growth by budding.

Yeast has been called the *E. coli* of eukaryotes because of the ease of forward and reverse mutant analysis. To isolate mutants using a forward genetic approach, hap-loid cells are mutagenized (with X rays, for example) and screened on plates for mutant phenotypes. This procedure is usually done by first plating cells on a rich medium on which all cells grow and by copying, or replica plating, the colonies from this master plate onto replica plates containing selec-tive media or special growth conditions. For example, tem-perature-sensitive mutants will grow on the master plate at the permissive temperature but not on a replica plate at a restrictive temperature. Comparison of the colonies on mas-ter and replica plates will reveal the temperature-sensitive mutants. Using reverse genetics, scientists can replace any yeast gene of known or unknown function with a mutant ver-sion to understand the nature of the gene product.

KEY CONCEPT Eukaryotic transcription factors are modular, having separable domains for DNA binding, activation/repression, dimerization, and ligand-binding.

Regulation of Gal4

Gal4 activity is regulated by the Gal80 and Gal3 proteins (**Figure 12-8**). In *GAL80* mutants, the *GAL* structural genes (*GAL1, GAL2, GAL7,* and *GAL10*) are transcriptionally active even in the absence of galactose. This suggests that the normal function of Gal80 is to inhibit *GAL* gene transcrip-tion. Conversely, in *GAL3* mutants, the *GAL* structural genes are not active in the presence of galactose, suggesting that Gal3 normally promotes transcription of the *GAL* genes.

Extensive biochemical analyses revealed that Gal80 is a corepressor of Gal4. Gal80 binds Gal4 with high affinity and directly inhibits Gal4 activity. Specifically, Gal80 binds within the Gal4 activation domain, blocking its ability to promote transcription. Gal80 is expressed continuously, so it is always acting to repress transcription of the *GAL* structural genes unless stopped.

The role of Gal3 is to release the *GAL* structural genes from their repression by Gal80 when galactose is present. Gal3 is thus both a sensor and inducer. When Gal3 binds galactose and ATP, it undergoes a conformational change that promotes binding to Gal80, which in turn causes Gal80 to be released from Gal4, which is then able to interact with coactivators and RNA polymerase II to activate transcrip-tion. Thus, Gal80, Gal3, and Gal4 are all part of a switch, whose state is determined by the presence or absence of galactose (Figure 12-8). In this switch, DNA binding by the transcription factor is not the physiologically regulated step

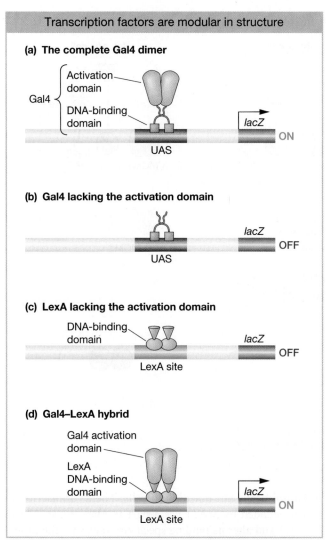

Transcription factors are modular in structure

(a) The complete Gal4 dimer

Gal4 { Activation domain / DNA-binding domain

lacZ   ON

UAS

(b) Gal4 lacking the activation domain

lacZ   OFF

UAS

(c) LexA lacking the activation domain

DNA-binding domain

lacZ   OFF

LexA site

(d) Gal4–LexA hybrid

Gal4 activation domain

LexA DNA-binding domain

lacZ   ON

LexA site

FIGURE 12-7 Transcription factors have multiple, separable domains. (a) Full-length Gal4 has three domains and activates transcription from UAS sites. (b) Removal of the Gal4 activation domain shows that dimerization and DNA binding is not sufficient for transcription activation. (c) Similarly, the LexA DNA-binding domain cannot activate transcription, but (d) when fused to the Gal4 activation domain, it can activate transcription through LexA-binding sites.

ANIMATED ART   **🌀 SaplingPlus**

Gal4 modularity and gene induction in non-yeast

(as is the case in the *lac* operon and bacteriophage λ); rather, the ability of the activation domain to perform its function is regulated.

KEY CONCEPT Environmental signals such as galactose alter the activity of eukaryotic transcription factors by controlling their interactions with other proteins.

Combinatorial control of transcription: lessons from yeast mating type

Thus far, we have focused on transcription regulation of single genes or a few genes in one pathway. In multicellular organisms, distinct cell types differ in the transcription of hundreds

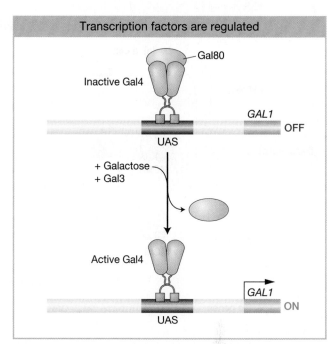

Transcription factors are regulated

Gal80

Inactive Gal4

GAL1   OFF

UAS

+ Galactose
+ Gal3

Active Gal4

GAL1   ON

UAS

FIGURE 12-8 Gal4 activity is regulated by Gal80 and Gal3. (*Top*) In the absence of galactose, Gal4 is inactive, even though it can bind UAS elements upstream of *GAL* genes such as *GAL1*. Gal4 activity is repressed by the binding of Gal80. (*Bottom*) In the presence of galactose, Gal3 induces a conformation change in Gal80, releasing it from Gal4, which can then activate *GAL* gene transcription.

ANIMATED ART   **🌀 SaplingPlus**

Gal4 regulation of galactose-metabolizing enzymes

of genes. The activation or repression of sets of genes must therefore be coordinated in the making of particular cell types. One of the best-understood examples of cell type-specific regulation in eukaryotes is the regulation of mating type in yeast. This regulatory system has been dissected by a combination of genetics, molecular biology, and biochemistry. Mating type serves as an excellent model for understanding the logic of transcription regulation in multicellular animals.

The yeast *Saccharomyces cerevisiae* can exist in any of three different cell types known as **a**, α(alpha), and **a**/α. The two cell types **a** and α are haploid and contain only one copy of each chromosome. The **a**/α cell is diploid and contains two copies of each chromosome. Although the two haploid cell types cannot be distinguished by their appearance, they can be differentiated by a number of specific cellular characteristics, principally their mating type (see the Model Organism box on page 404). An α cell mates only with an **a** cell, and an **a** cell mates only with an α cell. An α cell secretes an oligopeptide pheromone, or sex hormone, called α factor that arrests **a** cells in the cell cycle. Similarly, an **a** cell secretes a pheromone, called **a** factor, that arrests α cells. Cell arrest of both participants is necessary for successful mating. The diploid **a**/α cell does not mate, is larger than **a** and α cells, and does not respond to mating hormones.

Genetic analysis of mutants defective in mating has shown that cell type is controlled by a single genetic locus, the mating-type locus, *MAT*. There are two alleles of the *MAT* locus: haploid **a** cells have the *MATa* allele, and

Transcription factors work in combination to control cell type-specific transcription

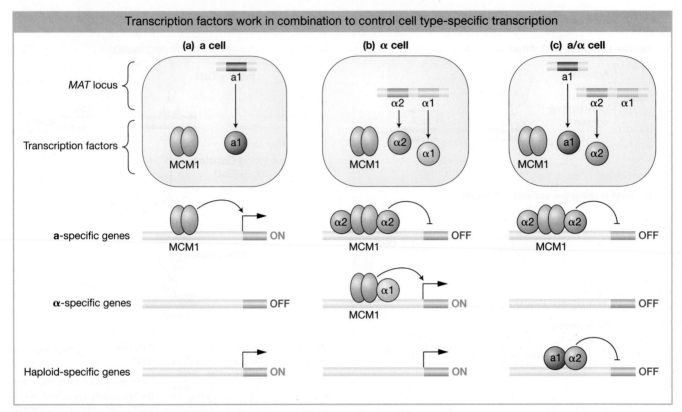

FIGURE 12-9 Control of cell type-specific transcription in yeast. The three cell types of *S. cerevisiae* are determined by differential expression of the transcription factors **a**1, α1, and α2, which regulate different subsets of target genes. The MCM1 transcription factor is expressed in all three cell types, but its function is changed by interactions with α1 and α2.

haploid α cells have the *MAT*α allele. The a/α diploid has both alleles. Yeast can switch mating type by a homologous recombination event that replaces one *MAT* allele with the other *MAT* allele. These two alleles activate different sets of genes because they encode different transcription factors. In addition, a transcription factor not encoded by the *MAT* locus, called MCM1, plays a key role in regulating cell type.

The simplest case is the **a** cell type (**Figure 12-9a**). The *MAT*a locus encodes a single transcription factor, a1. However, a1 has no effect in haploid cells, only in diploid cells. In a haploid **a** cell, the transcription factor MCM1 turns on the expression of the structural genes needed by an **a** cell by binding enhancers for **a**-specific genes.

In an α cell, the α-specific structural genes must be transcribed, but, in addition, MCM1 must be prevented from activating the **a**-specific genes (Figure 12-9b). The DNA sequence of the *MAT*α allele encodes two transcription factors, α1 and α2, that are produced by separate genes. These two proteins have different regulatory roles in the cell. The α1 protein is an activator of α-specific transcription. It binds in concert with the MCM1 protein to an enhancer that controls α-specific genes. The α2 protein represses transcription of the **a**-specific genes by binding with MCM1 to enhancers upstream of **a**-specific genes.

In a diploid yeast cell, transcription factors encoded by each *MAT* locus are expressed (Figure 12-9c). This results in repression of all genes involved in cell mating and a separate set of genes, called haploid specific, that are expressed

in haploid cells but not diploid cells. The a1 transcription factor encoded by *MAT*a has a part to play at last. a1 can bind α2 and alter its binding specificity such that the a1–α2 complex binds enhancers found upstream of haploid-specific genes and silences these genes. In diploid cells, then, the α2 protein exists in two forms: (1) as an α2–MCM1 complex that represses **a**-specific genes, and (2) in a complex with the a1 protein that represses haploid-specific genes. Moreover, the a1–α2 complex also represses expression of the α1 gene, which is thus no longer present to turn on α-specific genes. Thus, cell type-specific transcription of genes that control mating type in yeast is achieved by multiple transcription factors working in different combinations.

KEY CONCEPT The control of yeast mating type is an example of how cell type-specific patterns of transcription in eukaryotes can be governed by different combinations of interacting transcription factors.

12.2 CHROMATIN STRUCTURE

LO 12.2   Draw a segment of chromatin, labeling each histone, a nucleosome, and the structural features that are important to their function in transcription.

In eukaryotic cells, DNA is packaged with proteins to create **chromatin**. In the cell's nucleus, DNA in chromatin is compacted over 10,000-fold compared to its linear form. The structure of chromatin serves to fit DNA into the nucleus, and it also serves as a substrate for reversible changes in protein-DNA and protein-protein interactions that regulate transcription. In this section, we describe the structure of **histones** (the major protein components of chromatin), **nucleosomes** (the basic structural units of chromatin), and higher-order chromatin structures (three-dimensional assemblies of nucleosomes with one another). Because higher-order chromatin structures in eukaryotic cells can make DNA inaccessible to binding by transcription factors, an understanding of chromatin structure is essential for understanding how transcription is regulated.

Histones

Eukaryotic cells express five types of histone proteins: H1, H2A, H2B, H3, and H4. Histones H2A, H2B, H3, and H4 are known as **core histones** because they form a core complex around which DNA is wrapped to form nucleosomes. Histone H1 is known as a **linker histone** because it binds the DNA that links adjacent nucleosomes. In addition to these **canonical histones** that package the newly replicated genome, there are **variant histones** that are incorporated into nucleosomes in a DNA-replication independent manner. As an example, the histone variant H2A-Z is 60 percent identical in sequence to canonical histone H2A. H2A-Z replaces H2A in nucleosomes at promoters of both transcriptionally active and silent genes. In contrast, another H2A variant called H2A-X is incorporated into nucleosomes at sites of DNA damage (Chapter 15). Variants of H1, H2B, and H3 also play specialized roles, but, as yet, no variants of H4 have been identified.

Histone proteins have unusual features that are relevant to their roles as structural and regulatory components of chromatin. They are extremely abundant. In mammalian cells, histones constitute approximately 70 percent of the protein complement of chromatin. There are about 10 million copies of each core histone per cell and about half this amount of histone H1. Core histone proteins are small (11–15 kDa), unusually basic (at least 20 percent of their amino acids are lysine or arginine), and positively charged at neutral pH. Electrostatic interactions between positively charged amino acids and the negatively charged phosphate backbone of DNA play an important role in determining the structure of chromatin. The sequences of core histone proteins are among the most highly conserved in evolution. From yeast to humans, both H2A and H2B sequences are more than 70 percent identical, and both H3 and H4 are more than 90 percent identical. Because of this conservation, studies of histones in genetically controllable organisms such as yeast and *Drosophila* have provided considerable insights into the function of histones in higher eukaryotic organisms such as humans.

Core histone proteins have three types of structural domains: **histone folds, histone-fold extensions,** and **flexible tails** (**Figure 12-10a**). Histone folds are located in the central region of histone proteins. They are approximately 70 amino acids in length and are made up of three α-helices separated by loops (Figures 12-10b and c). Hydrophobic contacts between α-helices of histone folds are critical for the specific pairing of H2A with H2B and H3 with H4. Histone-fold extensions also make a significant contribution to the specificity of histone pairing. Finally, as the name suggests, flexible tails are located at the ends of histone proteins (Figure 12-10d). The tails are largely unstructured and are involved in interactions with non-histone proteins as well as neighboring nucleosomes.

The structure of linker histone H1 is substantially different than that of core histones. It is larger (~21 kDa) and it has much greater sequence and structural diversity. For example, in humans, histone H1 has three domains, a central domain of approximately 80 amino acids flanked by unstructured N- and C-terminal tails of approximately 20 and 100 amino acids, respectively; whereas in yeast, histone H1 has only a single, unstructured domain.

Nucleosomes

Nucleosomes are the basic structural units of chromatin. They contain 146 base pairs of DNA that wrap about 1.7 times around a **histone octamer** (eight proteins) consisting of two copies of each of the four core histones H2A, H2B, H3, and H4 (only one copy is shown in Figure 12-10c). The stability of nucleosomes is due to many protein-protein interactions within the histone octamer and electrostatic and hydrogen bonds between histones and DNA. The flexible tails of histones extend away from the nucleosomal DNA and are involved in interactions with adjacent nucleosomes and numerous nuclear factors (Figure 12-10d).

Molecular machines that assemble and disassemble nucleosomes play important roles in regulating transcription. During DNA replication, nucleosome formation begins with assembly of an H3/H4 tetramer (two H3/H4 dimers joined together) on DNA followed by sequential addition of two H2A/H2B dimers. Binding of histone H1 to nucleosomes organizes an additional 20 base pairs of linker DNA to form a complete nucleosome. Removal of histones from DNA occurs in the reverse order, beginning with sequential removal of the H2A/H2B dimers, followed by removal of the H3/H4 tetramer. Neighboring nucleosomes are separated from one another by ~20–75 base pairs of linker DNA.

> **KEY CONCEPT** In eukaryotes, DNA is packaged with histones in chromatin. Nucleosomes, the units of chromatin, contain two copies of each of the core histones (H2A, H2B, H3, and H4) around which is wrapped 146 base pairs of DNA. Complete nucleosomes also contain histone H1 and linker DNA of variable length.

Chromatin folding

Wrapping of DNA around histone octamers forms a structure of ~11 nanometers (nm) in diameter and compacts

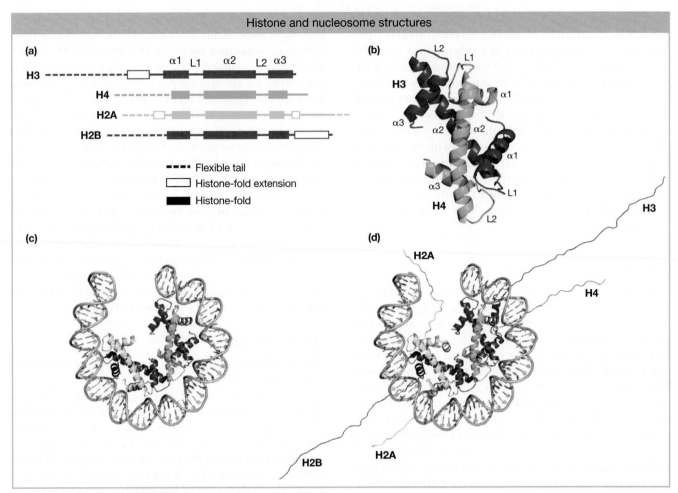

FIGURE 12-10 (a) Primary structures of the four core histones. α1, α2, and α3 are alpha-helices and L1 and L2 are loops. (b) Interactions between the histone-fold structures of histones H3 and H4. (c) The structure of a nucleosome, showing the histone-fold regions of one copy of each of the four core histones wrapped by DNA (gray). (d) The same view of the nucleosome as in (c) with the addition of H2A and H3 histone-fold extensions and flexible tails for all four core histones that extend beyond the wrapped DNA. [PDB ID 5y0c and 1aoi.]

DNA ~sixfold (**Figure 12-11**). This does not come close to the 10,000-fold compaction that occurs in eukaryotic cells. To achieve higher levels of compaction, nucleosomes fold upon themselves. The next order of chromatin folding produces the 30-nm fiber, a structure ~30 nm in diameter, and there are even more compact structures.

During the cell cycle, chromosomes vary in their level of compaction. Nucleosomes in mitosis are much more highly compacted than in interphase (see Appendix 2-1). Even in interphase, regions of chromosomes vary in their level of compaction. More compacted regions are called **heterochromatin**, and less compacted regions are called **euchromatin**. Heterochromatin constitutes a significant fraction of some eukaryotic genomes—approximately 20 percent for humans and 30 percent for *Drosophila*—but very little of others—less than 1 percent for the yeast. Chromatin that remains heterochromatic throughout the cell cycle is called **constitutive heterochromatin**, is concentrated at centromeres and telomeres, and is rich in repetitive sequences such as transposons but poor in genes. In contrast, **facultative heterochromatin** is spread along

chromosome arms, is gene-rich, and, through mechanisms described later in this chapter, can lose its compact structure and become transcriptionally active euchromatin.

KEY CONCEPT Regions of the genome with few genes, such as centromeres and telomeres, are compacted into heterochromatin throughout the cell cycle, whereas regions that are gene-rich vary in their level of chromatin compaction. Typically, genes are transcriptionally silent when compacted into heterochromatin, and they can be transcriptionally active when less compacted into euchromatin.

Over the past 20 years, new technologies such as chromatin immunoprecipitation (ChIP, Chapter 14) have made it possible to determine the genome-wide distribution of nucleosomes. As an example, in the yeast *S. cerevisiae*, Frank Pugh and colleagues found that approximately 70,000 nucleosomes occupy 81 percent of the genome and are typically separated by an 18-base-pair linker. Furthermore, nucleosomes are not equally distributed in the genome; they cover 87 percent of transcribed regions, but only 53 percent

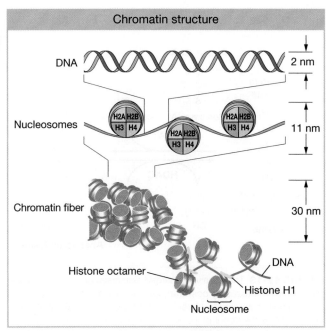

FIGURE 12-11 Chromatin is made up of 11 nm nucleosomes that fold upon one another into a compact filament of 30 nm.

ANIMATED ART 🐧 Sapling Plus

Three-dimensional structure of nuclear chromosomes

of intergenic regions. Focusing in on individual genes, transcription start sites are often located within a 150-base-pair **nucleosome free region (NFR)** that contains the promoter and is flanked by positioned nucleosomes, termed the −1 and +1 nucleosomes (**Figure 12-12a**). The precise positioning of nucleosomes is gradually reduced further upstream and downstream of the promoter. Enhancers are also flanked by a pair of nucleosomes. Enhancers of transcriptionally repressed genes can have nucleosomes positioned at the binding sites for transcription factors, but these nucleosomes are eliminated upon transcription activation.

The three-dimensional organization of chromatin in the nucleus is not random. Individual chromosomes occupy distinct territories, with gene-dense chromosomes located near the center of the nucleus and gene-poor chromosomes located near the nuclear periphery. Within and between chromosomes, large domains of transcriptionally active chromatin associate with one another. Similarly, inactive chromatin domains associate with one another. Smaller regions of chromatin are organized into **topologically associating domains (TADs)** whose DNA sequences preferentially contact one another. For example, interactions between gene enhancers and promoters are mostly limited to within a TAD (Figure 12-12b). Anchor points for the looping out of chromatin in TADs are

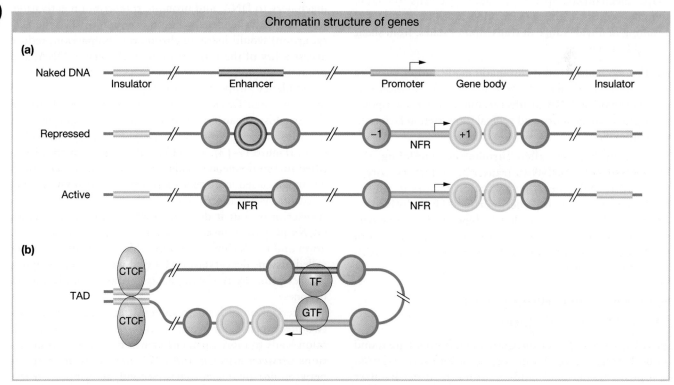

FIGURE 12-12 Genes contain insulator, enhancer, and promoter elements that regulate transcription. (a) Nucleosomes, indicated by purple circles, cover most of the transcribed region of genes (i.e., the gene bodies) but are excluded from enhancers and promoters of active genes to create nucleosome free regions (NFRs). (b) In chromosomes, groups of genes are segregated from one another by topologically associating domains (TADs) that are formed by the interaction of insulator binding proteins such as CTCF. Within TADs, enhancers bound by transcription factors (TFs) are positioned to act on general transcription factors (GTFs) at particular promoters.

defined by specialized regulatory sequences called **insulators** or **boundaries** that interact with one another or possibly the nuclear envelope through their associated proteins. In mammals, most of the known insulator sequences are bound by a zinc-finger DNA-binding protein called CTCF (CCCTC-binding factor). Therefore, insulators divide chromosomes into precisely defined loops that determine which enhancer-promoter interactions are allowed and which are prevented.

> **KEY CONCEPT** The wrapping of DNA enhancer elements into nucleosomes can prevent binding by transcription factors. Insulators prevent enhancers and their associated transcription factors from activating the transcription of genes outside a TAD.

12.3 CHROMATIN REGULATES TRANSCRIPTION

> **LO 12.3** Compare and contrast how chromatin modifying and chromatin remodeling mechanisms contribute to gene-specific transcription regulation.

The packaging of eukaryotic DNA into chromatin means that much of DNA is not readily accessible to the transcription machinery. Thus, eukaryotic genes are generally inaccessible and transcriptionally silent unless activated. Two major mechanisms operate in eukaryotic cells to enable dynamic access of the transcription machinery to DNA, resulting in a wide range of transcription states, from silent to highly active.

1. In a mechanism called **chromatin modification**, enzymes alter the chemical structure of amino acids in histones or nucleotides in DNA to affect recruitment of transcription factors, coregulators, and general transcription factors to chromatin.

2. In a mechanism called **chromatin remodeling**, the accessibility of DNA to transcription factors, coregulators, and general transcription factors is altered by enzymes that use energy from ATP hydrolysis to remodel nucleosomes; that is, reposition histone octamers along the DNA, remove histone octamers from DNA, or replace canonical histones in octamers with variant histones.

Histone modification: a type of chromatin modification

In 1964, Vincent Allfrey discovered that histones are found in both acetylated and non-acetylated forms. Acetylation is a **post-translational modification** (i.e., it occurs after translation) and consists of addition of an acetyl group to the amino group of a lysine amino acid side chain (Figure 12-13). Allfrey hypothesized that histone acetylation affects transcription. His thinking was that acetylation neutralizes the positive charge of lysine and thereby decreases the affinity of lysine for the negatively charged phosphate

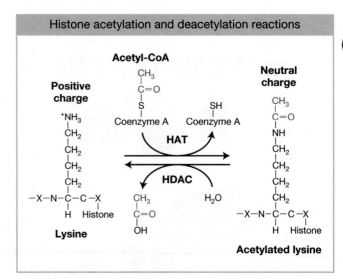

FIGURE 12-13 The positively charged side chain of lysine residues in histones is neutralized by post-translational addition of an acetyl group. Acetylation is catalyzed by histone acetyltransferases (HATs) that use acetyl-CoA as the donor of the acetyl group, indicated in red. The reverse reaction, deacetylation, is catalyzed by histone deacetylases (HDACs).

backbone of DNA. As a consequence, lysine acetylation (i.e., addition of an acetyl group) would reduce chromatin compaction, increase accessibility of the transcription machinery to DNA, and promote transcription activation. Conversely, lysine deacetylation (i.e., removal of the acetyl group) would increase chromatin compaction, reduce accessibility of the transcription machinery to DNA, and promote transcription repression.

Evidence supporting this hypothesis was uncovered in 1996 when David Allis and his colleagues identified the first **histone acetyltransferase (HAT)**, an enzyme that transfers an acetyl group from acetyl-CoA to lysines in histones (Figure 12-13). The HAT called p55 that they identified in Tetrahymena (a ciliated protozoan) turned out to be similar in sequence to a yeast protein called GCN5 that functions as a transcription coactivator. GCN5 promotes transcription but it does not directly bind DNA. Thus, GCN5 provides a mechanistic link between histone acetylation and transcription activation. Subsequently, enzymes called **histone deacetylases (HDACs)** were found to repress transcription by removing acetyl groups from lysines in histones.

Acetylation of lysine residues affects transcription by two mechanisms. First, as hypothesized by Allfrey, acetylation leads to more open chromatin by loosening interactions between histones and DNA as well as interactions between nearby nucleosomes. Second, acetylation creates a binding site for a protein motif called a bromodomain. Several transcription regulatory factors, including the TAF1 subunit of the general transcription factor TFIID, contain bromodomains that increase the affinity of the factor for particular genes by binding acetylated histones. Using the nomenclature introduced in Chapters 8 and 9, HATs are writers, HDACs are erasers, and bromodomains are readers of histone acetylation (**Figure 12-14**).

KEY CONCEPT Acetylation of lysines in histones by HATs (1) loosens interactions within and between nucleosomes and (2) creates a binding site for bromodomains, found in some transcription coregulators.

In summary, histone acetylation plays a crucial role in stepwise mechanisms that activate transcription: a transcription factor binds an enhancer, a HAT such as GCN5 binds the transcription factor, the HAT acetylates histones in nucleosomes at the promoter, a bromodomain protein such as TAF1 binds acetylated histones, and RNA polymerase II is recruited either directly or indirectly by the bromodomain protein. Similarly, an activated gene is turned off by transcription factor-mediated binding of an HDAC, which deacetylates histones and blocks recruitment of bromodomain-containing proteins. Acetylation affects transcription initiation, and elongation, by being targeted to nucleosomes positioned in different regions of genes and thereby affecting recruitment of bromodomain-containing initiation and elongation factors (Chapter 8).

Acetylation is one of many **histone modifications** that affect transcription. Other abundant modifications include methylation of lysine and arginine residues; phosphorylation of serine, threonine, and tyrosine residues; and ubiquitination of lysine residues. Furthermore, lysine can be methylated one, two, or three times (monomethyllysine, dimethyllysine, and trimethylysine, respectively); while arginine can be methylated one time (monomethylarginine) or two times in symmetric or asymmetric configurations (dimethylarginine) (**Figure 12-15**). Methylation is controlled by writers (histone methyltransferases, HMTs), erasers (histone demethylases, HDMs), and readers (proteins that contain a chromodomain or a plant homeodomain (PHD) finger). There are also writers, erasers, and readers for phosphorylation, ubiquitination, and other modifications, and both histones and DNA can be modified.

KEY CONCEPT Transcription is regulated by chemical modifications of amino acids in histones and nucleotides in DNA. Modifications are added by writer enzymes, removed by eraser enzymes, and bound by reader proteins.

The histone code hypothesis

Post-translational modifications occur in all parts of histone proteins but are concentrated in the tails. They are experimentally detected in vivo and in vitro using modification-specific antibodies, and in vitro by mass spectrometry of histones purified from cells. Unfortunately, these methods are largely unable to detect the extent to which modifications coexist on an individual histone protein. This information may be very important because different combinations of histone modifications may convey the information to bring about different transcription outputs. In 2000, this idea was formalized by David Allis and Thomas Jenuwein in the **histone code** hypothesis, which proposes that multiple histone modifications, acting

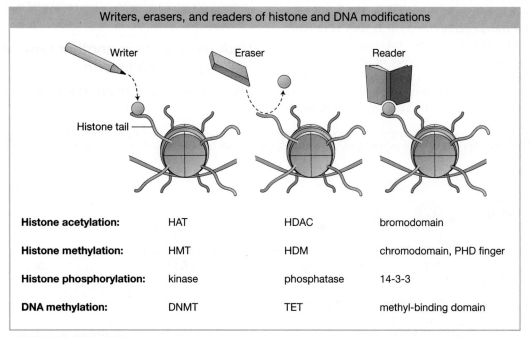

Writers, erasers, and readers of histone and DNA modifications			
	Writer	Eraser	Reader
Histone acetylation:	HAT	HDAC	bromodomain
Histone methylation:	HMT	HDM	chromodomain, PHD finger
Histone phosphorylation:	kinase	phosphatase	14-3-3
DNA methylation:	DNMT	TET	methyl-binding domain

FIGURE 12-14 The information content of histones and DNA is changed by writer enzymes that add and eraser enzymes that remove chemical modifications. Reader proteins interpret the information by binding chemical modifications. The text provides detailed descriptions of writer, eraser, and reader proteins for histone acetylation, methylation, and phosphorylation as well as DNA methylation.

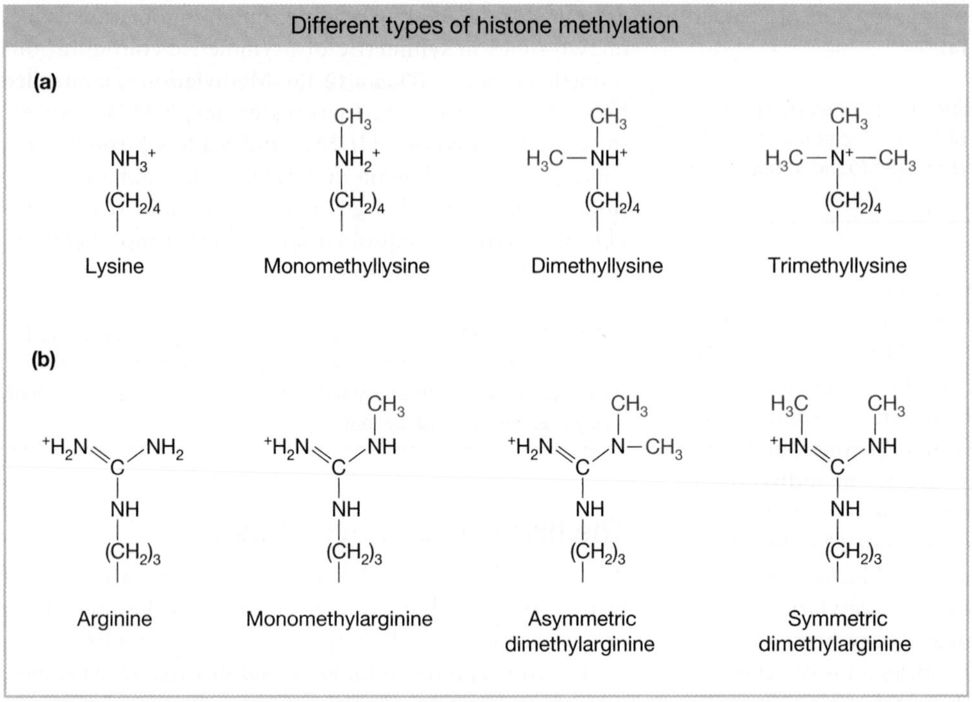

Different types of histone methylation

(a)

Lysine — Monomethyllysine — Dimethyllysine — Trimethyllysine

(b)

Arginine — Monomethylarginine — Asymmetric dimethylarginine — Symmetric dimethylarginine

FIGURE 12-15 There are several types of lysine and arginine methylation, each of which conveys different instructions to the transcription regulatory machinery. (a) Different types of lysine methylation. (b) Different types of arginine methylation. Only the amino acid side chain is drawn, with methyl groups shown in red.

sequentially or in combination on one or several histone tails, specify unique transcription outcomes.

In support of the histone code hypothesis, nucleosomes at promoters of transcriptionally active genes are commonly trimethylated (me3) on lysine (K) 4 of histone H3 (H3K4me3), whereas promoters of transcriptionally repressed genes are trimethylated on H3K9 (H3K9me3) (**Figure 12-16**). H3K4me3 activates transcription by serving as a binding site for transcription coactivators such as the PHD finger-containing TAF3 subunit of TFIID. In contrast, H3K9me3 represses transcription by serving as a binding site for transcription corepressors such as the chromodomain-containing protein heterochromatin protein 1 (HP1), which promotes the formation of heterochromatin. Combinations of modifications are also hallmarks of transcription activity—the combination of phosphorylation (P) of serine (S) 10 on histone H3 (H3S10P), which is bound by

14-3-3 proteins, and acetylation of lysine 14 on histone H3 (H3K14ac), which is bound by bromodomain proteins, signals transcription activation. The potential information content of histone modifications is enormous. For example, there are more than two million possible combinations of modifications that can occur on the N-terminal tail of histone H3.

KEY CONCEPT The histone code hypothesis posits that different combinations of histone modifications create unique binding sites that can be read by transcription coregulators, thereby conferring a variety of transcriptional outcomes.

DNA modification: another type of chromatin modification

Like histone modifications, **DNA modifications** affect transcription. In vertebrates, the predominant DNA

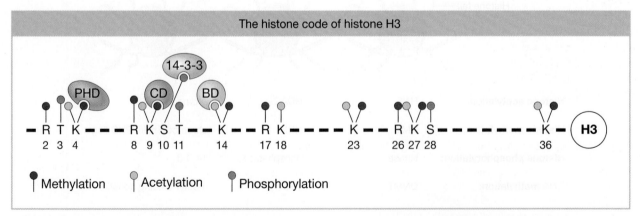

FIGURE 12-16 The N-terminal flexible tail of histone H3 contains lysine (K), arginine (R), serine (S), and threonine (T) amino acids that are post-translationally acetylated, methylated, or phosphorylated. Different combinations of modifications, termed a histone code, are thought to alter the level of transcription to different extents by conveying different information to the transcription machinery. The information is read by bromodomains (BD), chromodomains (CD), PHD fingers (PHD), 14-3-3 proteins, and other proteins and protein domains.

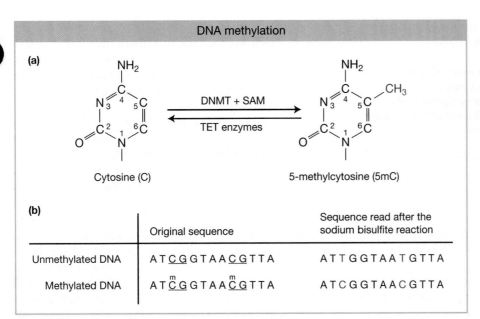

FIGURE 12-17 (a) Methylation of the fifth carbon position of cytosine (5mC) is catalyzed by DNA methyltransferases (DNMTs) using S-adenosylmethionine (SAM) as the methyl donor. TET enzymes reverse this reaction. (b) Sites of CpG methylation are detected by the sodium bisulfite reaction. CpGs are underlined, and differences in sequence after the sodium bisulfite reaction are indicated in red.

modification is 5-methylcytosine (5mC), where methylation by a DNA methyltransferase (DNMT) occurs at the fifth carbon in the cytosine ring of the dinucleotide CpG (cytidine-phosphodiester bond-guanosine) (**Figure 12-17a**). DNMTs use S-adenosyl methionine (SAM) as the methyl donor. In contrast to vertebrates, in plants, 5mC occurs in the dinucleotide CpG as well as other nucleotide contexts, and in *Drosophila*, *C. elegans*, and *S. cerevisiae*, little or no DNA methylation has yet been detected.

CpG methylation is reversed by enzymes in the TET (ten-eleven translocation) family (Figure 12-17a). Reversal comprises three steps, with each step catalyzed by a TET enzyme; the second and third intermediates are rapidly excised by a mechanism described in Chapter 15. In contrast, the first intermediate is more stable and is particularly abundant in embryonic stem cells and adult neurons where it is bound by reader proteins that regulate genes involved in development and tumorigenesis.

5mC is detected in the lab by the sodium bisulfite reaction. DNA isolated from cells is treated with sodium bisulfite, which in single-stranded DNA efficiently converts cytosine to uracil but inefficiently converts 5mC to thymine. After the sodium bisulfite reaction, the DNA is sequenced, and cytosines that are converted to uracil are read as thymine (T), whereas 5mC that are unchanged are read as cytosine (C) (Figure 12-17b).

In vertebrates, CpGs occur much less frequently than would be expected based on the C + G content of genomes. This is due to widespread methylation of CpGs and subsequent conversion by deamination over evolutionary time to TpG. Approximately 85 percent of CpGs are methylated and are scattered throughout the genome (**Figure 12-18**). The remaining CpGs are unmethylated, and many of these are highly clustered in 200- to 4000-base-pair regions called **CpG islands**. Approximately half of all CpG islands are located in gene promoters, and the remaining half are roughly equally divided between intragenic and intergenic locations. The majority of gene promoters are associated with a CpG island.

Unmethylated CpG islands at promoters are generally correlated with open chromatin and active transcription, whereas methylated islands are associated with closed chromatin and repressed transcription. These effects on transcription are mediated by proteins that distinctly bind unmethylated or methylated CpGs. In humans, the protein Cfp1 (CxxC finger protein 1) binds unmethylated CpGs and recruits a histone methyltransferase (HMT) that produces the transcription activating histone modification H3K4me3. In contrast, methylation of CpGs represses transcription by interfering with transcription factor binding to enhancers and by serving as a binding site for methyl binding domain (MBD) proteins that recruit transcription repressors such as HDACs that deacetylate lysine or HMTs that produce H3K9me3. Thus, DNMTs

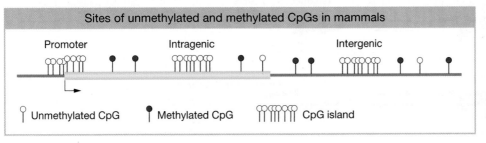

FIGURE 12-18 Mammalian genomes contain clusters of CpGs called CpG islands as well as dispersed CpGs that can be unmethylated or methylated. A representative gene is indicated by the green bar.

are writers, TETs are erasers, and a variety of proteins are readers of unmethylated and methylated CpGs.

KEY CONCEPT Methylation of cytosine in CpG islands at gene promoters is correlated with the repression of transcription. Like modifications of histone proteins, CpG methylation of DNA represses transcription by altering the affinity of transcription factors, coregulators, and general transcription factors for chromatin.

Chromatin remodeling

Chromatin remodeling is the workhorse in the process of altering chromatin structure to regulate transcription. Remodeling means changing histone-DNA interactions in nucleosomes to render DNA either more or less accessible to transcription regulators (**Figure 12-19**). To permit factors to bind enhancer and promoter elements, remodeling moves nucleosomes to new locations. On moderately active genes, remodeling displaces H2A/H2B dimers in front of RNA polymerase II and replaces them behind the polymerase during every round of transcription. H3/H4 tetramers may only be displaced and replaced on highly active genes. In contrast, eviction of histones from DNA may not be necessary to permit RNA polymerase II passage through genes being transcribed at a low level.

Chromatin remodeling complexes use energy from ATP hydrolysis to disrupt non-covalent histone-DNA interactions. Eukaryotic organisms contain four families of ATP-dependent chromatin remodeling complexes, which can be characterized as being associated with transcription activation or repression.

Two genetic screens in yeast for mutants in seemingly unrelated processes led to the discovery of a chromatin remodeling complex. In one study, mutagenized yeast cells were screened for the lack of growth on sucrose (sugar *nonf*ermenting mutants, *snf*, pronounced "sniff"). In the other study, mutagenized yeast cells were screened for defective switching of mating type (*swi*tch mutants, *swi*, pronounced "switch"). Many mutants for different loci were recovered in each screen, but one mutant gene was found to cause both phenotypes. Mutants at the *swi2/snf2* locus could neither use sucrose effectively nor switch mating type because the transcription of specific genes was blocked. The protein encoded in the *swi2/snf2* locus was found to be the ATPase subunit of the multisubunit SWI/SNF ("switch-sniff") chromatin remodeling complex.

SWI/SNF affects transcription by remodeling nucleosomes in two steps, initially removing an H2A/H2B dimer from DNA, followed by removal of the rest of the histone octamer. The gene specificity of the SWI/SNF complex is provided by binding to transcription factors such as Gal4 and through binding of a bromodomain-containing subunit of the complex to acetylated lysine. Thus, transcription factors, histone modifying enzymes, and chromatin remodeling factors function in concert to regulate transcription.

In contrast to the other chromatin remodeling complexes that slide, eject, or replace histone octamers on DNA, the SWR1 (pronounced "swur one") complex remodels chromatin by assembling the variant histone H2A-Z into chromatin. SWR1 does this by exchanging H2A-Z/H2B dimers for H2A/H2B dimers in histone octamers. This activity is targeted to enhancers and promoters of specific genes by the bromodomain-containing subunit of SWR1, which binds specific acetylated lysines on histone tails. Nucleosomes that contain H2A-Z are particularly prone to disassembly by other chromatin remodeling complexes, leading to increased access of transcription regulators to DNA.

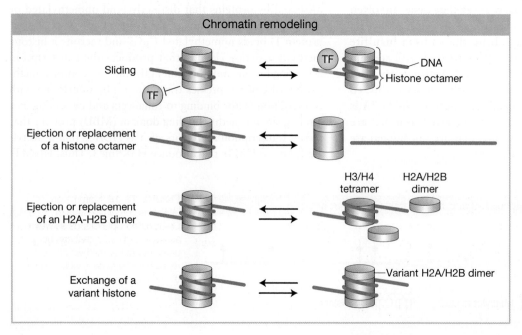

Chromatin remodeling

Sliding

Ejection or replacement of a histone octamer

Ejection or replacement of an H2A-H2B dimer

Exchange of a variant histone

TF — DNA
Histone octamer

H3/H4 tetramer   H2A/H2B dimer

Variant H2A/H2B dimer

FIGURE 12-19 Chromatin remodeling complexes use energy produced by ATP hydrolysis to slide, eject, or replace histone octamers on DNA or exchange variant histones for canonical histones in octamers. TF indicates a transcription factor that binds the green enhancer element.

ANIMATED ART   SaplingPlus

Chromatin remodeling

KEY CONCEPT Chromatin is dynamic; nucleosomes are not necessarily in fixed positions on the chromosome. Chromatin remodeling complexes change nucleosome density, position, and subunit composition to control access of the transcription machinery to DNA.

Connecting chromatin structure to transcription: lessons from the *interferon-β* gene

The human *interferon-beta* (*IFN-β*) gene, which encodes an antiviral protein, is one of the best-characterized genes in eukaryotes. Its transcription is normally switched off but, upon viral infection, is activated to very high levels. A central feature of activation of this gene is assembly of multiple, different transcription factors into an **enhanceosome** about 100 base pairs upstream of the TATA box promoter element and transcription start site.

A study of the *IFN-β* gene illustrates how regulated changes in chromatin structure affect transcription. Dimitris Thanos and colleagues used the chromatin immunoprecipitation (ChIP) technique (Chapter 14) to identify histone modifications and other events that occur at the *IFN-β* promoter as the gene shifts from transcriptionally inactive to active and back to inactive over a 24-hour period in response to virus infection of human cells (**Figure 12-20a**). They also used reverse transcription-PCR (RT-PCR) to quantify *IFN-β* mRNA levels. The following bullet points walk through the temporal pathway of molecular events uncovered in the study and they highlight the general mechanisms by which chromatin structure affects transcription:

- *IFN-β* mRNA is first detected 6 hours after viral infection, but histone modifications are detected as early as 3 hours after viral infection (Figure 12-20a). This illustrates that histone modifications occur prior to the first transcription initiation event to generate a chromatin structure that is conducive to transcription initiation and elongation.

- The earliest event is cooperative binding of a suite of transcription factors to the *IFN-β* proximal enhancer (Figure 12-20b) to form an enhanceosome (Figure 12-20c). The histone acetyltransferase GCN5 then binds the assembled enhanceosome and acetylates H4K8 and H3K9 and a kinase binds and phosphorylates H3S10 in nucleosomes near the *IFN-β* promoter (Figure 12-20d). This illustrates that transcription factors recruit histone modifying enzymes to generate specific histone modifications in particular nucleosomes.

- H4K8ac and H3S10P are last detected at 8 hours (see Figure 12-20a), even though transcription occurs up to 15 hours. In contrast, H3K9ac is last detected at 19 hours. Unknown HDACs and phosphatases are involved in removing these histone marks. This illustrates that both writing and erasing of histone modifications as well as the relative timing of histone modifications are important for transcription regulation.

- H3S10P peaks at 6 hours, the time at which H3K14ac is first detected (see Figure 12-20a), suggesting that H3S10P is required for GCN5 to acetylate H3K14 (Figure 12-20e). This illustrates that modification of one amino acid can promote or inhibit modification of other amino acids, a process called crosstalk.

- TBP binding is first detected at 6 hours, which is the same time that *IFN-β* transcription starts (see Figure 12-20a). The TFIID complex is recruited through direct binding of TBP to the TATA promoter element as well as binding of TAF1 bromodomains to H3K9ac and H3K14ac (Figure 12-20e). In addition, the SWI/SNF chromatin remodeling complex is recruited by binding to H4K8ac as well as interactions with another HAT called CBP (CREB-binding proteins), which replaces GCN5 at the enhanceosome. To initiate *IFN-β* transcription, TFIID recruits RNA polymerase II to the promoter and SWI/SNF remodels nucleosomes to allow the polymerase to initiate transcription. This illustrates that histone modifications aid in assembly of the transcription preinitiation complex as well as factors that remodel chromatin structure.

- Last, transcription is turned off at 24 hours, at which time activating histone modifications have mostly been removed by eraser enzymes (Figure 12-20g). This illustrates the rapid reversibility of chromatin-mediated control of transcription.

Lessons learned from the *IFN-β* gene are generalizable. The molecular mechanisms that alter chromatin structure turn transcription of particular genes on or off in response to developmental and environmental signals by permitting or preventing transcription regulatory proteins access to DNA. The information that controls transcription consists of DNA regulatory sequences (e.g., enhancers, promoters, and insulators), the histone code (e.g., various chemical modifications of histone amino acids), and DNA modifications (e.g., unmodified CpG and 5mC) (**Figure 12-21a**). Proteins read the information through physical interactions with the information-containing elements (Figure 12-21b). DNA sequences are bound by transcription factors via their DNA-binding reader domain. Modified histones and modified DNA are bound by other coregulators that contain a variety of reader domains (e.g., bromodomain, chromodomain, and methyl-binding domain). Once bound to chromatin, reader proteins serve as scaffolds for assembly of enzymes that change chromatin structure (Figure 12-21c). Reader proteins are either part of a stable complex that contains enzymes (e.g., SWI/SNF) or they recruit enzymes through protein-protein interactions. The enzymes then either edit the information at the gene by modifying histones or DNA, or they change the chromatin structure by sliding nucleosomes, ejecting or replacing histone octamers or parts of octamers, or exchanging variant histones for canonical histones (Figure 12-21d). Collectively, these events alter the access to DNA by RNA polymerase II, transcription initiation factors (e.g., TFIID), and transcription elongation factors (e.g., P-TEFb) (Figure 12-21e). Lastly,

Chromatin-mediated control of *IFN-β* transcription

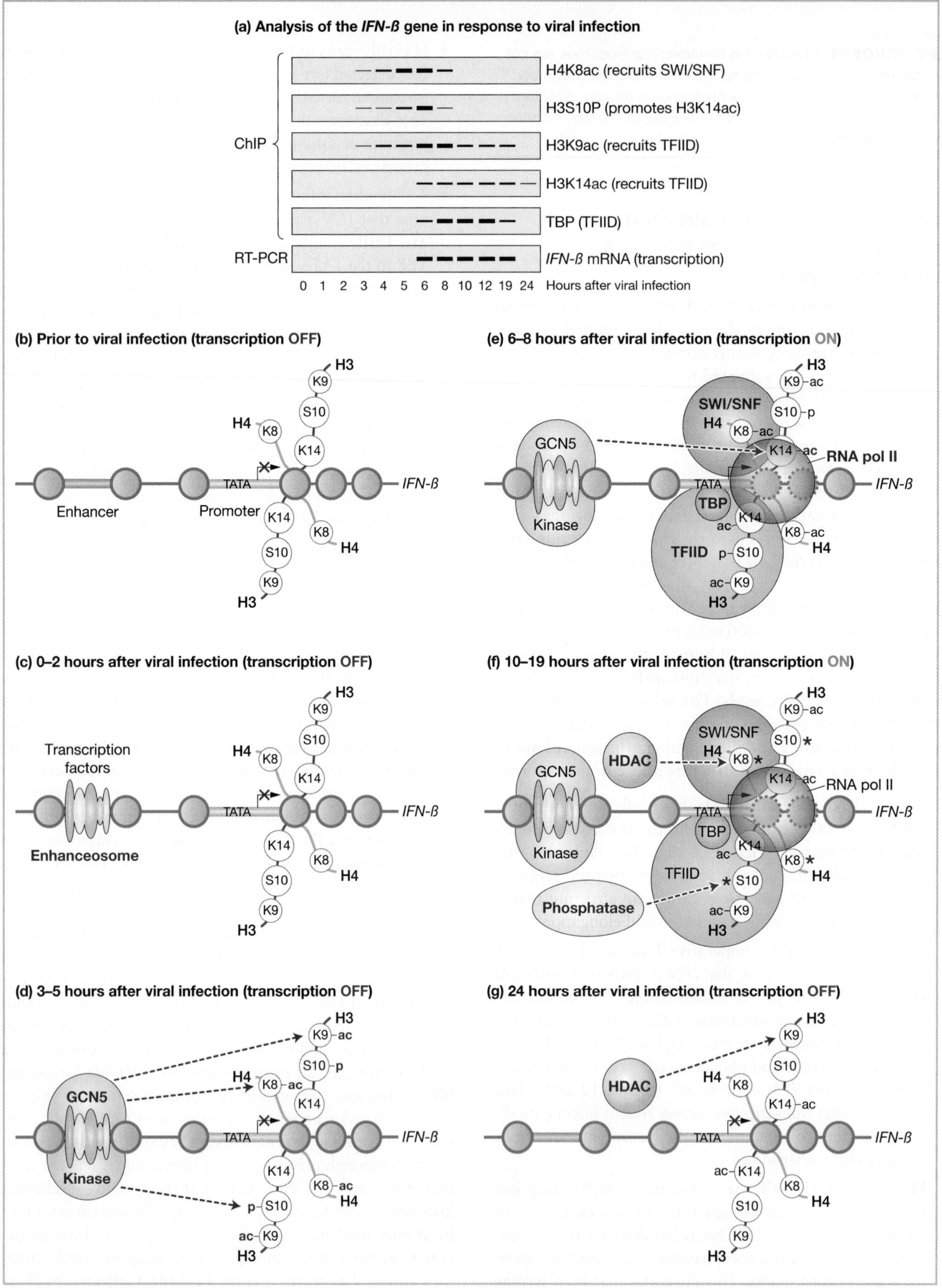

FIGURE 12-20 (a) ChIP and RT-PCR analysis of the *IFN-β* gene. For ChIP, the intensity of the black bands is proportional to the amount of modified histone or TBP associated with the *IFN-β* promoter. For RT-PCR, the intensity of the black bands indicates the amount of *IFN-β* mRNA. Descriptions in parentheses indicate the functional consequence of the experimental outcome. (b–g) Models that illustrate the sequential molecular interactions and enzymatic events that occur over time after viral infection. Purple circles indicate nucleosomes. Dotted purple circles indicate nucleosomes altered by chromatin remodeling. Note that transcription is on only 6 to 19 hours after viral infection. [*Data from Agalioti et al., "Deciphering the Transcriptional Histone Acetylation Code for a Human Gene,"* Cell *111, 2002, 381–392.*]

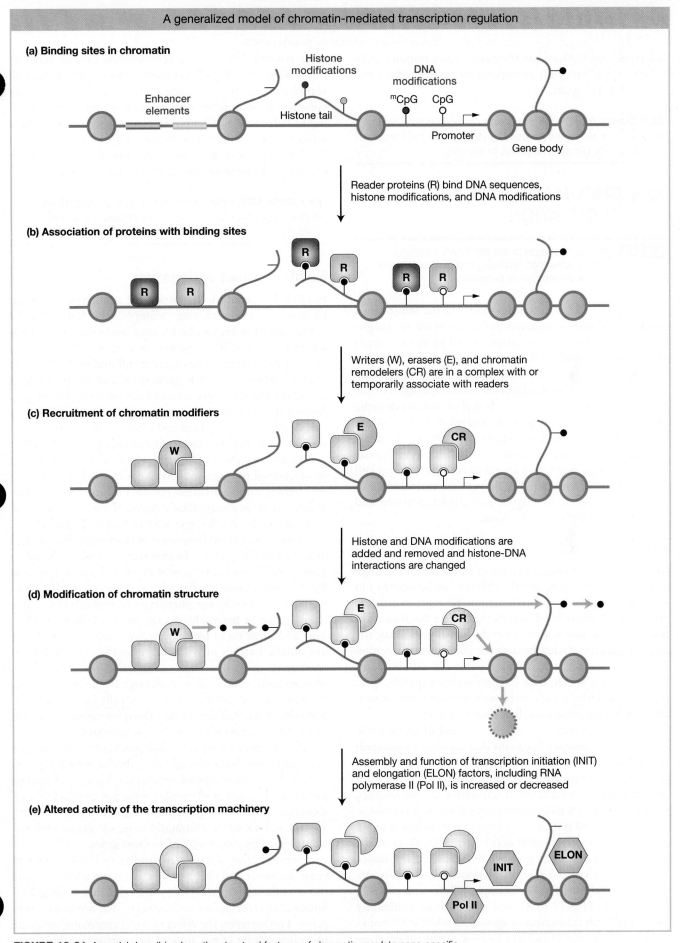

FIGURE 12-21 A model describing how the structural features of chromatin regulate gene-specific transcription in response to signals. Red and green indicate factors that, respectively, activate or repress transcription, and purple circles indicate nucleosomes.

since these mechanisms are reversible, transcription can be rapidly turned on or off in response to developmental and environmental signals.

KEY CONCEPT *IFN-β* transcription exemplifies how chromatin regulatory strategies are used by cells to alter the transcription of specific genes in response to signals.

12.4 CHROMATIN IN EPIGENETIC REGULATION

LO 12.4 Provide examples of the chromatin-based mechanisms that maintain gene expression over cellular or organismal generations.

When cells divide, information stored in the sequence of DNA is faithfully replicated and transferred to daughter cells. Similarly, information stored in the structure of chromatin is inherited through cell divisions. This form of inheritance is given a special name—**epigenetic inheritance**—because it affects the traits of daughter cells without altering DNA sequence. In this section, we describe four examples of the epigenetic control of transcription: cellular memory, position-effect variegation, genomic imprinting, and X-chromosome inactivation. In each case, the collection of genes that are transcribed in a parent cell is reproduced in daughter cells through the maintenance of chromatin structure by histone and DNA modifying and chromatin remodeling mechanisms.

Cellular memory

Unlike DNA sequence, chromatin structure can change during the life of a cell, and the changes can be inherited in successive generations of cell division. Changes in cell fate are based on short-lived signals that affect the transcription of specific genes. Even after the signal goes away, the cell fate does not change because the effect on transcription stays. For instance, once an embryonic cell differentiates into an intestinal cell, with its intestinal cell-specific spectrum of transcriptionally active and inactive genes, it usually remains an intestinal cell as long as it lives.

Studies that were initially performed in *Drosophila* identified two groups of proteins that function to maintain the cellular memory of transcription, *Polycomb* group proteins and *Trithorax* group proteins. Polycomb and Trithorax proteins often function in opposition to one another, with Polycomb proteins maintaining genes in a transcriptionally *repressed* state and Trithorax proteins maintaining genes in a transcriptionally *active* state. Members of the Polycomb and Trithorax groups are components of multiprotein complexes that post-translationally modify histones and remodel chromatin. For example, a Polycomb complex *trimethylates* H3K27 (a histone modification commonly associated with transcription silencing), while a Trithorax complex *acetylates* H3K27 (a histone modification commonly associated with transcription activation). Note that

since trimethyl-lysine and acetyl-lysine cannot occur at the same time on H3K27, trimethylation by the Polycomb complex blocks the activating acetylation by the Trithorax complex. As with other transcription coregulators, targeting of Polycomb and Trithorax complexes to chromatin is influenced by transcription factors, histone modifications, DNA methylation, and long noncoding RNAs (for example, see X-chromosome inactivation later in this section).

KEY CONCEPT Polycomb and Trithorax group proteins work in opposition to maintain the repressed and active transcription states of parent cells in daughter cells.

Position-effect variegation

In 1930, Hermann Muller discovered an interesting genetic phenomenon while studying *Drosophila*. He found that the expression of genes can be silenced when they are experimentally "relocated" to another region of a chromosome. In these experiments, flies were irradiated with X rays to induce mutations in their germ cells, and the progeny of the irradiated flies were screened for unusual phenotypes. Among the collection of mutants, Muller found flies with eyes that had patches of red and white color. This is unusual because wild-type flies have uniform red eyes, and flies that are mutant for the *white* gene, which is required for the production of red pigment, have uniform white eyes.

Cytological examination revealed a chromosomal rearrangement in the mutant flies: a region of the X-chromosome containing the *white* gene was inverted (**Figure 12-22**). Inversions and other chromosomal rearrangements will be discussed in Chapter 17. In this rearrangement, the *white* gene, which is normally located in a euchromatic region of the X-chromosome, now is near the heterochromatic centromere. The patchy eye phenotype of Muller's flies is due to spreading of heterochromatin into the wild-type *white* gene and silencing of *white* transcription in some cells but not others. Patches of white tissue in the eye are derived from descendants of a single cell in which the *white* gene is silenced and remains silenced through future cell divisions. In contrast, red patches arise from cells in which heterochromatin has not spread into the *white* gene, and so the *white* gene remains active in all its descendants.

The existence of red and white patches of cells in the eye of a single organism dramatically illustrates two features of epigenetic transcription regulation. First, as described earlier, differences in chromatin structure across chromosomes can be inherited from one cell generation to the next. Second, differences in chromatin structure across chromosomes affect the expression of resident genes.

Findings from subsequent studies in *Drosophila* and yeast demonstrated that many active genes are silenced in this mosaic fashion when they are relocated to neighborhoods near centromeres or telomeres that are heterochromatic. Furthermore, the effect of local chromatin structure on transcription is not limited to centromeres and telomeres. In mouse cells the degree of chromatin compaction

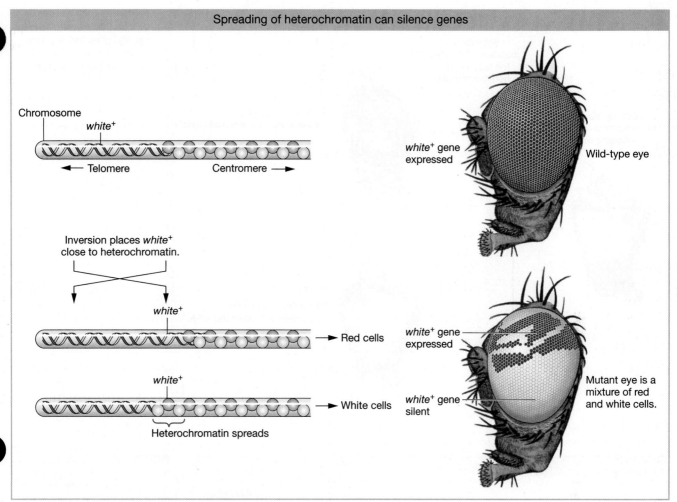

FIGURE 12-22 Chromosomal rearrangement produces position-effect variegation (PEV). A chromosomal inversion places the wild-type *white* gene close to heterochromatin, and spreading of heterochromatin into *white* silences transcription. Wherever *white* is silenced, cells are white because they do not make red pigment.

at the site of integration of a transgene correlates with the level of transcription of the transgene and accounts for about a 1000-fold variation in transcription level across the genome. This phenomenon has been called **position-effect variegation (PEV)**. It provides powerful evidence that chromatin structure is able to regulate the expression of genes—in this case, determining whether genes with identical DNA sequence will be active or silent.

Geneticists reasoned that PEV could be exploited to identify the proteins necessary for forming heterochromatin. To this end, they isolated mutations that either suppressed or enhanced the variegated pattern (**Figure 12-23**). A Suppressor of variegation (*Su(var)*) is a gene that when mutated reduces the spread of heterochromatin, meaning that the wild-type product of this gene is required for spreading. In contrast, an Enhancer of variegation (*E(var)*) is a gene that when mutated increases the spread of heterochromatin and normally functions to block spreading. *Su(var)* and *E(var)* genes have proved to be a treasure trove for scientists interested in the proteins that are required to establish and maintain the heterochromatic state.

Among more than 300 *Drosophila* mutants identified by these screens was *Su(var)2-5*, which encodes a histone reader protein heterochromatin protein 1 (HP1), and *Su(var)3-9*, which encodes a histone methyltransferase (**Figure 12-24**). HP1 contains a chromodomain that binds H3K9me3 and a chromoshadow domain involved in dimerization of HP1 proteins and recruitment of a variety of chromatin-modifying factors. On the other hand, *Su(var)3-9* trimethylates H3K9. HP1 and *Su(var)3-9* interact with one another to create a feed-forward loop that spreads heterochromatin. HP1 binds H3K9me3 and dimerizes with another HP1 molecule; the dimer recruits *Su(var)3-9*, which generates H3K9me3; and HP1 binds H3K9me3 to continue the process.

In the absence of any barriers, heterochromatin might spread into adjoining regions and inactivate genes in some cells but not in others. One can imagine that the spreading of heterochromatin into active gene regions could be disastrous for an organism because active genes would be silenced as they are converted into heterochromatin. To avert this potential disaster, boundary/insulator elements,

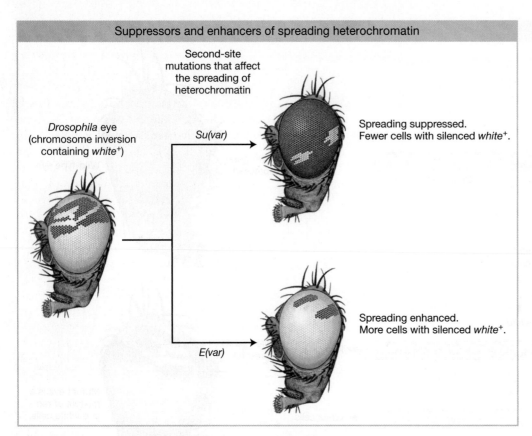

Suppressors and enhancers of spreading heterochromatin

FIGURE 12-23 Forward genetic screens were used to identify genes that suppress, *Su(var)*, or enhance, *E(var)*, position-effect variegation.

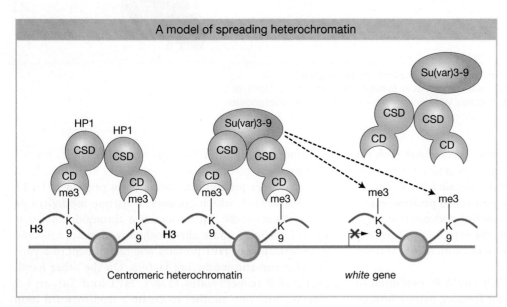

A model of spreading heterochromatin

FIGURE 12-24 The coordinated activities of HP1 and *Su(var)3-9* spread heterochromatin from the centromere into the repositioned *white* gene. The dotted arrows indicate that *Su(var)3-9* trimethylates H3K9 on adjacent nucleosomes, recruiting HP1 and silencing transcription of the *white* gene. CD and CSD indicate the HP1 chromodomain and chromoshadow domain, respectively, and purple circles indicate nucleosomes.

which were discussed earlier in the context of topologically associating domains (TADs, Figure 12-12) prevent the spreading of heterochromatin by creating a local environment that is not favorable to heterochromatin formation. Insulator-binding proteins may block the spread of heterochromatin by recruiting activating enzymes such as histone acetyltransferases, H3K4 methyltransferases, and SWI/SNF chromatin remodelers, or they may block access to histones by directly binding them.

KEY CONCEPT Proteins involved in the spread of heterochromatin include writers, readers, and erasers of histone modifications.

Genomic imprinting

The phenomenon of **genomic imprinting** was discovered about 35 years ago in mammals. In genomic imprinting, certain autosomal genes are expressed in a

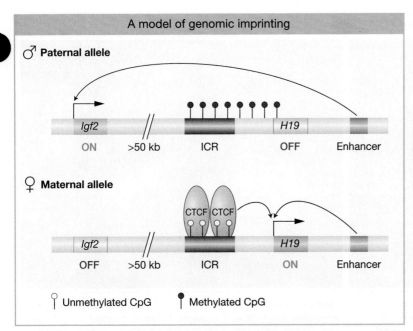

A model of genomic imprinting

♂ **Paternal allele**

Igf2 — ON   >50 kb   ICR   H19 — OFF   Enhancer

♀ **Maternal allele**

CTCF CTCF

Igf2 — OFF   >50 kb   ICR   H19 — ON   Enhancer

♀ Unmethylated CpG   ● Methylated CpG

FIGURE 12-25 Genomic imprinting in the mouse. The imprinting control region (ICR) is methylated in male gametes, blocking CTCF binding and directing the enhancer to activate transcription of *Igf2*. The unmethylated ICR in female gametes binds CTCF, forming an insulator that blocks enhancer activation of *Igf2* and directs activation of *H19*.

parent-of-origin-specific manner. For example, transcripts from the *Igf2* (*insulin-like growth factor 2*) gene in mammals come exclusively from the father's (i.e., paternal) allele because the mother's (i.e., maternal) allele is silenced—an example of **maternal imprinting** because the copy of the gene derived from the mother is transcriptionally inactive. Conversely, *H19* transcripts come exclusively from the mother's allele; *H19* is an example of **paternal imprinting** because the paternal copy is transcriptionally inactive. The consequence of parental imprinting is that imprinted genes are expressed as if only one copy of the gene is present in the cell even though there are two. Importantly, no changes are observed in the DNA sequences of imprinted genes; that is, the identical gene can be active or inactive in the progeny, depending on whether it was inherited from the mother or father. Imprinted genes are controlled by DNA regulatory elements called imprinting control regions (ICRs) that have parent-specific chromatin modifications. This then represents an epigenetic phenomenon.

Let's turn again to the mouse *Igf2* and *H19* genes to see how imprinting works at the molecular level. These two genes are located in a cluster of imprinted genes on mouse chromosome 7. There are an estimated 100 imprinted genes in the mouse, and most are found in clusters containing 3–12 imprinted genes that are spread out over 20 kilobases to 3.7 megabases of DNA. Humans have most of the same clustered imprinted genes as mice. In all cases examined, there is a specific pattern of DNA methylation and histone modification at the ICR for each parental copy of an imprinted gene. For the *Igf2–H19* cluster, the ICR DNA that lies between the two genes is methylated

in male germ cells and unmethylated in female germ cells (**Figure 12-25**). Thus, methylation of the ICR leads to *Igf2* being transcriptionally active and *H19* being inactive, whereas the lack of methylation leads to the reverse. This difference is due to the fact that only the unmethylated (female) ICR can be bound by CTCF, the same protein that binds insulator elements in TADs (Figure 12-12). When bound, CTCF acts as an enhancer-blocking insulator that prevents enhancer activation of *Igf2* transcription. However, the enhancer in females can still activate *H19* transcription. In males, CTCF cannot bind to the ICR, and the enhancer can activate *Igf2* transcription (recall that enhancers can act at great distances). However, the enhancer cannot activate *H19* because the methylated region extends into the *H19* promoter. Epigenetic marks such as DNA methylation that cause genes to be expressed in a parent-of-origin manner are established in germ cells (sperm and eggs) and, as organisms develop, are maintained through mitotic cell division of somatic cells.

KEY CONCEPT For most diploid organisms, both alleles of a gene are expressed independently; however, a few genes in mammals undergo genomic imprinting. Through this mechanism, epigenetic marks made in germline cells are retained throughout development of offspring, silencing one allele and allowing expression of the other.

Note that parental imprinting can greatly affect disease inheritance. For most diploid genes, mutation of the copy inherited from one parent does not produce a disease phenotype because there is an additional copy from the other parent. However, imprinted genes are essentially haploid because only one of the two copies is expressed. Thus, as you might expect, diseases occur due to mutations in the non-imprinted, transcriptionally active, copy of imprinted genes. Prader–Willi syndrome and Angelman syndrome are examples of imprinting diseases derived from loss of non-imprinted paternal and maternal genes, respectively. These diseases occur in about 1 in 15,000 births and are associated with distinct neurodevelopmental phenotypes. Prader–Willi syndrome is associated with severe obesity owing to an involuntary urge to eat constantly. Features of Angelman syndrome include severe mental retardation, seizures, and characteristic abnormal behaviors such as a happy, excitable demeanor.

X-chromosome inactivation

Epigenetic regulation of transcription can occur at specific genes, or it can be more global, as in the case of **dosage compensation** in animals. In mammals, females have two X chromosomes and males have only one, creating a potential imbalance in the transcription of genes residing on the X chromosome. This imbalance is corrected by transcriptional

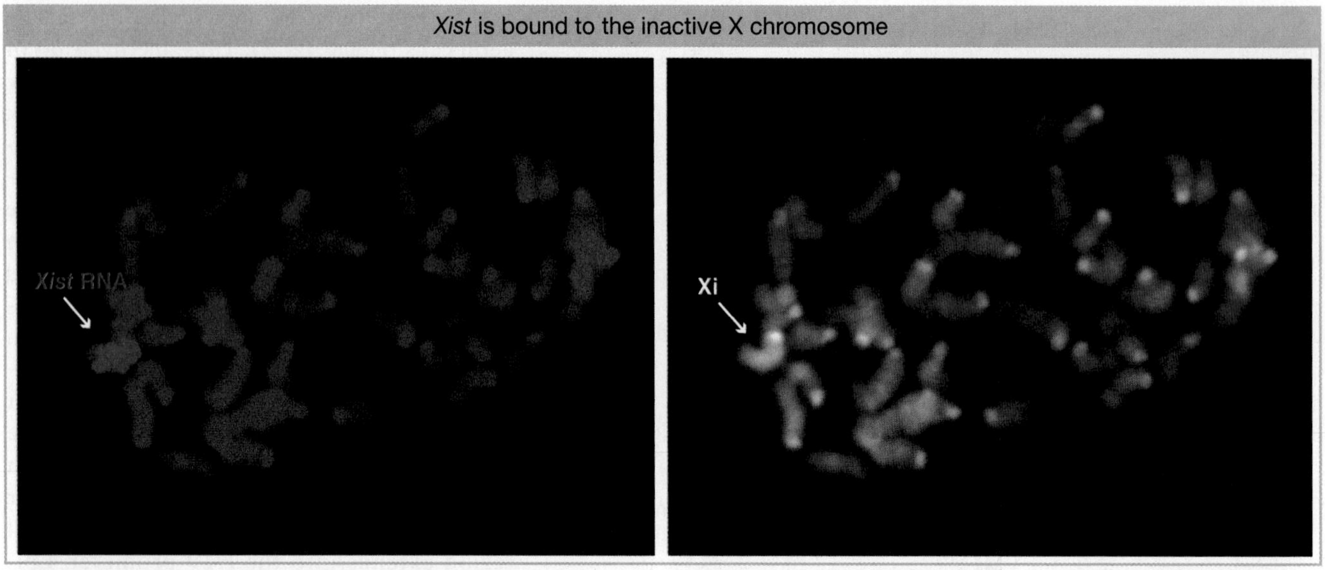

Xist is bound to the inactive X chromosome

Xist RNA

Xi

FIGURE 12-26 RNA fluorescent *in situ* hybridization (FISH; see Chapter 10) for *Xist* RNA performed on a metaphase chromosome spread of a female fibroblast cell. *Xist* (labeled with a red fluorescent dye) covers one of the two X chromosomes. DNA (blue) is visualized with DAPI. Binding of *Xist* is part of the X-inactivation mechanism that silences transcription. [*From: J.T. Lee et al., "Lessons from X-chromosome inactivation: long ncRNA as guides and tethers to the epigenome," Genes Dev., 23 (16), 2009, 1831–1842, Fig. 2 © Cold Spring Harbor Laboratory Press. Photography by Jeannie Lee.*]

silencing of one of the two X chromosomes in females through a process called **X-chromosome inactivation** or X-inactivation, for short. The inactivated X chromosome, called a **Barr body**, can be seen in the nucleus as a darkly staining, highly condensed, heterochromatic structure (**Figure 12-26**). This is a classic example of epigenetic regulation because the two X chromosomes in female cells are nearly identical in sequence; however, one is transcriptionally active and the other is silenced by the formation of heterochromatin. In human cells,

the choice of whether to inactivate the maternal or paternal X chromosome is random, but once an X chromosome is inactivated, it will remain inactive for the lifetime of the cell and its daughter cells.

A 17-kilobase-long noncoding RNA (lncRNA) called *Xist* (*X-inactive specific transcript*) plays a central role in initiating silencing of one of the X chromosomes, as does an antisense transcript *Tsix* ("Xist" spelled backward) from the same locus (**Figure 12-27**). Early on in development of the

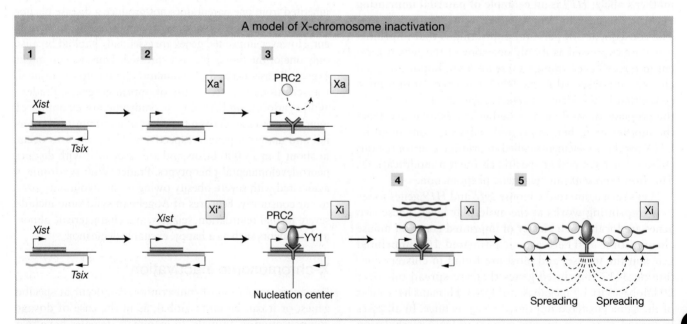

A model of X-chromosome inactivation

FIGURE 12-27 A model showing how the *Xist* lncRNA, YY1, and PRC2 act to inactivate one X chromosome by forming heterochromatin. The five steps illustrate the progression of events that begin early in embryonic development to silence the transcription of one X chromosome (Xi) and maintain transcription on the other (Xa). Spreading of *Xist* across Xa initiates at a nucleation center at the *Xist* locus. Xi* and Xa* indicate the future inactive and active X chromosomes, respectively.

embryo, when both female X chromosomes are transcriptionally active, *Tsix* is expressed from both alleles. At the beginning of X-chromosome inactivation, transient pairing of the X chromosomes represses the transcription of *Tsix* from one allele, establishing the future inactive X chromosome (Xi). Transcription that persists from the other allele blocks activation of *Xist* transcription and this establishes the future active X chromosome (Xa). *Xist* lncRNA spreads along the future Xi and induces silencing. Spreading is nucleated at the *Xist* locus by the YY1 (Yin-Yang 1) protein, which binds both *Xist* lncRNA and DNA. As it spreads, *Xist* recruits PRC2 (Polycomb Repressive Complex 2), which catalyzes the heterochromatin-associated modification H2K27me3. Other lncRNAs and structural changes in chromatin occur to establish and/or maintain X-chromosome inactivation, including H3K9 methylation, histone deacetylation, DNA CpG island methylation, and incorporation of the histone H2A variant macroH2A into nucleosomes.

Dosage compensation also takes place in *Drosophila*, but in this case, the X chromosome is transcriptionally upregulated two-fold in males to equal the transcription of the two X chromosomes in females. As in mammals, dosage compensation in *Drosophila* involves lncRNAs (*roX1* and *roX2*) that are transcribed from the X chromosome in males as well as alterations in chromatin structure that include acetylation of H4K16, phosphorylation of H3S10, and nucleosome remodeling by a chromatin remodeling complex. Twofold transcription upregulation of X-chromosome genes in males is probably achieved by precisely balancing activities that condense and decondense the X chromosome.

KEY CONCEPT In X-inactivation, epigenetic mechanisms enacted early in embryonic development silence an entire chromosome.

SUMMARY

This chapter has focused on the roles that transcription factors and chromatin structure play in directing cell type-specific transcription of eukaryotic genes. Transcription factors bind distal and proximal enhancers and alter transcription by recruiting coactivators and corepressors. A single transcription factor such as Gal4 can affect the transcription of multiple genes by binding an enhancer shared by the genes. In addition, as demonstrated by MCM1 in mating-type specification, one transcription factor can affect the activation/repression activity and transcriptional targets of other transcription factors. Transcription factors are spurred into action by environment signals such as the presence of galactose, hormones, and viruses or developmental signals such as those that specify the intestinal cell fate. Furthermore, cells control transcription by regulating the expression, cellular localization, stability, and activity (DNA binding, dimerization, ligand-binding, and interactions) of transcription factors.

In eukaryotic cells, chromatin, not naked DNA, is the substrate for transcription factors, general transcription factors, and RNA polymerase II. The wrapping of DNA around an octamer of core histones (H2A, H2B, H3, and H4) and the binding of linker histones (H1) to form nucleosomes as well as the gathering of nucleosomes into compact structures is generally repressive to transcription. Reversible chromatin modifying and chromatin remodeling activities work together to overcome the repressive effects of chromatin by changing its structure.

Chromatin modification encompasses modifications to histone amino acids and DNA nucleotides. Modification systems are made up of coactivator and corepressor proteins that add (write), remove (erase), and bind (read) modifications. Protein complexes containing one or several

of these activities have affinity for specific genes because of interactions with transcription factors and chromatin modifications themselves. Histone modifications are concentrated in the flexible tails of core histones. There are many different types of modifications, including acetylation, methylation, and phosphorylation, and an enormous number of different combinations of modifications can occur on single histone tails and within a histone octamer. Thus, histone modifications have the capacity to convey complex instructions, dubbed the histone code, for how the transcription machinery should operate. Similarly, DNA modifications, mainly cytosine methylation in CpG dinucleotides, provide an additional set of instructions to the transcription machinery in some eukaryotic organisms, including plants and mammals. Histone and DNA modifications control recruitment of transcription initiation factors such as TFIID and elongation factors such as P-TEFb to specific genes. In addition, they recruit ATP-dependent chromatin remodeling complexes such as SWI/SNF that change factor accessibility to DNA. Chromatin remodeling complexes make DNA more or less accessible by sliding, ejecting, replacing, or exchanging histones on DNA. Studies such as those of the *IFN-β* gene demonstrate how transcription factors, chromatin modifying factors, and chromatin remodeling factors function cooperatively to adjust the level of transcription of particular genes in response to a signal. In the end, the level of transcription is determined by the balance of activating and repressive mechanisms that act on a gene.

DNA replication faithfully copies both DNA sequence and chromatin structure from parent to daughter cells. Newly formed cells inherit both genetic information, inherent in the nucleotide sequence of DNA, and epigenetic information, built into histone and DNA modifications. Cellular memory, position-effect variegation, genomic imprinting, and X-chromosome inactivation are examples of epigenetic phenomenon where the transcription state of single genes, multiple genes, and even whole chromosomes is inherited without changing the sequence of DNA. Epigenetic inheritance mechanisms involve Polycomb group and Trithorax group factors, suppressor of variegation (*Su(var)*) and enhancer of variegation (*E(var)*) factors, insulator elements, and long noncoding RNAs (lncRNAs). Thus, the nucleotide sequence of genomes is not sufficient for understanding the inheritance of normal and disease states of transcription.

KEY TERMS

activation domain (p. 401)
Barr body (p. 422)
boundary (p. 410)
canonical histone (p. 407)
chromatin (p. 406)
chromatin modification (p. 410)
chromatin remodeling (p. 410)
coactivator (p. 400)
constitutive heterochromatin (p. 408)
core histone (p. 407)
core promoter (p. 400)
corepressor (p. 400)
CpG island (p. 413)
dimerization domain (p. 401)
distal enhancer (p. 400)
DNA-binding domain (p. 401)
DNA modification (p. 412)
dosage compensation (p. 421)
enhanceosome (p. 415)

enhancer (p. 400)
epigenetic inheritance (p. 418)
euchromatin (p. 408)
facultative heterochromatin (p. 408)
flexible tail (p. 407)
genomic imprinting (p. 420)
general transcription factor
    (GTF) (p. 400)
heterochromatin (p. 408)
histone (p. 407)
histone acetyltransferase (HAT) (p. 410)
histone code (p. 411)
histone deacetylase (HDAC) (p. 410)
histone fold (p. 407)
histone-fold extension (p. 407)
histone octamer (p. 407)
histone modification (p. 411)
insulator (p. 410)
ligand-binding domain (p. 402)

linker histone (p. 407)
maternal imprinting (p. 421)
nucleosome (p. 407)
nucleosome free region (NFR)
    (p. 409)
paternal imprinting (p. 421)
position-effect variegation (PEV)
    (p. 419)
post-translational modification (p. 410)
proximal enhancer (p. 400)
proximal promoter (p. 400)
repression domain (p. 401)
topologically associating domain
    (TAD) (p. 409)
transcription factor (p. 400)
upstream activation sequence (UAS)
    (p. 402)
variant histone (p. 407)
X-chromosome inactivation (p. 422)

PROBLEMS

Visit SaplingPlus for supplemental content. Problems with the icon are available for review/grading.

WORKING WITH THE FIGURES

(The first 27 questions require inspection of text figures.)

1. In Figure 12-1, name the cis-acting sequence elements and trans-acting proteins that regulate transcription.

2. In Figure 12-2, how might a corepressor block transcription by RNA polymerase II?

3. In Figure 12-3, what proteins bind the GGCCAATC and the ATATAA sequences?

4. In Figure 12-4, what is the function of each transcription factor domain?

5. In Figure 12-5, how does Gal4 regulate the transcription of four different *GAL* genes at the same time?

6. In Figure 12-6, how many individual Gal4 proteins can bind the DNA between the *GAL10* and *GAL1* genes?

7. In Figure 12-7, what effect would a Gal4 protein that lacks the DNA-binding domain have on transcription of the UAS-*lacZ* reporter gene, and why?

8. In Figure 12-8, is Gal3 a transcription factor, coactivator, corepressor, or none of these?

9. In Figure 12-9, hypothesize why MCM1 does not bind and activate α-specific genes in **a** cells and **a**/α cells.

10. In Figure 12-10, several protein subunits of the TFIID general transcription factor contain a histone-fold domain. Based on the function of the histone fold in histones, propose a function for the histone fold in TFIID proteins.

11. In Figure 12-11, how might the structure of chromatin bring enhancer and promoter elements close together that are far apart in linear DNA?

12. In Figure 12-12, what features of chromatin structure are shared between enhancers and promoters?

13. In Figure 12-13, what effect might reduced acetyl-CoA levels have on transcription?

14. Lysines in histone tails can be propionylated. The propionyl group is similar in structure to an acetyl group. Using the categories shown in Figure 12-14, what would you call enzymes that regulate propionyl addition and removal?

15. In Figure 12-15, what are the implications to the histone code of the different lysine and arginine methylation types?

16. In Figure 12-16, how many different codes could be produced on the histone H3 tail just by phosphorylation? 〰

17. Based on Figure 12-17, what sequence would be read after the sodium bisulfite reaction, if all of the CpGs in the sequence 5'-GGCGTCGAAGTCGAA-3' were methylated? 〰

18. In Figure 12-18, how might a CpG island function differently than an isolated CpG?

19. In Figure 12-19, what steps would need to occur to exchange a variant H2A for a canonical H2A in a nucleosome? 〰

20. In Figure 12-20, describe two ways in which the HDAC might be recruited to the *IFN-β* gene.

21. In Figure 12-21a, which of the transcription instructions, in the form of binding sites in chromatin, are reversible?

22. In Figure 12-22, will all flies with a *white* gene inversion have the same pattern of white and red cells as the eye shown at the bottom? Why or why not?

23. In Figure 12-23, name a type of gene that might be an *E(var)* and explain your answer.

24. In Figure 12-24, how is this mechanism similar to the mechanism by which transcription factors regulate transcription (for example, as in Figure 12-2)?

25. In Figure 12-25, what mechanisms might position the enhancer of the paternal allele to act on the *Igf2* promoter that is >50 kilobases away, and why might this not happen for the maternal allele?

26. In Figure 12-26, why is it specified in the figure legend that this is a female cell?

27. In Figure 12-27, what histone modification is expected to be enriched on the inactive X chromosome relative to the active X chromosome, and why?

BASIC PROBLEMS

28. Do all nucleosomes have the same eight core histones? Why or why not?

29. Why might binding of a transcription factor to DNA be inhibited for DNA that is part of a nucleosome?

30. The Lugar and Richmond crystal structure of the nucleosome used *Xenopus laevis* (toad) histones. Why is the structure thought to be a good representation of human nucleosomes?

31. Why are histone tails not visible in the crystal structure of the nucleosome core particle?

32. How might higher-order structures of chromatin activate, rather than repress, transcription?

33. What are the two general mechanisms by which histone acetylation affects transcription? 〰

34. What functions might be served by modifications of amino acids in the histone-fold domain?

35. How is the function of histone tails similar to that of the C-terminal domain (CTD) of RNA polymerase II?

36. By what two mechanisms could histone acetylation levels increase at a gene promoter?

37. Explain how phosphorylation of histone H3 serine 10 (H3S10P) might increase acetylation of histone H4 lysine 16 (H4K16ac)?

38. Vertebrate histone H1 can be phosphorylated on many amino acids in the C-terminal unstructured domain. What effect would you expect histone H1 phosphorylation to have on chromatin structure?

39. What type of factors would you expect to be involved in the regulation of histone H1 phosphorylation?

40. What molecular interactions must by broken by chromatin remodeling complexes to remove a histone octamer from DNA?

41. Why is the order of assembly and disassembly of nucleosomes important for understanding transcription regulation?

42. Why might insertion of a transgene at different places in the *Drosophila* genome cause the transgene to be transcribed at different levels?

43. How would you modify a transgene so that its expression was not affected by position-effect variegation (PEV)?

44. What purpose might be served by the long half-life of core histones?

45. What is meant by the term epigenetic inheritance? Describe two examples of such inheritance. 〰

46. Give three functions of insulator elements.

47. How many nucleosomes would be needed to cover the human genome (3×10^9 base pairs), if the average linker distance between nucleosomes was 50 base pairs? 〰

48. Why might the concentration of ATP in cells affect the structure of chromatin?

49. Why is acid used to extract histones from cell nuclei in experiments performed in vitro?

50. A researcher has identified a mutant cell line that has reduced transcription of gene *X* relative to the parental cell line. The mutant cell line has a single point mutation in the entire genome. Describe five possible mechanisms by which the point mutation could reduce the transcription of gene *X*.

51. For position-effect variegation to have been discovered, why is it critical that the *white* gene is on the X chromosome?

52. How might DNA methylation at a promoter lead to H3K9me3 at nearby nucleosomes?

53. Overexpression of a transcription factor changes the transcription of different genes in different cell types. Why?

54. Draw the pattern of H3K4 trimethylation expected at *IFN-β* promoter during the 24 hours following viral infection (Figure 12-20a).

55. Can a transcription factor both activate and repress transcription? Explain your answer.

56. To understand the inheritance of diseases, researchers are mapping genomes and epigenomes (i.e., genome-wide chemical modifications to histones and DNA). Describe the information that might be contained in an epigenome map.

GENETICS AND SOCIETY

Accumulating evidence suggests that epigenetic effects can be inherited across multiple generations. For example, the effects of a pregnant woman smoking on her child might also affect their children and their children's children. Do you think that this adds to the moral responsibility of a mother?

Gene expression in a developing fruit-fly embryo. The seven magenta stripes mark the cells expressing the mRNA of a gene encoding a regulatory protein that controls segment number in the *Drosophila* embryo. The spatial regulation of gene expression is central to the control of animal development. [*Dave Kosman, Ethan Bier, and Bill McGinnis.*]

427

Of all the phenomena in biology, few if any inspire more awe than the formation of a complex animal from a single-celled egg. In this spectacular transformation, unseen forces organize the dividing mass of cells into a form with a distinct head and tail, various appendages, and many organs. The great geneticist Thomas Hunt Morgan was not immune to its aesthetic appeal:

> A transparent egg as it develops is one of the most fascinating objects in the world of living beings. The continuous change in form that takes place from hour to hour puzzles us by its very simplicity. The geometric patterns that present themselves at every turn invite mathematical analysis. . . . This pageant makes an irresistible appeal to the emotional and artistic sides of our nature.[1]

Yet, for all its beauty and fascination, biologists were stumped for many decades concerning how biological form is generated during development. Morgan also said that "if the mystery that surrounds embryology is ever to come within our comprehension, we must . . . have recourse to other means than description of the passing show."

The long drought in embryology lasted well beyond Morgan's heyday in the 1910s and 1920s, but it was eventually broken by geneticists working very much in the tradition of Morgan-style genetics and with his favorite, most productive genetic model, the fruit fly *Drosophila melanogaster*.

The key catalysts to understanding the making of animal forms were the discoveries of genetic "monsters"—mutant fruit flies with dramatic alterations of body structures (**Figure 13-1**). In the early days of *Drosophila* genetics, rare mutants arose spontaneously or as by-products of other experiments with spectacular transformations of body parts. In 1915, Calvin Bridges, then Morgan's student, isolated a fly having a mutation that caused the tiny hind wings (halteres) of the fruit fly to resemble the large forewings. He dubbed the mutant *bithorax*. The transformation in *bithorax* mutants is called *homeotic* (Greek *homeos*, meaning same or similar) because one part of the body (the hind wing) is transformed to resemble another (the forewing), as shown in Figure 13-1b. Subsequently, several more homeotic mutants were identified in *Drosophila*, such as the dramatic

Homeotic mutants of *Drosophila melanogaster*

(a)

(b)

(c)

FIGURE 13-1 In homeotic mutants, the identity of one body structure has been changed into another. (a) Normal fly with one pair of forewings on the second thoracic segment and one pair of small hind wings on the third thoracic segment. (b) Mutations in the *Ultrabithorax* gene lead to loss of *Ubx* function in the posterior thorax, which causes the development of forewings in place of the hind wings. (c) *Antennapedia* mutant in which the antennae are transformed into legs. [*Sean Carroll.*]

Antennapedia mutant in which legs develop in place of the antennae (Figure 13-1c).

The spectacular effects of homeotic mutants inspired what would become a revolution in embryology, once the tools of molecular biology became available to understand what homeotic genes encoded and how they exerted such enormous influence on the development of entire body parts. Surprisingly, these strange fruit-fly genes turned out to be a passport to the study of the entire animal kingdom, as counterparts to these genes were discovered that played similar roles in almost all animals. Furthermore, the same

[1] T. H. Morgan, *Experimental Embryology*. Columbia University Press, 1927.

regulatory logic that underpins development in animals is also used to control development in plants.

The study of animal and plant development is a very large and still-growing discipline. As such, we do not attempt a comprehensive overview. Rather, in this chapter, we will focus on a few general concepts that illustrate the logic of the genetic control of animal development. We will explore how the information for building complex structures is encoded in the genome. In contrast to the control of gene regulation in single bacterial or eukaryotic cells, the genetic control of body formation and body patterning is fundamentally a matter of gene regulation in three-dimensional *space* and over *time*. Yet we will see that the principles governing the genetic control of development are connected to those already presented in Chapters 11 and 12, governing the physiological control of gene expression in bacteria and single-celled eukaryotes.

13.1 THE GENETIC APPROACH TO DEVELOPMENT

LO 13.1 Outline experimental approaches to identify and characterize members of the genetic toolkit for development in different animal phyla.

For many decades, the study of embryonic development largely entailed the physical manipulation of embryos, cells, and tissues. Several key concepts were established about the properties of developing embryos through experiments in which one part of an embryo was transplanted into another part of the embryo. For example, the transplantation of a part of a developing amphibian embryo to another site in a recipient embryo was shown to induce the surrounding tissue to form a second complete body axis (**Figure 13-2a**). Similarly, transplantation of the posterior part of a developing chick limb bud to the anterior could induce extra digits, but with reversed polarity with respect to the

normal digits (Figure 13-2b). These transplanted regions of the amphibian embryo and chick limb bud were termed **organizers** because of their remarkable ability to organize the development of surrounding tissues. The cells in the organizers were postulated to produce **morphogens**, molecules that induced various responses in surrounding tissue in a concentration-dependent manner.

KEY CONCEPT Organizers are groups of cells in an embryo that have the remarkable ability to instruct the development of other cells in an embryo via the production of morphogens, which are molecules that act in a concentration-dependent manner. Cells in close proximity to the organizer are exposed to high concentrations of morphogens and therefore develop into different structures from cells located further from the organizer.

Although these experimental results were spectacular and fascinating, further progress in understanding the nature of organizers and morphogens stalled after their discovery in the first half of the 1900s. It was essentially impossible to isolate the molecules responsible for these activities by using biochemical separation techniques. Embryonic cells make thousands of substances—proteins, glycolipids, hormones, and so forth. A morphogen could be any one of these molecules but would be present in minuscule quantities—one needle in a haystack of cellular products.

The long impasse in defining embryology in molecular terms was broken by genetic approaches—mainly the systematic isolation of mutants with discrete defects in development and the subsequent characterization and study of the gene products that they encoded. The genetic approach to studying development presented many advantages over alternative, biochemical strategies. First, the geneticist need not make any assumptions about the number or nature of molecules required for a process. Second, the (limited) quantity of a gene product is no impediment: all genes can

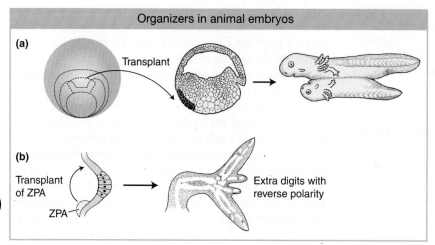

FIGURE 13-2 Transplantation experiments played a central role in early embryology and demonstrated the long-range organizing activity of embryonic tissues. (a) The Spemann-Mangold organizer. The dorsal blastopore "lip" of an early amphibian embryo can induce a second embryonic axis and embryo when transplanted to the ventral region of a recipient embryo. (b) In the developing chick limb bud, the zone of polarizing activity (ZPA) organizes pattern along the anteroposterior axis. Transplantation of the ZPA from a donor limb bud to the anterior position in a recipient limb bud induces extra digits with reverse polarity.

Mutational Analysis of Early *Drosophila* Development

The initial insights into the genetic control of pattern formation emerged from studies of the fruit fly *Drosophila melanogaster. Drosophila* development has proved to be a gold mine to researchers because developmental problems can be approached by the use of genetic and molecular techniques simultaneously.

The *Drosophila* embryo has been especially important in understanding the formation of the basic animal body plan. One important reason is that an abnormality in the body plan of a mutant is easily identified in the larval exoskeleton in the *Drosophila* embryo. The larval exoskeleton is a noncellular structure, made of a polysaccharide polymer called chitin that is produced as a secretion of the epidermal cells of the embryo. Each structure of the exoskeleton is formed from epidermal cells or cells immediately underlying that structure. With its intricate pattern of hairs, indentations, and other structures, the exoskeleton provides numerous landmarks to serve as indicators of the fates assigned to the many epidermal cells (see Figure 13-13). In particular, there are many distinct anatomical structures along the anteroposterior (A–P) and dorsoventral (D–V) axes (see the figure above). Furthermore, because all the nutrients necessary to develop to the larval stage are prepackaged in the egg, mutant embryos in which the A–P or D–V cell fates are drastically altered can nonetheless develop to the end of embryogenesis and produce a mutant larva in about 1 day (see the figure on the next page). The exoskeleton of such a mutant larva mirrors the mutant fates assigned to subsets of the epidermal cells and can thus identify genes worthy of detailed analysis.

The development of the *Drosophila* adult body pattern takes a little more than a week (see the figure on the next page). Small populations of cells set aside during embryogenesis proliferate during three larval stages (instars) and differentiate in the pupal stage into adult structures. These set-aside cells include

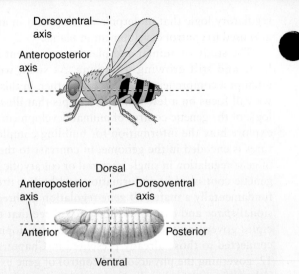

The relationship between adult and embryonic body axes. Note that most images of *Drosophila* embryos in this chapter are oriented so that anterior is to the left, and dorsal is at the top.

the *imaginal disks,* which are disk-shaped regions that give rise to specific appendages and tissues in each segment as the leg, wing, eye, and antennal disks. Imaginal disks are easy to remove for analysis of gene expression (see Figure 13-7).

Once a mutant with an effect on the *Drosophila* body plan has been identified, the underlying gene can be cloned and characterized at the molecular level with ease. The analysis of the cloned genes often provides valuable information on the function of the protein product—usually by identifying close relatives in amino acid sequence of the encoded polypeptide through comparisons with all the protein sequences stored in public databases. In addition, one can investigate the spatial and temporal patterns of expression of (1) an mRNA, by using histochemically tagged single-stranded DNA sequences complementary to the mRNA to perform RNA in situ hybridization, or (2) a protein, by using histochemically tagged antibodies that bind specifically to that protein (see Figure 13-5).

be mutated regardless of the amount of product made by a gene. And, third, the genetic approach can uncover phenomena for which there is no biochemical or other bioassay.

From the genetic viewpoint, there are four key questions concerning the number, identity, and function of genes taking part in development:

1. Which genes are important in development?

2. Where in the developing organism and at what times are these genes active?

3. How is the expression of developmental genes regulated?

4. Through what molecular mechanisms do gene products affect development?

To address these questions, strategies had to be devised to identify, catalog, and analyze genes that control development. One of the first considerations in the genetic analysis of animal development was which animal to study. Of the millions of living species, which offered the most promise? The fruit fly *Drosophila melanogaster* emerged as the leading genetic model of animal development because its ease of rearing, rapid life cycle, cytogenetics, and decades of classical genetic analysis (including the isolation of many very dramatic mutants) provided important experimental advantages (see the Model Organism box on *Drosophila melanogaster* above). The nematode worm *Caenorhabditis elegans* also presented many attractive features, most particularly its simple construction and well-studied cell

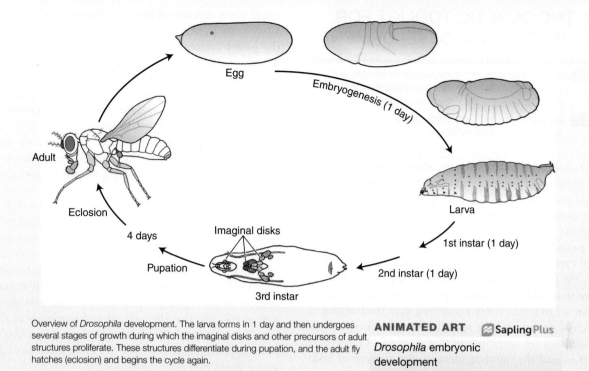

Overview of *Drosophila* development. The larva forms in 1 day and then undergoes several stages of growth during which the imaginal disks and other precursors of adult structures proliferate. These structures differentiate during pupation, and the adult fly hatches (eclosion) and begins the cycle again.

ANIMATED ART   Sapling Plus

Drosophila embryonic development

Using Knowledge from One Model Organism to Fast-Track Developmental Gene Discovery in Others

With the discovery of genes that regulate development within the *Drosophila* genome, similarities among the DNA sequences of these genes could be exploited in treasure hunts for other members of the gene family. These hunts depend on DNA base-pair complementarity. For this purpose, DNA hybridizations were carried out under *moderate stringency conditions,* in which there could be some mismatch of bases between the hybridizing strands without disrupting the proper hydrogen bonding of nearby base pairs. Some of these treasure hunts were carried out in the *Drosophila* genome itself, in looking for more family members. Others searched for similar genes in other animals, by means of *zoo blots* (Southern blots of restriction-enzyme-digested DNA from different animals), by using radioactive *Drosophila* DNA as the probe (see Chapter 10). This approach led to the discovery of homologous gene sequences in many different animals, including humans and mice. Now homologous genes are typically identified by computational searches of genome sequences (see Chapter 14).

lineages (see the Model Organism box on *Caenorhabditis elegans* on page 451). Among vertebrates, the development of targeted gene disruption techniques opened up the laboratory mouse *Mus musculus* to more systematic genetic study, and the zebrafish *Danio rerio* has recently become a favorite model owing to the transparency of the embryo and to advances in its genetic study. Among plants, *Arabidopsis thaliana* has played a similar role as *Drosophila* in illuminating fundamental mechanisms in plant development. More information about the most common model organisms can be found in "A Brief Guide to Model Organisms" at the end of this book.

Through systematic and targeted genetic analysis, as well as comparative genomic studies, much of the **genetic toolkit**—the set of genes that control the development of the bodies, body parts, and cell types of several different animal species—has been defined. We will first focus on the genetic toolkit of *Drosophila melanogaster* because its identification was a source of major insights into the genetic control of development; its discovery catalyzed the identification of the genetic toolkit of other animals, including humans.

KEY CONCEPT Genetic model organisms, particularly *Drosophila melanogaster*, have played a key role in the identification of the genetic toolkit for development. Remarkably, many of the toolkit genes discovered in model organisms play fundamental roles in human development and disease.

13.2 THE GENETIC TOOLKIT FOR *DROSOPHILA* DEVELOPMENT

LO 13.1 Outline experimental approaches to identify and characterize members of the genetic toolkit for development in different animal phyla.

LO 13.2 Differentiate members of the genetic toolkit for development from other genes.

Animal genomes typically contain about 13,000 to 22,000 genes. Many of these genes encode proteins that function in essential processes in all cells of the body (for example, in cellular metabolism or the biosynthesis of macromolecules). Such genes are often referred to as **housekeeping genes**. Other genes encode proteins that carry out the specialized tasks of various organ systems, tissues, and cells of the body such as the globin proteins in oxygen transport or antibody proteins that mediate immunity. Here, we are interested in a different set of genes, those concerned with the building of organs and tissues and the specification of cell types—the genetic toolkit for development that determines the overall body plan and the number, identity, and pattern of body parts.

Toolkit genes of the fruit fly have generally been identified through the monstrosities or catastrophes that arise when they are mutated. Toolkit-gene mutations from two sources have yielded most of our knowledge. The first source consists of spontaneous mutations that arise in laboratory populations, such as those found in the Morgan lab. The second source comprises mutations induced at random by treatment with mutagens (such as chemicals or radiation) that greatly increase the frequency of damaged genes throughout the genome. Elegant refinements of the latter approach have made possible systematic searches, called **genetic screens**, in which organisms are treated with a mutagen and allowed to reproduce, and then the offspring are examined for visible defects in a phenotype of interest. Such screens have identified many members of the fly's genetic toolkit. The members of this toolkit constitute only a small fraction, perhaps several hundred genes, of the roughly 14,000 genes in the fly genome.

KEY CONCEPT The genetic toolkit for animal development is composed of a small fraction of all genes. Only a small subset of the entire complement of genes in the genome affect development in discrete ways.

Classification of genes by developmental function

One of the first tasks following the execution of a genetic screen for mutations is to sort out those of interest. Many mutations are lethal when hemi- or homozygous because cells cannot survive without products affected by these mutations. The more interesting mutations are those that cause some discrete defect in either the embryonic or the adult body pattern, or both. It has proved useful to group the genes affected by mutations into several categories based on the nature of their mutant phenotypes. Many toolkit genes can be classified according to their function in controlling the identity of body parts (for example, of different segments or appendages), the formation of body parts (for example, of organs or appendages), the number of body parts, the formation of cell types, and the organization of the primary body axes (the anteroposterior, or A–P, and dorsoventral, or D–V, axes; see the Model Organism Box on page 430).

We will begin our inventory of the *Drosophila* toolkit by examining the genes that control the identity of segments and appendages. We do so for both historical and conceptual purposes. The genes controlling segmental and appendage identity were among the very first toolkit genes identified. Subsequent discoveries about their nature were sources of profound insights into not just how their products work, but also the content and workings of the toolkits of most animals. Furthermore, their spectacular mutant phenotypes indicate that they are among the most globally acting genes that affect animal form.

Homeotic genes and segmental identity

Among the most fascinating abnormalities to be described in animals are those in which one normal body part is replaced by another. Such **homeotic transformations** have been observed in many species in nature, including sawflies in which a leg forms in place of an antenna and frogs in which a thoracic vertebra forms in place of a cervical vertebra (**Figure 13-3**). Whereas only one member of a bilateral pair of structures is commonly altered in many naturally occurring variants, both members of a bilateral pair of structures are altered in homeotic mutants of fruit flies (see Figure 13-1). In the former case, the alteration is not heritable, but homeotic mutants breed true from generation to generation.

The scientific fascination with homeotic mutants stems from three properties. First, it is amazing that a single gene mutation can alter a developmental pathway so dramatically. Second, it is striking that the structure formed in the mutant is a well-developed likeness of another body part. And, third, it is important to note that homeotic mutations transform the identity of **serially reiterated structures**. Insect and many animal bodies are made of repeating parts of similar structure, like building blocks, arranged in a series. The forewings and hind wings, the segments, and the antennae, legs, and mouthparts of insects are sets of serially reiterated body parts. Homeotic mutations transform identities within these sets.

A mutation may cause a loss of homeotic gene function where the gene normally acts, or it may cause a gain of homeotic function where the homeotic gene does

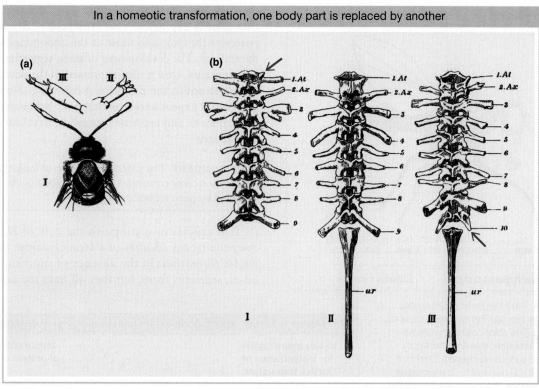

In a homeotic transformation, one body part is replaced by another

FIGURE 13-3 A late-nineteenth-century drawing from one of the first studies of homeotic transformations in nature. (a) Homeosis in a sawfly, with the left antenna transformed into a leg. (b) Homeosis in a frog. The middle specimen is normal. The specimen on the left has extra structures growing out of the top of the vertebral column, transforming a cervical vertebra into a thoracic vertebra (red arrow). The specimen on the right has an extra set of vertebrae (red arrow). [*From W. Bateson, Material for the Study of Variation. Macmillan, 1894.*]

not normally act. For example, the *Ultrabithorax* (*Ubx*) gene acts in the developing hind wing to promote hind-wing development and to repress forewing development. **Loss-of-function mutations** in *Ubx* transform the hind wing into a forewing (see Figure 13-1). Dominant **gain-of-function mutations** in *Ubx* transform the forewing into a hind wing. Similarly, the antenna-to-leg transformations of *Antennapedia* (*Antp*) mutants are caused by the dominant gain of *Antp* function in the antenna (see Figure 13-1). In addition to these transformations in appendage identity, homeotic mutations can transform segment identity, causing one body segment of the adult or larva to resemble another.

Although homeotic genes were first identified through spontaneous mutations affecting adult flies, they are required throughout most of a fly's development. Systematic searches for homeotic genes have led to the identification of eight loci, now referred to as **Hox genes,** that affect the identity of segments and their associated appendages in *Drosophila*. Generally, the complete loss of any *Hox*-gene function is lethal in early development. The dominant mutations that transform adults are viable in heterozygotes because the wild-type allele provides normal gene function to the developing animal.

Organization and expression of *Hox* genes

A most intriguing feature of *Hox* genes is that they are clustered together in two **gene complexes** that are located on the third chromosome of *Drosophila*. The *Bithorax* complex contains three *Hox* genes, and the *Antennapedia* complex contains five *Hox* genes. Moreover, the order of the genes in the complexes and on the chromosome corresponds to the order of body regions, from head to tail, that are influenced by each *Hox* gene (**Figure 13-4**).

The relation between the structure of the *Hox*-gene complexes and the phenotypes of *Hox*-gene mutants was illuminated by the molecular characterization of the genes. Molecular cloning of the sequences encompassing each *Hox* locus provided the means to analyze where in the developing animal each gene is expressed. These spatial aspects of gene expression and gene regulation are crucial to understanding the logic of the genetic control of development. In regard to the *Hox* genes and other toolkit genes, the development of technology that made possible the visualization of gene and protein expression was crucial to understanding the relation among gene organization, gene function, and mutant phenotypes.

Two principal technologies for the visualization of gene expression in embryos or other tissues are (1) the

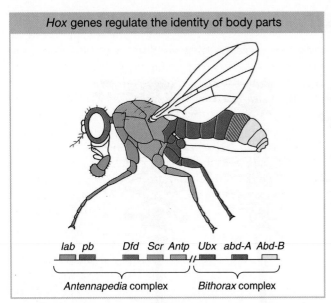

Hox genes regulate the identity of body parts

lab pb Dfd Scr Antp Ubx abd-A Abd-B

Antennapedia complex   *Bithorax* complex

FIGURE 13-4 The *Hox* genes of *Drosophila*. Eight *Hox* genes regulate the identity of regions within the adult. The color coding identifies the segments and structures that are affected by mutations in the various *Hox* genes. [*Data from S. B. Carroll, J. K. Grenier, and S. D. Weatherbee, From DNA to Diversity: Molecular Genetics and the Evolution of Animal Design 2e, Blackwell, 2005.*]

expression of RNA transcripts visualized by in situ hybridization and (2) the expression of proteins visualized by immunological methods. Each technology depends on the isolation of cDNA clones representing the mature mRNA transcript and protein (**Figure 13-5**).

In the developing embryo, the *Hox* genes are expressed in spatially restricted, sometimes overlapping domains within the embryo (**Figure 13-6**). The genes are also expressed in the larval and pupal tissues that will give rise to the adult body parts.

The patterns of *Hox*-gene expression (and other toolkit genes) generally correlate with the regions of the animal affected by gene mutations. For example,

FIGURE 13-5 The two principal technologies for visualizing where a gene is transcribed or where the protein that it encodes is expressed are (*left*) in situ hybridization of complementary RNA probe to mRNA and (*right*) immunolocalization of protein expression. The procedures for each method are outlined. Expression patterns may be visualized as the product of an enzymatic reaction or of a chromogenic substrate or with fluorescently labeled compounds.

the dark blue shading in Figure 13-6 indicates where the *Ubx* gene is expressed. This *Hox* gene is expressed in the posterior thoracic and most of the abdominal segments of the embryo. The development of these segments is altered in *Ubx* mutants. *Ubx* is also expressed in the developing hind wing but not in the developing forewing (**Figure 13-7**), as one would expect knowing that *Ubx* promotes hind-wing development and represses forewing development in this appendage.

> **KEY CONCEPT** The spatial expression of toolkit genes is usually closely correlated with the regions of the animal affected by gene mutations.

It is crucial to distinguish the role of *Hox* genes in determining the *identity* of a structure from that governing its *formation*. In the absence of function of all *Hox* genes, segments form, but they all have the same identity;

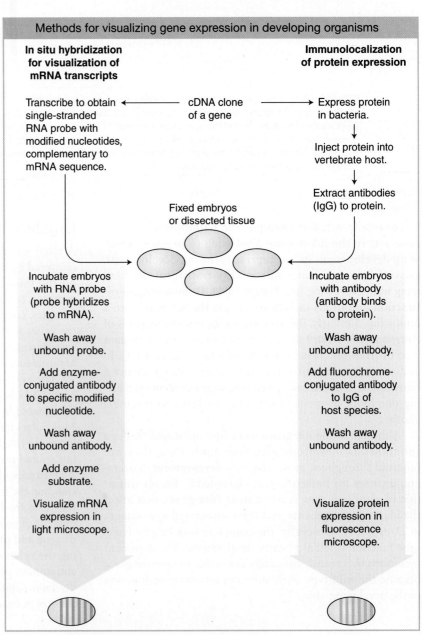

Methods for visualizing gene expression in developing organisms

In situ hybridization for visualization of mRNA transcripts

Immunolocalization of protein expression

cDNA clone of a gene

Transcribe to obtain single-stranded RNA probe with modified nucleotides, complementary to mRNA sequence.

Express protein in bacteria.

Inject protein into vertebrate host.

Extract antibodies (IgG) to protein.

Fixed embryos or dissected tissue

Incubate embryos with RNA probe (probe hybridizes to mRNA).

Wash away unbound probe.

Add enzyme-conjugated antibody to specific modified nucleotide.

Wash away unbound antibody.

Add enzyme substrate.

Visualize mRNA expression in light microscope.

Incubate embryos with antibody (antibody binds to protein).

Wash away unbound antibody.

Add fluorochrome-conjugated antibody to IgG of host species.

Wash away unbound antibody.

Visualize protein expression in fluorescence microscope.

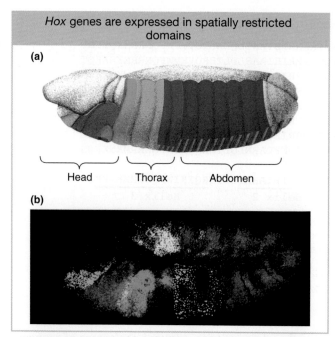

Hox genes are expressed in spatially restricted domains

(a)

Head   Thorax   Abdomen

(b)

FIGURE 13-6 Expression of *Hox* genes in the *Drosophila* embryo. (a) Schematic representation of *Drosophila* embryo showing regions where eight individual *Hox* genes are expressed. (b) Actual image of the expression of seven *Hox* genes visualized by in situ hybridization. Colors indicate expression of *labial* (turquoise), *Deformed* (lavender), *Sex combs reduced* (green), *Antennapedia* (orange), *Ultrabithorax* (dark blue), *Abdominal-A* (red), and *Abdominal-B* (yellow). The embryo is folded so that the posterior end (yellow) appears near the top center. [*(b) Dave Kosman, Ethan Bier, and Bill McGinnis.*]

limbs also can form, but they have antennal identity; and, similarly, wings can form, but they have forewing identity. Other genes control the formation of segments, limbs, and wings and will be described later. First, we must understand how *Hox* genes exert their dramatic effects on fly development.

The homeobox

Because *Hox* genes have large effects on the identities of entire segments and other body structures, the nature and function of the proteins that they encode are of special interest. Edward Lewis, a pioneer in the study of homeotic genes, noted early on that the clustering of *Bithorax* complex genes suggested that the multiple loci had arisen by tandem duplication of an ancestral gene. This idea led researchers to search for similarities in the DNA sequences of *Hox* genes. They found that all eight *Hox* genes of the two complexes have a short region of sequence similarity, 180 bp in length. Because this stretch of DNA sequence similarity is present in homeotic genes, it was dubbed the **homeobox**. The homeobox encodes a protein domain, the **homeodomain**, containing 60 amino acids. The amino acid sequence of the homeodomain is very similar among the Hox proteins (**Figure 13-8**).

Although the discovery of a common protein motif in each of the Hox proteins was very exciting, further analysis of the structure of the homeodomain revealed that it forms a helix-turn-helix motif—the structure common to the Lac repressor, the λ repressor, Cro, and the α2 and a1 regulatory proteins of the yeast mating-type loci! This similarity suggested immediately (and it was subsequently borne out) that Hox proteins are sequence-specific DNA-binding proteins and that they exert their effects by controlling the expression of genes within developing segments and appendages. Thus, the products of these remarkable genes function through principles that are already familiar from Chapters 11 and 12—by binding to regulatory elements of other genes to activate or repress their expression. We will see that it is also true of many other toolkit genes: a significant fraction of these genes encode transcription factors that control the expression of other genes.

KEY CONCEPT Homeotic transformations result from mutations in *Hox* genes, which are genes that contain a conserved sequence called the homeobox. This sequence encodes a protein domain called the homeodomain, which is similar to the helix-turn-helix motif found in many other transcription factors.

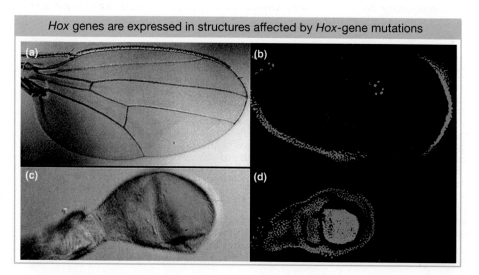

Hox genes are expressed in structures affected by Hox-gene mutations

(a)

(b)

(c)

(d)

FIGURE 13-7 An example of *Hox*-gene expression. (a) The adult forewing of *D. melanogaster*. (b) Ubx protein is not expressed in cells of the developing imaginal disk that will form the forewing. Cells enriched in Hox proteins are stained green; in this image, the green-stained cells are cells that do *not* form the wing. (c) The adult hind wing (haltere). (d) The Ubx protein is expressed at high levels in all cells of the developing hind-wing imaginal disk. [*Scott Weatherbee.*]

	Hox proteins have a sequence in common
lab	NNSGRTNFTNKQLTELEKEFHFNRYLTRARRIEIANTLQLNETQVKIWFQNRRMKQKKRV
pb	PRRLRTAYTNTQLLELEKEFHFNKYLCRPRRIEIAASLDLTERQVKVWFQNRRMKHKRQT
Dfd	PKRQRTAYTRHQILELEKEFHYNRYLTRRRRIEIAHTLVLSERQIKIWFQNRRMKWKKDN
Scr	TKRQRTSYTRYQTLELEKEFHFNRYLTRRRRIEIAHALCLTERQIKIWFQNRRMKWKKEH
Antp	RKRGRQTYTRYQTLELEKEFHFNRYLTRRRRIEIAHALCLTERQIKIWFQNRRMKWKKEN
Ubx	RRRGRQTYTRYQTLELEKEFHTNHYLTRRRRIEMAHALCLTERQIKIWFQNRRMKLKKEI
abd-A	RRRGRQTYTRFQTLELEKEFHNHYLTRRRRIEIAHALCLTERQIKIWFQNRRMKLKKEL
Abd-B	VRKKRKPYSKFQTLELEKEFLFNAYVSKQKRWELARNLQLTERQVKIWFQNRRMKNKKNS
Consensus sequence	-RRGRT-YTR-QTLELEKEFHFNRYLTRRRRIEIAHALCLTERQIKIWFQNRRMK-KKE- 　　　　　Helix 1　　　　　　　Helix 2　　　　　　　Helix 3

FIGURE 13-8 Sequences of fly homeodomains. All eight *Drosophila Hox* genes encode proteins containing a highly conserved 60-amino-acid domain, the homeodomain, composed of three α helices. Helices 2 and 3 form a helix-turn-helix motif similarly to the Lac repressor, Cro, and other DNA-binding proteins. Residues common to the *Hox* genes are shaded in yellow; divergent residues are shaded in red; those common to subsets of proteins are shaded in blue or green. [*Data from S. B. Carroll, J. K. Grenier, and S. D. Weatherbee,* From DNA to Diversity: Molecular Genetics and the Evolution of Animal Design *2e, Blackwell, 2005.*]

We will examine how Hox proteins and other toolkit proteins orchestrate gene expression in development a little later. First, there is one more huge discovery to describe, which revealed that what we learn from fly *Hox* genes has very general implications for the animal kingdom.

Clusters of *Hox* genes control development in most animals

When the homeobox was discovered in fly *Hox* genes, it raised the question whether this feature was some peculiarity of these bizarre fly genes or was more widely distributed, in other insects or segmented animals, for example. To address this possibility, researchers searched for homeoboxes in the genomes of other insects, as well as earthworms, frogs, cows, and even humans. They found many homeoboxes in each of these animal genomes.

The similarities in the homeobox sequences from different species were astounding. Over the 60 amino acids of the homeodomain, some mouse and fish Hox proteins were identical with the fly sequences at as many as 54 of the 60 positions (**Figure 13-9**). In light of the vast evolutionary distances between these animals, more than 500 million years

since their last common ancestor, the extent of sequence similarity indicates very strong pressure to maintain the sequence of the homeodomain.

The existence of *Hox* genes with homeoboxes throughout the animal kingdom was entirely unexpected. Why different types of animals would possess the same regulatory genes was not obvious, which is why biologists were further surprised by the results when the organization and expression of *Hox* genes was examined in other animals. In vertebrates, such as the laboratory mouse, the *Hox* genes also are clustered together in four large gene complexes on four different chromosomes. Furthermore, the order of the genes in the mouse *Hox* complexes parallels the order of their most related counterparts in the fly *Hox* complexes, as well as in each of the other mouse *Hox* clusters (**Figure 13-10a**). This correspondence indicates that the *Hox* complexes of insects and vertebrates are related and that some form of *Hox* complex existed in their distant common ancestor. The four *Hox* complexes in the mouse arose by duplications of entire *Hox* complexes (perhaps of entire chromosomes) in vertebrate ancestors.

Why would such different animals have these sets of genes in common? Their deep, common ancestry indicates

	Drosophila and vertebrate Hox proteins show striking similarities	
Fly *Dfd*	PKRQRTAYTRHQILELEKEFHYNRYLTRRRRIEIAHTLVLSERQIKIWFQNRRMKWKKDN	KLPNTKNVR
Amphibian *Hox4*	TKRSRTAYTRQQVLELEKEFHFNRYLTRRRRIEIAHSLGLTERQIKIWFQNRRMKWKKDN	RLPNTKTRS
Mouse *HoxB4*	PKRSRTAYTRQQVLELEKEFHYNRYLTRRRRVEIAHALCLSERQIKIWFQNRRMKWKKDH	RLPNTKIRS
Human *HoxB4*	PKRSRTAYTRQQVLELEKEFHYNRYLTRRRRVEIAHALCLSERQIKIWFQNRRMKWKKDH	RLPNTKIRS
Chick *HoxB4*	PKRSRTAYTRQQVLELEKEFHYNRYLTRRRRVEIAHSLCLSERQIKIWFQNRRMKWKKDH	RLPNTKIRS
Frog *HoxB4*	AKRSRTAYTRQQVLELEKEFHYNRYLTRRRRVEIAHTLRLSERQIKIWFQNRRMKWKKDH	KLPNTKIKS
Fugu *HoxB4*	PKRSRTAYTRQQVLELEKEFHYNRYLTRRRRVEIAHTLCLSERQIKIWFQNRRMKWKKDH	KLPNTKVRS
Zebrafish *HoxB4*	AKRSRTAYTRQQVLELEKEFHYNRYLTRRRRVEIAHTLRLSERQIKIWFQNRRMKWKKDH	KLPNTKIKS

FIGURE 13-9 The sequences of the *Drosophila* Deformed protein homeodomain and of several members of the vertebrate *Hox* group 4 genes are strikingly similar. Residues in common are shaded in yellow; divergent residues are shaded in red; residues common to subsets of proteins are shaded in blue. The very similar C-terminal flanking regions outside of the homeodomain are shaded in green. [*Data from S. B. Carroll, J. K. Grenier, and S. D. Weatherbee,* From DNA to Diversity: Molecular Genetics and the Evolution of Animal Design *2e, Blackwell, 2005.*]

The order of *Hox* genes parallels the order of body parts in which they are expressed

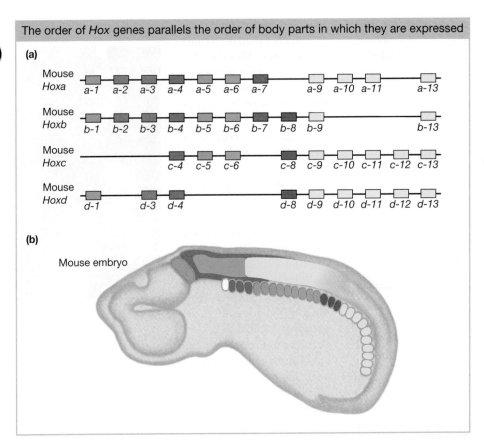

(a)

Mouse
Hoxa   a-1 a-2 a-3 a-4 a-5 a-6 a-7   a-9 a-10 a-11   a-13

Mouse
Hoxb   b-1 b-2 b-3 b-4 b-5 b-6 b-7 b-8 b-9   b-13

Mouse
Hoxc   c-4 c-5 c-6   c-8 c-9 c-10 c-11 c-12 c-13

Mouse
Hoxd   d-1 d-3 d-4   d-8 d-9 d-10 d-11 d-12 d-13

(b)

Mouse embryo

FIGURE 13-10 Like those of the fruit fly, vertebrate *Hox* genes are organized in clusters and expressed along the anteroposterior axis. (a) In the mouse, four complexes of *Hox* genes, comprising 39 genes in all, are present on four different chromosomes. Not every gene is represented in each complex; some have been lost in the course of evolution. (b) The *Hox* genes are expressed in distinct domains along the anteroposterior axis of the mouse embryo. The color shading represents the different groups of genes shown in part *a*. [*S. B. Carroll, "Homeotic Genes and the Evolution of Arthropods and Chordates,"* Nature *376, 1995, 479–485.*]

that *Hox* genes play some fundamental role in the development of most animals. That role is apparent from analyses of how the *Hox* genes are expressed in different animals. In vertebrate embryos, adjacent *Hox* genes also are expressed in adjacent or partly overlapping domains along the anteroposterior body axis. Furthermore, the order of the *Hox* genes in the complexes corresponds to the head-to-tail order of body regions in which the genes are expressed (Figure 13-10b).

The *Hox*-gene expression patterns of vertebrates suggested that they also specify the identity of body regions, and subsequent analyses of *Hox*-gene mutants have borne this suggestion out. For example, mutations in the *Hoxa11* and *Hoxd11* genes cause the homeotic transformation of sacral vertebrae to lumbar vertebrae (**Figure 13-11**). Thus, as in the fly, the loss or gain of function of *Hox* genes in vertebrates causes transformation of the identity of serially repeated structures. Such results have been obtained in

Hox genes regulate the identity of serially repeated structures in vertebrates

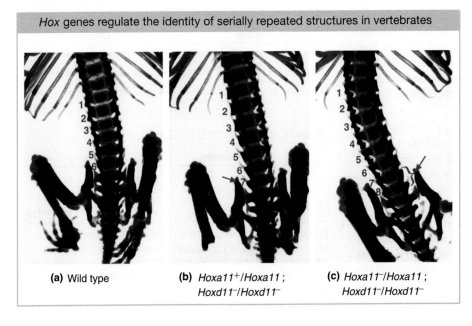

(a) Wild type

(b) *Hoxa11⁺/Hoxa11 ; Hoxd11⁻/Hoxd11⁻*

(c) *Hoxa11⁻/Hoxa11 ; Hoxd11⁻/Hoxd11⁻*

FIGURE 13-11 The morphologies of different regions of the vertebral column are regulated by *Hox* genes. (a) In the mouse, six lumbar vertebrae (numbers in red) form just anterior to the sacral vertebrae. (b) In mice lacking the function of the posteriorly acting *Hoxd11* gene and possessing one functional copy of the *Hoxa11* gene, seven lumbar vertebrae form and one sacral vertebra is lost. (c) In mice lacking both *Hoxa11* and *Hoxd11* function, eight lumbar vertebrae form and two sacral vertebrae are lost. [*Photographs courtesy of Dr. Ann Boulet, HHMI, University of Utah; from S. B. Carroll, J. K. Grenier, S. D. Weatherbee, from DNA to Diversity; Molecular Genetics and the Evolution of Animal Design, 2nd ed. Blackwell, 2005.*]

several classes, including mammals, birds, amphibians, and fish. Furthermore, clusters of *Hox* genes have been shown to govern the patterning of other insects and to be deployed in regions along the anteroposterior axis in annelids, molluscs, nematodes, various arthropods, primitive chordates, flatworms, and other animals. Therefore, despite enormous differences in anatomy, the possession of one or more clusters of *Hox* genes that are deployed in regions along the main body axis is a common, fundamental feature of at least all bilateral animals. Indeed, the surprising lessons from the *Hox* genes portended what turned out to be a general trend among toolkit genes; that is, most toolkit genes are common to different animals.

> **KEY CONCEPT** Despite great differences in anatomy, a broad array of different animal phyla have many toolkit genes in common.

Now let's take an inventory of the rest of the toolkit to see what other general principles emerge.

13.3 DEFINING THE ENTIRE TOOLKIT

LO 13.1 Outline experimental approaches to identify and characterize members of the genetic toolkit for development in different animal phyla.

LO 13.2 Differentiate members of the genetic toolkit for development from other genes.

LO 13.3 Predict both the phenotypic effects of mutations in toolkit genes based on their expression during development as well as the expression patterns of toolkit genes based on the phenotypic effects of mutations in toolkit genes.

The *Hox* genes are perhaps the best-known members of the toolkit, but they are just a small family in a much larger group of genes required for the development of the proper numbers, shapes, sizes, and kinds of body parts. Little was known about the rest of the toolkit until the late 1970s and early 1980s, when Christiane Nüsslein-Volhard and Eric Wieschaus, working at the European Molecular Biology Laboratory in Heidelberg, Germany, set out to find the genes required for the formation of the segmental organization of the *Drosophila* embryo and larva.

Until their efforts, most work on fly development focused on viable adult phenotypes and not the embryo. Nüsslein-Volhard and Wieschaus realized that the sorts of genes that they were looking for were probably lethal to embryos or larvae in homozygous mutants. So, they came up with a scheme to search for genes that were required in the **zygote** (the product of fertilization; **Figure 13-12**, bottom). They also developed genetic screens to identify those genes with products that function in the egg, before the zygotic genome is active, and that are required for the proper patterning of the embryo. Genes with products

provided by the female to the egg are called **maternal-effect genes**. Mutant phenotypes of strict maternal-effect genes depend only on the genotype of the mother (Figure 13-12, top).

In these screens, genes were identified that were necessary to make the proper number and pattern of larval segments, to make its three tissue layers (ectoderm, mesoderm, and endoderm), and to pattern the fine details of an animal's anatomy. The power of the genetic screens was their systematic nature. By saturating each of a fly's chromosomes (except the small fourth chromosome)

Christiane Nüsslein-Volhard and Eric Wieschaus at the European Molecular Biology Laboratory. [*Christiane Nüsslein-Volhard*.]

with chemically induced mutations, the researchers were able to identify most genes that were required for the building of the fly. For their pioneering efforts, Nüsslein-Volhard, Wieschaus, and Lewis shared the 1995 Nobel Prize in Physiology or Medicine.

The most striking and telling features of the newly identified mutants were that they showed dramatic but discrete defects in embryo organization or patterning. That is, the dead larva was not an amorphous carcass but exhibited specific, often striking patterning defects. The *Drosophila* larval body has various features whose number, position, or pattern can serve as landmarks to diagnose or classify

---

Genetic screens for maternally and zygotically required toolkit genes

MATERNALLY REQUIRED GENES

Parents		Offspring	
$m/+$ ♂ × $m/+$ ♀	→	m/m, $m/+$, $+/+$	all normal
m/m ♂ × $m/+$ ♀	→	m/m, $m/+$	all normal
$+/+$, $m/+$, or m/m ♂ × m/m ♀	→	$m/+$, m/m	all mutant phenotype

ZYGOTICALLY REQUIRED GENES

Parents		Offspring	
$m/+$ ♂ × $m/+$ ♀	→	$m/+$, $+/+$	normal
		m/m	mutant phenotype

FIGURE 13-12 Genetic screens identify whether a gene product functions in the egg or in the zygote. The phenotypes of offspring depend on either (*top*) the maternal genotype for maternal-effect genes or (*bottom*) the offspring (zygotic) genotype for zygotically required genes (*m*, mutant; +, wild type).

the abnormalities in mutant animals. Each locus could thus be classified according to the body axis that it affected and the pattern of defects caused by mutations. Each class of genes appeared to represent different steps in the progressive refinement of the embryonic body plan—from those that affect large regions of the embryo to those with more limited realms of influence.

KEY CONCEPT Genetic screens are a powerful and unbiased approach to systematically identify genes that affect a biological process, such as embryonic development.

For any toolkit gene, three pieces of information are key toward understanding gene function: (1) the mutant phenotype, (2) the pattern of gene expression, and (3) the nature of the gene product. Extensive study of a few dozen genes has led to a fairly detailed picture of how each body axis is established and subdivided into segments or germ layers.

The anteroposterior axis

To illustrate the principles of toolkit genes, we will focus on the anteroposterior body axis in *Drosophila*. However, the same principles apply to the making of the dorsoventral body axis of *Drosophila*, and indeed to the establishment of body axes in both animals and plants. Genetic screens have shown that only a few dozen genes are required for proper organization of the anteroposterior body axis of the fly embryo. The genes are grouped into five classes on the basis of their realm of influence on embryonic pattern.

KEY CONCEPT Toolkit genes can be classified by their roles in development; that is, where and when they function during the development of an organism.

- The first class sets up the anteroposterior axis and consists of the maternal-effect genes. A key member of this class is the *Bicoid* gene. Embryos from *Bicoid* mutant mothers are missing the anterior region of the embryo (**Figure 13-13**), telling us that the gene is required for the development of that region.

The next three classes are zygotically active genes required for the development of the segments of the embryo.

- The second class contains the **gap genes**. Each of these genes affects the formation of a contiguous block of segments; mutations in gap genes lead to large gaps in segmentation (**Figure 13-14**, left).
- The third class comprises the **pair-rule genes**, which act at a double-segment periodicity. Pair-rule mutants are missing part of each pair of segments, but different pair-rule genes affect different parts of each double segment. For example, the *even-skipped* gene affects one set of segmental boundaries, and the *odd-skipped* gene affects the complementary set of boundaries (Figure 13-14, middle).

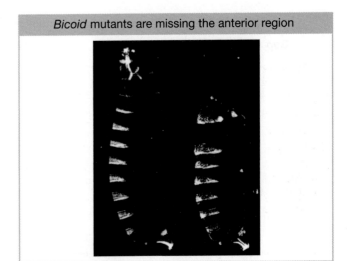

Bicoid mutants are missing the anterior region

FIGURE 13-13 The *Bicoid* (*bcd*) maternal-effect gene affects the anterior part of the developing larva. These photomicrographs are of *Drosophila* larvae that have been prepared to show their hard exoskeletons. Dense structures, such as the segmental denticle bands, appear white. (*Left*) A normal larva. (*Right*) A larva from a homozygous *bcd* mutant female. Head and anterior thoracic structures are missing. [*Republished with permission of the American Association for the Advancement of Science, from C.H. Nüsslein-Volhard, G. Frohnhofer, and R. Lehmann, "Determination of anteroposterior polarity in Drosophila" Science Vol. 238, Issue 4834 (1987) 1678, Figure 4. Permission conveyed through Copyright Clearance Center, Inc.]*

- The fourth class consists of the **segment-polarity genes**, which affect patterning within each segment. Mutants of this class display defects in segment polarity and number (Figure 13-14, right).

The fifth class of genes determines the fate of each segment.

- The fifth class includes the *Hox* genes already discussed; *Hox* mutants do not affect segment number, but they alter the appearance of one or more segments.

Expression of toolkit genes

To understand the relation between genes and mutant phenotype, we must know the timing and location of gene-expression patterns and the molecular nature of the gene products. The patterns of expression of the toolkit genes turn out to vividly correspond to their phenotypes, inasmuch as they are often precisely correlated with the parts of the developing body that are altered in mutants. Each gene is expressed in a region that can be mapped to specific coordinates along either axis of the embryo. For example, the maternal-effect Bicoid protein is expressed in a graded pattern emanating from the anterior pole of the early embryo, the section of the embryo missing in mutants (**Figure 13-15a**). Similarly, the gap proteins are expressed in blocks of cells that correspond to the future positions of the segments that are missing in respective gap-gene mutants (Figure 13-15b). The pair-rule proteins are expressed in striking striped patterns: one transverse stripe is expressed per every 2 segments, in a total of 7 stripes covering the 14 future body segments (the position and periodicity of the

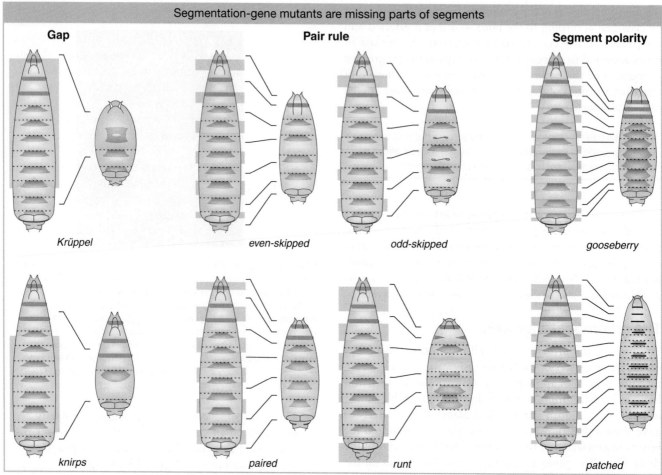

FIGURE 13-14 Classes of *Drosophila* segmentation-gene mutants. These diagrams depict representative gap, pair-rule, and segment-polarity mutants. The red trapezoids are the dense bands of exoskeleton seen in Figure 13-13. The boundary of each segment is indicated by a dotted line. The left-hand diagram of each pair depicts a wild-type larva, and the right-hand diagram depicts the pattern formed in a given mutant. The shaded light orange regions on the wild-type diagrams indicate the domains of the larva that are missing or affected in the mutant.

ANIMATED ART   SaplingPlus

Drosophila embryonic development

stripes correspond to the periodicity of defects in mutant larvae), as shown in Figure 13-15c. Many segment-polarity genes are expressed in stripes of cells within each segment, 14 stripes in all corresponding to 14 body segments (Figure 13-15d). Note that the domains of gene expression become progressively more refined as development proceeds: genes are expressed first in large regions (gap proteins), then in stripes from three to four cells wide (pair-rule proteins), and then in stripes from one to two cells wide (segment-polarity proteins).

In addition to what we have learned from the spatial patterns of toolkit-gene expression, the order of toolkit-gene expression over time is logical. The maternal-effect Bicoid protein appears before the zygotic gap proteins, which are expressed before the 7-striped patterns of pair-rule proteins appear, which in turn precede the 14-striped patterns of segment-polarity proteins. The order of gene expression and the progressive refinement of domains within the embryo reveal that the

making of the body plan is a step-by-step process, with major subdivisions of the body outlined first and then refined until a fine-grain pattern is established. The order of gene action further suggests that the expression of one set of genes might govern the expression of the succeeding set of genes.

One clue that this progression is indeed the case comes from analyzing the effects of mutations in toolkit genes on the expression of other toolkit genes. For example, in embryos from *Bicoid* mutant mothers, the expression of several gap genes is altered, as well as that of pair-rule and segment-polarity genes. This finding suggests that the Bicoid protein somehow (directly or indirectly) influences the regulation of gap genes.

Another clue that the expression of one set of genes might govern the expression of the succeeding set of genes comes from examining the protein products. Inspection of the Bicoid protein sequence reveals that it contains a homeodomain, related to but distinct from those of

Expression of anteroposterior-axis-patterning proteins

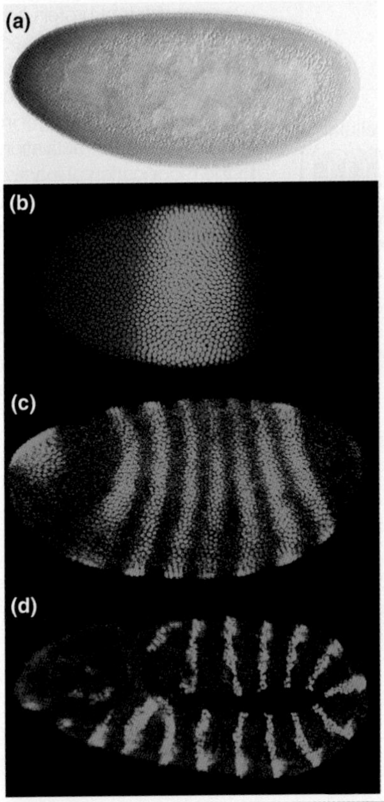

(a)

(b)

(c)

(d)

FIGURE 13-15 Patterns of toolkit-gene expression correspond to mutant phenotypes. *Drosophila* embryos have been stained with antibodies to the (a) maternally derived Bicoid protein, (b) Krüppel gap protein, (c) Hairy pair-rule protein, and (d) Engrailed segment-polarity protein and visualized by immunoenzymatic (staining is brown) (a) or immunofluorescence (staining is green) (b–d) methods. Each protein is localized to nuclei in regions of the embryo that are affected by mutations in the respective genes. [(a) *Photomicrographs courtesy of Ruth Lehmann, (b), (c), (d) Photomicrographs courtesy of James A. Langeland.*]

Hox proteins. Thus, Bicoid has the properties of a DNA-binding transcription factor. Each gap gene also encodes a transcription factor, as does each pair-rule gene, several segment-polarity genes, and, as described earlier, all *Hox* genes. These transcription factors include representatives of most known families of sequence-specific DNA-binding proteins; so, although there is no restriction concerning to which family they may belong, many early-acting toolkit proteins are transcription factors. Those that are not transcription factors tend to be components of signaling pathways (**Table 13-1**). These pathways, shown in generic form in **Figure 13-16**, mediate ligand-induced signaling processes between cells, and their output generally leads to gene activation or repression. Thus, most toolkit proteins either directly (as transcription factors) or indirectly (as components of signaling pathways) affect gene regulation.

> **KEY CONCEPT** Most toolkit proteins are transcription factors that regulate the expression of other genes or components of ligand-mediated signal-transduction pathways.

The genetic control of development, then, is fundamentally a matter of gene regulation in space and over time. How does the turning on and off of toolkit genes build animal form? And how is it choreographed during development? To answer these questions, we will examine the interactions among fly toolkit proteins and genes in more detail. The mechanisms that we will see for controlling toolkit-gene expression in the *Drosophila* embryo have emerged as models for the spatial regulation of gene expression in animal and plant development in general.

TABLE 13-1	Examples of *Drosophila* A–P Axis Genes That Contribute to Pattern Formation		
Gene symbol	**Gene Name**	**Protein function**	**Role(s) in early development**
bcd	*Bicoid*	Transcription factor—homeodomain protein	Maternal-effect gene
hb-z	*hunchback-zygotic*	Transcription factor—zinc-finger protein	Gap gene
Kr	*Krüppel*	Transcription factor—zinc-finger protein	Gap gene
kni	*knirps*	Transcription factor—steroid receptor-type protein	Gap gene
eve	*even-skipped*	Transcription factor—homeodomain protein	Pair-rule gene
ftz	*fushi tarazu*	Transcription factor—homeodomain protein	Pair-rule gene
opa	*odd-paired*	Transcription factor—zinc-finger protein	Pair-rule gene
prd	*paired*	Transcription factor—paired class homeodomain protein	Pair-rule gene
en	*engrailed*	Transcription factor—homeodomain protein	Segment-polarity gene
wg	*wingless*	Signaling protein-secreted ligand	Segment-polarity gene
hh	*hedgehog*	Signaling protein-secreted ligand	Segment-polarity gene
ptc	*patched*	Signaling protein-transmembrane receptor	Segment-polarity gene
lab	*labial*	Transcription factor—homeodomain protein	Segment-identity gene
Dfd	*Deformed*	Transcription factor—homeodomain protein	Segment-identity gene
Antp	*Antennapedia*	Transcription factor—homeodomain protein	Segment-identity gene
Ubx	*Ultrabithorax*	Transcription factor—homeodomain protein	Segment-identity gene

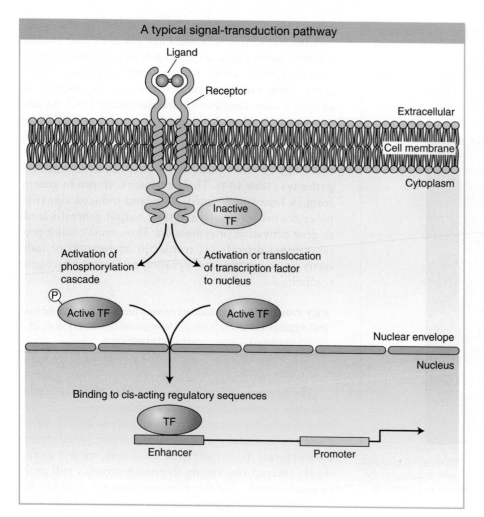

A typical signal-transduction pathway

Ligand

Receptor

Extracellular

Cell membrane

Cytoplasm

Inactive TF

Activation of phosphorylation cascade

Activation or translocation of transcription factor to nucleus

Ⓟ Active TF

Active TF

Nuclear envelope

Nucleus

Binding to cis-acting regulatory sequences

TF

Enhancer

Promoter

FIGURE 13-16 Most signaling pathways operate through similar logic but have different protein components and signal-transduction mechanisms. Signaling begins when a ligand binds to a membrane-bound receptor, leading to the release or activation of intracellular proteins. Receptor activation often leads to the modification of inactive transcription factors (TF). The modified transcription factors are translocated to the cell nucleus, where they bind to cis-acting regulatory DNA sequences or to DNA-binding proteins and regulate the level of target-gene transcription. [*Data from S. B. Carroll, J. K. Grenier, and S. D. Weatherbee,* From DNA to Diversity: Molecular Genetics and the Evolution of Animal Design *2e, Blackwell, 2005.*]

13.4 SPATIAL REGULATION OF GENE EXPRESSION IN DEVELOPMENT

LO 13.4 Infer how spatially and temporally restricted patterns of gene expression are generated during development from analyses of genetic mutations.

We have seen that toolkit genes are expressed in reference to coordinates in the embryo. But how are the spatial coordinates of the developing embryo conveyed as instructions to genes, to turn them on and off in precise patterns? As described in Chapters 11 and 12, the physiological control of gene expression in bacteria and simple eukaryotes is ultimately governed by sequence-specific DNA-binding proteins acting on cis-acting regulatory elements (for example, operators and upstream-activation-sequence, or UAS, elements). Similarly, the spatial control of gene expression during development is largely governed by the interaction of transcription factors with cis-acting regulatory elements. However, the spatial and temporal control of gene regulation in the development of a three-dimensional multicellular embryo requires the action of more transcription factors on more numerous and more complex cis-acting regulatory elements.

To define a position in an embryo, regulatory information must exist that distinguishes that position from adjacent regions. If we picture a three-dimensional embryo as a globe, then **positional information** must be specified that indicates longitude (location along the anteroposterior axis), latitude (location along the dorsoventral axis), and altitude or depth (position in the germ layers). We will illustrate the general principles of how the positions of gene expression are specified with three examples. These examples should be thought of as just a few snapshots of the vast number of regulatory interactions that govern fly and animal development. Development is a continuum in which every pattern of gene activity has a preceding causal basis. The entire process includes tens of thousands of regulatory interactions and outputs.

We will focus on a few connections between genes in different levels of the hierarchies that lay out the basic segmental body plan and on *nodal points* where key genes integrate multiple regulatory inputs and respond by producing simpler gene-expression outputs.

Maternal gradients and gene activation

The Bicoid protein is a homeodomain-type transcription factor that is translated from maternally derived mRNA that is deposited in the egg and localized at the anterior pole. Because the early *Drosophila* embryo is a *syncytium* with all nuclei in one cytoplasm, and lacks any cell membranes that would impede the diffusion of protein molecules, the Bicoid protein can diffuse through the cytoplasm. This diffusion establishes a protein concentration gradient (**Figure 13-17a**): the Bicoid protein is highly concentrated at the anterior end, and this concentration gradually decreases as distance from that end increases, until there is very little Bicoid protein beyond the middle of the embryo. This concentration gradient provides positional information about the location along the anteroposterior axis. A high concentration means anterior end, a lower concentration means middle, and so on. Thus, a way to ensure that a gene is activated in only one location along the axis is to link gene expression to the concentration level. A case in point is the gap genes, which must be activated in specific regions along the axis.

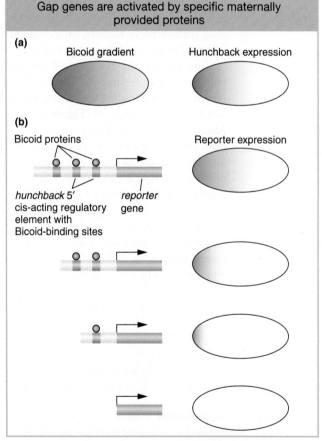

Gap genes are activated by specific maternally provided proteins

(a)

Bicoid gradient   Hunchback expression

(b)

Bicoid proteins   Reporter expression

hunchback 5′ cis-acting regulatory element with Bicoid-binding sites   *reporter gene*

FIGURE 13-17 The Bicoid protein activates zygotic expression of the *hunchback* gene. (a) Bicoid protein expression is graded along the anteroposterior axis. The *hunchback* gap gene is expressed in the anterior half of the zygote. (b) The Bicoid protein (blue) binds to three sites 5′ of the *hunchback* gene. When this 5′ DNA is placed upstream of a reporter gene, reporter-gene *expression* recapitulates the pattern of *hunchback* expression (*top right*). However, progressive deletion of one, two, or all three Bicoid-binding sites either leads to more restricted expression of the reporter gene or abolishes it altogether. These observations show that the level and pattern of *hunchback* expression are controlled by Bicoid through its binding to *hunchback* DNA regulatory sequences.

Several zygotic genes, including gap genes, are regulated by different levels of the Bicoid protein. For example, the *hunchback* gene is a gap gene activated in the zygote in the anterior half of the embryo. This activation is through direct binding of the Bicoid protein to three sites 5′ of the promoter of the *hunchback* gene. Bicoid binds to these sites *cooperatively*; that is, the binding of one Bicoid protein molecule to one site facilitates the binding of other Bicoid molecules to nearby sites.

In vivo experiments can demonstrate that the activation of *hunchback* depends on the concentration gradient. These tests require linking gene regulatory sequences to a reporter gene (an enzyme-encoding gene such as the *LacZ* gene or the green fluorescent protein of jellyfish; see Chapter 10), introducing the DNA construct into the fly germ line, and monitoring reporter expression in the embryo offspring of transgenic flies (a general overview of the method is shown in **Figure 13-18**). The wild-type sequences 5′ of the *hunchback* gene are sufficient to drive reporter expression in the anterior half of the embryo. Importantly, deletions of Bicoid-binding sites in this cis-acting regulatory element reduce or abolish reporter expression (Figure 13-17b). More than one Bicoid site must be occupied to generate a sharp boundary of reporter expression, which indicates that a threshold concentration of Bicoid protein is required to occupy multiple sites before gene expression is activated. A gap gene with fewer binding sites will not be activated at locations with lower concentration of Bicoid protein.

Each gap gene contains cis-acting regulatory elements with different arrangements of binding sites, and these binding sites may have different affinities for the Bicoid protein. Consequently, each gap gene is expressed in a unique distinct domain in the embryo, in response to different levels of Bicoid and other transcription-factor gradients. A similar theme is found in the patterning of the dorsoventral axis: cis-acting regulatory elements contain different numbers and arrangements of binding sites for the maternally supplied Dorsal protein and other zygotic transcription factors. Consequently, genes are activated in discrete domains along the dorsoventral axis.

KEY CONCEPT The concentration-dependent response of genes to graded inputs is a crucial feature of gene regulation in the early *Drosophila* embryo. The cis-acting regulatory elements governing distinct responses contain different numbers and arrangements of transcription-factor-binding sites.

Drawing stripes: integration of gap-protein inputs

The expression of each pair-rule gene in seven stripes is the first sign of the periodic organization of the embryo and future animal. How are such periodic patterns generated from prior aperiodic information? Before the molecular analysis of pair-rule-gene regulation, several models were put forth to explain stripe formation. Every one of these ideas viewed all seven stripes as identical outputs in

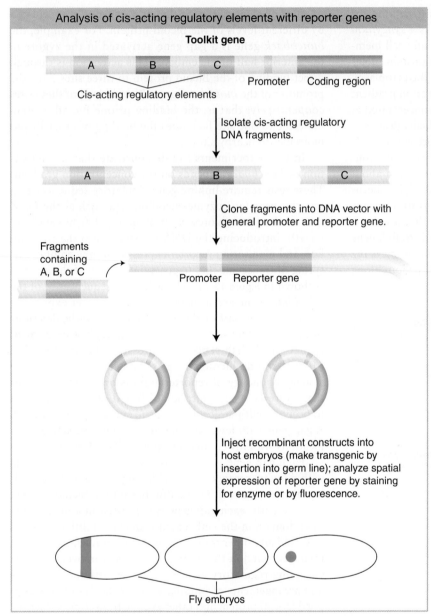

Analysis of cis-acting regulatory elements with reporter genes

Toolkit gene

A   B   C   Promoter   Coding region

Cis-acting regulatory elements

Isolate cis-acting regulatory DNA fragments.

A   B   C

Clone fragments into DNA vector with general promoter and reporter gene.

Fragments containing A, B, or C

Promoter   Reporter gene

Inject recombinant constructs into host embryos (make transgenic by insertion into germ line); analyze spatial expression of reporter gene by staining for enzyme or by fluorescence.

Fly embryos

FIGURE 13-18 Toolkit loci (such as *hunchback*, as described in the text) often contain multiple independent cis-acting regulatory elements that control gene expression in different places or at different times during development or both (for example, A, B, C, here). These elements are identified by their ability, when placed in cis to a reporter gene and inserted back into a host genome, to control the pattern, timing, or level, or all three, of reporter-gene expression. In this example, each element drives a different pattern of gene expression in a fly embryo. Most reporter genes encode enzymes or fluorescent proteins that can be easily visualized.

second stripe expressed by the *even-skipped* gene (**Figure 13-19a**). This stripe lies within the broad region of *hunchback* expression and on the edges of the regions of expression of two other gap proteins, Giant and Krüppel (Figure 13-19b). Thus, within the area of the future stripe, there will be large amounts of Hunchback protein and small amounts of Giant protein and Krüppel protein. There will also be a certain concentration of the maternal-effect Bicoid protein. No other stripe of the embryo will contain these proteins in these proportions. The formation of stripe 2 is controlled by a specific cis-acting regulatory element, an **enhancer**, that contains a number of binding sites for these four proteins (Figure 13-19c). Detailed analysis of the *eve* stripe 2 cis-acting regulatory element revealed that the position of this "simple" stripe is controlled by the binding of these four aperiodically distributed transcription factors, including one maternal protein and three gap proteins.

Specifically, the *eve* stripe 2 element contains multiple sites for the maternal Bicoid protein and the Hunchback, Giant, and Krüppel gap proteins (Figure 13-19d). Mutational analyses of different combinations of binding sites revealed that Bicoid and Hunchback activate the expression of the *eve* stripe 2 element over a broad region. The Giant and Krüppel proteins are repressors that sharpen the boundaries of the stripe to just a few cells wide. The *eve* stripe 2 element acts, then, as a genetic switch, integrating multiple regulatory protein activities to produce one stripe from three to four cells wide in the embryo.

The entire seven-striped periodic pattern of *even-skipped* expression is the sum of different sets of inputs into separate cis-acting regulatory elements. The enhancers for other stripes contain different combinations of protein binding sites.

response to identical inputs. However, the actual way in which the patterns of a few key pair-rule genes are encoded and generated is one stripe at a time. The solution to the mystery of stripe generation highlights one of the most important concepts concerning the spatial control of gene regulation in developing animals; namely, the distinct cis-acting regulatory elements of individual genes are controlled independently.

The key discovery was that each of the seven stripes that make up the expression patterns of the *even-skipped* and *hairy* pair-rule genes is controlled independently. Consider the

KEY CONCEPT The regulation of cis-acting regulatory elements by combinations of activators and repressors is a common theme in the spatial regulation of gene expression. Complex patterns of inputs are often integrated to produce simpler patterns of outputs.

Making segments different: integration of Hox inputs

The combined and sequential activity of the maternal-effect, gap, pair-rule, and segment-polarity proteins establishes the basic segmented body plan of the embryo and

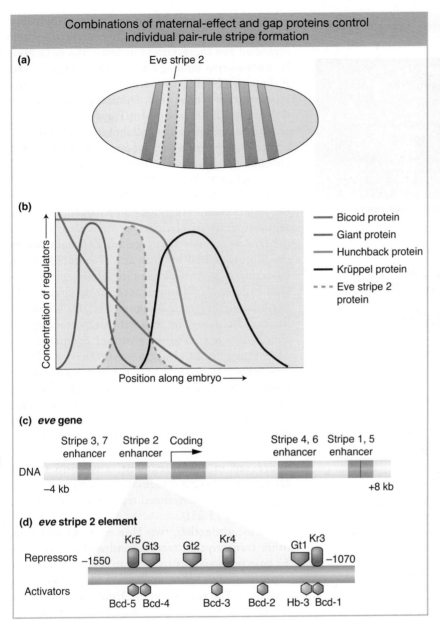

(a) Combinations of maternal-effect and gap proteins control individual pair-rule stripe formation

Eve stripe 2

(b)

Concentration of regulators →

Position along embryo →

— Bicoid protein
— Giant protein
— Hunchback protein
— Krüppel protein
--- Eve stripe 2 protein

(c) *eve* gene

Stripe 3, 7 enhancer   Stripe 2 enhancer   Coding   Stripe 4, 6 enhancer   Stripe 1, 5 enhancer

DNA

−4 kb   +8 kb

(d) *eve* stripe 2 element

Repressors   Kr5 Gt3   Gt2   Kr4   Gt1 Kr3
−1550   −1070

Activators   Bcd-5 Bcd-4   Bcd-3   Bcd-2   Hb-3 Bcd-1

FIGURE 13-19 Regulation of a pair-rule stripe: combinatorial control of an independent cis-acting regulatory element. (a) The regulation of the *eve* stripe 2 cis-acting regulatory element controls the formation of the second stripe of *eve* expression in the early embryo, just one of seven stripes of *eve* expression. (b) The stripe forms within the domains of the Bicoid (Bcd) and Hunchback (Hb) proteins and at the edge of the Giant (Gt) and Krüppel (Kr) gap proteins. Bcd and Hb are activators, Gt and Kr are repressors of the stripe. (c) The *eve* stripe 2 element is just one of several cis-acting regulatory elements of the *eve* gene, each of which controls different parts of *eve* expression. The *eve* stripe 2 element spans from about 1 to 1.7 kb upstream of the *eve* transcription unit. (d) Within the *eve* stripe 2 element, several binding sites exist for each transcription factor (repressors are shown above the element, activators below). The net output of this combination of activators and repressors is expression of the narrow *eve* stripe.

larva. How are the different segmental identities established by Hox proteins? This process has two aspects. First, the *Hox* genes are expressed in different domains along the anteroposterior axis. *Hox*-gene expression is largely controlled by segmentation proteins, especially gap proteins, through mechanisms that are similar to those already described herein for *hunchback* and *eve* stripe 2 (as well as some cross-regulation by Hox proteins of other *Hox* genes). The regulation of *Hox* genes will not be considered in depth here. The second aspect of Hox control of segmental identity is the regulation of target genes by Hox proteins. We will examine one example that nicely illustrates how a major feature of the fruit fly's body plan is controlled through the integration of many inputs by a single cis-acting regulatory element.

The paired limbs, mouthparts, and antennae of *Drosophila* each develop from initially small populations of

about 20 cells in different segments. Different structures develop from the different segments of the head and thorax, whereas the abdomen is limbless. The first sign of the development of these structures is the activation of regulatory genes within small clusters of cells, which are called the appendage *primordia*. The expression of the *Distal-less* (*Dll*) gene marks the start of the development of the appendages. This gene is one of the key targets of the *Hox* genes, and its function is required for the subsequent development of the distal parts of each of these appendages. The small clusters of cells expressing *Distal-less* arise in several head segments and in each of the three thoracic segments, but not in the abdomen (**Figure 13-20a**).

How is *Distal-less* expression restricted to the more anterior segments? Several lines of evidence have revealed that the *Distal-less* gene is repressed in the abdomen by two Hox proteins—the Ultrabithorax and Abdominal-A

Hox proteins repress appendage formation in the abdomen

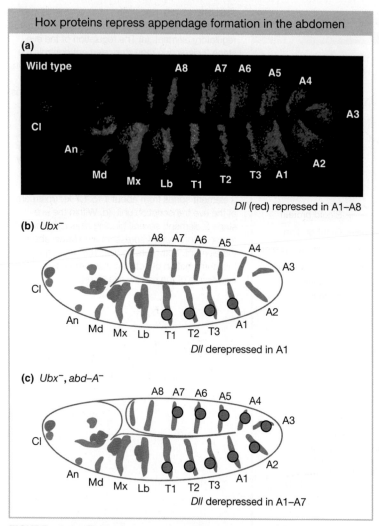

FIGURE 13-20 The absence of limbs in the abdomen is controlled by *Hox* genes. (a) The expression of the *Distal-less* (*Dll*) gene (red) marks the position of future appendages, expression of the *Hox* gene *Ultrabithorax* (purple) marks the position of the abdominal segments A1 through A7, and expression of the *engrailed* gene (blue) marks the posterior of each segment. (b) Schematic representation of *Ubx*⁻ embryo showing that *Dll* expression (red circles) is derepressed in segment A1. (c) Schematic representation of *Ubx*⁻ *abd-A*⁻ embryo showing that *Dll* expression (red circles) is derepressed in the first seven abdominal segments. [*(a) Photomicrograph by Dave Kosman, Ethan Bier, and Bill McGinnis; (b and c) Data from B. Gebelein, D. J. McKay, and R. S. Mann, "Direct Integration of Hox and Segmentation Gene Inputs During Drosophila Development," Nature 431, 2004, 653–659.*]

proteins—working in collaboration with two segmentation proteins. Notice in Figure 13-6 that Ultrabithorax is expressed in abdominal segments one through seven, and Abdominal-A is expressed in abdominal segments two through seven, overlapping with all but the first segment covered by Ultrabithorax. In *Ultrabithorax* mutant embryos, *Distal-less* expression expands to the first abdominal segment (Figure 13-20b), and in *Ultrabithorax/Abdominal-A* double-mutant embryos, *Distal-less* expression extends through the first seven abdominal segments (Figure 13-20c), indicating that both proteins are required for the repression of *Distal-less* expression in the abdomen.

The cis-acting regulatory element responsible for *Distal-less* expression in the embryo has been identified and characterized in detail (**Figure 13-21a**). It contains two binding sites for the Hox proteins. If these two binding sites are mutated such that the Hox proteins cannot bind, *Distal-less* expression is derepressed in the abdomen (Figure 13-21b). Several additional proteins collaborate with the Hox proteins in repressing *Distal-less*. Two are proteins encoded by segment-polarity genes, *Sloppy-paired* (*Slp*) and *engrailed* (*en*). The Sloppy-paired and Engrailed proteins are expressed in stripes that mark the anterior and posterior compartments of each segment, respectively. Each protein also binds to the *Distal-less* cis-acting regulatory element. When the Sloppy-paired-binding site is mutated in the cis-acting regulatory element, reporter-gene expression is derepressed in the anterior compartments of abdominal segments (Figure 13-21c). When the Engrailed-binding site is mutated, reporter expression is derepressed in the posterior compartments of each abdominal segment (Figure 13-21d). And when the binding sites for both proteins are mutated, reporter-gene expression is derepressed in both compartments of each abdominal segment, just as when the Hox-binding sites are mutated (Figure 13-21e). Two other proteins, called Extradenticle and Homothorax, which are broadly expressed in every segment, also bind to the *Distal-less* cis-acting regulatory element and are required for transcriptional repression in the abdomen (Figure 13-21f).

Thus, altogether, two Hox proteins and four other transcription factors bind within a span of 57 base pairs and act together to repress *Distal-less* expression and, hence, appendage formation in the abdomen. The repression of *Distal-less* expression is a clear demonstration of how Hox proteins regulate segment identity and the number of reiterated body structures. It is also a good illustration of how diverse regulatory inputs act combinatorially on cis-acting regulatory elements. In this instance, the presence of Hox-binding sites is not sufficient for transcriptional repression: collaborative and cooperative interactions are required among several proteins to fully repress gene expression in the abdomen.

KEY CONCEPT Combinatorial and cooperative regulation of gene transcription imposes greater specificity on spatial patterns of gene expression and allows for their greater diversity.

Although evolutionary diversity has not been explicitly addressed in this chapter, the presence of multiple independent cis-acting regulatory elements for each toolkit gene has profound implications for the evolution of form. Specifically, the modularity of these elements allows for changes in one aspect of gene expression independent of other gene

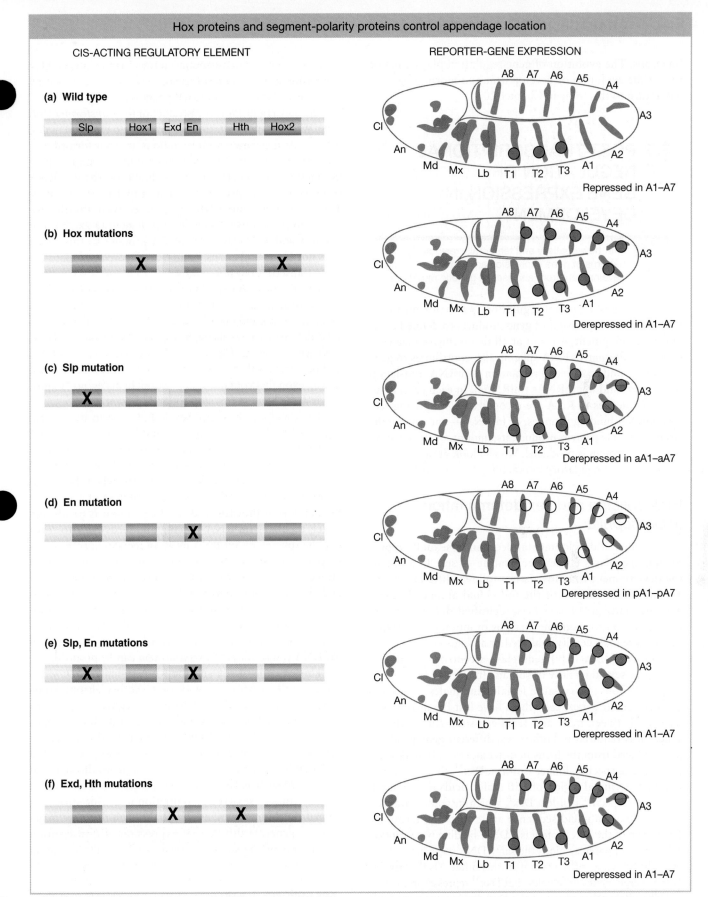

FIGURE 13-21 Integration of Hox and segmentation-protein inputs by a cis-acting regulatory element. (a) *Left:* A cis-acting regulatory element of the *Dll* gene governs the repression of *Dll* expression in the abdomen by a set of transcription factors. (a) *Right: Dll* expression (red) extends to the thorax but not into the abdomen in a wild-type embryo. (b–f) Mutations in the respective binding sites shown derepress *Dll* expression in various patterns in the abdomen. Binding sites are: Slp, Sloppy-paired; Hox1 and Hox2, Ultrabithorax and Abdominal-A; Exd, Extradenticle; En, Engrailed; Hth, Homothorax. [*Data from B. Gebelein, D. J. McKay, and R. S. Mann, "Direct Integration of Hox and Segmentation Gene Inputs During Drosophila Development," Nature 431, 2004, 653–659.*]

functions. The evolution of gene regulation plays a major role in the evolution of development and morphology. We will return to this topic in Chapter 20.

13.5 POST-TRANSCRIPTIONAL REGULATION OF GENE EXPRESSION IN DEVELOPMENT

LO 13.4 Infer how spatially and temporally restricted patterns of gene expression are generated during development from analyses of genetic mutations.

Although transcriptional regulation is a major means of restricting the expression of gene products to defined areas during development, it is not at all the exclusive means of doing so. Alternative RNA splicing also contributes to gene regulation, and so does the regulation of mRNA translation by proteins and microRNAs (miRNAs). In each case, regulatory sequences in RNA are recognized—by splicing factors, mRNA-binding proteins, or miRNAs—and govern the structure of the protein product, its amount, or the location where the protein is produced. We will look at one example of each type of regulatory interaction at the RNA level.

RNA splicing and sex determination in *Drosophila*

A fundamental developmental decision in sexually reproducing organisms is the specification of sex. In animals, the development of many tissues follows different paths, depending on the sex of the individual animal. In *Drosophila*, many genes have been identified that govern *sex determination* through the analysis of mutant phenotypes in which sexual identity is altered or ambiguous.

The *doublesex* (*dsx*) gene plays a central role in governing the sexual identity of somatic (non-germ-line) tissue. Null mutations in *dsx* cause females and males to develop as intermediate *intersexes*, which have lost the distinct differences between male and female tissues. Although *dsx* function is required in both sexes, different gene products are produced from the locus in different sexes. In males, the product is a specific, longer isoform, Dsx^M, that contains a unique C-terminal region of 150 amino acids not found in the female-specific isoform Dsx^F, which instead contains a unique 30-amino-acid sequence at its carboxyl terminus. Each form of the Dsx protein is a DNA-binding transcription factor that apparently binds the same DNA sequences. However, the activities of the two isoforms differ: Dsx^F activates certain target genes in females that Dsx^M represses in males.

The alternative forms of the Dsx protein are generated by alternative splicing of the primary *dsx* RNA transcript. Thus, in this case, the choice of splice sites must be regulated to produce mature mRNAs that encode different

proteins. The various genetic factors that influence Dsx expression and sex determination have been identified by mutations that affect the sexual phenotype.

One key regulator is the product of the *transformer* (*tra*) gene. Whereas null mutations in *tra* have no effect on males, XX female flies bearing *tra* mutations are transformed into the male phenotype. The Tra protein is an alternative splicing factor that affects the splice choices in the *dsx* RNA transcript. In the presence of Tra (and a related protein Tra2), a splice occurs that incorporates exon 4 of the *dsx* gene into the mature *dsx*^{F} transcript (**Figure 13-22**), but not exons 5 and 6. Males lack the Tra protein; so this splice does not occur, and exons 5 and 6 are incorporated into the *dsx*^{M} transcript, but not exon 4.

The Tra protein explains how alternative forms of Dsx are expressed, but how is Tra expression itself regulated to differ in females and males? The *tra* RNA itself is alternatively spliced. In females, a splicing factor encoded by the *Sex-lethal* (*Sxl*) gene is present. This splicing factor binds to the *tra* RNA and prevents a splicing event that would otherwise incorporate an exon that contains a stop codon. In males, no Tra protein is made because this stop codon is present.

The production of the Sex-lethal protein is, in turn, regulated both by RNA splicing and by factors that alter the level of transcription. The level of *Sxl* transcription is initially governed by activators on the X chromosome and repressors on the autosomes. In females, which have two X chromosomes and therefore a double dose of activators, *Sxl* activation prevails and the Sxl protein is produced, which regulates *tra* RNA splicing and feeds back to regulate the splicing of *Sxl* RNA itself. In females, a stop codon is spliced out so that Sxl protein production can continue. However, in males, which have only one X chromosome and therefore only half the dose of X-linked activators, transcription of *Sxl* is initially repressed. Later, *Sxl* transcription is activated in males, but the absence of Sxl protein means that the stop codon is still present in unspliced *Sxl* RNA transcript and no Sxl protein can be produced.

This cascade of sex-specific RNA splicing in *D. melanogaster* illustrates one way that the sex-chromosome genotype leads to different forms of regulatory proteins being expressed in one sex and not the other. Interestingly, the genetic regulation of sex determination differs greatly between animal species, in that sexual genotype can lead to differential expression of regulatory genes through distinctly different paths. However, proteins related to Dsx do play roles in sexual differentiation in a wide variety of animals, including humans. Thus, although there are many ways to generate differential expression of transcription factors, a family of similar proteins plays conserved roles in sexual differentiation across a diversity of species.

KEY CONCEPT The sex determination pathway in *D. melanogaster* is an example of how the spatial and temporal expression of genes involved in developmental pathways can be regulated by differential splicing.

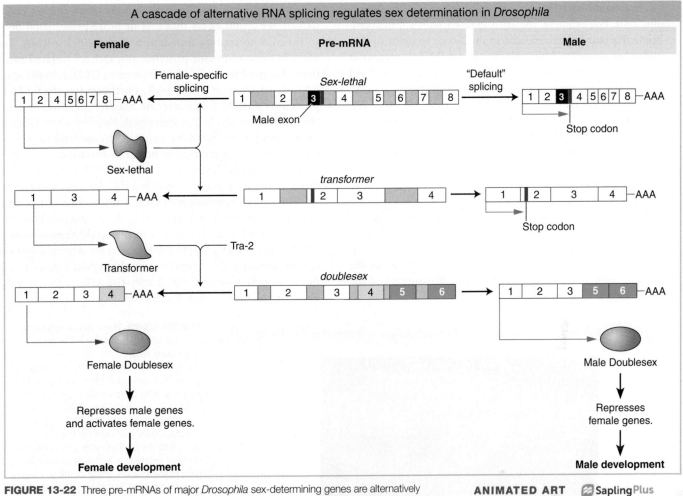

FIGURE 13-22 Three pre-mRNAs of major *Drosophila* sex-determining genes are alternatively spliced. The female-specific pathway is shown on the left and the male-specific pathway shown on the right. The pre-mRNAs are identical in both sexes and shown in the middle. In the male *Sex-lethal* and *transformer* mRNAs, there are stop codons that terminate translation. These sequences are removed by splicing to produce functional proteins in the female. The Transformer and Tra-2 proteins then splice the female *doublesex* pre-mRNA to produce the female-specific isoform of the Dsx protein, which differs from the male-specific isoform by the alternative splicing of several exons.

ANIMATED ART   Sapling Plus

Sex determination in *Drosophila*

Regulation of mRNA translation and cell lineage in *C. elegans*

In many animal species, the early development of the embryo entails the partitioning of cells or groups of cells into discrete lineages that will give rise to distinct tissues in the adult. This process is best understood in the nematode worm *C. elegans*, in which the adult animal is composed of just about 1000 somatic cells (a third of which are nerve cells) and a similar number of germ cells in the gonad. The simple construction, rapid life cycle, and transparency of *C. elegans* has made it a powerful model for developmental analysis (see the Model Organism box on *C. elegans* on page 451). All of this animal's cell lineages were mapped out in a series of elegant studies led by John Sulston at the Medical Research Council (MRC) Laboratory of Molecular Biology in Cambridge, England. Systematic genetic screens for mutations that disrupt or extend cell lineages have provided a bounty of information about the genetic control of lineage decisions. *C. elegans* genetics has been especially

important in understanding the role of post-transcriptional regulation at the RNA level, and we will examine two mechanisms here: (1) control of translation by mRNA-binding proteins, and (2) miRNA control of gene expression.

Translational control in the early embryo

We first look at how a cell lineage begins. After two cell divisions, the *C. elegans* embryo contains four cells, called blastomeres. Each cell will begin a distinct lineage, and the descendants of the separate lineages will have different fates. Already at this stage, differences are observed in the proteins present in the four blastomeres. However, the mRNAs encoding some worm toolkit proteins are present in *all* cells of the early embryo, and post-transcriptional regulation determines which of these mRNAs will be translated into proteins. Thus, in the *C. elegans* embryo, post-transcriptional regulation is critical for the proper specification of early cell fates. During the very first cell division, polarity within the zygote leads to the partitioning of regulatory molecules to specific embry-onic cells. For example, the *glp-1* gene encodes a transmem-brane receptor protein (related to the Notch receptor of flies and other animals). Although the *glp-1* mRNA is present in all cells at the four-cell stage, the GLP-1 protein is translated only in the two anterior cells ABa and ABp (**Figure 13-23a**). This localized expression of GLP-1 is critical for establish-ing distinct fates. Mutations that abolish *glp-1* function at the four-cell stage alter the fates of ABp and ABa descendants.

GLP-1 is localized to the anterior cells by repressing its translation in the posterior cells. The repression of GLP-1 translation requires sequences in the 3′ UTR of the *glp-1* mRNA—specifically, a 61-nucleotide region called the spa-tial control region (SCR). The importance of the SCR has been demonstrated by linking mRNA transcribed from reporter genes to different variants of the SCR. Deletion of this region or mutation of key sites within it causes the reporter gene to be expressed in all four blastomeres of the early embryo (Figure 13-23c).

On the basis of how we have seen transcription con-trolled, we might guess that one or more proteins bind(s) to the SCR to repress translation of the *glp-1* mRNA. To identify these repressor proteins, researchers isolated pro-teins that bind to the SCR. One protein, GLD-1, binds spe-cifically to a region of the SCR. Furthermore, the GLD-1 protein is enriched in posterior blastomeres, just where the expression of *glp-1* is repressed. Finally, when GLD-1 expression is inhibited by using RNA interference, the GLP-1 protein is expressed in posterior blastomeres (Figure 13-23d). This evidence suggests that GLD-1 is a transla-tional repressor protein controlling the expression of *glp-1*.

The spatial regulation of GLP-1 translation is but one example of translational control in development. Transla-tional control is also important in the establishment of the anteroposterior axis in *Drosophila* and the development of sperm in mammals. Again, we see that genetic analysis in model organisms can reveal deeply conserved mechanisms for the regulation of gene expression.

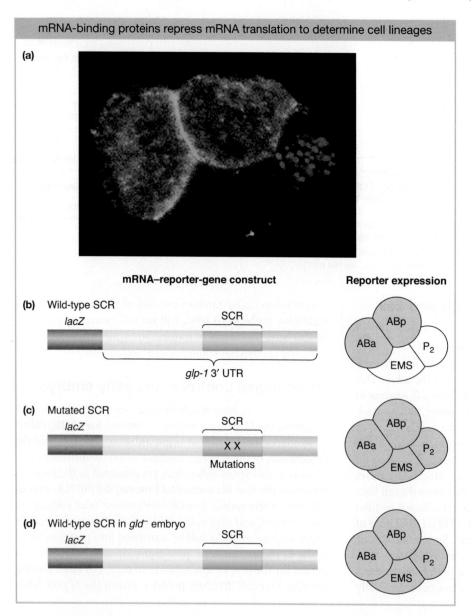

mRNA-binding proteins repress mRNA translation to determine cell lineages

(a)

mRNA–reporter-gene construct

Reporter expression

(b) Wild-type SCR

lacZ

SCR

glp-1 3′ UTR

(c) Mutated SCR

lacZ

SCR

X X

Mutations

(d) Wild-type SCR in *gld⁻* embryo

lacZ

SCR

FIGURE 13-23 Translational regulation and cell-lineage decisions in the early *C. elegans* embryo. (a) At the four-cell stage of the *C. elegans* embryo, the GLP-1 protein is expressed in two anterior cells ABa and ABp (bright green), but not in the EMS or P_2 cell (red). Translation of the *glp-1* mRNA is regulated by the GLD-1 protein in posterior cells. (b) Fusion of the *glp-1* 3′ UTR to the *lacZ* reporter gene leads to reporter expression in the ABa and ABp cells of the four-cell stage of the *C. elegans* embryo (shaded, *right*). (c) Mutations in GLD-1-binding sites in the spatial control region (SCR) cause derepression of translation in the EMS and P_2 lineages, as does (d) loss of *gld* function. [*(a) Courtesy of Thomas C. Evans, University of Colorado Anschutz Medical Campus.*]

The Nematode *Caenorhabditis elegans* as a Model for Cell-Lineage-Fate Decisions

In the past 20 years, studies of the nematode worm *Caenorhabditis elegans* (see the upper figure) have greatly advanced our understanding of the genetic control of cell-lineage decisions.

The transparency and simple construction of this animal led Sydney Brenner to establish its use as a model organism. The adult worm contains about 1000 somatic cells, and researchers, led by John Sulston, have carefully mapped out the entire series of somatic-cell decisions that produce the adult animal.

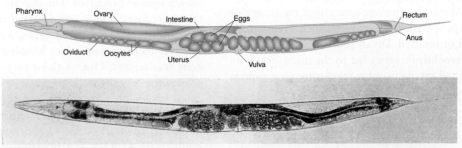

An adult hermaphrodite *Caenorhabditis elegans*, showing various organs. [*Republished with permission of Elsevier, from J. E. Sulston and H. R. Horvitz, "Post-embryonic Cell Lineages of the Nematode, Caenorhabditis elegans," Developmental Biology, 1977, March; 56(1):110–56, Figure 1. Permission conveyed through Copyright Clearance Center, Inc.*]

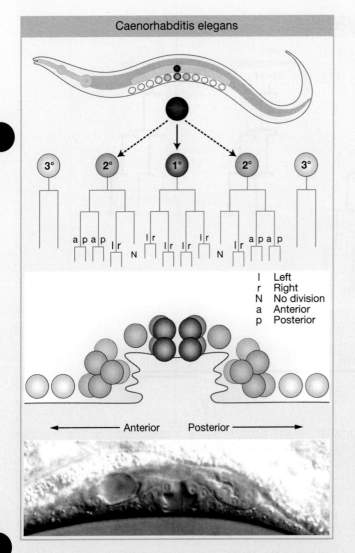

Some of the lineage decisions, such as the formation of the vulva (the opening from which eggs are laid), have been key models of so-called *inductive interactions* in development, where signaling between cells induces cell-fate changes and organ formation (see the lower figure). Exhaustive genetic screens have identified many components participating in signaling and signal transduction involved in the specification of the different cell types that form the vulva.

For some of the embryonic and larval cell divisions, particularly those that will contribute to a worm's nervous system, a progenitor cell gives rise to two progeny cells, one of which then undergoes programmed cell death. Analysis of mutants in which programmed cell death is aberrant, led by Robert Horvitz, has revealed many components of programmed-cell-death pathways common to most animals. Sydney Brenner, John Sulston, and Robert Horvitz shared the 2002 Nobel Prize in Physiology or Medicine for their pioneering work based on *C. elegans*.

Production of the vulval-cell lineages. Parts of the vulval anatomy are occupied by so-called primary (1°), secondary (2°), and tertiary (3°) cells. The lineages or pedigrees of the primary, secondary, and tertiary cells are distinguished by their cell-division patterns and will give rise to different parts of the vulva in the adult worm, as shown in the bottom image. [*Republished with permission of Jennifer L. Green, Takao Inoue and Paul W. Sternberg, Development and The Company of Biologists, The C. elegans ROR receptor tyrosine kinase, CAM-1, nonautonomously inhibits the Wnt pathway, of Jennifer L. Green, Takao Inoue and Paul W. Sternberg, Development 134, 4053–4062 (2007), and The Company of Biologists; permission conveyed through Copyright Clearance Center, Inc.*]

miRNA control of developmental timing in *C. elegans* and other species

Development is a temporally as well as spatially ordered process. When events take place is just as important as where. Mutations in the **heterochronic genes** of *C. elegans* have been sources of insight into the control of developmental timing. Mutations in these genes alter the timing of events in cell-fate specification, causing such events to be either reiterated or omitted. Detailed investigation into the products of heterochronic genes led to the discovery of an entirely unexpected mechanism for regulating gene expression, through microRNAs (see Chapter 9).

Among the first members of this class of regulatory molecules to be discovered in *C. elegans* is RNA produced by the *let-7* gene. The *let-7* gene regulates the transition from late-larval to adult cell fates. In *let-7* mutants, for example, larval cell fates are reiterated in the adult stage (Figure 13-24a). Conversely, increased *let-7* gene dosage causes the precocious specification of adult fates in larval stages.

The *let-7* gene does not encode a protein. Instead, it encodes a temporally regulated mature 22-nucleotide RNA that is processed from an approximately 70-nucleotide precursor. The mature RNA is complementary to sequences in 3′ untranslated regions of a variety of developmentally regulated genes, and the binding of the miRNA to these sequences hinders translation of these gene transcripts. One of these target genes, *lin-41*, also affects the larval-to-adult transition. The *lin-41* mutants cause precocious specification of adult cell fates, suggesting that the effect of *let-7* overexpression is due at least in part to an effect on *lin-41* expression. The *let-7* mRNA binds to *lin-41* RNA in vitro at several imperfect complementary sites (Figure 13-24b).

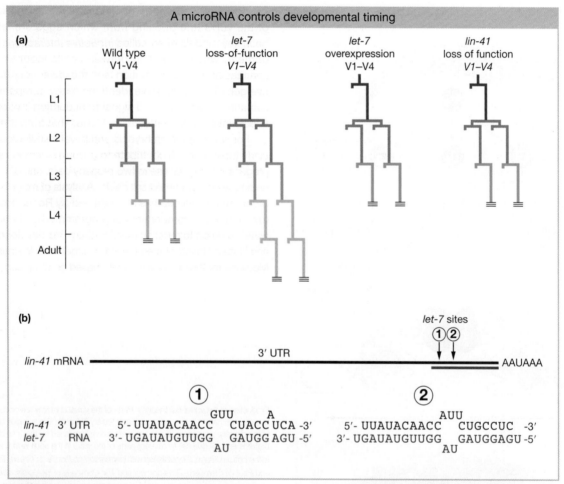

A microRNA controls developmental timing

(a)

FIGURE 13-24 Normally, *C. elegans* develops into an adult after four larval stages, and hypodermal cell lineages conclude their development at L4 (hatched lines at ends of V1–V4 lineages). (a) In *let-7* mutants, the transition from the L4 larval stage to adult is delayed and the cell lineages of lateral hypodermal cells (V) are reiterated. Conversely, in *lin-41* mutants, there is precocious development of adult cell fates in the L3 larval stage. (b) *let-7* encodes an miRNA that is complementary to sequences at two sites in the 3′ UTR of *lin-41* mRNA. [*Data from A. E. Rougvie, "Intrinsic and Extrinsic Regulators of Developmental Timing: From miRNAs to Nutritional Cues," Development 132, 2005, 3787–3798; and from D. M. Eisenmann, "Wnt signaling" (June 25, 2005), WormBook, ed. The C. elegans Research Community, WormBook, doi/10.1895/wormbook.1.7.1, http://www.wormbook.org.*]

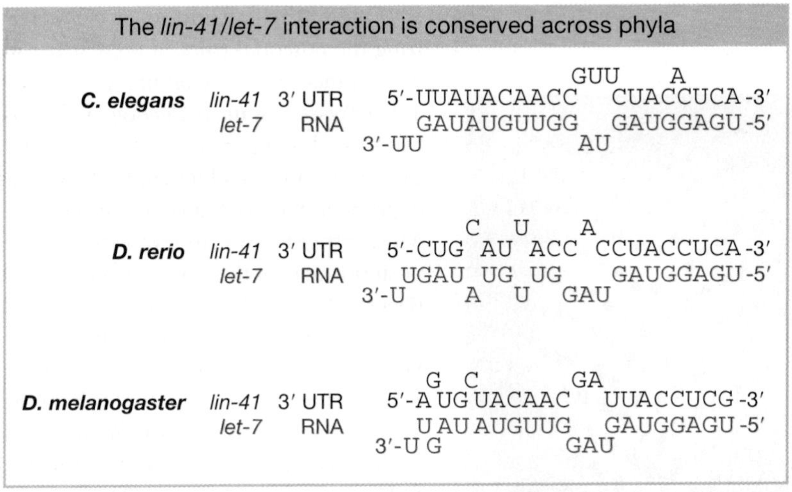

The *lin-41/let-7* interaction is conserved across phyla

			GUU	A
C. elegans	*lin-41*	3′ UTR	5′-UUAUACAACC	CUACCUCA -3′
	let-7	RNA	GAUAUGUUGG	GAUGGAGU -5′
		3′-UU		AU

			C U	A
D. rerio	*lin-41*	3′ UTR	5′-CUG AU ACC	CCUACCUCA -3′
	let-7	RNA	UGAU UG UG	GAUGGAGU -5′
		3′-U	A U GAU	

			G C	GA
D. melanogaster	*lin-41*	3′ UTR	5′-A UG UACAAC	UUACCUCG -3′
	let-7	RNA	U AU AUGUUG	GAUGGAGU -5′
		3′-U G		GAU

FIGURE 13-25 The sequences of both the *let-7* miRNA and its binding site in the 3′ UTR of the *lin-41* mRNA are conserved across *C. elegans*, *D. rerio* (zebrafish), and *D. melanogaster*. [Data from A. E. Pasquinelli et al., "Conservation of the Sequence and Temporal Expression of let-7 Heterochronic Regulatory RNA," *Nature* 408, 2000, 86–89.]

The role of miRNAs in *C. elegans* development extends far beyond *let-7*. Several hundred miRNAs have been identified, and many target genes have been shown to be miRNA regulated. Moreover, the discovery of this class of regulatory RNAs prompted the search for such genes in other genomes, and, in general, hundreds of candidate miRNA genes have been detected in both plant and animal genomes, including those of humans.

Quite surprisingly, the *let-7* miRNA gene is widely conserved and found in *Drosophila*, ascidian, mollusc, annelid, and vertebrate (including human) genomes. The *lin-41* gene also is conserved, and evidence suggests that the *let-7–lin-41* regulatory interaction also controls the timing of events in the development of other species, such as mouse and zebrafish (**Figure 13-25**).

The discoveries of miRNA regulation of developmental genes and of the scope of the miRNA repertoire are fairly recent. Geneticists and other biologists are quite excited about the roles of this class of regulatory molecules in normal development, as well as in the pathology and treatment of disease, leading to a very vigorous, fast-paced area of new research.

KEY CONCEPT Sequence-specific RNA-binding proteins and micro RNAs act through cis-acting sequences in the 3′ untranslated regions of mRNAs to regulate the spatial and temporal pattern of protein translation.

13.6 FROM FLIES TO FINGERS, FEATHERS, AND FLOOR PLATES: THE MANY ROLES OF INDIVIDUAL TOOLKIT GENES

LO 13.1 Outline experimental approaches to identify and characterize members of the genetic toolkit for development in different animal phyla.

LO 13.5 Summarize the evidence that the genetic toolkit for development is conserved across animal phyla.

We have seen that toolkit proteins and regulatory RNAs have multiple roles in development. For example, recall that the Ultrabithorax protein represses limb formation in the fly abdomen and promotes hind-wing development in the fly thorax. Similarly, Sloppy-paired and Engrailed participate in the generation of the basic segmental organization of the embryo and collaborate with Hox proteins to suppress limb formation. These roles are just a few of the many roles played by these toolkit genes in the entire course of fly development. Most toolkit genes function at more than one time and place, and most may influence the formation or patterning of many different structures that are formed in different parts of the larval or adult body. Those that regulate gene expression may directly regulate scores to hundreds of different genes. The function of an individual toolkit protein (or RNA) is almost always context dependent, which is why the toolkit analogy is perhaps so fitting. As with a carpenter's toolkit, a common set of tools can be used to fashion many structures.

To illustrate this principle more vividly, we will look at the role of one toolkit protein in the development of many vertebrate features, including features present in humans. This toolkit protein is the vertebrate homolog of the *Drosophila hedgehog* gene. The *hedgehog* gene was first identified by Nüsslein-Volhard and Wieschaus as a segment-polarity gene. It has been characterized as encoding a signaling protein secreted from cells in *Drosophila*.

As the evidence grew that toolkit genes are common to different animal phyla, the discovery and characterization of fly toolkit genes such as *hedgehog* became a common springboard to the characterization of genes in other taxa, particularly vertebrates. The identification of homologous genes based on sequence similarity was a fast track to the identification of vertebrate toolkit genes. The application of this strategy to the *hedgehog* gene illustrates the power and payoffs of using homology to discover important genes. Several distinct homologs of *hedgehog* were isolated from vertebrates including zebrafish, mice, chickens, and humans. In the whimsical spirit of the *Drosophila* gene nomenclature, the three vertebrate homologs were named *Sonic hedgehog* (after the video-game character), *Indian hedgehog*, and *Desert hedgehog*.

The *Sonic hedgehog* toolkit gene has multiple roles

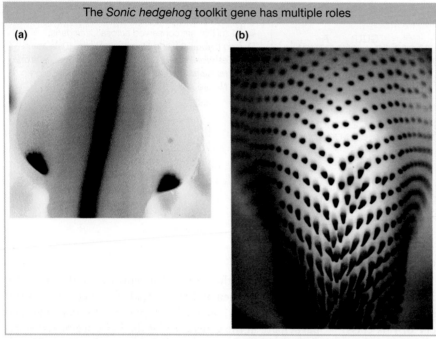

FIGURE 13-26 The *Shh* gene is expressed in many different parts of the developing chick embryo (indicated by dark staining), including (a) the zone of polarizing activity in each of the two developing limb buds and the long neural tube, and (b) the developing feather buds. *Shh* mRNA is visualized by in situ hybridization. [*Photomicrographs courtesy of (a) Cliff Tabin and (b) Photomicrographs courtesy Dr. John Fallon, University of Wisconsin / Matthew Harris, Harvard Medical School, Department of Genetics.*]

One of the first means of characterizing the potential roles of these genes in development was to examine where they are expressed. *Sonic hedgehog* (*Shh*) was found to be expressed in several parts of the developing chicken, with similar expression patterns in other vertebrates. Most intriguing was its expression in the posterior part of the developing limb bud (**Figure 13-26a**). This part of the limb bud was known for decades to be the *zone of polarizing activity* (ZPA) because it is an organizer responsible for establishing the anteroposterior polarity of the limb and its digits (see Figure 13-2). To test whether *Shh* might play a role in ZPA function, Cliff Tabin and his colleagues at Harvard Medical School caused the Shh protein to be expressed in the *anterior* region of developing chick limb buds. They observed the same effect as transplantation of the ZPA—the induction of extra digits with reversed polarity. Their results were stunning evidence that Shh was the long-sought morphogen produced by the ZPA (**Figure 13-27**).

FIGURE 13-27 (a) A normal chicken limb with a single organizer, the zone of polarizing activity (ZPA), has three digits (II-III-IV). (b) Transplantation of the ZPA from a donor limb bud to the anterior position in a recipient limb bud induces extra digits with reverse polarity (IV-III-II-II-III-IV). (c) Similarly, ectopic expression of the *Shh* gene (dark staining) in the anterior limb bud results in a mirror image duplication of the digits (IV-III-II-II-III-IV). [*Figures 6 and 9 republished with permission of Elsevier, from Robert D. Riddle, Randy L. Johnson, Ed Laufer, Cliff Tabin, "Sonic hedgehog mediates the polarizing activity of the ZPA," Cell, 1993, 31 December; 75 (7): 1401–1416. https://doi.org/10.1016/0092-8674(93)90626-2. Permission conveyed through Copyright Clearance Center, Inc.*]

The *Sonic hedgehog* gene is the morphogen produced by the limb organizer

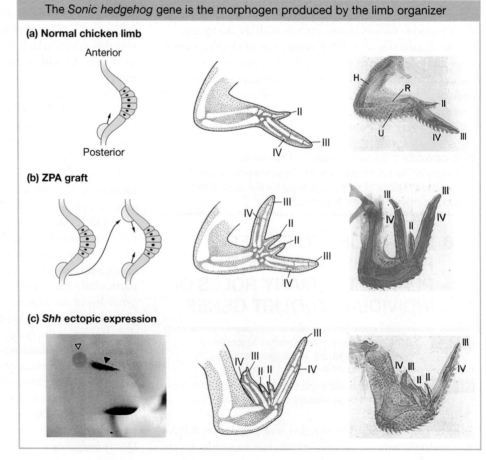

Shh is also expressed in other intriguing patterns in the chicken and other vertebrates. For example, Shh is expressed in developing feather buds, where it plays a role in establishing the pattern and polarity of feather formation (Figure 13-26b). Shh is also expressed in the developing neural tube of vertebrate embryos, in a region called the *floor plate* (Figure 13-26a). Subsequent experiments have shown that Shh signaling from these floor-plate cells is critical for the subdivision of the brain hemispheres and the subdivision of the developing eye into the left and right sides. When the function of the *Shh* gene is eliminated by mutation in the mouse using techniques described in Chapters 10 and 14, these hemispheres and eye regions do not separate, and the resulting embryo is *cyclopic*, with one central eye and a single forebrain (it also lacks limb structures).

Shh is just one striking example of the dramatic and diverse roles played by toolkit genes at different places and times in development. The outcomes of Shh signaling are different in each case: the Shh signaling pathway will induce the expression of one set of genes in the developing limb, a different set in the feather bud, and yet another set in the floor plate. How are different cell types and tissues able to respond differently to the same signaling molecule? Just as we learned from investigating the genetic control of patterning in the *Drosophila* embryo, the outcome of Shh signaling depends on the integration with the signals provided by other toolkit genes that are acting at the same time and in the same place.

> **KEY CONCEPT** Most toolkit genes have multiple roles in different tissues and cell types. The specificity of their action is determined by the context provided by the other toolkit genes that act in combination with them.

13.7 DEVELOPMENT AND DISEASE

LO 13.5 Summarize the evidence that the genetic toolkit for development is conserved across animal phyla.

The discovery that the fly genetic toolkit for development is largely conserved in vertebrates has also had a profound effect on the study of the genetic basis of human diseases, particularly of birth defects and cancer. A large number of toolkit-gene mutations have been identified that affect human development and health. We will focus here on just a few examples that illustrate how understanding gene function and regulation in model animals has translated into better understanding of human biology.

Polydactyly

A fairly common syndrome in humans is the development of extra partial or complete digits on the hands and feet. This condition, called *polydactyly*, arises in about 5 to 17 of every 10,000 live births. In the most dramatic cases, the condition is present on both hands and feet (**Figure 13-28**). Polydactyly occurs widely throughout vertebrates—in cats, chickens, mice, and other species.

The discovery of the role of Shh in digit patterning led geneticists to investigate whether the *Shh* gene was altered in polydactylous humans and other species. In fact, some cases of polydactyly in humans (and also in cats) result from mutations of the *Shh* gene. Importantly, the mutations are not in the coding region of the *Shh* gene; rather, they lie in a cis-acting regulatory element, far from the coding region, that controls *Shh* expression in the developing limb bud. The extra digits are induced by the expression of *Shh* in a part of the limb where the gene is not normally expressed. Mutations in cis-acting regulatory elements have two important properties that are distinct from mutations in coding regions. First, because they affect regulation in cis, the phenotypes are often dominant. Second, because only one of several cis-acting regulatory elements may be affected, other gene functions may be completely normal. Polydactyly can occur without any collateral developmental problems that would be expected given the multiple roles of *Shh* in development. For similar reasons, we will see in Chapter 20 that mutations in cis-acting regulatory elements of toolkit genes also play key roles during the evolution of morphological differences among species. Coding mutations in *Shh*, however, tell a different story, as we will see in the next section.

Holoprosencephaly

Mutations in the human *Shh* coding region also have been identified. The consequent alterations in the Shh protein are

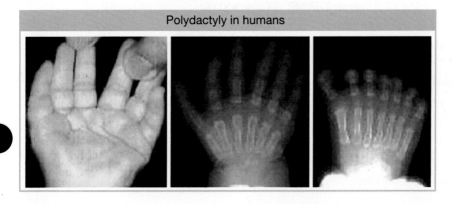

Polydactyly in humans

FIGURE 13-28 This person has six fingers on each hand and seven toes on each foot owing to a regulatory mutation in the *Sonic hedgehog* gene. [*Courtesy of Dr. Robert Hill, MRC Human Genetics Unit, Edinburgh, Scotland; from L. A. Lettice et al., "Disruption of a Long-Range Cis-Acting Regulator for Shh Causes Preaxial Polydactyly,"* Proc. Natl. Acad. Sci. USA *99, 7548. Copyright (2002) National Academy of Sciences, U.S.A.*]

associated with a syndrome termed *holoprosencephaly*, in which abnormalities occur in brain size, in the formation of the nose, and in other midline structures. These abnormalities appear to be less severe counterparts of the developmental defects observed in homozygous *Shh* mutant mice. Indeed, the affected children seen in clinics are heterozygous. One copy of a normal *Shh* gene appears to be insufficient for normal midline development (the gene is *haploinsufficient*). Human fetuses homozygous for loss-of-function *Shh* mutations very likely die in gestation with more severe defects.

Holoprosencephaly is not caused exclusively by *Shh* mutations. Shh is a ligand in a signal-transduction pathway. As might be expected, mutations in genes encoding other components of the pathway affect the efficiency of Shh signaling and are also associated with holoprosencephaly. Several components of the human Shh pathway were first identified as homologs of members of the fly pathway, demonstrating once again both the conservation of the genetic toolkit and the power of model systems for biomedical discovery.

Cancer as a developmental disease

In long-lived animals, such as ourselves and other mammals, development does not cease at birth or at the end of adolescence. Tissues and various cell types are constantly being replenished. The maintenance of many organ functions depends on the controlled growth and differentiation of cells that replace those that are sloughed off or otherwise die. Tissue and organ maintenance is generally controlled by signaling pathways. Inherited or spontaneous mutations in genes encoding components of these pathways can disrupt tissue organization and contribute to the loss of control of cell proliferation. Because unchecked cell proliferation is a characteristic of cancer, the formation of cancers may be a consequence. Cancer, then, is a developmental disease, a product of normal developmental processes gone awry.

Some of the genes associated with types of human cancers are shared members of the animal toolkit. For example, the *patched* gene encodes a receptor for the Hedgehog

signaling proteins. In addition to causing inherited developmental disorders such as polydactyly and holoprosencephaly, mutations in the human *patched* gene are associated with the formation of a variety of cancers. About 30 to 40 percent of patients with a dominant genetic disorder called *basal cell nevus syndrome* (BCNS) carry *patched* mutations. These persons are strongly disposed to develop a type of skin cancer called basal-cell carcinoma. They also have a greatly increased incidence of medulloblastoma, a very deadly form of brain tumor. A growing list of cancers are now associated with disruptions of signal-transduction pathways—pathways that were first elucidated by these early systematic genetic screens for patterning mutants in fruit flies (**Table 13-2**).

The discoveries of links between mutations of signal-transduction-pathway genes and human cancer have greatly facilitated the study of the biology of cancer and the development of new therapies. For example, about 30 percent of mice heterozygous for a targeted mutation in the *patched* gene develop medulloblastoma. These mice therefore serve as an excellent model for the biology of human disease and a testing platform for therapy.

One promising avenue for the development of new cancer therapies is to identify drugs that can specifically target and kill cancer cells without affecting normal cells. These so-called targeted therapies are already employed today for the treatment of some cancers. For example, Herceptin is a drug used to treat breast cancers with overexpression of the human *epidermal growth factor receptor 2* (*HER2* or *ERBB2*), a homolog of the *Drosophila torpedo* gene (Table 13-2). Much current research is focused on identifying additional drugs to specifically target the signal-transduction pathways that are disrupted in different types of tumors and that were often first identified in genetic screens in flies and worms.

It is fair to say that even the most optimistic and far-sighted researchers did not expect that the discovery of the genetic toolkit for building a fly would have such far-ranging effects on understanding human development and disease.

TABLE 13-2 Some Toolkit Genes Having Roles in Cancer

	Fly gene	Mammalian gene	Cancer type
Signaling-Pathway Components			
Wingless	*armadillo*	*β-catenin* (*CTNNB*)	Colon and skin
	TCF/pangolin	*TCF/LEF*	Colon
Hedgehog	*cubitus interruptus*	*GLI1*	Basal-cell carcinoma
	patched	*PTCH*	Basal-cell carcinoma, medulloblastoma
	smoothened	*SMO*	Basal-cell carcinoma
Notch	*Notch*	*NOTCH1*	T-cell leukemia, lymphoma, breast
EGF receptor	*torpedo*	*ERBB2*	Breast and colon
Decapentaplegic/TGF-β	*Medea*	*SMAD4*	Pancreatic and colon
Toll	*dorsal*	*NF-κB*	Lymphoma
Homeobox	*extradenticle*	*PBX1*	Acute pre-B-cell leukemia

But such huge unforeseen dividends are familiar in the recent history of basic genetic research. The advent of genetically engineered medicines, monoclonal antibodies for diagnosis and therapy, and forensic DNA testing all had similar origins in seemingly unrelated investigations.

KEY CONCEPT Investigation into the genetic control of development in model organisms such as *Drosophila* and *C. elegans* has led to unexpected and far-reaching consequences for the understanding and treatment of human disease.

SUMMARY

In Chapter 11, we mentioned the quip from Jacques Monod and François Jacob that "anything found to be true of *E. coli* must also be true of Elephants."[2] Now that we have seen the regulatory processes that build worms, flies, chickens, humans, and elephants, would we say that they were right? If Monod and Jacob were referring to the principle that gene transcription is controlled by sequence-specific regulatory proteins, we have seen that the bacterial Lac repressor and the fly Hox proteins do indeed act similarly. Moreover, their DNA-binding proteins have the same type of motif. The fundamental insights that Jacob and Monod had concerning the central role of the control of gene transcription in bacterial physiology and that they expected would apply to cell differentiation and development in complex multicellular organisms have been borne out in many respects in the genetic control of animal development.

Many features in single-celled and multicellular eukaryotes, however, are not found in bacteria and their viruses. Geneticists and molecular biologists have discovered the functions of introns, RNA splicing, distant and multiple cis-acting regulatory elements, chromatin, alternative splicing, and, more recently, miRNAs. Still, central to the genetic control of development is the control of differential gene expression.

This chapter has presented an overview of the logic and mechanisms for the control of gene expression and development in fruit flies and a few other model species. We have concentrated on the toolkit of animal genes for developmental processes and the mechanisms that control the organization of major features of the body plan—the establishment of body axes, segmentation, and segment identity. Although we explored only a modest number of regulatory mechanisms in depth, and just a few species, similarities in regulatory logic and mechanisms allow us to identify some general themes concerning the genetic control of development.

1. *Despite vast differences in appearance and anatomy, animals have in common a conserved toolkit of genes that govern development.* This toolkit is a small fraction of all genes in the genome, and most of these toolkit genes control transcription factors and components of signal-transduction pathways. Individual toolkit genes typically have multiple functions and affect the development of different structures at different stages.

2. *The development of the growing embryo and its body parts takes place in a spatially and temporally ordered progression.* Domains within the embryo are established by the expression of toolkit genes that mark out progressively finer subdivisions along both embryonic axes.

3. *Spatially restricted patterns of gene expression are products of combinatorial regulation.* Each pattern of gene expression has a preceding causal basis. New patterns are generated by the combined inputs of preceding patterns. In the examples presented in this chapter, the positioning of pair-rule stripes and the restriction of appendage-regulatory-gene expression to individual segments requires the integration of numerous positive and negative regulatory inputs by cis-acting regulatory elements.

Post-transcriptional regulation at the RNA level adds another layer of specificity to the control of gene expression. Alternative RNA splicing and translational control by proteins and miRNAs also contribute to the spatial and temporal control of toolkit-gene expression.

Combinatorial control is key to both the *specificity* and the *diversity* of gene expression and toolkit-gene function. In regard to specificity, combinatorial mechanisms provide the means to localize gene expression to discrete cell populations by using inputs that are not specific to cell type or tissue type. The actions of toolkit proteins can thus be quite specific in different contexts. In regard to diversity, combinatorial mechanisms provide the means to generate a virtually limitless variety of gene-expression patterns.

4. *The modularity of cis-acting regulatory elements allows for independent spatial and temporal control of toolkit-gene expression and function.* Just as the operators and UAS elements of bacteria and simple eukaryotes act as switches in the physiological control of gene expression, the cis-acting regulatory elements of toolkit genes act as switches in the developmental control of gene expression. The distinguishing feature of toolkit genes is the typical presence of numerous independent cis-acting regulatory elements that govern gene expression in different spatial domains and at different stages of development. The independent spatial and temporal regulation of gene expression enables individual toolkit genes to have different but specific functions in different contexts. In this light, it is not adequate or accurate to describe a given toolkit-gene function solely in relation to the protein (or miRNA) that it encodes because the function of the gene product almost always depends on the context in which it is expressed.

[2] F. Jacob and J. Monod, *Cold Spring Harbor Quant. Symp. Biol.* 26, 1963, 393.

KEY TERMS

enhancer (p. 444)
gain-of-function mutation (p. 433)
gap gene (p. 439)
gene complex (p. 433)
genetic screen (p. 432)
genetic toolkit (p. 431)
heterochronic gene (p. 452)
homeobox (p. 435)

homeodomain (p. 435)
homeotic transformation (p. 432)
housekeeping gene (p. 432)
Hox gene (p. 433)
loss-of-function mutation (p. 433)
maternal-effect gene (p. 438)
morphogen (p. 429)
organizer (p. 429)

pair-rule gene (p. 439)
positional information (p. 442)
segment-polarity gene (p. 439)
serially reiterated structure (p. 432)
zygote (p. 438)

SOLVED PROBLEMS

SOLVED PROBLEM 1

The *Bicoid* gene (*bcd*) is a maternal-effect gene required for the development of the *Drosophila* anterior region. A mother heterozygous for a *bcd* deletion has only one copy of the *bcd* gene. With the use of *P* elements to insert copies of the cloned *bcd*+ gene into the genome by transformation, it is possible to produce mothers with extra copies of the gene. The early *Drosophila* embryo develops an indentation called the cephalic furrow that is more or less perpendicular to the longitudinal, anteroposterior (A–P) body axis. In the progeny of mothers with only a single copy of *bcd*+, this furrow is very close to the anterior tip, lying at a position one-sixth of the distance from the anterior to the posterior tip. In the progeny of standard wild-type diploids (having two copies of *bcd*+), the cephalic furrow arises more posteriorly, at a position one-fifth of the distance from the anterior to the posterior tip of the embryo. In the progeny of

mothers with three copies of *bcd*+, it is even more posterior. As additional gene doses are added, the cephalic furrow moves more and more posteriorly, until, in the progeny of mothers with six copies of *bcd*+, it is midway along the A–P axis of the embryo. Explain the gene-dosage effect of *bcd*+ on the formation of the cephalic furrow in light of the contribution that *bcd*+ makes to A–P pattern formation.

SOLUTION

The determination of anterior–posterior parts of the embryo is governed by a concentration gradient of Bicoid protein, which is therefore a morphogen. The furrow develops at a critical concentration of *bcd*. As *bcd*+ gene dosage (and, therefore, Bicoid protein concentration) decreases, the furrow shifts anteriorly; as the gene dosage increases, the furrow shifts posteriorly.

PROBLEMS

Visit SaplingPlus for supplemental content. Problems with the 🐟 icon are available for review/grading.

WORKING WITH THE FIGURES

(The first 16 questions require inspection of text figures.)

1. In Figure 13-2, the transplantation of certain regions of embryonic tissue induces the development of structures in new places. What are these special regions called, and what are the substances they are proposed to produce?

2. In Figure 13-5, two different methods are illustrated for visualizing gene expression in developing animals. Which method would allow one to detect where within a cell a protein is localized?

3. Figure 13-7 illustrates the expression of the Ultra-bithorax (Ubx) Hox protein in developing flight appendages. What is the relationship between where

the protein is expressed and the phenotype resulting from the loss of its expression (shown in Figure 13-1)?

4. Why might there be more differences among the sequences of all the Hox proteins within *Drosophila* (shown in Figure 13-8) than there are among the sequences of the Hox group 4 proteins in *Drosophila* and different vertebrate species (shown in Figure 13-9)?

5. In Figure 13-11, what is the evidence that vertebrate *Hox* genes govern the identity of serially repeated structures?

6. As shown in Figure 13-14, what is the fundamental distinction between a pair-rule gene and a segment-polarity gene?

7. In Table 13-1, what is the most common function of proteins that contribute to pattern formation? Why is this the case?

8. Based on the information provided in Figure 13-17 and Figure 13-19, do you predict that there are many or few Bicoid-binding sites in the regulatory elements that control expression of the *Giant* gene?

9. In Figure 13-19, which gap protein regulates the posterior boundary of *eve* stripe 2? Describe how it does so in molecular terms.

10. In Figure 13-20, the *Ultrabithorax* (*Ubx*) gene is expressed in abdominal segments one through seven, and the *Distal-less* (*Dll*) gene is expressed in the head and thoracic segments. What do you predict would happen to *Dll* expression if *Ubx* were expressed in thoracic segments one through three?

11. Figure 13-21 shows a cis-acting regulatory element of the *Distal-less* (*Dll*) gene.

    a. How many different transcription factors govern where the *Dll* gene will be expressed?

    b. Are there any combinations of mutations that would lead to expression of the *Dll* gene in abdominal segment 8? 🔁

12. Examine the *Drosophila* sex determination cascade shown in Figure 13-22.

    a. Which isoform of the *doublesex* transcript would be found in males that express the Sex-lethal protein?

    b. Which isoform of the *doublesex* transcript would be found in males that have a loss-of-function mutation in the *Sex-lethal* gene?

    c. Which isoform of the *doublesex* transcript would be found in females that have a loss-of-function mutation in the *Sex-lethal* gene? 🔁

13. What do you predict would happen to expression of the *lacZ* reporter gene in Figure 13-23 if the GLD-1 protein was expressed in all four cells of the early *C. elegans* embryo?

14. In Figure 13-24, we see that overexpression of the *let-7* gene has the same phenotype as a loss-of-function mutation in the *lin-41* gene. Explain this result based on the molecular function of the *let-7* gene.

15. As shown in Figure 13-26, the *Sonic hedgehog* gene is expressed in many places in a developing chicken. Is the identical Sonic hedgehog protein expressed in each tissue? If so, how do the tissues develop into different structures? If not, how are different Sonic hedgehog proteins produced?

16. Mutations in a cis-acting regulatory element of the *Sonic hedgehog* gene lead to polydactyly in humans, as shown in Figure 13-28. Based on Figure 13-27, where do you think the *Sonic hedgehog* gene in expressed in humans with this mutation during limb development?

BASIC PROBLEMS

17. *Engrailed, even-skipped, hunchback,* and *Antennapedia*. To a *Drosophila* geneticist, what are they? How do they differ? 🔁

18. Describe the expression pattern of the *Drosophila* gene *eve* in the early embryo, and the phenotypic effects of mutations in the *eve* gene. 🔁

19. Contrast the function of homeotic genes with that of pair-rule genes. 🔁

20. When an embryo is homozygous mutant for the gap gene *Kr*, the fourth and fifth stripes of the pair-rule gene *ftz* (counting from the anterior end) do not form normally. When the gap gene *kni* is mutant, the fifth and sixth *ftz* stripes do not form normally. Explain these results in regard to how segment number is established in the embryo.

21. Some of the mammalian *Hox* genes have been shown to be more similar to one of the insect *Hox* genes than to the others. Design an experimental approach that would enable you to demonstrate this finding in a functional test in living flies.

22. The three homeodomain proteins Abd-B, Abd-A, and Ubx are encoded by genes within the *Bithorax* complex of *Drosophila*. In wild-type embryos, the *Abd-B* gene is expressed in the posterior abdominal segments, *Abd-A* in the middle abdominal segments, and *Ubx* in the anterior abdominal and posterior thoracic segments. When the *Abd-B* gene is deleted, *Abd-A* is expressed in both the middle and the posterior abdominal segments. When *Abd-A* is deleted, *Ubx* is expressed in the posterior thorax and in the anterior and middle abdominal segments. When *Ubx* is deleted, the patterns of *Abd-A* and *Abd-B* expression are unchanged from wild type. When both *Abd-A* and *Abd-B* are deleted, *Ubx* is expressed in all segments from the posterior thorax to the posterior end of the embryo. Explain these observations, taking into consideration the fact that the gap genes control the initial expression patterns of the homeotic genes.

23. What genetic tests allow you to tell if a gene is required zygotically or if it has a maternal effect? 🔁

24. In considering the formation of the A–P and D–V axes in *Drosophila*, we noted that, for mutations such as *bcd*, homozygous mutant mothers uniformly produce mutant offspring with segmentation defects. This outcome is always true regardless of whether the offspring themselves are *bcd*⁺/*bcd* or *bcd*/*bcd*. Some other maternal-effect lethal mutations are different, in that the mutant phenotype can be "rescued" by introducing a wild-type allele of the gene from the father. In other words, for such rescuable maternal-effect lethals, *mut*⁺/*mut* animals are normal, whereas *mut*/*mut* animals have the mutant defect. Explain the difference between rescuable and nonrescuable maternal-effect lethal mutations.

25. Suppose you isolate a mutation affecting A–P patterning of the *Drosophila* embryo in which every other segment of the developing mutant larva is missing.

    a. Would you consider this mutation to be a mutation in a gap gene, a pair-rule gene, a segment-polarity gene, or a segment-identity gene?

    b. You have cloned a piece of DNA that contains four genes. How could you use the spatial-expression pattern of their mRNA in a wild-type embryo to identify which represents a candidate gene for the mutation described?

    c. Assume that you have identified the candidate gene. If you now examine the spatial-expression pattern of its mRNA in an embryo that is homozygous mutant for the gap gene *Krüppel*, would you expect to see a normal expression pattern? Explain.

26. In an embryo from a homozygous *Bicoid* mutant female, which class(es) of gene expression is (are) abnormal?

    a. Gap genes

    b. Pair-rule genes

    c. Segment-polarity genes

    d. *Hox* genes

    e. All answer options are correct.

27. The Hunchback protein in normally expressed in the anterior half of the *Drosophila* embryo. You find a mutation in the 3′ untranslated region of the *hunchback* gene that results in expression of Hunchback protein throughout the entire embryo. Provide a molecular explanation for this result.

28. During mouse development, a homolog of the *Drosophila wingless* gene called *Wnt7a* is expressed in the developing limbs and the female reproductive tract. What phenotypes would be predicted to occur in mice with a mutation in the coding region of *Wnt7a*?

29. Mutations in the *Wnt7a* gene have been associated with a human syndrome in which there are abnormalities of both the limbs and genitalia. Do you predict that these mutations are in the coding sequence or in a cis-acting regulatory element of the *Wnt7a* gene?

CHALLENGING PROBLEMS

30. Which of the proteins involved in *Drosophila* development can be classified as a morphogen?

31. You are interested in the genes that control the development of the eyes in *Drosophila*.

    a. Outline the steps you would take to identify and characterize these genes.

    b. Outline the steps you would take to determine whether the genes you find in *Drosophila* are found in other species.

    c. Summarize the advantages of your experimental approaches.

32. The *eyeless* gene is required for eye formation in *Drosophila*. It encodes a homeodomain.

    a. What would you predict about the biochemical function of the Eyeless protein?

    b. Where would you predict that the *eyeless* gene is expressed in development? How would you test your prediction?

    c. The *Small eye* and *Aniridia* genes of mice and humans, respectively, encode proteins with very strong sequence similarity to the fly Eyeless protein, and they are named for their effects on eye development. Devise one test to examine whether the mouse and human genes are functionally equivalent to the fly *eyeless* gene.

33. Gene *X* is expressed in the developing brain, heart, and lungs of mice. Mutations that selectively affect gene *X* function in these three tissues map to three different regions (A, B, and C, respectively) 5′ of the *X* coding region.

    a. Explain the nature of these mutations.

    b. Draw a map of the *X* locus consistent with the preceding information.

    c. How would you test the function of the A, B, and C regions?

34. Why are regulatory mutations at the mouse *Sonic hedgehog* gene dominant and viable? Why do coding mutations cause more widespread defects?

35. A mutation occurs in the *Drosophila doublesex* gene that prevents Tra from binding to the *dsx* RNA transcript. What would be the consequences of this mutation for Dsx protein expression in males? In females?

36. You isolate a *glp-1* mutation of *C. elegans* and discover that the DNA region encoding the spatial control region (SCR) has been deleted. What will the GLP-1 protein expression pattern be in a four-cell embryo in mutant heterozygotes? In mutant homozygotes?

37. Assess the validity of Monod and Jacob's remark that "anything found to be true of *E. coli* must also be true of Elephants."

    a. Compare the structures and mechanisms of action of animal Hox proteins and the Lac repressor. In what ways are they similar? In what ways are they different?

    b. Compare the structure and function of the *lac* operator with the even-skipped stripe 2 enhancer (*eve* stripe 2). How is the control of these "genetic switches" similar or different?

GENETICS AND SOCIETY

Justify the genetic study of development in model organisms such as *Drosophila* and *C. elegans* to understand human development and disease.

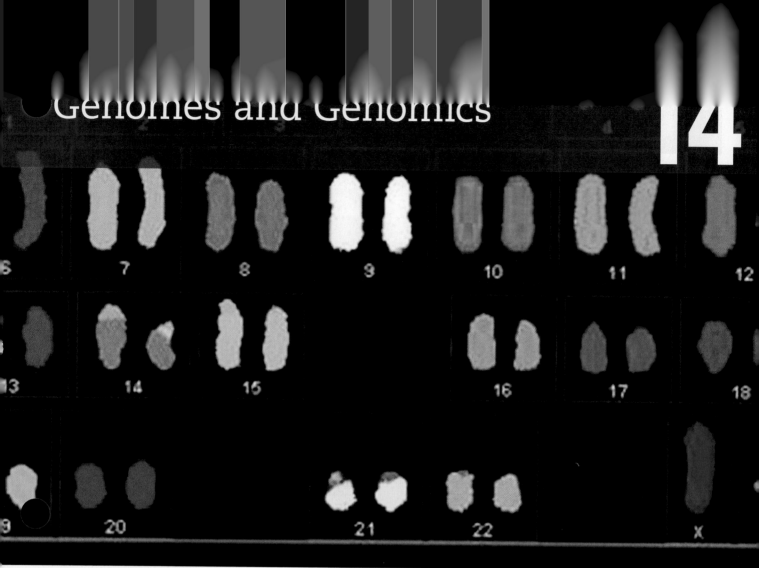

Genomes and Genomics

14

The human nuclear genome viewed as a set of labeled DNA. The DNA of each chromosome has been labeled with a dye that emits fluorescence at one specific wavelength (producing a specific color). [*Nallasivam Palanisamy, MSc., MPhil., PhD., Associate Professor of Pathology, Michigan Center for Translational Pathology, University of Michigan.*]

<table>
<tr><td>

CHAPTER OBJECTIVE

</td><td>

In this chapter, we will see that the ability to sequence whole genomes has revolutionized the field of genetics. Our broad objective is to learn how a combination of experimental and computational methods are used to sequence genomes and to identify functional elements within those genomes.

</td></tr>
</table>

In the summer of 2009, Dr. Alan Mayer, a pediatrician at Children's Hospital of Wisconsin in Milwaukee, wrote to a colleague about the heartbreaking and baffling case of a four-year-old patient of his (**Figure 14-1**). For two years, little Nicholas Volker had endured over 100 trips to the operating room as doctors tried to manage a mysterious disease that was destroying his intestines, leaving him vulnerable to dangerous infections, severely underweight, and often unable to eat.

Neither Mayer nor any other doctors had ever seen a disease like Nicholas's; they were unable to diagnose it, or to stem its ravages by any medical, surgical, or nutritional treatment. It was difficult to treat a disease that no one could identify. So, Dr. Mayer asked his colleague, Dr. Howard Jacob at the Medical College of Wisconsin, "if there is some way we can get his genome sequenced. There is a good chance Nicholas has a genetic defect, and it is likely to be a new disease. Furthermore, a diagnosis soon could save his life and truly showcase personalized genomic medicine."[1]

Dr. Jacob knew that it would be a longshot. Finding a single mutation responsible for a disease would require sifting through thousands of variations in Nicholas's DNA. One key decision was to narrow the search to just the exon sequences in Nicholas's DNA. The rationale was that if the causal mutation was a protein-coding change, then it could be identified by sequencing all of the exons, or Nicholas's *exome*, which comprise a little over 1 percent of the entire human genome. Still, it would be an expensive search—the sequencing would cost about $75,000 with the technology available at the time. Nevertheless, the money was raised from donors, and Jacob and a team of collaborators undertook the task.

As Jacob expected, they found more than 16,000 possible candidate variations in Nicholas's DNA. They narrowed this long list by focusing on those mutations that had not been previously identified in humans, and that caused amino acid replacements that were not found in other species. Eventually, they identified a single base substitution in a gene called the *X-linked inhibitor of apoptosis* (*XIAP*) that changed one amino acid at position 203 of the protein—an amino acid that was invariant among mammals, fish, and even the fruit-fly counterparts of the *XIAP* gene.

Fortunately, the identification of Nicholas's *XIAP* mutation suggested a therapeutic approach. The *XIAP* gene

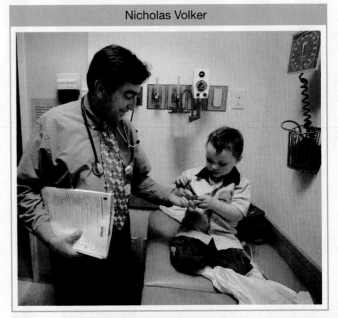

Nicholas Volker

FIGURE 14-1 DNA sequencing of all the exons of Nicholas Volker's genome revealed a single mutation responsible for his debilitating, but previously unidentified, disease. [*Gary Porter/Tribune News Service/WAUWATOSA/WI/USA/Newscom.*]

was previously known to have a role in the inflammatory response, and mutations in the gene were associated with a very rare but potentially fatal immune disorder (although not Nicholas's intestinal symptoms). Based on that knowledge, Nicholas's doctors boosted his immune system with an infusion of umbilical-cord blood from a well-matched donor. Over the next several months, Nicholas's health improved to the point where he was able to eat steak and other foods. And over the next two years, Nicholas did not require any further intestinal surgeries.

The diagnosis and treatment of Nicholas Volker illustrate the dramatic advances in the technology and impact of **genomics**—the study of genomes in their entirety. The long-awaited promise that genomics would shape clinical medicine is now very much a reality. The technological and biological progress from what started as a trickle of data in the 1990s has been astounding. In 1995, the 1.8-Mb (1.8-megabase) genome of the bacterium *Haemophilus influenzae* was the first genome of a free-living organism to be sequenced. In 1996 came the 12-Mb genome of *Saccharomyces cerevisiae*; in 1998, the 100-Mb genome of *Caenorhabditis elegans*; in 2000, the 180-Mb genome of *Drosophila melanogaster*; in 2001, the first draft of the 3000-Mb human genome; and, in 2005, the first draft of

---

[1] M. Johnson and K. Gallagher, "A Baffling Illness," *Milwaukee Journal Sentinel*. Published Dec. 10, 2010. Accessed Mar. 5, 2014.

our closest living relative, the chimpanzee. These species are just a small sample. By the end of 2017, over 130,000 bacterial genomes, and nearly 5500 eukaryotic genomes (including protists, fungi, plants, and animals) had been sequenced. At the beginning of 2018, the Earth BioGenome Project announced its bold intention to sequence all of the approximately 1.5 million known species of eukaryotes in the next 10 years.

It is no hyperbole to say that genomics has revolutionized how genetic analysis is performed and has opened avenues of inquiry that were not conceivable just a few years ago. Most of the genetic analyses that we have so far considered employ a **forward genetics** approach to analyzing genetic and biological processes. That is, the analysis begins by first screening for mutants that affect some observable phenotype, and the characterization of these mutants eventually leads to the identification of the gene and the function of DNA, RNA, and protein sequences. In contrast, having the entire DNA sequences of an organism's genome allows geneticists to work in both directions—forward from phenotype to gene, and in reverse from gene to phenotype (**Figure 14-2**). Without exception, genome sequences reveal many genes that were not detected from classical mutational analysis. Using so-called **reverse genetics**, geneticists can now systematically study the roles of such formerly unidentified genes. Moreover, a lack of prior classical genetic study is no longer an impediment to the genetic investigation of organisms. The frontiers of experimental analysis are growing far beyond the bounds of the very modest number of long-explored model organisms (for more, see the *Beyond Model Organisms* section of *A Brief Guide to Model Organisms*, at the back of this book).

Analyses of whole genomes now contribute to every corner of biological research. In human genetics, genomics is providing new ways to locate genes that contribute to many genetic diseases, like Nicholas's, which had previously eluded investigators. The day is soon approaching when a person's genome sequence is a standard part of his or her medical record. The availability of genome sequences for long-studied model organisms and their relatives has dramatically accelerated gene identification, the analysis of gene function, and the characterization

of noncoding elements of the genome. New technologies for the global, genome-wide analysis of the physiological role of all gene products are driving the development of the new field called *systems biology*. From an evolutionary perspective, genomics provides a detailed view of how genomes and organisms have diverged and adapted over geological time.

The DNA sequence of the genome is the starting point for a whole new set of analyses aimed at understanding the structure, function, and evolution of the genome and its components. In this chapter, we will focus on three major aspects of genomic analysis:

- *Bioinformatics*, the analysis of the information content of entire genomes. This information includes the numbers and types of genes and gene products as well as the location, number, and types of binding sites on DNA and RNA that allow functional products to be produced at the correct time and place.

- *Comparative genomics*, which considers the genomes of closely and distantly related species for evolutionary insight.

- *Functional genomics*, the use of an expanding variety of methods, including reverse genetics, to understand gene and protein function in biological processes.

14.1 THE GENOMICS REVOLUTION

After the development of recombinant DNA technology in the 1970s, research laboratories typically undertook the cloning and sequencing of one gene at a time (see Chapter 10), and then only after having had first found out something interesting about that gene from a classic mutational analysis. The steps in proceeding from the classical genetic map of a locus to isolating the DNA encoding a gene (*cloning*) to determining its sequence were often numerous and time consuming. In the 1980s, some scientists realized that a large team of researchers making a concerted effort could clone and sequence the *entire* genome of a selected organism. Such **genome projects** would then make the clones and the sequence publicly available resources. One appeal of having these resources available is that, when researchers become interested in a gene of a species whose genome has been sequenced, they need only find out where that gene is located on the map of the genome to be able to zero in on its sequence and potentially its function. By this means, a gene could be characterized much more rapidly than by cloning and sequencing it from scratch, a project that at the time could take several years to carry out. This quicker approach is now a reality for all model organisms.

Similarly, the Human Genome Project aimed to revolutionize the field of human genetics. The availability of human genome sequences, and the ability to sequence the

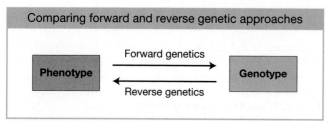

FIGURE 14-2 Forward genetics is phenotype driven, and asks *what genes underlie a particular phenotype*, while reverse genetics is genotype driven and asks *what phenotypes are associated with a particular gene*.

genomes of patients and their relatives, has greatly aided the identification of disease-causing genes. Furthermore, the ability to determine gene sequences in normal and diseased tissues (for example, cancers) has been a great catalyst to the understanding of disease processes, and pointed the way to new therapies.

From a broader perspective, the genome projects had the appeal that they could provide some glimmer of the principles on which genomes are built. The human genome contains 3 billion base pairs of DNA. Having the entire sequence raised questions such as: How many genes does it contain? How are they distributed, and why? What fraction of the genome is coding sequence? What fraction is regulatory sequence? How is our genome similar to or different from other animals? Although we might convince ourselves that we understand a single gene of interest, the major challenge of genomics today is genomic literacy: How do we read the storehouse of information enciphered in the sequence of complete genomes?

The basic techniques needed for sequencing entire genomes were already available in the 1980s. But the scale that was needed to sequence a complex genome was, as an engineering project, far beyond the capacity of the research community then. Genomics in the late 1980s and the 1990s evolved out of large research centers that could integrate these elemental technologies into an industrial-level production line. These centers developed robotics and automation to carry out the many thousands of cloning steps and millions of sequencing reactions necessary to assemble the sequence of a complex organism. Just as important,

advances in information technology aided the analysis of the resulting data.

The first successes in genome sequencing set off waves of innovation that led to faster and much less expensive sequencing technologies. Now, individual machines can produce as much sequence in a day as centers used to accomplish in months. New technologies can now obtain more than 1×10^{12} bases of sequence in a working day on a single instrument. This represents an approximately *1 million-fold* increase in throughput over earlier instruments used to obtain the first human genome sequences (**Figure 14-3**).

Genomics, aided by the explosive growth in information technology, has encouraged researchers to develop ways of experimenting on the genome as a whole rather than simply one gene at a time. Genomics has also demonstrated the value of collecting large-scale data sets in advance so that they can be used later to address specific research problems. In the last sections of this chapter, we will explore some ways that genomics now drives basic and applied genetics research. In subsequent chapters, we will see how genomics is catalyzing advances in understanding the dynamics of mutation, recombination, and evolution.

KEY CONCEPT Characterizing whole genomes is fundamental to understanding the entire body of genetic information underlying the physiology, development, and evolution of living organisms, and to the discovery of new genes such as those having roles in human genetic disease.

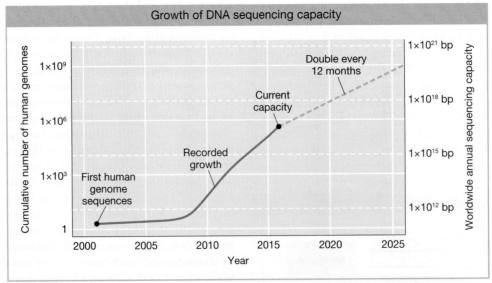

FIGURE 14-3 This plot shows the increase in DNA sequencing capacity from the publication of the first human genome sequences in 2001 until the end of 2015 (solid line), along with the projected growth if capacity doubles every year (dashed line). The total number of human genomes sequenced is indicated on the y-axis on the left, and the total worldwide annual sequencing capacity is indicated on the y-axis on the right. [*Data from Z. D. Stephens et al., "Big Data: Astronomical or Genomical?" PLoS Biology, 13(7), 2015, e1002195, https://doi.org/10.1371/journal.pbio.1002195.*]

14.2 OBTAINING THE SEQUENCE OF A GENOME

LO 14.1 Describe the combinations of strategies typically necessary for obtaining and assembling the complete DNA sequences of organisms.

When people encounter new territory, one of their first activities is to create a map. This practice has been true for explorers, geographers, oceanographers, and astronomers, and it is equally true for geneticists. Geneticists use many kinds of maps to explore the terrain of a genome. Examples are linkage maps based on inheritance patterns of gene alleles and cytogenetic maps based on the location of microscopically visible features such as rearrangement break points (see Chapters 4 and 17).

The highest-resolution map is the complete DNA sequence of the genome—that is, the complete sequence of nucleotides A, T, C, and G of each double helix in the genome. Because obtaining the complete sequence of a genome is such a massive undertaking of a sort not seen before in biology, new strategies must be used, all based on automation.

Turning sequence reads into an assembled sequence

You've probably seen a magic act in which the magician cuts up a newspaper page into a great many pieces, mixes it in his hat, says a few magic words, and *voila!* an intact newspaper page reappears. Basically, that's how genomic sequences are obtained. The approach is to (1) break the DNA molecules of a genome up into thousands to millions of more or less random, overlapping small segments; (2) read the sequence of each small segment; (3) computationally find the overlap among the small segments where their sequences are identical; and (4) continue overlapping ever larger pieces until all the small segments are linked (**Figure 14-4**). At that point, the sequence of a genome is assembled.

Why does this process require automation? To understand why, let's consider the human genome, which contains about 3×10^9 bp of DNA, or 3 billion base pairs (3 gigabase pairs = 3 Gbp). Suppose we could purify the DNA intact from each of the 24 human chromosomes (the 22 autosomes, plus the X and the Y sex chromosomes), separately put each of these 24 DNA samples into a sequencing machine, and read their sequences directly from one telomere to the other. Obtaining a complete sequence would be utterly straightforward, like reading a book with 24 chapters—albeit a very, very long book with 3 billion characters (about the length of 3000 novels). Unfortunately, such a sequencing machine does not yet exist.

Rather, automated sequencing is the current state of the art in DNA sequencing technology. Initially based on the pioneering dideoxy chain-termination sequencing method developed by Fred Sanger (discussed in Chapter 10; see Figure 10-18), automated sequencing now employs a variety of chemistries and optical-detection methods. The methods now available vary in the length of DNA sequence obtained, the bases determined per second, and raw accuracy. For large-scale sequencing projects that seek to analyze large individual genomes or the genomes of many different individuals or species, choosing a method requires balancing speed, cost, and accuracy.

Individual sequencing reactions (called *sequencing reads*) provide letter strings that, depending on the sequencing technique employed, range on average from about 100 to 15,000 bases long. Such lengths are tiny compared with the DNA of a single chromosome. For example, an individual read of 300 bases is only 0.0001 percent of the longest human chromosome (about 3×10^8 bp of DNA) and only about 0.00001 percent of the entire human genome. Thus, one major challenge facing a genome project is **sequence assembly**—that is, building up all of the individual reads into a **consensus sequence**, a sequence for which there is consensus (or agreement) that it is an authentic representation of the sequence for each of the DNA molecules in that genome.

Let's look at these numbers in a somewhat different way to understand the scale of the problem. As with any experimental observation, automated sequencing machines do not always give perfectly accurate sequence reads. Indeed, newer, higher-throughput sequencing technologies generate a *greater* frequency of errors than older methods; the error rate may range from less than 1 percent to as much as 15 percent, depending upon the technology. Thus, to ensure accuracy, genome projects conventionally obtain many independent sequence reads of each base pair in a genome. Many-fold coverage ensures that chance errors in the reads do not give a false reconstruction of the consensus sequence.

Given a sequence read of about 100 bases of DNA and a human genome of 3 billion base pairs, 300 million independent reads are required to give 10-fold average coverage of each base pair. However, not all sequences are represented equally, and so the number of reads required is even larger. Typically, 30-fold average coverage is desired when sequencing a genome. The amount of information to be tracked is enormous. Thus, genome sequencing has required many advances in automation and information technology.

What are the goals of sequencing a genome? First, we strive to produce a consensus sequence that is a true and accurate representation of the genome, starting with one individual organism or standard strain from which the DNA was obtained. This sequence will then serve as a reference sequence for the species. We now know that there are many differences in DNA sequence between different individuals within a species and even between the maternally and paternally contributed genomes within a single diploid individual. Thus, no single genome sequence truly represents the genome of the entire species. Nonetheless, the genome sequence serves as a standard or reference with which other sequences can be compared, and it can be analyzed to determine the information encoded within the DNA, such as the inventory of encoded RNAs and polypeptides.

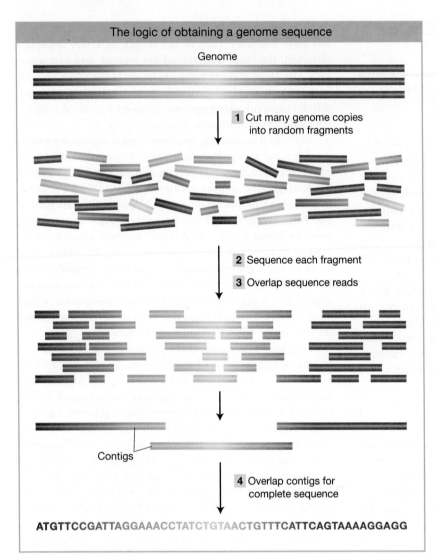

The logic of obtaining a genome sequence

Genome

1 Cut many genome copies into random fragments

2 Sequence each fragment

3 Overlap sequence reads

Contigs

4 Overlap contigs for complete sequence

ATGTTCCGATTAGGAAACCTATCTGTAACTGTTTCATTCAGTAAAAGGAGG

FIGURE 14-4 To obtain a genome sequence, multiple copies of the genome are cut into small pieces that are sequenced. The resulting sequence reads are overlapped by matching identical sequences in different fragments until a consensus sequence of each DNA double helix in the genome is produced.

Like written manuscripts, genome sequences can range from *draft* quality (the general outline is there, but there are typographical errors, grammatical errors, gaps, sections that need rearranging, and so forth), to *finished* quality (a very low rate of typographical errors, some missing sections but everything that is currently possible has been done to fill in these sections), to truly *complete* (no typographical errors, every base pair absolutely correct from telomere to telomere). Although complete assemblies have been obtained for organisms with small genomes, such as bacteria and yeast, this is currently not possible for large and complex eukaryotic genomes, including human. In the following sections, we will examine the strategy and some methods for producing draft and finished genome-sequence assemblies. We will also encounter some of the features of genomes that challenge genome-sequencing projects.

Whole-genome sequencing

The current general strategy for obtaining and assembling the sequence of a genome is called **whole-genome shotgun (WGS) sequencing**. This approach is based on determining the sequence of many segments of genomic DNA that have been

generated by breaking the long chromosomes of DNA into many short segments. Two approaches to WGS sequencing are responsible for most genome sequences obtained to date. The fundamental differences between them are in how the short segments of DNA are obtained and prepared for sequencing and the sequencing chemistry employed. The first method, used to sequence the first human genome, relied on the cloning of DNA in microbial cells and employed the dideoxy sequencing technique. We will refer to this approach as "traditional WGS sequencing." Methods in the second group are generally cell-free methods that employ new techniques for sequencing and are designed for very high throughput (referring to the number of reads per machine per unit time). We will refer to this group of methods as "next-generation WGS sequencing."

Traditional WGS sequencing

The traditional WGS approach begins with the construction of genomic libraries, which are collections of these short segments of DNA, representing the entire genome. The short DNA segments in such a library have been inserted into one of a number of types of *accessory* chromosomes (nonessential elements such as plasmids, modified bacterial viruses, or artificial chromosomes) and propagated in microbes, usually bacteria or yeast. These accessory chromosomes carrying DNA inserts are called vectors (see Chapter 10).

To generate a genomic library, a researcher first uses restriction enzymes, which cleave DNA at specific sequences, to cut up purified genomic DNA. Some enzymes cut the DNA at many places, whereas others cut it at fewer places; so the researcher can control whether the DNA is cut, on average, into longer or shorter pieces. The resulting fragments have short single strands of DNA at both ends. Each fragment is then joined to the DNA molecule of the accessory chromosome, which also has been cut with a restriction enzyme and which has ends that are complementary to those of the genomic fragments. In order for the entire genome to be represented, multiple copies of the genomic DNA are cut into fragments. By this means, thousands to millions of different fragment-vector recombinant molecules are generated.

As discussed in Chapter 10, the resulting pool of recombinant DNA molecules is then propagated, typically by introducing the molecules into bacterial cells. Each cell takes up one recombinant molecule. Then each recombinant molecule is replicated in the normal growth and division of its host so that many identical copies of the inserted fragment are produced for use in analyzing the fragment's DNA sequence. Because each recombinant molecule is amplified from an individual cell, each cell is a distinct *clone*. The resulting library of clones is called a *shotgun library* because sequence reads are obtained from clones randomly selected from the whole-genome library without any information on where these clones map in the genome.

Next, the genome fragments in clones from the shotgun library are partially sequenced. The sequencing reaction must start from a primer of known sequence. Because the sequence of a cloned insert is not known (and is the goal of the exercise), primers are based on the sequence of adjacent vector DNA. These primers are used to guide the sequencing reaction into the insert. Hence, short regions at one or both ends of the genomic inserts can be sequenced (**Figure 14-5**). After

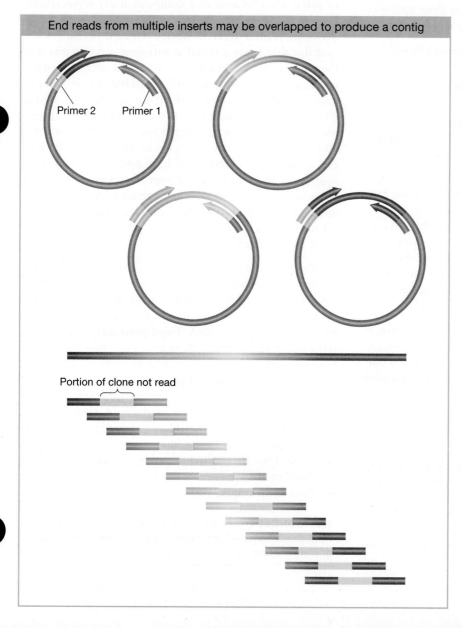

End reads from multiple inserts may be overlapped to produce a contig

Primer 2   Primer 1

Portion of clone not read

FIGURE 14-5 Sequencing reads are taken only of the ends of cloned inserts. The use of two different sequence-priming sites, one at each end of the vector, makes possible the sequencing of as many as 600 base pairs at each end of the genomic insert. If both ends of the same clone are sequenced, the two resulting sequence reads are called *paired-end reads*. When paired-end reads from many different clones are obtained, they can be assembled into a sequence contig even though the sequence from the middle of each single clone is missing (gray bars).

sequencing, the output is a large collection of random short sequences, some of them overlapping. These sequence reads are assembled into a consensus sequence covering the whole genome by matching homologous sequences shared by reads from overlapping clones. The sequences of overlapping reads are assembled into units called **sequence contigs**, which are sequences that are contiguous, or touching.

KEY CONCEPT Whole genomes can be assembled from sequencing many short segments of DNA.

Next-generation WGS sequencing

The goal of next-generation WGS is the same as that of traditional WGS—to obtain a large number of overlapping sequence reads that can be assembled into contigs. However, the methodologies used differ in several substantial ways from traditional WGS. A few different systems have been developed that, while they differ in their sequencing chemistry and machine design, each employ three strategies that have dramatically increased throughput:

1. DNA molecules are prepared for sequencing in *cell-free* reactions, without cloning in microbial hosts.

2. Millions of individual DNA fragments are isolated and sequenced in parallel during each machine run.

3. Advanced fluid-handling technologies, cameras, and software make it possible to detect the products of sequencing reactions in extremely small reaction volumes.

Since the field of genomic technology is evolving rapidly, we will not describe every next-generation system. Here, we will examine *Illumina sequencing*, which is currently the most widely used approach that employs all of these features. The Illumina approach illustrates the gains that have been made in throughput and what such gains enable geneticists to do. The approach can be considered to have three stages:

Stage 1. A **DNA sequencing library** of DNA molecules is constructed. After genomic DNA is isolated from an organism of interest, it is fragmented into smaller pieces of a uniform size. Then, short sequences called *adaptors* are added to both ends of the DNA fragments. There are two adaptor sequences; one sequence is added to one end of the DNA fragment, and the other sequence is added to the other end of the DNA fragment (**Figure 14-6a**).

Stage 2. The DNA fragments are bound to a sequencing *flow cell*. This is a glass slide with small channels that are coated with oligonucleotides containing sequences complementary to both adaptor sequences (Figure 14-6b, inset). A single DNA molecule will bind to a unique location in the flow cell due to hybridization between the adaptor sequence at one end of the DNA molecule and

the oligonucleotide on the flow cell. Then, the adaptor on the other end of the DNA molecule will bind to its complementary oligonucleotide, which is called *bridge formation*. Once immobilized, each DNA molecule is amplified across this bridge by the polymerase chain reaction (PCR; see Chapter 10). After one round of *PCR bridge amplification*, there will be two DNA molecules with complementary sequence on the same location on the flow cell. One end of each of the two DNA molecules will be dissociated from the flow cell. This *dissociation* allows for another round of bridge formation and PCR bridge amplification to take place. Repeating this process many times will generate *clusters*. Each cluster contains thousands of copies of the same DNA fragment in a tiny spot (Figure 14-6b). Each channel of the flow cell contains millions to billions of these clusters.

Stage 3. The sequencing of each cluster is performed using a novel "sequencing-by-synthesis" approach (Figure 14-6c). DNA polymerase and a primer are added to the flow cell to prime the synthesis of a complementary DNA strand. Each of the four deoxyribonucleotide triphosphates, dATP, dGTP, dTTP, and dCTP, is labeled with a different fluorescent dye that emits a signal at different wavelengths (and therefore appears as a different color). In each sequencing cycle, a single nucleotide will be added that is complementary to the next base in the template strand in a given cluster. When the nucleotide is incorporated, the reaction emits a unique wavelength depending upon which base was added. After each sequencing cycle, an image of the flow cell is taken. Each cluster will have added only one of the four bases and will therefore appear as a spot of a single color in the image. The reaction is repeated for at least 100 and up to 300 cycles, and the signals from each cluster over all of the cycles are integrated to generate the sequence reads from each cluster.

The pace of development of next-generation sequencing technologies has been astonishing and is continuing at a dizzying rate. Recently, so-called "third-generation" sequencing technologies have been developed to enable sequencing of single molecules of DNA. Third-generation methods like those developed by Pacific Biosciences (PacBio) and Oxford Nanopore Technologies provide a number of advantages over second-generation sequencing methods such as Illumina. These include the ability to generate very long sequence reads, which greatly enables the assembly of whole genomes, as detailed in the next section. However, these newer sequencing methods currently have a lower throughput and a higher error rate. Thus, the method chosen by researchers depends a great deal on the application, and these choices will continue to evolve in the coming years.

KEY CONCEPT Next-generation WGS sequencing methods have already enabled enormous gains in sequencing output and are continuing to evolve at a rapid rate.

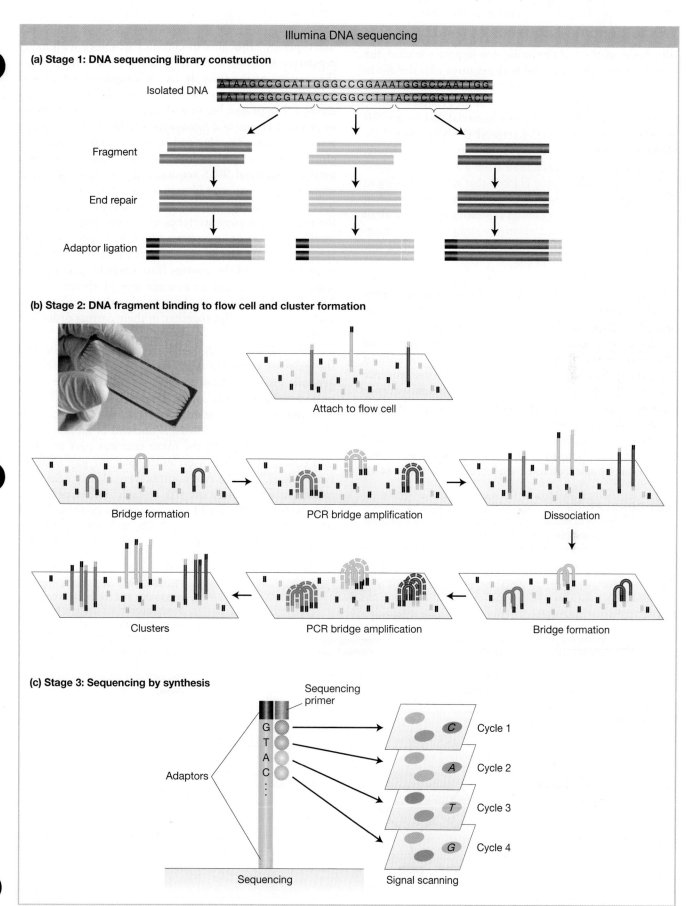

FIGURE 14-6 Illumina DNA sequencing consists of three stages: (a) DNA sequencing library construction; (b) DNA fragment binding to flow cell and cluster formation; and (c) sequencing by synthesis. See text for details. [*Bainscou, Wikimedia Commons, Creative Commons Attribution 3.0 Unported license, https://commons.wikimedia.org/wiki/File:Next_generation_sequencing_slide.jpg#filehistory.*]

ANIMATED ART SaplingPlus

Next-generation sequencing

Whole-genome-sequence assembly

Whichever method of obtaining raw sequence is used, the challenge remains to assemble the contigs into the entire genome sequence. The difficulty of that process depends strongly on the size and complexity of the genome.

For instance, the genomes of bacterial species are relatively easy to assemble. Bacterial DNA is essentially *single-copy* DNA, with no repeating sequences. Therefore, any given DNA sequence read from a bacterial genome will come from one unique place in that genome. Owing to these properties, contigs within bacterial genomes can often be assembled into larger contigs representing most or all of the genome sequence in a relatively straightforward manner. In addition, a typical bacterial genome is only a few megabase pairs of DNA in size.

For eukaryotes, genome assembly often presents some difficulties. A big stumbling block is the existence of numerous classes of repeated sequences, some arranged in tandem and others dispersed (see Chapter 16). Why are they a problem for genome sequencing? In short, because a sequencing read of repetitive DNA fits into many places in the draft of the genome. Not infrequently, a tandem repetitive sequence is in total longer than the length of a maximum sequence read. In that case, there is no way to bridge the gap between adjacent unique sequences. Dispersed repetitive elements can cause reads from different chromosomes or different parts of the same chromosome to be mistakenly assembled together in a single, collapsed sequence contig (**Figure 14-7**).

KEY CONCEPT The landscape of eukaryotic chromosomes includes a variety of repetitive DNA segments. These segments are difficult to assemble as sequence reads.

WGS sequencing is particularly good at producing draft-quality sequences of complex genomes with many repetitive sequences. As an example, we will consider the genome of the fruit fly *D. melanogaster*, which was initially sequenced by the traditional WGS sequencing method. The project began with the sequencing of libraries of genomic clones of different sizes (2 kb, 10 kb, 150 kb). Sequence reads were obtained from *both* ends of genomic-clone inserts and aligned by a logic identical to that used for bacterial WGS sequencing. Through this logic, sequence overlaps were identified and clones were placed in order, producing sequence contigs—consensus sequences for these single-copy stretches of the genome. However, unlike the situation in bacteria, the contigs eventually ran into a repetitive DNA segment that prevented unambiguous assembly of the contigs into a whole genome. The sequence contigs had an average size of about 150 kb. The challenge, then, was how to glue the thousands of such sequence contigs together in their correct order and orientation.

The solution to this problem was to make use of the pairs of sequence reads from opposite ends of the genomic inserts in the same clone—these reads are called **paired-end reads**. The idea was to find paired-end reads that spanned the gaps between two sequence contigs (**Figure 14-8**). In other words, if one end of an insert was part of one contig and the other end was part of a second contig, then this insert must span the gap between two contigs, and the two contigs were clearly near each other. Indeed, because the size of each clone was known (that is, it came from a library containing genomic inserts of uniform size, either the 2-kb, 100-kb, or 150-kb library), the distance between the end reads was known.

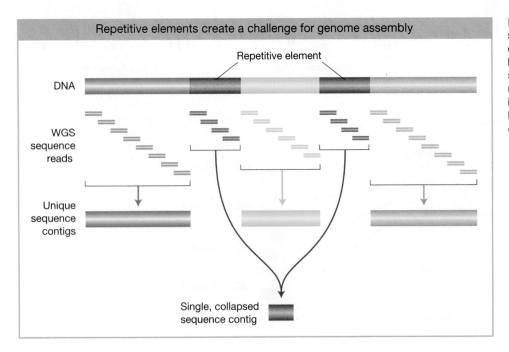

Repetitive elements create a challenge for genome assembly

DNA

Repetitive element

WGS sequence reads

Unique sequence contigs

Single, collapsed sequence contig

FIGURE 14-7 WGS reads from sequences that are found in only one location in the genome can be assembled into many unique sequence contigs. By contrast, WGS reads from repetitive elements found in many locations in the genome will be collapsed into a single sequence contig.

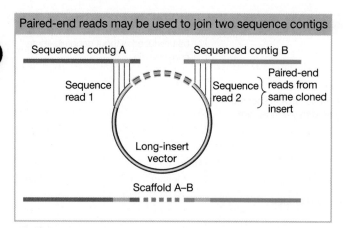

Paired-end reads may be used to join two sequence contigs

Sequenced contig A

Sequenced contig B

Sequence read 1

Sequence read 2

Paired-end reads from same cloned insert

Long-insert vector

Scaffold A–B

FIGURE 14-8 Paired-end reads can be used to join two sequence contigs into a single ordered and oriented scaffold.

next-generation WGS researchers had to devise a way to bridge these gaps without building genomic libraries in vectors. One solution was to build a library of circularized genomic DNA fragments of desired sizes. The circularization allows for short segments of previously distant sequences located at the ends of each fragment to be juxtaposed. Shearing of these circular molecules and amplification and sequencing of fragments containing the junction produces paired-end reads equivalent to those obtained from sequencing of traditional genomic-library inserts (**Figure 14-10**).

KEY CONCEPT Paired-end reads are crucial for assembling genomes from both traditional and next-generation WGS sequencing data.

Further, aligning the sequences of the two contigs by using paired-end reads automatically determines the relative orientation of the two contigs. In this manner, single-copy contigs could be joined together, albeit with gaps where the repetitive elements reside. These gapped collections of joined-together sequence contigs are called **scaffolds** (sometimes also referred to as **supercontigs**). Because most *Drosophila* repeats are large (3–8 kb) and widely spaced (one repeat approximately every 150 kb), this technique was extremely effective at producing a correctly assembled draft sequence of the single-copy DNA. A summary of the logic of this approach is shown in **Figure 14-9**.

Next-generation WGS sequencing does not circumvent the problem of repetitive sequences and gaps. Since this approach is intended to circumvent the construction of libraries, which would otherwise facilitate the bridging of gaps between contigs via paired-end reads,

In both traditional and next-generation WGS sequencing, some gaps usually remain. Specific procedures targeted to individual gaps must be used to fill the missing data in the sequence assemblies. If the gaps are short, missing fragments can be generated by using the known sequences at the ends of the assemblies as primers to amplify and analyze the genomic sequence in between. If the gaps are longer, attempts can be made to isolate the missing sequences as parts of larger inserts that have been cloned into a vector, and then to sequence the inserts. In the future, the longer sequencing reads generated by third-generation sequencing methods will also contribute to filling the gaps in sequence assemblies, particularly in regions of the genome that contain many repeat sequences.

Whether a genome is sequenced to "draft" or "finished" standards is a cost–benefit judgment. Currently, it is relatively straightforward to create a draft but very hard to complete a finished sequence.

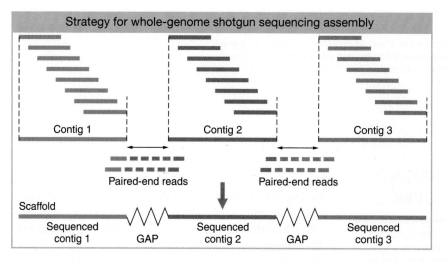

Strategy for whole-genome shotgun sequencing assembly

Contig 1

Contig 2

Contig 3

Paired-end reads

Paired-end reads

Scaffold

Sequenced contig 1

GAP

Sequenced contig 2

GAP

Sequenced contig 3

FIGURE 14-9 In whole-genome shotgun sequencing, first, the unique sequence overlaps between sequence reads are used to build contigs. Paired-end reads are then used to span gaps and to order and orient the contigs into larger units, called *scaffolds*.

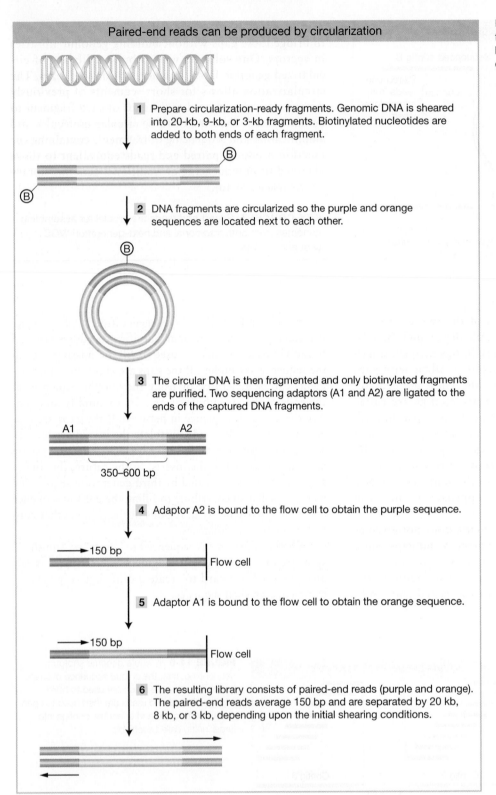

FIGURE 14-10 Paired-end reads for high-throughput sequencing can be produced without genomic-library construction.

Paired-end reads can be produced by circularization

1 Prepare circularization-ready fragments. Genomic DNA is sheared into 20-kb, 9-kb, or 3-kb fragments. Biotinylated nucleotides are added to both ends of each fragment.

2 DNA fragments are circularized so the purple and orange sequences are located next to each other.

3 The circular DNA is then fragmented and only biotinylated fragments are purified. Two sequencing adaptors (A1 and A2) are ligated to the ends of the captured DNA fragments.

A1   A2

350–600 bp

4 Adaptor A2 is bound to the flow cell to obtain the purple sequence.

150 bp   Flow cell

5 Adaptor A1 is bound to the flow cell to obtain the orange sequence.

150 bp   Flow cell

6 The resulting library consists of paired-end reads (purple and orange). The paired-end reads average 150 bp and are separated by 20 kb, 8 kb, or 3 kb, depending upon the initial shearing conditions.

14.3 BIOINFORMATICS: MEANING FROM GENOMIC SEQUENCE

LO 14.2 Explain the role of various functional elements within genomes, and differentiate between computational and experimental methods used to identify these elements.

The genomic sequence is a highly encrypted code containing the raw information for building and operation of organisms. The study of the information content of genomes is called **bioinformatics**. We are far from being able to read this information from beginning to end in the way that we would read a book. Even though we know which triplets encode which amino acids in the protein-coding segments, much of the information contained in a genome is not decipherable from mere inspection.

The nature of the information content of DNA

DNA contains information, but in what way is it encoded? Conventionally, the information is thought of as the sum of all the gene products, both proteins and RNAs. However, the information content of the genome is more complex than that. The genome also contains binding sites for different proteins and RNAs. Many proteins bind to sites located in the DNA itself, whereas other proteins and RNAs bind to sites located in mRNA (**Figure 14-11**). The sequence and relative positions of those sites permit genes to be transcribed, spliced, and translated properly, at the appropriate time in the appropriate tissue. For example, regulatory protein-binding sites determine when, where, and at what level a gene will be expressed. At the RNA level in eukaryotes, the locations of binding sites for the RNAs and proteins of spliceosomes will determine the 5′ and 3′ splice sites where introns are removed. Regardless of whether a binding site actually functions as such in DNA or RNA, the site must be encoded in the DNA. The information in the genome can be thought of as the sum of all the sequences that encode proteins and RNAs, plus the binding sites that govern the time and place of their actions. As a genome draft continues to be improved, the principal objective is the identification of all of the functional elements of the genome. This process is referred to as **annotation**.

KEY CONCEPT The functional elements of the genome include the sequences that encode proteins and RNAs, as well as the binding sites for the proteins and RNAs that regulate gene expression.

Deducing the protein-encoding genes from genomic sequence

Because the proteins present in a cell largely determine its morphology and physiological properties, one of the first orders of business in genome analysis and annotation is to try to determine an inventory of all of the polypeptides encoded by an organism's genome. This inventory is referred to as the organism's *proteome*. It can be considered a "parts list" for the cell. To determine the list of polypeptides, the sequence of each mRNA encoded by the genome must be deduced. Because of intron splicing, this task is particularly challenging in multicellular eukaryotes, where introns are the norm. In humans, for example, an average gene has about 10 exons. Furthermore, many genes encode alternative exons; that is, some exons are included in some versions of a processed mRNA but are not included in others (see Chapter 8). The alternatively processed mRNAs can encode polypeptides having much, but not all, of their amino acid sequences in common. Even though we have a great many examples of completely sequenced genes and mRNAs, we cannot yet identify 5′ and 3′ splice sites merely from DNA sequence with a high degree of accuracy. Therefore, we cannot be certain which sequences are introns. Predictions of alternatively used exons are even more error prone. For such reasons, deducing the total polypeptide parts list in higher eukaryotes is a large problem. Some approaches follow.

ORF detection The main approach to producing a polypeptide list is to use the computational analysis of the genome sequence to predict mRNA and polypeptide sequences, an important part of bioinformatics. The procedure is to look for sequences that have the characteristics of genes. These sequences would be gene-size and composed of sense codons after possible introns had been removed. The appropriate 5′- and 3′-end sequences would be present, such as start and stop codons. Sequences with these characteristics typical of genes are called **open reading frames** (ORFs). To find candidate ORFs, computer programs scan the DNA sequence on both strands in each reading frame. Because there are three possible reading frames on each strand, there are six possible reading frames in all.

Direct evidence from cDNA sequences Another means of identifying ORFs and exons is through the analysis of mRNA expression. This analysis can be done in

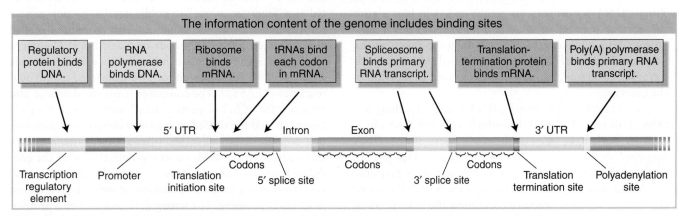

FIGURE 14-11 A gene within DNA may be viewed as a series of binding sites for proteins and RNAs.

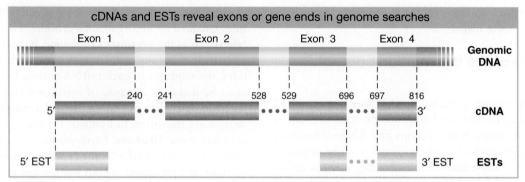

FIGURE 14-12 Alignment of fully sequenced complementary DNAs (cDNAs) and expressed sequence tags (ESTs) with genomic DNA. The dashed lines indicate regions of alignment; for the cDNA, these regions are the exons of the gene. The dots between segments of cDNA or ESTs indicate regions in the genomic DNA that do not align with cDNA or EST sequences; these regions are the locations of the introns. The numbers above the cDNA line indicate the base coordinates of the cDNA sequence, where base 1 is the 5′-most base and base 816 is the 3′-most base of the cDNA. For the ESTs, only a short sequence read is obtained from each end (5′ and 3′) of the corresponding cDNA. These sequence reads establish the boundaries of the transcription unit, but they are not informative about the internal structure of the transcript unless the EST sequences cross an intron (as is true for the 3′ EST depicted here).

two ways. Both methods involve the synthesis of libraries of DNA molecules that are complementary to mRNA sequences, called cDNA (see Chapter 10). The longest established method entails the cloning and amplification of these cDNA molecules in a vector. However, the next-generation sequencing technologies described in the previous section also allow for the direct sequencing of short cDNA molecules without the cloning step, called **RNA sequencing** or "**RNA-seq**" for short (this technique will be described in more detail later in Section 14.7). Whichever method is utilized, complementary DNA sequences are extremely valuable in two ways. First, they are direct evidence that a given segment of the genome is expressed and may thus encode a gene. Second, because the cDNA is complementary to the mature mRNA, the introns of the primary transcript have been removed, which greatly facilitates the identification of the exons and introns of a gene (**Figure 14-12**).

The alignment of cDNAs with their corresponding genomic sequence clearly delineates the exons, and hence introns are revealed as the regions falling between the exons. In the assembled cDNA sequence, the ORF should be continuous from initiation codon through stop codon. Thus, cDNA sequences can greatly assist in identifying the correct reading frame, including the initiation and stop codons. Full-length cDNA evidence is taken as the gold-standard proof that one has identified the sequence of a transcription unit, including its exons and its location in the genome.

In addition to full-length cDNA sequences, there are large data sets of cDNAs for which only the 5′ or the 3′ ends or both have been sequenced. These short cDNA sequence reads are called **expressed sequence tags (ESTs)**. Expressed sequence tags can be aligned with genomic DNA and thereby used to determine the 5′ and 3′ ends of transcripts—in other words, to determine the boundaries of the transcript as shown in Figure 14-12.

Predictions of binding sites As already discussed, a gene consists of a segment of DNA that encodes a transcript as well as the regulatory signals that determine when, where, and how much of that transcript is made. In turn, that transcript has the signals necessary to determine its splicing into mRNA and the translation of that mRNA into a polypeptide (**Figure 14-13**). There are now statistical "gene-finding" computer programs that search for the predicted sequences of the various binding sites used for promoters, for transcription start sites, for 3′ and 5′ splice sites, and for translation initiation codons within genomic DNA. These predictions are based on consensus motifs for such known sequences, but they are not perfect.

Using polypeptide and DNA similarity Because organisms have common ancestors, they also have many genes with similar sequences in common. Hence, a gene will likely have relatives among the genes isolated and sequenced in other organisms, especially in the closely related ones. Candidate genes predicted by the preceding techniques can often be verified by comparing them with all the other gene sequences that have ever been found. A candidate sequence is submitted as a "query sequence" to public databases containing a record of all known gene sequences. This procedure is called a BLAST search (BLAST stands for Basic Local Alignment Search Tool). The sequence can be submitted as a nucleotide sequence (a BLASTn search) or as a translated amino acid sequence (BLASTp). The computer scans the database and returns a list of full or partial "hits," starting with the closest matches. If the candidate sequence closely resembles that of a gene previously identified from another organism, then this resemblance provides a strong indication that the candidate gene is a real gene. Less-close matches are still useful. For example, an amino acid identity of only 35 percent, but at identical positions, is a strong indicator that two proteins have a common three-dimensional structure.

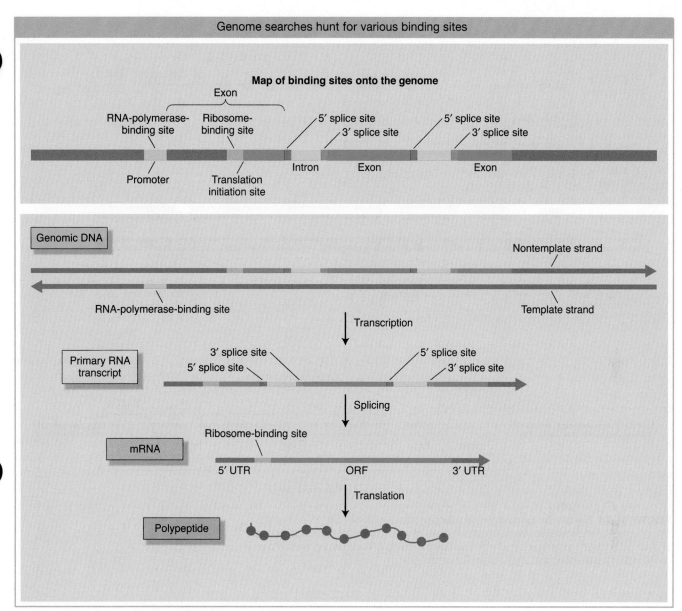

FIGURE 14-13 Eukaryotic information transfer from gene to polypeptide chain. Note the DNA and RNA "binding sites" that are bound by protein complexes to initiate the events of transcription, splicing, and translation.

BLAST searches are used in many other ways, but always the goal is to find out more about some identified sequence of interest.

Predictions based on codon bias Recall from Chapter 9 that the triplet code for amino acids is degenerate; that is, most amino acids are encoded by two or more codons (see Figure 9-8). The multiple codons for a single amino acid are termed *synonymous codons*. In a given species, not all synonymous codons for an amino acid are used with equal frequency. Rather, certain codons are present much more frequently in mRNAs (and hence in the DNA that encodes them). For example, in *D. melanogaster*, of the two codons for cysteine, UGC is used 73 percent of the time, whereas UGU is used 27 percent. This usage is

a diagnostic for *Drosophila* because, in other organisms, this "codon bias" pattern is quite different. Codon biases are thought to be due to the relative abundance of the tRNAs complementary to these various codons in a given species. If the codon usage of a predicted ORF matches that species' known pattern of codon usage, then this match is supporting evidence that the proposed ORF is genuine.

Putting it all together A summary of how different sources of information are combined to create the best-possible mRNA and gene predictions is depicted in **Figure 14-14**. These different kinds of evidence are complementary and can cross-validate one another. For example, the structure of a gene may be inferred from evidence

Many forms of evidence are integrated to make gene predictions

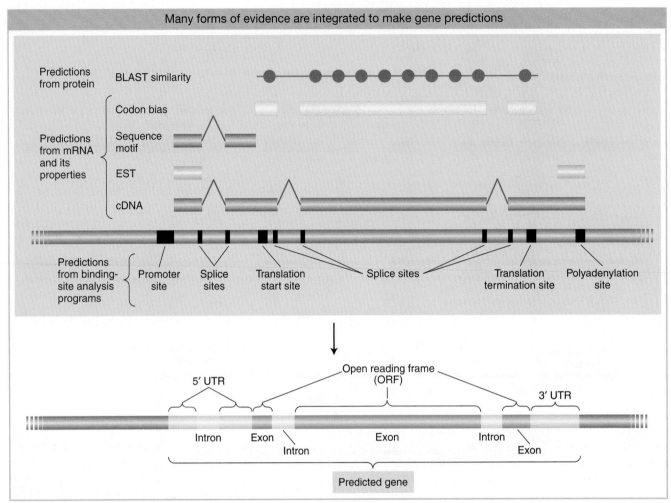

FIGURE 14-14 The different forms of gene-product evidence—cDNAs, ESTs, BLAST-similarity hits, codon bias, and motif hits—are integrated to make gene predictions. Where multiple classes of evidence are found to be associated with a particular genomic DNA sequence, there is greater confidence in the likelihood that a gene prediction is accurate.

of protein similarity within a region of genomic DNA bounded by 5' and 3' ESTs. Useful predictions are possible even without a cDNA sequence or evidence of protein similarities. A binding-site-prediction program can propose a hypothetical ORF, and proper codon bias would be supporting evidence.

KEY CONCEPT Predictions of mRNA and polypeptide structure from genomic DNA sequence depend on the integration of information from cDNA and EST sequence, binding-site predictions, polypeptide similarities, and codon bias.

Let's consider some of the insights from our first view of the overall genome structures and global parts lists of a few species whose genomes have been sequenced. We will start with ourselves. What can we learn by looking at the human genome by itself? Then we will see what we can learn by comparing our genome with others.

14.4 THE STRUCTURE OF THE HUMAN GENOME

LO 14.2 Explain the role of various functional elements within genomes, and differentiate between computational and experimental methods used to identify these elements.

LO 14.5 Outline reverse genetic approaches to analyze the function of genes and genetic elements identified by genome sequencing and comparative genomics.

In describing the overall structure of the human genome, we must first confront its repeat structure. A considerable fraction of the human genome, about 45 percent, is repetitive. Much of this repetitive DNA is composed of copies of transposable elements (discussed in Chapter 16). Indeed, even within the remaining single-copy DNA, a fraction has sequences suggesting that they might be descended from ancient transposable elements that are now immobile and have accumulated random mutations, causing them to diverge in sequence from the ancestral transposable elements. Thus, much of the human genome appears to be composed of genetic "hitchhikers."

Only a small part of the human genome encodes polypeptides; that is, somewhat less than 3 percent of it encodes exons of mRNAs. Exons are typically small (about 150 bases), whereas introns are large, many extending more than 1000 bases and some extending more than 100,000 bases. Transcripts are composed of an average of 10 exons, although many have substantially more. Finally, introns may be spliced out of the same gene in locations that vary. This variation in the location of splice sites generates considerable added diversity in mRNA and polypeptide sequence. On the basis of current cDNA and EST data, at least 60 percent of human protein-coding genes are likely to have two or more splice variants. On average, there are several splice variants per gene. Hence, the number of distinct proteins encoded by the human genome is several-fold greater than the number of recognized genes.

> **KEY CONCEPT** Only a small proportion of the human genome consists of protein-coding genes.

The number of genes in the human genome has not been easy to pin down. In the initial draft of the human genome, there were an estimated 30,000 to 40,000 protein-coding genes. However, the complex architecture of these genes and the genome can make annotation difficult. Some sequences scored as genes may actually be exons of larger genes. In addition, there are approximately 15,000 **pseudogenes**, which are ORFs or partial ORFs that may at first appear to be genes but are either nonfunctional or inactive due to the manner of their origin or to mutations. So-called **processed pseudogenes** are DNA sequences that

have been reverse-transcribed from RNA and randomly inserted into the genome. Seventy percent or so of human pseudogenes appear to be of this type. Most of the other pseudogenes in the human genome appear to have arisen from gene duplication events in which one of the duplicates has acquired one or more ORF-disrupting mutations in the course of evolution. As the challenges in annotation have been overcome, the estimated number of genes in the human genome has dropped steadily. A recent estimate is that there are about 20,000 protein-coding genes.

The annotation of the human genome progressed as the sequences of each chromosome were finished one by one. These sequences then became the searching ground in the hunt for candidate genes for human diseases. An example of gene predictions for a chromosome from the human genome is shown in **Figure 14-15**. Such predictions are being revised continually as new data become available. The current state of the predictions can be viewed at many Web sites, most notably at the public DNA databases in the United States and Europe (see Appendix B). These predictions are the current best inferences of the protein-coding genes present in the sequenced species and, as such, are works in progress.

Noncoding functional elements in the genome

The discussion thus far has focused exclusively on the protein-coding regions of the genome. This emphasis is due more to analytical ease than to biological importance. Because of the simplicity and universality of the genetic code, and the ability to synthesize cDNA from mRNA, the detection of ORFs and exons is much easier than the detection of functional noncoding sequences. As stated earlier, only 3 percent of the human genome encodes exons of mRNAs, and fewer than half of these exon sequences, a

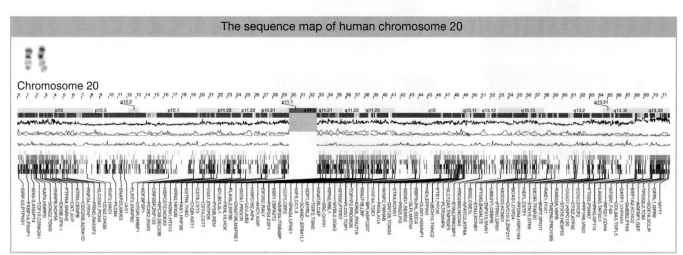

The sequence map of human chromosome 20

Chromosome 20

FIGURE 14-15 Numerous genes have been identified on human chromosome 20. The recombinational and cytogenetic map coordinates are shown in the top lines of the figure. Various graphics depicting gene density and different DNA properties are shown in the middle sections. The identifiers of the predicted genes are shown at the bottom of the panel. [*Reprinted with permission from Macmillan Publishers Ltd: from Jim Kent, Ewan Birney, Darryl Leja, and Francis Collins, After the International Human Genome Sequencing Consortium, "Initial Sequencing and Analysis of the Human Genome," Nature, 2001, February 15; 409, 860–921, Figure 9, pdf 1. Permission conveyed through Copyright Clearance Center, Inc.*]

little over 1 percent of the total genome DNA, encode protein sequences. So, nearly 99 percent of our genome does not encode proteins. How do we identify other functional parts of the genome?

Introns and 5′ and 3′ untranslated sequences are readily annotated by analysis of gene transcripts, while gene promoters are usually identified by their proximity to transcription units and signature DNA sequences. However, other regulatory sequences such as enhancers are not identifiable by mere inspection of DNA sequences, and other sequences that encode various kinds of RNA transcripts (microRNAs, small interfering RNAs, piwi-interacting RNAs, long noncoding RNAs; see Chapter 8) require detection and annotation of their transcripts. While many such noncoding elements have been identified in the course of the study of human molecular genetics, the potentially vast number of such elements warrants a more systematic approach. The Encyclopedia of DNA Elements (ENCODE) project was thus launched with the ambitious goal of identifying all functional elements within the human genome.

This large-scale collaborative endeavor has employed a diverse array of techniques to detect sequences potentially involved in the control of gene transcription, as well as all transcribed regions. Because such sequences are expected to be active in only individual or subsets of cell types, researchers studied 147 human cell types. By searching for regions that were associated with the binding of transcription factors, the ENCODE project estimated that there are approximately 500,000 potential enhancers and promoters associated with known genes. The project also detected transcripts emanating from nearly 80 percent of the human genome.

This is a much larger fraction of the genome than was expected. After all, as stated earlier, only a little over 1 percent of the genome is protein-coding sequence. However, the production of a transcript does not necessarily mean that the transcript contributes to human biology. It is possible that some proportion of these transcripts represent "noise" in the cell—transcripts that have no biological function, but also do no harm. It is not sound to ascribe function to a sequence without some form of additional data, so what kinds of additional data can be used to resolve questions of function?

Evolutionary conservation of sequences has proven to be a good indicator of biological function. Sequences will not be preserved over evolutionary time unless mutations that alter them are weeded out by natural selection. One way to locate potentially functional noncoding elements then is to look for conserved sequences, which have not changed much over millions of years of evolution.

For example, one can search for very highly conserved sequences of modest length among a few species or for less perfectly conserved sequences of greater length among a larger number of species. Comparisons of the human, rat, and mouse genomes have led to the identification of so-called *ultraconserved elements*, which are sequences that are perfectly conserved among the three species. Searches of these genomes have found more than 5000 sequences of more than 100 bp and 481 sequences of more than 200 bp that are absolutely conserved. Nearly all of these elements were highly conserved in the chicken genome, and about two-thirds were also conserved in a fish genome. Although many of these elements are found in gene-poor regions, they are most richly concentrated near regulatory genes important for development. The majority of highly conserved noncoding elements may largely take part in regulating the expression of the genetic toolkit for the development of mammals and other vertebrates (see Chapter 13).

How can we verify that such conserved elements play a role in gene regulation? These elements can be tested in the same manner as the transcriptional cis-acting regulatory elements examined in earlier chapters, with the use of reporter genes (see Figure 13-18). A researcher places candidate regulatory regions adjacent to a promoter and reporter gene and introduces the reporter gene into a host species. One such example is shown in **Figure 14-16**. An element that is highly conserved among

Testing the role of a conserved element in gene regulation

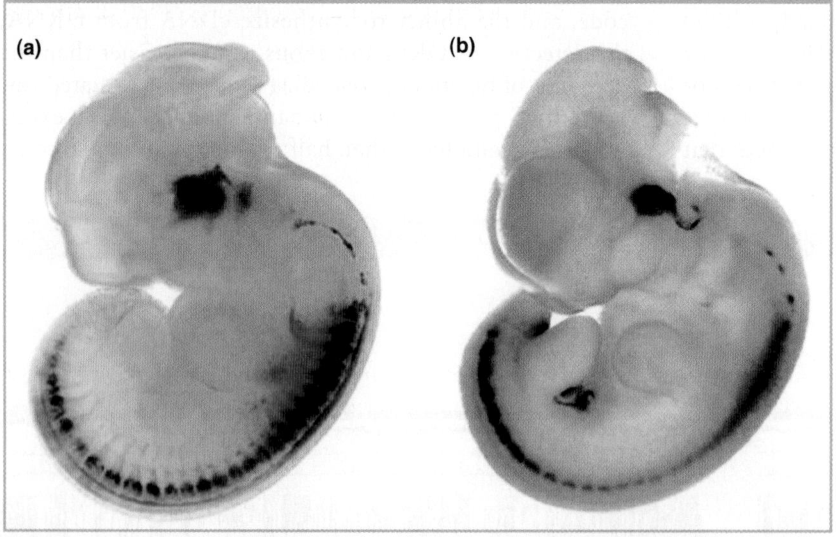

FIGURE 14-16 A transcriptional cis-acting regulatory element is identified in an ultraconserved element of the human genome. An ultraconserved element lying near the human *ISL1* gene was coupled to a reporter gene and injected into fertilized mouse oocytes. The regions where the gene is expressed are stained dark blue or black. (a) The reporter gene is expressed in the head and spinal cord of a transgenic mouse, as seen here on day 11.5 of gestation. This expression pattern corresponds to (b) the native pattern of expression of the mouse *ISL1* gene on day 11.5 of gestation. This experiment demonstrates how functional noncoding elements can be identified by comparative genomics and tested in a model organism. [*Reprinted with permission from Macmillan Publishers Ltd: from G. Bejerono et al. "A distal enhancer and an ultraconserved exon are derived from a novel retroposon" Nature, 2006, April 16; 441: 87–90. Figure 3. Permission conveyed through Copyright Clearance Center, Inc.*]

mammalian, chicken, and a frog species lies 488 kb from the 3′ end of the human *ISL1* gene, which encodes a protein required for motor-neuron differentiation. This element was placed upstream of a promoter and the β-galactosidase (*lacZ*) reporter gene, and the construct was injected into the pronuclei of fertilized mouse oocytes (see Figure 10-25). The reporter protein is expressed along the spinal cord and in the head, as one would expect for the location of future motor neurons (see Figure 14-16). Most significantly, the expression pattern corresponds to part of the expression pattern of the native mouse *ISL1* gene (presumably other noncoding elements control the other features of *ISL1* expression). The expression pattern strongly suggests that the conserved element is a regulatory region for the *ISL1* gene in each species. The success of this approach suggests that many additional human noncoding regulatory elements will likely be identified on the basis of sequence conservation and the activity of those elements in reporter assays.

> **KEY CONCEPT** Noncoding regulatory elements can be identified through a combination of computational approaches and reporter gene assays.

14.5 THE COMPARATIVE GENOMICS OF HUMANS WITH OTHER SPECIES

LO 14.3 Infer the evolutionary direction of genomic changes among species based on their phylogenetic relationships.

LO 14.5 Outline reverse genetic approaches to analyze the function of genes and genetic elements identified by genome sequencing and comparative genomics.

Fundamentally, much of the science of genomics entails a comparative approach. For instance, most of what we know about the function of human proteins is based on the function of those proteins as analyzed in model species. And many of the questions that may be addressed through genomics are comparative. For example, we often want to know, as in the case of Nicholas Volker, how an individual with a trait or disease differs genetically from those without it.

Comparative genomics also has the potential to reveal how species diverge. Species evolve and traits change through changes in DNA sequence. The genome thus contains a record of the evolutionary history of a species. Comparisons among species' genomes can reveal events unique to particular lineages that may contribute to differences in physiology, behavior, or anatomy. Such events could include, for example, the gain and loss of individual genes or groups of genes. Here, we will explore the key principles underlying comparative genomics and look at a few examples of how comparisons reveal what is similar and different among humans and other species. In the next section

we will examine how differences are identified among individual humans.

Phylogenetic inference

The first step in comparing species' genomes is to decide which species to compare. In order for comparisons to be informative, it is crucial to understand the evolutionary relationships among the species to be compared. The evolutionary history of a group is called an evolutionary tree, or a **phylogeny**. Phylogenies are useful because they allow us to infer how species' genomes have changed over time.

The second step in comparing genomes is the identification of the most closely related genes, called **homologous genes** (**Figure 14-17**). These genes can be recognized by similarities in their DNA sequences and in the amino acid sequences of the proteins they encode. It is important to distinguish here two classes of homologous genes. Some homologs are genes at the same genetic locus in different species. These genes would have been inherited from a common ancestor and are referred to as **orthologs**. However, many homologous genes belong to families that have expanded (and contracted) in number in the course of evolution. These homologous genes are at different genetic loci in the same organism. They arose when genes within a genome were duplicated. Genes that are related by gene-duplication events in a genome are called **paralogs**. The history of gene families can be quite revealing about the evolutionary history of a group.

For example, suppose we would like to know how the mammalian genome has evolved over the history of the group. We would like to know whether mammals as

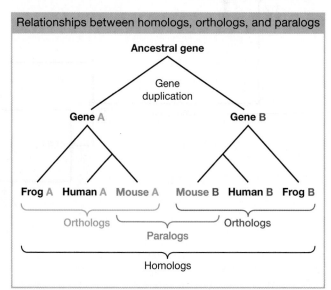

FIGURE 14-17 A gene in the common ancestor of a particular group of species (here frogs, mice, and humans) is duplicated, creating the A and the B genes, which are known as homologs. The A gene present in the frog genome is an ortholog of the A gene present in the mouse or human genome. Similarly, the B gene present in the frog genome is an ortholog of the B gene present in the mouse or human genome. The A gene present in the mouse genome is a paralog of the B gene in the mouse genome.

a group might have acquired some unique genes, whether mammals with different lifestyles might possess different sets of genes, and what the fate was of genes that existed in mammalian ancestors.

Fortunately, we now have a large and rapidly expanding set of mammal genome sequences to compare that includes representatives of the three main branches of mammals—monotremes (for example, platypus), marsupials (for example, wallaby, opossum), and eutherian mammals (for example, human, chimpanzee, dog, mouse). The relationships between these groups, some members within these groups, and other amniote vertebrates (amniotes are mostly land-dwelling vertebrates that have a terrestrially adapted egg) are shown in **Figure 14-18**.

To illustrate the importance of understanding phylogenies and how to utilize them, we consider the platypus genome. Monotremes differ from other mammals in that they lay eggs. Inspection of the platypus genome revealed

that it contains one egg-yolk gene called vitellogenin. Analyses of marsupial and eutherian genomes revealed no such functional yolk genes. The presence of vitellogenin in the platypus and its absence from other mammals could be explained in one of two ways: (1) vitellogenin is a novel invention of the platypus, or (2) vitellogenin existed in a common ancestor of monotremes, marsupials, and eutherians but was subsequently lost from marsupials and eutherians. The direction of evolutionary change is opposite in these two alternatives.

A simple pair-wise comparison between the platypus and another mammal does not distinguish between these alternatives. To do that, first we have to infer whether vitellogenin was likely to be present in the last common ancestor of the platypus, marsupials, and eutherians. We make this **phylogenetic inference** by examining whether vitellogenin is found in taxa outside of this entire group of mammals, what is referred to as an evolutionary **outgroup**.

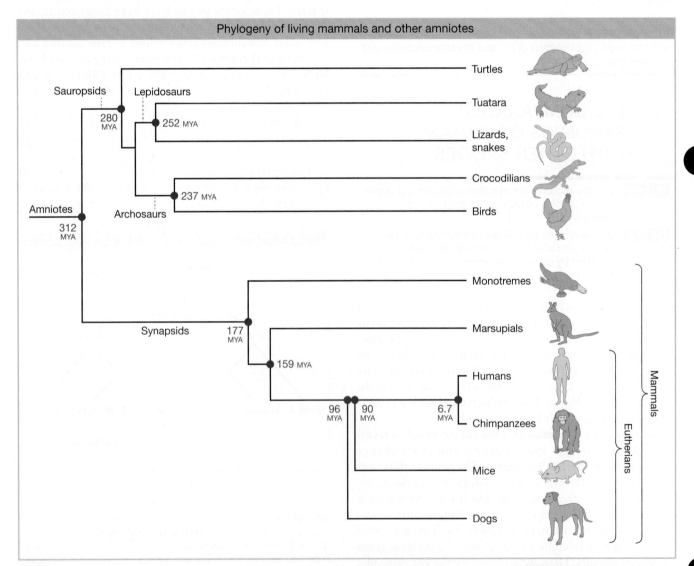

FIGURE 14-18 The phylogenetic tree depicts the evolutionary relationships among the three major groups of mammals (monotremes, marsupials, and eutherians) and other amniotes, including birds and various reptiles. By mapping the presence or absence of genes in particular groups onto known phylogenies, one can infer the direction of evolutionary change (gain or loss) in particular lineages.

INTERACTIVE RESOURCE ⧉ SaplingPlus

Understanding evolutionary trees

Indeed, three homologous vitellogenin genes exist in the chicken. Next, we consider the relationship of the chicken to mammals. Chickens belong to another major branch of the amniotes. Looking at the evolutionary tree in Figure 14-18, we can explain the presence of vitellogenins in chickens and the platypus as the result of two independent acquisitions (in the platypus lineage and the chicken lineage, respectively) or as the result of just one acquisition in a common ancestor of the platypus and chicken (which, based on the tree, would be a common ancestor of all amniotes) followed by the loss of vitellogenin genes in marsupials and eutherians.

How do we decide between these alternatives? When studying infrequent events such as the invention of a gene, evolutionary biologists prefer to rely on the principle of **parsimony,** that is, to favor the simplest explanation involving the smallest number of evolutionary changes. Therefore, the preferred explanation for the pattern of vitellogenin evolution in mammals is that this egg-yolk protein and corresponding gene were present in some egg-laying amniote ancestor and were retained in the egg-laying platypus and lost from non-egg-laying mammals.

As it turns out, there is one additional and very compelling piece of evidence that supports this inference. While inspection of eutherian genomes does not reveal any intact, functional vitellogenin genes, traces of vitellogenin gene sequences are detectable in the human and dog genomes at positions that are in the same position as (syntenic to) the vitellogenin genes of the platypus and chicken (**Figure 14-19**). These sequences are molecular relics of our egg-laying ancestors. As our mammalian ancestors shifted away from yolky eggs, natural selection was relaxed on the vitellogenin gene sequences such that they have been nearly eroded away by mutations over tens of millions of years. Our genome contains numerous relics of genes that once functioned in our ancestors, and as we will see again in

this section, the identities of those pseudogenes reflect how human biology has diverged from that of our ancestors.

Of course, evolution is also about the acquisition of new traits. For example, milk production is a shared trait among all mammals. A family of genes encoding the casein milk proteins are unique to mammals and tightly clustered together in their genomes, including that of the platypus. Just this brief glance at a few mammalian genomes informs us that, indeed, some mammals have genes that others do not, some genes are shared by all mammals, and the presence or absence of certain genes correlates with mammals' lifestyle. The latter is a pervasive finding in comparative genomics.

> **KEY CONCEPT** Determining which genomic elements have been gained or lost during evolution requires knowledge of the phylogeny of the species being compared. The presence or absence of genes often correlates with organism lifestyles.

Let's look at a few more examples that illuminate the evolutionary history of our genome and how we are different from, and similar to, other mammals.

Of mice and humans

The sequence of the mouse genome has been particularly informative for understanding the human genome because of the mouse's long-standing role as a model genetic species, the vast knowledge of its classical genetics, and the mouse's evolutionary relationship to humans. The mouse and human lineages diverged approximately 90 million years ago, which is sufficient time for mutations to cause their genomes to differ, on average, at about one of every two nucleotides. Thus, sequences common to the mouse and human genomes are likely to indicate common functions.

Homologs are identified because they have similar DNA sequences. Analysis of the mouse genome indicates that the number of protein-coding genes that it contains is similar to

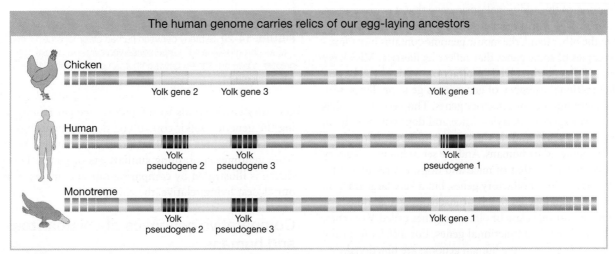

FIGURE 14-19 Strings of genes along chicken chromosome 8 and human chromosome 1 and in the platypus are in the same relative order (boxes). Whereas the chicken genome has three genes that encode egg-yolk proteins, the egg-laying platypus has one functional gene and two pseudogenes, and humans have fragmented, very short remnants of the yolk genes.

that of the human genome. Further inspection of the mouse genes reveals that at least 99 percent of all mouse genes have some homolog in the human genome and that at least 99 percent of all human genes have some homolog in the mouse genome. Thus, the kinds of proteins encoded in each genome are essentially the same. Furthermore, about 80 percent of all mouse and human genes are clearly identifiable orthologs.

The similarities between the genomes extend well beyond the inventory of protein-coding genes to overall genome organization. More than 90 percent of the mouse and human genomes can be partitioned into corresponding regions of conserved **synteny**, where the order of genes within variously sized blocks is the same as their order in the most recent common ancestor of the two species. This synteny is very helpful in relating the maps of the two genomes. For example, human chromosome 17 is orthologous to a single mouse chromosome (chromosome 11). Although there have been extensive intrachromosomal rearrangements in the human chromosome, there are 26 segments of collinear sequences more than 100 kb in size (**Figure 14-20**).

> **KEY CONCEPT** The mouse and human genomes contain similar sets of genes, often arranged in similar order. This conserved gene order between species is known as synteny.

There are some detectable differences between the inventories of mouse and human genes. In one family of genes involved in color vision, the opsins, humans possess one additional paralog. The presence of this opsin has equipped humans with so-called trichromatic vision, so that we can perceive colors across the entire spectrum of visible light—violet, blue, green, red—whereas mice cannot. But again, the presence of this additional paralog in humans and its absence in mice does not alone tell us whether it was gained in the human lineage or lost in the mouse lineage. Analysis of other primate and mammalian genomes has revealed that Old World primates such as chimpanzees, gorillas, and the colobus monkey possess this gene, but that all nonprimate mammals lack it. We can safely infer from this phylogenetic distribution of the additional opsin gene that it evolved in an ancestor of Old World primates (that includes humans).

On the other hand, the mouse genome contains more functional copies of some genes that reflect its lifestyle. Mice have about 1400 genes involved in olfaction—this is the largest single functional category of genes in its genome. Dogs, too, have a large number of olfactory genes. This certainly makes sense for the species' lifestyles. Mice and dogs rely heavily on their sense of smell, and they encounter different odors from those encountered by humans. And the set of human olfactory genes, compared to that of mice and dogs, is strikingly inferior. We have a lot of olfactory genes, but a very large fraction of them are pseudogenes that bear inactivating mutations. For example, in just one class of olfactory genes called *V1r* genes, mice have about 160 functional genes, but just 5 out of the 200 or so *V1r* genes in the human genome are functional.

Still, these differences in gene content are relatively modest in light of the vast differences in anatomy and behavior. The overall similarity in the mouse and human

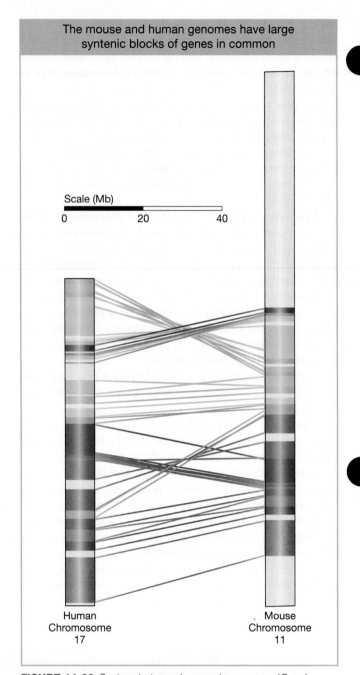

FIGURE 14-20 Synteny between human chromosome 17 and mouse chromosome 11. Large conserved syntenic blocks 100 kb or greater in size are shown between human chromosome 17 and mouse chromosome 11.

genomes corresponds to the picture we get from examining the genetic toolkit controlling development in different taxa (see Chapter 13)—that great differences can evolve from genomes containing similar sets of genes. This same theme is illustrated by comparing our genome with that of our closest living relative, the chimpanzee.

Comparative genomics of chimpanzees and humans

Chimpanzees and humans last had a common ancestor about 5–7 million years ago. Since that time, genetic differences have accumulated by mutations that have occurred in each lineage. Genome sequencing has revealed that there are about 35 million single-nucleotide differences between chimpanzees and humans, corresponding to about a 1.06 percent degree of divergence. In addition, about 5 million insertions and deletions, ranging in length from just a single nucleotide to more than 15 kb, contribute a total of about 90 Mb of divergent DNA sequence (about 3 percent of the overall genome). Most of these insertions or deletions lie outside of coding regions.

Overall, the proteins encoded by the human and chimpanzee genomes are extremely similar. Twenty-nine percent of all orthologous proteins are *identical* in sequence. Most proteins that differ do so by only about two amino acid replacements. There are a few detectable differences between chimpanzees and humans in the sets of functional genes. About 80 or so genes that were functional in their common ancestor are no longer functional in humans, owing to their deletion or to the accumulation of mutations. Some of these changes may contribute to differences in physiology.

In addition to changes in particular genes, duplications of chromosome segments in a single lineage have contributed to genome divergence. More than 170 genes in the human genome and more than 90 genes in the chimpanzee genome are present in large duplicated segments. These duplications are responsible for a greater amount of the total genome divergence than all single-nucleotide mutations combined. Intriguingly, duplications unique to the human genome are enriched for genes that are predicted to play a role in brain development. It has been suggested that at least some of these gene duplications were involved in the expansion of the neocortex in humans relative to other primates. However, whether these duplicated genes contribute to major phenotypic differences between humans and our closest relatives is not yet clear.

Despite the existence of these few differences in gene content between chimpanzees and humans, we have seen that the vast majority of genes are highly conserved, with very few changes in protein-coding regions. How, then, can we explain the dramatic differences in morphology, behavior, and physiology between chimpanzees and humans? In 1975, well before the advent of whole-genome sequencing, Mary-Claire King and Allan Wilson boldly proposed that most of the phenotypic differences between humans and chimpanzees result from mutations that affect gene regulation. Comparative genomics has now provided a tool to identify regulatory mutations that might be responsible for the phenotypic differences between chimpanzees and humans.

> **KEY CONCEPT** Great phenotypic differences can evolve from genomes containing similar sets of genes. Many of the phenotypic differences between species are likely due to genetic changes that affect gene regulation.

Here, we will discuss just one example of the many approaches used to identify putative cis-acting regulatory elements that differ between chimpanzees and humans. In this case, researchers searched for noncoding sequences that were highly conserved in the genomes of chimpanzee, macaque, and other mammals but missing in the human genome. There were 510 such deletions in the human genome, and these deletions were enriched near genes with neural function as well as steroid hormone signaling. One of these deletions is near the androgen receptor gene, which encodes a protein necessary for responses to circulating androgens such as testosterone. Using the previously introduced reporter gene assays in transgenic mice (see Chapter 10), the researchers showed that both the mouse and chimpanzee sequence drove expression in the developing sensory vibrissae (or whiskers) as well as the penile spines, which are both androgen-responsive structures that are present in most mammals but have been lost in humans (**Figure 14-21**). Testing the functions of other putative cis-acting regulatory elements missing in the human genome will likely uncover additional insight into the genetic changes that underlie differences between humans and our closest relatives.

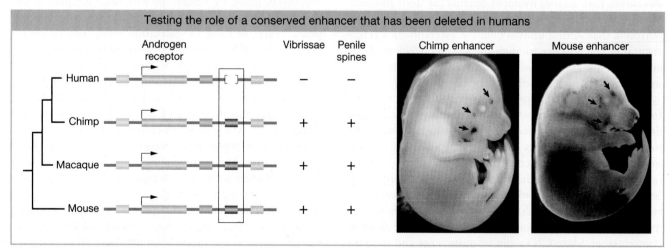

Testing the role of a conserved enhancer that has been deleted in humans

FIGURE 14-21 The androgen receptor coding sequence (pink) is present in the human, chimp, macaque, and mouse genomes. Although some conserved, noncoding sequences near the androgen receptor gene are present in the genomes of all four species (yellow, green, and light pink rectangles), one conserved, noncoding sequence is present in the chimp, macaque, and mouse genomes but absent from the human genome (red rectangle). Using either the chimp or mouse sequence in a reporter gene assay in mice shows that this sequence is a cis-acting enhancer that drives expression in the sensory vibrissae (arrows) and penile spines (not shown). [*Reprinted by permission from Macmillan Publishers Ltd. from McLean et al., "Human-specific loss of regulatory DNA and the evolution of human-specific traits," Nature, 2011, March 10; 471, 216–219, Figure 2. Permission conveyed through Copyright Clearance Center, Inc.*]

KEY CONCEPT Genetic changes that underlie phenotypic differences between humans and our closest relatives can be identified using a combination of computational approaches and reporter gene assays.

Of course, all genetic differences between species originate as variations within species. The sequencing of the human genome and the advent of faster and less expensive high-throughput sequencing methods have opened the door to the detailed analysis of human genetic variation.

14.6 COMPARATIVE GENOMICS AND HUMAN MEDICINE

LO 14.4 Compare genomic methods used to identify mutations that have been associated with human disease thus far.

The human species, *Homo sapiens*, originated in Africa approximately 200,000 years ago. Sometime between 50,000 and 100,000 years ago, populations left Africa and migrated across the world, eventually populating five additional continents. These migrating populations encountered different climates, adopted different diets, and combated different pathogens in different parts of the world. Much of the recent evolutionary history of our species is recorded in our genomes, as are the genetic differences that make individuals or populations more or less susceptible to disease.

Overall, any two unrelated humans' genomes are 99.9 percent identical. That difference of just 0.1 percent still corresponds to roughly 3 million bases. The challenge today is to decipher which of those base differences are meaningful with respect to physiology, development, or disease.

Once the sequence of the first human genome was advanced, that accomplishment opened the door to much more rapid and less costly analysis of other individuals. The reason is that with a known genome assembly as a reference, it is much easier to align the raw sequence reads of additional individuals, and to design approaches to studying and comparing parts of the genome.

One of the first and greatest surprises that has emerged from comparing individual human genomes is that humans differ not merely at one base in a thousand, but also in the number of copies of parts of individual genes, entire genes, or sets of genes. These **copy number variations (CNVs)** include repeats and duplications that increase copy number and deletions that reduce copy number. Between any two unrelated individuals, there may be hundreds of segments of DNA greater than 1000 bp in length that differ in copy number. Some CNVs can be quite large and span up to 5 million base pairs. Together, CNVs account for more sequence variation among humans than all the 3 million single base pair changes combined.

How such copy numbers may play a role in human evolution and disease is of intense interest. One case where increased copy number appears to have been adaptive concerns diet. People with high-starch diets have, on average, more copies of a salivary amylase (an enzyme that breaks down starch) gene than people with traditionally low-starch diets. In other cases, copy number variations have been associated with human diseases. For example, it now appears that at least 15 percent of human neurodevelopmental diseases are due to changes in copy number that are found at a very low frequency in human populations. Copy number polymorphisms that are relatively common in human populations have also been associated with immune-related diseases such as Crohn's disease, psoriasis, and lupus.

The evolutionary history of human disease genes

One might ask when and where the mutations that cause human disease originated, and why some of these disease alleles are maintained at a relatively high frequency in human populations. Although we are still a long way from answering these questions, some insight has come from analyzing the genome sequences of ancient hominins, including those of our own species *Homo sapiens* as well as now-extinct, archaic hominin lineages like Neanderthals. Advances in sequencing and other technologies have made it possible to extract and sequence whole genomes from ancient DNA samples, even when very small amounts of tissue are found. For example, whole genome sequencing of ancient DNA from a single finger bone and three teeth found in the Denisova Cave in Siberia revealed the existence of an archaic hominin lineage, now called Denisovans, which are genetically very distinct from Neanderthals, diverging approximately 640,000 years ago.

Analyses of these archaic genomes has revealed that as anatomically modern humans left Africa and spread around the globe, they interbred with other hominin species that had already been living in Eurasia for over 200,000 years. Traces of these hybridization events can be seen in the genomes of humans living today. Reflecting the migratory paths of modern humans, all non-African individuals sequenced to date have between 1 percent and 4 percent Neanderthal ancestry, while indigenous Australians and Melanesians also have up to 6 percent Denisovan ancestry (**Figure 14-22**). Many direct-to-consumer genetic testing services will now report what percentage of a person's DNA has been inherited from their archaic human ancestors. **Box 14-1** discusses the various types of direct-to-consumer genetic testing options available today, as well as some important ethical and social implications of such services.

Remarkably, some of the Neanderthal and Denisovan alleles present in modern humans have effects on physiology. For example, gene variants that cause lighter skin in northern Eurasians were present in Neanderthals, and one of the gene variants that has enabled Tibetans to live at high altitudes (see Chapter 1) is Denisovan in origin. However, Neanderthal-derived alleles of some genes are associated with

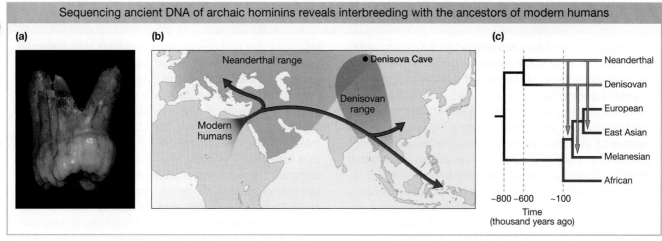

FIGURE 14-22 (a) Side view of a molar found in the Denisova Cave in Siberia. (b) Anatomically modern humans migrated out of Africa into Eurasia through the ranges of both Neanderthals and Denisovans. (c) Genome sequencing provides evidence for early hybridization between Neanderthals and the ancestors of modern Melanesians, East Asians, and Europeans, as well as later hybridization between Neanderthals and the ancestors of modern East Asians (blue arrows). There is also evidence for hybridization between Denisovans and the ancestors of modern Melanesians (green arrow). [*Courtesy of the Max Planck Institute for Evolutionary Anthropology.*]

the risk of diseases in modern humans. For example, modern humans with a Neanderthal allele at a gene involved in blood coagulation have a higher risk of blood clots and stroke. Rapid clotting might have been an advantage in early hominids, who were hunting dangerous animals and at risk of excessive bleeding during childbirth. However, in our modern times, these risks are lessened, and humans live much longer. Thus, fast clotting is no longer an advantage and leads to an increased risk of stroke and blood clots. Genetic variants of Neanderthal origin are also linked to increased risk of neurological, immunological, and skin diseases in modern humans. These examples serve as a reminder that our genetic susceptibility to disease has been shaped by our evolutionary history. Genomics has provided a tool that allows us to explore this evolutionary history in ways that could not be imagined before.

KEY CONCEPT The ability to sequence whole genomes of modern and archaic humans provides a tool to uncover the evolutionary history of humans and to identify mutations associated with disease.

The exome and personalized genomics

Advances in sequencing technologies have reduced the cost of sequencing individual genomes from about $300 million in 2000, to $1 million in 2008, to about $1000 in 2017. But for many large-scale studies, that figure is still prohibitive. For some applications, it is more practical and cost effective, and can be just as informative, to sequence only part of the genome. For example, since many disease-causing mutations occur in coding sequences, strategies have been designed to sequence all of the exons, or the **exome**, of individuals, as was done in the case of Nicholas Volker.

The strategy for exome sequencing involves generating a library of genomic DNA that is enriched for

exon sequences (**Figure 14-23**). The DNA is prepared by (1) shearing genomic DNA into short, single-stranded pieces, (2) hybridizing the single-stranded pieces to biotin-labeled probes complementary to the known exonic regions of the human genome and purifying the biotin-labeled duplexes, (3) amplifying the exon-rich duplexes, and (4) sequencing the exon-rich duplexes. In this manner, 30–60 megabases of the human genome is targeted for sequencing, as opposed to the 3200 megabases of total sequence.

As of late 2017, the exomes of more than 120,000 individuals have been sequenced, at the current cost of only a few hundred dollars per exome. One particularly important power of exome sequencing is to identify de novo mutations in individuals (mutations that are not present in either parent). Such mutations are responsible for many spontaneously appearing genetic diseases whose origins would not be revealed by traditional pedigree-based studies. As such, whole-exome sequencing is now a rapidly spreading clinical diagnostic tool, particularly for neurodevelopmental and other disorders in pediatric populations.

And just as exome sequencing can be used to identify genetic differences between individuals, it can also be used to identify differences between normal and abnormal cells, such as cancer cells. Cancer is a suite of genetic diseases in which combinations of gene mutations typically contribute to the loss of growth control and metastasis. Understanding what genetic changes are common to particular cancers, or to subsets of cancers, will not only further our understanding of cancer, but also promises to impact diagnosis and treatment in powerful ways. Researchers across the world have recently completed an "atlas" of cancer genomes that has uncovered the extraordinary genetic heterogeneity present in cancer cells and provided a framework for classifying tumor subtypes based on the underlying genomic

BOX 14-1   Direct-to-Consumer Genetic Testing

The genomics revolution has also led to the democratization of access to personal genetic information. The Human Genome Project was started in part due to the promise of **personal genomics**, exemplified by the case of Nicholas Volker at the beginning of this chapter. Thus, shortly after the first draft of the human genome was completed, a number of so-called "direct-to-consumer genetic testing" companies began to emerge with the goal of fulfilling that promise. Currently, a handful of companies offer direct-to-consumer genetic testing. For the cost of approximately $100 to $200, a consumer can provide a saliva sample or a cheek swab to a company. Their DNA will then be genotyped at roughly 700,000 of the 3 million sites in the genome that are known to vary among humans. The consumer can then retrieve the results of their genome analyses via a Web site or an app. Currently, the services provided by these companies fall into three main categories: medical testing, genetic genealogy, and personal ancestry.

Medical testing   The first direct-to-consumer genetic testing companies popped up quickly around 2005–2006, promising to provide individuals with personalized information about their genetic risk for common diseases like diabetes or cancer. Just as quickly, concerns over these tests emerged. For example, it was unknown whether consumers would understand their personal genetic risks without the help of a health care professional or whether they would stop taking preventative health measures based on the results of these tests. Furthermore, there were concerns over privacy and the potential for misuse of the data. Based on these and other concerns, at the end of 2013 the U.S. Food and Drug Administration (FDA) served "cease and desist" letters to these companies, requiring them to obtain FDA authorization for their tests. As of early 2018, only one company, 23andMe, has been authorized by the FDA to provide direct-to-consumer testing for genetic risk factors associated with a limited number of diseases, such as breast cancer, Parkinson's, and Alzheimer's. Consumers can currently also use their services to assess carrier status for over 40 inherited diseases, such as cystic fibrosis and sickle cell anemia.

Genetic genealogy   The second most common hobby in the United States is genealogy, or the tracing of family lineages and history. This popularity is reflected in the fact that the companies that offer direct-to-consumer genetic testing for the purpose of genealogy have now collected genetic data for over 15 million people. When an individual submits their DNA sample, their relationship to every other individual in the database is estimated from genetic data. For example, if that individual had a monozygotic twin in the database, they would show up as a perfect match, while a parent, child, or full sibling would show up as a first-degree relative. Most matches in the database comprise second, third, or fourth cousins. Customers can use this information to identify and contact possible relatives to fill in their family tree. Of course, these genetic matches may reveal unexpected relatives or relationships that were not previously known, and customers must be aware of the repercussions, both positive and negative, of this knowledge. The International Society of Genetic Genealogy has compiled a chart comparing features of the top five companies that offer autosomal DNA testing; this can be found at https://isogg.org/wiki/Autosomal_DNA_testing_comparison_chart.

Personal ancestry   Where did we come from? Humans have been asking this universal question for millennia. Direct-to-consumer genetic testing promises to answer this question by providing consumers with information about their genetic ancestry, including the percentage of ancestry derived from archaic humans such as Neanderthals and Denisovans. It is important to note, however, that the ability to assign ancestry is reliant upon the other data in the database. For example, if the database of a company is comprised mostly of people of European descent, it is more difficult to determine the ancestry of a person of Asian or African descent. Thus, the results provided by any one company about an individual's ancestry should be interpreted as a rough estimate that is likely to evolve over time as more and more people decide to submit their own DNA samples for ancestry testing.

Ethical, legal, and social implications   The ethical, legal, and social implications of direct-to-consumer genetic testing are far-reaching and need to be carefully considered. Thus, the future of direct-to-consumer genetic testing is not currently clear. However, discussions among a wide variety of stakeholders, including geneticists, ethicists, medical providers, companies, regulators, and consumers, are ongoing to ensure that the avalanche of personal genetic information that is now upon us is used for the benefit of individuals and society.

---

alterations. This knowledge opens up new opportunities to develop therapies that specifically target the genetic changes found in a particular tumor rather than treating cancer as a homogeneous disease. (See http://cancergenome.nih.gov/ for further information.)

KEY CONCEPT   Exome sequencing is a powerful approach to cheaply and rapidly identify mutations associated with human disease.

The ability to rapidly analyze organisms' genomes is also impacting other dimensions of medicine. We will look at one such case next.

Comparative genomics of nonpathogenic and pathogenic *E. coli*

Escherichia coli are found in our mouths and intestinal tracts in vast numbers, and this species is generally a benign

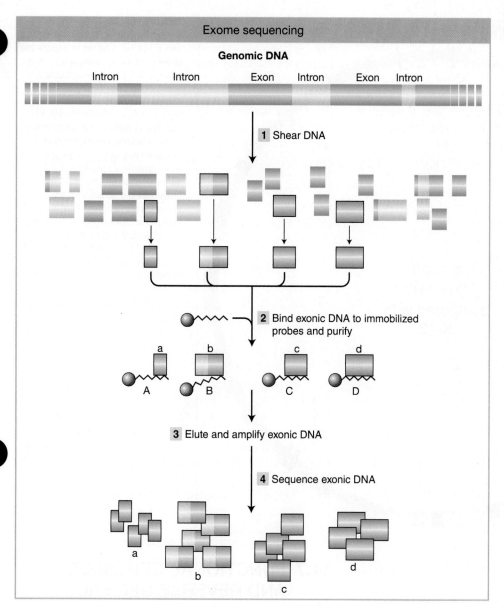

FIGURE 14-23 In order to sequence just the exon fraction of the genome, genomic DNA is fragmented and denatured, and exon-containing fragments are hybridized with biotin-labeled probes that are complementary to the known exon sequences in the genome. Duplexes containing annealed probes are then purified and prepared for sequencing.

symbiont. Because of its central role in genetics research, it was one of the first bacterial genomes sequenced. The *E. coli* genome is about 4.6 Mb in size and contains 4405 genes. However, calling it "the *E. coli* genome" is really not accurate. The first genome sequenced was derived from the common laboratory *E. coli* strain K-12. Many other *E. coli* strains exist, including several important to human health.

In 1982, a multistate outbreak of human disease was traced to the consumption of undercooked ground beef. The *E. coli* strain O157:H7 was identified as the culprit, and it has since been associated with a number of large-scale outbreaks of infection. In fact, there are an estimated 75,000 cases of *E. coli* infection annually in the United States. Although most people recover from the infection, a fraction develop hemolytic uremia syndrome, a potentially life-threatening kidney disease.

To understand the genetic bases of pathogenicity, the genome of an *E. coli* O157:H7 strain has been sequenced. The O157 and K-12 strains have a backbone of 3574 protein-coding genes in common, and the average nucleotide identity among orthologous genes is 98.4 percent, comparable to that of human and chimpanzee orthologs. About 25 percent of the *E. coli* orthologs encode identical proteins, similar to the 29 percent for human and chimpanzee orthologs.

Despite the similarities in many proteins, the genomes and proteomes differ enormously in content. The *E. coli* O157 genome encodes 5416 genes, whereas the *E. coli* K-12 genome encodes 4405 genes. The *E. coli* O157 genome contains 1387 genes that are not found in the K-12 genome, and the K-12 genome contains 528 genes not found in the O157 genome. Comparison of the genome maps reveals that the backbones common to the two strains

Two *E. coli* strains contain islands of genes specific to each strain

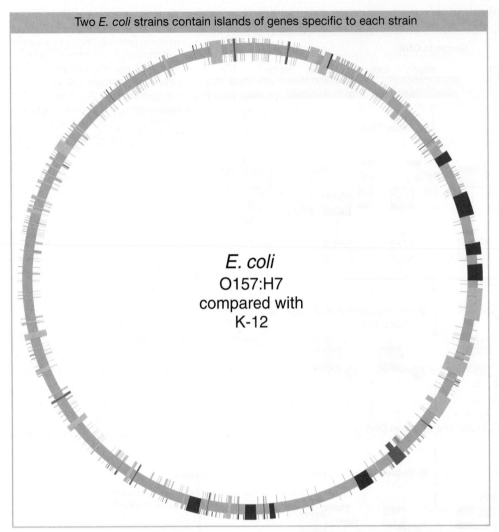

E. coli
O157:H7
compared with
K-12

FIGURE 14-24 The circular genome maps of *E. coli* strains K-12 and O157:H7. The circle depicts the distribution of sequences specific to each strain. The colinear backbone common to both strains is shown in blue. The positions of O157:H7-specific sequences are shown in red. The positions of K-12-specific sequences are shown in green. The positions of O157:H7- and K-12-specific sequences at the same location are shown in tan. Hypervariable sequences are shown in purple. [*Data from N. T. Perna et al., "Genome Sequence of Enterohaemorrhagic Escherichia coli O157:H7," Nature 409, 2001, 529–533. Courtesy of Guy Plunkett III and Frederick Blattner.*]

are interspersed with islands of genes specific to either K-12 or O157 (**Figure 14-24**).

Among the 1387 genes specific to *E. coli* O157 are many genes that are suspected to encode virulence factors, including toxins, cell-invasion proteins, adherence proteins, and secretion systems for toxins, as well as possible metabolic genes that may be required for nutrient transport, antibiotic resistance, and other activities that may confer the ability to survive in different hosts. Most of these genes were not known before sequencing and would not be known today had researchers relied solely on *E. coli* K-12 as a guide to all *E. coli*.

The surprising level of diversity between two members of the same species shows how dynamic genome evolution can be. Most new genes in *E. coli* strains are thought to have been introduced by horizontal transfer from the genomes of viruses and other bacteria (see Chapter 6). Differences can also evolve owing to gene deletion. Other pathogenic *E. coli* and bacterial species also exhibit many differences in gene content from their nonpathogenic cousins. The identification of genes that may contribute directly to pathogenicity opens new avenues to the understanding, prevention, and treatment of infectious disease.

14.7 FUNCTIONAL GENOMICS AND REVERSE GENETICS

LO 14.2 Explain the role of various functional elements within genomes, and differentiate between computational and experimental methods used to identify these elements.

LO 14.5 Outline reverse genetic approaches to analyze the function of genes and genetic elements identified by genome sequencing and comparative genomics.

Geneticists have been studying the expression and interactions of individual gene products for the past several decades. With the advent of genomics, we have an opportunity to expand these studies to a global level by using genome-wide approaches to study most or all gene products systematically and simultaneously, and in species that are not previously established experimental models (see the *Beyond Model Organisms* section of *A Brief Guide to Model Organisms*, at the back of this book). This global approach to the study of the function, expression, and interaction of gene products is termed **functional genomics**.

"'Omics"

In addition to the genome, other global data sets are of interest. Following the example of the term *genome*, for which "gene" plus "-ome" becomes a word for "all genes," genomics researchers have coined a number of terms to describe other global data sets on which they are working. This *'ome* wish list includes

> The **transcriptome**. The sequence and expression patterns of all RNA transcripts (which kinds, where in tissues, when, how much).
>
> The **proteome**. The sequence and expression patterns of all proteins (where, when, how much).
>
> The **interactome**. The complete set of physical interactions between proteins and DNA segments, between proteins and RNA segments, and between proteins.

We will not consider all of these *'omes* in this section but will focus on some of the global techniques that are beginning to be exploited to obtain these data sets.

Using RNA-seq to study the transcriptome Suppose we want to answer the question, what genes are active in a particular cell under certain conditions? Those conditions could be one or more stages in development, or they could be the presence or absence of a pathogen or a hormone. Active genes are transcribed into RNA, and so the set of RNA transcripts present in the cell can tell us what genes are active. Here, the application of next-generation sequencing technologies has been extremely powerful by permitting the assay of RNA transcripts for all genes simultaneously in a single experiment. Let's see how this process works in more detail.

The first step is to isolate the total set of RNA molecules from cells of interest. For example, one set might be extracted from a particular cell type grown under typical conditions. A second set might be made from RNA extracted from cells grown under some experimental condition. Although methods exist for capturing and sequencing different types of RNAs in the cell, we will focus here on the sequencing of mRNA, which is the fraction of the RNA that encodes proteins. The mRNA can be captured from total RNA using an oligo-dT primer, which is complementary to the 3' poly(A) tail of the mRNA. Then, the mRNA is subjected to reverse transcription to transform it into cDNA (see Chapter 10), which can then be used as a substrate for next-generation sequencing libraries just as for genomic DNA (see Figure 14-6). The sequencing reads are then mapped to the genome, where they align to the transcribed regions of genes. The number of reads present for a particular transcript should reflect its levels of expression in the cell; genes expressed at a low level in a particular cell type will have few reads, and genes expressed at a high level in a particular cell type will have many reads (**Figure 14-25**). In this manner, genes whose levels of expression are increased or decreased under the given experimental condition are identified. Similarly, genes

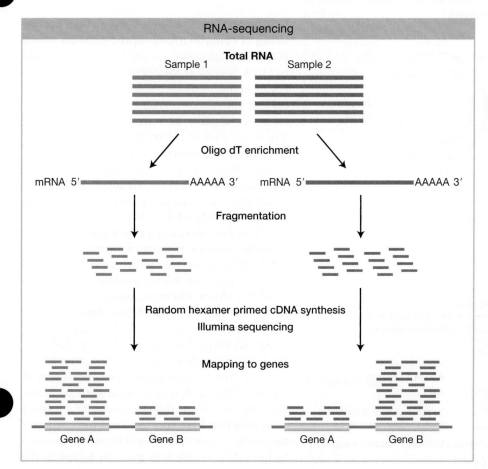

RNA-sequencing

FIGURE 14-25 Total RNA is isolated from cells in two different conditions, followed by mRNA enrichment and cDNA synthesis. The resulting cDNA is sequenced using a next-generation sequencing method. The resulting sequencing reads are aligned to the exonic sequences in the genome, and the number of reads mapping to genes in the different conditions is compared.

ANIMATED ART SaplingPlus

RNA-seq

that are active in a given cell type or at a given stage of development can be identified.

With an understanding of which genes are active or inactive at a given developmental stage, in a particular cell type, or in various environmental conditions, the sets of genes that may respond to similar regulatory inputs can be identified. Furthermore, gene-expression profiles can paint a picture of the differences between normal and diseased cells. By identifying genes whose expression is altered by mutations, in cancer cells, or by a pathogen, researchers may be able to devise new therapeutic strategies.

Using the two-hybrid test to study the protein–protein interactome
One of the most important activities of proteins is their interaction with other proteins. Because of the large number of proteins in any cell, biologists have sought ways of systematically studying all of the interactions of individual proteins in a cell. One of the most common ways of studying the interactome uses an engineered system in yeast cells called the **two-hybrid test**, which detects physical interactions between two proteins. The basis for the test is the transcriptional activator encoded by the yeast *GAL4* gene (see Chapter 12).

Recall that this protein has two domains: (1) a DNA-binding domain that binds to the transcriptional start site and (2) an activation domain that will activate transcription but cannot itself bind to DNA. Thus, the two domains must be in close proximity in order for transcriptional activation to take place. Suppose that you are investigating whether two proteins interact. The strategy of the two-hybrid system is to separate the two domains of the activator encoded by *GAL4*, making activation of a reporter gene impossible. Each domain is connected to a different protein. If the two proteins interact, they will join the two domains together. The activator will become active and start transcription of the reporter gene.

How is this scheme implemented in practice? The *GAL4* gene is divided between two plasmids so that one plasmid contains the part encoding the DNA-binding domain and the other plasmid contains the part encoding the activation domain. On one plasmid, a gene for one protein under investigation is spliced next to the DNA-binding domain, and this fusion protein acts as "bait." On the other plasmid, a gene for another protein under investigation is spliced next to the activation domain, and this fusion protein is said to be the "target" (**Figure 14-26**). The two hybrid plasmids are then introduced into the same yeast cell—perhaps by mating haploid cells containing bait and target plasmids. The final step is to look for activation of transcription by a *GAL4*-regulated reporter gene construct, which would be proof that bait and target bind to each other. The two-hybrid system can be automated to make it possible to hunt for protein interactions throughout the proteome.

Studying the protein–DNA interactome using chromatin immunoprecipitation assay (ChIP)
The sequence-specific binding of proteins to DNA is critical for correct gene expression. For example, regulatory proteins bind to promoters and activate or repress transcription in both bacteria and eukaryotes (see Chapters 11, 12, and 13). In the case of eukaryotes, chromosomes are organized into chromatin, in which the fundamental unit, the nucleosome, contains DNA wrapped around histones. Post-translational modification of histones often dictates what proteins bind and where (see Chapter 12). A variety of technologies have been developed that allow researchers to isolate specific regions of chromatin so that DNA and its associated proteins can be analyzed together. The most widely used method is called **ChIP** (for **chromatin immunoprecipitation**), and its application is described below (**Figure 14-27**).

Let's say that you have isolated a gene from yeast and suspect that it encodes a protein that binds to DNA when yeast is grown at high temperature. You want to know whether this protein binds to DNA and, if so, to what yeast sequence. One way to address this question is first to treat yeast cells that have been grown at high temperature with a chemical that will cross-link proteins to the DNA. In this way proteins bound to the

Studying protein interactions with the use of the yeast two-hybrid system

Yeast two-hybrid vectors

Cam^R   2 μ ori   Gal4-binding domain (BD)   amp^R   2 μ ori   Gal4-activation domain (AD)

"Bait" protein   "Target" protein

Trp 1⁺   Leu 2⁺

Unite

Interaction
Target   Bait
Gal4 AD
Gal4 BD

Transcription

GAL promoter   Reporter *lacZ*

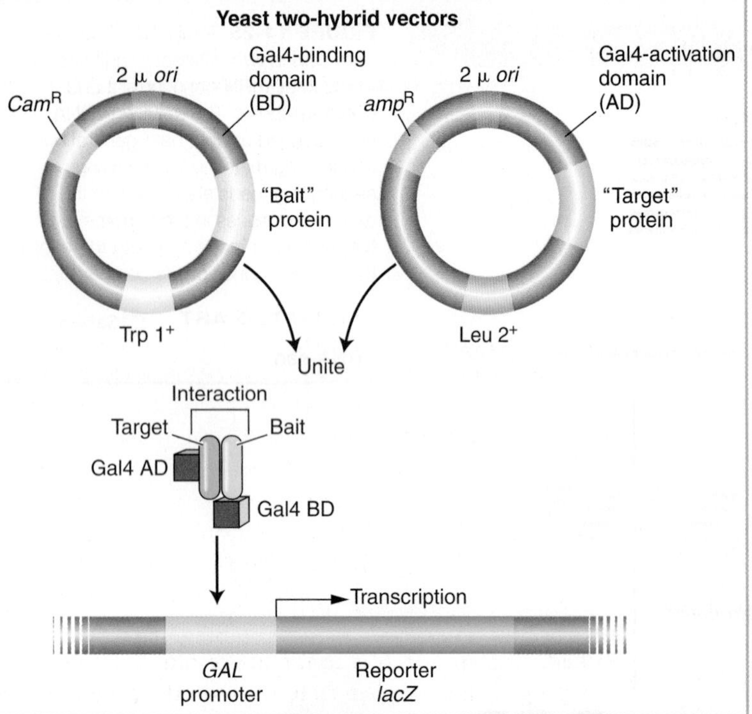

FIGURE 14-26 The system uses the binding of two proteins, a "bait" protein and a "target" protein, to restore the function of the Gal4 protein, which activates a reporter gene. *Cam*, Trp, and Leu are components of the selection systems for moving the plasmids around between cells. The reporter gene is *lacZ*, which resides on a yeast chromosome (shown in blue).

ANIMATED ART 🔷 SaplingPlus

Yeast two-hybrid systems

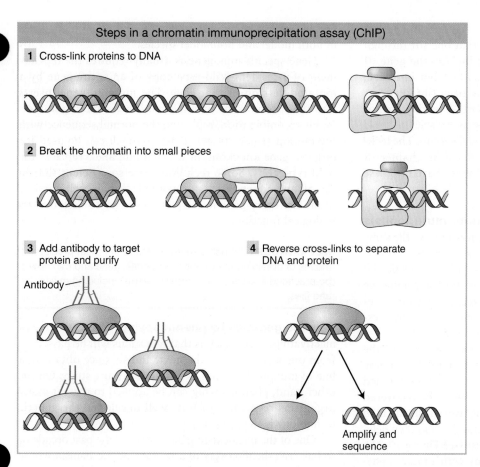

Steps in a chromatin immunoprecipitation assay (ChIP)

1 Cross-link proteins to DNA

2 Break the chromatin into small pieces

3 Add antibody to target protein and purify

Antibody

4 Reverse cross-links to separate DNA and protein

Amplify and sequence

FIGURE 14-27 ChIP is a technique for isolating the DNA and its associated proteins in a specific region of chromatin so that both can be analyzed together.

ANIMATED ART 🔁 SaplingPlus

ChIP

DNA at the time of chromatin isolation will remain bound through subsequent treatments. The next step is to break the chromatin into small pieces. To separate the fragment containing your protein–DNA complex from others, you use an antibody that reacts specifically with the encoded protein. You add your antibody to the mixture so that it forms an immune complex that can be purified. The DNA bound in the immune complex can be analyzed after cross-linking is reversed. DNA bound by the protein may be amplified into many copies by PCR to prepare for DNA sequencing, or the DNA may be sequenced directly.

As we saw in Chapter 12, regulatory proteins often activate transcription of many genes simultaneously by binding to several promoter regions. A variation of the ChIP procedure, called **ChIP-seq**, has been devised to identify all the binding sites of a protein in a sequenced genome. Proteins that bind to many genomic regions are immunoprecipitated as described previously. Then, after cross-linking is reversed, the DNA fragments are subjected to DNA sequencing using a next-generation method such as Illumina sequencing. The sequencing reads are mapped to the genome to reveal the locations where the regulatory protein binds in a particular cell type, environmental condition, or disease state.

KEY CONCEPT Advances in genomic technologies have made it possible to catalog the transcripts and proteins as well and protein–DNA and protein–protein interactions found in normal and diseased cells.

Reverse genetics

The kinds of data obtained from RNA-seq, ChIP-seq, and protein-interaction screens are suggestive of interactions within the genome and proteome, but they do not allow one to draw firm conclusions about gene functions and interactions in vivo. For example, finding out that the expression of certain genes is lost in some cancers is not proof of cause and effect. The gold standard for establishing the function of a gene or genetic element is to disrupt its function and to understand phenotypes in native conditions. Starting from available gene sequences, researchers can now use a variety of methods to disrupt the function of a specific gene. These methods are referred to as reverse genetics. Reverse-genetic analysis starts with a known molecule—a DNA sequence, an mRNA, or a protein—and then attempts to disrupt this molecule to assess the role of the normal gene product in the biology of the organism (see Figure 14-2).

There are several approaches to reverse genetics, and new technologies are constantly being developed and refined. One approach is to introduce random mutations into the genome, but then to hone in on the gene of interest by molecular identification of mutations in the gene. A second approach is to conduct a targeted mutagenesis that produces mutations directly in the gene of interest. A third approach is to create *phenocopies*—effects comparable to mutant phenotypes—usually by treatment with agents that interfere with the mRNA transcript of the gene.

Each approach has its advantages. Random mutagenesis is well established, but it requires that one sift through all the mutations to find those that include the gene of interest. Targeted mutagenesis can also be labor intensive, but, after the targeted mutation has been obtained, its characterization is more straightforward. Creating phenocopies can be very efficient, especially as libraries of tools have been developed for particular model species. The technical details of these methods are covered in Chapters 8 and 10, so we will here consider examples of each of these approaches.

Reverse genetics through random mutagenesis

Random mutagenesis for reverse genetics employs the same kinds of general mutagens that are used for forward genetics: chemical agents, radiation, or transposable genetic elements (see Figure 6-38). However, instead of screening the genome at large for mutations that exert a particular phenotypic effect, reverse genetics focuses on the gene in question, which can be done in one of two general ways.

One approach is to focus on the map location of the gene. Only mutations falling in the region of the genome where the gene is located are retained for further detailed molecular analysis. Thus, in this approach, the recovered mutations must be mapped. One straightforward way is to cross a new mutant with a mutant containing a known deletion or mutation of the gene of interest (see Figure 17-21). Only the pairings that result in progeny with a mutant phenotype (showing lack of complementation) are saved for study.

In another approach, the gene of interest is identified in the mutagenized genome and checked for the presence of mutations. For example, if the mutagen causes small deletions, then, after PCR amplification of gene fragments, genes from the parental and mutagenized genomes can be compared, looking for a mutagenized genome in which the gene of interest is reduced in size. Similarly, transposable-element insertions into the gene of interest can be readily detected because they increase its size. As the ability to rapidly and cheaply sequence whole genomes improves, it is becoming feasible to identify mutations in genes of interest, including single–base-pair substitutions, by simply sequencing the parental and mutagenized genomes. In these ways, a set of genomes containing random mutations can be effectively screened to identify the small fraction of mutations in a gene of interest to a researcher.

Reverse genetics by targeted mutagenesis

For most of the twentieth century, researchers viewed the ability to direct mutations to a specific gene as the unattainable "holy grail" of genetics. However, now several such techniques are available. After a gene has been inactivated in an individual, geneticists can evaluate the phenotype exhibited for clues to the gene's function. While the tools for targeted gene mutations were first developed using genetic techniques for model organisms, new technologies, particularly those that are CRISPR-based (see Chapter 10), are revolutionizing the ability to disrupt and manipulate genes in both model and nonmodel species.

Gene-specific mutagenesis usually requires the replacement of a resident wild-type copy of an entire gene by a mutated version of that gene. The mutated gene inserts into the chromosome by a mechanism resembling homologous recombination, replacing the normal sequence with the mutant (**Figure 14-28**). This approach can be used for targeted gene knockout, in which a null allele replaces the wild-type copy. Some techniques are so efficient that in *E. coli* and *S. cerevisiae*, for example, it has been possible to mutate every gene in the genome to try to ascertain its biological function.

KEY CONCEPT Targeted mutagenesis is the most precise means of obtaining mutations in a specific gene and can now be practiced in a variety of model systems, including mice and flies.

Reverse genetics by phenocopying

The advantage of inactivating a gene itself is that mutations will be passed on from one generation to the next, and so, once obtained, a line of mutants is always available for future study. On the other hand, phenocopying can be applied to a great many organisms regardless of how well developed the genetic technology is for a given species.

One of the most exciting discoveries of the past decade or so has been the discovery of a widespread mechanism whose natural function seems to be to protect a cell from foreign DNA. This mechanism is called **RNA interference (RNAi)**, described in Chapter 8. Researchers have capitalized on this

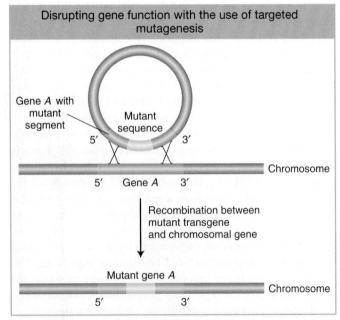

FIGURE 14-28 The basic molecular event in targeted gene replacement. A transgene containing sequences from two ends of a gene but with a selectable segment of DNA in between is introduced into a cell. Double recombination between the transgene and a normal chromosomal gene produces a recombinant chromosomal gene that has incorporated the abnormal segment.

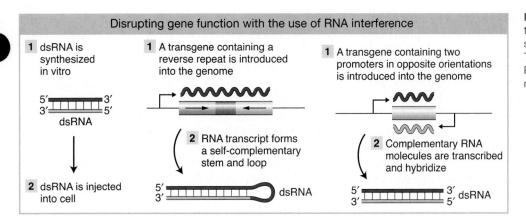

Disrupting gene function with the use of RNA interference

1 dsRNA is synthesized in vitro

5′ |||||||| 3′
3′ |||||||| 5′
dsRNA

2 dsRNA is injected into cell

1 A transgene containing a reverse repeat is introduced into the genome

2 RNA transcript forms a self-complementary stem and loop

5′ ||||||||
3′ |||||||| dsRNA

1 A transgene containing two promoters in opposite orientations is introduced into the genome

2 Complementary RNA molecules are transcribed and hybridize

5′ |||||||| 3′
3′ |||||||| 5′   dsRNA

FIGURE 14-29 Three ways to create and introduce double-stranded RNA (dsRNA) into a cell. The dsRNA will then stimulate RNAi, degrading sequences that match those in the dsRNA.

cellular mechanism to make a powerful method for inactivating specific genes. The inactivation is achieved as follows. A double-stranded RNA is made with sequences homologous to part of the gene under study and is introduced into a cell (Figure 14-29). The RNA-induced silencing complex, or RISC, then degrades native mRNA that is complementary to the double-stranded RNA. The net result is a complete or considerable reduction of mRNA levels that lasts for hours or days, thereby nullifying expression of that gene. Because the RISC complex is found in most eukaryotes, the technique has been widely applied in model systems such as *C. elegans*, *Drosophila*, zebrafish, and several plant species.

But what makes RNAi especially powerful is that it can be applied to nonmodel organisms. First, target genes of interest can be identified by comparative genomics. Then RNAi sequences are produced to target the inhibition of the specific target genes. This technique has been applied, for example, to a mosquito that carries malaria (*Anopheles gambiae*). Using these techniques, scientists can better understand the biological mechanisms relating to the medical or economic effect of such species. The genes that control the complicated life cycle of the malaria parasite, partly inside a mosquito host and partly inside the human body, can be better understood, revealing new ways to control the single most common infectious disease in the world.

KEY CONCEPT RNAi-based methods provide general ways of experimentally interfering with the function of a specific gene without changing its DNA sequence (generally called *phenocopying*).

Functional genomics with nonmodel organisms

Much of our consideration of mutational dissection and phenocopying has focused on genetic model organisms. One current focus of many geneticists is the broader application of these techniques to other species, including those that have negative effects on human society, such as parasites, disease carriers, or agricultural pests, or those species that are of interest to evolutionary biologists and ecologists (see Chapter 20). Classical genetic techniques are not readily applicable to most of these species, but whole-genome sequencing and functional genomics can now be conducted in any species for which DNA and tissue can be obtained. Furthermore, the roles of specific genes can be assessed through the generation of phenocopies by RNAi and targeted mutagenesis. In particular, the recently developed CRISPR-based methods for genome engineering are already being used in a number of nonmodel systems and promise to enable reverse genetic approaches in nearly any species (see Chapter 10).

KEY CONCEPT Reverse genetic methods are the gold standard to test the functions of genes and genetic elements discovered through genomic approaches. Recent technological advances mean that these methods can now be practiced in a variety of model and nonmodel systems.

SUMMARY

Genomic analysis takes the approaches of genetic analysis and applies them to the collection of global data sets to fulfill goals such as the mapping and sequencing of whole genomes and the characterization of all transcripts and proteins. Genomic techniques require the rapid processing of large sets of experimental material, all dependent on extensive automation.

The key problem in compiling an accurate sequence of a genome is to take short sequence reads and relate them to one another by sequence identity to build up a consensus sequence of an entire genome. This can be done in a straightforward way for bacterial or archaeal genomes by aligning overlapping sequences from different sequence reads to compile the entire genome, because few or no DNA segments are present in more than one copy in such organisms. The problem is that complex genomes of plants and animals are replete with such repetitive sequences. These repetitive sequences interfere with accurate sequence-contig production. The problem is resolved in whole-genome shotgun (WGS) sequencing with the use of paired-end reads.

Having a genomic sequence map provides the raw, encrypted text of the genome. The job of bioinformatics is to interpret this encrypted information. For the analysis of gene products, computational techniques are used to identify ORFs and noncoding RNAs, then to integrate these results with available experimental evidence for mRNA transcript structures (cDNA sequences), protein similarities, and knowledge of characteristic sequence motifs.

One of the most powerful means to advance the analysis and annotation of genomes is by comparing with the genomes of related species. Conservation of sequences among species is a reliable guide to identifying functional sequences in the complex genomes of many animals and plants. Comparative genomics can also reveal how genomes have changed in the course of evolution and how these changes may relate to differences in physiology, anatomy, or behavior among species. Comparisons of modern and archaic human genomes are accelerating the discovery of rare disease mutations. In bacterial genomics, comparisons of pathogenic and nonpathogenic strains have revealed many differences in gene content that contribute to pathogenicity.

Functional genomics attempts to understand the working of the genome as a whole system. Two key elements are the transcriptome, the set of all transcripts produced, and the interactome, the set of interacting gene products and other molecules that together enable a cell to be produced and to function. The function of individual genes and gene products for which classical mutations are not available can be tested through reverse genetics—by targeted mutation or phenocopying.

KEY TERMS

annotation (p. 473)
bioinformatics (p. 472)
ChIP (chromatin immunoprecipitation) (p. 490)
ChIP-seq (p. 491)
comparative genomics (p. 479)
consensus sequence (p. 465)
copy number variation (CNV) (p. 484)
DNA sequencing library (p. 468)
exome (p. 485)
expressed sequence tag (EST) (p. 474)
forward genetics (p. 463)
functional genomics (p. 488)

genome project (p. 463)
genomics (p. 462)
homologous gene (p. 479)
interactome (p. 489)
open reading frame (ORF) (p. 473)
ortholog (p. 479)
outgroup (p. 480)
paired-end read (p. 470)
paralog (p. 479)
parsimony (p. 481)
personal genomics (p. 486)
phylogeny (p. 479)
phylogenetic inference (p. 480)
processed pseudogene (p. 477)
proteome (p. 489)

pseudogene (p. 477)
reverse genetics (p. 463)
RNA interference (RNAi) (p. 492)
RNA sequencing (RNA-seq) (p. 474)
scaffold (p. 471)
sequence assembly (p. 465)
sequence contig (p. 468)
supercontig (p. 471)
synteny (p. 482)
transcriptome (p. 489)
two-hybrid test (p. 490)
whole-genome shotgun (WGS) sequencing (p. 466)

SOLVED PROBLEMS

SOLVED PROBLEM 1

You want to study the development of the olfactory (smell-reception) system in the mouse. You know that the cells that sense specific chemical odors (odorants) are located in the lining of the nasal passages of the mouse. Describe some approaches for using functional genomics and reverse genetics to study olfaction.

SOLUTION

Many approaches can be imagined. For reverse genetics, you would want to first identify candidate genes that are expressed in the lining of the nasal passages. Given the techniques of functional genomics, this identification could be accomplished by purifying RNA from isolated nasal-passage-lining cells and using this RNA for an RNA-seq experiment. For example, you may choose to first examine mRNAs that are expressed in the nasal-passage lining but nowhere else in the mouse as important candidates for a specific role in olfaction. (Many of the important molecules may also have other jobs elsewhere in the body, but you have to start somewhere.) Alternatively, you may choose to start with those genes whose protein products are candidate proteins for binding the odorants themselves. Regardless of your choice, the next step would be to engineer a targeted knockout of the gene that encodes each mRNA or protein of interest or to use RNA interference to attempt to phenocopy the loss-of-function phenotype of each of the candidate genes.

PROBLEMS

Visit SaplingPlus for supplemental content. Problems with the ⚌ icon are available for review/grading.

WORKING WITH THE FIGURES

(The first 18 questions require inspection of text figures.)

1. You have identified a noncoding sequence that is conserved across all mammals, except for primates. You decided to engineer a targeted knockout of this sequence in mice. Based on Figure 14-2, is this a forward or reverse genetic experiment?

2. Based on the projection shown in Figure 14-3, what is the approximate number of human genomes that will be sequenced by 2025? How many basepairs will this represent? ⚌

3. Based on Figure 14-4, why must the DNA fragments sequenced overlap in order to obtain a genome sequence?

4. In Figure 14-6, the color pink indicates the base T, the color orange indicates the base A, the color yellow indicates the base G, and the color purple indicates the base C. What is the scanned sequence of the middle cluster in this figure? What is the scanned sequence of the cluster on the left? ⚌

5. Filling gaps in draft genome sequences is a major challenge. Based on Figures 14-8 and 14-9, can paired-end reads from a library of 2-kb fragments fill a 10-kb gap?

6. In Figure 14-11, how are the positions of codons determined?

7. In Figure 14-11, how are the positions of transcriptional regulatory elements determined?

8. In Figure 14-12, expressed sequence tags (ESTs) are aligned with genomic sequence. How are ESTs helpful in genome annotation?

9. In Figure 14-12, cDNA sequences are aligned with genomic sequence. How are cDNA sequences helpful in genome annotation? Are cDNAs more important for bacterial or eukaryotic genome annotations?

10. Based on Figure 14-16 and the features of ultraconserved elements, what would you predict you would observe if you injected a reporter-gene construct of the rat ortholog of the *ISL1* ultraconserved element into fertilized mouse oocytes and examined reporter gene expression in the developing embryo?

11. Based on Figure 14-17, did the duplication that created the A and B genes occur before or after speciation of the common ancestor of frogs, humans, and mice?

12. Based on Figure 14-18, are humans more closely related to mice or to dogs?

13. Figure 14-20 shows syntenic regions of mouse chromosome 11 and human chromosome 17. What do these syntenic regions reveal about the genome of the last common ancestor of mice and humans?

14. Based on Figure 14-22, what percent of Denisovan ancestry do you predict would be found in modern Western Europeans?

15. In Figure 14-23, what key step enables exome sequencing and distinguishes it from whole-genome sequencing? ⚌

16. The genomes of two *E. coli* strains are compared in Figure 14-24. Would you expect any third strain to contain more of the blue, tan, or red regions shown in Figure 14-24? Explain.

17. In Figure 14-25, why do the mRNA-sequencing reads map only to parts of the genome? Which gene is more highly expressed in sample 1?

18. Figure 14-26 depicts the Gal4-based two-hybrid system. Why do the "bait" proteins fused to the Gal4 DNA-binding protein not activate reporter-gene expression?

BASIC PROBLEMS

19. Explain the approach that you would apply to sequencing the genome of a newly discovered bacterial species.

20. Terminal-sequencing reads of clone inserts are a routine part of genome sequencing. How is the central part of the clone insert ever obtained?

21. What is the difference between a contig and a scaffold?

22. Two particular contigs are suspected to be adjacent, possibly separated by repetitive DNA. In an attempt to link them, end sequences are used as primers to try to bridge the gap. Is this approach reasonable? In what situation will it not work?

23. In a genomic analysis looking for a specific disease gene, one candidate gene was found to have a single–base-pair substitution resulting in a nonsynonymous amino acid change. What would you have to check before concluding that you had identified the disease-causing gene?

24. Is a bacterial operator a binding site?

25. A sequenced fragment of DNA in *Drosophila* was used in a BLAST search. The best (closest) match was to a kinase gene from *Neurospora*. Does this match mean that the *Drosophila* sequence contains a kinase gene?

26. In a two-hybrid test, a certain gene A gave positive results with two clones, M and N. When M was used, it gave positives with three clones, A, S, and Q. Clone N gave only one positive (with A). Develop a tentative interpretation of these results. 🅜

27. You have the following sequence reads from a genomic clone of the *Drosophila melanogaster* genome:

    Read 1: TGGCCGTGATGGGCAGTTCCGGTG

    Read 2: TTCCGGTGCCGGAAAGA

    Read 3: CTATCCGGGCGAACTTTTGGCCG

    Read 4: CGTGATGGGCAGTTCCGGTG

    Read 5: TTGGCCGTGATGGGCAGTT

    Read 6: CGAACTTTTGGCCGTGATGGGCAGTTCC

    Use these six sequence reads to create a sequence contig of this part of the *D. melanogaster* genome. 🅜

28. Sometimes, cDNAs turn out to be "chimeras"; that is, fusions of DNA copies of two different mRNAs accidentally inserted adjacently to each other in the same clone. You suspect that a cDNA clone from the nematode *Caenorhabditis elegans* is such a chimera because the sequence of the cDNA insert predicts a protein with two structural domains not normally observed in the same protein. How would you use the availability of the entire genomic sequence to assess if this cDNA clone is a chimera or not?

29. In browsing through the human genome sequence, you identify a gene that has an apparently long coding region, but there is a two–base-pair deletion that disrupts the reading frame.

    a. How would you determine whether the deletion was correct or an error in the sequencing?

    b. You find that the exact same deletion exists in the chimpanzee homolog of the gene but that the gorilla gene reading frame is intact. Given the phylogeny of great apes in the figure below, what can you conclude about when in ape evolution the mutation occurred?

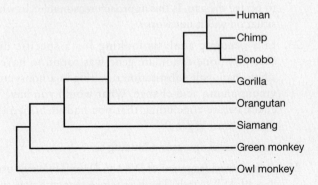

30. In browsing through the chimpanzee genome, you find that it has three homologs of a particular gene, whereas humans have only two.

    a. What are two alternative explanations for this observation?

    b. How could you distinguish between these two possibilities?

31. The platypus is one of the few venomous mammals. The male platypus has a spur on the hind foot through which it can deliver a mixture of venom proteins. Looking at the phylogeny in Figure 14-18, how would you go about determining whether these venom proteins are unique to the platypus? 🅜

32. You have sequenced the genome of the bacterium *Salmonella typhimurium*, and you are using BLAST analysis to identify similarities within the *S. typhimurium* genome to known proteins. You find a protein that is 100 percent identical in the bacterium *Escherichia coli*. When you compare nucleotide sequences of the *S. typhimurium* and *E. coli* genes, you find that their nucleotide sequences are only 87 percent identical. 🅜

    a. Explain this observation.

    b. What do these observations tell you about the merits of nucleotide- versus protein-similarity searches in identifying related genes?

33. If you sequenced the genomes of any two unrelated humans, what types of sequence changes would you expect to find, and how many total base pairs would be affected by each type of sequence change?

34. You have access to both normal cells and cancerous cells taken from a biopsy from a patient with liver cancer. Describe the genomic approaches you would use to characterize the differences between these cells.

35. To inactivate a gene by RNAi, what information do you need? Do you need the map position of the target gene? 🅜

36. What is the purpose of generating a phenocopy?

37. What is the difference between forward and reverse genetics?

38. Why might exome sequencing fail to identify a disease-causing mutation in an affected person?

39. You have identified a noncoding sequence that is conserved in all mammals. Can you conclude that it is functional?

CHALLENGING PROBLEMS

40. You have the following sequence reads from a genomic clone of the *Homo sapiens* genome:

    Read 1: ATGCGATCTGTGAGCCGAGTCTTTA

    Read 2: AACAAAAATGTTGTTATTTTTATTTCAGATG

    Read 3: TTCAGATGCGATCTGTGAGCCGAG

    Read 4: TGTCTGCCATTCTTAAAAACAAAAATGT

    Read 5: TGTTATTTTTATTTCAGATGCGA

Read 6: AACAAAAATGTTGTTATT

a. Use these six sequence reads to create a sequence contig of this part of the *H. sapiens* genome.

b. Translate the sequence contig in all possible reading frames.

c. Go to the BLAST page of the National Center for Biotechnology Information, or NCBI (http://www.ncbi.nlm.nih.gov/BLAST/, Appendix B) and see if you can identify the gene of which this sequence is a part by using each of the reading frames as a query for protein–protein comparison (BLASTp).

41. Some sizable regions of different chromosomes of the human genome are more than 99 percent nucleotide identical with one another. These regions were overlooked in the production of the draft genome sequence of the human genome because of their high level of similarity. Of the techniques discussed in this chapter, which would allow genome researchers to identify the existence of such duplicate regions?

42. Some exons in the human genome are quite small (less than 75 bp long). Identification of such "microexons" is difficult because these distances are too short to reliably use ORF identification or codon bias to determine if small genomic sequences are truly part of an mRNA and a polypeptide. What techniques of "gene finding" can be used to try to assess if a given region of 75 bp constitutes an exon?

43. A certain cDNA of size 2 kb hybridized to eight genomic fragments of total size 30 kb and contained two short ESTs. The ESTs were also found in two of the genomic fragments each of size 2 kb. Sketch a possible explanation for these results.

44. You are studying proteins having roles in translation in the mouse. By BLAST analysis of the predicted proteins of the mouse genome, you identify a set of mouse genes that encode proteins with sequences similar to those of known eukaryotic translation-initiation factors. You are interested in determining the phenotypes associated with loss-of-function mutations of these genes.

a. Would you use forward- or reverse-genetics approaches to identify these mutations?

b. Briefly outline two different approaches that you might use to look for loss-of-function phenotypes in one of these genes.

45. You are interested in identifying genetic changes that might contribute to behavioral differences between two species of mice: one is promiscuous and the other is monogamous.

a. Would you conduct whole-genome sequencing or exome sequencing of these two species? Defend your decision.

b. What additional functional genomics experiments would you do to identify differences between these two species?

c. How would you show that the genetic differences you identify actually contribute to behavioral differences between the species?

46. Different strains of *E. coli* are responsible for enterohemorrhagic and urinary tract infections. Based on the differences between the benign K-12 strain and the enterohemorrhagic O157:H7 strain, would you predict that there are obvious genomic differences

a. between K-12 and uropathogenic strains?

b. between O157:H7 and uropathogenic strains?

c. What might explain the observed pair-by-pair differences in genome content?

d. How might the function of strain-specific genes be tested?

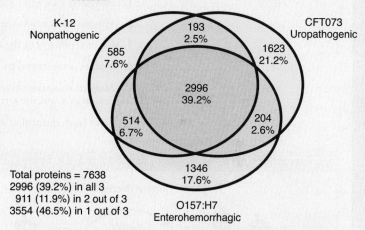

Total proteins = 7638
2996 (39.2%) in all 3
911 (11.9%) in 2 out of 3
3554 (46.5%) in 1 out of 3

GENETICS AND SOCIETY

1. You decide to submit samples to two different "direct-to-consumer genetic testing" companies to learn more about your genetic ancestry. However, the results provided by the two companies give you very different estimates of your genetic origins. What might explain these differences?

2. What advice would you give to a friend who was considering doing "direct-to-consumer genetic testing" to learn more about their genealogy?

PART 3

Core Principles in Mutation, Variation, and Evolution

I magine for a moment that the first living cell contained within it a single DNA molecule or chromosome and that this molecule was replicated without error each time the cell and its descendants divided. Had life on Earth been conceived this way, our biological world would be monochromatic. There would be a single species within which all individuals are exactly the same. What makes life as we know it remarkable is not the fidelity of DNA replication, but rather its imprecision. If mutation generates imperfections, then life flaunts its imperfections.

Part 1 of this text introduced the roles of constancy and variation in heredity, and Part 2 elucidated how that constancy is maintained and regulated through the transfer and expression of genetic information. In this final section of the text, you will learn how the process of *mutation* generates heritable *variation* within populations that provides the raw material for the evolution of the diversity of life on Earth. In the pipeline that runs from mutation to variation to evolution, there are checkpoints. A mutation created by an error in DNA replication or cellular damage can be lost or retained by *chance*, or affected by *natural selection* if it alters the *fitness* of the cell or individual that carries it.

These six core principles can be seen as life's essential ingredients:

1. *Mutation* generates variation in populations.

2. *Variation* occurs at the level of DNA and phenotype.

3. *Chance* influences the fate of variation within populations.

4. *Fitness* is measured by the number of offspring contributed to the next generation.

5. *Natural selection* occurs when individuals with a particular variation have higher fitness in a specific environment.

6. Over time, natural selection results in Darwinian *evolution*.

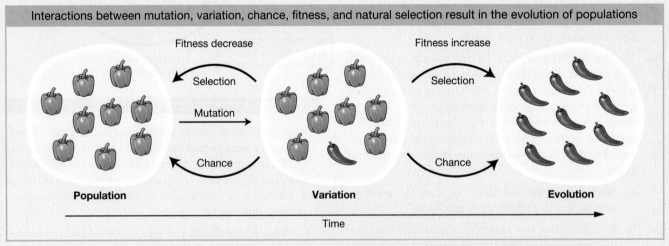

Interactions between mutation, variation, chance, fitness, and natural selection result in the evolution of populations

As an example, a population of peppers may evolve over time from a population of round and green peppers to a population of long and red peppers due to chance and/or selection acting on variation produced by mutation.

CORE PRINCIPLES OF EVOLUTION

1. Mutation generates variation in populations.

Mutation is a change in the nucleotide sequence of a cell, which can result from either an error in intrinsic cellular processes, like DNA replication, or extrinsic factors, like exposure to ionizing radiation or UV light. Mutations occur in many forms, ranging from changes at a single nucleotide, to movement of transposable elements within or between genomes, to changes in chromosome number and structure.

We are often accustomed to thinking about the deleterious effects that mutations can have on the individuals that carry them. Indeed, a mutation can have negative impacts on the health, fertility, or viability of an individual. However, many, and perhaps most, mutations have no effect on the phenotype of the individual. Furthermore, mutations can sometimes result in phenotypic changes that are beneficial. Whether a particular mutation is detrimental, neutral, or beneficial can depend greatly upon the environment, and this will affect whether the mutation is lost, maintained, or fixed within a population or a species.

2. Variation occurs at the level of DNA and phenotype.

Because mutation produces change in the genomes of some individuals while other individuals retain a wild-type genome, mutation produces genetic differences among individuals; that is, mutation generates genetic variation within the population. The amount of variation in a population can be measured in different ways. For example, one can simply count the number of different alleles at a locus as a measure of variation. A genetic locus with 10 alleles found within a population would be more variable than a locus with just two alleles. The frequency of the alleles at a locus in the population is another useful metric. For a locus with two alleles, the amount of variation depends on the allele frequencies in the population. A population of 100 individuals in which 50 individuals are wild type and 50 carry the mutant allele is more variable than one in which 99 are wild type and just one carries the mutant allele.

Variation at the level of DNA can create variation at the phenotypic level. Some mutant alleles create new phenotypes that differ from wild type, although other mutant alleles may have no effect on phenotype and thus are considered neutral. Populations with much genetic variation tend to have much phenotypic variation as well. When there are many genes that affect a trait, the variation in phenotype can be continuous, such as for height in humans. The evolutionary potential of a population to respond to a changing environment depends on the amount of heritable phenotypic variation within the population. Populations with more heritable variation have a greater repertoire of potential solutions to a changing environment.

3. Chance influences the fate of variation within populations.

When a DNA mutation first arises in a diploid species, it will exist in a single individual who would be heterozygous for the mutant and wild-type alleles—*A/a*. Just by chance alone, that individual might transmit only its wild-type allele to its offspring. Thus, the mutation can be lost from the population in one generation by chance. Similarly, the individual might transmit only the mutant allele to its offspring, in which case the mutant allele could be present in multiple individuals of the next generation, rising in frequency in the population. Over long periods of time, chance can even cause the mutant allele to completely replace the wild-type allele in the population.

The importance of chance in determining the fate of a mutant allele depends on the number of individuals in the population or population size. In a small population, a mutant allele has a greater chance of replacing the wild-type allele than it does in a large population. If you flip a coin twice (a small population of trials), you have pretty good odds of getting all tails and no heads; but if you flip a coin 1000 times (a large population of trials), the chance of getting all tails is diminishingly small. When mutations have no effect on phenotype, chance is the only force that affects their fate. When mutations have either detrimental or beneficial effects on reproductive success (i.e., fitness), then their fate can be governed both by chance and the action of natural selection.

4. Fitness is measured by the number of offspring contributed to the next generation.

There are two components of fitness: survival and reproduction. An individual that does not survive will not reproduce. However, the ultimate measure of fitness is reproductive success; that is, the number of offspring that an individual contributes to the next generation. It is possible to measure fitness at the level of alleles, genotypes, phenotypes, and/or individuals in a population.

Take the example shown in the figure: there is a population with two types of peppers, round-green and long-red. In this population, we can ask how many offspring round-green peppers have on average and how many offspring long-red peppers have on average. If long-red peppers tend to have more offspring than round-green peppers, then the long-red phenotype has a higher fitness. If we knew that the long-red phenotype was due to genotype *a/a* at a particular locus, and the round-green phenotype was due to genotype *A/A* or *A/a* at that locus, then we could also say that the *a/a* genotype has a higher fitness than the *A/A* or *A/a* genotype. It is also important to remember that fitness is always relative; we are interested not in the absolute number of offspring produced by a particular genotype or phenotype, but rather in knowing if one produces a higher number of offspring relative to the other.

5. Natural selection occurs when individuals with a particular variation have higher fitness in a specific environment.

For natural selection to occur within a population, four conditions must be met: (1) individuals within a population must be variable for the phenotype of interest; (2) phenotypic variation among individuals must be heritable; (3) there must be variation in reproductive success within the population such that not all individuals within a population survive and reproduce, but some are more successful than others; and (4) survival and reproductive success is not random; rather, those individuals that possess the most favorable variations in a phenotype of interest will survive and reproduce.

When these conditions are met, and a mutation occurs that has an effect on a phenotype, there are three possible outcomes. First, the new phenotype may be neutral; that is, individuals with the new phenotype have the same reproductive success as individuals with the wild-type phenotype. In this case, only chance governs the fate of this mutation in the population. Second, individuals with the new phenotype might have lower reproductive success than individuals with the wild-type phenotype. In this case, there is natural selection against this phenotype and the underlying mutation. Third, individuals with the new phenotype might have a higher reproductive success than individuals with the wild-type phenotype. In this case, there is natural selection for the new phenotype and the underlying mutation. If, in our example, the peppers with the long-red phenotype (*a/a* genotype) have higher reproductive success than the round-green peppers in a population, then natural selection will cause the population to evolve, and over time it will contain only long-red peppers with the *a/a* genotype.

6. Over time, natural selection results in Darwinian evolution.

Evolution can be simply defined as a change in the frequencies of genotypes or phenotypes in a population over time. Charles Darwin and Alfred Russel Wallace independently proposed the process of natural selection as a mechanism to explain evolution. Although natural selection is often equated with evolution, natural selection is but one mechanism by which evolution can occur. As discussed previously, other mechanisms such as chance can also lead to changes in the frequencies of genotypes or phenotypes within populations and therefore evolution over time. This is why evolution caused by natural selection is often referred to as Darwinian evolution.

Crucially, it is sometimes possible to experimentally distinguish between the effects of chance and natural selection on evolution. For example, the population shown in the figure has evolved from containing mostly round-green peppers to containing only long-red peppers. If evolution resulted from selection, we would find that long-red peppers have a higher fitness than round-green peppers in the current environment of the population. Importantly, signatures of Darwinian selection can also be observed at the DNA-sequence level. In modern evolutionary genetics, it is therefore now possible to identify phenotypes that have evolved due to selection and to link these phenotypes with their underlying genotypes.

In Part 1 of this book, you learned of the discoveries of Mendel and others that established the basic rules of inheritance, which are the foundation for the science of genetics. In Part 2, you learned of the discoveries of many other geneticists and biochemists that elucidated the details of how information encoded in DNA controls the metabolism of cells and the growth and development of whole organisms. The final part of the book builds on these breakthroughs and covers: (1) the generation of genetic variation through mutation; (2) the rules governing the transmission of genetic variation from one generation to the next in whole populations; (3) the theory of how genetic variation in many genes working together can give rise to continuous trait variation among individuals, such as difference in height among people; and (4) the synthesis of Mendel's theory of inheritance with Darwin's theory of evolution. When combined, these two theories provide a powerful paradigm for how the diversity of life on Earth evolved.

DNA Damage, Repair, and Mutation

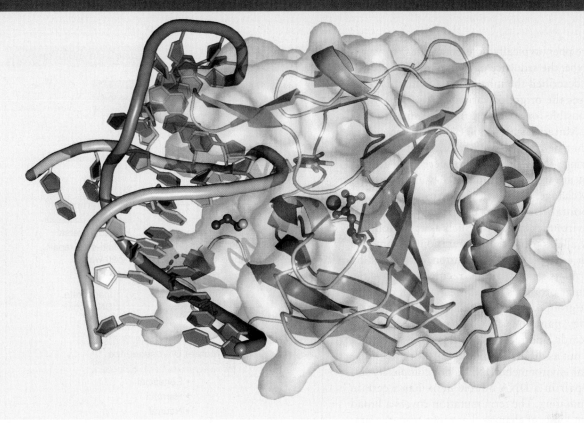

Molecular model of a DNA repair enzyme alpha-ketoglutarate-dependent dioxygenase, which removes methyl groups and larger alkyl groups from purine and pyrimidine bases.

CHAPTER OUTLINE AND LEARNING OBJECTIVES

15.1 MOLECULAR CONSEQUENCES OF POINT MUTATIONS

LO 15.1 Explain the effects of point mutations on the sequence and expression of RNAs and proteins.

15.2 MOLECULAR BASIS OF SPONTANEOUS MUTATIONS

LO 15.2 Summarize the causes of spontaneous DNA damage that lead to mutations following DNA replication.

15.3 MOLECULAR BASIS OF INDUCED MUTATIONS

LO 15.3 Tabulate the causes of induced DNA damage that result in mutations following DNA replication.

15.4 DNA REPAIR MECHANISMS

LO 15.4 Illustrate the molecular mechanisms that repair distinct types of DNA damage.

LO 15.5 Using examples, describe how different types of mutations in different genes lead to particular human genetic diseases.

501

Building upon all of the preceding chapters that explain how the sequence of genomic DNA determines the characteristics of organisms, the broad objective for this chapter is to explain how damage to DNA that is not correctly repaired leads to mutations in DNA sequence that can alter the characteristics of organisms.

Individuals are phenotypically different because of variation in genotype, the sequence of genomic DNA. Preceding chapters described the inheritance of variants. This chapter addresses the origin of variants. Two major processes are responsible for genetic variation: mutation and recombination. **Mutations** are changes in the sequence of DNA that cannot be repaired. Therefore, mutations are transmitted during DNA replication to succeeding generations. Mutations are significant because they are the source of evolutionary change; new alleles arise in all organisms, some occur spontaneously, and others are induced by exposure to environmental factors such as radiation and chemicals. New alleles produced by mutations become the raw material for a second level of variation, carried out by recombination. As its name suggests, **recombination** is the outcome of cellular processes that cause alleles of different genes to become grouped in new combinations (Chapter 4). To use an analogy, mutations change the identities of individual playing cards, and recombination shuffles the cards and deals them out as different hands.

In the cellular environment, DNA is not completely stable: each base pair in a DNA double helix has a certain probability of mutating. The term mutation covers a broad array of different kinds of changes that range from the simple swapping of one base pair for another, to the elimination of an entire chromosome (**Figure 15-1**). Mutations arise from **DNA damage** (also called a lesion), which is a physical or chemical abnormality in the structure of DNA. Types of DNA damage include abasic sites, base mismatches, modified bases, interstrand and intrastrand crosslinks, and strand breaks. Chapter 17 considers mutational changes that affect entire chromosomes or large pieces of chromosomes, while this chapter focuses on mutations that occur within individual genes.

Cells have evolved sophisticated systems to detect and repair damaged DNA, thereby preventing the occurrence of most, but not all, mutations. DNA can be viewed as being subjected to a dynamic tug-of-war between chemical processes that damage DNA and lead to new mutations and cellular repair processes that constantly monitor DNA for such damage and correct it. However, this tug-of-war is not straightforward. As already mentioned, mutations provide the raw material for evolution, and, thus, the introduction of a low level of mutation must be tolerated. In fact, DNA replication and DNA repair systems can actually introduce mutations.

In unicellular organisms such as *E. coli* (bacteria) and *S. cerevisiae* (yeast), mutations are passed from parent to

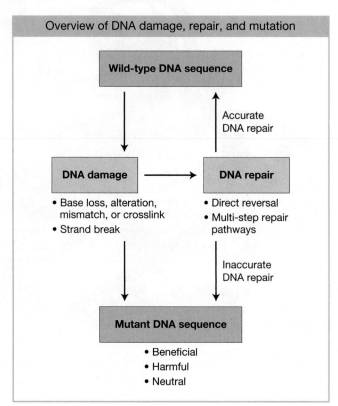

FIGURE 15-1 DNA damage (also called a lesion) occurs through spontaneous and induced mechanisms that alter bases and the phosphodiester backbone. A variety of DNA repair mechanisms can detect and correct DNA damage. However, DNA damage that escapes repair, or is caused during the process of repair, leads to mutations (i.e., DNA sequence changes) that can have beneficial, harmful, or neutral effects on organisms.

daughter cells when cells divide by mitosis. In contrast, in multicellular eukaryotic organisms, there are two general types of mutational inheritance because there are two general types of cells, somatic and germ-line. **Somatic mutations** arise in single cells such as human skin or liver cells during an organism's life and are passed on to daughter cells but are not inherited by offspring. In contrast, **germ-line mutations** in gametes such as human sperm and eggs are inherited by offspring and are present in all of their cells (i.e., both somatic and germ-line cells). In this chapter, we discuss both somatic and germ-line mutations in the context of human diseases. However, keep in mind that not all mutations are detrimental. Some mutations are beneficial to an organism, while others are neutral and do not help or harm the organism.

KEY CONCEPT DNA damage and mutation do not mean the same thing. DNA damage is a physical or chemical defect in the structure of DNA that can often be fixed by repair mechanisms in cells. In contrast, mutations are changes in the base sequence of both strands of DNA that cannot be repaired. Therefore, mutations are transmitted during DNA replication to succeeding generations.

KEY CONCEPT In multicellular organisms, there are two types of mutations. Germ-line mutations occur in gametes and are passed on to offspring, whereas somatic mutations occur in all other cell types and are not passed on to offspring.

15.1 MOLECULAR CONSEQUENCES OF POINT MUTATIONS

LO 15.1 Explain the effects of point mutations on the sequence and expression of RNAs and proteins.

LO 15.5 Using examples, describe how different types of mutations in different genes lead to particular human genetic diseases.

The term **point mutation** refers to a single–base-pair change in DNA sequence. In this section, we will consider the effects on gene expression of point mutations in the protein-coding regions and the noncoding regions of genes.

The types of point mutations

There are three types of point mutations in DNA: base substitutions, base insertions, and base deletions. **Base substitutions** are mutations in which one base pair is replaced by another. They can be divided into two subtypes: transitions and transversions. A **transition** replaces a purine with a purine (from A to G or G to A) or a pyrimidine with a pyrimidine (from C to T or T to C). A **transversion** replaces a pyrimidine with a purine (from C to A, C to G, T to A, or T to G) or a purine with a pyrimidine (from A to C, A to T, G to C, or G to T). Thus, there are four possible transitions and eight possible transversions. In describing the changes in double-stranded DNA, an example of a transition is $G \cdot C \rightarrow A \cdot T$ and a transversion is $G \cdot C \rightarrow T \cdot A$. A **base insertion** is the addition of one base pair, and a **base deletion** is the removal of one base pair. Collectively, base insertions and deletions are termed **indel mutations** (for *in*sertion-*del*etion).

The molecular consequences of a point mutation in an open reading frame

Figure 15-2 shows the three types of point mutations and their effects when they occur within the open reading frame (ORF, the protein-coding region) of a gene. The variety of outcomes is a direct consequence of features of the genetic code: mRNA codons are read as triplets, the code is degenerate (multiple codons can encode the same amino acid), codons for amino acids with similar chemical properties and size often differ by one nucleotide, and 3 of the 64 codons signal translation termination (i.e., stop) (the genetic code is shown in Figure 9-8). New mRNA codons produced by base substitutions in DNA may code for the same amino acid (synonymous mutations), a different amino acid (missense mutations), or a translation stop (nonsense mutations) (Figure 15-2a).

- **Synonymous mutations** (also called **silent mutations**) change the sequence of a codon but not the encoded amino acid. As an example, for 32 of the 61 codons that encode an amino acid, mutating the third position of the codon to any other base does not change the encoded amino acid. For instance, GU<u>A</u>, GU<u>C</u>, GU<u>G</u>, and GU<u>U</u> all encode valine (see Figure 9-8, bottom left).

- **Missense mutations** (also called nonsynonymous mutations) change the sequence of a codon to one that codes for a different amino acid. Missense mutations can result in an amino acid being replaced with a chemically similar amino acid. This is called a **conservative mutation** because it may not significantly affect the protein's structure and function. A lysine-to-arginine change is an example of a conservative mutation because both amino acids are positively charged, and they are similar in size (see Figure 9-2 to compare their chemical structures). Alternatively, an amino acid may be replaced by a chemically different amino acid. This is called a **nonconservative mutation** because it is likely to produce a change in the protein's structure and function. A lysine-to-threonine change is an example of a nonconservative mutation because lysine is positively charged and has a long hydrocarbon chain, whereas threonine is uncharged and has a short hydrocarbon chain.

- **Nonsense mutations** change the sequence of a codon that codes for an amino acid into one that stops translation (i.e., UAA, UAG, or UGA). The closer a nonsense mutation is to the 3′ end of the ORF, the more likely that the resulting protein will retain its biological activity. However, nonsense mutations often produce proteins that are completely inactive. In addition, in eukaryotes, nonsense mutations can completely block protein production by triggering decay of the mRNA by nonsense-mediated decay (NMD) (Chapter 8).

KEY CONCEPT A point mutation in the open reading frame of a gene changes the sequence of a single codon and has three potential consequences on the translated amino acid: (1) no change, (2) a change to another amino acid, or (3) a change to a translation stop.

Because the genetic code consists of triplet nucleotides, the other two types of point mutations—single base pair insertions and base pair deletions in the DNA sequence—lead to a frameshift, that is, a change in the translation

Consequences of point mutations in open reading frames

(a)

		Single base substitution			
	No mutation	Silent mutation (synonymous)	Missense mutation (nonsynonymous) Conservative	Missense mutation (nonsynonymous) Nonconservative	Nonsense mutation
DNA	CAT AAG CAG AGT ACT GTA TTC GTC TCA TGA	CAT AAA CAG AGT ACT GTA TTT GTC TCA TGA	CAT AGG CAG AGT ACT GTA TCC GTC TCA TGA	CAT ACG CAG AGT ACT GTA TGC GTC TCA TGA	CAT TAG CAG AGT ACT GTA ATC GTC TCA TGA
mRNA	CAU AAG CAG AGU ACU	CAU AAA CAG AGU ACU	CAU AGG CAG AGU ACU	CAU ACG CAG AGU ACU	CAU UAG CAG AGU ACU
Protein	His Lys Gln Ser Thr	His Lys Gln Ser Thr	His Arg Gln Ser Thr	His Thr Gln Ser Thr	His Stop

(b)

		Single base insertion or base deletion (indel)	
	No mutation	Insertion (frameshift)	Deletion (frameshift)
DNA	CAT TGC GAC AAG GAT AGT ACT CCT GTA ACG CTG TTC CTA TCA TGA GGA	CAT GTG CGA CAA GGA TAG TAC TCC T GTA CAC GCT GTT CCT ATC ATG AGG A	T CAT GCG ACA AGG ATA GTA CTC CT GTA CGC TGT TCC TAT CAT GAG GA A
mRNA	CAU UGC GAC AAG GAU AGU ACU CCU	CAU GUG CGA CAA GGA UAG UAC UCC U	CAU GCG ACA AGG AUA GUA CUC CU
Protein	His Cys Asp Lys Val Ser Thr Pro	His Val Arg Gln Gly Stop	His Ala Thr Arg Ile Val Leu

FIGURE 15-2 (a) Mutations that change a single base pair in an open reading frame can be silent, missense, or nonsense, with regard to encoded amino acids. (b) Mutations that insert or delete a single base pair change the codon reading frame downstream of the mutation, which often generates a stop codon, as shown in the insertion example.

reading frame for all codons downstream of the mutation (Figure 15-2b). Hence, single base insertions and base deletions are also called **frameshift mutations**. The string of wrong amino acids that is encoded after the frameshift typically does not continue to the end of the original open reading frame in the mRNA because there is a high probability of encountering a stop codon. Thus, the effect of a frameshift on a protein's normal structure and function can vary depending on where the insertion or deletion occurs in the ORF and what sequence is appended downstream of the frameshift. As described in Figure 9-6, insertions or deletions of two base pairs also cause frameshifts, but insertions or deletions of three base pairs or any multiple of three do not.

KEY CONCEPT A single–base-pair insertion or deletion in the open reading frame of a gene shifts the reading frame of all of the downstream codons, which changes the encoded amino acids and often leads to a translation stop.

Point mutations in ORFs can have major phenotypic consequences, as illustrated by a missense mutation in the human *ras* gene. The mutation in question changes a glycine at position 12 of the Ras protein to a valine (G12V) (**Figure 15-3a**). Glycine is a unique amino acid in that it contains a hydrogen as its side chain, which gives it much more conformational flexibility relative to other amino acids, which all have carbon side chains. This means that changing a glycine to any other amino acid can have severe effects on protein function. The Ras protein is a GTPase that cycles between active and inactive states when bound to GTP and GDP, respectively (Figure 15-3b). In response to signals received by cells, GTP exchange factors (GEFs) activate Ras by exchanging its GDP for GTP, and GTPase activating proteins (GAPs) inactivate Ras by activating its intrinsic GTPase activity that converts GTP to GDP. The G12V mutation blocks the GTPase activity of Ras, inappropriately locking it into its active form. This matters because when Ras is active, it affects cells in a number of

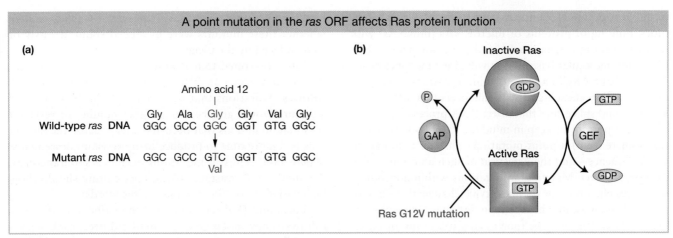

A point mutation in the *ras* ORF affects Ras protein function

(a)

Amino acid 12

		Gly	Ala	Gly	Gly	Val	Gly
Wild-type *ras* DNA		GGC	GCC	GGC	GGT	GTG	GGC
Mutant *ras* DNA		GGC	GCC	GTC	GGT	GTG	GGC
				Val			

(b)

FIGURE 15-3 (a) A G-to-T transversion in codon 12 of the *ras* gene changes a glycine (gly, G) to a valine (val, V) in the Ras protein. (b) The G12V mutation blocks conversion of Ras from the active to the inactive state. Abbreviations: GAP, GTPase activating protein; GEF, GTP exchange factor.

ways, including promoting cell proliferation. The uncontrolled cell proliferation resulting from the mutant Ras protein can lead to **cancer**. In fact, approximately 30 percent of human cancers are driven by mutations in Ras.

The molecular consequences of a point mutation in a noncoding region

Point mutations do not need to occur within a protein-coding region to affect the phenotype of cells and organisms. Previous chapters have drawn attention to the importance of noncoding DNA sequence elements for the regulation of DNA replication and transcription. In addition, some DNA sequences that are copied into RNA function as sequence

elements for the regulation of RNA processing, stability, localization, translation, and function. In general, DNA and RNA regulatory elements are short and serve as binding sites for proteins and RNAs that scaffold or catalyze molecular processes. Thus, point mutations in regulatory elements can block binding and disrupt molecular processes.

As shown in **Figure 15-4**, point mutations in transcription enhancer or promoter elements that block binding of transcription factors or general transcription factors, respectively, will affect transcription activation of the associated gene (Chapters 8 and 12). Point mutations in eukaryotic pre-mRNA splice sites that block complementary base pairing of small nuclear RNAs (snRNAs) will affect the removal of introns (Chapter 8). Point mutations

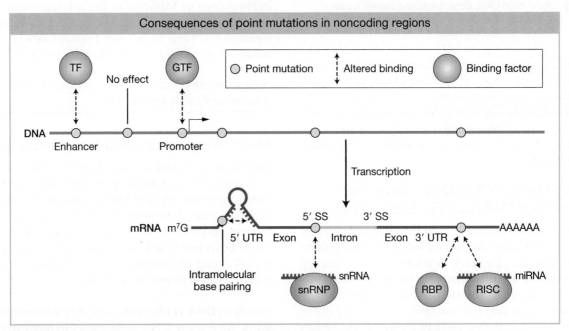

FIGURE 15-4 Point mutations in noncoding regions affect gene expression by altering the binding of proteins and RNAs to DNA and RNA regulatory elements. Abbreviations: TF, transcription factor; GTF, general transcription factor; snRNP, small nuclear ribonucleoprotein; snRNA, small nuclear RNA; RBP, RNA-binding protein; RISC, RNA-induced silencing complex; miRNA, microRNA, SS, splice site; UTR, untranslated region; m^{7}G, 7-methylguanosine cap; and AAAAAA, poly(A) tail.

in eukaryotic mRNA untranslated regions (UTRs) that block binding of proteins or microRNAs (miRNAs) will affect translation (Chapter 9), and point mutations that block intramolecular base pairing will affect the function of RNAs (Chapter 8). In addition, point mutations can create new regulatory elements such as enhancers that affect transcription and splice sites that affect RNA processing.

It is important to keep in mind the distinction between the occurrence of a point mutation—that is, a change in DNA sequence—and the detection of such an event at the phenotypic level. Many point mutations within noncoding sequences elicit little or no phenotypic change; these mutations are often located between binding sites for regulatory factors. Such sites may be functionally irrelevant, or other sites may duplicate their function.

> **KEY CONCEPT** A point mutation in a noncoding region of a gene can affect expression of the encoded RNA by altering binding sites for proteins and RNAs that regulate transcription, post-transcription, and translation events.

> **KEY CONCEPT** Point mutations in coding or noncoding regions of genes can have phenotypic consequences.

15.2 MOLECULAR BASIS OF SPONTANEOUS MUTATIONS

LO 15.2 Summarize the causes of spontaneous DNA damage that lead to mutations following DNA replication.

LO 15.5 Using examples, describe how different types of mutations in different genes lead to particular human genetic diseases.

Mutations can arise spontaneously, or they can be induced. **Spontaneous mutations** occur naturally and arise in all cells. **Induced mutations** arise through the action of external agents called **mutagens** that increase the rate at which mutations occur. In this section, we consider the nature of spontaneous mutations. Induced mutations are the subject of the next section.

Evidence for spontaneous mutations: the Luria and Delbrück fluctuation test

The causes of mutations are of considerable interest because they are the basis of evolution and disease. In 1943, Salvador Luria and Max Delbrück developed an experimental approach to test whether mutations occur spontaneously as a result of cellular processes that act on DNA. It was known at the time that if *E. coli* are spread on a plate of nutrient medium in the presence of phage T1, the phages infect and kill the bacteria. Rarely, but regularly, bacterial colonies were seen that were resistant to phage attack; the resistance phenotype was heritable, and so it appeared to

be due to genuine mutations. However, it was not known whether these mutants were produced spontaneously or were induced by the phage.

Luria reasoned that if mutations were spontaneous, they would occur at different times in different *E. coli* cultures. Mutations that occurred early in the growth of a culture would give rise to a higher number of resistant cells than mutations that occurred later because the mutant cells had more time to produce many resistant descendants (**Figure 15-5**). Thus, if mutations occurred spontaneously, the numbers of resistant colonies per culture should show high variation (or "fluctuation," in his words).

Luria and Delbrück designed their *fluctuation test* as follows. They inoculated 20 small cultures, each with a few cells, and incubated them until there were 10^8 cells per milliliter. At the same time, a larger culture was inoculated and incubated until there were 10^8 cells per milliliter. The 20 individual cultures and 20 aliquots (samples) from the large culture were plated in the presence of phage. The 20 individual cultures showed high variation in the number of phage-resistant colonies: 11 plates had no resistant colonies, and the remainder had 1, 1, 3, 5, 5, 6, 35, 64, and 107 per plate. In contrast, the 20 aliquots from the large culture showed much less variation from plate to plate, all in the range of 14 to 26. If the phage were inducing mutations, there was no reason why fluctuation should be higher for the individual cultures than the aliquots because they were all similarly exposed to phage. This result led to the reigning "paradigm" of mutation; that is, whether in bacteria or eukaryotes, mutations can occur in any cell at any time and their occurrence is random. For this and other work, Luria and Delbrück were awarded the Nobel Prize in Physiology or Medicine in 1969. Interestingly, this was after Luria's first graduate student, James Watson, was awarded his Nobel Prize (with Francis Crick in 1964) for the discovery of the DNA double-helix structure.

> **KEY CONCEPT** Mutations can occur spontaneously, that is, independently of external agents.

Mechanisms of spontaneous mutations

Spontaneous mutations arise from a variety of sources. One source is DNA replication. Although DNA replication is a remarkably accurate process, mistakes are made in the copying of the millions to billions of base pairs in a genome. Spontaneous mutations also arise because of damage to DNA by the cellular environment. Lastly, as we will see in Chapter 16, spontaneous mutations can be caused by insertion of a transposable element.

Errors in DNA replication cause spontaneous mutations An error in DNA replication can result when a mismatched base pair forms, leading to a base substitution that may be either a transition or a transversion. The generation of a transition by a DNA replication error involves the

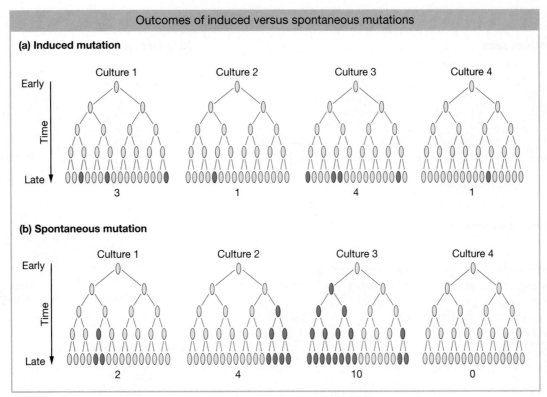

FIGURE 15-5 Cell pedigrees illustrate the expected outcomes of induced versus spontaneous mutation as the cause of resistant cells in the Luria and Delbrück fluctuation experiment. Dark green cells are resistant to phage infection.

pairing of a pyrimidine with the wrong purine (e.g., C·A, instead of T·A, where A is the template) or a purine with the wrong pyrimidine (e.g., G·T, instead of A·T, where T is the template). Other errors may insert or delete base pairs.

Mismatching occurs because of **tautomerization** and **ionization** of bases (**Figure 15-6**). Each of the bases in DNA can reside in one of several forms, called tautomers, which are isomers that differ in the positions of their atoms and in the bonds between the atoms. The forms are in equilibrium. The keto form of each base is normally present in DNA, but in rare instances a base may shift to the imino or enol form. The imino and enol forms may pair with the wrong base, forming a mispair. For example, when a C shifts to its rare imino form, the DNA polymerase incorporates an A rather than a G to pair with it (Figure 15-6b). Similarly, ionization of bases brought about by proton exchange between water and hydrogen bonds can allow DNA polymerase to insert a mismatch (Figure 15-6c) that resembles a wobble base pair (Figure 15-6d). Fortunately, mismatch errors due to tautomerization and ionization of bases are usually detected and removed by the 3′-to-5′ exonuclease activity (proofreading activity) of DNA polymerase (Figure 7-18). If proofreading does not occur, the mismatches lead to transition mutations, in which a purine substitutes for a purine or a pyrimidine for a pyrimidine. Repair systems (described later in this chapter) correct many of the mismatched bases that escape correction by DNA polymerases.

Base insertions and deletions (indels) are also caused by DNA replication errors. The prevailing model (**Figure 15-7**) proposes that indels arise when loops in single-stranded regions of DNA are stabilized by the "slipped mispairing" of repeated sequences in the course of DNA replication. This mechanism is sometimes called **replication slippage** or slipped-strand mispairing.

> **KEY CONCEPT** Spontaneous mutations are generated by errors in DNA replication.

DNA replication slippage and other mechanisms lead to expansion of three base-pair repeats that are responsible for more than 40 human neurodegenerative diseases, collectively called **trinucleotide-repeat diseases**. Fragile X syndrome is an example of a trinucleotide-repeat disease. It is the most common form of inherited intellectual disability, affecting about 1 of 4000 males and 1 of 8000 females. Fragile X syndrome results from an increase in the number of CGG repeats in the 5′ UTR of the *FMR1* (*fragile X mental retardation*) gene. Normal individuals have fewer than 45 CGG repeats, but Fragile X syndrome occurs when the number of CGG repeats expands to at least 200 (**Figure 15-8**). Sometimes, unaffected parents and grandparents contain increased numbers of repeats, but only ranging from 55 to 200. For this reason, these ancestors are said to

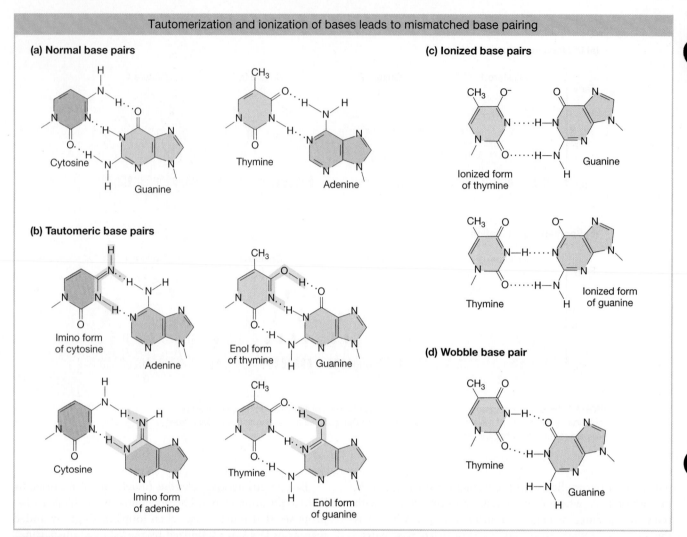

FIGURE 15-6 Normal base pairing compared with mismatched bases. (a) Pairing between the normal (keto) forms of bases. (b) Rare tautomeric forms of bases (imino and enol forms) result in mismatches. (c) Ionization of bases results in mismatches that resemble wobble base pairs (d).

carry premutations. The repeats in these premutation alleles are not sufficient to cause the disease phenotype, but they are much more prone to expansion than normal alleles, and so they lead to even greater expansion in their offspring. Expansion over generations to the full mutation allele with more than 200 repeats results in silencing of *FMR1* transcription by DNA methylation of CpGs in an island (Chapter 12) at the *FMR1* promoter, in surrounding regulatory regions, and in the C̲GG repeats (the CpG is underlined). Because no *FMR1* mRNA is produced, no FMR1 protein is produced, which leads to impaired cognitive development.

Other trinucleotide-repeat diseases are associated with expansion of trinucleotide CAG codons that code for glutamine (Q in the single letter amino acid nomenclature). Huntington's disease (Chapter 2) is one of nine diseases caused by polyglutamine (polyQ) expansion of a protein. In normal individuals, the N-terminus of the Huntingtin protein has about 20 glutamines encoded by 20 CAG repeats, but when the number of glutamines is expanded to more than about 40, the protein becomes abnormally folded.

It aggregates and causes disease of the central nervous system at some point during a normal human life span. The age of onset of the disease correlates with the number of repeat copies, that is, more repeats lead to earlier onset. Because the *Huntingtin* disease allele is autosomal dominant, it means that an individual needs to inherit only one copy of the expanded gene to develop the disorder.

KEY CONCEPT Trinucleotide-repeat diseases arise through expansion of the number of copies of a three–base-pair sequence normally present in several copies within coding or noncoding regions of a gene.

The cellular environment causes spontaneous mutations
An important source of point mutations is damage to DNA by water and reactive oxygen species that naturally reside in cells. Chemical reactions of DNA with water (i.e., hydrolysis) lead to depurination and deamination, and reactions with reactive molecules that contain oxygen lead to several types of DNA damage.

converts cytosine to uracil, adenine to hypoxanthine, and guanine to xanthine (Figure 15-9b). Uracil base pairs with adenine in replication, converting a C·G base pair into a T·A base pair, and hypoxanthine base pairs with cytosine, converting an A·T base pair into a G·C base pair. On the other hand, conversion of guanine to xanthine is less harmful because xanthine still base pairs with cytosine but with only two hydrogen bonds rather than three. Lastly, as discussed in Chapter 12, deamination of 5-methylcytosine (5mC) produces thymine, which converts a C·G base pair into a T·A base pair. Although T·G mispairs that result from deamination of 5mC can be repaired (discussed later in this chapter), CpG dinucleotides, the major sites of cytosine methylation in eukaryotic genomes, remain hotspots for mutations.

Oxidative damage represents a third type of spontaneous lesion that generates mutations. Reactive oxygen species (ROS) such as superoxide radicals ($\cdot O_2^-$), hydrogen peroxide (H_2O_2), and hydroxyl radicals ($\cdot OH$) are byproducts of normal aerobic metabolism of molecular oxygen by mitochondria. Over 100 different types of oxidative DNA modifications have been identified in mammalian genomes. For example, thymine is converted to thymine glycol, which cannot base pair with any nucleotide and blocks DNA replication, and guanine is converted to 8-oxoguanine (8-oxoG), which mispairs with A, resulting in G·C to T·A transversions (Figure 15-9c).

> **KEY CONCEPT** Spontaneous mutations are generated by chemical reactions of DNA with water and reactive oxygen species in cells.

15.3 MOLECULAR BASIS OF INDUCED MUTATIONS

LO 15.3 Tabulate the causes of induced DNA damage that result in mutations following DNA replication.

While some mutations are spontaneously produced by reactive molecules within cells, other mutations are induced by agents present in the external environment. These exogenous agents, called *mutagens*, can be present in air, food, or water. Mutagens can be chemical agents such as reactive oxygen species (ROS), alkylating agents, DNA adducts, base analogs, and intercalating agents, or they can be physical agents such as ultraviolet (UV) light and ionizing radiation (IR). Mutagens induce mutations by at least three different mechanisms. They can *replace* a base in DNA, *alter* a base so that it mispairs with another base, or *damage* a base so that it can no longer base pair with any base.

Mechanisms of induced mutagenesis

Base modification by alkylating agents Alkylation is the addition of an alkyl group (C_nH_{2n+1}, e.g., a methyl group (CH_3) or an ethyl group (C_2H_5)) to a nucleotide base.

Base alkylation can be mutagenic by preventing base pairing. Certain alkylating agents, including ethylmethanesulfonate (EMS) and methylnitronitrosoguanidine (MNNG), operate by adding alkyl groups to many positions on all four bases. As an example, EMS adds an ethyl group to the oxygen at position 6 of guanine to create O-6-ethylguanine (**Figure 15-10a**). This addition leads to base pairing with thymine and results in G·C → A·T transitions at the next round of DNA replication. Similarly, MNNG adds a methyl group to the oxygen at position 4 of thymine to produce O-4-methylthymine, which base pairs with guanine, resulting in T·A → C·G transitions. One of the primary experimental strategies used by geneticists to understand the relationship between genotype and phenotype is to use exogenous agents such as EMS to induce mutations in genes and observe the phenotypic consequences.

Base damage by bulky adducts Aflatoxin B_1 causes mutations by attaching to guanine at the N-7 position (Figure 15-10b). Formation of this addition product leads to breakage of the glycosidic bond between the base and the sugar, thereby liberating the base and generating an apurinic site. When covalently bound to DNA, aflatoxin B_1 is called a DNA adduct. Other compounds that form DNA adducts include the diol epoxides of benzo(a)pyrene, a compound produced by internal combustion engines. All compounds of this class induce mutations, although the mechanisms are not always clear.

Incorporation of base analogs Some chemical compounds are sufficiently similar to the normal nitrogenous bases of DNA that they are occasionally incorporated into DNA in place of normal bases; such compounds are called **base analogs**. To be mutagenic, a base analog must mispair more often than the normal base it replaces. The base analog exists in only a single strand, but it can cause a base-pair substitution that is replicated in all DNA copies descended from the original strand. A base analog widely used in research is 2-aminopurine (2-AP). This analog of adenine base pairs with thymine but also mispairs with cytosine when protonated, as shown in Figure 15-10c. Therefore, when 2-AP is incorporated into DNA by base pairing with thymine, it can generate A·T → G·C transitions by mispairing with cytosine in subsequent rounds of DNA replication. Alternatively, if 2-AP is incorporated by mispairing with cytosine, then G·C → A·T transitions will result when it base pairs with thymine. Genetic studies have shown that 2-AP almost exclusively causes transitions.

Binding of intercalating agents A group of compounds called **intercalating agents** are planar molecules that mimic base pairs and are able to slip themselves in (intercalate) between the stacked nitrogenous bases inside the DNA double helix (Figure 15-10d). Intercalating compounds such as proflavin and acridine orange differ from other mutagenic compounds in that they distort the DNA duplex "fooling" DNA polymerase into inserting extra bases or skipping

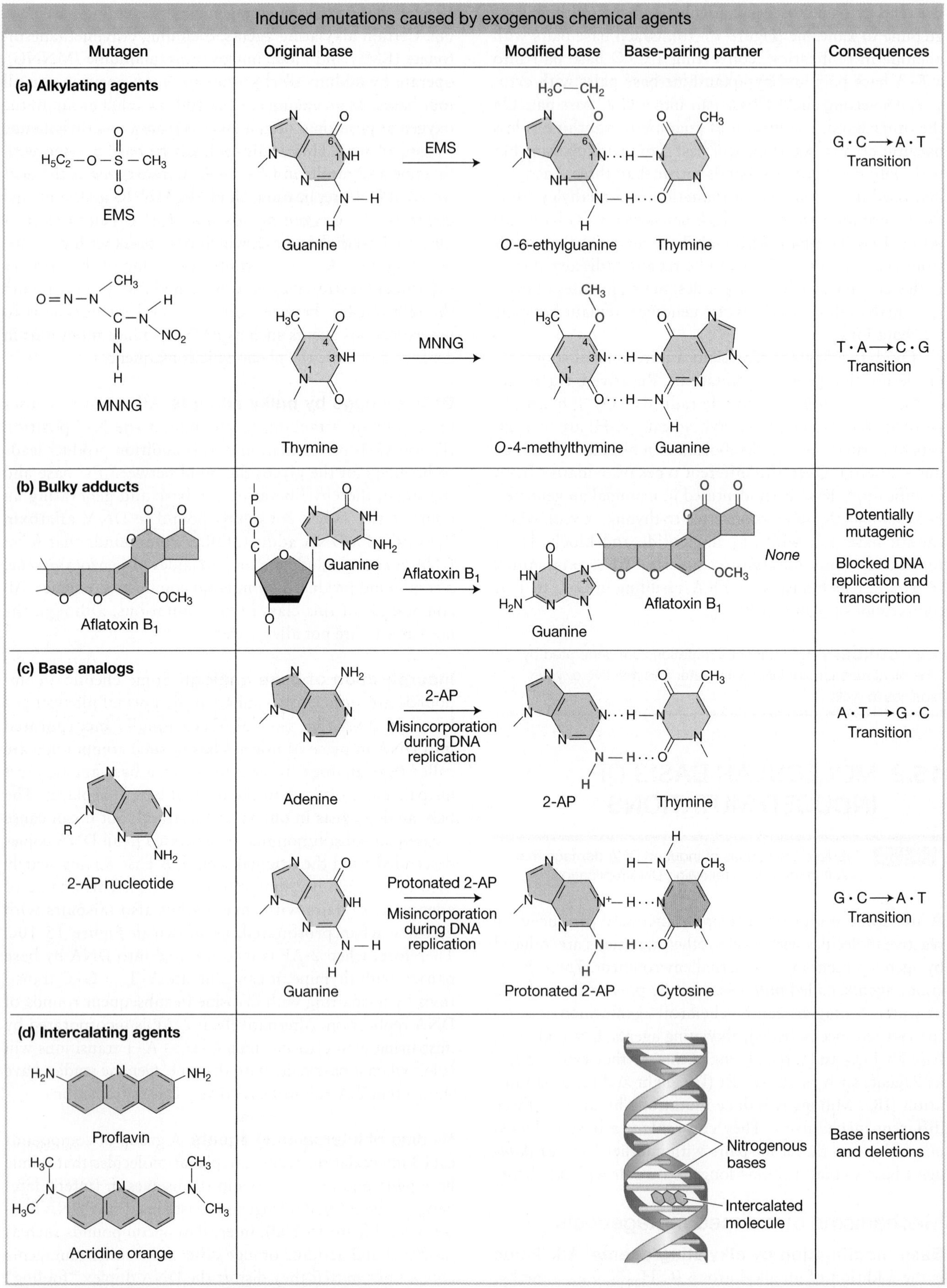

Induced mutations caused by exogenous chemical agents				
Mutagen	Original base	Modified base	Base-pairing partner	Consequences
(a) Alkylating agents	Guanine	O-6-ethylguanine	Thymine	$G \cdot C \longrightarrow A \cdot T$ Transition
EMS	Thymine	O-4-methylthymine	Guanine	$T \cdot A \longrightarrow C \cdot G$ Transition
MNNG				
(b) Bulky adducts	Guanine	Guanine / Aflatoxin B_1	None	Potentially mutagenic / Blocked DNA replication and transcription
Aflatoxin B_1				
(c) Base analogs	Adenine	2-AP	Thymine	$A \cdot T \longrightarrow G \cdot C$ Transition
2-AP nucleotide	Guanine	Protonated 2-AP	Cytosine	$G \cdot C \longrightarrow A \cdot T$ Transition
(d) Intercalating agents		Nitrogenous bases / Intercalated molecule		Base insertions and deletions
Proflavin / Acridine orange				

FIGURE 15-10 Examples of induced mutations caused by (a) alkylating agents, EMS (ethylmethanesulfonate) and MNNG (methylnitronitrosoguanidine); (b) bulky adducts, aflatoxin B_1; (c) base analogs, 2-AP (2-aminopurine); and (d) intercalating agents, proflavin and acridine orange. Atoms shown in red are altered by the chemical reaction.

templated bases, leading to base insertions and deletions, respectively, rather than base substitutions.

KEY CONCEPT Chemical agents cause damage to DNA by adding chemical groups to bases, mimicking bases, or altering the structure of DNA. In doing so, these agents increase the frequency of mutations due to DNA replication errors.

Base damage by ultraviolet light Ultraviolet light (UV) of wavelengths around 300 nanometers can form various types of **pyrimidine dimers** in DNA, which are covalently bonded adjacent pyrimidines on the same DNA strand. Two common types of pyrimidine dimers are cyclobutane pyrimidine dimers, which are characterized by a cyclobutane ring involving carbons 5 and 6 of adjacent pyrimidines, and 6-4 photoproducts, which contain linked carbons 6 and 4 of adjacent pyrimidines. Figure 15-11a illustrates these two types of UV-induced thymine dimers. During DNA replication, pyrimidine dimers stall DNA polymerases because the bases cannot specify a complementary partner by hydrogen bonding. Mechanisms that repair pyrimidine dimers frequently introduce mutations such as $T \cdot T \rightarrow T \cdot C$ transitions (see

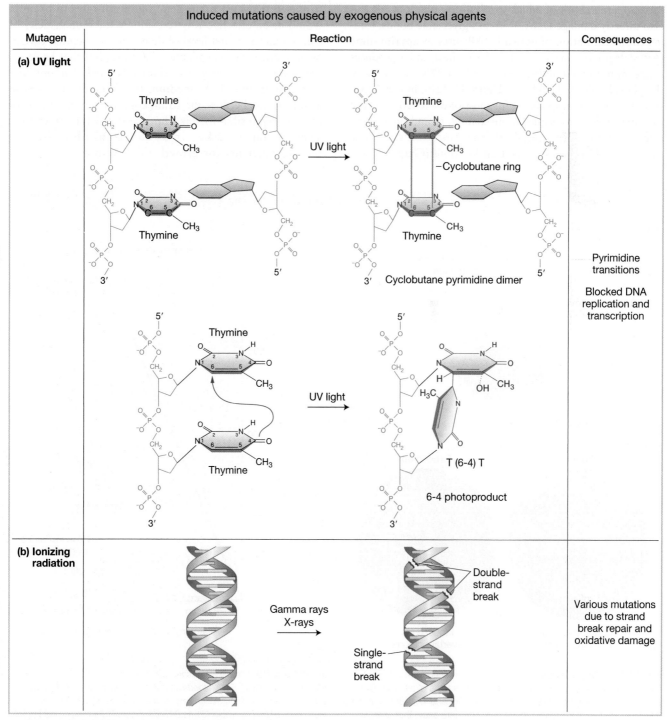

FIGURE 15-11 Examples of mutations caused by (a) UV light and (b) ionizing radiation. Atoms shown in red are altered by the chemical reaction.

the section on base excision repair). In addition, unrepaired pyrimidine dimers can cause transcription errors.

Base damage and modification and DNA strand breaks by ionizing radiation

Ionizing radiation (IR) from gamma rays and X rays causes mutations by generating reactive oxygen species (ROS) through radiolysis of water and disruption of mitochondrial functions. IR produces many different types of ROS, but the most damaging to DNA bases are superoxide radicals ($\cdot O_2^-$), hydrogen peroxide (H_2O_2), and hydroxyl radicals ($\cdot OH$). As illustrated in Figure 15-8c, ROS leads to the generation of thymine glycol and 8-oxoguanine, both of which can lead to mutations. IR can also directly damage DNA. It breaks glycosidic bonds, leading to the formation of **apurinic (AP) sites** or **apyrimidinic (AP) sites** (more generally called **abasic sites**), and it produces single-strand and double-strand breaks in DNA by severing the phosphodiester backbone (Figure 15-11b). In fact, DNA strand breaks are responsible for most of the lethal effects of ionizing radiation because if unrepaired, strand breaks can direct a cell to undergo cell death, and if incorrectly repaired, they can lead to chromosomal translocations (Chapter 17).

> **KEY CONCEPT** Physical agents such as UV light and ionizing radiation cause damage to DNA by inducing intra-strand crosslinks or strand breaks and in doing so increase the frequency of mutations due to DNA replication errors.

Identifying mutagens in the environment: the Ames test

In the 1970s, Bruce Ames recognized that there is a strong correlation between the ability of chemical compounds to cause cancer and their ability to cause mutations. He surmised that measurement of mutation rates in bacteria would be an effective way to evaluate the mutagenicity of compounds as a first level of detection of potential carcinogens (cancer-causing agents). The **Ames test** that he developed uses strains of the bacterium *Salmonella typhimurium* that can grow only in medium that includes the amino acid histidine because the strain contains mutant alleles of a gene responsible for histidine synthesis (**Figure 15-12a**). These mutants are known as **auxotrophs** because they require nutrients for growth that are not needed by a

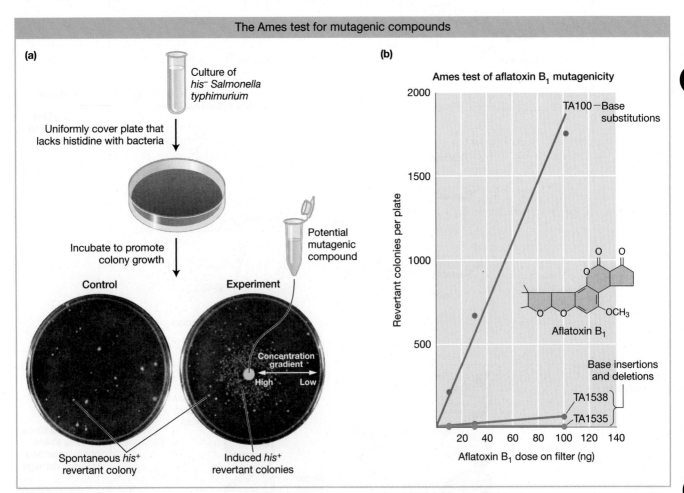

The Ames test for mutagenic compounds

FIGURE 15-12 (a) Summary of the procedure used for the Ames test for mutagenic compounds. (b) Results of Ames tests of aflatoxin B_1 with three strains of *S. typhimurium* (TA100, TA1538, and TA1535) containing different histidine auxotrophic mutations. The data show that aflatoxin B_1 is a potent mutation that causes base substitutions but not base insertions or deletions. [*Republished with permission of Elsevier, Ames BN, McCann J, Yamasaki E., "Methods for detecting carcinogens and mutagens with the* Salmonella/*mammalian-microsome mutagenicity test," Mutation Research, 1975, December; 31 (6):347–64, Figure 2. Permission conveyed through Copyright Clearance Center, Inc.*]

wild-type strain. Additionally, the *S. typhimurium* mutants were known to "revert"—that is, grow in histidine-free medium—only by certain kinds of additional mutations. For example, an allele called TA100 could be reverted to wild type only by a base substitution mutation, whereas TA1538 and TA1535 could be reverted only by base insertion or base deletion mutations that result in a protein frameshift. Thus, as illustrated by a study of the mutagen aflatoxin B_1, these strains can be used in the Ames test not only to identify mutagens but also to determine the types of mutations they induce (Figure 15-12b).

The Ames test works as follows. A potential mutagenic compound such as aflatoxin B_1 is absorbed onto a filter disc, which is placed in the center of a plate that has *S. typhimurium* uniformly spread over medium that lacks histidine (Figure 15-12a). The compound diffuses from the disc into the surrounding medium, creating a gradient of concentrations, with the highest concentration closest to the disc. The absence of histidine ensures that only revertant bacteria containing the appropriate base substitution or frameshift mutation will grow and form colonies. After incubation, the number of colonies on each plate and the total number of bacteria tested are determined, allowing a determination of the frequency of reversion (Figure 15-12b).

While this works well, it became clear that not all compounds were directly mutagenic; rather, the actual mutagenic agent is sometimes a metabolite of the compound that is produced in the body. Typically, these metabolites are produced in the liver, and the enzymatic reactions that convert the compound into the bioactive metabolites does not take place in bacteria. Ames realized that he could overcome this problem by pre-treating compounds with extracts of rat livers containing the metabolic enzymes. Treated compounds that yielded elevated levels of reversion relative to the compound or liver extract alone would be defined as mutagenic and possibly carcinogenic. Therefore, the Ames test provides an important way of screening thousands of compounds and evaluating one aspect of their risk to health and the environment. The Ames test is still in use today as an important tool for evaluating the safety of chemical compounds.

KEY CONCEPT The Ames test is used to determine the mutagenic activity of chemicals by testing whether they increase the frequency of mutations in bacteria.

15.4 DNA REPAIR MECHANISMS

LO 15.4 Illustrate the molecular mechanisms that repair distinct types of DNA damage.

LO 15.5 Using examples, describe how different types of mutations in different genes lead to particular human genetic diseases.

After surveying the numerous ways that DNA can be damaged—from sources both inside the cell (DNA replication, water, and reactive oxygen species) and outside (chemical and physical agents)—you might be wondering how life has managed to survive and thrive for billions of years. The fact is that organisms ranging from bacteria to plants to humans can efficiently repair their DNA. All organisms make use of a variety of repair mechanisms. The major DNA repair pathways are base excision repair (BER), nucleotide excision repair (NER), mismatch repair (MMR), translesion synthesis (TLS), homologous recombination (HR), and nonhomologous end joining (NHEJ) (**Figure 15-13**). Each of these pathways involves numerous

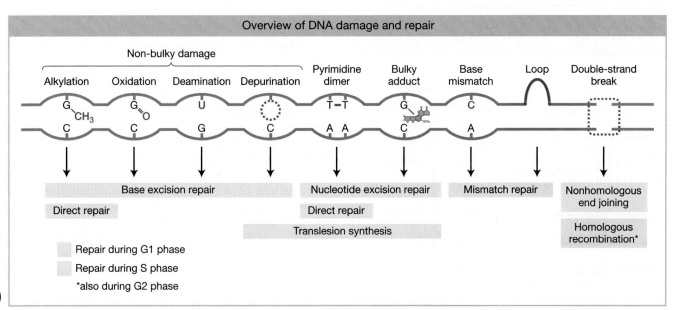

FIGURE 15-13 DNA repair mechanisms are paired with the types of DNA damage they act on. DNA damage is indicated in red. DNA repair mechanisms highlighted in tan function during G1 phase of the cell cycle, and those highlighted in blue function during S phase of the cell cycle. Homologous recombination also functions in G2 phase of the cell cycle.

proteins. In addition, pyrimidine dimers and O-alkylated bases can also be directly repaired by single enzymes. As you will see, failure of any of these mechanisms is a significant cause of inherited human diseases.

Direct repair of damaged DNA

The most straightforward way to repair a base lesion is to directly reverse it, thereby regenerating the normal base. Although most types of DNA damage are essentially irreversible and require removal and replacement of the damaged area in order to be repaired, a few can be directly reversed. One case is a mutagenic pyrimidine dimer caused by UV light (Figure 15-11a). Cyclobutane pyrimidine dimers (CPDs) are repaired by an enzyme called CPD photolyase (**Figure 15-14a**). The enzyme binds pyrimidine dimers and breaks the covalent bonds to regenerate the original bases. This repair mechanism is called photoreactivation because the enzyme requires light as an energy source for repair. Note that placental mammals, including humans, lack a functional CPD photolyase gene and rely on other mechanisms such as nucleotide excision repair to fix CPDs.

Base alkylation (Figure 15-10a) is reversed by enzymes known as alkyltransferases. For example, a type of alkyltransferase, O^6-methylguanine DNA methyltransferase (MGMT), repairs O^6-methylguanine back to guanine by transferring the methyl group from O^6 (oxygen 6) to a cysteine residue in the enzyme's active site (Figure 15-14b). As a consequence, MGMT is irreversibly inactivated, and thus is known as a "suicide enzyme."

> **KEY CONCEPT** DNA damage caused by pyrimidine dimers or base alkylation can be directly repaired by enzymes that recognize the damage and reverse the chemical process that created the damage.

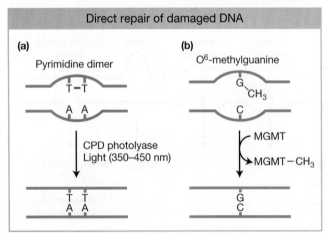

Direct repair of damaged DNA

(a) Pyrimidine dimer

(b) O^6-methylguanine

FIGURE 15-14 (a) The enzyme CPD photolyase splits a UV light-induced cyclobutane pyrimidine dimer to repair this damage.
(b) The enzyme O^6-methylguanine DNA methyltransferase (MGMT) reverts O^6-methylguanine to guanine.

ANIMATED ART 📶 SaplingPlus

UV-induced photodimers and excision repair

Base excision repair

An overarching principle that guides genetic systems is the power of nucleotide sequence complementarity. Repair systems exploit the properties of antiparallel complementarity to restore damaged DNA segments to their initial, undamaged state. In these systems, damaged nucleotides on one strand of DNA are removed and replaced with a newly synthesized nucleotide segment that is complementary to the undamaged strand. Unlike the examples of direct repair of damage described in the preceding section, these pathways involve the removal and replacement of one or more bases.

The first repair system of this type is **base excision repair (BER)**. The main target of BER is non-bulky damage to bases produced by alkylation, oxidation, and deamination (Figure 15-13). In bacteria, the damaged base is detected by a DNA glycosylase, which cleaves the glycosidic bond between the base and sugar, thereby creating an apurinic or an apyrimidinic (AP) site (**Figure 15-15a**). An enzyme called AP endonuclease nicks the damaged strand upstream of the AP site, and DNA phosphodiesterase removes the AP sugar-phosphate. Then DNA polymerase fills the gap with a complementary nucleotide, and DNA ligase seals the remaining nick in the backbone.

In eukaryotes, the first two steps of BER are the same as in bacteria; however, the third step is carried out by DNA polymerase β (Pol β) rather than DNA phosphodiesterase (Figure 15-15b). Pol β can insert a single nucleotide before excising the AP residue; this is called short patch BER. Alternatively, in long patch BER, Pol β typically inserts 2–10 nucleotides, which prevents excision of the AP residue by Pol β. Instead, Pol β generates a "flap," a single-stranded region of DNA, that is removed by Flap endonuclease. DNA ligase completes the repair in both branches of the pathway.

Cells contain several DNA glycosylases, each of which recognizes one or several types of damaged bases. For example, uracil-DNA glycosylase removes uracil from DNA. Uracil residues, which result from spontaneous deamination of cytosine (Figure 15-9b), can lead to a C-to-T transition after DNA replication. One advantage of having thymine (5-methyluracil) rather than uracil as the natural pairing partner of adenine in DNA is that spontaneous cytosine deamination events can be recognized as abnormal and then excised and repaired. If uracil were a normal constituent of DNA, such repair would not occur.

However, deamination does pose problems for both bacteria and eukaryotes. By analyzing a large number of mutations in the *lacI* gene, Jeffrey Miller identified places in the gene where one or more bases were prone to frequent mutation. Miller found that these so-called mutational hotspots corresponded to deaminations at cytosine residues. DNA sequence analysis of $C \cdot G \rightarrow T \cdot A$ transition hotspots in the *lacI* gene showed that 5-methylcytosine residues were present at each hotspot. As described in Chapter 12, in some eukaryotic organisms, cytosine methylation regulates transcription (Figure 12-17). Similarly, in *E. coli* and other bacteria, DNA is methylated at cytosines but for different

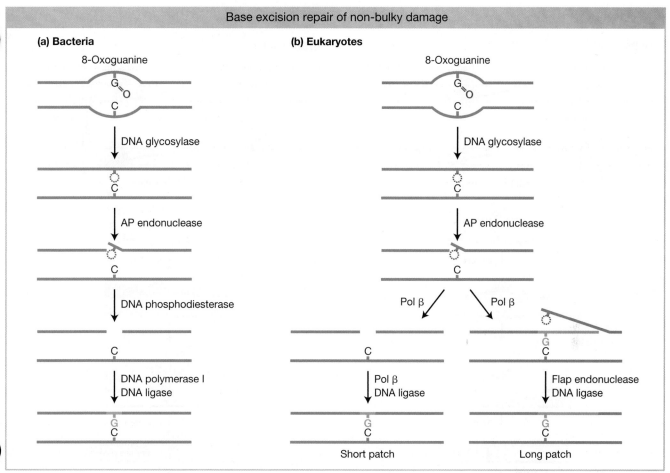

FIGURE 15-15 In base excision repair in both bacteria (a) and eukaryotes (b), damaged bases are removed and repaired in a stepwise manner through the sequential action of enzymes. DNA damage is indicted in red, abasic sites are indicated by dotted circles, and replacement DNA is indicated in green.

ANIMATED ART   SaplingPlus

DNA repair mechanisms: BER, NER, and mismatch repair

purposes. 5-Methylcytosines are hotspots for mutations because deamination of 5-methylcytosine generates thymine (5-methyluracil) (Figure 15-9b), which is not recognized by uracil-DNA glycosylase and thus is not repaired. A consequence of the frequent mutation of 5-methylcytosine to thymine is that methylated regions of the genome are converted, over evolutionary time, to AT-rich regions.

KEY CONCEPT In base excision repair, non-bulky damage to DNA is detected by one of several DNA glycosylases that cleave the base–sugar bond, releasing the damaged base. Repair consists of removal of the abasic site and insertion of the correct nucleotide as guided by the complementary nucleotide in the undamaged strand.

Nucleotide excision repair

Although the vast majority of DNA damage sustained by an organism is minor base damage that can be handled by BER, this mechanism cannot repair bulky adducts that distort the DNA helix (Figure 15-10b), pyrimidine dimers caused by UV

light (Figure 15-11a), or damage to more than one base. If left unrepaired, these lesions can severely affect cellular physiology because they obstruct the progress of DNA and RNA polymerases, resulting in DNA replication and transcription blocks, respectively. **Nucleotide excision repair (NER)** pathways relieve replication and transcription blocks by repairing the damaged DNA. In bacteria and eukaryotes, NER pathways are comprised of a common set of steps: damage detection, strand separation, incision (i.e., cleavage), excision (i.e., removal), polymerization, and ligation.

In *E. coli*, bulky adducts and pyrimidine dimers are either detected by UvrA acting along with UvrB or by RNA polymerase, which stalls during transcription at the damaged site and recruits UvrA-UvrB. The pathways are respectively called **global genome nucleotide excision repair (GG-NER)** and **transcription-coupled nucleotide excision repair (TC-NER)** (Figure 15-16a). In both cases, the DNA damage site is handed from UvrA to UvrB, which is a helicase that separates the DNA strands and promotes release of UvrA. UvrB then serves as a scaffold for recruitment of UvrC, which uses separate endonuclease domains to cleave phosphodiester bonds 8 nucleotides 5′ and 4–5 nucleotides 3′

Nucleotide excision repair of pyrimidine dimers and bulky adducts

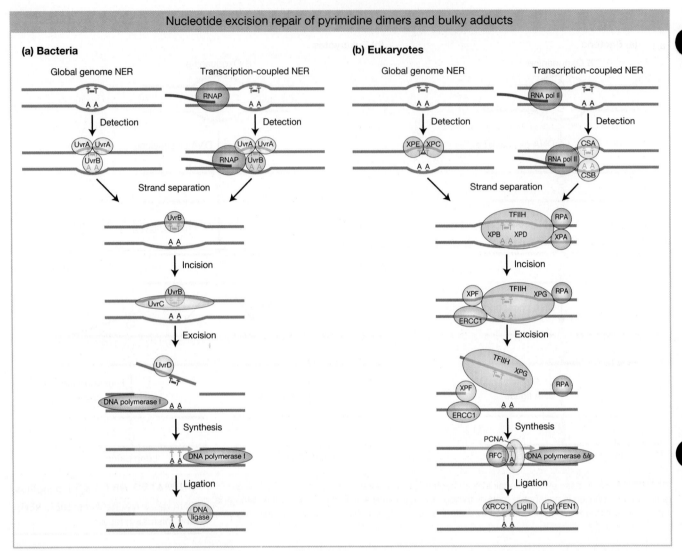

FIGURE 15-16 NER pathways in bacteria (a) and eukaryotes (b) are comprised of the same basic steps but involve different protein factors. In both bacteria and eukaryotes, pyrimidine dimers are detected by global genome and transcription-coupled mechanisms that feed into a common repair pathway consisting of strand separation, incision, excision, polymerization, and ligation. DNA damage is indicated in red, and repaired DNA is indicated in green.

ANIMATED ART ⬡ Sapling Plus
UV-induced photodimers and excision repair

ANIMATED ART ⬡ Sapling Plus
DNA repair mechanisms: BER, NER, and mismatch repair

of the DNA damage site. Next, the DNA helicase UvrD and DNA polymerase I (Pol I) work together to excise the 12- to 13-nucleotide region between the two incision sites and fill in the gap with complementary nucleotides. Lastly, DNA ligase seals the nick left by Pol I.

In eukaryotes, NER pathways follow the same steps but involve a distinct set of enzymes and excise a larger, about 27-nucleotide, region spanning the site of DNA damage. Much of what is known about NER in humans has come from studies of human diseases, including Cockayne syndrome (CS), xeroderma pigmentosum (XP), and trichothiodystrophy (TTD) that are caused by autosomal recessive mutations in genes that encode protein components of the NER machinery. Patients with CS have mutations in genes encoding the CSA and CSB proteins, which recognize stalled transcription complexes in TC-NER. Patients with XP carry mutations in genes

encoding components of the GG-NER pathway (e.g., XPC and XPE) and shared components of the GG-NER and TC-NER pathways (e.g., XPB, XPD, XPF, and XPG). Lastly, patients with TTD also carry mutations in genes encoding components of the shared pathway (e.g., TTD-A, XPB, and XPA). Because they have reduced capacity to repair DNA damage caused by exposure to sunlight (i.e., UV light), XP patients are thousands of times more likely to develop certain types of skin cancer. In contrast, CS is characterized by developmental and neurological symptoms, and TTD patients have brittle hair and ichthiosis (a dry skin disorder). Furthermore, some individuals present with symptoms of XP and CS or XP and TTD.

Similar to the process in bacteria, in eukaryotes, GG-NER is initiated when XPC and XPE complexes detect damaged DNA, whereas TC-NER is initiated when RNA polymerase II stalls at the site of DNA damage and recruits CSA and CSB

(Figure 15-16b). After lesion detection, the GG-NER and TC-NER pathways mostly use the same proteins to remove and repair the damaged DNA because both XPC-XPE and CSA-CSB recruit the multiprotein TFIIH complex, which also functions in RNA polymerase II transcription (Chapter 8). Two of the 11 TFIIH subunits, XPB and XPD, are helicases (3′-to-5′ and 5′-to-3′, respectively) that separate the strands of DNA around the site of DNA damage. XPA along with RPA (replication protein A) further expand the DNA bubble around the damage. Then the XPF-ERCC1 endonuclease cleaves a phosphodiester bond 5′ of the damage and the TFIIH endonuclease XPG cleaves 3′ of the damage, excising an approximately 27-nucleotide region. The gap is filled by DNA replication factors (Table 7-2): RPC (replication factor C), PCNA (proliferating cell nuclear antigen), and DNA polymerases δ and ε. Lastly, the nick is either sealed by DNA ligase III-XRCC1 (X-ray repair cross-complementing protein 1) or by DNA ligase I-FEN1 (Flap endonuclease 1).

> **KEY CONCEPT** Nucleotide excision repair corrects DNA damage due to UV light and bulky adducts by global genome and transcription-coupled mechanisms that each involve six steps (detection, strand separation, incision, excision, synthesis, and ligation) and, in doing so, relieves stalled DNA replication and transcription.

> **KEY CONCEPT** Individuals that carry mutations in genes that encode components of NER pathways are less able to repair DNA damage caused by sunlight, resulting in mutations that lead to certain types of skin cancer.

Mismatch repair

In *E. coli*, DNA polymerase III, the main replication enzyme for both the leading and lagging strands, has proofreading activity, which reduces the error rate to about 10^{-7} (i.e., one error in about 10,000,000 base pairs). The major pathway that corrects the remaining replication errors is called **mismatch repair (MMR)**. This repair pathway reduces the error rate to less than 10^{-9} by recognizing and repairing mismatched bases and small loops caused by insertions and deletions of nucleotides (indels) that occur in the course of DNA replication (Figure 15-13). Thus, loss of the MMR pathway increases the mutation frequency up to 100-fold and, like defects in other repair pathways, is associated with specific types of cancer.

MMR systems must detect mismatched base pairs, identify the new DNA strand that contains the mismatched base, excise the mismatched base, and carry out repair synthesis (**Figure 15-17**). Much of what is known about the

FIGURE 15-17 A model for mismatch repair in *E. coli*. During DNA replication, mismatch errors are incorporated into the newly synthesized strand. To initiate repair, MutS detects a base mismatch and MutH determines which of the bases is incorrect by binding a nearby GATC sequence and distinguishing the parental from the newly synthesized strand by the presence of a methylated adenosine. The base mismatch is then repaired through a series of enzymatic steps that removes a region of DNA containing the incorrect nucleotide and synthesizing the correct sequence using the parental strand as a template.

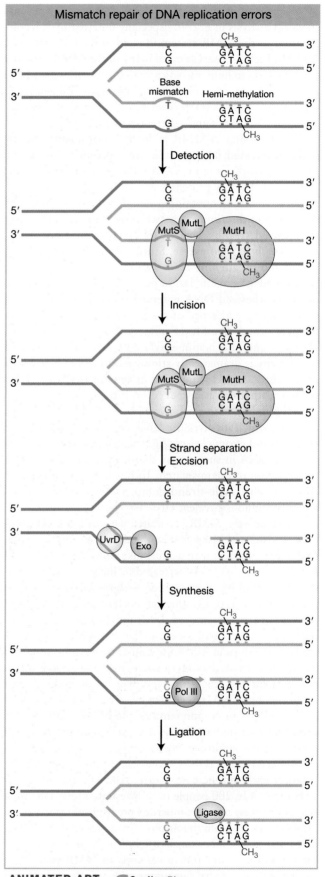

Mismatch repair of DNA replication errors

ANIMATED ART Sapling Plus

DNA repair mechanisms: BER, NER, and mismatch repair

MMR mechanism comes from decades of genetic and biochemical analysis in *E. coli*. Especially noteworthy is reconstitution of the MMR system in vitro in the laboratory of Paul Modrich. Conservation of many of the MMR proteins from bacteria to humans indicates that this pathway is both ancient and important in all living organisms. We will focus on the very well-characterized *E. coli* system and briefly highlight similarities and differences in eukaryotes.

The first step in MMR is detection of mismatches in newly replicated DNA by the MutS protein. Binding of MutS to distortions in the DNA double helix recruits MutL and MutH. The key protein is MutH, which cuts the newly synthesized strand containing the incorrect base. Without the ability to discriminate between the parental and newly synthesized strands, the MMR system could not determine which base to excise. Strand recognition by MutH is directed by adenine methylation at GATC sequences. Because adenine methylation occurs after DNA synthesis, newly synthesized DNA is temporarily unmodified, and this temporary absence of methylation directs repair to the new strand. The MutH endonuclease cuts the unmethylated strand at a hemimethylated GATC sequence, and this activity is dramatically stimulated by interactions with MutS-MutL. MutH is targeted by hemimethylated GATC sequences that can be either 5′ or 3′ to the mismatch and several hundred base pairs away. MutS-MutL also activates excision, which involves the DNA helicase UvrD and four single-strand exonucleases. After incision, UvrD is loaded onto the DNA in such a way that it unwinds the DNA in the direction of the mismatch. To finish, DNA polymerase III repairs the single-stranded gap, and DNA ligase generates a continuous covalent DNA backbone.

In eukaryotes, MMR is initiated when a MutS-like complex of Msh2 and Msh6 proteins binds to a mismatch. Msh2-Msh6 then interacts with a MutL-like complex of Mlh1 and Pms1 proteins, which is activated to incise the newly synthesized strand by an interaction with the DNA replication β-clamp PCNA (proliferating cell nuclear antigen). After the replication error is removed, DNA polymerases δ and ε synthesize the correct sequence, and ligation follows. So, while many of the steps in MMR are similar between bacteria and eukaryotes, the mechanism of strand discrimination is different. In eukaryotes, the strand-specificity of the endonuclease is directed by PCNA, rather than by DNA methylation.

Mutations in components of the MMR pathway are responsible for human diseases, especially cancers. A case in point is Lynch Syndrome, often called hereditary nonpolyposis colorectal cancer (HNPCC), which, despite its name, is not a cancer itself but increases cancer risk. The disease affects as many as 1 in 200 people in the Western world, making it one of the most common inherited predispositions to cancer. HNPCC results from mutations in the *Msh2*, *Msh6*, *Mlh1*, and *Pms1* genes. Inheritance of HNPCC is autosomal dominant. Cells with one functional copy of MMR genes have normal MMR activity, but tumor cells arise from cells that have lost the one functional copy and are thus deficient for MMR. As the mutant cells replicate their DNA and divide,

they accumulate DNA damage that can lead to uncontrolled cell growth and cancer. NHPCC is frequent because an important target of the human MMR system is short repeat sequences that can be expanded or deleted during DNA replication by the slipped-mispairing mechanism described previously (Figure 15-7). There are thousands of short repeats located throughout the human genome (Chapter 4). Although most are located in noncoding regions because most of the genome is noncoding, a few are located in genes that are critical for normal growth and development.

KEY CONCEPT The mismatch repair system corrects errors in DNA replication that are not corrected by the proofreading function of replicative DNA polymerases. Repair is restricted to the newly synthesized strand, which is identified in bacteria by the lack of DNA methylation and in eukaryotes by a DNA replication factor.

KEY CONCEPT Individuals that carry mutations in genes that encode components of MMR pathway are at increased risk of cancer because dividing cells accumulate DNA damage.

Translesion synthesis

Despite the fact that mechanisms exist to repair lesions (i.e., DNA damage) that stall DNA replication forks, some lesions persist and, by blocking DNA replication, they can cause severe consequences to cells, including death. To avoid these consequences, both bacterial and eukaryotic cells use a variety of DNA polymerases to replicate past lesions and permit the completion of genome duplication. This process, called **translesion synthesis (TLS)**, provides additional time for other mechanisms to repair the lesion before the replicative DNA polymerase returns to finish synthesizing the genome.

The TLS mechanism is conserved from *E. coli* to humans (**Figure 15-18**). It is initiated by stalled DNA polymerase, which triggers recruitment of a **translesion (TLS) polymerase** that synthesizes past the lesion. In some cases, the TLS polymerase extends beyond the lesion and in other cases extension is carried out by another DNA polymerase. Once extension passes the lesion, the TLS polymerase is replaced by the replicative DNA polymerase.

More specifically, replicative DNA polymerases, *E. coli* Pol III and eukaryotic Pol ε, stall at sites of depurination (Figure 15-9a), bulky adducts (Figure 15-10b), and pyrimidine dimers (Figure 15-11a). TLS polymerases are then recruited by an interaction with the replicative β-clamp in *E. coli* and PCNA (the β-clamp equivalent) in eukaryotes. *E. coli* has three TLS polymerases (Pol II, Pol IV, and Pol V) and humans have at least five (Pol η, Pol ι, Pol κ, Rev1, and Pol ζ). TLS polymerases differ from replicative polymerases in three important respects. First, whereas replicative polymerases stall because damaged bases do not fit into their active site, TLS polymerases have much larger active sites that can accommodate damaged bases. Second, depending on the type of lesion, TLS polymerases can be error-prone,

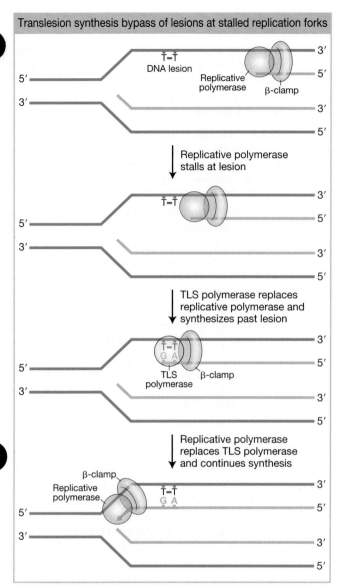

Translesion synthesis bypass of lesions at stalled replication forks

FIGURE 15-18 A generalized model for translesion synthesis in *E. coli* and humans. In the course of DNA replication, replicative polymerase stalls at lesions and is temporarily replaced by a TLS polymerase that can continue replicating past the lesion. TLS polymerases are error prone, as shown by the G·T mismatch. Newly synthesized DNA is indicated in green. Specific *E. coli* and human factors are described in the text.

in part because they lack 3′-to-5′ proofreading activity. Third, TLS polymerases have low processivity; that is, they can add only a few nucleotides before falling off the DNA template. Thus, TLS polymerases can unblock the replication fork but cannot synthesize long stretches of DNA.

In *E. coli*, TLS is activated by the **SOS response**. Analogous to its common use, Save Our Ship, the name SOS comes from the idea that this system is induced as an emergency response to prevent cell death in the presence of significant DNA damage. The SOS pathway regulates the production of DNA repair and damage tolerance proteins, including TLS polymerases. As such, SOS induction is a mechanism of last resort, a form of damage tolerance

that allows the cell to trade death for a certain level of mutagenesis.

> **KEY CONCEPT** In translesion synthesis, TLS polymerases are recruited to replication forks that have stalled because of damage in the template strand. TLS polymerases may introduce errors in the course of synthesis that either persist as mutations or are corrected by other mechanisms such as mismatch repair.

Repair of double-strand breaks

Many of the DNA damage repair systems covered thus far exploit DNA complementarity to make error-free repairs. Such error-free repair is characterized by two events: (1) removal of the damaged bases, perhaps along with nearby DNA, from one strand of the double helix; and (2) use of the other strand as a template for DNA synthesis to fill the single-strand gap. However, DNA complementarity cannot be exploited to repair some types of DNA damage. For example, exposure to X rays often causes both strands of the double helix to break at sites that are close together. This type of damage is called a **double-strand break (DSB)**. If left unrepaired, DSBs can cause a variety of chromosomal aberrations resulting in cancer or cell death. Interestingly, the generation of DSBs is an integral feature of normal cellular processes that require DNA rearrangements. One example is meiotic recombination (Chapter 3). Cells use many of the same proteins and pathways to repair DSBs as they do to carry out meiotic recombination.

DSBs can arise spontaneously (for example, in response to reactive oxygen species produced as a by-product of cellular metabolism) or they can be induced (for example, by ionizing radiation). There are two primary pathways used to repair DSBs in higher eukaryotes such as mammals: **nonhomologous end joining (NHEJ)** and **homologous recombination (HR)**. NHEJ joins DNA ends independent of sequence complementarity, whereas HR uses complementary sequence on a homologous chromosome as a template to extend DNA ends past a break point. Because of mechanistic differences between NHEJ and HR, NHEJ is more prone to incorporating errors at the break-point junction than HR. Furthermore, NHEJ can function in both dividing and non-dividing cells, whereas HR occurs primarily in S (DNA replication) and G2 phases of the cell cycle and is thus restricted to cells that are actively dividing. As described in Chapter 10, NHEJ and HR pathways are used to repair DSBs that are introduced by CRISPR-Cas9 for the purpose of genome engineering. Repair of the DNA break by NHEJ often results in inactivation of the target gene, whereas repair by HR introduces a precise change in the sequence of the target gene.

Nonhomologous end joining One way that higher eukaryotes repair DSBs is by NHEJ, which processes and then rejoins DSB ends, frequently generating small insertions and deletions at the break site (**Figure 15-19**). Like in

Nonhomologous end joining repair of double-strand breaks

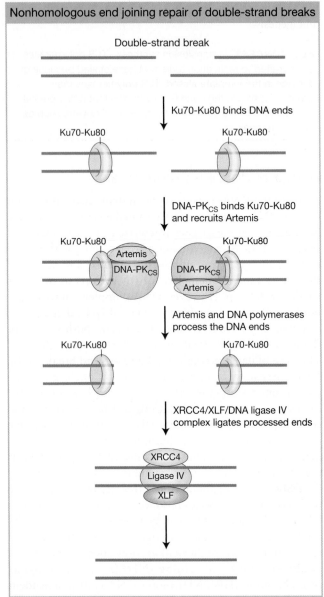

Double-strand break

Ku70-Ku80 binds DNA ends

Ku70-Ku80   Ku70-Ku80

DNA-PK$_{CS}$ binds Ku70-Ku80
and recruits Artemis

Ku70-Ku80   Ku70-Ku80
Artemis   DNA-PK$_{CS}$
DNA-PK$_{CS}$ Artemis

Artemis and DNA polymerases
process the DNA ends

Ku70-Ku80   Ku70-Ku80

XRCC4/XLF/DNA ligase IV
complex ligates processed ends

XRCC4
Ligase IV
XLF

FIGURE 15-19 A model for nonhomologous end joining in higher eukaryotes. As in other repair pathways, nonhomologous end joining involves proteins that detect DNA damage, remove and synthesize DNA at the site of DNA damage, and ligate DNA strands. Details are provided in the text. DNA polymerases that fill in small gaps are not shown.

ANIMATED ART   Sapling Plus

Double-strand break repair

other repair mechanisms, the first step in NHEJ is detection of the damage. The NHEJ pathway is initiated when Ku70 and Ku80 proteins form a heterodimer that binds to each broken end. Binding of Ku70-Ku80 prevents further damage to the ends and recruits DNA-PK$_{CS}$ (DNA-protein kinase, catalytic subunit) and the nuclease Artemis. DNA-PK$_{CS}$ then phosphorylates Artemis, activating its endonuclease and 5'-exonuclease activities that remove 5'- and 3'-single-stranded DNA overhangs and hairpins. Small gaps in the DNA are filled in by DNA polymerase μ or DNA polymerase λ, which leave 5'-phosphate and 3'-hydroxyl ends that are necessary for ligation by a complex of XRCC4, XLF (XRCC4-like factor), and DNA ligase IV.

As might be expected, cells deficient for NHEJ proteins are more sensitive to ionizing radiation.

It is important to note that NHEJ does not join telomeres, the double-stranded ends of eukaryotic chromosomes. This is because multiple mechanisms cooperate to inhibit NHEJ at telomeres, including many proteins that specifically bind telomere sequences (Figure 7-25a).

KEY CONCEPT Nonhomologous end joining is an error-prone pathway that repairs double-strand breaks in higher eukaryotes by ligating the free DNA ends back together, independently of extensive sequence homology.

Homologous recombination The repair of DSBs by homologous recombination requires an undamaged homologous double-stranded DNA template. During DNA replication the template can be the sister chromatid, and in diploid cells the template can be the second chromosomal copy. The two main pathways used to repair DSBs by homologous recombination are the **double-strand break repair (DSBR)** pathway and the **synthesis-dependent strand annealing (SDSA)** pathway. The initial steps of these pathways are shared (**Figure 15-20a**). Following DSB formation, the broken DNA ends are processed by exonucleases to generate 3' single-strand DNA overhangs that participate in strand exchange with homologous double-stranded DNA. Through the action of a class of enzymes called recombinases, the invading 3' overhang displaces one strand of the homologous DNA and base pairs to the other. This creates a structure called a displacement loop (D-loop). The invading strand is then extended by DNA synthesis using the homologous strand as a template. At this point, the DSBR and SDSA pathways diverge. For DSBR, the other 3' overhang invades, creating a four-branched, double cross-over intermediate called a double **Holliday junction (HJ)**, which is named for Robin Holliday who first proposed the recombination model (Figure 15-20b). Next, in a process called HJ resolution, endonucleases called resolvases cleave the Holliday junctions to yield either noncrossover or crossover DNA segments. Lastly, gaps are filled by DNA polymerases, and DNA ligase seals the remaining nicks. For SDSA, DNA helicases displace the extended invading strand, followed by annealing of the original broken chromosome pieces, DNA synthesis, and ligation (Figure 15-20c).

Repair of DSBs is also critical during meiosis, where it contributes to the formation of chiasmata that are required for chromosome pairing, exchange, and segregation (Chapter 4). In meiosis, homologous recombination is initiated by the introduction of DSBs at multiple chromosomal sites followed by exonuclease-mediated generation of 3' single-strand DNA overhangs and strand invasion of the 3' overhangs, as in Figure 15-20a. Crossovers are formed by resolution of double HJs through the DSBR mechanism (Figure 15-20b), and noncrossovers are primarily produced by the SDSA mechanism (Figure 15-20c).

KEY CONCEPT Homologous recombination accurately repairs double-strand breaks in DNA using homologous chromosomes. The repair mechanisms are also critical for chromosome pairing, exchange, and segregation during meiosis.

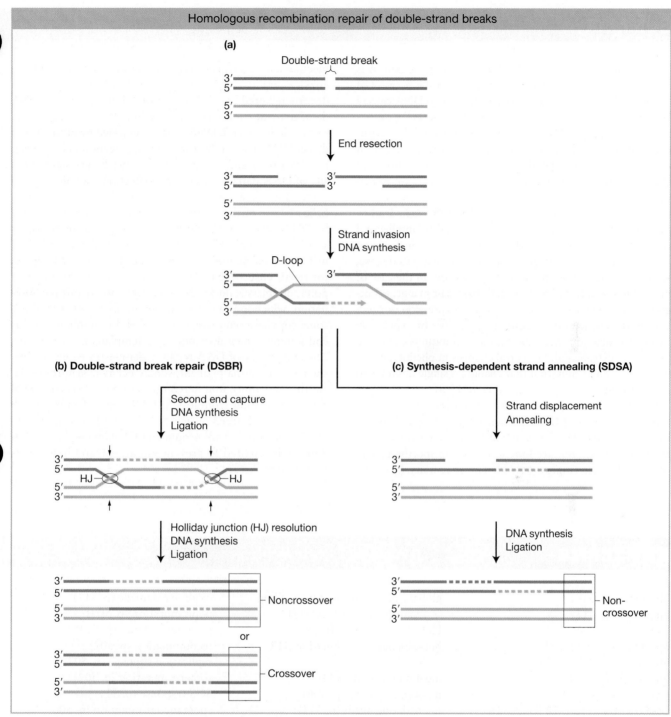

FIGURE 15-20 A model for homologous recombination repair of DSBs. (a) Both the DSBR and SDSA pathways are initiated in the same way by resection of DSB ends and strand invasion into a homologous sequence. Blue strands indicate the broken chromosome and orange strands indicate the homologous chromosome. (b) In the DSBR pathway, the double Holliday junction structure is either resolved in a crossover or noncrossover manner (small arrows indicate potential sites of crossover). The black boxes highlight the chromosomal differences that result from the two mechanisms. (c) In the SDSA pathway, strand displacement followed by annealing leads exclusively to noncrossover homologous recombination repair.

ANIMATED ART 📀 SaplingPlus

Double-strand break repair

SUMMARY

The genome sequence in individual cells of an organism is subject to change during the organism's lifetime because of errors in DNA replication as well as damage to DNA caused by intrinsic (i.e., cellular) factors and extrinsic (i.e., environmental) factors. Mutations that result from DNA replication errors and DNA damage can have a wide variety of phenotypic effects, depending on the type of mutation and the location of the mutation in the genome. As an example, DNA replication errors that cause trinucleotide-repeat expansion in a noncoding region of the *FMR1* gene lead to Fragile X syndrome, which is characterized by intellectual disability, while a trinucleotide-repeat expansion in a coding region of the *Huntingtin* gene leads to Huntington's disease, which is characterized by neurodegeneration. In the case of point mutations that change a single base pair in a gene, outcomes can range from no effect on gene expression or function of the encoded protein to complete loss of expression or protein function. Accordingly, point mutations can cause beneficial, detrimental, or neutral changes in phenotype. For example, a point mutation that changes a single amino acid in the Ras protein causes cancer (Figure 15-3).

Both bacterial and eukaryotic cells reduce the potential mutagenic effects of DNA damage by detecting and repairing the damage. DNA repair mechanisms act on specific types of damage and are functional during specific phases of the cell cycle (Figure 15-13). The specificity of repair mechanisms is largely driven by unique activities of the enzymes involved. A few types of DNA damage can be directly repaired by single enzymes, but most types of DNA damage require multiple enzymes and DNA-binding factors for repair. In general, DNA repair is initiated by proteins that detect DNA damage. In some cases, detection occurs when the DNA replication or transcription machinery stalls at the site of DNA damage. Following detection, damaged nucleotides and often surrounding nucleotides are removed by helicases and nucleases and the correct sequence is replaced by DNA polymerases, which fills in the gap using the other strand as a template, and DNA ligase, which seals the final nick in the phosphodiester backbone. The use of a template for repair ensures that the repair is error-free. As a case in point, double-strand break repair by homologous recombination, which uses a sister chromatid or the other chromosome copy as a template, is error-free, but nonhomologous end joining, which does not use a template, is error-prone. The importance of DNA repair is demonstrated by diseases such as xeroderma pigmentosum, Cockayne syndrome, trichothiodystrophy, and hereditary nonpolyposis colorectal cancer that result from loss of DNA repair factors. Much remains to be learned by geneticists about how cellular and environmental factors produce DNA damage, how DNA damage is repaired to prevent mutations, and how mutations lead to disease.

KEY TERMS

abasic site (p. 514)
alkylation (p. 511)
Ames test (p. 514)
apurinic (AP) site (p. 514)
apyrimidinic (AP) site (p. 514)
auxotroph (p. 514)
base analog (p. 511)
base excision repair (BER) (p. 516)
base deletion (p. 503)
base insertion (p. 503)
base substitution (p. 503)
cancer (p. 505)
conservative mutation (p. 503)
deamination (p. 509)
depurination (p. 509)
DNA damage (p. 502)
double-strand break (DSB) (p. 521)
double-strand break repair (DSBR) (p. 522)
frameshift mutation (p. 504)
germ-line mutation (p. 502)

global genomic nucleotide excision repair (GG-NER) (p. 517)
Holliday junction (HJ) (p. 522)
homologous recombination (HR) (p. 521)
indel mutation (p. 503)
induced mutation (p. 506)
intercalating agent (p. 511)
ionization (p. 507)
ionizing radiation (IR) (p. 514)
mismatch repair (MMR) (p. 519)
missense (nonsynonymous) mutation (p. 503)
mutagen (p. 506)
mutation (p. 502)
nonconservative mutation (p. 503)
nonhomologous end joining (NHEJ) (p. 521)
nonsense mutation (p. 503)
nucleotide excision repair (NER) (p. 517)

oxidative damage (p. 511)
point mutation (p. 503)
pyrimidine dimer (p. 513)
recombination (p. 502)
replication slippage (p. 507)
somatic mutation (p. 502)
SOS response (p. 521)
spontaneous mutation (p. 506)
synonymous (silent) mutation (p. 503)
synthesis-dependent strand annealing (SDSA) (p. 522)
tautomerization (p. 507)
transcription-coupled nucleotide excision repair (TC-NER) (p. 517)
transition (p. 503)
translesion (TLS) polymerase (p. 520)
translesion synthesis (TLS) (p. 520)
transversion (p. 503)
trinucleotide-repeat disease (p. 507)

PROBLEMS

Visit SaplingPlus for supplemental content. Problems with the 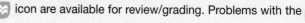 icon are available for review/grading. Problems with the icon have an Unpacking the Problem exercise.

WORKING WITH THE FIGURES

(The first 20 questions require inspection of text figures.)

1. In Figure 15-1, describe a mutation that is neutral.

2. In Figure 15-2, using an example, explain why it is essential to know the reading frame to determine how a mutation in an open reading frame affects the encoded protein.

3. In Figure 15-3, are there any other single base changes that would convert a GGC glycine codon into a valine codon? Consult Figure 9-8.

4. In Figure 15-4, what are the two main types of molecular interactions that are affected by point mutations in noncoding regions?

5. In Figure 15-5, if the mutation occurred spontaneously in the original cell, what percentage of the cells in the fifth generation would contain the mutation?

6. In Figure 15-6, when a G shifts to its rare enol form, what nucleotide can the DNA polymerase add rather than a C? Also, following additional rounds of DNA replication, wild-type cells have a G·C base pair; what base pair will mutant cells have?

7. In Figure 15-7, write a DNA template sequence that might lead to a two-base insertion due to slippage during DNA replication.

8. In Figure 15-8, how might DNA methylation inhibit transcription of *FMR1* (see Section 12.3)?

9. In Figure 15-9, oxidation of guanine to 8-oxoguanine leads to a G·C → T·A transversion after DNA replication. Write out the DNA replication steps that lead to the base transversion.

10. In Figure 15-10, the reaction of EMS with guanine generates O-6-ethylguanine, which leads to a G·C → A·T transition. Write out the DNA replication steps that lead to the base transition.

11. In Figure 15-11, draw a 6-4 photoproduct between thymine and cytosine.

12. In Figure 15-12, what types of DNA damage are likely to have led to mutations in spontaneous *his*⁺ revertant colonies?

13. In Figure 15-13, which of the types of DNA damage are caused by DNA replication errors?

14. In Figure 15-14, do cells need to undergo DNA replication for these repair mechanisms to work? Explain your answer.

15. In Figure 15-15, draw the chemical reaction between nucleotides that is catalyzed by DNA ligase.

16. In Figure 15-16, why might the repair of pyrimidine dimers be faster and more efficient on the coding strand of transcriptionally active genes than on other parts of the genome?

17. In Figure 15-17, draw the steps of mismatch repair with a hemimethylated GATC sequence located 5′ of the base mismatch. What type of exonuclease, 5′-to-3′ or 3′-to-5′ is required in this case?

18. In Figure 15-18, why does repair by translesion synthesis not take place in non-dividing cells such as neurons?

19. In Figure 15-19, draw three different DNA end structures that might be formed by a double-strand break and would be substrates for repair by nonhomologous end joining. Based on these structures, explain why the proteins involved in nonhomologous end joining are described as being distinctive in their versatility.

20. In Figure 15-20, how are the noncrossover products of DSBR and SDSA different?

BASIC PROBLEMS

21. What is the difference between a DNA lesion and a DNA mutation?

22. Consider the following wild-type and mutant sequences:

    Wild-type   ...CTTGC<u>A</u>AGCGAATC...

    Mutant   ...CTTGC<u>T</u>AGCGAATC...

    The substitution shown seems to have created a stop codon. What additional information do you need to be confident that it has done so?

23. Can a missense mutation of proline to histidine be made with a G·C → A·T transition-causing mutagen? What about a proline-to-serine missense mutation? Refer to the genetic code in Figure 9-8.

24. By base-pair substitution, what are the synonymous and nonsynonymous changes that can occur starting with the codon CGG? Refer to Figure 9-8.

25. A mutational lesion results in a sequence containing a mismatched base pair:

    5′ AGCT G CCTT 3′

    3′ TCG <u>ATG</u> GAA 5′

    codon

    If mismatch repair occurs in either direction, which amino acids could be found at this site? Refer to Figure 9-8.

26. Why does acridine orange commonly produce null alleles?

27. Defend the statement "Cancer is a genetic disease."

28. Where within a gene might trinucleotide repeat expansion occur, and how might expansion at those sites lead to disease?

29. Do you think that mutations in the base excision repair factor DNA polymerase β increase the risk of cancer? Why or why not?

30. In mismatch repair, only mismatches in the newly synthesized strand are corrected. How are bacteria and eukaryotes able to detect the newly synthesized strand?

31. Why are many chemicals that test positive by the Ames test also classified as carcinogens?

32. Differentiate between the elements of the following pairs: 🌊

    a. Transitions and transversions

    b. Synonymous and neutral mutations

    c. Missense and nonsense mutations

    d. Frameshift and nonsense mutations

33. Describe two spontaneous lesions that can lead to mutations.

34. What are translesion polymerases? How do they differ from the replicative polymerases? How do their special features facilitate their role in DNA repair? 🌊

35. A certain compound that is an analog of the base cytosine can become incorporated into DNA. It normally hydrogen bonds just as cytosine does, but it quite often isomerizes to a form that hydrogen bonds as thymine does. Do you expect this compound to be mutagenic, and, if so, what types of changes might it induce in DNA?

36. In cells that have stopped dividing, what types of DNA repair systems are possible?

37. Two pathways, homologous recombination (HR) and nonhomologous end joining (NHEJ), can repair double-strand breaks in DNA. If HR is an error-free pathway whereas NHEJ is not always error free, why is NHEJ used most of the time in eukaryotes?

38. Which repair pathways detect DNA damage during transcription? 🌊

39. Which of the following is not possible?

    a. A nonsynonymous mutation in an intron

    b. A nonsynonymous mutation in an exon

    c. An indel mutation in an intron

    d. An indel mutation in an exon

40. Which of the following statements best describe the mismatch repair pathway?

    a. It is part of the 3′-to-5′ proofreading function of DNA polymerases.

    b. It acts after DNA replication by recognizing mismatched base pairs.

    c. It is activated by stalled replication forks.

    d. It is coupled to transcription.

CHALLENGING PROBLEMS

41. Hydroxylamine (HA) only causes $G \cdot C \rightarrow A \cdot T$ transitions in DNA. Will HA produce nonsense mutations? Will HA revert nonsense mutations? 🌊

42. You are using methylnitronitrosoguanidine (MNNG) to "revert" mutant *nic-2* (nicotinamide-requiring) alleles in *Neurospora*. You treat cells, plate them on a medium without nicotinamide, and look for prototrophic colonies (i.e., colonies that grow on minimal media). You obtain the following results for two mutant alleles. Explain these results at the molecular level, and indicate how you would test your hypotheses. 🧬

    a. With *nic-2* allele 1, you obtain no prototrophs at all.

    b. With *nic-2* allele 2, you obtain three prototrophic colonies A, B, and C, and you cross each separately with a wild-type strain. From the cross prototroph A × wild type, you obtain 100 progeny, all of which are prototrophic. From the cross prototroph B × wild type, you obtain 100 progeny, of which 78 are prototrophic and 22 are nicotinamide requiring. From the cross prototroph C × wild type, you obtain 1000 progeny, of which 996 are prototrophic and 4 are nicotinamide requiring.

GENETICS AND SOCIETY

Despite considerable data showing that sunscreen provides protection from DNA damage following exposure to UV irradiation, many individuals are unwilling to protect their skin against the sun. Based on what you now know about the causes and consequences of DNA damage and the mechanisms of DNA repair, what arguments would you make in support of public health programs that provide guidance for sun protection, particularly in younger individuals?

The Dynamic Genome: Transposable Elements

Barbara McClintock, shown here at the ceremony to receive the Nobel Prize in Physiology or Medicine. McClintock made several seminal breakthroughs in genetics, most notably the discovery of transposable elements, for which she was awarded the Nobel Prize. [*Keystone/Getty Images.*]

CHAPTER OUTLINE AND LEARNING OBJECTIVES

16.1 DISCOVERY OF TRANSPOSABLE ELEMENTS IN MAIZE

LO 16.1 Design and interpret experiments related to the excision, insertion, recombination, repression, and transcription of transposable elements.

16.2 TRANSPOSABLE ELEMENTS IN BACTERIA

LO 16.2 Determine the class of a transposon from its DNA structure and the proteins it encodes and predict its behavior based on its class.

16.3 TRANSPOSABLE ELEMENTS IN EUKARYOTES

LO 16.3 Compare the structures of transposons and viruses.

LO 16.4 Use transposons as tools to clone genes and create transgenic organisms.

16.4 THE DYNAMIC GENOME: MORE TRANSPOSABLE ELEMENTS THAN EVER IMAGINED

LO 16.5 Predict the short-term and evolutionary fate of transposable elements in a species.

16.5 REGULATION OF TRANSPOSABLE ELEMENT MOVEMENT BY THE HOST

Transposable elements (transposons) are genetic elements with the ability to move from one location in the genome to another. Through their movement, they have the ability to create new mutations, and researchers have harnessed this ability for use in the laboratory. The broad objective for this chapter is to describe transposon genetics and behavior in different groups of organisms such as bacteria, plants, animals, and humans.

In the 1940s, one of the most remarkable discoveries in the history of genetics was made. Namely, there are genetic loci that can move from one location in the genome to another. This special class of loci became known as **transposable elements**, or **transposons** for short. A particular transposon might be on one chromosome in an individual but appear on a different chromosome in its offspring. In the 1940s, the idea that there are mobile genetic loci was heretical, and so it was viewed skeptically and never broadly accepted until demonstrated at the level of DNA decades later.

Perhaps more remarkable than the discovery of transposons themselves was the woman who discovered them—Barbara McClintock. McClintock was recognized by her colleagues as an exceptional, indeed profoundly gifted, scientist from the time she was a graduate student at Cornell University. She earned the distinct honor of being elected to the U.S. National Academy of Sciences at the relatively youthful age of 41. The same year she was elected the first woman president of the Genetics Society of America. Among the numerous honors she won in science, she was awarded the Nobel Prize in Physiology or Medicine in 1983, the first woman to win the Nobel Prize unshared. In 2005, the United States Postal Service issued a stamp in her honor.

McClintock began her career in maize genetics in the laboratory of Rollins Emerson at Cornell University in the 1920s (Figure 16-1). Remarkably, at that time, George Beadle, who also won a Nobel Prize for genetics research (the one-gene, one-enzyme hypothesis; Chapters 6 and 9) was another member of the Emerson lab group. McClintock and Beadle are the only two contemporary graduate students in a single genetics lab who both went on to win separate Nobel Prizes. The Emerson group also included Marcus Rhoades and Charles Burnham, two leading maize geneticists of this era. Nevertheless, even in this august company, McClintock stood out. George Beadle recognized that McClintock's skills in cytogenetics exceeded his own. The story goes that one day, Beadle prepared some chromosome spreads on microscope slides that he knew would take him some time to understand. On leaving the lab that day, he instructed the others, "Don't show these slides to Barbara," as he knew she would solve the puzzle before he could.

McClintock had an exceptional talent for discerning underlying genetic mechanisms by combining cytological observations, the progeny ratios from genetic crosses, and her knowledge of maize. A self-described introvert with "a capacity to be alone," she had a deeply creative mind that saw well beyond the limits of knowledge of her time. In

Barbara McClintock and colleagues at Cornell University

FIGURE 16-1 Rollins A. Emerson laboratory members at Cornell University, 1929. Standing from left to right: Charles Burnham, Marcus Rhoades, Rollins Emerson, and Barbara McClintock. Kneeling is George Beadle. Both McClintock and Beadle were awarded Nobel Prizes. [*Department of Plant Breeding, Cornell University.*]

addition to her Nobel Prize-winning work on transposons, she made important contributions to the understanding of cytological crossing-over and genetic recombination (Chapter 4) and telomeres (Chapter 2 and 7). She defined the morphology of maize chromosomes and published the first genetic map for maize. She discovered a cycle in some dividing cells by which a broken chromosome is repaired and then breaks again with each successive cell division, generating a series of large-scale somatic mutations. This cycle remains important in cancer research today.

In this chapter, you will learn about transposable elements, segments of DNA that can move from one location in the genome to another. We will begin by going through a few of McClintock's experiments and the logic she used to infer that some loci are mobile. We will see that transposons are found in virtually all organisms across the tree of life and that they can comprise a substantial portion of an organism's genome. There are two classes of transposons with different biological properties, and the host species regulates their movement. Finally, transposons are important sources of new mutations and can be used as tools for reverse genetics (Chapter 14) and producing transgenic organisms.

16.1 DISCOVERY OF TRANSPOSABLE ELEMENTS IN MAIZE

LO 16.1 Design and interpret experiments related to the excision, insertion, recombination, repression, and transcription of transposable elements.

To understand McClintock's experiments, let's begin with a little biology of maize. Hundreds of kernels are borne on each maize ear (Figure 16-2a). Each kernel is a mature or ripened fruit that has an embryo and a nutritive tissue surrounding the embryo called the endosperm. Each kernel develops after a pollen grain lands on the single silk (style) attached to the immature kernel, germinates, and then the pollen tube grows through the silk into the immature kernel (ovary) to complete fertilization. Each kernel contains a single offspring (embryo) of the parent plant on which the ear is borne.

Like most other flowering plants, the maize life cycle includes a process called double fertilization. Each pollen grain possesses two haploid sperm cells—one sperm cell combines with the haploid egg cell in the ovary to form

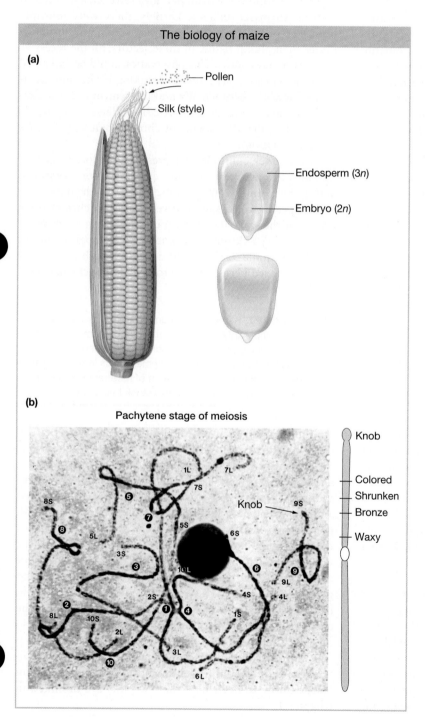

The biology of maize

(a)

Pollen
Silk (style)
Endosperm (3n)
Embryo (2n)

(b)

Pachytene stage of meiosis

Knob
Colored
Shrunken
Bronze
Waxy

FIGURE 16-2 Maize ear, kernel, and chromosomes. (a) An ear of maize contains hundreds of kernels, each one representing a single offspring of the mother plant on which the ear is borne. The offspring embryo (2n) is located on one side of each kernel and is surrounded by the endosperm (3n), a nutritive tissue. The back side of each kernel shows only the endosperm. (b) Meiotic prophase chromosomes of maize with the long (L) and short (S) arms of the 10 chromosomes labeled. A knob marks the end of the short arm of chromosome 9. A genetic map of the loci on chromosome arm 9S used by McClintock is shown. [*Part (b) Photo kindly supplied by Ron Phillips, from a photo taken by John Stout. William Sheridan, University of North Dakota.*]

the diploid embryo, and a second sperm cell combines with a diploid cell (the central cell) in the ovary to form a triploid ($3n$) cell that divides mitotically to form the triploid endosperm that surrounds and nourishes the embryo (see Figure 16-2a). The two sperm cells are genetically identical, as are the sets of chromosomes in the egg cell and diploid central cell. The endosperm makes up the bulk of the starchy tissue that comprises the maize kernel. Importantly, since the endosperm carries sets of chromosomes from both the pollen (male) and ear (female) parents, it expresses the genotype of the next generation.

A strength of maize for McClintock's work is that it has relatively large chromosomes that vary in size and bear distinctive features (*knobs*) that allowed her to identify individual chromosomes on her microscope slides (Figure 16-2b). The *knobs* are dense segments of heterochromatin that produce bulges at specific and diagnostic locations on the different chromosomes. Maize has a haploid number of 10 chromosomes that are numbered 1 to 10 in order of largest to smallest. Each chromosome has a cytologically defined short (S) and long (L) arm. Finally, McClintock took advantage of several linked genes on the short arm of chromosome 9 that are all expressed in the developing kernel and have visible mutant phenotypes in the kernel. *Waxy* (*Wx*) has a recessive mutant allele (*wx*) that converts normal starch to waxy starch; *Bronze* (*Bz*) has a recessive mutant (*bz*) that converts a blue pigment to a bronze color; *Shrunken* (*Sh*) has a recessive mutant (*sh*) that gives shriveled rather than plump kernels. The *Colorless* (*C*) locus has three alleles: *C* conditions a blue pigment in kernels; *c* does not make blue pigment; and C^I is a dominant inhibitor allele that represses pigment production; it is dominant to *C*. The order of dominance is $C^I > C > c$.

McClintock's experiments: the *Ds* element

In the 1930s, McClintock was working with X-ray-induced mutants of maize that cause frequent chromosomal breaks. At this time, she encountered a plant whose progeny exhibited a wide variety of cytological abnormalities on chromosome 9 including chromosome losses, translocations, and inversions that she could observe cytologically. Among the descendants of this special plant, there was one plant carrying C^I/C^I that was expected to produce all heterozygous C^I/C colorless kernels (lack blue pigment) when used as the pollen parent with a *C/C* ear parent. However, when she made this cross, she received a surprise: some of the kernels had multiple blue sectors (**Figure 16-3a**).

McClintock suspected that the C^I allele was being lost in the sectors with blue cells during kernel development so that it would no longer inhibit the *C* allele and its ability to make the blue pigment. An obvious way for this loss to happen would be by breakage of chromosome 9, which she had been studying. If chromosome 9 broke between the *C* locus and the centromere on the chromosome carrying C^I, then the C^I allele would be on an acentric fragment and be lost during mitosis in the descendant cells, allowing the

C allele to produce the blue pigment (Figure 16-3a). Consistent with this interpretation, this special stock of maize showed a high frequency of breakage of chromosome 9 when examined cytologically.

McClintock made crosses between this special stock as the male parent and a stock carrying additional marker loci on chromosome 9—the recessive *waxy*, the recessive *bronze*, and recessive *shrunken*—as well as *C* for colored (blue) kernels (Figure 16-3b). McClintock expected the kernels to lack blue pigment; have normal, not waxy starch; and be plump rather than shrunken since the dominant alleles from the special stock should obscure the recessive alleles from the female parent. However, she received another surprise: on some kernels, there were multiple bronze-colored sectors that had waxy starch and shrunken rather than plump tissue. In these sectors, all the recessive alleles were uncovered. This observation could be explained if the chromosome was always breaking in the same location somewhere between *Wx* and the centromere, simultaneously uncovering all four recessive alleles—*wx*, *bz*, *sh*, and *C*. She called this locus for chromosome breaking **Ds** (for **Dissociation**).

McClintock confirmed the presence and location of *Ds* on chromosome 9 through her cytological observations. When she examined the meiotic prophase chromosomes of the special stock, she observed recurrent chromosome breaks at the same location on the short arm of chromosome 9, using a chromosome knob as a cytological marker for chromosome 9. She could detect *Ds* both genetically by the sectored kernels and cytologically by chromosome breaks.

> **KEY CONCEPT** Kernel sectors uncovering the recessive alleles at multiple linked loci on maize chromosome 9 indicated that the chromosome broke (dissociated) repeatedly in the same position during kernel development. This inference was confirmed by cytological observation of recurrent breaks on chromosome 9 of the special maize stock being used.

Another observation McClintock made was that, in some backcross families with the special stock, about 50 percent of the kernels were colorless with no sectors, and about 50 percent had blue sectors on a colorless background. The 1:1 ratio of these types of kernels suggested another Mendelian factor was required for *Ds* to break the chromosome. If the pollen parent was heterozygous for this additional factor, then half of the kernels on the ear would inherit it, and it would "activate" the breakage of chromosome 9 at the *Ds* locus, uncovering the *C* allele and giving a blue sector. The other half of the kernels would not inherit this factor, so *Ds* would not be activated, the chromosome would not break, and C^I would be present in all cells, inhibiting the formation of blue-colored sectors. McClintock called this second factor *Ac* (for **Activator**). She knew *Ac* was unlinked from *Ds* the same way you learned how to determine that two loci are unlinked in Chapter 4.

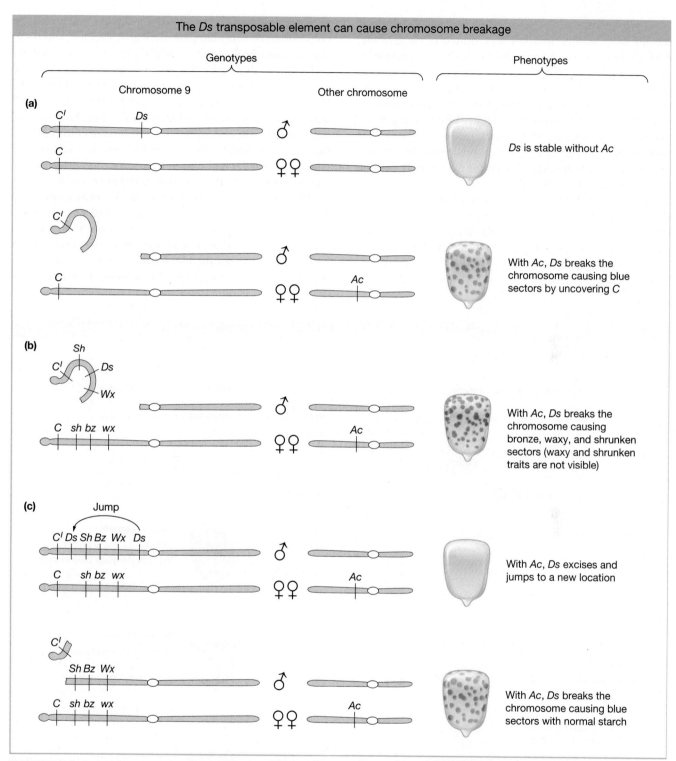

FIGURE 16-3 Chromosome 9 of maize breaks at the *Ds* locus, where the *Ds* transposable element has inserted. (a) Chromosome composition of the endosperm (3n) of a kernel with the *Ds* near the centromere on the copy of chromosome 9 contributed by the pollen parent. The endosperm is heterozygous *C/C/C^I*, with two copies of *C* from the female parent. When *Activator* (*Ac*) is not present (top), the *Ds* element is stable. When *Ac* is present on a separate chromosome (bottom), the *Ds* element breaks the chromosome, producing an acentric fragment with the *C^I* allele and sectors of cells lacking *C^I* in which blue pigment is produced. (b) Like panel (a) except recessive alleles at *shrunken*, *bronze*, and *waxy* are incorporated on the chromosomes. When *C*, *sh*, *bz*, and *wx* are all uncovered by a chromosome break, bronze sectors with waxy and shrunken tissue are produced. (c) Like panel (b) except *Ds* has moved to a new location between *C* and *Sh* (top). When *Ds* breaks the chromosome between *Sh* and *C*, only *C* is uncovered, so the sectors are blue but have normal, not waxy starch, and they are plump, not shriveled (bottom).

Having hypothesized another factor, *Ac*, McClintock wanted to determine where it was located in the genome, that is, she wanted to map it. She did this by crossing plants of stocks with *Ac* and *Ds* to other stocks with marker genes on different chromosomes. If *Ac* was near one of the marker genes on another chromosome, she would observe linkage of *Ac* and the marker locus. When doing these mapping experiments, McClintock received still another surprise: *Ac* mapped to different chromosomes in different crosses. These observations raised the question: Is *Ac* able to move around the genome?

McClintock found another piece of the puzzle when she discovered another derivative of her special stock in which *Ds* seemed to have moved. With this derivative, chromosome 9 no longer broke between *Wx* and the centromere; rather, it broke between *Sh* and *C* (Figure 16-3c). She could observe this change cytologically by the change

in the position of the breaks along the chromosome. She could observe this change genetically, because when the chromosome broke, only C^I was on the acentric fragment and lost. The dominant *Wx*, *Sh*, and *Bz* alleles were on the centromere side of the breakpoint. Thus, the sectors in the kernels with the broken chromosome were blue since C^I was lost, but they were plump, not shrunken; had normal, not waxy starch; and were blue, not bronze color because the dominant *Wx*, *Sh*, and *Bz* alleles were all retained. She concluded that *Ds* had moved from a location between *Wx* and the centromere to a location between *Sh* and *C*.

Here is one last piece of evidence that convinced McClintock that loci can move around the genome. McClintock made the cross shown in **Figure 16-4**. For this cross, when *Ac* is not present, we expect the chromosome to remain intact and the kernels to be entirely blue. When *Ac* is present, we expect the chromosome to break at *Ds*

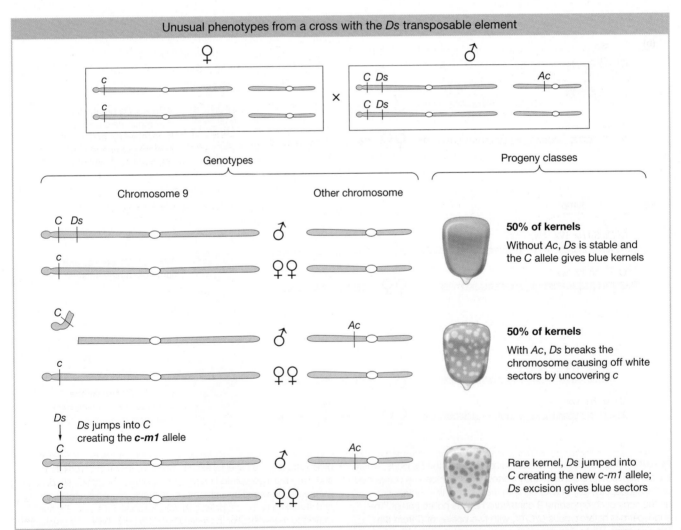

FIGURE 16-4 A female parent homozygous for the recessive *c* allele is pollinated with a male parent that is homozygous for the dominant *C* allele that makes blue pigment and for *Ds*, but that is heterozygous for *Ac* on another chromosome. Fifty percent of the offspring kernels are solid blue because *Ac* is absent, so chromosome 9 remains intact and all cells in the kernel are heterozygous—*C/c*. Fifty percent of the offspring kernels are blue with colorless sectors because *Ac* is present

and breaks the chromosome in the sectors, so that the *C* allele is lost on the acentric fragment and only the recessive *c* allele that does not make blue pigment is present in the sectors. In one rare kernel, *Ds* jumps into the *C* locus, creating a new unstable mutant allele (*c-m1*), giving a *c/c-m1* heterozygote with a colorless kernel. When *Ds* jumps out of *c-m1* in the presence of *Ac*, blue sectors are formed as *c-m1* reverts to *C*.

and the dominant *C* allele for blue color to be lost, so one should see colorless sectors on an otherwise blue kernel. The result should be a 1 : 1 ratio of blue and sectored kernels.

When McClintock made the cross in Figure 16-4, she did see a 1 : 1 ratio of blue to sectored kernels. However, among about 4000 kernels examined, she observed one kernel that had the reversed type of sectoring—blue sectors on a colorless background. Subsequent experiments revealed that in this kernel, *Ds* had moved right into the *C* locus, creating a new recessive allele called *c-m1* by McClintock. The presence of *Ds* in the *C* locus disrupted the ability of the *c-m1* allele to make the blue pigment. However, in the presence of *Ac*, *Ds* jumps out of *c-m1* to restore a functional dominant *C* allele in sectors of the kernel, giving blue spots on a white background. The "*m*" in *c-m1* stands for *mutable*, and it is a mutable or **unstable allele**. The leaving *Ds* element is said to **excise** from the chromosome or **transpose**. This was further evidence that *Ds* was a mobile locus.

> **KEY CONCEPT** *Ds* is a genetic element that can jump around the genome or break chromosomes in the presence of *Ac*. *Ac* also has the ability to jump from one genomic location to another itself, and *Ac* provides functionality that enables *Ds* to move.

In summary, three observations from her experiments convinced McClintock that there are loci that can move around the genome, called "jumping genes" or transposons.

- First, *Ds* moved from near *Wx* to between *Sh* and *C*.

- Second, *Ac* mapped to different chromosomal locations in different crosses, suggesting it could move around the genome.

- Third, when *Ds* was inserted in *C* and created a recessive mutant allele (*c-m1*), it could subsequently jump out of *C* to restore a functional wild-type allele that confers blue color.

Ac (*Activator*) and *Ds* (*Dissociation*) today

Later in this chapter, we will examine how McClintock's ground-breaking discovery of transposons was confirmed at the DNA level when transposons were cloned, their molecular makeup determined, and the mechanism by which they move discerned. At this juncture, a few features of *Ac* and *Ds* learned after McClintock's initial work are worth noting to tie up the story of their discovery.

First, *Ac* encodes an enzyme called **transposase** that catalyzes the excision of *Ds* from a chromosome and its insertion in a new location. *Ds* does not encode transposase, and this is why *Ds* relies upon *Ac* to activate it. The transposase encoded by *Ac* can cut the *Ds* element out of one place in the genome and paste it into another. Thus, the transposase could cut *Ds* out of the location near *Wx* and paste into a location between *Sh* and *C*. Because *Ds* relies on *Ac*, it is said to be a **nonautonomous transposable element**.

Second, the transposase that *Ac* encodes not only can activate the transposition of *Ds*, but it can activate its own transposition. *Ac* can move from one location to another just like *Ds*. As noted previously, McClintock discovered that *Ac* mapped to different chromosomes in different stocks of maize, suggesting it could move. As further proof that *Ac* transposes, McClintock found an additional unstable allele of the *C* gene into which *Ac* was inserted. This allele is called *c-m(Ac)*. Since *Ac* activates the movement of transposons, the *Ac* in *c-m(Ac)* could activate its own transposition out of *c-m(Ac)* to restore a functional allele of *C* that conferred blue sectors in the kernels. Because *Ac* can move on its own, it is said to be an **autonomous transposable element**.

Third, *Ds* elements are not all the same but come in different forms with different properties. The *Ds* element that McClintock first discovered near *Wx* had a structure that causes it to break the chromosome when it excises such that the two ends of the broken chromosome are not joined back together by the transposase. Other *Ds* elements have a different structure such that when they excise, the transposase joins the broken ends of the chromosome back together. This type of *Ds* is found in the *c-m1* allele. When the *Ds* in *c-m1* excises, the two ends of the chromosome are joined together, restoring a functional dominant *C* allele.

> **KEY CONCEPT** Transposable elements in maize can inactivate a gene in which they reside, cause chromosome breaks, and transpose to new locations within the genome. Autonomous elements can perform these functions unaided; nonautonomous elements can transpose only with the help of an autonomous element elsewhere in the genome.

Transposable elements: only in maize?

McClintock was a highly respected geneticist, but the relevance of transposition to other organisms was questioned by some who argued that maize is not a natural organism: it is a crop plant that is the product of human selection and domestication. The existence of transposons in all organisms would imply that genomes are inherently unstable and dynamic. This view was inconsistent with the fact that the genetic maps of members of the same species were the same. After all, if genes can be genetically mapped to a precise chromosomal location, this observation would appear to indicate that they are not moving around.

Skepticism regarding the importance of transposable elements remained until the 1960s, when the first transposable elements were isolated from the *E. coli* genome and studied at the DNA-sequence level. Transposable elements were subsequently isolated from the genomes of many organisms, including yeast and *Drosophila* (**Figure 16-5**). It was only after it became apparent that transposable elements are a significant component of the genomes of most and perhaps all organisms, that Barbara McClintock was recognized for her seminal discovery by being awarded the 1983 Nobel Prize in Medicine or Physiology.

Transposable element in *Drosophila*

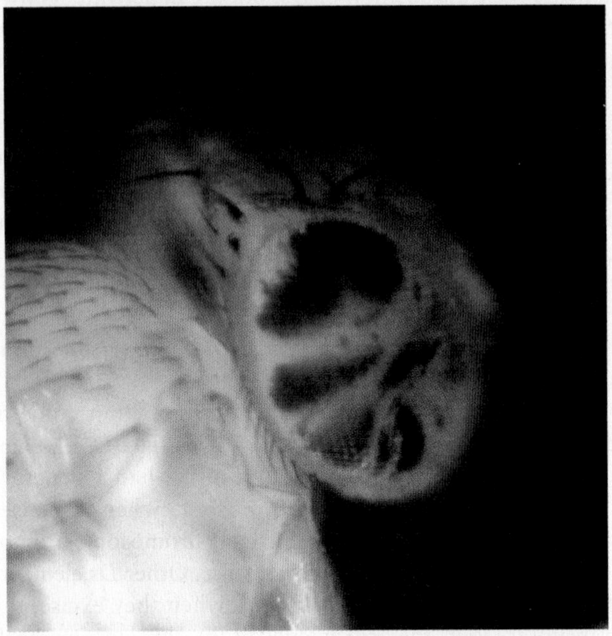

FIGURE 16-5 Excision of the *mariner* transposable element from the *white* gene causes mosaicism in the eye of a *Drosophila*. In the red sectors, *mariner* has been excised from the *white* gene, restoring a functional allele and wild-type red eye color. [*Courtesy of Emilie Robillard, CNRS.*]

16.2 TRANSPOSABLE ELEMENTS IN BACTERIA

LO 16.2 Determine the class of a transposon from its DNA structure and the proteins it encodes and predict its behavior based on its class.

The molecular nature of transposable elements was first understood in a bacterium, *E. coli*. The discovery of transposons in *E. coli* parallels McClintock's work in several ways—there were unstable mutants that could revert to wild type, and a "locus" could appear at several locations around the genome. However, the array of molecular tools that could be applied in bacteria, along with the small genome size of bacteria, enabled resolving these elements to the DNA level.

Evidence for transposable elements in bacteria

The story begins with the isolation of some new mutants in the *E. coli gal* operon—a cluster of three genes that encode the enzymes required for use of galactose as an energy source (see Chapter 11). These mutants cannot grow on a medium in which galactose is the energy source. Like the *c-m1* allele of McClintock, these *gal* mutants could revert to wild type spontaneously. The ability to revert suggested that they were not simple deletions of the operon. Furthermore, adding a chemical mutagen to the media did not increase the frequency of reversion. If the mutants were single–base-pair

nonsense or missense nucleotide substitutions and the revertants were being restored to the original nucleotide, then a chemical mutagen that increased the single nucleotide mutation rate should increase the rate of reversion. It did not, and so something else must have been happening.

The scientists took advantage of the ability of λ phage to pick up the *gal* operon from *E. coli*, inserting it into the phage DNA and thereby into the resultant phage particles. Such λ phage with *gal* could be made from wild-type (λd*gal*$^+$) or mutant (λd*gal*$^-$) *E. coli*. The "d" is for "defective" because when *gal* inserts into the phage, a part of the phage's own chromosome is deleted. When a mixture of λd*gal*$^+$ and λd*gal*$^-$ phage was subjected to density centrifugation (see Chapter 7), the two types of phage particles could be separated because they had different buoyant densities (**Figure 16-6a**). The λd*gal*$^-$ had a higher density or molecular weight (more DNA) than the λd*gal*$^+$. Why should the mutant have more DNA?

The next experiment took advantage of the ability to hybridize the DNA from λd*gal*$^+$ and λd*gal*$^-$ to one another. The DNAs from these two phages are mixed, then denatured, and then allowed to anneal to one another. When the annealed heteroduplexes (double-stranded DNA molecules composed of one λd*gal*$^+$ and one λd*gal*$^-$ strand) are examined under an electron microscope, one observes a loop of single-stranded DNA where the λd*gal*$^+$ has no complement to the λd*gal*$^-$ strand (Figure 16-6b). The size of the loop was estimated to be about 1000 bp. When this experiment was done with *gal* mutants that are caused by point mutations, there was no loop in the heteroduplex. This experiment showed that the *gal*$^-$ mutants contained extra, inserted DNA relative to wild type.

Finally, other experiments showed the extra DNA (the insertion) in the *gal*$^-$ strain hybridizes to different regions of the *E. coli* genome in different *E. coli* strains, that is, it was moving around like *Ac* and *Ds*. The small-phage DNAs could be readily sequenced and the sequences of the λd*gal*$^+$ and λd*gal*$^-$ compared. This work provided the first DNA sequence of a transposable element. The sequences revealed that the inserted element contains a single open reading frame that encodes a transposase flanked by short **inverted repeat** (**IR**) **sequences** of about 20 bp in length (**Figure 16-7a**). The first such insertion element isolated from the *gal* operon was called IS1 for **insertion-sequence** (**IS**) **element** 1. Subsequently, different *E. coli* strains were found to harbor other similar elements named IS2, IS3, etc. The genome of the standard wild-type *E. coli* is rich in IS elements: it contains eight copies of IS1, five copies of IS2, and copies of other less well-studied IS types.

To summarize, some mutants of the *gal* operon of *E. coli* contain transposable elements:

- The mutants arose spontaneously, and they can revert spontaneously to wild type.

- The mutants contained an extra segment of DNA inserted into the *gal* operon.

- The inserted DNA could be found in different locations in the *E. coli* genome; that is, it moves.

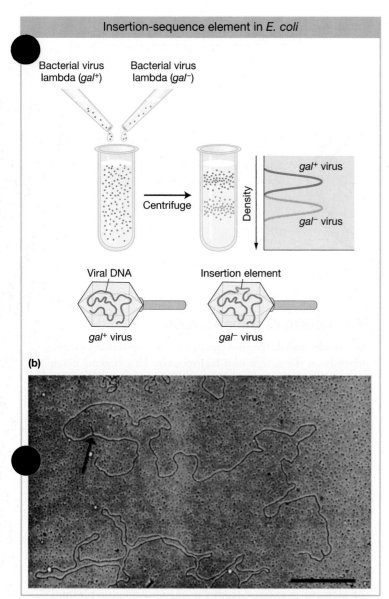

Insertion-sequence element in *E. coli*

Bacterial virus lambda (*gal*⁺) Bacterial virus lambda (*gal*⁻)

Centrifuge

Density

gal⁺ virus

gal⁻ virus

Viral DNA   Insertion element

gal⁺ virus   *gal*⁻ virus

(b)

FIGURE 16-6 Evidence that *E. coli* has a type of transposon called an insertion sequence (IS). (a) A mixture of wild-type (λd*gal*⁺) or mutant (λd*gal*⁻) phage particles are subjected to density centrifugation and separate into two bands with low (λd*gal*⁺) and high (λd*gal*⁻) density, indicating that the latter contains a longer DNA molecule. (b) Electron micrograph of a λd*gal*⁺/λd*gal*⁻ heteroduplex. The arrow indicates a single-stranded loop caused by the presence of the IS element in λd*gal*⁻. [*Part (b) Republished with permission of Springer Science+Business Media, from A. Ahmed and D. Scraba, "The nature of the gal3 mutation of Escherichia coli,"* Molecular and General Genetics MGG, *1975, September; 136 (3) 233–242, Figure 2. Permission conveyed through Copyright Clearance Center, Inc.*]

- DNA sequencing of the insertion revealed that the inserted DNA encodes a transposase flanked by inverted repeats.

KEY CONCEPT The bacterial genome contains segments of DNA, termed IS elements, that can move from one position on the chromosome to another.

Simple and composite transposons

In Chapter 6, you learned about **R plasmids**, which carry genes that encode resistance to several antibiotics. These

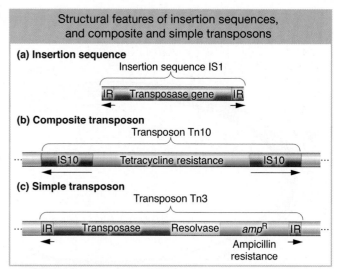

Structural features of insertion sequences, and composite and simple transposons

(a) Insertion sequence

Insertion sequence IS1

IR   Transposase gene   IR

(b) Composite transposon

Transposon Tn10

IS10   Tetracycline resistance   IS10

(c) Simple transposon

Transposon Tn3

IR   Transposase   Resolvase   *amp*ᴿ   IR

Ampicillin resistance

FIGURE 16-7 (a) Structure of IS1 including the transposase gene and terminal inverted repeats (IRs). (b) Tn10, an example of a composite transposon. The IS elements are inserted in opposite orientation and form IRs. Each IS element carries a transposase, but only one is usually functional. (c) Tn3, an example of a simple transposon that encodes its own transposase. The resolvase is a protein that promotes recombination (see Figure 16-10).

R plasmids (for resistance), also known as R factors, are transferred rapidly from one bacterial cell to another during conjugation, much like the F factor in *E. coli*. The R factors proved to be just the first of many similar F-like factors to be discovered. R factors have been found to carry many different kinds of genes in bacteria. In particular, R factors pick up genes conferring resistance to different antibiotics. How do they acquire their new genetic abilities?

It turns out that the drug-resistance genes can reside in transposons. There are two types of bacterial transposons. Let's first discuss one type called **composite transposons**, which can contain a variety of genes that reside between two nearly identical IS elements that are oriented in the opposite direction (Figure 16-7b). The two IS elements form an inverted repeat sequence. Transposase encoded by one of the two IS elements is necessary to catalyze the movement of the entire composite transposon. Figure 16-7b shows a composite transposon (Tn10) that carries a gene conferring resistance to the antibiotic tetracycline flanked by two IS10 elements in opposite orientation. The IS elements that make up composite transposons are not capable of transposing on their own because of mutations in their short inverted repeats.

Simple transposons are composed of short (<50 bp) inverted repeat sequences that can encompass bacterial genes. The mobility of simple transposons is catalyzed by a transposase that is encoded within the transposon itself rather than in an IS element. Simple transposons also encode resolvase, an enzyme that promotes site-specific recombination. An example of a simple transposon is Tn3, shown in Figure 16-7c.

To review, IS elements are short mobile sequences that encode only those proteins necessary for their mobility.

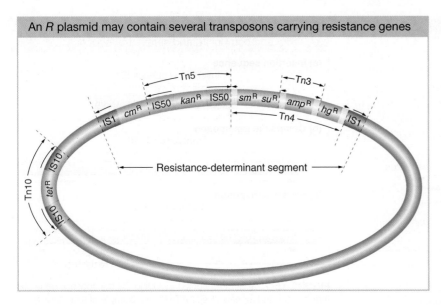

An *R* plasmid may contain several transposons carrying resistance genes

FIGURE 16-8 A schematic map of a plasmid with several insertions of simple and composite transposons carrying resistance genes. Plasmid sequences are in blue. Genes encoding resistance to the antibiotics tetracycline (*tet*^R), kanamycin (*kan*^R), streptomycin (*sm*^R), sulfonamide (*su*^R), and ampicillin (*amp*^R) and to mercury (*hg*^R) are shown. The resistance-determinant segment can move as a cluster of resistance genes. Tn3 is within Tn4. Each transposon can be transferred independently. [*Data from S. N. Cohen and J. A. Shapiro, "Transposable Genetic Elements." Copyright 1980 by Scientific American, Inc. All rights reserved.*]

Composite transposons and simple transposons contain additional genes that confer new functions to bacterial cells. Whether composite or simple, transposons are usually just called transposons, and different transposons are designated Tn1, Tn2, Tn505, and so forth.

A transposon can jump from a plasmid to a bacterial chromosome or from one plasmid to another plasmid. In this manner, multiple-drug-resistant plasmids are generated. **Figure 16-8** shows a composite diagram of an R factor, indicating the various places at which transposons can be located. We next consider the question of how such **transposition** or mobilization events occur.

KEY CONCEPT Some bacterial transposons were detected as mobile genetic elements that confer drug resistance. These elements can consist of two IS elements flanking a gene that encodes drug resistance. This organization promotes the spread of drug-resistant bacteria by facilitating movement of the resistance gene from the chromosome of a resistant bacterium to a plasmid that can be conjugated into another (susceptible) bacterial strain.

Mechanism of transposition

As already stated, the movement of a transposable element depends on the action of a transposase. This enzyme plays key roles in the two stages of transposition: excision (leaving) from the original location, and insertion into the new location.

Excision from the original location Most transposable elements in bacteria (and in eukaryotes) employ one of two mechanisms of transposition, called **replicative transposition** and **conservative** (or nonreplicative) **transposition**, as illustrated in **Figure 16-9**. In the replicative pathway (as shown for Tn3), a new copy of the transposable element is generated during the transposition event. The results of the transposition are that one copy appears at the new site and one copy remains at the old site. In the conservative pathway (as shown for Tn10), there is no replication. Instead, the element is excised from the chromosome or plasmid and is integrated into the new site.

Replicative transposition Because this mechanism is complicated, it will be described here in detail. As Figure 16-9

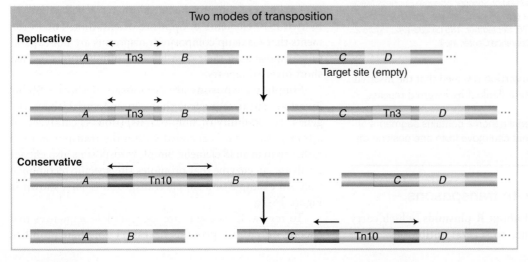

Two modes of transposition

Replicative

Conservative

Target site (empty)

FIGURE 16-9 Mobile-element transposition may be either replicative or conservative. See text for details.

illustrates, one copy of Tn3 is produced from an initial single copy, yielding two copies of Tn3 altogether. **Figure 16-10** shows the details of the intermediates in the transposition of Tn3 from one plasmid (the donor) to another plasmid (the target). During transposition, the donor and recipient plasmids are temporarily fused together to form a double plasmid. The formation of this intermediate is catalyzed by Tn3-encoded transposase, which makes single-strand cuts at the two ends of Tn3 and staggered cuts at the target

sequence and then joins the free ends together, forming a fused circle called a **cointegrate**. The transposable element is duplicated in the fusion event. The cointegrate then resolves by a recombination-like event catalyzed by the resolvase (see Figure 16-7) that turns a cointegrate into two smaller circles, leaving one copy of the transposable element in each plasmid. Because a new copy of the element is created, this mechanism is called "copy and paste."

Conservative transposition Some transposons, such as Tn10, excise from the chromosome and integrate into the target DNA. In these cases, the DNA of the element is not replicated, and the element is lost from the site of the original chromosome (see Figure 16-9). This mechanism is called "cut and paste." Like replicative transposition, this reaction is initiated by the element-encoded transposase, which cuts at the ends of the transposon. However, in contrast with replicative transposition, the transposase cuts the element out of the donor site by making a double-stranded cut. (These double-stranded breaks create the opportunity for a chromosome break, as seen by McClintock, if the cell fails to ligate the ends at the donor site back together.) It then makes a cut at a target site and inserts the element into the target site. We will revisit this mechanism in greater detail in a discussion of the transposition of eukaryotic transposable elements.

Insertion into a new location Let's now look a little closer at how the transposase catalyzes the insertion into the target site. In one of the first steps of insertion, the transposase makes a staggered cut in the target-site DNA (not unlike the staggered breaks catalyzed by restriction endonucleases in the sugar–phosphate backbone of DNA). **Figure 16-11** shows the steps in the insertion of a generic transposable element. In this case, the transposase makes a five–base-pair staggered cut. The transposable element inserts between the staggered ends, and the host DNA repair machinery (see Chapter 15) fills in the gap opposite each single-strand overhang by using the bases in the overhang as a template. There are now two duplicate sequences, each five base pairs in length, at the sites of the former overhangs. These sequences are called a **target-site duplication**. Virtually all transposable elements (in both bacteria and eukaryotes) are flanked by a target-site duplication, indicating that all use a mechanism of insertion similar to that shown in Figure 16-11. What differs is the length of the duplication; a particular type of transposable element has a characteristic length for its target-site duplication—as small as two base pairs for some elements. It is important to keep in mind that the transposable elements have *inverted repeats* at their ends and that the inverted repeats are flanked by the target-site duplication—which is a *direct repeat*.

> **KEY CONCEPT** In bacteria, transposition occurs by at least two different mechanisms. Some transposable elements can replicate a copy of the element into a target site, leaving one copy behind at the original site. In other cases, transposition consists of the excision of the element and its reinsertion into a new site.

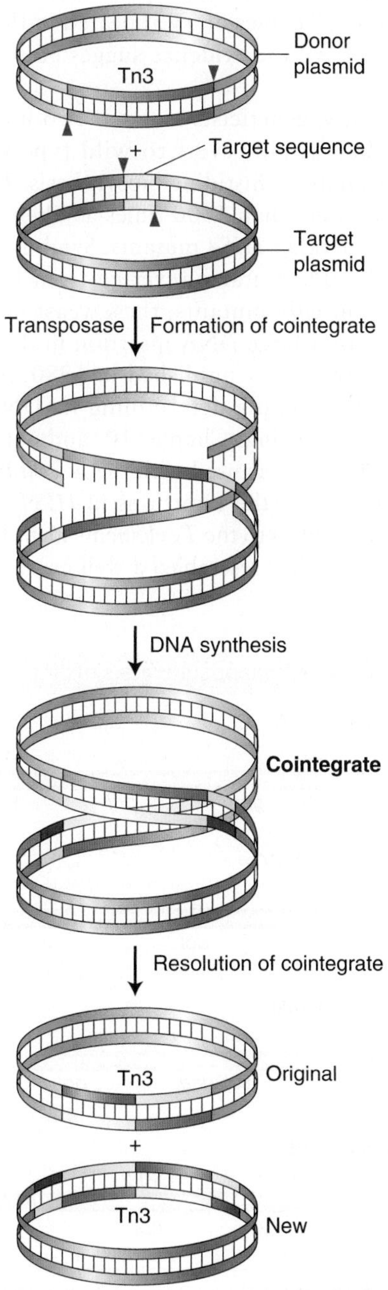

FIGURE 16-10 Replicative transposition of Tn3 takes place through a cointegrate intermediate.

ANIMATED ART SaplingPlus

Replicative transposition

An inserted element is flanked by a short repeat

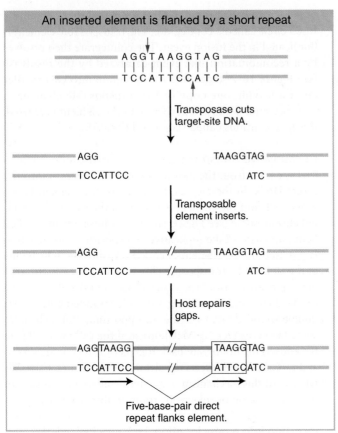

FIGURE 16-11 A short sequence of DNA is duplicated at the transposon insertion site. The recipient DNA is cleaved at staggered sites (a 5-bp staggered cut is shown), leading to the production of two copies of the five–base-pair sequence flanking the inserted element.

16.3 TRANSPOSABLE ELEMENTS IN EUKARYOTES

LO 16.2	Determine the class of a transposon from its DNA structure and the proteins it encodes and predict its behavior based on its class.
LO 16.3	Compare the structures of transposons and viruses.
LO 16.4	Use transposons as tools to clone genes and create transgenic organisms.

Although transposable elements were first discovered in maize, the first eukaryotic elements to be characterized at the molecular level were isolated from mutant yeast and *Drosophila* genes. Eukaryotic transposable elements fall into two classes: class 1 retrotransposons and class 2 DNA transposons. The first class to be isolated, the retrotransposons, are not at all like the bacterial transposable elements.

Class 1: retrotransposons

In the late 1970s, yeast geneticists discovered a 5.6-kb-long repeated sequence in the yeast genome that they called *Ty1*, the founding member of the **Ty element** family of

transposons. This element occurred in about 35 copies distributed on the 16 yeast chromosomes. These 5.6-kb elements had **long terminal repeat (LTR)** sequences of about 350 bp in length (**Figure 16-12a**). The geneticists also found that there were many more LTRs than the 70 that formed part of the 35 full-length elements, suggesting that some LTRs existed on their own within the genome. One technology used to make these discoveries was standard Southern blotting (Chapter 10). Using a cloned DNA probe for *Ty*, one can visualize about 35 bands on the Southern blot. Later, it was learned that *Ty* elements are flanked by a 5-bp direct repeat, much like bacterial transposons. Finally, comparison of parent and derivative yeast strains revealed that new, "transposed" copies of *Ty* appeared in the genome in the derivatives. All the evidence suggested that *Ty* was a transposon.

Subsequently, geneticists isolated two unstable yeast mutations that would revert to wild type in the *HIS4* gene, which catalyzes histidine biosynthesis. The unstable mutants were more than 1000 times as likely to revert to wild type as the other *HIS4* mutants. Symbolically, we say that these unstable mutants reverted from *His⁻* to *His⁺*. Like the *E. coli gal⁻* mutants, these yeast mutants were found to harbor a large DNA insertion in the *HIS4* gene. However, for this work published in 1980, the insertion was observed by simply determining its DNA sequence using Sanger sequencing (Chapter 10) and not by buoyant density centrifugation as used with the *E. coli* IS elements.

The sequences of the insertions in *HIS4* showed that they were homologous to the *Ty* elements identified a decade earlier. Moreover, they resembled a well-characterized class

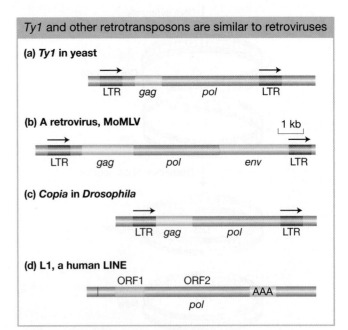

FIGURE 16-12 Structural comparison of *Ty*, other retrotransposons, and a retrovirus. (a) A retrotransposon, *Ty1*, in yeast. (b) A retrovirus, Moloney murine leukemia virus (MoMLV), of mice. (c) A retrotransposon, *copia*, in *Drosophila*. (d) A long interspersed element (LINE) in humans. Abbreviations: LTR, long terminal repeat; ORF, open reading frame.

of animal viruses called retroviruses (Figure 16-12b). A **retrovirus** is a single-stranded RNA virus that employs a double-stranded DNA intermediate for replication. The RNA is copied into DNA by the enzyme **reverse transcriptase**. The double-stranded DNA is integrated into host chromosomes, from which it is transcribed to produce the RNA viral genome and proteins that form new viral particles. When integrated into host chromosomes as double-stranded DNA, the double-stranded DNA copy of the retroviral genome is called a **provirus**. The life cycle of a typical retrovirus is shown in **Figure 16-13**. Some retroviruses, such as mouse mammary tumor virus (MMTV) and Rous sarcoma virus (RSV), are responsible for the induction of cancerous tumors. For MMTV, this happens when it inserts randomly into the genome next to a gene whose altered expression leads to cancer.

Figure 16-12 shows the similarity in structure and gene content of a retrovirus and the *Ty* element isolated from the *HIS4* mutants. Both are flanked by LTR sequences that are several hundred base pairs long. Retroviruses encode at least three proteins that take part in viral replication: the products of the *gag*, *pol*, and *env* genes. The *gag*-encoded protein has a role in the maturation of the RNA genome,

pol encodes the all-important reverse transcriptase, and *env* encodes a protein that is embedded in the viral membrane. This protein is necessary for the virus to leave the cell to infect other cells. Interestingly, *Ty* elements have genes related to *gag* and *pol* but not *env*. These features led to the hypothesis that, like retroviruses, *Ty* elements are transcribed into RNA transcripts that are copied into double-stranded DNA by the reverse transcriptase. However, unlike retroviruses, *Ty* elements cannot leave the cell because they do not encode *env*. Instead, the double-stranded DNA copies are inserted back into the genome of the same cell. These steps are diagrammed in **Figure 16-14**.

In 1985, scientists showed that, like retroviruses, *Ty* elements do in fact transpose through an RNA intermediate. **Figure 16-15** diagrams their experimental design. They began by altering a yeast *Ty* element, cloned on a plasmid. First, near one end of an element, they inserted a promoter that can be activated by the addition of galactose to the medium. This enabled them to control expression of the element by adding galactose. Second, they introduced an intron from another yeast gene into the coding region of the *Ty* transposon as a reporter. If *Ty* was transcribed and the mRNA processed, then the intron would be spliced out.

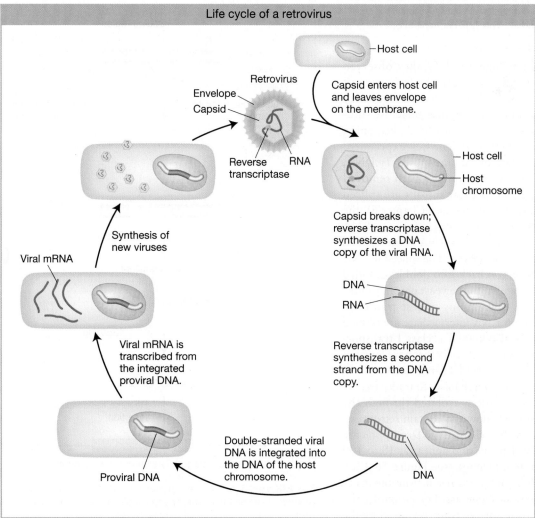

FIGURE 16-13 The retrovirus RNA genome undergoes reverse transcription into double-stranded DNA inside the host cell.

ANIMATED ART   Sapling Plus

Life cycle of a retrovirus

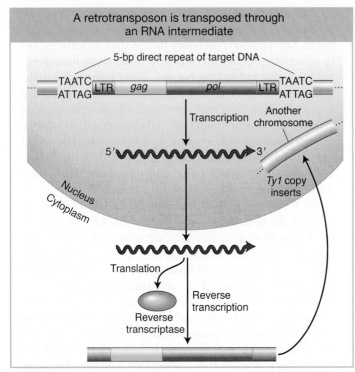

FIGURE 16-14 An RNA transcript from the retrotransposon undergoes reverse transcription into DNA, by a reverse transcriptase encoded by the retrotransposon. The DNA copy is inserted at a new location in the genome.
ANIMATED ART 🦋 SaplingPlus The *Ty1* mechanism of retrotransposition

When they assayed the engineered *Ty*, they observed that adding galactose greatly increased the frequency of transposition of the element. This increased frequency suggests the participation of RNA because galactose stimulates the transcription of *Ty* DNA into RNA, beginning at the galactose-sensitive promoter. The key experimental result, however, is the fate of the transposed *Ty* DNA. The researchers found that the intron had been removed from the transposed *Ty* DNA. Because introns are spliced only in the course of RNA processing (see Chapter 8), the transposed *Ty* DNA must have been copied from an RNA intermediate. The conclusion was that RNA is transcribed from the altered *Ty* element and spliced. The spliced mRNA undergoes reverse transcription back into double-stranded DNA, which is then integrated into the yeast chromosome.

Transposable elements that employ reverse transcriptase to transpose through an RNA intermediate are termed **retrotransposons**. They are also known as **class 1 transposable elements**. Retrotransposons such as *Ty* that have *long terminal repeats* at their ends are called **LTR-retrotransposons**, and they use a "copy and paste" mechanism to transpose.

Several spontaneous mutations isolated through the years in *Drosophila* also were shown to contain retrotransposon insertions. The *copia*-like elements of *Drosophila* are structurally similar to *Ty* elements and appear at 10 to 100 positions in the *Drosophila* genome (see Figure 16-12c). Certain classic *Drosophila* mutations result from the insertion of *copia*-like and other elements. For example, the *white-apricot* (w^a) mutation for eye color is caused by the

insertion of an element of the *copia* family into the *white* locus. The insertion of LTR-retrotransposons into plant genes (including maize) also has been shown to contribute to spontaneous mutations in this kingdom.

Before we leave retrotransposons, one question needs to be answered. Recall that the first LTR-retrotransposon was discovered in an unstable *His⁻* strain of yeast that reverted frequently to *His⁺*. However, we have just seen that LTR-retrotransposons, unlike most DNA transposable elements, do not excise when they transpose. What, then, is responsible for this allele's ~1000-fold increase in reversion frequency when compared to other *His⁻* alleles? The answer is shown in **Figure 16-16**, which shows that the *Ty* element in the *His⁻* allele is located in the promoter region of the *His* gene, where it prevents gene transcription. In contrast, the revertants contain a single copy of the LTR, called a **solo LTR**. This much smaller insertion does not interfere with the transcription of the *His* gene.

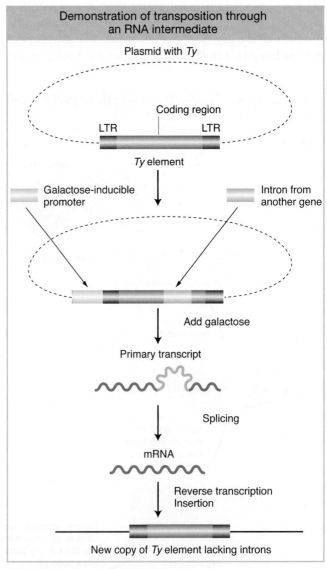

FIGURE 16-15 A *Ty* element is altered by adding an intron and a promoter that can be activated by the addition of galactose. The intron sequences are spliced before reverse transcription.

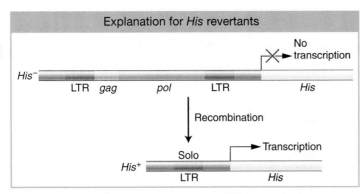

FIGURE 16-16 *His*⁺ revertants contain a solo LTR that results from recombination between the identical DNA sequences in the two LTRs of the LTR-retrotransposon in the *His* promoter.

The solo LTR is the product of recombination between the identical LTRs, which results in the deletion of the rest of the element (see Chapters 4 and 15 for more on recombination). Solo LTRs are a very common feature in the genomes of virtually all eukaryotes, indicating the importance of this process. The sequenced yeast genome contains more than fivefold as many solo LTRs as complete *Ty* elements.

KEY CONCEPT Retrotransposons, also known as class 1 transposable elements, encode a reverse transcriptase that produces a double-stranded DNA copy (from an RNA intermediate) that is capable of integrating at a new position in the genome.

Class 2: DNA transposons

Like IS elements, some eukaryotic mobile elements use a "cut and paste" mechanism and physically move to a new position in the genome after they excise. Elements that transpose in this manner are called **class 2 elements**, or **DNA transposons**. The first transposable elements discovered by McClintock in maize are now known to be DNA transposons. However, the first DNA transposons to be molecularly characterized were the P elements in *Drosophila*.

P elements Of all the transposable elements in *Drosophila*, the most intriguing and useful to geneticists are the **P elements**. The full-size P element resembles the simple transposons of bacteria in that its ends are short (31-bp) inverted repeats and it encodes a single protein—the transposase—that is responsible for its mobilization (**Figure 16-17**). P elements vary in size, ranging from 0.5 to

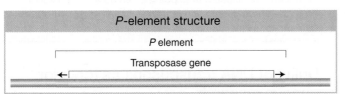

FIGURE 16-17 DNA sequence analysis of the 2.9-kb autonomous P element reveals a gene that encodes transposase. A perfect 31-bp inverted repeat resides at each of the element's termini.

2.9 kb in length. This size difference is due to the presence of many defective P elements from which parts of the middle of the element—encoding the transposase gene—have been deleted, rendering them nonautonomous elements.

The discovery of P elements traces back to the 1970s and Margaret Kidwell, a professor at Brown University who was studying natural (wild) populations of *Drosophila melanogaster* and crossing wild and laboratory strains. When she crossed a wild-strain female with a lab-strain male, the F_1's were normal. However, when she crossed a lab-strain female with a wild male, she noticed the F_1's were sterile. Indeed, the F_1's showed a range of surprising phenotypes in the germ-line cells, including a high mutation rate and a high frequency of chromosomal aberration and nondisjunction (**Figure 16-18**). The defects in these F_1 hybrid progeny made them *dysgenic*, that is, incapable of breeding; hence, this phenomenon was called **hybrid dysgenesis**. It occurs only when females from lab strains are mated with males derived from natural populations.

One observation about the dysgenic flies was that a large percentage of the induced mutations were unstable; that is, they revert to wild type at a high frequency. The unstable *Drosophila* mutants had similarities to the unstable maize mutants characterized by McClintock. Investigators hypothesized that these unstable mutations were caused by the insertion of transposable elements into specific genes, thereby rendering them inactive. Reversion would occur when the transposons excised from the genes. These observations suggested that hybrid dysgenesis was somehow linked to transposable elements.

The role of transposable elements in hybrid dysgenesis was confirmed when an active family of elements was discovered in and molecularly isolated from the dysgenic flies. They called these P elements. Interestingly, P elements are found in wild fly populations, but they are absent from laboratory strains. Thus, wild strains have become known as P strains and are said to have **P cytotype** (cell type). The lab strains are called M strains and said to have **M cytotype**. M stands for "maternal" and P for "paternal," symbolizing that dysgenesis occurs when the lab (M) strain is maternal and the wild (P) strain is paternal.

KEY CONCEPT P elements are DNA transposons found in wild strains of *D. melanogaster*. They were identified to cause hybrid dysgenesis when wild males were mated to lab-strain females, but not when lab-strain males were mated to wild females.

Why are P elements found only in wild strains? The answer to this question requires a bit of history. *Drosophila melanogaster* is native to the Old World, and it was brought to the Americas only in the post-Columbian era. Another species, *Drosophila willistoni*, is native to the Americas, and critically it contains P elements. In the early 1900s, *D. melanogaster* was brought out of the wild and into the laboratory for use in genetic research by T. H. Morgan. At that time, it was expected that P elements did not exist in wild populations of

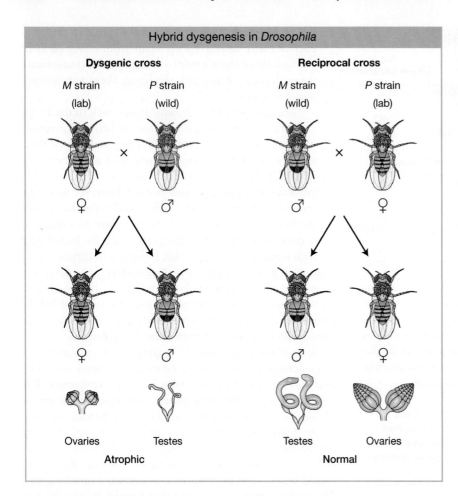

FIGURE 16-18 Dysgenesis and fertile crosses between a laboratory and wild stock of *Drosophila*. When the lab strain is the female, the F_1's are dysgenic (sterile). When the lab strain is the male, the F_1's are normal (fertile). See text for details.

D. melanogaster at all. As the lab strains were maintained in captivity over the years and shared among researchers, they remained genetically isolated from their wild relatives. Meanwhile in nature, *D. melanogaster* populations were evolving separately from their captive cousins. The hypothesis is that the *P* element was transferred horizontally from *D. willistoni* into wild *D. melanogaster*, and then it spread rapidly throughout the global population. The transfer may have been accomplished by a virus that infected *D. willistoni*, picked up the *P* element, and then infected *D. melanogaster*.

When the *P* element was first introduced into wild *D. melanogaster*, it likely caused hybrid dysgenesis. However, over time, natural populations evolved a mechanism to repress *P* transposition in the germ line. The mechanism to repress *P* elements never evolved in the laboratory strains. Later in this chapter, we will review the mechanism for *P* element repression and why hybrid dysgenesis appears only when an M female (no *P* elements) is mated with a P male (*P* elements), but not in the reciprocal cross.

Maize transposable elements revisited Although the causative agent responsible for unstable mutants was first shown genetically to be transposable elements in maize, it was almost 50 years before the maize *Ac* and *Ds* elements were isolated and shown to be related to DNA transposons in bacteria and in other eukaryotes. Like the *P* element of *Drosophila*, *Ac* has terminal inverted repeats and encodes a single protein, a transposase. The nonautonomous *Ds* element does not encode transposase and thus cannot transpose on its own. When *Ac* is in the genome, the transposase it encodes can bind to the ends of *Ac* or *Ds* elements and promote their transposition (**Figure 16-19**).

Ac and *Ds* are members of a single transposon family, and there are other families of transposable elements in maize. Each family contains autonomous elements encoding a transposase that can mobilize elements in the same family but cannot mobilize elements in other families because the transposase can bind only to the specific inverted repeat DNA sequence at the end of the elements of its family members. Although some organisms such as yeast have no DNA transposons, elements structurally similar to the *P* or *Ac* elements have been isolated from many plant and animal species.

> **KEY CONCEPT** DNA transposons structurally resemble DNA bacterial IS elements and are found in many eukaryotes. DNA transposons encode a transposase that cuts the transposon from the chromosome and catalyzes its reinsertion at other chromosomal locations.

Utility of DNA transposons as tools for genetic research

Quite apart from their interest as a genetic phenomenon, DNA transposons have become important tools for

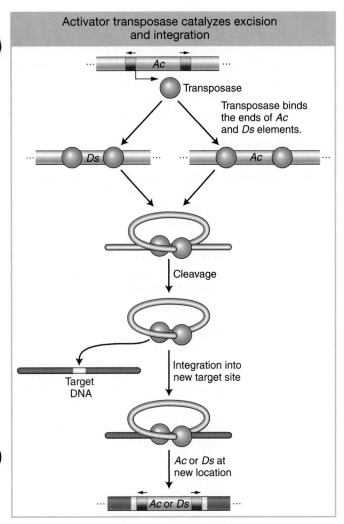

Activator transposase catalyzes excision and integration

Ac

Transposase

Transposase binds the ends of *Ac* and *Ds* elements.

Ds   *Ac*

Cleavage

Target DNA

Integration into new target site

Ac or *Ds* at new location

Ac or *Ds*

FIGURE 16-19 The *Ac* element in maize encodes a transposase that binds its own ends or those of a *Ds* element, excising the element, cleaving the target site, and allowing the element to insert elsewhere in the genome.

geneticists working with a variety of organisms. Their mobility has been exploited to generate new mutations in genes and to insert transgenes into genomes.

> **KEY CONCEPT** A knowledge of the genetics and molecular biology of transposons has enabled scientists to harness them as tools to clone genes and create transgenic organisms.

Using transposons in reverse genetics The complete sequence of the genomes of all model organisms are now available, revealing that eukaryotes typically contain 20,000 to 40,000 genes (Chapter 14). The function of the vast majority of these genes is unknown, and mutant alleles of them are unavailable. Creating mutant alleles by transposon insertions offers a powerful means to interrogate the functions of all the genes in a genome. But how can this be done?

In maize, researchers have used a DNA transposon called *Mutator* (*Mu*) to create insertion alleles in many genes. About 40 nonautonomous *Mu* elements are in the genome

of typical maize inbred lines. The active autonomous *Mu* element, called *MuDr*, is only in a few special maize stocks. By crossing a stock with the active *MuDr* to one that lacks *MuDr*, the nonautonomous elements are activated in the F$_1$. Activation can be confirmed using an allele of *Bronze* locus (*bz-mum*) that has a *Mu* insertion (**Figure 16-20a**). When *Mu*

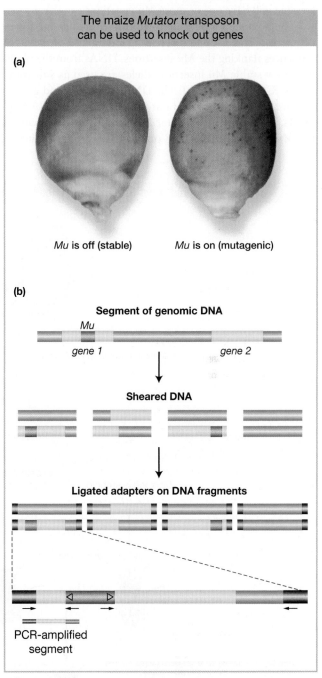

The maize *Mutator* transposon can be used to knock out genes

(a)

Mu is off (stable)   *Mu* is on (mutagenic)

(b)

Segment of genomic DNA

Mu

gene 1   gene 2

Sheared DNA

Ligated adapters on DNA fragments

PCR-amplified segment

FIGURE 16-20 Use of *Mutator* (*Mu*) transposon to create insertion alleles to tag and clone maize genes. (a) *Mu* is inserted in the *Bronze* locus so kernels have a stable bronze color when the autonomous *MuDr* element is not present (*left*). When *MuDr* is present, *Mu* is active, excises from *Bronze*, producing blue sectors (*right*). (b) Flow chart for the steps needed to clone genes into which Mu has inserted. See text for details. [*Part (a) John Doebley.*]

excises from *Bronze*, there are small blue sectors on a bronze background in the kernels. Once activated, the nonautonomous *Mu* elements cause new mutations in many genes in the F₁'s. When the F₁'s are selfed, 3/4 of the progeny will have *MuDr* and sectored kernels, but ¼ will lack *MuDr* and have stable bronze kernels. Because *MuDr* was segregated away in the unsectored bronze kernels, the new insertion alleles that they carry will be stable mutant alleles. If one creates a large enough population, then potentially every gene in the genome will have one or more *Mu* insertions.

The next step is to find a *Mu* insertion allele in a gene of interest. This is done by creating a database of the DNA sequences flanking the *Mu* insertions. DNAs from the plants with new stable *Mu* insertion alleles are used as substrates for PCR (Figure 16-20b). The DNA is sheared and a short adapter sequence is ligated onto the ends of the DNA fragments. Then, PCR is performed with a primer that matches the *Mu* terminal repeat and one for the adapter sequence.

NextGen sequencing (Chapter 10) is applied to pools of the PCR products that are tagged with "DNA barcodes" so that the individual sequence reads can be traced back to specific plants. The DNA sequences are collected into a database that can be queried by BLAST (Chapter 14) with the sequence of a known gene from the maize genome sequence. The barcode is used to trace the sequence back to a specific parent plant for which selfed seed has been saved and which will segregate for the *Mu* insertion allele in the gene of interest.

Using *P* elements to insert genes Geneticists have also shown that *P*-element DNA can be an effective vehicle for transferring donor genes into the germ line of a recipient fly; that is, to make a transgenic fly. They devised the following experimental procedure (**Figure 16-21**). Suppose the goal is to transfer the wild-type allele of *rosy* (*ry*⁺), which confers a characteristic eye color, into the fly genome. The recipient genotype is homozygous for the *rosy* (*ry*⁻) mutation. From this strain, embryos are

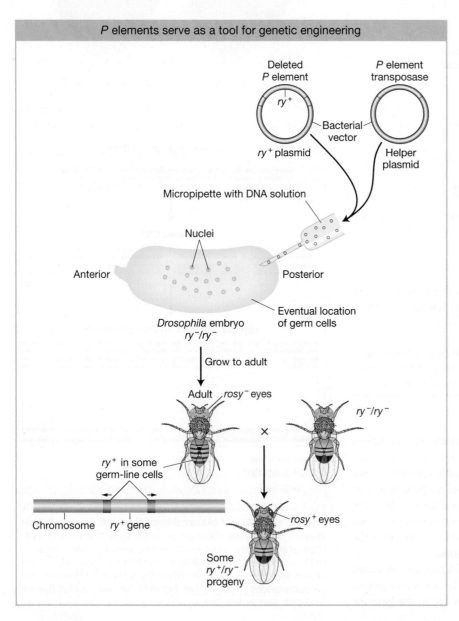

P elements serve as a tool for genetic engineering

FIGURE 16-21 *P*-element-mediated gene transfer in *Drosophila*. The *rosy*⁺ (*ry*⁺) eye-color gene is engineered into a deleted *P* element carried on a bacterial vector. At the same time, a helper plasmid bearing an intact *P* element transposase is used. Both are injected into an *ry*⁻/*ry*⁻ embryo, where *ry*⁺ transposes with the *P* element terminal repeats into the chromosomes of the germ-line cells.

collected at the completion of about nine nuclear divisions. At this stage, the embryo is one multinucleate cell, and the nuclei destined to form the germ cells are clustered at one end. Two types of DNA are injected into embryos of this type. The first is a bacterial plasmid carrying a defective *P* element into which the *ry*+ gene has been inserted. The defective *P* element does not encode transposase but still has the ends that bind transposase and allow transposition. This defective element is not able to transpose, and so, a second (helper) plasmid encoding transposase but without the terminal repeats (so it cannot transpose) also is injected. Flies developing from these embryos are phenotypically still *rosy* mutants, but their offspring include a large proportion of *ry*+ flies. Follow-up experiments show that the *ry*+ gene within the deleted *P* element was inserted into one of several distinct chromosome locations. These new *ry*+ genes are stable and inherited in a Mendelian fashion.

KEY CONCEPT DNA transposons have been used by scientists in two important ways: (1) to make new mutant alleles with transposon insertions; and (2) to serve as vectors that can introduce foreign genes into a chromosome.

16.4 THE DYNAMIC GENOME: MORE TRANSPOSABLE ELEMENTS THAN EVER IMAGINED

LO 16.5 Predict the short-term and evolutionary fate of transposable elements in a species.

Once geneticists had discovered transposable elements and determined the DNA sequences and molecular structures, new sets of questions could be addressed. How prevalent are transposable elements in genomes? Are there other families of transposable elements in the genome that have remained unknown because they had not caused a mutation that could be studied in the laboratory? Do all organisms have transposable elements in their genomes? In this section, we will review how these questions were addressed and the answers that were found.

Large genomes are largely transposable elements

Long before the advent of DNA-sequencing projects, scientists using a variety of molecular techniques discovered that genome size varied dramatically among eukaryotic species and did not correlate with an organism's biological complexity. For example, the genomes of salamanders are 20 times as large as the human genome, whereas the genome of barley is more than 10 times as large as the genome of rice. In addition, the genomes of all eukaryotes contain repetitive DNA elements. Specifically, there are some short DNA sequences (hundreds to a few thousand bp in length) that are repeated thousands, even hundreds of thousands, of times in a genome,

and these repetitive sequences can make up a large fraction (up to 90 percent) of eukaryotic genomes.

Thanks to the many projects to sequence the complete genomes of a wide variety of organisms (including *Drosophila*, humans, the mouse, *Arabidopsis*, maize, and rice), we now know that there are many classes of repetitive sequences in the genomes of higher organisms and that some of these repeat elements are similar to the DNA transposons and retrotransposons discussed in this chapter. Most remarkably, these repetitive sequences make up most of the DNA in the genomes of most multicellular eukaryotes. The portion of the genomes that are composed of repetitive sequences varies widely among species.

We now see that variation in genome size results from variation in the numbers of repetitive sequences and not the numbers of genes. Barley and rice both have about 40,000 genes. The 10-fold larger genome of barley is due to a much greater amount of repetitive DNA, most of which is either transposable elements or decayed transposable elements. Rather than correlating with the numbers and sizes of gene in a genome, genome size frequently correlates with the amount of DNA in the genome that is derived from transposable elements. Organisms with big genomes have lots of sequences that resemble transposable elements, whereas organisms with small genomes have many fewer. Two examples, one from the human genome and the other from a comparison of the plant genomes, illustrate this point.

KEY CONCEPT Genes make up only a small proportion of the genomes of multicellular organisms. Genome size usually correlates with the amount of transposable-element sequences and not gene number.

Transposable elements in the human genome

Almost half of the human genome is derived from transposable elements. The vast majority of these transposable elements are two types of retrotransposons called **long interspersed elements**, or **LINEs**, and **short interspersed elements**, or **SINEs** (**Figure 16-22**). LINEs move like a retrotransposon with the help of an element-encoded reverse transcriptase but lack some structural features of retrovirus-like elements, including LTRs (see Figure 16-12d). SINEs can be best described as nonautonomous LINEs because they have the structural features of LINEs but do not encode their own reverse transcriptase. Presumably, they are mobilized by reverse transcriptase enzymes that are encoded by LINEs residing in the genome.

The most abundant SINE in humans is called *Alu*, so named because it contains a target site for the *Alu* restriction enzyme. The human genome contains more than 1 million whole and partial *Alu* sequences, scattered between genes and within introns. These *Alu* sequences make up more than 10 percent of the human genome. The full *Alu* sequence is about 300 nucleotides long and bears a remarkable resemblance to 7SL RNA, an RNA that is part of a complex by which newly synthesized

Types of transposable elements in the human genome					
Element	Transposition	Structure	Length	Copy number	Fraction of genome
LINEs	Autonomous	ORF1 ORF2 *(pol)* AAA	1–5 kb	20,000–40,000	21%
SINEs	Nonautonomous	AAA	100–300 bp	1,500,000	13%
DNA transposons	Autonomous	← transposase →	2–3 kb	300,000	3%
	Nonautonomous	← →	80–3000 bp		

FIGURE 16-22 Several general classes of transposable elements are found in the human genome. [*Data from* Nature *409, 880 (15 February 2001), "Initial Sequencing and Analysis of the Human Genome," The International Human Genome Sequencing Consortium.*]

polypeptides are targeted to the endoplasmic reticulum (see Figure 9-25). Presumably, the *Alu* sequences originated as reverse transcripts of these RNA molecules.

There is about 20 times as much DNA in the human genome derived from transposable elements as there is DNA encoding all human proteins. **Figure 16-23** illustrates how diverse types of transposons are distributed within and between genes in the human genome. *Alu* elements are often found within introns.

If such a large fraction of the human and other eukaryotic genomes is composed of intact or decayed transposable elements, an obvious question arises: how do animals and

plants survive and thrive with so much mobile DNA in their genomes? Several factors come into play. As we will see in Section 16.5, organisms suppress transposon activity so that the elements are inactive most of the time and do not cause new mutations. When transposons do move, they may insert into exons, introns, or noncoding regions of the genome between genes. If a transposon inserts into an exon, it will disrupt the coding sequence, which is apt to destroy protein function. Such a deleterious insertion will be removed from the population by Darwinian selection (Chapters 19 and 20). If they insert into an intron, the mRNA produced by the gene will not include any sequences from transposable elements because they will have

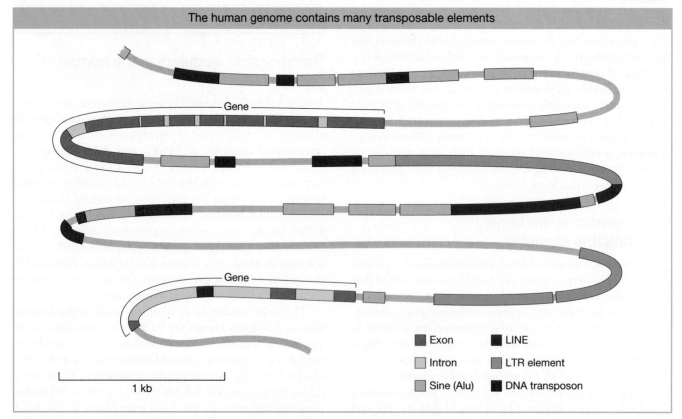

The human genome contains many transposable elements

Gene

Gene

1 kb

■ Exon		■ LINE
▢ Intron		▨ LTR element
▨ Sine (Alu)		■ DNA transposon

FIGURE 16-23 Distribution of transposable elements in a typical segment of the human genome. Transposons are found in introns and intergenic regions, but not in exons of functional alleles.

been spliced out of the pre-mRNA with the surrounding intron. So, many of the transposons that survive in genomes are safely hidden away in introns. Similarly, if they insert in the noncoding regions between genes, they may not affect gene function unless they insert into a regulatory element. Thus, transposons accumulate over evolutionary time in parts of the genome where they do not affect gene function.

When transposons do insert into critical regions of the genome, they can cause severe mutations that remain in the population at least transiently. Three separate insertions of LINEs have disrupted the coagulation factor VIII gene, causing hemophilia A. At least 11 *Alu* insertions into human genes have been shown to cause several diseases, including hemophilia B (in the coagulation factor IX gene), neurofibromatosis (in the *NF1* gene), and breast cancer (in the *BRCA2* gene).

KEY CONCEPT Transposable elements compose the largest fraction of the human genome, with LINEs and SINEs being the most abundant. The vast majority of transposable elements can no longer move or increase their copy number. A few elements remain active, and their movement into genes can cause disease.

Plants: LTR-retrotransposons thrive in large genomes

In plants, differences in the genome sizes of different species have been shown to correlate primarily with the number of one class of elements, the LTR-retrotransposons. Plants share a common biology and homologous organs, including roots, stems, leaves, and flowers, and as such, their genomes are similar with respect to gene content. Despite these similarities, the genome size of plants varies widely, from about 125 Mbp for *Arabidopsis thaliana* (mustard weed) to 5100 Mbp for barley (**Figure 16-24**). For plants like *Arabidopsis* with small genomes, LTR-retrotransposons comprise less than 10 percent of the genome, but for plants like barley with large genomes, this class of transposon comprises about 80 percent of the genome. Most of the expansion in plant genome size is due to growth in the numbers of LTR-retrotransposons.

Safe havens

The abundance of transposable elements in the genomes of multicellular organisms led some investigators to postulate that successful transposable elements (those that attain very high

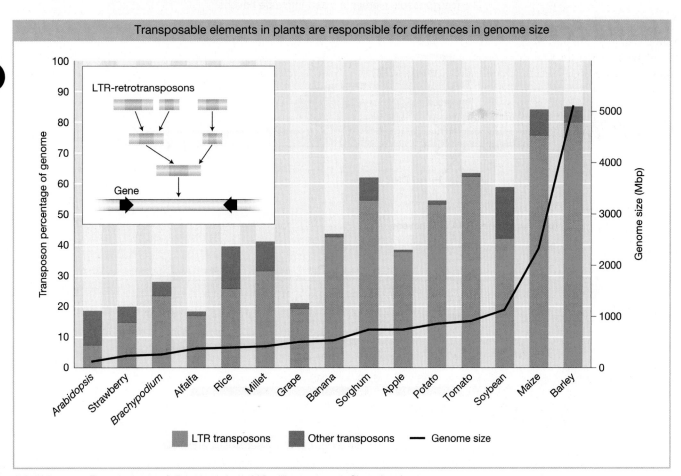

FIGURE 16-24 Flowering plants first evolved about 200 million years ago. Since that time, transposable elements have accumulated to different levels in each species. Maize and barley have genomes with large numbers of LTR-retrotransposons. *Arabidopsis* has a small genome with fewer LTR-retrotransposons. The inset shows how LTR-retrotransposons preferentially insert into other LTR-retrotransposons (safe havens) rather than critical genes.

copy numbers) have evolved mechanisms to prevent harm to their hosts by not inserting into host genes. Instead, successful transposable elements insert into so-called **safe havens** in the genome. For the grasses, a safe haven for new insertions appears to be into other retrotransposons that are located in the regions between genes (see Figure 16-24, insert). Another safe haven is the heterochromatin of centromeres, where there are very few genes but lots of repetitive DNA (see Chapter 12 for more on heterochromatin). Many classes of transposable elements in both plant and animal species tend to insert into the centromeric heterochromatin.

In contrast to the genomes of multicellular eukaryotes, the genome of unicellular yeast is very compact, with closely spaced genes and very few introns. Because almost 70 percent of its genome consists of exons, there is a high probability that new insertions of transposable elements will disrupt a coding sequence. Yet, as we have seen earlier in this chapter, the yeast genome supports a collection of LTR-retrotransposons called *Ty* elements.

How are transposable elements able to spread to new sites in genomes with few safe havens? Investigators have identified hundreds of *Ty* elements in the sequenced yeast genome and have determined that they are not randomly distributed. Instead, each family of *Ty* elements inserts into a specific genomic region. For example, the *Ty3* family inserts almost exclusively near but not in tRNA genes, at sites where they do not interfere with the production of tRNAs and, presumably, do not harm their hosts. This region-specific integration is made possible by a mechanism that evolved in *Ty* elements: the proteins necessary for integration interact with specific yeast proteins bound to genomic DNA. *Ty3* proteins, for example, recognize and bind to subunits of the RNA polymerase complex that have assembled at tRNA promoters (**Figure 16-25a**).

The ability of some transposons to insert preferentially into certain sequences or genomic regions is called **targeting**. A remarkable example of targeting is illustrated by the *R1* and *R2* elements of arthropods, including *Drosophila*. *R1* and *R2*

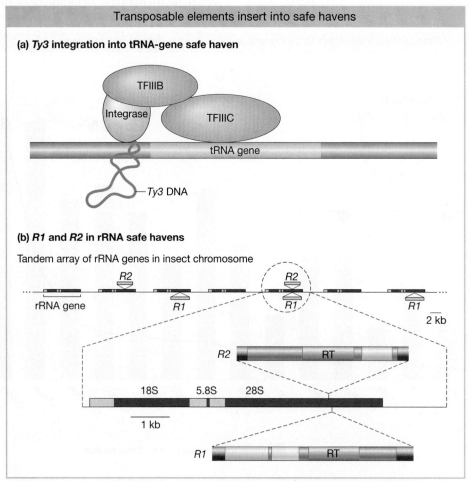

FIGURE 16-25 Some transposable elements are targeted to specific safe havens. (a) The yeast *Ty3* retrotransposon inserts near transfer RNA genes. (b) The *Drosophila R1* and *R2* non-LTR-retrotransposons (LINEs) insert into the genes encoding ribosomal RNA that are found in long tandem arrays on the chromosome. Only the reverse transcriptase (RT) genes of *R1* and *R2* are noted.

are LINEs (see Figure 16-22) that insert only into the genes that produce ribosomal RNA. In arthropods, several hundred rRNA genes are organized in tandem arrays (Figure 16-25b). With so many genes encoding the same product, the host tolerates insertion into a subset. However, too many insertions of *R1* and *R2* have been shown to decrease insect viability, presumably by interfering with ribosome assembly.

KEY CONCEPT A successful transposable element increases copy number without harming its host. One way in which an element safely increases copy number is to target new insertions into safe havens, regions of the genome where there are few genes.

16.5 REGULATION OF TRANSPOSABLE ELEMENT MOVEMENT BY THE HOST

LO 16.5 Predict the short-term and evolutionary fate of transposable elements in a species.

The repression of transposable elements was first investigated in the late 1990s, using the model organism *Caenorhabditis elegans* (a nematode; see the Model Organism box in Chapter 13). This story starts with the observation of a striking difference between the mobility of a transposable element called *Tc*1 in two different cell types of this model organism. *Tc*1 is a DNA transposon that, like the *Ac* element of maize, can lead to an unstable mutant phenotype when it excises from a gene with a visible phenotype. There are 32 *Tc*1 elements in the sequenced genome of the common laboratory strain of *C. elegans*. Significantly, *Tc*1 transposes in somatic but not in germ-line cells. That observation suggested that transposition is repressed in the germ line by the host. Evidently, germ-line repression results from the silencing of the transposase genes of all 32 *Tc*1 copies in germ-line cells.

RNAi silencing of transposable elements

Researchers set out to identify *C. elegans* genes responsible for silencing the transposase gene. They began with a *C. elegans* strain that had *Tc*1 inserted in the *unc-22* gene (designated *unc-22/Tc*1; **Figure 16-26**). Whereas wild-type *C. elegans* glides smoothly on the surface of the agar in a petri dish (as illustrated by horizontal arrows in Figure 16-26), worms with the mutant *unc-22/Tc*1 gene have a twitching movement (as illustrated by vertical arrows in Figure 16-26) that can be easily observed with a microscope. Because *Tc*1 cannot normally transpose in the germ line, it remains inserted in the *unc-22* gene and continues to disrupt its function. Thus, the strain with the mutant *unc-22/Tc*1 gene should express a twitching phenotype from generation to generation. However, researchers reasoned that mutations that inactivated *C. elegans* genes required for repression would allow *Tc*1

to excise from the *unc-22/Tc*1 allele in the germ line and revert the twitching phenotype to wild type (*unc-22*). To this end, they exposed the mutant *unc-22/Tc*1 strain to a chemical (called a mutagen; see Chapter 16) that greatly increased the frequency of mutation and examined their progeny under a microscope, searching for rare worms that no longer twitched.

This and subsequent genetic screens identified over 25 *C. elegans* genes that, when mutated, allowed the host to excise *Tc*1 in the germ line. Significantly, many of the products of these genes are integral components of the RNAi silencing pathway, including proteins found in Dicer and RISC (see Chapters 8 and 12). Recall from Chapter 8 that Dicer binds to long dsRNAs and cleaves them into small dsRNA fragments. These fragments are then unwound so that one strand, the siRNA, can target RISC to chop up complementary mRNAs (see Figure 8-29).

Beginning with this elegant genetic screen, many years of experimentation have led to the following model for the repression of transposable elements in the germ line of *C. elegans*. With 32 *Tc*1 elements scattered throughout the *C. elegans* genome, a few elements near genes are transcribed along with the nearby gene (**Figure 16-27**). Because the ends of *Tc*1 are 54-bp terminal inverted repeats, the *Tc*1 RNA spontaneously forms dsRNA. Like all dsRNAs produced in most eukaryotes, this RNA is recognized by Dicer

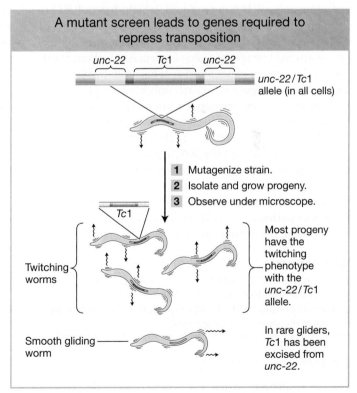

FIGURE 16-26 Experimental design used to identify genes required to repress transposition. Investigators look for mutants that have regained normal movement because mutations in these individuals would have disabled the repression mechanism that prevents the transposition of the *Tc*1 element from the *unc-22* gene.

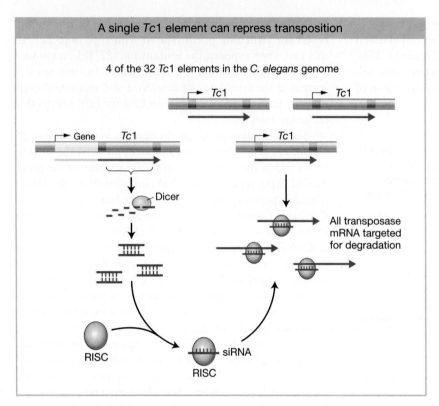

A single *Tc*1 element can repress transposition

4 of the 32 *Tc*1 elements in the *C. elegans* genome

*Tc*1

*Tc*1

Gene *Tc*1

*Tc*1

Dicer

All transposase
mRNA targeted
for degradation

RISC

siRNA

RISC

FIGURE 16-27 The production of dsRNA from only a single *Tc*1 element is sufficient to silence all of the *Tc*1 transposase genes and thereby repress transposition in the germ line. The siRNA derived from *Tc*1 dsRNA is bound to RISC and targets all complementary RNA for degradation.

and ultimately siRNA is produced, which directs RISC to chop up complementary *Tc*1 transcripts. Because all *Tc*1 RNA is efficiently chopped up in the germ line, the element-encoded transposase gene is silenced. Without transposase, the element cannot excise. It has been hypothesized that *Tc*1 can transpose in somatic cells because RNAi is not as efficient and some transposase can be produced.

Over the past decade, numerous laboratories working with both plants and animals have discovered that mutations that disrupt the RNAi pathway often lead to the activation of transposable elements in their respective genomes. Because of the abundance of transposable elements in eukaryotic genomes, it has been suggested that the RNAi pathway evolved to maintain genome stability by repressing the movement of transposable elements.

KEY CONCEPT Eukaryotes use RNAi to repress the expression of active transposable elements in their genomes. In this way, a single element that inserts near a gene can be transcribed to produce dsRNA that will trigger the silencing of all copies of the element in the genome.

Genome surveillance

The RNAi silencing pathway is akin to radar in that enables the host to detect new insertions of transposons into the genome if they generate antisense RNA. The host then responds by producing siRNAs that target the transposase mRNA, silencing the gene and preventing the movement of

all transposable-element (TE) family members. Two other types of genome radar (also called **genome surveillance**) have been described that utilize different classes of small noncoding RNAs to target "invasive" nucleic acids including transposons and viruses (see Chapter 6). These mechanisms are presented here because they illustrate how different solutions evolve to solve similar biological problems.

piRNAs in animals In the germ lines of animal species including *Drosophila*, active transposons are repressed through the action of **piRNAs** (short for Piwi-interacting RNAs). Animal genomes contain several long (often >100 kb) loci called **pi-clusters** that serve as traps that ensnare active transposons as they insert randomly around the genome. A pi-cluster can contain remnants of several different transposon families that represent a historical record of prior insertions of active transposons into it. Long RNAs are transcribed from the pi-clusters and then processed into short single-stranded piRNAs of 23–30 nt in length. piRNAs form an RNA-protein complex with the protein Piwi-Argonaute. The piRNA-Piwi complex has two effects. First, the piRNA-Piwi complex guides the degradation of mRNAs complementary to the piRNA, that is, mRNAs from the transposons that compose the pi-cluster (**Figure 16-28**). By degrading the mRNAs complementary to transposases, any transposons represented in a pi-cluster are silenced. Second, the piRNA-Piwi complex directs the placement of histone marks on the pi-cluster chromatin to promote the transcription of the long RNAs from it.

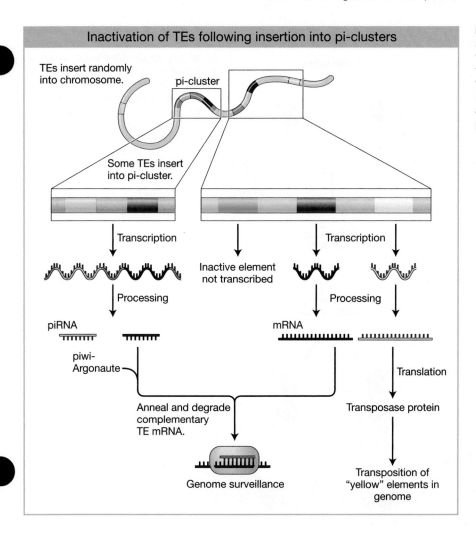

Inactivation of TEs following insertion into pi-clusters

TEs insert randomly into chromosome.

pi-cluster

Some TEs insert into pi-cluster.

Transcription

Transcription

Inactive element not transcribed

Processing

Processing

piRNA

mRNA

piwi-Argonaute

Anneal and degrade complementary TE mRNA.

Translation

Transposase protein

Genome surveillance

Transposition of "yellow" elements in genome

FIGURE 16-28 Insertion of the green and purple transposons into a pi-cluster in the genome results in the degradation of transcripts from these two transposons by the steps shown and described in the text. In contrast, the yellow transposon will remain active until copies insert by chance into a pi-cluster.

This latter step enables the production of the piRNAs needed for the first step.

piRNAs and hybrid dysgenesis in *Drosophila* Now, let's return to *Drosophila P* elements. As discussed earlier, *P* elements entered the genome of wild *D. melanogaster* a little over 100 years ago. At first, they were probably active and jumping, causing many mutations. As the elements inserted randomly around the genome, eventually one or more landed in a pi-cluster, after which *P* elements could be repressed in wild *Drosophila* by the piRNA-Piwi complex. Since the laboratory strains of *D. melanogaster* lacked *P* elements, so too did their pi-clusters. Thus, there would be no means for the piRNA-Piwi complex to repress *P* element activity in lab stocks.

Next, let's consider why hybrid dysgenesis does not occur when a *P* cytotype female is mated with an M cytotype male but does occur in the reciprocal mating. The *P* elements of *Drosophila* are repressed in the germ line by piRNA-Piwi early in embryonic development (**Figure 16-29**, right). Repression is initially activated by maternally produced piRNAs from the P female that are loaded into the embryo and

deposited in cells of the primordial germ line. Here, these piRNAs direct histone marks to be placed on the pi-clusters, allowing them to be transcribed to make piRNAs for the piRNA-Piwi complex. Essentially, the maternal piRNAs from the *P* female jump-start the system to repress *P* elements and allow healthy $P(\text{female}) \times M(\text{male})F_1$'s to develop.

In the reciprocal cross, an M female has no P elements in her pi-clusters, and so no maternal piRNAs complementary to *P* elements are loaded into her embryos (**Figure 16-29**, left). Sperm from a *P* stock do not deposit piRNAs into the embryo. The result is that the F_1 has *P* elements from the male parent but no maternally loaded piRNAs to activate the piRNA-Piwi complex to repress their movement. As development proceeds, the *P* elements are mobilized, cause genomic havoc in the germ line, and the $M(\text{female}) \times P(\text{male})F_1$'s are dysgenic.

KEY CONCEPT Like siRNAs, piRNAs in animals interact with protein complexes and guide them to degrade complementary sequences in transposons. These small noncoding RNAs have their origin in long RNAs transcribed from pi-clusters that capture fragments of invasive DNA.

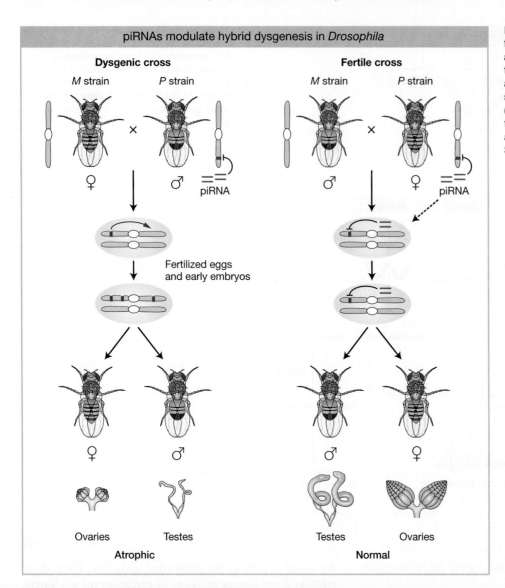

FIGURE 16-29 Dysgenesis and fertile crosses between a laboratory and wild stock of *Drosophila*. When the lab strain is the female, the F₁'s are dysgenic (sterile). When the lab strain is the male, the F₁'s are normal (fertile). piRNAs produced in *P* cytotype females are deposited in her embryos and repress *P* element transposition. See text for further details.

SUMMARY

Transposable elements were discovered in maize by Barbara McClintock as an astounding feature of the genome, stretches of DNA (loci) that could move from one location to another. McClintock determined that some transposons, like *Ac*, are autonomous and able to move on their own; but others, like *Ds*, are nonautonomous and require the presence of an autonomous element for mobilization. Transposons that insert into genes can produce unstable mutant alleles that revert to wild type when the element excises.

Bacterial insertion-sequence elements were the first transposable elements isolated molecularly. There are many different types of IS elements in *E. coli* strains, and they are usually present in at least several copies. Composite transposons contain IS elements flanking one or more genes, such as genes conferring resistance to antibiotics. Transposons with resistance genes can insert into plasmids and can then be transferred by conjugation to nonresistant bacteria.

There are two major groups of transposable elements in eukaryotes: class 1 elements (retrotransposons) and class 2

elements (DNA transposons). The *P* element of *Drosophila* was the first eukaryotic class 2 DNA transposon to be isolated molecularly. It was isolated from unstable mutations that were induced in crosses between P and M cytotype strains. When *P* elements are mobilized in P × M crosses, the F₁'s exhibit hybrid dysgenesis. *P* elements have been developed into vectors for the introduction of foreign DNA into *Drosophila* germ cells.

Ac, *Ds*, and *P* are examples of DNA transposons, so named because the transposition intermediate is the DNA element itself. Autonomous elements such as *Ac* encode a transposase that binds to the ends of autonomous and nonautonomous elements and catalyzes excision of the element from the donor site and reinsertion into a new target site elsewhere in the genome.

Retrotransposons were first molecularly isolated from yeast mutants, and their resemblance to retroviruses was immediately apparent. Retrotransposons are class 1 elements, as are all transposable elements that use RNA as their transposition intermediate.

DNA sequencing of whole genomes, including the human genome, has led to the remarkable finding that almost half of the human genome is derived from transposable elements. For some plants like maize, transposable elements compose fully 80 percent of the DNA in the genome. Despite having so many transposable elements, eukaryotic genomes are extremely stable, as transposition is relatively rare because of two factors. First, most of the transposable elements in eukaryotic genomes cannot move because inactivating mutations prevent the production of normal transposase and reverse transcriptase. Second, expression of the vast majority of the remaining elements is silenced by the RNAi and/or the piRNA pathways.

KEY TERMS

Activator (*Ac*) element (p. 530)
Alu (p. 545)
autonomous transposable element (p. 533)
class 1 element (retrotransposon) (p. 540)
class 2 element (DNA transposon) (p. 541)
cointegrate (p. 537)
composite transposon (p. 535)
conservative transposition (p. 536)
copia-like element (p. 540)
Dissociation (*Ds*) element (p. 530)
DNA transposon (p. 541)
excise (p. 533)
genome surveillance (p. 550)
hybrid dysgenesis (p. 541)

insertion-sequence (IS) element (p. 534)
inverted repeat (IR) sequence (p. 534)
long interspersed element (LINE) (p. 545)
long terminal repeat (LTR) (p. 538)
LTR-retrotransposon (p. 540)
M cytotype (p. 541)
nonautonomous transposable element (p. 533)
P cytotype (p. 541)
P element (p. 541)
pi-cluster (p. 550)
piRNAs (p. 550)
provirus (p. 539)
replicative transposition (p. 536)
retrotransposon (p. 540)

retrovirus (p. 539)
reverse transcriptase (p. 539)
R plasmid (p. 535)
safe haven (p. 548)
short interspersed element (SINE) (p. 545)
simple transposon (p. 535)
solo LTR (p. 540)
targeting (p. 548)
target-site duplication (p. 537)
transposable element (p. 528)
transposase (p. 533)
transpose (p. 533)
transposition (p. 536)
transposon (p. 528)
Ty element (p. 538)
unstable allele (p. 533)

PROBLEMS

SOLVED PROBLEM 1

Transposable elements have been referred to as "jumping genes" because they appear to jump from one position to another, leaving the old locus and appearing at a new locus. In light of what we now know concerning the mechanism of transposition, how appropriate is the term "jumping genes" for bacterial transposable elements?

SOLUTION

In bacteria, transposition takes place by two different modes. The conservative mode results in true jumping genes because, in this case, the transposable element excises from its original position and inserts at a new position. The other mode is the replicative mode. In this pathway, a transposable element moves to a new location by replicating into the target DNA, leaving behind a copy of the transposable element at the original site. When operating by the replicative mode, transposable elements are not really jumping genes because a copy does remain at the original site.

SOLVED PROBLEM 2

Following from Solved Problem 1, in light of what we now know concerning the mechanism of transposition, how appropriate is the term "jumping genes" for the vast majority of transposable elements in the human genome and in the genomes of most other mammals?

SOLUTION

The vast majority of transposable elements in the characterized mammalian genomes are retrotransposons. In humans, two retrotransposons (the LINE called *L1* and the SINE called *Alu*) account for fully one-third of the entire genome. Retrotransposons do not excise from the original site, so they are not really jumping genes. Instead, the element serves as a template for the transcription of RNAs that can be reverse-transcribed by the enzyme reverse transcriptase as a step toward synthesis of a double-stranded DNA complementary to the retrotransposon. Each double-stranded DNA can potentially insert into target sites throughout the genome.

PROBLEMS

Visit SaplingPlus for supplemental content. Problems with the icon are available for review/grading. Problems with the ⟩ icon have an Unpacking the Problem exercise.

WORKING WITH THE FIGURES

(The first 8 questions require inspection of text figures.)

1. For Figure 16-3b, if the *Ds* element were located between *C* and *Sh*, what phenotype would you expect in the sectors?

2. In Figure 16-5, the fly has a *mariner* transposon inserted in the *white* locus.

    a. The wild-type allele is w^+. Make up an allele designation for the *mariner* insertion allele.

    b. Referring back to what you learned in Chapter 2, what is the genotype for the *white* locus in the red and white sectors if the fly shown is male and if it is female?

3. For Figure 16-8, draw out a series of steps that could explain the origin of this large plasmid containing many transposable elements.

4. Draw a figure for the mode of transposition not shown in Figure 16-9, retrotransposition.

5. In Figure 16-11, show where the transposase would have to cut to generate a 6-bp target-site duplication. Also show the location of the cut to generate a 4-bp target-site duplication.

6. If the transposable element in Figure 16-15 were a DNA transposon that had an intron in its transposase gene, would the intron be removed during transposition? Justify your answer.

7. For Figure 16-23, draw the pre-mRNA that is transcribed from one of the genes shown and then draw its mRNA.

8. For Figure 16-29, how would you modify this figure for a P cytotype × P cytotype cross and an M cytotype × M cytotype cross?

BASIC PROBLEMS

9. At a garden shop, you notice that there are three varieties of petunia for sale: one with blue pigmented corollas, one with white corollas, and one with small patches (sectors) of blue on a white background. What hypotheses would you propose to explain these three phenotypes? Can you design any experiments to test your hypotheses?

10. Working with the petunias from Problem 9, you isolate DNA from leaves of each and design PCR primers to amplify the chalcone synthase gene, an enzyme involved in pigment biosynthesis. If sectored flowers are caused by a transposon in the chalcone synthase gene, how would this be manifest by the results of the PCR with the three leaf DNA samples? The PCR with leaf DNA indicates there is a transposon in the chalcone synthase gene. How could you investigate this hypothesis further using isolated DNA from the corollas of the sectored flowers?

11. Propose a model for the generation of a multiple-drug-resistant plasmid.

12. Propose an experiment to prove that the transposition of the *Ty* element in yeast takes place through an RNA intermediate.

13. Explain how the properties of *P* elements in *Drosophila* make gene-transfer experiments possible in this organism.

14. Although DNA transposons are abundant in the genomes of multicellular eukaryotes, class 1 elements usually make up the largest fraction of very large genomes. Given what you know about class 1 and class 2 elements, what is it about their distinct mechanisms of transposition that would account for this consistent difference in abundance?

15. As you saw in Figure 16-23, the genes of multicellular eukaryotes often contain many transposable elements. Why do most of these elements not affect the expression of genes?

16. What are safe havens? Are there any places in the much more compact bacterial genomes that might be a safe haven for insertion elements?

17. Nobel Prizes are usually awarded many years after the actual discovery. For example, James Watson, Francis Crick, and Maurice Wilkens were awarded the Nobel Prize in Medicine or Physiology in 1962, almost a decade after their discovery of the double-helical structure of DNA. However, Barbara McClintock was awarded the Nobel Prize in 1983, almost four decades after her discovery of transposable elements in maize. Why do you think it took this long for the significance of her discovery to be recognized in this manner?

18. Why can't retrotransposons move from one cell to another like retroviruses?

    a. Because they do not encode the Env protein.

    b. Because they are nonautonomous elements.

    c. Because they require reverse transcriptase.

    d. Both a and b are true.

19. Which of the following is true of reverse transcriptase?

   a. It is required for the movement of DNA transposons.

   b. It catalyzes the synthesis of DNA from RNA.

   c. It is required for the transposition of retrotransposons.

   d. Both b and c are correct.

20. What is the major reason why the barley genome is much larger than the rice genome?

   a. Barley has more genes than rice.

   b. Rice has more genes than maize.

   c. Barley has more DNA transposons than rice.

   d. Barley has more retrotransposons than rice.

21. If 80% or even 90% of an organism's genome is composed of largely decayed retrotransposons and other classes of transposons, what cost, if any, would you expect this to exert on the organism? Why do you suppose the organism does not simply delete all of this "junk" DNA?

CHALLENGING PROBLEMS

22. The insertion of transposable elements into genes can alter the normal pattern of expression. In the following situations, describe the possible consequences on gene expression.

   a. A LINE inserts into an enhancer of a human gene.

   b. A transposable element contains a binding site for a transcriptional repressor and inserts adjacent to a promoter.

   c. An *Alu* element inserts into the 3′ splice (AG) site of an intron in a human gene.

   d. A *Ds* element that was inserted into the exon of a maize gene excises imperfectly and leaves three base pairs behind in the exon.

   e. Another excision by that same *Ds* element leaves two base pairs behind in the exon.

   f. A *Ds* element that was inserted into the middle of an intron excises imperfectly and leaves five base pairs behind in the intron.

23. Before the integration of a transposon, its transposase makes a staggered cut in the host target DNA. If the staggered cut is at the sites of the arrows below, draw what the sequence of the host DNA will be after the transposon has been inserted. Represent the transposon as a rectangle.

$$\downarrow$$
AATTTGGCCTAGTACTAATTGGTTGG
TTAAACCGGATCATGATTAACCAACC
$$\uparrow$$

24. In *Drosophila*, there is a *singed* allele (*sn^{cm}*) with some unusual characteristics. Females homozygous for this X-linked allele have singed bristles, but they have numerous patches of *sn^+* (wild-type) bristles on their heads, thoraxes, and abdomens. When these flies are mated with *sn^-* males, some females give only singed progeny, but others give both singed and wild-type progeny in variable proportions. Explain these results.

25. Consider two maize plants:

   a. Genotype *C/c^m*; *Ac/Ac^+*, where *c^m* is an unstable allele caused by a *Ds* insertion

   b. Genotype *C/c^m*, where *c^m* is an unstable allele caused by *Ac* insertion

   What kernel phenotypes would be produced and in what proportions when (1) each plant is crossed with a stable recessive loss of function mutant *c/c* and (2) the plant in part *a* is crossed with the plant in part *b*? Assume that *Ac* and *c* are unlinked, that the chromosome-breakage frequency is negligible, and that mutant *c/c* is *Ac^+*.

26. The yeast *His^-* mutant with the *Ty* retroelement insertion is unstable and undergoes a high rate of reversion to wild-type *His^+*. The high rate of reversion is notable because retroelements have no means for excision. The mechanism of reversion is recombination between the direct repeats to leave only a solo LTR, while the rest of the element is deleted. Diagram the recombination event that would give this outcome.

27. DNA transposons use a "cut and paste" mechanism by which they are excised from one genomic location and pasted into another. This is a conservative or nonreplicative process. Yet, DNA elements, like the *Drosophila P* element, can enter a genome as a single copy and rise to have 30 or more copies over time. Describe several mechanisms by which the copy number of such elements increases over time in a species.

28. The evidence indicates that *Drosophila P* elements first entered the *D. melanogaster* genome in the Americas in the early 1900s and then spread worldwide throughout all *D. melanogaster* populations by the 1980s such that today there are no longer wild flies that lack *P* elements. How can a fundamentally parasitic DNA element like *P* spread so rapidly and pervasively when its effect is to cause deleterious mutations by inserting into genes?

29. You are leading the bioinformatics team analyzing a newly discovered species of nematode related to *C. elegans*. You have its complete genome DNA sequence and a database of all the RNAs expressed in this species. How would you proceed to (1) identify all class 1 and class 2 transposons, (2) distinguish between autonomous and nonautonomous elements, and (3) determine if any of the family of elements is active?

30. The yeast genome has class 1 elements (*Ty1*, *Ty2*, and so forth) but no class 2 elements. What is a possible reason why DNA elements have not been successful in the yeast genome?

31. In addition to *Tc*1, the *C. elegans* genome contains other families of DNA transposons such as *Tc*2, *Tc*3, *Tc*4, and *Tc*5. Like *Tc*1, their transposition is repressed in the germ line but not in somatic cells. Predict the behavior of these elements in the mutant strains where *Tc*1 is no longer repressed due to mutations in the RNAi pathway. Justify your answer.

32. Based on the mechanism of gene silencing, what features of transposable elements does the RNAi pathway exploit to ensure that the host's own genes are not also silenced? 🔵

33. What are the similarities and differences between retroviruses and retrotransposons? It has been hypothesized that retroviruses evolved from retrotransposons. Do you agree with this model? Justify your answer.

34. You have isolated a transposable element from the human genome and have determined its DNA sequence. How would you use this sequence to determine the copy number of the element in the human genome based on analytical methods you learned in Chapters 10 and 14? 🔵

35. Devise a genetic screen to identify the molecular components of the mechanism that represses *P* element transposition in *Drosophila*. Describe the genetic stocks you would use, how you would treat them, any crosses you would make, and the results that you expect to obtain.

36. When a typical maize solo *Ds* element excises, the two ends of the chromosome at the excision site are normally ligated back together. Thus, when *Ds* is in a gene, the gene can revert to wild type when *Ds* excises. The first *Ds* element that McClintock discovered was unusual in that it induced a high frequency of chromosome breakage because the broken ends of the chromosome were not ligated back together. Suppose two *Ds* elements are close neighbors on a chromosome. Diagram a mechanism by which the transposase could cause an excision event involving these neighbor elements that promotes the failure of the two ends of the chromosome to be ligated back together.

37. McClintock mapped *Ac* and found that it mapped to different locations in different crosses. Let's consider a cross that would allow us to map *Ac* relative to the *Sugary* (*Su*) locus on chromosome 4 which has a

recessive allele (*su*) that conditions kernels that accumulate sugar but not starch (*su* is the sweet corn allele). We have two parent stocks—one stock is homozygous *c-m1/c-m1* on chromosome 9 and homozygous *Su/Su* for starchy kernels on chromosome 4; the other stock is homozygous *C/C* and *su/su*, and homozygous for an *Ac* element at an unknown location. Remember, *c-m1* is an unstable allele containing a *Ds* element. You cross these two stocks and self-pollinate their F_1.

a. What are the phenotypes of the two parent stocks?

b. Make a table showing the genotypes and phenotypes and their proportions among the F_2's, given that the *Ac* is so tightly linked to *su* that there are no crossovers between them.

c. Make the same table, given that the *Ac* is on a chromosome other than chromosomes 4 and 9.

38. You are using nitrosoguanidine to "revert" mutant nic-2 (nicotinamide-requiring) alleles in *Neurospora*. You treat cells, plate them on a medium without nicotinamide, and look for prototrophic colonies. You obtain the following results from two mutant alleles.

a. With nic-2 allele 1, you obtain no prototrophs at all.

b. With nic-2 allele 2, you obtain three prototrophic colonies, A, B, and C, and you cross each separately with a wild-type strain of *Neurospora*. From the cross prototroph A × wildtype, you obtain 100 progeny, all of which are prototrophic. From the cross prototroph B × wildtype, you obtain 100 progeny, of which 78 are prototrophic and 22 are nicotinamide requiring. From the cross prototroph C × wildtype, you obtain 1000 progeny, of which 996 are prototrophic and 4 are nicotinamide requiring. Explain these results at the molecular level. 🧬

GENETICS AND SOCIETY

Researchers in France have developed a technique to transiently mobilize transposable elements in crop plants to rapidly generate large numbers of new mutations with the expectation that some of these mutants will improve the crop. Plant breeders would screen the mutants and identify those that are beneficial. In the past, plant breeders have used X-ray mutagenesis to create new mutations in crop plants. The technique does not involve "genetic engineering," the insertion of foreign genes into a species. How might consumers or governmental agencies respond to this new technology? How would you guide their decision-making process based on your knowledge of transposons?

Large-Scale Chromosomal Changes

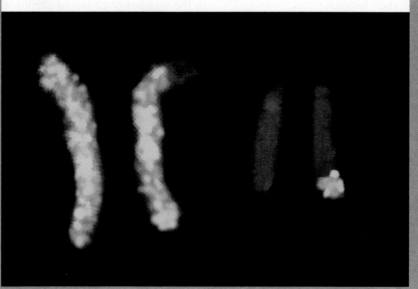

A reciprocal translocation demonstrated by chromosome painting. A suspension of chromosomes from many cells is passed through an electronic device that sorts the chromosomes by size. DNA is extracted from individual chromosomes, denatured, bound to one of several fluorescent dyes, and then added to partly denatured chromosomes on a slide. The fluorescent DNA "finds" its own chromosome and binds along its length by base complementarity, thus "painting" it. In this example, a red and a green dye have been used to paint different chromosomes. The figure shows unpainted (above) and painted (below) preparations. The painted preparation shows one normal green chromosome, one normal red, and two that have exchanged segments. [SPL/Science Source.]

The broad objective for this chapter is to distinguish between the major types of chromosomal mutations using cytological and genetic data, and to predict the effects of chromosomal mutations on organismal phenotypes.

A young couple is planning to have children. The husband knows that his grandmother had a child with Down syndrome by a second marriage. Down syndrome is a set of physical and mental disorders caused by the presence of an extra chromosome 21 (**Figure 17-1**). No records of the birth, which occurred early in the twentieth century, are available, but the couple knows of no other cases of Down syndrome in their families.

The couple has heard that Down syndrome results from a rare chance mistake in egg production and therefore decide that they stand only a low chance of having such a child. They decide to have children. Their first child is unaffected, but the next conception aborts spontaneously (a miscarriage), and their second child is born with Down syndrome. Was their having a Down syndrome child a coincidence, or did a connection between the genetic makeup of the child's father and that of his grandmother lead to their both having Down syndrome children? Was the spontaneous abortion significant? What tests might be necessary to investigate this situation? The analysis of such questions is the topic of this chapter.

We have seen throughout the book that gene mutations are an important source of change in the genomic sequence. However, the genome can also be remodeled on a larger scale by alterations to chromosome structure or by changes in the number of copies of chromosomes in a cell. These large-scale variations are termed **chromosome mutations** to distinguish them from gene mutations. Broadly speaking, gene mutations are defined as changes that take place within a gene, whereas chromosome mutations are changes in a chromosome region encompassing multiple genes. Gene mutations are never detectable microscopically; a chromosome bearing a gene mutation looks the same under the microscope as one carrying the wild-type allele. In contrast, many chromosome mutations can be detected by microscopy, by genetic or molecular analysis, or by a combination of these techniques. Chromosome mutations have been best characterized in eukaryotes, and all the examples in this chapter are from that group.

Chromosome mutations are important to scientists and clinicians for several reasons. First, they can be sources of insight into how genes act in concert on a genomic scale. Second, they reveal several important features of meiosis and chromosome architecture. Third, they constitute useful tools for experimental genomic manipulation. Fourth, they are sources of insight into evolutionary processes. Fifth, chromosomal mutations are regularly found in humans, and some of these mutations cause genetic disease.

Many chromosome mutations cause abnormalities in cell and organismal function. Most of these abnormalities stem from changes in *gene number* or *gene position*. In some cases, a chromosome mutation results from chromosome breakage. If the break occurs within a gene, the result is functional *disruption* of that gene.

For our purposes, we will divide chromosome mutations into two groups: changes in chromosome *number* and changes in chromosome *structure*. These two groups represent two fundamentally different kinds of events. Changes in chromosome number are not associated with structural alterations of any of the DNA molecules of the cell. Rather, it is the *number* of these DNA molecules that is changed, and this change in number is the basis of their genetic effects. Changes in chromosome structure, on the other hand, result in novel sequence arrangements within one or more DNA double helices. These two types of chromosome mutations are illustrated in **Figure 17-2**, which is a summary of the topics of this chapter. We begin by exploring the nature and consequences of changes in chromosome number.

Child with Down syndrome

FIGURE 17-1 Down syndrome results from having an extra copy of chromosome 21. [*Terry Harris/Shutterstock.*]

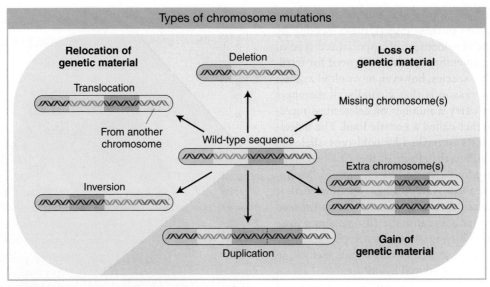

FIGURE 17-2 The illustration is divided into three colored regions to depict the main types of chromosome mutations that can occur: the loss, gain, or relocation of entire chromosomes or chromosome segments. The wild-type chromosome is shown in the center.

17.1 CHANGES IN CHROMOSOME NUMBER

LO 17.1	Distinguish among the different types of polyploidy and predict their effects on meiosis and subsequent progeny.
LO 17.2	Predict the outcome of first and second division meiotic nondisjunction.

In genetics as a whole, few topics impinge on human affairs quite so directly as that of changes in the number of chromosomes present in our cells. Foremost is the fact that a group of common genetic disorders results from the presence of an abnormal number of chromosomes. Although this group of disorders is small, these disorders are the leading genetic cause of miscarriage, birth defects, and developmental disabilities in humans. Also of relevance to humans is the role of chromosome mutations in plant breeding: plant breeders have routinely manipulated chromosome number to improve commercially important agricultural crops.

Changes in chromosome number are of two basic types: changes in *whole* chromosome sets, resulting in a condition called *aberrant euploidy*, and changes in *parts* of chromosome sets, resulting in a condition called *aneuploidy*.

Aberrant euploidy

Organisms with multiples of the basic chromosome set (genome) are referred to as **euploid**. You learned in earlier chapters that familiar eukaryotes such as plants, animals, and fungi carry in their cells either one chromosome set (haploidy) or two chromosome sets (diploidy). In these species, both the **haploid** and the **diploid** states are cases of normal euploidy. Organisms that have more or fewer than the normal number of sets are aberrant euploids. **Polyploids**

are individual organisms that have more than two chromosome sets. They can be represented by 3n (**triploid**), 4n (**tetraploid**), 5n (**pentaploid**), 6n (**hexaploid**), and so forth. The number of chromosome sets is called the ploidy or ploidy level. The number of chromosomes in a set is called the **haploid chromosome number**. An individual member of a normally diploid species that has only one chromosome set (n) is called a **monoploid** to distinguish it from an individual member of a normally haploid species (also n). Examples of these conditions are shown in **Table 17-1**.

TABLE 17-1 Chromosome Constitutions in a Normally Diploid Organism with Three Chromosomes (Identified as A, B, and C) in the Basic Set*

Name	Designation	Constitution	Number of chromosomes
Normal Euploid			
Diploid	2n	AA BB CC	6
Aberrant Euploids			
Monoploid	n	A B C	3
Triploid	3n	AAA BBB CCC	9
Tetraploid	4n	AAAA BBBB CCCC	12
Aneuploids			
Monosomic	2n − 1	A BB CC	5
		AA B CC	5
		AA BB C	5
Trisomic	2n + 1	AAA BB CC	7
		AA BBB CC	7
		AA BB CCC	7

*In the case shown, the number of chromosomes in the basic set (the haploid chromosome number) is three.

Monoploids Male bees, wasps, and ants are monoploid. In the normal life cycles of these insects, males develop by **parthenogenesis** (the development of a specialized type of unfertilized egg into an embryo without the need for fertilization). In most other species, however, monoploid zygotes fail to develop. The reason is that virtually all members of a diploid species carry a number of deleterious recessive mutations, together called a **genetic load**. The deleterious recessive alleles are masked by wild-type alleles in the diploid condition, but the effects of these alleles can be observed in a monoploid derived from a diploid. Monoploids that do develop to advanced stages are abnormal. If they survive to adulthood, their germ cells cannot proceed through meiosis normally because the chromosomes have no pairing partners. Thus, monoploids are characteristically sterile. (Male bees, wasps, and ants bypass meiosis; in these groups, gametes are produced by *mitosis*.)

Polyploids Polyploidy is very common in plants but rarer in animals. The reasons for this difference are currently unknown. Nonetheless, it is clear that an increase in the number of chromosome sets has been an important factor in the origin of new plant species. The evidence for this benefit is that above a haploid chromosome number of about 12, even numbers of chromosomes are much more common than odd numbers. This pattern is a consequence of the polyploid origin of many plant species, because doubling and redoubling of a number can give rise only to even numbers. Animal species do not show such a distribution, owing to the relative rarity of polyploid animals.

In aberrant euploids, there is often a correlation between the number of copies of the chromosome set and the size of the organism. A tetraploid organism, for example, typically looks very similar to its diploid counterpart in its proportions, except that the tetraploid is bigger, both as a whole and in its component parts. The higher the ploidy level, the larger the size of the organism (**Figure 17-3**).

> **KEY CONCEPT** Polyploids are often larger and have larger component parts than their diploid relatives.

In the realm of polyploids, we must distinguish between **autopolyploids**, which have multiple chromosome sets originating from within one species, and **allopolyploids**, which have sets from two or more different species. Allopolyploids form only between closely related species; however, the different chromosome sets are only **homeologous** (partly homologous), not fully homologous as they are in autopolyploids.

Autopolyploids Triploids (3*n*) are usually autopolyploids. They arise spontaneously in nature, but they can be constructed by geneticists from the cross of a 4*n* (tetraploid) and a 2*n* (diploid). The 2*n* and the *n* gametes produced by the tetraploid and the diploid, respectively, unite to form a 3*n* triploid. Triploids are characteristically sterile. For example, the bananas that are widely available commercially are

Higher ploidy produces larger size

FIGURE 17-3 A frog species (*Xenopus laevis*) with a tetraploid genome (one individual on top) is larger than a closely related species (*X. tropicalis*) with a diploid genome (two individuals on bottom). [*Courtesy of Atsushi Suzuki, Amphibian Research Center, Hiroshima University.*]

sterile triploids with 11 chromosomes in each set (3*n* = 33). The most obvious expression of the sterility of bananas is the absence of seeds in the fruit that we eat. (The black specks in bananas are ovules, not seeds; banana seeds are rock hard—real tooth breakers.) Seedless watermelons are another example of the commercial exploitation of triploidy in plants.

The problem (which is also true of monoploids) lies in the presence of unpaired chromosomes at meiosis. The molecular mechanisms for synapsis (see page 39), or true pairing, dictate that, in a triploid, pairing can take place between only two of the three chromosomes (**Figure 17-4**). Paired homologs (**bivalents**) segregate to opposite poles, but the unpaired homologs (**univalents**) pass to either pole randomly. In a **trivalent**, a paired group of three, the paired centromeres segregate as a bivalent and the unpaired one as a univalent. These segregations take place for every chromosome threesome; so, for each of the independent chromosome types (e.g., A, B, and C in an organism with a haploid chromosome number of three), the gamete could receive either one or two chromosomes. It is unlikely that a gamete will receive *two* for *every* chromosomal type or that it will receive *one* for *every* chromosomal type. For example, a gamete might receive two copies of chromosome A, one copy of chromosome B, and two copies of chromosome C. Hence, the likelihood is that gametes will have chromosome numbers intermediate between the haploid number and the diploid number; such genomes are of a type called **aneuploid** ("not euploid").

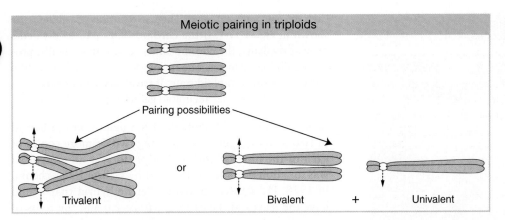

FIGURE 17-4 The three homologous chromosomes of a triploid may pair in two ways at meiosis I, as a trivalent or as a bivalent plus a univalent.

Aneuploid gametes do not generally give rise to viable offspring. In plants, aneuploid pollen grains are generally inviable and hence unable to fertilize the female gamete. In any organism, zygotes that might arise from the fusion of a haploid and an aneuploid gamete will themselves be aneuploid, and typically these zygotes also are inviable. We will examine the underlying reason for the inviability of aneuploids when we consider gene balance later in the chapter.

KEY CONCEPT Polyploids with odd numbers of chromosome sets, such as triploids, are sterile or highly infertile because their gametes and offspring are aneuploid.

Autotetraploids arise by the doubling of a $2n$ complement to $4n$. This doubling can occur spontaneously, but it can also be induced artificially by applying chemical agents that disrupt microtubule polymerization. As stated in Chapter 2, chromosome segregation is powered by spindle fibers, which are polymers of the protein tubulin. Hence, disruption of microtubule polymerization blocks chromosome segregation. The chemical treatment is normally applied to somatic tissue during the formation of spindle fibers in cells undergoing division. The resulting polyploid tissue (such as a polyploid branch of a plant) can be detected by examining stained chromosomes from the tissue under a microscope. Such a branch can be removed and used as a cutting to generate a polyploid plant or allowed to produce flowers, which, when selfed, would produce polyploid offspring. A commonly used antitubulin agent is colchicine, an alkaloid extracted from the autumn crocus. In colchicine-treated cells, the S phase of the cell cycle takes place, but chromosome segregation or cell division does not. As the treated cell enters telophase, a nuclear membrane forms around the entire doubled set of chromosomes. Thus, treating diploid ($2n$) cells with colchicine for one cell cycle leads to tetraploids ($4n$) with exactly four copies of each type of chromosome (**Figure 17-5**). Note that all alleles in the genotype are doubled. Therefore, if a diploid cell of genotype A/a ; B/b is doubled, the resulting autotetraploid will be of genotype $A/A/a/a$; $B/B/b/b$. Treatment for an additional cell cycle produces octoploids ($8n$), and so forth. This method works in both plant and animal cells, but, generally, plants seem to be much more tolerant of polyploidy. Many natural autotetraploid plants, such as potatoes,

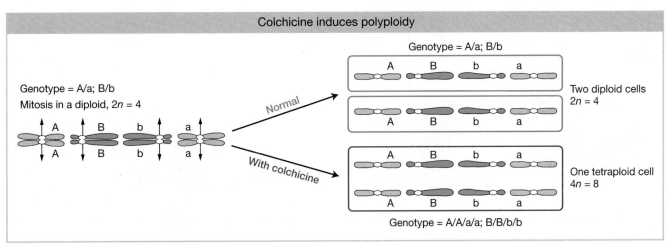

FIGURE 17-5 Colchicine may be applied to generate a tetraploid from a diploid. Colchicine added to mitotic cells during metaphase and anaphase disrupts spindle-fiber formation, preventing the migration of sister chromatids after the centromere has split. A single cell is created that contains pairs of identical chromosomes that are homozygous at all loci.

FIGURE 17-6 Diploid (*left*) and tetraploid (*right*) watermelon leaves and flowers. [*Michael E. Compton, University of Wisconsin—Platteville.*]

peanuts, and coffee, and induced autotetraploid plants such as watermelon have been developed as commercial crops to take advantage of their increased size (**Figure 17-6**). Large fruits and flowers are particularly favored.

Because four is an even number, autotetraploids can have a regular meiosis, although this result is by no means always the case. The crucial factor is how the four chromosomes of

each set pair and segregate. There are several possibilities, as shown in **Figure 17-7**. If the chromosomes pair as bivalents or quadrivalents, the chromosomes segregate normally, producing diploid gametes. The fusion of gametes at fertilization regenerates the tetraploid state. If trivalents form, segregation leads to nonfunctional aneuploid gametes and, hence, sterility.

Allopolyploids   An allopolyploid is a plant that is a hybrid of two or more species, containing two or more copies of each of the input genomes. The prototypic allopolyploid was an allotetraploid synthesized by Georgi Karpechenko in 1928. He wanted to make a fertile hybrid that would have the leaves of the cabbage (*Brassica*) and the roots of the radish (*Raphanus*), because they were the agriculturally important parts of each plant. Each of these two species has 18 chromosomes, and so $2n_1 = 2n_2 = 18$, and $n_1 = n_2 = 9$. The species are related closely enough to allow intercrossing. Fusion of an *n*1 and an *n*2 gamete produced a viable hybrid progeny individual of constitution $n_1 + n_2 = 18$. However, this hybrid was functionally sterile because the 9 chromosomes from the cabbage parent were different enough from the radish chromosomes that pairs did not synapse and segregate normally at meiosis, and thus the hybrid could not produce functional gametes.

Eventually, one part of the hybrid plant produced some seeds. On planting, these seeds produced fertile individuals with 36 chromosomes. All these individuals were allopolyploids. They had apparently been derived from spontaneous, accidental chromosome doubling to $2n_1 + 2n_2$ in one region of the sterile hybrid, presumably in tissue that eventually became a flower and underwent meiosis to produce gametes. In $2n_1 + 2n_2$ tissue, there is a pairing partner for each

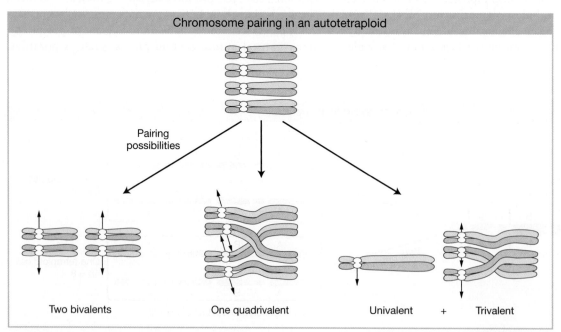

FIGURE 17-7 There are three different pairing possibilities at meiosis I in tetraploids. The four homologous chromosomes may pair as two bivalents or as a quadrivalent, and each can yield functional gametes. A third possibility, a trivalent plus a univalent, yields nonfunctional gametes.

ANIMATED ART   *Sapling*Plus

Segregation outcomes of autotetraploid meiosis

chromosome, and functional gametes of the type $n_1 + n_2$ are produced. These gametes fuse to give $2n_1 + 2n_2$ allopolyploid progeny, which also are fertile. This kind of allopolyploid is sometimes called an **amphidiploid**, or doubled diploid (**Figure 17-8**). Treating a sterile hybrid with colchicine greatly increases the chances that the chromosome sets will double. Amphidiploids are now synthesized routinely in this manner.

When Karpechenko's allopolyploid was crossed with either parental species—the cabbage or the radish—sterile offspring resulted. The offspring of the cross with cabbage were $2n_1 + n_2$, constituted from an $n_1 + n_2$ gamete from the allopolyploid and an n_1 gamete from the cabbage. The n_2 chromosomes had no pairing partners; hence, a normal meiosis could not take place, and the offspring were sterile. Thus, Karpechenko had effectively created a new species, with no possibility of gene exchange with either cabbage or radish. He called his new plant *Raphanobrassica*.

In nature, allopolyploidy seems to have been a major force in the evolution of new plant species. One convincing example is shown by the genus *Brassica*, as illustrated in **Figure 17-9**. Here, three different parent species have

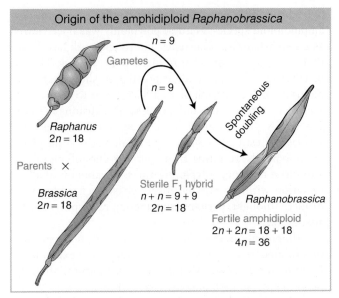

FIGURE 17-8 In the progeny of a cross of cabbage (*Brassica*) and radish (*Raphanus*), the fertile amphidiploid arose from spontaneous doubling in the $2n = 18$ sterile hybrid. The sketches depict the seedpods of each parental species and the hybrids.

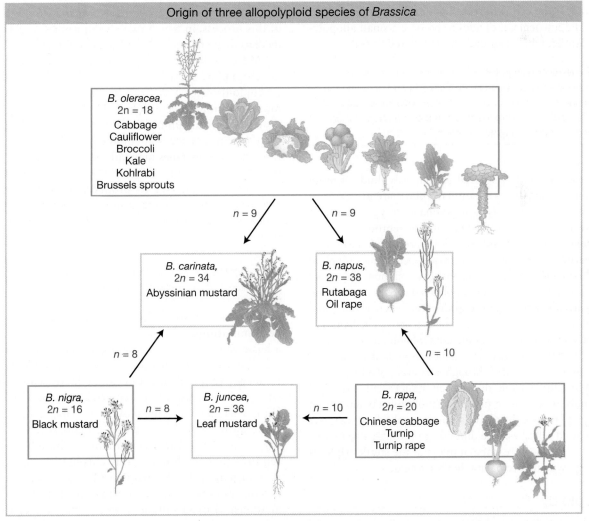

FIGURE 17-9 Allopolyploidy is important in the production of new species. In the example shown, three diploid species of *Brassica* (light green boxes) were crossed in different combinations to produce their allopolyploids (tan boxes). Some of the agricultural derivatives of some of the species are shown within the boxes.

hybridized in all possible pair combinations to form new amphidiploid species. Natural polyploidy was once viewed as a somewhat rare occurrence, but recent work has shown that it is a recurrent event in many plant species. The use of DNA markers has made it possible to show that polyploids in any population or area that appear to be the same are the result of many independent past fusions between genetically distinct individuals of the same two parental species. An estimated 50 percent of all angiosperm plants are polyploids, resulting from auto- or allopolyploidy. As a result of multiple polyploidizations, the amount of allelic variation within a polyploid species is much higher than formerly thought, perhaps contributing to its potential for adaptation.

Allopolyploidy has also been important in the production of modern crop plants. In addition to the examples of *Brassica* species, New World cotton is a natural allopolyploid that arose spontaneously, as is wheat.

Allopolyploid plant cells can also be produced artificially by fusing diploid cells from different species. First, the walls of two diploid cells are removed by treatment with an enzyme, and the membranes of the two cells fuse and become one. The nuclei often fuse, too, resulting in the polyploid. If the cell is nurtured with the appropriate hormones and nutrients, it divides to become a small allopolyploid plantlet, which can then be transferred to soil.

KEY CONCEPT Allopolyploid plants can arise in nature or be synthesized by crossing related species and doubling the chromosomes of the hybrid or by fusing diploid cells. Allopolyploidy has played an important role in the formation of many natural and agricultural plant species.

Polyploid animals As noted earlier, polyploidy is more common in plants than in animals, but there are cases of naturally occurring polyploid animals. Polyploid species of flatworms, leeches, and brine shrimps reproduce by parthenogenesis. Triploid and tetraploid *Drosophila* have been synthesized experimentally. Naturally occurring polyploid amphibians and reptiles are surprisingly common. They have several modes of reproduction: polyploid species of frogs and toads participate in sexual reproduction, whereas polyploid salamanders and lizards are parthenogenetic. The Salmonidae (the family of fishes that includes salmon and trout) provide a familiar example of the numerous animal species that appear to have originated through ancestral polyploidy.

The sterility of triploids has been commercially exploited in animals as well as in plants. Triploid oysters have been developed because they have a commercial advantage over their diploid relatives. The diploids go through a spawning season, when they are unpalatable, but the sterile triploids do not spawn and are palatable year-round.

Aneuploidy

Aneuploidy is the second major category of chromosomal aberrations in which the chromosome number is abnormal.

An aneuploid is an individual organism whose chromosome number differs from the wild type by part of a chromosome set. Generally, the aneuploid chromosome set differs from the wild type by only one chromosome or by a small number of chromosomes. An aneuploid can have a chromosome number either greater or smaller than that of the wild type. Aneuploid nomenclature (see Table 17-1) is based on the number of copies of the specific chromosome in the aneuploid state. For autosomes in diploid organisms, the aneuploid $2n + 1$ is **trisomic**, $2n - 1$ is **monosomic**, and $2n - 2$ (the "-2" represents the loss of both homologs of a chromosome) is **nullisomic**. In haploids, $n + 1$ is **disomic**. Special notation is used to describe sex-chromosome aneuploids because it must deal with the two different chromosomes. The notation merely lists the copies of each sex chromosome, such as XXY, XYY, XXX, or XO (the "O" stands for absence of a chromosome and is included to show that the single X symbol is not a typographical error).

Nondisjunction The cause of most aneuploidy is **nondisjunction** in the course of meiosis or mitosis. *Disjunction* is another word for the normal segregation of homologous chromosomes or chromatids to opposite poles at meiotic or mitotic divisions. *Nondisjunction* is a failure of this process, in which two chromosomes or chromatids incorrectly go to one pole and none to the other.

Mitotic nondisjunction can occur as cells divide during development. Sections of the body will be aneuploid (aneuploid *sectors*) as a result. *Meiotic* nondisjunction is more commonly encountered. In this case, the products of meiosis are aneuploid, leading to descendants in which the entire organism is aneuploid. In meiotic nondisjunction, the chromosomes may fail to disjoin at either the first or the second meiotic division (**Figure 17-10**). Either way, $n - 1$ and $n + 1$ gametes are produced. If an $n - 1$ gamete is fertilized by an n gamete, a monosomic ($2n - 1$) zygote is produced. The fusion of an $n + 1$ and an n gamete yields a trisomic $2n + 1$.

KEY CONCEPT Aneuploid organisms result mainly from nondisjunction in a parental meiosis.

Nondisjunction occurs spontaneously. Like most gene mutations, it is an example of a chance failure of a basic cellular process. The precise molecular processes that fail are not known, but in experimental systems, the frequency of nondisjunction can be increased by interference with microtubule polymerization, thereby inhibiting normal chromosome movement. Disjunction appears to be more likely to go awry in meiosis I. This failure is not surprising, because normal anaphase I disjunction requires that the homologous chromatids of the tetrad remain paired during prophase I and metaphase I, and it requires crossovers. In contrast, proper disjunction at anaphase II or at mitosis requires that the centromere split properly but does not require chromosome pairing or crossing over.

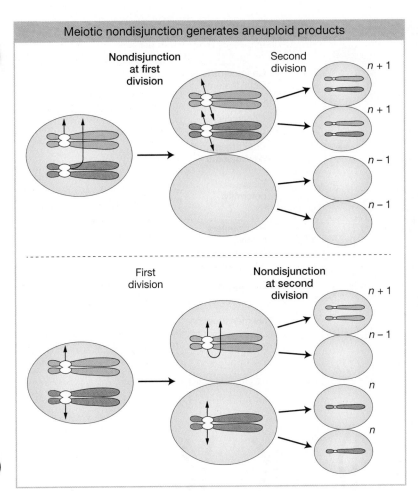

FIGURE 17-10 Aneuploid products of meiosis (that is, gametes) are produced by nondisjunction at the first or second meiotic division. Note that all other chromosomes are present in normal number, including in the cells in which no chromosomes are shown.

ANIMATED ART   🌀 Sapling*Plus*

Nondisjunctional meiosis I

ANIMATED ART   🌀 Sapling*Plus*

Nondisjunctional meiosis II

Crossovers are a necessary component of the normal disjunction process. Somehow the formation of a chiasma helps to hold a bivalent together and ensures that the two dyads will go to opposite poles. In most organisms, the amount of crossing over is sufficient to ensure that all bivalents will have at least one chiasma per meiosis. In *Drosophila*, many of the nondisjunctional chromosomes seen in disomic ($n + 1$) gametes are nonrecombinant, showing that they arise from meioses in which there is no crossing over on that chromosome. Similar observations have been made in human trisomies. In addition, in several different experimental organisms, mutations that interfere with recombination have the effect of massively increasing the frequency of meiosis I nondisjunction. All these observations provide evidence for the role of crossing over in maintaining chromosome pairing; in the absence of these associations, chromosomes are vulnerable to anaphase I nondisjunction.

KEY CONCEPT Crossovers are needed to keep bivalents paired until anaphase I. If crossing over fails for some reason, first-division nondisjunction occurs.

Monosomics ($2n - 1$) Monosomics are missing one copy of a chromosome. In most diploid organisms, the absence of one chromosome copy from a pair is deleterious. In humans, monosomics for any of the autosomes die in utero. Many X-chromosome monosomics also die in utero, but some are viable. A human chromosome complement of 44 autosomes plus a single X produces a condition known as **Turner syndrome**, represented as XO. Affected persons have a characteristic phenotype: they are sterile females, short in stature, and often have a web of skin extending between the neck and shoulders (**Figure 17-11**). Although their intelligence is near normal, some of their specific cognitive functions are defective. About 1 in 5000 female births show Turner syndrome.

Geneticists have used viable plant monosomics to map newly discovered recessive mutant alleles to a specific chromosome. For example, one can make a set of monosomic lines, each known to lack a different chromosome. Homozygotes for the new mutant allele are crossed with each monosomic line, and the progeny of each cross are inspected for the recessive phenotype. The appearance of the recessive phenotype identifies the chromosome that has one copy missing as the one on which the gene is normally located. The test works because half the gametes of a fertile $2n - 1$ monosomic

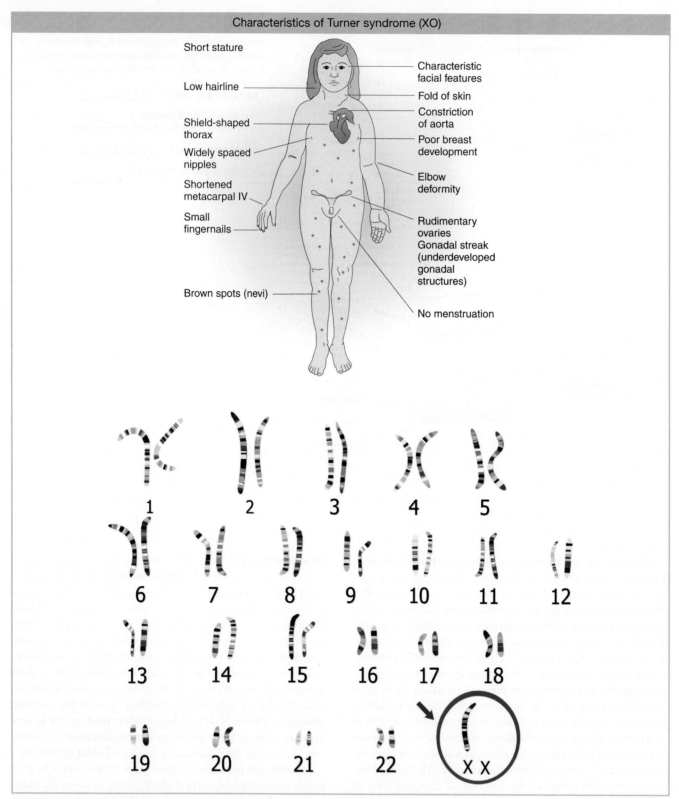

FIGURE 17-11 Turner syndrome results from the presence of a single X chromosome (XO), which can be seen in the image of chromosomes from an individual with Turner syndrome. [*Zuzanae/Shutterstock.com.*]

will be $n-1$, and, when an $n-1$ gamete is fertilized by a gamete bearing a new mutation on the homologous chromosome, the mutant allele will be the only allele of that gene present and hence its phenotype can be observed.

As an illustration, let's assume that a gene A with a mutant allele a is on chromosome 2. Crosses of a/a and monosomics for chromosome 1 and chromosome 2 are predicted to produce different results (chromosome is abbreviated chr):

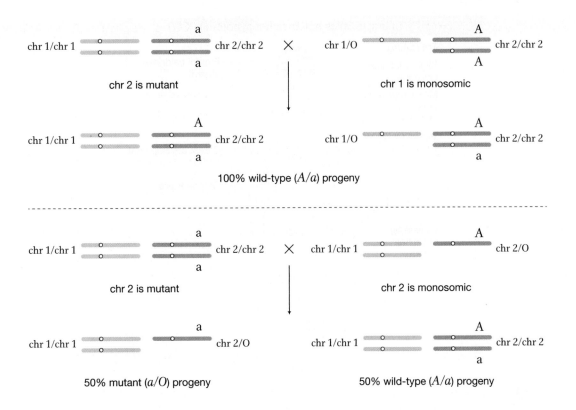

Trisomics (2n + 1) Trisomics contain an extra copy of one chromosome. In diploid organisms generally, the chromosomal imbalance from the trisomic condition can result in abnormality or death. However, there are many examples of viable trisomics. Furthermore, trisomics can be fertile. When cells from some trisomic organisms are observed under the microscope at the time of meiotic chromosome pairing, the trisomic chromosomes are seen to form an associated group of three (a trivalent), whereas the other chromosomes form regular bivalents.

What genetic ratios might we expect for genes on the trisomic chromosome? Let's consider a gene A that is close to the centromere on that chromosome so that the probability of recombination with the centromere is low, and let's assume that the genotype is A/a/a. Furthermore, let's postulate that, at anaphase I, the two paired centromeres in the trivalent pass to opposite poles and that the other centromere passes randomly to either pole. Then we can predict the three equally frequent segregations shown in **Figure 17-12**. These segregations result in an overall gametic ratio as shown in the six compartments of Figure 17-12; that is,

$$\frac{1}{6}\,A$$
$$\frac{2}{6}\,a$$
$$\frac{2}{6}\,A/a$$
$$\frac{1}{6}\,a/a$$

There are several examples of viable human trisomies. Several types of sex-chromosome trisomics can live

FIGURE 17-12 Three equally likely segregations may take place in the meiosis of an A/a/a trisomic, yielding the genotypes shown.

to adulthood. Each of these types is found at a frequency of about 1 in 1000 births of the relevant sex. (In considering human sex-chromosome trisomies, recall that mammalian sex is determined by the presence or absence of the Y chromosome.) The combination XXY results in **Klinefelter syndrome**. Persons with this syndrome are males who have lanky builds and a mildly impaired IQ and are sterile (**Figure 17-13**). Another abnormal combination, XYY, has a controversial history. Attempts have been made to link the XYY condition with a predisposition toward violence. However, it is now clear that an XYY condition in no way guarantees such behavior. Most males with XYY are fertile. Meioses show normal pairing of the X with one of the Y's; the other

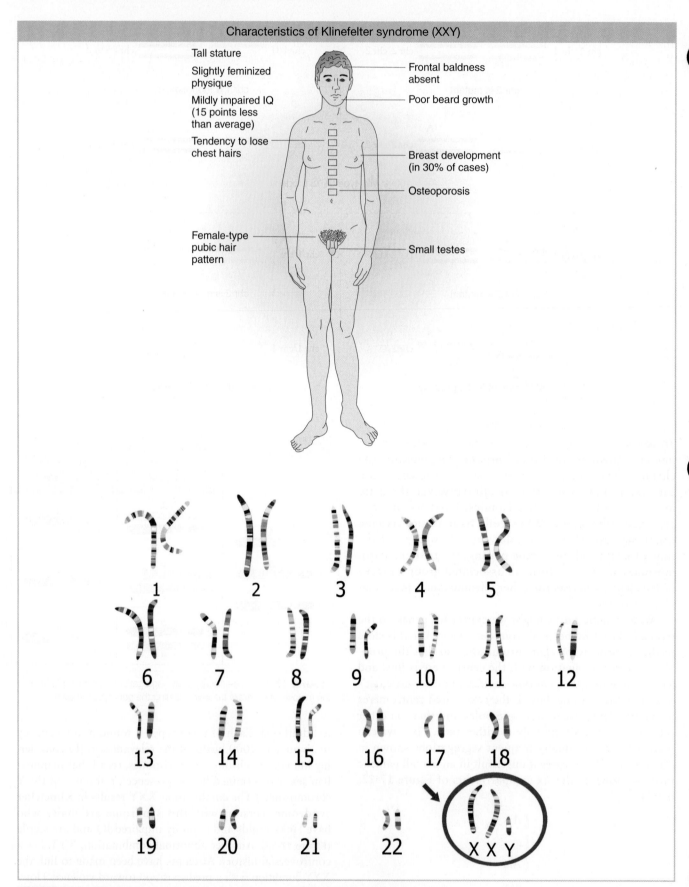

FIGURE 17-13 Klinefelter syndrome results from the presence of two X chromosomes and a Y chromosome, which can be seen in the image of chromosomes from an individual with Klinefelter syndrome. [*Zuzanae/Shutterstock.com*.]

Y does not pair and is not transmitted to gametes. Therefore, the gametes contain either X or Y, never YY or XY. Triplo-X trisomics (XXX) are phenotypically normal and fertile females. Meiosis shows pairing of only two X chromosomes; the third does not pair. Hence, eggs bear only one X and, like that of XYY males, the condition is not passed on to progeny.

Of human trisomies, the most familiar type is **Down syndrome** (**Figure 17-14**), discussed briefly at the beginning of the chapter. The frequency of Down syndrome is about 0.15 percent of all live births. Most affected persons have an extra copy of chromosome 21 caused by nondisjunction of chromosome 21 in a parent who is chromosomally

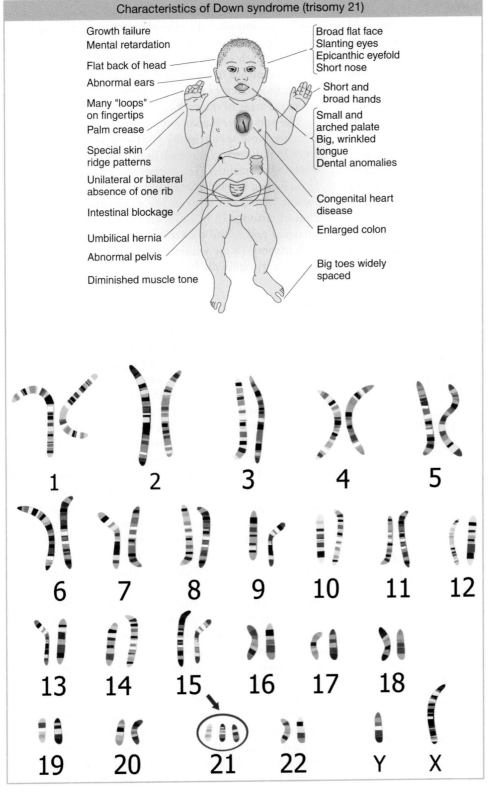

FIGURE 17-14 Down syndrome results from the presence of an extra copy of chromosome 21, which can be seen in the image of chromosomes from an individual with Down syndrome. [*Zuzana Egertova/ Alamy.*]

normal. In this *sporadic* type of Down syndrome, there is no family history of aneuploidy. Some rarer types of Down syndrome arise from translocations (a type of chromosomal rearrangement discussed later in the chapter); in these cases, as we will see, Down syndrome recurs in the pedigree because the translocation may be transmitted from parent to child.

The combined phenotypes that make up Down syndrome include mental retardation (with an IQ in the 20 to 50 range); a broad, flat face; eyes with an epicanthic fold (a skin fold of the upper eyelid that covers the inner corner of the eye); short stature; short hands with a crease across the middle; and a large, wrinkled tongue. Females may be fertile and may produce normal or trisomic progeny, but males are sterile with very few exceptions. Due to improvements in the care of individuals with Down syndrome, the average life expectancy is now 60 years.

The incidence of Down syndrome is related to maternal age: older mothers run a greatly elevated risk of having a child with Down syndrome (**Figure 17-15**). For this reason, fetal chromosome analysis (by cell-free fetal DNA testing, amniocentesis, or chorionic villus sampling) is now recommended for older expectant mothers. Although paternal age has not been linked to Down syndrome, a paternal-age effect on point mutations also has been found (see Chapter 1).

Even though the maternal-age effect has been known for many years, the causes are still not known. Nonetheless, there are some interesting biological correlations. With age, possibly the chromosome bivalent is less likely to stay together during prophase I of meiosis. Meiotic arrest of oocytes (female meiocytes) in late prophase I is a common phenomenon in many animals. In female humans, all oocytes are arrested at diplotene before birth. Meiosis resumes at each menstrual period, which means that the chromosomes in the bivalent must remain properly associated for as long as five or more decades. If we speculate that these associations have an increasing probability of breaking down by accident as time passes, we can envision a mechanism contributing to increased maternal nondisjunction with age. Consistent with this speculation, most nondisjunction related to the effect of maternal age is due to nondisjunction at anaphase I, not anaphase II. However, recent research suggests that additional factors contribute to the prevalence of aneuploidy in older mothers, and identifying those factors as well as preventative interventions are active areas of research.

The only other human autosomal trisomics to survive to birth are those with trisomy 13 (Patau syndrome) and trisomy 18 (Edwards syndrome). Both have severe physical and mental abnormalities. The phenotypic syndrome of trisomy 13 includes a harelip, a small and malformed head, "rocker-bottom" feet, and a mean life expectancy of 130 days. That of trisomy 18 includes "faunlike" ears, a small jaw, a narrow pelvis, and rocker-bottom feet; almost all babies with trisomy 18 die within the first few weeks after birth. All other trisomics die in utero.

The concept of gene balance

In considering aberrant euploidy, we noted that an increase in the number of full chromosome sets correlates with increased organism size but that the general shape and proportions of the organism remain very much the same. In contrast, autosomal aneuploidy typically alters the organism's shape and proportions in characteristic ways.

Plants tend to be somewhat more tolerant of aneuploidy than are animals. Studies in jimsonweed (*Datura stramonium*) provide a classic example of the effects of aneuploidy and polyploidy. In jimsonweed, the haploid chromosome number is 12. As expected, the polyploid jimsonweed is proportioned like the normal diploid, only larger. In contrast, each of the 12 possible trisomics is disproportionate but in ways different from one another, as exemplified by changes in the shape of the seed capsule (**Figure 17-16**). The 12 different trisomies lead to 12 different and characteristic shape changes in the capsule. Indeed, these characteristics and others of the individual trisomics are so reliable that the phenotypic syndrome can be used to identify plants carrying a particular trisomy. Similarly, the 12 monosomics are themselves different from one another and from each of the trisomics. In general, a monosomic for a particular chromosome is more severely abnormal than is the corresponding trisomic.

We see similar trends in aneuploid animals. In the fruit fly *Drosophila*, the only autosomal aneuploids that survive to adulthood are trisomics and monosomics for chromosome 4, which is the smallest *Drosophila* chromosome, representing only about 1 to 2 percent of the genome. Trisomics for chromosome 4 are only very mildly affected and

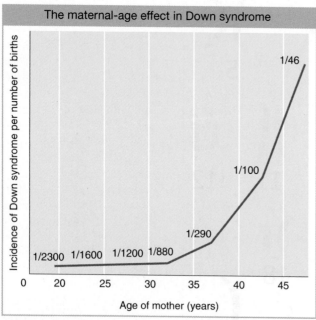

FIGURE 17-15 Older mothers have a higher proportion of babies with Down syndrome than younger mothers do. [*Data from L. S. Penrose and G. F. Smith,* Down's Anomaly. *Little, Brown and Company, 1966.*]

The trisomics of *Datura*

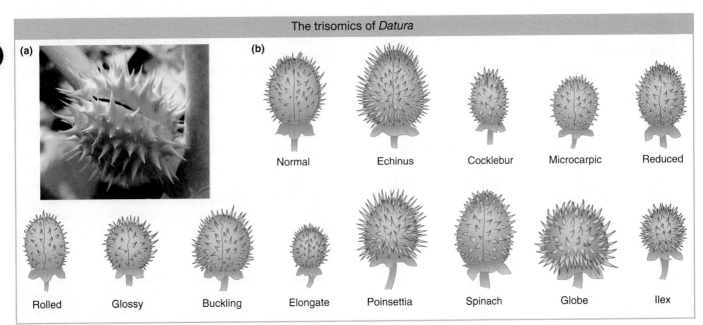

FIGURE 17-16 Each of the 12 possible trisomics of *Datura* is disproportionate in a different way. (a) *Datura* fruit. (b) The drawings show the fruit of a normal *Datura* (2*n*) or different trisomic *Datura* (2*n* + 1), each of which has been named. [*(a) Konrad Lange/Getty Images.*]

are much less abnormal than are monosomics for chromosome 4. In humans, no autosomal monosomic survives to birth, but, as already stated, three types of autosomal trisomics can do so. As is true of aneuploid jimsonweed, each of these three trisomics shows unique phenotypic syndromes.

Why does aneuploidy for each chromosome have its own characteristic phenotypic effects? Why are monosomics typically more severely affected than are the corresponding trisomics? And why are aneuploids generally so much more abnormal than polyploids? The answers seem certain to be a matter of **gene balance**. In a euploid, the ratio of genes on any one chromosome to the genes on other chromosomes is always 1:1, regardless of whether we are considering a monoploid, diploid, triploid, or tetraploid. For example, in a tetraploid, for gene *A* on chromosome 1 and gene *B* on chromosome 2, the ratio is 4 *A*:4 *B*, or 1:1. In contrast, in an aneuploid, the ratio of genes on the aneuploid chromosome to genes on the other chromosomes differs from the wild type up or down by 50 percent: 50 percent for monosomics and 150 percent for trisomics. For example, in a trisomic for chromosome 2 with gene *A* on chromosome 1 and gene *B* on chromosome 2, we find that the ratio of the *A* and *B* genes is 2 *A*:3 *B*. Thus, the aneuploid genes are out of balance. How does their being out of balance help us answer the questions raised at the opening of this paragraph?

In general, the amount of transcript produced by a gene is directly proportional to the number of copies of that gene in a cell. That is, for a given gene, the rate of transcription is directly related to the number of DNA templates available. Thus, the more copies of the gene, the more transcripts are produced and the more of the corresponding protein

product is made (**Figure 17-17**). This relation between the number of copies of a gene and the amount of the gene's product made is called a **gene-dosage effect**.

We can infer that normal physiology in a cell depends on the proper ratio of gene products in the euploid cell. This ratio is the normal gene balance. If the relative dosage of certain genes changes—for example, because of the removal of one of the two copies of a chromosome (or even a segment thereof)—physiological imbalances in cellular pathways can arise. These imbalances are the reason that aneuploids are generally much more abnormal than polyploids.

In some cases, the imbalances of aneuploidy result from the effects of a few "major" genes whose dosage has changed, rather than from changes in the dosage of all the genes on a chromosome. Such genes can be viewed as *haplo-abnormal* (resulting in an abnormal phenotype if present only in one copy) or *triplo-abnormal* (resulting in an abnormal phenotype if present in three copies). They contribute significantly to the aneuploid phenotypic syndromes. For example, the study of persons trisomic for only part of chromosome 21 has made it possible to localize genes contributing to Down syndrome to various regions of chromosome 21; the results hint that some aspects of the phenotype might be due to triplo-abnormality for a specific gene in these chromosome regions. However, other aspects of aneuploid syndromes are likely to result from the cumulative effects of aneuploidy for numerous genes whose products are all out of balance. This is supported by the fact that the only trisomy that survives to adulthood in humans is for chromosome 21, which is the smallest human chromosome. Trisomies of larger chromosomes with even more genes out of balance do not survive. Thus, the characteristic

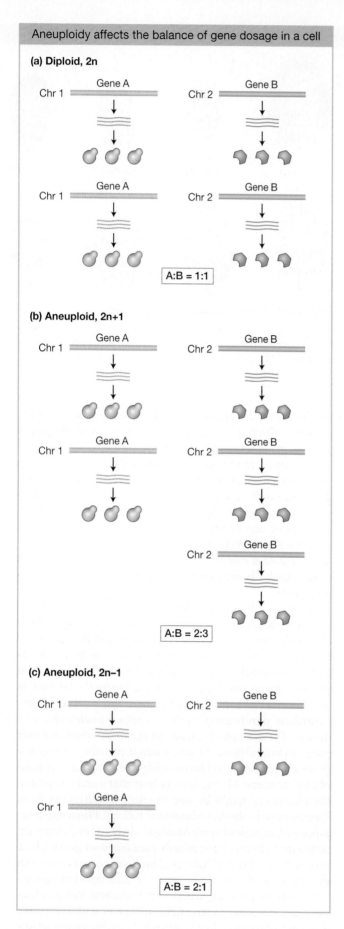

Aneuploidy affects the balance of gene dosage in a cell

(a) Diploid, 2n

Chr 1 — Gene A   Chr 2 — Gene B

A:B = 1:1

(b) Aneuploid, 2n+1

Chr 1 — Gene A   Chr 2 — Gene B

Chr 1 — Gene A   Chr 2 — Gene B

Chr 2 — Gene B

A:B = 2:3

(c) Aneuploid, 2n–1

Chr 1 — Gene A   Chr 2 — Gene B

Chr 1 — Gene A

A:B = 2:1

phenotypic effects associated with aneuploidy for each different chromosome, as seen in jimsonweed or humans, likely result from a combination of the imbalance of genes with very specific phenotypic effects and the total number of genes that are out of balance.

However, the concept of gene balance does not tell us why having too few gene products (monosomy) is much worse for an organism than having too many gene products (trisomy). In a parallel manner, we can ask why there are many more haplo-abnormal genes than triplo-abnormal ones. A key to explaining the extreme abnormality of monosomics relative to trisomics is that the phenotypes of any deleterious recessive alleles present on a monosomic autosome will be automatically observed due to the absence of the wild-type allele.

Sex chromosome gene balance How do we apply the idea of gene balance to cases of sex-chromosome aneuploidy? Gene balance holds for sex chromosomes as well, but we also have to take into account the special properties of the sex chromosomes. In many organisms with XY sex determination, such as mammals, the Y chromosome seems to be a degenerate version of the X chromosome in which there are very few functional genes other than some concerned with sex determination itself, sperm production, or both (**Figure 17-18**). The X chromosome, on the other hand, contains many genes concerned with basic cellular processes ("housekeeping genes") that just happen to reside on the chromosome that eventually evolved into the X chromosome. XY sex-determination mechanisms have likely evolved independently at least a hundred times in different taxonomic groups. For example, there appears to be one sex-determination mechanism for all mammals, but it is completely different from the mechanism governing sex determination in fruit flies (see Chapter 13).

In a sense, X chromosomes are naturally aneuploid. In species with an XY sex-determination system, females have two X chromosomes, whereas males have only one. Nonetheless, the X chromosome's housekeeping genes are expressed to approximately equal extents per cell in females and in males. In other words, there is **dosage compensation**. How is this compensation accomplished? The answer depends very much on the organism. In fruit flies, the male's X chromosome appears to be hyperactivated, allowing it to be transcribed at twice the

FIGURE 17-17 Gene A is located on chromosome 1 and produces a protein that physically interacts with the protein produced by gene B, which is located on chromosome 2. (a) In a diploid cell, the number of transcripts and proteins produced by gene A and gene B are equal, and there is normal gene balance. (b) In an aneuploid ($2n + 1$) cell with a trisomy for chromosome 2, there is more of protein B produced, leading to an imbalance in the ratio of proteins A and B. (c) In an aneuploid ($2n - 1$) cell with a monosomy for chromosome 2, there is more of protein A produced, leading to an imbalance in the ratio of proteins A and B.

Human X and Y chromosome pair

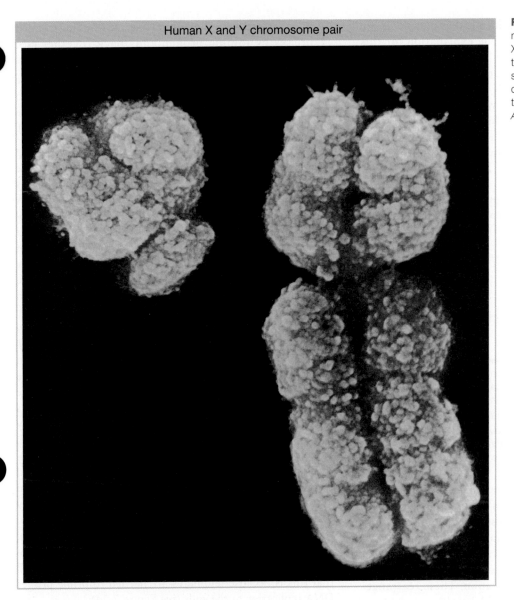

FIGURE 17-18 Scanning electron micrograph of a human Y (*left*) and X (*right*) chromosome pair shows that the Y chromosome is much smaller than the X chromosome, due to the loss of most genes on the Y chromosome. [*BIOPHOTO ASSOCIATES/Getty Images*.]

rate of either X chromosome in the female. As a result, the XY male *Drosophila* has an X gene dosage equivalent to that of an XX female. In mammals, in contrast, the rule is that no matter how many X chromosomes are present, there is only one transcriptionally active X chromosome in each somatic cell. This rule gives the XX female mammal an X gene dosage equivalent to that of an XY male. Dosage compensation in mammals is achieved by X-chromosome inactivation. A female with two X chromosomes, for example, is a mosaic of two cell types in which one or the other X chromosome is active. We examined this phenomenon in Chapter 12. Thus, XY and XX individuals produce the same amounts of X-chromosome housekeeping-gene products. X-chromosome inactivation also explains why triplo-X humans are phenotypically normal: only one of the three X chromosomes is transcriptionally active in a given cell. Similarly, an XXY male is only moderately affected because only one of his two X chromosomes is active in each cell.

Why are XXY individuals abnormal at all, given that triplo-X individuals are phenotypically normal? It turns out that a few genes scattered throughout an "inactive X" are still transcriptionally active. In XXY males, these genes are transcribed at twice the level that they are in XY males. In XXX females, on the other hand, the few transcribed genes are active at only 1.5 times the level that they are in XX females. This lower level of "functional aneuploidy" in XXX than in XXY, plus the fact that the active X genes appear to lead to feminization, may explain the feminized phenotype of XXY males. The severity of Turner syndrome (XO) may be due to the deleterious effects of monosomy and to the lower activity of the transcribed genes of the X (compared with XX) females. As is usually observed for aneuploids, monosomy for the X chromosome produces a more abnormal phenotype than does having an extra copy of the same chromosome (triplo-X females or XXY males).

Gene dosage is also important in the phenotypes of polyploids. Human polyploid zygotes do arise through various kinds of mistakes in cell division. Most die in utero. Occasionally, triploid babies are born, but none survive.

This fact seems to violate the principle that polyploids are more normal than aneuploids. The explanation for this contradiction seems to lie with X-chromosome dosage compensation. Part of the rule for gene balance in organisms that have a single active X seems to be that there must be one active X for every two copies of the autosomal chromosome complement. Thus, some cells in triploid mammals are found to have one active X, whereas others, surprisingly, have two. Neither situation is in balance with autosomal genes.

> **KEY CONCEPT** Aneuploidy is nearly always deleterious because of gene imbalance: the ratio of gene products is different from that in euploids, and this difference interferes with the normal function of the genome.

17.2 CHANGES IN CHROMOSOME STRUCTURE

> **LO 17.3** Distinguish among the major types of chromosome rearrangements (translocations, inversions, deletions, duplications) and diagnose their presence in progeny analysis.
>
> **LO 17.4** In a cross involving a known chromosome rearrangement, predict the inheritance of genes linked and unlinked to the rearrangement.

Changes in chromosome structure, called **rearrangements**, encompass several major classes of events (see Figure 17-2). A chromosome segment can be lost, resulting in a **deletion**, or doubled, to form a **duplication**. The orientation of a segment within the chromosome can be reversed, called an **inversion**, or a segment can be moved to a different chromosome, constituting a **translocation**. These rearrangements can be caused by either DNA breakage or crossing over between repetitive DNA (**Figure 17-19**).

DNA breakage is a major cause of each of these chromosomal rearrangements. Both DNA strands must break at two different locations, followed by a rejoining of the broken ends to produce a new chromosomal arrangement (Figure 17-19, left side). Chromosomal rearrangements by breakage can be induced artificially by using ionizing radiation. This kind of radiation, particularly X rays and gamma rays, is highly energetic and causes numerous double-stranded breaks in DNA. To understand how chromosomal rearrangements are produced by breakage, several points should be kept in mind:

1. Each chromosome is a single double-stranded DNA molecule.

2. The first event in the production of a chromosomal rearrangement is the generation of two or more double-stranded breaks in the chromosomes of a cell (see Figure 17-19, top row at left).

3. Double-stranded breaks are potentially lethal, unless they are repaired.

4. Repair systems in the cell correct the double-stranded breaks by joining broken ends back together (see Chapter 15 for a detailed discussion of DNA repair).

5. If the two ends of the same break are rejoined, the original DNA order is restored. If the ends of two different breaks are joined, however, one result is some type of chromosomal rearrangement (see Figure 17-19, left side).

Another important cause of rearrangements is crossing over between repetitive (duplicated) DNA segments. This type of unequal crossing over is termed **nonallelic homologous recombination (NAHR)**. In organisms with repeated DNA sequences within one chromosome or on different chromosomes, there is ambiguity about which of the repeats will pair with each other at meiosis. If sequences pair up that are not in the same relative positions on the homologs, crossing over can produce aberrant chromosomes. Deletions, duplications, inversions, and translocations can all be produced by such crossing over (see Figure 17-19, right side).

Regardless of the mechanism of formation, the only chromosomal rearrangements that survive meiosis are those that produce DNA molecules that have one centromere and two telomeres. If a rearrangement produces a chromosome that lacks a centromere, such an **acentric chromosome** will not be dragged to either pole at anaphase of mitosis or meiosis and will not be incorporated into either progeny nucleus. Therefore, acentric chromosomes are not inherited. If a rearrangement produces a chromosome with two centromeres (a **dicentric chromosome**), it will often be pulled simultaneously to opposite poles at anaphase, forming an **anaphase bridge**. Anaphase-bridge chromosomes typically will not be incorporated into either progeny cell. If a chromosome break produces a chromosome lacking a telomere, DNA will be progressively lost from the end of the chromosome with every round of replication. Recall from Chapter 7 that this is because telomeres are needed to prime proper DNA replication at the ends (see Figure 7-24).

There are two general types of rearrangements: unbalanced and balanced. **Unbalanced rearrangements** change the gene dosage of a chromosome segment. As with aneuploidy for whole chromosomes, the loss of one copy of a segment or the addition of an extra copy can disrupt normal gene balance. The larger the segment that is lost or duplicated, the more likely it is that gene imbalance will cause phenotypic abnormalities. The two simple classes of unbalanced rearrangements are deletions and duplications. A *deletion* is the loss of a segment within one chromosome arm and the juxtaposition of the two segments on either side of the deleted segment, as in this example, which shows loss of segment C–D:

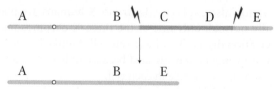

A *duplication* is the repetition of a segment of a chromosome arm. In the simplest type of duplication, the two

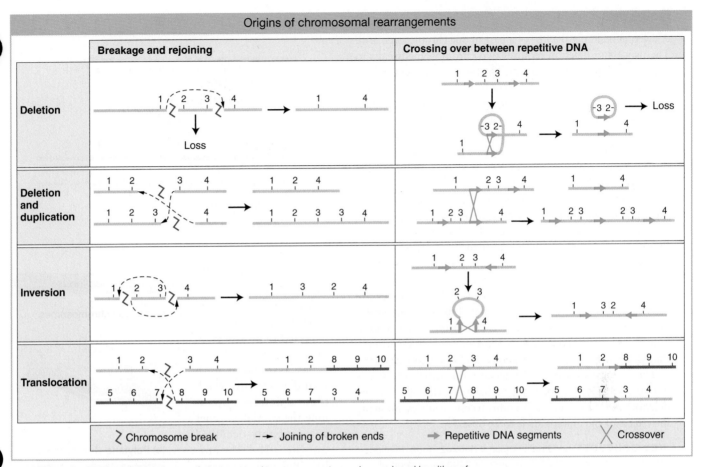

FIGURE 17-19 Each of the four types of chromosomal rearrangements can be produced by either of two basic mechanisms: chromosome breakage and rejoining or crossing over between repetitive DNA. Chromosome regions are numbered 1 through 10. Homologous chromosomes are the same color.

segments are adjacent to each other (a tandem duplication), as in this duplication of segment C:

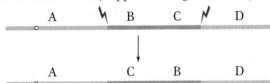

However, the duplicate segment can end up at a different position on the same chromosome or even on a different chromosome.

Balanced rearrangements change the chromosomal gene order but do not remove or duplicate any DNA. The two simple classes of balanced rearrangements are inversions and reciprocal translocations. An *inversion* is a rearrangement in which an internal segment of a chromosome has been broken twice, flipped 180 degrees, and rejoined.

A *reciprocal translocation* is a rearrangement in which two nonhomologous chromosomes are each broken once, creating acentric fragments, which then trade places:

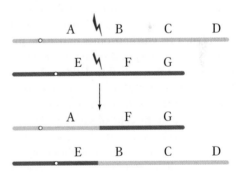

Sometimes the DNA breaks that precede the formation of a rearrangement occur *within* genes. When they do, they disrupt gene function because part of the gene moves to a new location and no complete transcript can be made. In addition, the DNA sequences on either side of the rejoined ends of a rearranged chromosome are sequences that are not normally juxtaposed. Sometimes the junction occurs in such a way that fusion produces a hybrid gene composed of parts of two other genes. Usually such hybrid genes are nonfunctional, but occasionally, the hybrid gene can acquire a new function. Later in this chapter, we will see an example in which the formation of a hybrid gene can result in cancer.

The following sections consider the properties of these balanced and unbalanced rearrangements.

Deletions

A deletion is simply the loss of a part of one chromosome arm. The process of deletion requires two chromosome breaks to cut out the intervening segment. The deleted fragment has no centromere; consequently, it cannot be pulled to a spindle pole in cell division and is lost. The effects of deletions depend on their size. A small deletion *within* a gene, called an **intragenic deletion**, inactivates the gene and has the same effect as that of other null mutations of that gene. If the homozygous null phenotype is viable (as, for example, in human albinism), the homozygous deletion also will be viable.

For most of this section, we will be dealing with **multigenic deletions**, in which several to many genes are missing. The consequences of these deletions are more severe than those of intragenic deletions. If such a deletion is made homozygous by inbreeding (that is, if both homologs have the same deletion), the combination is nearly always lethal. This fact suggests that all regions of the chromosomes are essential for normal viability and that complete elimination of any segment from the genome is deleterious. Even an individual organism heterozygous for a multigenic deletion—that is, having one normal homolog and one that carries the deletion—may not survive. Principally, this lethal outcome is due to disruption of normal gene balance. Alternatively, the deletion may "uncover" deleterious recessive alleles, allowing the effects of the mutant allele to be observed.

KEY CONCEPT The lethality of large heterozygous deletions can be explained by gene imbalance and the expression of deleterious recessive alleles.

Small deletions are sometimes viable in combination with a normal homolog. Such deletions may be identified by examining meiotic chromosomes under the microscope. The failure of the corresponding segment on the normal homolog to pair creates a visible **deletion loop** in meiotic chromosomes (**Figure 17-20a**). In *Drosophila*, deletion loops are also visible in the **polytene chromosomes**. These chromosomes are found in the cells of salivary glands and other specific tissues of certain insects. In these cells, the homologs pair and replicate many times without separating, and so each homologous chromosome pair is represented by a thick bundle of replicates. These polytene chromosomes provide a rare opportunity to visualize interphase chromosomes under light microscopy. Each chromosome has a set of dark-staining bands of fixed position and number. These bands act as useful chromosomal landmarks. An example of a polytene chromosome in which one original homolog carried a deletion is shown in Figure 17-20b. A deletion can be assigned to a specific chromosome location by examining polytene chromosomes microscopically and determining the position of the deletion loop.

Another clue to the presence of a deletion is that the loss of a segment on one homolog sometimes unmasks recessive alleles present on the other homolog, leading to the appearance of phenotypes associated with those mutations. Consider, for example, the deletion shown in the following diagram:

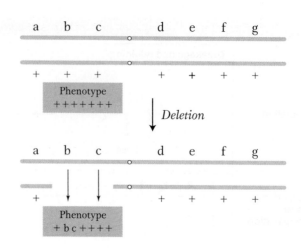

If there is no deletion, none of the phenotypes associated with the seven recessive alleles are expected to be observed; however, if the phenotypes associated with *b* and *c* are observed, then a deletion spanning the b^+ and c^+ genes has probably occurred on the other homolog. Because recessive alleles seem to be showing dominance in such cases, the effect is called **pseudodominance**.

In the reverse case—if we already know the location of the deletion—we can apply the pseudodominance effect in the opposite direction to map the positions of mutant alleles. This procedure, called **deletion mapping**, pairs mutations against a set of defined overlapping deletions. An example from *Drosophila* is shown in **Figure 17-21**. In this diagram, the recombination map is shown at the top, marked with distances in map units from the left end. The location of a specific deletion is identified by the presence of deletion loops in polytene chromosomes, as described earlier. The horizontal bars below the chromosome show the extent of the deletions listed at the left. Each deletion is paired with each mutation under test, and the phenotype is examined to see if the mutation is pseudodominant. The mutation *pn* (prune), for example, shows pseudodominance only with deletion 264-38, and

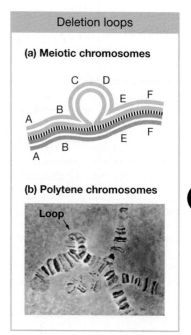

FIGURE 17-20 (a) Schematic of meiotic pairing in a deletion heterozygote. The normal homolog forms a loop because the genes in this loop have no alleles with which to synapse. (b) A deletion loop is visible in *Drosophila* polytene chromosomes where the normal chromosome homolog is unable to align with the deletion chromosome homolog. [*William M. Gelbart, Harvard University.*]

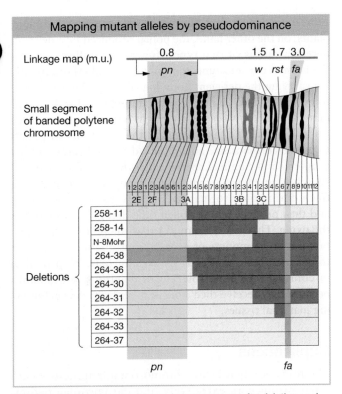

Mapping mutant alleles by pseudodominance

FIGURE 17-21 A *Drosophila* strain heterozygous for deletion and normal chromosomes may be used to map mutant alleles. The red bars show the extent of the deleted segments in 10 deletions. When an individual is heterozygous for a chromosome with a recessive allele and the homologous chromosome with a deletion covering the location of the wild-type allele, the mutant phenotype will be observed.

this result determines its location in the 2E-1 to 3A-2 region. However, *fa* (facet) shows pseudodominance with all but two deletions (258-11 and 258-14); so its position can be pinpointed to band 3C-7, which is the region that all but two deletions have in common.

KEY CONCEPT Deletions can be recognized by deletion loops and pseudodominance.

Deletions in human chromosomes are not uncommon. Although these deletions are usually small, they still may encompass many genes. In this case, a deletion will likely have adverse effects, even when heterozygous, due to the disruption of normal gene balance for many genes. Deletions of specific human chromosome regions cause unique syndromes of phenotypic abnormalities. One example is cri du chat syndrome, caused by a heterozygous deletion of the tip of the short arm of chromosome 5 (**Figure 17-22**). The specific bands deleted in cri du chat syndrome are 5p15.2 and 5p15.3, the two most distal bands identifiable on 5p. (The short and long arms of human chromosomes are traditionally called p and q, respectively.) The most characteristic phenotype in the syndrome is the one that gives it its name, the distinctive catlike mewing cries made by affected infants. Other manifestations of the syndrome

are microencephaly (abnormally small head) and a moonlike face. Like syndromes caused by other deletions, cri du chat syndrome includes mental retardation. Fatality rates are low, and many persons with this deletion reach adulthood.

Another instructive example is Williams syndrome. This syndrome is autosomal dominant and is characterized by unusual development of the nervous system and certain external features. Williams syndrome is found at a frequency of about 1 in 10,000 people. Patients often have pronounced musical or singing ability, as well as hypersociality. The syndrome is almost always caused by a 1.5-Mb deletion on one homolog of chromosome 7, specifically at band 7q11.23. Sequence analysis has shown that this deletion encompasses between 26 and 28 genes of known and unknown function. It is likely that haploinsufficiency of different genes within the deletion contribute to different phenotypes associated with the syndrome. Sequence analysis also reveals the origin of this deletion because the normal sequence is bounded by repeated copies of a gene called *PMS*, which happens to encode a DNA-repair protein. As we have seen, repeated sequences can act as substrates for

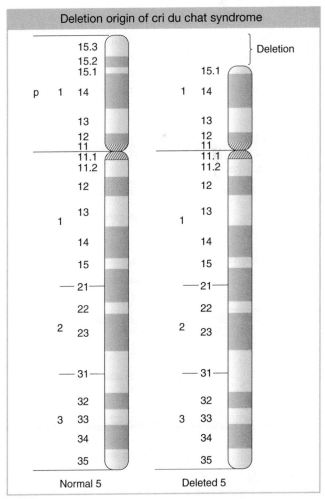

FIGURE 17-22 Cri du chat syndrome is caused by the loss of the tip of the short arm of one of the homologs of chromosome 5.

unequal crossing over. A crossover between flanking copies of *PMS* on opposite ends of the chromosomal segment leads to a Williams syndrome deletion and a duplication, as shown in **Figure 17-23**. Although less is currently known about the traits associated with having a duplication of the 7q11.23 region, affected individuals seem to have traits that are the opposite of those individuals with a deletion of this region of chromosome 7. This finding is consistent with the concept of gene balance that we considered in the previous section when comparing the effects of whole-chromosome monosomies and trisomies.

Most human deletions, such as those that we have just considered, arise spontaneously in the gonads of a normal parent of an affected person; thus, no signs of the deletions are usually found in the chromosomes of the parents. Less commonly, deletion-bearing individuals appear among the offspring of an individual having an undetected balanced rearrangement of chromosomes. For example, cri du chat syndrome can result from a parent heterozygous for a reciprocal translocation, because (as we will see) segregation produces deletions. Deletions may also result from recombination within a heterozygote having a pericentric inversion (an inversion spanning the centromere) on one chromosome. Both mechanisms will be detailed later in the chapter.

Animals and plants show differences in the survival of gametes or offspring that bear deletions. A male animal with a deletion in one chromosome produces sperm carrying one or the other of the two chromosomes in approximately equal numbers. These sperm seem to function to

some extent regardless of their genetic content. In diploid plants, on the other hand, the pollen produced by a deletion heterozygote is of two types: functional pollen carrying the normal chromosome and nonfunctional (aborted) pollen carrying the deficient homolog. Thus, pollen cells seem to be sensitive to changes in the amount of chromosomal material, and this sensitivity might act to weed out deletions. This effect is analogous to the sensitivity of pollen to whole-chromosome aneuploidy, described earlier in this chapter. Unlike animal sperm cells, whose metabolic activity relies on enzymes that have already been deposited in them during their formation, pollen cells must germinate and then produce a long pollen tube that grows to fertilize the ovule. This growth requires that the pollen cell manufacture large amounts of protein, thus making it sensitive to genetic abnormalities in its own nucleus. Plant ovules, in contrast, are quite tolerant of deletions, presumably because they receive their nourishment from the surrounding maternal tissues.

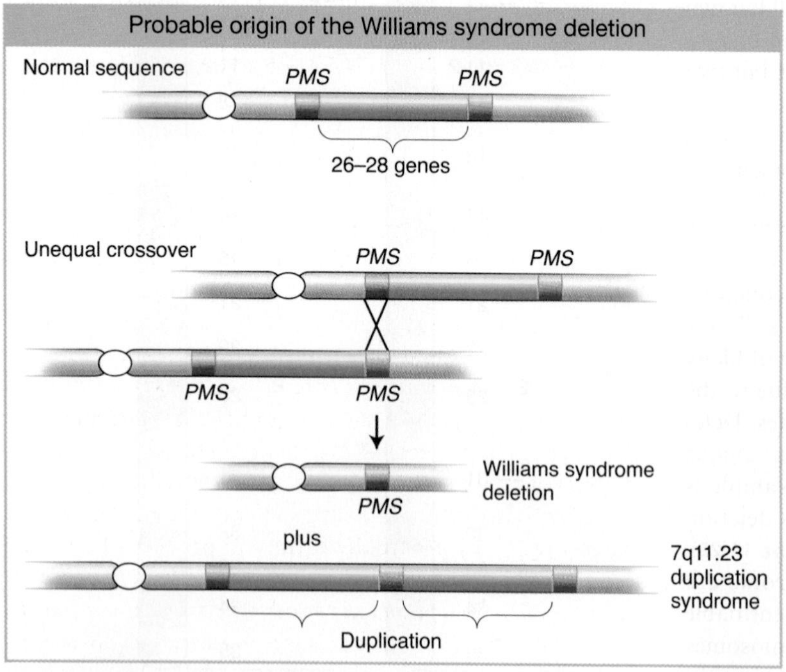

Duplications

The processes that cause chromosomal mutations sometimes produce an extra copy of a chromosome region. The duplicate regions can be located adjacent to each other, called a **tandem duplication**; or the extra copy can be located elsewhere in the genome, called an **insertional duplication**. A diploid cell containing a duplication will have three copies of the chromosome region in question: two in one chromosome set and one in the other. This is known as a duplication heterozygote. In meiotic prophase, tandem-duplication heterozygotes are seen as a loop consisting of the unpaired extra region.

Analyses of genome DNA sequences have revealed a high level of duplications in humans and in most model organisms. Simple sequence repeats, which are extensive throughout the genome and useful as molecular markers in mapping, were discussed in earlier chapters. However, another class of duplications is based on duplicated units that are much bigger than simple sequence repeats. Duplications in this class are termed **segmental duplications**. The duplicated units in segmental duplications generally range from 10 to 50 kilobases in length and encompass whole genes and the regions in between. An example of segmental duplications on human chromosome 7 is shown in **Figure 17-24**. Most of the duplications are dispersed within chromosome 7, but there are some tandem duplications, and even some duplicated segments from chromosome 7 that are found on other chromosomes. One of the tandem segmental duplications overlaps with the previously discussed chromosomal rearrangement associated with Williams syndrome.

FIGURE 17-23 A crossover between left and right repetitive flanking genes results in two reciprocal rearrangements, one corresponding to the Williams syndrome deletion, and the other to the 7q11.23 duplication syndrome.

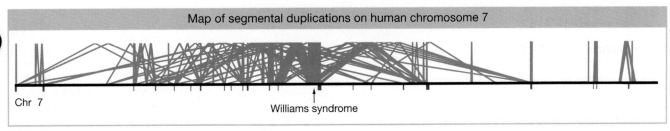

FIGURE 17-24 A map of human chromosome 7 shows the positions of duplications greater than 10 kilobases in size. Blue connecting lines show intrachromosomal duplications (the great majority). Interchromosomal duplications are shown with red bars. The location where recombination between duplications has given rise to the Williams syndrome deletion is indicated. [*Data from J. A. Bailey et al., "Recent Segmental Duplications in the Human Genome," Science 297, 2002, 1003–1007.*]

Duplicated units dispersed within the same chromosome are called intrachromosomal, while duplicated segments found on other chromosomes are known as interchromosomal. Every chromosome in the human genome contains numerous intra- and interchromosomal segmental duplications.

Segmental duplications are thought to have an important role as substrates for nonallelic homologous recombination (NAHR). As shown in Figure 17-19, crossing over between segmental duplications can lead to various chromosomal rearrangements. Such rearrangements seem to have been important in human evolution. For example, major inversions that are key differences between human and ape sequences have almost certainly come from NAHR mediated by segmental duplications. It also seems likely that NAHR is responsible for rearrangements that cause some human diseases. In addition to the previously discussed association between a segmental duplication on chromosome 7 with Williams syndrome (Figure 17-23), segmental duplications are found more frequently than expected in regions of the genome that have been associated with dozens of other human genetic disorders, including neurofibromatosis, hemophilia A, and red-green color blindness.

KEY CONCEPT Crossing over between segmental duplications can lead to other chromosomal rearrangements.

Inversions

We have seen that, to create an inversion, a segment of a chromosome is cut out, flipped, and reinserted. Inversions are of two basic types. If the centromere is outside the inversion, the inversion is said to be a **paracentric inversion**. Inversions spanning the centromere are **pericentric inversions**.

A	B		C	D	E	F

Normal sequence

A	B		C	E	D	F

Paracentric

A	D	C		B	E	F

Pericentric

ANIMATED ART   SaplingPlus

Formation of paracentric inversions

Because inversions are balanced rearrangements, they do not change the overall amount of genetic material, and so they do not result in gene imbalance. Individuals with inversions are generally normal, if there are no breaks within genes. A break that disrupts a gene produces a mutation that may be detectable as an abnormal phenotype. If the gene has an essential function, then the break point acts as a lethal mutation linked to the inversion. In such a case, the inversion cannot be bred to homozygosity. However, many inversions can be made homozygous, and furthermore, inversions can be detected in haploid organisms. In these cases, the break points of the inversion are clearly not in essential regions. Some of the possible consequences of inversion at the DNA level are shown in **Figure 17-25**.

Most analyses of inversions are carried out on diploid cells that contain one normal chromosome set plus one set carrying the inversion. This type of cell is called an **inversion heterozygote**, but note that this designation does not imply that any gene locus is heterozygous; rather, it means that for the chromosome pair with the inversion, there is one normal chromosome and one inverted chromosome present in the cell. The location of the inverted segment can often be detected microscopically. In meiosis, one chromosome twists once at the ends of the inversion to pair with its untwisted homolog; in this way, the paired homologs form a visible **inversion loop** (**Figure 17-26**).

In a *paracentric* inversion, crossing over within the inversion loop at meiosis connects homologous centromeres in a **dicentric bridge** while also producing an **acentric fragment** (**Figure 17-27**). Then, as the chromosomes separate in anaphase I, the centromeres remain linked by the bridge. The acentric fragment cannot align itself or move; consequently, it is lost. Tension eventually breaks the dicentric bridge, forming two chromosomes with terminal deletions. Either the gametes containing such chromosomes or the zygotes that they eventually form will probably be inviable. Hence, a crossover event, which normally generates the recombinant class of meiotic products, is instead lethal to those products. The overall result is a drastically lower frequency of viable recombinants. In fact, for genes within the inversion, the recombinant frequency is close to zero. (It is not exactly zero because rare double crossovers between only two chromatids are viable.) For genes flanking the inversion,

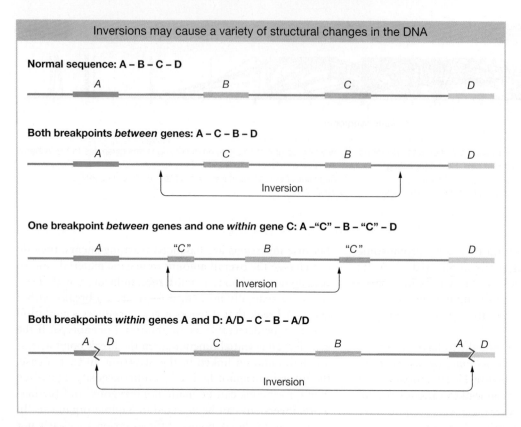

FIGURE 17-25 An inversion may have no effect on genes, may disrupt a gene, or may fuse parts of two genes, depending on the location of the break points. Genes are represented by *A, B, C,* and *D*. Arrows indicate the positions of the break points.

the recombination frequency (RF) is reduced in proportion to the size of the inversion because, for a longer inversion, there is a greater probability of a crossover occurring within it and producing an inviable meiotic product.

In a heterozygous *pericentric* inversion, the net genetic effect is the same as that of a paracentric inversion—crossover products are not recovered—but the reasons are different. In a pericentric inversion, the centromeres are contained within the inverted region. Consequently, the chromosomes that have engaged in crossing over separate in the normal fashion, without the creation of a bridge (**Figure 17-28**). However, the crossover produces chromatids that contain a duplication and a deletion for different parts of the chromosome. In this case, if a gamete carrying a crossover chromosome is fertilized, the zygote dies because of gene imbalance. Again, the result is that only noncrossover chromatids are present in viable progeny.

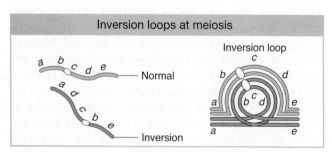

FIGURE 17-26 The chromosomes of inversion heterozygotes pair in a loop at meiosis.

Hence, the RF value of genes within a pericentric inversion also is zero.

Inversions affect recombination in another way, too. Inversion heterozygotes often have mechanical pairing problems in the region of the inversion. The inversion loop causes a large distortion that can extend beyond the loop itself. This distortion reduces the opportunity for crossing over in the neighboring regions.

Let us consider an example of the effects of an inversion on recombinant frequency. A wild-type *Drosophila* specimen from a natural population is crossed with a homozygous recessive laboratory stock *dp cn/dp cn*. (The *dp* allele encodes dumpy wings and *cn* encodes cinnabar eyes. The two genes are known to be 45 map units apart on chromosome 2.) The F_1 generation is wild type. When an F_1 female is crossed with the recessive parent, the progeny are

250	wild type	$+ +/dp\ cn$
246	dumpy cinnabar	$dp\ cn/dp\ cn$
5	Dumpy	$dp\ +/dp\ cn$
7	Cinnabar	$+ cn/dp\ cn$

In this cross, which is effectively a dihybrid testcross, 45 percent of the progeny are expected to be either dumpy or cinnabar (they constitute the crossover classes), but only 12 of 508, about 2 percent, are obtained. Something is reducing crossing over in this region, and a likely explanation is an inversion spanning most of the *dp–cn* region.

Paracentric inversions can lead to deletion products

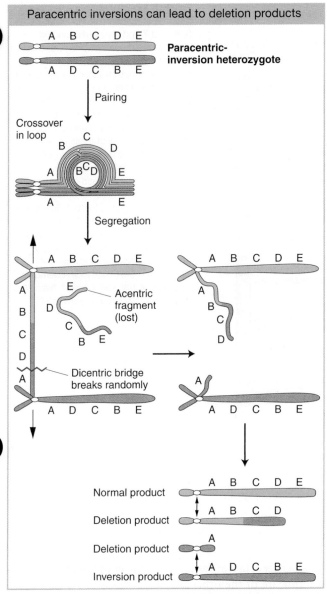

FIGURE 17-27 A crossover in the loop of a paracentric-inversion heterozygote gives rise to chromosomes containing deletions.

ANIMATED ART 📊 SaplingPlus

Meiotic behavior of paracentric inversions

Because the expected RF was based on measurements made on laboratory strains, the wild-type fly from nature was the most likely source of the inverted chromosome. Hence, chromosome 2 in the F_1 can be represented as follows:

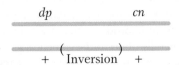

KEY CONCEPT The main diagnostic features of heterozygous inversions are inversion loops, reduced recombinant frequency, and reduced fertility because of unbalanced or deleted meiotic products.

Pericentric inversions can lead to duplication-and-deletion products

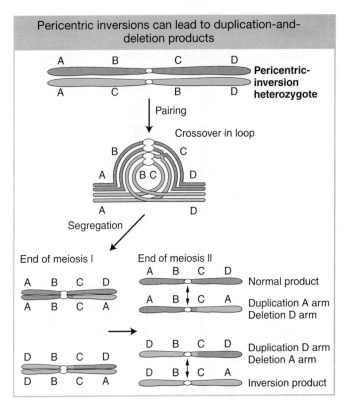

FIGURE 17-28 A crossover in the loop of a pericentric-inversion heterozygote gives rise to chromosomes containing duplications and deletions.

Pericentric inversions also can be detected microscopically through new arm ratios. Consider the following pericentric inversion:

Normal ————o———— Arm ratio, long : short ~ 4:1

Inversion —(o)—— Arm ratio, long : short ~ 1:1

Note that the length ratio of the long arm to the short arm has been changed from about 4:1 to about 1:1 by the inversion. Paracentric inversions do not alter the arm ratio, but they may be detected microscopically by observing changes in banding or other chromosomal landmarks, if available. The ability to sequence whole genomes (see Chapter 14) has also enabled the discovery of inversions that are not cytologically visible and revealed that many inversions exist both within and between species. For example, a comparison of the chimpanzee and human genomes revealed that there are over 1500 inversions between these two species, including 33 that encompass more than 100 kb (see **Figure 17-29**). Previously, only nine pericentric inversions between these two species had been identified using cytogenetic methods. The phenotypic consequences of these inversions are not known, but such inversions have been proposed to play key roles in forming and maintaining reproductive barriers between species (see Chapter 20).

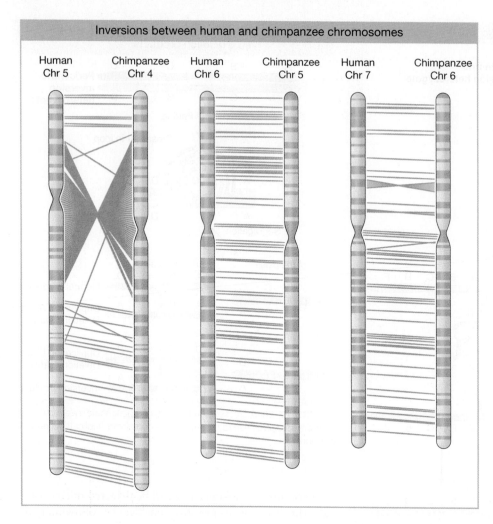

Inversions between human and chimpanzee chromosomes

Human Chr 5   Chimpanzee Chr 4   Human Chr 6   Chimpanzee Chr 5   Human Chr 7   Chimpanzee Chr 6

FIGURE 17-29 Cytological banding patterns of human chromosomes 5, 6, and 7 and the corresponding chimpanzee chromosomes 4, 5, and 6 are shown. Crossing lines indicate that the homologous sequences are found in an inverted orientation on the human and chimpanzee chromosomes. Except for the large pericentric inversion between human chromosome 5 and chimpanzee chromosome 4, all of these inversions were discovered by sequencing the genomes of these two species. Note the difference in the ratio of arm lengths between human chromosome 5 and chimpanzee chromosome 4. [*Information from Feuk L, MacDonald JR, Tang T, Carson AR, Li M, Rao G, et al. (2005) Discovery of Human Inversion Polymorphisms by Comparative Analysis of Human and Chimpanzee DNA Sequence Assemblies. PLoS Genet 1(4): e56, Figure 1. https://doi .org/10.1371/journal.pgen.0010056.*]

Reciprocal translocations

ANIMATED ART   SaplingPlus

Formation of reciprocal translocations

There are several types of translocations, but here we consider only reciprocal translocations, the simplest type. Recall that, to form a reciprocal translocation, two nonhomologous chromosomes trade acentric fragments created by two simultaneous chromosome breaks (see the Chapter 17 opening photo on page 557). As with other rearrangements, meiosis in heterozygotes having two translocated chromosomes and their normal counterparts produces characteristic configurations. **Figure 17-30** illustrates meiosis in an individual that is heterozygous for a reciprocal translocation. Note the cross-shaped pairing configuration. Because the law of independent assortment is still in force, there are two common patterns of segregation. Let us use N_1 and N_2 to represent the normal chromosomes and T_1 and T_2 as the translocated chromosomes. The segregation of each of the structurally normal chromosomes with one of the translocated ones ($T_1 + N_2$ and $T_2 + N_1$) is called **adjacent-1 segregation**. Each of the two meiotic

products is deficient for a different arm of the cross and has a duplicate of the other. These products are inviable. On the other hand, the two normal chromosomes may segregate together, as will the reciprocal parts of the translocated ones, to produce $N_1 + N_2$ and $T_1 + T_2$ products. This segregation pattern is called **alternate segregation**. These products are both balanced and viable.

Adjacent-1 and alternate segregations are equal in number, and so half the overall population of gametes will be nonfunctional, a condition known as **semisterility** or "half-sterility." Semisterility is an important diagnostic tool for identifying translocation heterozygotes. However, semisterility is defined differently for plants and animals. In plants, the 50 percent of meiotic products that are from the adjacent-1 segregation generally abort at the gametic stage (**Figure 17-31**). In animals, these products are viable as gametes but lethal to the zygotes that they produce on fertilization.

Remember that heterozygotes for inversions also may show some reduction in fertility but by an amount dependent on the size of the affected region. Thus, the precise 50 percent reduction in viable gametes or zygotes is usually a reliable diagnostic clue for the presence of a translocation.

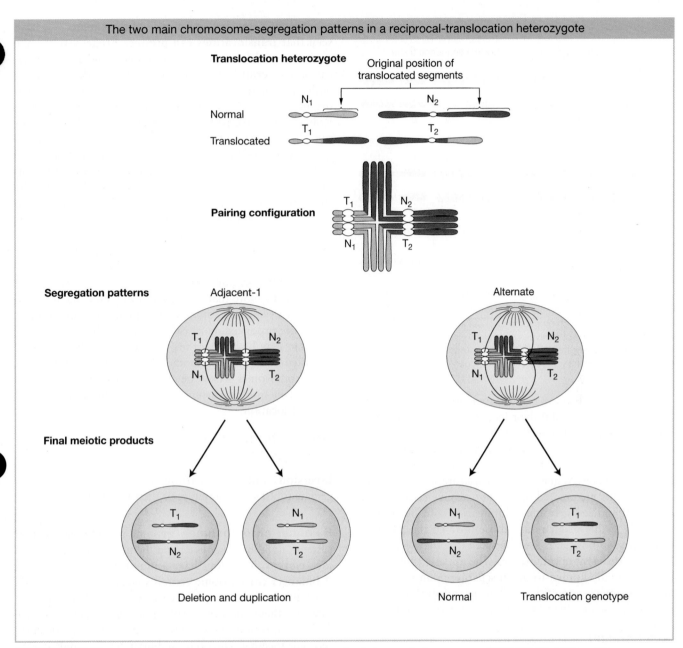

FIGURE 17-30 The segregating chromosomes of a reciprocal-translocation heterozygote form a cross-shaped pairing configuration. The two most commonly encountered segregation patterns that result are the often inviable "adjacent-1" and the viable "alternate." N_1 and N_2, normal nonhomologous chromosomes; T_1 and T_2, translocated chromosomes.

ANIMATED ART   SaplingPlus

Meiotic behavior of reciprocal translocations

Normal and aborted pollen of a translocation heterozygote

FIGURE 17-31 Pollen of a semisterile corn plant. The clear pollen grains contain chromosomally unbalanced meiotic products of a reciprocal-translocation heterozygote. The opaque pollen grains, which contain either the complete translocation genotype or normal chromosomes, are functional in fertilization and development. [*William Sheridan.*]

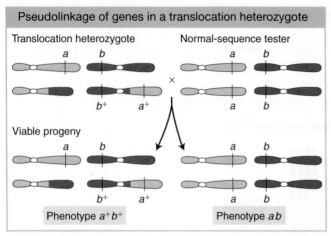

Pseudolinkage of genes in a translocation heterozygote

| Translocation heterozygote | Normal-sequence tester |

Viable progeny

Phenotype a^+b^+ Phenotype ab

FIGURE 17-32 When a translocated fragment carries a marker gene, this marker can show linkage to genes on the other chromosome. All viable progeny that inherit the translocation chromosome will show the wild-type phenotypes associated with genes *a* and *b*, while all viable progeny that do not inherit the translocation will show the mutant phenotypes associated with genes *a* and *b*.

ANIMATED 🖥 SaplingPlus **ART**

Pseudolinkage of genes by reciprocal translocations

Genetically, genes on translocated chromosomes act as though they are linked if their loci are close to the translocation break point. **Figure 17-32** shows a translocation heterozygote that has been established by crossing an *a/a*;*b/b* individual with a translocation homozygote bearing the wild-type alleles. When the heterozygote is testcrossed, recombinants are created but do not survive because they carry unbalanced genomes (duplication-and-deletions). The only viable progeny are those bearing the parental genotypes; so linkage is seen between loci that were originally on different chromosomes. The apparent linkage of genes normally known to be on separate nonhomologous chromosomes—sometimes called **pseudolinkage**—is a genetic diagnostic clue to the presence of a translocation.

KEY CONCEPT Heterozygous reciprocal translocations are diagnosed genetically by semisterility and by the apparent linkage of genes whose normal loci are on separate chromosomes.

Robertsonian translocations

Let's return to the family with the Down syndrome child, introduced at the beginning of the chapter. The birth of two children with Down syndrome in the family can indeed be a coincidence—after all, coincidences do happen. However, the miscarriage gives a clue that something else might be going on. A large proportion of spontaneous abortions carry chromosomal abnormalities, so perhaps that is the case in this example. If so, the couple may have had two conceptions with chromosome mutations, which would be very unlikely unless there was a common cause. However, a small proportion of Down syndrome cases are known to

result from a translocation in one of the parents. We have seen that translocations can produce progeny that have extra material from part of the genome, and so a translocation concerning chromosome 21 can produce progeny with extra material from that chromosome. In Down syndrome, the translocation responsible is of a type called a **Robertsonian translocation**, which involves breakage of two chromosomes at or near their centromeres, and subsequent fusions of the long arms of the chromosomes as well as loss of the short arms of the chromosomes. Note that in humans, Robertsonian translocations usually involve the five chromosomes (13, 14, 15, 20, and 21) with almost no unique genes on their short arms. Thus, loss of these short arms can be tolerated. In the case of a Robertsonian translocation involving chromosome 21, the progeny therefore will carry an almost complete extra copy of chromosome 21. The translocation and its segregation are illustrated in **Figure 17-33**. Note that, in addition to complements causing Down syndrome, other aberrant chromosome complements are produced, most of which abort. In our example, the man may have this translocation, which he may have inherited from his grandmother. To confirm this possibility, his chromosomes would be checked. His unaffected child might have normal chromosomes or might have inherited his translocation.

Applications of inversions and translocations

Inversions and translocations have proven to be useful genetic tools; some examples of their uses follow.

Balancer chromosomes In some model experimental systems, notably *Drosophila* and the nematode *Caenorhabditis elegans*, inversions have a practical use as balancers. A **balancer chromosome** contains *multiple* inversions; so, when it is combined with the corresponding wild-type chromosome, there can be no viable crossover products. In some analyses, it is important to keep all the alleles on one chromosome together with no recombination between them. The geneticist creates individuals having genotypes that combine such a chromosome of interest with a balancer. This combination eliminates progeny with crossovers, so only parental combinations appear in the progeny. For convenience, balancer chromosomes are marked with a dominant morphological mutation. The marker allows the geneticist to track the segregation of the entire balancer or its normal homolog by noting the presence or absence of the marker.

Gene mapping Inversions and translocations are useful for mapping and subsequent isolation of specific genes. The gene for human neurofibromatosis was isolated in this way. The critical information came from people who not only had the disease, but also carried chromosomal translocations. All the translocations had one break point in common, in a band close to the centromere of chromosome 17. Hence, this band appeared to be the locus of the

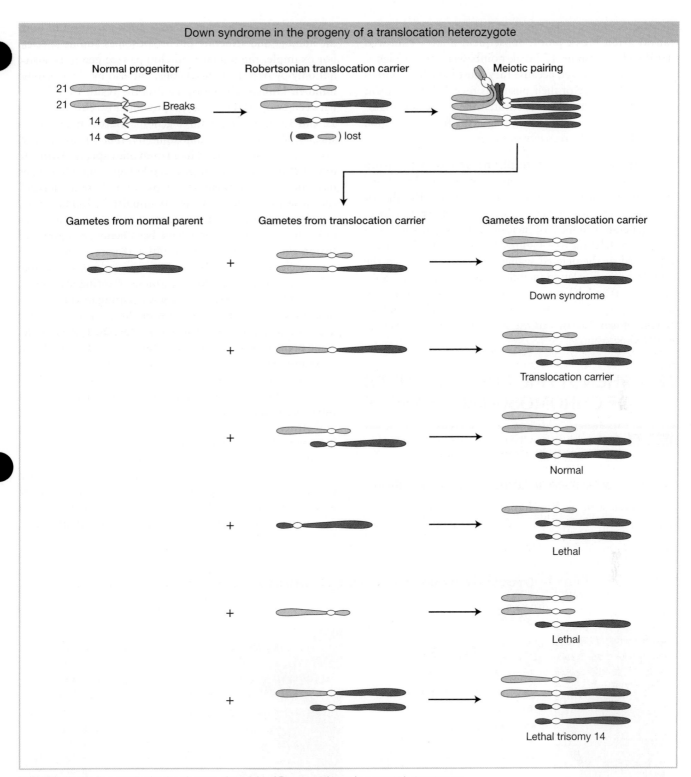

Down syndrome in the progeny of a translocation heterozygote

FIGURE 17-33 In a small minority of cases, the origin of Down syndrome is a parent heterozygous for a Robertsonian translocation concerning chromosome 21. Meiotic segregation results in some gametes carrying a chromosome with a large additional segment of chromosome 21. In combination with a normal chromosome 21 provided by the gamete from the opposite sex, the symptoms of Down syndrome are produced even though there is not full trisomy 21.

neurofibromatosis gene, which had been disrupted by the translocation break point. Subsequent analysis showed that the chromosome 17 break points were not at identical positions; however, because they must have been within the gene, the range of their positions revealed the segment of the chromosome that constituted the neurofibromatosis gene. The isolation of DNA fragments from this region eventually led to the recovery of the gene itself.

Synthesizing specific duplications or deletions

Translocations and inversions are routinely used to delete or duplicate specific chromosome segments. Recall, for example, that pericentric inversions as well as translocations generate products of meiosis that contain a duplication *and* a deletion (see Figures 17-28 and 17-30). If the duplicated or the deleted segment is very small, then the duplication-and-deletion meiotic products are often viable. Duplications and deletions are useful for a variety of experimental applications, including the mapping of genes and the varying of gene dosage for the study of regulation, as seen in preceding sections.

17.3 PHENOTYPIC CONSEQUENCES OF CHROMOSOMAL CHANGES

LO 17.5 Distinguish among the main human syndromes resulting from chromosomal changes.

Chromosome rearrangements and evolution

As we have seen in this chapter, chromosomal changes within species can have many detrimental effects. Thus, it is perhaps surprising that there is great variation in both the number and structure of chromosomes among species. For example, the human and chimpanzee genomes differ not only by over 1500 inversions, but also by chromosome number. Human chromosome 2 is the result of a fusion between two chromosomes, represented by chromosomes 12 and 13 in the chimpanzee genome. Across mammalian species, the diploid chromosome number varies greatly, ranging from 6 to 102. The mammalian species with the lowest chromosome number is the Indian muntjac: females have only 6 chromosomes (3 pairs) and males have 7 (because the Y chromosome is unpaired). Remarkably, the closely related Chinese muntjac has 46 chromosomes (23 pairs), suggesting there have been many chromosomal fusion events in the Indian muntjac (**Figure 17-34**). These two species have been observed to mate and produce viable offspring in captivity. However, these offspring are sterile, consistent with problems in meiotic pairing of these chromosomes. Dramatic differences in chromosome number and structure can also occur within species. For example, some populations of house mice have a reduced number of chromosomes due to the presence of multiple Robertsonian translocations.

These mammalian examples are just a few to highlight the remarkable diversity of chromosome number and structure found in nature. Currently, almost nothing is known about the mechanisms that evolved to overcome the deleterious effects of these chromosome rearrangements. However, many scientists have hypothesized that chromosome rearrangements, such as inversions, might actually facilitate evolutionary change. Recent research in a number of different systems, including plants, ants,

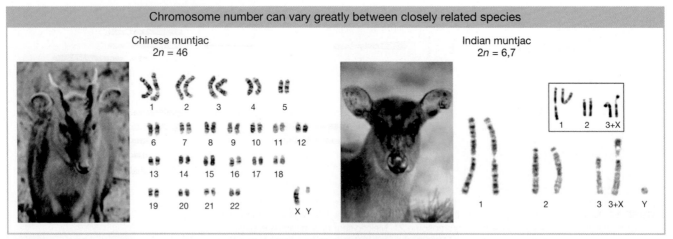

FIGURE 17-34 Chinese muntjacs have $2n = 46$ chromosomes, including an XY sex chromosome pair in males, as shown in the chromosome image on the left. Indian muntjacs have $2n = 7$ chromosomes in males and $2n = 6$ chromosomes in females, as shown in the chromosome images on the right. [*Republished with permission of Springer Science+Business Media, from Markus O. Scheuermann, Andrea E. Merman, et al., "Characterization of nuclear compartments identified by ectopic markers in mammalian cells with distinctly different karyotype," Chromosoma, 2005, May; 114 (1) 39–53, Figure 1. Permission conveyed through Copyright Clearance Center, Inc.*]

butterflies, and birds, supports this hypothesis. A classic example is Müllerian mimicry in butterflies, in which two distasteful species mimic each other's wing patterns. In this co-mimicry, predators more easily learn to avoid a particular wing pattern, providing a benefit for both species. It has been observed that the genes controlling different aspects of the wing pattern are tightly linked to each other, creating a so-called *supergene*. Any recombination between these genes would create hybrid wing patterns that might no longer be recognized as distasteful by predators, and thus increase the risk that the butterfly would be eaten. As we have seen, recombination between genes is rare within inversions, and it was therefore predicted that these wing pattern supergenes would be found within inversions. Indeed, there are now examples of butterfly species in which wing pattern supergenes are found in chromosomal inversions (**Figure 17-35**). Similar evidence is accumulating in other systems, which emphasizes that chromosome rearrangements are an important substrate for evolution.

Chromosome rearrangements and cancer

Cancer is a disease of abnormal cell proliferation. As a result of some insult inflicted on it, a cell of the body divides out of control to form a population of cells called a cancer. A localized knot of proliferated cells is called a tumor, whereas cancers of mobile cells such as blood cells disperse throughout the body. Cancer is most often caused by mutations in the coding or regulatory sequence of genes whose normal function is to regulate cell division. Such genes are called *proto-oncogenes* before a cancer-causing mutation occurs, and *oncogenes* after a cancer-causing mutation occurs. Chromosomal rearrangements, especially translocations, can interfere with the normal function of such proto-oncogenes.

There are two basic ways in which translocations can alter the function of proto-oncogenes. In the first mechanism, the translocation relocates a proto-oncogene next to a new regulatory element. A good example is provided by Burkitt lymphoma. The proto-oncogene in this cancer encodes the protein MYC, a transcription factor that activates genes required for cell proliferation. Normally, the *MYC* gene is transcribed only when a cell needs to undergo proliferation, but in cancerous cells, the proto-oncogene *MYC* is relocated next to the regulatory region of immunoglobulin (Ig) genes (**Figure 17-36a**). These immunoglobulin genes are constitutively transcribed; that is, they are on all the time. Consequently, the *MYC* gene is transcribed at all times, and the cell-proliferation genes are continuously activated.

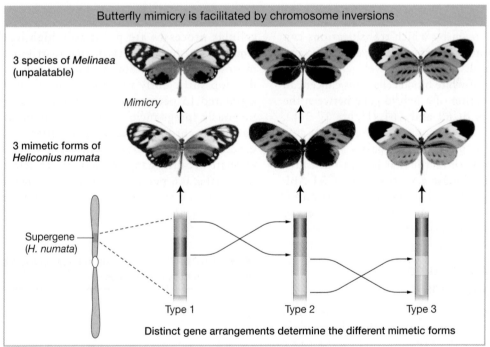

Butterfly mimicry is facilitated by chromosome inversions

3 species of *Melinaea* (unpalatable)

Mimicry

3 mimetic forms of *Heliconius numata*

Supergene (*H. numata*)

Type 1   Type 2   Type 3

Distinct gene arrangements determine the different mimetic forms

FIGURE 17-35 *Heliconius numata* is a distasteful butterfly species that is a Müllerian mimic with different species of the butterfly genus *Melinaea*. When the two species co-occur, *H. numata* has a wing pattern that mimics the local *Melinaea* species. However, these wing patterns differ between Melinaea species, and so *H. numata* wing patterns differ from location to location. The genes controlling these differences in wing pattern in *H. numata* are found in a supergene on chromosome 15. Distinct wing patterns are associated with different chromosomal inversions around the supergene. [*Courtesy of Mathieu Joron.*]

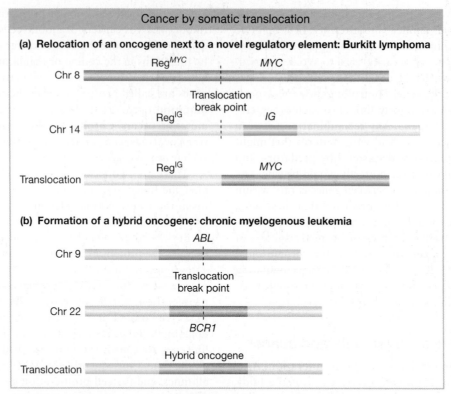

FIGURE 17-36 The two main ways that translocations can cause cancer in a body (somatic) cell are illustrated by the cancers Burkitt lymphoma (a) and chronic myelogenous leukemia (b). The genes *MYC*, *BCR1*, and *ABL* are proto-oncogenes.

The other mechanism by which translocations can cause cancer is the formation of a hybrid gene. An example is provided by the disease chronic myelogenous leuke-mia (CML), a cancer of white blood cells. This cancer can result from the formation of a hybrid gene between the two proto-oncogenes *BCR1* and *ABL* (Figure 17-36b). The *ABL* proto-oncogene encodes a protein kinase in a signaling pathway. The protein kinase passes along a sig-nal initiated by a growth factor that leads to cell pro-liferation. The protein kinase activity of the BCR1-ABL fusion protein is always on. The fusion protein contin-ually propagates its growth signal onward, regardless of whether the initiating signal is present.

Overall incidence of human chromosome mutations

Chromosome mutations arise surprisingly frequently in human sexual reproduction, showing that the relevant cellular processes are prone to a high level of error. **Figure 17-37** shows the estimated distribution of chro-mosome mutations among human conceptions that develop sufficiently to implant in the uterus. Of the estimated 15 percent of conceptions that abort spon-taneously (pregnancies that terminate naturally), fully half show chromosomal abnormalities. Some medical geneticists believe that even this high level is an under-estimate because many cases are never detected. Among live births, 0.6 percent have chromosomal abnormali-ties, resulting from both aneuploidy and chromosomal rearrangements.

KEY CONCEPT Chromosome mutations have a large impact on human fertility and disease.

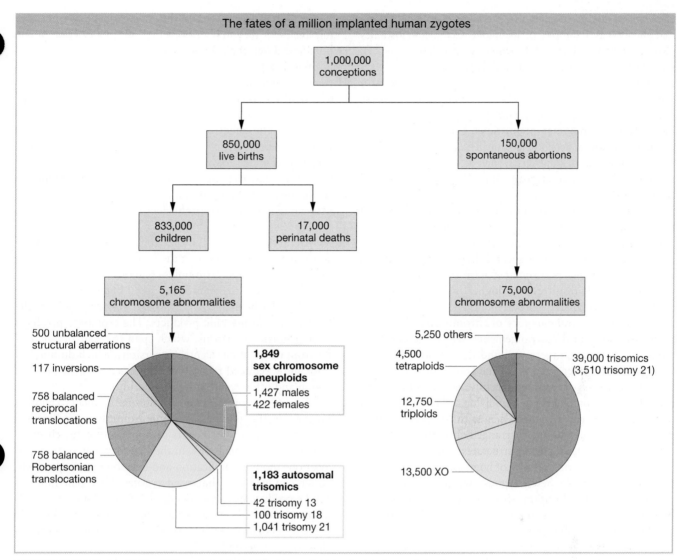

FIGURE 17-37 The proportion of chromosomal mutations is much higher in spontaneous abortions. [*Data from K. Sankaranarayanan, Mutat. Res. 61, 1979, 249–257.*]

SUMMARY

Polyploidy is an abnormal condition in which there is a larger-than-normal number of chromosome sets. Polyploids such as triploids ($3n$) and tetraploids ($4n$) are common among plants and are represented even among animals. Organisms with an odd number of chromosome sets are sterile because not every chromosome has a partner at meiosis. Unpaired chromosomes segregate randomly to the poles of the cell in meiosis, leading to unbalanced sets of chromosomes in the resulting gametes. Such unbalanced gametes do not yield viable progeny because the normal balance of gene dosage is disrupted. In polyploids with an even number of sets, each chromosome has a potential pairing partner and

hence can produce balanced gametes and progeny. Polyploidy can result in an organism of larger dimensions; this discovery has permitted important advances in horticulture and in crop breeding.

In plants, allopolyploids (polyploids formed by combining chromosome sets from different species) can be made by crossing two related species and then doubling the progeny chromosomes through the use of colchicine. These techniques have potential applications in crop breeding because allopolyploids combine the features of the two parental species.

When cellular accidents change parts of chromosome sets, aneuploids result. Aneuploidy itself usually results in unbalanced dosage of gene products, which leads to abnormal phenotypes. Examples of aneuploids include monosomics $(2n - 1)$ and trisomics $(2n + 1)$. Down syndrome (trisomy 21), Klinefelter syndrome (XXY), and Turner syndrome (XO) are well-documented examples of aneuploid conditions in humans. The spontaneous level of aneuploidy in humans is quite high and accounts for a large proportion of genetically based disorders in human populations. The phenotype of an aneuploid organism depends very much on the particular chromosome affected. In some cases, such as human trisomy 21, there is a highly characteristic constellation of associated phenotypes.

Most instances of aneuploidy result from accidental chromosome missegregation at meiosis (nondisjunction). The error is spontaneous and can occur in any particular meiocyte at the first or second division. In humans, a maternal-age effect is associated with nondisjunction of chromosome 21, resulting in a higher incidence of Down syndrome in the children of older mothers.

The other general category of chromosome mutations comprises structural rearrangements, which include deletions, duplications, inversions, and translocations. These changes result either from breakage and incorrect reunion or from crossing over between repetitive elements (nonallelic homologous recombination). In individuals heterozygous for a chromosome rearrangement (i.e., with one normal chromosome homolog and one rearranged chromosome homolog), there are unusual pairing structures at meiosis resulting from the strong pairing affinity of homologous chromosome regions. For example, heterozygous inversions show loops, and reciprocal translocations show cross-shaped structures. Segregation of these structures results in abnormal meiotic products unique to the rearrangement.

A deletion is the loss of a section of chromosome, either because of chromosome breaks followed by loss of the intervening segment or because of segregation in heterozygous translocations or inversions. If the genes removed in a deletion are essential to life, a homozygous deletion is lethal. Heterozygous deletions may be lethal because of gene dosage imbalance or because they uncover recessive deleterious alleles, or they may be nonlethal. When a deletion in one homolog allows the phenotypic expression of recessive alleles in the other, the unmasking of the recessive alleles is called pseudodominance.

Duplications are generally produced from other rearrangements or by aberrant crossing over. They also unbalance the genetic material, producing a deleterious phenotypic effect or death of the organism. Segmental duplications are also a substrate for additional chromosomal rearrangements due to nonallelic homologous recombination. Many human chromosome disorders are associated with regions of the genome that harbor segmental duplications.

An inversion is a 180-degree turn of a part of a chromosome. In the homozygous state, inversions may cause little problem for an organism unless one of the breaks disrupts a gene. On the other hand, inversion heterozygotes show inversion loops at meiosis, and crossing over within the loop results in inviable products. The crossover products of pericentric inversions, which span the centromere, differ from those of paracentric inversions, which do not, but both show reduced recombinant frequency in the affected region and often result in reduced fertility.

A translocation moves a chromosome segment to another position in the genome. A simple example is a reciprocal translocation, in which parts of nonhomologous chromosomes exchange positions. In the heterozygous state, translocations produce duplication-and-deletion meiotic products, which can lead to unbalanced zygotes. New gene linkages can be produced by translocations. The random segregation of centromeres in a translocation heterozygote results in 50 percent unbalanced meiotic products and, hence, 50 percent sterility (semisterility).

Chromosomal rearrangements are an important cause of sterility, birth defects, and disease in human populations. However, they are also substrates for evolution and useful in engineering special strains of organisms for experimental and applied genetics.

KEY TERMS

acentric chromosome (p. 574)
acentric fragment (p. 579)
adjacent-1 segregation (p. 582)
allopolyploid (p. 560)
alternate segregation (p. 582)
amphidiploid (p. 563)
anaphase bridge (p. 574)
aneuploid (p. 560)
autopolyploid (p. 560)
balanced rearrangement (p. 575)
balancer chromosome (p. 584)
bivalent (p. 560)

chromosome mutation (p. 558)
deletion (p. 574)
deletion loop (p. 576)
deletion mapping (p. 576)
dicentric bridge (p. 579)
dicentric chromosome (p. 574)
diploid (p. 559)
disomic (p. 564)
dosage compensation (p. 572)
Down syndrome (p. 569)
duplication (p. 574)
euploid (p. 559)

gene balance (p. 571)
gene-dosage effect (p. 571)
genetic load (p. 560)
haploid (p. 559)
haploid chromosome number (p. 559)
hexaploid (p. 559)
homeologous chromosomes (p. 560)
insertional duplication (p. 578)
intragenic deletion (p. 576)
inversion (p. 574)
inversion heterozygote (p. 579)
inversion loop (p. 579)

SOLVED PROBLEMS

SOLVED PROBLEM 1

A corn plant is heterozygous for a reciprocal translocation and is therefore semisterile. This plant is crossed with a chromosomally normal strain that is homozygous for the recessive allele brachytic (b), located on chromosome 2. A semisterile F_1 plant is then backcrossed to the homozygous brachytic strain. The progeny obtained show the following phenotypes:

Nonbrachytic		Brachytic	
Semisterile	Fertile	Semisterile	Fertile
334	27	42	279

a. What ratio would you expect to result if the chromosome carrying the brachytic allele does not take part in the translocation?

b. Do you think that chromosome 2 takes part in the translocation? Explain your answer, showing the conformation of the relevant chromosomes of the semisterile F_1 and the reason for the specific numbers obtained.

SOLUTION

a. We should start with the methodical approach and simply restate the data in the form of a diagram, where

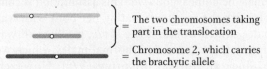

= The two chromosomes taking part in the translocation

= Chromosome 2, which carries the brachytic allele

To simplify the diagram, we do not show the chromosomes divided into chromatids (although they would be at this stage of meiosis). We then diagram the first cross:

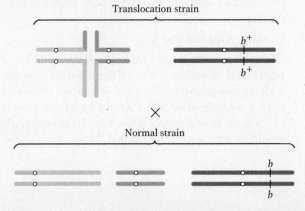

All the progeny from this cross will be heterozygous for the chromosome carrying the brachytic allele, but what about the chromosomes taking part in the translocation? In this chapter, we have seen that only alternate-segregation products survive and that half of these survivors will be chromosomally normal and half will carry the two rearranged chromosomes. The rearranged combination will regenerate a translocation heterozygote when it combines with the chromosomally normal complement from the normal parent. These latter types—the semisterile F_1's—are diagrammed as part of the backcross to the parental brachytic strain:

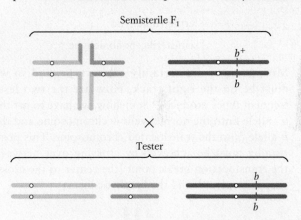

In calculating the expected ratio of phenotypes from this cross, we can treat the behavior of the translocated chromosomes independently of the behavior of chromosome 2. Hence, we can predict that the progeny will be

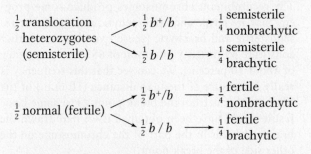

This predicted 1:1:1:1 ratio is quite different from that obtained in the actual cross.

b. Because we observe a departure from the expected ratio based on the independence of the brachytic phenotype and semisterility, chromosome 2 likely *does* take part in the translocation. Let's assume that the brachytic locus (*b*) is on the orange chromosome. But where? For the purpose of the diagram, it does not matter where we put it, but it does matter genetically because the position of the *b* locus affects the ratios in the progeny. If we assume that the *b* locus is near the tip of the piece that is translocated, we can redraw the pedigree:

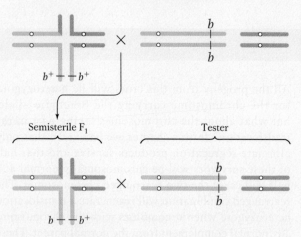

If the chromosomes of the semisterile F₁ segregate as diagrammed here, we could then predict

$$\frac{1}{2}\text{ fertile, brachytic}$$

$$\frac{1}{2}\text{ semisterile, nonbrachytic}$$

Most progeny are certainly of this type, and so we must be on the right track. How are the two less-frequent types produced? Somehow, we have to get the b^+ allele onto the normal yellow chromosome and the *b* allele onto the translocated chromosome. This positioning must be achieved by crossing over between the translocation break point (the center of the cross-shaped structure) and the brachytic locus:

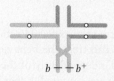

The recombinant chromosomes produce some progeny that are fertile and nonbrachytic and some that are semisterile and brachytic (these two classes together constitute 69 progeny of a total of 682, or a frequency of about 10 percent). We can see that this frequency is really a measure of the map distance (10 m.u.) of the brachytic locus from the break point. (The same basic result would have been obtained if we had drawn the brachytic locus in the part of the chromosome on the other side of the break point.)

We have lines of mice that breed true for two alternative behavioral phenotypes that we know are determined by two alleles at a single locus: *v* causes a mouse to move with a "waltzing" gait, whereas *V* determines a normal gait. After crossing the true-breeding waltzers and normals, we observe that most of the F₁ is normal, but, unexpectedly, there is one waltzer female. We mate the F₁ waltzer with two different waltzer males and note that she produces only waltzer progeny. When we mate her with normal males, she produces normal progeny and no waltzers. We mate three of her normal female progeny with two of their brothers, and these mice produce 60 progeny, all normal. When, however, we mate one of these same three females with a third brother, we get six normals and two waltzers in a litter of eight. By thinking about the parents of the F₁ waltzer, we can consider some possible explanations of these results:

a. A dominant allele may have mutated to a recessive allele in her normal parent.

b. In one parent, there may have been a dominant mutation in a second gene to create an epistatic allele (see Chapter 5) that acts to prevent the expression of *V*, leading to waltzing.

c. Meiotic nondisjunction of the chromosome carrying *V* in her normal parent may have given a viable aneuploid.

d. There may have been a viable deletion spanning *V* in the meiocyte from her normal parent.

Which of these explanations are possible, and which are eliminated by the genetic analysis? Explain in detail.

SOLUTION

The best way to answer the question is to take the explanations one at a time and see if each fits the results given.

a. Mutation *V* to *v*

This hypothesis requires that the exceptional waltzer female be homozygous *v/v*. This assumption is compatible with the results of mating her both with waltzer males, which would, if she is *v/v*, produce all waltzer offspring (*v/v*), and with normal males, which would produce all normal offspring (*V/v*). However, all brother–sister matings within this normal progeny should then produce a 3:1 normal-to-waltzer ratio. Because some of the brother–sister matings actually produced no waltzers, this hypothesis does not explain the data.

b. Epistatic mutation *s* to *S*

Here the parents would be $V/V \cdot s/s$ and $v/v \cdot s/s$, and a germinal mutation in one of them would give the F₁ waltzer the genotype $V/v \cdot S/s$. When we crossed her with a waltzer male, who would be of the genotype $v/v \cdot s/s$, we would expect some $V/v \cdot S/s$ progeny, which would be phenotypically normal. However, we saw no normal progeny from this cross, and so the hypothesis is already overthrown. Linkage could save the hypothesis temporarily if we assumed that the mutation was in the normal parent, giving a gamete $V\ S$. Then the F_1 waltzer would be $V\ S/v\ s$, and, if linkage were tight enough, few or no $V\ s$ gametes would be produced, the type that are necessary to combine with the $v\ s$ gamete from the male to give $V\ s/v\ s$ normals. However, if the linkage hypothesis were true, the cross with the normal males would be $V\ S/v\ s \times V\ s/V\ s$, and this would give a high percentage of $V\ S/V\ s$ progeny, which would be waltzers, none of which were seen.

c. Nondisjunction in the normal parent

This explanation would give a nullisomic gamete that would combine with v to give the F_1 waltzer the hemizygous genotype v. The subsequent matings would be

- $v \times v/v$, which gives v/v and v progeny, all waltzers. This fits.

- $v \times V/V$, which gives V/v and V progeny, all normals. This also fits.

- First intercrosses of normal progeny: $V \times V$. These intercrosses give V and V/V, which are normal. This fits. Note that this intercross would also produce progeny that are homozygous for the loss of the chromosome, but these progeny would not survive.

- Second intercrosses of normal progeny: $V \times V/v$. These intercrosses give 25 percent each of V/V, V/v, V (all normals), and v (waltzers). This also fits.

This hypothesis is therefore consistent with the data.

d. Deletion of V in normal parent

Let's call the deletion D. The F_1 waltzer would be D/v, and the subsequent matings would be

- $D/v \times v/v$, which gives v/v and D/v, which are waltzers. This fits.

- $D/v \times V/V$, which gives V/v and D/V, which are normal. This fits.

- First intercrosses of normal progeny: $D/V \times D/V$, which give D/V and V/V, all normal. This fits. Again, note that the D/D progeny produced by this intercross would likely not survive.

- Second intercrosses of normal progeny: $D/V \times V/v$, which give 25 percent each of V/V, V/v, D/V (all normals), and D/v (waltzers). This also fits.

Once again, the hypothesis fits the data provided; so we are left with two hypotheses that are compatible with the results, and further experiments are necessary to distinguish them. One way of doing so would be to examine the chromosomes of the exceptional female under the microscope: aneuploidy should be easy to distinguish from deletion.

PROBLEMS

Visit SaplingPlus for supplemental content. Problems with the 🔄 icon are available for review/grading. Problems with the 🔲 icon have a Problem Solving Video. Problems with the 🧬 icon have an Unpacking the Problem exercise.

WORKING WITH THE FIGURES

(The first 20 questions require inspection of text figures.)

1. Based on Table 17-1, how would you categorize the following genomes? (Letters H through J stand for four different chromosomes.) 🔄

    HH II J KK

    HH II JJ KKK

    HHHH IIII JJJJ KKKK

2. Based on Figure 17-4, how many chromatids are in a trivalent?

3. Based on Figure 17-5, if colchicine is used on a plant in which $2n = 18$, how many chromosomes would be in the abnormal product? 🔄

4. Basing your work on Figure 17-8, use colored pens to represent the chromosomes of the fertile amphidiploid.

5. In Figure 17-9, we can designate the genome of *B. oleracea* as "A," the genome of *B. nigra* as "B," and the genome of *B. rapa* as "C." Which species shown in this figure would have an "AC" genome? 🔄

6. In Figure 17-10, what would be the constitution of an individual formed from the union of a monosomic from a first-division nondisjunction in a female and a disomic from a second-division nondisjunction in a male, assuming the gametes were functional?

7. In Figure 17-12, what would be the expected percentage of each type of segregation?

8. In Figure 17-19, is there any difference between the inversion products formed from breakage and those formed from crossing over?

9. Referring to Figure 17-19, draw a diagram showing the process whereby an inversion formed from crossing over could generate a normal sequence.

10. In Figure 17-21, would the phenotype associated with the recessive *fa* allele be visible when paired with deletion 264-32? With 258-11?

11. Look at Figure 17-22 and state which bands are missing in the cri du chat deletion.

12. Explain why the phenotypes associated with the Williams syndrome deletion and the 7q11.23 duplication shown in Figure 17-23 are not the same.

13. Referring to Figure 17-25, draw the product if breaks occurred within genes *A* and *B*.

14. In Figure 17-27, what would be the consequence of a crossover between the centromere and locus A?

15. In Figure 17-29, which of the three pairs of homologous human and chimpanzee chromosomes shown differ in the ratio of their arm length?

16. Based on Figure 17-30, are normal gametes ever formed from an adjacent-1 segregation?

17. Referring to Figure 17-32, draw an *inviable* product from the same meiosis.

18. Based on Figure 17-33, what fraction of progeny would be phenotypically normal in a cross between a translocation heterozygote and a normal parent?

19. Based on Figure 17-36, write a sentence stating how translocation can lead to cancer. Can you think of another genetic cause of cancer?

20. Using Figure 17-37, calculate what percentage of conceptions are triploid. The same figure shows XO in the spontaneous-abortion category; however, we know that many XO individuals are viable. In which of the viable categories would XO be grouped?

BASIC PROBLEMS

21. In keeping with the style of Table 17-1, what would you call organisms that are MM N OO; MM NN OO; MMM NN PP?

22. A large plant arose in a natural population. Qualitatively, it looked just the same as the others, except much larger. Is it more likely to be an allopolyploid or an autopolyploid? How would you test that it was a polyploid and not just growing in rich soil?

23. Is a trisomic an aneuploid or a polyploid?

24. Seedless watermelons are sterile triploids. However, sometimes viable seeds are found in these watermelons. What is an explanation for this finding?

25. How many different types of gametes can be produced by a triploid organism with a haploid number of three (i.e., chromosomes A, B, C)?

26. In a tetraploid *B/B/b/b*, how many quadrivalent possible pairings are there? Draw them (see Figure 17-5).

27. Someone tells you that cauliflower is an amphidiploid. Do you agree? Explain.

28. Why is *Raphanobrassica* fertile, whereas its progenitor was not?

29. A disomic product of meiosis is obtained. What is its likely origin? What other genotypes would you expect among the products of that meiosis under your hypothesis?

30. Can a trisomic *A/A/a* ever produce a gamete of genotype *a*?

31. Which, if any, of the following sex-chromosome aneuploids in humans are fertile: XXX, XXY, XYY, XO?

32. If you observed a dicentric bridge at meiosis, what rearrangement would you predict had taken place?

33. Why do acentric fragments get lost?

34. Diagram a translocation arising from repetitive DNA. Repeat for a deletion.

35. From a large stock of *Neurospora* rearrangements available from the fungal genetics stock center, what type would you choose to synthesize a strain that had a duplication of the right arm of chromosome 3 and a deletion for the tip of chromosome 4?

36. You observe a very large pairing loop at meiosis. Is it more likely to be from a heterozygous inversion or a heterozygous deletion? Explain.

37. A new recessive mutant allele does not show pseudodominance with any of the deletions that span *Drosophila* chromosome 2. What might be the explanation?

38. Compare and contrast the origins of Turner syndrome, Williams syndrome, cri du chat syndrome, and Down syndrome. (Why are they called *syndromes*?)

39. List the diagnostic features (genetic or cytological) that are used to identify these chromosomal alterations:

   a. Deletions

   b. Duplications

   c. Inversions

   d. Reciprocal translocations

40. The normal sequence of nine genes on a certain *Drosophila* chromosome is 123·456789, where the dot represents the centromere. Some fruit flies were found to have aberrant chromosomes with the following structures:

   a. 123·476589   c. 123·46789

   b. 1654·32789   d. 123·4566789

   Name each type of chromosomal rearrangement, and draw diagrams to show how each would synapse with the normal chromosome.

41. The two loci *P* and *Bz* are normally 36 m.u. apart on the same arm of a certain plant chromosome. A paracentric inversion spans about one-fourth of this region but does not include either of the loci. What approximate recombinant frequency between *P* and *Bz* would you predict in plants that are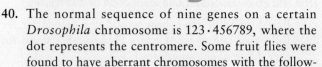

a. heterozygous for the paracentric inversion?

b. homozygous for the paracentric inversion?

42. As stated in Solved Problem 2, recessive mutation in certain mice called *waltzers* causes them to execute bizarre steps. W. H. Gates crossed waltzers with pure-breeding normal mice and found, among several hundred normal progeny, a single waltzing female mouse. This mouse was mated with a waltzing male, and her offspring were waltzers. When mated with a homozygous normal male, all her progeny were normal. Some of these normal males and females were intercrossed, and, unexpectedly, none of their progeny were waltzers. T. S. Painter examined the chromosomes of some of Gates's waltzing mice that showed a breeding behavior similar to that of the original, unusual waltzing female. He found that these mice had the normal number of 40 chromosomes. In the unusual waltzers, however, one member of a chromosome pair was abnormally short. Interpret these observations as completely as possible, both genetically and cytologically.

43. A salivary-gland chromosome of *Drosophila* has six bands as shown in the following illustration. Below the chromosome are shown the extent of five deletions (Del 1 to Del 5):

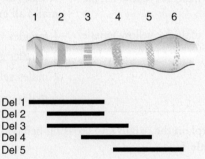

Recessive alleles *a*, *b*, *c*, *d*, *e*, and *f* are known to be in the region, but their order is unknown. When the deletions are combined with each allele, the following results are obtained:

	a	b	c	d	e	f
Del 1	−	−	−	+	+	+
Del 2	−	+	−	+	+	+
Del 3	−	+	−	+	−	+
Del 4	+	+	−	−	−	+
Del 5	+	+	+	−	−	−

In this table, a minus sign means that the deletion uncovers the recessive allele (the recessive phenotype is observed), and a plus sign means that the corresponding wild-type allele is still present. Match each salivary band with a gene.

44. A fruit fly was found to be heterozygous for a paracentric inversion. However, obtaining flies that were homozygous for the inversion was impossible even after many attempts. What is the most likely explanation for this inability to produce a homozygous inversion?

45. Orangutans are now recognized as a group of three endangered species in their natural environments (one species on the island of Borneo and two species on the island of Sumatra). Before the distinction between the three species was clear, a captive-breeding program was established using orangutans held in zoos throughout the world. One component of this program is research into orangutan cytogenetics. This research has shown that all orangutans from Borneo carry one form of chromosome 2, as shown in the accompanying diagram, and all orangutans from Sumatra carry the other form. Before this cytogenetic difference became known, some matings were carried out between animals from different islands, and 14 hybrid progeny are now being raised in captivity.

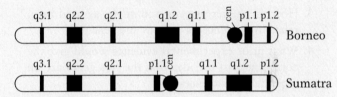

a. What term or terms describe the differences between these chromosomes?

b. Draw the chromosomes 2, paired in the first meiotic prophase, of such a hybrid orangutan. Be sure to show all the landmarks indicated in the accompanying diagram, and label all parts of your drawing.

c. In 30 percent of meioses, there will be a crossover somewhere in the region between bands p1.1 and q1.2. Draw the gamete chromosomes 2 that would result from a meiosis in which a single crossover occurred within band q1.1.

d. What fraction of the gametes produced by a hybrid orangutan will give rise to viable progeny, if these chromosomes are the only ones that differ between the parents? (Problem 45 is from Rosemary Redfield.)

46. In corn, the genes for tassel length (alleles *T* and *t*) and rust resistance (alleles *R* and *r*) are known to be on separate chromosomes. In the course of making routine crosses, a breeder noticed that one *T/t*; *R/r* plant gave unusual results in a testcross with the double-recessive pollen parent *t/t*; *r/r*. The results were

Progeny:	*T/t*; *R/r*	98
	t/t; *r/r*	104
	T/t; *r/r*	3
	t/t; *R/r*	5

Corncobs: Only about half as many seeds as usual

a. What key features of the data are different from the expected results?

b. State a concise hypothesis that explains the results.

c. Show genotypes of parents and progeny.

d. Draw a diagram showing the arrangement of alleles on the chromosomes.

e. Explain the origin of the two classes of progeny having three and five members.

UNPACKING PROBLEM 46

Before attempting a solution to this problem, try answering the following questions:

1. What do a "gene for tassel length" and a "gene for rust resistance" mean?

2. Does it matter that the precise meaning of the allelic symbols *T*, *t*, *R*, and *r* is not given? Why or why not?

3. How do the terms *gene* and *allele*, as used here, relate to the concepts of locus and gene pair?

4. What prior experimental evidence would give the corn geneticist the idea that the two genes are on separate chromosomes?

5. What do you imagine "routine crosses" are to a corn breeder?

6. What term is used to describe genotypes of the type *T/t*; *R/r*?

7. What is a "pollen parent"?

8. What are testcrosses, and why do geneticists find them so useful?

9. What progeny types and frequencies might the breeder have been expecting from the testcross?

10. Describe how the observed progeny differ from expectations.

11. What does the approximate equality of the first two progeny classes tell you?

12. What does the approximate equality of the second two progeny classes tell you?

13. What were the gametes from the unusual plant, and what were their proportions?

14. Which gametes were in the majority?

15. Which gametes were in the minority?

16. Which of the progeny types seem to be recombinant?

17. Which allelic combinations appear to be linked in some way?

18. How can there be linkage of genes supposedly on separate chromosomes?

19. What do these majority and minority classes tell us about the genotypes of the parents of the unusual plant?

20. What effect could lead to the absence of half the seeds?

21. Did half the seeds die? If so, was the female or the male parent the reason for the deaths?

Now try to solve the problem. If you are unable to do so, try to identify the obstacle and write a sentence or two describing your difficulty. Then go back to the expansion questions and see if any of them relate to your difficulty. If this approach does not work, inspect the Learning Objectives and Key Concepts of this chapter and ask yourself which might be relevant to your difficulty.

47. A yellow body in *Drosophila* is caused by a mutant allele *y* of a gene located at the tip of the X chromosome (the wild-type allele causes a gray body). In a radiation experiment, a wild-type male was irradiated with X rays and then crossed with a yellow-bodied female. Most of the male progeny were yellow, as expected, but the scanning of thousands of flies revealed two gray-bodied (phenotypically wild-type) males. These gray-bodied males were crossed with yellow-bodied females, with the following results:

	Progeny
gray male 1 × yellow female	females all yellow males all gray
gray male 2 × yellow female	$\frac{1}{2}$ females yellow
	$\frac{1}{2}$ females gray
	$\frac{1}{2}$ males yellow
	$\frac{1}{2}$ males gray

a. Explain the origin and crossing behavior of gray male 1.

b. Explain the origin and crossing behavior of gray male 2.

48. In corn, the allele *Pr* stands for green stems, *pr* for purple stems. A corn plant of genotype *pr/pr* that has standard chromosomes is crossed with a *Pr/Pr* plant that is homozygous for a reciprocal translocation between chromosomes 2 and 5. The F$_1$ is semisterile and phenotypically Pr. A backcross with the parent with standard chromosomes gives 764 semisterile Pr, 145 semisterile pr, 186 normal Pr, and 727 normal pr. What is the map distance between the *Pr* locus and the translocation point?

49. Distinguish among Klinefelter, Down, and Turner syndromes. Which syndromes are found in both sexes?

50. Show how you could make an allotetraploid between two related diploid plant species, both of which are $2n = 28$.

51. In *Drosophila*, trisomics and monosomics for the tiny chromosome 4 are viable, but nullisomics and tetrasomics are not. The *b* locus is on this chromosome.

Deduce the phenotypic proportions in the progeny of the following crosses of trisomics.

a. $b^+/b/b \times b/b$

b. $b^+/b^+/b \times b/b$

c. $b^+/b^+/b \times b^+/b$

52. A woman with Turner syndrome is found to be color-blind (an X-linked recessive phenotype). Both her mother and her father have normal vision.

   a. Explain the simultaneous origin of Turner syndrome and color blindness by the abnormal behavior of chromosomes at meiosis.

   b. Can your explanation distinguish whether the abnormal chromosome behavior occurred in the father or in the mother?

   c. Can your explanation distinguish whether the abnormal chromosome behavior occurred at the first or second division of meiosis?

   d. Now assume that a color-blind Klinefelter man has parents with normal vision, and answer parts *a*, *b*, and *c*.

53. a. How would you synthesize a pentaploid?

   b. How would you synthesize a triploid of genotype *A/a/a*?

   c. You have just obtained a rare recessive mutation *a** in a diploid plant, which Mendelian analysis tells you is *A/a**. From this plant, how would you synthesize a tetraploid ($4n$) of genotype *A/A/a*/a**?

   d. How would you synthesize a tetraploid of genotype *A/a/a/a*?

54. Suppose you have a line of mice that has cytologically distinct forms of chromosome 4. The tip of the chromosome can have a knob (called 4^K) or a satellite (4^S) or neither (4). Here are sketches of the three types:

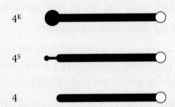

You cross a $4^K/4^S$ female with a 4/4 male and find that most of the progeny are $4^K/4$ or $4^S/4$, as expected. However, you occasionally find some rare types as follows (all other chromosomes are normal):

a. $4^K/4^K/4$

b. $4^K/4^S/4$

c. 4^K

Explain the rare types that you have found. Give, as precisely as possible, the stages at which they originate, and state whether they originate in the male parent, the female parent, or the zygote. (Give brief reasons.)

55. A cross is made in tomatoes between a female plant that is trisomic for chromosome 6 and a normal diploid male plant that is homozygous for the recessive allele for potato leaf (*p/p*). A trisomic F_1 plant is backcrossed to the potato-leaved male.

   a. What is the ratio of normal-leaved plants to potato-leaved plants when you assume that *p* is located on chromosome 6?

   b. What is the ratio of normal-leaved to potato-leaved plants when you assume that *p* is not located on chromosome 6?

56. A tomato geneticist attempts to assign five recessive mutations to specific chromosomes by using trisomics. She crosses each homozygous mutant ($2n$) with each of three trisomics, in which chromosomes 1, 7, and 10 take part. From these crosses, the geneticist selects trisomic progeny (which are less vigorous) and backcrosses them to the appropriate homozygous recessive. The *diploid* progeny from these crosses are examined. Her results, in which the ratios are wild type:mutant, are as follows:

Trisomic chromosome	Mutation				
	d	*y*	*c*	*h*	*cot*
1	48:55	72:29	56:50	53:54	32:28
7	52:56	52:48	52:51	58:56	81:40
10	45:42	36:33	28:32	96:50	20:17

Which of the mutations can the geneticist assign to which chromosomes? (Explain your answer fully.)

57. A petunia is heterozygous for the following autosomal homologs:

A	B	C	D	E	F	G	H	I
a	b	c	d	h	g	f	e	i

   a. Draw the pairing configuration that you would see at metaphase I, and identify all parts of your diagram. Number the chromatids sequentially from top to bottom of the page.

   b. A three-strand double crossover occurs, with one crossover between the *C* and *D* loci on chromatids 1 and 3, and the second crossover between the *G* and *H* loci on chromatids 2 and 3. Diagram the results of these recombination events as you would see them at anaphase I, and identify all parts of your diagram.

   c. Draw the chromosome pattern that you would see at anaphase II after the crossovers described in part *b*.

   d. Give the genotypes of the gametes from this meiosis that will lead to the formation of viable progeny. Assume that all gametes are fertilized by pollen that has the gene order *A B C D E F G H I*.

58. Two groups of geneticists, in California and in Chile, begin work to develop a linkage map of the medfly (see Chapter 4). They both independently find that the loci for body color (B = black, b = gray) and eye shape (R = round, r = star) are linked 28 m.u. apart. They send strains to each other and perform crosses; a summary of all their findings is shown here:

Cross	F_1	Progeny of $F_1 \times$ any $b\ r/b\ r$	
$B\ R/B\ R$ (Calif.)	$B\ R/b\ r$	$B\ R/b\ r$	36%
$\times\ b\ r/b\ r$ (Calif.)		$b\ r/b\ r$	36
		$B\ r/b\ r$	14
		$b\ R/b\ r$	14
$B\ R/B\ R$ (Chile)	$B\ R/b\ r$	$B\ R/b\ r$	36
$\times\ b\ r/b\ r$ (Chile)		$b\ r/b\ r$	36
		$B\ r/b\ r$	14
		$b\ R/b\ r$	14
$B\ R/B\ R$ (Calif.)	$B\ R/b\ r$	$B\ R/b\ r$	48
$\times b\ r/b\ r$ (Chile) or		$b\ r/b\ r$	48
$b\ r/b\ r$ (Calif.)		$B\ r/b\ r$	2
$\times B\ R/B\ R$ (Chile)		$b\ R/b\ r$	2

a. Provide a genetic hypothesis that explains the three sets of testcross results.

b. Draw the key chromosomal features of meiosis in the F_1 from a cross of the Californian and Chilean lines.

59. An aberrant corn plant gives the following RF values when testcrossed:

	Interval				
	d–f	*f–b*	*b–x*	*x–y*	*y–p*
Control	5	18	23	12	6
Aberrant plant	5	2	2	0	6

(The locus order is centromere-*d–f–b–x–y–p*.) The aberrant plant is a healthy plant, but it produces far fewer normal ovules and pollen than does the control plant.

a. Propose a hypothesis to account for the abnormal recombination values and the reduced fertility in the aberrant plant.

b. Use diagrams to explain the origin of the recombinants according to your hypothesis.

60. The following corn loci are on one arm of chromosome 9 in the order indicated (the distances between them are shown in map units):

$$c–bz–wx–sh–d–\text{centromere}$$
$$12 \quad 8 \quad 10 \quad 20 \quad 10$$

C gives colored aleurone; c, white aleurone.

Bz gives green leaves; bz, bronze leaves.

Wx gives starchy seeds; wx, waxy seeds.

Sh gives smooth seeds; sh, shrunken seeds.

D gives tall plants; d, dwarf.

A plant from a standard stock that is homozygous for all five recessive alleles is crossed with a wild-type plant from Mexico that is homozygous for all five dominant alleles. The F_1 plants express all the dominant alleles and, when backcrossed to the recessive parent, give the following progeny phenotypes:

colored, green, starchy, smooth, tall	360
white, bronze, waxy, shrunk, dwarf	355
colored, bronze, waxy, shrunk, dwarf	40
white, green, starchy, smooth, tall	46
colored, green, starchy, smooth, dwarf	85
white, bronze, waxy, shrunk, tall	84
colored, bronze, waxy, shrunk, tall	8
white, green, starchy, smooth, dwarf	9
colored, green, waxy, smooth, tall	7
white, bronze, starchy, shrunk, dwarf	6

Propose a hypothesis to explain these results. Include

a. a general statement of your hypothesis, with diagrams if necessary.

b. why there are 10 classes.

c. an account of the origin of each class, including its frequency.

d. at least one test of your hypothesis.

61. Chromosomally normal corn plants have a p locus on chromosome 1 and an s locus on chromosome 5.

P gives dark green leaves; p, pale green leaves.

S gives large ears; s, shrunken ears.

An original plant of genotype P/p;S/s has the expected phenotype (dark green, large ears) but gives unexpected results in crosses as follows:

- On selfing, fertility is normal, but the frequency of p/p;s/s types is 1/4 (not 1/16 as expected).

- When crossed with a normal tester of genotype p/p;s/s, the F_1 progeny are $\frac{1}{2}P/p$;S/s and $\frac{1}{2}p/p$;s/s; fertility is normal.

- When an F_1 P/p;S/s plant is crossed with a normal p/p;s/s tester, it proves to be semisterile, but, again, the progeny are; $\frac{1}{2}P/p$;S/s and $\frac{1}{2}p/p$;s/s.

Explain these results, showing the full genotypes of the original plant, the tester, and the F_1 plants. How would you test your hypothesis?

	Embryos (mean number)			
Mating	Implanted in the uterine wall	Degeneration after implantation	Normal	Degeneration (%)
exceptional ♂ × normal ♀	8.7	5.0	3.7	37.5
normal ♂ × normal ♀	9.5	0.6	8.9	6.5

62. A male rat that is phenotypically normal shows reproductive anomalies when compared with normal male rats, as shown in the table above. Propose a genetic explanation of these unusual results, and indicate how your idea could be tested.

63. A tomato geneticist working on *Fr*, a dominant mutant allele that causes rapid fruit ripening, decides to find out which chromosome contains this gene by using a set of lines of which each is trisomic for one chromosome. To do so, she crosses a homozygous diploid mutant with each of the wild-type trisomic lines.

   a. A trisomic F_1 plant is crossed with a diploid wild-type plant. What is the ratio of fast- to slow-ripening plants in the diploid progeny of this second cross if *Fr* is on the trisomic chromosome? Use diagrams to explain.

   b. What is the ratio of fast- to slow-ripening plants in the diploid progeny of this second cross if *Fr* is not located on the trisomic chromosome? Use diagrams to explain.

   c. Here are the results of the crosses. On which chromosome is *Fr*, and why?

Trisomic chromosome	Fast ripening : slow ripening in diploid progeny
1	45:47
2	33:34
3	55:52
4	26:30
5	31:32
6	37:41
7	44:79
8	49:53
9	34:34
10	37:39

   (Problem 63 is from Tamara Western.)

CHALLENGING PROBLEMS

64. The *Neurospora un-3* locus is near the centromere on chromosome 1, and crossovers between *un-3* and the centromere are very rare. The *ad-3* locus is on the other side of the centromere of the same chromosome, and crossovers occur between *ad-3* and the centromere in about 20 percent of meioses (no multiple crossovers occur).

   a. What types of linear asci (see Chapter 3) do you predict, and in what frequencies, in a normal cross of *un-3 ad-3* × wildtype? (Specify genotypes of spores in the asci.)

   b. Most of the time such crosses behave predictably; but in one case, a standard *un-3 ad-3* strain was crossed with a wild type isolated from a field of sugarcane in Hawaii. The results follow:

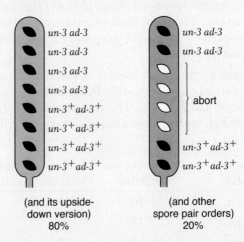

un-3 ad-3	un-3 ad-3
un-3 ad-3	un-3 ad-3
un-3 ad-3	
un-3 ad-3	} abort
un-3⁺ad-3⁺	
un-3⁺ad-3⁺	
un-3⁺ad-3⁺	un-3⁺ad-3⁺
un-3⁺ad-3⁺	un-3⁺ad-3⁺

   (and its upside-down version) 80% | (and other spore pair orders) 20%

   Explain these results, and state how you could test your idea. (**Note:** In *Neurospora*, ascospores with extra chromosomal material survive and are the normal black color, whereas ascospores lacking any chromosome region are white and inviable.)

65. Two mutations in *Neurospora*, *ad-3* and *pan-2*, are located on chromosomes 1 and 6, respectively. An unusual *ad-3* line arises in the laboratory, giving the results shown in the table below. Explain all three results with the aid of clearly labeled diagrams. (**Note:** In *Neurospora*, ascospores with extra chromosomal material survive and are the normal black color, whereas ascospores lacking any chromosome region are white and inviable.)

	Ascospore appearance	RF between *ad-3* and *pan-2*
1. Normal *ad-3* × normal *pan-2*	All black	50%
2. Abnormal *ad-3* × normal *pan-2*	About $\frac{1}{2}$ black and $\frac{1}{2}$ white (inviable)	1%

   3. Of the black spores from cross 2, about half were completely normal and half repeated the same behavior as the original abnormal *ad-3* strain.

66. The New World cotton species *Gossypium hirsutum* has a 2*n* chromosome number of 52. The Old World species *G. thurberi* and *G. herbaceum* each have a 2*n* number of 26. When these species are crossed, the resulting hybrids show the following chromosome pairing arrangements at meiosis:

Hybrid	Pairing arrangement
G. hirsutum	13 small bivalents
× *G. thurberi*	+13 large univalents
G. hirsutum	13 large bivalents
× *G. herbaceum*	+13 small univalents
G. thurberi	13 large univalents
× *G. herbaceum*	+13 small univalents

Interpret these observations phylogenetically, using diagrams. Clearly indicate the relationships between the species. How would you prove that your interpretation is correct?

67. There are six main species in the *Brassica* genus: *B. carinata*, *B. campestris*, *B. nigra*, *B. oleracea*, *B. juncea*, and *B. napus*. You can deduce the interrelationships among these six species from the following table:

Species or F₁ hybrid	Chromosome number	Number of bivalents	Number of univalents
B. juncea	36	18	0
B. carinata	34	17	0
B. napus	38	19	0
B. juncea × *B. nigra*	26	8	10
B. napus × *B. campestris*	29	10	9
B. carinata × *B. oleracea*	26	9	8
B. juncea × *B. oleracea*	27	0	27
B. carinata × *B. campestris*	27	0	27
B. napus × *B. nigra*	27	0	27

a. Deduce the chromosome number of *B. campestris*, *B. nigra*, and *B. oleracea*.

b. Show clearly any evolutionary relationships between the six species that you can deduce at the chromosomal level.

68. Several kinds of sexual mosaicism are well documented in humans. Suggest how each of the following examples may have arisen by nondisjunction at *mitosis*:

a. XX/XO (that is, there are two cell types in the body, XX and XO)

b. XX/XXYY

c. XO/XXX

d. XX/XY

e. XO/XX/XXX

69. In *Drosophila*, a cross (cross 1) was made between two mutant flies, one homozygous for the recessive mutation bent wing (*b*) and the other homozygous for the recessive mutation eyeless (*e*). The mutations *e* and *b* are alleles of two different genes that are known to be very closely linked on the tiny autosomal chromosome 4. All the progeny had a wild-type phenotype. One of the female progeny was crossed with a male of genotype *b e*/*b e*; we will call this cross 2. Most of the progeny of cross 2 were of the expected types, but there was also one rare female of wild-type phenotype.

a. Explain what the common progeny are expected to be from cross 2.

b. Could the rare wild-type female have arisen by (1) crossing over or (2) nondisjunction? Explain.

c. The rare wild-type female was testcrossed to a male of genotype *b e* / *b e* (cross 3). The progeny were

$\frac{1}{6}$ wild type $\frac{1}{3}$ bent

$\frac{1}{6}$ bent, eyeless $\frac{1}{3}$ eyeless

Which of the explanations in part *b* is compatible with this result? Explain the genotypes and phenotypes of the progeny of cross 3 and their proportions.

UNPACKING PROBLEM 69

Before attempting a solution to this problem, try answering the following questions:

1. Define *homozygous*, *mutation*, *allele*, *closely linked*, *recessive*, *wild type*, *crossing over*, *nondisjunction*, *testcross*, *phenotype*, and *genotype*.

2. Does this problem concern sex linkage? Explain.

3. How many chromosomes does *Drosophila* have?

4. Draw a clear pedigree summarizing the results of crosses 1, 2, and 3.

5. Draw the gametes produced by both parents in cross 1.

6. Draw the chromosome 4 constitution of the progeny of cross 1.

7. Is it surprising that the progeny of cross 1 are wild-type phenotype? What does this outcome tell you?

8. Draw the chromosome 4 constitution of the male tester used in cross 2 and the gametes that he can produce.

9. With respect to chromosome 4, what gametes can the female parent in cross 2 produce in the absence of nondisjunction? Which would be common, and which rare?

10. Draw first- and second-division meiotic nondisjunction in the female parent of cross 2, as well as in the resulting gametes.

11. Are any of the gametes from part 10 aneuploid?

12. Would you expect aneuploid gametes to give rise to viable progeny? Would these progeny be nullisomic, monosomic, disomic, or trisomic?

13. What progeny phenotypes would be produced by the various gametes considered in parts 9 and 10?

14. Consider the phenotypic ratio in the progeny of cross 3. Many genetic ratios are based on halves and quarters, but this ratio is based on thirds and sixths. To what might this ratio point?

15. Could there be any significance to the fact that the crosses concern genes on a very small chromosome? When is chromosome size relevant in genetics?

16. Draw the progeny expected from cross 3 under the two hypotheses, and give some idea of relative proportions.

Now try to solve the problem. If you are unable to do so, try to identify the obstacle and write a sentence or two describing your difficulty. Then go back to the expansion questions and see if any of them relate to your difficulty. If this approach does not work, inspect the Learning Objectives and Key Concepts of this chapter and ask yourself which might be relevant to your difficulty.

70. In the fungus *Ascobolus* (similar to *Neurospora*), ascospores are normally black. The mutation *f*, producing fawn-colored ascospores, is in a gene just to the right of the centromere on chromosome 6, whereas mutation *b*, producing beige ascospores, is in a gene just to the left of the same centromere. In a cross of fawn and beige parents ($+f \times b+$), most octads showed four fawn and four beige ascospores, but three rare exceptional octads were found, as shown in the accompanying illustration. In the sketch, black is the wild-type phenotype, a vertical line is fawn, a horizontal line is beige, and an empty circle represents an aborted (dead) ascospore.

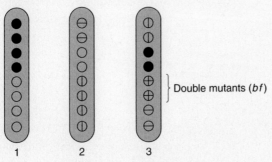

a. Provide reasonable explanations for these three exceptional octads.

b. Diagram the meiosis that gave rise to octad 2.

71. The life cycle of the haploid fungus *Ascobolus* is similar to that of *Neurospora*. A mutational treatment produced two mutant strains, 1 and 2, both of which when crossed with wild type gave unordered tetrads, all of the following type (fawn is a light brown color; normally, crosses produce all black ascospores):

spore pair 1 black	spore pair 3 fawn
spore pair 2 black	spore pair 4 fawn

a. What does this result show? Explain.

The two mutant strains were crossed. Most of the unordered tetrads were of the following type:

spore pair 1 fawn	spore pair 3 fawn
spore pair 2 fawn	spore pair 4 fawn

b. What does this result suggest? Explain.

When large numbers of unordered tetrads were screened under the microscope, some rare ones that contained black spores were found. Four cases are shown here:

	Case A	Case B	Case C	Case D
spore pair 1	black	black	black	black
spore pair 2	black	fawn	black	abort
spore pair 3	fawn	fawn	abort	fawn
spore pair 4	fawn	fawn	abort	fawn

(**Note:** Ascospores with extra genetic material survive, but those with less than a haploid genome abort.)

c. Propose reasonable genetic explanations for each of these four rare cases.

d. Do you think the mutations in the two original mutant strains were in one single gene? Explain.

GENETICS AND SOCIETY

Based on the data shown in Figures 17-15 and 17-37, at what age would you recommend screening of expectant mothers for fetal chromosome abnormalities? And, for which chromosomal abnormalities? What aspects of an expectant mother's health history would influence this judgment?

Population Genetics

Breeds of dogs vary greatly in size, shape, and coat color, demonstrating the considerable genetic variation in the species. [*CAROLYN McKEONE/Science Source.*]

In 2009, Sean Hodgson was released from a British prison after serving 27 years behind bars for the 1979 murder of Teresa De Simone, a clerk and part-time barmaid. Hodgson, who suffers from mental illness, initially confessed to the crime but withdrew his confession during the trial. Throughout his years in prison, he maintained his innocence. More than two decades after the crime, the courts analyzed DNA of the assailant found at the crime scene and determined that it did not come from Mr. Hodgson. Hodgson's conviction was overturned, and the police subsequently identified David Lace as the likely murderer. Strangely, Lace had come forward in 1983 and confessed, but with Hodgson already convicted, the police refused to believe Lace. In 1988, Lace committed suicide long before the DNA evidence implicated him. As you will learn in this chapter, the DNA-based analysis used to exonerate Mr. Hodgson and hundreds of other wrongly convicted prisoners was dependent on population genetic analysis.

The principles of population genetics are at the heart of many questions facing society today. What are the risks that a couple will have a child with a genetic disease? Have the practices of plant and animal breeding caused a loss of genetic diversity on the farm, and does this loss of diversity place our food supply at risk? As the human population continues to expand and wildlife retreats into smaller and smaller parts of the earth, will wildlife species be able to avoid inbreeding and survive? The principles of population genetics are also fundamental to understanding many historical and evolutionary questions. How are human populations from different regions of the world related to one another? How has the human genome responded as humans have spread out across the globe and become adapted to different environments and lifestyles? How do populations and species evolve over time?

A **population** is a group of individuals of the same species. **Population genetics** analyzes the amount and distribution of genetic variation in populations and the forces that control this variation. It has its roots in the early 1900s, when geneticists began to study how Mendel's laws could be extended to understand genetic variation within whole populations of organisms. While Mendel's laws explain how genes are passed from parent to offspring in known pedigrees, these laws are insufficient to understand the transmission of genes from one generation to the next in natural populations, in which not all individuals produce

offspring and not all offspring survive. When geneticists began developing the principles of population genetics, they had rather limited tools to actually measure genetic variation. With the development of DNA-based technologies over the past three decades, geneticists now have the ability to observe directly differences between the DNA sequences of individuals throughout their genomes, and they can measure these differences in large samples of individuals in many species. The result has been a revolution in our understanding of genetic variation in populations.

In this chapter, we will consider the concept of the gene pool and how geneticists estimate allele and genotype frequencies in populations. Next, we will examine the impact that mating systems have on the frequencies of genotypes in a population. We will also discuss how geneticists measure variation using DNA-based technologies. We will then discuss the forces that modulate the levels of genetic variation within populations. Finally, we will look at some case studies involving the application of population genetics to questions of interest to society.

18.1 DETECTING GENETIC VARIATION

LO 18.1 Describe and analyze data to determine how much genetic variation exists within populations.

The methods of population genetics can be used to analyze any variable or polymorphic locus in the DNA sequences of a population of organisms. Over the past several decades, multiple technologies, such as DNA sequencing, DNA microarrays, and PCR (see Chapters 10 and 14), have been developed that allow geneticists to observe differences in the DNA sequences among large samples of individuals.

In population genetics, a **locus** is simply a location in the genome; it can be a single nucleotide site or a stretch of many nucleotides. The simplest form of variation one might observe among individuals at a locus is a difference in the nucleotide present at a single nucleotide site, whether adenine, cytosine, guanine, or thymine. These types of variants are called **single nucleotide polymorphisms (SNPs)**, and they are the most widely studied variants in human population genetics (**Figure 18-1**; see also Chapter 4). Population

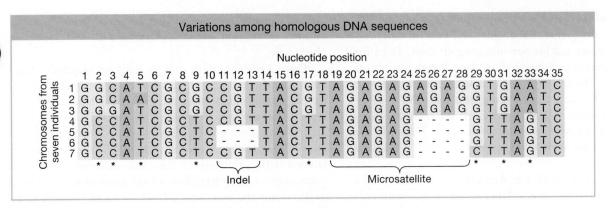

Variations among homologous DNA sequences

FIGURE 18-1 Variation in the aligned DNA sequences of seven chromosomes from different people. The asterisks show the location of SNPs. The location of an indel (insertion/deletion of a string of nucleotide pairs) and a microsatellite are also indicated.

genetics also makes extensive use of **microsatellite** loci (see Chapter 4). These loci have a short sequence motif, 2 to 6 base-pairs long, that is repeated multiple times with different alleles having different numbers of repeats. For example, the 2-bp-sequence motif AG at a locus might be tandemly repeated five times in one allele (AGAGAGAGAG) but three times in another (AGAGAG) (see Figure 18-1).

Single nucleotide polymorphisms (SNPs)

SNPs are the most prevalent types of polymorphism in most genomes. Most SNPs have just two alleles—for example, A and C. SNPs are usually considered **common SNPs** in a population if the less common allele occurs at a frequency of about 5 percent or greater. SNPs for which the less common allele occurs at a frequency below 5 percent are considered **rare SNPs**. For humans, there is a common SNP about every 300 to 1000 bp in the genome. Of course, there are a far greater number of rare SNPs.

SNPs occur within genes, including within exons, introns, and regulatory regions. SNPs within protein-coding regions can be classified into one of three groups: *synonymous* if the different alleles encode the same amino acid, *nonsynonymous* if the two alleles encode different amino acids, and *nonsense* if one allele encodes a stop codon and the other an amino acid. Thus, it is sometimes possible to associate a SNP with functional variation in proteins and an associated change in phenotype. SNPs located outside of coding sequences are called *noncoding* SNPs (ncSNPs). If ncSNPs have no effect on gene function and phenotype, they are called *silent*.

To study SNP variation in a population, one can first determine which nucleotide sites in the genome are variable—that is, constitute a SNP. This first step is called SNP discovery. SNPs are often discovered by sequencing the genomes of a small sample of individuals of a species, then comparing these sequences. For example, SNP discovery in humans began by partially sequencing the genomes of a **discovery panel** of 48 individuals from around the world. Variable nucleotide sites were discovered by comparing the

partial genome sequences of these 48 individuals with one another. This initial effort led to the discovery of more than 1 million SNPs.

Once SNPs have been discovered, the genotype (allelic composition) of different individuals in the population at each SNP can be determined. DNA microarrays are a widely used technology for this purpose (**Figure 18-2**). The microarrays used for SNP assays can contain thousands of probes corresponding to known SNPs. Biotechnologists have developed several different methods to detect SNP variants using microarrays. In one method, DNA from

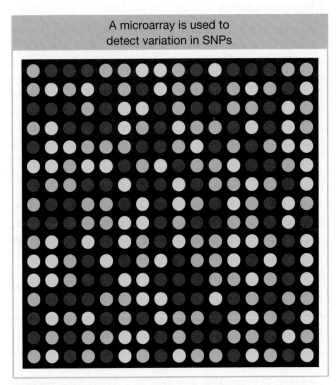

A microarray is used to detect variation in SNPs

FIGURE 18-2 Detecting variation in DNA: SNPs. View of a small portion of a microarray used to scan a single individual's genome. Each dot represents one SNP, with red and green for the homozygous classes and yellow for heterozygous.

an individual is labeled with fluorescent tags and hybridized to the microarray. Each spot (SNP) on the microarray will fluoresce red for one homozygous class, green for the other homozygote, and yellow for a heterozygote (see Figure 18-2). The entire procedure has been enhanced with robotics to allow rapid *genotyping*, or assignment of genotypes (for example, A/A versus A/C) on a large-scale basis. Direct-to-consumer genotyping services like 23andMe use SNP microarrays, as do some large-scale medical research projects.

More recently with the dramatic reduction in the cost of DNA sequencing, it has become possible to collect the genome sequences of many individuals in a species and then compare the SNP difference between the multiple genome sequences. For many questions in population genetics, a relatively small sample of genome sequences of 20 to 100 individuals is adequate. In other cases, thousands or even tens of thousands of genome sequences of individuals are being gathered. Such large samples are used when researchers are attempting to identify regions of the genome that control a phenotype such as the risk of having a disease, like type 2 diabetes. This type of analysis, called a genome-wide association study, is covered in Chapter 19. For our own species, the first large-scale multiple genome project was called the *1000 Genomes Project*, and it collected that many genome sequences of people from around the world (www.1000genomes.org).

Microsatellites

Microsatellites are powerful loci for population genetic analysis for several reasons. First, unlike SNPs, which typically have only two alleles per locus and can never have more than four alleles, the number of alleles at a microsatellite is often very large (20 or more). Second, they have a high mutation rate, typically in the range of 10^{-3} to 10^{-4} mutations per locus per generation as compared to 10^{-8} to 10^{-9} mutations per site per generation for SNPs. The high mutation rate means that levels of variation are higher: more alleles per locus and a greater chance that any two individuals will have different genotypes. Third, microsatellites are very abundant in most genomes. Humans have over a million microsatellites. Discovering microsatellite loci in the genome of a species is done by performing a computer search of its complete genomic sequence.

Microsatellites are found throughout the genomes of most organisms and may be present in exons, introns, regulatory regions, and nonfunctional DNA sequences. Microsatellites with trinucleotide repeats are found in the coding sequences of some genes; these encode strings of a single amino acid. The Huntington disease gene (*HD*) (see Chapter 15) contains a repeat of CAG, which encodes a string of glutamines. Individuals carrying alleles with more than 30 glutamines are predisposed to develop the disease. In general, however, most microsatellites are located outside of coding sequences, and variation in the number of repeats is not associated with differences in phenotype.

Once a microsatellite and its flanking sequences have been identified, DNA samples from a set of individuals in the population can be analyzed to determine the number of repeats that are present in each individual. To carry out the analysis, oligonucleotide primers are designed that match the flanking sequences for use in PCR. If the primers are labeled with a fluorescent tag, then the sizes of the PCR products can be determined on the same apparatus used to determine the sequence of DNA molecules (**Figure 18-3**). These sizes reveal the number of repeats in a microsatellite allele. For example, the PCR product of a microsatellite allele containing seven AG repeats will be 8 bp longer than an allele containing three AG repeats. Heterozygous individuals will possess products of two different sizes. Since PCR, the sizing of PCR products, and scoring of the alleles can all be automated, it is possible to determine the genotypes of large samples of individuals for large numbers of microsatellites relatively rapidly.

Haplotypes

For some questions in population genetics, it is important to consider the genotypes of linked loci as a group rather than individually. Geneticists use the term **haplotype** to refer to the combination of alleles at multiple loci on the same chromosomal homolog. Two homologous chromosomes that share the same allele at each of the loci under consideration have the same haplotype. If two chromosomes have different genotypes at even one of the loci in question, then they

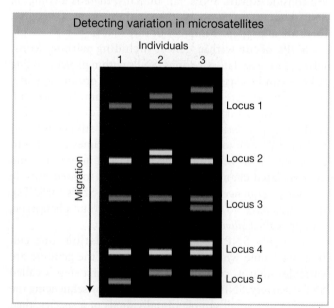

FIGURE 18-3 Detecting variation in DNA: microsatellites. Schematic drawing of a gel image of the loci for five microsatellites scored simultaneously. The three vertical lanes correspond to three individuals. Notice that there are three alleles present for Locus 1 and that individuals 2 and 3 are both heterozygous for this locus.

have different haplotypes. If the *A* locus with alleles *A* and *a* is linked to the *B* locus with alleles *B* and *b*, then there are four possible haplotypes for the chromosomal segment on which these two loci are located:

A	*B*
A	*b*
a	*B*
a	*b*

A more complex, but more realistic, example is shown in **Figure 18-4**. In Figure 18-4a, there are seven chromosome segments but only six haplotypes because chromosome segments 5 and 6 have the same haplotype (E).

Haplotypes are most often used in population genetics for loci that are physically close. For example, the variable-nucleotide sites in a single gene can be used to define haplotypes for that gene. However, the haplotype concept works for larger regions when there is little or no recombination over the region. It can even be applied to the human Y chromosome, most of which does not experience recombination with the X chromosome. Finally, it is sometimes useful to group haplotypes into classes. As shown in Figure 18-4a, there are two major classes of haplotypes (I and II) that differ at five nucleotide sites plus a microsatellite. However, each class contains several subtypes

(I-a, I-b, . . .). The **haplotype network** shows the relationships among the haplotypes, placing each mutation on one of the branches (Figure 18-4b).

What insights can we gain from haplotype analysis? Population geneticists studying the human Y chromosome among Asian men discovered one highly prevalent haplotype, termed the "star-cluster" haplotype (**Figure 18-5a**). Typically, most men have a rare Y chromosome haplotype, but the "star-cluster" haplotype is present in 8 percent of Asian men. Using the known mutation rate, the researchers estimated that this common haplotype arose between 700 and 1300 years ago. (Later in this chapter, we will discuss mutation rates and their use in population genetics.) This haplotype is most common in Mongolia, suggesting that it arose there. The researchers inferred that the "star-cluster" haplotype traces back to one man in Mongolia about 1000 years ago. Remarkably, the present-day distribution of this haplotype follows the geographic boundaries of the Mongolian Empire established by Genghis Khan about 1200 years ago (Figure 18-5b). It appears that contemporary men with this haplotype are all descendants of Genghis Khan (or his male-lineage relatives).

Y-chromosome star clusters, like the one linked to Genghis Khan, are a common feature in the genetic history of our species. Notably, there are multiple star clusters dating to

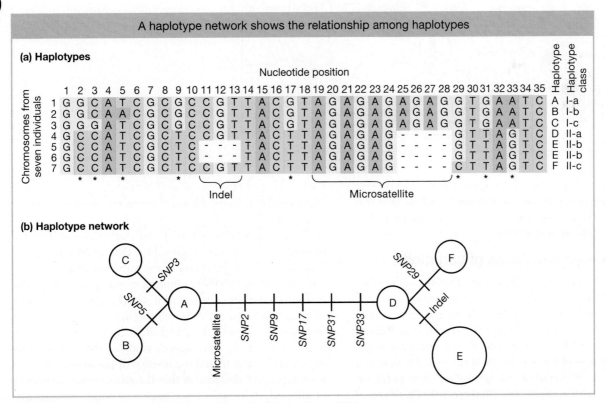

FIGURE 18-4 (a) There are a total of six haplotypes (A–F) in the aligned DNA sequences from seven individual chromosomes from different people. (b) These six haplotypes are joined in a haplotype network showing the relationships among the haplotypes. Each circle represents one of the six haplotypes. Any two haplotypes differ at the loci noted on all of the branches connecting them. The asterisks show the location of SNPs.

CHAPTER 18 Population Genetics

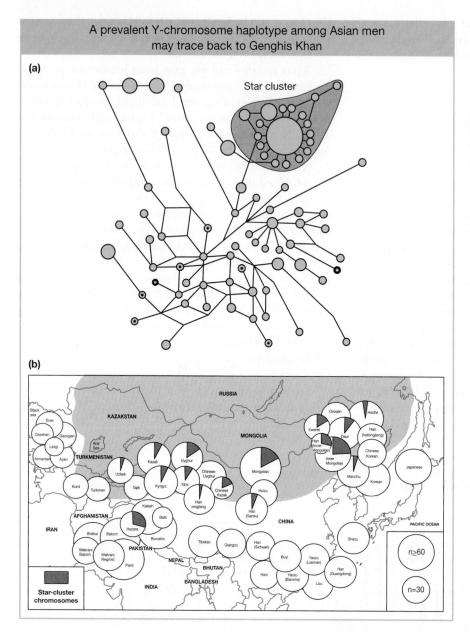

A prevalent Y-chromosome haplotype among Asian men may trace back to Genghis Khan

(a)

Star cluster

(b)

FIGURE 18-5 (a) Haplotype network for the Y chromosomes of Asian men showing the predominance of the star-cluster haplotype thought to trace back to Genghis Khan. The area of the circle is proportional to the number of individuals with the specific haplotype that the circle represents. (b) Geographical distribution of the star-cluster haplotype. Populations are shown as circles with an area proportional to sample size; the proportion of individuals in the sample carrying star-cluster chromosomes is indicated by green sectors. No star-cluster chromosomes were found in populations having no green sector in the circle. The shaded area represents the extent of Genghis Khan's empire. [*Data from T. Zerjal et al., Am. J. Hum. Genet. 72, 2003, 717–721.*]

the time when agriculture was invented, suggest that this cultural innovation is associated with differential reproductive success of the men of that time.

Other sources and forms of variation

Beyond SNP and microsatellites, any variation in the DNA sequence of the chromosomes in a population is amenable to population genetic analysis. Variations that can be analyzed include inversions, translocations, deletions or duplications, and the presence or absence of a transposable element at a particular locus in the genome. Another common form of variation is insertion-deletion polymorphism, or *indel* for short (see Chapter 15). This type of polymorphism involves the presence or absence of one or more nucleotides at a locus in one allele relative to another. In Figure 18-1, chromosome segments 5 and 6 differ from the other five segments by a 3-bp indel. Unlike

microsatellites, indels do not contain repeat motifs such as AGAGAGAG.

Thus far, our discussion of SNP and microsatellites has focused on the nuclear genome. However, interesting genetic variation can also be found in the mitochondrial (mtDNA) and chloroplast (cpDNA) genomes of eukaryotes. Both SNP and microsatellites are found in these organelle genomes. Since mtDNA and cpDNA are usually maternally inherited, their analysis can be used to follow the history of female lineages. In 1987, a prominent study of the human mitochondrial lineage traced the history of the human mtDNA haplotypes and determined that the mitochondrial genomes of all modern humans trace back to a single woman who lived in Africa about 150,000 years ago (**Figure 18-6**). She was dubbed the "mitochondrial Eve" in the popular press. This study of mtDNA was the first thorough genetic analysis to suggest that all modern humans came from Africa.

Mitochondrial haplotypes can be used to trace human origins to Africa

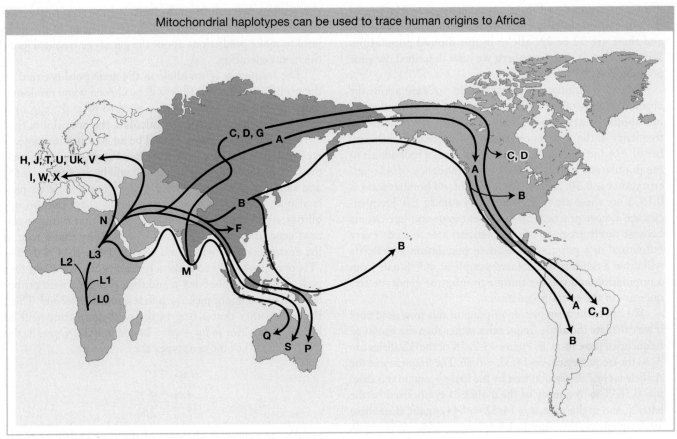

FIGURE 18-6 The haplotype network for human mtDNA haplotype groups drawn onto a world map. The ancestral L haplotype group appears in Africa, and the derived groups (A, B, and so on) are dispersed throughout the world. This haplotype network is like the one shown in Figure 18-4, except here the SNPs are not labeled on the branches. [*Data from www.mitomap.org*.]

KEY CONCEPT Genomes are replete with diverse types of variation suitable for population genetic analysis. SNPs and microsatellites are the two most commonly studied types of polymorphism in population genetics. High-throughput technologies allow hundreds of thousands of polymorphisms to be scored in tens of thousands of individuals.

18.2 THE GENE-POOL CONCEPT AND THE HARDY–WEINBERG LAW

LO 18.2 Apply the Hardy–Weinberg formula to calculate expected allele and genotype frequencies.

Perhaps you have watched someone performing a death-defying stunt and thought that they were at risk of eliminating themselves from the "gene pool." If so, you were using a concept, the gene pool, that comes straight out of population genetics and has worked its way into popular culture. The gene-pool concept is a basic tool for thinking about genetic variation in populations. We can define the **gene pool** as the sum total of all alleles in the breeding members of a population at a given time. For example, **Figure 18-7**

The gene pool is the sum total of alleles in a population

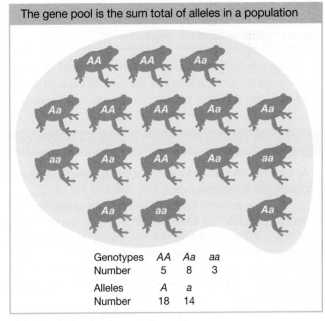

Genotypes	*AA*	*Aa*	*aa*
Number	5	8	3
Alleles	*A*	*a*	
Number	18	14	

FIGURE 18-7 A frog gene pool.

shows a population of 16 frogs, each of which carries two alleles at the autosomal locus *A*. By simple counting, we can determine that there are five *A/A* homozygotes, eight

A/a heterozygotes, and three *a/a* homozygotes. The size of the population, usually symbolized by the letter *N*, is 16, and there are 32 or *2N* alleles in this diploid population. With this simple set of numbers, we have described the gene pool with regard to the *A* locus.

Typically, population geneticists do not care about the absolute counts of the different genotypes in a population, but about the **genotype frequencies**. We can calculate the frequency of the *A/A* genotype simply by dividing the number of *A/A* individuals by the total number of individuals in the population (*N*) to get 0.31. The frequency of *A/a* heterozygotes is 0.50, and the frequency of *a/a* homozygotes is 0.19. Since these are frequencies, they sum to 1.0. Frequencies are a more practical measurement than absolute counts because rarely are population geneticists able to study every individual in a population. Rather, population geneticists will draw a random or unbiased sample of individuals from a population and use the sample to infer the genotype frequencies in the entire population.

We can make a simpler description of this frog gene pool if we calculate the **allele frequencies** rather than the genotype frequencies (**Box 18-1**). In Figure 18-7, 18 of the 32 alleles are *A*, so the frequency of *A* is 18/32 = 0.56. The frequency of the *A* allele is typically symbolized by the letter *p*, and in this case, *p* = 0.56. The frequency of the *a* allele is symbolized by the letter *q*, and in this case, *q* = 14/32 = 0.44. Again, since these are frequencies, they sum to 1.0 : *p* + *q* = 0.56 + 0.44 = 1.0. We now have a description of our frog gene pool using only two numbers, *p* and *q*.

> **KEY CONCEPT** The gene pool is a fundamental concept for the study of genetic variation in populations: it is the sum total of all alleles in the breeding members of a population at a given time. We can describe the variation in a population in terms of genotype and allele frequencies.

As mentioned previously, an important goal of population genetics is to understand the transmission of alleles from one generation to the next in natural populations. In this section, we will begin to look at how this works. We will see how we can use the allele frequencies in the gene pool to make predictions about the genotype frequencies in the next generation.

The frequency of an allele in the gene pool is equal to the probability that the allele will be chosen when randomly picking an allele from the gene pool to form an egg or a sperm. Knowing this, we can calculate the probability that a frog in the next generation will be an *A/A* homozygote. If we reach into the frog gene pool (see Figure 18-7) and pick the first allele, the probability that it will be an *A* is *p* = 0.56, and similarly, the probability that the second allele we pick is also an *A* is *p* = 0.56. The product of these two probabilities, or $p^2 = 0.3136$, is the probability that a frog in the next generation will be *A/A*. The probability that a frog in the next generation will be *a/a* is $q^2 = 0.44 \times 0.44 = 0.1936$. There are two ways to make a heterozygote. We might first pick an *A* with probability *p* and then pick an *a* with probability *q*, or we might pick the *a* first and the *A* second. Thus, the probability that a frog in the next generation will be heterozygous *A/a* is *pq* + *qp* = 2*pq* = 0.4928. Overall, the frequencies (*f*) of the genotypes are

$$f_{A/A} = p^2$$
$$f_{a/a} = q^2$$
$$f_{A/a} = 2pq$$

Finally, as expected, the sum of the probability of being *A/A* plus the probability of being *A/a* plus the probability of being *a/a* is 1.0:

$$p^2 + 2pq + q^2 = 1.0$$

This simple equation is the **Hardy–Weinberg law**, and it is part of the foundation for the theory of population genetics.

The process of reaching into the gene pool to pick an allele is called *sampling* the gene pool. Since any individual that contributes to the gene pool can produce many eggs or sperm that carry exactly the same copy of an allele, it is possible to pick a particular copy and then reach back into

---

BOX 18-1 Calculation of Allele Frequencies

At a locus with two alleles *A* and *a*, let's define the frequencies of the three genotypes *A/A*, *A/a*, and *a/a* as $f_{A/A}$, $f_{A/a}$, and $f_{a/a}$, respectively. We can use these genotype frequencies to calculate the allele frequencies: *p* is the frequency of the *A* allele, and *q* is the frequency of the *a* allele. Because each homozygote *A/A* consists only of *A* alleles, and because half the alleles of each heterozygote *A/a* are *A* alleles, the total frequency *p* of *A* alleles in the population is calculated as

$$p = f_{A/A} + \tfrac{1}{2}f_{A/a} = \text{frequency of } A$$

Similarly, the frequency *q* of the *a* allele is given by

$$q = f_{a/a} + \tfrac{1}{2}f_{A/a} = \text{frequency of } a$$

Therefore,

$$p + q = f_{A/A} + f_{A/a} + f_{a/a} = 1.0$$

and

$$q = 1 - p$$

If there are more than two different allelic forms, the frequency for each allele is simply the frequency of its homozygote plus half the sum of the frequencies for all the heterozygotes in which it appears.

the gene pool and pick exactly the same copy again. There is also an element of chance involved when sampling the gene pool. Just by chance, some copies may be picked more than once and other copies may not be picked at all. Later in the chapter, we will look at how these properties of sampling the gene pool can lead to changes in the gene pool over time.

We used the Hardy–Weinberg law to calculate genotype frequencies in the next generation from the allele frequencies in the current generation. We can also use the Hardy–Weinberg law to calculate allele frequencies from the genotype frequencies within a single generation. For example, some forms of albinism in humans are due to recessive alleles at the *OCA2* locus. In Africa, a form of albinism called brown oculocutaneous albinism results from a recessive allele of *OCA2* (**Figure 18-8**). Individuals with this condition, who have two recessive alleles, are present at frequencies as high as 1 in 1100 among some ethnic groups in Africa. We can use the Hardy–Weinberg law to calculate the allele frequencies:

$$f_{a/a} = q^2 = 1/1100 = 0.0009$$

so

$$q = \sqrt{0.0009} = 0.03$$

and

$$p = 1 - q = 0.97$$

Using the allele frequencies, we can also calculate the frequency of heterozygotes in the population as

$$2pq = 2 \times 0.97 \times 0.03 = 0.06$$

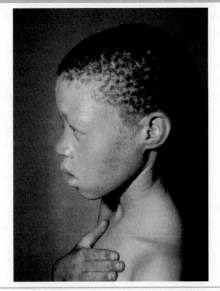

A form of albinism common among some African ethnic groups

FIGURE 18-8 Individual of African ancestry with brown oculocutaneous albinism (BOCA), a condition defined by light tan skin and beige to light brown hair. [*Dr. Michele Ramsay Department of Human Genetics, School of Pathology, the National Health Laboratory Service University of Witwatersrand.*]

The latter number predicts that about 6 percent of this population are heterozygotes, or carriers of the recessive allele at *OCA2*.

When we use the Hardy–Weinberg law to calculate allele or genotype frequencies, we make some critical assumptions.

- *First*, we assume that mating is random in the population with respect to the gene in question. Deviation from random mating violates this assumption, making it inappropriate to apply Hardy–Weinberg. For example, a tendency for individuals who are phenotypically similar to mate with each other violates the Hardy–Weinberg law. If albinos mated more frequently with other albinos than with non-albinos, then the Hardy–Weinberg law would overestimate the frequency of the recessive allele.

- *Second*, if one of the genotypes has reduced viability such that some individuals with that genotype die before the genotype frequencies are counted, then the estimate of the gene frequencies will be inaccurate.

- *Third*, for the Hardy–Weinberg law to apply, the population must not be divided into subpopulations that are partially or fully genetically isolated. If there are separate subpopulations, alleles may be present at different frequencies in the different subpopulations. If so, using genotypic counts from the overall population may not give an accurate estimate of the overall allele frequencies.

- *Finally*, the Hardy–Weinberg law strictly applies only to infinitely large populations. For finite populations, there will be deviations from the frequencies predicted by the Hardy–Weinberg law due to chance when sampling the gene pool to produce the next generation.

We have seen how we can use the Hardy–Weinberg law and the gene frequencies in the current generation (t_0) to calculate genotype frequencies in the next generation (t_1) by randomly sampling the gene pool for the production of eggs and sperm. Similarly, the predicted genotype frequencies for generation t_1 can be used in turn to calculate gene frequencies for the next generation (t_2). The gene frequencies in generation t_2 will remain the same as in generation t_1. Under the Hardy–Weinberg law, neither gene nor genotype frequencies change from one generation to the next when an infinitely large population is randomly sampled for the formation of eggs and sperm. Thus, an important lesson from the Hardy–Weinberg law is that, in large populations, genetic variation is neither created nor destroyed by the process of transmitting genes from one generation to the next. Populations that adhere to this principle are said to be at **Hardy–Weinberg equilibrium**.

Generation	Genotype frequencies			Gene frequencies	
	A/A	*A/a*	*a/a*	*A*	*a*
t_0	0.64	0.32	0.04	0.8	0.2
t_1	0.64	0.32	0.04	0.8	0.2
⋮	⋮	⋮	⋮	⋮	⋮
t_n	0.64	0.32	0.04	0.8	0.2

Here are a few more points about the Hardy–Weinberg law.

1. For any allele that exists at a very low frequency, homozygous individuals will only very rarely be found. If allele *a* has a frequency of 1 in a thousand ($q = 0.001$), then only 1 in a million (q^2) individuals will be homozygous for that allele. As a consequence, recessive alleles for genetic disorders can occur in the heterozygous state in many more individuals than there are individuals that actually express the genetic disorder in question.

2. The Hardy–Weinberg law still applies where there are more than two alleles per locus. If there are *n* alleles, $A_1, A_2, \ldots A_n$ with frequencies $p_1, p_2, \ldots p_n$, then the sum of all the individual frequencies equals 1.0. The frequencies of each of the homozygous genotypes are simply the square of the frequencies of the alleles, and the frequencies of the different heterozygous classes are two times the product of the frequencies of the first and second allele. **Table 18-1** gives an example with $p_1 = 0.5$, $p_2 = 0.3$, and $p_3 = 0.2$.

3. Hardy–Weinberg logic applies to X-linked loci as well. Males are hemizygous for X-linked genes, meaning that a male has a single copy of these genes. Thus, for X-linked genes in males, the genotype frequencies are equal to the allele frequencies. For females, genotype frequencies for X-linked genes follow normal Hardy–Weinberg expectations.

    Male pattern baldness is an X-linked trait (**Figure 18-9**). *AR* (for androgen receptor) is an X-linked gene involved in male development. There is an *AR* haplotype called *Eur-H1* that is strongly associated with pattern baldness. Male pattern baldness is common in Europe, where the *Eur-H1* haplotype occurs at a frequency of 0.71, meaning that 71 percent of European men carry it. Using the Hardy–Weinberg law, we can calculate that 50 percent of European women are *Eur-H1* homozygotes and 41 percent are heterozygous. The inheritance of baldness is complex and is affected by multiple genes, and so not all men who have *Eur-H1* go bald.

Male pattern baldness

FIGURE 18-9 Individual showing male pattern baldness, an X-chromosome-linked condition. [*B2M Productions/Getty Images.*]

4. One can test whether the observed genotype frequencies at a locus fit Hardy–Weinberg predictions using the chi-square test (see Chapter 3). An example is provided by the human leukocyte antigen gene, *HLA-DQA1*, of the major histocompatibility complex (MHC). MHC is a cluster of genes on chromosome 6 that play roles in the immune system. **Table 18-2** has genotype frequencies for an SNP (rs9272426) in the *HLA-DQA1* for 84 residents of Tuscany, Italy. This SNP has alleles *A* and *G*. From the genotype frequencies in Table 18-2, we can calculate the allele frequencies: $f(A) = p = 0.53$ and $f(G) = q = 0.47$. Next, we can calculate expected genotype frequencies under the Hardy–Weinberg law: $p^2 = 0.281$, $2pq = 0.498$, and $q^2 = 0.221$. Multiplying the expected genotype frequencies times the sample size ($N = 84$) gives us the expected number of individuals for each genotype. Now we can calculate the χ^2 statistic to be 8.29. Using Table 3-1, we see that the probability under the null hypothesis that the observed data fit Hardy–Weinberg predictions is $P < 0.005$ with df = 1. [We have only one degree of freedom because we have three genotypic categories and we used two numbers from the data (N and p) to calculate the expected values ($3 - 2$ leaves 1 degree of freedom). We did not need to use q since $q = p - 1$.] This analysis makes us strongly suspect that Tuscans do not conform to Hardy–Weinberg expectations with regard to *HLA-DQA1*.

    The Hardy–Weinberg law is part of the foundation of population genetics. It applies to an idealized population that is infinite in size and in which mating is random. It also assumes that all genotypes are equally fit—that is, that they are all equally viable and have the same success at reproduction. Real populations deviate from this idealized one. In the rest of the chapter, we will examine how factors such as nonrandom mating, finite population size, and the unequal fitness of different genotypes cause deviations from Hardy–Weinberg expectations.

TABLE 18-1	Hardy–Weinberg Genotype Frequencies for a Locus with Three Alleles A_1, A_2, and A_3 with Frequencies 0.5, 0.3, and 0.2, Respectively	
Genotype	**Expectation**	**Frequency**
A_1A_1	p_1^2	0.25
A_2A_2	p_2^2	0.09
A_3A_3	p_3^2	0.04
A_1A_2	$2p_1p_2$	0.30
A_1A_3	$2p_1p_3$	0.20
A_2A_3	$2p_2p_3$	0.12
Sum		1.00

TABLE 18-2 Frequencies of SNP rs9272426 Genotypes in *HLA-DQA1* of the MHC Locus for People from Tuscany, Italy

	Genotypes			
	A/A	**A/G**	**G/G**	**Sum**
Observed number	17	55	12	84
Observed frequency	0.202	0.655	0.143	1
Expected frequency	0.281	0.498	0.221	1
Expected number	23.574	41.851	18.574	84
$(\text{Observed}-\text{expected})^2/\text{expected}$	1.833	4.131	2.327	8.29

Source: International HapMap Project.

KEY CONCEPT The Hardy–Weinberg law describes the relationship between allele and genotype frequencies. This law informs us that genetic variation is neither created nor destroyed by the process of transmitting genes from one generation to the next. The Hardy–Weinberg law strictly applies only in infinitely large and randomly mating populations.

18.3 MATING SYSTEMS

LO 18.3 Quantify the effect of inbreeding in a population.

Random mating is a critical assumption of the Hardy–Weinberg law. The assumption of random mating is met if all individuals in the population are equally likely as a choice when a mate is chosen. However, if a relative, a neighbor, or a phenotypically similar individual is a more likely mate than a random individual, then the assumption of random mating has been violated. Populations that are not random mating will not exhibit exact Hardy–Weinberg proportions for the genotypes at some or all genes. Three types of bias in mate choice that violate the assumption of random mating are assortative mating, isolation by distance, and inbreeding.

Assortative mating

Assortative mating occurs if individuals choose mates based on resemblance or non-resemblance to themselves. **Positive assortative mating** occurs when similar types mate; for example, if tall individuals preferentially mate with other tall individuals and short individuals mate with other short individuals. In these cases, genes controlling the difference in height will not follow the Hardy–Weinberg law. Rather, we would expect to see an excess of homozygotes for the "tall" alleles among the progeny of tall mating pairs and an excess of homozygotes for "short" alleles among the progeny of short mating pairs. In humans, there is positive assortative mating for height.

Negative assortative or **disassortative mating** occurs when unlike individuals mate—that is, when opposites attract. One example of negative assortative mating is provided by the self-incompatibility, or *S*, locus in plants such as *Brassica* (broccoli and its relatives). There are numerous alleles at the *S* locus, S_1, S_2, S_3, and so forth. The stigma of

a plant will not be receptive to pollen that carries either of its own two alleles (**Figure 18-10**). For example, the stigma of an S_1/S_2 heterozygote will not allow pollen grains carrying either an S_1 or S_2 allele to germinate and fertilize its ovules, although pollen grains carrying the S_3 or S_4 alleles can do so. This mechanism blocks self-fertilization, thereby enforcing cross-pollination. The *S* locus violates the Hardy–Weinberg law since homozygous genotypes at *S* are not formed.

A second example of negative assortative mating is provided by the major histocompatibility complex (MHC), which is known to influence mate choice in vertebrates. MHC affects body odor in mice and rats, providing a basis for mate choice. In what are known as the "sweaty T-shirt experiments," researchers asked a group of men to wear T-shirts for two days. Then they asked a group of women to smell the T-shirts and rate them for "pleasantness." Women preferred the scent of men whose MHC haplotypes were different from their own. Data

Self-incompatibility leads to disassortative mating in *Brassica*

(a) Pollen inhibition

(b) Pollen-tube growth

FIGURE 18-10 Disassortative mating caused by the self-incompatibility locus (*S*) of the flowering plant genus *Brassica*. (a) A self-pollinated S_1/S_2 stigma shows no pollen-tube growth. (b) There is pollen-tube growth for an S_1/S_2 stigma cross-pollinated with pollen from an S_3/S_4 heterozygote. [*June Bowman Nasrallah.*]

from the human HapMap project have since confirmed that American couples are significantly more heterozygous at the MHC than expected by chance. The MHC plays a central role in our immune response to pathogens, and heterozygotes may be more resistant to pathogens. Therefore, our offspring benefit if we mate disassortatively with respect to our MHC genotype. This mechanism may explain why the SNP in the MHC gene *HLA-DQA1* that we discussed earlier does not follow the Hardy–Weinberg law among residents of Tuscany. Look back at Table 18-2 and you will notice that there are more heterozygotes than expected, 55 versus 42. Tuscans appear to be practicing disassortative mating with respect to this SNP.

Isolation by distance

Another form of bias in mate choice arises from the amount of geographic distance between individuals. Individuals are more apt to mate with a neighbor than another member of their species on the opposite side of the continent—that is, individuals can show **isolation by distance**. As a consequence, allele and genotype frequencies often differ between fish in separate lakes or between pine trees in different regions of a continent. Species or populations exhibiting such patterning of genetic variation are said to show **population structure**. A species can be divided into a series of subpopulations such as frogs in different ponds or people in different cities.

If a species has population structure, the proportion of homozygotes will be greater species-wide than expected under the Hardy–Weinberg law. Consider a hypothetical example of a species of wild sunflowers distributed across Kansas with a gradient in the frequency of the *A* allele from 0.9 near Kansas City to 0.1 near Elkhart (**Figure 18-11a**). We sample 100 sunflower plants from each of these two cities plus 100 from Hutchinson, in the middle of the state, and we calculate allele frequencies. Each city represents a subpopulation of the sunflowers. For any of the three cities, the Hardy–Weinberg law works fine. For example, in Elkhart, we expect $Nq^2 = 100 \times (0.9)^2 = 81$ *a/a* homozygotes, and that is what we observe. However, statewide, we would predict $Nq^2 = 300 \times (0.5)^2 = 75$ *a/a* homozygotes, yet we observed 107. Because of population structure, there are more homozygous sunflower plants than expected.

	Number of individuals					
	N	**A/A**	**A/a**	**a/a**	**p**	**q**
Kansas City	100	81	18	1	0.90	0.10
Hutchinson	100	25	50	25	0.50	0.50
Elkhart	100	1	18	81	0.10	0.90
Statewide (observed)	300	107	86	107	0.50	0.50
Statewide (expected)	300	75	150	75	–	–

Here is a real example of population structure from our own species. In Africa, the *FY*^{null} allele of the Duffy blood group shows a gradient with a low frequency in eastern and northern Africa, moderate frequency in southern Africa, and high frequency across central Africa (Figure 18-11b).

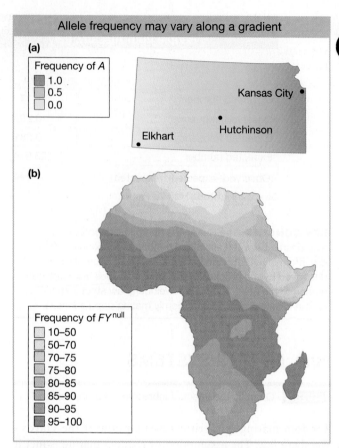

FIGURE 18-11 (a) Allele frequency variation across Kansas for a hypothetical species of wild sunflower. (b) Frequency variation for the *FY*^{null} allele of the Duffy blood group locus in Africa. [*Data from P. C. Sabeti et al., Science 312, 2006, 1614–1620.*]

This allele is rare outside of Africa. Because of this gradient, we cannot use overall allele frequencies in Africa to calculate genotype frequencies using the Hardy–Weinberg law. Later in the chapter and in Chapter 20, we will discuss the relationship between *Fy*^{null} and malaria.

KEY CONCEPT Assortative mating and isolation by distance violate the Hardy–Weinberg law and can cause genotype frequencies to deviate from Hardy–Weinberg expectations.

Inbreeding

The third type of bias in mating is **inbreeding**, or mating between relatives. Long before anyone knew about deleterious recessive alleles, some societies recognized that disorders such as muteness, deafness, and blindness were more frequent among the children of marriages between relatives. Accordingly, brother–sister and first-cousin marriages were either outlawed or discouraged. Nevertheless, many famous individuals have married a cousin, including Charles Darwin, Albert Einstein, J. S. Bach, Edgar Allan Poe, Jesse James, and Queen Victoria. As we will see, the offspring of marriages between relatives are at higher risk of having an inherited disorder.

Progeny of inbreeding are more likely to be homozygous at any locus than progeny of non-inbred matings. Thus, they are more likely to be homozygous for deleterious recessive alleles. For this reason, inbreeding can lead to a reduction in vigor and reproductive success called **inbreeding depression**. However, inbreeding can have advantages, too. Many plant species are highly self-pollinating and highly inbred. These include the cereal crops rice and wheat, and the model plant *Arabidopsis*, a successful weed. Since most plant species bear male and female organs on the same individual, self-pollination can be accomplished more easily than outcrossing. Another advantage of self-pollination is that when a single seed is dispersed to a new location, the plant that grows from the seed has a ready mate—itself, enabling a new population to be established from a single seed. Finally, if an individual plant has a beneficial combination of alleles at different loci, then inbreeding preserves that combination. In selfing plant species, benefits such as these offer advantages that outweigh the cost associated with inbreeding depression.

> **KEY CONCEPT** Inbreeding increases the frequency of homozygotes in a population, and can result in a higher frequency of recessive genetic disorders.

The inbreeding coefficient

Inbreeding increases the risk that an individual will be homozygous for a recessive deleterious allele and exhibit a genetic disease. The amount that risk increases depends on two factors: (1) the frequency of the deleterious allele in the population and (2) the degree of inbreeding. To measure the degree of inbreeding, geneticists use the **inbreeding coefficient** (F), which is the probability that two alleles in an individual trace back to the same copy in a common ancestor. Let's first consider how to calculate F using pedigrees and then examine how F can be used to determine the increase in risk of inheriting a recessive disease condition.

Consider a simple pedigree for a mating between half-sibs, individuals who have one parent in common (**Figure 18-12a**). In the figure, B and C are half-sibs who have the same mother, A, but different fathers; B and C have a daughter, I. Notice that there is a closed loop from I through B and A and back to I through C. The presence of a closed loop in the pedigree informs us that I is inbred. The two copies of the gene in A are colored blue and pink—the blue from A's father and pink from her mother. As drawn, I has inherited the pink copy both through her father (B) and her mother (C). Since I's two copies of the gene trace back to the same copy in her grandmother, her two copies are **identical by descent** (IBD). More generally, if the two copies of a gene in an individual trace back to the same copy in an ancestor, then the copies are IBD. We would like a way to calculate the probability that I's two alleles will be IBD. This probability is the inbreeding coefficient for I, which in symbol form is F_I.

First, since we are interested only in tracing the path of IBD alleles, we can simplify the pedigree to contain only the individuals in the closed loop and still follow the transmission of any IBD alleles (Figure 18-12b). Also, since the sex of the individual does not matter, we use circles for both sexes. The alleles transmitted with each mating are labeled w, x,

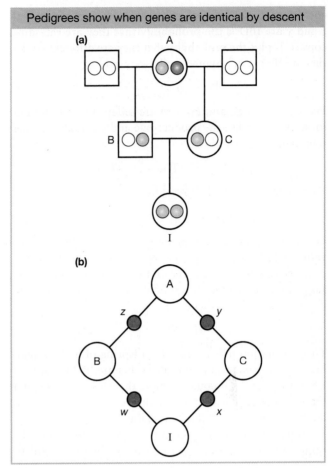

FIGURE 18-12 (a) Pedigree for a half-sib mating drawn in the standard format. Small colored balls represent a single copy of a gene. Within individual A, the pink and blue copies represent the copies of the gene that she inherited from her mother and father, respectively. (b) Pedigree for a half-sib mating drawn in the simplified format used for the analysis of inbreeding. Only lines connecting parent to offspring are drawn, and only individuals in the "closed inbreeding loop" are included. w, x, y, and z are symbols for the allele transmitted from parent to offspring.

y, and z. We use "~" to symbolize IBD. We would like to calculate the probability that w and x are IBD, but let's take this calculation step by step. First, what is the probability that x and y are IBD or, symbolically, what is $P(x \sim y)$? This is the probability that C transmits the copy inherited from A to I, which is 1/2, or $P(x \sim y) = 1/2$. Similarly, the probability that B transmits the copy inherited from A to I is 1/2, or $P(w \sim z) = 1/2$.

Now we need to calculate the probability that z and y are IBD. There are two ways that z and y can be IBD. The first way is when z and y are both the same copy (both pink or both blue). This happens 1/2 of the time, since 1/4 of the time they are both blue and 1/4 of the time both pink. The second way is when z and y are different copies (one pink and the other blue), but individual A was inbred. If individual A is inbred, then there is a probability that her two copies of the gene are IBD. The probability that A's two copies are IBD is the inbreeding coefficient of A, F_A. The probability that z and y are different copies (one pink, the other blue) is 1/2. So, the probability that z and y are different copies that are IBD is 1/2 multiplied by the inbreeding coefficient (F_A) to give $\frac{1}{2}F_A$. Altogether, the probability that z and y are IBD is the probability that they are the same copy (1/2) plus the probability that they are different copies that are IBD $\left(\frac{1}{2}F_A\right)$. Symbolically, we write

$$P(z \sim y) = \frac{1}{2} + \frac{1}{2}F_A$$

$P(x \sim y)$, $P(w \sim z)$, and $P(z \sim y)$ are independent probabilities, so we can use the product rule and put it all together to obtain

$$F_I = P(x \sim y) \times P(w \sim z) \times P(z \sim y)$$
$$= \frac{1}{2} \times \frac{1}{2} \times \left(\frac{1}{2} + \frac{1}{2}F_A\right)$$
$$= \left(\frac{1}{2}\right)^3 (1 + F_A)$$

In the analysis of inbred pedigrees, we can substitute the value of F_A into the equation above if it is known. Otherwise, we can assume F_A is zero if there is no information to suggest that individual A is inbred. In the current example, if we assume $F_A = 0$, then

$$F_I = \left(\frac{1}{2}\right)^3 = \frac{1}{8}$$

This calculation tells us that the offspring of half-sib matings will be homozygous for alleles that are IBD for at least 1/8 of their genes. It could be more than 1/8 if F_A is greater than zero. Additional inbred pedigrees and a general formula for calculating F can be found in **Box 18-2**.

When there is inbreeding in a population, the random-mating assumption of Hardy–Weinberg will be violated. However, Hardy–Weinberg can be modified to correct the predicted genotypic proportions for different degrees of inbreeding using F, the mean inbreeding coefficient for the population. The modified Hardy–Weinberg frequencies are

$$f_{A/A} = p^2 + pqF$$
$$f_{A/a} = 2pq - 2pqF$$
$$f_{a/a} = q^2 + pqF$$

These modified Hardy–Weinberg proportions make intuitive sense, showing how inbreeding reduces the frequency of heterozygotes by $2pqF$ and adds half this amount to each of the homozygous classes. With these modified Hardy–Weinberg equations, you'll also notice that when there is no inbreeding ($F = 0$), you regain standard Hardy–Weinberg genotypic frequencies, and when there is complete inbreeding ($F = 1$), you get $f_{A/A} = p$ and $f_{a/a} = q$.

How much does inbreeding increase the risk that offspring will exhibit a recessive disease condition? **Table 18-3** shows the inbreeding coefficients for offspring of some different inbred matings and the predicted number of homozygous recessives for different frequencies (q) of the recessive allele. When $q = 0.01$, there is a 7-fold (7.19/1.0) increase in homozygous recessive offspring for first-cousin matings as compared to matings between unrelated individuals. The increase in risk jumps 13-fold (3.36/0.25) when $q = 0.005$ and 63-fold (0.63/0.01) when $q = 0.001$. In other words, the degree of risk jumps dramatically for rare alleles. Brother–sister and parent–offspring matings are the riskiest: when

---

BOX 18-2 Calculating Inbreeding Coefficients from Pedigrees

In the main text, we saw that the inbreeding coefficient (F_I) for the offspring of a mating between half-sibs is

$$F_I = \left(\frac{1}{2}\right)^3 (1 + F_A)$$

where F_A is the inbreeding coefficient of the ancestor. This expression includes the term 1/2 to the third power, $\left(\frac{1}{2}\right)^3$. In Figure 18-12, you'll see there are three individuals in the inbreeding loop, not counting I. The general formula for computing inbreeding coefficients from pedigrees is

$$F_I = \left(\frac{1}{2}\right)^n (1 + F_A)$$

where n is the number of individuals in the inbreeding loop not counting I. Let's look at another pedigree, one in which the grandparents of I are half-sibs:

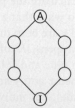

There are five individuals in the inbreeding loop other than I, so if we assume that the ancestor was not inbred ($F_A = 0$), then

$$F_I = \left(\frac{1}{2}\right)^5 (1 + F_A) = 0.03125$$

In some pedigrees, there is more than one inbreeding loop. Here's a pedigree in which I is the offspring of a mating between full sibs:

For pedigrees with multiple inbreeding loops, you sum the contribution over all of the loops where F_A is the inbreeding coefficient of the ancestor (A) of the given loop:

$$F_I = \sum_{\text{loops}} \left(\frac{1}{2}\right)^n (1 + F_A)$$

Thus, for the pedigree where I is the offspring of a mating between full sibs, we get

$$F_I = \left(\frac{1}{2}\right)^3 (1 + F_{A_1}) + \left(\frac{1}{2}\right)^3 (1 + F_{A_2}) = \frac{1}{4}$$

assuming that the inbreeding coefficients for both ancestors are 0.

TABLE 18-3	Number of Homozygous Recessives per 10,000 Individuals for Different Allele Frequencies (*q*)			
Mating	**F**	**q = 0.01**	**q = 0.005**	**q = 0.001**
Unrelated parents	0.0	1.00	0.25	0.01
Parent–offspring or brother–sister	1/4	25.75	12.69	2.51
Half-sib	1/8	13.38	6.47	1.26
First cousin	1/16	7.19	3.36	0.63
Second cousin	1/64	2.55	1.03	0.17

$q = 0.001$, they show a 250-fold (2.51/0.01) greater risk compared to matings between unrelated individuals.

The impact of inbreeding on the frequency of genetic disorders in human populations can be seen in **Figure 18-13**. Children of marriages of first cousins show about a two-fold higher frequency of disorders as compared to children of unrelated parents. Historical records suggest that the risks of inbreeding were understood long before the field of genetics existed.

KEY CONCEPT The inbreeding coefficient (*F*) is the probability that two alleles in an individual trace back to the same copy in a common ancestor.

Population size and inbreeding

Population size is a major factor contributing to the level of inbreeding in populations. In small populations, individuals are more likely to mate with a relative than in large ones. The phenomenon is seen in small human populations such

as the one on the Tristan de Cunha Islands in the South Atlantic, which has fewer than 300 people. Let's look at the effect of population size on the overall level of inbreeding in a population as measured by F.

Consider a population with F_t being the level of inbreeding at generation t. To form an individual in the next generation $t + 1$, we select the first allele from the gene pool. Suppose the population size is N. After the first allele is selected, the probability that the second allele we pick will be exactly the same copy is $1/2N$ and the inbreeding coefficient for this individual is 1.0. The probability that the second allele we pick will be a different copy from the first allele is $1 - 1/2N$ and the level of inbreeding for the resulting individual would be F_t, the average inbreeding coefficient for the initial population at generation t. The level of inbreeding in the next generation is the sum of these two possible outcomes, or

$$F_{t+1} = \left(\frac{1}{2N}\right)1 + \left(1 - \frac{1}{2N}\right)F_t$$

This equation informs us that F will increase over time as a function of population size. When N is large, F increases slowly over time. When N is small, F increases rapidly over time. For example, suppose F_t in the initial population is 0.1 and $N = 10,000$. Then F_{t+1} would be 0.10005, just a slightly higher value. However, if $N = 10$, then F_{t+1} would be 0.145, a much higher value. We can also use this equation recursively to calculate F_{t+2} by using F_{t+1} in place of F_t on the right side. The result with $N = 10$ and $F_t = 0.1$ would be $F_{t+2} = 0.188$. The effects of population size on inbreeding in populations are further explored in **Box 18-3**.

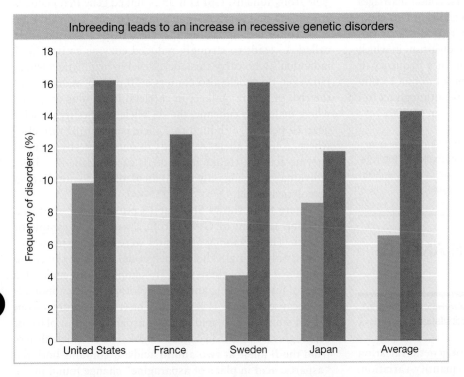

FIGURE 18-13 Frequency of genetic disorders among children of unrelated parents (blue columns) compared to that of children of parents who are first cousins (red columns). [*Data from C. Stern*, Principles of Human Genetics, *W. H. Freeman, 1973.*]

BOX 18-3   Inbreeding in Finite Populations

In the main text, we derived the formula for the increase in inbreeding between generations in finite populations as

$$F_{t+1} = \left(\frac{1}{2N}\right)1 + \left(1 - \frac{1}{2N}\right)F_t$$

which can be rewritten as

$$(1 - F_{t+1}) = \left(1 - \frac{1}{2N}\right)(1 - F_t)$$

We also presented the formula for the frequency of heterozygotes (H) with inbreeding as

$$H = f_{A/a} = 2pq - 2pqF$$

which can be rewritten as

$$(1 - F) = H/2pq$$

Combining these two equations, we obtain

$$H_{t+1}/2pq = \left(1 - \frac{1}{2N}\right)H_t/2pq$$

and then

$$H_{t+1} = \left(1 - \frac{1}{2N}\right)H_t$$

Thus, for each generation, the level of heterozygosity is reduced by the fraction $(1 - 1/2N)$. The reduction in H over t generations is

$$H_t = \left(1 - \frac{1}{2N}\right)^t H_0$$

and the change in F over t generations is given by

$$F_t = 1 - \left(1 - \frac{1}{2N}\right)^t (1 - F_0)$$

As shown in the figure below, inbreeding will increase with time in a finite population even when there is no inbreeding in the initial population.

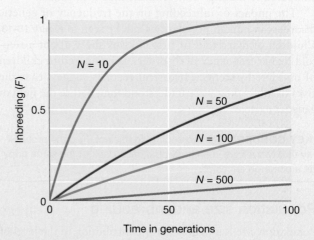

Increase in inbreeding (F) over time for several different population sizes.

A consequence of the increased inbreeding is that individuals in small populations are more likely to be homozygous for deleterious alleles just as the offspring of first-cousin marriages are more likely to be homozygous for such alleles. This effect is seen in ethnic groups that live in small, reproductively isolated communities. For example, a form of dwarfism in which affected individuals have six fingers occurs at a frequency of more than 1 in 200 among a population of about 13,000 Amish in Lancaster County, Pennsylvania, although its frequency in the general U.S. population is only 1 in 60,000.

KEY CONCEPT The inbreeding coefficient (F) increases over time as a function of population size (N). For smaller N, the rate at which F grows is faster than for larger values of N. In smaller populations, there is a greater risk of homozygosity for deleterious recessive alleles.

18.4 GENETIC VARIATION AND ITS MEASUREMENT

LO 18.1 Describe and analyze data to determine how much genetic variation exists within populations.

To study the amount and distribution of genetic variation in populations, we need some ways to quantify variation.

To describe how we can quantify variation, we will use data for the glucose-6-phosphate dehydrogenase (G6PD) gene from humans. G6PD is an X-linked gene that encodes an enzyme that catalyzes a step in glycolysis. The wild-type allele (B) of G6PD has full enzyme activity. A second allele called A^- leads to strongly reduced enzyme activity, and individuals who carry this allele develop hemolytic anemia. However, this allele also confers a 50 percent reduction in the risk of severe malaria in carriers. In regions of Africa where malaria is endemic, the A^- allele reaches frequencies near 20 percent, although this allele is absent or rare elsewhere. Another allele (A^+) leads to only modestly reduced enzyme activity. Unlike individuals carrying the A^- allele, individuals carrying only the A^+ or B alleles do not develop hemolytic anemia.

Figure 18-14 shows SNPs at 18 polymorphic sites in a 5102-bp segment of G6PD from a worldwide sample of 47 men. The remaining 5084 sites were **fixed**, or invariant: that is, only a single allele (nucleotide) exists in the entire sample for each of these sites. By sampling only males, we observe just one allele and one haplotype for each individual because the gene is X linked. The A^+ allele differs from B by a single amino acid change (aspartic acid in place of asparagine) at SNP3 in Figure 18-14. The A^- allele differs from the B allele at two amino acids: it contains both the "aspartic acid in place of asparagine" change found in the

Nucleotide variation at the *G6PD* gene in humans

Individual	Origin	Allele	1	2	3	4	5	6	7	8	9	10	11	12	13	14	15	16	17	18	Haplotype
			A	G	A	C	C	G	C	C	C	C	C	G	G	C	T	C	A	C	
1	Southern African	A−	G	A	G	·	G	·	·	T	·	T	·	·	·	·	C	·	G	·	1
2	Central African	A−	G	A	G	·	G	·	·	T	·	T	·	·	·	·	C	·	G	·	1
3	Central African	A−	G	A	G	·	G	·	·	T	·	T	·	·	·	·	C	·	G	·	1
4	African American	A−	G	A	G	·	G	·	·	T	·	T	·	·	·	·	C	·	G	·	1
5	African American	A−	G	A	G	·	G	·	·	T	·	T	·	·	·	·	C	·	G	·	1
6	Central African	A−	G	A	G	·	G	·	·	T	·	T	·	·	·	·	C	·	G	·	1
7	Central African	A+	G	·	G	·	·	·	·	·	·	T	·	·	·	·	C	·	G	·	2
8	Central African	A+	G	·	G	·	·	·	·	·	·	T	·	·	·	·	C	·	G	·	2
9	Central African	B	·	·	·	·	·	·	·	·	·	·	·	·	·	·	C	·	G	·	3
10	Southern African	B	·	·	·	·	·	·	·	·	·	·	·	A	·	·	C	T	G	·	4
11	Southern African	B	·	·	·	·	·	·	·	·	·	·	·	A	·	·	C	T	G	·	4
12	Southern African	B	·	·	T	·	·	·	·	·	·	·	·	·	·	·	C	T	G	·	5
13	Southern African	B	·	·	·	·	·	·	·	·	·	·	·	·	·	·	C	T	G	·	6
14	Southern African	B	·	·	·	·	·	·	·	·	·	·	T	·	A	·	C	·	G	·	7
15	Central African	B	·	·	·	·	·	·	·	·	·	·	·	·	T	·	C	·	G	·	8
16	European	B	·	·	·	·	·	·	·	·	·	·	·	·	T	·	C	·	G	·	8
17	European	B	·	·	·	·	·	·	·	·	·	·	·	·	T	·	C	·	G	·	8
18	European	B	·	·	·	·	·	·	·	·	·	·	·	·	T	·	C	·	G	·	8
19	Southwest Asian	B	·	·	·	·	·	·	·	·	·	·	·	·	T	·	C	·	G	·	8
20	East Asian	B	·	·	·	·	·	·	·	·	·	·	·	·	·	·	C	·	G	·	3
21	Native American	B	·	·	·	·	·	·	A	T	·	·	·	·	·	·	C	·	G	·	9
22	Southern African	B	·	·	·	·	·	·	·	·	·	·	·	·	·	·	·	·	·	·	10
23	Native American	B	·	·	·	·	·	·	·	·	·	·	·	·	·	·	·	·	·	·	10
24	Native American	B	·	·	·	·	·	·	·	·	·	·	·	·	·	·	·	·	·	·	10
25	Native American	B	·	·	·	·	·	·	·	·	·	·	·	·	·	·	·	·	·	·	10
26	Native American	B	·	·	·	·	·	·	·	·	·	·	·	·	·	·	·	·	·	·	10
27	Native American	B	·	·	·	·	·	·	·	·	·	·	·	·	·	·	·	·	·	·	10
28	Native American	B	·	·	·	·	·	·	·	·	·	·	·	·	·	·	·	·	·	·	10
29	Native American	B	·	·	·	·	·	·	·	·	·	·	·	·	·	·	·	·	·	·	10
30	Native American	B	·	·	·	·	·	·	·	·	·	·	·	·	·	·	·	·	·	·	10
31	Native American	B	·	·	·	·	·	·	·	·	·	·	·	·	·	·	·	·	·	·	10
32	European	B	·	·	·	·	·	·	·	·	·	·	·	·	·	·	·	·	·	·	10
33	European	B	·	·	·	·	·	·	·	·	·	·	·	·	·	·	·	·	·	·	10
34	European	B	·	·	·	·	·	·	·	·	·	·	·	·	·	·	·	·	·	·	10
35	European	B	·	·	·	·	·	·	·	·	·	·	·	·	·	·	·	·	·	·	10
36	European	B	·	·	·	·	·	·	·	·	·	·	·	·	·	·	·	·	·	·	10
37	European	B	·	·	·	·	·	·	·	·	·	·	·	·	·	·	·	·	·	·	10
38	Southwest Asian	B	·	·	·	·	·	·	·	·	·	·	·	·	·	·	·	·	·	·	10
39	East Asian	B	·	·	·	·	·	·	·	·	·	·	·	·	·	·	·	·	·	·	10
40	East Asian	B	·	·	·	·	·	·	·	·	·	·	·	·	·	·	·	·	·	·	10
41	East Asian	B	·	·	·	·	·	·	·	·	·	·	·	·	·	·	·	·	·	·	10
42	East Asian	B	·	·	·	·	·	·	·	·	·	·	·	·	·	·	·	·	·	·	10
43	East Asian	B	·	·	·	·	·	·	·	·	·	·	·	·	·	·	·	·	·	·	10
44	East Asian	B	·	·	·	·	·	·	·	·	·	·	·	·	·	·	·	·	·	·	10
45	East Asian	B	·	·	·	·	·	·	·	·	·	·	·	·	·	·	·	·	·	·	10
46	Pacific Islander	B	·	·	·	·	·	·	·	·	T	·	·	·	·	·	·	·	·	·	11
47	East Asian	B	·	·	·	·	·	·	·	·	·	·	·	·	·	·	·	·	T	·	12

FIGURE 18-14 Nucleotide variation for 5102 bp of the *G6PD* gene for a worldwide sample of 47 men. Only the 18 variable sites are shown. The functional allele class (*A*⁻, *A*⁺, or *B*) is shown for each sequence. *SNP2* is a nonsynonymous SNP that causes a valine-to-methionine change that underlies differences in enzyme activity associated with the *A*⁻ allele. *SNP3* is a nonsynonymous SNP that causes an aspartic-acid-to-asparagine amino acid change. [*Data from M. A. Saunders et al., Genetics 162, 2002, 1849–1861.*]

A^+ allele and a second amino acid difference (methionine in place of valine) at *SNP2*.

How can we quantify variation at the *G6PD* locus? One simple measure is the number of polymorphic or **segregating sites** (*S*). For the *G6PD* data, *S* is 18 for the total sample, 14 for the African sample, and 7 for the non-African sample. Africans contain twice the number of segregating sites despite the fact that our sample has fewer Africans. Another simple measure is the **number of haplotypes** (*NH*). The value of *NH* is 12 for the total sample, 9 for the African sample, and 6 for the non-African sample. Again, the African sample has greater variation. One shortcoming of measures such as *S* and *NH* is that the values we observe depend heavily on sample size. If one samples more individuals, then the values of *S* and *NH* are apt to increase. For example, our sample has 16 Africans compared to 31 non-Africans. Although *S* is twice as large in Africans as in non-Africans, the difference would likely be even greater if we had an equal number (31) of Africans and non-Africans.

In place of *S* and *NH*, we can calculate allele frequencies, which are not biased by differences in sample size. For the *G6PD* data, *B*, A^-, and A^+ have worldwide frequencies of 0.83, 0.13, and 0.04, respectively. However, you will note that A^- has a frequency of 0.0 outside of Africa and 0.38 in our African sample, which is a substantial difference. We can use allele frequency data to calculate a statistic called **gene diversity** (*GD*), which is the probability that two alleles drawn at random from the gene pool will be different. The probability of drawing two different alleles is equal to 1 minus the probability of drawing two copies of the same allele summed over all alleles at the locus. Thus,

$$GD = 1 - \sum p_i^2$$
$$= 1 - (p_1^2 + p_2^2 + p_3^2 + \cdots p_n^2)$$

where p_i is the frequency of the *i*th allele and Σ is the summation sign, indicating that we add the squares of all *n* observed values of *p* for *i* = 1, 2, through the *n*th allele. The value of *GD* can vary from 0 to 1. It will approach 1 when there is a large number of alleles of roughly equal frequencies. It is 0 when there is a single allele, and it is near 0 whenever there is a single very common allele with a frequency of 0.99 or higher. **Table 18-4** shows that gene diversity is quite high in Africans (0.47). Since non-Africans have only the *B* allele, gene diversity is 0.0.

The value of *GD* is equal to the expected proportion of heterozygotes under Hardy–Weinberg equilibrium, which is **heterozygosity** (*H*). However, *H* as a concept applies only to diploids, and it would not apply to X-linked loci in males. Thus, conceptually gene diversity (*GD*) is more appropriate even if it is mathematically the same quantity as *H* for populations of diploids under Hardy–Weinberg equilibrium.

Gene diversity can be calculated for a single nucleotide site. It can be averaged over all the nucleotide sites in a gene, in which case it is referred to as **nucleotide diversity**. Since the vast majority of nucleotides in any two copies of a gene from a species are typically the same, values

TABLE 18-4	Diversity Data for Glucose-6-Phosphate Dehydrogenase (*G6PD*) in Humans		
	Total sample	Africans	Non-Africans
Sample size	47	16	31
Number of segregating sites	18	14	7
Number of haplotypes	12	9	6
Gene diversity (*GD*) at *SNP2*	0.22	0.47	0.00
Nucleotide diversity	0.0006	0.0008	0.0002

for nucleotide diversity for genes are typically very small. For *G6PD*, there are only 18 polymorphic nucleotide sites but 5084 invariant sites. The average nucleotide diversity for the entire *G6PD* gene sequence is 0.0008 in Africans, 0.0002 in non-Africans, and 0.0006 for the entire sample. These values tell us that Africans have four times as much nucleotide diversity at *G6PD* as non-Africans.

Figure 18-15 shows the level of nucleotide diversity in several organisms. Unicellular eukaryotes are the most diverse, followed by plants and then invertebrates. Vertebrates are the least diverse group; however, most vertebrates still possess a lot of nucleotide diversity. For humans,

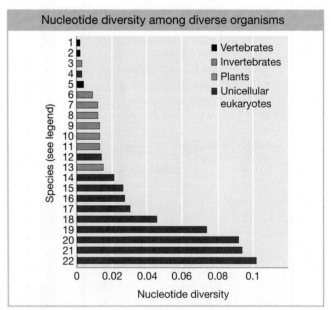

FIGURE 18-15 Levels of nucleotide diversity at synonymous and silent sites in some different organisms. (1) *Mus musculus*, (2) *Homo sapiens*, (3) *Oryza sativa*, (4) *Plasmodium falciparum*, (5) *Fugu rubripes*, (6) *Strongylocentrotus purpuratus*, (7) *Anopheles gambiae*, (8) *Ciona intestinalis*, (9) *Arabidopsis thaliana*, (10) *Caenorhabditis elegans*, (11) *Zea mays*, (12) *Encephalitozoon cuniculi*, (13) *Drosophila melanogaster*, (14) *Leishmania major*, (15) *Trypanosoma* species, (16) *Toxoplasma gondii*, (17) *Giardia lamblia*, (18) *Neurospora crassa*, (19) *Dictyostelium discoideum*, (20) *Saccharomyces cerevisiae*, (21) *Cryptosporidium parvum*, (22) *Cryptococcus neoformans*. [Data from M. Lynch and J. S. Conery, *Science* 302, 2003, 1401–1404.]

nucleotide diversity is about 0.001, meaning that two randomly chosen human chromosomes will differ at about 1 bp per thousand. With 3 billion bp in our genome, that adds up to a total of about 3 million differences between the set of chromosomes inherited from a person's mother and the set inherited from a person's father for non-inbred individuals.

KEY CONCEPT The rich genetic variation in biological populations can be quantified by different statistics, such as number of segregating sites, number of haplotypes, gene diversity, and nucleotide diversity, to compare levels of variation among populations and species.

18.5 THE MODULATION OF GENETIC VARIATION

LO 18.4	Explain how new alleles enter a population by mutation and migration.
LO 18.5	Measure the degree of linkage disequilibrium between two loci.
LO 18.6	Calculate the effect of genetic drift on gene frequencies within populations.
LO 18.7	Calculate the effect of selection on gene frequencies within populations.

What are the forces that modulate the amount of genetic variation in a population? How do new alleles enter the gene pool? What forces remove alleles from the gene pool? How can genetic variants be recombined to create novel combinations of alleles? Answers to these questions are at the heart of understanding the process of evolution. In this section, we will examine the roles of mutation, migration, recombination, genetic drift (chance), and selection in sculpting the genetic composition of populations.

New alleles enter the population: mutation and migration

Mutation is the ultimate source of all genetic variation. In Chapter 15, we discussed the molecular mechanisms that underlie small-scale mutations such as point mutations, indels, and changes in the number of repeat units in microsatellites. Population geneticists are particularly interested in the **mutation rate**, which is the probability that a copy of an allele changes to some other allelic form in one generation. The mutation rate is typically symbolized by the Greek letter μ. As we will see in this section, if we know the mutation rate and the number of nucleotide differences between two sequences, then we can estimate how long ago the two sequences diverged.

Geneticists can estimate mutation rates by starting with a single homozygous individual and following the pedigree of its descendants for several generations. Then they can compare the DNA sequence of the founding individual to the DNA sequences of the descendants several generations

later and record any new mutations that have occurred. The number of observed mutations per genome per generation provides an estimate of the rate. Because one is looking for rather rare events, it is necessary to sequence billions of nucleotides to find just a few SNP mutations. In 2009, the SNP mutation rate for a part of the human Y chromosome was estimated by this approach to be 3.0×10^{-8} mutations/nucleotide/generation, or about one mutation every 30 million bp. If we extrapolate to the entire human genome (3 billion bp), then each of us has inherited 100 new mutations from each of our parents. Luckily, the vast majority of mutations are not detrimental since they occur in regions of the genome that are not critical.

Table 18-5 lists the mutation rates for SNPs and microsatellites in several model organisms. The SNP mutation rate is several orders of magnitude lower than the microsatellite rate. Their higher mutation rate and greater variation make microsatellites particularly useful in population genetics and DNA forensics. The SNP mutation rate per generation appears to be lower for unicellular organisms than for large multicellular organisms. This difference can be explained at least partially by the number of cell divisions per generation. There are about 200 cell divisions from zygote to gamete in humans but only 1 in *E. coli*. If the human rate is divided by 200, then the rate per cell division in humans is remarkably close to the rate in *E. coli*.

Other than mutation, the only other means for new variation to enter a population is through **migration** or **gene flow**, the movement of individuals (or gametes) between populations. Most species are divided into a set of small local populations or subpopulations. Physical barriers such as oceans, rivers, or mountains may reduce gene flow between subpopulations, but often some degree of gene flow occurs despite such barriers. Within subpopulations, an individual may have a chance to mate with any other member of the opposite sex; however, individuals from different subpopulations cannot mate unless there is migration.

TABLE 18-5	Approximate Mutation Rates per Generation per Haploid Genome	
Organism	**SNP mutations (per bp)**	**Microsatellite**
Arabidopsis	7×10^{-9}	9×10^{-4}
Maize	3×10^{-8}	8×10^{-4}
E. coli	5×10^{-10}	–
Yeast	5×10^{-10}	4×10^{-5}
C. elegans	3×10^{-9}	4×10^{-3}
Drosophila	4×10^{-9}	9×10^{-6}
Mouse	4×10^{-9}	3×10^{-4}
Human	3×10^{-8}	6×10^{-4}

Note: Microsatellite rate is for di- or trinucleotide repeat microsatellites.
Source: Data from multiple published studies.

Isolated subpopulations tend to diverge as each accumulates its own unique mutations. Gene flow limits genetic divergence between subpopulations. One of the genetic consequences of migration is **genetic admixture**, the mix of genes that results when individuals have ancestry from more than one subpopulation. This phenomenon is common in human populations. It is readily observed in South Africa, where migrants from around the world were brought together. As shown in **Figure 18-16**, the genomes of South Africans of mixed ancestry are complex and include parts from the indigenous people of southern Africa plus contributions of migrants from western Africa, Europe, India, East Asia, and other regions.

KEY CONCEPT Mutation is the ultimate source of all genetic variation. Migration can add genetic variation to a population via gene flow from another population of the same species.

Recombination and linkage disequilibrium

Recombination is a critical force sculpting patterns of genetic variation in populations. In this case, alleles are not

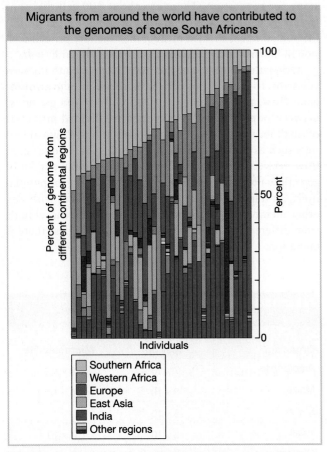

Migrants from around the world have contributed to the genomes of some South Africans

Percent of genome from different continental regions

Individuals

- Southern Africa
- Western Africa
- Europe
- East Asia
- India
- Other regions

FIGURE 18-16 Graphical representation of genetic admixture for 39 people of mixed ancestry from South Africa. Each column represents one person's genome, and the colors represent the parts of their genome contributed by their ancestors, who came from many regions of the world. The figure is based on the population genetic analysis of over 800 microsatellites and 500 indels that were scored for nearly 4000 people from around the world, including the 39 of mixed ancestry from South Africa. [*Data from S. A. Tishkoff et al.*, Science 324, 2009, 1035–1044.]

gained or lost; rather, recombination creates new haplotypes. Let's look at how this works. Consider linked loci A and B. There could be a population in which only two haplotypes are found at generation t_0: AB and ab. Suppose an individual in this population is heterozygous for these two haplotypes:

$$\frac{A \qquad\qquad B}{a \qquad\qquad b}$$

If a crossover occurs during meiosis in this individual, then gametes with two new haplotypes, Ab and aB, could be formed and enter the population in generation t_1.

$$\underline{A \qquad\qquad b} \quad \underline{a \qquad\qquad B}$$

Thus, recombination can create variation that takes the form of new haplotypes. The new haplotypes can have unique properties that alter protein function. For example, suppose an amino acid variant in a protein on one haplotype increases the enzyme activity of the protein twofold and a second amino acid variant on another haplotype also increases activity twofold. A recombination event that combines these two variants would yield a protein with fourfold higher activity.

Let's now consider the observed and expected frequencies of the four possible haplotypes for two loci, each with two alleles. Linked loci, A and B, have alleles A and a and B and b with frequencies p_A, p_a, p_B, and p_b, respectively. The four possible haplotypes are AB, Ab, aB, and ab with observed frequencies P_{AB}, P_{Ab}, P_{aB}, and P_{ab}. At what frequency do we expect to find each of these four haplotypes? If there is a random relationship between the alleles at the two loci, then the frequency of any haplotype will be the product of the frequencies of the two alleles that compose that haplotype:

$$P_{AB} = p_A \times p_B$$
$$P_{Ab} = p_A \times p_b$$
$$P_{aB} = p_a \times p_B$$
$$P_{ab} = p_a \times p_b$$

For example, suppose that the frequency of each of the alleles is 0.5; that is, $p_A = p_a = p_B = p_b = 0.5$. When we sample the gene pool, the probability of drawing a chromosome with an A allele is 0.5. If the relationship between the alleles at locus A and the alleles at locus B is random, then the probability that the selected chromosome has the B allele is also 0.5. Thus, the probability that we draw a chromosome with the AB haplotype is

$$P_{AB} = p_A \times p_B = 0.5 \times 0.5 = 0.25$$

If the association between the alleles at two loci is random as just described, then the two loci are said to be at **linkage equilibrium**. In this case, the observed and expected frequencies will be the same. **Figure 18-17a** diagrams a case of two loci at linkage equilibrium.

If the association between the alleles at two loci is nonrandom, then the loci are said to be in

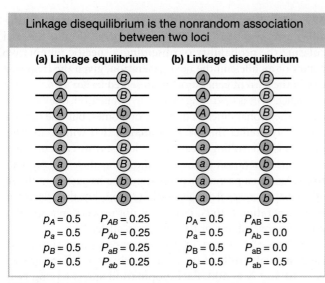

FIGURE 18-17 (a) Linkage equilibrium and (b) linkage disequilibrium for two loci (*A* and *B*).

linkage disequilibrium (LD). In this case, a specific allele at the first locus is associated with a specific allele at the second locus more often than expected by chance. Figure 18-17b diagrams a case of complete LD between two loci. The *A* allele is always associated with the *B* allele, while the *a* allele is always associated with the *b* allele. There are no chromosomes with haplotypes *Ab* or *aB*. In this case, the observed and expected frequencies will not be the same.

We can quantify the level of LD between two loci as the difference (*D*) between the observed frequency of a haplotype and the expected frequency given a random association among alleles at the two loci. If both loci involved have just two alleles, then

$$D = P_{AB} - p_A p_B$$

In Figure 18-17a, $D = 0$ since there is no LD, and in Figure 18-17b, $D = 0.25$, which is greater than 0, indicating the presence of LD.

How does LD arise? Whenever a new mutation occurs at a locus, the mutation appears on a single specific chromosome and so it is instantly linked to (or associated with) the specific alleles at any neighboring loci on that chromosome. Consider a population in which there are just two haplotypes: *AB* and *Ab*. If a new mutation (*a*) arises at the *A* locus on a chromosome that already possesses the *b* allele at the *B* locus, then a new *ab* haplotype would be formed. Over time, this new *ab* haplotype might rise in frequency in the population. Other chromosomes in the population would possess the *AB* or *Ab* haplotypes at these two loci, but no chromosomes would possess *aB*. Thus, the loci would be in LD. Migration can also cause LD when one subpopulation possesses only the *AB* haplotype and another only the *ab* haplotype. Any migrants between the subpopulations would give rise to LD within the subpopulation that receives the migrants.

LD between two loci will decline over time as crossovers between them randomize the relationship between

their alleles. The rate of decline in LD depends on the rate at which crossing over occurs. The frequency of recombinants (RF) between the two loci among the gametes that form the next generation (see Chapter 4) provides an estimate of recombination rate, which in population genetics is symbolized by the lowercase letter *r*. If D_0 is the value for linkage disequilibrium between two loci in the current generation, then the value in the next generation (D_1) is given by this equation:

$$D_1 = D_0(1 - r)$$

In other words, linkage disequilibrium as measured by *D* declines at a rate of $(1 - r)$ per generation. When *r* is small, *D* declines slowly over time. When *r* is at its maximum (0.5), then *D* declines by 1/2 each generation.

Since LD decays as a function of time and the recombination fraction, population geneticists can use the level of LD between a mutation and the loci surrounding it to estimate the time in generations since the mutation first arose in the population. Older mutations have little LD with neighboring loci, while recent mutations show a high level of LD with neighboring loci. If you look again at Figure 18-14, you will notice that there is considerable LD between *SNP2* in *G6PD* and the neighboring SNPs. *SNP2* encodes the amino acid change of valine to methionine in the A^- allele that confers resistance to malaria. Population geneticists have used LD at *G6PD* to estimate that the A^- allele arose about 10,000 years ago. Malaria is not thought to have been prevalent in Africa until then. Thus, the A^- arose by random mutation but was maintained in the population because it provided protection against malaria.

> **KEY CONCEPT** Linkage disequilibrium is the outcome of the fact that new mutations arise on a single haplotype. Linkage disequilibrium will decay over time because of recombination.

Genetic drift and population size

The Hardy–Weinberg law tells us that allele frequencies remain the same from one generation to the next in an *infinitely large population*. However, actual populations of organisms in nature are *finite* rather than infinite in size. In finite populations, allele frequencies may change from one generation to the next as the result of chance (sampling error) when gametes are drawn from the gene pool to form the next generation. Change in allele frequencies between generations due to sampling error is called **random genetic drift**, or just drift for short.

Let's consider a simple but extreme case—a population composed of a single heterozygous (*A/a*) individual ($N = 1$) at generation t_0. We will allow self-fertilization. In this case, the gene pool can be described as having two alleles, *A* and *a*, each present at a frequency of $p = q = 0.5$. The size of the population remains the same, $N = 1$, in the subsequent generation, t_1. What is the probability that the allele frequencies will change ("drift") to $p = 1$ and $q = 0$ at generation t_1? In other words, what is the probability that the population will become fixed for the *A* allele, so that it consists of a single homozygous *A/A* individual? Since $N = 1$, we need to draw just two gametes from the gene pool to form a single individual. The probability of drawing two *A*'s is $p^2 = 0.5^2 = 0.25$. Thus, 25 percent of the time this population will "drift" away from the initial allele frequencies and become fixed for the *A* allele after just one generation.

What happens if we increase the population size to $N = 2$ and the initial gene pool still has $p = q = 0.5$? The allele frequencies will change to $p = 1$ and $q = 0$ in the next generation only if the population consists of two *A/A* individuals. For this to happen, we need to draw four *A* alleles, each with a probability of $p = 0.5$, so the probability that the next generation will have $p = 1$ and $q = 0.0$ is $p^4 = (0.5)^4 = 0.0625$, or just over 6 percent. Thus, an $N = 2$ population is less likely to drift to fixation of the *A* allele than an $N = 1$ population. More generally, the probability of a population drifting to the fixation of the *A* allele in a single generation is p^{2N}, and thus this probability gets progressively smaller as the population size (N) gets larger. Drift is a weaker force in large populations.

Drift means any change in allele frequencies due to sampling error, not just loss or fixation of an allele. In a population of $N = 500$ with two alleles at a frequency of $p = q = 0.5$, there are 500 copies of *A* and 500 copies of *a*. If the next generation has 501 copies of *A* ($p = 0.501$) and

499 copies of *a* ($q = 0.499$), then there has been genetic drift, albeit a very modest level of drift. A general formula for calculating the probability of observing a specific number of copies of an allele in the next generation, given the frequencies in the current generation, is presented in **Box 18-4**.

When drift is operating in a finite population, one can calculate the probabilities of different outcomes, but one cannot accurately predict the specific outcome that will occur. The process is like rolling dice. At any locus, drift can continue from one generation to the next until one allele has become fixed. Also, in a particular population, the frequency of the *A* allele may increase from generation t_0 to t_1 but then decrease from generation t_1 to t_2. Drift does not proceed in a specific direction toward loss or fixation of an allele.

Figures 18-18a and **18-18b** show computer-simulated random trials (rolls of the dice) for six populations of size $N = 10$ and $N = 500$. Each population starts having two alleles at a frequency of $p = q = 0.5$, then the random trials proceed for 30 generations. First, notice the randomness of the process from one generation to the next. For example, the frequency of *A* in the population depicted by the yellow line in Figure 18-18a bounces up and down from one generation to the next, hitting a low of $p = 0.15$ at t_{16} but then rebounding to $p = 0.75$ at t_{30}. Second, whether $N = 10$ or $N = 500$, notice that no two populations have exactly the same trajectory. Drift is a random process, and we are not likely to observe exactly the same outcome with different

---

BOX 18-4 Allele Frequency Changes Under Drift

Consider a population of *N* diploid individuals segregating for two alleles *A* and *a* at the *A* locus with frequencies *p* and *q*, respectively. The population undergoes random mating, and the size of the population remains the same (*N*) in each generation. When the gene pool is sampled to create the next generation, the exact number of copies of the *A* allele that are drawn cannot be strictly predicted because of sampling error. However, the probability that a specific number of copies of *A* will be drawn can be calculated using the binomial formula. Let *k* be a specific number of copies of the *A* allele. The probability of drawing *k* copies is

$$\text{Prob}(k) = \left(\frac{2N!}{k!(2N-k)!} \right) p^k q^{(2N-k)}$$

If we set $N = 10$ and $p = q = 0.5$, then the probability of drawing 10 copies of the *A* allele is

$$\text{Prob}(10) = \left(\frac{20!}{10!(20-10)!} \right) 0.5^{10} 0.5^{(20-10)} = 0.176$$

Thus, only 17.6 percent of the time will the next generation have the same frequency of *A* and *a* as the original generation. We can use this formula to calculate the outcomes for all possible values of *k* and obtain a probability distribution, shown in the figure to the right.

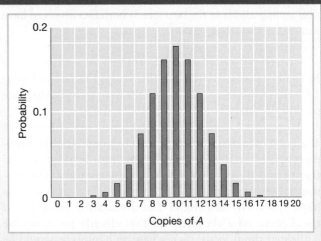

Probability distribution showing the likelihood that different numbers of *A* will be present after one generation.

The most probable single outcome is no drift, with $k = 10$ and a probability of 0.176. However, the other outcomes all involve some drift, and so the probability that the population will experience some drift is 0.824.

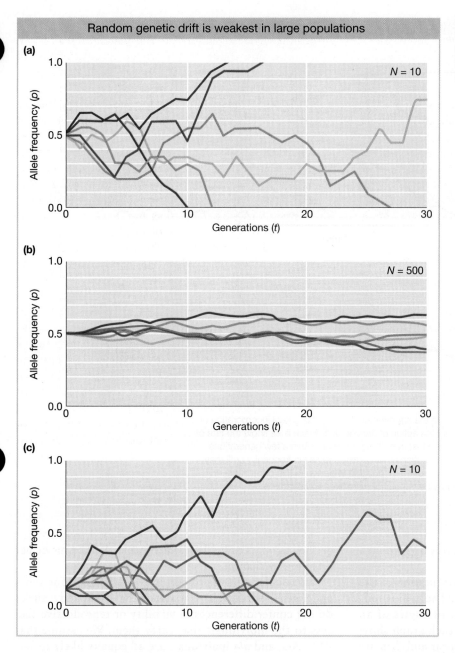

Random genetic drift is weakest in large populations

(a) $N = 10$

(b) $N = 500$

(c) $N = 10$

FIGURE 18-18 Computer simulations of random genetic drift. Each colored line represents a simulated population over 30 generations. (a) $N = 10, p = q = 0.5$. (b) $N = 500, p = q = 0.5$. (c) $N = 10, p = 0.1, q = 0.9$.

populations over many generations except when N is very small. Third, notice that when $N = 10$, the populations became fixed (either $p = 1$ or $p = 0$) before generation 30 in five of the six trials. However, when $N = 500$, the populations retained both alleles in all six trials even after 30 generations.

In addition to population size, the fate of an allele is determined by its frequency in the population. Specifically, the probability that an allele will drift to fixation in a future generation is equal to its frequency in the present generation. An allele that is at a frequency of 0.5 has a 50:50 chance of fixation or loss from the population in a future generation. You can see the effect of allele frequency on the fate of an allele in Figure 18-18c. For 10 populations with an initial frequency of $p = 0.1$, eight populations experienced the loss of the A allele, one its fixation, and one population retained both alleles after 30 generations. That

is very close to the expectation that A will go to fixation 10 percent of the time when $p = 0.1$.

The fact that the frequency of an allele is equal to its probability of fixation means that most newly arising mutations will ultimately be lost from a population because of drift. The initial frequency of a new mutation in the gene pool is

$$\frac{1}{2N}$$

If N is even modestly large, such as 10,000, then the probability that a new mutation will ultimately reach fixation is extremely small: $1/2N = 1/20,000 = 5 \times 10^{-5}$. The probability that a new mutation will ultimately be lost from the population is

$$\frac{2N-1}{2N} = 1 - \frac{1}{2N}$$

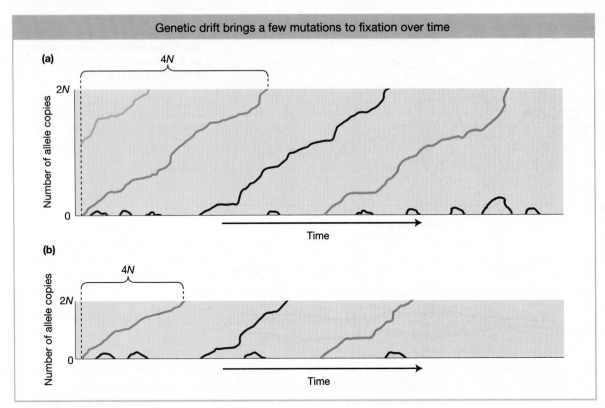

Genetic drift brings a few mutations to fixation over time

FIGURE 18-19 (a) Graphical representation of the appearance, loss, and eventual incorporation of new mutations in a population over time under the action of genetic drift. Black lines show the fate of most new mutations, which appear and then are lost from the population within a few generations. Colored lines show the fate of the few "lucky" mutations that continue to rise in frequency until they reach fixation. (b) A population that is 1/2 the size of the population in part *a*. In this population, 4*N* generations are about 1/2 as long and the lucky new mutations are fixed more rapidly.

which is close to 1.0 in large populations. It is 0.99995 in a population of 10,000.

Figure 18-19a shows a graphical representation of the fate of new mutations in a population. The x-axis represents time and the y-axis the number of copies of an allele. The black lines at the bottom of each graph show the fate of most new mutations. They appear and then are soon lost from the population. The colored lines show the few "lucky" new mutations that become fixed. From population genetic theory, it can be shown that the average time required for a lucky mutation to become fixed is 4*N* generations. Figure 18-19b shows a population that is 1/2 the size of the population in Figure 18-19a. Thus, 4*N* generations are about 1/2 as long and the lucky new mutations are fixed more rapidly.

An important consequence of drift is that slightly deleterious alleles can be brought to fixation or advantageous alleles lost by this random process. Consider a new allele that arises in a population and endows the individual carrying it with a stronger immune system. This individual can pass the advantageous allele to his or her offspring, but those offspring might die before reproducing because of a random event such as being struck by lightning. Or if the individual carrying the favorable allele is heterozygous, he

or she may pass only the less favorable allele to his or her offspring by chance.

In calculating the probabilities of different outcomes under genetic drift, we are assuming that the A and a alleles do not confer differences in viability or reproductive success to the individuals that carry them. We assume that *A/A*, *A/a*, and *a/a* individuals are all equally likely to survive and reproduce. In this case, A and a would be termed **neutral alleles** (or variants) relative to each other. Change in the frequencies of neutral alleles over time due to drift is called **neutral evolution**. The process of neutral evolution is the foundation for the **molecular clock**, the constant rate of substitution of newly arising allelic variants for preexisting ones over long periods (Box 18-5). Neutral evolution is distinct from Darwinian evolution, in which favorable alleles rise in frequency because the individuals that carry them leave more offspring. We will discuss Darwinian evolution in the next section of this chapter and in Chapter 20.

Up until now, we have been considering drift in the context of populations that remain the same size from one generation to the next. In reality, populations often contract or expand in size over time. For example, a new population of much smaller size can suddenly form when a relatively small number of the members of a population migrate to a

BOX 18-5 The Molecular Clock

As species diverge over time, their DNA sequences become increasingly different as mutations arise and become fixed in the population. At what rate do sequences diverge? To answer this question, consider a population at generation t_0. The number of mutations that will appear in generation t_1 is the product of the number of copies of the sequence in the gene pool ($2N$) times the rate at which they mutate (μ), that is, $2N\mu$. If a mutation is neutral, then the probability that it drifts to fixation is $1/2N$. So each generation, $2N\mu$ new mutations enter the gene pool, and $1/2N$ of these will become fixed. The product of these two numbers is the rate (k) at which sequences evolve:

$$k = 2N\mu \times \frac{1}{2N} = \mu$$

The value k is called the substitution rate, and it is equal to the mutation rate for neutral mutations. If the mutation rate remains constant over time, then the substitution rate will "tick" regularly like a clock, the molecular clock.

Consider two species A and B and their common ancestor. Let's define d (divergence) as the number of neutral substitutions at nucleotide sites in the DNA sequence of a gene that have occurred since the divergence of A and B from their ancestor.

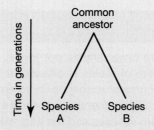

The expected value for d will be the product of the rate (k) at which substitutions occur and two times the time in generations ($2t$) during which substitution accumulated. The 2 is required because there are two lineages leading away from the common ancestor. Thus, we have

$$d = 2tk$$

This equation can be rewritten as

$$t = \frac{d}{2k}$$

showing how we can calculate the time in generations since the divergence of two species if we know d and k. The SNP mutation rate per generation (μ) is known for many groups of organisms (see Table 18-5), and it is the same as the substitution rate (k) for neutral mutations. One can sequence one or more genes from two species and determine the proportion of silent (neutral) nucleotide sites at which they differ and use this proportion as an estimate for d. Thus, one can calculate the time since two sequences (two species) diverged using the molecular clock. Between humans and chimps, there are about 0.018 base differences at synonymous sites in coding sequences. The SNP mutation rate for humans is 3×10^{-8}, and the generation time is about 20 years. Using these values and the equation above, the estimated divergence time for humans and chimps is 6.0 million years ago. These calculations assume that the substitutions are neutral and that the rate of substitution has been constant over time.

---

new location and establish a new population. The migrants, or "founders," of the new population may not carry all the alleles present in the original population, or they may carry the same alleles but at different frequencies. Genetic drift caused by random sampling of the original population to create the new population is known as the **founder effect**. One of many founder events in human history occurred when people crossed the Bering land bridge from Asia to the Americas during the ice age about 15,000 to 30,000 years ago. As a result, genetic diversity among Native Americans is lower than among people in other regions of the world (**Figure 18-20**).

Population size can also change within a single location. A period of one or several consecutive generations of contraction in population size is known as a population **bottleneck**. Bottlenecks occur in natural populations because of environmental fluctuations such as a reduction in the food supply or increase in predation. The gray wolf, American bison, bald eagle, California condor, whooping crane, and many whale species are some familiar examples

of species that have experienced recent bottlenecks because of hunting by humans or encroachment by humans on their habitat. The reduction in population size during a bottleneck increases the level of drift in a population. As explained earlier in the chapter, the level of inbreeding in populations is also dependent on population size. Thus, bottlenecks also cause an increase in the level of inbreeding.

The California condor presents a remarkable example of a bottleneck. This species was once wide ranging but in the 1980s declined to a breeding population of only 14 captive birds. The population is now about 450 individuals, but the average heterozygosity in the genome decreased by 8 percent during the initial bottleneck. Furthermore, a deleterious recessive allele for a lethal form of dwarfism occurs at a frequency of about 9 percent among the surviving animals, presumably as a result of drift from a lower frequency in the pre-bottleneck population. To manage these problems, conservation biologists set up matings of captive animals to minimize further inbreeding and to purge deleterious alleles from the population.

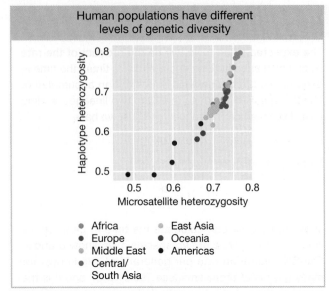

Human populations have different levels of genetic diversity

- Africa
- Europe
- Middle East
- Central/South Asia
- East Asia
- Oceania
- Americas

FIGURE 18-20 Plot of haplotype heterozygosity versus microsatellite heterozygosity shows genetic diversity for different geographical groups of humans. Genetic diversity is lowest for Native Americans because of the founder effect. [*Data from D. F. Conrad et al.*, Nat. Genet. *38, 2006, 1251–1260.*]

Box 18-6 discusses the well-characterized bottleneck that occurred during the domestication of crop species. This

bottleneck explains why our crop plants possess much less genetic diversity than their wild ancestors.

KEY CONCEPT Population size is a key factor affecting genetic variation in populations. Genetic drift is a stronger force in small populations than in large ones. The probability that an allele will become fixed in (or lost from) a population by drift is a function of its frequency in the population and population size. Most new neutral mutations are lost from populations by drift.

Selection

So far, we have considered how new alleles enter a population through mutation and migration and how these alleles can become fixed in (or lost from) a population by random drift. But mutation, migration, and drift cannot explain why organisms seem so well adapted to their environments. They cannot explain **adaptations**, features of an organism's form or physiology that allow it to cope with the environmental conditions under which it lives. To explain the origin of adaptations, Charles Darwin, in 1859 in his historic book *The Origin of Species*, proposed that adaptations arise through the action of another process, which he called "natural selection." In this section, we will

BOX 18-6 The Domestication Bottleneck

Before 10,000 years ago, our ancestors around the world provided for themselves by hunting wild animals and collecting wild plant foods. At about that time, human societies began to develop farming. People took local wild plants and animals and bred them into crop plants and domesticated animals. Some of the major crops that were domesticated at this time include wheat in the Middle East, rice in Asia, sorghum in Africa, and maize in Mexico.

When the first farmers collected seeds from the wild to begin domestication, they drew a sample of the wild gene pool. This sample possessed only a subset of the genetic variation found in the wild. The domesticated populations were put through a bottleneck. As a consequence, crop plants and domesticated animals typically have less genetic variation than their wild progenitors.

Modern scientific plant breeding aimed at crop improvement has created a second bottleneck. By sampling the gene pool of the traditional crop varieties, modern plant breeders have created elite varieties with traits of commercial value such as high yield and suitability for mechanical harvesting and processing. As a consequence, elite or modern varieties have even less genetic variation than traditional varieties.

The loss of genetic variation resulting from the domestication and improvement bottlenecks can pose a threat.

Since there are fewer alleles per locus, crops have a smaller repertoire of alleles at disease-resistance genes and potentially greater susceptibility to emerging pathogens. To reduce this vulnerability, breeders make crosses between modern varieties and the wild relatives (or traditional varieties) to reintroduce critically important alleles into modern crops.

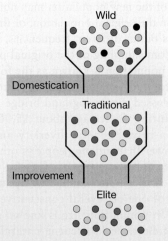

Crop domestication and improvement bottlenecks. Colored dots represent different alleles. [*Data from M. Yamasaki et al.*, Plant Cell *17, 2005, 2859–2872.*]

explore the role of natural selection in modulating genetic variation within populations. Later, in Chapter 20, we will consider the effects of natural selection on the evolution of genes and traits over extended periods.

Let's define **natural selection** as the process by which individuals with certain heritable features are more likely to survive and reproduce than are other individuals that lack these features. As outlined by Darwin, the process works like this. In each generation, more offspring are produced than can survive and reproduce in the environment. Nature has a mechanism (mutation) to generate new heritable forms or variants. Individuals with particular variants of some features are more likely to survive and reproduce. Individuals with features that enhance their ability to survive and reproduce will transmit these features to their offspring. Over time, these features will rise in frequency in the population. Thus, populations will change over time (*evolve*) as the environment (*nature*) favors (*selects*) features that enhance the ability to survive and reproduce. This is Darwin's theory of evolution by means of natural selection.

Darwinian evolution is often described using the phrase *"survival of the fittest."* This phrase can be misleading. An individual who is physically strong, resistant to disease, and lives a long life but has no offspring is not fit in the Darwinian sense. **Darwinian fitness** refers to the ability to survive *and reproduce*. It considers both viability and fecundity. One measure of Darwinian fitness is simply the number of offspring that an individual has. This measure is called **absolute fitness**, and we will symbolize it with an uppercase W. For an individual with no offspring, W equals 0, for an individual with one offspring, W equals 1, for an individual with two offspring, W equals 2, and so forth. W is also the number of alleles at a locus that an individual contributes to the gene pool.

Absolute fitness confounds population size and differences in reproductive success among individuals. Population geneticists are primarily interested in the latter, and so they use a measure called **relative fitness** (symbolized by a lowercase w), which is the fitness of an individual relative to some other individual, usually the most fit individual in the population. If individual X has 2 offspring and the most fit individual, Y, has 10 offspring, then the relative fitness of X is $w = 2/10 = 0.2$. The relative fitness of Y is $w = 10/10 = 1$. For every 10 alleles Y contributes to the next generation, X will contribute 2.

The concept of fitness applies to genotypes as well as to individuals. The absolute fitness for the *A/A* genotype ($W_{A/A}$) is the average number of offspring left by individuals with that genotype. If we know the absolute fitnesses for all genotypes at a locus, we can calculate the relative fitnesses for each of the genotypes.

Let's now look at how allele frequencies can change over time when different genotypes have different fitnesses; that is, when natural selection is at work. Below are the fitnesses and genotype frequencies for the three genotypes at the *A* locus in a population. In this case, *A* is a favored dominant allele since the fitnesses of the *A/A* and *A/a*

individuals are the same and superior to the fitness of the *a/a* individuals. We are assuming that this population follows the Hardy–Weinberg law, with $p = 0.1$ and $q = 0.9$.

	A/A	**A/a**	**a/a**
Average number of offspring (*W*)	10	10	5
Relative fitness (*w*)	1.0	1.0	0.5
Genotype frequency	0.01	0.18	0.81

The relative contribution of each genotype to the gene pool is determined by the product of its fitness and its frequency. The more fit and the higher the frequency of a genotype, the more it contributes.

Genotype	*A/A*	*A/a*	*a/a*	Sum
Relative contribution	1×0.01 $= 0.01$	1×0.18 $= 0.18$	0.5×0.81 $= 0.405$	0.595

The relative contributions do not sum to 1, so we need to rescale them by dividing each by the sum of all three (0.595) to get the expected frequencies of the genotypes that contribute to the gene pool.

Genotype	*A/A*	*A/a*	*a/a*	Sum
Genotype frequencies	0.02	0.30	0.68	1.0

Using these expected genotype frequencies and the Hardy–Weinberg law, we can calculate the frequencies of the alleles in the next generation:

$$p' = 0.02 + \left(\tfrac{1}{2} \times 0.3\right) = 0.17$$

and

$$q' = 0.68 + \left(\tfrac{1}{2} \times 0.3\right) = 0.83$$

The difference between p' and p ($\Delta p = p' - p$) is $0.17 - 0.1 = 0.07$, so we conclude that the *A* allele has climbed 7 percent in one generation due to natural selection. **Box 18-7** presents the standard equations for calculating changes in allele frequencies over time due to natural selection.

We could go through this process recursively, using the allele frequencies from the first generation to calculate those in the second generation, then using those from the second to calculate the third, and so forth. If we then plotted p by time measured in number of generations (t), we would have a picture of the tempo with which allele frequencies change under the force of natural selection. **Figure 18-21** shows such a plot for both a favored dominant and a favored recessive allele. The dominant allele rises rapidly to start but then hits a plateau and only slowly approaches fixation. Once the favored dominant allele is at a high frequency, the unfavored recessive allele occurs mostly in heterozygotes and rarely as homozygotes with reduced fitness, so selection is ineffective at purging it from the population. The favored recessive behaves in the opposite manner—it rises slowly in frequency at first since *a/a* homozygotes with enhanced fitness are rare but proceeds more rapidly to fixation later.

BOX 18-7 The Effect of Selection on Allele Frequencies

Selection causes change in allele frequencies between generations because some genotypes contribute more alleles to the gene pool than others. Let's describe a set of equations to predict gene frequencies in the next generation when selection is operating. The genotype frequencies and absolute fitnesses are symbolized as follows:

genotype	A/A	A/a	a/a
frequency	p^2	$2pq$	q^2
absolute fitness	$W_{A/A}$	$W_{A/a}$	$W_{a/a}$

The average number of alleles contributed by individuals of a given genotype is the frequency of the genotype times the absolute fitness. If N is the population size, the total number of alleles contributed by all individuals of a given genotype is N multiplied by the average number of alleles contributed by individuals of a given genotype:

average number	$p^2 W_{A/A}$	$2pq W_{A/a}$	$q^2 W_{a/a}$
total number	$N(p^2)W_{A/A}$	$N(2pq)W_{A/a}$	$N(q^2)W_{a/a}$

Thus, the gene pool will have

number of A alleles $= N(p^2)W_{A/A} + \frac{1}{2}[N(2pq)W_{A/a}]$

number of a alleles $= N(q^2)W_{a/a} + \frac{1}{2}[N(2pq)W_{A/a}]$

The mean fitness of the population is

$$\overline{W} = p^2 W_{A/A} + 2pq W_{A/a} + q^2 W_{a/a}$$

which is the average number of alleles contributed to the gene pool by an individual. $N\overline{W}$ is the total number of alleles in the gene pool.

We can now calculate the proportion of A alleles in the gene pool for the next generation as

$$p' = \frac{Np^2 W_{A/A} + Npq W_{A/a}}{N\overline{W}}$$

This equation reduces to

$$p' = p\frac{pW_{A/A} + qW_{A/a}}{\overline{W}}$$

Notice the expression $pW_{A/A} + qW_{A/a}$. This is called the allelic fitness or mean fitness of A alleles (W_A):

$$W_A = pW_{A/A} + qW_{A/a}$$

From the Hardy–Weinberg law, we know that a proportion p of all A alleles are present in homozygotes with another A, in which case they have a fitness of $W_{A/A}$, whereas a proportion q of all the A alleles are present in heterozygotes with a and have a fitness of $W_{A/a}$. Substituting W_A into the equation at lower left, we obtain

$$p' = p\frac{W_A}{\overline{W}}$$

This equation can be used to calculate the frequency of A in the next generation and used recursively to follow the change in p over time.

Although we derived these formulas using absolute fitness, generally we are not interested in population size, so we use forms of these equations with relative fitness:

$$\overline{w} = p^2 w_{A/A} + 2pq w_{A/a} + q^2 w_{a/a}$$
$$w_A = p w_{A/A} + q w_{A/a}$$
$$p' = p\frac{w_A}{\overline{w}}$$

Finally, we can express change in allele frequency between generations as

$$\Delta p = p' - p = p\frac{w_A}{\overline{w}} - p$$
$$= \frac{p(w_A - \overline{w})}{\overline{w}}$$

But $\overline{w}$, the mean relative fitness of the population, is the average of w_A and w_a, which are the allelic fitnesses of the A and a alleles, respectively:

$$\overline{w} = pw_A + qw_a$$

Substituting this expression for $\overline{w}$ in the formula for Δp and remembering that $q = 1 - p$, we obtain

$$\Delta p = \frac{pq(w_A - w_a)}{\overline{w}}$$

Since the heterozygous class has reduced fitness, the unfavored dominant allele can eventually be purged from the population.

Forms of selection

Natural selection can operate in several different ways. **Directional selection**, which we have been discussing, moves the frequency of an allele in one direction until it reaches fixation or loss. Directional selection can be either *positive* or *purifying*. **Positive selection** works to bring a new, favorable mutation or allele to a higher frequency. This type of selection is at work when new adaptations evolve. A *selective sweep* occurs when a favorable allele reaches fixation. Directional selection can also work to remove deleterious mutations from the population. This form of selection is called **purifying selection**, and it prevents existing adaptive features from being degraded or lost. Selection does not always proceed directionally until loss or fixation of an allele. If the heterozygous class has a higher fitness than either of the homozygous classes, then natural selection will favor the maintenance of both alleles

Allele frequencies change under the force of natural selection

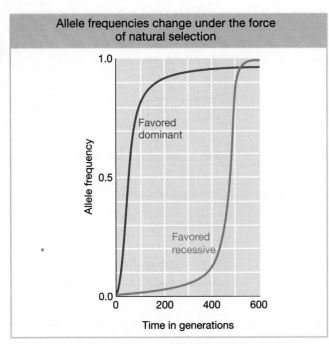

FIGURE 18-21 Change in allele frequency of a favored dominant allele (red) and a favored recessive allele (blue) driven by natural selection over the course of 600 generations.

in the population. In this case, the locus is under **balancing selection**, and natural selection will move the population to an equilibrium point at which both alleles are maintained in the population (see Chapter 20).

The different forms of selection each leave a distinct signature on the DNA sequence near the target locus in a population. For example, positive selection can be detected in DNA sequences by its effects on genetic diversity and linkage disequilibrium. Figure 18-22 shows schematic haplotypes before and after an episode of positive selection. In the panel showing the haplotypes before selection, the bracketed region has many polymorphisms and multiple haplotypes. However, after selection, there is only a single

Positive selection leaves a distinct signature

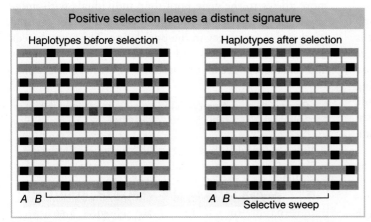

FIGURE 18-22 Schematic of haplotypes found in a population before and after a favored allele (red) is swept to fixation. There are 11 loci altogether. There are two alleles (red and gray) at the locus that was the target of selection. There are two alleles (black and gray) at each locus that is linked to the target locus. After selection, the target and some neighboring sites have all been swept to fixation.

In Europe, a selective sweep caused a loss of all diversity at the SLC24A5 locus

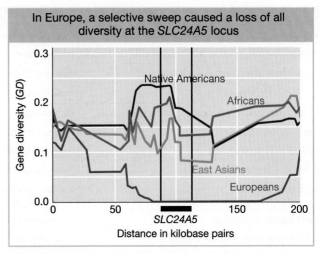

FIGURE 18-23 Gene diversity in human continental groups along a 2-million-bp segment of human chromosome 15 surrounding the *SLC24A5* gene. [*Data from Human Genome Diversity Project, www.hagsc.org/hgdp.*]

haplotype in this region and thus no polymorphism. When selection is applied to the target site (shown in red), the target and neighboring sites can all be swept to fixation before recombination breaks up the haplotype in which the favorable mutation first occurred. The result is lower diversity and higher LD near the target. As distance from the target increases, there is more opportunity for recombination, and so diversity goes gradually back up.

Figure 18-23 shows the pattern of diversity in the region surrounding the *SLC24A5* gene in humans. This gene influences the deposition of melanin in the skin. When people migrated from Africa to Europe, a selective sweep at *SLC24A5* caused a loss of all diversity at this locus (gene diversity $\cong 0.0$). As a consequence, there is a single allele and a single haplotype at this locus in Europe. The single allele that was selected for in Europe produces lighter skin color. Moving away from the gene in either direction, gene diversity rises in European populations since recombination disrupted the linkage disequilibrium between *SLC24A5* and more distance sites. Light skin may be adaptive in northern latitudes. People are able to synthesize vitamin D, but to do so they need to absorb UV radiation through the skin. In the equatorial latitudes, people are exposed to high levels of UV light and can synthesize vitamin D even with heavily pigmented skin. At greater distance from the equator, people are exposed to less UV light, and lighter skin color may facilitate vitamin D synthesis at these latitudes.

Table 18-6 lists a few of the genes that show evidence for natural selection in modern humans. These genes fall into a few basic categories. One group strengthens resistance to pathogens. The genes *G6PD*, *FY*[null], and *Hb* (*hemoglobin β*, the sickle-cell-anemia gene) all help humans adapt to the threat of malaria. Figure 18-11b shows that the frequency of *FY*[null] is highest in central Africa. Central Africa also has the highest prevalence of malaria, suggesting that selection has driven *FY*[null] to

TABLE 18-6   Some Genes Showing Evidence for Natural Selection in Specific Human Populations

Gene	Presumed Trait	Population
EDA2R (ectodysplasin A2 receptor)	Male pattern baldness	Europeans
EDAR (ectodysplasin A receptor)	Hair morphology	East Asians
FY^{null} (Duffy antigen)	Resistance to malaria	Africans
G6PD (glucose-6-phosphate dehydrogenase)	Resistance to malaria	Africans
Hb (hemoglobin β)	Resistance to malaria	Africans
KITLG (KIT ligand)	Skin pigmentation	East Asians and Europeans
LARGE (glycosyltransferase)	Resistance to Lassa fever	Africans
LCT (lactase)	Lactase persistence; ability to digest milk sugar as an adult	Africans, Europeans
LPR (leptin receptor)	Processing of dietary fats	East Asians
MC1R (melanocortin receptor)	Hair and skin pigmentation	East Asians
MHC (major histocompatibility complex)	Infectious disease resistance	Multiple populations
OCA2 (oculocutaneous albinism)	Skin pigmentation and eye color	Europeans
PPARD (peroxisome proliferator-activated receptor delta)	Processing of dietary fats	Europeans
SI (sucrase-isomaltase)	Sucrose metabolism	East Asians
SLC24A5 (solute carrier family 24)	Skin pigmentation	Europeans and West Asians
TYRP1 (tyrosinase-related protein 1)	Skin pigmentation	Europeans

Sources: P. C. Sabeti et al., *Science* 312, 2006, 1614–1620; P. C. Sabeti et al., *Nature* 449, 2007, 913–919; B. F. Voight et al., *PLoS Biology* 4, 2006, 446–458; J. K. Pickrell et al., *Genome Research* 19, 2009, 826–837.

its highest frequency in the region where selection pressure is greatest. In the 1990s, medical geneticists discovered an allele of the gene *CCR5* (*chemokine receptor 5*) that provides resistance to AIDS. This allele is now a target of natural selection. As long as there are pathogens, natural selection will continue to operate in human populations.

Another group of selected genes in Table 18-6 adapts people to regional diets. Before 10,000 years ago, all humans were hunter–gatherers. More recently, most humans switched to agricultural foods, but there are regional differences in diet. In northern Europe and parts of Africa, milk products are a substantial part of the diet. In most populations, the lactase enzyme for digesting milk sugar (lactose) is expressed during childhood but is switched off in adults. In parts of Europe and Africa where adults drink milk, however, special alleles of the *lactase* gene that continue to express the lactase enzyme during adulthood have risen in frequency due to natural selection. Finally, Table 18-6 includes some genes for physiological adaptations to climate. Among these are the genes for skin pigmentation such as *SLC24A5*, discussed earlier.

Whereas directional selection causes a loss of genetic variation in the region surrounding the target locus, balancing selection can prevent the loss of diversity by random genetic drift, leading to regions of unusually high genetic diversity in the genome. One region of high genetic diversity surrounds the major histocompatibility complex (MHC) gene complex on chromosome 6. **Figure 18-24** shows a distinct spike in the number of SNPs at the MHC. This complex includes the human leukocyte antigen (HLA)

genes, which are involved in immune system recognition of (and response to) pathogens. Balancing selection is one hypothesis proposed to explain the high diversity observed at the MHC. Since heterozygotes have two alleles, they may be resistant to a greater repertoire of pathogen types, giving heterozygotes a fitness advantage.

Finally, selection can be imposed by an agent other than nature. Humans have imposed selection in the process of domesticating and improving cultivated plants and animals. This form of selection is called **artificial selection**. In this case, individuals with traits that humans prefer contribute more alleles to the gene pool than individuals with unfavored traits. Over time, the alleles that confer the favored traits rise in frequency in the population. The many breeds of dogs and dairy cows and varieties of garden vegetables and cereal crops are all the products of artificial selection.

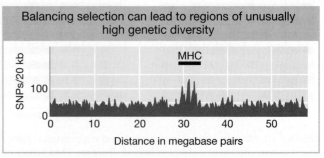

FIGURE 18-24 Number of segregating sites (S) or SNPs in 20-kilobase-pair windows along the short arm of human chromosome 6. There is a spike of high diversity at the MHC locus. [*Data from the 1000 Genomes Project, www.internationalgenome.org.*]

Balance between mutation and drift

We have considered the forces that regulate variation in populations individually. Let's now consider the opposing effects of mutation and drift, the former adding variation and the latter removing it from populations. When these two forces are in balance, a population can reach an equilibrium at which the loss and gain of variation are equal. We will use heterozygosity (H) as a measure of variation. Remember that H will be near 0 when a population is near fixation for a single allele (low variation), and H approaches 1 when there are many alleles of equal frequency (high variation).

Let's use H with a "hat," $\hat{H}$, as the symbol for the equilibrium value of H. To find $\hat{H}$, we start with two mathematical equations: one equation that relates change in H to population size (drift) and another equation that relates change in H to the mutation rate. We can then set these equations equal to each other and solve for $\hat{H}$.

First, we need an equation for the decline in variation (H) between generations as a function of population size (drift). We developed such an equation in Box 18-3 when discussing inbreeding:

$$H' = \left(1 - \frac{1}{2N}\right)H$$

This equation applies to the effects of drift as well as those of inbreeding. From this equation, it follows that the change in H between generations due to drift is

$$\Delta H = H - H' = \frac{1}{2N}H$$

Second, we need an equation for the increase in variation, as measured by H, between generations due to mutation. Any new mutation will increase heterozygosity at a rate proportional to the frequency of homozygotes in the population $(1 - H)$ times the rate at which mutation converts them to heterozygotes (2μ). (The 2 is necessary because there are two alleles that could mutate in a diploid.) Thus, the change in H between generations due to mutation is

$$\Delta H = 2\mu(1 - H)$$

When the population reaches an equilibrium, the loss of heterozygosity by drift will equal the gain from mutation. Thus, we have

$$\frac{1}{2N}\hat{H} = 2\mu(1 - \hat{H})$$

which can be rewritten as

$$\hat{H} = \frac{4N\mu}{4N\mu + 1}$$

This equation gives the equilibrium value of $\hat{H}$ when the loss by drift and gain by mutation are balanced. This equation applies only to neutral variation; that is, we are assuming selection is not at work. We are also assuming that each new mutation yields a unique allele.

Expressions such as this are useful when we have estimates for two of the variables and would like to know the third. For example, nucleotide diversity (H at the nucleotide level) for noncoding sequences, which are largely neutral, is about 0.0013 in humans, and μ for humans is 3×10^{-8} (see Table 18-5). Using these values and solving the equation above for N yields an estimate of the human population size of 10,498 humans. This estimate is far below the 7.6 billion of us alive today. What's up? This is an estimate for the equilibrium value. Modern humans are a young group, only about 150,000 years old. Over the last 150,000 years, our population has grown dramatically as we filled the globe, but mutation is a slow process, so genetic diversity has not kept up and the human population is not at equilibrium. The population size of 10,498 represents an estimate of our historical size, or how many breeding members there were about 150,000 years ago.

Balance between mutation and selection

Allelic frequencies may also reach a stable equilibrium when the introduction of new alleles by repeated mutation is balanced by their removal by natural selection. This balance probably explains the persistence of genetic diseases as low-level polymorphisms in human populations. New deleterious mutations are constantly arising spontaneously. These mutations may be completely recessive or partly dominant. Selection removes them from the population, but there is an equilibrium between their appearance and removal.

Let's begin with the simplest case—the frequency for a deleterious recessive when an equilibrium is reached between mutation and selection. For this purpose, it is convenient to express the relative fitnesses in terms of the **selection coefficient** (s), which is the selective disadvantage of (or loss of fitness in) a genotype:

$W_{A/A}$	$W_{A/a}$	$w_{a/a}$
1	1	$1 - s$

Then, as shown in **Box 18-8**, the equation for equilibrium frequency of a deleterious recessive allele is

$$\hat{q} = \sqrt{\frac{\mu}{s}}$$

This equation shows that the frequency at equilibrium depends on the ratio μ/s. When the mutation rate for $A \rightarrow a$ gets larger and the selective disadvantage smaller, then the equilibrium frequency ($\hat{q}$) of a recessive deleterious allele will rise. As an example, a recessive lethal allele ($s = 1$) that arises by mutation from the wild-type allele at the rate of $\mu = 10^{-6}$ will have an equilibrium frequency of 10^{-3}.

BOX 18-8 The Balance Between Selection and Mutation

If we let q be the frequency of the deleterious allele a and $p = 1 - q$ be the frequency of the normal allele A, then the change in allele frequency due to the mutation rate μ is

$$\Delta q_{mut} = \mu p$$

A simple way to express the fitnesses of the genotypes in the case of a recessive deleterious allele a is $w_{A/A} = w_{A/a} = 1.0$ and $w_{a/a} = 1 - s$, where s, the selection coefficient, is the loss of fitness in the recessive homozygotes. We now can substitute these fitnesses in our general expression for allele frequency change (see Box 18-7) and obtain

$$\Delta q_{sel} = \frac{-pq(sq)}{1 - sq^2} = \frac{-spq^2}{1 - sq^2}$$

Equilibrium means that the increase in the allele frequency due to mutation exactly balances the decrease in the allele frequency due to selection, so

$$\mu \hat{p} = \frac{-s\hat{p}\hat{q}^2}{1 - s\hat{q}^2}$$

The frequency of a recessive deleterious allele ($\hat{q}$) at equilibrium will be quite small, so $1 - s\hat{q}^2 \approx 1$, and we have

$$\mu \hat{p} = -s\hat{p}\hat{q}^2$$

$$\hat{q} = \sqrt{\frac{\mu}{s}}$$

at equilibrium.

Let's consider the equilibrium between selection and mutation for the slightly more complicated case of a partially dominant deleterious allele—that is, an allele with some deleterious effect in heterozygotes as well as its effect in homozygotes. We will define h as the degree of dominance of the deleterious allele. When h is 1, the deleterious allele is fully dominant, and when h is 0, the deleterious allele is fully recessive. Then, the fitnesses are

$W_{A/A}$	$W_{A/a}$	$w_{a/a}$
1	$1 - hs$	$1 - s$

where a is a partially dominant deleterious allele. A derivation similar to the one in Box 18-8 gives us

$$\hat{q} = \frac{\mu}{hs}$$

Here is an example. If $\mu = 10^{-6}$ and the lethal allele is not totally recessive but causes a 5 percent reduction in fitness in heterozygotes ($s = 1.0$, $h = 0.05$), then

$$\hat{q} = \frac{\mu}{hs} = 2 \times 10^{-5}$$

This result is smaller by two orders of magnitude than the equilibrium frequency for the purely recessive case described earlier. In general, then, we can expect deleterious, completely recessive alleles to have frequencies much higher than those of partly dominant alleles because the recessive alleles are protected in heterozygotes.

KEY CONCEPTS The amount of genetic variation in populations represents a balance between opposing forces: mutation and migration, which add new variation, versus drift and selection, which remove variation. Balancing selection also serves to maintain variation in populations. As a result of these processes, allele frequencies can reach equilibrium values, explaining why populations often maintain high levels of genetic variation.

18.6 BIOLOGICAL AND SOCIAL APPLICATIONS

LO 18.8 Explain how population genetics informs many issues facing modern societies.

Just as the principles of physics guide engineers who design bridges and jet airliners, so the principles of population genetics touch all of our lives in many, if unseen, ways. In Chapter 19, you'll see how population genetics figures prominently in the search for genes that contribute to disease risk in people, using concepts such as linkage disequilibrium, described in this chapter. In this final section of the chapter, we will examine three other areas in which the principles of population genetics are being to applied to issues affecting modern societies.

Conservation genetics

Conservation biologists attempting to save endangered wild species, and zookeepers attempting to maintain small populations of captive animals, often perform population genetic analyses. Earlier in this chapter, we discussed how a genetic bottleneck caused a loss of genetic variation in the California condor and an increase in the frequency of a lethal form of dwarfism. Bottlenecks may also increase the level of inbreeding in a population, perhaps leading to a decline in fitness through inbreeding depression. The issue is complex, however, because inbreeding is not always associated with a decline in fitness. Inbreeding can sometimes help purge deleterious recessive alleles from a population. Purifying selection is more effective at eliminating deleterious recessive alleles with inbreeding since the homozygous recessive class becomes more frequent in inbred populations. Thus, conservation biologists have debated whether they should attempt to maximize genetic diversity and minimize inbreeding or deliberately subject zoo populations to inbreeding with the goal of purging deleterious alleles.

To help address this question, researchers looked for evidence of successful purging among zoo populations. Let's define inbreeding depression as delta (δ)

$$\delta = 1 - \frac{w_f}{w_0}$$

where w_f is the fitness of inbred individuals and w_0 the fitness of non-inbred individuals. The value of δ will be positive when there is a decline in fitness with inbreeding but negative when fitness improves with inbreeding. Researchers calculated δ for 119 zoo populations, including 88 species, and they found evidence that purging had improved fitness (negative values for δ) in 14 populations. Still, it is not clear that deliberate inbreeding of zoo animals is advisable. For one thing, although 14 of the 119 populations improved, the majority of the populations declined in fitness when inbred. Thus, if one starts with a small zoo population and purposely inbreeds the animals, a decline in fitness is the most likely outcome.

Calculating disease risks

In Chapter 2, we saw how alleles for genetic disorders could be traced in pedigrees and we discussed how to calculate the risk that a couple will have a child who inherits such a disorder. Population genetic principles allow us to extend this type of analysis. We will consider two examples.

The disease allele for cystic fibrosis (CF) occurs at a frequency of about 0.025 in Caucasians. In the pedigree for a Caucasian family below, individual II-2 has a first cousin (II-1) with cystic fibrosis. II-2 is married to an unrelated Caucasian (II-3), and they are planning to have a child. What is the chance that the child (III-1) will have cystic fibrosis?

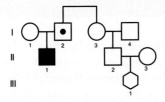

One of II-2's maternal grandparents must have been a carrier. We begin by calculating the probability that III-1 will inherit this cystic fibrosis allele from this grandparent through his father, II-2, using methods already familiar from Chapter 2. The probability that this grandparent transmitted the disease allele to I-3 is 1/2. The probability that I-3 transmitted it to II-2 and that II-2 will transmit it to III-1 are also both 1/2. So the probability that III-1 inherits the same CF allele as II-2 is $\left(\frac{1}{2}\right)^3$, or 1/8. We now extend the calculation to determine the probability that III-1 could inherit the cystic fibrosis allele from his mother, II-3. Individual II-3 does not have CF, but we are not sure whether or not she is a carrier. If the frequency (q) of the disease allele in the population is 0.025, then the probability

that an unaffected individual such as II-3 is a carrier is $2pq/(1 - q^2) = 0.049$. If II-3 is a carrier, then there is a $\frac{1}{2}$ chance she will transmit the disease allele to III-1. These are all independent probabilities, so we can use the product rule. The probability that III-1 will have cystic fibrosis is

$$\frac{1}{8} \times \frac{1}{2} \times 0.049 = 0.003$$

The frequency of cystic fibrosis among Caucasians is $q^2 = (0.025)^2 = 0.000625$. These calculations tell us that individuals who have a first cousin with cystic fibrosis have a $0.003 \div 0.000625 = 4.9$-fold higher risk of having a child with the disease than members of the general population.

Here is another application of population genetics for assessing disease risk. Sickle-cell anemia, a recessive disease, has a frequency of about 0.25 percent, or 1 in 400, among African Americans (see Chapter 5). Applying the Hardy–Weinberg law, we estimate the frequency of the disease allele (Hb^S) as 0.05. What would be the expected frequency of this disease among the offspring of African Americans who are first cousins? Using the method described in Box 18-2, we calculate that the inbreeding coefficient (F) for the offspring of first-cousin marriages is 1/16. In the earlier section on inbreeding, we saw that the frequency of the homozygotes when there is inbreeding is increased, as shown by this equation:

$$f_{a/a} = q^2 + pqF$$

Using this equation, we obtain

$$f(Hb^S/Hb^S) = (0.05)^2 + (0.05 \times 0.95)(1/16) = 0.0055$$

This represents a 2.2-fold increase in the risk of having a child with the disease for first-cousin marriages compared to that in a marriage between unrelated individuals.

DNA forensics

Criminals can leave DNA evidence at the scene of a crime in the form of blood, semen, hairs, or even buccal cells from saliva on a cigarette butt. The polymerase chain reaction (PCR) enables forensic scientists to amplify very tiny amounts of DNA and determine the genotype of the individual who left the specimen. If the DNA found at the crime scene matches that of the suspect, then they "may be" the same individual. The key phrase here is "may be," and this is where population genetics comes into play. Let's see how this works.

Consider two microsatellite loci, each with multiple alleles: $A_1, A_2, \ldots A_n$ and $B_1, B_2, \ldots B_n$. Forensic scientists determine that a DNA specimen from a crime scene and the suspect are both $A_3/A_8\ B_1/B_7$. They have determined that there is a "match" between the evidence and the suspect. Does the match prove that the DNA evidence came from the suspect? Does it prove that that the suspect was at the crime scene?

What population geneticists do with this type of evidence is to test a specific hypothesis: *The evidence came from someone other than the suspect.* This is what statisticians call the "null hypothesis," or the hypothesis that is considered true unless the evidence shows that it is very unlikely (see Chapter 4). To perform the test, we calculate the probability of observing a match between the evidence and the suspect, given that the suspect and the person who left the evidence are different individuals. Symbolically, we write

$$\text{Prob(match} \mid \text{different individuals)}$$

where "|" means "given." If this probability is very small, then we can reject the null hypothesis and argue in favor of an alternative hypothesis: *The evidence was left by the suspect.* We never formally prove the suspect left the evidence since there could be alternative hypotheses such as *The evidence was left by the suspect's identical twin.*

To calculate the probability of observing a match between the evidence and the suspect if the evidence is from a different individual, we need to know the frequencies of the microsatellite alleles in the population.

A_4	0.03
A_6	0.05
B_1	0.01
B_7	0.12

Prob(match | different individuals) is the same as the probability that the evidence came from a randomly chosen individual. We can calculate this probability using the allele frequencies above. First, we will assume that the Hardy–Weinberg law applies and calculate the probability of being A_4/A_6 at the first locus and B_1/B_7 at the second:

$$\text{Prob}(A_4/A_6) = 2pq = 2 \times 0.03 \times 0.05 = 0.003$$
$$\text{Prob}(B_1/B_7) = 2 \times 0.01 \times 0.12 = 0.0024$$

To combine these two probabilities, we need to make one more assumption. We need to assume that the two loci are *independent;* that is, that the loci are at *linkage equilibrium.* By making this assumption, we can apply the *product rule* for independent events (see Chapter 2) and determine that

$$\text{Prob(match} \mid \text{different individuals)}$$
$$= \text{Prob}(A_4/A_6) \times \text{Prob}(B_1/B_7) = 7.2 \times 10^{-6}$$

Thus, the probability under the null hypothesis that the evidence came from someone other than the suspect is 7.2×10^{-6}, or about 7 in a million. That is a small probability, and so the null hypothesis seems unlikely in this case. However, if Prob(match | different individuals) were 0.1, then 10 percent of the population would be a match and could have left the evidence. In that case, we would not want to reject the null hypothesis.

Two microsatellites do not provide very much power to discriminate, so the FBI in the United States uses a set of 20 microsatellites. Microsatellite loci typically have large numbers of alleles (10 to 20 or more); therefore, the number of possible genotypes based on 20 microsatellites is astronomically large. With 10 alleles per locus, there are 55 possible genotypes at each locus and 55^{20}, or 6.4×10^{34}, possible multilocus genotypes for 20 loci. The FBI has also assembled a database called CODIS (Combined DNA Index System) that contains the frequencies of different alleles at these loci in the population, including data specific to different ethnic groups and regions of the country.

SUMMARY

Population genetics seeks to understand the laws that govern the amount of genetic variation within populations and changes in genetic variation over time. The concept of the gene pool provides a model for thinking about the transmission of genetic variation from one generation to the next for an entire population. Basic population genetic theory starts with an idealized population that is infinite in size and in which mating is random. In such a population, the Hardy–Weinberg law defines the relationship between allele frequencies in the gene pool and genotype frequencies in the population.

Real populations deviate to various degrees from the Hardy–Weinberg model. One source of deviation comes in the form of nonrandom or assortative mating. If individuals preferentially mate with others who share a similar phenotype, then there will be an excess of homozygotes at genes controlling that phenotype compared to Hardy–Weinberg expectations. When individuals mate more frequently with relatives than expected by chance, then there will be an excess of homozygous genotypes throughout the entire genome and the population becomes inbred. Even when local populations of a species conform to Hardy–Weinberg expectations, those populations are apt to be isolated from other populations at distant locations. Thus, a species often consists of a series of genetically distinct subpopulations; that is, species show population genetic structure.

Several forces can add new variation to a population or remove existing variation from it. Mutation is the ultimate source of all genetic variation. Population geneticists have determined reasonably precise estimates of the rate at which new mutations arise in populations. Migration can also bring new variation into a population. Migration results in some individuals who are genetically admixed, having ancestry from multiple populations. Genetic recombination can also add variation to populations by recombining alleles into new haplotypes.

Two forces control the fate of genetic variation in populations. First, genetic drift is a random force that can lead to the loss or fixation of an allele as a result of sampling error in finite populations. Drift is a strong force in small populations and a weak force in large ones. Second, natural selection drives changes in allele frequencies in populations over time. Alleles that enhance the fitness of the individuals that carry them will rise in frequency and can become fixed, while deleterious alleles that reduce fitness will be purged from the population.

The fundamental goal of population genetics is to understand the relative contributions made by mating systems, mutation, migration, recombination, drift, and natural selection to the amount and distribution of genetic variation in populations. In this chapter, we have seen how research in population genetics has both developed the basic theory and collected a vast amount of data to achieve this goal. Our understanding of the population genetics of our own species is remarkably detailed.

Finally, the methods and results of population genetics both inform us about the evolutionary process and have practical applications to issues facing modern societies. Population genetic theory and analyses play important roles in the management of endangered species, the identification of perpetrators of crimes, plant and animal breeding, and assessing the risks that a couple will have a child with a disease condition.

KEY TERMS

absolute fitness (p. 629)
adaptation (p. 628)
allele frequency (p. 610)
artificial selection (p. 632)
balancing selection (p. 631)
bottleneck (p. 627)
common SNP (p. 605)
Darwinian fitness (p. 629)
directional selection (p. 630)
disassortative mating (p. 613)
discovery panel (p. 605)
fixed (p. 618)
founder effect (p. 627)
gene diversity (GD) (p. 620)
gene flow (p. 621)
gene pool (p. 609)
genetic admixture (p. 622)
genotype frequency (p. 610)
haplotype (p. 606)

haplotype network (p. 607)
Hardy–Weinberg equilibrium (p. 611)
Hardy–Weinberg law (p. 610)
heterozygosity (H) (p. 620)
identical by descent (IBD) (p. 615)
inbreeding (p. 614)
inbreeding coefficient (F) (p. 615)
inbreeding depression (p. 615)
isolation by distance (p. 614)
linkage disequilibrium (LD) (p. 623)
linkage equilibrium (p. 622)
locus (p. 604)
microsatellite (p. 605)
migration (p. 621)
molecular clock (p. 626)
mutation rate (μ) (p. 621)
natural selection (p. 629)
negative assortative mating (p. 613)

neutral allele (p. 626)
neutral evolution (p. 626)
nucleotide diversity (p. 620)
number of haplotypes (NH) (p. 620)
population (p. 604)
population genetics (p. 604)
population structure (p. 614)
positive assortative mating (p. 613)
positive selection (p. 630)
purifying selection (p. 630)
random genetic drift (p. 623)
rare SNP (p. 605)
relative fitness (p. 629)
segregating sites (S) (p. 620)
selection coefficient (s) (p. 633)
single nucleotide polymorphism (SNP) (p. 604)

SOLVED PROBLEMS

SOLVED PROBLEM 1

About 70 percent of all Caucasians can taste the chemical phenylthiocarbamide, and the remainder cannot. The ability to taste this chemical is determined by the dominant allele T, and the inability to taste is determined by the recessive allele t. If the population is assumed to be in Hardy–Weinberg equilibrium, what are the genotype and allele frequencies in this population?

SOLUTION

Because 70 percent are tasters (T/T and T/t), 30 percent must be nontasters (t/t). This homozygous recessive frequency is equal to q^2; so, to obtain q, we simply take the square root of 0.30:

$$q = \sqrt{0.30} = 0.55$$

Because $p + q = 1$, we can write $p = 1 - q = 1 - 0.55 = 0.45$. Now we can calculate

$$p^2 = (0.45)^2 = 0.20, \quad \text{the frequency of } T/T$$
$$2pq = 2 \times 0.45 \times 0.55 = 0.50, \quad \text{the frequency of } T/t$$
$$q^2 = 0.3, \quad \text{the frequency of } t/t$$

SOLVED PROBLEM 2

In a large experimental *Drosophila* population, the relative fitness of a recessive phenotype is calculated to be 0.90, and the mutation rate to the recessive allele is 5×10^{-5}. If the population is allowed to come to equilibrium, what allele frequencies can be predicted?

SOLUTION

Here, mutation and selection are working in opposite directions, and so an equilibrium is predicted. Such an equilibrium is described by the formula

$$\hat{q} = \sqrt{\frac{\mu}{s}}$$

In the present question,

$$\mu = 5 \times 10^{-5} \text{ and } s = 1 - w = 1 - 0.9 = 0.1$$

Hence,

$$\hat{q} = \sqrt{\frac{5 \times 10^{-5}}{0.1}} = 0.022$$

$$\hat{p} = 1 - 0.022 = 0.978$$

SOLVED PROBLEM 3

A colony of 50 horned puffins (*Fratercula corniculata*) is established at a zoo and maintained there for 30 generations.

a. If the inbreeding coefficient of the founding members was zero ($F = 0.0$), what is the expected inbreeding co-efficient for this population at present?

b. For a deleterious disease allele with a frequency of 0.001 in the wild, what is the predicted frequency of homozygous affected birds in the wild and in the zoo population at present?

SOLUTION

a. In Box 18-3, we saw that inbreeding will increase as a function of population size (N) over time (t) as measured in generations according to the following equation:

$$F_t = 1 - \left(1 - \frac{1}{2N}\right)^t (1 - F_0)$$

Substituting in $N = 50$, $t = 30$, and $F_0 = 0$, we obtain

$$F_{30} = 1 - \left(1 - \frac{1}{2 \times 50}\right)^{30} (1 - 0) = 0.26$$

b. If the frequency of a recessive disease allele (q) in the wild is 0.001, then by applying the Hardy–Weinberg law, we predict that the frequency of homozygous affected individuals in the wild will be $q^2 = 10^{-6}$. For the zoo population, the frequency of homozygotes will be higher because of inbreeding according to the following equation:

$$f_{a/a} = q^2 + pqF$$

Substituting in $q = 0.001$, $p = 0.999$, and $F = 0.26$, we obtain

$$f_{a/a} = 10^{-6} + (0.001 \times 0.999 \times 0.26) = 2.61 \times 10^{-4}$$

The ratio of 2.61×10^{-4} to 10^{-6} shows us that there is a 261-fold increase in the expected frequency of affected individuals in the current zoo population compared to the ancestral wild population.

SOLVED PROBLEM 4

At a criminal trial, the prosecutor presents genotypes for three microsatellite loci from the FBI CODIS set. He reports that a DNA sample from the crime scene and one from the suspect both have the genotype FGA_1/FGA_4, $TPOX_1/TPOX_3$, VWA_2/VWA_7 at these three microsatellites. He also presents the allelic frequencies for the general population to which the suspect belongs (see the table that follows). What is the probability that the genotype of the DNA evidence would match that of the suspect, given that the person who committed the crime and the suspect are different individuals? What assumptions do you make when calculating this probability?

Allele	Frequency
FGA_1	0.30
FGA_4	0.26
$TPOX_1$	0.32
$TPOX_3$	0.65
VWA_2	0.23
VWA_7	0.59

SOLUTION

The probability that the genotype of the DNA evidence matches that of the suspect given that the person who committed the crime and the suspect are different individuals is the same as the probability that a randomly chosen member of the population would have the same genotype as the DNA evidence. The probability of a randomly chosen person being $FGA_1/FGA_4 = 2pq = 2(0.30)(0.26) = 0.156$ and, similarly, the probability of a random person being $TPOX_1/TPOX_3 = 0.416$ and $VWA_2/VWA_7 = 0.2714$. Applying the multiplicative rule, the probability of a random member of the population being FGA_1/FGA_4, $TPOX_1/TPOX_3$, $VWA_2/VWA_7 = 0.156 \times 0.416 \times 0.2714 = 0.0176$. In calculating this probability, we have assumed that the population is at Hardy–Weinberg equilibrium and that the three loci in question are at linkage equilibrium with one another.

PROBLEMS

Visit SaplingPlus for supplemental content. Problems with the 🜨 icon are available for review/grading.

WORKING WITH THE FIGURES

(The first 6 questions require inspection of text figures.)

1. Which individual in Figure 18-3 has the most heterozygous loci, and which individual has the fewest?

2. Suppose that the seven chromosomes in Figure 18-4a represent a random sample of chromosomes from a population.

   a. Calculate gene diversity (*GD*) separately for the indel, the microsatellite locus, and the SNP at position 3.

   b. If the sequence was shortened so that you had data only for positions 1 through 24, how many haplotypes would there be?

   c. Calculate the linkage disequilibrium parameter (*D*) between the SNPs at positions 29 and 33.

3. Looking at Figure 18-6, can you count how many mitochondrial haplotypes were carried from Asia into the Americas?

4. In Figure 18-13, the "unrelated" (blue) column for Japan is higher than the "unrelated" column for France. What does this tell you?

5. In Figure 18-14, some individuals have unique SNP alleles—for example, the T allele at SNP4 occurs only in individual 12. Can you identify two individuals each of whom have unique alleles at two SNPs?

6. Looking at Figure 18-20, do people of the Middle East tend to have higher or lower levels of heterozygosity compared to the people of East Asia? Why might this be the case?

BASIC PROBLEMS

7. What are the forces that can change the frequency of an allele in a population, and what effect does each have on variation in a population?

8. What assumptions are made when using the Hardy–Weinberg formula to estimate genotypic frequencies from allele frequencies?

9. In a population of mice, there are two alleles of the *A* locus (A_1 and A_2). Tests showed that, in this population, there are 384 mice of genotype A_1/A_1, 210 of A_1/A_2, and 260 of A_2/A_2. What are the frequencies of the two alleles in the population?

10. In a natural population of *Drosophila melanogaster*, the alcohol dehydrogenase gene has two alleles called *F* (fast) and *S* (slow) with frequencies of *Adh-F* at 0.75 and *Adh-S* at 0.25. In a sample of 480 flies from this population, how many individuals of each genotypic class would you expect to observe under Hardy–Weinberg equilibrium?

11. In a randomly mating laboratory population of *Drosophila*, 4 percent of the flies have black bodies (encoded by the autosomal recessive *b*), and 96 percent have brown bodies (the wild type, encoded by *B*). If this population is assumed to be in Hardy–Weinberg equilibrium, what are the allele frequencies of *B* and *b* and the genotypic frequencies of *B/B* and *B/b*?

12. In a population of a beetle species, you notice that there is a 3:1 ratio of shiny to dull wing covers. Does this ratio prove that the *shiny* allele is dominant? (Assume that the two states are caused by two alleles of one gene.) If not, what does it prove? How would you elucidate the situation?

13. Cystic fibrosis (CF) is an autosomal recessive disorder that occurs relatively frequently among people of European descent. In an Amish community in Ohio, medical researchers reported the occurrence of cystic fibrosis (CF) as being 1/569 live births. Using the Hardy–Weinberg rule, estimate the frequency of carriers of the disease allele in this Amish population.

14. The relative fitness values of three genotypes are $w_{A/A} = 1.0$, $w_{A/a} = 1.0$, and $w_{a/a} = 0.7$.

    a. If the population starts at the allele frequency $p = 0.5$, what is the value of *p* in the next generation?

    b. What is the predicted equilibrium allele frequency if the rate of mutation of *A* to *a* is 2×10^{-5}?

15. *A/A* and *A/a* individuals are equally fertile. If 0.1 percent of the population is *a/a*, what selection pressure exists against *a/a* if the $A \rightarrow a$ mutation rate is 10^{-5}? Assume that the frequencies of the alleles are at their equilibrium values.

16. When alleles at a locus act in a semidominant fashion on fitness, the relative fitness of the heterozygote is midway between the two homozygous classes. For example, genotypes with semidominance at the *A* locus might have these relative fitnesses: $w_{A/A} = 1.0$, $w_{A/a} = 0.9$, and $w_{a/a} = 0.8$.

    a. Change one of these fitness values so that *a* becomes a deleterious recessive allele.

    b. Change one of these fitness values so that *A* becomes a favored dominant allele.

17. If the recessive allele for an X-linked recessive disease in humans has a frequency of 0.02 in the population, what proportion of individuals in the population will

have the disease? Assume that the population is 50:50 male:female.

18. Red-green color blindness is an X-linked recessive disorder in humans caused by mutations in one of the genes that encodes the light-sensitive protein, opsin. If the mutant allele has a frequency of 0.08 in the population, what proportion of females will be carriers? Assume that the population is 50:50 male:female.

19. Is a new neutral mutation more likely to reach fixation in a large or small population?

20. It seems clear that inbreeding causes a reduction in fitness. Can you explain why?

21. In a population of 50,000 diploid individuals, what is the probability that a new neutral mutation will ultimately reach fixation? What is the probability that it will ultimately be lost from the population?

22. Inbreeding in a population causes a deviation from Hardy–Weinberg expectations such that there are more homozygotes than expected. For a locus with a rare deleterious allele at a frequency of 0.04, what would be the frequency of homozygotes for the deleterious allele in populations with inbreeding coefficients of $F = 0.0$ and $F = 0.125$?

23. Sickle-cell anemia is a recessive autosomal disorder that is caused by an amino acid substitution in the β-hemoglobin protein. The DNA mutation underlying this substitution is a SNP that alters a GAG codon for the amino acid glutamate to a GTG that codes a valine. The frequency of sickle-cell anemia among African Americans is about 1/400. What is the frequency of this GTG codon in the β-hemoglobin gene among African Americans?

24. You have a sample of 10 DNA sequences of 100 bp in length from a section of highly conserved gene from 10 individuals of a species. The 10 sequences are almost entirely identical; however, each sequence carries one unique SNP not found in any of the others. What is the nucleotide diversity for this sample of sequences?

CHALLENGING PROBLEMS

25. Figure 18-14 presents haplotype data for the *G6PD* gene in a worldwide sample of people.

    a. Draw a haplotype network for these haplotypes. Label the branches on which each SNP occurs.

    b. Which of the haplotypes has the most connections to other haplotypes?

    c. On what continents is this haplotype found?

    d. Counting the number of SNPs along the branches of your network, how many differences are there between haplotypes 1 and 12?

26. Figure 18-12 shows a pedigree for the offspring of a half-sib mating.

    a. If the inbreeding coefficient for the common ancestor (A) in Figure 18-12 is 1/2, what is the inbreeding coefficient of I?

    b. If the inbreeding coefficient of individual I in Figure 18-12 is 1/8, what is the inbreeding coefficient of the common ancestor, A?

27. Consider 10 populations that have the genotype frequencies shown in the following table:

Population	A/A	A/a	a/a
1	1.0	0.0	0.0
2	0.0	1.0	0.0
3	0.0	0.0	1.0
4	0.50	0.25	0.25
5	0.25	0.25	0.50
6	0.25	0.50	0.25
7	0.33	0.33	0.33
8	0.04	0.32	0.64
9	0.64	0.32	0.04
10	0.986049	0.013902	0.000049

    a. Which of the populations are in Hardy–Weinberg equilibrium?

    b. What are p and q in each population?

    c. In population 10, the $A \rightarrow a$ mutation rate is discovered to be 5×10^{-6}. What must be the fitness of the a/a phenotype if the population is at equilibrium?

    d. In population 6, the a allele is deleterious; furthermore, the A allele is incompletely dominant; so A/A is perfectly fit, A/a has a fitness of 0.8, and a/a has a fitness of 0.6. If there is no mutation, what will p and q be in the next generation?

28. The hemoglobin β gene (*Hb*) has a common allele (*A*) of a SNP (*rs334*) that encodes the Hb^A form of (adult) hemoglobin and a rare allele (*T*) that encodes the sickling form of hemoglobin, Hb^S. Among 571 residents of a village in Nigeria, 440 were *A/A* and 129 were *A/T*, and 2 were *T/T* individuals were observed. Use the chi-square test to determine whether these observed genotypic frequencies fit Hardy–Weinberg expectations.

29. A population has the following gametic frequencies at two loci: $AB = 0.4$, $Ab = 0.1$, $aB = 0.1$, and $ab = 0.4$. If the population is allowed to mate at random until linkage equilibrium is achieved, what will be the expected frequency of individuals that are heterozygous at both loci?

30. Two species of palm trees differ by 50 bp in a 5000-bp stretch of DNA that is thought to be neutral. The mutation rate for these species is 2×10^{-8} substitutions per site per generation. The generation time for these

species is five years. Estimate the time since these species had a common ancestor.

31. Color blindness in humans is caused by an X-linked recessive allele. Ten percent of the males of a large and randomly mating population are color-blind. A representative group of 1000 people from this population migrates to a South Pacific island, where there are already 1000 inhabitants and where 30 percent of the males are color-blind. Assuming that Hardy–Weinberg equilibrium applies throughout (in the two original populations before the migration and in the mixed population immediately after the migration), what fraction of males and females can be expected to be color-blind in the generation immediately after the arrival of the migrants?

32. Using pedigree diagrams, calculate the inbreeding coefficient (F) for the offspring of (a) parent–offspring matings; (b) first-cousin matings; (c) aunt–nephew or uncle–niece matings; (d) self-fertilization of a hermaphrodite.

33. A group of 50 men and 50 women establish a colony on a remote island. After 50 generations of random mating, how frequent would a recessive trait be if it were at a frequency of 1/500 back on the mainland? The population remains the same size over the 50 generations, and the trait has no effect on fitness.

34. Figure 18-22 shows 10 haplotypes from a population before a selective sweep and another 10 haplotypes many generations later after a selective sweep has occurred for this chromosomal region. There are 11 loci defining each haplotype, including one with a red allele that was the target of selection. In the figure, two loci are designated as A and B. These loci each have two alleles: one black and the other gray. Calculate the linkage disequilibrium parameter (D) between A and B, both before and after the selective sweep. What effect has the selective sweep had on the level of linkage disequilibrium?

35. The recombination rate (r) between linked loci A and B is 0.10. In a population, we observe the following haplotypic frequencies:

AB	0.40
aB	0.10
Ab	0.10
Ab	0.40

   a. What is the level of linkage disequilibrium as measured by D in the present generation?

   b. What will D be in the next generation?

   c. What is the expected frequency of the Ab haplotype in the next generation?

   d. Using a spreadsheet computer software program, make a graph of the decline in D over 10 generations.

36. Allele B is a deleterious autosomal dominant. The frequency of affected individuals is 4.0×10^{-6}. The reproductive capacity of these individuals is about 30 percent that of normal individuals. Estimate μ, the rate at which b mutates to its deleterious allele B. Assume that the frequencies of the alleles are at their equilibrium values.

37. What is the equilibrium heterozygosity for a SNP in a population of 50,000 when the mutation rate is 3×10^{-8}?

38. Of 31 children born of father–daughter matings, 6 died in infancy, 12 were very abnormal and died in childhood, and 13 were normal. From this information, calculate roughly how many recessive lethal genes we have, on average, in our human genomes. (**Hint:** If the answer were 1, then a daughter would stand a 50 percent chance of carrying the lethal allele, and the probability of the union's producing a lethal combination would be $1/2 \times 1/4 = 1/8$. So, 1 is not the answer.) Consider also the possibility of undetected fatalities in utero in such matings. How would they affect your result?

39. The B locus has two alleles B and b with frequencies of 0.95 and 0.05, respectively, in a population in the current generation. The genotypic fitnesses at this locus are $w_{B/B} = 1.0, w_{B/b} = 1.0$, and $w_{b/b} = 0.0$.

   a. What will the frequency of the b allele be in two generations?

   b. What will the frequency of the b allele be in two generations if the fitnesses were $w_{B/B} = 1.0$, $w_{B/b} = 0.0$, and $w_{b/b} = 0.0$?

   c. Explain why there is a difference in the rate of change for the frequency of the b allele under parts a and b of this problem.

40. The sd gene causes a lethal disease of infancy in humans when homozygous. One in 100,000 newborns die each year of this disease. The mutation rate from Sd to sd is 2×10^{-4}. What must the fitness of the heterozygote be to explain the observed gene frequency in view of the mutation rate? Assign a relative fitness of 1.0 to Sd/Sd homozygotes. Assume that the population is at equilibrium with respect to the frequency of sd.

41. If we define the *total selection cost* to a population of deleterious recessive genes as the loss of fitness per individual affected (s) multiplied by the frequency of affected individuals (q^2), then selection cost = sq^2.

   a. Suppose that a population is at equilibrium between mutation and selection for a deleterious recessive allele, where $s = 0.5$ and $\mu = 10^{-5}$. What is the equilibrium frequency of the allele? What is the selection cost?

b. Suppose that we start irradiating individual members of the population so that the mutation rate doubles. What is the new equilibrium frequency of the allele? What is the new selection cost?

c. If we do not change the mutation rate but we lower the selection coefficient to 0.3 instead, what happens to the equilibrium frequency and the selection cost?

42. Balancing selection acts to maintain genetic diversity at a locus since the heterozygous class has a greater fitness than the homozygous classes. Under this form of selection, the allele frequencies in the population approach an equilibrium point somewhere between 0 and 1. Consider a locus with two alleles A and a with frequencies p and q, respectively. The relative genotypic fitnesses are shown below, where s and g are the selective disadvantages of the two homozygous classes.

Genotype	A/A	A/a	a/a
Relative fitness	$1 - s$	1	$1 - g$

a. At equilibrium, the mean fitness of the A alleles (w_A) will be equal to the mean fitness of the a alleles (w_a) (see Box 18-7). Set the mean fitness of the A alleles (w_A) equal to the mean fitness of the a alleles (w_a). Solve the resulting equation for the frequency of the A allele. This is the expression for the equilibrium frequency of A ($\hat{p}$).

b. Using the expression that you just derived, find $\hat{p}$ when $s = 0.2$ and $g = 0.8$.

GENETICS AND SOCIETY

Genome-wide SNP data from companies like 23andMe allow people to compare their genotypes to other people's genotypes to identify distant relatives such as third cousins. A free Web site (GEDmatch.com) allows you to upload your SNP genotypes and search for relatives in a public database. In 2018, police collected SNP genotypes from DNA left by the "Golden State Killer" at one of his crime scenes. Using GEDmatch, police identified relatives of the person who left the DNA at the crime scene that led them to identify and arrest Joseph DeAngelo as the Golden State Killer. Do you see any ethical issues in use of public databases to track criminals? Do you think people might be reluctant to upload their personal data to GEDmatch if they thought it might lead to the arrest of a relative? Are there ways in which making your own DNA data public could cause unjustified harm to you or your relatives?

The Inheritance of Complex Traits

Former basketball star Kareem Abdul-Jabbar (7 feet, 2 inches tall) and former renowned jockey Willie Shoemaker (4 feet, 11 inches tall) show some of the extremes in human height—a quantitative trait. [*RF/AP Images*.]

CHAPTER OUTLINE AND LEARNING OBJECTIVES

19.1 MEASURING QUANTITATIVE VARIATION

LO 19.1 Understand how quantitative genetics uses mathematical models and statistics to investigate complex traits.

LO 19.2 Analyze data to assess the amount and distribution of trait variation in populations.

19.2 A SIMPLE GENETIC MODEL FOR QUANTITATIVE TRAITS

LO 19.3 Assess the relative contributions of genetic and environmental factors to phenotypic traits.

19.3 BROAD-SENSE HERITABILITY: NATURE VERSUS NURTURE

LO 19.4 Calculate and interpret broad-sense heritability.

19.4 NARROW-SENSE HERITABILITY: PREDICTING PHENOTYPES

LO 19.5 Calculate and interpret narrow-sense heritability.

LO 19.6 Use knowledge of parental phenotypes to predict the phenotype of offspring.

19.5 MAPPING QTL IN POPULATIONS WITH KNOWN PEDIGREES

LO 19.7 Determine how many genes contribute to the genetic variation for a trait.

19.6 ASSOCIATION MAPPING IN RANDOM-MATING POPULATIONS

LO 19.8 Design and analyze experiments to identify the loci controlling quantitative traits in populations.

Complex traits, also known as quantitative traits, are ones that do not behave in simple Mendelian fashion, but instead have a continuous range of variation. These phenotypes are the result of a set of interactions among multiple genes and various environmental factors. Geneticists utilize mathematical models and statistical methods to analyze complex traits. Understanding the factors and identifying the genes that control complex traits is of great importance to plant and animal breeders and to medical clinicians.

Look at almost any large group of men or women and you'll notice a considerable range in their heights—some are short, some tall, and some about average. Kareem Abdul-Jabbar, a star basketball center of the 1970s and 1980s, was a towering 7 feet, 2 inches tall, whereas Willie Shoemaker, a renowned jockey who won the Kentucky Derby four times, was a mere 4 feet, 11 inches. You might also have noticed that in some families, the parents and their adult children are all on the tall side, whereas in other families, the parents and adult children are all fairly short. Such observations suggest that genes play a role in determining our heights. Still, people do not segregate cleanly into tall and short categories as we saw for Mendel's pea plants. At first inspection, continuous traits, such as height, do not appear to follow Mendel's laws despite the fact that they are heritable.

Traits such as height that show a continuous range of variation and do not behave in a simple Mendelian fashion are known as **quantitative** or **complex traits**. The term *complex trait* is often preferred because variation for such traits is governed by a "complex" of genetic and environmental factors. How tall you are is partly explained by the genes you inherited from your parents and partly by environmental factors such as how well you were nourished as a child. Teasing apart the genetic and environmental contributions to an individual phenotype is a substantial challenge, but geneticists have a powerful set of tools to meet it.

In the early 1900s, when Mendel's laws were rediscovered, controversy arose about whether these laws were applicable to continuous traits. A group known as the biometricians discovered that there are correlations between relatives for continuous traits such that tall parents tend to have tall children. However, the biometricians saw no evidence that such traits followed Mendel's laws. Some biometricians concluded that Mendelian loci do not control continuous traits. On the other hand, some adherents of Mendelism thought continuous variation was unimportant and could be ignored when studying inheritance. By 1920, this controversy was resolved with the formulation of the **multifactorial hypothesis**. This hypothesis proposed that continuous traits are governed by a combination of multiple Mendelian loci, each with a small effect on the trait, and environmental factors. The multifactorial hypothesis brought quantitative traits into the realm of Mendelian genetics.

Although the multifactorial hypothesis provided a sensible explanation for continuous variation, classic Mendelian analysis is inadequate for the study of complex traits. If progeny cannot be sorted into categories with expected ratios, then the Mendelian approach has little utility for the analysis of complex traits. In response to this problem, geneticists developed a set of mathematical models and statistical methods for the analysis of complex traits. Through the application of these analytical methods, geneticists have made great strides in understanding complex traits. The subfield of genetics that develops and applies these methods to understand the inheritance of complex traits is called **quantitative genetics**.

At the heart of the field of quantitative genetics is the goal of defining the **genetic architecture** of complex traits. Genetic architecture is a description of all of the genetic factors that influence a trait. It includes the number of genes affecting the trait and the relative contribution of each gene. Some genes may have a large effect on the trait, while others have only a small effect. As we will see in this chapter, genetic architecture is the property of a specific population and can vary among populations of a species. For example, the genetic architecture of a trait such as systolic blood pressure in humans differs among different populations. This is because different alleles segregate in different populations and different populations experience different environments; therefore, different populations are apt to have different architectures for many traits.

Understanding the inheritance of complex traits is one of the most important challenges facing geneticists in the twenty-first century. Complex traits are of paramount importance in medical and agricultural genetics. For humans, blood pressure, body weight, susceptibility to depression, serum cholesterol levels, and the risk of developing cancer or other disorders are all complex traits. For crop plants, yield, resistance to pathogens, ability to tolerate drought stress, efficiency of fertilizer uptake, and even flavor are all complex traits. For livestock, milk production in dairy cows, muscle mass in beef cattle, litter size in pigs, and egg production in chickens are all complex traits. Despite the importance of such traits, we know far less about their inheritance than we do about the inheritance of simply inherited traits such as cystic fibrosis or sickle-cell anemia.

In this chapter, we will explore the inheritance of complex traits. We will begin with a review of some basic statistical concepts. Next, we will develop the mathematical model used to connect the action of genes inside the cell with the phenotypes we observe at the level of the whole organism. Using this model, we will then show how quantitative geneticists partition the phenotypic variation in a population into the parts that are due to genetic and environmental factors. We will review the methods used by plant and animal breeders to predict the phenotype of offspring from the phenotype of their parents. Finally, we will see how a combination of the statistical analysis and molecular markers can be used to identify the specific genes that control quantitative traits.

19.1 MEASURING QUANTITATIVE VARIATION

LO 19.1 Understand how quantitative genetics uses mathematical models and statistics to investigate complex traits.

LO 19.2 Analyze data to assess the amount and distribution of trait variation in populations.

To study the inheritance of quantitative traits, we need some basic statistical tools. In this section, we will introduce the mean (or average), which can be used to describe differences between groups, and the variance, which can be used to quantify the amount of variation that exists within a group. We will also discuss the normal distribution, which is central to understanding quantitative variation in populations. But before discussing the statistical tools, let's define the different types of complex trait variation that can occur in a population.

Types of traits and inheritance

A **continuous trait** is one that can take on a potentially infinite number of states over a continuous range. Height in humans is a good example. People can range from about 140 cm to 230 cm in height. If we measured height precisely, then the number of possible heights is infinite. For example, a person might be 170 cm tall, or 170.2 cm, or 170.02 cm. Continuous traits typically have **complex inheritance** involving multiple genes plus environmental factors.

For some traits, the individuals in a population can be sorted into discrete groups or categories. Such traits are known as **categorical traits**. Examples include purple versus white flowers or tall versus short stems for Mendel's pea plants, as seen in Chapter 2. Categorical traits often exhibit **simple inheritance** such that the progeny of crosses segregate into standard Mendelian ratios such as 3:1 for a single gene or 15:1 for two genes. The inheritance is simple because only one or two genes are involved and the environment has little or no effect on the phenotype.

Some categorical traits do not show simple inheritance. These include many disease conditions in humans. In medical genetics, individuals can be classified into the categories "affected" or "not affected" by a disease. For example, an

individual may or may not have type 2 diabetes. However, type 2 diabetes does not follow simple Mendelian rules or produce Mendelian ratios in pedigrees. Rather, there are multiple genetic and environmental factors that place someone at risk of developing this disease. Individuals who have a certain number of risk factors will exceed a threshold and develop the disease. Type 2 diabetes is a form of a categorical trait called a **threshold trait**. Type 2 diabetes has complex inheritance.

Another type of trait is a **meristic trait,** or counting trait, which takes on a range of discrete values. An example would be clutch size in birds. A bird can lay 1, 2, 3, or more eggs, but it cannot lay 2.49 eggs. Meristic traits are quantitative, but they are restricted to certain discrete values. They do not take on a continuous range of values. Meristic traits usually have complex inheritance.

Quantitative geneticists seek to understand the inheritance of traits that show complex inheritance resulting from a mix of genetic and environmental factors. They may investigate traits that are categorical, meristic, or continuous. The emphasis is on the type of inheritance—complex. For this reason, the term *complex trait* is often preferred to continuous or quantitative trait because it includes all the types of traits with which quantitative genetics is concerned. Any biological phenomenon for which variation exists may show complex inheritance and can be studied as a complex trait. Thus, size and shape of structures, enzyme kinetics, mRNA levels, circadian rhythms, and bird songs can all be treated as complex traits.

KEY CONCEPT A complex trait is any trait that does not show simple Mendelian inheritance. A complex trait can be either a categorical trait such as the presence or absence of a disease condition, or a continuously variable trait such as height in humans.

The mean

When quantitative geneticists study the inheritance of a trait, they work with a particular group of individuals, or **population**. For example, we might be interested in the inheritance of height for the population of adult men in Shanghai, China. Here, we are using "population" to denote a group that shares certain features in common, such as age, sex, ethnicity, or geographic origin. Since there are more than 5 million adult men in Shanghai, determining each of their heights would be a herculean task. Therefore, quantitative geneticists typically study just a subset or **sample** of the full population. The sample should be *randomly* chosen such that each of the 5 million men has an equal chance of being included in the sample. If the sample meets this criterion, then we can use measurements made on the sample to make inferences about the entire population.

Using the example of height for men from Shanghai, we can describe the population using the **mean** or average value for the trait. We select a random sample of 100 men from the population and measure their heights. Some of the men might be 166 cm tall, others 172 cm tall, and so forth. To calculate the mean, we simply sum all the individual measurements and divide the sum by the size of the sample (*n*), which in this case is 100. For the data in Table 19-1, the result would be 170 cm, or 5 feet, 7 inches. Since we have a random sample, we can infer that the average height in the entire population is 170 cm.

Height is a *random variable*, which means it can take on different values, and when we select someone at random from the population, the value we observe is governed by an element of chance. Random variables are usually represented by the letter *X* in statistics. We have measurements for $X_1, X_2, X_3, \ldots X_{100}$ for the $n = 100$ men in the sample. Symbolically, we can express the mean as

$$\bar{X} = \frac{1}{n}\sum_{i=1}^{n} X_i$$

where $\bar{X}$ represents the sample mean. The uppercase Greek letter sigma (Σ) is the summation sign, indicating that we add all *n* observed values of *X* for *i* = 1, 2, through *n*. (Often, the *n* above Σ and the *i* = 1 below Σ are omitted to simplify the appearance of equations.)

There is a distinction made between the mean of a sample ($\bar{X}$) and the true mean of the population. To learn the true mean for the height of men in Shanghai, we would need to determine the height of each and every man. The true mean is symbolized by the Greek letter μ, so that we have different symbols for the sample and population means.

Here is another way to calculate the mean, which is often quite useful. We can add the products of each class of values of *X* in the data set times the frequency of that class in the data set. This operation is symbolized as

$$\bar{X} = \sum_{i=1}^{k} f_i X_i$$

where f_i is the frequency of the *i*th class of observations, X_i is the value of the *i*th class, and there are a total of *k* classes. For the data in Table 19-1, one man of the 100 ($f = 0.01$) is 156 cm tall, two men ($f = 0.02$) are 157 cm tall, and so forth, so we can calculate the sample mean as

$$\bar{X} = (0.01 \times 156) + (0.02 \times 157) + \ldots + (0.02 \times 184) = 170$$

The mean is useful for both describing populations and comparing differences between populations. For example, men in urban areas of China are on average 170 cm tall, while men in rural areas of China are 166 cm tall. These values were calculated using samples drawn from each region. One question that a quantitative geneticist might ask about the observed difference in height between rural and urban Chinese men is the following: Is the difference due to genetic factors, or is it due to differences in nutrition, health care, or other environmental factors? Later in the chapter, we will see how quantitative geneticists tease apart genetic versus environmental contributions to a trait.

Lastly, here is another helpful notation from statistics that can be used to define the mean. The mean of a random variable, *X*, is the *expectation* or *expected value* of that random variable. The expected value is the average of all the values we would observe if we measured *X* many times. The expectation is symbolized by *E*, and we write *E(X)* to signify "the expected value of *X*." Symbolically, we write

$$E(X) = \bar{X}$$

We will use the notation of expectation in several places in this chapter.

The variance

Besides the mean, we also need a measure of how much variation exists in populations. We can create a visual representation of the variation by plotting the count or frequency of each height class. Figure 19-1 shows such a plot for our simulated height data for 100 men from Shanghai. The *x*-axis shows different height classes, and the y-axis shows the count or frequency of each class. In this figure,

TABLE 19-1	Simulated Data for the Heights of 100 Men from Shanghai, China	
Height (cm)	**Count**	**Frequency × Height**
156	1	1.56
157	2	3.14
158	1	1.58
159	2	3.18
160	1	1.60
161	1	1.61
162	2	3.24
164	7	11.48
165	7	11.55
166	1	1.66
167	6	10.02
168	9	15.12
169	7	11.83
170	9	15.30
171	5	8.55
172	5	8.60
173	6	10.38
174	5	8.70
175	6	10.50
176	3	5.28
177	4	7.08
178	2	3.56
179	2	3.58
180	2	3.60
181	2	3.62
184	2	3.68
Sum	100	170.00

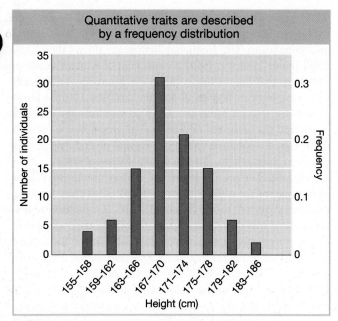

FIGURE 19-1 Frequency histogram of simulated data for the height of adult men from Shanghai, China.

the men were binned into 4-cm groups, for example, from 155 to 158 cm. This type of graph is called a **frequency histogram**. If the values are clustered tightly around the mean, then there is less variation, and if the values are spread out along the x-axis, there is greater variation.

We can quantify the amount of variation in a population using a statistical measure called the **variance**. The variance measures the extent to which individuals in the population deviate from the population mean. If all 100 men in our sample had heights very close to the mean, then the variance would be small. If their heights deviated greatly from the mean, the variance would be large.

Since the variance is a measure of **deviation** from the mean, let's define deviation mathematically. Knowing the mean value for the random variable X, we can calculate the deviation of each individual from the mean by subtracting $\bar{X}$ from the individual observations. We will represent the deviations by a lowercase x:

$$x = X - \bar{X}$$

Some individuals will have X values above the mean, and they will have a positive deviation. Others will have X values below the mean, and they will have a negative deviation. For the population overall, the expected value of x is 0, or $E(x) = 0$.

To measure the amount of variation for X in the population, we use the variance, which is the mean of the squared deviations. First, we calculate the sum of the squared deviations (or *sum of squares*, for short) as

$$\text{sum of squares} = \sum_i (X_i - \bar{X})^2$$

$$= \sum_i (x_i)^2$$

Since deviations with negative values form positive squares, both negative and positive deviations will contribute positively to the sum of squares. The variance is the mean of the squared deviations (or the sum of squares divided by n). Symbolically, we express the population variance as

$$V_X = \frac{1}{n} \sum_i (X_i - \bar{X})^2$$

$$= \frac{1}{n} \sum_i (x_i)^2$$

where V_X denotes the variance of X. The population variance is sometimes symbolized using the lowercase Greek letter sigma squared (σ^2). In statistics, there is also a distinction made between the population variance (σ^2) and the sample variance (s^2). The latter is calculated by dividing the sums of squares by $n - 1$ rather than n to correct a bias caused by small sample size. For simplicity, we will use the population variance and the formula above throughout this chapter.

There are several points to understand about the variance. First, the variance provides a measure of dispersion about the mean. When the variance is high, the individual values are spread farther apart from the mean; when it is low, then the individual values cluster closer to the mean. Second, the variance is measured in squared units such that if we measure human height in centimeters, then the variance would be in centimeters². Third, the variance can range from 0.0 to infinity. Fourth, the variance is equal to the expected value of the squared deviation (x^2) or $E(x^2)$.

The variance of quantitative traits is measured in squared units. These squared units have desirable mathematical properties as we will see below; however, they do not make intuitive sense. If we measure weight in kilograms, then the variance would be in kilograms², which has no clear meaning. Therefore, another statistic used to quantify the extent of deviation from the mean in a population is the **standard deviation** (σ), which is the square root of the variance:

$$\sigma = \sqrt{\sigma^2}$$

The standard deviation is expressed in the same units as the trait itself, so its meaning is more intuitive. We will use the standard deviation in the description of traits below.

The normal distribution

Even if you have never taken a statistics course, you likely have heard of the **normal distribution**, also known as the "bell curve" in popular culture. The normal distribution is remarkably useful in biology in general and quantitative genetics in particular because the frequency distribution for many biological traits approximates a normal curve. For this reason, geneticists can take advantage of several features of the normal distribution to describe quantitative traits and dissect the underlying genetics.

The normal distribution is a continuous frequency distribution similar to the frequency histogram shown in

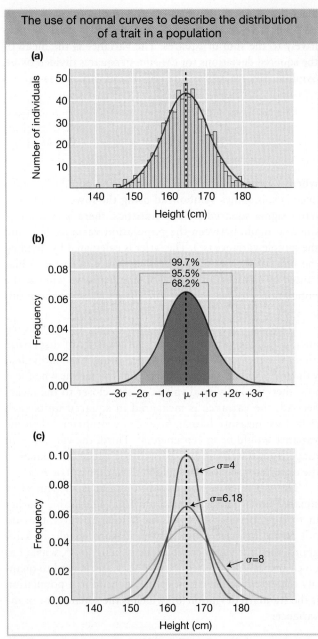

The use of normal curves to describe the distribution of a trait in a population

(a)

(Frequency histogram with y-axis "Number of individuals" ranging from 10 to 50, and x-axis "Height (cm)" ranging from 140 to 180.)

(b)

(Normal curve with y-axis "Frequency" from 0.00 to 0.08, and x-axis marked −3σ −2σ −1σ μ +1σ +2σ +3σ, showing 99.7%, 95.5%, 68.2%.)

(c)

(Normal curves with y-axis "Frequency" from 0.00 to 0.10, and x-axis "Height (cm)" from 140 to 180, showing σ=4, σ=6.18, and σ=8.)

FIGURE 19-2 (a) Frequency histogram of actual data for the height of adult women from the United States. The red line represents the normal curve fit to these data with a mean of 164.4 cm and standard deviation of 6.18 cm. (b) Normal curve for the height of U.S. women showing the predicted percentages of women who will fall within different numbers of standard deviations from the mean. (c) Normal curves with the same mean (164.4 cm) but different standard deviations, showing the effect of the standard deviation on the shape of the curve.

Figure 19-1. The normal distribution applies to continuous traits. As mentioned in the previous section, continuous traits can take on an infinite number of values. A person might be 170 cm tall, or 170.2 cm, or 170.02 cm, and so forth. For such traits, the expected frequency of different trait values is better represented by a curve than by a frequency histogram. For the normal distribution, the shape of the curve is determined by two factors—the mean and the standard deviation.

Here is an example using height data for 660 women from the United States collected by the Centers for Disease Control and Prevention. The frequency histogram shows the classic "bell curve" shape with the peak near the mean value of 164.4 cm and the off-mean values distributed symmetrically around the mean (**Figure 19-2a**). We can fit a normal curve to this distribution using just two pieces of information—the mean and the standard deviation. The shape of the curve is defined by an equation called the normal probability density function, into which the mean and the standard deviation are plugged. The normal distribution allows us to predict the percentage of the observations that will fall within a certain distance from the mean (Figure 19-2b). If we measure distance along the x-axis in standard deviations, then 68 percent of the observations are expected to fall within 1 standard deviation (σ) of the mean and 95.5 percent within 2 standard deviations. For the height data for U.S. women, 71 percent (449 women) fall within 1 standard deviation of the mean and 96 percent (633 women) within 2 standard deviations. These values are very close to the predictions of 68.2 percent and 95.5 percent based on the normal curve.

If we know just the mean and the standard deviation for a trait, we can predict the shape of the distribution of the trait in the population, and we can predict how likely we are to observe certain values when sampling the population. For example, if the mean height for U.S. women is 164.4 cm (5 feet, 5 inches) and the standard deviation is 6.18 cm, we can predict that only 2 percent of women will be more than 177 cm tall, or 5 feet, 10 inches. As shown in Figure 19-2c, if the standard deviation is greater (for example, 8), then the curve would be flatter and a greater percentage would fall above 177 cm. However, it would still be true that only 2 percent would be more than 2σ above the mean, or 180.4 cm [(164.4 + (2 × 8)].

KEY CONCEPT The field of quantitative genetics studies the inheritance of complex traits using some basic statistical tools including the mean, variance, standard deviation, and normal distribution.

19.2 A SIMPLE GENETIC MODEL FOR QUANTITATIVE TRAITS

LO 19.3 Assess the relative contributions of genetic and environmental factors to phenotypic traits.

A mathematical model is a simplified representation of a complex phenomenon. Models allow us to describe a phenomenon in terms of the variables that influence it and then to use the model to make predictions about the state of the phenomenon under different values for these variables. In this section, we will define the mathematical model used by quantitative geneticists to study complex traits.

The exceptional height of Yao Ming

FIGURE 19-3 Former basketball star center Yao Ming, who stands 7 feet, 6 inches, talks with retired golf star Gary Player, who is 5 feet, 6 inches. [*Power Sport Images/Getty Images*.]

Genetic and environmental deviations

We will now examine how phenotypes can be decomposed into their genetic and environmental contributions, using as an example the height of Yao Ming, the former center for the Houston Rockets basketball team. Yao Ming stands out at 229 cm, or 7 feet, 6 inches (**Figure 19-3**). That's right: Yao Ming is nearly two feet taller than the average man from Shanghai, which happens to be Yao Ming's hometown. As for all of us, Yao Ming's height is the combined result of his genotype and the environment in which he was raised. Let's do an imaginary experiment and see how we can tease apart the genetic and environmental contributions to Yao Ming's exceptional height.

First, we will define a simple mathematical model that can be applied to any quantitative trait. The value of a trait (X) for an individual member of a population can be expressed in terms of the population mean and deviations from the mean due to genetic (g) and environmental (e) factors.

$$X = \bar{X} + g + e$$

We are using lowercase g and e for the genetic and environmental deviations, just as we used a lowercase x for the deviation of X from the mean. Thus, in Yao Ming's case, his height can be expressed as the mean value for men from Shanghai (170 cm) plus his specific genetic and environmental deviations ($g + e = 59$ cm). We can simplify the equation above by subtracting $\bar{X}$ from both sides to obtain

$$x = g + e$$

where x represents the individual's phenotypic deviation. For Yao Ming's height, $x = g + e = 59$ cm.

How can we determine the values of g and e for Yao Ming? One way would be if we had clones of Yao Ming (clones are genetically identical individuals). Let's imagine that we cloned Yao Ming and distributed these clones (as newborns) to a set of randomly chosen households in

Shanghai. Twenty-one years later, we locate this army of Yao Ming clones, measure their heights, and determine that their average height is 212 cm. The expectation of e over the many environments in which the Yao Ming clones were reared is 0. In some households, the clones get a positive environment ($+e$) and in others a negative environment ($-e$). Overall, $E(e) = 0$. Thus, the mean for the clones minus the population mean equals Yao Ming's genotypic deviation, or $g = (212 - 170) = 42$ cm. The remaining 17 cm of his remarkable 59-cm phenotypic deviation is e for the specific environment in which the real Yao Ming was raised. Plugging these values into the equation, we obtain

$$229 = 170 + 42 + 17$$

We conclude that Yao Ming's exceptional height is mostly due to exceptional genetics, but he also experienced an environment that boosted his height.

Although our imaginary experiment of cloning Yao Ming is far-fetched, many plant species and some animal species can be clonally propagated with ease. For example, one can use "cuttings" of an individual plant to produce multiple genetically identical individuals. Another way of creating genetically identical individuals is by producing **inbred lines** or **stocks** that are homozygous at all loci throughout their entire genomes. Like clones, all individuals of an inbred line are genetically identical to each. By using clones or inbred lines, geneticists can estimate the genetic and environmental contributions to a trait by rearing the clones in randomly assigned environments. Here is an example.

Table 19-2 (experiment I) shows simulated data for 10 inbred strains of maize that were grown in three different environments and scored for the number of days between planting and the time that the plants first shed pollen. The overall mean is 70 days. Let's consider line A when grown in environment 1. The mean for all lines in environment 1 is 68, or 2 less than the overall mean, so e for environment 1 is −2. The mean line A over all three environments is 64, or 6 less than the overall mean, so g for line A is −6. Putting these two values together, we decompose the phenotype of line A when grown in environment 1 as

$$62 = 70 + (-6) + (-2)$$

We could do the same calculations for the other nine inbred lines, and then we would have a complete description of all the phenotypes in each environment in terms of the extent to which their deviation from the overall mean is due to genetic and environmental factors.

Genetic and environmental variances

We can use the simple model $x = g + e$ to think further about the variance of quantitative traits. Recall that the variance is a way to measure how much individuals deviate from the population mean. Under this model, the trait variance can be partitioned into the genetic and the environmental variances:

$$V_X = V_g + V_e$$

TABLE 19-2 Simulated Data for Days to Pollen Shed for 10 Inbred Lines of Maize Grown in Two Experiments

Experiment I

Inbred lines	A	B	C	D	E	F	G	H	I	J	Mean
Environment 1	62	64	66	66	68	68	70	70	72	74	68
Environment 2	64	66	68	68	70	70	72	72	74	76	70
Environment 3	66	68	70	70	72	72	74	74	76	78	72
Mean	64	66	68	68	70	70	72	72	74	76	70

Experiment II

Inbred lines	A	B	C	D	E	F	G	H	I	J	Mean
Environment 4	58	60	62	62	64	64	66	66	68	70	64
Environment 5	64	66	68	68	70	70	72	72	74	76	70
Environment 6	70	72	74	74	76	76	78	78	80	82	76
Mean	64	66	68	68	70	70	72	72	74	76	70

This simple equation tells us that the trait or phenotypic variation (V_X) is the sum of two components—the genetic (V_g) variance and the environmental (V_e) variance. As noted in **Box 19-1**, there is an important assumption behind this equation; namely, that genotype and environment are not correlated—that is, they are independent. If the best genotypes are placed in the best environments and the worst genotypes in the worst environments, then this equation gives inaccurate results. We will discuss this important assumption later in the chapter.

We can use the data in Table 19-2 (experiment I) to explore the equation for variances. First, let's use all 30 phenotypic values for the 10 lines in the three environments to calculate the variance. The result is $V_X = 14.67$ days^2. Now, to estimate V_g, we calculate the variance of the means among the 10 inbred lines. The result is $V_g = 12.0$ days^2.

BOX 19-1 Genetic and Environmental Variances

To better understand the basic equation $V_X = V_g + V_e$, we need to introduce a new concept from statistics— **covariance**. The covariance provides a measure of association between traits. For two random variables X and Y, their covariance is

$$COV_{X,Y} = \frac{1}{n}\sum_i (X_i - \bar{X})(Y_i - \bar{Y})$$

$$= \frac{1}{n}\sum_i (x_i y_i)$$

where x and y are the deviations of X and Y from their respective means as described in the main text. The term $(X_i - \bar{X})(Y_i - \bar{Y})$, or $(x_i y_i)$, is referred to as the *cross product*. The covariance is obtained by summing all the cross products together and dividing by n. The covariance is the average or expected value, $E(xy)$, of the cross products. The covariance can vary from negative infinity to positive infinity. If large values of X are associated with large values of Y, then the covariance will be positive. If large values of X are associated with small values of Y, then the covariance will be negative. If there is no association between X and Y, then the covariance will be zero. For independent traits, the covariance will be zero.

In the main text, we saw that the variance is the expected value of the squared deviations:

$$V_X = E(x^2)$$

Since the phenotypic deviation (x) is the sum of the genotypic (g) and environmental (e) deviations, we can substitute ($g + e$) for x and obtain

$$V_X = E[(g + e)^2]$$
$$= E[(g^2 + e^2 + 2ge]$$
$$= E(g^2) + E(e^2) + E(2ge)$$

The first term $E(g^2)$ is the **genetic variance**, the middle term $[E(e^2)]$ is the **environmental variance**, and the last term is twice the covariance between genotype and environment.

In controlled experiments, different genotypes are placed into different environments at random. In other words, genotype and environment are independent. If genotype and environment are independent, then the covariance between genotype and environment $E(ge) = 0$, and the equation reduces to

$$V_X = E(g^2) + E(e^2)$$
$$= V_g + V_e$$

Thus, the phenotypic variance is the sum of the variance due to the different genotypes in the population and the variance due to the different environments within which the organisms are reared.

Finally, to estimate V_e, we calculate the variance of the means among the three environments. The result is $V_e = 2.67$ days^2. Thus, the phenotypic variance (14.67) is equal to the genetic variance (12.0) plus the environmental variance (2.67). The equation works for these data because genotype and environment are not correlated.

If we calculate the standard deviations for the data in Table 19-2 (experiment I), we observe that the phenotypic standard deviation (3.83) is not the sum of the genetic (3.46) and environmental (1.63) standard deviations. Variances can be decomposed into difference sources. Standard deviations cannot be decomposed in this manner. In Section 19.3, we will see how this property of the variance is helpful for quantifying the extent to which trait variation is heritable versus environmental.

Finally, let's look at what would happen to the variances if genotype and environment are correlated. To do this, imagine that we knew the genetic deviations (g) for nine Thoroughbred horses for the time it takes them to run the Kentucky Derby. We also know the environmental deviations (e) that their trainers contribute to the time it takes each horse to run this race. We will suppose that besides training, there are no other sources of environmental variation. The population mean for this set of Thoroughbreds is 123 seconds to run the Derby. We assign the best horses to the best trainers and the worst horses to the worst trainers. By doing this, we have created a nonrandom relationship or correlation between horses (genotypes) and trainers (environments).

Table 19-3 shows the data for this imaginary experiment. You'll notice that V_X (6.67) is not equal to the sum of V_g (2.22) and V_e (1.33). Because genotype and environment are correlated, we violated the assumption of the equation that states $V_X = V_g + V_e$. The equation works only when genotype and environment are uncorrelated.

Correlation between variables

Let's look a little more closely at the concept of **correlation**, the existence of a relationship between two variables. This is a critical concept in quantitative genetics, as we will see throughout this chapter.

To visualize the degree of correlation between two variables, we can construct scatter plots, or scatter diagrams. **Figure 19-4** shows the scatter plots that we would see under several different strengths of correlation between two variables. These plots use simulated data for the heights of imaginary sets of identical adult male twins. The top panel of the figure shows a perfect correlation, which is what we would see if the height of one twin was exactly the same as that of the other twin for all sets of twins. The middle panel shows a strong but not perfect correlation. Here, when one twin is short, the other also tends to be short, and when one is tall, the other tends to be tall. The bottom panel shows the relationship we would see if the height of one twin was uncorrelated with that of the other twin of the set. Here, the height of one twin of each set is random with respect to the other twin of the set. In the next section, we will see that the data for real twins would look something like the middle panel.

In statistics, there is a specific measure of correlation called the **correlation coefficient**, which is symbolized by a lowercase r. It is a measure of association between two variables. The correlation coefficient is related to the covariance, which was introduced in Box 19-1; however, it is scaled to vary between -1 and $+1$. If we symbolize one random variable by X and the other by Y, then the correlation coefficient between X and Y is

$$r_{X,Y} = \frac{COV_{X,Y}}{\sqrt{V_X V_Y}}$$

The term $\sqrt{V_X V_Y}$ is used to scale the covariance to vary between -1 and $+1$. The expanded equation for the correlation coefficient is

$$r_{X,Y} = \frac{\Sigma(X_i - \bar{X})(Y_i - \bar{Y})}{\sqrt{\Sigma(X_i - \bar{X})^2 \Sigma(Y_i - \bar{Y})^2}}$$

The equation is cumbersome, and in practice, the calculation of correlation coefficients is done with the aid of

TABLE 19-3	Simulated Data for Time in Seconds (X) that Horses Run the Kentucky Derby Decomposed into the Genetic (g) and Environmental (e) Deviations from the Population Mean					
Horse	**Population mean**	**g**	**Trainer**	**e**	**x**	**X**
Secretariat	123	−2	Lucien	−2	−4	119
Decidedly	123	−2	Horatio	−1	−3	120
Barbaro	123	−1	Mike	−1	−2	121
Unbridled	123	−1	Carl	0	−1	122
Ferdinand	123	0	Charlie	0	0	123
Cavalcade	123	1	Bob	0	1	124
Meridian	123	1	Albert	1	2	125
Whiskery	123	2	Fred	1	3	126
Gallant Fox	123	2	Jim	2	4	127
Mean (sec)	123	0		0	0	123
Variance (sec^2)		2.22		1.33	6.67	6.67

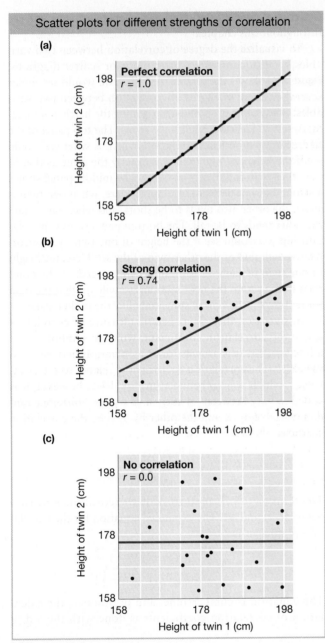

Scatter plots for different strengths of correlation

(a) Perfect correlation $r = 1.0$

(b) Strong correlation $r = 0.74$

(c) No correlation $r = 0.0$

FIGURE 19-4 Scatter plots for the case of a perfect correlation (a), strong correlation (b), and no correlation (c). Red lines have a slope that is equal to the correlation coefficient.

computers. For two variables that are perfectly correlated, $r = +1.0$ if as one variable gets larger, the other gets larger, or $r = -1.0$ if as one gets larger, the other gets smaller. For completely independent variables, $r = 0.0$.

In Figure 19-4, the correlation coefficient is shown on each panel. It is 1.0 in the top panel for a perfect positive correlation, 0.74 in the middle panel for a strong correlation, and 0.0 in the bottom panel for no correlation (X and Y are independent). The slope of the red line in each panel is equal to the correlation coefficient and provides a visual indicator of the strength of the correlation.

As an exercise, use the data in Table 19-3 to construct a scatter diagram and calculate the correlation coefficient.

This would best be done with a computer and spreadsheet software. Use the genetic deviations (g) for the x-axis and the environmental deviations (e) for the y-axis. Then calculate the correlation coefficient between g and e. The scatter diagram will be similar to the one in Figure 19-4 (middle panel), and the correlation coefficient will be 0.90. Thus, when the best horses are placed with the best trainers, genetics and environment are correlated and the $V_X = V_g + V_e$ model cannot be used.

KEY CONCEPT An individual's phenotype for a trait can be expressed in terms of its deviation from the population mean. The phenotypic deviation (x) of an individual is composed of the sum of its genetic deviation (g) and its environmental deviation (e). The phenotypic variation in a population for a trait (V_X) can be decomposed into the genetic (V_g) and the environmental (V_e) variances.

19.3 BROAD-SENSE HERITABILITY: NATURE VERSUS NURTURE

LO 19.4 Calculate and interpret broad-sense heritability.

A key question in genetics is, how much of the variation in a population is due to genetic factors and how much is due to environmental factors? In the popular press, this question is often phrased in terms of *nature* versus *nurture*—that is, what is the influence of innate (genetic) factors compared to external (environmental) factors? Answers to some nature-versus-nurture questions are of practical importance. If high blood pressure is primarily due to lifestyle choices (environment), then changes in diet or exercise habits would be most appropriate. However, if high blood pressure is largely predetermined by our genes, then drug therapy may be recommended.

Quantitative geneticists have developed the statistical tools needed to estimate the extent to which variation in complex traits is due to genes versus the environment. Below, we will describe these tools. At the end of this section, we will discuss the assumptions underlying these estimates and the limits to their utility.

Let's begin by defining **broad-sense heritability** (H^2) as the proportion of the phenotypic variance that is due to genetic differences among individuals in a population. Mathematically, we write this as the ratio of the genetic variance to the total variance in the population:

$$H^2 = \frac{V_g}{V_X}$$

The H is squared because it is the ratio of two variances, which are measured in squared units. H^2 can vary from 0 to 1.0. When all of the variation in a population is due to environmental sources and there is no genetic variation, then H^2 is 0. When all of the variation in a population is due to genetic sources, then V_g equals V_X and H^2 is

1.0. H^2 is called "broad sense" because it encompasses several different ways by which genes contribute to variation. For example, some of the variation will be due to the contributions of individual genes. Additional genetic variation can be contributed by the way genes work together, the interactions between genes known as epistasis.

In Section 19.2, we showed how we can calculate the genetic and environmental variances when we have inbred lines or clones. For the imaginary example of days to pollen shed for maize inbred lines in Table 19-2 (experiment I), we saw that V_g is 12.0 and V_X is 14.67. Using these values, the heritability of the trait is 12.0/14.67 = 0.82, or 82 percent. This estimate of H^2 tells us that genes contribute most of the variation and environmental factors contribute a more modest share of the variation. Thus, we might conclude that days to pollen shed is a highly heritable trait in maize.

Let's look at the data for experiment II in Table 19-2. The genotypes are exactly the same as in experiment I; these are the genotypes of the inbred lines A through J. In this case, however, the lines are reared in more extreme environments. If we calculate the variance for the means of the inbred line in experiment II, V_g will be 12.0 days^2 as in experiment I. Since the genotypes are the same in both experiments, the genetic variance is the same. If we calculate the variance for the means of the different environments (V_e) in experiment II, we will obtain 24.0 days^2, which is much larger than the value for V_e in experiment I (2.67). Since the environments are more extreme, the environmental variance is larger. Finally, if we calculate H^2 for experiment II, we obtain

$$H^2 = \frac{V_g}{V_g + V_e} = \frac{12}{12 + 24} = 0.33$$

The estimate of H^2 for experiment II is on the small side—closer to 0 than to 1. Thus, we might conclude that days to pollen shed is *not* a highly heritable trait in maize.

The contrast between the estimates of the heritability for the same set of maize inbred lines reared in different environments highlights the point that *heritability is the proportion of the phenotypic variance* (V_X) *due to genetics*. Since $V_X = V_g + V_e$, as V_e increases, then V_g will represent a smaller part of V_X and H^2 will go down. Similarly, if the environmental variance is kept to a minimum, then V_g will represent a larger part of V_X and H^2 will go up. H^2 is a moving target, and results from one study may not apply to another.

Measuring heritability in humans using twin studies

How can we measure heritability in humans? Although we don't have inbred lines for humans, we do have genetically identical individuals—monozygotic or identical twins (**Figure 19-5**). In most cases, identical twins are raised in the same household and so experience a similar environment. When individuals with the same genotypes are reared in the same environments, we have violated the assumption of our genetic model that genes and environment are independent.

So, to estimate heritability in humans, we need to use sets of identical twins who were separated shortly after birth and reared apart by unrelated adoptive parents.

The equation for estimating H^2 in studies of identical twins who are reared apart is relatively simple. It makes use of the statistical measure called the covariance, which was introduced in Box 19-1. As explained in **Box 19-2**, the covariance between identical twins who are reared apart is equal to the genetic variance (V_g). Thus, we can estimate H^2 in humans by using this covariance as the numerator and the trait variance (V_X) as the denominator:

Monozygotic twins are genetically identical

FIGURE 19-5 A set of identical twins. [*Barbara Penoyar/Getty Images.*]

$$H^2 = \frac{COV_{X',X''}}{V_X}$$

Here is how it's done. For each set of twins, let's designate the trait value for one twin as X' and the other as X''. If we have n sets of twins, then the trait values for the n sets could be designated $X_1' \, X_1''$, $X_2' \, X_2''$, ... $X_n' \, X_n''$.

Suppose we had IQ measurements for five sets of twins as follows:

	Twin	
	X'	**X''**
1	100	110
2	125	118
3	97	90
4	92	104
5	86	89

Using these data and the formula for the covariance from Box 19-1, we calculate that $COV_{X',X''}$ is 119.2 points^2. Using the formula for trait variance, we would calculate that the value of V_X is 154.3 points^2. Thus, we obtain

$$H^2 = \frac{119.2 \text{ points}^2}{154.3 \text{ points}^2} = 0.77$$

The points^2 in the numerator and denominator cancel out, and we are left with a unitless measure that is the proportion of the total variance that is due to genetics.

Box 19-2 provides some additional details about estimating H^2 from twin data, including the derivation of the formula we just used. It also discusses the relationship between the ratio $COV_{X',X''} / V_X$ and the correlation coefficient. Quantitative geneticists have developed several means for estimating heritability using the correlation among relatives. Identical twins share 100 percent of their genes, while brothers, sisters, and dizygotic twins share 50 percent of their genes. The strength of the correlation

BOX 19-2 Estimating Heritability from Human Twin Studies

If we had many sets of identical twins who were reared apart, how could we use them to measure H^2? Let's represent the trait value for one member of each pair of twins as X' and the trait value for the other as X''. We have many (n) sets of twins: $X_1' X_1'', X_2' X_2'', \ldots X_n' X_n''$. We can express the phenotypic deviations for one set of twins as the sum of their genetic and environmental deviations,

$$x' = g + e' \text{ and } x'' = g + e''$$

using x' as the deviation for one twin and x'' for the other twin. Notice that g is the same because the twins are genetically identical, but e' and e'' are different because the twins were reared in separate households. Next, we develop an expression for the covariance between the twins. In Box 19-1, we saw that the covariance is the average or expected value of the cross products $E(xy)$. Using our notation for twins, x' and x'', in place of x and y, we get

$$COV_{X',X''} = E(x'x'')$$

We can substitute $(g + e')$ for x' and $(g + e'')$ for x'', giving us

$$
\begin{aligned}
COV_{X',X''} &= E[(g \times e')(g \times e'')] \\
&= E(g^2 + ge' + ge'' + e'e'') \\
&= E(g^2) + E(ge') + E(ge'') + E(e'e'')
\end{aligned}
$$

Let's consider the last three terms of this expression. Under our model, the twins are assigned randomly to households, and thus there should be no correlation between the environments to which the X' and X'' twin of each pair are assigned. Accordingly, the covariance

between the environments $[E(e'e'')]$ will be 0.0. Similarly, because the assignment of twins to households is random, we expect no correlation between the genetic deviation of twins (g) and the household to which they are assigned, so $E(ge')$ and $E(ge'')$ will be 0.0. Therefore, the equation for the covariance among twins reduces to

$$COV_{X',X''} = E(g^2) = V_g$$

In other words, the covariance among identical twins reared apart is equal to the genetic variance. If we have a large set of identical twins who were reared apart, we can use the covariance between the twins to estimate the amount of genetic variation for a trait in the population. If we divide this covariance by the phenotypic variance, then we have an estimate of H^2:

$$H^2 = \frac{COV_{X',X''}}{V_X}$$

This equation is essentially the correlation coefficient between the twins. The variance for the twin of each set designated X' and that for the twin designated X'' are expected to be the same over a large sample. Thus, we can rewrite the denominator of the equation as follows:

$$r_{X',X''} = \frac{COV_{X',X''}}{\sqrt{V_{X'}V_{X''}}} = H^2$$

and we will see that H^2 is equivalent to the correlation between twins.

---

between different types of relatives can be scaled for the proportion of their genes that they share and the results used to estimate the genetic and environmental contributions to trait variation.

Over the last 100 years, there have been extensive genetic studies of twins and other sets of relatives. A great deal has been learned about heritable variation in humans from these studies. Table 19-4 lists some results from twin studies. It may or may not be surprising to you, but there is a genetic contribution to the variance for many different traits, including physique, physiology, personality attributes, psychiatric disorders, and even our social attitudes and political beliefs. We readily observe that traits such as hair and eye color run in families, and we know these traits are the manifestation of genetically controlled biochemical, developmental processes. In this context, it is not so surprising that other aspects of who we are as people also have a genetic influence.

Twin studies and the estimates of heritability that they provide can easily be over- or misinterpreted. Here are a few important points to keep in mind. First, H^2 is a property of a particular population and environment. For this reason, estimates of H^2 can differ widely among different

populations and environments. We saw this phenomenon above in the case of the days to pollen shed for maize inbred lines. Second, the twin sets used in many studies were separated at birth and placed into adoptive homes. Adoption agencies do not assign babies randomly to the full range of households in a society; rather, they place babies in economically, socially, and emotionally stable households. As a result, V_e is smaller than in the general population, and the estimate of H^2 will be inflated. Accordingly, the published estimates likely lead us to underestimate the importance of environment and overestimate the importance of genetics. Third, for twins, prenatal effects could cause a positive correlation between genotype and environment. As we saw in the earlier case of Thoroughbreds and jockeys, such a correlation violates our model and will bias H^2 upward.

Finally, heritability is not useful for interpreting differences between groups. Table 19-4 shows that the heritability for height in humans can be very high: 0.88. However, this high value for heritability does not tell us anything about whether groups with different heights differ because of genetics or the environment. For example, men in the Netherlands today average 184 cm in height, while around 1800, men in the Netherlands were about 168 cm tall on

TABLE 19-4	Broad-Sense Heritability for Some Traits in Humans as Determined by Twin Studies	
Trait		H^2
Physical attributes		
Height		0.88
Chest circumference		0.61
Waist circumference		0.25
Fingerprint ridge count		0.97
Systolic blood pressure		0.64
Heart rate		0.49
Mental attributes		
IQ		0.69
Speed of spatial processing		0.36
Speed of information acquisition		0.20
Speed of information processing		0.56
Personality attributes		
Extraversion		0.54
Conscientiousness		0.49
Neuroticism		0.48
Positive emotionality		0.50
Antisocial behavior in adults		0.41
Psychiatric disorders		
Autism		0.90
Schizophrenia		0.80
Major depression		0.37
Anxiety disorder		0.30
Alcoholism		0.50–0.60
Beliefs and political attitudes		
Religiosity among adults		0.30–0.45
Conservatism among adults		0.45–0.65
Views on school prayer		0.41
Views on pacifism		0.38

Sources: J. R. Alford et al., *American Political Science Review* 99, 2005, 1-15; T. Bouchard et al., *Science* 250, 1990, 223–228; T. Bouchard, *Curr. Dir. Psych. Sci.* 13, 2004, 148–151; P. J. Clark, *Am. J. Hum. Genet.* 7, 1956, 49–54; C. M. Freitag, *Mol. Psychiatry* 12, 2007, 2–22.

average, a 16-cm difference. The gene pool of the Netherlands has probably not changed appreciably over that time, so genetics cannot explain the huge difference in height between the current population and the one of 200 years ago. Rather, improvements in health and nutrition are the likely cause. Thus, even though height is highly heritable and the past and present Dutch populations differ greatly in height, the difference has an environmental basis.

KEY CONCEPT Broad-sense heritability (H^2) is the ratio of the genetic (V_g) to the phenotypic (V_X) variance. H^2 provides a measure of the extent to which differences among individuals within a population are due to genetic versus environmental factors. Estimates of H^2 apply only to the population and environment in which they were made. H^2 is not useful for interpreting differences in trait means among populations.

19.4 NARROW-SENSE HERITABILITY: PREDICTING PHENOTYPES

LO 19.5 Calculate and interpret narrow-sense heritability.
LO 19.6 Use knowledge of parental phenotypes to predict the phenotype of offspring.

Broad-sense heritability tells us the proportion of the variance in a population that is due to genetic factors. Broad-sense heritability expresses the degree to which the differences in the phenotypes among individuals in a population are determined by differences in their genotypes. However, even when there is genetic variation in a population as measured by broad-sense heritability, it may not be transmissible to the next generation in a predictable way. In this section, we will explore how genetic variation comes in two forms—additive and dominance (nonadditive) variation. Whereas additive variation is predictably transmitted from parent to offspring, dominance variation is not. We will also define another form of heritability called **narrow-sense heritability** (h^2), which is the ratio of the additive variance to the phenotypic variance. Narrow-sense heritability provides a measure of the degree to which the genetic constitution of individuals determines the phenotypes of their offspring.

The different modes of **gene action** (interaction among alleles at a locus) are at the heart of understanding narrow-sense heritability, so we will briefly review them. Consider a locus, B, that controls the number of flowers on a plant. The locus has two alleles, B_1 and B_2, and three genotypes—B_1/B_1, B_1/B_2, and B_2/B_2. As diagrammed in **Figure 19-6a**, plants with the B_1/B_1 genotype have 1 flower,

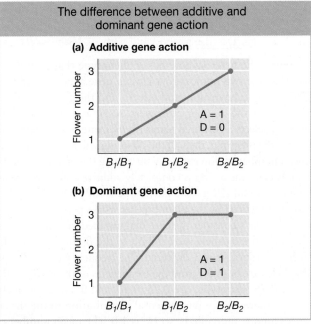

FIGURE 19-6 Plot of genotype (x-axis) by phenotype (y-axis) for a hypothetical locus, B, that regulates number of flowers per plant. (a) Additive gene action. (b) Dominant gene action.

B_1/B_2 plants have 2 flowers, and B_2/B_2 plants have 3 flowers. In a case like this, when the heterozygote's trait value is midway between those of the two homozygous classes, gene action is defined as **additive gene action**. In Figure 19-6b, the heterozygote has 3 flowers, the same as the B_2/B_2 homozygote. Here, the B_2 allele is dominant to the B_1 allele. In this case, the gene action is defined as **dominant gene action**. (We could also define this gene action as recessive with the B_1 allele being recessive to the B_2 allele.) Gene action need not be purely additive or dominant but can show **partial dominance**. For example, if B_1/B_2 heterozygotes had 2.5 flowers on average, then we would say that the B_2 allele shows partial dominance.

Gene action and the transmission of genetic variation

Let's work through a simple example to show how the mode of gene action influences heritability. Suppose a plant breeder wants to create an improved plant population with more flowers per plant. Flower number is controlled by the B locus, which has two alleles, B_1 and B_2, as diagrammed in Figure 19-6a. The frequencies of the B_1 and B_2 alleles are both 0.5, and the frequencies of the B_1/B_1, B_1/B_2, and B_2/B_2 genotypes are 0.25, 0.50, and 0.25, respectively. Plants with the B_1/B_1 genotype have 1 flower, B_1/B_2 plants have 2 flowers, and B_2/B_2 plants have 3 flowers. The mean number of flowers per plant in the population is 2.0. (Remember that we can calculate the mean as the sum of the products of frequency of each class times the value for that class.)

Genotype	Frequency	Trait value (no of flowers)	Contribution to the mean (frequency × value)
B_1/B_1	0.25	1	0.25
B_1/B_2	0.50	2	1.0
B_2/B_2	0.25	3	0.75
			Mean = 2.0

Since the heterozygote has a phenotype that is midway between the two homozygous classes, gene action is additive. There are no environmental effects, and the genotype alone determines the number of flowers, so H^2 is 1.0. If the plant breeder selects 3-flowered plants (B_2/B_2), intermates them, and grows the offspring, then all the offspring will be $B_2 B_2$, and the mean number of flowers per plant among the offspring will be 3.0. When gene action is completely additive and there are no environmental effects, the phenotype is fully heritable. Selection as practiced by the plant breeder works perfectly.

Now let's consider the case diagrammed in Figure 19-6b, in which the B_2 allele is dominant to the B_1. In this case, the $B_1 B_2$ heterozygote is 3-flowered. The frequency of the B_1 and B_2 alleles are both 0.5, and the frequencies of the B_1/B_1, B_1/B_2, and B_2/B_2 genotypes are 0.25, 0.50, and 0.25, respectively. Again, there is no environmental contribution to the differences among individuals, so H^2 is 1.0. The mean number of flowers per plant in the starting population is 2.5.

Genotype	Frequency	Phenotype	Contribution to the mean (frequency × value)
B_1/B_1	0.25	1	0.25
B_1/B_2	0.50	3	1.5
B_2/B_2	0.25	3	0.75
			Mean = 2.5

If the plant breeder selects a group of 3-flowered plants, 2/3 will be B_1/B_2 and $1/3 B_2/B_2$. When the breeder intermates the selected plants, 0.44 ($2/3 \times 2/3$) of the crosses would be between heterozygotes, and 1/4 of the offspring from these crosses would be B_1/B_1 and thus 1-flowered. The remainder of the offspring would be either B_1/B_2 or B_2/B_2 and thus 3-flowered. The overall mean for the offspring would be 2.78, although the mean of their parents was 3.0. Hence, when there is dominance, the phenotype is not fully heritable. Selection as practiced by the plant breeder worked, but not perfectly, because some of the differences among individuals are due to dominance.

In conclusion, when there is dominance, we cannot strictly predict the offspring's phenotypes from the parents' phenotypes. Some of the differences (variation) among the individuals in the parental generation are due to the dominance interactions between alleles. Since parents transmit their genes but not their genotypes to their offspring, these dominance interactions are not transmitted to the offspring.

The additive and dominance effects

As described earlier, traits controlled by genes with additive gene action will respond very differently to selection than those with dominance. Thus, geneticists need to quantify the degree of dominance and additivity. In this section, we will see how this is done. Let's again consider the B locus that controls the number of flowers on a plant (see Figure 19-6). The **additive effect** (**A**) provides a measure of the degree of change in the phenotype that occurs with the substitution of one B_2 allele for one B_1 allele. The additive effect is calculated as the difference between the two homozygous classes divided by 2. For example, as shown in Figure 19-6a, if the trait value of the B_1/B_1 genotype is 1 and the trait value of the B_2/B_2 genotype is 3, then

$$A = \frac{X_{B_2 B_2} - X_{B_1 B_1}}{2} = \frac{3 - 1}{2} = 1$$

The **dominance effect** (**D**) is the deviation of the heterozygote (B_1/B_2) from the midpoint of the two homozygous classes. As shown in Figure 19-6b, if the trait value of the B_1/B_1 genotype is 1, of the B_1/B_2 genotype, 3, and of the B_2/B_2 genotype, 3, then

$$D = X_{B_1 B_2} - \left(\frac{X_{B_2 B_2} + X_{B_1 B_1}}{2} \right) = 3 - 2 = 1$$

If you calculate D for the situation depicted in Figure 19-6a, you'll find $D = 0$; that is, no dominance.

The ratio of D/A provides a measure of the degree of dominance. For Figure 19-6a, D/A = 0.0, indicating pure additivity or no dominance. For Figure 19-6b, D/A = 1.0, indicating complete dominance. A D/A ratio of −1 would indicate a complete recessive. (The distinction between dominance and recessivity depends on how the phenotypes are coded and is in this sense arbitrary.) Values that are greater than 0 and less than 1 represent partial dominance, and values that are less than 0 and greater than −1 represent partial recessivity.

Here is an example of calculating additive and dominance effects at a single locus. Three-spined sticklebacks (*Gasterosteus aculeatus*) have marine populations with long pelvic spines and populations that live near the bottoms of freshwater lakes with highly reduced pelvic spines (**Figure 19-7a**). The spines are thought to play a role in defense against predation. The bottom-dwelling freshwater populations are derived from the ancestral marine populations. A change in predation between the marine and freshwater environments may explain the loss of spines in the freshwater environments (see Chapter 20).

Pitx1 is one of several genes that contribute to pelvic-spine length in sticklebacks. This gene encodes a transcription factor that regulates the development of the pelvis in vertebrates, including the growth of pelvic spines in sticklebacks. Michael Shapiro and his colleagues at Stanford University measured the pelvic-spine length in an F_2 population that segregated for the marine or long (*l*) allele and freshwater or short (*s*) allele of *Pitx1*. They recorded the following mean values (in units of proportion of body length) for pelvic-spine length for the three genotypic classes:

s/s	s/l	l/l
0.068	0.132	0.148

Using these values and the formulas above, we can calculate the additive and dominance effects. The additive effect (A) is

$$(0.148 - 0.068)/2 = 0.04$$

or 4 percent of body length. The dominance effect (D) is

$$0.132 - [(0.148 + 0.068)/2] = 0.024$$

The dominance/additivity ratio is

$$0.024/0.04 = 0.6$$

The 0.6 value for the ratio indicates that the long (*l*) allele of *Pitx1* is partially dominant to the short (*s*) allele.

One can also calculate additive and dominance effects averaged over all the genes in the genome that affect the trait. Here is an example using cave fish (*Astyanax mexicanus*) and their surface relatives (Figure 19-7b). The cave populations have highly reduced (small-diameter) eyes compared to the surface populations. Populations colonizing lightless caves do not benefit from having eyes. Since there are physiological and neurological costs to forming

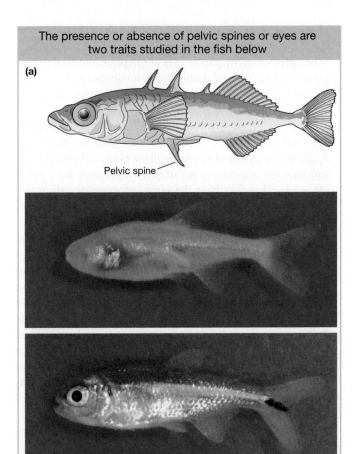

The presence or absence of pelvic spines or eyes are two traits studied in the fish below

(a)

Pelvic spine

FIGURE 19-7 (a) Three-spined stickleback (*Gasterosteus aculeatus*). (b) Blind cave fish (*Astyanax mexicanus*) (*top*) and its sighted, surface relative (*bottom*). [(a) *Masato Yoshizawa and William Jeffery, University of Maryland; (b) Masato Yoshizawa and William Jeffery, University of Maryland.*]

and maintaining eyes, evolution may have favored a reduction in the size of the eye in cave populations.

Horst Wilkins at the University of Hamburg measured mean eye diameter (in mm) for the cave and surface populations and their F_1 hybrid:

Cave	F_1	Surface
2.10	5.09	7.05

Using the formulas above, we calculate that A = 2.48, D = 0.52, and D/A = 0.21. In this case, gene action is closer to a purely additive state, although the surface genome is slightly dominant.

KEY CONCEPT When the trait value for the heterozygous class is midway between the two homozygous classes, gene action is called additive. Any deviation of the heterozygote from the midpoint between the two homozygous classes indicates a degree of dominance of one allele.

A model with additivity and dominance

The example provided earlier with the *B* locus and flower number shows that we cannot accurately predict offspring phenotypes from parental phenotypes when there is dominance, although we can do so in cases of pure additivity. When predicting the phenotypes of offspring, we need to separate the additive and dominance contributions. To do this, we need to modify the simple model introduced in Section 19.2, $x = g + e$.

Let's begin by looking more closely at the situation depicted in Figure 19-6b. Individuals with the B_1/B_2 and B_2/B_2 genotypes have the same phenotype, 3 flowers. If we subtract the population mean (2.5) from their trait value (3), we see that they have the same genotypic deviation (g):

$$g_{B_1B_2} = g_{B_2B_2} = 0.5$$

Now let's calculate the mean phenotypes of their offspring. If we self-pollinate a B_1/B_2 individual, the offspring will be $\frac{1}{4} B_1/B_1$, $\frac{1}{2} B_1/B_2$, and $\frac{1}{4} B_2/B_2$, and the mean trait value of these offspring would be 2.75. However, if we self-pollinate a B_2/B_2 individual, the offspring will all be B_2/B_2, and the mean trait value of these offspring would be 3.0. Even though the B_1/B_2 and B_2/B_2 individuals have the same trait value and the same value for their genotypic deviation (g), they do not produce the equivalent offspring because the underlying basis of their phenotypes is different. The phenotype of the B_1/B_2 individual depends on the dominance effect (D), while that of the B_2/B_2 individual does not involve dominance.

We can expand the simple model ($x = g + e$) to incorporate the additive and dominance contributions. The genotypic deviation (g) is the sum of two components: a, the additive deviation, which is transmitted to offspring; and d, the dominance deviation, which is not transmitted to offspring. We can rewrite the simple model and separate out these two components as follows:

$$x = g + e$$
$$x = a + d + e$$

The additive deviation is transmitted from parent to offspring in a predictable way. The dominance deviation is not transmitted from parent to offspring since new genotypes and thus new interactions between alleles are created each generation.

Let's look at how the genetic deviation is decomposed into the additive and dominance deviations for the case shown in Figure 19-6b.

	B_1B_1	B_1B_2	B_2B_2
Trait value	1	3	3
Genetic deviation (g)	−1.5	0.5	0.5
Additive deviation (a)	−1	0	1
Dominance deviation (d)	−0.5	0.5	−0.5

The genotypic deviations (g) are simply calculated by subtracting the population mean (2.5) from the trait value for each genotype. Each genotypic deviation is then decomposed into the additive (a) and dominance (d) deviations using formulas that are beyond the scope of this book. These formulas include the additive (A) and dominance (D) effects as well as

the frequencies of the B_1 and B_2 allele in the population. You will notice that $a + d$ sum to g. The additive (a) and dominance (d) deviations are dependent on the allele frequencies because the phenotype of an offspring receiving a B_1 allele from one parent will depend on whether that allele combines with a B_1 or B_2 allele from the other parent, and that outcome depends on the frequencies of the alleles in the population.

The additive deviation (a) has an important meaning in plant and animal breeding. It is the **breeding value**, or the part of an individual's deviation from the population mean that is due to additive effects. This is the part that is transmitted to its progeny. Thus, if we wanted to increase the number of flowers per plant in the population, the B_2/B_2 individuals have the highest breeding value. Breeding values can also be calculated for the genome overall for an individual. Animal breeders estimate the genomic breeding values of individual animals, and these estimates can determine the economic value of the animal.

We have partitioned the genetic deviation (g) into the additive (a) and dominance (d) deviations. Using algebra similar to that described in Box 19-1, we can also partition the genetic variance into the additive and dominance variances as follows:

$$V_g = V_a + V_d$$

where V_a is the **additive genetic variance** and V_d is the dominance variance. V_a is the variance of the additive deviations or the variance of the breeding values. It is the part of the genetic variation that is transmitted from parents to their offspring. V_d is the variance of the dominance deviations. Finally, we can substitute these terms in the equation for the phenotypic variance presented earlier in the chapter:

$$V_X = V_g + V_e$$
$$V_X = V_a + V_d + V_e$$

where V_e is the environmental variance. This equation assumes that the additive and dominance components are not correlated with the environmental effects. This assumption will be true in experiments in which individuals are randomly assigned to environments.

Thus far, we have described models with genetic, environmental, additive, and dominance deviations and variances. In quantitative genetics, the models can get even more complex. In particular, the models can be expanded to include interaction between factors. If one factor alters the effect of another factor, then there is an interaction. **Box 19-3** briefly reviews how interactions are factored into quantitative genetic models.

KEY CONCEPT The genetic deviation (g) of an individual from the population mean is composed of the additive deviation (a) and dominance deviation (d). The additive deviation is known as the breeding value, and it represents the component of an individual's phenotype that is transmitted to its offspring. The genetic variation for a trait in a population (V_g) can be decomposed into the additive (V_a) and the dominance (V_d) variances.

BOX 19-3 Interaction Effects

The simple model for decomposing traits into genetic and environmental deviations, $x = g + e$, assumes that there is no genotype–environment interaction. By this statement, we mean that the differences between genotypes do not change across environments. In other words, a *genotype–environment interaction* occurs when the performance of different genotypes is unequally affected by a change in the environment. Here is an example. Consider two inbred lines, IL1 and IL2, that have different genotypes. We rear both of these inbred lines in two environments, E1 or E2. We can visualize the performance of these two lines in the two environments using a graph (below). This type of graph, which shows the pattern of trait values of different genotypes across two or more environments, is called a *reaction norm.*

If there is no interaction, then the difference in trait value between the inbred lines will be the same in both environments, as shown by the graph on the left.

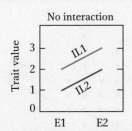
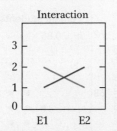

With no interaction, the difference between the two inbreds is 1.0 in both environments, and so the difference between the lines averaged over the two environments is 1.0.

$$\text{Environment 1: IL1} - \text{IL2} = 2 - 1 = 1.0$$
$$\text{Environment 2: IL1} - \text{IL2} = 3 - 2 = 1.0$$

The difference in the overall mean shows that the lines are genetically different. The mean over both environments is 2.5 for IL1 and 1.5 for IL2.

The graph on the right shows a case of an interaction between genotype and environment. IL1 does well in Environment 1 but poorly in Environment 2. The opposite is true for IL2. The difference in the trait value between the two lines is +1.0 in Environment 1 but −1.0 in Environment 2.

$$\text{Environment 1: IL1} - \text{IL2} = 2 - 1 = +1.0$$
$$\text{Environment 2: IL1} - \text{IL2} = 1 - 2 = -1.0$$

The difference between the lines averaged over the two environments is 0.0, so we might incorrectly conclude that these inbreds are genetically equivalent if we looked just at the overall mean.

The simple model can be expanded to include a genotype–environment interaction term $(g \times e)$:

$$x = g + e + g \times e$$

and

$$V_X = V_g + V_e + V_{g \times e}$$

where $V_{g \times e}$ is the variance of the genotype–environment interaction. If the interaction term is not included in the model, then there is an implicit assumption that there are no genotype–environment interactions.

Interactions can also occur between the alleles at separate genes. This type of interaction is called epistasis. Let's look at how epistatic interactions affect variation in quantitative traits.

Consider two genes, A with alleles A_1 and A_2, and B with alleles B_1 and B_2. The left side of the table below shows the case of no interaction between these genes. Starting with the A_1/A_1; B_1/B_1 genotype, whenever you substitute an A_2 allele for an A_1 allele, the trait value goes up by 1 regardless of the genotype at the B locus. The same is true when substituting alleles at the B locus. The effects of alleles at the A locus are independent of those at the B locus and vice versa. There is no interaction or epistasis.

	No interaction				Interaction		
	B_1/B_1	B_1/B_2	B_2/B_2		B_1/B_1	B_1/B_2	B_2/B_2
A_1/A_1	0	1	2	A_1/A_1	0	1	2
A_1/A_2	1	2	3	A_1/A_2	0	1	3
A_2/A_2	2	3	4	A_2/A_2	0	1	4

Now look at the right side of the table. Starting with the A_1/A_1; B_1/B_1 genotype, substituting an A_2 allele for an A_1 allele has an effect on the trait value only when the genotype at the B locus is B_2/B_2. The effects of alleles at the A locus are *dependent* of those at the B locus. There is an interaction or epistasis between the genes.

The genetic model can be expanded to include an epistatic or interaction term (i):

$$x = a + d + i + e$$

and

$$V_X = V_a + V_d + V_i + V_e$$

where V_i is the interaction or epistatic variance.

If the interaction term is not included in the model, then there is an implicit assumption that the genes work independently; that is, there is no epistasis. The interaction variance (V_i), like the dominance variance, is not transmitted from parents to their offspring, since new genotypes and thus new epistatic relationships are formed with each generation.

Narrow-sense heritability

We can now define *narrow-sense heritability*, which is symbolized by a lowercase h squared (h^2), as the ratio of the additive variance to the total phenotypic variance:

$$h^2 = \frac{V_a}{V_X} = \frac{V_a}{V_a + V_d + V_e}$$

This form of heritability measures the extent to which variation among individuals in a population is predictably transmitted to their offspring. Narrow-sense heritability is the form of heritability of interest to plant and animal breeders because it provides a measure of how well a trait will respond to selective breeding.

To estimate h^2, we need to measure V_a, but how can this be accomplished? Using algebra and logic similar to what we used to show that V_g can be estimated using the covariance between monozygotic twins reared separately (see Box 19-2), it can be shown that the covariance between a parent and its offspring is equal to one-half the additive variance:

$$COV_{P,O} = \frac{1}{2}V_a$$

The parent–offspring covariance is one-half of V_a because the offspring inherits only one-half of its genes from the parent. Combining this formula with the one for h^2, we get

$$h^2 = \frac{V_a}{V_X} = \frac{2COV_{P,O}}{V_X}$$

To estimate V_a using the covariance between parents and offspring requires controlling environmental factors in experiments. This can be a challenge because parents and offspring are necessarily reared at different times. V_a can also be estimated using the covariance between half-sibs, in which case all individuals in the experiment can be reared at the same time in the same environment. Half-sibs share one-fourth of their genes, so V_a equals $4 \times$ the covariance between half-sibs.

If you compare the equation for h^2 to the one for H^2 (see Box 19-2), you will see that both involve the ratio of a covariance to a variance. The correlation coefficient introduced earlier in the chapter is also the ratio of a covariance to a variance. We are using the degree of correlation among relatives to infer the extent to which traits are heritable.

Here is an exercise that your class can try. Have each student submit his or her height and the height of their same-sex parent. Using these data and spreadsheet computer software, calculate the covariance between parents and their offspring (the students). Then estimate h^2 as two times the covariance divided by the phenotypic variance. For the total phenotypic variance (V_X) in the denominator of the equation, you can use the variance among the parents. Data for male and female students should be analyzed separately.

Typically, values for narrow-sense heritability of height in humans are about 0.8, meaning that about 80 percent of the variance is additive, or transmissible, from parent

to offspring. The results for your class could deviate from this value for several reasons. First, if your class is small, sampling error can affect the accuracy of your estimate of h^2. Second, you will not be conducting a randomized experiment. If parents re-create in their households the growth-promoting (or growth-limiting) environments that they experienced as children, then there will be a correlation between the environments of the parents and their offspring. This correlation of environments violates an assumption of the analysis. Third, the population of students in your class may not be representative of the population in which the 0.8 value was obtained.

Figure 19-8 is a scatter plot with the height data for male and female students and their parents. There is a clear correlation between the heights of the students and their same-sex parent. These data give estimates of narrow-sense heritability of 0.86 for mother–daughter and 0.82 for father–son.

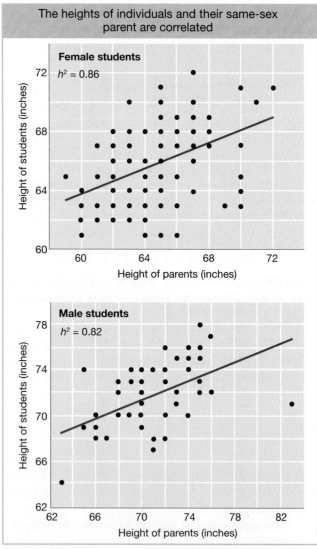

FIGURE 19-8 Scatter diagrams for height in inches of female (*top*) and male (*bottom*) students and their same-sex parent. The plots show positive correlations between the heights of the students and their parents. The slope of the diagonal line is equal to the correlation coefficient.

The results are close to the value of h^2 equals 0.8 obtained from studies in which the children were separated at birth from their parents and reared in adoptive households.

Here are a few more points about narrow-sense heritability. First, when $h^2 = 1.0$ ($V_a = V_X$), the expected value for an offspring's phenotype will equal the mid-parent value. All the variation in the population is additive and heritable in the narrow sense. Second, when $h^2 = 0.0$ ($V_a = 0$), the expected value of any offspring's phenotype will be the population mean. All the variation in the population is due either to dominance or to environmental factors, and thus it is not transmissible to offspring. Finally, as with broad-sense heritability (H^2), narrow-sense heritability is the property of the specific environment and population in which it was measured. An estimate from one population and environment may not be meaningful for another population or environment.

Narrow-sense heritability is an important concept both in plant and animal breeding and in evolution. For a breeder, h^2 indicates which traits can be improved by artificial selection. For an evolutionary biologist, h^2 is critical to understanding how populations will change in response to natural selection imposed by a changing environment. **Table 19-5** lists estimates of narrow-sense heritability for some traits and organisms.

Predicting offspring phenotypes

In order to efficiently improve crops and livestock for traits of agronomic importance, the breeder must be able to predict an offspring's phenotype from its parents' phenotypes. Such predictions are made using the breeder's knowledge of narrow-sense heritability. An individual's phenotypic deviation (x) from the population mean is the sum of the additive, dominance, and environmental deviations:

$$x = a + d + e$$

TABLE 19-5	Narrow-Sense Heritability for Some Traits in Several Different Species
Trait	**h^2 (%)**
Agronomic species	
Body weight in cattle	65
Milk yield in cattle	35
Back-fat thickness in pig	70
Litter size in pig	5
Body weight in chicken	55
Egg weight in chicken	50
Natural species	
Bill length in Darwin's finch	65
Flight duration in milkweed bug	20
Plant height in jewelweed	8
Fecundity in red deer	46
Life span in collared flycatchers	15

Source: D. F. Falconer and T. F. C. Mackay, *Introduction to Quantitative Genetics*, Longman, 1996; J. C. Conner and D. L. Hartl, *A Primer in Ecological Genetics*, Sinauer, 2004.

The additive part is the heritable part that is transmitted to the offspring. Let's look at a set of parents with phenotypic deviations x' for the mother and x'' for the father. The parents' dominance deviations (d' and d'') are not transmitted to their offspring since new genotypes and new dominance interactions are created with each generation. Similarly, the parents do not transmit their environmental deviations (e' and e'') to their offspring.

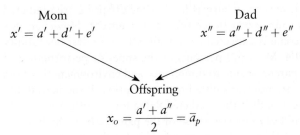

Thus, the only factors that parents transmit to their offspring are their additive deviations (a' and a''). Accordingly, we can estimate the offspring's phenotypic deviation (x_o) as the mean of the additive deviations of its parents $\bar{a}_p$.

So, to predict the offspring's phenotype, we need to know its parents' additive deviations. We cannot directly observe the parents' additive deviations, but we can estimate them. The additive deviation of an individual is the heritable part of its phenotypic deviation; that is,

$$\hat{a} = h^2 x$$

where $\hat{a}$ signifies an estimate of the additive deviation or breeding value. Thus, we can estimate the mean of the parents' additive deviations as the product of h^2 times the mean of their phenotypic deviation and this product will be an estimate of the phenotypic deviation of the offspring ($\hat{x}_o$):

$$\hat{x}_o = h^2 \left(\frac{x' + x''}{2} \right)$$

or

$$\hat{x}_o = h^2 \bar{x}_p$$

The offspring will have its own dominance and environmental deviations. However, these cannot be predicted. Since they are deviations, they will be zero on average over a large number of offspring.

Here is an example. Icelandic sheep are prized for the quality of their fleece. The average adult sheep in a particular population produces 6 lb of fleece per year. A sire that produces 6.5 lb per year is mated with a dam that produces 7.0 lb per year. The narrow-sense heritability of fleece production in this population is 0.4. What is the predicted fleece production for offspring of this mating? First, calculate the phenotypic deviations for the parents by subtracting the population mean from their phenotypic values:

Sire	$6.5 - 6.0 = 0.5$
Dam	$7.0 - 6.0 = 1.0$
Parent mean ($\bar{x}_p$)	$(0.5 + 1.0)/2 = 0.75$

Now multiply h^2 by $\bar{x}_p$ to determine $\hat{x}_0$, the estimated phenotypic deviation of the offspring:

$$0.4 \times 0.75 = 0.3$$

Finally, add the population mean (6.0) to the predicted phenotypic deviation of the offspring (0.3) and obtain the result that the predicted phenotype of the offspring is 6.3 lb of fleece per year.

It may seem surprising that the offspring are predicted to produce less fleece than either parent. However, this outcome is expected for a trait with a modest heritability of 0.4. Most (60 percent) of the superior performance of the parents is due to dominance and environmental factors that are not transmitted to the offspring. If the heritability were 1.0, then the predicted value for the offspring would be midway between those of the parents. If the heritability were 0.0, then the predicted value for the offspring would be at the population mean since all the variation would be due to nonheritable factors.

Selection on complex traits

Our final topic regarding narrow-sense heritability is the application of selection over the long term to improve the performance of a population for a complex trait. By applying selection, plant breeders over the past 10,000 years transformed a host of wild plant species into the remarkable array of fruit, vegetable, cereal, and spice crops that we enjoy today. Similarly, animal breeders applied selection to domesticate many wild species, transforming wolves into dogs, jungle fowl into chickens, and wild boar into pigs.

Selection is a process by which only individuals with certain features contribute to the gene pool that forms the next generation (see Chapters 18 and 20). Selection applied by humans to improve a crop or livestock population is termed *artificial selection* to distinguish it from natural selection. Let's look at an example of how artificial selection works.

Provitamin A is a precursor in the biosynthesis of vitamin A, an important nutrient for healthy eyes and a well-functioning immune system. Plant products are an important source of provitamin A for humans; however, people in many areas of the globe have too little provitamin A in their diets. To solve this problem, a plant breeder seeks to increase the provitamin A content of a maize population used in parts of Latin America where vitamin A deficiency is common. At present, this population produces 1.25 μg of provitamin A per gram of kernels. The variance for the population is 0.06 μg^2 (**Figure 19-9**). To improve the population, the breeder selects a group of plants that produce 1.5 μg or more of provitamin A per gram of kernels. The mean for the selected group is 1.63 μg. The breeder randomly intermates the selected plants and grows the offspring to produce the next generation, which has a mean of 1.44 μg per gram of kernels.

If the narrow-sense heritability of a trait is not known before performing an artificial selection experiment, one

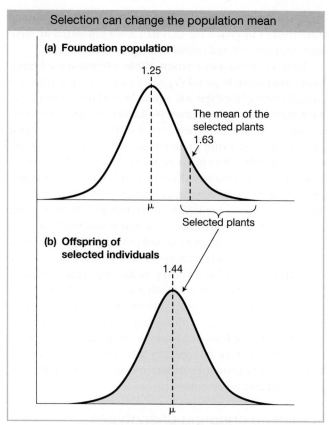

FIGURE 19-9 Distribution of trait values for provitamin A in maize kernels in a starting population (a) and offspring population (b) after one generation of selection. The starting population had a mean of 1.25 μg/g, the selected individuals a mean of 1.63 μg/g, and the offspring population a mean of 1.44 μg/g.

can use the results of such experiments to estimate it. Here is an example using the case of provitamin A in maize. Let's start with the equation from above:

$$\hat{x}_0 = h^2 \bar{x}_p$$

and rewrite it as

$$h^2 = \frac{\bar{x}_o}{\bar{x}_p}$$

$\bar{x}_p$ is the mean deviation of the parents (the selected plants) from the population mean. This is known as the **selection differential (S)**, the difference between the mean of the selected group and that of the base population. For our example,

$$\bar{x}_p = 1.63 - 1.25 = 0.38$$

$\bar{x}_p$ is the mean deviation of the offspring from the population mean. This is known as the **selection response (R)**, the difference between the mean of the offspring and that of the base population. For our example,

$$\bar{x}_o = 1.44 - 1.25 = 0.19$$

Now we can calculate the narrow-sense heritability for this trait in this population as

$$h^2 = \frac{R}{S} = \frac{\bar{x}_o}{\bar{x}_p} = \frac{0.19}{0.38} = 0.5$$

The underlying logic of this calculation is that the response represents the heritable or additive part of the selection differential.

Over the last century, quantitative geneticists have conducted a large number of selection experiments like this. Typically, these experiments are performed over many generations and are referred to as long-term selection studies. Each generation, the best individuals are selected to produce the subsequent generation. Such studies have been performed in economically important species such as crop plants and livestock and in many model organisms such as *Drosophila*, mice, and nematodes. This work has shown that virtually any species will respond to selection for virtually any trait. Populations contain deep pools of additive genetic variation.

Here are two examples of long-term selection experiments. In the first experiment, fruit flies were selected for increased flight speed over a period of 100 generations (**Figure 19-10a**). Each generation, the speediest flies were selected and bred to form the next generation. Over the 100 generations, the average flight speed of the flies in the population increased from 2 to 170 cm/sec, and neither the flies nor the gains made by selection showed any signs of slowing down after 100 generations. In the second experiment, mice were selected over 10 generations for the amount of "wheel running" they did per day (Figure 19-10b). There was a 75 percent increase over just 10 generations. These studies and many more like them demonstrate the tremendous power of artificial selection.

KEY CONCEPT Narrow-sense heritability (h^2) is the proportion of the phenotypic variance that is attributable to additive effects. This form of heritability measures the extent to which variation among individuals in a population is predictably transmitted to their offspring. Narrow-sense heritability is an important quantity in plant and animal breeding since it provides a measure of how well a trait will respond to selective breeding.

19.5 MAPPING QTL IN POPULATIONS WITH KNOWN PEDIGREES

LO 19.7 Determine how many genes contribute to the genetic variation for a trait.

The genes that control variation in quantitative (or complex) traits are known as **quantitative trait loci**, or **QTL** for short. As we will see below, QTL are genes just like any others that you have learned about in this book. They may encode metabolic enzymes, cell-surface proteins, DNA-repair enzymes, transcription factors, or any of many other classes of genes. What is of interest here is that QTL have allelic variants that typically make relatively small, quantitative contributions to the phenotype.

We can visualize the contributions of the alleles at a QTL to the trait value by looking at the frequency distributions associated with each genotype at a QTL as shown in **Figure 19-11**. The QTL locus is *B* and the genotypic classes are *B/B*, *B/b*, and *b/b*. The *B/B* individuals tend to have higher trait values, *B/b* intermediate values, and *b/b* small values. However, their distributions overlap, and we cannot determine genotype simply by looking at an individual's phenotype as we can for genes that segregate in Mendelian ratios. In Figure 19-11, an individual with an intermediate trait value could be *B/B*, *B/b*, or *b/b*.

Because of this property of QTL, we need special tools to determine their location in the genome and characterize their effects on traits. One way to do this is by a form of analysis called **QTL mapping**. The fundamental idea behind QTL mapping is that the location of QTL in the genome can be identified using marker loci linked to a QTL. Here is how the method works. Suppose you make a

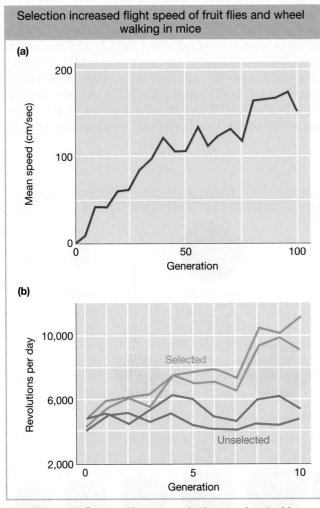

FIGURE 19-10 Results of long-term selection experiments. (a) Selection for an increase in flight speed of fruit flies. Speed was tested in a wind tunnel in which flies flew against the wind to reach a light source. (b) Selection for an increase in the amount of voluntary wheel walking done by mice. [(a) Data from K. E. Weber, Genetics 144, 1996, 205–213, (b) Data from J. G. Swallow et al., Behav. Genet. 28, 1998, 227–237.]

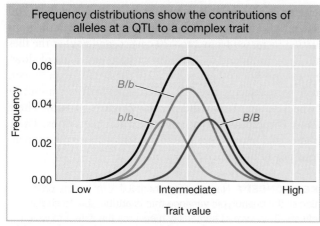

Frequency distributions show the contributions of alleles at a QTL to a complex trait

FIGURE 19-11 Frequency distributions showing how the distributions for the different genotypic classes at QTL locus *B* relate to the overall distribution for the population (black line).

cross between two inbred strains—parent one (P₁) with a high trait value and parent two (P₂) with a low trait value. The F₁ can be backcrossed to P₁ to create a BC₁ population in which the alleles at all the genes in the two parental genomes will segregate. Marker loci such as single nucleotide polymorphisms (SNPs) or microsatellites (also called simple sequence repeats, or SSRs) can be scored unambiguously as homozygous P₁ or heterozygous for each BC₁ individual. If there is a QTL linked to the marker locus, then the mean trait value for individuals that are homozygous P₁ at the marker locus will be different from the mean trait value for the heterozygous individuals. Based on such evidence, one can infer that a QTL is located near the marker locus.

The basic method for QTL mapping

Let's look at a simple experimental design used in QTL mapping experiments. We have two inbred lines of tomato that differ in fruit weight—Beefmaster with fruits of 230 g in weight, and Sungold with fruits of 10 g in weight (**Figure 19-12**). We cross the two lines to produce an F₁ hybrid and then backcross the F₁ to the Beefmaster line to produce a BC₁ generation. We grow several hundred BC₁ plants to maturity and measure the weight of the fruit on each. We also extract DNA from each of the BC₁ plants. We use these DNA samples to determine the genotype of each plant at a set of marker loci (SNPs or SSRs) that are distributed across all of the chromosomes such that we have a marker locus about every 5 centimorgans. (Recall that a centimorgan is a genetic map unit defined as the distance between genes for which 1 product of meiosis per 100 is recombinant.)

From this process, we would assemble a data set for several hundred plants and 200 or more marker loci distributed around the genome.

Table 19-6 shows part of such a data set for just 20 plants and 5 marker loci that are linked on a single chromosome. For each BC₁ plant, we have the weight of its fruit and the genotypes at the marker loci. Notice that trait values for the BC₁ plants are intermediate between the two parents as expected but closer to the Beefmaster value because this is a BC₁ population and Beefmaster was the backcross parent. Also, since this is a backcross population, the genotypes at each marker locus are either homozygous for the Beefmaster allele (*B/B*) or heterozygous (*B/S*). In Table 19-6, you can see the positions of crossovers between the marker loci that occurred during meiosis in the F₁ parent. For example, plant BC₁-001 has a recombinant chromosome with a crossover between marker loci *M3* and *M4*.

The overall mean fruit weight for the BC₁ population is 175.7. We can also calculate the mean for the two genotypic classes at each marker locus as shown in Table 19-6. For marker *M1*, the means for the *B/B* (176.3) and *B/S* (175.3) genotypic classes are very close to the overall mean (175.7). This is the expectation if there is no QTL affecting fruit weight near *M1*. For marker *M3*, the means for the *B/B* (180.7) and *B/S* (169.6) genotypic classes are quite different from the overall mean (175.7) and from each other. This is the expectation if there is a QTL affecting fruit weight near *M3*. Thus, we have evidence for a QTL affecting fruit weight near marker *M3*. Also notice that the *B/B* class has heavier fruit than the *B/S* class of *M3*. Plants that inherited the *S* allele from the small-fruited Sungold

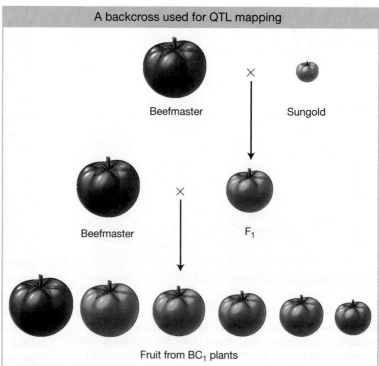

A backcross used for QTL mapping

FIGURE 19-12 Breeding scheme for a backcross population between Beefmaster and Sungold tomatoes. In the BC₁ generation, there is a continuous range of fruit sizes.

		Markers				
Plant	**Fruit wt. (g)**	**M1**	**M2**	**M3**	**M4**	**M5**
Beefmaster	230	B/B	B/B	B/B	B/B	B/B
Sungold	10	S/S	S/S	S/S	S/S	S/S
BC$_1$-001	183	B/B	B/B	B/B	B/S	B/S
BC$_1$-002	176	B/S	B/S	B/B	B/B	B/B
BC$_1$-003	170	B/B	B/S	B/S	B/S	B/S
BC$_1$-004	185	B/B	B/B	B/B	B/S	B/S
BC$_1$-005	182	B/B	B/B	B/B	B/B	B/B
BC$_1$-006	170	B/S	B/S	B/S	B/S	B/B
BC$_1$-007	170	B/B	B/S	B/S	B/S	B/S
BC$_1$-008	174	B/S	B/S	B/S	B/S	B/S
BC$_1$-009	171	B/S	B/S	B/S	B/B	B/B
BC$_1$-010	180	B/S	B/S	B/S	B/B	B/B
BC$_1$-011	185	B/S	B/B	B/B	B/S	B/S
BC$_1$-012	169	B/S	B/S	B/S	B/S	B/S
BC$_1$-013	165	B/B	B/B	B/S	B/S	B/S
BC$_1$-014	181	B/S	B/S	B/B	B/B	B/S
BC$_1$-015	169	B/S	B/S	B/S	B/B	B/B
BC$_1$-016	182	B/B	B/B	B/B	B/S	B/S
BC$_1$-017	179	B/S	B/S	B/B	B/B	B/B
BC$_1$-018	182	B/S	B/B	B/B	B/B	B/B
BC$_1$-019	168	B/S	B/S	B/S	B/B	B/B
BC$_1$-020	173	B/B	B/B	B/B	B/B	B/B
Mean of B/B	—	176.3	179.6	180.7	176.1	175.0
Mean of B/S	—	175.3	173.1	169.6	175.3	176.4
Overall mean	175.7					

TABLE 19-6 Simulated Fruit Weight and Marker-Locus Data for a Backcross Population between Two Tomato Inbred Lines—Beefmaster and Sungold

line have smaller fruits than those that inherited the *B* allele from the Beefmaster line.

Figure 19-13 is a graphical representation of QTL-mapping data for many plants along one chromosome using values from Table 19-6. The phenotypic data for the *B/B* and *B/S* genotypic classes are represented as frequency distributions so we can see the distributions of the trait values. At marker *M1*, the distributions are fully overlapping and the means for the *B/B* and *B/S* distributions are very close. It appears that the *B/B* and *B/S* classes have the same underlying distribution. At marker *M3*, the distributions are only partially overlapping and the means for the *B/B* and *B/S* distributions are quite different. The *B/B* and *B/S* classes around M3 have different underlying distributions similar to the situation in Figure 19-11. Again, we have evidence for a QTL near *M3*.

As shown in Figure 19-13, the trait means for the *B/B* and *B/S* groups at some markers are nearly the same. At other markers, these means are rather different. How

different do they need to be before we declare that a QTL is located near a marker? The statistical details for answering this question are beyond the scope of this text. However, let's review the basic logic behind the statistics. The statistical analysis involves calculating the probability of observing the data (the specific fruit weights and marker-locus genotypes for all the plants) given that there is a QTL near the marker locus and the probability of observing the data given that there is *not* a QTL near the marker locus. The ratio of these two probabilities is called the "odds":

$$\text{odds} = \frac{\text{Prob(data|QTL)}}{\text{Prob(data|no QTL)}}$$

The vertical line | means "given," and the term Prob(data|QTL) reads "the probability of observing the data given that there is a QTL." If the probability of the data when there is a QTL is 0.1 and the probability of the data when there is no QTL is 0.001, then the odds are 0.1/0.001 = 100. That is, the odds are 100 to 1 in favor of

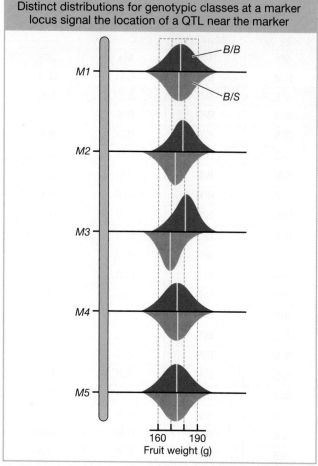

Distinct distributions for genotypic classes at a marker locus signal the location of a QTL near the marker

M1
B/B
B/S
M2
M3
M4
M5

160   190
Fruit weight (g)

FIGURE 19-13 A tomato chromosomal segment with marker loci *M1* through *M5*. At each marker locus, the frequency distributions for fruit weight from a BC_1 population of a Beefmaster × Sungold cross are shown. The red distributions are for the homozygous Beefmaster (*B/B*) genotypic class at the marker; the gray distributions are for the heterozygous (*B/S*) genotypic class. Yellow lines represent the mean of each distribution.

peaks of various heights as well as stretches that are relatively flat. The peaks represent putative QTL, but how high does a peak need to be before we declare that it represents a QTL? As discussed in Chapters 4 and 18, we can set a statistical threshold for rejecting the "null hypothesis." In this case, the null hypothesis is that "there is *not* a QTL at a specific position along the chromosome." Where the Lod score exceeds the threshold value, then we reject the null hypothesis in favor of the alternative hypothesis that a QTL is located at that position. In Figure 19-14, the Lod score exceeds the threshold value (red line) near marker locus *M3*. We conclude that a QTL is located near *M3*.

In addition to backcross populations, QTL mapping can be done with F_2 populations and other breeding designs. An advantage of using an F_2 population is that one gets estimates of the mean trait values for all three QTL genotypes: homozygous parent-1, homozygous parent-2, and heterozygous. With these data, one can get estimates of the additive (A) and dominance (D) effects of the QTL as discussed earlier in this chapter. Thus, QTL mapping enables us to learn about gene action, whether dominant or additive, for each QTL.

Here is an example. Suppose we studied an F_2 population from a cross of Beefmaster and Sungold tomatoes and we identified two QTL for fruit weight. The mean fruit weights for the different genotypic classes at the QTL might look something like this:

	Fruit weights			Effects	
	B/B	*B/S*	*S/S*	*A*	*D*
QTL 1	180	170	160	10	0
QTL 2	200	185	110	45	30

We can use these fruit weight values for the QTL to calculate the additive and dominance effects. QTL 1 is purely additive (D = 0), but QTL 2 has a large dominance effect. Also, notice that the additive effect of QTL 2 is more than 4 times that of QTL 1 (45 versus 10). Some QTL have large effects, and others have rather small effects.

there being a QTL. Researchers report the $\log_{10}$ of the odds, or the Lod score. So, if the odds ratio is 100, then the $\log_{10}$ of 100, or Lod score, is 2.0.

In addition to testing for QTL at the marker loci where the genotypes are known, Lod scores can be calculated for points between the markers. This can be done by using the genotypes of the flanking markers to infer the genotypes at points between the markers. For example, in Table 19-6, plant BC_1-001 is *B/B* at markers *M1* and *M2*, and so it has a high probability of being *B/B* at all points in between. Plant BC_1-003 is *B/B* at marker *M1* but *B/S* at *M2*, and so the plant might be either *B/B* or *B/S* at points in between. The odds equation incorporates this uncertainty when one calculates the Lod score at points between the markers.

The Lod scores can be plotted along the chromosome as shown by the blue line in Figure 19-14. Such plots typically show some

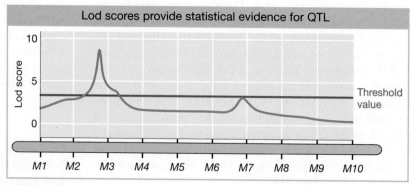

Lod scores provide statistical evidence for QTL

10

Lod score

5

0

M1   M2   M3   M4   M5   M6   M7   M8   M9   M10

Threshold value

FIGURE 19-14 Plot of Lod scores from a QTL-mapping experiment along a chromosome with 10 marker loci. The blue line shows the value of the Lod score at each position. Where the Lod score exceeds the threshold value, there is statistical evidence for a QTL.

What can be learned from QTL mapping? With the most powerful QTL-mapping designs, geneticists can estimate (1) the number of QTL (genes) affecting a trait, (2) the genomic locations of these genes, (3) the size of the effects of each QTL, (4) the mode of gene action for the QTL (dominant versus additive), and (5) whether one QTL affects the action of another QTL (epistatic interaction). In other words, one can get a rather complete description of the genetic architecture for the trait.

Much has been learned about genetic architecture from QTL-mapping studies in diverse organisms. Here are two examples. First, flowering time in maize is a classic quantitative or continuous trait. Flowering time is a trait of critical importance in maize breeding since the plants must flower and mature before the end of the growing season. Maize from Canada is adapted to flower within 45 days after planting, while maize from Mexico can require 120 days or longer. QTL mapping has shown that the genetic architecture for flowering time in maize involves more than 50 genes. Results from one experiment are shown in **Figure 19-15a**; these results show evidence for 15 QTL. QTL for maize flowering time generally have a small effect, such that substituting one allele for another at a QTL alters flowering time by only one day or less. Thus, the difference in flowering time between tropical and temperate maize involves many QTL.

Second, mice have been used to map QTL for many disease-susceptible traits. What one learns about disease-susceptibility genes in mice is often true in humans as well. Figure 19-15b shows the results of a genomic scan in mice for QTL for bone mineral density (BMD), the trait underlying osteoporosis. This scan identified two QTL, one on chromosome 9 and one on chromosome 12. From studies such as this, researchers have identified over 80 QTL in mice that may contribute to susceptibility to osteoporosis.

Similar studies have been done on dozens of other disease conditions.

From QTL to gene

QTL mapping does not typically reveal the identity of the gene(s) at the QTL. At its best, the resolution of QTL mapping is on the order of 1 to 10 cM, the size of a region that can contain 100 or more genes. To go from QTL to a single gene requires additional experiments to **fine-map** a QTL. To do this, the researcher creates a set of genetic homozygous stocks (also called lines), each with a crossover near the QTL. These stocks or lines differ from one another near the QTL, but they are identical to one another (**isogenic**) throughout the rest of their genomes. Lines that are identical throughout their genomes except for a small region of interest are called **congenic lines** or **nearly isogenic lines**. The isolation of QTL in an isogenic background is critical because only the single QTL region differs between the congenic lines. Thus, the use of congenic lines eliminates the complications caused by having multiple QTL segregate at the same time.

Using the tomato fruit weight example provided earlier, the chromosome region for a set of such congenic lines is shown in **Figure 19-16**. The genes (*flc*, *arf4*, . . .) are shown at the top, and the location for each crossover is indicated by the switch in color from red (Beefmaster genotype) to yellow (Sungold genotype). The mean fruit weight for the congenic lines carrying these recombinant chromosomes is indicated on the right. By inspection of Figure 19-16, you will notice that all lines with the Beefmaster allele of *kin1* (a kinase gene) have fruit of ~ 180 g, while those with the Sungold allele of *kin1* have fruit of about ~ 170 g. None of the other genes are associated with fruit weight in this way. If confirmed by appropriate statistical tests, this result allows us to identify *kin1* as the gene underlying this QTL.

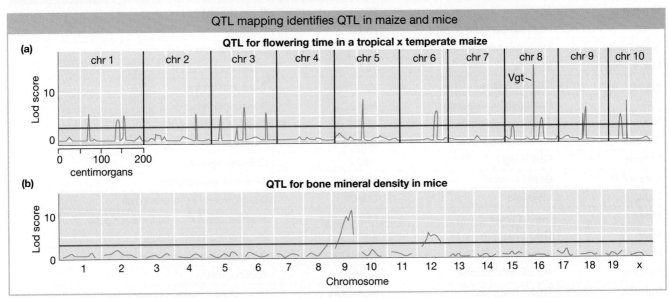

FIGURE 19-15 Plot of Lod scores from genomic scans for QTL. (a) Results from a scan for flowering time QTL in maize. (b) Results from a scan for bone-mineral-density QTL in mice. [*(a) Data from E. S. Buckler et al.,* Science 325, *2009, 714–718; (b) Data from N. Ishimori et al.,* J. Bone Min. Res. 23, *2008, 1529–1537.*]

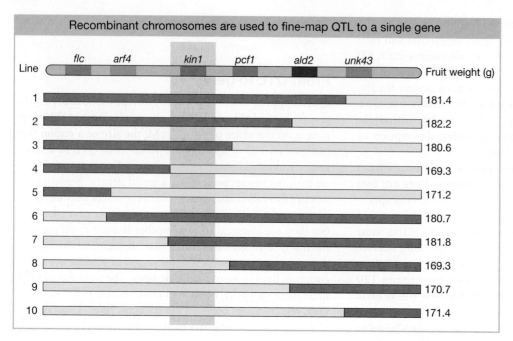

FIGURE 19-16 A tomato chromosomal segment for a set of 10 congenic lines that have crossovers near a QTL for fruit weight. Red chromosomal segments are derived from the Beefmaster line and yellow segments from the Sungold line. Differences in fruit weight among the lines make it possible to identify the *kin1* gene as the gene underlying this QTL.

Table 19-7 lists a small sample of the hundreds of genes or QTL affecting quantitative variation from different species that have been identified. The list includes the gene for maize flowering time, *Vgt*, that underlies one of the Lod peaks in Figure 19-15a. One notable aspect of this list is the diversity of gene functions. There does not appear to be a rule that only particular types of genes can be a QTL. Most, if not all, genes in the genomes of organisms are likely to contribute to quantitative variation in populations.

KEY CONCEPT Quantitative trait locus (QTL) mapping is a procedure for identifying the genomic locations of the genes (QTL) that control variation for complex traits. QTL mapping evaluates the progeny of controlled crosses for their genotypes at molecular markers and for their trait values. If the different genotypes at a marker locus have different mean values for the trait, then there is evidence for a QTL near the marker.

TABLE 19-7 Some Genes Contributing to Quantitative Variation That Were First Identified Using QTL Mapping

Organism	Trait	Gene	Gene function
Yeast	High-temperature growth	RHO2	GTPase
Arabidopsis	Flowering time	CRY2	Cryptochrome
Maize	Branching	Tb1	Transcription factor
Maize	Flowering time	Vgt	Transcription factor
Rice	Photoperiod sensitivity	Hd1	Transcription factor
Rice	Photoperiod sensitivity	CK2α	Casein kinase subunit
Tomato	Fruit-sugar content	Brix9-2-5	Invertase
Tomato	Fruit weight	Fw2.2	Cell-cell signaling
Drosophila	Bristle number	Scabrous	Secreted glycoprotein
Cattle	Milk yield	DGAT1	Diacylglycerol acyltransferase
Mice	Colon cancer	Mom1	Modifier of a tumor-suppressor gene
Mice	Type 1 diabetes	I-Aβ	Histocompatibility antigen
Humans	Asthma	ADAM33	Metalloproteinase-domain-containing protein
Humans	Alzheimer's disease	ApoE	Apolipoprotein
Humans	Type 1 diabetes	HLA-DQA	MHC class II surface glycoprotein

Source: A. M. Glazier et al., *Science* 298, 2002, 2345–2349.

19.6 ASSOCIATION MAPPING IN RANDOM-MATING POPULATIONS

LO 19.8 Design and analyze experiments to identify the loci controlling quantitative traits in populations.

Association mapping is a method for finding QTL in the genome based on naturally occurring linkage disequilibrium (see Chapter 18) between a marker locus and the QTL in a random-mating population. As we will see, this method often allows researchers to directly identify the specific genes that control the differences in phenotype among members of a population.

The basic idea behind association mapping has been around for decades. Here is an example from the 1990s for the *ApoE* gene in humans, a gene involved in lipoprotein (lipid-protein-complex) metabolism. Because of its role in lipoprotein metabolism, *ApoE* was considered a **candidate gene** for a causative role in cardiovascular disease, the accumulation of fatty (lipid) deposits in the arteries. Researchers looked for statistical associations between the alleles of *ApoE* that people carry and whether they had cardiovascular disease. They found an association between the *e4* allele of this gene and the disease—people carrying the *e4* allele were 42 percent more likely to have the disease than those who carried other alleles. Although this type of study was successful, it required that a candidate gene suspected to affect the trait be known in advance.

Over the past two decades, association mapping has been revolutionized by the development of genome-wide SNP maps and high-throughput DNA sequencing methods that allow the genomes of thousands of individuals to be sequenced (see Chapter 18). Association mapping is now routinely used to scan the entire genome for genes contributing to quantitative variation. This type of study is known as a **genome-wide association** study (**GWA** study, or **GWAS**). A major advantage of GWA studies is that candidate genes are not required since one is scanning every gene in the genome.

Association mapping offers several advantages over QTL mapping. First, since it is performed with random-mating populations, there is no need to make controlled crosses or work with human families with known parent–offspring relationships. Second, it tests many alleles at a locus at once. In QTL-mapping studies, there are two parents (Beefmaster and Sungold tomatoes in the example in the previous section), and so only two alleles are being compared. With association mapping, all the alleles in the population are being assayed at the same time. Finally, association mapping can lead to the direct identification of the genes at the QTL without the need for subsequent fine-mapping studies. This is possible because the SNPs in the gene that influences the trait will show stronger associations with the trait than SNPs in other linked genes.

The basic method for GWAS

Let's begin by looking at how genetic variation is patterned across the genome in a population. In Chapter 18, we discussed linkage disequilibrium (LD), or the nonrandom association of alleles at two loci. **Figure 19-17** shows how LD could appear among a sample of chromosomes from 18 different individuals. SNPs (or other polymorphisms) that are close to each other tend to be in strong disequilibrium, while those that are farther apart are in weak or no disequilibrium. Genomes also tend to have recombination hotspots, points where crossing over occurs at a high frequency. Hotspots disrupt linkage disequilibrium such that SNPs on either side of the hotspot are in equilibrium with each other. SNPs that are not separated by a hotspot form a haplotype block of strongly correlated SNPs.

Suppose SNP8 in Figure 19-17 is a SNP in a gene that causes a difference in phenotype such that individuals with the A/A genotype have a different phenotype than those with either A/G or G/G. SNP8 could affect phenotype by causing an amino acid change or affecting gene expression. SNP8 or any SNPs that directly affect a phenotype are called functional SNPs. Since SNP8 is in strong disequilibrium with other SNPs in the haplotype block (SNPs 6, 7, 9, and 10), any of these other SNPs can serve as a proxy for the functional SNP8. Individuals who are T/T at SNP7 will have the same phenotype as those who are A/A at SNP8 because SNP7 and SNP8 are in LD. When the SNP genotypes are correlated (in disequilibrium), then the trait values will be correlated. For this reason, GWA studies do not need to survey the actual functional SNPs, but they do need to have SNPs in every haplotype block.

To conduct a GWA study for a disease condition in humans, we might survey 50,000 individuals with a disorder such as adult-onset, or type 2, diabetes. We would also select another 50,000 control individuals who do not have this disorder. Each of the 100,000 participants would donate blood from which their DNA would be extracted. The DNA samples would be genotyped for a set of 500,000 SNPs that are distributed across the entire genome or their entire genomes could be sequenced using next-generation sequencing technology (Chapter 14). We want a sufficient number of SNPs so that each of the haplotype blocks in the genome is marked by one or more SNPs (see Figure 19-17). The resulting data set would be enormous—consisting of 500,000 genotypes in 100,000 individuals, a total of 50 billion data points. A small part of such a data set is shown in **Table 19-8**.

Once the data are assembled, the researcher performs a statistical test on each SNP to determine whether one of its alleles is more frequently associated with diabetes than expected by chance. In the case of a categorical trait such as being "affected" or "not affected" by diabetes, statistical tests similar to the chi-square (χ^2) test (see Chapter 3) can be used. A statistical test is performed separately on each SNP and the P values plotted along the chromosome.

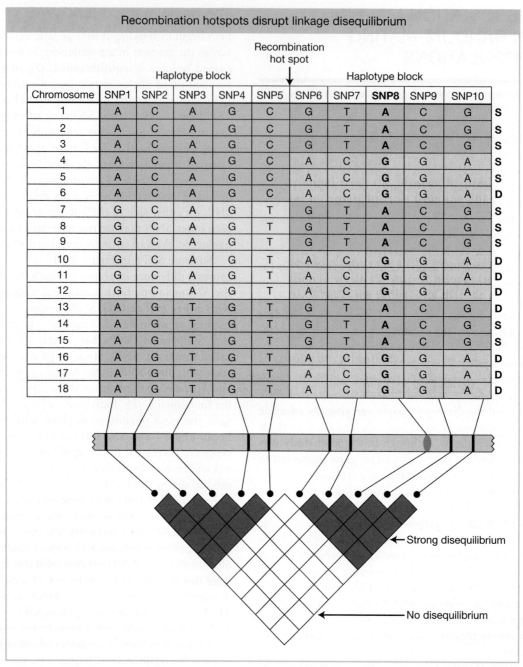

FIGURE 19-17 (*top*) Diagram of the distribution of SNPs and haplotypes for a chromosomal segment (golden bar) from 18 individuals. Haplotypes often occur in blocks (regions of lower recombination) separated from one another by recombination hotspots (different colors indicate haplotype blocks). (The column of S's and D's at the right are for Problem 19-4.) SNP8 (bold) controls a difference in trait values. (*bottom*) You can tell whether two SNPs show disequilibrium by noting the color of the square where the rows for the markers intersect. Within a haplotype block, SNPs show strong disequilibrium (red). SNPs in different haplotype blocks show weak or no disequilibrium (white). [*Data from David Altshuler et al.*, Science 322, *2008, 881–888.*]

The null hypothesis is that the SNP is not associated with the trait. If the P value for a SNP falls below 0.05, then the evidence for the null hypothesis is weak and we will favor the alternative hypothesis that the different genotypes at the SNP are associated with different phenotypes for the trait. Association mapping does not actually prove that a gene or a SNP within a gene affects a trait. It only provides statistical evidence for an association between the SNP and

the trait. Proof requires molecular characterization of the gene and its different alleles.

Figure 19-18a shows the results of an association-mapping study for body size in dogs. Each dot plotted along the chromosomes (x-axis) represents the P value (y-axis) for a test of association between body size and a SNP. The P values are plotted using an inverse scale such that the higher up the y-axis, the smaller the value. On

TABLE 19-8   Part of a Simulated Data Set for an Association-Mapping Experiment

Individual	SNP1	SNP2	SNP3	Type 2 diabetes	Height (cm)
1	C/C	A/G	T/T	yes	173
2	C/C	A/A	C/C	yes	170
3	C/G	G/G	T/T	no	183
4	C/G	G/G	C/T	no	180
5	C/C	G/G	C/T	no	173
6	G/G	A/G	C/T	yes	178
7	G/G	A/G	C/T	no	163
8	C/G	G/G	C/T	no	168
9	C/G	A/G	C/T	yes	165
10	G/G	A/A	C/C	yes	157

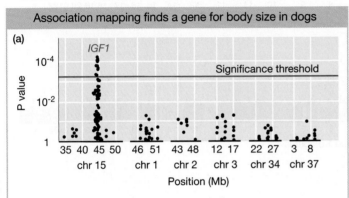

FIGURE 19-18 (a) Results from an association-mapping experiment for body size in dogs. Each dot in the plot represents the P value for a test of association between a SNP and body size. Dots above the "threshold line" show evidence for a statistically significant association. (b) Examples of a small and a large breed of dog. [*(b)Tetra Images/Getty Images*.]

chromosome 15, there is a cluster of SNPs above the threshold line, indicating that the null hypothesis of no association can be rejected for these SNPs in favor of the alternative hypothesis that a gene affecting body size in dogs is located at this position. The strong peak on chromosome 15 involves SNPs in the *insulin-like growth factor-1* (*IGF1*) gene, a gene that encodes a hormone involved in juvenile growth in mammals. This gene is the major contributor to the difference in size between small and large breeds of dogs (Figure 19-18b).

GWA, genes, disease, and heritability

A large number of GWA studies have been performed, and much has been learned from them about heritable variation in humans and other species. Let's look at one of the largest studies, which was a search for disease-risk genes in a group of 17,000 people using 500,000 SNPs. **Figure 19-19** shows plots of the P values for associations between SNPs and several common diseases. Green dots are the statistically significant associations. Notice the spike of green dots on chromosome 6 for rheumatoid arthritis and type 1 (juvenile) diabetes. These are two autoimmune diseases, and this spike is positioned over a human leukocyte antigen (*HLA*) gene of the major histocompatibility complex (MHC) of genes that regulates immune response in humans and other vertebrates. Thus, genes active in the normal immune response are implicated as a cause of autoimmune diseases. The gene *PTPN22* is also associated with risk for type 1 diabetes. *PTPN22* encodes the protein tyrosine phosphatase, which is expressed in lymphoid cells of the immune system. For coronary artery disease, there is a significant association with the *ApoE* gene, confirming an earlier study mentioned previously.

GWA studies have identified thousands of risk genes for hundreds of diseases, and the numbers are growing. You can explore a compilation of the results at the GWAS Catalog Web site (www.ebi.ac.uk/gwas/).

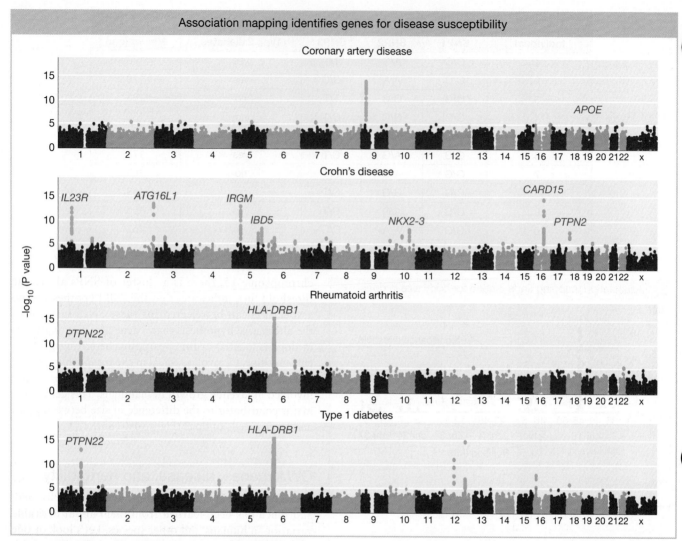

Association mapping identifies genes for disease susceptibility

FIGURE 19-19 Results from a genome-wide association study of common diseases in humans. The 23 human chromosomes are arrayed from left to right. The y-axis shows the P value for the statistical test of an association between the disease and each SNP. Significant test results are shown as green dots. The names of some genes identified by this analysis are shown in red. [*Reprinted by permission from Macmillan Publishers Ltd. from The Wellcome Trust Case Control Consortium, "Genome-wide association study of 14,000 cases of seven common diseases and 3,000 shared controls," Nature 447, 2007, 661–678, Figure 4. Permission conveyed through Copyright Clearance Center, Inc.*]

These data are ushering in a new era of **precision medicine**, in which an individual can have his or her genome scanned to determine their genotype at genes known to increase disease risk. Although this science is relatively young, it is possible to identify individuals who have a 10-fold higher risk for certain diseases than other members of the population. Such information can be used to initiate preventative measures and changes in lifestyle (environment) that contribute to disease risk. Some companies offer direct-to-consumer genotyping service including information on whether you carry risk alleles for diseases like breast cancer, Alzheimer's disease, macular degeneration, and more.

Since height in humans is a classic quantitative trait, quantitative geneticists had great interest in performing GWA studies for this trait. GWA studies have identified over 180 genes affecting height. Each of these genes has a small additive effect (~1 to 4 mm), as expected for a trait

governed by many genes. However, a perplexing result was that the 180 genes accounted for only 10 percent of the genetic variance in height. This falls far short of the roughly 80 percent value for broad-sense heritability for height. The difference between 10 percent and 80 percent has been dubbed the missing heritability. For disease risk, there is also much missing heritability. For example, GWA studies have succeeded in explaining only 10 percent of the genetic variation for Crohn's disease and only 5 percent of the genetic variation for type 2 diabetes.

It has come as a surprise to many geneticists that GWA studies with hundreds of thousands of SNPs blanketing the genome and samples of over 10,000 individuals should be able to account for only a tiny fraction of the heritable variation. Currently, it is unknown why this is the case, although there are some possible explanations. For example, even with tens of thousands of individuals,

GWA studies do not have enough statistical power to detect genes of very small effect. Thus, one hypothesis is that susceptibility for many common diseases (or height variation) is largely controlled by very many genes but each with a very small effect.

Despite the inability of GWA studies to explain all the heritable variation for traits, this approach has provided a major advance in understanding quantitative genetic variation. Hundreds of new genes contributing to quantitative variation for disease risk have been identified. These genes are now targets for the development of new therapies. Beyond humans, GWA studies have advanced our understanding of the inheritance of quantitative traits in *Arabidopsis*, *Drosophila*, yeast, and maize.

KEY CONCEPT GWAS is a method for identifying statistical associations between molecular markers and phenotypic variation for complex traits. Linkage disequilibrium in a population between the marker locus and a functional variant in a gene can cause the association.

SUMMARY

Quantitative genetics seeks to understand the inheritance of complex traits—traits that are influenced by a mix of genetic and environmental factors and do not segregate in simple Mendelian ratios. Complex traits can be categorical traits, threshold traits, counting (meristic) traits, or continuously variable traits.

The genetic architecture of a trait is the full description of the number of genes affecting the trait, their relative contributions to the phenotype, the contribution of environmental factors to the phenotype, and an understanding of how the genes interact with one another and with environmental factors. To decipher the genetic architecture of complex traits, quantitative geneticists have developed a simple mathematical model that decomposes the phenotypes of individuals into differences that are due to genetic factors (g) and those that are due to environmental factors (e).

The differences in trait values among members of a population can be summarized by the variance, a statistical measure of the extent to which individuals deviate from the population mean. The variance for a trait can be partitioned into a part that is due to genetic factors (the genetic variance) and a part that is due to environmental factors (the environmental variance). A key assumption behind partitioning the trait variance into genetic and environmental components is that genetic and environmental factors are uncorrelated or independent.

The degree to which variation for a trait in a population is explained by genetic factors is measured by the broad-sense heritability (H^2) of the trait. H^2 is the ratio of the genetic variance to the phenotypic variance. Broad-sense heritability expresses the degree to which the differences in phenotype among the individuals in a population are determined by differences in their genotypes. The measurement of H^2 in humans has revealed that most traits have some genetic influences.

Parents transmit genes but not genotypes to their offspring. At each generation, new dominance interactions between the alleles at a locus are created. To incorporate this phenomenon into the mathematical model for quantitative variation, the genetic deviation (g) is decomposed into the additive (a) and dominance (d) deviations. Only the additive deviation is transmitted from parents to offspring. The additive deviation represents the heritable part of the phenotype in the narrow sense. The additive part of the variance in a population is the heritable part of the variance. Narrow-sense heritability (h^2) is the ratio of the additive variance to the phenotypic variance. Narrow-sense heritability provides a measure of the degree to which the phenotypes of individuals are determined by the genes they inherit from their parents.

A knowledge of the narrow-sense heritability of a trait is fundamental to understanding how a trait will respond to selective breeding or the force of natural selection. Plant and animal breeders use their knowledge of narrow-sense heritability for traits of interest to guide plant and animal improvement programs. Narrow-sense heritability is used to predict the phenotypes of offspring and estimate the breeding value of individual members of the breeding population.

The genetic loci underlying variation in complex traits are known as quantitative trait loci, or QTL for short. There are two experimental methods for characterizing QTL and determining their locations in the genome. First, QTL mapping looks for statistical correlations between the genotypes at marker loci and trait values in populations with known pedigrees such as a BC_1 population. Subsequently, the QTL can be fine-mapped to the underlying gene. Second, association mapping looks for statistical correlations between the genotypes at marker loci and trait values in random-mating populations. Genome-wide association (GWA) studies can directly identify the gene that underlie a QTL.

Most traits of importance in medicine, agriculture, and evolutionary biology show complex inheritance. Examples include disease risk in humans, yield in soybeans, milk production in dairy cows, and the full spectrum of phenotypes that differentiate all the species of plants, animals, and microbes on earth. Quantitative genetic analyses are at the forefront of understanding the genetic basis of these critical traits.

KEY TERMS

additive effect (A) (p. 656)
additive gene action (p. 656)
additive genetic variance (p. 658)
association mapping (p. 669)
breeding value (p. 658)
broad-sense heritability (H^2)
  (p. 652)
candidate gene (p. 669)
categorical trait (p. 645)
complex inheritance (p. 645)
complex trait (p. 644)
congenic line (p. 667)
continuous trait (p. 645)
correlation (p. 651)
correlation coefficient (p. 651)
covariance (p. 650)
deviation (p. 647)

dominance effect (D) (p. 656)
dominant gene action (p. 656)
environmental variance (p. 650)
fine-map (p. 667)
frequency histogram (p. 647)
gene action (p. 655)
genetic architecture (p. 644)
genetic variance (p. 650)
genome-wide association (GWA or
  GWAS) (p. 669)
inbred line or stock (p. 649)
isogenic (p. 667)
mean (p. 645)
meristic trait (p. 645)
multifactorial hypothesis (p. 644)
narrow-sense heritability (h^2)
  (p. 655)

nearly isogenic line (p. 667)
normal distribution (p. 647)
partial dominance (p. 656)
precision medicine (p. 672)
population (p. 645)
QTL mapping (p. 663)
quantitative genetics (p. 644)
quantitative trait (p. 644)
quantitative trait loci (QTL)
  (p. 663)
sample (p. 645)
selection differential (S) (p. 662)
selection response (R) (p. 662)
simple inheritance (p. 645)
standard deviation (p. 647)
threshold trait (p. 645)
variance (p. 647)

SOLVED PROBLEMS

SOLVED PROBLEM 1

In a flock of 100 broiler chickens, the mean weight is 700 g and the standard deviation is 100 g. Assume the trait values follow the normal distribution.

a. How many of the chickens are expected to weigh more than 700 g?

b. How many of the chickens are expected to weigh more than 900 g?

c. If H^2 is 1.0, what is the genetic variance for this population?

SOLUTION

a. Since the normal distribution is symmetrical about the mean, 50 percent of the population will have a trait value above the mean and the other 50 percent will have a trait value below the mean. In this case, 50 of the 100 chickens are expected to weigh more than 700 g.

b. The value of 900 g is 2 standard deviations greater than the mean. Under the normal distribution, 95.5 percent of the population will fall within 2 standard deviations of the mean and the remaining 4.5 percent will lie more than 2 standard deviations from the mean. Of this 4.5 percent, one-half (2.25 percent) will be more than 2 standard deviations less than the mean, and the other half (2.25 percent) will be more than 2 standard deviations greater than the mean. Thus, we expect about 2.25 percent of the 100 chickens (or roughly 2 chickens) to weigh more than 900 g.

c. When H^2 is 1.0, then all of the variance is genetic. We know that the standard deviation is 100, and the variance is the square of the standard deviation.

$$\text{Variance} = \sigma^2$$

Thus, the genetic variance would be $(100)^2 = 10,000\ g^2$.

SOLVED PROBLEM 2

Two inbred lines of beans are intercrossed. In the F_1, the variance in bean weight is measured at $15\ g^2$. The F_1 is selfed; in the F_2, the variance in bean weight is $61\ g^2$. Estimate the broad heritability of bean weight in the F_2 population of this experiment.

SOLUTION

The key here is to recognize that all the variance in the F_1 population must be environmental because all individuals have the same genotype. Furthermore, the F_2 variance must be a combination of environmental and genetic components, because all the genes that are heterozygous in the F_1 will segregate in the F_2 to give an array of different genotypes that relate to bean weight. Hence, we can estimate

$$V_e = 15\ g^2$$
$$V_g + V_e = 61\ g^2$$

Therefore,

$$V_g = 61 - 15 = 46\ g^2$$

and broad heritability is

$$H^2 = \frac{46}{61} = 0.75 \ (75\%)$$

SOLVED PROBLEM 3

In an experimental population of *Tribolium* (flour beetles), body length shows a continuous distribution with a mean of 6 mm. A group of males and females with a mean body length of 9 mm are removed and interbred. The body lengths of their offspring average 7.2 mm. From these data, calculate the heritability in the narrow sense for body length in this population.

SOLUTION

The selection differential (S) is $9 - 6 = 3$ mm, and the selection response (R) is $7.2 - 6 = 1.2$ mm. Therefore, the heritability in the narrow sense is

$$h^2 = \frac{R}{S} = \frac{1.2}{3.0} = 0.4 (40\%)$$

SOLVED PROBLEM 4

One research team reports that the broad-sense heritability for height in humans is 0.5 based on a study of identical twins reared apart in Iceland. Another team reports that the narrow-sense heritability for human height is 0.8 based on a study of parent–offspring correlation in the United States.

What seems unexpected about these results? How could the unexpected results be explained?

SOLUTION

Broad-sense heritability is the ratio of the total genetic variance (V_g) to the phenotypic variance (V_X). The total genetic variance includes both the additive (V_a) and the dominance (V_d) variance

$$h^2 = \frac{V_g}{V_X} = \frac{V_a + V_d}{V_X}$$

Narrow-sense heritability is the ratio of the additive variance (V_a) to the phenotypic variance (V_X).

$$h^2 = \frac{V_a}{V_X}$$

Thus, all other variables being equal, H^2 should be greater than or equal to h^2. It will be equal to h^2 when V_d is 0.0. It is unexpected that h^2 should be greater than H^2. However, the two research teams studied different populations—in Iceland and in the United States. Estimates of heritability apply only to the population and environment in which they were measured. Estimates made in one population can be different from those made in another population because the two populations may segregate for different alleles at numerous genes and the two populations experience different environments.

PROBLEMS

Visit SaplingPlus for supplemental content. Problems with the ⚏ icon are available for review/grading.

WORKING WITH THE FIGURES

(The first 6 questions require inspection of text figures.)

1. Figure 19-9 shows the trait distributions before and after a cycle of artificial selection. Does the variance of the trait appear to have changed as a result of selection? Explain.

2. Figure 19-11 shows the expected distributions for the three genotypic classes if the *B* locus is a QTL affecting the trait value.

   a. As drawn, what is the dominance/additive (D/A) ratio?

   b. How would you redraw this figure if the *B* locus had no effect on the trait value?

   c. How would the positions along the x-axis of the curves for the different genotypic classes of the *B* locus change if D/A = 1.0? ⚏

3. Figure 19-16 shows the results of a QTL fine-mapping experiment. Which gene would be implicated as controlling fruit weight if the mean fruit weight for each line was as follows?

Line	Fruit weight (g)
1	181.4
2	169.3
3	170.7
4	171.2
5	171.4
6	182.2
7	180.6
8	180.7
9	181.8
10	169.3

4. Figure 19-17 shows a set of haplotypes. Suppose these are haplotypes for a chromosomal segment from 18 haploid yeast strains. On the right edge of the figure, the S and D indicate whether the strain survives (S) or dies (D) at high temperature (40°C). Using the χ^2 test (see Chapter 3) and Table 3-1, does either SNP1

or SNP6 show evidence for an association with the growth phenotype? Explain. 🐟

5. Figure 19-18a shows a plot of P values (represented by the dots) along the chromosomes of the dog genome. Each P value is the result of a statistical test of association between a SNP and body size. Other than the cluster of small P values near *IGF1*, do you see any chromosomal regions with evidence for a significant association between a SNP and body size? Explain. 🐟

6. Figure 19-19 shows plots of P values (represented by the dots) along the chromosomes of the human genome. Each P value is the result of a statistical test of association between a SNP and a disease condition. There is a cluster, or spike, of statistically significant P values (green dots) at the gene *HLA-DRB1* for two diseases. Why might this particular gene contribute to susceptibility for the autoimmune diseases rheumatoid arthritis and type 1 diabetes?

BASIC PROBLEMS

7. Distinguish between continuous and discontinuous variation in a population, and give some examples of each. 🐟

8. What are the central assumptions of the multifactorial hypothesis?

9. The table below shows a distribution of bristle number in a *Drosophila* population. Calculate the mean, variance, and standard deviation for these data.

Bristle number	Number of individuals
1	1
2	4
3	7
4	31
5	56
6	17
7	4

10. Suppose that the mean IQ in the United States is roughly 100 and the standard deviation is 15 points. People with IQs of 145 or higher are considered "geniuses" on some scales of measurement. What percentage of the population is expected to have an IQ of 145 or higher? In a country with 300 million people, how many geniuses are there expected to be?

11. In a sample of adult women from the United States, the average height was 164.4 cm and the standard deviation was 6.2 cm. Women who are more than 2 standard deviations above the mean are considered very tall, and women who are more than 2 standard deviations below the mean are considered very short. Height in women is normally distributed.

a. What are the heights of very tall and very short women?

b. In a population of 10,000 women, how many are expected to be very tall and how many very short?

12. A bean breeder is working with a population in which the mean number of pods per plant is 50 and the variance is 10 pods^2. The broad-sense heritability is known to be 0.8. Given this information, can the breeder be assured that the population will respond to selection for an increase in the number of pods per plant in the next generation?

13. The table below shows the number of piglets per litter for a group of 60 sows. What is the mean number of piglets per litter? What is the relative frequency of litters with at least 12 piglets? 🐟

Number of litters	Piglets/litter
1	6
3	7
7	8
12	9
18	10
20	11
17	12
14	13
6	14
2	15

14. A chicken breeder is working with a population in which the mean number of eggs laid per hen in one month is 28 and the variance is 5 eggs^2. The narrow-sense heritability is known to be 0.8. Given this information, can the breeder expect that the population will respond to selection for an increase in the number of eggs per hen in the next generation?

a. No, applying selection is always risky and a breeder never knows what to expect.

b. No, a breeder needs to know the broad-sense heritability to know what to expect.

c. Yes, since the narrow-sense heritability is close to 1(0.8), then we would expect selective breeding could lead to increased egg production in the next generation.

d. Yes, since the variance is greater than 0.

e. Both *c* and *d* are correct.

15. The narrow-sense heritability of the number of peas per pod in a population of sugar snap peas is 0.5. The mean of the population is 6.2 peas per pod. A plant breeder selects one plant with 6.8 peas per pod and crosses with a second plant that has 8.0 peas per pod. What is the expected numbers of peas per pod among the offspring of this cross?

16. QTL mapping and GWA (association) mapping are two different methods used to identify genes that affect complex traits. For each of the following statements, choose whether it applies to QTL mapping, association mapping, or both.

Statement	QTL	GWA	Both
This method requires that the experimenter make crosses between different strains to produce a mapping population.			
This method can scan the entire genome to find QTL for a trait.			
This method can often identify the specific genes that represent the QTL.			
This method may sample a large number of individuals from a random-mating population that has variation for the trait being studied.			
This method typically tests two alleles that differ between the two parents of the mapping population.			

CHALLENGING PROBLEMS

17. In a large herd of cattle, three different characters showing continuous distribution are measured, and the variances in the following table are calculated:

	Characters		
Variance	Shank length	Neck length	Fat content
Phenotypic	310.2	730.4	106.0
Environmental	248.1	292.2	53.0
Additive genetic	46.5	73.0	42.4
Dominance genetic	15.6	365.2	10.6

  a. Calculate the broad- *and* narrow-sense heritabilities for each character.

  b. In the population of animals studied, which character would respond best to selection? Why?

  c. A project is undertaken to decrease mean fat content in the herd. The mean fat content is currently 10.5 percent. Animals with a mean of 6.5 percent fat content are interbred as parents of the next generation. What mean fat content can be expected in the descendants of these animals?

18. In a species of the Darwin's finches (*Geospiza fortis*), the narrow-sense heritability of bill depth has been estimated to be 0.79. Bill depth is correlated with the ability of the finches to eat large seeds. The mean bill depth for the population is 9.6 mm. A male with a bill depth of 10.8 mm is mated with a female with a bill depth of 9.8 mm. What is the expected value for bill depth for the offspring of this mating pair?

19. Two inbred lines of laboratory mice are intercrossed. In the F_1 (which have identical genotypes at all loci), the variance in adult weight is measured at 3 g^2. The F_1 animals are intercrossed to create an F_2 in which the variance in adult weight is 16 g^2. Estimate the broad-sense heritability of adult weight in the F_2 population of this experiment. (The environments in which the F_1 and F_2 animals were reared were equivalent.)

20. The table below shows the weights of 100 individual mice of the same inbred strain reared on different diets. For an individual mouse that weights 27 g, how much of its weight is due to its genetics and how much to the specific diet it was fed (environment)? (Other than diet, the mice were reared in equivalent environments.)

Number of mice	Weight (g)
5	21
13	22
18	23
21	24
22	25
16	26
5	27

21. The table below contains measurements of total serum cholesterol (mg/dl) for 10 sets of monozygotic twins who were reared apart. Calculate the following: overall mean, overall variance, covariance between the twins, and broad-sense heritability (H^2).

X'	X''
228	222
186	152
204	220
142	185
226	210
217	190
207	226
185	213
179	159
170	129

22. The table on the next page contains the height in centimeters for 10 sets of adult women twins. Calculate the correlation coefficient (*r*) between the heights of the sisters for the twin pairs.

Twin 1	Twin 2
158	163
156	150
172	173
156	154
160	163
159	153
170	174
177	174
165	168
172	165

23. Population A consists of 100 hens that are fully isogenic and that are reared in a uniform environment. The average weight of the eggs they lay is 52 g, and the variance is 3.5 g^2. Population B consists of 100 genetically variable hens that produce eggs with a mean weight of 52 g and a variance of 21.0 g^2. Population B is raised in an environment that is equivalent to that of Population A. What is the environmental variance (V_e) for egg weight? What is the genetic variance in Population B? What is the broad-sense heritability in Population B? 🌊

24. Maize plants in a population are on average 180 cm tall. Narrow-sense heritability for plant height in this population is 0.5. A breeder selects plants that are 10 cm taller on average than the population mean to produce the next generation, and the breeder continues applying this level of selection for eight generations. What will be the average height of the plants after eight generations of selection? Assume that h^2 remains 0.5 and V_e does not change over the course of the experiment.

25. In a population of *Drosophila melanogaster* reared in the laboratory, the mean wing length is 0.55 mm and the range is 0.35 to 0.65. A geneticist selects a female with wings that are 0.42 mm in length and mates her with a male that has wings that are 0.56 mm in length.

    a. What is the expected wing length of their offspring if wing length has a narrow-sense heritability of 1.0?

    b. What is the expected wing length of their offspring if wing length has a narrow-sense heritability of 0.0?

26. Different species of crickets have distinct songs, and they use these songs for mate recognition. Researchers crossed two species of Hawaiian crickets (*Laupala paranigra* and *L. kohalensis*) whose songs are distinguished by pulse rate (the number of pulses per second; Shaw et al., *Molecular Ecology* 16, 2007, 2879–2892). Then, they mapped QTL in the F$_2$ population derived from this cross. Six autosomal QTL were detected.

The mean trait values (pulses per second) at the three genotypic classes in the F$_2$ for each QTL are shown in the table below, where P indicates the *L. paranigra* allele and K indicates the *L. kohalensis* allele. 🌊

QTL	P/P	P/K	K/K
1	1.54	1.89	2.10
2	1.75	1.87	1.94
3	1.72	1.88	1.92
4	1.70	1.82	2.02
5	1.67	1.80	2.13
6	1.57	1.88	2.19

a. Calculate the additive (A) and dominance (D) effects and the D/A ratio for each of the six QTL.

b. Which of these QTL shows the greatest amount of dominance?

c. Which of these has the largest additive effect?

d. The mean pulse rate for *L. kohalensis* is 3.72, and it is 0.71 for *L. paranigra*. Do all six QTL act in the expected direction with the *L. kohalensis* allele conferring a higher pulse rate than the *L. paranigra* allele?

27. Question 26 refers to QTL on the cricket autosomes. For the sex chromosomes, female crickets are XX and male crickets are XO, having just one X chromosome but no Y chromosome. Can QTL for pulse rate be mapped on cricket X chromosomes? If the song is sung only by male crickets, can the dominance effects of QTL on the X be estimated?

28. GWA studies reveal statistical correlations between the genotypes at marker loci in genes and complex traits. Do GWA studies prove that allelic variation in a gene actually causes the variation in the trait? If not, what experiments could prove that allelic variants in a gene in a population are responsible for variation in a trait?

29. The *ocular albinism-2* (*OCA2*) gene and the *melanocortin-1-receptor* (*MC1R*) gene are both involved in melanin metabolism in skin cells in humans. To test whether variation at these genes contributes to sun sensitivity and the associated risk of being afflicted with skin cancer, you perform association analyses. A sample of 1000 people from Iceland were asked to classify themselves as having tanning or burning (nontanning) skin when exposed to the sun. The individuals were also genotyped for a SNP in each gene (rs7495174 and rs1805007). The table shows the number of individuals in each class.

	OCA2 (rs7495174)			MC1R (rs1805007)		
	A/A	A/G	G/G	C/C	C/T	T/T
Burning	245	56	1	192	89	21
Tanning	555	134	9	448	231	19

a. What are the frequencies of tanning and burning phenotypes in Iceland?

b. What are the allelic frequencies at each locus (SNP)?

c. Using the χ^2 test (see Chapter 3) and Table 3-1, test the null hypothesis that there is no association between these SNPs and sun-sensitive skin. Does either SNP show evidence for an association?

d. If you find evidence for an association between the gene and the trait, what is the mode of gene action?

e. If the P value is greater than 0.05, does that prove that the gene does not contribute to variation for sun sensitivity? Why?

GENETICS AND SOCIETY

Bioethicists have expressed concern that consumers are not prepared to properly evaluate the results from direct-to-consumer genetic testing for disease risk alleles because consumers lack adequate knowledge to interpret the results appropriately and may react to test results by taking an action that is damaging to their health. Should society outlaw such direct-to-consumer testing, require that a genetic counselor be consulted, or allow such testing?

Evolution of Genes, Traits, and Species

The theory of evolution by natural selection was developed independently by two intrepid British naturalists, Charles Darwin (1809–1882) and Alfred Russel Wallace (1823–1913), in the course of their respective long voyages. [*Left: Charles Darwin, 1840 (w/c on paper) (for pair see 369470)/Richmond, George (1809-96)/HISTORIC ENGLAND ARCHIVE/Down House, Downe, Kent, UK/ Bridgeman Images; Right: Pictorial Press Ltd/Alamy.*]

CHAPTER OUTLINE AND LEARNING OBJECTIVES

In this chapter, we will see that both natural selection and neutral evolutionary processes can lead to changes in DNA sequences, which can then impact the evolution of phenotypic variation within and between species. Our broad objective is to identify and distinguish the genetic mechanisms that underlie the evolution of genes, genomes, traits, and species.

Charles Darwin (1809–1882) arrived in the Galápagos Islands in 1835, well into the fourth year of what was supposed to be a two-year voyage. One might think that these islands, now inextricably linked with Darwin's name, were the young naturalist's paradise. Far from it. Darwin found the islands hellishly hot, their broken black volcanic rock scorching under the hot sun. In his diary he observed that "the stunted trees show little signs of life . . . the plants also smell unpleasantly. . . . The black lava rocks on the beach are frequented by large (2–3 ft.) most disgusting clumsy lizards. . . . They assuredly well become the land they inhabit."[1] Other than the lizards and the tortoises, the animal life on the islands was scant and unimpressive. He could not wait to leave the place. The 26-year-old explorer did not know that his five weeks in the Galápagos would inspire a series of radical ideas that, some 24 years later with the publication of his *On the Origin of Species* (1859), would change our perception of the world and our place in it.

Several months after leaving the islands, on the last leg of the voyage home to England, Darwin had his first flash of insight. He had begun to organize his copious field notes from his nearly five years of exploration and collecting. His plan was for experts back in England to lead the study of his collections of fossils, plants, animals, and rocks. Turning to his observations on the birds of the Galápagos, he recalled that he had found slightly different forms of mockingbirds on three different islands. Now, there was a puzzle. The prevailing view of the origin of species in 1835, held by most of Darwin's teachers and much of the scientific establishment, was that species were specially created by God in their present form, unchangeable, and placed in the habitat to which they were best suited. Why, then, would there be slightly different birds on such similar islands? Darwin jotted in his ornithology notebook:

> When I see these Islands in sight of each other and possessed of but a scanty stock of animals, tenanted by these birds but slightly differing in structure filling the same place in Nature, I must suspect they are only varieties. . . . If there is the slightest foundation for these remarks, the zoology of Archipelagoes will be well worth examining; *for such facts would undermine the stability of species* [emphasis added].[2]

[1]C. Darwin, *Charles Darwin's* Beagle *Diary*, Ed. R. D. Keynes, Cambridge University Press, 2001.
[2]C. Darwin, *Charles Darwin's* Beagle *Diary*, Ed. R. D. Keynes, Cambridge University Press, 2001.

Darwin's insight was that species might change. This was not what he had learned at Cambridge University. This was heresy. Although Darwin decided to keep such dangerous thoughts to himself, he was gripped by the idea. After arriving home in England, he filled a series of notebooks with thoughts about species changing. Within a year he had convinced himself that species arise naturally from preexisting species, as naturally as children are born from parents and parents from grandparents. He then pondered *how* species change and adapt to their particular circumstances. In 1838, just two years after the conclusion of his voyage and before he had yet turned 30, he conceived his answer—**natural selection**. In this competitive process, individuals bearing some relative advantage over others live longer and produce more offspring, which in turn inherit the advantage.

Darwin knew that to convince others of these two ideas—the descent of species from ancestors and natural selection—he would need more evidence. He spent the next two decades marshaling all of the facts he could from botany, zoology, embryology, and the fossil record.

He received crucial information from experts who helped to sort out and characterize his collections. For example, ornithologist John Gould pointed out to Darwin that what the young naturalist thought were blackbirds, grosbeaks, and finches from the Galápagos were actually 13 new and distinct species of ground finches (**Figure 20-1**). We now know that the Galápagos species, though clearly finches, exhibit an immense variation in feeding behavior and in the bill shape that corresponds to their food sources. For example, the vegetarian tree finch uses its heavy bill to eat fruits and leaves, the insectivorous finch has a bill with a biting tip for eating large insects, and, most remarkable of all, the woodpecker finch grasps a twig in its bill and uses it to obtain insect prey by probing holes in trees.

This diversity of species, Darwin deduced, must have arisen from an original population of finch that arrived in the Galápagos from the mainland of South America and populated the islands. The descendants of the original colonizers spread to the different islands and formed local populations that diverged from one another and eventually formed different species.

The finches illustrate the process of **adaptation**, in which the characteristics of a species change over time if those traits increase the chance of survival and reproduction in the environment in which the species lives. A measure of an organism's ability to survive and reproduce is called its **fitness**. Darwin provided one level of explanation

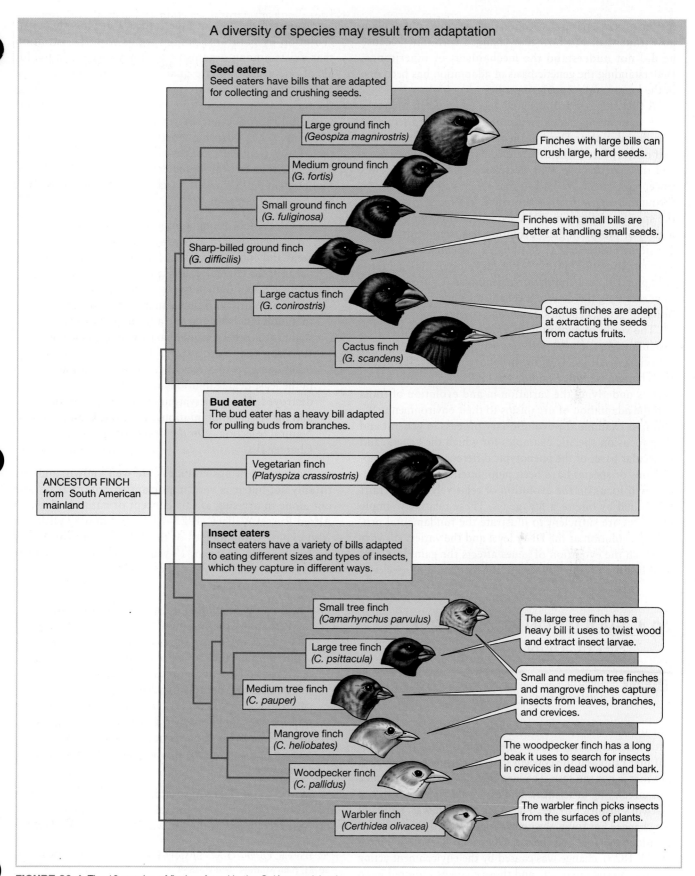

A diversity of species may result from adaptation

Seed eaters
Seed eaters have bills that are adapted for collecting and crushing seeds.

Large ground finch
(*Geospiza magnirostris*)

Medium ground finch
(*G. fortis*)

Small ground finch
(*G. fuliginosa*)

Sharp-billed ground finch
(*G. difficilis*)

Large cactus finch
(*G. conirostris*)

Cactus finch
(*G. scandens*)

Finches with large bills can crush large, hard seeds.

Finches with small bills are better at handling small seeds.

Cactus finches are adept at extracting the seeds from cactus fruits.

Bud eater
The bud eater has a heavy bill adapted for pulling buds from branches.

Vegetarian finch
(*Platyspiza crassirostris*)

ANCESTOR FINCH
from South American mainland

Insect eaters
Insect eaters have a variety of bills adapted to eating different sizes and types of insects, which they capture in different ways.

Small tree finch
(*Camarhynchus parvulus*)

Large tree finch
(*C. psittacula*)

Medium tree finch
(*C. pauper*)

Mangrove finch
(*C. heliobates*)

Woodpecker finch
(*C. pallidus*)

Warbler finch
(*Certhidea olivacea*)

The large tree finch has a heavy bill it uses to twist wood and extract insect larvae.

Small and medium tree finches and mangrove finches capture insects from leaves, branches, and crevices.

The woodpecker finch has a long beak it uses to search for insects in crevices in dead wood and bark.

The warbler finch picks insects from the surfaces of plants.

FIGURE 20-1 The 13 species of finches found in the Galápagos Islands.

for the process, natural selection, but he could not explain how traits varied or how they changed with time because he did not understand the mechanisms of inheritance. Understanding the genetic basis of adaptation has been one of the long-standing goals of evolutionary biology.

A first step toward this goal was taken when Mendel's work pointing to the existence of genes was rediscovered two decades after Darwin died. Another key emerged a half century later, when the molecular basis of inheritance and the genetic code were deciphered. For many decades since, biologists have known that species and traits evolve through changes in DNA sequence. However, elucidating the specific changes in DNA sequence underlying physiological or morphological evolution has been fraught with considerable technical challenges. Advances in molecular genetics (Chapters 7–12), developmental genetics (Chapter 13), and comparative genomics (Chapter 14) are now revealing the diverse mechanisms underlying the evolution of genes, genomes, traits, and species.

The study of evolution is a very large and expanding discipline. As such, we will not attempt a comprehensive overview of all facets of evolutionary analysis. Rather, in this chapter, we will examine the molecular genetic mechanisms underlying the variation in and evolution of traits and the adaptation of organisms to their environments. We will first examine the evolutionary process in general and then focus on specific examples for which the genetic and molecular bases of the phenotypic differences between populations or species have been pinpointed. All of the examples will focus on the evolution of relatively simple traits controlled by one or a few genes. These relatively simple examples are sufficient to illustrate the fundamental process of evolution at the DNA level and the variety of ways in which the evolution of genes affects the gain, loss, and modification of traits, as well as the formation of species.

20.1 EVOLUTION BY NATURAL SELECTION

LO 20.1 Identify and explain the principles of evolution by natural selection.

The modern theory of evolution is so completely identified with Darwin's name that many people think Darwin himself first proposed the concept that organisms have evolved, but that is not the case. The idea that life changed over time was circulating in scientific circles for many decades before Darwin's historic voyage. The great question was, *How* did life change? For some, the explanation was a series of special creations by God. To others, such as Jean-Baptiste Lamarck (1744–1829), change was caused by the environment acting directly on the organism, and those changes acquired in an organism's lifetime were passed on to its offspring.

What Darwin provided was a detailed explanation of the mechanism of the evolutionary process that correctly incorporated the role of inheritance. Darwin's theory of evolution by natural selection begins with the variation that exists among organisms within a species. Individuals of one generation are qualitatively different from one another. Evolution of the species as a whole results from the fact that the various types differ in their rates of survival and reproduction. Better-adapted types leave more offspring. When offspring inherit the type of their parents, the relative frequencies of the types will change over time. Thus, the four critical ingredients to evolutionary change that Darwin put forth were variation, selection, inheritance, and time. Darwin said:

> Can it, then, be thought improbable . . . that variations useful in some way to each being in the great and complex battle of life, should sometimes occur in the course of thousands of generations? . . . Can we doubt (remembering that many more individuals are born than can possibly survive) that individuals having any advantage, however slight, over others, would have the best chance of surviving and of procreating their kind? On the other hand, we may feel sure that any variation in the least degree injurious would be rigidly destroyed. This preservation of favorable variations and the rejection of injurious variations I call Natural Selection. (*On the Origin of Species*, Chapter IV)[3]

Darwin's writings and ideas are well known, and justifiably so, but it is very important to note that he was not alone in arriving at this concept of natural selection. Alfred Russel Wallace (1823–1913), a fellow Englishman who explored the jungles of the Amazon and the Malay Archipelago for a total of 12 years, reached a very similar conclusion in a paper that was copublished with an excerpt from Darwin in 1858:

> The life of wild animals is a struggle for existence. . . . Perhaps all the variations from the typical form of a species must have some definite effect, however slight, on the habits or capacities of the individuals. . . . It is also evident that most changes would affect, either favourably or adversely, the powers of prolonging existence. . . . If, on the other hand, any species should produce a variety having slightly increased powers of preserving existence, that variety must inevitably in time acquire a superiority in numbers.[4]

While today Darwin's name tends to be exclusively linked to evolution by natural selection, in their day, the theory was recognized as the Darwin-Wallace theory. Perhaps

[3]C. Darwin, *On the Origin of Species by Means of Natural Selection*, 80–81. John Murray, London, 1859.
[4]C. Darwin and A. Wallace, "On the Tendency of Species to Form Varieties; and on the Perpetuation of Varieties and Species by Natural Means of Selection," *Journal of the Proceedings of the Linnean Society of London. Zoology* 3, 1858, 45–50.

the current perception is at least in part due to Wallace himself, who was always deferential to Darwin and referred to the emergent theory of evolution as "Darwinism."

KEY CONCEPT Darwin and Wallace proposed a new explanation to account for the phenomenon of evolution. They understood that the population of a given species at a given time includes individuals of varying characteristics. They realized that the population of succeeding generations will contain a higher frequency of those types that most successfully survive and reproduce under the existing environmental conditions. Thus, the frequencies of various types within the species will change over time.

There is an obvious similarity between the process of evolution as Darwin and Wallace described it and the process by which the plant or animal breeder improves a domestic stock. The plant breeder selects the highest-yielding plants from the current population and uses them as the parents of the next generation. If the characteristics causing the higher yield are heritable, then the next generation should produce a higher yield. It was no accident that Darwin chose the term *natural selection* to describe his model of evolution through differences in the rates of reproduction shown by different variants in the population. As a model for this evolutionary process in the wild, he had in mind the selection that breeders exercise on successive generations of domestic plants and animals.

We can summarize the theory of evolution by natural selection in three principles:

1. *Principle of variation.* Among individuals within any population, there is variation in morphology, physiology, and behavior.

2. *Principle of heredity.* Offspring resemble their parents more than they resemble unrelated individuals.

3. *Principle of selection.* Some forms are more successful at surviving and reproducing than other forms in a given environment.

A selective process can produce change in the population composition only if there are some variations among which to select. If all individuals are identical, no differences in the reproductive rates of individuals, no matter how extreme, will alter the composition of the population. Furthermore, the variation must be in some part heritable if these differences in reproductive rates are to alter the population's genetic composition. If large animals within a population have more offspring than do small ones, but their offspring are no larger on average than those of small animals, then there will be no change in population composition from one generation to another. Finally, if all variant types leave, on average, the same number of offspring, then we can expect the population to remain unchanged.

KEY CONCEPT The principles of variation, heredity, and selection must all apply for evolution to take place through natural selection.

Heritable variation provides the raw material for successive changes within a species and for the multiplication of new species. The basic mechanisms of those changes (as discussed in Chapter 18) are the origin of new genetic and phenotypic variation in a population by mutation or migration and the change in frequency of alleles and phenotypes within populations by selective and random processes (**Figure 20-2**). From those basic mechanisms, a set of principles governing changes in the genetic composition of

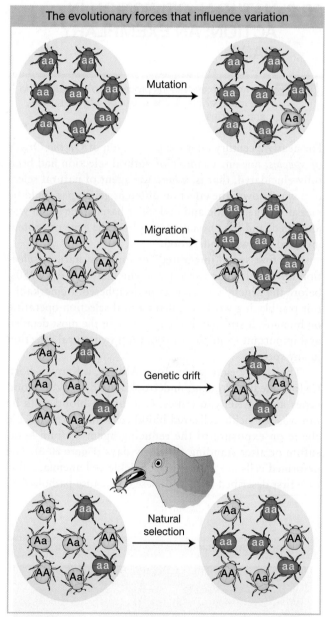

FIGURE 20-2 The effects of various forces of evolution on genetic and phenotypic variation in populations. Bugs with a yellow phenotype have genotype AA or Aa, and bugs with a green phenotype have genotype aa. In a population comprised of green bugs with genotype aa, the *A* allele and yellow phenotype can appear either due to mutation or migration from a yellow population. In a population with both genetic and phenotypic variation, a change in the allele and phenotype frequencies can occur due to genetic drift (e.g., after a bottleneck event) or due to natural selection (e.g., more yellow bugs are eaten by birds).

populations can be derived. The application of these principles of population genetics provides a genetic theory of evolution.

KEY CONCEPT Evolution, the change in populations or species over time, is the conversion of heritable variation between individuals within populations into heritable differences between populations in time and in space by population genetic mechanisms.

20.2 NATURAL SELECTION IN ACTION: AN EXEMPLARY CASE

LO 20.2 Summarize the various processes of evolution and the role they play in the evolution of genes, traits, and species.

For nearly a century after the publication of *On the Origin of Species*, not one example of natural selection had been fully elucidated, that is, where the agent of natural selection was known, the effect on different genotypes could be measured, the genetic and molecular basis of variation was identified, and the physiological role of the gene or protein involved was well understood.

The first such "integrated" example of natural selection on a molecular variant was elucidated in the 1950s, before the genetic code was even deciphered. Remarkably, this trailblazing work revealed natural selection operating on humans. It still stands today as one of the most detailed and important examples of evolution by natural selection in any species.

The story began when Tony Allison, a Kenyan-born Oxford medical student, undertook a field study of blood types among Kenyan tribes. One of the blood tests he ran was for sickle cells, red blood cells that form a sickle shape on exposure to the reducing agent sodium beta-sulfite or after standing for a few days (**Figure 20-3**). The deformed cells are a hallmark of sickle-cell anemia, a disease first described in 1910. These cells cause pathological

complications by occluding blood vessels and lead to early mortality.

In 1949, the very year Allison went into the field, Linus Pauling's research group demonstrated that patients with sickle-cell anemia had a hemoglobin protein with an abnormal charge (Hemoglobin S, or HbS) in their blood, compared with the hemoglobin of unaffected individuals (Hemoglobin A, or HbA). This was the first demonstration of a molecular abnormality linked to a complex disease. It was generally understood at the time that carriers of sickle cell were heterozygous and thus had a mixture of HbA and HbS (denoted *AS*), whereas affected individuals were homozygous for the Hb^S allele (denoted *SS*).

Allison collected blood specimens from members of the Kikuyu, Masai, Luo, and other tribes across the very diverse geography of Kenya. While he did not see any particularly striking association between ABO or MN blood types among the tribes, he measured remarkably different frequencies of Hb^S. In tribes living in arid central Kenya or in the highlands, the frequency of Hb^S was less than 1 percent; however, in tribes living on the coast or near Lake Victoria, the frequency of Hb^S often exceeded 10 percent and approached 30 percent in some locations (**Table 20-1**).

The allele frequencies were surprising for two reasons. First, since sickle-cell anemia was usually lethal, why were the frequencies of the Hb^S allele so high? And second, given the relatively short distances between regions, why was the Hb^S frequency high in some places and not in others?

Allison's familiarity with the terrain, tribes, and tropical diseases of Kenya led him to the crucial explanation. Allison realized that the Hb^S allele was at high frequency in low-lying humid regions with very high levels of malaria, and nearly absent at high altitudes such as around Nairobi. Carried by mosquitoes, the intracellular parasite *Plasmodium falciparum*, which causes malaria, multiplies inside red blood cells (**Figure 20-4**). Mosquitoes and the disease are prevalent throughout sub-Saharan Africa in humid, low-lying regions near bodies of water where the mosquitoes reproduce. Allison surmised that the Hb^S allele might, by altering red blood cells, confer some degree of resistance to malarial infection.

The selective advantage of Hb^S

To test this idea, Allison carried out a much larger survey of Hb^S frequencies across eastern Africa, including Uganda, Tanzania, and Kenya. He examined about 5000 individuals representing more than 30 different tribes. Again, he

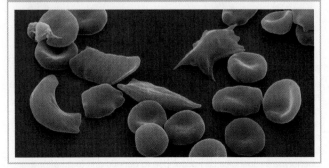

Red blood cells in someone with sickle-cell trait

FIGURE 20-3 A colorized electron micrograph showing sickle cells among normal red blood cells. [*Eye of Science/Science Source.*]

TABLE 20-1	Frequency of Hb^S in Particular Kenyan Tribes		
Tribe	**Ethnic affinity**	**District/region**	**%Hb^S**
Luo	Nilotic	Kisumu (Lake Victoria)	25.7
Suba	Bantu	Rusingo Island	27.7
Kikuyu	Bantu	Nairobi	0.4

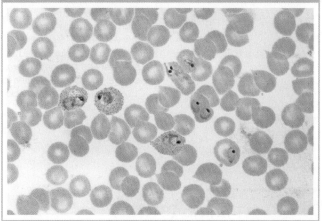

Malarial parasites live within red blood cells

FIGURE 20-4 A blood smear of an individual infected with malarial parasites. A red blood cell sample was treated with Giemsa stain to reveal parasites within cells (red dots). [*CDC/Dr. Mae Melvin.*]

found Hb^S frequencies of up to 40 percent in malarial areas and frequencies as low as 0 percent where malaria was absent.

The link suggested that the Hb^S allele might affect parasite levels, so Allison also undertook a study of the level of parasites in the blood of heterozygous *AS* children versus wild-type *AA* children. In a study of nearly 300 children, he found the incidence of malarial parasites was indeed lower in *AS* children (27.9 percent) than in *AA* children (45.7 percent) and that parasite density was also lower in *AS* children. The results indicated that *AS* children had a lower incidence and severity of malarial infection and would thus have a selective advantage in areas where malaria was prevalent.

The advantage to *AS* heterozygotes was especially striking in light of the disease suffered by *SS* homozygotes. Allison noted:

> The proportion of individuals with sickle cells in any population . . . will be the result of a balance between two factors: the severity of malaria, which will tend to increase the frequency of the gene, and the rate of elimination of the sickle-cell genes in individuals dying of sickle-cell anaemia.

The sickle-cell mutation was under *balancing selection* (see Chapter 18) in areas where malaria was present. Natural selection operating in favor of *AS* individuals is

balanced by natural selection operating against *AA* individuals susceptible to malaria and *SS* individuals who would succumb to sickle-cell anemia. In other words, the heterozygote has an advantage over either of the homozygotes.

How much of an advantage do *AS* individuals experience? This can be calculated by measuring the frequency of the Hb^S allele in populations and examining how these frequencies differ from the frequencies expected under the assumptions of the Hardy–Weinberg equation (see Chapter 18). A large-scale survey of 12,387 West Africans revealed an Hb^S allele frequency (q) of 0.123 (see Box 18.1 for assistance in calculating allele frequencies). The frequencies calculated from the Hardy–Weinberg equation (expected phenotype frequencies in **Table 20-2**) are higher than what was actually observed for the homozygous phenotypes and lower for the heterozygous phenotype (Table 20-2). If it is assumed that the *AS* heterozygote has a fitness of 1.0, then the relative fitness of the two homozygous genotypes can be estimated. For example, the relative fitness of the homozygous *AA* genotype compared to the *AS* genotype is 0.88. The *AS* genotype thus has a selective advantage of 1.136, or approximately 14 percent, over the *AA* genotype.

This selective advantage has been well documented by long-term survival studies of *AA*, *AS*, and *SS* children in Kenya. These studies have found that *AS* individuals have a pronounced survival advantage over *AA* and *SS* individuals in the first few years of life (**Figure 20-5**).

KEY CONCEPT The sickle-cell hemoglobin allele, Hb^S, is under balancing selection in malarial zones and conveys a large survival advantage in heterozygotes over the first few years of life.

The molecular origins of Hb^S

After Allison's discovery, there was keen interest in determining the molecular basis of the difference(s) between Hb^S and Hb^A. Protein sequencing determined that Hb^S differs from Hb^A by just one amino acid, a valine in place of a glutamic acid residue. This single amino acid change alters the charge of hemoglobin and causes it to aggregate into long rodlike structures within red blood cells. Once the genetic code was deciphered and methods for sequencing DNA were developed, Hb^S was determined to be caused by a single point mutation (CTC → CAC) in the glutamic acid

TABLE 20-2   The Fitness Advantage of Sickle-Cell Heterozygotes

Genotype	Observed phenotype frequency	Expected phenotype frequency	Ratio of observed/expected	*w* (relative fitness)	Selective advantage
SS	29	187.4	0.155	0.155/1.12 = 0.14	
AS	2993	2672.4	1.12	1.12/1.12 = 1.00	**1.0/0.88 = 1.136**
AA	9365	9527.2	0.983	0.983/1.12 = 0.88	
Total	12,387	12,387			

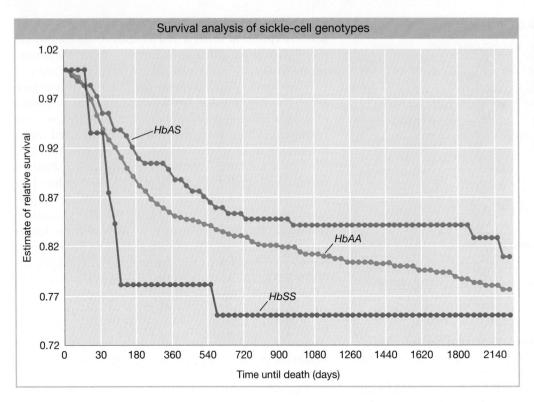

FIGURE 20-5 The relative survival of approximately 1000 children from Kisumu is plotted from birth through the first few years of life. Sickle-cell heterozygotes experienced a significant advantage in overall survival from ages 2 months to 16 months. [*Data from M. Aidoo et al., The Lancet 359, 2002, 1311–1312.*]

codon encoding the sixth amino acid of the β-globin subunit within the hemoglobin protein.

Interestingly, Allison also noted a high incidence of Hb^S outside of Africa, including in Italy, Greece, and India. Other blood-type markers did not indicate strong genetic relationships among these populations. Rather, Allison observed that these were also areas with a high incidence of malaria. The correlation between Hb^S frequency and the incidence of malaria held across not only East Africa, but the African continent, southern Europe, and the Indian subcontinent.

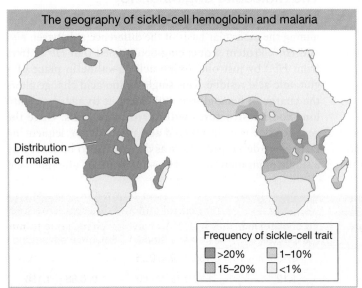

FIGURE 20-6 These maps show the close correspondence between the distribution of malaria (*left*) and the frequency of the sickle-cell trait (*right*) across Africa. [*Data from A. C. Allison, Genetics 66, 2004, 1591; redrawn by Leanne Olds.*]

Allison composed maps showing these striking correlations (**Figure 20-6**) and inferred that the Hb^S alleles in different regions arose independently, rather than through spreading by migration. Indeed, with the advent of tools for DNA genotyping, it is clear that the Hb^S mutation has arisen independently at least five times and then increased to high frequency in particular regions. Based on the limited genetic diversity of malarial populations, it is believed that Hb^S mutations arose in just the past several thousand years, once populations began living around bodies of water as part of the advent of agriculture.

Elucidating the role of Hb^S in conferring resistance to malaria illustrated three important facets of the evolutionary process:

1. *Evolution can and does repeat itself.* The multiple independent origins and expansions of the Hb^S mutation demonstrate that the same mutations can arise and spread repeatedly. Many examples are now known of the precise, independent repetition of the evolution of adaptive mutations, and we will encounter several more in this chapter.

2. *Fitness is a relative, conditional status.* Whether a mutation is advantageous, disadvantageous, or neither depends very much on environmental conditions. In the absence of malaria, Hb^S is very rare and disfavored. Where malaria is present, Hb^S can reach high frequencies despite the disadvantages imparted to *SS* homozygotes. In African Americans, the frequency of Hb^S is declining because there is selection against the allele in the absence of malaria in North America.

3. *Natural selection acts on whatever variation is available, and not necessarily by the best means imaginable.* The Hb^S mutation, while protective against malaria (in the heterozygous state), also causes a life-threatening condition (in the homozygous state). Over 40 percent of the world's population lives in areas where malaria is prevalent. In these places, the imperative of combating malaria counterbalances the deleterious effect of the sickle-cell mutation.

20.3 MOLECULAR EVOLUTION

LO 20.3 Distinguish among the signatures of neutral evolution, positive selection, and purifying selection in DNA and protein sequences.

Darwin and Wallace conceived of evolution largely as "changes in organisms brought about by natural selection." Indeed, this is what most people think of as the meaning of "evolution." However, a century after Darwin's theory, as molecular biologists began to confront evolution at the level of proteins and DNA molecules, they encountered and identified another dimension of the evolutionary process, neutral molecular evolution, which did not involve natural selection. This led to the realization that an understanding of when molecular evolution occurs due to neutral or selective processes is crucial to grasping how organisms change over time.

The development of the neutral theory of evolution

In the 1950s and early 1960s, methods were developed that enabled biologists to determine the amino acid sequences of proteins. This new capability raised the prospect that the fundamental basis of evolutionary change was finally at hand. However, as the sequences of proteins from a variety of species were deciphered, a paradox emerged. The sequences of globins and cytochrome c, for example, typically differ between any two species at a number of amino acids, and that number increases with the time elapsed since their divergence from a common ancestor (**Figure 20-7**). Yet, the function of these proteins is the same in different species—to carry and deliver oxygen to tissues in the case of hemoglobin and to shuttle electrons during cellular respiration in the case of cytochrome c.

The puzzle then was whether the amino acid replacements between species reflected changes in protein function and adaptations to selective conditions. Biochemists Linus Pauling and Emile Zuckerkandl did not think so. They observed that many substitutions were of one amino acid for another with similar properties. They concluded that most amino acid substitutions were "neutral" or "nearly neutral" and did not change the function of a protein whatsoever.

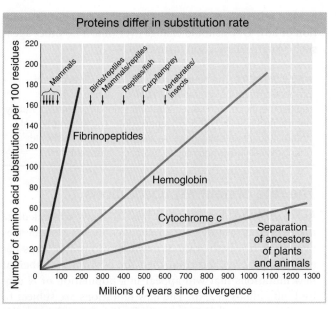

FIGURE 20-7 Number of amino acid substitutions in the evolution of the vertebrates as a function of time since divergence. The three proteins—fibrinopeptides, hemoglobin, and cytochrome c—differ in substitution rate because different proportions of their amino acid substitutions are selectively neutral.

This line of reasoning was rejected at first by many evolutionary biologists, who at the time viewed all evolutionary changes as the result of natural selection and adaptation. Paleontologist George Gaylord Simpson argued that "there is a strong consensus that completely neutral genes or alleles must be rare if they exist at all. To an evolutionary biologist it therefore seems highly improbable that proteins . . . should change in a regular but non-adaptive way."[5]

Zuckerkandl and Pauling asserted that the similarity or differences among organisms need not be reflected at the level of protein—that molecular change and visible change were not necessarily linked or proportional.

The debate over whether molecular changes could be neutral was resolved by an onslaught of empirical data and the deciphering of the genetic code. Because multiple codons encode the same amino acid, a mutation that changes, say, CAG to CAC does not change the amino acid encoded. Therefore, variation can exist at the DNA level that has no effect on protein sequences, and thus neutral alleles do exist. But even more important for population genetics was the development of the "neutral theory of molecular evolution" by Motoo Kimura, Jack L. King, and Thomas Jukes. These authors proposed that most, but not all, mutations that are invariant or *fixed* between species are neutral or nearly neutral and any differences between species at such sites in DNA evolve by chance due to *random genetic drift*.

The "neutral theory" marked a profound conceptual shift away from a view of evolution as always guided by

[5]G. G. Simpson, "Organisms and Molecules in Evolution," *Science* 146, 1964, 1535–1538.

natural selection. Moreover, it provided a baseline assumption of how DNA should change over time if no other agent such as natural selection intervened.

> **KEY CONCEPT** The neutral theory of molecular evolution proposed that most mutations in DNA or amino acid replacements between species are functionally neutral or nearly neutral and fixed by random genetic drift. The assumption of neutrality offers a baseline expectation of how DNA should change over time when natural selection is absent.

The rate of neutral substitutions

As we saw in Chapter 18 (see Box 18-5), we can calculate the expected rate of neutral changes in DNA sequences over time. If μ is the rate of new mutations at a locus per gene copy per generation, then the absolute number of new mutations that will appear in a population of N diploid individuals is $2N\mu$. The new mutations are subject to random genetic drift: most will be lost from the population, while a few will become fixed and replace the previous allele. If a newly arisen mutation is neutral, then there is a probability of $1/(2N)$ that it will replace the previous allele because of random genetic drift. Each one of the $2N\mu$ new mutations that will appear in a population has a probability of $1/(2N)$ of eventually taking over that population. Thus, the absolute substitution rate k is the mutation rate multiplied by the probability that any one mutation will eventually take over by drift:

$$k = \text{rate of neutral substitution} = 2N\mu \times 1/(2N) = \mu$$

That is, we expect that, in every generation, there will be μ substitutions in the population, purely from the genetic drift of neutral mutations.

> **KEY CONCEPT** The rate of substitutions in DNA in evolution resulting from the random genetic drift of neutral mutations is equal to the mutation rate to such alleles, μ.

The signature of purifying selection on DNA sequences

When measurements of molecular change deviate from what is expected for neutral changes, that is an important signal—a signal that selection has intervened. That signal may reveal that selection has favored some specific change or that it has rejected others. We have seen, in the case of the Hb^S mutation, how natural selection favors the mutation in the presence of the malarial parasite, but rejects it where malaria is absent. The most pervasive influence of natural selection on DNA is, in fact, to conserve gene function and sequence.

All classes of DNA sequences, including exons, introns, regulatory sequences, and sequences in between genes, show nucleotide diversity among individuals within populations and between species. The constant rate of neutral substitutions predicts that, if the number of nucleotide differences between two species were plotted against the time

since their divergence from a common ancestor, the result should be a straight line with slope equal to μ. That is, evolution should proceed according to a molecular clock (see Box 18-5) that is ticking at the rate μ. **Figure 20-8** shows such a plot for the β-globin gene. The results are quite consistent with the claim that nucleotide substitutions in this gene have been neutral in the past 500 million years. Two sorts of neutral nucleotide substitutions are plotted: **synonymous substitutions**, which are from one codon to another, making no change in the amino acid, and **nonsynonymous substitutions**, which result in an amino acid change. Figure 20-8 shows a much lower slope for nonsynonymous substitutions than for synonymous changes, which means that the substitution rate of neutral nonsynonymous substitutions is much lower than that of synonymous neutral substitutions.

This outcome is precisely what we expect under natural selection. Mutations that cause an amino acid substitution should have a deleterious effect more often than synonymous substitutions, which do not change the protein. Such deleterious variants will be removed from populations by *purifying selection* (see Chapter 18). A lower-than-expected ratio of nonsynonymous to synonymous changes is a signature of purifying selection. It is important to note that these observations do not show that synonymous substitutions have no selective constraints on them; rather, they show that these constraints are, on the average, not as strong as those for mutations that change amino acids. So, a synonymous change, although it has no effect on the amino acid sequence, does change the mRNA for that sequence and thus may affect mRNA stability or efficiency at which the mRNA is translated.

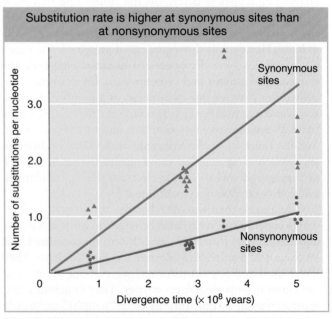

FIGURE 20-8 The amount of nucleotide divergence at synonymous sites is greater than the amount of divergence at nonsynonymous sites of the β-globin gene.

Purifying selection is the most widespread, but often underappreciated, facet of natural selection. The "rejection of injurious variations," as Darwin termed it, is pervasive. Purifying selection explains why we find many protein sequences that are unchanged or nearly unchanged over vast spans of evolutionary time. For example, there are several dozen genes that exist in all domains of life—Archaea, Bacteria, and Eukarya (fungi, plants, protists, and animals)—and encode proteins whose sequences have been largely conserved over 3 billion years of evolution. To preserve such sequences, variants that have arisen at random in billions of individuals in tens of millions of species have been rejected by selection over and over again.

KEY CONCEPT Purifying selection is a pervasive aspect of natural selection that reduces genetic variation and preserves DNA and protein sequences over eons of time.

Another prediction of the theory of neutral evolution is that different proteins will have different clock rates because the metabolic functions of some proteins will be much more sensitive to changes in their amino acid sequences. Proteins that are more sensitive to amino acid changes will have a lower rate of neutral mutation because a smaller proportion of their mutations will be neutral compared with proteins that are more tolerant of substitution. Figure 20-7 shows a comparison of the clocks for fibrinopeptides, hemoglobin, and cytochrome c. That fibrinopeptides have a much higher proportion of neutral mutations is reasonable because these peptides are not known to have a function after they are cut out of fibrinogen to activate the blood-clotting reaction. It is not obvious why hemoglobins are less sensitive to amino acid changes than is cytochrome c.

KEY CONCEPT The rate of neutral evolution for the amino acid sequence of a protein depends on the sensitivity of the protein's function to amino acid changes.

Because so much sequence evolution is neutral, there is no simple relation between the amount of change in a gene's DNA sequence and the amount of change, if any, in the encoded protein's function. At one extreme, almost the entire amino acid sequence of a protein can be replaced while maintaining the original function if those amino acids that are substituted maintain the enzyme's three-dimensional structure. This is the case for the fibrinopeptides, in which nearly any amino acid can be substituted at any position of the peptide.

In contrast, the function of an enzyme can be changed by a single amino acid substitution. The sheep blowfly, *Lucilia cuprina*, has evolved resistance to organophosphate insecticides used widely to control it. This resistance is the consequence of a single substitution of an aspartic acid for a glycine residue in the active site of a blowfly enzyme that is ordinarily a carboxylesterase (splits a carboxyl ester, R–COO–R, into an alcohol and a carboxylate). The mutation causes complete loss of the carboxylesterase activity and its replacement by esterase activity (splits any ester, R–O–R, into an acid and an alcohol). Three-dimensional modeling of the molecule indicates that the protein with esterase activity gains the ability to bind a water molecule close to the site of attachment of the organophosphate. The water molecule then reacts with the organophosphate, splitting it in two.

KEY CONCEPT There is no proportionate relation between how much DNA change takes place in evolution and how much change in function results.

The signature of positive selection on DNA sequences

Evidence for the molecular clock supports the idea that most nucleotide substitutions that have occurred in evolution were neutral. However, it does not tell us how much of molecular evolution has been adaptive change driven by *positive selection*. One way of detecting the adaptive evolution of a protein is by comparing the synonymous and nonsynonymous nucleotide polymorphisms within species with the synonymous and nonsynonymous nucleotide changes between species. If all mutations are neutral, the ratio of nonsynonymous to synonymous nucleotide polymorphisms within a species should be the same as the ratio of nonsynonymous to synonymous nucleotide substitutions between species. On the other hand, if the amino acid changes between species have been driven by positive selection, there ought to be an excess of nonsynonymous changes between species.

One test for detecting positive selection on DNA sequences was developed by John McDonald and Martin Kreitman. This test involves several logical but simple steps:

1. The DNA sequence of a gene is obtained from a number of separate individuals from each of two species. Sequences from ten or more individuals of each species would be desirable. The fixed nucleotide differences between species are then classified into nonsynonymous (*a* in the table below) and synonymous (*b* in the table below) differences.

2. The nucleotide differences among individuals within each species (polymorphisms) are then tabulated, and classified as either those that result in amino acid changes (nonsynonymous polymorphisms; *c* in the table below) or those that do not change the amino acid (synonymous polymorphisms; *d* in the table below).

3. If the divergence between the species is purely the result of random genetic drift, then we expect *a/b* to be equal to *c/d*. If, on the other hand, there has been selective divergence, there should be an excess of fixed nonsynonymous differences, and so *a/b* should be greater than *c/d*.

	Fixed species differences	Polymorphisms
Nonsynonymous	*a*	*c*
Synonymous	*b*	*d*
Ratio	*a/b*	*c/d*

TABLE 20-3   Synonymous and Nonsynonymous Polymorphisms and Species Differences for Alcohol Dehydrogenase in Three Species of *Drosophila*

	Species differences	Polymorphisms
Nonsynonymous	7	2
Synonymous	17	42
Ratio	7/17 = 0.41	2/42 = 0.05

Data from J. McDonald and M. Kreitman, "Adaptive Protein Evolution at the *Adh* locus in *Drosophila*," *Nature* 351, 1991, 652–654.

Table 20-3 shows an application of this principle to the alcohol dehydrogenase gene in three closely related species of *Drosophila*. Clearly, there is an excess of amino acid replacements between species over what is expected. Therefore, we conclude that some of the amino acid replacements in the enzyme were adaptive changes driven by natural selection.

KEY CONCEPT It is possible to detect signatures of adaptive evolution at the level of DNA sequences.

20.4 EVOLUTION OF GENES AND GENOMES

LO 20.4 Identify evidence for gene and genome duplications, and assess the role of gene duplication in the evolution of protein function, traits, and species.

Evolution consists of more than just substitutions in the amino acid sequences of protein-coding genes. A large fraction of protein-coding and RNA-encoding genes belong to **gene families**, groups of genes that are related in sequence and typically in biochemical function as well. For example, there are over 1000 genes encoding structurally related olfactory receptors in a mouse and, in humans, there are three structurally related opsin genes that encode proteins necessary for color vision. Within families such as these, new functions have evolved that have made possible new capabilities. These new functions may be expansions of existing capabilities. In the examples shown earlier, new receptors appeared in mice with the ability to detect new chemicals in the environment. In the case of humans and their Old World primate relatives, new opsin proteins appeared that enabled their owners to detect wavelengths of light that other mammals cannot. In other cases, the evolution of gene families may lead to novel functions that open up new ways of living, such as the acquisition of additional globin genes in placental mammals. In this section, we seek answers to the questions: Where does the DNA for new genes come from? What are the fates of new genes? What are the evolutionary consequences when whole genomes are duplicated?

Expanding gene number

There are several genetic mechanisms that can expand the number of genes or parts of genomes in an organism. One large-scale process for the expansion of gene number is the formation of *polyploids*, individuals with more than two chromosome sets (see Chapter 17). Polyploids result from the duplication of the entire genome. We will examine the consequences of whole genome duplications later in this section.

A second mechanism that can increase gene number is **gene duplication**. Misreplication of DNA during meiosis can cause segments of DNA to be duplicated. The lengths of the segments duplicated can range from just one or two nucleotides up to substantial segments of chromosomes containing scores or even hundreds of genes. Detailed analyses of variation in the human genome has revealed that individual humans commonly carry small duplications that result in variation in gene-copy number (see Chapter 14).

A third mechanism that can generate gene duplications is *transposition*. Sometimes, when a transposable element moves to another part of the genome, it may carry along additional host genetic material and insert a copy of some part of the genome into another location (see Chapter 16).

A fourth mechanism that can expand gene number is **retrotransposition**. Many animal genomes harbor retroviral-like genetic elements (see Chapter 16) that encode reverse-transcriptase activity. Retrotransposons themselves make up approximately 40 percent of the human genome. Occasionally, a host genome mRNA transcript is reverse transcribed into cDNA and inserted back in the genome, producing an intronless gene duplicate.

The fate of duplicated genes

It was once thought that because the ancestral function is provided by the original gene, duplicate genes are essentially spare genetic elements that are free to evolve new functions (termed **neofunctionalization**), and that would be a common fate. However, the detailed analysis of genomes and population-genetic considerations has led to a better understanding of the alternative fates of new gene duplicates, with the evolution of new function being just one pathway.

For simplicity's sake, let's consider a duplication event that results in the duplication of the entire coding and regulatory region of a gene (**Figure 20-9a**). Many different outcomes can unfold from such a duplication. The simplest result is that the allele bearing the duplicate is lost from the population before it rises to any significant frequency, as is the fate of many new mutations (see Chapter 18). But let's consider next the more interesting scenarios: suppose the duplication survives and new mutations begin to occur within the duplicate gene pair. Keeping in mind that the original and duplicated genes are initially exact copies and therefore redundant, once new mutations arise, the following fates are possible:

1. An inactivating mutation may occur in the coding region of either duplicate. The inactivated *paralog* is called a

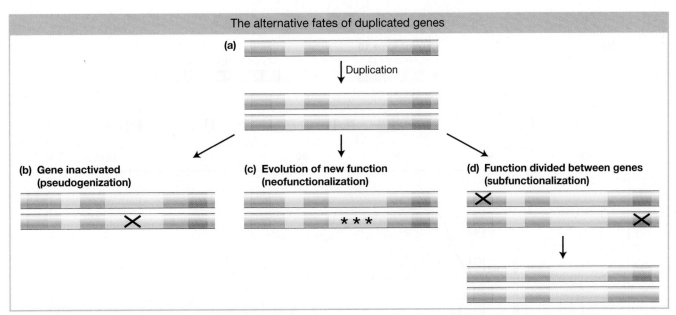

The alternative fates of duplicated genes

(a)

Duplication

(b) Gene inactivated (pseudogenization)

(c) Evolution of new function (neofunctionalization)

★ ★ ★

(d) Function divided between genes (subfunctionalization)

FIGURE 20-9 The alternative fates of duplicated genes. (a) The duplication of a gene. The orange, green, and purple boxes denote cis-regulatory elements; the beige box denotes the coding region. After duplication, several fates of the duplicates are possible: (b) any inactivating mutation in a coding region will render that duplicate into a pseudogene, and purifying selection will then operate on the remaining paralog; (c) mutations may arise that alter the function of a protein and may be favored by positive selection; (d) mutations may affect a subfunction of either duplicate, and so long as the two paralogs together provide the ancestral functions, different subfunctions may be retained, resulting in the evolution of two complementary loci.

pseudogene and will generally be invisible to natural selection. Thus, it will accumulate more mutations and evolve by random genetic drift, while natural selection will maintain the functional paralog (Figure 20-9b).

2. Mutations may occur that alter the regulation of one duplicate or the activity of one encoded protein. These alleles may then become subject to positive selection and acquire a new function (neofunctionalization) (Figure 20-9c).

3. In cases where the ancestral gene has more than one function and more than one regulatory element, as for most "toolkit" genes involved in development (see Chapter 13), a third possible outcome is that initial mutations inactivate or alter one regulatory element in each duplicate. The original gene function is now divided between the duplicates, which complement each other. To preserve the ancestral function, natural selection will maintain the integrity of both gene-coding regions. Loci that follow this path of duplication and mutation that produce complementary paralogs are said to be **subfunctionalized** (Figure 20-9d).

KEY CONCEPT Once a duplicated gene is fixed within a population and begins to acquire mutations, it can follow one of the following alternative evolutionary trajectories: pseudogenization, neofunctionalization, or subfunctionalization.

Some of these alternative fates of gene duplicates are illustrated in the history of the evolution of human globin genes. The evolution of our lineage, from fish ancestors to terrestrial amniotes that laid eggs to placental mammals, has required a series of innovations in tissue oxygenation. These include the evolution of additional globin genes with novel patterns of regulation and the evolution of hemoglobin proteins with distinct oxygen-binding properties.

Adult hemoglobin is a tetramer consisting of two α polypeptide chains and two β chains, each with its bound heme molecule. The gene encoding the adult α chain is on chromosome 16, and the gene encoding the β chain is on chromosome 11. The two chains are about 49 percent identical in their amino acid sequences; this similarity reflects their common origin from an ancestral globin gene deep in evolutionary time. The α chain gene resides in a cluster of five related genes (α and ζ) on chromosome 16, while the β chain resides in a cluster of six related genes on chromosome 11 (ε, β, δ, and γ) (**Figure 20-10**). Each cluster contains a pseudogene, ψ_α and ψ_β, respectively, that has accumulated random, inactivating mutations.

Each cluster also contains genes that have evolved distinct expression profiles, distinct functions, or both. Of greatest interest for this example are the two γ genes in the β cluster. These genes are expressed during the last seven months of fetal development to produce fetal hemoglobin (also known as hemoglobin F), which is composed of two α chains and two γ chains. Fetal hemoglobin has a greater affinity for oxygen than does adult hemoglobin, which allows the fetus to extract oxygen from the mother's circulation via the placenta. At birth, up to 95 percent of hemoglobin is the fetal type, then expression of the adult

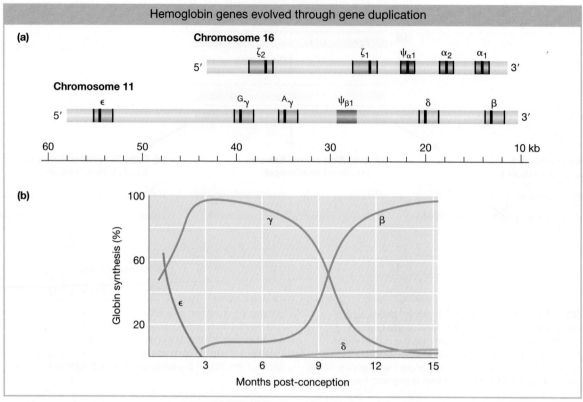

FIGURE 20-10 (a) Chromosomal distribution of the genes for the α family of globins on chromosome 16 and the β family of globins on chromosome 11 in humans. Gene structure is shown by black bars (exons) and colored bars (introns). Some duplicates of the hemoglobin genes evolved into nonfunctional pseudogenes (ψ_α and ψ_β). (b) The relative expression levels of the β globin genes during embryonic and fetal development and after birth are indicated.

β form replaces γ, and a small amount of δ globin is also produced. The order of appearance of globin chains during development is orchestrated by a complex set of cis-acting regulatory sequences and, remarkably, follows the order of the genes on each chromosome.

The γ genes are restricted to placental mammals. Their distinct developmental regulation and protein products mean that these duplicates have evolved differences in function that have contributed to the evolution of the placental lifestyle. Interestingly, regulatory variants of these genes are known that cause expression of the fetal hemoglobin to persist into childhood and adulthood. These naturally occurring variants appear to moderate the severity of sickle-cell anemia by suppressing the levels of Hb^S produced. One widespread treatment of sickle-cell anemia is to administer drugs that stimulate the reactivation of fetal-hemoglobin expression.

The fate of duplicated genomes

We have seen that, in some organisms such as polyploids, the present-day genome evolved as a result of a **whole genome duplication (WGD)** event (see Chapter 17). Approximately 25 to 30 percent of flowering plants are currently polyploids, and the formation of polyploids

has played a major role in the evolution of plant species. Consider the frequency distribution of haploid chromosome numbers among the dicotyledonous plant species shown in **Figure 20-11**. Above a chromosome number of about 12, even numbers are much more common than odd numbers—a consequence of frequent polyploidy. Although rarer in animals, hundreds of insect and vertebrate species, particularly fish and amphibians, are also polyploids. Many polyploidy events occur at the tips of evolutionary trees, suggesting that polyploid lineages are short-lived and prone to extinction. This is consistent with the problems in mitosis and meiosis that can be caused by changes in chromosome number discussed in Chapter 17.

However, there is emerging evidence from whole genome sequencing studies that some lineages have undergone one or more rounds of WGD at some point in their evolutionary history. The first case identified was in baker's yeast, *Saccharomyces cerevisiae*. The evolution of this genome was analyzed by comparing the whole-genome sequence of *S. cerevisiae* with that of another yeast, *Kluyveromyces*, whose genome is similar to that of the ancestral genome of yeast. Apparently, in the course of the evolution of *Saccharomyces*, the *Kluyveromyces*-like ancestral genome doubled, and so there were two sets, each containing the whole genome. After doubling occurred, many gene copies were

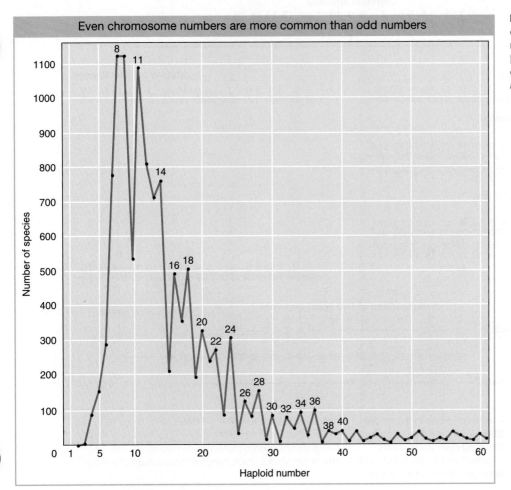

Even chromosome numbers are more common than odd numbers

FIGURE 20-11 Frequency distribution of haploid chromosome numbers in dicotyledonous plants. [*Data from Verne Grant*, The Origin of Adaptations. *Columbia University Press, 1963.*]

lost from one set or the other, and the remaining sets were rearranged, resulting in the present *Saccharomyces* genome (**Figure 20-12**). Similar lines of evidence suggest that one WGD occurred at the base of the lineage leading to flowering plants (angiosperms), with many independent WGDs occurring later in different angiosperm lineages. The human genome is the result of two WGDs that occurred at the base of the vertebrate lineage. After the ancestors of fishes and land vertebrates diverged, yet another WGD occurred in the lineage leading to teleost fishes.

KEY CONCEPT Many modern plant, animal, and fungal genomes are the result of ancestral whole genome duplication events.

What are the evolutionary consequences of these WGDs? The fact that every gene in the genome would be duplicated led Susumu Ohno to hypothesize that genome duplications might facilitate the evolution of novel traits, particularly **key innovations** that would allow a lineage to take advantage of a novel environment and rapidly diversify. Incredibly, several decades before their existence was revealed by analyses of genome sequences, Ohno predicted that vertebrates had undergone two rounds of genome

duplications. Consistent with Ohno's hypothesis, the emergence of key innovations such as flowers in angiosperms and hinged jaws in vertebrates are associated with the timing of ancient WGDs. As further support of this hypothesis, the set of genes that are retained as duplicates in modern genomes is not random. Key toolkit genes known to play a role in development (see Chapter 13), such as transcriptional regulators and signaling proteins, are preferentially retained after WGD in plants, vertebrates, fish, and yeast. For example, recall from Chapter 13 that the *Hox* genes are key regulators of segmental identity found in gene complexes, with multiple *Hox* genes in a cluster on the same chromosome. In vertebrates, all four *Hox* clusters have been retained from the two rounds of ancient WGD, and up to eight *Hox* gene clusters are retained from the additional round of WGD in teleost fish (**Figure 20-13**).

The retention of duplicate genes over long evolutionary times implies that they are maintained by natural selection and not by drift. Nonetheless, the function of most duplicate genes has not been directly tested. Thus, the evidence in support of the hypothesis that whole genome duplications have facilitated the evolution of novel traits remains mostly correlative. To test this hypothesis, we must identify the specific genetic changes that underlie the evolution of traits, a topic to which we now turn.

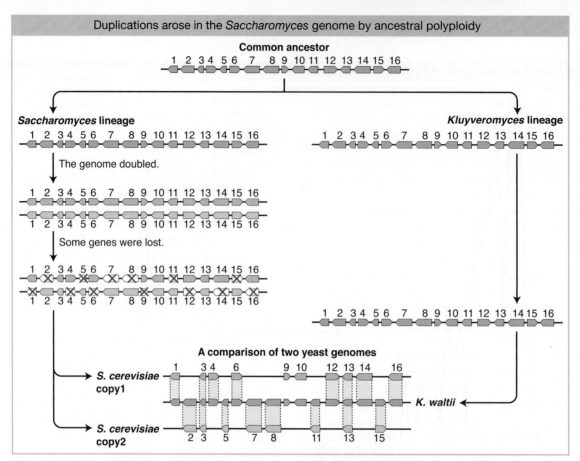

FIGURE 20-12 A common ancestor similar to the modern *Kluyveromyces* yeast duplicated its genome. For simplicity, 16 genes on one chromosome are shown. Some genes were lost. Duplicate genes such as 3 and 13 are in the same relative order. The bottom panel compares the two modern genomes. [*Data from Figure 1, Manolis Kellis, Bruce W. Birren, and Eric S. Lander, "Proof and Evolutionary Analysis of Ancient Genome Duplication in the Yeast Saccharomyces cerevisiae," Nature 428, April 8, 2004, copyright Nature Publishing Group.*]

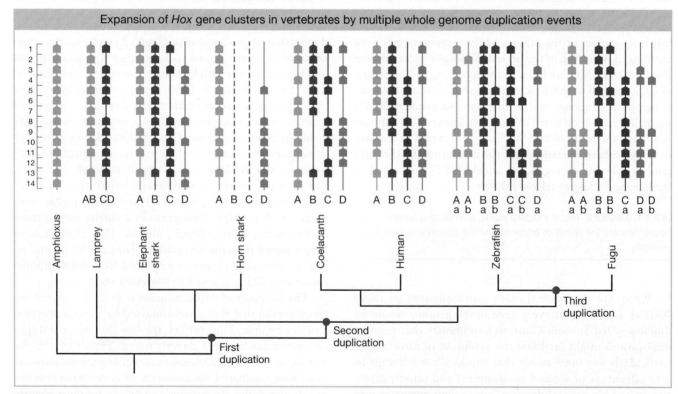

FIGURE 20-13 Three rounds of whole genome duplication have occurred in the vertebrate lineage. The first occurred after the split between chordate invertebrates (represented by Amphioxus) and jawless vertebrates (represented by Lamprey) leading to two *Hox* clusters (AB and CD). The second occurred in the lineage leading to jawed vertebrates (represented by Elephant shark, Horn shark, Coelacanth, and Humans), leading to four *Hox* clusters (A, B, C, and D). The third occurred in the lineage leading to telost fish (represented by Fugu, Zebrafish), leading to eight *Hox* clusters (Aa, Ab, Ba, Bb, Ca, Cb, Da, and Db), with subsequent loss of the Cb cluster in Fugu and the Db cluster in Zebrafish. The ancestral chordate *Hox* cluster had 14 genes (numbered 1–14), but there has been independent loss of different genes in each of the lineages.

20.5 EVOLUTION OF TRAITS

LO 20.5 Explain the critical role of regulatory sequences in the evolution of morphological traits.

One of the most apparent and interesting categories of evolving traits is that of organism morphology. Among animals, for example, there is great diversity in the number, kind, size, shape, and color of body parts. Since adult form is the product of embryonic development, changes in form must be the result of changes in what happens during development. Recent advances in understanding the genetic control of development (see Chapter 13) have enabled researchers to investigate the genetic and molecular bases of the evolution of animal form. We will see that some dramatic changes in animal form can have a relatively simple genetic and molecular basis, and that the evolution of traits governed by many toolkit genes involves molecular mechanisms that are distinct from those we have examined thus far. We will examine cases in which coding substitutions, gene inactivation, and regulatory sequence evolution underlie morphological divergence.

Adaptive changes in a pigment-regulating protein

Some of the most striking and best-understood examples of morphological divergence are found in animal body-color patterns. Mammalian-coat, bird-plumage, fish-scale, and insect-wing color schemes are wonderfully diverse. Investigators have made much progress in understanding the genetic control of color formation and its role in the evolution of color differences within and between species.

In the Pinacate region of southwestern Arizona, dark rocky outcrops are surrounded by lighter-colored sandy granite (**Figure 20-14**). The rock pocket mouse, *Chaetodipus intermedius*, inhabits the Pinacate as well as other rocky areas of the Southwest. The mice found on the lava outcrops are typically dark in color, whereas those found in surrounding areas of sandy-colored granite or on the desert floor are usually light colored (**Figure 20-15**). Field studies suggest that mice whose coat color closely matches their environment are seen less often by predators.

The rock pocket mice are an example of *melanism*—the occurrence of a dark form within a population or species. Melanism is one of the most common types of phenotypic variation in animals. The dark color of the fur is due to heavy deposition of the pigment *melanin*, the most widespread pigment in the animal kingdom. In mammals, two types of melanin are produced in melanocytes (the pigment cells of the epidermis and hair follicles): eumelanin, which forms black or brown pigments, and phaeomelanin, which forms yellow or red pigments. The relative amounts of eumelanin and phaeomelanin are controlled by the products of several genes. Two key proteins are the melanocortin 1 receptor (MC1R) and the agouti protein. During the hair-growth cycle, the α-melanocyte-stimulating hormone (α-MSH) binds to the MC1R protein, which triggers the induction of pigment-producing enzymes. The agouti protein blocks MC1R activation and inhibits the production of eumelanin.

When scientists examined the DNA sequences of the *Mc1r* genes of light- and dark-colored pocket mice, they found four mutations in the *Mc1r* gene in dark mice that cause the MC1R protein to differ at four amino acid residues from the corresponding protein in light mice. Findings from biochemical studies suggest that such mutations cause the MC1R protein to be constitutively active (active at all times), bypassing the regulation of receptor activity by the agouti protein. Indeed, mutations in *Mc1r* are associated with melanism in all sorts of wild and domesticated

FIGURE 20-14 Lava flows in the Pinacate desert have produced outcrops of black-colored rock adjacent to sandy-colored substrates. [*Michael Nachman, University of Arizona.*]

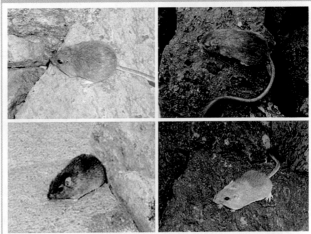

Melanism in the rock pocket mouse

FIGURE 20-15 Light- and dark-colored *Chaetodipus intermedius* from the Pinacate region of Arizona are shown on sandy-colored and dark lava-rock backgrounds. [*Michael Nachman, from M. W. Nachman et al., "The Genetic Basis of Adaptive Melanism in Pocket Mice,"* Proc. Natl. Acad. Sci. USA *100, 2003, 5268–5273. Copyright (2003) National Academy of Sciences, U.S.A.*]

vertebrates. Many of these mutations alter residues in the same part of the MC1R protein, and the same mutations have occurred independently in some species (**Figure 20-16**).

In many ways, we can think of these dark mice as analogs of Darwin's finches and the lava outcrops as new "island" habitats produced by the same volcanic activity that produced the Galápagos Islands. The sandy-colored form of the mouse appears to be the ancestral type, akin to the continental ancestral finch that colonized the Galápagos. The advantage of being less visible to predators resulted in natural selection for coat color, and the invasion of the lava-rock islands by the mice led to the spread of an allele that was favored on the black-rock background and selected against on the sandy-colored background. New mutations in the *Mc1r* gene were essential to this adaptation to the differing landscapes.

The evolution of melanism in the pocket mice illustrates how fitness depends on the conditions in which an organism lives. The new black mutation was favored on the lava outcrops but disfavored in the ancestral population living on sandy-colored terrain.

KEY CONCEPT The relative fitness of a new variant depends on the immediate selective conditions. A mutation that may be beneficial in one population may be deleterious in another.

Gene inactivation

It has long been noted that cave-dwelling animals are often blind and uncolored. Darwin noted in *On the Origin of Species* that "several animals belonging to the most different classes, which inhabit the caves of Carniola [in Slovenia] and Kentucky, are blind. As it is difficult to imagine that eyes, though useless, could be in any way injurious to animals living in darkness, I attribute their loss wholly to disuse."[6]

Many species of fish that live in caves have lost their eyes and body color. Because these species belong to many different families that include surface-dwelling, eye-bearing species, the loss of eyes and pigmentation has clearly occurred repeatedly. For example, the Mexican blind cave fish (*Astyanax mexicanus*) belongs to the same order as the piranha and the colorful neon tetra. About 30 cave populations of fish in Mexico have lost the body color of their surface-dwelling relatives (**Figure 20-17**).

Genetic studies have indicated that albinism in the Pachón cave fish population is due to a single recessive mutation. Furthermore, a complementation test (see

[6]C. Darwin, *On the Origin of Species by Means of Natural Selection*, 137. John Murray, London, 1859.

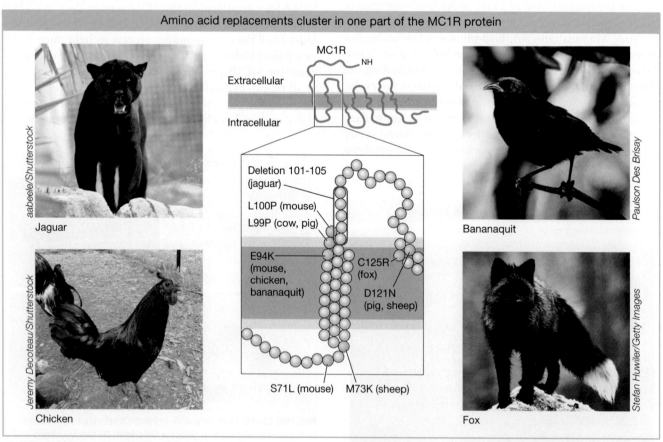

Amino acid replacements cluster in one part of the MC1R protein

Jaguar — *aabeele/Shutterstock*

Chicken — *Jeremy Decoteau/Shutterstock*

Bananaquit — *Paulson Des Brisay*

Fox — *Stefan Huwiler/Getty Images*

MC1R
NH
Extracellular
Intracellular

Deletion 101-105 (jaguar)
L100P (mouse)
L99P (cow, pig)
E94K (mouse, chicken, bananaquit)
C125R (fox)
D121N (pig, sheep)
S71L (mouse)   M73K (sheep)

FIGURE 20-16 Amino acid replacements (orange circles) and deletions (red oval) associated with melanism vary slightly in location in different species but are located in the same part of the MC1R protein. The upper part of the figure shows the general topology of the MC1R protein. The region in which replacements or deletions are located is enlarged in the lower part of the figure. (Note that L100P means leucine at position 100 has been replaced by proline. For single-letter amino acid abbreviations, see Figure 9-2.) [*Data from E. Eizirik et al., "Molecular Genetics and Evolution in the Cat Family," Curr. Biol. 13, 2003, 448–453.*]

Evolution of albinism in blind cave fish

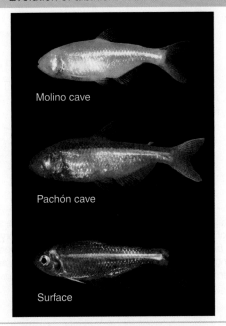

FIGURE 20-17 Surface forms of the fish *Astyanax mexicanus* appear normal, but cave populations, such as those from the Molino and Pachón caves in Mexico, have repeatedly evolved blindness and albinism. [*Courtesy of Richard Borowsky.*]

Chapter 5) was conducted by making a cross between a Molino cave individual and a Pachón cave individual. This cross produced only albino offspring, suggesting that the albinism in the two populations is due to the same genetic locus. To identify the gene responsible for albinism in the fish, Meredith Protas and colleagues performed quantitative trait locus (QTL) mapping (see Chapter 19). In offspring that were created either by a backcross between Molino and Molino/surface F_1 progeny or by an intercross between Pachón/surface F_1 progeny, they found that there was an association between the phenotype of albinism and the genotype of the *Oca2* gene, which is known to cause albinism in mice and humans.

Further inspection of the *Oca2* gene revealed that the Pachón population was homozygous for a deletion that extended from an intron through most of an exon and that the Molino population was homozygous for the deletion of a different exon. Functional analyses proved that each deletion in the *Oca2* gene caused loss of *Oca2* function.

The identification of different lesions in the *Oca2* gene of the two cave populations indicates that albinism evolved separately in the two cave populations. There is also evidence that a third cave population carries yet a third, distinct *Oca2* mutation. It is known from other vertebrates that albinism can evolve through mutations in other genes. What might account for the repeated inactivation of the *Oca2* gene? There are two likely explanations. First, *Oca2* mutations appear to cause no serious collateral defects other than loss of pigmentation and vision. Some other pigmentation genes, when mutated in fish, cause

dramatic reductions in viability. The effects of *Oca2* mutations appear, then, to be less *pleiotropic* and have effects on overall fitness that are less harmful than those of mutations in other fish pigmentation genes. Second, the *Oca2* locus is very large, spanning some 345 kb in humans and containing 24 exons. It presents a very large target for random mutations that would disrupt gene function; *Oca2* mutations are therefore more likely to arise than are mutations at smaller loci.

Loss of gene function is not what we usually think about when we think about evolution. But gene inactivation is certainly what we should predict to happen when selective conditions change or when populations or species shift their habitats or lifestyles and certain gene functions are no longer necessary.

KEY CONCEPT Gene-inactivating mutations may occur and rise to high frequency when habitat or lifestyle changes relax natural selection on traits and underlying gene functions.

Regulatory-sequence evolution

As discussed earlier, a major constraint on gene evolution is the potential for harmful side effects caused by mutations in coding regions that alter protein function. These effects can be circumvented by mutations in regulatory sequences, which play a major role in the evolution of gene regulation and body form.

The examples of body-coloration evolution we have looked at thus far have the coat or scale pattern changing over the entire body. The evolution of solid black or entirely unpigmented body coloration can arise through mutations in pigmentation genes. However, many color schemes are often made up of two or more colors in some spatial pattern. In such cases, the expression of pigmentation genes must differ in areas of the body that will be of different colors. In different populations or species, the regulation of pigmentation genes must evolve by some mechanism that does not disrupt the function of pigmentation proteins.

The species of the fruit-fly genus *Drosophila* display extensive diversity of body and wing markings. A common pattern is the presence of a black spot near the tip of the wing in males (**Figure 20-18**). The production of the black spots requires enzymes that synthesize melanin, the same pigment made in pocket

Wing spots on fruit flies

FIGURE 20-18 *Drosophila melanogaster* males lack wing spots (*top*), whereas *Drosophila biarmipes* males (*bottom*) have dark wing spots that are displayed in a courtship ritual. This simple morphological difference is due to differences in the regulation of pigmentation genes. [*Nicolas Gompel.*]

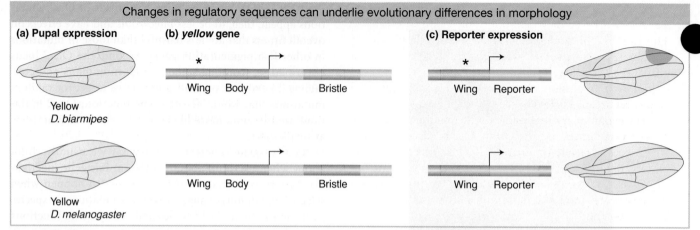

FIGURE 20-19 The evolution of gene regulation and morphology due to differences in cis-acting regulatory sequences. (a) In spotted fruit flies such as *D. biarmipes*, the Yellow pigmentation protein is expressed at high levels in cells that will produce large amounts of melanin. (b) The *yellow* locus of *Drosophila* species contains several discrete cis-acting regulatory elements (red) that govern *yellow* transcription in different body parts. Exons are shown in gold. Arrows indicate the start and direction of transcription. (c) The wing cis-acting regulatory element from a *D. biarmipes* drives reporter-gene expression (green) in a spot on the wing, suggesting that this sequence (indicated by an asterisk) must differ from the homologous sequence in *D. melanogaster*.

mice. Many genes controlling the melanin synthesis pathway have been well studied in the model organism *Drosophila melanogaster*. One gene is named *yellow* because mutations in the gene cause darkly pigmented areas of the body to appear yellowish or tan. The *yellow* gene plays a central role in the development of divergent melanin patterns. In species with spots, the Yellow protein is expressed at high levels in wing cells that will produce the black spot, whereas in species without spots, Yellow is expressed at a low level throughout the wing blade (**Figure 20-19a**).

The difference in Yellow expression between spotted and unspotted species could be due to differences in how the *yellow* gene is regulated in the two species. Either or both of two possible mechanisms could be at play: the species could differ in the spatial deployment of transcription factors that regulate *yellow* (that is, changes in trans-acting sequences to the *yellow* gene), or they could differ in cis-acting regulatory sequences that govern how the *yellow* gene is regulated. To examine which mechanisms are involved, investigators examined the activity of *yellow* cis-acting regulatory sequences from different species by placing them upstream of a reporter gene (see Figure 13-18) and introducing them into *D. melanogaster*.

The *yellow* gene is regulated by an array of cis-acting regulatory sequences that govern gene transcription in different tissues and cell types and at different times during development (Figure 20-19b). These regulatory sequences control transcription in larval mouthparts, the pupal thorax and abdomen, and the developing wing blade (labeled as bristle, body, and wing in Figure 20-19b).

Results of the reporter gene experiments showed that the "wing" regulatory element from *D. biarmipes* drives reporter-gene expression in a spot pattern in the developing wing, whereas the homologous element from the unspotted *D. melanogaster* does not drive a spot pattern of reporter expression (Figure 20-19c). This difference in wing cis-acting regulatory elements demonstrates that sequence changes in the cis-acting regulatory element underlie differences in Yellow expression and pigmentation between the two species.

Thus, changes in cis-acting regulatory sequences play a critical role in the evolution of body form. Changes in the regulatory sequence rather than the coding sequence of this toolkit gene exemplify an important way in which evolution can occur. In this instance, the *yellow* gene is highly pleiotropic: it is required for the pigmentation of many structures and functions in the nervous system as well. A mutation in the coding sequence that alters Yellow protein activity would alter Yellow activity in *all* tissues, which might be detrimental for fitness. However, because individual cis-acting regulatory sequences usually affect gene expression in only one tissue or cell type or at one developmental timepoint, mutations in these sequences provide a mechanism for altering gene expression at a specific time or a specific location while preserving the role of protein products in other developmental processes.

KEY CONCEPT Evolutionary changes in cis-acting regulatory sequences play a critical role in the evolution of gene expression. They circumvent the pleiotropic effects of mutations in the coding sequences of genes that have multiple roles in development.

Loss of characters through regulatory-sequence evolution

Morphological characters may be lost as well as gained as the result of adaptive changes in cis-acting regulatory sequences. If there is no selective pressure to maintain a character, it can be lost over time. But some losses are beneficial because they facilitate some change in lifestyle. Hind limbs, for example, have been lost many times in vertebrates—in snakes, lizards, whales, and manatees—as these organisms adapted to different habitats and means of locomotion. Evolutionary changes in cis-acting regulatory sequences are also linked to these dramatic changes.

The evolutionary forerunners of the hind limbs of four-legged vertebrates are the pelvic fins of fish. Dramatic differences in pelvic-fin anatomy have evolved in closely related fish populations. The threespine stickleback fish occurs in two forms in many lakes in North America—an open-water form that has a full spiny pelvis and a shallow-water, bottom-dwelling form with a dramatically reduced pelvis and spines. In open water, the long spines

help protect the fish from being swallowed by larger predators. But on the lake bottom, those spines are thought to be a liability, because dragonfly larvae that feed on the young fish can grasp the spines (**Figure 20-20**).

The differences in pelvic morphology have evolved repeatedly in just the past 10,000 years, since the recession of the glaciers of the last ice age. Many separate lakes were colonized by long-spined oceanic sticklebacks, and forms with reduced pelvic spines evolved independently several times. Because the fish are so closely related and interbreed in the laboratory, geneticists can map the genes involved in the reduction of the pelvis. Genetic linkage mapping was used to identify one major factor involved in pelvic differences to the *Pitx1* gene, which encodes a transcription factor. Like most other developmental toolkit genes, the

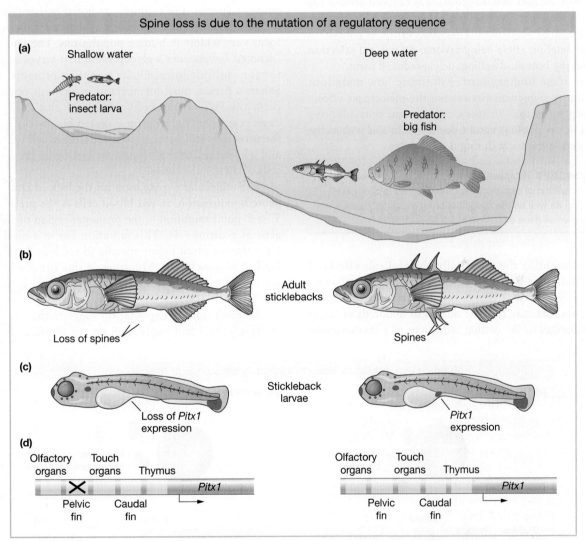

FIGURE 20-20 Deletions within a *Pitx1* cis-regulatory element underlie the adaptive evolution of the pelvic skeleton of stickleback fish. (a) One form of the threespine stickleback fish inhabits shallow water, and a different form inhabits open water. (b) The shallow-water form has a reduced pelvic skeleton (*left*) relative to the open-water form (*right*). (c) This reduction is due to the selective loss of expression of the *Pitx1* gene (orange) from the pelvic fin bud during development of the stickleback larvae (compare left and right stickleback larvae). (d) The loss of *Pitx1* expression in turn is due to the mutation of an enhancer of the *Pitx1* gene specific to the pelvic fin (X marks the mutated enhancer). Other enhancers of the *Pitx1* gene, which control expression of the gene elsewhere in the developing body, are unaffected and function similarly in both forms of the fish.

Pitx1 gene has several distinct functions in fish development. However, in the stickleback with a reduced pelvis, *Pitx1* expression is lost from the area of the developing fish embryo that will give rise to the pelvic-fin bud and spines (see Figure 20-20c).

The fact that the difference in pelvic morphology between the two forms mapped to the *Pitx1* locus and was associated with the loss of gene expression suggested that changes in *Pitx1* regulatory sequences were responsible for the difference in phenotypes. Like most pleiotropic toolkit genes, the expression of the *Pitx1* gene in different parts of the developing fish is controlled by separate cis-acting regulatory elements. The regulatory element that controls *Pitx1* expression in the developing pelvis has been inactivated by deletion mutations in multiple, independent populations of pelvic-reduced fish (Figure 20-20d). Furthermore, it was observed that heterozygosity was reduced around the cis-acting sequences controlling pelvic expression relative to other nearby sequences. This observation is consistent with the deletion allele being favored by natural selection acting on the bottom-dwelling, pelvic-reduced form.

Thus, these findings further illustrate how mutations in regulatory sequences circumvent the pleiotropic effects of coding mutations in toolkit genes and that adaptive changes in morphology can be due to the loss as well as the gain of gene expression during development.

> **KEY CONCEPT** Adaptive changes in morphology can result from inactivation of regulatory sequences and loss of gene expression as well as the modification of regulatory sequences and the gain of gene expression.

Circumventing the potentially harmful side effects of coding mutations is a very important factor in explaining why evolution acts by generating new roles for transcription factors that may regulate dozens to hundreds of target genes. Changes in the coding sequences of a transcription factor—for example, the DNA-binding domain—may affect all target genes, with catastrophic consequences for the animal. The constraint on the coding sequences of highly pleiotropic proteins, with many functions, explains the extraordinary conservation of the DNA-binding domains of Hox proteins (see Figure 13-9) and many other transcription factors over vast expanses of evolutionary time. But, although the proteins' biochemical functions are constrained, their regulation does diverge. The evolution of the expression patterns of *Hox* and other toolkit genes plays a major role in the evolution of body form.

Regulatory evolution in humans

Regulatory evolution is not limited to genes affecting development. The level, timing, or spatial pattern of the expression of any gene may vary within populations or diverge between species. For example, as noted earlier (see Chapter 18), the frequencies of alleles at the *Duffy* blood-group locus vary widely in human populations. The *Duffy* locus (denoted *Fy*) encodes a glycoprotein that serves as a receptor for multiple intercellular signaling proteins. In sub-Saharan Africa, most members of indigenous populations carry the *Fy*[null] allele. Individuals with this allele do not express any Duffy glycoprotein on red blood cells, although the protein is still being produced in other cell types. How and why is the Duffy glycoprotein lacking on the red blood cells of these individuals?

The molecular explanation for the lack of Duffy glycoprotein expression on red blood cells is the presence of a T → C point mutation in the promoter region of the *Duffy* gene at position −46. This mutation lies in a binding site for a transcription factor specific to red blood cells called GATA1 (**Figure 20-21**). Mutation of this site abolishes the activity of a *Duffy* gene enhancer in reporter-gene assays (see Figure 13-18).

An evolutionary explanation suggests that the lack of Duffy glycoprotein expression on red blood cells among

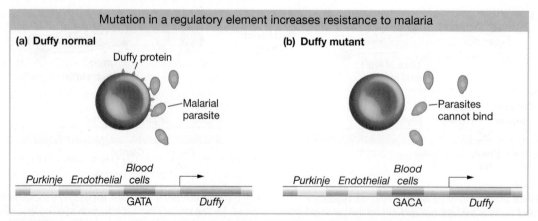

FIGURE 20-21 A regulatory mutation in a human *Duffy* gene enhancer is associated with resistance to malaria. (a) The Duffy protein (dark blue) is typically expressed on blood cells as well as on Purkinje cells in the brain and endothelial cells. (b) A high proportion of West Africans lack Duffy expression on their red blood cells due to a mutation in a blood-cell enhancer (the GATA sequence is mutated to GACA). Since the Duffy protein is part of the receptor for the *P. vivax* malarial parasite (orange), individuals with the regulatory mutation are resistant to infection but have normal Duffy expression elsewhere in the body.

Africans is the result of natural selection favoring resistance to malarial infection. The malarial parasite *Plasmodium vivax* is the second most prevalent form of malarial parasite (after the *Plasmodium falciparum* parasite discussed in Section 20.2) in most tropical and subtropical regions of the world but is currently absent from sub-Saharan Africa. The parasite gains entry to red blood cells and red-blood-cell precursors by binding to the Duffy glycoprotein (see Figure 20-21). The very high frequency of Fy^{null} homozygotes in Africa prevents *P. vivax* from being common there. Moreover, if we suppose that *P. vivax* was common in Africa in the past, then there would have been selection for the Fy^{null} allele.

The complete absence of Duffy protein on the red blood cells of a large subpopulation raises the question of whether the Duffy protein has any necessary function, because it is apparently dispensable. But it is not the case that these individuals lack Duffy protein expression altogether. The protein is expressed on endothelial cells of the vascular system and the Purkinje cells of the cerebellum. As with the evolution of *Yellow* expression in wing-spotted fruit flies and of *Pitx* expression in stickleback fish, the regulatory mutation at the *Fy* locus allows one aspect of gene expression (in red blood cells) to change without disrupting others (see Figure 20-21).

Modifications to coding and regulatory sequences are common means to evolutionary change within species. We will now examine whether similar genetic mechanisms lead to divergence between species.

20.6 EVOLUTION OF SPECIES

LO 20.6 Compare examples of genetic mechanisms that contribute to reproductive isolating barriers between species.

Darwin begins *On the Origin of Species* by stating that his observations while a naturalist on the HMS *Beagle* "seemed to me to throw some light on the origin of species—that mystery of mysteries."[7] Indeed, in the one figure present in *On the Origin of Species*, Darwin illustrates his idea that species are formed from the successive branching of lineages from a common ancestor. Despite putting forward this revolutionary idea, Darwin himself had remarkably little to say about how species arise through the processes of natural selection and evolution. Since that time, however, we have gained a much greater understanding of the evolution of new species, particularly at the genetic level.

Species concepts

One challenge in the study of speciation is that species are surprisingly hard to define. We often recognize discrete clusters in nature, but different biologists might cluster

different groups together as a "species" based on different criteria. For example, humans and chimpanzees are recognized as distinct species, with many differences in morphological, behavioral, and physiological traits. Yet, recall from Chapter 14 that the genomes of chimpanzees and humans are very similar, with *single nucleotide polymorphisms* at only 1 percent of all the sites in the genome. However, similar or even greater levels of polymorphism can be found within groups that are recognized as the same species. Thus, the recognition of discrete clusters very much depends upon perspective. Because different biological questions require different perspectives, many species definitions, or species concepts, have been proposed and are in use today. However, the species concept favored by most evolutionary geneticists is the **biological species concept,** as defined by Ernst Mayr: "species are groups of actually or potentially interbreeding natural populations, which are reproductively isolated from other such groups."[8] Here, the key focus is on **reproductive isolation**. With this focus, the biological species concept is useful for empirical studies of speciation because reproductive isolation can be measured as the amount of *gene flow*, or movement of gametes (see Chapter 18), between two populations or species. A major focus of speciation genetics is therefore to identify both the traits and genes that contribute to reproductive isolation between species.

KEY CONCEPT The biological species concept places the focus of speciation studies on the evolution of reproductive isolation.

Mechanisms of reproductive isolation

Many types of **reproductive isolating barriers** can prevent gene flow between species. These are broadly classified by the time at which the barriers to the exchange of gametes between species act (**Table 20-4**). Hybrid formation may be reduced or completely absent due to **premating isolating barriers,** which act before mating occurs. These barriers may involve differences in where species breed (*habitat isolation*) or when species breed (*temporal isolation*). Even if a species can move into the breeding habitat of another species, it might have reduced viability in that habitat and therefore have a reduced ability to breed (*immigrant inviability*). In animals, differences in mating preferences between the species can lead to *sexual isolation*, while in flowering plants, differences in pollinator preferences for the species can lead to *pollinator isolation*. Even when species meet and mate, **postmating, prezygotic isolating barriers** can prevent the formation of hybrid zygotes. There may be incompatibilities between the mating structures of the two species (*mechanical isolation*), the mating behaviors of the two species (*copulatory isolation*), or between the sperm and eggs (*gametic isolation*). Finally, hybrids may be formed but have reduced fitness; i.e., viability or fertility,

[7]C. Darwin, *On the Origin of Species by Means of Natural Selection*, 1. John Murray, London, 1859.

[8]E. Mayr, *Systematics and the Origin of Species*, Columbia University Press, New York, 1942.

TABLE 20-4 Reproductive Isolating Barriers

Premating isolating barriers	Postmating, prezygotic isolating barriers	Postzygotic isolating barriers

Ecological isolation: species do not meet
 	
   Habitat isolation: species mate in different habitats
 	
   Temporal isolation: species mate at different times
 	
   Immigrant inviability: immigrants do not survive to breed in habitat of other species
	Mechanical isolation: mating structures of species do not fit together	Extrinsic: hybrids have reduced fitness that is dependent on the environment
		
		   Behavioral sterility: hybrids have lower success in finding mates
		
		   Ecological inviability: hybrids have reduced viability in parental environments

Behavioral isolation: species meet but do not mate
 	
   Sexual isolation (animals): individuals prefer to mate with their own species
 	
   Pollinator isolation (plants): pollinators do not transfer pollen between species
	Copulatory isolation: mating behavior of one species does not stimulate the other species	Intrinsic: hybrids have reduced fitness, which is independent of the environment
		
		   Hybrid sterility: hybrids produce fewer functional gametes
		
		   Hybrid inviability: hybrids have reduced viability

	Gametic isolation: fertilization does not occur	

due to **postzygotic isolating barriers**. In many cases, hybrid inviability or sterility is independent of the environment (*intrinsic isolation*). In other cases, hybrids are viable and fertile when created in the laboratory or a zoo, but in nature they suffer reduced viability or fertility that is dependent upon environmental conditions (*extrinsic isolation*).

It is important to recognize that reproductive isolation between species does not need to be complete under the biological species concept. This is useful because much insight about how new species evolve can be gained by studying pairs of species that are at different stages of reproductive isolation. For example, researchers have compared the isolating barriers present between pairs of species that have diverged more recently (i.e., high gene flow and low genetic divergence) with the isolating barriers present between pairs of species that diverged more anciently (i.e., low gene flow and high genetic divergence). Studies across many *Drosophila* species suggest that premating isolating barriers evolve before postmating, prezygotic isolating barriers, and that intrinsic postzygotic isolating mechanisms like hybrid sterility and hybrid inviability evolve even later. Consistent with this observation, many closely related species of butterflies, birds, fish, frogs, and plants can still form hybrids that do not suffer from intrinsic sterility or inviability.

KEY CONCEPT Under the biological species concept, reproductive isolation between species does not need to be complete. Studying species pairs along a continuum ranging from little reproductive isolation to complete reproductive isolation provides insight into the mechanisms that promote speciation.

Nonetheless, overall reproductive isolation between closely related species can still be quite strong because multiple isolating barriers often work together to prevent hybridization between the species. For example, the reproductive isolating barriers between two species of closely related monkeyflowers from western North America have been analyzed in detail. One species, *Mimulus lewisii*, is pollinated by bees and has pink flowers with a wide corolla, while the other species, *M. cardinalis*, is pollinated by hummingbirds and has red flowers with a narrow corolla tube (**Figure 20-22**). Overall, reproductive isolation is almost complete between the species. Nearly 60 percent of the reproductive isolation between the species can be attributed to habitat isolation; the plants live mostly at different altitudes. Where the species do overlap in habitat, differences in pollinator preferences maintain the species as distinct; only 2 of the 2336 plants examined at one site of overlap were F_1 hybrids. Hybrids do exhibit reduced germination and fertility when made in the laboratory, but these postzygotic isolating barriers are less important than premating isolating barriers in the maintenance of these two species in the wild. Identifying the reproductive isolating barriers that maintain species in the wild is a challenging endeavor that requires careful observation and experimentation in the field and in the lab. Nonetheless, identifying them is necessary in order to identify the genetic changes that underlie speciation.

Genetics of reproductive isolation

Identifying the genetic changes that underlie the formation of new species has been a major question in evolutionary genetics. However, using genetic mapping approaches to address this question has been challenging because species are by definition reproductively isolated. This challenge has been circumvented in some cases by focusing on closely related species pairs in which reproductive isolation is not yet complete. Coupled with advances in sequencing technologies that have enabled genetic and genomic studies of non-model systems (see Beyond Model Organisms, pp. 728–730), there has been progress in identifying the

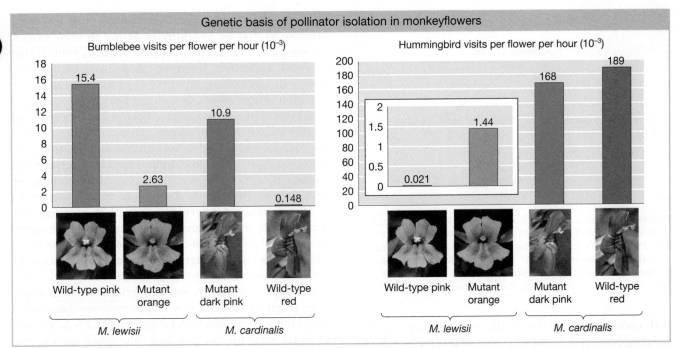

Genetic basis of pollinator isolation in monkeyflowers

Bumblebee visits per flower per hour (10^{-3})

15.4

2.63

10.9

0.148

Wild-type pink | Mutant orange | Mutant dark pink | Wild-type red

M. lewisii | *M. cardinalis*

Hummingbird visits per flower per hour (10^{-3})

189

168

0.021 | 1.44

Wild-type pink | Mutant orange | Mutant dark pink | Wild-type red

M. lewisii | *M. cardinalis*

FIGURE 20-22 Flower color is important for reproduction isolation between *M. lewisii* and *M. cardinalis*. Wild-type *M. lewisii* has pink flowers, and wild-type *M. cardinalis* has red flowers. A mutation that substitutes the *YUP* gene from *M. cardinalis* into *M. lewisii* results in orange flowers, and a mutation that substitutes the *YUP* gene from *M. lewisii* into *M. cardinalis* results in dark pink flowers. Bumblebees (*left panel*) visit pink *M. lewisii* 100 times more often than red *M. cardinalis*, while hummingbirds (*right panel*) visit red *M. cardinalis* over 9000 times more than pink *M. lewisii*. Manipulation of flower color by the *YUP* gene changes these preferences. Bumblebees (*left panel*) now visit the dark pink *M. cardinalis* 4.1 times more often than the orange *M. lewisii*. Although hummingbirds still visit *M. cardinalis* more often than *M. lewisii*, they visit orange *M. lewisii* nearly 70 times more than pink *M. lewisii*. [*Reprinted with permission from Macmillan Publishers Ltd: from H. D. Bradshaw Jr., Douglas W. Schemske, "Allele substitution at a flower colour locus produces a pollinator shift in monkey flowers." Nature, 426, 2003, November 13, 176–178, Figure 1. Permission conveyed through Copyright Clearance Center, Inc.*]

genetic basis of traits that underlie isolating barriers between species. Here, we will examine a few exemplary cases in which the genetic basis of traits involved in premating isolation or postzygotic isolation has been identified.

Genetic basis of premating isolation A diversity of isolating barriers contribute to premating isolation in plants and animals (see Table 20-4), and genetic studies of these different barriers are still quite rare. One exception is the monkeyflower system discussed earlier, in which pollinator isolation is a major component of reproductive isolation between *M. lewisii* and *M. cardinalis*. Performing QTL analyses in an F$_2$ intercross between the two species identified a locus called *YELLOW UPPER* (*YUP*) that is responsible for the species difference in the deposition of yellow carotenoid pigment on the petals. The pink flowers of *M. lewisii* have no carotenoid pigment and have only pink anthocyanin pigments, while the red flowers of *M. cardinalis* result from the deposition of both anthocyanin and carotenoid pigments. Researchers created *nearly isogenic lines* in which the *M. lewisii* version of the *YUP* locus was substituted into the *M. cardinalis* genome, and vice versa. These genetic manipulations create *M. cardinalis* with dark pink flowers and *M. lewisii* with orange flowers (Figure 20-22). Remarkably, changing this single genetic locus results in a shift in pollinator visitation in the wild. Bumblebees visited the dark pink *M. cardinalis* flowers

nearly 75 times more than the wild-type red *M. cardinalis* flowers, and hummingbirds visited the orange *M. lewisii* flowers nearly 70 times more than the wild-type pink flowers (Figure 20-22). Although the specific gene that underlies the *YUP* locus is not yet known, this study demonstrates that identifying the genetic basis of a trait enables researchers to use genetic manipulations to specifically test the contribution of that trait to reproductive isolation.

Genetic basis of postzygotic isolation Until recently, most of the genetic studies of speciation have focused on the intrinsic postzygotic isolating barriers of hybrid sterility and inviability, which are widespread across taxa. Yet, these types of isolating barriers posed a conundrum for Darwin and many who followed him: how can such incompatibilities evolve *between* species without causing problems *within* species? This evolutionary conundrum was solved by the geneticists Theodosius Dobzhansky and Hermann Muller. Although their solution is commonly referred to as the **Dobzhansky–Muller model**, it was later discovered that William Bateson had independently arrived at the same solution 25 years earlier. Regardless of the name, the key to the model is *epistasis*, or interactions between two or more loci. The simplest version of the Dobzhansky–Muller model with two interacting loci (*A* and *B*) is shown in **Figure 20-23**. Imagine that an ancestral species with genotype *aabb* is split into two species that are no longer able to reproduce

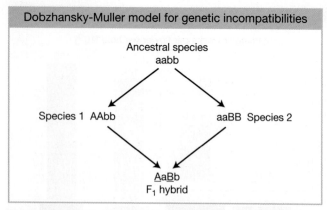

Dobzhansky-Muller model for genetic incompatibilities

FIGURE 20-23 Two interacting loci (A and B) are fixed for genotypes *aa* and *bb* in an ancestral species. After a speciation event, fixation of a new allele (A) at locus A occurs in species 1, and fixation of a new allele (B) at locus B occurs in species 2. In an F₁ hybrid, the A allele and the B allele (underlined) are now in the same organism for the first time. When these alleles do not interact properly, hybrid sterility or inviability can result.

with each other, either due to the evolution of prezygotic isolating barriers or to the formation of geographical barriers. Each species can therefore accumulate mutations independently of the other. One species might fix a new allele (B) at locus B, while the other species fixes a new allele (A) at locus A. If these two species come back together and form an F₁ hybrid, the alleles A and B have fixed in different populations and never been tested together in combination. If these two loci interact, there could be a genetic incompatibility between alleles A and B that results in reduced hybrid fertility or viability.

KEY CONCEPT The Dobzhansky–Muller model explains how hybrid incompatibilities *between* species can evolve *within* species.

There is now strong empirical support for the Dobzhansky–Muller model: many studies have shown that genetic incompatibilities between two or more loci do indeed underlie hybrid sterility and inviability across a number of taxonomic groups. In a handful of cases, the genes that underlie Dobzhansky–Muller incompatibilities have been identified. These studies have revealed two general mechanisms that contribute to genetic incompatibilities between species: gene duplication and genetic conflict. We will now discuss an example of each of these mechanisms.

Recall that one evolutionary fate of duplicate genes is loss of one copy of the gene (see Figure 20-9b). It was predicted that loss of different gene copies in different species would provide a plausible molecular mechanism for the Dobzhansky–Muller model. Specifically, if one copy of a duplicated gene is lost in one species, and the other copy of a duplicated gene is lost in the other species, hybrid progeny could result that completely lack a functional copy of the gene. This mechanism has now been observed to underlie hybrid sterility in *Drosophila* and hybrid inviability in plants.

As one example, we will focus on two species of monkeyflowers, *Mimulus nasutus* and *M. guttatus* (these are

different species of monkeyflowers from those discussed previously) that do hybridize in nature to some extent. Crossing the two species in the lab yields F₁ hybrids that are fertile and viable. However, intercrossing the F₁ plants reveals that one-sixteenth of the F₂ progeny die after germination (**Figure 20-24**). By genetic linkage mapping and positional cloning (see Chapter 10), Andrea Sweigart and Matthew Zuellig identified two genetic loci, one on chromosome 13 and the other on chromosome 14, that contribute to F₂ lethality. On chromosome 13, *M. nasutus* has a gene called *pTAC14*, which is essential for chloroplast development. This gene is also present on chromosome 13 in *M. guttatus*, but it harbors a frameshift mutation, rendering the gene non-functional. However, there is a duplicate copy of *pTAC14* on chromosome 14 in *M. guttatus*. This duplicate gene is not present at the chromosome 14 locus in *M. nasutus*. Thus, each species has a single functional copy of *pTAC14*, and F₁ hybrids have both duplicates (Figure 20-24a, *bottom*). However, one-sixteenth of F₂ hybrids have no functional copy of *pTAC14*. As a result, these are white because they lack chlorophyll and die after germination (Figure 20-24b). In this case, gene duplication did not lead to the evolution of novel traits (as suggested in Section 20.4), but rather contributed to reproductive isolation between species.

It is not clear whether the duplication and subsequent degeneration of one duplicate copy of *pTAC14* in *M. guttatus* was due to genetic drift or to the action of natural selection. Both drift and selection are plausible mechanisms for the differential fixation of mutations in different species under the Dobzhansky–Muller model. Nonetheless, there is now strong evidence that many genes found to underlie Dobzhansky–Muller incompatibilities have sequences that are rapidly evolving under positive selection *within* species (see Section 20.3). In these cases, it seems that genetic incompatibilities *between* species are simply a by-product of rapid evolution within species. Why might these genes be rapidly evolving?

One possible mechanism underlying rapid evolution within species is genetic conflict between a host and its pathogens or parasites. These pathogens can be from an external source, such as viruses or bacteria. But parasites can also be embedded in the genome of the host, such as transposable elements (see Chapter 16). Another type of genomic parasite is a **selfish genetic element**, which is an allele that violates the Mendelian rules of segregation and is transmitted to more than 50 percent of the gametes. Such alleles can rapidly increase in frequency in a population, although they are often are associated with reductions in fertility. In all of these cases, the host is engaged in an evolutionary arms race, in which the pathogen or parasite is evolving rapidly to escape the host's defense mechanism, and the host is evolving rapidly in response to reductions in fitness caused by the pathogen or parasite.

There are now several cases in which an evolutionary arms race within a species might have led to reproductive isolation between species. One example was identified in crosses between two closely related species of *Drosophila*.

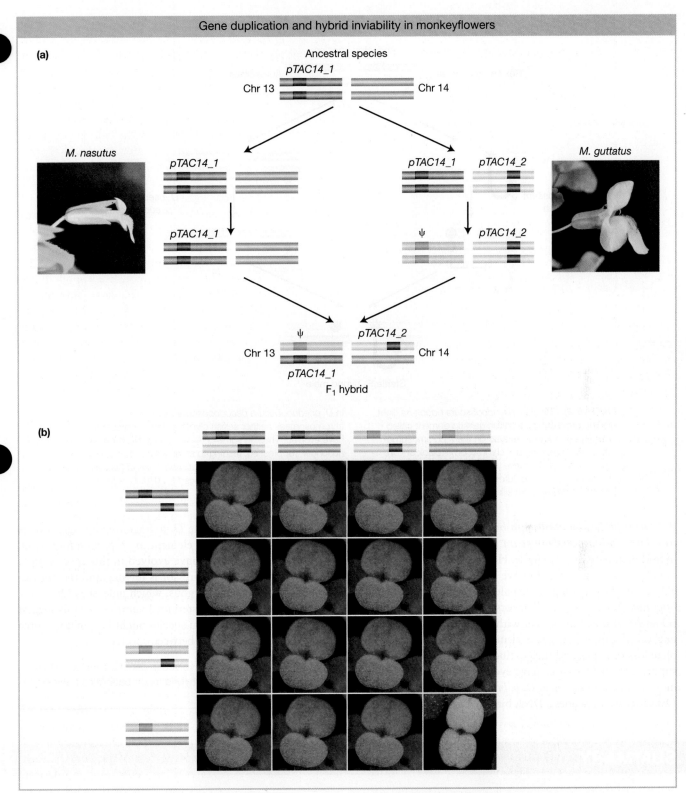

FIGURE 20-24 (a) There is a single copy of *pTAC14_1* (black bar) on chromosome 13 (dark red) and no *pTAC14* gene on chromosome 14 (dark blue) in the hypothesized ancestral species and in *M. nasutus*. There have been two genetic changes in *M. guttatus*: a mutation that creates a pseudogene (ψ) in the *pTAC14_1* gene (gray bar) on chromosome 13 (light red), and a duplication creating the *pTAC14_2* gene (black bar) on chromosome 14 (light blue). (b) Segregation of the *M. nasutus* and *M. guttatus* versions of chromosomes 13 and 14 when F₁ hybrids are intercrossed to create F₂ seedlings. Most seedlings are green because they carry at least one functional copy of *pTAC14*. One-sixteenth of the seedlings are white because they are homozygous for the *M. guttatus* version of chromosome 13 (light red) and the *M. nasutus* version of chromosome 14 (dark blue) and therefore have no functional copy of *pTAC14*. [*Top photos courtesy of Andrea L. Sweigart; bottom photo courtesy of Adam Bewick.*]

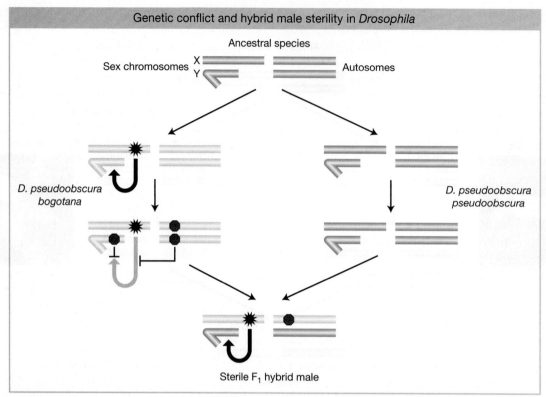

FIGURE 20-25 In *D. pseudoobscura bogotana* (light blue chromosomes), a selfish genetic element called *Overdrive* (black star) located on the X chromosome kills sperm harboring a Y chromosome (black arrow). Suppressors of *Overdrive* have evolved on the Y chromosome and the autosomes (red octagons) and prevent *Overdrive* from killing the Y-bearing sperm.

In *D. pseduoobscura pseudoobscura* (dark blue chromosomes), neither selfish elements nor suppressors have evolved. Thus, in some F₁ hybrid males, *Overdrive* is no longer suppressed, leading to sterility. [*Data from Daven C. Presgraves, "The Molecular Evolutionary Basis of Species Formation," Nature Reviews: Genetics 11, 2010, 175–180.*]

If a *Drosophila pseudoobscura bogotana* female is crossed to a *Drosophila pseudoobscura pseudoobscura* male, the hybrid males are mostly sterile. However, when these males are allowed to age, they do have a few offspring. Remarkably, all of the offspring are female. This is because the *D. p. bogotana* X chromosome harbors a selfish genetic element (*Overdrive*) that kills sperm with a Y chromosome. If left unchecked, this selfish genetic element would result in a complete loss of males and thus extinction of the species. Thus, suppressors of *Overdrive* have evolved on both the Y chromosome and autosomes in *D. p. bogotana* (**Figure 20-25**). The *Overdrive* gene encodes a DNA binding protein that has been

rapidly evolving within the *D. p. bogotana* lineage. It does not appear to be a selfish element in *D. p. pseudoobscura*, and thus no suppressors have evolved in this species. In F₁ hybrids, the Y-linked suppressor is missing and the autosomal suppressor is heterozygous, which unleashes *Overdrive* and results in hybrid male sterility. Examples like this suggest that genetic conflicts within species might be a major source of hybrid incompatibilities between species.

KEY CONCEPT Genes involved in genetic conflicts within species often contribute to hybrid incompatibilities between species.

SUMMARY

The theory of evolution by natural selection explains the changes that take place in populations of organisms as the result of changes in the relative frequencies of different variants in the population. If there is no variation within a species for some trait, there can be no evolution. Moreover, that variation must be influenced by genetic differences. If differences are not heritable, they cannot evolve, because the reproductive advantage of a variant will not carry

across generational lines. It is crucial to understand that the mutational processes that generate variation within the genome act at random, but that the selective process that sorts out the advantageous and disadvantageous variants is not random.

The ability to study evolution at the level of DNA and proteins has transformed our understanding of the evolutionary process. Before we had the ability to study evolution at the molecular level, there was no inkling that much of evolution was in fact a result of genetic drift and not natural selection. A great deal of molecular evolution seems to be the replacement of one protein sequence by another one of equivalent function. Among the evidence for the prevalence of neutral evolution is that the number of amino acid differences between two different species in some molecule—for example, hemoglobin—is directly proportional to the number of generations since their divergence from a common ancestor in the evolutionary past. We would not expect such a "molecular clock" with a constant rate of change to exist if the selection of differences were dependent on particular changes in the environment.

So much sequence evolution is neutral that there is no simple relation between the amount of change in a gene's DNA sequence and the amount of change, if any, in the encoded protein's function. However, it is possible to use sequence data to determine whether a gene is evolving neutrally or under the influence of natural selection. Most genes in a genome are evolving under purifying selection to maintain sequence conservation, but some genes show evidence of positive selection for adaptive sequence changes.

New protein functions often arise through the duplication of genes and subsequent mutation. New DNA may arise by duplication of the entire genome (polyploidy), a frequent occurrence in plants, or by various mechanisms that produce duplicates of individual genes or sets of genes. The fate of duplicate genes depends a great deal on the nature of mutations acquired after duplication. Possible fates are the inactivation of one duplicate, the splitting of function between two duplicates, or the gain of new functions.

Before the advent of molecular genetics, it was not possible to know whether independent evolutionary events might have given rise to the same adaptation multiple times. By pinpointing the genes and exact mutations involved in changes in function, we now appreciate that evolution can and does repeat itself by acting on the same genes to produce similar results in independent cases. For example, changes to the same genes are responsible for independently arising cases of melanism and albinism in some vertebrates and for the loss of pelvic spines in different stickleback-fish populations. Evolution may repeat itself by altering the very same nucleotide in independently arising sickle-cell mutations that lead to adaptive resistance to malaria.

An important constraint on the evolution of coding sequences is the potentially harmful side effects of mutations. If a protein serves multiple functions in different tissues, as is the case for many genes involved in the regulation of developmental processes, mutations in coding sequences may affect all functions and decrease fitness. The potential pleiotropic effects of coding mutations can be circumvented by mutations in noncoding regulatory sequences. Mutations in these sequences may selectively change gene expression in only one tissue or body part and not others. The evolution of cis-acting regulatory sequences is central to the evolution of morphological traits and the expression of toolkit genes that control development.

Molecular genetics has also begun to reveal the genetic changes that underlie the evolution of reproductive isolation between species. The same principles (variation, heredity, selection, and drift) that lead to the evolution of genes, genomes, and traits within species also contribute to the evolution of new species. In particular, differential gene duplication and rapid sequence evolution due to genetic conflict *within* species seem to be common mechanisms of intrinsic reproductive incompatibilities *between* species.

Overall, genetic evolution is subject to historical contingency and chance, but it is constrained by the necessity of organisms to survive and reproduce in a constantly changing world. "The fittest" is a conditional status, subject to change as the planet and habitats change.

KEY TERMS

adaptation (p. 682)
biological species concept (p. 703)
Dobzhansky–Muller model (p. 705)
fitness (p. 682)
gene duplication (p. 692)
gene families (p. 692)
key innovation (p. 695)
natural selection (p. 682)
neofunctionalization (p. 692)

nonsynonymous substitution (p. 690)
postmating, prezygotic isolating barriers (p. 703)
postzygotic isolating barriers (p. 704)
premating isolating barriers (p. 703)
pseudogene (p. 693)
reproductive isolation (p. 703)
reproductive isolating barriers (p. 703)

retrotransposition (p. 692)
selfish genetic element (p. 706)
subfunctionalization (p. 693)
synonymous substitution (p. 690)
whole genome duplication (WGD) (p. 694)

PROBLEMS

Visit SaplingPlus for supplemental content. Problems with the icon are available for review/grading. Problems with the icon have an Unpacking the Problem exercise.

WORKING WITH THE FIGURES

(The first 15 questions require inspection of text figures.)

1. The frequency of allele *A* in a population is 0.5. Based on Figure 20-2, what is the frequency of this allele after a bottleneck has occurred? Will the frequency of the *A* allele increase, decrease, or stay the same if there is natural selection against yellow bugs? What happens to the frequency of the *A* allele if there is natural selection against green bugs?

2. In Table 20-2, what is the selective advantage of the *AS* genotype relative to the *SS* genotype? What is the selective advantage of the *AA* genotype relative to the *SS* genotype?

3. In Figure 20-5, note that the difference in survival rates between *AS* and *AA* genotypes declines as children get older. Offer one possible explanation for this observation.

4. As shown in Figure 20-7, why is the substitution rate in fibrinopeptides higher than the substitution rate in cytochrome c?

5. Examining Figure 20-8, explain why the rate of substitution at nonsynonymous sites is lower. Do you expect this to be true only of globin genes or of most genes?

6. Examining Table 20-3, how would the interpretation of the McDonald–Kreitman test results differ if the number of nonsynonymous observed species differences was 1 instead of 7?

7. In Figure 20-9c, mutations in the coding region of the gene are responsible for neofunctionalization. Propose a mechanism by which changes to a cis-regulatory element might lead to neofunctionalization of the gene.

8. In Figure 20-11, what is the evidence that polyploid formation has been important in plant evolution?

9. Looking at Figure 20-12, which genes would be predicted to have undergone neofunctionalization or subfunctionalization in *S. cerevisiae*?

10. Looking at Figure 20-13, provide an explanation for why clusters B and C are missing in the horn shark.

11. Based on Figure 20-20d, do you predict that sticklebacks with mutations in the coding sequence of *Pitx1* would be viable in the wild?

12. Using Figure 20-21, explain how the mutation in the GATA sequence of the *Duffy* gene imparts resistance to *P. vivax* infection.

13. Looking at Figure 20-22, can all of the differences in the preference of bumblebees for *M. lewisii* be explained by differences in the *YUP* locus?

14. Based on Figure 20-24, why are F₁ hybrids between *M. guttatus* and *M. nasutus* viable? What fraction of F₂ progeny have four copies of *pTAC14*?

15. Based on Figure 20-25, there is still one autosomal repressor segregating in the F₁ hybrid males. Why are these males still infertile?

BASIC PROBLEMS

16. Compare Darwin's description of natural selection as quoted on page 684 with Wallace's description of the tendency of varieties to depart from the original type quoted below it, on the same page. What ideas do they have in common?

17. What are the three principles of the theory of evolution by natural selection?

18. Explain why the case of sickle-cell anemia exemplifies the three principles of the theory of evolution by natural selection.

19. Why was the neutral theory of molecular evolution a revolutionary idea?

20. What would you predict to be the relative rate of synonymous and nonsynonymous substitutions in a globin pseudogene?

21. Are *AS* heterozygotes completely resistant to malarial infection? Explain the evidence for your answer.

22. What are the four mechanisms that can lead to gene duplication?

23. What are three alternative fates of a new gene duplicate?

24. Summarize the evidence that supports the occurrence of a whole genome duplication in a specific lineage.

25. Mutations in over 100 genes are known to affect pigmentation in laboratory mice. Yet, mutations in only a few of these genes have been found to underlie differences in pigmentation in wild populations of mice. Provide one possible explanation for this finding.

26. Why can mutations in cis-regulatory elements lead to both loss and gain of structures?

27. According to the biological species concept, what is the key to determining whether two groups can be classified as species?

28. Which type of reproductive isolating barriers generally evolve later than others?

29. Summarize how the Dobzhansky–Muller model can explain the evolution of genetic incompatibilities between species.

30. What are two molecular mechanisms that have been implicated in the evolution of Dobzhansky–Muller incompatibilities between species? 🖥

CHALLENGING PROBLEMS

31. If the mutation rate to a new allele is 10^{-5}, assuming there is no migration, how large must isolated populations be to prevent chance differentiation among them in the frequency of this allele? 🖥

32. Glucose-6-phosphate dehydrogenase (G6PD) is a critical enzyme involved in the metabolism of glucose, especially in red blood cells. Deficiencies in the enzyme are the most common human enzyme defect and occur at a high frequency in certain populations of East African children.

    a. Offer one hypothesis for the high incidence of G6PD mutations in East African children.

    b. How would you test your hypothesis further?

    c. Scores of different G6PD mutations affecting enzyme function have been found in human populations. Offer one explanation for the abundance of different G6PD mutations.

33. Large differences in Hb^S frequencies among Kenyan and Ugandan tribes had been noted in surveys conducted by researchers other than Tony Allison. These researchers offered alternative explanations different from the malarial linkage proposed by Allison. Offer one counterargument to, or experimental test for, the following alternative hypotheses:

    a. The mutation rate is higher in certain tribes.

    b. There is a low degree of genetic mixing among tribes, so the allele rose to high frequency through inbreeding in certain tribes.

34. How would you determine whether a gene had duplicated by whole genome duplication, transposition or retrotransposition? 🖥

35. The *MC1R* gene affects skin and hair color in humans. There are at least 13 polymorphisms of the gene in European and Asian populations, 10 of which are nonsynonymous. In Africans, there are at least 5 polymorphisms of the gene, none of which are nonsynonymous. What might be one explanation for the differences in *MC1R* variation between Africans and non-Africans?

36. Opsin proteins detect light in photoreceptor cells of the eye and are required for color vision. The nocturnal owl monkey, the nocturnal bush baby, and the subterranean blind mole rat have different mutations in an opsin gene that render it nonfunctional. Explain why all three species can tolerate mutations in this gene that operates in most other mammals.

37. Full or partial limblessness has evolved many times in vertebrates (snakes, lizards, manatees, whales). Do you expect the mutations that occurred in the evolution of limblessness to be in the coding or noncoding sequences of toolkit genes? Why?

38. Several *Drosophila* species with unspotted wings are descended from a spotted ancestor. Would you predict the loss of spot formation to entail coding or noncoding changes in pigmentation genes? How would you test which is the case?

39. Several independent populations of sticklebacks have evolved the loss of pelvic spines. Without knowing the identity of the gene that causes the loss of pelvic spines, how could you genetically test whether the loss of pelvic spines is due to mutations at the same locus?

40. It has been claimed that "evolution repeats itself." What is the evidence for this claim from

    a. the analysis of Hb^S alleles?

    b. the analysis of *Oca2* mutations in cave fish?

    c. the analysis of stickleback *Pitx1* loci?

    d. the analysis of *Mc1r* mutations in melanistic mammals and birds?

41. Support the claim that fitness is a relative, conditional status from the following examples:

    a. The evolution of sickle-cell anemia in Africa.

    b. The evolution of pigmentation differences among pocket mice.

    c. The evolution of blind cave fish.

42. What is the molecular evidence that natural selection includes the "rejection of injurious change"?

43. Defend the statement that "Natural selection acts on whatever variation is available, and not necessarily by the best means imaginable."

44. What is the evidence that gene duplication has been the source of the α and β gene families for human hemoglobin?

45. DNA-sequencing studies for a gene in two closely related species produce the following numbers of sites that vary:

Synonymous polymorphisms	50
Nonsynonymous polymorphisms	20
Synonymous species differences	18
Nonsynonymous species differences	2

    Does this result support neutral evolution of the gene? Does it support an adaptive replacement of amino acids? What explanation would you offer for the observations? 🖥

46. In humans, two genes encoding the opsin visual pigments that are sensitive to green and red wavelengths of light are found adjacent to one another on the X chromosome. They encode proteins that are 96 percent identical. Nonprimate mammals possess just one gene encoding an opsin sensitive to the red/green wavelength.

   a. Offer one explanation for the presence of the two opsin genes on the human X chromosome.

   b. How would you test your explanation further and pinpoint when in evolutionary history the second gene arose?

47. About 9 percent of Caucasian males are color-blind and cannot distinguish red-colored from green-colored objects.

   a. Offer one genetic model for color blindness.

   b. Explain why and how color blindness has reached a frequency of 9 percent in this population.

48. Within some lakes, there are two types of stickleback fish: those with pelvic spines and those without pelvic spines. These two types of sticklebacks are found in different habitats within the lake. Furthermore, females without spines prefer to mate with males without spines, and females with spines prefer to mate with males with spines. Yet spined and spineless sticklebacks produce viable and fertile offspring in the laboratory. Would these two types of sticklebacks be classified as species under the biological species concept? Why or why not?

49. What would you predict to be the ratio of nonsynonymous to synonymous substitutions between species relative to the ratio of nonsynonymous to synonymous polymorphisms within species in a gene involved in a Dobzhansky-Muller incompatibility?

GENETICS AND SOCIETY

Defend the relevance of the theory of evolution by natural selection for understanding human disease.

A BRIEF GUIDE TO MODEL ORGANISMS

- Escherichia coli
- Saccharomyces cerevisiae
- Neurospora crassa
- Arabidopsis thaliana
- Caenorhabditis elegans
- Drosophila melanogaster
- Mus musculus
- Beyond Model Organisms

This brief guide collects in one place the main features of model organisms as they relate to genetics. Each of seven model organisms is given its own two-page spread; the format is consistent, allowing readers to compare and contrast the features of model organisms. Each treatment focuses on the special features of the organism that have made it useful as a model; the special techniques that have been developed for studying the organism; and the main contributions that studies of the organism have made to our understanding of genetics. Although many differences will be apparent, the general approaches of genetic analysis are similar but have to be tailored to take account of the individual life cycle, ploidy level, size and shape, and genomic properties, such as the presence of natural plasmids and transposons.

Model organisms have always been at the forefront of genetics (see Table 1 on page 731). Initially, in the historical development of a model organism, a researcher selects the organism because of some feature that lends itself particularly well to the study of a genetic process in which the researcher is interested. The advice of the past hundred years has been, "Choose your organism well." For example, the ascomycete fungi, such as *Saccharomyces cerevisiae* and *Neurospora crassa*, are well suited to the study of meiotic processes, such as crossing over, because their unique feature, the ascus, holds together the products of a single meiosis.

Different species tend to show remarkably similar processes, even across the members of large groups, such as the eukaryotes. Hence, we can reasonably expect that what is learned in one species can be at least partly applied to others. In particular, geneticists have kept an eye open for new research findings that may apply to our own species. Compared with other species, humans are relatively difficult to study at the genetic level, and so advances in human genetics owe a great deal to more than a century of work on model organisms.

All model organisms have far more than one useful feature for genetic or other biological study. Hence, after a model organism has been developed by a few people with specific interests, it then acts as a nucleus for the development of a research community—a group of researchers with an interest in various features of one particular model organism. There are organized research communities for all the model organisms mentioned in this summary. The people in these communities are in touch with one another regularly, share their mutant strains, and often meet at least annually at conferences that may attract thousands of people. Such a community makes possible the provision of important services, such as databases of research information, techniques, genetic stocks, clones, DNA libraries, and genomic sequences.

Another advantage to individual researchers in belonging to such a community is that they may develop "a feeling for the organism" (a phrase of maize geneticist and Nobel laureate Barbara McClintock). This idea is difficult to convey, but it implies an understanding of the general ways of an organism. No living process takes place in isolation, and so knowing the general ways of an organism is often beneficial in trying to understand one process and to interpret it in its proper context.

As the database for each model organism expands (which it currently is doing at a great pace thanks to genomics), geneticists are increasingly able to take a holistic view, encompassing the integrated workings of all parts of the organism's makeup. In this way, model organisms become not only models for isolated processes but also models of integrated life processes. The term *systems biology* is used to describe this holistic approach.

Recently, the advent of whole genome sequencing and new tools for genetic manipulations has led to the development of many new "non-traditional" model organisms, which have been developed to address specific questions that were not accessible with the more traditional model organisms presented in this guide. For example, genetic analyses of development in model systems like *Drosophila melanogaster*, *Caenorhabditis elegans*, and *Mus musculus* has revealed that the genetic toolkit for development has been highly conserved across metazoans. However, this conservation cannot explain why flies, worms, and mice look very different from each other. To address questions about how the genetic toolkit is modified to create the dramatic differences in phenotypes that we see in nature, additional model systems are needed. In the section "Beyond Model Organisms," some of these new model systems and research questions are introduced.

Escherichia coli

GENETIC "VITAL STATISTICS"

Genome size:	4.6 Mb
Chromosomes:	1, circular
Number of genes:	4000
Percentage with human homologs:	8%
Average gene size:	1 kb, no introns
Transposons:	Strain specific, ~60 copies per genome
Genome sequenced in:	1997

The unicellular bacterium *Escherichia coli* is widely known as a disease-causing pathogen, a source of food poisoning and intestinal disease. However, this negative reputation is undeserved. Although some strains of *E. coli* are harmful, others are natural and essential residents of the human gut. As model organisms, strains of *E. coli* play an indispensable role in genetic analyses. In the 1940s, several groups began investigating the genetics of *E. coli*. The need was for a simple organism that could be cultured inexpensively to produce large numbers of individual bacteria to be able to find and analyze rare genetic events. Because *E. coli* can be obtained from the human gut and is small and easy to culture, it was a natural choice. Work on *E. coli* defined the beginning of "black box" reasoning in genetics: through the selection and analysis of mutants, the workings of cellular processes could be deduced even though an individual cell was too small to be seen.

E. coli genome

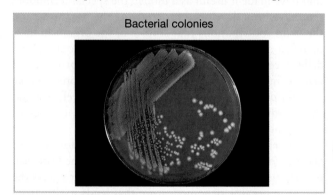

Electron micrograph of the genome of the bacterium *E. coli,* released from the cell by osmotic shock. [*SPL/Science Source.*]

SPECIAL FEATURES

Much of *E. coli*'s success as a model organism can be attributed to two statistics: its 1-µm cell size and a 20-minute generation time. (Replication of the chromosome takes 40 minutes, but multiple replication forks allow the cell to divide in 20 minutes.) Consequently, this bacteria can be grown in staggering numbers—a feature that allows geneticists to identify mutations and other rare genetic events such as intragenic recombinants. *E. coli* is also remarkably easy to culture. When cells are spread on plates of nutrient medium, each cell divides in situ and forms a visible colony. Alternatively, batches of cells can be grown in liquid shake culture. Phenotypes such as colony size, drug resistance, ability to obtain energy from particular carbon sources, and colored dye production take the place of the morphological phenotypes of eukaryotic genetics.

Geneticists have also taken advantage of some unique genetic elements associated with *E. coli*. Bacterial plasmids and phages are used as vectors to clone the genes of other organisms within *E. coli*. Transposable elements from *E. coli* are harnessed to disrupt genes in cloned eukaryotic DNA. Such bacterial elements are key players in recombinant DNA technology.

Bacterial colonies

[*BIOPHOTO ASSOC./Science Source.*]

GENETIC ANALYSIS

Spontaneous *E. coli* mutants show a variety of DNA changes, ranging from simple base substitutions to the insertion of transposable elements. The study of rare spontaneous mutations in *E. coli* is feasible because large populations can be screened. However, mutagens also are used to increase mutation frequencies.

To obtain specific mutant phenotypes that might represent defects in a process under study, screens or selections must be designed. For example, nutritional mutations and mutations conferring resistance to drugs or phages can be obtained on plates supplemented with specific chemicals, drugs, or phages. Null mutations of any essential gene will result in no growth; these mutations can be selected by adding penicillin (an antibacterial drug isolated from a fungus), which kills dividing cells but not the nongrowing mutants. For conditional lethal mutations, replica plating can be used: mutated colonies on a master plate are transferred by a felt pad to other plates that are then subjected to some toxic environment. Mutations affecting the expression of a specific gene of interest can be screened by fusing it to a reporter gene such as the *lacZ* gene, whose protein product can make a blue dye, or the *GFP* gene, whose product fluoresces when exposed to light of a particular wavelength.

After a set of mutants affecting the process of interest has been obtained, the mutations are sorted into their genes by recombination and complementation analyses. These genes are cloned and sequenced to obtain clues to function. Targeted mutagenesis can be used to tailor mutational changes at specific protein positions (see page 492).

LIFE CYCLE

Escherichia coli reproduces asexually by simple cell fission; its haploid genome replicates and partitions with the dividing cell. In the 1940s, Joshua Lederberg and Edward Tatum discovered that *E. coli* also has a type of sexual cycle in which cells of genetically differentiated "sexes" fuse and exchange some or all of their genomes, sometimes leading to recombination (see Chapter 6). "Males" can convert "females" into males by the transmission of a particular plasmid. This circular extragenomic 100-kb DNA plasmid, called F, determines a type of "maleness." F$^+$ cells acting as male "donors" transmit a copy of the F plasmid to a recipient cell. The F plasmid can integrate into the chromosome to form an Hfr cell type, which transmits the chromosome linearly into F$^-$ recipients. Other plasmids are found in *E. coli* in nature. Some carry genes whose functions equip the cell for life in specific environments; R plasmids that carry drug-resistance genes are examples.

Length of life cycle: 20 minutes

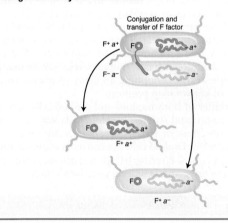

In *E. coli*, crosses are used to map mutations and to produce specific cell genotypes (see Chapter 6). Recombinants are made by mixing Hfr cells (having an integrated F plasmid) and F cells. Generally, an Hfr donor transmits part of the bacterial genome, forming a temporary merozygote in which recombination takes place. Hfr crosses can be used to perform mapping by time-of-marker entry or by recombinant frequency. By transfer of F′ derivatives carrying donor genes to F$^-$, it is possible to make stable partial diploids to study gene interaction or dominance.

TECHNIQUES OF GENETIC MODIFICATION

Standard mutagenesis:

Chemicals and radiation	Random somatic mutations
Transposons	Random somatic insertions

Transgenesis:

On plasmid vector	Free or integrated
On phage vector	Free or integrated
Transformation	Integrated

Targeted gene knockouts:

Null allele on vector	Gene replacement by recombination
Engineered allele on vector	Site-directed mutagenesis by gene replacement

GENETIC ENGINEERING

Transgenesis. *E. coli* plays a key role in introducing transgenes to other organisms (see Chapter 10). It is the standard organism used for cloning genes of any organism. *E. coli* plasmids or bacteriophages are used as vectors, carrying the DNA sequence to be cloned. These vectors are introduced into a bacterial cell by transformation, if a plasmid, or by transduction, if a phage, where they replicate in the cytoplasm. Vectors are specially modified to include unique cloning sites that can be cut by a variety of restriction enzymes. Other "shuttle" vectors are designed to move DNA fragments from yeast ("the eukaryotic *E. coli*") into *E. coli*, for its greater ease of genetic manipulation, and then back into yeast for phenotypic assessment.

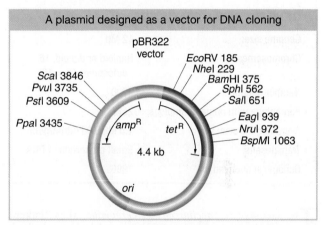

A plasmid designed as a vector for DNA cloning

Successful insertion of a foreign gene into the plasmid is detected by inactivation of either drug-resistance gene (*tet*^{R} or *amp*^{R}). Restriction sites are identified.

Targeted gene knockouts. A complete set of gene knockouts has been generated. In one procedure, a kanamycin-resistance transposon is introduced into a cloned gene in vitro (by using a transposase). The construct is transformed in, and resistant colonies are knockouts produced by homologous recombination.

MAIN CONTRIBUTIONS

Pioneering studies for genetics as a whole were carried out in *E. coli*. Perhaps the greatest triumph was the elucidation of the universal 64-codon genetic code, but this achievement is far from alone on the list of accomplishments attributable to this organism. Other fundamentals of genetics that were first demonstrated in *E. coli* include the spontaneous nature of mutation (the fluctuation test, page 506), the various types of base changes that cause mutations, and the semiconservative replication of DNA (the Meselson and Stahl experiment, page 249). This bacterium helped open up whole new areas of genetics, such as gene regulation (the *lac* operon, Chapter 11) and DNA transposition (IS elements, page 534). Last but not least, recombinant DNA technology was invented in *E. coli*, and the organism still plays a central role in this technology today.

OTHER AREAS OF CONTRIBUTION

- Cell metabolism
- Nonsense suppressors
- Colinearity of gene and polypeptide
- The operon
- Plasmid-based drug resistance
- Active transport

Saccharomyces cerevisiae

GENETIC "VITAL STATISTICS"

Genome size:	12 Mb
Chromosomes:	Haploid or diploid, 16 autosomes ($n = 16$)
Number of genes:	6000
Percentage with human homologs:	25%
Average gene size:	1.5 kb, 0.03 intron/gene
Transposons:	Small proportion of DNA
Genome sequenced in:	1996

The ascomycete *Saccharomyces cerevisiae*, alias "baker's yeast," "budding yeast," or simply "yeast," has been the basis of the baking and brewing industries since antiquity. In nature, it probably grows on the surfaces of plants, using exudates as nutrients, although its precise niche is still a mystery. Although laboratory strains are mostly haploid, cells in nature can be diploid or polyploid. In approximately 70 years of genetic research, yeast has become "the *E. coli* of the eukaryotes." Because it is haploid and unicellular, and forms compact colonies on plates, it can be treated in much the same way as a bacterium. However, it has eukaryotic meiosis, cell cycle, and mitochondria, and these features have been at the center of the yeast success story.

Yeast cells, *Saccharomyces cerevisiae*

[SCIMAT/Science Source.]

SPECIAL FEATURES

As a model organism, yeast combines the best of two worlds: it has much of the convenience of a bacterium, but with the key features of a eukaryote. Yeast cells are small (10 µm) and complete their cell cycle in just 90 minutes, allowing them to be produced in huge numbers in a short time. Like bacteria,

yeast can be grown in large batches in a liquid medium that is continuously shaken. And, like bacteria, yeast produces visible colonies when plated on agar medium, can be screened for mutations, and can be replica plated. In typical eukaryotic manner, yeast has a mitotic cell-division cycle, undergoes meiosis, and contains mitochondria housing a small unique genome. Yeast cells can respire anaerobically by using the fermentation cycle and hence can do without mitochondria, allowing mitochondrial mutants to be viable.

GENETIC ANALYSIS

Performing crosses in yeast is quite straightforward. Strains of opposite mating type (*MAT*a and *MAT*α) are simply mixed on an appropriate medium. The resulting a/α diploids are induced to undergo meiosis by using a special sporulation medium. Investigators can isolate ascospores from a single tetrad by using a machine called a micromanipulator. They also have the option of synthesizing a/a or α/α diploids for special purposes or creating partial diploids by using specially engineered plasmids.

Because a huge array of yeast mutants and DNA constructs are available within the research community, special-purpose strains for screens and selections can be built by crossing various yeast types. Additionally, new mutant alleles can be mapped by crossing with strains containing an array of phenotypic or DNA markers of known map position.

The availability of both haploid and diploid cells provides flexibility for mutational studies. Haploid cells are convenient for large-scale selections or screens because mutant phenotypes are expressed directly. Diploid cells are convenient for obtaining dominant mutations, sheltering lethal mutations, performing complementation tests, and exploring gene interaction.

LIFE CYCLE

Yeast is a unicellular species with a very simple life cycle consisting of sexual and asexual phases. The asexual phase can be haploid or diploid. A cell divides asexually by budding: a mother cell throws off a bud into which is passed one of the nuclei that result from mitosis. For sexual reproduction, there are two mating types, determined by the alleles *MAT*α and *MAT*a. When haploid cells of different mating type unite, they form a diploid cell, which can divide mitotically or undergo meiotic division. The products of meiosis are a nonlinear tetrad of four ascospores.

Total length of life cycle: 90 minutes to complete cell cycle

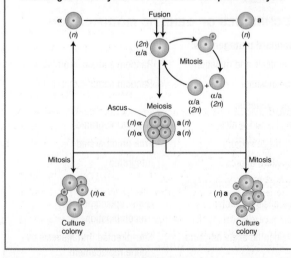

TECHNIQUES OF GENETIC MANIPULATION

Standard mutagenesis:

Chemicals and radiation	Random somatic mutations
Transposons	Random somatic insertions

Transgenesis:

Integrative plasmid	Inserts by homologous recombination
Replicative plasmid	Can replicate autonomously (2μ or ARS origin of replication)
Yeast artificial chromosome	Replicates and segregates as a chromosome
Shuttle vector	Can replicate in yeast or *E. coli*

Targeted gene knockouts:

Gene replacement	Homologous recombination replaces wild-type allele with null copy

GENETIC ENGINEERING

Transgenesis. Budding yeast provides more opportunities for genetic manipulation than any other eukaryote (see Chapter 10). Exogenous DNA is taken up easily by cells whose cell walls have been partly removed by enzyme digestion or abrasion. Various types of vectors are available. For a plasmid to replicate free of the chromosomes, it must contain a normal yeast replication origin (ARS) or a replication origin from a 2-μm plasmid found in certain yeast isolates. The most elaborate vector, the yeast artificial chromosome (YAC), consists of an ARS, a yeast centromere, and two telomeres. A YAC can carry large transgenic inserts, which are then inherited in the same way as Mendelian chromosomes. YACs have been important vectors in cloning and sequencing large genomes such as the human genome.

A simple yeast vector

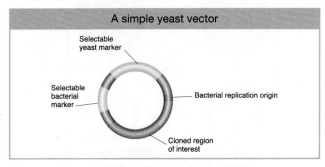

This type of vector is called a yeast integrative plasmid (YIp).

Targeted knockouts. Transposon mutagenesis (transposon tagging) can be accomplished by introducing yeast DNA into *E. coli* on a shuttle vector; the bacterial transposons integrate into the yeast DNA, knocking out gene function. The shuttle vector is then transferred back into yeast, and the tagged mutants replace wild-type copies by homologous recombination. Gene knockouts can also be accomplished by replacing wild-type alleles with an engineered null copy through homologous recombination. By using these techniques, researchers have systematically constructed a complete set of yeast knockout strains (each carrying a different knockout) to assess null function of each gene at the phenotypic level.

Cell-cycle mutants

(a) Mutants that elongate without dividing. (b) Mutants that arrest without budding. [*Reprinted with permission from Macmillan Publishers Ltd: from Susan L. Forsburg, The Salk Institute. "The Art and Design of Genetic Screens: Yeast," Nature Reviews: Genetics, 2001, September; 2 (9) 659–668, Figure 2. Permission conveyed through Copyright Clearance Center, Inc.*]

MAIN CONTRIBUTIONS

Thanks to a combination of good genetics and good biochemistry, yeast studies have made substantial contributions to our understanding of the genetic control of cell processes.

Cell cycle. The identification of cell-division genes through their temperature-sensitive mutants (*cdc* mutants) has led to a powerful model for the genetic control of cell division. The different Cdc phenotypes reveal the components of the machinery required to execute specific steps in the progression of the cell cycle. This work has been useful for understanding the abnormal cell-division controls that can lead to human cancer.

Recombination. Many of the key ideas for the current molecular models of crossing over (such as the double-strand-break model) are based on tetrad analysis of gene conversion in yeast (see page 118). Gene conversion (aberrant allele ratios such as 3:1) is quite common in yeast genes, providing an appropriately large data set for quantifying the key features of this process.

Gene interactions. Yeast has led the way in the study of gene interactions. The techniques of traditional genetics have been used to reveal patterns of epistasis and suppression, which suggest gene interactions (see Chapter 5). The two-hybrid plasmid system for finding protein interactions was developed in yeast and has generated complex interaction maps that represent the beginnings of systems biology (see Chapter 14). Synthetic lethals—lethal double mutants created by intercrossing two viable single mutants—also are used to plot networks of interaction (see page 173).

Mitochondrial genetics. Mutants with defective mitochondria are recognizable as very small colonies called "petites." The availability of these petites and other mitochondrial mutants enabled the first detailed analysis of mitochondrial genome structure and function in any organism.

Genetics of mating type. Yeast *MAT* alleles were the first mating-type genes to be characterized at the molecular level. Interestingly, yeast undergoes spontaneous switching from one mating type to the other. A silent "spare" copy of the opposite *MAT* allele, residing elsewhere in the genome, enters into the mating-type locus, replacing the resident allele by homologous recombination. Yeast has provided one of the central models for signal transduction during detection and response to mating hormones from the opposite mating type.

OTHER AREAS OF CONTRIBUTION

- Genetics of switching between yeast-like and filamentous growth
- Genetics of aging

Neurospora crassa

GENETIC "VITAL STATISTICS"

Genome size:	43 Mb
Chromosomes:	Haploid, 7 autosomes ($n = 7$)
Number of genes:	10,000
Percentage with human homologs:	6%
Average gene size:	1.7 kb, 1.7 introns/gene
Transposons:	rare
Genome sequenced in:	2003

*N*eurospora crassa, the orange bread mold, was one of the first eukaryotic microbes to be adopted by geneticists as a model organism. Like yeast, it was originally chosen because of its haploidy, its simple and rapid life cycle, and the ease with which it can be cultured. Of particular significance was the fact that it will grow on a medium with a defined set of nutrients, making it possible to study the genetic control of cellular chemistry. In nature, it is found in many parts of the world growing on dead vegetation. Because fire activates its dormant ascospores, it is most easily collected after burns—for example, under the bark of burnt trees and in fields of crops such as sugar cane that are routinely burned before harvesting.

Neurospora crassa growing on sugarcane

[David J. Jacobson, Ph.D.]

SPECIAL FEATURES

Neurospora holds the speed record for fungi because each hypha grows more than 10 cm per day. This rapid growth, combined with its haploid life cycle and ability to grow on defined medium, has made it an organism of choice for studying biochemical genetics of nutrition and nutrient uptake.

Another unique feature of *Neurospora* (and related fungi) allows geneticists to trace the steps of single meioses. The four haploid products of one meiosis stay together in a sac called an ascus. Each of the four products of meiosis undergoes a further mitotic division, resulting in a linear octad of eight ascospores (see page 91). This feature makes *Neurospora* an ideal system in which to study crossing over, gene conversion, chromosomal rearrangements, meiotic nondisjunction, and the genetic control of meiosis itself. Chromosomes, although small, are easily visible, and so meiotic processes can be studied at both the genetic and the chromosomal levels. Hence, in *Neurospora*, fundamental studies have been carried out on the mechanisms underlying these processes (see page 118).

LIFE CYCLE

N. crassa has a haploid eukaryotic life cycle. A haploid asexual spore (called a conidium) germinates to produce a germ tube that extends at its tip. Progressive tip growth and branching produce a mass of branched threads (called *hyphae*), which forms a compact colony on growth medium. Because hyphae have no cross walls, a colony is essentially one cell containing many haploid nuclei. The colony buds off millions of asexual spores, which can disperse in air and repeat the asexual cycle.

In *N. crassa*'s sexual cycle, there are two identical-looking mating types MAT-*A* and MAT-*a*, which can be viewed as simple "sexes." As in yeast, the two mating types are determined by two alleles of one gene. When colonies of different mating type come into contact, their cell walls and nuclei fuse. Many transient diploid nuclei arise, each of which undergoes meiosis, producing an octad of ascospores. The ascospores germinate and produce colonies exactly like those produced by asexual spores.

Length of life cycle: 4 weeks for sexual cycle

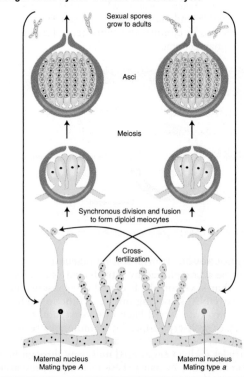

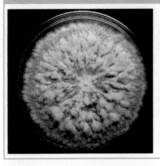
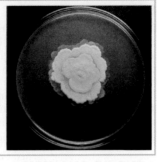

Wild-type (*left*) and mutant (*right*) *Neurospora* grown in a petri dish

[*Courtesy of Anthony Griffiths/Olivera Gavric.*]

GENETIC ANALYSIS

Genetic analysis is straightforward (see page 91). Stock centers provide a wide range of mutants affecting all aspects of the biology of the fungus. *Neurospora* genes can be mapped easily by crossing them with a bank of strains with known mutant loci or known polymorphic alleles. Strains of opposite mating type are crossed simply by growing them together. A geneticist with a handheld needle can pick out a single ascospore for study. Hence, analyses in which either complete asci or random ascospores are used are rapid and straightforward.

Because *Neurospora* is haploid, newly obtained mutant phenotypes are easily detected with the use of various types of screens and selections. A favorite system for study of the mechanism of mutation is the *ad-3* gene, because *ad-3* mutants are purple and easily detected.

Although vegetative diploids of *Neurospora* are not readily obtainable, geneticists are able to create a "mimic diploid," useful for complementation tests and other analyses requiring the presence of two copies of a gene (see page 166). Namely, the fusion of two different strains produces a heterokaryon, an individual containing two different nuclear types in a common cytoplasm. Heterokaryons also enable the use of a version of the specific-locus test, a way to recover mutations in a specific recessive allele. (Cells from a *+/m* heterokaryon are plated and *m/m* colonies are sought.)

TECHNIQUES OF GENETIC MANIPULATION

Standard mutagenesis:

Chemicals and radiation	Random somatic mutations
Transposon mutagenesis	Not available

Transgenesis:

Plasmid-mediated transformation	Random insertion

Targeted gene knockouts:

RIP	GC → AT mutations in transgenic duplicate segments before a cross
Quelling	Somatic posttranscriptional inactivation of transgenes

---

GENETIC ENGINEERING

Transgenesis. The first eukaryotic transformation was accomplished in *Neurospora*. Today, *Neurospora* is easily transformed with the use of bacterial plasmids carrying the desired transgene, plus a selectable marker such as hygromycin resistance to show that the plasmid has entered. No plasmids

replicate in *Neurospora*, and so a transgene is inherited only if it integrates into a chromosome.

Targeted knockouts. In special strains of *Neurospora*, transgenes frequently integrate by homologous recombination. Hence, a transgenic strain normally has the resident gene plus the homologous transgene, inserted at a random ectopic location. Because of this duplication of material, if the strain is crossed, it is subject to RIP, a genetic process that is unique to *Neurospora*. RIP is a premeiotic mechanism that introduces many GC-to-AT transitions into both duplicate copies, effectively disrupting the gene. RIP can therefore be harnessed as a convenient way of deliberately knocking out a specific gene.

MAIN CONTRIBUTIONS

George Beadle and Edward Tatum used *Neurospora* as the model organism in their pioneering studies on gene–enzyme relations, in which they were able to determine the enzymatic steps in the synthesis of arginine (see page 166). Their work with *Neurospora* established the beginning of molecular genetics. Many comparable studies on the genetics of cell metabolism with the use of *Neurospora* followed.

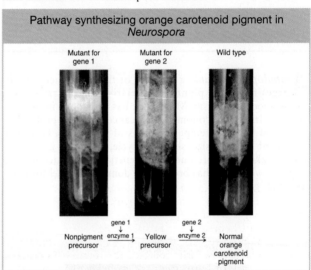

Pathway synthesizing orange carotenoid pigment in *Neurospora*

[*Anthony Griffiths.*]

Pioneering work has been done on the genetics of meiotic processes, such as crossing over and disjunction, and on sporulation rhythms. Continuously growing cultures show a daily rhythm of spore formation. The results of pioneering studies using mutations that alter this rhythm have contributed to a general model for the genetics of circadian rhythms.

Neurospora serves as a model for the multitude of pathogenic filamentous fungi affecting crops and humans because these fungi are often difficult to culture and manipulate genetically. It is even used as a simple eukaryotic test system for mutagenic and carcinogenic chemicals in the human environment.

Because crosses can be made by using one parent as female, the cycle is convenient for the study of mitochondrial genetics and nucleus–mitochondria interaction. A wide range of linear and circular mitochondrial plasmids have been discovered in natural isolates. Some of them are retroelements that are thought to be intermediates in the evolution of viruses.

OTHER AREAS OF CONTRIBUTION

- Fungal diversity and adaptation
- Cytogenetics (chromosomal basis of genetics)
- Mating-type genes
- Heterokaryon-compatibility genes (a model for the genetics of self and nonself recognition)

Arabidopsis thaliana

GENETIC "VITAL STATISTICS"

Genome size:	125 Mb
Chromosomes:	Diploid, 5 autosomes ($2n = 10$)
Number of genes:	25,000
Percentage with human homologs:	18%
Average gene size:	2 kb, 4 introns/gene
Transposons:	10% of the genome
Genome sequenced in:	2000

*A*rabidopsis thaliana, a member of the Brassicaceae (cabbage) family of plants, is a relatively late arrival as a genetic model organism. Most work has been done in the past 25 years. It has no economic significance: it grows prolifically as a weed in many temperate parts of the world. However, because of its small size, short life cycle, and small genome, it has overtaken the more traditional genetic plant models such as corn and wheat and has become the dominant model for plant molecular genetics.

Arabidopsis thaliana growing in the wild

The versions grown in the laboratory are smaller. [FloralImages/Alamy.]

SPECIAL FEATURES

In comparison with other plants, *Arabidopsis* is small in regard to both its physical size and its genome size—features that are advantageous for a model organism. *Arabidopsis* grows to a height of less than 10 cm under appropriate conditions; hence, it can be grown in large numbers, permitting large-scale mutant screens and progeny analyses. Its total genome size of 125 Mb made the genome relatively easy to sequence compared with other plant model organism genomes, such as the maize genome (2500 Mb) and the wheat genome (16,000 Mb).

GENETIC ANALYSIS

The analysis of *Arabidopsis* mutations through crossing relies on tried and true methods—essentially those used by Mendel. Plant stocks carrying useful mutations relevant to the experiment in hand are obtained from public stock centers. Lines can be manually crossed with each other or self-fertilized. Although the flowers are small, cross-pollination is easily accomplished by removing undehisced (i.e., unopened) anthers (which are sometimes eaten by the experimenter as a convenient means of disposal). Each pollinated flower then produces a long pod containing a large number of seeds. This abundant production of offspring (thousands of seeds per plant) is a boon to geneticists searching for rare mutants or other rare events. If a plant carries a new recessive mutation in the germ line, selfing allows progeny homozygous for the recessive mutation to be recovered in the plant's immediate descendants.

LIFE CYCLE

Arabidopsis has the familiar plant life cycle, with a dominant diploid stage. A plant bears several flowers, each of which produces many seeds. Like many annual weeds, its life cycle is rapid: it takes only about 6 weeks for a planted seed to produce a new crop of seeds.

Total length of life cycle: 6 weeks

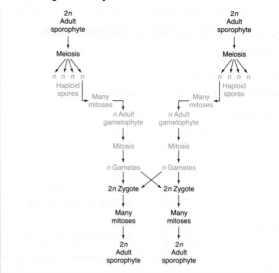

Arabidopisis mutants

(*Left*) Wild-type flower of *Arabidopsis*. (*Middle*) The *agamous* mutation (*ag*), which results in flowers with only petals and sepals (no reproductive structures). (*Right*) A double-mutant *ap1, cal*, which makes a flower that looks like a cauliflower. (Similar mutations in cabbage are probably the cause of real cauliflowers.) [*Courtesy of George Haughn.*]

TECHNIQUES OF GENETIC MODIFICATION

Standard mutagenesis:

Chemicals and radiation	Random germ-line or somatic mutations
T-DNA itself or transposons	Random tagged insertions

Transgenesis:

T-DNA carries the transgene	Random insertion

Targeted gene knockouts:

T-DNA or transposon-mediated mutagenesis	Random insertion; mutagenesis knockouts selected with PCR
RNAi	Mimics targeted knockout

GENETIC ENGINEERING

Transgenesis. *Agrobacterium* T-DNA is a convenient vector for introducing transgenes (see Chapter 10). The vector–transgene construct inserts randomly throughout the genome. Transgenesis offers an effective way to study gene regulation. The transgene is spliced to a reporter gene such as GUS, which produces a blue dye at whatever positions in the plant the gene is active.

Targeted knockouts. Because homologous recombination is rare in *Arabidopsis*, specific genes cannot be easily knocked out by homologous replacement with a transgene. Hence, in *Arabidopsis*, genes are knocked out by the random insertion of a T-DNA vector or transposon (maize transposons such as *Ac-Ds* are used), and then specific gene knockouts are selected by applying PCR analysis to DNA from large pools of plants. The PCR uses a sequence in the T-DNA or in the transposon as one primer and a sequence in the gene of interest as the other primer. Thus, PCR amplifies only copies of the gene of interest that carry an insertion. Subdividing the pool and repeating the process lead to the specific plant carrying the knockout. Alternatively, RNAi may be used to inactivate a specific gene.

Large collections of T-DNA insertion mutants are available; they have the flanking plant sequences listed in public databases, so if you are interested in a specific gene, you can see if the collection contains a plant that has an insertion in that gene. A convenient feature of knockout populations in plants is that they can be easily and inexpensively maintained as collections of seeds for many years, perhaps even decades. This feature is not possible for most populations of animal models. The worm

Caenorhabditis elegans can be preserved as a frozen animal, but fruit flies (*Drosophila melanogaster*) cannot be frozen and revived. Thus, lines of fruit-fly mutants must be maintained as living organisms.

MAIN CONTRIBUTIONS

As the first plant genome to be sequenced, *Arabidopsis* has provided an important model for plant genome architecture and evolution. In addition, studies of *Arabidopsis* have made key contributions to our understanding of the genetic control of plant development. Geneticists have isolated homeotic mutations affecting flower development, for example. In such mutants, one type of floral part is replaced by another. Integration of the action of these mutants has led to an elegant model of flower-whorl determination based on overlapping patterns of regulatory-gene expression in the flower meristem. *Arabidopsis* has also contributed broadly to the genetic basis of plant physiology, gene regulation, and the interaction of plants and the environment (including the genetics of disease resistance). Because *Arabidopsis* is a natural plant of worldwide distribution, it has great potential for the study of evolutionary diversification and adaptation.

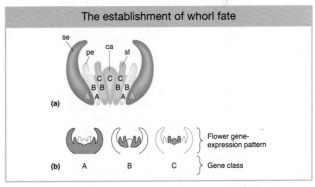

The establishment of whorl fate

(a) Patterns of gene expression corresponding to the different whorl fates. From outermost to innermost, the fates are sepal (se), petal (pe), stamen (st), and carpel (ca). (b) The shaded regions of the cross-sectional diagrams of the developing flower indicate the gene-expression patterns for the genes of the A, B, and C classes.

OTHER AREAS OF CONTRIBUTION

- Environmental-stress response
- Hormone control systems

Caenorhabditis elegans

KEY ORGANISM FOR STUDYING:

- Development
- Behavior
- Nerves and muscles
- Aging

GENETIC "VITAL STATISTICS"

Genome size:	97 Mb
Chromosomes:	Diploid, 5 autosomes, X chromosome ($2n = 12$)
Number of genes:	19,000
Percentage with human homologs:	25%
Average gene size:	5 kb, 5 exons/gene
Transposons:	Several types, active in some strains
Genome sequenced in:	1998

*C*aenorhabditis elegans may not look like much under a microscope, and, indeed, this 1-mm-long soil-dwelling roundworm (a nematode) is relatively simple as animals go. But that simplicity is part of what makes *C. elegans* a good model organism. Its small size, rapid growth, ability to self, transparency, and low number of body cells have made it an ideal choice for the study of the genetics of eukaryotic development.

SPECIAL FEATURES

Geneticists can see right through *C. elegans*. Unlike other multicellular model organisms, such as fruit flies or *Arabidopsis*, this tiny worm is transparent, making it efficient to screen large populations for interesting mutations affecting virtually any aspect of anatomy or behavior. Transparency also lends itself well to studies of development: researchers can directly observe all stages of development simply by watching the worms under a light microscope (see the Model Organism Box on p. 451). The results of such studies have shown that *C. elegans*'s development is tightly programmed and that each worm has a surprisingly small and consistent number of cells (959 in hermaphrodites and 1031 in males). In fact, biologists have tracked the fates of specific cells as the worm develops and have determined the exact pattern of cell division leading to each adult organ. This effort has yielded a lineage pedigree for every adult cell (see page 451).

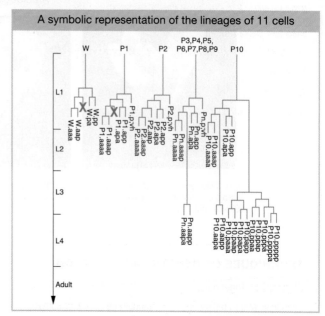

A symbolic representation of the lineages of 11 cells

A cell that undergoes programmed cell death is indicated by a blue X at the end of a branch of a lineage.

GENETIC ANALYSIS

Because the worms are small and reproduce quickly and prolifically (selfing produces about 300 progeny and crossing yields about 1000), they produce large populations of progeny that can be screened for rare genetic events. Moreover, because hermaphroditism in *C. elegans* makes selfing possible, individual worms with homozygous recessive mutations can be recovered quickly by selfing the progeny of treated individual worms. In contrast, other animal models, such as fruit flies or mice, require matings between siblings and take more generations to recover recessive mutations.

LIFE CYCLE

C. elegans is unique among the major model animals in that one of the two sexes is hermaphrodite (XX). The other is male (XO). The two sexes can be distinguished by the greater size of the hermaphrodites and by differences in their sex organs. Hermaphrodites produce both eggs and sperm, and so they can be selfed. The progeny of a selfed hermaphrodite also are hermaphrodites, except when a rare nondisjunction leads to an XO male. If hermaphrodites and males are mixed, the sexes copulate, and many of the resulting zygotes will have been fertilized by the males' amoeboid sperm. Fertilization and embryo production take place within the hermaphrodite, which then lays the eggs. The eggs finish their development externally.

Total length of life cycle: 3½ days

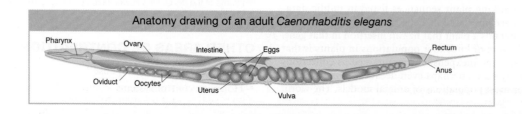

Anatomy drawing of an adult *Caenorhabditis elegans*

TECHNIQUES OF GENETIC MODIFICATION

Standard mutagenesis:

Chemical (EMS) and radiation	Random germ-line mutations
Transposons	Random germ-line insertions

Transgenesis:

Transgene injection of gonad	Unintegrated transgene array; occasional integration

Targeted gene knockouts:

Transposon-mediated mutagenesis	Knockouts selected with PCR
RNAi	Mimics targeted knockout
Laser ablation	Knockout of one cell

GENETIC ENGINEERING

Transgenesis. The introduction of transgenes into the germ line is made possible by a special property of *C. elegans* gonads. The gonads of the worm are syncytial, meaning that there are many nuclei in a common cytoplasm. The nuclei do not become incorporated into cells until meiosis, when formation of the individual egg or sperm begins. Thus, a solution of DNA containing the transgene injected into the gonad of a hermaphrodite exposes more than 100 germ-cell precursor nuclei to the transgene. By chance, a few of these nuclei will incorporate the DNA (see Chapter 10).

Transgenes recombine to form multicopy tandem arrays. In an egg, the arrays do not integrate into a chromosome, but transgenes from the arrays are still expressed. Hence, the gene carried on a wild-type DNA clone can be identified by introducing it into a specific recessive recipient strain (functional complementation). In some but not all cases, the transgenic arrays are passed on to progeny. To increase the chance of inheritance, worms are exposed to ionizing radiation, which can induce the integration of an array into an ectopic chromosomal position, and, in this site, the array is reliably transmitted to progeny.

Targeted knockouts. In strains with active transposons, the transposons themselves become agents of mutation by inserting into random locations in the genome, knocking out the interrupted genes. If we can identify organisms with insertions into a specific gene of interest, we can isolate a targeted gene knockout. Inserts into specific genes can be detected by using PCR if one PCR primer is based on the transposon sequence and another one is based on the sequence of the gene of interest. Alternatively, RNAi can be used to knockout the function of specific genes. As an alternative to mutation, individual cells can be killed by a laser beam to observe the effect on worm function or development (laser ablation).

MAIN CONTRIBUTIONS

C. elegans has become a favorite model organism for the study of various aspects of development because of its small and invariant number of cells. One example is programmed cell death, a crucial aspect of normal development. Some cells are genetically programmed to die in the course of development (a process called apoptosis). The results of studies of *C. elegans* have contributed a useful general model for apoptosis, which is also known to be a feature of human development.

Another model system is the development of the vulva, the opening to the outside of the reproductive tract. Hermaphrodites with defective vulvas still produce progeny, which in screens are easily visible clustered within the body. The results of studies of hermaphrodites with no vulva or with too many have revealed how cells that start off completely equivalent can become differentiated into different cell types (see page 451).

Behavior also has been the subject of genetic dissection. *C. elegans* offers an advantage in that worms with defective behavior can often still live and reproduce. The worm's nervous and muscular systems have been genetically dissected, allowing behaviors to be linked to specific genes.

OTHER AREAS OF CONTRIBUTION

- Cell-to-cell signaling
- RNA interference

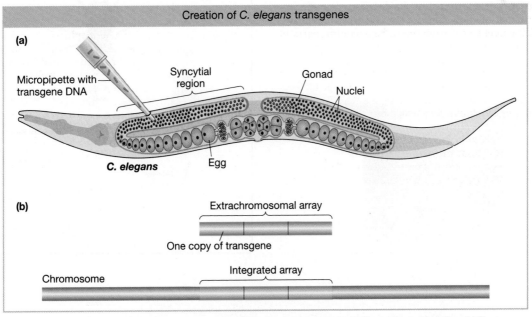

Transgenic *C. elegans* are created by injecting transgene DNA directly into a gonad. (a) The method of injection. (b) The two main types of transgenic results: extrachromosomal arrays and arrays integrated in ectopic chromosomal locations.

Drosophila melanogaster

KEY ORGANISM FOR STUDYING:

- Transmission genetics
- Cytogenetics
- Development
- Population genetics
- Evolution

GENETIC "VITAL STATISTICS"

Genome size:	180 Mb
Chromosomes:	Diploid, 3 autosomes, X and Y ($2n = 8$)
Number of genes:	13,000
Percentage with human homologs:	~50%
Average gene size:	3 kb, 4 exons/gene
Transposons:	*P* elements, among others
Genome sequenced in:	2000

The fruit fly *Drosophila melanogaster* was one of the first model organisms to be used in genetics. It was chosen in part because it is readily available in nature, has a short life cycle, has a diploid genome, and is simple to culture in the laboratory. Early genetic analysis showed that its inheritance mechanisms have strong similarities to those of other eukaryotes, underlining its role as a model organism. Its popularity as a model organism went into decline during the years when *E. coli*, yeast, and other microorganisms were being developed as molecular tools. However, *Drosophila* has experienced a renaissance because it lends itself so well to the study of the genetic basis of development, one of the central questions of biology (Chapter 13). *Drosophila*'s importance as a model for human genetics is demonstrated by the discovery that approximately 60 percent of known disease-causing genes in humans, as well as 70 percent of cancer genes, have counterparts in *Drosophila*.

Polytene chromosomes

[*William M. Gelbart, Harvard University.*]

SPECIAL FEATURES

Drosophila came into vogue as an experimental organism in the early twentieth century because of features common to most model organisms. It is small (3 mm long), simple to raise (originally, in milk bottles), quick to reproduce (only 10 days from egg to adult), and easy to obtain (just leave out some rotting fruit). It proved easy to amass a large range of interesting mutant alleles that were used to lay the ground rules of transmission genetics. Early researchers also took advantage of a feature unique to the fruit fly: polytene chromosomes (see page 576). In salivary glands and certain other tissues, these "giant chromosomes" are produced by multiple rounds of DNA replication without chromosomal segregation. Each polytene chromosome displays a unique banding pattern, providing geneticists with landmarks that could be used to correlate recombination-based maps with actual chromosomes (see Chapter 17). The momentum provided by these early advances, along with the large amount of accumulated knowledge about the organism, made *Drosophila* an attractive genetic model.

GENETIC ANALYSIS

Crosses in *Drosophila* can be performed quite easily. Wild-type or mutant male flies are cultured with wild-type or mutant virgin female flies, the male and female flies mate, and 10 days later F_1 progeny are produced.

LIFE CYCLE

Drosophila has a short life cycle that lends itself well to genetic analysis. After hatching from an egg, the fly develops through several larval stages and a pupal stage, during which it undergoes metamorphosis and emerges as an adult, and soon becomes sexually mature. The number of X chromosomes in relation to the number of autosomes determines sex (see page 48). XX is female and XY is male.

Total length of life cycle: 10 days from egg to adult

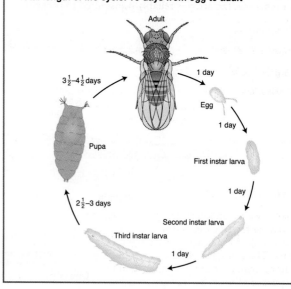

To perform a cross, males and females are placed together in a vial, and the females lay eggs in semisolid food covering the vial's bottom. After emergence from the pupae, offspring can be anesthetized to permit counting members of phenotypic classes and to distinguish males and females (by their different abdominal stripe patterns). However, because female progeny stay virgin for only a few hours after emergence from the pupae, they must immediately be isolated if they are to be used to make controlled crosses. Crosses designed to build specific gene combinations must be carefully planned, because crossing over does not take place in *Drosophila* males. Hence, in the male, linked alleles will not recombine to help create new combinations.

For obtaining new recessive mutations, special breeding programs provide convenient screening systems. In these tests, mutagenized flies are crossed with a stock having a balancer chromosome (see page 584). Recessive mutations are eventually brought to homozygosity by inbreeding for one or two generations, starting with single F_1 flies.

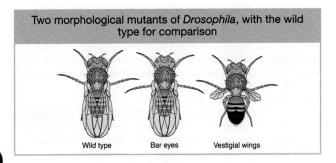

Two morphological mutants of *Drosophila*, with the wild type for comparison

Wild type   Bar eyes   Vestigial wings

TECHNIQUES OF GENETIC MODIFICATION

Standard mutagenesis:

Chemical (EMS) and radiation	Random germ-line and somatic mutations

Transgenesis:

P element mediated	Random insertion

Targeted gene knockouts:

Induced replacement	Null ectopic allele exits and recombines with wild-type allele
RNAi	Mimics targeted knockout

GENETIC ENGINEERING

Transgenesis. Building transgenic flies requires the help of a *Drosophila* transposon called the *P* element. Geneticists construct a vector that carries a transgene flanked by *P*-element repeats. The transgene vector is then injected into a fertilized egg along with a helper plasmid containing a transposase. The transposase allows the transgene to jump randomly into the genome in germ cells of the embryo (see Chapter 16).

Targeted knockouts. Targeted gene knockdown is accomplished by RNAi and targeted knockouts are generated by CRISPR-Cas9 gene engineering.

MAIN CONTRIBUTIONS

Much of the early development of the chromosome theory of heredity was based on the results of *Drosophila* studies. Geneticists working with *Drosophila* made key advances in developing techniques for gene mapping, in understanding the origin and nature of gene mutation, and in documenting the nature and behavior of chromosomal rearrangements (see pages 115 and 119). Their discoveries opened the door to other pioneering studies:

- Early studies on the kinetics of mutation induction and the measurement of mutation rates were performed with the use of *Drosophila* due to convenient screening methods for recessive mutations.

- Chromosomal rearrangements that move genes adjacent to heterochromatin were used to discover and study position-effect variegation.

- In the last part of the twentieth century, after the identification of certain key mutational classes such as homeotic and maternal-effect mutations, *Drosophila* assumed a central role in the genetics of development, a role that continues today (see Chapter 13). Maternal-effect mutations that affect the development of embryos, for example, have been crucial in the elucidation of the genetic determination of the *Drosophila* body plan; these mutations are identified by screening for abnormal developmental phenotypes in the embryos from a specific female. Techniques such as enhancer trap screens have enabled the discovery of new regulatory regions in the genome that affect development. Through these methods and others, *Drosophila* biologists have made important advances in understanding the determination of segmentation and of the body axes. Some of the key genes discovered, such as the homeotic genes, have widespread relevance in animals generally.

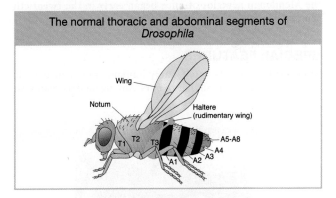

The normal thoracic and abdominal segments of *Drosophila*

Wing
Notum
Haltere (rudimentary wing)
A5-A8
A4
T1 T2 T3
A1 A2 A3

OTHER AREAS OF CONTRIBUTION

- Innate immunity
- Behavioral genetics
- Neurobiology
- Circadian rhythms

Mus musculus

KEY ORGANISM FOR STUDYING:

- Human disease
- Mutation
- Development
- Coat color
- Immunology

GENETIC "VITAL STATISTICS"

Genome size:	2600 Mb
Chromosomes:	Diploid, 19 autosomes, X and Y ($2n = 40$)
Number of genes:	30,000
Percentage with human homologs:	99%
Average gene size:	40 kb, 8.3 exons/gene
Transposons:	38% of genome
Genome sequenced in:	2002

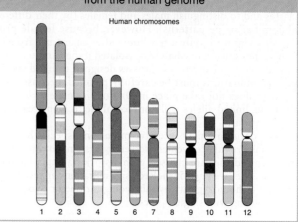

A mouse–human synteny map of 12 chromosomes from the human genome

Human chromosomes

1 2 3 4 5 6 7 8 9 10 11 12

Color coding is used to depict the regional matches of each block of the human genome to the corresponding sections of the mouse genome. Each color represents a different mouse chromosome.

Because humans and most domesticated animals are mammals, the genetics of mammals is of great interest. However, mammals are not ideal for genetics: they are relatively large in size compared with other model organisms, thereby taking up large and expensive facilities; their life cycles are long; and their genomes are large and complex. Compared with other mammals, however, mice (*Mus musculus*) are relatively small, have short life cycles, and are easily obtained, making them an excellent choice for a mammal model. In addition, mice had a head start in genetics because mouse "fanciers" had already developed many different interesting lines of mice that provided a source of variants for genetic analysis. Research on the Mendelian genetics of mice began early in the twentieth century.

SPECIAL FEATURES

Mice are not exactly small, furry humans, but their genetic makeup is remarkably similar to ours. Among model organisms, the mouse is the one whose genome most closely resembles the human genome. The mouse genome is about 14 percent smaller than that of humans (the human genome is 3000 Mb), but it has approximately the same number of genes (current estimates are just under 30,000). A surprising 99 percent of mouse genes seem to have homologs in humans. Furthermore, a large proportion of the genome is syntenic with that of humans; that is, there are large blocks containing the same genes in the same relative positions (see page 482). Such genetic similarities are the key to the mouse's success as a model organism; these similarities allow mice to be treated as "stand-ins" for their human counterparts in many ways. Potential mutagens and carcinogens that we suspect of causing damage to humans, for example, are tested on mice, and mouse models are essential in studying a wide array of human genetic diseases.

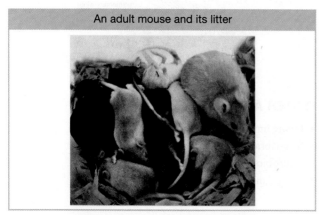

An adult mouse and its litter

[*Anthony Griffiths.*]

LIFE CYCLE

Mice have a familiar diploid life cycle, with an XY sex-determination system similar to that of humans. Litters are from 5 to 10 pups; however, the fecundity of females declines after about 9 months, and they rarely have more than five litters.

Total length of life cycle: 10 weeks from birth to giving birth, in most laboratory strains

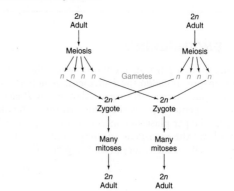

GENETIC ANALYSIS

Mutant and "wild-type" (though not actually from the wild) mice are easy to come by: they can be ordered from large stock centers that provide mice suitable for crosses and various other types of experiments. Many of these lines are derived from mice bred in past centuries by mouse fanciers. Controlled crosses can be performed simply by pairing a male with a nonpregnant female.

Most of the standard estimates of mammalian mutation rates (including those of humans) are based on measurements in mice. Indeed, mice provide the final test of agents suspected of causing mutations in humans. Mutation rates in the germ line are measured with the use of the specific-locus test: mutagenize $+/+$ gonads, cross to m/m (m is a known recessive mutation at the locus under study), and look for $m*/m$ progeny ($m*$ is a new mutation). The procedure is repeated for seven sample loci. The measurement of somatic mutation rates uses a similar setup, but the mutagen is injected into the fetus. Mice have been used extensively to study the type of somatic mutation that gives rise to cancer.

TECHNIQUES OF GENETIC MODIFICATION

Standard mutagenesis:

Chemicals and radiation	Germ-line and somatic mutations

Transgenesis:

Transgene injection into zygote	Random and homologous insertion
Transgene uptake by stem cells	Random and homologous insertion

Targeted gene knockouts:

Null transgene uptake by stem cells	Targeted knockout stem cells selected

GENETIC ENGINEERING

Transgenesis. The creation of transgenic mice is straightforward but requires the careful manipulation of a fertilized egg (see Chapter 10). First, mouse genomic DNA is cloned in *E. coli* with the use of bacterial or phage vectors. The DNA is then injected into a fertilized egg, where it integrates at ectopic (random) locations in the genome or, less commonly, at the normal locus. The activity of the transgene's protein can be monitored by fusing the transgene with a reporter gene such as *GFP* before the gene is injected. With the use of a similar method, the somatic cells of mice also can be modified by transgene insertion: specific fragments of DNA are inserted into individual somatic cells and these cells are, in turn, inserted into mouse embryos.

Targeted knockouts. Knockouts of specific genes for genetic dissection can be accomplished by introducing a transgene containing a defective allele and two drug-resistance markers into a wild-type embryonic stem cell (see Chapter 10). The markers are used to select those specific transformant cells in which the defective allele has replaced the homologous wild-type allele. The transgenic cells are then introduced into mouse embryos. A similar method can be used to replace wild-type alleles with a functional transgene (gene therapy).

MAIN CONTRIBUTIONS

Early in the mouse's career as a model organism, geneticists used mice to elucidate the genes that control coat color and pattern, providing a model for all fur-bearing mammals, including cats, dogs, horses, and cattle (see page 159). More recently, studies of mouse genetics have made an array of contributions with direct bearing on human health:

- A large proportion of human genetic diseases have a mouse counterpart—called a "mouse model"—useful for experimental study.
- Mice serve as models for the mechanisms of mammalian mutation.
- Studies on the genetic mechanisms of cancer are performed on mice.
- Many potential carcinogens are tested on mice.
- Mice have been important models for the study of mammalian developmental genetics (see page 436). For example, they provide a model system for the study of genes affecting cleft lip and cleft palate, a common human developmental disorder.
- Cell lines that are fusion hybrids of mouse and human genomes played an important role in the assignment of human genes to specific human chromosomes. There is a tendency for human chromosomes to be lost from such hybrids, and so loss of specific chromosomes can be correlated with loss of specific human alleles.

OTHER AREAS OF CONTRIBUTION

- Behavioral genetics
- Quantitative genetics
- Evolutionary genetics

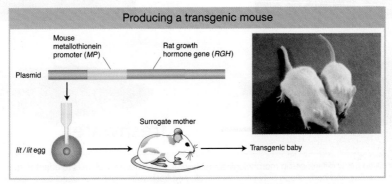

Producing a transgenic mouse

The transgene, a rat growth-hormone gene joined to a mouse promoter, is injected into a mouse egg homozygous for dwarfism (*lit/lit*). [R. L. Brinster, School of Veterinary Medicine, University of Pennsylvania.]

Beyond Model Organisms

Although most genetic research today is carried out in the model organisms described on the previous pages, this has not always been the case. Many key discoveries in genetics were made using different organisms (see Table 1 on page 731). For example, Gregor Mendel uncovered the principles of inheritance by crossing peas (*Pisum sativum*) (see Chapter 1), Barbara McClintock discovered transposable elements in maize (*Zea mays*) (see Chapter 16), and Elizabeth Blackburn and Carol Greider discovered telomeres and telomerase in a ciliated protozoan (*Tetrahymena thermophila*) (see Chapter 7). In recent years, the ability to sequence the genome of any species (see Chapter 14) has enabled the use of a diverse array of "non-traditional model organisms" to address fundamental questions in genetics. Even more recently, the development of the CRISPR/Cas9 technology for genome engineering (see Chapter 10) means that many more systems are amenable to genetic manipulations such as manipulating expression of a gene or creating a targeted knockout of a gene. These advances mean that the choice of model organism is no longer limited to those systems in which genetic and genomic resources have already been developed. Rather a researcher can choose an organism that is best suited to address a specific biological question.

Monkeyflowers (*Mimulus* species)

There is extensive phenotypic diversity across the approximately 170 species of *Mimulus*. These species are distributed across the world, with the highest species diversity in western North America and in Australia. Across their range, *Mimulus* species have evolved striking differences in morphology, life history, and physiology to adapt to varied and extreme habitats. Geneticists have identified genes that underlie some of these phenotypic differences between species, including the size, shape, and color of the flowers, which are highlighted here. [*Courtesy of Dena Grossenbacher.*]

Stickleback fish (*Gasterosteus* species)

marine — freshwater —

Stickleback fish have repeatedly evolved many differences in morphological, behavioral, and physiological traits to adapt to marine and freshwater habitats. For example, differences in color, body size, and body shape between marine (*left*) and freshwater (*right*) sticklebacks are seen in the upper photos. The genetic basis of some of these differences has been elucidated, including the identification of the gene responsible for the difference in bony lateral plates (highlighted by the red stain in the lower photos) between marine and freshwater sticklebacks. [(top left): *courtesy of Seiichi Mori and R. Uchiyama;* (top right): *courtesy of Seiichi Mori and Y. Kano;* (bottom left and right): *courtesy of Jun Kitano.*]

Heliconius butterflies

Heliconius erato (top row) and *H. melpomene* (bottom row), from different sites in Ecuador and Northern Peru. Within any site, the two species are Müllerian co-mimics and have evolved very similar color patterns. However, major geographic differences in color patterns have evolved within each species, and these have been the focus of extensive genetic analyses. [*Courtesy of James Mallet, Biological Laboratories, Harvard University.*]

Deer mice (*Peromyscus* species)

Species in the genus *Peromyscus* are widespread in North America, where they have adapted to a variety of ecological habitats. One of the most obvious differences among populations and species is coat color, which is often well matched to the substrate on which the mice live and provides camouflage from visual predators. For example, (a) *Peromyscus polionotus* that live on light sand beaches display a light coat color, while (b) *Peromyscus maniculatus* that live on darker substrates have a darker body color. In addition, these species differ in burrowing and parental behaviors. Genetic studies have led to the identification of genes that underlie both morphological and behavioral differences between these species. [*PureStock/Alamy; Thomas Kitchin & Victoria Hurst/Design Pics Inc/Alamy.*]

SPECIAL FEATURES

One key feature of many non-traditional model organisms is that they are specifically chosen for the genetic and phenotypic diversity that they harbor in the wild. This is in stark contrast to traditional model organisms, which have been inbred in the laboratory for many hundreds of generations to create genetically homogeneous backgrounds. Although inbred lines are powerful for the genetic dissection of conserved molecular and developmental processes, they are less useful for the genetic dissection of complex traits, which are affected by multiple genetic factors and environmental factors. As many agricultural traits and human diseases are complex traits, identifying the genetic architecture of traits in outbred, non-traditional model systems is of fundamental importance.

GENETIC ANALYSIS

In many cases, researchers studying non-traditional model organisms are interested in the natural genetic and phenotypic variation present within or between species. For example, if a researcher wants to identify the genetic changes that underlie the evolution of morphological diversity, then the researcher will look for populations or species that differ in a particular morphological trait, but that can still be crossed. In these cases, nature has already done the genetic screen for mutations that affect the morphological trait of interest.

The next step to identify the genetic changes that underlie a trait of interest is often to cross the populations or species that differ and then perform linkage mapping to identify regions of the genome that are associated with differences in the trait. In rare cases, there may be a single locus associated with phenotypic variation. However, in the vast majority of cases, many loci underlie variation in the trait. In these cases, quantitative trait locus or QTL mapping is used (see Chapter 19). Of course, this approach is limited to those systems in which we have the ability to cross populations in the laboratory or greenhouse. As in humans, it is not always possible to perform controlled crosses in organisms that are large and long-lived. In these cases, association mapping in natural populations can also be used to identify genetic loci that underlie phenotypic variation (see Chapter 19).

GENETIC ENGINEERING

Transgenesis. The generation of transgenic plants and animals is now possible in many systems (see Chapter 10), but the particular method used will depend very much on the specific biology of the organism. In plants, transgenesis is generally accomplished by T-DNA mediated insertion of a transgene into the host genome (see Chapter 10). In animals, transgenesis is generally accomplished by injecting DNA containing the transgene into gonads, as in *C. elegans*, or into fertilized eggs, as in *D. melanogaster* or *M. musculus* (see Chapter 10). The ability to inject fertilized eggs is facilitated in species with external fertilization, such as many fish, amphibians, and marine invertebrates.

Nonetheless, methods for harvesting and re-implanting fertilized eggs have been developed in *M. musculus* and other mammals of economic significance, and these methods could also be adapted for non-traditional mammalian systems.

Targeted knockouts. Even in many model organisms, creating specific mutations in a gene of interest has been challenging. In both model and non-traditional model systems, RNAi (see Chapter 14) has been used to decrease or completely ablate the expression of a gene. However, this method will knock out gene expression only for a few days and is therefore less useful for studying the effects of gene knockouts on phenotypes that appear later in development or in adults. However, the CRISPR/Cas9 gene editing system (see Chapter 10) appears to be a method that can be universally applied to eukaryotic systems to create targeted mutations. Thus far, CRISPR/Cas9 has been successfully used in both plants and animals, including traditional genetic model systems (*M. musculus*, *D. melanogaster*, *C. elegans*, *A. thaliana*), many emerging model systems in evolutionary genetics (stickleback fish, cavefish, butterflies), as well as species of importance for agriculture or human health (rice, wheat, maize, tomato, pig, goat, cow, salmon, silkworm, mosquito).

MAIN CONTRIBUTIONS

Because the main reason for developing new model organisms is to address diverse questions in genetics, only a few key areas of research are highlighted here:

Genetics of morphological evolution. A diversity of plant and animal systems, including species of monkeyflowers (*Mimulus*), butterflies (*Heliconius*), fruit flies (*Drosophila*), deer mice (*Peromyscus*), stickleback fish (*Gasterosteus*) and cave fish (*Astyanax*), have been used to identify genetic changes that underlie dramatic morphological differences seen in nature. These studies have revealed that mutations that underlie morphological evolution are largely those that avoid having pleiotropic effects on multiple phenotypes (see Chapter 20). In some cases, mutations occur in genes that have few pleiotropic effects, and in other cases, the mutations occur in the cis-acting regulatory elements of toolkit genes, which normally affect many developmental processes (see Chapter 13).

Genetics of speciation. Much of the classic research on the genetic basis of speciation was done using different species of *Drosophila*. Since then, an expanding array of systems has been used to investigate the genetic basis of both pre- and postzygotic isolating barriers between species. For example, genetic studies of premating isolating barriers like mate choice in animals such as stickleback fish (*Gasterosteus*), crickets (*Laupala*), and butterflies (*Heliconius*) have revealed that there is often linkage between genetic loci that underlie mate choice and loci that underlie traits important for adaptation to habitat differences between the species. Genetic studies of hybrid sterility and inviability in both plants (*Arabidopsis*, *Mimulus*) and animals (*Drosophila*, *Xiphophorus* (platyfish), and *Mus*) have uncovered roles for both gene duplication and genetic conflicts in mediating hybrid incompatibilities (see Chapter 20).

Agricultural genetics. Although animals and plants of agricultural importance are not traditionally considered genetic model organisms, selective breeding of plants and animals to improve food yield and quality has driven many foundational discoveries in genetics. Perhaps the most famous is Barbara McClintock's discovery of transposable elements, but the entire field of quantitative genetics has its foundations in plant and animal breeding (see Chapter 19).

OTHER AREAS OF CONTRIBUTION

- Behavioral genetics
- Quantitative genetics
- Genetics of aging

Year	Prize	Person	Model Organism	Discovery
		TABLE 1 List of Nobel Prizes awarded for research conducted at least in part in genetic model organisms		
1933	Physiology or Medicine	Thomas H. Morgan	*Drosophila melanogaster*	The role played by the chromosome in heredity
1946	Physiology or Medicine	Hermann J. Muller	*Drosophila melanogaster*	The production of mutations by means of X-ray irradiation
1958	Physiology or Medicine	George W. Beadle Edward L. Tatum Joshua Lederberg	*Neurospora crassa; Escherichia coli*	Genes act by regulating definite chemical events; genetic recombination and the organization of the genetic material of bacteria
1965	Physiology or Medicine	François Jacob Jacques Monod André Lwoff	*Escherichia coli*	Genetic control of enzyme and virus synthesis
1969	Physiology or Medicine	Max Delbrück Alfred D. Hershey Salvador E. Luria	*Escherichia coli*	The replication mechanism and the genetic structure of viruses
1983	Physiology or Medicine	Barbara McClintock	*Zea mays*	Mobile genetic elements
1995	Physiology or Medicine	Edward B. Lewis Christiane Nüsslein-Vollhard Eric F. Wieschaus	*Drosophila melanogaster*	The genetic control of early embryonic development
2001	Physiology or Medicine	Leland H. Hartwell Tim Hunt Paul M. Nurse	*Saccharomyces cerevisiae*	Key regulators of the cell cycle
2002	Physiology or Medicine	Sydney Brenner H. Robert Horvitz John E. Sulston	*Caenorhabditis elegans*	Genetic regulation of organ development and programmed cell death
2004	Physiology or Medicine	Richard Axel Linda B. Buck	*Drosophila melanogaster*	Odorant receptors and the organization of the olfactory system
2006	Physiology or Medicine	Andrew Z. Fire Craig C. Mello	*Caenorhabditis elegans*	RNA interference—gene silencing by double-stranded RNA
2007	Physiology or Medicine	Mario R. Capecchi Martin J. Evans Oliver Smithies	*Mus musculus*	Principles for introducing specific gene modifications in mice by the use of embryonic stem cells
2008	Chemistry	Osamu Shimamura Martin Chalfie Roger Y. Tsien	*Caenorhabditis elegans*	Development of the green fluorescent protein, GFP
2009	Physiology or Medicine	Elizabeth H. Blackburn Carol W. Greider Jack W. Szostak	*Tetrahymena thermophila*	How chromosomes are protected by telomeres and the enzyme telomerase
2011	Physiology or Medicine	Jules A. Hoffmann Bruce A. Beutler Ralph M. Steinman	*Drosophila melanogaster*	The activation of innate immunity; the dendritic cell and its role in adaptive immunity
2017	Physiology or Medicine	Jeffrey C. Hall Michael Rosbash Michael W. Young	*Drosophila melanogaster*	Molecular mechanisms controlling the circadian rhythm

Genetic Nomenclature

There is no universally accepted set of rules for naming genes, alleles, protein products, and associated phenotypes. At first, individual geneticists developed their own symbols for recording their work. Later, groups of people working on any given organism met and decided on a set of conventions that all would use. Because *Drosophila* was one of the first organisms to be used extensively by geneticists, most of the current systems are variants of the *Drosophila* system. However, there has been considerable divergence. Some scientists now advocate a standardization of this symbolism, but standardization has not been achieved. Indeed, the situation has been made more complex by the advent of DNA technology. Whereas most genes previously had been named for the phenotypes produced by mutations within them, the new technology has shown the precise nature of the products of many of these genes. Hence, it seems more appropriate to refer to them by their cellular function. However, the old names are still in the literature, so many genes have two parallel sets of nomenclature.

The following examples by no means cover all the organisms used in genetics, but most of the nomenclature systems follow one of these types.

Drosophila melanogaster (insect)

ry	A gene, *rosy*, that when mutated causes rosy eyes
ry^{S02}	A specific recessive mutant allele producing rosy eyes in homozygotes
ry$^+$	The wild-type allele of *rosy*
ry	The rosy mutant phenotype
ry$^+$	The wild-type phenotype (red eyes)
Ry, Rosy	The protein product of the *rosy* gene
Xdh	Xanthine dehydrogenase, an alternative description of the protein product of the *rosy* gene; named for the enzyme that it encodes
D	A gene, *Dichaete*, that when mutated causes a loss of certain bristles and wings to be held out laterally in heterozygotes and causes lethality in homozygotes
D^3	A specific mutant allele of the *Dichaete* gene
D$^+$	The wild-type allele of *Dichaete*
D	The Dichaete mutant phenotype
D$^+$	The wild-type phenotype
D	(Depending on context) the protein product of the *Dichaete* gene (a DNA-binding protein)

Neurospora crassa (fungus)

arg	A gene that when mutated causes arginine requirement
arg-1	One specific *arg* gene
arg-1	An unspecified mutant allele of the *arg-1* gene
arg-1 (1)	A specific mutant allele of the *arg-1* gene
arg-1$^+$	The wild-type allele
arg-1	The protein product of the *arg-1*$^+$ gene
Arg$^+$	A strain not requiring arginine
Arg$^-$	A strain requiring arginine

Saccharomyces cerevisiae (fungus)

ARG	A gene that when mutated causes arginine requirement
ARG1	One specific *ARG* gene
arg1	An unspecified mutant allele of the *ARG1* gene
arg1-1	A specific mutant allele of the *ARG1* gene
ARG1$^+$	The wild-type allele
ARG1p	The protein product of the *ARG1*$^+$ gene
Arg$^+$	A strain not requiring arginine
Arg$^-$	A strain requiring arginine

Homo sapiens (mammal)

ACH	A gene that when mutated causes achondroplasia
ACH1	A mutant allele of the *ACH* gene
ACH	Protein product of the *ACH* gene; nature unknown
FGFR3	Recent name for gene for achondroplasia
FGFR3^1 or *FGFR3* *1 or *FGFR3* <1>	Mutant allele of *FGFR3*
FGFR3	The protein product of the *FGFR3* gene

Mus musculus (mammal)

Tyr	A gene for tyrosinase
Tyr$^+$	The wild-type allele of the *Tyr* gene
Tyr^{c-ch}	A mutant allele causing chinchilla color
TYR	The protein product of the *Tyr* gene
Tyr$^+$	The wild-type phenotype
Tyr^{c-ch}	The chinchilla phenotype

Escherichia coli (bacterium)

lacZ	A gene for utilizing lactose
lacZ$^+$	The wild-type allele
lacZ1	A mutant allele of the *lacZ* gene
LacZ	The protein product of the *lacZ* gene
Lac$^+$	A strain able to use lactose (phenotype)
Lac$^-$	A strain unable to use lactose (phenotype)

Arabidopsis thaliana (plant)

YGR	A gene that when mutant produces yellow-green leaves
YGR1	A specific *YGR* gene
YGR1	The wild-type *YGR* allele
ygr1-1	A specific recessive mutant allele of *YGR1*
ygr1-2D	A specific dominant (D) mutant allele of *YGR1*
YGR1	The protein product of *YGR1*
Ygr$^-$	Yellow-green phenotype
Ygr$^+$	Wild-type phenotype

"You certainly usually find something, if you look, but it is not always quite the something you were after."
—*The Hobbit,* J. R. R. Tolkien

The field of bioinformatics encompasses the use of computational tools to distill complex data sets. Genetic and genomic data are so diverse that it has become a considerable challenge to identify the authoritative site(s) for a specific type of information. Furthermore, the landscape of Web-accessible software for analyzing this information is constantly changing as new and more powerful tools are developed. This appendix is intended to provide *some* valuable starting points for exploring the rapidly expanding universe of online resources for genetics and genomics.

1. FINDING GENETIC AND GENOMIC WEB SITES

Here are listed several central resources that contain large lists of relevant Web sites:

- The scientific journal called *Nucleic Acids Research* (NAR) publishes a special issue every January listing a wide variety of online database resources at https://academic.oup.com/nar.

- The Virtual Library has Model Organisms and Genetics subdivisions with rich arrays of Internet resources at http://ceolas.org/VL/mo/.

- The National Human Genome Research Institute (NHGRI) provides information about the Human Genome Project at https://www.genome.gov/10001772/.

- The Swiss Institute of Bioinformatics has an extensive list of databases at https://www.expasy.org/links.html.

2. GENERAL DATABASES

Nucleic Acid and Protein Sequence Databases By international agreement, three groups collaborate to house the primary DNA and mRNA sequences of all species: the National Center for Biotechnology Information (NCBI) houses GenBank, the European Bioinformatics Institute (EBI) houses the European Molecular Biology Laboratory (EMBL) Data Library, and the National Institute of Genetics in Japan houses the DNA Data-Base of Japan (DDBJ).

Primary DNA sequence records, called accessions, are submitted by individual research groups. In addition to providing access to these DNA sequence records, these sites provide many other data sets. For example, NCBI also houses RefSeq, a summary synthesis of information on the DNA sequences of fully sequenced genomes and the gene products that are encoded by these sequences.

Many other important features can be found at the NCBI, EBI, and DDBJ sites. Home pages and some other key Web sites are

- **NCBI** https://www.ncbi.nlm.nih.gov/
- **NCBI-Genomes** https://www.ncbi.nlm.nih.gov/genome
- **NCBI-RefSeq** https://www.ncbi.nlm.nih.gov/refseq/
- **EBI** https://www.ebi.ac.uk
- **DDBJ** https://www.ddbj.nig.ac.jp
- **UniProt** https://www.uniprot.org

Because protein sequence predictions are a natural part of the analysis of DNA and mRNA sequences, these same sites provide access to a variety of protein databases. One important protein database is UniProt, which contains two sections: SwissProt and TrEMBL. TrEMBL sequences are automatically predicted from DNA and/or mRNA sequences. SwissProt sequences are curated, meaning that an expert scientist reviews the output of computational analysis and makes expert decisions about which results to accept or reject. In addition to the primary protein sequence records, SwissProt also offers databases of protein domains and protein signatures (amino acid sequence strings that are characteristic of proteins of a particular type).

Protein Domain Databases The functional units within proteins are thought to be local folding regions called domains. Prediction of domains within newly discovered proteins is one way to guess at their function. Numerous protein domain databases have emerged that predict domains in somewhat different ways. Some of the individual domain databases are Pfam, PROSITE, PRINTS, SMART, ProDom, TIGRFAMs, BLOCKS, and CDD. InterPro allows querying of multiple protein domain databases simultaneously and presents the combined results. Web sites for some domain databases are

- **InterPro** http://www.ebi.ac.uk/interpro/
- **Pfam** http://pfam.xfam.org
- **PROSITE** https://prosite.expasy.org
- **PRINTS** http://www.bioinf.man.ac.uk/dbbrowser/PRINTS/
- **SMART** http://smart.embl-heidelberg.de/
- **ProDom** http://prodom.prabi.fr/
- **TIGRFAMs** http://tigrfams.jcvi.org/cgi-bin/index.cgi
- **BLOCKS** http://130.88.97.239/bioactivity/newblocksrch.html
- **CDD** http://www.ncbi.nlm.nih.gov/Structure/cdd/cdd.shtml

Protein Structure Databases The representation of three-dimensional protein structures has become an important aspect of global molecular analysis. Three-dimensional structure databases are available from the major DNA/protein sequence database sites and from independent protein structure databases, notably the Protein DataBase (PDB). NCBI has an application called Cn3D that helps in viewing PDB data.

- **PDB** http://www.rcsb.org/pdb/
- **Cn3D** http://www.ncbi.nlm.nih.gov/Structure/CN3D/cn3d.shtml

Genome Sequencing Databases Both of the following outstanding sites contain the reference sequence and working draft assemblies for a large collection of genomes and a number of tools for exploring those genomes. The UCSC Genome Browser zooms and scrolls over chromosomes, showing the work of annotators worldwide. The UCSC Gene Sorter shows expression, homology, and other information on groups of genes that can be related in many ways. The UCSC Table Browser provides access to the underlying database. Blat is an alignment tool that quickly maps sequences to the genome.

- **Ensembl Genomes** http://ensemblgenomes.org/
- **The UCSC Genome Bioinformatics Site** http://genome.ucsc.edu/

The harsh reality is that, with so much biological information, the goal of making these online resources "transparent" to the user is not fully achieved. Thus, exploration of these sites will entail familiarizing yourself with the contents of each site and exploring some of the ways the site helps you to focus your queries so you get the right answer(s). For one example of the power of these sites, consider a search for a nucleotide sequence at NCBI. Databases typically store information in separate bins called "fields." By using queries that limit the search to the appropriate field, a more directed question can be asked. Using the "Limits" option, a query phrase can be used to identify or locate a specific species, type of sequence (genomic or mRNA), gene symbol, or any of several other data fields. Query engines usually support the ability to join multiple query statements together. For example: retrieve all DNA sequence records that are from the species *Caenorhabditis elegans* AND that were published after January 1, 2000. Using the "History" option, the results of multiple queries can be joined together, so that only those hits common to multiple queries will be retrieved. By proper use of the available query options on a site, a great many false positives can be computationally eliminated while not discarding any of the relevant hits.

3. SPECIALIZED DATABASES

Organism-Specific Genetic Databases To mass some classes of genetic and genomic information, especially phenotypic information, expert knowledge of a particular species is required. Thus, MODs (model organism databases) have emerged to fulfill this role for the major genetic systems. These include databases for *Saccharomyces cerevisiae* (SGD), *Caenorhabditis elegans* (WormBase), *Drosophila melanogaster* (FlyBase), the zebrafish *Danio rerio* (ZFIN), the mouse *Mus musculus* (MGI), the rat *Rattus norvegicus* (RGD), *Zea mays* (MaizeGDB), and *Arabidopsis thaliana* (TAIR). Home pages for these MODs can be found at

- **SGD** https://www.yeastgenome.org/
- **WormBase** http://www.wormbase.org/
- **FlyBase** http://flybase.org/
- **ZFIN** http://zfin.org/
- **MGI** http://www.informatics.jax.org/
- **RGD** http://rgd.mcw.edu/
- **MaizeGDB** http://www.maizegdb.org
- **TAIR** http://www.arabidopsis.org/

Human Genetics and Genomics Databases Because of the importance of human genetics in clinical as well as basic research, a diverse set of human genetic databases has emerged. This set includes a human genetic disease database called Online Mendelian Inheritance in Man (OMIM), a database of brief descriptions of human genes called GeneCards, a compilation of all known mutations in human genes called Human Gene Mutation Database (HGMD), and some links to human genetic disease databases:

- **OMIM** http://www.omim.org
- **GeneCards** http://genecards.org
- **HGMD** http://www.hgmd.org
- **Disease InfoSearch** http://diseaseinfosearch.org
- **The Encyclopedia of Genome Elements (ENCODE)** http://encodeproject.org
- **GENCODE** https://www.gencodegenes.org/
- **The Cancer Genome Atlas Program** http://cancergenome.nih.gov/

Genome Project Databases The individual genome projects also have Web sites, where they display their results, often including information that does not appear on any other Web site in the world. The largest of the publicly funded genome centers include

- **Broad Institute** https://www.broadinstitute.org/
- **Washington University School of Medicine McDonnell Genome Institute** http://genome.wustl.edu/
- **Baylor College of Medicine Human Genome Sequencing Center** http://www.hgsc.bcm.tmc.edu
- **Sanger Institute** http://www.sanger.ac.uk/
- **DOE Joint Genome Institute** http://www.jgi.doe.gov/
- **International Genome Sample Resource** http://www.internationalgenome.org/

4. RELATIONSHIPS OF GENES WITHIN AND BETWEEN DATABASES

Gene products may be related by virtue of sharing a common evolutionary origin, sharing a common function, or participating in the same pathway.

BLAST: Identification of Sequence Similarities Evidence for a common evolutionary origin comes from the identification of sequence similarities between two or more sequences. One of the most important tools for identifying such similarities is BLAST (Basic Local Alignment Search Tool), which was developed by NCBI. BLAST is really a suite of related programs and databases in which local matches between long stretches of sequence can be identified and ranked. A query for similar DNA or protein sequences through BLAST is one of the first things that a researcher does with a newly sequenced gene. Different sequence databases can be accessed and organized by type of sequence (reference genome, recent updates, nonredundant, ESTs, etc.), and a particular species or taxonomic group can be specified. One BLAST routine matches a query nucleotide sequence translated in all six frames to a protein sequence database. Another matches a protein query sequence to the six-frame translation of a nucleotide sequence database. Other BLAST routines are customized to identify short sequence pattern matches or pair-wise alignments, to screen genome-sized DNA segments, and so forth, and can be accessed through the same top-level page:

- **NCBI-BLAST** http://www.ncbi.nlm.nih.gov/BLAST/

Functional Ontology Databases Another approach to developing relationships among gene products is by assigning these products to functional roles based on experimental evidence or prediction. Having a common way of describing these roles, regardless of the experimental system, is then of great importance. A group of scientists from different databases are working together to develop a common set of hierarchically arranged terms—an ontology—for *function* (biochemical event), *process* (the cellular event to which a protein contributes), and *subcellular location* (where a product is located in a cell) as a way of describing the activities of a gene product. This particular ontology is called the Gene Ontology (GO), and many different databases of gene products now incorporate GO terms. A full description can be found at

- **The Gene Ontology Resource** http://www.geneontology.org/

Pathway Databases Still another way to relate products to one another is by assigning them to steps in biochemical or cellular pathways. Pathway diagrams can be used as organized ways of presenting relationships of these products to one another. The Kyoto Encyclopedia of Genes and Genomes (KEGG) produces such a pathway database.

- **KEGG** http://www.genome.ad.jp/kegg/

Glossary

3′ splice site A sequence that spans the intron-exon boundary at the 3′ end of an intron and defines where cleavage occurs in the second catalytic step of splicing.

3′ untranslated region (3′ UTR) The part of an mRNA located downstream of (i.e., after) the translation stop codon.

5′ splice site A sequence that spans the exon-intron boundary at the 5′ end of an intron and defines where cleavage occurs in the first catalytic step of splicing.

5′ untranslated region (5′ UTR) The part of an mRNA located upstream of (i.e., before) the translation initiation codon.

7-methylguanosine (m⁷G) *See* **cap.**

A *See* **additive effect; adenine.**

abasic site A site in DNA that has neither a purine nor a pyrimidine base.

absolute fitness The number of offspring an individual has.

***Ac* element** *See Activator* element.

acentric chromosome A chromosome having no centromere.

acentric fragment A chromosome fragment having no centromere.

activation domain A part of a transcription factor required for the activation of target-gene transcription; it may bind to components of the transcriptional machinery or may recruit proteins that modify chromatin structure or both.

activator A protein that, when bound to a cis-acting regulatory DNA element such as an operator or an enhancer, activates transcription from an adjacent promoter.

***Activator (Ac)* element** A class 2 DNA transposable element so named by its discoverer, Barbara McClintock, because it is required to activate chromosome breakage at the *Dissociation (Ds)* locus.

active site The part of a protein that must be maintained in a specific shape if the protein is to be functional—for example, in an enzyme, the part to which the substrate binds.

adaptation In the evolutionary sense, some heritable feature of an individual's phenotype that improves its chances of survival and reproduction in the existing environment.

additive effect (A) Half the difference between the mean of the phenotypic values for the homozygous genotypic classes at a QTL.

additive gene action When the trait value for the heterozygous class at a QTL is exactly intermediate between the trait values for the two homozygous classes.

additive genetic variance The part of the genetic variance for a trait in a population that is predictably transmitted from parent to offspring.

adenine (A) A purine base that pairs with thymine in the DNA double helix.

adjacent-1 segregation In a reciprocal translocation, the passage of a translocated and a normal chromosome to each of the poles.

alkylation The transfer of an alkyl group to DNA.

allele One of the different forms of a gene that can exist at a single locus.

allele frequency A measure of the commonness of an allele in a population; the proportion of all alleles of that gene in the population that are of this specific type.

allelic series The set of known alleles of one gene. *See also* **multiple alleles.**

allopolyploid *See* **amphidiploid.**

allosteric effector A small molecule that binds to an allosteric site.

allosteric site A site on a protein to which a small molecule binds, causing a change in the conformation of the protein that modifies the activity of its active site.

allosteric termination model A model proposing that transcription termination by RNA pol II is caused by a conformational change within its active site.

allosteric transition A change from one conformation of a protein to another.

alternate segregation In a reciprocal translocation, the passage of both normal chromosomes to one pole and both translocated chromosomes to the other pole.

alternative splicing A process by which different messenger RNAs are produced from the same primary transcript, through variations in the splicing pattern of the transcript. Multiple mRNA "isoforms" can be produced in a single cell, or the different isoforms can display different tissue-specific patterns of expression. If the alternative exons fall within the open reading frames of the mRNA isoforms, different proteins will be produced by the alternative mRNAs.

Alu A short transposable element that makes up more than 10 percent of the human genome. *Alu* elements are retroelements that do not encode proteins and as such are nonautonomous elements.

Ames test A way to test whether a chemical compound is mutagenic by exposing special mutant bacterial strains to the product formed by that compound's digestion by liver extract and then counting the number of colonies. Only new mutations, presumably produced by the compound, can produce revertants to wild type able to form colonies.

amino acid A peptide; the basic building block of proteins (or polypeptides).

aminoacyl-tRNA binding site (A site) A site in the ribosome that binds incoming aminoacyl-tRNAs. The anticodon of each incoming aminoacyl-tRNA matches the codon of the mRNA.

aminoacyl-tRNA synthetase An enzyme that attaches an amino acid to a tRNA before its use in translation. There are 20 different aminoacyl-tRNA synthetases, one for each amino acid.

amino group A functional group on amino acids consisting of a nitrogen atom attached to hydrogen atoms.

amphidiploid An allopolyploid; a polyploid formed from the union of two separate chromosome sets and their subsequent doubling.

anaphase bridge In a dicentric chromosome, the segment between centromeres being drawn to opposite poles at nuclear division.

aneuploid A genome having a chromosome number that differs from the normal chromosome number for the species by a small number of chromosomes.

annotation The identification of all the functional elements of a particular genome.

antibody A protein (immunoglobulin) molecule, produced by the immune system, that recognizes a particular substance (antigen) and binds to it.

anticodon A nucleotide triplet in a tRNA molecule that aligns with a particular codon in mRNA under the influence of the ribosome; the amino acid carried by the tRNA is inserted into a growing protein chain.

antiparallel A term used to describe the opposite orientations of the two strands of a DNA double helix; the 5′ end of one strand aligns with the 3′ end of the other strand.

antiterminator A protein that promotes the continuation of transcription by preventing the termination of transcription at specific sites on DNA.

apurinic (AP) site A DNA site that lacks a purine base.

apyrimidinic (AP) site A site in DNA that lacks a pyrimidine base.

artificial selection Breeding of successive generations by the deliberate human selection of certain phenotypes or genotypes as the parents of each generation.

A site *See* **aminoacyl-tRNA binding site.**

association mapping A method for locating quantitative trait loci in the genome based on linkage disequilibrium between a marker locus and the quantitative trait locus in a random-mating population.

attenuation A regulatory mechanism in which the level of transcription of an operon (such as *trp*) is reduced when the end product of a pathway (for example, tryptophan) is plentiful; the regulated step is after the initiation of transcription.

attenuator A region of RNA sequence that forms alternative secondary structures that govern the level of transcription of attenuated operons.

autonomous transposable element A transposable element that encodes the protein(s)—for example, transposase or reverse transcriptase—necessary for its transposition and for the transposition of nonautonomous elements in the same family.

autopolyploid Polyploid formed from the doubling of a single genome.

autoradiography The use of X-ray film to detect radioactive materials, including nucleic acids.

auxotroph A strain of microorganisms that will proliferate only when the medium is supplemented with a specific substance not required by wild-type organisms (*compare* **prototroph**).

BAC *See* **bacterial artificial chromosome.**

bacterial artificial chromosome (BAC) An F plasmid engineered to act as a cloning vector that can carry large inserts.

bacteriophage (phage) A virus that infects bacteria.

balanced rearrangement A change in the chromosomal gene order that does not remove or duplicate any DNA. The two classes of balanced rearrangements are *inversions* and reciprocal *translocations*.

balancer chromosome A chromosome with multiple inversions, used to retain favorable allele combinations in the uninverted homolog.

balancing selection Natural selection that results in an equilibrium with intermediate allele frequencies.

Barr body A densely staining mass that represents an inactivated X chromosome.

base *See* **nucleotide base.**

base analog A chemical whose molecular structure mimics that of a DNA base; because of the mimicry, the analog may act as a mutagen.

base deletion A mutation in which a base pair is removed from a DNA sequence.

base excision repair (BER) One of several excision-repair pathways. In this pathway, subtle base-pair distortions are repaired by the creation of abasic sites followed by repair synthesis.

base insertion A mutation in which a base pair is added to a DNA sequence.

base substitution A mutation that exchanges one base pair for another in a DNA sequence.

BER *See* **base excision repair.**

β (beta) clamp A protein that encircles the DNA, keeping the enzyme DNA pol III attached to the DNA molecule at the replication fork, converting it from a distributive to a processive enzyme.

bioinformatics Computational information systems and analytical methods applied to biological problems such as genomic analysis.

biological species concept Groups of individuals that actually or potentially interbreed with each other and that are reproductively isolated from other such groups.

bivalents Two homologous chromosomes paired at meiosis.

boundary *See* **insulator.**

bottleneck A period of one or several consecutive generations of contraction in population size.

branch point An intron sequence near the 3′ splice site that contains an adenosine residue involved in the first catalytic step of splicing.

breeding value The part of an individual's deviation from the population mean that is due to additive effects and is transmitted to its progeny.

broad-sense heritability (H^2) The proportion of total phenotypic variance at the population level that is contributed by genetic variance.

C *See* **cytosine.**

cAMP *See* **cyclic adenosine monophosphate.**

cancer A class of disease characterized by the rapid and uncontrolled proliferation of cells within a tissue of a multitissue eukaryote. Cancers are generally thought to be genetic diseases of somatic cells, arising through sequential mutations that create oncogenes and inactivate tumor-suppressor genes.

candidate gene A gene that, because of its chromosomal position or some other property, becomes a candidate for a particular function such as disease risk.

canonical histone Histones encoded by replication-dependent genes that are expressed in high levels during S phase.

CAP *See* **catabolite activator protein.**

cap A special structure, consisting of a 7-methylguanosine residue linked to the transcript by three phosphate groups, that is added in the nucleus to the 5′ end of eukaryotic mRNA. The cap protects an mRNA from degradation and is required for translation of the mRNA in the cytoplasm.

carboxyl end The end of a protein having a free carboxyl group. The carboxyl end is encoded by the 3′ end of the mRNA and is the last part of the protein to be synthesized in translation.

carboxyl group A functional group on amino acids consisting of a carbon atom attached to two oxygen atoms.

carboxy-terminal domain (CTD) The protein tail of the β subunit of RNA polymerase II; it coordinates the processing of eukaryotic pre-mRNAs including capping, splicing, and termination.

Cas protein *See* **CRISPR-associated protein (Cas).**

catabolite A product of catabolism, the breakdown of larger molecules into smaller units.

catabolite activator protein (CAP) A protein that unites with cAMP at low glucose concentrations and binds to the *lac* promoter to facilitate RNA polymerase action.

catabolite repression The inactivation of an operon caused by the presence of large amounts of the metabolic end product of the operon.

categorical trait A trait for which individuals can be sorted into discrete or discontinuous groupings, such as tall versus short stems for Mendel's pea plants.

cDNA *See* **complementary DNA.**

cDNA library A collection of cDNAs derived from particular cells, tissues, or organisms.

cell clone Members of a colony that have a single genetic ancestor.

cell cycle The series of events that lead to division of a mother cell into two daughter cells.

centimorgan (cM) *See* **map unit.**

centromere A specialized region of DNA on each eukaryotic chromosome that acts as a site for the binding of kinetochore proteins.

chaperone A protein that assists in the folding of other proteins.

character An attribute of individual members of a species for which various heritable differences can be defined.

charged tRNA A transfer RNA molecule with an amino acid attached to its 3′ end. Also called aminoacyl-tRNA.

chimera (chimeric) An organism with cells of different genotype.

ChIP *See* **chromatin immunoprecipitation.**

ChIP-seq A variation of the ChIP procedure in which all of the DNA binding sites of a regulatory protein in a genome are identified by sequencing.

chi-square (χ^2) test A statistical test used to determine the probability of obtaining observed proportions by chance, under a specific hypothesis.

chloroplast DNA (cpDNA) The small genomic component found in the chloroplasts of plants, concerned with photosynthesis and other functions taking place within that organelle.

chromatid One of the two side-by-side replicas produced by chromosome division.

chromatin A complex of DNA and proteins of which eukaryotic chromosomes are composed.

chromatin immunoprecipitation (ChIP) The use of antibodies to isolate specific regions of chromatin and to identify the regions of DNA to which regulatory proteins are bound.

chromatin modification An enzyme catalyzed alteration to the chemical structure of an amino acid in histones.

chromatin remodeling Enzymatic processes that alter interactions between histones and DNA in chromatin.

chromosome A linear end-to-end arrangement of genes and other DNA, sometimes with associated protein and RNA.

chromosome map A representation of all chromosomes in the genome as lines, marked with the positions of genes known from their mutant phenotypes, plus molecular markers. Based on analysis of recombinant frequency.

chromosome mutation Any type of change in chromosome structure or number.

chromosome theory Inheritance of traits is controlled by genes on chromosomes.

chromosome walk A method for the dissection of large segments of DNA, in which a cloned segment of DNA, usually eukaryotic, is used to screen recombinant DNA clones from the same genome bank for other clones containing neighboring sequences.

cis-acting Refers to a site on a DNA (or RNA) molecule that functions as a binding site for a sequence-specific DNA- (or RNA-) binding protein. The term cis-acting indicates that protein binding to this site affects only nearby DNA (or RNA) sequences on the same molecule.

cis conformation In a heterozygote having two mutant sites within a gene or within a gene cluster, the arrangement A_1A_2/a_1a_2.

class 1 element A transposable element that moves through an RNA intermediate. Also called a retrotransposon.

class 2 element A transposable element that moves directly from one site in the genome to another. Also called a DNA transposon.

cloning vector In cloning, the plasmid or phage chromosome used to carry the cloned DNA segment.

clustered, regularly interspaced short palindromic repeats (CRISPRs) Sequences in bacteria that function together with Cas enzymes to provide immunity against pathogens.

CNV *See* **copy number variation.**

coactivator A special class of eukaryotic regulatory complex that serves as a bridge to bring together regulatory proteins and RNA polymerase II.

c.o.c. *See* **coefficient of coincidence.**

coding strand The nontemplate strand of a DNA molecule having the same sequence as that in the RNA transcript.

codominance A situation in which a heterozygote shows the phenotypic effects of both alleles equally.

codon A section of RNA (three nucleotides in length) that encodes a single amino acid.

coefficient of coincidence (c.o.c.) The ratio of the observed number of double recombinants to the expected number.

cointegrate The product of the fusion of two circular transposable elements to form a single, larger circle in replicative transposition.

colony A visible clone of cells.

common SNP A single nucleotide polymorphism (SNP) for which the less common allele occurs at a frequency of about 5 percent or greater.

comparative genomics Analysis of the relations of the genome sequences of two or more species.

complementary (base pairs) Refers to specific pairing between adenine and thymine, guanine and cytosine, and adenine and uracil.

complementary bases *See* **complementary (base pairs)**.

complementary DNA (cDNA) DNA synthesized from a messenger RNA template through the action of the enzyme reverse transcriptase.

complementation Production of wild-type phenotype when two full or partial haploid genomes are united in the same cell.

complementation test A test for determining whether two mutations are in different genes (they complement) or the same gene (they do not complement).

complete dominance Describes an allele that expresses itself the same in single copy (heterozygote) as in double copy (homozygote).

complex inheritance The type of inheritance exhibited by traits affected by a mix of genetic and environmental factors. Continuous traits, such as height, typically have complex inheritance.

complex trait A trait exhibiting complex inheritance.

composite transposon A type of bacterial transposable element containing a variety of genes that reside between two nearly identical insertion-sequence (IS) elements.

congenic lines Strains or stocks of a species that are identical throughout their genomes except for a small region of interest.

conjugation The union of two bacterial cells during which chromosomal material is transferred from the donor to the recipient cell.

consensus sequence The nucleotide or amino acid sequence of a nucleic acid or protein, respectively, that is derived by aligning similar sequences (either from the same or different organisms) and determining the most common nucleotide or amino acid at each position.

conservative mutation Nucleotide-pair substitution within a protein-coding region that leads to the replacement of an amino acid by one having similar chemical properties.

conservative replication A disproved model of DNA synthesis suggesting that one-half of the daughter DNA molecules should have both strands composed of newly polymerized nucleotides.

conservative transposition A mechanism of transposition that moves a mobile element to a new location in the genome as it removes that element from its previous location.

constitutive Describes a gene that is continuously transcribed or an RNA transcript that is continuously produced.

constitutive expression Refers to genes that are expressed continuously regardless of biological conditions.

constitutive heterochromatin Chromosomal regions of permanently condensed chromatin, usually around the telomeres and centromeres.

constitutive mutation A change in a DNA sequence that causes a gene that is repressed at times to be expressed continuously, or "constitutively."

continuous trait A trait that can take on a potentially infinite number of states over a continuous range, such as height in humans.

coordinately controlled genes Genes whose products are simultaneously activated or repressed in parallel.

***copia*-like element** A transposable element (retrotransposon) of *Drosophila* that is flanked by long terminal repeats and typically encodes a reverse transcriptase.

copy number variation (CNV) Variation for a large DNA segment among homologous chromosomes caused by differences in the numbers of tandem copies of a single or multiple genes.

core histone Histones H2A, H2B, H3, and H4 that make up the nucleosome octamer.

core promoter The region of a eukaryotic promoter that contains the transcription start site, often defined as -50 to $+50$ relative to the transcription start site.

corepressor A repressor that facilitates gene repression but is not itself a DNA-binding repressor.

correlation The tendency of one variable to vary in proportion to another variable, either positively or negatively.

correlation coefficient A statistical measure of association that signifies the extent to which two variables vary together.

cosuppression An epigenetic phenomenon whereby a transgene becomes reversibly inactivated along with the gene copy in the chromosome.

cotranscriptional processing The simultaneous transcription and processing of eukaryotic pre-mRNA.

cotransductants Two donor alleles that simultaneously transduce a bacterial cell; their frequency is used as a measure of closeness of the donor genes on the chromosome map.

covariance A statistical measure of the extent to which two variables change together. It is used in computing the correlation coefficient between two variables.

cpDNA Chloroplast DNA.

CpG island Unmethylated or methylated CG dinucleotides found in clusters near gene promoters.

CRISPR-associated protein (Cas) A DNA endonuclease from bacteria that functions together with CRISPR sequences to provide immunity against pathogens.

CRISPR loci Regions in bacterial chromosomes containing *c*lustered, *r*egularly *i*nterspaced *s*hort *p*alindromic *r*epeats involved in immunity to viruses.

cross The deliberate mating of two parental types of organisms in genetic analysis.

crossing over The exchange of corresponding chromosome parts between homologs by breakage and reunion.

crossover product Meiotic product cell with chromosomes that have engaged in a crossover.

crRNA RNA transcribed from CRISPR loci that guide a protein complex to degrade complementary invading viral nucleic acid.

CTD *See* **carboxy-terminal domain**.

cumulative selection The situation when natural selection promotes multiple substitutions that alter the function of a protein or regulatory element through repeated rounds of mutation and selection.

C-value The DNA content of a haploid genome.

C-value paradox The discrepancy (or lack of correlation) between the DNA content of an organism and its biological complexity.

cyclic adenosine monophosphate (cAMP) A molecule containing a diester bond between the 3′ and 5′ carbon atoms of the ribose part of the nucleotide. This modified nucleotide cannot be incorporated into DNA or RNA. It plays a key role as an intracellular signal in the regulation of various processes.

cytoplasmic segregation Segregation in which genetically different daughter cells arise from a progenitor that is a cytohet.

cytosine (C) A pyrimidine base that pairs with guanine.

Darwinian fitness The relative probability of survival and reproduction for a genotype.

daughter molecule One of the two products of DNA replication composed of one template strand and one newly synthesized strand.

deadenylase An ribonuclease that specifically removes the poly(A) tail from mRNAs.

deamination The removal of an amino group from the DNA bases cytosine, adenine, or guanine.

decapping enzyme A protein that removes the m^7G cap from the 5′ end of mRNAs.

decay The process by which RNA molecules are enzymatically degraded.

decoding center The region in the small ribosomal subunit where the decision is made whether an aminoacyl-tRNA can bind in the A site. This decision is based on complementarity between the anticodon of the tRNA and the codon of the mRNA.

degenerate code A genetic code in which some amino acids are encoded by more than one codon each.

deletion The removal of a stretch of nucleotides from a gene or chromosome.

deletion loop The loop formed at meiosis by the pairing of a normal chromosome and a deletion-containing chromosome.

deletion mapping The use of a set of known deletions to map new recessive mutations by pseudodominance.

deoxynucleotide A component of DNA consisting of a deoxyribose sugar, a phosphate, and a base.

deoxyribonucleic acid *See* **DNA.**

deoxyribose The pentose sugar in the DNA backbone.

depurination The process of losing a purine base from a site in DNA.

deviation Difference of an individual trait value from the mean trait value for the population.

dicentric bridge In a dicentric chromosome, the segment between centromeres being drawn to opposite poles at nuclear division.

dicentric chromosome A chromosome with two centromeres.

dideoxy (Sanger) sequencing A DNA sequencing method that uses dideoxynucleoside triphosphates mixed with standard nucleoside triphosphates to produce a ladder of DNA strands that end at each nucleotide in the DNA being sequenced. This method has been incorporated into automated DNA-synthesis machines. Also called Sanger sequencing after its inventor, Frederick Sanger.

dihybrid A double heterozygote such as $A/a \cdot B/b$.

dihybrid cross A cross between two individuals identically heterozygous at two loci—for example, $A\ B/a\ b \times A\ B/a\ b$.

dimerization domain A protein region that permits binding between two identical or similar proteins.

dimorphism A polymorphism with only two forms.

dioecious species A plant species in which male and female organs are on separate plants.

diploid A cell having two chromosome sets or an individual organism having two chromosome sets in each of its cells.

directional selection Selection that changes the frequency of an allele in a constant direction, either toward or away from fixation for that allele.

disassortative mating *See* **negative assortative mating.**

discovery panel A group of individuals used to detect variable nucleotide sites by comparing the partial genome sequences of these individuals with one another.

disomic An abnormal haploid carrying two copies of one chromosome.

dispersive replication A disproved model of DNA synthesis suggesting more or less random interspersion of parental and new segments in daughter DNA molecules.

Dissociation (Ds) element A nonautonomous transposable element named by Barbara McClintock for its ability to break chromosome 9 of maize but only in the presence of another element called *Activator (Ac)*.

distal enhancer An enhancer that is located far from the transcription site.

distributive enzyme An enzyme that can add only a limited number of nucleotides before falling off the DNA template.

DNA (deoxyribonucleic acid) A chain of linked nucleotides (having deoxyribose as their sugars). Two such chains in double-helical form are the fundamental substance of which genes are composed.

DNA adapter *See* **DNA linker.**

DNA amplification The production of multiple copies of a DNA sequence.

DNA assembly The joining together of multiple DNA fragments.

DNA-binding domain The site on a DNA-binding protein that directly interacts with specific DNA sequences.

DNA cloning The creation of recombinant DNA molecules that can be replicated in cells.

DNA damage A physical or chemical abnormality in the structure of DNA.

DNA gyrase A bacterial enzyme that belongs to a class of enzymes called topoisomerase that removes supercoils from DNA.

DNA ligase An enzyme involved in DNA replication and repair that seals the DNA backbone by catalyzing the formation of phosphodiester bonds.

DNA linker (DNA adapter) A short single-stranded or double-stranded piece of DNA, often containing a restriction site, that is ligated onto the end of another DNA molecule.

DNA modification A reversible chemical change to nucleotides in genomic DNA.

DNA polymerase A general term for any of the enzymes responsible for synthesizing DNA in bacteria or eukaryotes.

DNA polymerase I (DNA pol I) A bacterial enzyme that synthesizes DNA to connect Okazaki fragments during DNA replication.

DNA polymerase III (DNA pol III) A bacterial enzyme that is the primary DNA polymerase during DNA replication.

DNA polymerase III holoenzyme (DNA pol III holoenzyme) In *E. coli,* the large multisubunit complex at the replication fork consisting of two catalytic cores and many accessory proteins.

DNA replication The process of synthesizing two identical copies of a DNA molecule from one original copy.

DNA sequencing The process used to decipher the sequence of A's, C's, G's, and T's in a DNA molecule.

DNA sequencing library A set of DNA fragments that will be used for sequencing.

DNA technologies The collective techniques for obtaining, amplifying, and manipulating specific DNA fragments.

DNA transposon A class 2 transposable element found in both bacteria and eukaryotes and so named because the DNA element participates directly in transposition.

Dobzhansky-Muller model Explains the evolution of genetic incompatibilities between loci that result in hybrid sterility or hybrid inviability.

domain A region of a protein associated with a particular function. Some proteins contain more than one domain.

dominance effect The difference between the trait value for the heterozygous class at a QTL and the midpoint between the trait values of the two homozygous classes.

dominant The phenotype shown by a heterozygote. Also, an allele that expresses its phenotypic effect even when heterozygous with a recessive allele; thus, if *A* is dominant over *a*, then *A/A* and *A/a* have the same phenotype.

dominant gene action The situation when the trait value for the heterozygous class at a QTL is equal to the trait value for one of the two homozygous classes.

dominant negative mutation A mutant allele that in a single dose (a heterozygote) wipes out gene function by a spoiler effect on the protein.

donor Bacterial cell used in studies of unidirectional DNA transmission to other cells; examples are Hfr in conjugation and phage source in transduction.

donor DNA (insert DNA) Any DNA to be used in cloning or in DNA-mediated transformation.

dosage compensation The process in organisms using a chromosomal sex-determination mechanism (such as XX versus XY) that allows standard structural genes on the sex chromosome to be expressed at the same levels in females and males, regardless of the number of sex chromosomes. In mammals, dosage compensation operates by maintaining only a single active X chromosome in each cell; in *Drosophila*, it operates by hyperactivating the male X chromosome.

double helix The structure of DNA first proposed by James Watson and Francis Crick, with two interlocking helices joined by hydrogen bonds between paired bases.

double (mixed) infection Infection of a bacterium with two genetically different phages.

double mutant Genotype with mutant alleles of two different genes.

double-strand break (DSB) A DNA break cleaving the sugar–phosphate backbones of both strands of the DNA double helix.

double-strand break repair (DSBR) One of two main pathways for repair of double-strand DNA breaks by homologous recombination.

double-stranded RNA (dsRNA) An RNA molecule comprised of two complementary strands.

double transformation Simultaneous transformation by two different donor markers.

downstream A way to describe the relative location of a site in a DNA or RNA molecule. A downstream site is located closer to the 3′ end of a transcription unit.

Down syndrome An abnormal human phenotype, including mental retardation, due to a trisomy of chromosome 21; more common in babies born to older mothers.

drift *See* **random genetic drift.**

DSB *See* **double-strand break.**

DSBR *See* **double-strand break repair.**

dsRNA *See* **double-stranded RNA.**

duplication More than one copy of a particular chromosomal segment in a chromosome set.

E site *See* **exit site.**

ectopic (ectopically) In a transgenic organism, the insertion of an introduced gene at a site other than its usual locus.

elongation The stage of transcription that follows initiation and precedes termination.

elongation factor (EF) A protein involved in bringing aminoacyl-tRNAs to the ribosome or ribosome translocation during the elongation phase of protein synthesis.

endogenote *See* **merozygote.**

endonuclease An enzyme that internally cleaves RNA molecules into fragments.

enhanceosome The macromolecular assembly responsible for interaction between enhancer elements and the promoter regions of genes.

enhancer A cis-regulatory sequence in DNA that is bound by transcription factors.

environmental variance The part of the phenotypic variation among individuals in a population that is due to the different environments the individuals have experienced.

epigenetic inheritance Heritable modifications in gene function not due to changes in the base sequence of the DNA of the organism. Examples of epigenetic inheritance are paramutation, X-chromosome inactivation, and parental imprinting.

epistasis A situation in which the differential phenotypic expression of a genotype at one locus depends on the genotype at another locus; a mutation that exerts its expression while canceling the expression of the alleles of another gene.

epitope tag A short polypeptide sequence added to the end of a recombinant protein that can be used for the purpose of protein purification or analysis.

essential gene A gene without at least one copy of which the organism dies.

EST *See* **expressed sequence tag.**

euchromatin A less-condensed chromosomal region that is gene-rich.

euploid A cell having any number of complete chromosome sets or an individual organism composed of such cells.

excise Describes what a transposable element does when it leaves a chromosomal location. Also called transpose.

exconjugant A female bacterial cell that has just been in conjugation with a male and contains a fragment of male DNA.

exit site (E site) The site on the ribosome where the deacylated tRNA can be found.

exogenote *See* **merozygote.**

exome The sequence of all of the exons in a genome.

exon A segment of a pre-mRNA that is retained after the removal of introns by splicing.

exonuclease An enzyme that removes nucleotides successively from the end of RNA molecules.

expressed sequence tag (EST) A cDNA clone for which only the 5′ or the 3′ ends or both have been sequenced; used to identify transcript ends in genomic analysis.

expressivity The degree to which a particular genotype is expressed in the phenotype.

F⁻ (recipient) In *E. coli*, a cell having no fertility factor; a female cell.

F⁺ (donor) In *E. coli*, a cell having a free fertility factor; a male cell.

F factor *See* **fertility factor.**

F′ plasmid A fertility factor into which a part of the bacterial chromosome has been incorporated.

factor-independent termination A type of transcription termination mechanism in bacteria that occurs without the help of the protein Rho.

facultative heterochromatin Chromatin that may be packaged as heterochromatin in one cell type or under one condition, but euchromatin in another cell type or under another condition.

fertility factor (F factor) A bacterial episome whose presence confers donor ability (maleness).

fibrous protein A protein with a linear shape such as the components of hair and muscle.

fine mapping (*also* fine-map) Finding the genomic location of a gene of interest (or a functional region within a gene) with marker loci that are very tightly linked to it.

first-division segregation pattern (MI pattern) A linear pattern of spore phenotypes within an ascus for a particular allele pair, produced when the alleles go into separate nuclei at the first meiotic division, showing that no crossover has taken place between the allele pair and the centromere.

first filial generation (F1) The progeny individuals arising from a cross of two homozygous diploid lines.

FISH *See* **fluorescence** in situ **hybridization**

fitness Reproductive success, or the number of offspring that an allele, genotype, and/or individual contributes to the next generation.

fixed allele An allele for which all members of the population under study are homozygous, and so no other alleles for this locus exist in the population.

flexible tail The N-terminal ends of core histones that are structurally variable.

fluorescence in situ **hybridization (FISH)** A method that uses fluorescently labeled probes to detect a specific DNA sequence in vivo.

forward genetics The classical approach to genetic analysis, in which genes are first identified by mutant alleles and mutant phenotypes and later cloned and subjected to molecular analysis.

fosmid A vector that can carry a 35- to 45-kb insert of foreign DNA.

founder effect A random difference in the frequency of an allele or a genotype in a new colony as compared to the parental population that results from a small number of founders.

frameshift mutation The insertion or deletion of a nucleotide pair or pairs, causing a disruption of the translational reading frame.

frequency histogram A "step curve" in which the frequencies of various arbitrarily bounded classes are graphed.

full dominance *See* **complete dominance.**

functional genomics The study of the patterns of transcript and protein expression and of molecular interactions at a genome-wide level.

functional RNA An RNA type that plays a role without being translated.

G *See* **guanine.**

gain-of-function mutation A mutation that confers a new function to a gene. Gain-of-function mutations are often dominant.

gap gene In *Drosophila*, a class of cardinal genes that are activated in the zygote in response to the anterior–posterior gradients of positional information.

GD *See* **gene diversity.**

gel electrophoresis A method of molecular separation in which DNA, RNA, or proteins are separated in a gel matrix according to molecular size, with the use of an electrical field to draw the molecules through the gel in a predetermined direction.

gene The fundamental physical and functional unit of heredity, which carries information from one generation to the next; a segment of DNA composed of a transcribed region and a regulatory sequence that makes transcription possible.

gene action Interaction among alleles at a locus.

gene balance The idea that a normal phenotype requires a 1:1 relative proportion of genes in the genome.

gene complex A group of adjacent functionally and structurally related genes that typically arise by gene duplication in the course of evolution.

gene discovery The process whereby geneticists find a set of genes affecting some biological process of interest by the single-gene inheritance patterns of their mutant alleles or by genomic analysis.

gene diversity (GD) The probability that two alleles drawn at random from the gene pool will be different.

gene-dosage effect (1) Proportionality of the expression of some biological function to the number of copies of an allele present in the cell. (2) A change in phenotype caused by an abnormal number of wild-type alleles (observed in chromosomal mutations).

gene duplication The duplication of genes or segments of DNA through misreplication of DNA.

gene expression The process by which a gene's DNA sequence is transcribed into RNA and, for protein-coding genes, into a polypeptide.

gene family A set of genes in one genome, all descended from the same ancestral gene.

gene flow *See* **migration.**

gene knockout The inactivation of a gene by either a naturally occurring mutation or through the integration of a specially engineered introduced DNA fragment. In some systems, such inactivation is random, with the use of transgenic constructs that insert at many different locations in the genome. In other systems, it can be carried out in a directed fashion.

gene pool The sum total of all alleles in the breeding members of a population at a given time.

generalized transduction The ability of certain phages to transduce any gene in the bacterial chromosome.

general transcription factor (GTF) A eukaryotic protein or protein complex that does not take part in RNA synthesis but binds to the promoter region to attract and correctly position RNA polymerase II for transcription initiation.

gene replacement The insertion of a genetically engineered transgene in place of a resident gene; often achieved by a double crossover.

genetic admixture The mix of genes that results when individuals have ancestry from more than one subpopulation.

genetically modified organism (GMO) A popular term for a transgenic organism, especially applied to transgenic agricultural organisms.

genetic architecture All of the genetic and environmental factors that influence a trait.

genetic code A set of correspondences between nucleotide triplets in RNA and amino acids in protein.

genetic dissection The use of recombination and mutation to piece together the various components of a given biological function.

genetic engineering The process of producing modified DNA in a test tube and reintroducing that DNA into host organisms.

genetic load The total set of deleterious alleles in an individual genotype.

genetic map unit (m.u.) A distance on the chromosome map corresponding to 1 percent recombinant frequency.

genetic marker An allele used as an experimental probe to keep track of an individual organism, a tissue, a cell, a nucleus, a chromosome, or a gene.

genetics (1) The study of genes. (2) The study of inheritance.

genetic screen *See* **screen.**

genetic switch A segment of regulatory DNA and the regulatory protein(s) that binds to it that govern the transcriptional state of a gene or set of genes.

genetic toolkit The set of genes responsible for the regulation of animal development, largely comprised of members of cell-cell signaling pathways and transcription factors.

genetic variance The part of the phenotypic variation among individuals in a population that is due to the genetic differences among the individuals.

genome The entire complement of genetic material in a chromosome set.

genome project A large-scale, often multilaboratory effort required to sequence a complex genome.

genome surveillance A collection of mechanisms that recognize and destroy invading nucleic acids or active transposons.

genome-wide association (GWA or GWAS) Association mapping that uses marker loci throughout the entire genome.

genomic imprinting A phenomenon in which a gene inherited from one of the parents is not expressed, even though both gene copies are functional. Imprinted genes are methylated and inactivated in the formation of male or female gametes.

genomic library A library encompassing an entire genome.

genomics The cloning and molecular characterization of entire genomes.

genotype The allelic composition of an individual or of a cell—either of the entire genome or, more commonly, of a certain gene or a set of genes.

genotype frequency The proportion of individuals in a population having a particular genotype.

germ-line mutation A mutation in the DNA of germ cells, sperm, or oocyte.

GG-NER *See* **global genomic nucleotide excision repair.**

global genomic nucleotide excision repair (GG-NER) A DNA repair mechanism that acts on bulky adducts and pyrimidine dimers independently of transcription.

globular protein A protein with a compact structure, such as an enzyme or an antibody.

GMO *See* **genetically modified organism.**

guanine (G) A purine base that pairs with cytosine.

GWA *See* **genome-wide association.**

H *See* **heterozygosity.**

half-life The amount of time that it takes for half the pool of an RNA molecule to decay.

haploid A cell having one chromosome set or an organism composed of such cells.

haploid chromosome number The total number of chromosomes that comprises the basic chromosome set in an organism.

haploinsufficient Describes a gene that, in a diploid cell, is insufficient to promote wild-type function in only one copy (dose).

haplosufficient Describes a gene that, in a diploid cell, can promote wild-type function in only one copy (dose).

haplotype The type (or form) of a haploid segment of a chromosome as defined by the alleles present at the loci within that segment.

haplotype network A network that shows relationships among haplotypes and the positions of the mutations defining the haplotypes on the branches.

Hardy–Weinberg equilibrium The stable frequency distribution of genotypes A/A, A/a, and a/a, in the proportions of p^2, $2pq$, and q^2, respectively (where p and q are the frequencies of the alleles A and a), that is a consequence of random mating in the absence of mutation, migration, natural selection, or random drift.

Hardy–Weinberg law An equation used to describe the relationship between allelic and genotypic frequencies in a random-mating population.

helicase An enzyme that catalyzes the separation of duplex nucleic acids, DNA and RNA, into single strands.

hemizygous Refers to a gene present in only one copy in a diploid organism—for example, a gene found in the differential region of the X chromosome in a male mammal.

heterochromatin Highly-condensed regions of eukaryotic chromosomes.

heterochronic genes that regulate the timing of events in cell-fate specification. Mutations in these genes cause such events to be either repeated or omitted.

heteroduplex DNA DNA in which there is one or more mismatched nucleotide pairs in a gene under study.

heterogametic sex The sex that has heteromorphic sex chromosomes (for example, XY) and hence produces two different kinds of gametes with respect to the sex chromosomes.

heterokaryon A culture of cells composed of two different nuclear types in a common cytoplasm.

heterozygote An individual organism having a heterozygous gene pair.

heterozygous In diploids, refers to a gene pair consisting of two different alleles of that specific gene—for example, A/a or A^1/A^2.

heterozygosity (H) A measure of the genetic variation in a population; with respect to one locus, stated as the frequency of heterozygotes for that locus.

hexaploid A cell having six chromosome sets or an organism composed of such cells.

Hfr *See* **high frequency of recombination.**

high frequency of recombination (Hfr) In *E. coli*, a cell having its fertility factor integrated into the bacterial chromosome; a donor (male) cell.

histone An alkaline protein that forms the unit around which DNA is coiled in the nucleosomes of eukaryotic chromosomes.

histone acetyltransferase (HAT) An enzyme that adds an acetyl group to lysine amino acids in histones.

histone code Refers to the pattern of modification (for example, acetylation, methylation, phosphorylation) of the histone tails that carries information that regulates DNA-dependent events such as transcription.

histone deacetylase (HDAC) An enzyme that removes an acetyl group from a histone tail, which promotes the repression of gene transcription.

histone fold A structural element in core histones that facilitates heterodimerization, H2A with H2B and H3 with H4.

histone-fold extension A region in core histones that contributes to the specificity of histone pairing, H2A with H2B and H3 with H3.

histone modification Covalent alteration of one or more amino acid residues of the histone protein. Modifications include acetylation, phosphorylation, and methylation.

histone octamer The eight proteins found in the center of a nucleosome, two copies of H2A, H2B, H3, and H4.

Holliday junction (HJ) A four-branched structure that forms during DNA recombination.

homeobox (homeotic box) A family of quite similar 180-bp DNA sequences that encode a polypeptide sequence called a homeodomain, a sequence-specific DNA-binding sequence. Although the homeobox was first discovered in all homeotic genes, it is now known to encode a much more widespread DNA-binding motif.

homeodomain A highly conserved family of sequences, 60 amino acids in length and found within a large number of transcription factors, that can form helix-turn-helix structures and bind DNA in a sequence-specific manner.

homeologous chromosomes Partly homologous chromosomes, usually indicating some original ancestral homology.

homeotic transformation A mutation in which the identity of one normal body structure is changed into another.

homogametic sex The sex with homologous sex chromosomes (for example, XX).

homologous gene Genes that share a common ancestor. Homologous genes can be **paralogs** or **orthologs**.

homologous recombination (HR) A type of recombination in which DNA is exchanged between DNA molecules with similar or identical sequences; a DNA repair mechanism that relies upon homologous sequences to repair double-strand breaks.

homozygote An individual organism that is homozygous.

homozygous Refers to the state of carrying a pair of identical alleles at one locus.

homozygous dominant Refers to a genotype such as A/A.

homozygous recessive Refers to a genotype such as a/a.

horizontal transmission Inheritance of DNA from another member of the same generation.

housekeeping gene An informal term for a gene whose product is required in all cells and carries out a basic physiological function.

Hox **genes** Members of this gene class are the clustered homeobox-containing, homeotic genes that govern the identity of body parts along the anterior–posterior axis of most bilateral animals.

HR *See* **homologous recombination.**

hybrid dysgenesis A syndrome of effects including sterility, mutation, chromosome breakage, and male recombination in the hybrid progeny of crosses between certain laboratory and natural isolates of *Drosophila*.

hybridize (hybridization) (1) To form a hybrid by performing a cross. (2) To anneal complementary nucleic acid strands from different sources.

hybrid vigor A situation in which an F_1 is larger or healthier than its two different pure parental lines.

IBD *See* **identical by descent.**

identical by descent (IBD) When two copies of a gene in an individual trace back to the same copy in an ancestor.

immunofluorescence A microscopy technique that uses antibodies linked to fluorescent dyes.

inbred line or stock A stock consisting of genetically identical individuals that were fully inbred from a common parent(s).

inbreeding Mating between relatives.

inbreeding coefficient (F) The probability that the two alleles at a locus in an individual are identical by descent.

inbreeding depression A reduction in vigor and reproductive success from inbreeding.

incomplete dominance A situation in which a heterozygote shows a phenotype quantitatively (but not exactly) intermediate between the corresponding homozygote phenotypes. (Exact intermediacy means no dominance.)

indel mutation A mutation in which one or more nucleotide pairs is added or deleted.

independent assortment *See* **Mendel's second law.**

induced mutation A mutation that arises through the action of an agent that increases the rate at which mutations occur.

inducer An environmental agent that triggers transcription from an operon.

induction (1) The relief of repression of a gene or set of genes under negative control. (2) An interaction between two or more cells or tissues that is required for one of those cells or tissues to change its developmental fate.

infection Invasion of an organism and subsequent multiplication by microorganisms such as bacteria and viruses.

initiation The first stage of transcription or translation. Its main function in transcription is to correctly position RNA polymerase before the elongation stage, and in translation it is to correctly position the first aminoacyl-tRNA in the P site.

initiation codon The first codon in an open reading frame, which is commonly AUG and specifies fMet in bacteria and methionine in eukaryotes.

initiation factor (IF) A protein required for the correct initiation of translation.

initiator A region of DNA sequence where the binding of activator proteins initiates gene transcription.

initiator tRNA A special tRNA that inserts the first amino acid of a polypeptide chain into the ribosomal P site at the start of translation. The amino acid carried by the initiator in bacteria is N-formylmethionine.

insert DNA *See* **donor DNA.**

insertional duplication A duplication in which the extra copy is not adjacent to the normal one.

insertional mutagenesis The situation when a mutation arises by the interruption of a gene by foreign DNA, such as from a transgenic construct or a transposable element.

insertion-sequence (IS) element A mobile piece of bacterial DNA (several hundred nucleotide pairs in length) capable of inactivating a gene into which it inserts.

in situ hybridization (ISH) Detection of RNA or DNA in vivo by hybridization with radioactive single-stranded nucleic acid probes.

insulator A DNA sequence element that prevents inappropriate interactions between nearby chromatin domains.

interactome The entire set of molecular interactions within cells, including in particular protein–protein interactions.

intercalating agent A mutagen that can insert itself between the stacked bases at the center of the DNA double helix, causing an elevated rate of *indel mutations.*

interference A measure of the independence of crossovers from each other, calculated by subtracting the coefficient of coincidence from 1.

interrupted mating A technique used to map bacterial genes by determining the sequence in which donor genes enter recipient cells.

intragenic deletion A deletion within a gene.

intron A segment of a pre-mRNA that is removed by splicing.

inversion A chromosomal mutation consisting of the removal of a chromosome segment, its rotation through 180°, and its reinsertion in the same location.

inversion heterozygote A diploid with a normal and an inverted homolog.

inversion loop A loop formed by meiotic pairing of homologs in an inversion heterozygote.

inverted repeat (IR) sequence A sequence found in identical (but inverted) form—for example, at the opposite ends of a DNA transposon.

ionization The process by which an atom in DNA acquires a negative or positive charge.

ionizing radiation (IR) A form of energy that that causes ionization of DNA.

IR sequence *See* **inverted repeat sequence.**

IS element *See* **insertion-sequence element.**

ISH *See* **in situ hybridization.**

isoforms Members of a set of similar proteins that are produced from a single gene by alternative splicing or from a gene family.

isogenic Genetically identical.

isolation by distance A bias in mate choice that arises from the amount of geographic distance between individuals, causing individuals to be more apt to mate with a neighbor than another member of their species farther away.

key innovation A trait that allows a lineage to take advantage of a novel environment and rapidly diversify.

kinase An enzyme that transfers a phosphate group from ATP to a molecule.

Klinefelter syndrome An abnormal human male phenotype due to an extra X chromosome (XXY).

Kozak sequence A nucleotide sequence surrounding the AUG translation start codon in eukaryotic mRNAs.

lagging strand In DNA replication, the strand that is synthesized apparently in the 3′-to-5′ direction by the ligation of short fragments synthesized individually in the 5′-to-3′ direction.

λ (lambda) attachment site Where the λ prophage inserts in the *E. coli* chromosome.

large ribosomal subunit The part of the ribosome that contains the peptidyl transferase center, which catalyzes peptide bond formation and peptide release.

law of equal segregation (Mendel's first law) The production of equal numbers (50 percent) of each allele in the meiotic products (for example, gametes) of a heterozygous meiocyte.

law of independent assortment (Mendel's second law) Unlinked or distantly linked segregating gene pairs assort independently at meiosis.

leader sequence The sequence at the 5′ end of an mRNA that is not translated into protein *See also* **5′ UTR**.

leading strand In DNA replication, the strand that is made in the 5′-to-3′ direction by continuous polymerization at the 3′ growing tip.

leaky mutation A mutation that confers a mutant phenotype but still retains a low but detectable level of wild-type function.

lethal allele An allele whose expression results in the death of the individual organism expressing it.

LINE *See* **long interspersed element**.

ligase Enzymes that catalyze formation of a phosphodiester bond between a 3′-OH and 5′ monophosphate on separate strands of DNA or RNA.

ligand-binding domain A region in a transcription factor where small molecules, often hormones, bind and alter the activity of the transcription factor.

linkage disequilibrium (LD) Deviation in the frequencies of different haplotypes in a population from the frequencies expected if the alleles at the loci defining the haplotypes are associated at random.

linkage equilibrium A perfect fit of haplotype frequencies in a population to the frequencies expected if the alleles at the loci defining the haplotypes are associated at random.

linkage map A chromosome map; an abstract map of chromosomal loci that is based on recombinant frequencies.

linked The situation in which two genes are on the same chromosome as deduced by recombinant frequencies less than 50 percent.

linker histone Histone that binds the linker DNA that exits from nucleosome core particles.

lncRNA *See* **long noncoding RNA**.

locus (plural, **loci**) The specific place on a chromosome where a gene is located.

long interspersed element (LINE) A type of class 1 transposable element that encodes a reverse transcriptase. LINEs are also called non-LTR retrotransposons.

long noncoding RNA (lncRNA) Nonprotein-coding transcripts that are over approximately 200 nucleotides in length.

long terminal repeat (LTR) A direct repeat of DNA sequence at the 5′ and 3′ ends of retroviruses and retrotransposons.

loss-of-function mutation A mutation that disrupts the function of a gene. Loss-of-function mutations are often recessive.

LTR *See* **long terminal repeat**.

LTR-retrotransposon A type of class 1 transposable element that terminates in long terminal repeats and encodes several proteins including reverse transcriptase.

lysate Population of phage progeny.

lysis The rupture and death of a bacterial cell on the release of phage progeny.

lysogen *See* **lysogenic bacterium**.

lysogenic bacterium A bacterial cell containing an inert prophage integrated into, and that is replicated with, the host chromosome.

lysogenic cycle The life cycle of a normal bacterium when it is infected by a wild-type λ phage and the phage genome is integrated into the bacterial chromosome as an inert prophage.

lytic cycle The bacteriophage life cycle that leads to lysis of the host cell.

M$_I$ pattern *See* **first-division segregation pattern**.

M$_{II}$ pattern *See* **second-division segregation pattern**.

major groove The larger of the two grooves in the DNA double helix.

mapping function A formula expressing the relation between distance in a linkage map and recombinant frequency.

map unit (m.u.) The "distance" between two linked gene pairs where 1 percent of the products of meiosis are recombinant; a unit of distance in a linkage map.

maternal-effect gene A gene that produces an effect only when present in the mother.

maternal imprinting The expression of a gene only when inherited from the father, because the copy of the gene inherited from the mother is inactive due to methylation in the course of gamete formation.

maternal inheritance A type of uniparental inheritance in which all progeny have the genotype and phenotype of the parent acting as the female.

mating type In fungi, a simple form of sex. Only different mating types can unite and complete the sexual cycle. Usually determined by alleles of one or two genes.

M cytotype Laboratory stocks of *Drosophila melanogaster* that completely lack the *P* element transposon, which is found in stocks from the wild (P cytotype).

mean The arithmetic average.

meiocyte A cell in which meiosis takes place.

meiosis Two successive nuclear divisions (with corresponding cell divisions) that produce gametes (in animals) or sexual spores (in plants and fungi) that have one-half of the genetic material of the original cell.

meiotic recombination Recombination from assortment or crossing over at meiosis.

Mendel's first law (law of equal segregation) The two members of a gene pair segregate from each other in meiosis; each gamete has an equal probability of obtaining either member of the gene pair.

Mendel's second law (law of independent assortment) Unlinked or distantly linked segregating gene pairs assort independently at meiosis.

meristic trait A counting trait, taking on a range of discrete values.

merozygote A partly diploid *E. coli* cell formed from a complete chromosome (the endogenote) plus a fragment (the exogenote).

messenger RNA (mRNA) An RNA molecule transcribed from the DNA of a gene; a protein is translated from this RNA molecule by the action of ribosomes.

microRNA (miRNA) A class of functional RNA that regulates the amount of protein produced by a eukaryotic gene.

microsatellite A locus composed of several to many copies (repeats) of a short (about 2- to 6-bp) sequence motif. Different alleles have different numbers of repeats.

migration The movement of individuals (or gametes) between populations.

minimal medium Medium containing only inorganic salts, a carbon source, and water.

minor groove The smaller of the two grooves in the DNA double helix.

miRNA *See* **microRNA.**

mismatch repair (MMR) A system for repairing damage to DNA that has already been replicated.

missense (nonsynonymous) mutation Nucleotide-pair substitution within a protein-coding region that leads to the replacement of one amino acid by another amino acid.

mitochondrial DNA (mtDNA) The subset of the genome found in the mitochondrion, specializing in providing some of the organelle's functions.

mitosis A type of nuclear division (occurring at cell division) that produces two daughter nuclei identical with the parent nucleus.

mixed (double) infection The infection of a bacterial culture with two different phage genotypes.

MMR *See* **mismatch repair.**

model organism A non-human species used in experimental biology because of certain features such as a short generation time, small genome, and ease with which it can be reared in the laboratory. It is presumed that what is learned from the analysis of that species will hold true for other species, especially other closely related species.

modifier A mutation at a second locus that changes the degree of expression of a mutated gene at a first locus.

molecular clock The constant rate of substitution of amino acids in proteins or nucleotides in nucleic acids over a long evolutionary time.

molecular genetics The study of the molecular processes underlying gene structure and function.

molecular marker A DNA sequence variant that can be used to map an interesting phenotype to a specific region of DNA.

monohybrid A single-locus heterozygote of the type *A/a*.

monohybrid cross A cross between two individuals identically heterozygous at one gene pair—for example, *A/a* × *A/a*.

monoploid A cell having only one chromosome set (usually as an aberration) or an organism composed of such cells.

monosomic A cell or individual organism that is basically diploid but has only one copy of one particular chromosome type and thus has chromosome number $2n + 1$.

morph One form of a genetic polymorphism; the morph can be either a phenotype or a molecular sequence.

morphogen A molecule that induces a response in surrounding cells in a concentration-dependent manner.

mRNA *See* **messenger RNA.**

mtDNA *See* **mitochondrial DNA.**

m.u. *See* **map unit.**

multifactorial hypothesis A hypothesis that explains quantitative variation by proposing that traits are controlled by a large number of genes, each with a small effect on the trait.

multigenic deletion A deletion of several adjacent genes.

multiple alleles The set of forms of one gene, differing in their DNA sequence or expression or both.

multiple cloning site (MCS) (polylinker) A region of a vector that contains multiple restriction sites that occur only once in the vector.

mutagen An agent capable of increasing the mutation rate.

mutant An organism or cell carrying a mutation.

mutation (1) The process that produces a gene or a chromosome set differing from that of the wild type. (2) The gene or chromosome set that results from such a process.

mutation rate (μ) The probability that a copy of an allele changes to some other allelic form in one generation.

NAHR *See* **nonallelic homologous recombination.**

narrow-sense heritability (h^2) The proportion of phenotypic variance that can be attributed to additive genetic variance.

natural selection The differential rate of reproduction of different types in a population as the result of different physiological, anatomical, or behavioral characteristics of the types.

ncRNA *See* **non-coding RNA.**

nearly isogenic line *See* **congenic line.**

negative assortative mating Preferential mating between phenotypically unlike partners.

negative regulation Regulation mediated by factors that block or turn off transcription.

negative selection The elimination of a deleterious trait from a population by natural selection.

neofunctionalization The evolution of a new function by a gene.

NER *See* **nucleotide excision repair.**

neutral allele An allele that has no effect on the fitness of individuals that possess it.

neutral evolution Nonadaptive evolutionary changes due to random genetic drift.

NH *See* **number of haplotypes.**

NHEJ *See* **nonhomologous end joining.**

NLS *See* **nuclear localization sequence.**

nonallelic homologous recombination (NAHR) Crossing over between short homologous units found at different chromosomal loci. *See also* **unequal crossing over.**

nonautonomous transposable element A transposable element that relies on the protein products of autonomous elements for its mobility. *Dissociation (Ds)* is an example of a nonautonomous transposable element.

noncoding RNA (ncRNA) RNA that is not translated into protein.

noncoding strand *See* **template strand.**

nonconservative mutation Nucleotide-pair substitution within a protein-coding region that leads to the replacement of an amino acid by one having different chemical properties.

nondisjunction The failure of homologs (at meiosis) or sister chromatids (at mitosis) to separate properly to opposite poles.

nonhomologous end joining (NHEJ) A mechanism used by eukaryotes to repair double-strand breaks.

nonsense codon A codon that terminates translation (UAA, UAG, and UGA).

nonsense mutation Nucleotide-pair substitution within a protein-coding region that changes a codon for an amino acid into a termination (nonsense) codon.

nonsynonymous mutation *See* **missense mutation.**

nonsynonymous substitution *See* **missense mutation.**

non-template strand *See* **coding strand.**

normal distribution A continuous distribution defined by the normal density function with a specified mean and standard deviation showing the expected frequencies for different values of a random variable (the "bell curve").

Northern blotting A method to detect a particular RNA in a mixture of many RNAs by hybridization with a complementary nucleic acid probe.

nuclear localization sequence (NLS) Part of a protein required for its transport from the cytoplasm to the nucleus.

nuclease An enzyme that cuts DNA molecules in specific locations or degrades an entire DNA molecule into single nucleotides.

nucleolus A non-membrane bound region of the nucleus that functions as the site of ribosome biogenesis.

nucleoside A component of DNA or RNA consisting of a sugar and a base.

nucleosome The basic unit of eukaryotic chromosome structure; a ball of eight histone molecules that is wrapped by two coils of DNA.

nucleosome free region (NFR) Segments of DNA, often at promoters of active genes, that lack nucleosomes.

nucleotide A molecule composed of a nitrogen base, a sugar, and a phosphate group; the basic building block of nucleic acids.

nucleotide diversity Heterozygosity or gene diversity averaged over all the nucleotide sites in a gene or any other stretch of DNA.

nucleotide excision repair (NER) An excision-repair pathway that breaks the phosphodiester bonds on either side of a damaged base, removing that base and several on either side followed by repair replication.

null allele An allele whose effect is the absence either of normal gene product at the molecular level or of normal function at the phenotypic level.

null hypothesis In statistics, the hypothesis being tested that makes a prediction about the expected results of an experiment. If the probability of observing the results under the null hypothesis is less than 0.05, then the null hypothesis is rejected.

nullisomic Refers to a cell or individual organism with one chromosomal type missing, with a chromosome number such as $n - 1$ or $2n - 2$.

null mutation A mutation that results in complete absence of function for the gene.

number of haplotypes (NH) A simple count of the number of haplotypes at a locus in a population.

O *See* **origin of replication.**

octad An ascus containing eight ascospores, produced in species in which the tetrad normally undergoes a postmeiotic mitotic division.

Okazaki fragment A small segment of single-stranded DNA synthesized as part of the lagging strand in DNA replication.

one-gene–one-enzyme hypothesis *See* **one-gene-one-polypeptide hypothesis.**

one-gene–one-polypeptide hypothesis A mid-twentieth-century hypothesis that originally proposed that each gene (nucleotide sequence) encodes a polypeptide sequence; generally true, with the exception of untranslated functional RNA.

open reading frame (ORF) A stretch of nucleotide sequence that is not interrupted by a stop codon in a given reading frame.

operator A DNA region at one end of an operon that acts as the binding site for a repressor protein.

operon A set of adjacent structural genes whose mRNA is synthesized in one piece, plus the adjacent regulatory signals that affect transcription of the structural genes.

ORF *See* **open reading frame.**

origin (O) *See* **origin of replication.**

origin of replication (O) The point of a specific sequence at which DNA replication is initiated.

origin recognition complex (ORC) A eukaryotic complex that binds to origins of replication to initiate DNA replication.

organizer A group of cells in an embryo that has the ability to organize the development of surrounding cells.

orthologs Genes in different species that evolved from a common ancestral gene by speciation.

outgroup Taxa outside of a group of organisms among which evolutionary relationships are being determined.

oxidative damage Damage to DNA caused by reactive oxygen species.

paired-end reads In whole-genome shotgun sequence assembly, the DNA sequences corresponding to both ends of a genomic DNA insert in a recombinant clone.

pair-rule gene In *Drosophila*, a member of a class of zygotically expressed genes that act at an intermediary stage in the process of establishing the correct numbers of body segments. Pair-rule mutations have half the normal number of segments, owing to the loss of every other segment.

palindrome (palindromic) A DNA sequence that reads the same in the $5'$-to-$3'$ direction on both strands.

paracentric inversion An inversion not including the centromere.

paralogs Genes that are related by gene duplication in a genome.

parental generation (P) The two strains or individual organisms that constitute the start of a genetic breeding experiment; their progeny constitute the F_1 generation.

parental molecule The DNA in a cell that is replicated when the cell divides.

parsimony To favor the simplest explanation involving the smallest number of evolutionary changes.

parthenogenesis The production of offspring by a female with no genetic contribution from a male.

partial diploid *See* merozygote.

partial dominance Gene action under which the phenotype of heterozygotes is intermediate between the two homozygotes but more similar to that of one homozygote than the other.

paternal imprinting The expression of a gene only when inherited from the mother, because the allele of the gene inherited from the father is inactive due to methylation in the course of gamete formation.

PCNA *See* proliferating cell nuclear antigen.

PCR *See* polymerase chain reaction.

P cytotype Natural stocks of *Drosophila melanogaster* that contain 20 to 50 copies of the *P* element. Laboratory stocks have none. *See also* **M cytotype.**

pedigree analysis Deducing single-gene inheritance of human phenotypes by a study of the progeny of matings within a family, often stretching back several generations.

***P* element** A DNA transposable element in *Drosophila* that has been used as a tool for insertional mutagenesis and for germ-line transformation.

penetrance The proportion of individuals with a specific genotype that manifest that genotype at the phenotype level.

pentaploid An individual organism with five sets of chromosomes.

peptide bond A chemical bond between the carboxyl group of one amino acid and the amino group of another amino acid.

peptidyl site (P site) The site in the ribosome to which a tRNA with the growing polypeptide chain is bound.

peptidyltransferase center The site in the large ribosomal subunit at which the joining of two amino acids is catalyzed.

pericentric inversion An inversion that includes the centromere.

permissive temperature The temperature at which a temperature-sensitive mutant allele is expressed the same as the wild-type allele.

personal genomics The analysis of the genome of an individual to better understand his or her ancestry or the genetic basis of phenotypic traits such as his or her risk of developing a disease.

PEV *See* position-effect variegation.

phage *See* bacteriophage.

phage recombination The production of recombinant phage genotypes as a result of doubly infecting a bacterial cell with different "parental" phage genotypes.

phenotype (1) The form taken by some character (or group of characters) in a specific individual. (2) The detectable outward manifestations of a specific genotype.

phosphatase An enzyme that removes a phosphate group from a molecule.

phosphate An ion formed of four oxygen atoms attached to a phosphorus atom or the chemical group formed by the attachment of a phosphate ion to another chemical species by an ester bond.

phosphodiester bond A chemical linkage between successive nucleotide sugars in an RNA or DNA molecule.

phosphorylation The addition of a phosphate group to a molecule.

phylogenetic inference Determining the state of a character or the direction of change in a character based on the distribution of that character within a phylogeny of organisms.

phylogeny The evolutionary history of a group.

physical map The ordered and oriented map of cloned DNA fragments on the genome.

PIC *See* preinitiation complex.

pi-cluster Region in vertebrate and invertebrate genomes that codes for clusters of piRNAs.

piRNA *See* piwi-interacting RNA.

piwi-interacting RNA (piRNA) An RNA transcribed from pi-clusters that helps to protect the integrity of plant and animal genomes and to prevent the spread of transposable elements to other chromosomal loci. piRNAs restrain transposable elements in animals.

P site *See* peptidyl site.

plaque A clear area on a bacterial lawn, left by lysis of the bacteria through progressive infections by a phage and its descendants.

plasmid An autonomously replicating extrachromosomal DNA molecule.

plating Spreading the cells of a microorganism (bacteria, fungi) on a dish of nutritive medium to allow each cell to form a visible colony.

pleiotropic allele An allele that affects several different properties of an organism.

point mutation A mutation that alters a single base position in a DNA molecule by converting it to a different base or by the insert/deletion of a single base in a DNA molecule.

Poisson distribution A mathematical distribution giving the probability of observing various numbers of a particular event in a sample when the mean probability of an event on any one trial is very small.

pol III holoenzyme *See* DNA polymerase III holoenzyme.

poly(A) polymerase (PAP) An enzyme that catalyzes template-independent addition of AMP from ATP to the $3'$ end of mRNAs.

poly(A) tail A string of adenine nucleotides added to mRNA after transcription.

polyadenylation The process of adding a poly(A) tail to an mRNA.

polygene (quantitative trait locus) A gene whose alleles are capable of interacting additively with alleles at other loci to affect a phenotype (trait) showing continuous distribution.

polylinker *See* **multiple cloning site (MCS).**

polymerase chain reaction (PCR) An in vitro method for amplifying a specific DNA segment that uses two primers that hybridize to opposite ends of the segment in opposite polarity and, over successive cycles, prime exponential replication of that segment only.

polymerase III holoenzyme *See* **DNA polymerase III holoenzyme.**

polymorphism The occurrence in a population of multiple forms of a trait or multiple alleles at a genetic locus.

polypeptide A chain of linked amino acids; a protein.

polyploid A cell having three or more chromosome sets or an organism composed of such cells.

polytene chromosome A giant chromosome in specific tissues of some insects, produced by an endomitotic process in which the multiple DNA sets remain bound in a haploid number of chromosomes.

population (1) A group of individuals that mate with one another to produce the next generation. (2) A group of individuals from which a sample is drawn.

population genetics The study of genetic variation in populations and changes over time in the amount or patterning of that variation resulting from mutation, migration, recombination, random genetic drift, natural selection, and mating systems.

population structure The division of a species or population into multiple genetically distinct subpopulations.

positional information The process by which chemical cues that establish cell fate along a geographic axis are established in a developing embryo or tissue primordium.

position effect Describes a situation in which the phenotypic influence of a gene is altered by changes in the position of the gene within the genome.

position-effect variegation (PEV) Variegation caused by the inactivation of a gene in some cells through its abnormal juxtaposition with heterochromatin.

positive assortative mating A situation in which like phenotypes mate more commonly than expected by chance.

positive regulation Regulation mediated by a protein that is required for the activation of a transcription unit.

positive selection The process by which a favorable allele is brought to a higher frequency in a population because individuals carrying that allele have more viable offspring than other individuals.

postmating, prezygotic isolating barrier A reproductive barrier that acts after mating between species has occurred, but before hybrid zygotes have formed.

postzygotic isolating barrier A reproductive barrier between species that acts after hybrid zygotes have formed.

post-translational modification An alteration of amino acid residues that occurs after the protein has been translated.

precision medicine A form of medicine that uses information about a person's genes, proteins, and environment to prevent, diagnose, and treat disease.

precursor RNA (pre-RNA) A nascent RNA that has not yet been processed.

preinitiation complex (PIC) A very large eukaryotic protein complex comprising RNA polymerase II and the six general transcription factors (GTFs), each of which is a multiprotein complex.

premating isolating barrier A reproductive barrier that prevents mating between species.

primary structure (of a protein) The sequence of amino acids in the polypeptide chain.

primase An enzyme that makes RNA primers in DNA replication.

primer An RNA or DNA oligonucleotide that can serve as a substrate for extension by DNA polymerase when annealed to a longer DNA molecule.

primosome A protein complex at the replication fork whose central component is primase.

probe Labeled nucleic acid segment that can be used to identify specific DNA molecules bearing the complementary sequence, usually through autoradiography or fluorescence.

processed pseudogene A pseudogene that arose by the reverse transcription of a mature mRNA and its integration into the genome.

processive enzyme As used in Chapter 7, describes the behavior of DNA polymerase III, which can perform thousands of rounds of catalysis without dissociating from its substrate (the template DNA strand).

product rule The probability of two independent events occurring simultaneously is the product of the individual probabilities.

prokaryote A single-celled organism that lacks a membrane-bound nucleus and other membrane-bound organelles.

promoter A DNA sequence that defines where transcription by RNA polymerase will begin.

prophage A phage "chromosome" inserted as part of the linear structure of the DNA chromosome of a bacterium.

propositus In a human pedigree, the person who first came to the attention of the geneticist.

proteasome A protein complex containing proteases that degrades proteins.

proteome The complete set of proteins expressed in a cell, tissue, or organism.

prototroph A strain of organisms that will proliferate on minimal medium (*compare* **auxotroph**).

provirus The chromosomally inserted DNA genome of a retrovirus.

proximal enhancer An enhancer that is located close to the transcription start site.

proximal promoter *See* **proximal enhancer.**

pseudoautosomal regions 1 and 2 Small regions at the ends of the X and Y sex chromosomes; they are homologous and undergo pairing and crossing over at meiosis.

pseudodominance The sudden appearance of a recessive phenotype in a pedigree, due to the deletion of a masking dominant gene.

pseudogene A mutationally inactive gene for which no functional counterpart exists in wild-type populations.

pseudolinkage The appearance of linkage of two genes on translocated chromosomes.

pure line A population of individuals all bearing the identical fully homozygous genotype.

purifying selection Natural selection that removes deleterious variants of a DNA or protein sequence, thus reducing genetic diversity.

purine A type of nitrogen base; the purine bases in DNA are adenine and guanine.

pyrimidine A type of nitrogen base; the pyrimidine bases in DNA are cytosine and thymine.

pyrimidine dimer Covalent linkages between thymine or cytosine bases.

qPCR *See* **quantitative PCR.**

QTL *See* **quantitative trait locus.**

QTL mapping A method for locating QTL in the genome and characterizing the effects of QTL on trait variation.

quantitative genetics The subfield of genetics that studies the inheritance of complex or quantitative traits.

quantitative PCR (qPCR) A method to measure the amount of a specific DNA molecule in a sample.

quantitative trait Any trait exhibiting complex inheritance because it is controlled by a mix of genetic and/or environmental factors.

quantitative trait locus (QTL) A gene contributing to the phenotypic variation in a trait that shows complex inheritance, such as height and weight.

quantitative trait locus mapping *See* **QTL mapping.**

quaternary structure (of a protein) The multimeric constitution of a protein.

radioisotope An unstable form of an element that emits radiation.

random genetic drift Changes in allele frequency in a population resulting from chance differences in the actual numbers of offspring of different genotypes produced by different individual members.

rare SNP A single nucleotide polymorphism (SNP) for which the less common allele occurs at a frequency below 5 percent.

reactive group (R group) The unique function group of an amino acid.

reading frame The consecutive set of nonoverlapping nucleotide triplets that equate to amino acids or stop signals during translation.

rearrangement The production of abnormal chromosomes by the breakage and incorrect rejoining of chromosomal segments; examples are inversions, deletions, and translocations.

recessive Refers to an allele whose phenotypic effect is not expressed in a heterozygote.

recipient The bacterial cell that receives DNA in a unilateral transfer between cells; examples are F⁻ in a conjugation or the transduced cell in a phage-mediated transduction.

recombinant Refers to an individual organism or cell having a genotype produced by recombination.

recombinant DNA A DNA molecule generated in the laboratory that brings together pieces of DNA from multiple sources.

recombinant frequency (RF) The proportion (or percentage) of recombinant cells or individuals.

recombination (1) In general, any process in a diploid or partly diploid cell that generates new gene or chromosomal combinations not previously found in that cell or in its progenitors. (2) At meiosis, the process that generates a haploid product of meiosis whose genotype is different from either of the two haploid genotypes that constituted the meiotic diploid.

recombination map A chromosome map in which the positions of loci shown are based on recombinant frequencies.

regulatory element DNA sequence motif that influences the timing, cell, or tissue specificity, or level of expression of a gene.

regulon Genes that are transcribed in a manner that is coordinated by the same regulatory protein (for example, sigma factor).

relative fitness A measure of the fitness of an individual or genotype relative to some other individual or genotype, usually the most fit individual or genotype in the population.

release factor (RF) A protein that binds to the A site of the ribosome when a stop codon is in the mRNA.

replication fork The point at which the two strands of DNA are separated to allow the replication of each strand.

replication slippage A DNA replication mechanism that leads to indel mutations.

replicative transposition A mechanism of transposition that generates a new insertion element integrated elsewhere in the genome while leaving the original element at its original site of insertion.

replisome The molecular machine at the replication fork that coordinates the numerous reactions necessary for the rapid and accurate replication of DNA.

repression domain A region of a transcription factor that inhibits transcription.

repressor A protein that binds to a cis-acting element such as an operator or a silencer, thereby preventing transcription from an adjacent promoter.

reproductive isolating barrier Any mechanism that reduces gene flow between populations or species.

reproductive isolation A reduction in the amount of gene flow between populations or species.

resistant mutant A mutant that can grow in a normally toxic environment.

restriction enzyme An endonuclease that will recognize specific target nucleotide sequences in DNA and break the DNA chain at those points; a variety of these enzymes are known, and they are used in genetic engineering.

restriction fragment A DNA fragment resulting from cutting DNA with a restriction enzyme.

restriction map A map of the restriction sites in a piece of DNA.

restriction site The DNA sequence that is recognized and cut by a restriction enzyme.

restrictive temperature The temperature at which a temperature-sensitive mutation expresses the mutant phenotype.

retrotransposition A mechanism of transposition characterized by the reverse flow of information from RNA to DNA.

retrotransposon A transposable element that uses reverse transcriptase to transpose through an RNA intermediate. *See also* **class 1 element.**

retrovirus An RNA virus that replicates by first being converted into double-stranded DNA.

reverse genetics An experimental procedure that begins with a cloned segment of DNA or a protein sequence and uses it (through directed mutagenesis) to introduce programmed mutations back into the genome to investigate function.

reverse transcriptase An enzyme that catalyzes the synthesis of a DNA strand from an RNA template.

reverse transcription-PCR (RT-PCR) A method to amplify an RNA sequence.

revertant An allele with wild-type function arising by the mutation of a mutant allele; caused either by a complete reversal of the original event or by a compensatory second-site mutation.

RF *See* **recombinant frequency; release factor.**

Rho-dependent termination A type of transcription termination mechanism in bacteria that involves the protein Rho.

ribonucleic acid *See* **RNA.**

ribose The pentose sugar of RNA.

ribosomal RNA (rRNA) Several different noncoding RNAs that are components of the ribosome and are essential for protein synthesis.

ribosome A complex of RNAs and proteins that catalyzes the synthesis of proteins using mRNA as a template.

ribosome-binding site (RBS) A sequence in a bacterial mRNA responsible for recruitment of a ribosome upstream of the start codon to initiate translation.

ribozyme An RNA with enzymatic activity—for instance, the self-splicing RNA molecules in *Tetrahymena*.

RNA (ribonucleic acid) A nucleic acid similar to DNA but having ribose sugar rather than deoxyribose sugar and uracil rather than thymine as one of the bases.

RNA editing Molecular processes that modify specific nucleotides in synthesized RNA molecules.

RNAi *See* **RNA interference.**

RNA interference (RNAi) A system in some eukaryotes to control the expression of genes through the action of siRNAs. *See also* **gene silencing.**

RNA polymerase An enzyme that catalyzes the synthesis of an RNA strand from a DNA template. Eukaryotes possess several classes of RNA polymerase; structural genes encoding proteins are transcribed by RNA polymerase II.

RNA polymerase I An enzyme in eukaryotic cells that is specialized to synthesize rRNA.

RNA polymerase II An enzyme in eukaryotic cells that is specialized to synthesize mRNA and a variety of noncoding RNAs.

RNA polymerase III An enzyme in eukaryotic cells that is specialized to synthesize tRNAs, 5S rRNA, and a variety of noncoding RNAs.

RNA polymerase core enzyme Bacterial RNA polymerase without sigma (σ) factor.

RNA polymerase holoenzyme The bacterial multisubunit complex composed of the four subunits of the core enzyme plus the σ factor.

RNA processing The collective term for co-transcriptional and post-transcriptional modifications to RNA, such as mRNA capping and splicing in eukaryotes.

RNA-seq *See* **RNA sequencing.**

RNA sequencing (RNA-seq) A method used to determine the transcribed regions of a genome within some specific cell population, tissue sample, or organism.

Robertsonian translocation A chromosomal rearrangement that involves breakage of two chromosomes at or near their centromeres and subsequent fusions of the long arms of the two chromosomes.

rolling circle replication A mode of replication used by some circular DNA molecules in bacteria (such as plasmids) in which the circle seems to rotate as it reels out one continuous leading strand.

R plasmid A plasmid containing one or several transposons that bear resistance genes.

rRNA *See* **ribosomal RNA.**

RT-PCR *See* **reverse transcription-PCR.**

s See **selection coefficient.**

S See **segregating sites** or **selection differential.**

safe haven A site in the genome where the insertion of a transposable element is unlikely to cause a mutation, thus preventing harm to the host.

sample A small group of individual members or observations meant to be representative of a larger population from which the group has been taken.

Sanger sequencing *See* **dideoxy (Sanger) sequencing.**

scaffold (1) The central framework of a chromosome to which the DNA solenoid is attached as loops; composed largely of topoisomerase. (2) In genome projects, an ordered set of contigs in which there may be unsequenced gaps connected by paired-end sequence reads.

screen A mutagenesis procedure in which essentially all mutagenized progeny are recovered and are individually evaluated for mutant phenotype; often the desired phenotype is marked in some way to enable its detection.

SDSA *See* **synthesis-dependent strand annealing.**

secondary structure (of a protein) Local regions of protein folding into specific shapes such as an α-helix and β-sheet.

second-division segregation pattern (MII pattern) A pattern of ascospore genotypes for a gene pair showing that the two alleles separate into different nuclei only at the second meiotic division, as a result of a crossover between that gene pair and its centromere; can be detected only in a linear ascus.

second filial generation (F_2) The progeny of a cross between two individuals from the F_1 generation.

segmental duplication Presence of two or more large nontandem repeats.

segment-polarity gene In *Drosophila*, a member of a class of genes that contribute to the final aspects of establishing the correct number of segments. Segment-polarity mutations cause a loss of or change in a comparable part of each of the body segments.

segregating sites (S) The number of variable or polymorphic nucleotide sites in a set of homologous DNA sequences.

selection coefficient (s) The loss of fitness in (or selective disadvantage of) one genotype relative to another genotype.

selection differential (S) The difference between the mean of a population and the mean of the individual members selected to be parents of the next generation.

selection response (R) The amount of change in the average value of some phenotypic character between the parental generation and the offspring generation as a result of the selection of parents.

selective system A mutational selection technique that enriches the frequency of specific (usually rare) genotypes by establishing environmental conditions that prevent the growth or survival of other genotypes.

self To fertilize eggs with sperms from the same individual.

selfish genetic element Any sequence that enhances its own transmission to the next generation at the expense of other sequences in the genome, even if it has no positive or negative effects on organismal fitness.

semiconservative replication The established model of DNA replication in which each double-stranded molecule is composed of one parental strand and one newly polymerized strand.

semidiscontinuous A term used to describe the fact that DNA replication is continuous on one strand and discontinuous on the other.

semisterility The phenotype of an organism heterozygotic for certain types of chromosome aberration; expressed as a reduced number of viable gametes and hence reduced fertility.

sequence assembly The compilation of thousands or millions of independent DNA sequence reads into a set of contigs and scaffolds.

sequence contig A group of overlapping cloned segments.

serially reiterated structures Body parts that are members of repeated series, such as digits, ribs, teeth, limbs, and segments.

sex chromosome A chromosome whose presence or absence is correlated with the sex of the bearer; a chromosome that plays a role in sex determination.

sex linkage The location of a gene on a sex chromosome.

sexual cell division The two successive cell divisions of a meiocyte that produce sex cells; sexual cell division is accompanied by the two successive nuclear divisions called meiosis.

sgRNA *See* **single guide RNA.**

Shine–Dalgarno sequence A short sequence in bacterial mRNA that precedes the initiation AUG and base pairs to the 3′ end of the 16S RNA in the 30S ribosomal subunit. This positions the AUG codon in the P site of the ribosome.

short interspersed element (SINE) A type of class 1 transposable element that does not encode reverse transcriptase but is thought to use the reverse transcriptase encoded by LINEs. *See also* **Alu.**

side chain *See* **reactive group.**

sigma factor (σ) A bacterial protein that, as part of the RNA polymerase holoenzyme, recognizes the −10 and −35 regions of bacterial promoters, thus positioning the holoenzyme to initiate transcription correctly at the start site. The σ factor dissociates from the holoenzyme before RNA synthesis.

signal sequence The amino-terminal sequence of a secreted protein; it is required for the transport of the protein through the cell membrane.

silent mutation *See* **synonymous (silent) mutation.**

simple inheritance A form of inheritance in which only one (or a few) genes are involved and the environment has little or no effect on the phenotype; categorical traits often exhibit simple inheritance.

simple transposon A type of bacterial transposable element containing a variety of genes that reside between short inverted repeat sequences.

SINE *See* **short interspersed element.**

single guide RNA (sgRNA) A designed small RNA that directs DNA cleavage by the CRISPR-Cas9 editing technology.

single nucleotide polymorphism (SNP) (snip) A nucleotide-pair difference at a given location in the genomes of two or more naturally occurring individuals.

single-strand DNA-binding (SSB) protein A protein that binds to DNA single strands and prevents the duplex from re-forming before replication.

siRNA *See* **small interfering RNA.**

sliding clamp *See* **β (beta) clamp.**

small interfering RNA (siRNA) Short double-stranded RNAs produced by the cleavage of long double-stranded RNAs by Dicer.

small nuclear ribonucleoprotein (snRNP) A complex containing an snRNA and proteins that resides in the nucleus.

small nuclear RNA (snRNA) Any of several short noncoding RNAs found in the eukaryotic nucleus, where they assist in RNA processing events.

small ribosomal subunit The part of the ribosome that contains the decoding center where tRNAs base pair with mRNA.

SNP *See* **single nucleotide polymorphism.**

snRNA *See* **small nuclear RNA.**

snRNP *See* **small nuclear ribonucleoprotein.**

solo LTR A single copy of an *LTR*.

somatic cell division The division of a body (somatic) cell into two daughter cells. Somatic cell division is accompanied by the nuclear division called mitosis.

somatic mutation A mutation in a cell that can be passed on to daughter cells in the course of cell division.

SOS response A global response induced when bacteria are exposed to stresses that cause substantial DNA damage.

Southern blotting A method to detect a particular fragment of DNA in a mixture of many DNA fragments by hybridization with a complementary nucleic acid probe.

specialized transduction The situation in which a particular phage will transduce only specific regions of the bacterial chromosome.

spliceosome The ribonucleoprotein processing complex that removes introns from eukaryotic pre-mRNAs.

splicing A reaction that removes introns and joins together exons in pre-RNA.

spontaneous mutation A mutation occurring in the absence of exposure to mutagens.

***SRY* gene** The maleness gene, residing on the Y chromosome.

SSB *See* **single-strand-binding protein.**

standard deviation The square root of the variance.

stop codon A three-nucleotide sequence (UAA, UAG, and UGA) in mRNA that terminates translation.

subfunctionalization A path of gene duplication and mutation that produces *paralogs* with complementary functions.

subunit A single polypeptide in a protein containing multiple polypeptides.

sum rule The probability that one or the other of two mutually exclusive events will occur is the sum of their individual probabilities.

supercontig *See* **scaffold** (2).

suppressor A secondary mutation that can cancel the effect of a primary mutation, resulting in wild-type phenotype.

synonymous codon A codon that codes for the same amino acid as another codon.

synonymous (silent) mutation A change to the sequence of a codon that does not change the encoded amino acid.

synonymous substitution *See* **synonymous (silent) mutation.**

synteny A situation in which genes are arranged in similar blocks in different species.

synthesis-dependent strand annealing (SDSA) An error-free mechanism for correcting double-strand breaks that occur after the replication of a chromosomal region in a dividing cell.

synthetic lethal Refers to a double mutant that is lethal, whereas the component single mutations are not.

T *See* **thymine.**

t-loop *See* **telomeric loop.**

tandem duplication Adjacent identical chromosome segments.

targeting A feature of certain transposable elements that facilitates their insertion into regions of the genome where they are not likely to insert into a gene causing a mutation.

target-site duplication A short direct-repeat DNA sequence (typically from 2 to 10 bp in length) adjacent to the ends of a transposable element that was generated during the element's integration into the host chromosome.

TATA box A DNA sequence found in many eukaryotic genes that is located about 30 bp upstream of the transcription start site.

tautomerization The spontaneous isomerization of a nitrogen base from its normal keto form to an alternative hydrogen-bonding enol (or imino) form.

TC-NER *See* **transcription-coupled nucleotide excision repair.**

telomerase An enzyme that, with the use of a small RNA as a template, adds repetitive units to the ends of linear chromosomes to prevent shortening after replication.

telomere The tip, or end, of a chromosome.

telomeric loop (t-loop) A structure at the end of a telomere that masks the end from the factors that detect DNA breaks.

temperate phage A phage that can become a prophage.

temperature-sensitive (ts) mutation A conditional mutation that produces the mutant phenotype in one temperature range and the wild-type phenotype in another temperature range.

template A single-stranded DNA or RNA molecule that directs the synthesis of another molecule; for example, the nucleotide sequence of DNA acts as a template to determine the nucleotide sequence of RNA during transcription.

template strand (noncoding strand) The strand of DNA that is copied (i.e., transcribed) by RNA polymerase.

termination The last stage of transcription; it results in the release of the RNA and RNA polymerase from the DNA template.

termination codon *See* **stop codon.**

termination factor A protein involved in stopping translation and releasing the ribosome and newly synthesized protein.

terminus The end represented by the last added monomer in the unidirectional synthesis of a polymer such as RNA or a polypeptide.

tertiary structure (of a protein) The overall three-dimensional shape of a polypeptide.

testcross A cross of an individual organism of unknown genotype or a heterozygote (or a multiple heterozygote) with a tester.

tester An individual organism homozygous for one or more recessive alleles; used in a testcross.

tetraploid A cell having four chromosome sets; an organism composed of such cells.

three-point testcross (three-factor testcross) A testcross in which one parent has three heterozygous gene pairs.

threshold trait A categorical trait for which the expression of the different phenotypic states depends on a combination of multiple genetic and/or environmental factors that place an individual above or below a critical value for trait expression.

thymine (T) A pyrimidine base that pairs with adenine.

Ti plasmid A circular plasmid of *Agrobacterium tumifaciens* that enables the bacterium to infect plant cells and produce a tumor (crown gall tumor).

TLS *See* **translesion synthesis.**

TLS polymerase *See* **translesion polymerase.**

Tn *See* **transposon.**

topoisomerase An enzyme that can cut and re-form polynucleotide backbones in DNA to allow it to assume a more relaxed configuration.

topologically associating domain (TAD) Genomic regions that preferentially contact each other and are bound by proteins such as CTCF.

torpedo termination model A model proposing that transcription termination is caused by displacement of RNA polymerase II from DNA by an exonuclease.

trait More or less synonymous with character.

trans-acting Refers to a diffusible regulatory molecule (almost always a protein) that binds to a specific cis-acting element. The term trans-acting indicates that the factor can regulate all downstream genes, whether residing on the same or different DNA molecules.

trans conformation In a heterozygote with two mutant sites within a gene or gene cluster, the arrangement a_1 +/+ a_2.

transcript The RNA molecule copied from the DNA template strand by RNA polymerase.

transcription The synthesis of RNA from a DNA template.

transcription bubble The site at which the double helix is unwound so that RNA polymerase can use one of the DNA strands as a template for RNA synthesis.

transcription factor A protein that contains a sequence-specific DNA binding domain and an activation/repression domain that regulates transcription by binding enhancer elements and interacting with coregulators.

transcription-coupled nucleotide excision repair (TC-NER) A form of nucleotide excision repair that is activated by stalled transcription complexes and corrects DNA damage in transcribed regions of the genome.

transcription start site The location in a gene where transcription begins.

transcriptome The set of RNAs expressed in cells, tissues, or organisms.

transduction The movement of genes from a bacterial donor to a bacterial recipient with a phage as the vector.

transfer RNA (tRNA) A class of small RNA molecules that carry specific amino acids to the ribosome in the course of translation; an amino acid is inserted into the growing polypeptide chain when the anticodon of the corresponding tRNA pairs with a codon on the mRNA being translated.

transformation The directed modification of a genome by the external application of DNA from a cell of different genotype.

transgene A gene that has been modified by externally applied recombinant DNA techniques and reintroduced into the genome by germ-line transformation.

transgenic organism An organism whose genome has been modified by externally applied new DNA.

transition A type of nucleotide-pair substitution in which a purine replaces another purine or in which a pyrimidine replaces another pyrimidine—for example, G–C to A–T.

translation The ribosome- and tRNA-mediated production of a polypeptide whose amino acid sequence is derived from the codon sequence of an mRNA molecule.

translesion (TLS) polymerase A DNA polymerase that can continue to replicate DNA past a site of damage that would halt replication by the normal replicative polymerase. Translesion polymerases contribute to a damage-tolerance mechanism called translesion synthesis (TLS).

translesion synthesis (TLS) A damage-tolerance mechanism in eukaryotes that uses translesion polymerases to replicate DNA past a site of damage.

translocation The relocation of a chromosomal segment to a different position in the genome.

transposable element *See* **transposon.**

transposase An enzyme encoded by transposable elements that undergo conservative transposition.

transpose To move from one location in the genome to another; said of a mobile genetic element.

transposition A process by which mobile genetic elements move from one location in the genome to another.

transposon A mobile piece of DNA that is flanked by terminal repeat sequences and typically bears genes encoding transposition functions. Bacterial transposons can be simple or composite.

transversion A type of nucleotide-pair substitution in which a pyrimidine replaces a purine or vice versa—for example, G–C to T–A.

trinucleotide-repeat disease A disease caused by a kind of mutation where the number of three-nucleotide repeats is increased over a stable threshold.

triplet Three nucleotide pairs that compose a codon.

triploid A cell having three chromosome sets or an organism composed of such cells.

trisomic Basically a diploid with an extra chromosome of one type, producing a chromosome number of the form $2n + 1$.

trivalent Refers to the meiotic pairing arrangement of three homologs in a triploid or trisomic.

tRNA *See* **transfer RNA.**

ts mutation *See* **temperature-sensitive mutation.**

Turner syndrome An abnormal human female phenotype produced by the presence of only one X chromosome (XO).

two-hybrid test A method for detecting protein–protein interactions, typically performed in yeast.

Ty **element** A yeast LTR-retrotransposon; the first isolated from any organism.

U *See* **uracil.**

UAS *See* **upstream activation sequence.**

ubiquitin A protein that, when attached as a multicopy chain to another protein, targets that protein for degradation by a protease called the 26S proteasome. The addition of a single ubiquitin molecule to a protein can change protein–protein interactions, as in the case of PCNA and bypass polymerases.

ubiquitination The process of adding ubiquitin chains to a protein targeted for degradation.

unbalanced rearrangement A rearrangement in which chromosomal material is gained or lost in one chromosome set.

unequal crossing over Crossover involving improper synaptic pairing of tandemly duplicated DNA segments, resulting in one chromatid with one copy of the segment and another with three copies. *See also* **nonallelic homologous recombination.**

uniparental inheritance Inheritance pattern in which the progeny have the genotype and phenotype of one parent only; for example, inheritance of mitochrondrial genomes.

univalent A single unpaired meiotic chromosome, as is often found in trisomics and triploids.

unselected marker In a bacterial recombination experiment, an allele scored in progeny for the frequency of its cosegregation with a linked selected allele.

unstable allele an allele that can change states; for example, between producing or not producing a functional product. Often these are mutant alleles with a high rate of reversion to wild type.

upstream Refers to a DNA or RNA sequence located on the 5′ side of a point of reference.

upstream activation sequence (UAS) A DNA sequence of yeast located 5′ of the gene promoter; a transcription factor binds to the UAS to positively regulate gene expression.

uracil (U) A pyrimidine base in RNA in place of the thymine found in DNA.

UTR *See* 3′ untranslated region; 5′ untranslated region.

variance A statistical measure used to quantify the degree to which the trait values of individuals deviate from the population mean.

variant histone Non-allelic isoforms of canonical histones that are incorporated into chromatin outside of S-phase and perform specialized functions.

vector *See* cloning vector.

vertical transmission Inheritance of DNA from a member of a previous generation.

virulent phage A phage that cannot become a prophage; infection by such a phage always leads to lysis of the host cell.

virus A particle consisting of nucleic acid and protein that must infect a living cell to replicate and reproduce.

Western blotting A method to detect a particular protein in a mixture of many proteins by interaction with an antibody probe.

whole genome duplication The duplication of an entire genome leading to polyploidy.

whole-genome shotgun (WGS) sequencing A strategy for obtaining the sequence of a genome by sequencing and assembling many short segments of genomic DNA.

wild type The genotype or phenotype that is found in nature or in the standard laboratory stock for a given organism.

wobble Pairing between two RNA bases that does not follow the Watson-Crick base pair rules; wobble pairing often occurs between tRNA anticodon bases and mRNA codon bases during translation.

X chromosome One of a pair of sex chromosomes, distinguished from the Y chromosome.

X-chromosome inactivation The process by which one of the two X-chromosomes in female cells is transcriptionally silenced.

X linkage The inheritance pattern of genes found on the X chromosome but not on the Y chromosome.

Y chromosome One of a pair of sex chromosomes, distinguished from the X chromosome.

Y linkage The inheritance pattern of genes found on the Y chromosome but not on the X chromosome (rare).

zygote A cell formed by the fusion of an egg and a sperm; the unique diploid cell that will divide mitotically to create a differentiated diploid organism.

zygotic induction The sudden release of a lysogenic phage from an Hfr chromosome when the prophage enters the F⁻ cell followed by the subsequent lysis of the recipient cell.

Answers to Selected Problems

This section includes selected answers to Basic Problems and Challenging Problems. Answers to all the problems are available in the Solutions Manual.

CHAPTER 1

1. Approximately equal proportions of purple and white.
4. **a.** Blue spheres are atoms in the sugar phosphate backbone.
   b. Brown slabs are nucleotide bases.
6. Archaea.
11. The complementary strand reads CAAGCGCCGGCGCTTG. The two strands are both complementary and palindromic.
13. 68 cytosines, 68 guanines, and 32 thymines.

CHAPTER 2

1. *Ap3* has no petals. *Ag* has no stamens.
5. $651 \times 2/3 = 434$
8. At meiosis, the replicated homologs are paired. At mitosis, each homolog is by itself.
12. One possibility is that a mutation in a non-active site could affect the folding (and thus the function) of the protein.
16. Half the males would have red eyes and half would have white eyes. All females would have red eyes. Half of them would be homozygous for the wild-type allele, and half would be heterozygous.
18. (1) No.
   (2) No; small sample sizes often do not concur with expected frequencies.
19. *PARK7, HTT, PAH, EXA, ASPA, PRNP, NF2*
21. No.
22. The male III-2 may have been misdiagnosed, or in his genome the dominant allele may have been suppressed by other genes.
23. II-1 must be *Tt*; III-11 is most likely *TT*.
25. (c) The cross in generation V must have been $Aa \times Aa$, so the probability of having five affected children would be $\frac{1}{4} \times \frac{1}{4} \times \frac{1}{4} \times \frac{1}{4} \times \frac{1}{4} = 1/1024$. This result could be purely by chance, or there might be another, more subtle, biological reason.
27. Gel electrophoresis separates DNA molecules by size. When DNA is carefully isolated from *Neurospora* (which has seven different chromosomes), seven bands should be produced with the use of this technique. Similarly, the pea has seven different chromosomes and will produce seven bands (homologous chromosomes will comigrate as a single band).
30. The key function of mitosis is to generate two daughter cells genetically identical with the original parent cell.
34. As cells divide mitotically, each chromosome consists of identical sister chromatids that are separated to form genetically identical daughter cells. Although the second division of meiosis appears to be a similar process, the "sister" chromatids are likely to be different from each other. Recombination in earlier meiotic stages will have swapped regions of DNA between sister and nonsister chromosomes such that the two daughter cells of this division are typically not genetically identical.
38. Yes. Half of our genetic makeup is derived from each parent, half of each parent's genetic makeup is derived from half of each of their parents', etc.
42. (5) Chromosome pairing.
47. The progeny ratio is approximately 3:1, indicating classic heterozygous-by-heterozygous mating. Because Black (*B*) is dominant over white (*b*),

   Parents: $B/b \times B/b$
   Progeny: 3 black : 1 white ($1\,B/B : 2\,B/b : 1\,b/b$)

51. The fact that about half of the F₁ progeny are mutant suggests that the mutation that results in three cotyledons is dominant and the original mutant was heterozygous. If C = the mutant allele and c = the wild-type allele, the cross is as follows:

   P $C/c \times c/c$
   F₁ C/c three cotyledons
         c/c two cotyledons

56. p (child has galactosemia) = p (John is G/g) $\times p$ (Martha is G/g) $\times p$ (both parents passed g to the child) = $(2/3)(1/4)(1/4)$ = $2/48$ = $1/24$
62. **a.** The disorder appears to be dominant because all affected individuals have an affected parent. If the trait were recessive, then I-1, II-2, III-1, and III-8 would all have to be carriers (heterozygous for the rare allele).
   b. With the assumption of dominance, the genotypes are

   I: $d/d, D/d$
   II: $D/d, d/d, D/d, d/d$
   III: $d/d, D/d, d/d, D/d, d/d, d/d, D/d, d/d$
   IV: $D/d, d/d, D/d, d/d, d/d, d/d, d/d, D/d, d/d$

   c. The probability of an affected child (D/d) equals 1/2, and the probability of an unaffected child (d/d) equals 1/2. Therefore, the chance of having four unaffected children (since each is an independent event) is $(1/2) \times (1/2) \times (1/2) \times (1/2) = 1/16$.
68. **a.** Sons inherit the X chromosome from their mothers. The mother has free earlobes; the son has attached earlobes. If the allele for free earlobes is dominant and the allele for attached earlobes is recessive, then the mother could be heterozygous for this trait and the gene could be X linked.
   b. It is not possible from the data given to decide which allele is dominant. If attached earlobes is dominant, then the father would be heterozygous and the son would have a 50% chance of inheriting the dominant attached earlobes allele. If attached earlobes is recessive, then the trait could be autosomal or X linked, but, in either case, the mother would be heterozygous.
72. Let H = hypophosphatemia and h = normal. The cross is $H/Y \times h/h$, yielding Hh (females) and h/Y (males). The answer is 0%.
77. **a.** $X^C/X^c, X^c/X^c$
   b. p (color-blind) $\times p$ (male) = $(1/2)(1/2) = 1/4$
   c. The girls will be 1 normal (X^C/X^C) : 1 color-blind (X^c/X^c).
   d. The cross is $X^C/X^c \times X^c/Y$, yielding 1 normal:1 color-blind for both sexes.
85. **a.** The pedigree suggests that the allele causing red hair is recessive because most red-haired individuals are from parents without this trait.
   b. Observation of those around us makes the allele appear to be somewhat rare.
89. Note that only males are affected and that, in all but one case, the trait can be traced through the female side. However, there is one example of an affected male having affected sons. If the trait is X linked, this male's wife must be a carrier. Depending on how rare this trait is in the general population, that could be unlikely, suggesting that the disorder is caused by an autosomal dominant allele with expression limited to males.

CHAPTER 3

1. **a.** 0.05
   b. 0.01
   c. p values decrease as df decreases.
5. Anaphase I.
9. *AAaa*.
11. No, because all progeny would be one phenotype.
12. Left hand parent, right hand parent, and tester.

18. Trihybrid selfs have $\frac{1}{4} \times \frac{1}{4} \times \frac{1}{4}$ fully recessive progeny, which $= 1/64$.

21. Some green, some white, some variegated.

23. Affected fathers do not transmit the disease.

25. The genotype of the daughter cells will be identical with that of the original cell: (f) A/a ; B/b.

30. Mitosis produces daughter cells having the same genotype as that of the original cell: A/a ; B/b ; C/c.

33. His children will have to inherit the satellite-containing 4 (probability = 1/2), the abnormally staining 7 (probability = 1/2), and the Y chromosome (probability = 1/2). To inherit all three, the probability is $(1/2)(1/2)(1/2) = 1/8$.

38. With the assumption of independent assortment and simple dominant–recessive relations of all genes, the number of genotypic classes expected from selfing a plant heterozygous for n gene pairs is $3n$ and the number of phenotypic classes expected is $2n$.

41. a. and **b.** Cross 2 indicates that purple (G) is dominant over green (g), and cross 1 indicates that cut (P) is dominant over potato (p).

Cross 1: G/g ; $P/p \times g/g$; P/p	There are 3 cut : 1 potato, and 1 purple : 1 green.
Cross 2: G/g ; $P/p \times G/g$; p/p	There are 3 purple : 1 green, and 1 cut : 1 potato.
Cross 3: G/g ; $P/p \times g/g$; P/p	There is no green, and there are 3 cut : 1 potato.
Cross 4: G/g ; $P/P \times g/g$; p/p	There is no potato, and there is 1 purple : 1 green.
Cross 5: G/g ; $p/p \times g/g$; P/p	There is 1 cut : 1 potato, and there is 1 purple : 1 green.

46. The crosses are
Cross 1:   stop-start female × wild-type male →
                                   all stop-start progeny
As mitochondrial DNA is inherited only through the female in *Neurospora*, the progeny display the maternal mitochondrial phenotype.
Cross 2:   wild-type female × stop-start male →
                                   all wild-type progeny
The reciprocal will result in a different outcome, wild-type progeny. Again, mitochondria are inherited through the maternal parent and gamete.

53. a. There should be nine classes corresponding to 0, 1, 2, 3, 4, 5, 6, 7, 8 "doses."

b. There should be 13 classes corresponding to 0, 1, 2, 3, 4, 5, 6, 7, 8, 9, 10, 11, 12 "doses."

62. a. and **b.** Begin with any two of the three lines and cross them. If, for example, you began with a/a ; B/B ; $C/C \times A/A$; b/b ; C/C, all the progeny would be A/a; B/b ; C/C. Crossing two of them would yield

   9   $A/-$; $B/-$; C/C
   3   a/a ; $B/-$; C/C
   3   $A/-$; b/b ; C/C
   1   a/a ; b/b ; C/C

The a/a ; b/b ; C/C genotype has two of the genes in a homozygous recessive state and is found in 1/16 of the offspring. If that genotype were crossed with A/A ; B/B ; c/c, all the progeny would be A/a ; B/b ; C/c. Crossing two of them (or "selfing") would lead to a $27 : 9 : 9 : 9 : 3 : 3 : 3 : 1$ ratio, and 1/64 of the progeny would be the desired a/a ; b/b ; c/c.

There are several different routes to obtaining a/a ; b/b ; c/c, but the one just outlined requires only four crosses.

69. a. Let B = brachydactylous, b = normal, T = taster, and t = nontaster. The genotypes of the couple are B/b ; T/t for the male and b/b ; T/t for the female.

b. For all four children to be brachydactylous, $p = (1/2)^4 = 1/16$.

c. For none of the four children to be brachydactylous, $p = (1/2)^4 = 1/16$.

d. For all to be tasters, $p = (3/4)^4 = 81/256$.

e. For all to be nontasters, $p = (1/4)^4 = 1/256$.

f. For all to be brachydactylous tasters, $p = (1/2 \times 3/4)^4 = 81/4096$.

g. The probability of not being a brachydactylous taster is $1 -$ (the probability of being a brachydactylous taster), or $1 - (1/2 \times 3/4) = 5/8$. The probability that all four children are not brachydactylous tasters is $(5/8)^4 = 625/4096$.

h. The probability that at least one is a brachydactylous taster is $1 -$ (the probability of none being a brachydactylous taster), or $1 - (5/8)^4$.

CHAPTER 4

3.   Figure 4-3 is drawn to show how crossing over can produce new combinations of alleles, so only the crossover meiotic products are shown for simplicity. In addition to those, there would be two noncrossover products: *pr vg* designated with brown, and *pr+ vg+* designated yellow.

8.   **a.** The parental meiotic products are chromosomes that have descended intact from one of the two original parents (P) in the cross. In this case, the parent for the $A\ B/a\ b$ genotype would be the $A\ B/A\ B$ parent (homozygous brown chromosomes), and the parent for the $a\ b/a\ b$ genotype would be the $a\ b/a\ b$ (homozygous yellow chromosomes).

10.   Figure 4-10 shows four cells undergoing meiosis, depicting crossovers at various locations along the chromosome. One crossover occurs in the A–B region, producing 2 recombinant chromosomes out of a total of 16, for a frequency of 0.125. Four crossovers (two singles and a three-stranded double) occur in the B–C region, producing 6 recombinant chromosomes out of a total of 16, for a frequency of 0.375. (The strand in meiocyte 3 that is involved in two crossovers would appear to have had no crossovers.)

14.   a. Figure 4-14 represents a trihybrid testcross with linked genes, so there are two genetic intervals to consider. In a typical three-point testcross, those intervals will be different sizes with correspondingly different frequencies of SCOs. The colored boxes depicting SCOs are different sizes to reflect a difference in the number of single crossovers. Note that the two adjacent colors in each pair (green/brown, purple/gray) are about the same size, reflecting the roughly equal number of reciprocal products in each crossover.

b. Lines 3, 4, and 5; all different results of testing or selfing in dihybrid crosses.

c. Lines 6 and 7; trihybrids testcrossed for unlinked and linked genes.

19.   a. The heteroduplex DNA in the upper chromatid would produce a non-identical spore pair with GC (= A) on top and AT (= a) on bottom. The bottom chromatid would replicate to produce a normal identical spore pair. Thus, the final octad would be *A-A A-a a-a a-a*.

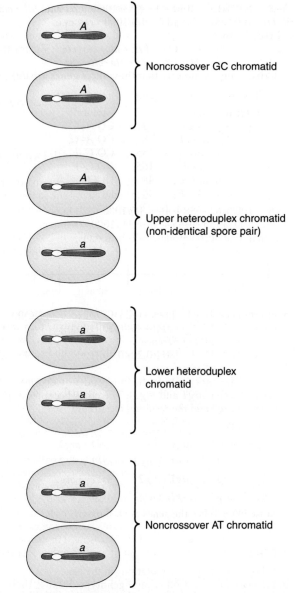

Noncrossover GC chromatid

Upper heteroduplex chromatid (non-identical spore pair)

Lower heteroduplex chromatid

Noncrossover AT chromatid

b. That the two loci show no non-Mendelian segregation means that both lie outside the heteroduplex region. If P and Q were in cis and positioned to the left and right, respectively, of the gene A crossover region, they would recombine due to the crossover. The final octad would be $PQ\ PQ\ Pq\ Pq\ pQ\ pQ\ pq\ pq$.

22. P $A\ d/A\ d \times a\ D/a\ D$
      F_1 $A\ d/a\ D$
      F_2 $1\ A\ d/A\ d$ phenotype: $A\ d$
             $2\ A\ d/a\ D$ phenotype: $A\ D$
             $1\ a\ D/a\ D$ phenotype: $a\ D$

25. Because only parental types are recovered, the two genes must be tightly linked and recombination must be very rare. Knowing how many progeny were looked at would give an indication of how close the genes are.

30. a. The three genes are linked.

b. A comparison of the parentals (most frequent) with the double crossovers (least frequent) reveals that the gene order is $v\ p\ b$. There were 2200 recombinants between v and p, and 1500 between p and b. The general formula for map units is

m.u. = 100%(number of recombinants)/total number of progeny

Therefore, the map units between v and p = 100%(2200)/ 10,000 = 22 m.u., and the map units between p and b = 100% (1500)/10,000 = 15 m.u. The map is

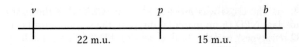

v p b
      22 m.u.   15 m.u.

c. I = 1 − observed double crossovers/expected double crossovers
= 1 − 132/(0.22)(0.15)(10,000)
= 1 − 0.4 = 0.6

36. a.

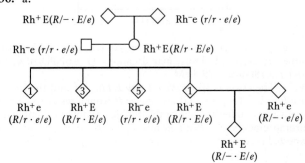

$Rh^+E\ (R/-\cdot E/e)$ $Rh^-e\ (r/r\cdot e/e)$

$Rh^-e\ (r/r\cdot e/e)$ $Rh^+E\ (R/r\cdot E/e)$

Rh^+e Rh^+E Rh^-e Rh^+E Rh^+e
$(R/r\cdot e/e)$ $(R/r\cdot E/e)$ $(r/r\cdot e/e)$ $(R/r\cdot E/e)$ $(R/-\cdot e/e)$

Rh^+E
$(R/-\cdot E/e)$

b. Yes.

c. Dominant.

d. As drawn, the pedigree hints at linkage. If unlinked, expect that the phenotypes of the 10 children should be in a 1:1: 1:1 ratio of Rh^+E, Rh^+e, Rh^-E, and Rh^-e. There are actually five Rh^-e, four Rh^+E, and one Rh^+e. If linked, this last phenotype would represent a recombinant, and the distance between the two genes would be 100%(1/10) = 10 m.u. However, there is just not enough data to strongly support that conclusion.

42. a. If the genes are unlinked, the cross is

P $hyg/hyg\ ;\ her/her \times hyg^+/hyg^+\ ;\ her^+/her^+$

F_1 $hyg^+/hyg\ ;\ her^+/her \times hyg^+/hyg\ ;\ her^+/her$

F_2 9/16 $hyg^+/-\ ;\ her^+/-$
       3/16 $hyg^+/-\ ;\ her/her$
       3/16 $hyg/hyg\ ;\ her^+/-$
       1/16 $hyg/hyg\ ;\ her/her$

So only 1/16 (or 6.25%) of the seeds are expected to germinate.

b. and **c.** No. More than twice the expected seeds germinated; so assume that the genes are linked. The cross, then, is

P $hyg\ her/hyg\ her \times hyg^+\ her^+/hyg^+\ her^+$

F_1 $hyg^+\ her^+/hyg\ her \times hyg^+\ her^+/hyg\ her$

F_2 13% $hyg\ her/hyg\ her$

Because this class represents the combination of two parental chromosomes, it is equal to

$$p(hyg\ her) \times p(hyg\ her) = (\tfrac{1}{2}\ parentals)^2 = 0.13$$

and

$$parentals = 0.72$$

So

$$recombinants = 1 - 0.72 = 0.28$$

Therefore, a testcross of $hyg^+\ hyg^+/hyg\ her$ will give

         36% $hyg^+\ her^+/hyg\ her$
         36% $hyg\ her/hyg\ her$
         14% $hyg^+\ her/hyg\ her$
         14% $hyg\ her^+/hyg\ her$

and 36% of the progeny will grow (the $hyg\ her/hyg\ her$ class).

46. The formula for this problem is $f(i) = e^{-m}m^i/i!$ where $m = 2$ and $i = 0, 1,$ or 2.

a. $f(0) = e^{-2}2^0/0! = e^{-2} = 0.135$, or 13.5%
b. $f(1) = e^{-2}2^1/1! = e^{-2}(2) = 0.27$, or 27%
c. $f(2) = e^{-2}2^2/2! = e^{-2}(2) = 0.27$, or 27%

52. a. The cross was *pro* + × + *his*, which makes the first tetrad class NPD (6 nonparental ditypes), the second tetrad class T (82 tetratypes), and the third tetrad class PD (112 parental ditypes). When PD >> NPD, you know that the two genes are linked.

b. Map distance can be calculated by using the formula RF = [NPD + (1/2)T]100%. In this case, the frequency of NPD is 6/200, or 3%, and the frequency of T is 82/200, or 41%. Map distance between these two loci is therefore 23.5 cM.

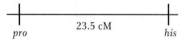

c. To correct for multiple crossovers, the Perkins formula can be used. Thus, map distance = (T + 6NPD)50%, or (0.41 + 0.18)50% = 29.5 cM.

56. a. The cross is *W e F/W e F* × *w E f/w E f* and the F₁ are *W e F/w E f*. Progeny that are *ww ee ff* from a testcross of this F₁ must have inherited one of the double-crossover recombinant chromosomes (*w e f*). With the assumption of no interference, the expected percentage of double crossovers is 8% × 24% = 1.92%, half of which is 0.96%.

b. To obtain a *ww ee ff* progeny from a self cross of this F₁ requires the independent inheritance of two doubly recombinant *w e f* chromosomes. Its chances of happening, based on the answer to part *a* of this problem, are 0.96 × 0.96 = 0.009%.

62. The short answer is that the results tell us little about linkage. Although the number of recombinants (3) is less than the number of parentals (5), one can have no confidence in the fact that the RF is < 50%. The main problem is that the sample size is small, so just one individual more or less in a genotypic class can dramatically affect the ratios. Even the chi-square test is unreliable at such small sample sizes. It *is* probably safe to say that there is not tight linkage because several recombinants were found in a relatively small sample. However, one cannot distinguish between more distant linkage and independent assortment. A larger sample size is required.

67. The data given for each of the three-point testcrosses can be used to determine the gene order when one realizes that the rarest recombinant classes are the result of double-crossover events. A comparison of these chromosomes with the "parental" types reveals that the alleles that have switched represent the gene in the middle.

For example, in data set 1, the most common phenotypes (+ + + and *a b c*) represent the parental-allele combinations. A comparison of these phenotypes with the rarest phenotypes of this data set (+ *b c* and *a* + +) indicates that the *a* gene is recombinant and must be in the middle. The gene order is *b a c*.

For data set 2, + *b c* and *a* + + (the parentals) should be compared with + + + and *a b c* (the rarest recombinants) to indicate that the *a* gene is in the middle. The gene order is *b a c*.

For data set 3, compare + *b* + and *a* + *c* with *a b* + and + + *c*, which gives the gene order *b a c*.

For data set 4, compare + + *c* and *a b* + with + + + and *a b c*, which gives the gene order *a c b*.

For data set 5, compare + + + and *a b c* with + + *c* and *a b* +, which gives the gene order *a c b*.

73. a. Cross 1 reduces to

P   *A/A · B/B · D/D* × *a/a · b/b · d/d*

F₁   *A/a · B/b · D/d* × *a/a · b/b · d/d*

The testcross progeny indicate that these three genes are linked (CO = crossover, DCO = double crossover).

Testcross	A B D	316	parental
Progeny	*a b d*	314	parental
	A B d	31	CO B–D
	a b D	39	CO B–D
	A b d	130	CO A–B
	a B D	140	CO A–B
	A b D	17	DCO
	a B d	13	DCO

A–B: 100%(130 + 140 + 17 + 13)/1000 = 30 m.u.
B–D: 100%(31 + 39 + 17 + 13)/1000 = 10 m.u.

Cross 2 reduces to

P   *A/A · C/C · E/E* × *a/a · c/c · e/e*

F₁   *A/a · C/c · E/e* × *a/a · c/c · e/e*

The testcross progeny indicate that these three genes are linked.

Testcross	A C E	243	parental
Progeny	*a c e*	237	parental
	A c e	62	CO A–C
	a C E	58	CO A–C
	A C e	155	CO C–E
	a c E	165	CO C–E
	a C e	46	DCO
	A c E	34	DCO

A–B: 100%(62 + 58 + 46 + 34)/1000 = 20 m.u.
B–D: 100%(155 + 165 + 46 + 34)/1000 = 40 m.u.

The map that accommodates all the data is

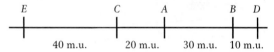

b. Interference (I) = 1 − [(observed DCO)/(expected DCO)]
For cross 1: I = 1 − {30/[(0.30)(0.10)(1000)]} = 1 − 1 = 0, no interference
For cross 2: I = 1 − {80/[(0.20)(0.40)(1000)]} = 1 − 1 = 0, no interference

78. a. and b. The data support the independent assortment of two genes (call them *arg1* and *arg2*). The cross becomes *arg1 ; arg2⁺* × *arg1⁺ ; arg2* and the resulting tetrads are

4 : 0 (PD)	3 : 1 (T)	2 : 2 (NPD)
arg1 ; arg2⁺	*arg1 ; arg2⁺*	*arg1 ; arg2*
arg1 ; arg2⁺	*arg1⁺ ; arg2*	*arg1 ; arg2*
arg1⁺ ; arg2	*arg1 ; arg2*	*arg1⁺ ; arg2⁺*
arg1⁺ ; arg2	*arg1⁺ ; arg2⁺*	*arg1⁺ ; arg2⁺*

Because PD = NPD, the genes are unlinked.

CHAPTER 5

1. a. Yellow stars represent mutations in the gene.

b. Heterozygous is a functional genotype because there is a sufficient amount of a gene product from the second (wild-type) gene on the second chromosome, which does not have a mutation.

c. The heterozygote would have black pigment.

d. The proteins could differ in size, shape, or general conformation.

4. a. The blue bands represent normal and mutant hemoglobin proteins migrating through the electrophoretic gel.

b. No, it is the opposite. Normal hemoglobin migrates faster than mutant hemoglobin.

c. Different migration rates could be due to differences in size or charge of the proteins.

7. This pedigree shows an incomplete penetrance of a dominant gene (*A*). Individual Q has a genotype *A/a* (if it is a rare dominant allele, there is a very low likelihood that another parent would harbor the same mutation, especially since it is a dominant trait). Two of the children in the last generation have complete penetrance of the dominant allele (with genotypes *A/a*), while female R might have the allele *A*, which is not showing in this case. Probability that R female is *A/a* is 1/2 and that she is *a/a* is 1/2.

17. Possible genotypes for the three flowers are:
left: *D/D ; W/W*
middle: *d/d ; w/w*
right: *D/D ; w/w*

21. With the assumption of homozygosity for the normal gene, the mating is $A/A \cdot b/b \times a/a \cdot B/B$. The children would be normal, $A/a \cdot B/b$.

24. a. Red.

b. Purple.

c.

9	$M_1/-$; $M_2/-$	purple
3	m_1/m_1 ; $M_2/-$	blue
3	$M_1/-$; m_2/m_2	red
1	m_1/m_1 ; m_2/m_2	white

d. The mutant alleles do not produce functional enzyme. However, enough functional enzyme must be produced by the single wild-type allele of each gene to synthesize normal levels of pigment.

28. a. The original cross was a dihybrid cross. Both oval and purple must represent an incomplete dominant phenotype.

b. A long, purple oval, purple cross is as follows:

P L/L ; $R/R' \times L/L'$; R/R''

F$_1$

$\frac{1}{2} L/L \times \frac{1}{2} R/R'$

$\frac{1}{4} R/R$ $\frac{1}{8}$ long, red

$\frac{1}{2} R/R'$ $\frac{1}{4}$ long, purple

$\frac{1}{4} R'/R'$ $\frac{1}{8}$ long, white

$\frac{1}{4} R/R$ $\frac{1}{8}$ oval, red

$\frac{1}{2} L/L' \times \frac{1}{2} R/R'$ $\frac{1}{4} R/R'$ $\frac{1}{4}$ oval, purple

$\frac{1}{2} R'/R'$ $\frac{1}{4}$ oval, white

31.

	Parents	Child
a.	AB × O	B
b.	A × O	A
c.	A × AB	AB
d.	O × O	O

35. a. The sex ratio is expected to be 1 : 1.

b. The female parent was heterozygous for an X-linked recessive lethal allele, which would result in 50% fewer males than females.

c. Half of the female progeny should be heterozygous for the lethal allele, and half should be homozygous for the nonlethal allele. Individually mate the F$_1$ females and determine the sex ratio of their progeny.

38. a. The mutations are in two different genes because the heterokaryon is prototrophic (the two mutations complemented each other).

b. $leu1^+$; $leu2^-$ and $leu1^-$; $leu2^+$

c. With independent assortment, expect

$\frac{1}{4} leu1^+$; $leu2^-$

$\frac{1}{4} leu1^-$; $leu2^+$

$\frac{1}{4} leu1^-$; $leu2^-$

$\frac{1}{4} leu1^+$; $leu2^+$

42. a. P A/a (frizzle) × A/a (frizzle)

F$_1$ 1 A/A (normal) : 2 A/a (frizzle) : 1 a/a (woolly)

b. If A/A (normal) is crossed with a/a (woolly), all offspring will be A/a (frizzle).

44. The production of black offspring from two pure-breeding recessive albino parents is possible if albinism results from mutations in two different genes. If the cross is designated

$$A/A ; b/b \times a/a ; B/B$$

all offspring would be

$$A/a ; B/b$$

and they would have a black phenotype because of complementation.

48. The purple parent can be either A/a ; b/b or a/a ; B/b for this answer. Assume that the purple parent is A/a ; b/b. The blue parent must be A/a ; B/b.

51. The cross is gray × yellow, or $A/- ; R/- \times A/- ; r/r$. The F$_1$ progeny are

$\frac{3}{8}$ yellow $\frac{1}{8}$ black $\frac{3}{8}$ gray $\frac{1}{8}$ white

For white progeny, both parents must carry an r and an a allele. Now the cross can be rewritten as A/a ; $R/r \times A/a$; r/r.

54. The original brown dog is w/w ; b/b and the original white dog is W/W ; B/B. The F$_1$ progeny are W/w ; B/b and the F$_2$ progeny are

9 $W/-$; $B/-$	white
3 w/w ; $B/-$	black
3 $W/-$; b/b	white
1 w/w ; b/b	brown

58. Pedigrees such as this one are quite common. They indicate lack of penetrance due to epistasis or environmental effects. Individual A must have the dominant autosomal gene.

60. a. Let W^O = oval, W^S = sickle, and W^R = round. The three crosses are

Cross 1: $W^S/W^S \times W^R/Y \to W^S/W^R$ and W^S/Y

Cross 2: $W^S/W^S \times W^S/Y \to W^S/W^R$ and W^R/Y

Cross 3: $W^S/W^S \times W^O/Y \to W^O/W^S$ and W^S/Y

b. $W^O/W^S \times W^R/Y$

$\frac{1}{4} W^O/W^R$	female oval
$\frac{1}{4} W^S/W^R$	female sickle
$\frac{1}{4} W^O/Y$	male oval
$\frac{1}{4} W^S/Y$	male sickle

63. a. The genotypes are

P B/B ; $i/i \times b/b$; I/I

F$_1$ B/b ; I/i hairless

F$_2$

9 $B/-$; $I/-$	hairless
3 $B/-$; i/i	straight
3 b/b ; $I/-$	hairless
1 b/b ; i/i	bent

b. The genotypes are B/b ; $I/i \times B/b$; i/i.

66. There are a total of 159 progeny that should be distributed in a 9 : 3 : 3 : 1 ratio if the two genes are assorting independently. You can see that

Observed	Expected
88 $P/-$; $Q/-$	90
32 $P/-$; q/q	30
25 p/p ; $Q/-$	30
14 p/p ; q/q	10

69. a. Cross-feeding is taking place, whereby a product made by one strain diffuses to another strain and allows growth of the second strain.

b. For cross-feeding to take place, the growing strain must have a block that is present earlier in the metabolic pathway than the block in the strain from which the growing strain is obtaining the product for growth.

c. The data suggest that the metabolic pathway is

$$trpE \to trpD \to trpB$$

d. Without some tryptophan, there would be no growth at all, and the cells would not have lived long enough to produce a product that could diffuse.

71. a. The best explanation is that Marfan's syndrome is inherited as a dominant autosomal trait.

b. The pedigree shows both pleiotropy (multiple affected traits) and variable expressivity (variable degree of expressed phenotype).

c. Pleiotropy indicates that the gene product is required in a number of different tissues, organs, or processes. When the gene is mutant, all tissues needing the gene product will be affected. Variable expressivity of a phenotype for a given genotype indicates modification by one or more other genes, random noise, or environmental effects.

74. a. This type of gene interaction is called *epistasis*. The phenotype of *e/e* is epistatic to the phenotypes of *B/–* or *b/b*.

b. The inferred genotypes are as follows:

I	1 (*B/b E/e*)	2 (*B/b E/e*)		
II	1 (*b/b E/e*)	2 (*B/b E/e*)	3 (*–/– e/e*)	4 (*b/b E/–*)
	5 (*B/b E/e*)	6 (*b/b E/e*)		
III	1 (*B/b E/–*)	2 (*–/b e/e*)	3 (*b/b E/–*)	4 (*B/b E/–*)
	5 (*b/b E/–*)	6 (*B/b E/–*)	7 (*–/b e/e*)	

77. a. A multiple-allelic series has been detected: superdouble > single > double.

b. Although the explanation for part *a* does rationalize all the crosses, it does not take into account either the female sterility or the origin of the superdouble plant from a double-flowered variety.

79. a. A trihybrid cross would give a 63 : 1 ratio. Therefore, there are three *R* loci segregating in this cross.

b.

P	R_1/R_1 ; R_2/R_2 ; $R_3/R_3 \times r_1/r_1$; r_2/r_2 ; r_3/r_3	
F_1	R_1/r_1 ; R_2/r_2 ; R_3/r_3	
F_2	27 $R_1/–$; $R_2/–$; $R_3/–$	red
	9 $R_1/–$; $R_2/–$; r_3/r_3	red
	9 $R_1/–$; r_2/r_2 ; $R_3/–$	red
	9 r_1/r_1 ; $R_2/–$; $R_3/–$	red
	3 $R_1/–$; r_2/r_2 ; r_3/r_3	red
	3 r_1/r_1 ; $R_2/–$; r_3/r_3	red
	3 r_1/r_1 ; r_2/r_2 ; $R_3/–$	red
	1 r_1/r_1 ; r_2/r_2 ; r_3/r_3	white

c. (1) To obtain a 1 : 1 ratio, only one of the genes can be heterozygous. A representative cross is R_1/r_1 ; r_2/r_2 ; $r_3/r_3 \times r_1/r_1$; r_2/r_2 ; r_3/r_3.
(2) To obtain a 3 red : 1 white ratio, two alleles must be segregating and they cannot be within the same gene. A representative cross is R_1/r_1 ; R_2/r_2 ; $r_3/r_3 \times r_1/r_1$; r_2/r_2 ; r_3/r_3.
(3) To obtain a 7 red : 1 white ratio, three alleles must be segregating, and they cannot be within the same gene. The cross is R_1/r_1 ; R_2/r_2 ; $R_3/r_3 \times r_1/r_1$; r_2/r_2 ; r_3/r_3.

d. The formula is $1 - \left(\frac{1}{4}\right)n$, where n = the number of loci that are segregating in the representative crosses in part *c*.

83. a. and **b.** Epistasis is implicated, and the homozygous recessive white genotype seems to block the production of color by a second gene.

Assume the following dominance relations: red > orange > yellow. Let the alleles be designated as follows:

red	A^R
orange	A^O
yellow	A^Y

Crosses 1 through 3 now become

P	$A^O/A^O \times A^Y/A^Y$	$A^R/A^R \times A^O/A^O$	$A^R/A^R \times A^Y/A^Y$
F_1	A^O/A^Y	A^R/A^O	A^R/A^Y
F_2	3 $A^O/–$: 1 A^Y/A^Y	3 $A^R/–$: 1 A^O/A^O	3 $A^R/–$: 1 A^Y/A^Y

Cross 4: To do this cross, you must add a second gene. You must also rewrite crosses 1 through 3 to include the second gene. Let *B* allow color expression and *b* block its expression, producing white. The first three crosses become

P	A^O/A^O ; $B/B \times A^Y/A^Y$; B/B
	A^R/A^R ; $B/B \times A^O/A^O$; B/B
	A^R/A^R ; $B/B \times A^Y/A^Y$; B/B
F_1	A^O/A^Y ; B/B
	A^R/A^O ; B/B
	A^R/A^Y ; B/B
F_2	3 $A^O/–$; B/B : 1 A^Y/A^Y ; B/B
	3 $A^R/–$; B/B : 1 A^O/A^O ; B/B
	3 $A^R/–$; B/B : 1 A^Y/A^Y ; B/B

The fourth cross is

P	A^R/A^R ; $B/B \times A^R/A^R$; b/b
F_1	A^R/A^R ; B/b
F_2	3 A^R/A^R ; $B/–$: 1 A^R/A^R ; b/b

Cross 5: To do this cross, note that there is no orange. Therefore, the two parents must carry the alleles for red and yellow, and the expression of red must be blocked.

P	A^Y/A^Y ; $B/B \times A^R/A^R$; b/b	
F_1	A^R/A^Y ; B/b	
F_1	9 $A^R/–$; $B/–$	red
	3 $A^R/–$; b/b	white
	3 A^Y/A^Y ; $B/–$	yellow
	1 A^Y/A^Y ; b/b	white

Cross 6: This cross is identical with cross 5 except that orange replaces yellow.

P	A^O/A^O ; $B/B \times A^R/A^R$; b/b	
F_1	A^R/A^O ; B/b	
F_1	9 $A^R/–$; $B/–$	red
	3 $A^R/–$; b/b	white
	3 A^O/A^O ; $B/–$	orange
	1 A^O/A^O ; b/b	white

Cross 7: In this cross, yellow is suppressed by *b/b*.

P	A^R/A^R ; $B/B \times A^Y/A^Y$; b/b	
F_1	A^R/A^Y ; B/b	
F_2	9 $A^R/–$; $B/–$	red
	3 $A^R/–$; b/b	white
	3 A^Y/A^Y ; $B/–$	yellow
	1 A^Y/A^Y ; b/b	white

85. a. Intercrossing mutant strains that all have a common recessive phenotype is the basis of the complementation test. This test is designed to identify the number of different genes that can mutate to a particular phenotype. In this problem, if the progeny of a given cross still express the wiggle phenotype, the mutations fail to complement and are considered alleles of the same gene; if the progeny are wild type, the mutations complement and the two strains carry mutant alleles of separate genes.

b. These data identify five complementation groups (genes).

c. mutant 1 : $a^1/a^1 \cdot b^+/b^+ \cdot c^+/c^+ \cdot d^+/d^+ \cdot e^+/e^+$ (although only the mutant alleles are usually listed)

mutant 2 : $a^+/a^+ \cdot b^2/b^2 \cdot c^+/c^+ \cdot d^+/d^+ \cdot e^+/e^+$

mutant 5 : $a^5/a^5 \cdot b^+/b^+ \cdot c^+/c^+ \cdot d^+/d^+ \cdot e^+/e^+$

$\frac{1}{3}$ hybrid : $a^1/a^5 \cdot b^+/b^+ \cdot c^+/c^+ \cdot d^+/d^+ \cdot e^+/e^+$

phenotype: wiggles

Conclusion: 1 and 5 are both mutant for gene *A*.

$\frac{2}{5}$ hybrid : $a^+/a^5 \cdot b^+/b^2 \cdot c^+/c^+ \cdot d^+/d^+ \cdot e^+/e^+$

phenotype: wild type

Conclusion: 2 and 5 are mutant for different genes.

CHAPTER 6

13. a. $F^+ \ a^-$ cells result from transfer of the F plasmid into the $F^- \ a^-$ strain.

b. $F^- \ a^-$ strains result when the recipient in an Hfr cross does not acquire the donor a^+ allele by recombination.

c. $F^- \ a^+$ strains result when the recipient in an Hfr cross acquires the donor a^+ allele by recombination.

d. $F^+ \ a^+$ strains result from resynthesis of the donated F plasmid in the donor strain.

17. Ten different bacterial species have donated sequences to the R plasmid.

20. No. In this figure all of the phage particles contain only phage DNA. Only five phages are shown; in a real-world phage infection, millions of phages would be involved and some would very likely carry DNA from the host and thus facilitate transduction.

22. In this example, the phages infecting the host cells are either h^-r^+ or h^+r^-. In bacteria infected with two phages simultaneously, recombination is possible and h^+r^+ or h^-r^- recombinant phages may be produced. The recombinant plaque types are small, clear plaques and large, cloudy plaques.

31. The gene is the one colored orange.

32. An Hfr strain has the fertility factor F integrated into the chromosome. An F^+ strain has the fertility factor free in the cytoplasm. An F^- strain lacks the fertility factor.

36. Although the interrupted-mating experiments will yield the gene order, it will be relative only to fairly distant markers. Thus, the mutation cannot be precisely located with this technique. Generalized transduction will yield information with regard to very close markers, which makes it a poor choice for the initial experiments because of the massive amount of screening that would have to be done. Together, the two techniques allow, first, for localization of the mutant (interrupted mating) and, second, for the precise determination of the location of the mutant (generalized transduction) within the general region.

41. The best explanation is that the integrated F factor of the Hfr looped out of the bacterial chromosome abnormally and is now an F′ that contains the *pro*⁺ gene. This F′ is rapidly transferred to F⁻ cells, converting them into *pro*⁺ (and F⁺).

46. The expected number of double recombinants is $(0.01)(0.002)(100,000) = 2$. Interference $= 1 - (\text{observed double crossover/expected double crossover}) = 1 - 5/2 = -1.5$. By definition, the interference is negative.

50. a. This process appears to be specialized transduction. It is characterized by the transduction of specific markers based on the position of the integration of the prophage. Only those genes near the integration site are possible candidates for misincorporation into phage particles that then deliver this DNA to recipient bacteria.

b. The only media that supported colony growth were those lacking either cysteine or leucine. These media selected for *cys*⁺ or *leu*⁺ transductants and indicate that the prophage is located in the *cys–leu* region.

55. No. Closely linked loci would be expected to be cotransduced; the greater the cotransduction frequency, the closer the loci are. Because only 1 of 858 *metE*⁺ was also *pyrD*⁺, the genes are not closely linked. The lone *metE*⁺ *pyrD*⁺ could be the result of cotransduction, it could be a spontaneous mutation of *pyrD* to *pyrD*⁺, or it could be the result of coinfection by two separate transducing phages.

60. a. To determine which genes are close, compare the frequencies of double transformants. Pair-by-pair testing gives low values whenever B is included but fairly high rates when any drug but B is included. This finding suggests that the gene for B resistance is not close to the other three genes.

b. To determine the relative order of genes for resistance to A, C, and D, compare the frequencies of double and triple transformants. The frequency of resistance to AC is approximately the same as that of resistance to ACD, which strongly suggests that D is in the middle. Additionally, the frequency of AD coresistance is higher than AC (suggesting that the gene for A resistance is closer to D than to C) and the frequency of CD is higher than AC (suggesting that C is closer to D than to A).

64. To isolate the specialized transducing particles of phage $\phi80$ that carried *lac*⁺, the researchers would have had to lysogenize the strain with $\phi80$, induce the phage with UV, and then use these lysates to transduce a Lac⁻ strain to Lac⁺. The Lac⁺ colonies would then have been used to make a new lysate, which should have been highly enriched for the *lac*⁺ transducing phage.

CHAPTER 7

1. If A = 20%, then T = 20%, G = 30%, and C = 30%.

4. Proteins could be degraded using assorted proteases, and RNAs could be destroyed with RNA nucleases (also called RNases).

7. That DNA is a long helical molecule made of two similar parts that run parallel to one another.

10. Each dot represents one hydrogen bond. So the two dots between an A–T base pair signify that A and T form two hydrogen bonds with each other. A G–C base pair forms three hydrogen bonds, hence there are three dots.

20. The lagging strand is shown looping around so that the replisome can coordinate the synthesis of both strands and move in the direction of the replication fork.

23. Initiation of replication requires the proteins Cdc6 and Cdt1. In yeast, these proteins are synthesized during late mitosis (M) and G1 and are destroyed at the beginning of S phase. Thus these initiator proteins aren't available during G2.

31. 5′-UAACCCUAA-3′

32. A characteristic of Werner syndrome is chromosome instability, which can produce mutations that lead to cancer.

36. Telomerase is not required in bacteria because the genome is circular and does not have ends.

38. (1) DNA polymerase I is too abundant. Only two molecules can be engaged on DNA at any one time. (2) DNA polymerase I is too slow to complete replication of the bacterial genome within the time it takes for one cell division. (3) DNA polymerase I is not processive. It cannot synthesize long stretches of DNA that are needed to produce both the leading strand and Okazaki fragments of the lagging strand.

44. If the GC content is 48%, the AT content must be the remaining 52% $(100 - 48 = 52)$. Since the amount of G equals the amount of C, the 48% GC is 24% G and 24% C. Likewise, since the amount of A equals the amount of T, the 52% AT is 26% A and 26% T.

45. Yes. DNA replication is also semiconservative in diploid eukaryotes.

47. 5′. . . .CCTTAAGACTAACTACTTACTGGGATC. . . .3′

48. Chargaff's rules are that A = T and G = C. Because these equalities are not observed, the most likely interpretation is that the DNA is single stranded. The phage would first have to synthesize a complementary strand before it could begin to make multiple copies of itself.

50. Continuous replication is unlikely to occur in any organism because it would require DNA and RNA polymerases that synthesize in the 3′-to-5′ direction. All of the characterized DNA and RNA polymerases from bacteria and eukaryotes synthesize in the 5′-to-3′ direction.

CHAPTER 8

2. After a longer period of time, the RNAs would no longer be detected in either the nucleus or the cytoplasm because they would have decayed and no new radioactive RNAs would be made.

8. 5′-GUUGAGACGA-3′

14. Mutating TFIIIB would block transcription of all three types of RNA polymerase III genes because it is required for transcription initiation of all three gene types.

17. Transcription can start with any nucleotide, so presumably capping is equally efficient regardless of the initiating nucleotide.

18. The AAUAAA element and the 5′ half of the poly(A) site are retained in the mRNA.

23. Exon 2a is mutually exclusive with 2b, and exon 6a is mutually exclusive with 6b (i.e., only one exon at a time is included in the spliced mRNA, not both).

27. In (a), the dsRNA source is annealed transcripts from the transgene that are synthesized in the petunia. In (b), the dsRNA source is annealed sense and anti-sense transcripts that are synthesized in vitro. In (c), the dsRNA source is annealed sense and anti-sense transcripts from the viral genome that are synthesized in the tobacco plant.

31. 5′-AAUGCCGGUAACGAUUAACGCCCGAUAUCCG-3′
                            3′-UUGCGGGCUAUA-5′

38. Non-template
   5′-GTTCACTGGGACTAAAGCCCGGGAACTAGG-3′
   Template
   3′-CAAGTGACCCTGATTTCGGGCCCTTGATCC-5′

41. The answer can be displayed three different ways (N is any nucleotide).

           UCGNNAGAUUCC
           C   GCC

          U/CCGNNAGA/GU/CU/CCC

        (U/C)CGNNAG(A/G)(U/C)(U/C)CC

45. A random 21-nucleotide sequence appears once every 4.4×10^{12} (i.e., 4^{21}) nucleotides. RNAi in the human genome is very specific because the genome is only 3.2×10^{9} nucleotides, which is much smaller than 4.4×10^{12}.

47. The stress might increase the process of mRNA transcription or reduce the process of mRNA decay.

52. There are many answers to this question. Here is one.
       5′-CGGCAAUGCGACCAAGUCGUA-3′
       3′-AGGCCGUUACGCUGGUUCAGC-5′

53. In bacteria, translation begins at the 5′ end while the 3′ end is still being synthesized. In eukaryotes, processing (capping, splicing) takes place at the 5′ end while the 3′ end is still being synthesized.

57. The RNA was likely a target of A-to-I editing. The inosine (I) is read as a guanine (G) in the process of sequencing because it base pairs with cytosine.

59. Because RNA has a hydroxyl group (OH) at the 2′ position of the ribose sugar, it is susceptible to base-catalyzed hydrolysis (in this context, base refers to a hydroxide ion). DNA does not have 2′-OH groups and thus is not susceptible to base-catalyzed hydrolysis.

CHAPTER 9

4. The hydrophobic amino acids are mostly buried in the middle of the protein and the hydrophilic amino acids are exposed on the surface of the protein.

7. CCA (Proline), CAC (Histidine), and ACC (Threonine).

15. Less than 10. The three loops in the figure have 4, 8, and 9 nucleotides.

20. RF1 is a protein.

21. Mutating the first (G) nucleotide in the tRNA^{Tyr} to a U.

25. The diagram explains the association of ribosomes with the endoplasmic reticulum.

27. a. and **b.** 5′-GCU UCC CAA-3′

    c. and **d.** With the assumption that the reading frame starts at the first base,

$$NH_3 - Ala - Ser - Gln - COOH$$

    For the top strand, the mRNA is 5′-UUG GGA AGC-3′ and, with the assumption that the reading frame starts at the first base, the corresponding amino acid chain is

$$NH_3 - Leu - Gly - Ser - COOH$$

31. Quaternary structure is due to the interactions of subunits of a protein. In this example, the enzyme activity being studied may be that of a protein consisting of two different subunits. The polypeptides of the subunits are encoded by separate and unlinked genes.

33. No. The enzyme may require post-translational modification to be active. Mutations in the enzymes required for these modifications would not map to the isocitrate lyase gene.

34. The suppressor mutation could be in tRNA for tryptophan such that its anticodon now recognizes UAG instead of UGG, allowing translation to continue but putting a tryptophan into a position that was another amino acid in the wild-type gene.

39. No. Translation would not be able to initiate without a 5′ cap.

41. Single amino acid changes throughout the enzyme can inactivate an enzyme by affecting its folding, targeting, or post-translational modifications.

CHAPTER 10

5. Probe 2 would detect a 9 kb *Pvu*II fragment. Probe 3 would detect a 15 kb *Nsi*I fragment.

11. Sample A had 8 times more DNA ($2^{-(24-27)}$).

17. Yes, more than one insert can ligate into a single vector, but it is a rare event because its occurrence is equal to the frequency of a single insertion squared.

20. Cutting the assembled plasmid with *Eco*RI and *Bam*HI will release the full-length insert from the vector.

26. Cells with a double crossover lack the marker gene and cannot grow in the presence of the corresponding drug. Selectable markers are often antibiotic-resistance genes such as kanamycin and ampicillin.

27. All of the cells should contain the T-DNA because the T-DNA is the source of kanamycin resistance.

32. Using DNA oligonucleotide primers that are complementary to the genomic DNA upstream and downstream of the site of cleavage, amplify the region by PCR, sequence the PCR product, and compare the sequence to the wild-type genomic DNA sequence.

35. Each cycle takes 5 minutes and doubles the DNA. In 1 hour, there would be 12 cycles; so the DNA would be amplified $2^{12} = 4096$-fold.

38. a. Digest the plasmid with two restriction enzymes that cut on either side of *Eco*RI within the vector polylinker. If these enzymes do not cut within the insert, the DNA fragment that is cut out of the vector will be progressively larger as the number of inserts increases. The size of the DNA fragment can be determined by agarose gel electrophoresis followed by ethidium bromide staining and exposure to UV light.

b. Using DNA oligonucleotide primers that are complementary to the sequences on either site of the vector polylinker, amplify the insert by PCR. The size of the PCR product can be determined by agarose gel electrophoresis followed by ethidium bromide staining and exposure to UV light.

c. Digest the plasmid with *Eco*RI, fractionate the products by agarose gel electrophoresis, and perform Southern blot analysis using a probe complementary to the insert. For a given amount of digested plasmid, the amount of signal on the Southern blot will be progressively greater as the number of inserts increases.

43. Ligase is an essential enzyme within all cells that seals breaks in the sugar–phosphate backbone of DNA. In DNA replication, ligase joins Okazaki fragments to create a continuous strand, and, in cloning, it is used to join the various DNA fragments with the vector. If it were not added, the vector and cloned DNA would simply fall apart.

44. The positive selection for *neo*^{R} ensures that the transgene is inserted somewhere in the genome, whereas the negative selection for *tk*$^-$ ensures that the transgene specifically inserted at the homologous site in the genome.

45. Test for presence of the transgene by Southern hybridization. Test for mRNA expression by Northern blot analysis; test for protein expression by Western blot analysis.

47. To knock out the *Ins1* gene in mice, inject mouse embryos with three plasmids: (1) a plasmid that encodes the Cas9 endonuclease, (2) a plasmid that encodes an sgRNAs that is complementary to sequences upstream of the *Ins1* transcription start site and is positioned appropriately relative to a PAM site, and (3) a plasmid that encodes an sgRNAs that is complementary to sequences downstream of the *Ins1* transcription start site and is positioned appropriately relative a PAM site. The DNA double-strand breaks will remove the *Ins1* gene and the DNA will be repaired by NHEJ.

50. a. The gel can be read from the bottom to the top in a 5′-to-3′ direction. The sequence is

5′-T T C G A A A G G T G A C C C C T G G A C C T T T A G A-3′

b. By complementarity, the template was

3′-A A G C T T T C C A C T G G G G A C C T G G A A A T C T-5′

c. The double helix is

5′-T T C G A A A G G T G A C C C C T G G A C C T T T A G A-3′
3′-A A G C T T T C C A C T G G G G A C C T G G A A A T C T-5′

53. The promoter and control regions of the plant gene of interest must be cloned and joined in the correct orientation with the glucuronidase gene, which places the reporter gene under the same transcriptional control as the gene of interest. The text describes the methodology used to create transgenic plants. Transform plant cells with the reporter gene construct, and, as discussed in the text, grow them into transgenic plants. The glucuronidase gene will now be expressed in the same developmental pattern as that of the gene of interest, and its expression can be easily monitored by bathing the plant in an X-Gluc solution and assaying for the blue reaction product.

CHAPTER 11

4. Because these are the regions of DNA that are bound by the sigma subunit of RNA polymerase.

6. The *lac* repressor binds to the operator sequence and blocks transcription.

10. No, because there is some variation among genes in the exact sequence of functional sequences at −35 and at −10.

13. O^C mutants are changes in the DNA sequence of the operator that impair the binding of the *lac* repressor. Because an operator controls only the genes on the same DNA strand, it is cis (on the same strand).

17. A gene is turned off or inactivated by the "modulator" (usually called a *repressor*) in negative regulation, and the repressor must be removed for transcription to take place. A gene is turned on by the "modulator" (usually called an *activator*) in positive regulation, and the activator must be added or converted into an active form for transcription to take place.

32. The S mutation is an alteration in *lacI* such that the repressor protein binds to the operator, regardless of whether inducer is present. In other words, it is a mutation that inactivates the allosteric site that binds to inducer but does not affect the ability of the repressor to bind to the operator site. The dominance of the S mutation is due to the binding of the mutant repressor, even under circumstances when normal repressor does not bind to DNA (i.e., in the presence of inducer). The constitutive reverse mutations that map to *lacI* are mutational events that inactivate the ability of this repressor to bind to the operator. The constitutive reverse mutations that map to the operator alter the operator DNA sequence such that it will not permit binding to any repressor molecules (wild-type or mutant repressor).

35. Mutations in *cI*, *cII*, and *cIII* would all affect lysogeny: *cI* encodes the repressor, *cII* encodes an activator of P_{RE}, and *cIII* encodes a protein that protects *cII* from degradation. Mutations in *N* (an antiterminator) also would affect lysogeny because its function is required for transcription of the *cII* and *cIII* genes, but it is also necessary for genes having roles in lysis. Mutations in the gene encoding the integrase (*int*) also would affect the ability of a mutant phage to lysogenize.

CHAPTER 12

2. A corepressor might bind a transcription factor and block its ability to interact with TFIID or RNA polymerase II.

3. The transcription factor C/EBP binds the sequence GGCCAATC (a CAAT box), and the general transcription factor TBP binds the sequence ATATAA (a TATA box).

7. The DNA-binding domain is essential for the activity of a transcription factor, so a Gal4 protein that lacks the DNA-binding domain would not be able to activate transcription of the UAS-*lacZ* reporter gene.

13. Acetyl-CoA is a substrate for histone acetylation reactions, which generally activate transcription. So, reduced acetyl-CoA levels might reduce histone acetylation and thus repress transcription.

21. Histone and DNA modifications are reversible. They are added by writer enzymes and removed by eraser enzymes.

22. Different flies will have different patterns of white and red cells because the process of establishing heterochromatin spreading in cells that make up the eye is random.

26. X-chromosome inactivation only occurs in female cells.

29. Interactions between DNA and histone proteins might obstruct sites in the DNA that are necessary for binding by a transcription factor.

37. H3S10P might serve as a binding site for a reader protein that interacts with a histone acetyltransferase (HAT) that acetylates H4K16.

42. Chromatin structure affects transcription and the structure of chromatin varies across the genome.

45. The term *epigenetic inheritance* is used to describe heritable alterations in which the DNA sequence itself is not changed. It can be defined operationally as the inheritance of chromatin states from one cell generation to the next. Genomic imprinting, X-chromosome inactivation, and position-effect variegation are several such examples.

46. Insulator elements serve as binding sites for proteins, block the spread of heterochromatin, and control interactions between enhancers and promoters.

49. Histones are soluble in acid because they contain a large number of basic amino acids (lysine and arginine).

55. Yes, a transcription factor can be both an activator and a repressor of transcription. Switching between these states can be controlled by post-translational modification or by binding ligands, which alter interactions with coregulators (i.e., coactivators and corepressors), or by interactions with adjacent DNA-bound transcription factors.

CHAPTER 13

1. Organizers, which produce morphogens.

2. Immunolocalization.

7. Most are transcription factors, which coordinately regulate the expression of multiple genes.

11a. Two Hox proteins and four other factors for a total of six transcription factors.

18. The primary pair-rule gene *eve* (*even-skipped*) would be expressed in seven stripes along the A–P axis of the late blastoderm.

22. If you diagram these results, you will see that the deletion of a gene that functions posteriorly allows the next most anterior segments to extend in a posterior direction. Deletion of an anterior gene does not allow extension of the next most posterior segment in an anterior direction. The gap genes activate *Ubx* in both thoracic and abdominal segments, whereas the *abd-A* and *Abd-B* genes are activated only in the middle and posterior abdominal segments. The functioning of the *abd-A* and *Abd-B* genes in those segments somehow prevents *Ubx* expression. However, if the *abd-A* and *Abd-B* genes are deleted, *Ubx* can be expressed in these regions.

25. a. A pair-rule gene.

 b. Look for expression of the mRNA from the candidate gene in a repeating pattern of seven stripes along the A–P axis of the developing embryo.

 c. No. An embryo mutant for the gap gene *Krüppel* would be missing many anterior segments. This effect would be epistatic to the expression of a pair-rule gene.

32. a. The homeodomain is a conserved protein domain containing 60 amino acids found in a significant number of transcription factors. Any protein that contains a functional homeodomain is almost certainly a sequence-specific DNA-binding transcription factor.

 b. The *eyeless* gene (named for its mutant phenotype) regulates eye development in *Drosophila*. You would expect that it is expressed only in those cells that will give rise to the eyes. To test this prediction, visualization of the location of *eyeless* mRNA expression by in situ hybridization and the Eyeless protein by immunological methods should be performed. Through genetic manipulation, the *eyeless* gene can be expressed in tissues in which it is not ordinarily expressed. For example, when *eyeless* is turned on in cells destined to form legs, eyes form on the legs.

 c. Transgenic experiments have shown that the mouse *Small eye* gene and the *Drosophila eyeless* gene are so similar that the mouse gene can substitute for *eyeless* when introduced into *Drosophila*. As in the answer to part *b*, when the mouse *Small eye* gene is expressed in *Drosophila*, even in cells destined to form legs, eyes form on the legs. (However, the "eyes" are not mouse eyes, because *Small eye* and *eyeless* act as master switches that turn on the entire cascade of genes needed to build the eye—in this case, the *Drosophila* set to build a *Drosophila* eye.)

36. GLP-1 protein is localized to the two anterior cells of the four-cell *C. elegans* embryo by repression of its translation in the two posterior cells. The repression of GLP-1 translation requires the 3′ UTR spatial control region (SCR). Deletion of the SCR will allow *glp-1* expression in both anterior and posterior cells. In both heterozygous and homozygous mutants, you would expect GLP-1 protein expression in all cells.

CHAPTER 14

1. This is a reverse genetic experiment.

4. The sequence of the middle cluster is ATAC. The sequence of the left cluster is TGCG.

6. Several methods for identifying protein-coding sequences are discussed in the textbook. Computer analysis of the nucleotide sequence to look for open reading frames can identify at least some of the exons. These searches can be refined to include the detection of codon bias, binding site predictions, and conservation of predicted amino acid sequences. However, comparison of cDNA sequences to the genomic sequence will directly reveal the sequences comprising the coding portion of the gene.

9. The cDNA defines the exons of a particular gene. The sequences in between, therefore, represent the introns. Consequently, cDNAs are more important for eukaryotic genome annotation because bacterial genes lack introns.

11. The duplication occurred before speciation because frogs, humans, and mice all have both the A and B genes, and the A gene of three species are more similar to each other than the A gene is to the B gene in the same species.

17. The RNA sequencing reads map only to the parts of the genome that are transcribed and present in mature mRNA transcripts like the 5′ and 3′ UTRs, and the exons. Gene A is more highly expressed than gene B in sample 1.

19. Because bacteria have small genomes (roughly 3-Mb pairs) and essentially no repeating sequences, the whole-genome shotgun approach would be used.

21. A scaffold is also called a supercontig. A contig is a sequence of overlapping reads assembled into a unit, and a scaffold is a collection of joined-together contigs.

24. Yes. The operator is the location at which repressor functionally binds through interactions between the DNA sequence and the repressor protein.

28. You can determine whether the cDNA clone is a chimera or not by the alignment of the cDNA sequence against the genomic sequence. (Computer programs for doing such alignments are available.) Is the sequence derived from two different sites? If the cDNA maps within one (gene-size) region in the genome, it is likely not a chimera. However, if the cDNA maps to two different regions of the genome, it is likely a chimera. Introns may complicate the matter.

32. a. Because the triplet code is redundant, changes in the DNA nucleotide sequence (especially at those nucleotides encoding the third position of a codon) can occur without changing its encoded protein.

 b. Protein sequences can be expected to evolve and diverge more slowly than do the genes that encode them.

41. The correct assembly of large and nearly identical regions is problematic with either method of genomic sequencing. However, the whole-genome shotgun method is less effective at finding these regions than the clone-based method. This method also has the added advantage of easy access to the suspect clone(s) for further analysis.

44. a. Forward genetics identifies heritable differences by their phenotypes and map locations and precedes the molecular analysis of the gene products. Reverse genetics starts with an identified protein or RNA and works toward mutating the gene that encodes it (and in the process, discovers the phenotype when the gene is mutated). Because you have known proteins and want to determine the phenotypes associated with loss-of-function mutations in the genes that encoded them, a reverse genetic approach is the answer.

b. The two general approaches would be either directed mutations in the gene of interest or using a method such as RNAi to inactivate the gene product rather than the gene itself.

CHAPTER 15

2. If the sequence ATGCCAGAGTCA was mutated to ATGTCAGAGTCA, the mutation would have the following effect on the encoded amino acid, depending on the reading frame:

	Wild type	Mutant
Reading frame 1	CCA (Pro)	UCA (Ser)
Reading frame 2	GCA (Ala)	GUA (Val)
Reading frame 3	UGC (Cys)	UGU (Cys)

4. Protein-DNA and RNA-RNA interactions can be affected by point mutations in noncoding regions.

9. (1) guanine (G) is converted to 8-oxoguanine by oxidation, (2) during DNA replication an adenine (A) is added opposite the 8-oxoguanine template, and (3) during the next round of replication a thymine (T) is added opposite the adenine (A), so the original G · C base pair is converted to a T · A base pair.

13. Base mismatches and loops can be caused by errors in DNA replication.

16. Transcription may enhance repair because it serves as a mechanism to detect the pyrimidine dimer.

22. You need to know the reading frame of the possible message.

24. Synonymous: changing the third base to an A, C, or T or the first base to an A. Nonsynonymous: changing the first base to a C or a T or changing the second base to an A, C, or T.

25. The mismatched T would be corrected to C, and the resulting ACG, after transcription, would be 5′ UGC 3′ and encode cysteine. Or, if the other strand were corrected, ATG would be transcribed to 5′ UAC 3′ and encode tyrosine.

27. The following list of observations argues, "cancer is a genetic disease":

(1) Certain cancers are inherited as highly penetrant simple Mendelian traits.

(2) Most carcinogenic agents are also mutagenic.

(3) Various oncogenes have been isolated from tumor viruses.

(4) A number of genes that lead to the susceptibility to particular types of cancer have been mapped, isolated, and studied.

(5) Dominant oncogenes have been isolated from tumor cells.

(6) Certain cancers are highly correlated to specific chromosomal rearrangements.

35. Yes, it is mutagenic. It will cause CG-to-TA transitions.

36. Many repair systems are available: direct reversal, excision repair, transcription-coupled repair, and nonhomologous end joining.

41. a. Yes. An example is 5′-UGC-3′, which codes for trp, to 5′-UAG-3′.

b. No. In the three stop codons, the only base that can be acted upon is G (in UAG, for instance). Replacing the G with an A would result in 5′-UAA-3′, a stop codon.

42. a. A lack of revertants suggests either a deletion or an inversion within the gene.

b. To understand these data, recall that half the progeny should come from the wild-type parent.

Prototroph A: Because 100% of the progeny are prototrophic, a reversion at the original mutant site may have occurred.

Prototroph B: Half the progeny are parental prototrophs, and the remaining prototrophs, 28%, are the result of the new mutation. Notice that 28% is approximately equal to the 22% auxotrophs. The suggestion is that an unlinked suppressor mutation occurred, yielding independent assortment with the *nic-2* mutant.

Prototroph C: There are 496 "revertant" prototrophs (the other 500 are parental prototrophs) and four auxotrophs. This suggests that a suppressor mutation occurred in a site very close to the original mutation and was infrequently separated from the original mutation by recombination [$100\% (4 \times 2)/1000 = 0.8$ m.u.].

CHAPTER 16

5. The cuts indicated by the arrows below would result in a 6-bp duplication.

The cuts indicated by the arrows below would result in a 4-bp duplication.

6. No, the intron would not be removed during transposition. DNA transposons jump by cutting the DNA from the source location and pasting it into the target location. There is no mRNA intermediate from which introns are removed.

16. Some transposable elements have evolved strategies to insert into safe havens, regions of the genome where they will do minimal harm. Safe havens include duplicate genes (such as tRNA or rRNA genes) and other transposable elements. Safe havens in bacterial genomes might be very specific sequences between genes or the repeated rRNA genes.

23. The staggered cut will lead to a 9-bp target-site duplication that flanks the inserted transposon.

CHAPTER 17

2. Six.

7. Each type is expected at 33.3%.

11. 15.3 and 15.2 at the tip of p5.

14. E B C D—o—A

21. MM N OO would be classified as $2n - 1$; MM NN OO would be classified as $2n$; and MMM NN PP would be classified as $2n + 1$.

26. There would be one possible quadrivalent.

30. Yes.

33. An acentric fragment cannot be attached to spindle fibers in meiosis (or mitosis) and is consequently lost.

36. Very large deletions tend to be lethal, likely owing to genomic imbalance or the unmasking of recessive lethal genes. Therefore, the observed very large pairing loop is more likely to be from a heterozygous inversion.

38. Williams syndrome is the result of a deletion of the 7q11.23 region of chromosome 7. Cri du chat syndrome is the result of a deletion of a significant part of the short arm of chromosome 5 (specifically bands 5p15.2 and 5p15.3). Both Turner syndrome (XO) and Down syndrome (trisomy 21) result from meiotic nondisjunction. The term *syndrome* is used to describe a set of phenotypes (often complex and varied) that are generally present together.

43. The order is *b a c e d f.*

Allele	Band
b	1
a	2
c	3
e	4
d	5
f	6

44. The data suggest that one or both break points of the inversion are located within an essential gene, causing a recessive lethal mutation.

47. a. When crossed with yellow females, the results would be

X^e/Y^{e+} gray males
X^e/X^e yellow females

b. If the e^+ allele was translocated to an autosome, the progeny would be as follows, where "A" indicates autosome:

P	$A^{e+}/A; X^e/Y \times A/A ; X^e/X^e$	
F_1	$A^{e+}/A ; X^e/X^e$	gray female
	$A^{e+}/A ; X^e/Y$	gray male
	$A/A ; X^e/X^e$	yellow female
	$A/A; X^e/Y$	yellow male

49.

Klinefelter syndrome	XXY male
Down syndrome	trisomy 21
Turner syndrome	XO female

53. a. If a hexaploid were crossed with a tetraploid, the result would be pentaploid.

b. Cross *A/A* with *a/a/a/a* to obtain *A/a/a/a.*

c. The easiest way is to expose the *A/a** plant cells to colchicine for one cell division, which will result in a doubling of chromosomes to yield *A/A/a*/a*.*

d. Cross a hexaploid (*a/a/a/a/a/a*) with a diploid (*A/A*) to obtain *A/a/a/a.*

55. a. The ratio of normal-leaved to potato-leaved plants will be 5 : 1.

b. If the gene is not on chromosome 6, there should be a 1 : 1 ratio of normal-leaved to potato-leaved plants.

59. a. The aberrant plant is semisterile, which suggests an inversion. Because the *d–f* and *y–p* frequencies of recombination in the aberrant plant are normal, the inversion must implicate *b* through *x.*

b. To obtain recombinant progeny when there has been an inversion requires the occurrence of either a double crossover within the inverted region or single crossovers between *f* and the inversion, which occurred someplace between *f* and *b.*

61. The original plant is homozygous for a translocation between chromosomes 1 and 5, with break points very close to genes *P* and *S.* Because of the close linkage, a ratio suggesting a monohybrid cross, instead of a dihybrid cross, was observed, both with selfing and with a testcross. All gametes are fertile because of homozygosity.

original plant:	*P S/p s*
tester:	*p s/p s*

F_1 progeny: heterozygous for the translocation:
The easiest way to test this hypothesis is to look at the chromosomes of heterozygotes in meiosis I.

66. The original parents must have had the following chromosome constitution:

G. hirsutum	26 large, 26 small
G. thurberi	26 small
G. herbaceum	26 large

G. hirsutum is a polyploid derivative of a cross between the two Old World species, which could easily be checked by looking at the chromosomes.

71. a. Each mutant is crossed with wild type, or

$$m \times m^+$$

The resulting tetrads (octads) show 1 : 1 segregation, indicating that each mutant is the result of a mutation in a single gene.

b. The results from crossing the two mutant strains indicate either that both strains are mutant for the same gene:

$$m_1 \times m_2$$

or that they are mutant in different but closely linked genes:

$$m_1 \ m_2^+ \times m_1^+ \ m_2$$

c. and **d.** Because phenotypically black offspring can result from nondisjunction (notice that, in cases C and D, black appears in conjunction with aborted spores), mutant 1 and mutant 2 are likely to be mutant in different but closely linked genes. The cross is therefore

$$m_1 \ m_2^+ \times m_1^+ \ m_2$$

Case A is an NPD tetrad and would be the result of a four-strand double crossover.

$m_1^+ \ m_2^+$	black
$m_1^+ \ m_2^+$	black
$m_1 \ m_2$	fawn
$m_1 \ m_2$	fawn

Case B is a tetratype and would be the result of a single crossover between one of the genes and the centromere.

$m_1^+ \ m_2^+$	black
$m_1^+ \ m_2$	fawn
$m_1 \ m_2^+$	fawn
$m_1 \ m_2$	fawn

Case C is the result of nondisjunction in meiosis I.

$m_1^+ \ m_2^+ ; m_1^+ \ m_2^+$	black
$m_1^+ \ m_2^+ ; m_1^+ \ m_2^+$	black
no chromosome	abort
no chromosome	abort

Case D is the result of recombination between one of the genes and the centromere followed by nondisjunction in meiosis II. For example,

$m_1^+ \ m_2 ; m_1 \ m_2^+$	black
no chromosome	abort
$m_1 \ m_2^+$	fawn
$m_1^+ \ m_2$	fawn

CHAPTER 18

3. Four mtDNA haplotypes.

4. The frequency of recessive disorders in children of unrelated parents is higher in Japan (~8.5%) than in France (~3.5%).

5. Individuals 14 and 21.

7. Selection and drift can cause an allele to either rise or fall in its frequency and possibly become fixed or lost from the population. Mutation and migration can bring new alleles into a population.

11. The frequency of b is $q = \sqrt{0.04} = 0.2$, and the frequency of B is $p = 1 - q = 0.8$. The frequency of B/B is $p^2 = 0.64$, and the frequency of B/b is $2pq = 0.32$.

14. a. $p' = 0.5[(0.5)(1.0) + 0.5(1.0)]/[(0.25)(1.0) + (0.5)(1.0)] + (0.25)(0.7)] = 0.54$
   b. 0.008
17. $\frac{1}{2}(0.02) + \frac{1}{2}(0.02)^2 = 0.0102$
21. Probability of fixation $= \dfrac{1}{2N} = \dfrac{1}{100,000}$;

probability of loss $= 1 - \dfrac{1}{2N} = \dfrac{99,999}{100,000}$

26. a. $F_I = (\frac{1}{2})^3 \times (1 + \frac{1}{2}) = 3/16$
   b. $1/8 = (\frac{1}{2})^3 \times (1 + F_A)$, so $F_A = 0$
29. $p_A = p_a = p_B = p_b = 0.5$. At equilibrium, the frequency of doubly heterozygous individuals is $2(p_A p_a) \times 2(p_B p_b) = 0.25$
31. Before migration, $q_A = 0.1$ and $q_B = 0.3$ in the two populations. Because the two populations are equal in number, immediately after migration $q_{A+B} = \frac{1}{2}(q_A + q_B) = \frac{1}{2}(0.1 + 0.3) = 0.2$. At the new equilibrium, the frequency of affected males is $q = 0.2$, and then frequency of affected females is $q^2 = (0.2)^2 = 0.04$. (Color blindness is an X-linked trait.)
33. $q^2 = 0.002$; $q = 0.045$. Assuming F in the founders is 0.0, the $F_{50} = 0.222$ (see Box 18-3). $f_{a/a} = q^2 + pqF = 0.012$.
37. $\hat{H} = [4 \times 50,000 \times (3 \times 10^{-8})]/[4 \times 50,000 \times (3 \times 10^{-8}) + 1]$
   $= 5.96 \times 10^{-3}$
41. a. $\hat{q} = \sqrt{1.0 \times 10^{-5}/0.5} = 4.47 \times 10^{-3}$
   Selection cost $= sq^2 = 0.5(4.47 \times 10^{-3})^2 = 10^{-5}$
   b. $\hat{q} = 6.32 \times 10^{-3}$
   Selection cost $= sq^2 = 0.5(6.32 \times 10^{-3})^2 = 2 \times 10^{-5}$
   c. Selection cost $= sq^2 = 0.3(5.77 \times 10^{-3})^2 = 10^{-5}$

CHAPTER 19

3. The *ald2* gene.
5. No, the interval near *IGF1* is the only place where the P-values for the SNPs rise above the threshold for statistical significance.
7. Many traits vary more or less continuously over a wide range. For example, height, weight, shape, color, reproductive rate, metabolic activity, etc., vary quantitatively rather than qualitatively. Continuous variation can often be represented by a bell-shaped curve, where the "average" phenotype is more common than the extremes. Discontinuous variation describes the easily classifiable, discrete phenotypes of simple Mendelian genetics: seed shape, auxotrophic mutants, sickle-cell anemia, etc. These traits often show a simple relation between genotype and phenotype, although discontinuous traits such as affected versus not-affected for a disease condition can also exhibit complex inheritance.
9. The mean is 4.7 bristles, the variance is 1.11 bristles^2, and the standard deviation is 1.05 bristles.
12. The breeder cannot be assured that this population will respond to selective breeding even though the broad-sense heritability is high. Broad-sense heritability is the ratio of the genetic variance to the phenotypic variance. The genetic variance is the sum of the additive and dominance variances. Only additive variance is transmitted from parent to offspring. Dominance variance is not transmitted from parent to offspring. If all the genetic variance in the population is dominance variance, then selective breeding will not succeed.
18. $\bar{x}_{par} = [(9.8 + 10.8)/2] - 9.6 = 0.7$ mm, $\bar{a}_{par} = 0.79 \times 0.7 = 0.55$ mm, $\hat{x}_{off} = 9.6 + 0.55 = 10.15$ mm
23. $V_e = 3.5\ g^2$, V_g in population B is $21.0 - 3.5 = 17.5\ g^2$, $H^2 = 17.5/21.0 = 0.83$.

CHAPTER 20

6. The ratio of nonsynonymous to synonymous species differences would be similar to the ratio of nonsynonymous to

synonymous polymorphisms, indicating that selection did not play a role and the differences are due to random genetic drift.
8. The high frequency of even (versus odd) haploid chromosome numbers.
12. The mutation prevents expression of Duffy on red blood cells. Thus, *P. vivax* cannot bind and infect red blood cells.
17. The three principles are (1) individuals within any one population vary from one another, (2) offspring resemble their parents more than they resemble unrelated individuals, and (3) some forms are more successful at surviving and reproducing than other forms in a given environment.
20. The relative rate of synonymous and nonsynonymous substitutions would not be higher than expected in a globin pseudogene because a pseudogene is inactive and has no function to be preserved.
23. A new gene duplicate can (1) evolve a new function, (2) become inactivated, or (3) perform part of the original function, thereby sharing full function with the original gene.
31. A population will not differentiate from other populations by local inbreeding if:

$$\mu \geq 1/N$$

and so

$$N \geq 1/\mu$$
$$N \geq 10^5$$

35. When amino acid changes have been driven by positive selection, there should be an excess of nonsynonymous changes. The *MC1R* gene (melanocortin 1 receptor) encodes a key protein controlling the amount of melanin in skin and hair. Asian and European populations appear to have experienced positive selection for more lightly pigmented skin relative to their African counterparts.
37. Noncoding sequences. A major constraint on gene evolution comprises the potential pleiotropic effects of mutations in coding regions. These effects can be circumvented by mutations in regulatory sequences, which play a major role in the evolution of body form. Changes in noncoding sequences provide a mechanism for altering one aspect of gene expression while preserving the role of pleiotropic proteins in other essential developmental processes.
40. a. The HbS mutation has arisen independently in five different haplotypes in different regions and then increased to high frequency.
   b. Three independent populations of blind cave fish have different mutations in the *Oca2* gene.
   c. Multiple independent populations of spineless sticklebacks have distinct deletions of a regulatory element of the *Pitx1* gene.
   d. Many species of melanistic birds and mammals have different mutations in the same region of the *Mc1r* gene.
45. For polymorphic sites with a species, let nonsynonymous $= a$ and synonymous $= b$. For polymorphic sites between the species, let nonsynonymous $= c$ and synonymous $= d$. If divergence is due to neutral evolution, then

$$a/b = c/d$$

If divergence is due to positive selection, then

$$a/b < c/d$$

However, in this example, $a/b = 20/50 > c/d = 2/18$, which fits neither expectation. Because the ratio of nonsynonymous to synonymous polymorphisms (a/b) is relatively high, the gene being studied may encode a protein tolerant of relatively fewer species differences. The relatively fewer species differences may suggest that speciation was a recent event so new polymorphisms have been fixed in one species that are not variants in the other.

Index

Note: Page numbers followed by f indicate figures; those followed by t indicate tables. **Boldface** page numbers indicate Key Terms.